Computational Stochastic Mechanics

FIRST INTERNATIONAL CONFERENCE
ON
COMPUTATIONAL STOCHASTIC METHODS

SCIENTIFIC COMMITTEE

Acknowledgement is made to F. Casciati *et al.* for the use of Figure 2 on p. 489, which appears on the front cover of this book.

Computational Stochastic Mechanics

Editors: P.D. Spanos, Rice University, U.S.A.
C.A. Brebbia, Wessex Institute of Technology, U.K.

Computational Mechanics Publications
Southampton Boston

Co-published with

Elsevier Applied Science
London New York

P.D. Spanos
Brown School of Engineering
Rice University
PO Box 1892
Houston
TX 77251
USA

C.A. Brebbia
Computational Mechanics Institute
Wessex Institute of Technology
Ashurst Lodge
Ashurst
Southampton SO4 2AA
UK

Co-published by

Computational Mechanics Publications
Ashurst Lodge, Ashurst, Southampton, UK

Computational Mechanics Publications Ltd
Sole Distributor in the USA and Canada:

Computational Mechanics Inc.
25 Bridge Street, Billerica, MA 01821, USA

and

Elsevier Science Publishers Ltd
Crown House, Linton Road, Barking, Essex IG11 8JU, UK

Elsevier's Sole Distributor in the USA and Canada:

Elsevier Science Publishing Company Inc.
655 Avenue of the Americas, New York, NY 10010, USA

British Library Cataloguing-in-Publication Data

A Catalogue record for this book is available
from the British Library

ISBN 1-85166-698-2 Elsevier Applied Science, London, New York
ISBN 1-85312-147-9 Computational Mechanics Publications, Southampton
ISBN 1-56252-074-1 Computational Mechanics Publications, Boston, USA

Library of Congress Catalog Card Number 91-75141

CONTENTS

SECTION 9: MATERIALS

SECTION 10: APPLIED RANDOM VIBRATIONS

SECTION 11: APPLIED STOCHASTIC FINITE ELEMENT ANALYSIS

Dedication to Dr. Masanobu Shinozuka
1st International Conference on Computational Stochastic Mechanics
17-19 September 1991
Corfu, Greece

Masanobu Shinozuka, Ph.D.
23.12.30 -

Since September, 1990, Dr. Masanobu Shinozuka has been director of the National Center for Earthquake Engineering Research (NCEER) with headquarters at the State University of New York at Buffalo. He is also visiting Capen Professor of Structural Engineering at the University at Buffalo, on leave from Princeton University where he is Sollenberger Professor of Civil Engineering. Dr. Shinozuka earned his Ph.D. from Columbia University in New York, and holds M.S. and B.S. degrees from Kyoto University in Japan.

An internationally renowned expert in structural dynamics, structural reliability, and lifeline earthquake engineering, Dr. Shinozuka also coordinates NCEER's research program on lifeline systems and the risk and reliability of structures. In this capacity, he heads a team of researchers from several institutions that is investigating water delivery systems in San Francisco and water and oil lifelines in the Memphis, Tennessee area. He is also a member of the U.S. Panel on Active Structural Control that is formulating an international cooperative research program with Japan.

A consultant to numerous government and private organizations, Dr. Shinozuka is an elected member of the National Academy of Engineering, and an elected fellow of the American Academy of Mechanics, and American Society of Mechanical Engineers. He is also Executive Vice-President of the International Association for Structural Safety and Reliability (IASSR), a member of the American Society of Civil Engineering, the American Institute of Aeronautics and Astronautics, the Architectural Institute of Japan, and the Japan Society of Civil Engineers (JSCE). He is the recipient of numerous awards including the American Society of Civil Engineers' Moisseiff Award (1988), Nathan M. Newmark Medal (1985), Alfred M. Freudenthal Medal (1978), and Walter L. Huber Civil Engineering Research Prize (1972). He is the Editor of the International Journal of Probabilistic Engineering Mechanics. He has authored more that 300 technical papers and journal articles and has served on numerous editorial review boards. Additionally, he has organized or co-chaired several international structural engineering conferences. Dr. Shinozuka is listed in *Who's Who in the World, Who's Who in America, Who's Who in American Men and Women of Science, Who's Who in Engineering*, and *Who's Who in Technology Today.*

PREFACE

Over a period of several years the field of probabilistic mechanics and computational mechanics have progressed vigorously, but independently. With the advent of powerful computational hardware and the development of novel mechanical techniques, the field of stochastic mechanics has progressed in such a manner that the inherent uncertainty of quite complicated systems can be addressed. The first International Conference on Computational Stochastic Mechanics was convened in Corfu in September 1991 in an effort to provide a forum for the exchanging of ideas on the current status of computational methods as applied to stochastic mechanics and for identifying needs for further research. The Conference covered both theoretical techniques and practical applications.

The Conference also celebrated the 60th anniversary of the birthday of Dr. Masanobu Shinozuka, the Sollenberger Professor of Civil Engineering at Princeton University, whose work has contributed in such a great measure to the development of Computational Stochastic Mechanics. A brief summary of his career and achievements are given in the Dedication.

This book comprises some of the papers presented at the meeting and covers sections on Theoretical Reliability Analysis; Damage Analysis; Applied Reliability Analysis; Theoretical Random Vibrations; Stochastic Finite Element Concept; Fatigue and Fracture; Monte Carlo Simulations; Earthquake Engineering Applications; Materials; Applied Random Vibrations; Applied Stochastic Finite Element Analysis, and Flow Related Applications and Chaotic Dynamics.

The Editors hope that the book will be a valuable contribution to the growing literature covering the field of Computational Stochastic Mechanics.

The Editors
Corfu
September 1991

Preface

[illegible] years [illegible] general [illegible] in the field of [illegible] mechanics and [illegible] [illegible] techniques have progressed [illegible] rapidly [illegible] of powerful computers [illegible] and [illegible] development of [illegible] statistical techniques [illegible] in [illegible] systems [illegible] be discussed. [illegible] the first International Conference on Computational Stochastic Mechanics was convened in Corfu in September 1991 in an effort to provide a forum for [illegible] of ideas [illegible] the analysis of computational stochastic mechanics [illegible] and for identifying needs for further research. The [illegible] covered both theoretical techniques and practical applications.

The Conference was dedicated to the [illegible] birthday of Dr Masanobu Shinozuka, the Sollenberger Professor of Civil Engineering at Princeton University, whose work has contributed in a great measure to the development of computational stochastic mechanics. A brief summary of his career and achievements are given in the Dedication.

This book comprises some of the papers presented at the meeting and covers sections on: Theoretical Reliability Analysis; Stochastic Methods; Applied Reliability Analysis; Theoretical Random Vibrations; Stochastic Finite Element Concepts; Fatigue and Fracture; Monte Carlo Simulation Techniques; Engineering Applications; Materials; Applied Random Vibrations; Applied Stochastic Finite Element Analysis; and Flow-Related Applications and Chaotic Dynamics.

The Editors hope that the book will be a valuable contribution to the growing literature covering the field of Computational Stochastic Mechanics.

The Editors
Corfu
September 1991

SECTION 1: THEORETICAL RELIABILITY ANALYSIS

On Excursions of Non-Homogeneous Vector-Valued Gaussian Random Fields

M.H. Faber (*), R. Rackwitz (**)
(*) RCP-DENMARK ApS, Teglgade 17p, 9550 Mariager, Denmark
(**) Technical University of Munich, Arcisstrasse 21, 8000 Munich, Germany

ABSTRACT

Extreme excursions of non–homogeneous vector–valued Gaussian random fields into non–linearly bounded failure domains are investigated based on suitable scalarizations of the failure boundary. The scalarization scheme is based on a second order expansion of the failure surface in the most likely excursion point. The expected number of excursions of the scalarized random field above a given threshold function in a certain domain are then determined by the corresponding expected number of Bolotin excursion characteristics for scalar fields. The numerical evaluation of the resulting multi–dimensional integrals is performed utilizing the asymptotic integral expansion technique due to Laplace. The scalarization scheme is illustrated by a numerical example.

INTRODUCTION

Many random phenomena can appropriately be modeled by Gaussian random fields. Examples include the strength properties of soils, the fracture toughness around a crack tip and the load effects in long span bridges. In reliability applications their extremes are of special interest. However, their probabilistic characteristics are difficult to assess. The difficulties arise from the fact that an excursion of a random field above a given threshold is not easy to describe – not even in the scalar case. Several excursion characteristics have been proposed in the literature. The simplest and most obvious characteristic is to associate the event of an excursion of a random field above a given threshold with the event of a local maxima above the threshold (Belyayev [1] and Hasofer [2]) leading to an asymptotic (high thresholds) expression for the expected number of local maxima of a homogeneous Gaussian field above a constant threshold function (see also Vanmarcke [3]). An alternative approach followed by Adler/Hasofer [4] is to describe the excursions of a random field in terms of a differential topology excursion characteristic based on the Euler characteristic. The expected number of this excursion characteristic can be derived in closed form for homogeneous Gaussian random fields and constant threshold functions (see Adler [5] for a full account of the results). Bolotin [6] described a similar excursion characteristic for which closed form results can also be obtained (Faber [7]). The non–homogeneous, scalar case has recently been studied by the authors in a sequence of papers. In Faber/Rackwitz [8] the local maxima of a suitably normalized field are taken as the excursion characteristics, in Faber et al. [9] Adler's excursion characteristic is used and in Faber/Rackwitz [10] the characteristic due to Bolotin has been applied. In all those cases the resulting integrals are approximated by the asymptotic integral expansion method due to Laplace (see e.g. Bleistein & Handelsman [11]). This method can be shown to be sufficiently accurate for most practical applications (see Faber [7] for some numerical comparisons).

In this paper the results for scalar, non–homogeneous Gaussian fields are generalized for vector–valued fields by making use of a special approximate scalarization scheme. Only the excursion characteristics based on local maxima and the excursion characteristic due to Bolotin are then applicable. In the following only the more powerful excursion characteristic due do Bolotin for scalar valued random fields is considered. It is shown how a non–homogeneous vector–valued Gaussian random field can be scalarized such that the results for scalar fields become applicable.

DEFINITIONS

Consider the probability $P_f(\mathbf{B})$ that a m–componental vector–valued, non–homogeneous Gaussian random field $\mathbf{X}(\mathbf{t})$, $\mathbf{t} \in \mathbf{B} \in \mathbb{R}^n$, has at least one realization $\mathbf{x}(\mathbf{t})$ in the failure domain $\Omega_f(\mathbf{t})$. The failure domain $\Omega_f(\mathbf{t}) = \{g(\mathbf{x}(\mathbf{t})) \leq 0\}$ is assumed to be bounded by an at least twice differentiable failure surface $\partial\Omega_f(\mathbf{t}) = \{g(\mathbf{x}(\mathbf{t})) = 0\}$. The field $\mathbf{X}(\mathbf{t})$ is assumed to be sufficiently mixing so that its excursions into the failure domain tend to follow a Poisson process for increasing threshold levels. Then, $P_f(\mathbf{B})$ can be written as

$$P_f(\mathbf{B}) = P(g(\mathbf{x}(\mathbf{t})) \leq 0 \,|\, \mathbf{t} \in \mathbf{B}) \approx 1 - \exp\{-E[N_{\mathbf{X}}(\mathbf{B})]\} \tag{1}$$

where $E[N_{\mathbf{X}}(\mathbf{B})]$ is the expected number of excursions of $\mathbf{X}(\mathbf{t})$ into the failure domain $\Omega_f(\mathbf{t})$ in the domain $\mathbf{B}$. The m elements of $\mathbf{X}(\mathbf{t})$, $X_i(\mathbf{t})$, i=1,2,..m are assumed to be non–homogeneous Gaussian random fields with at least twice differentiable sample paths $x_i(\mathbf{t})$, i=1,2,..m. The derivative fields $\dot{X}_i(\mathbf{t})$ and $\ddot{X}_i(\mathbf{t})$ are assumed to exist and are defined as

$$\dot{X}_i(\mathbf{t}) = \left(\frac{\partial X_i(\mathbf{t})}{\partial t_j}\,;\, j=1,2,...n\right)^T, \quad i=1,2,..m \tag{2}$$

$$\ddot{X}_i(\mathbf{t}) = \left\{\frac{\partial^2 X_i(\mathbf{t})}{\partial t_j\,\partial t_k}\,;\, j,k=1,2,...n\right\}, \quad i=1,2,..m \tag{3}$$

The mean values of the random field and its derivatives are defined as

$$\mathbf{m}_{X,\dot{X},\ddot{X}}(\mathbf{t}) = (\mathbf{m}_X(\mathbf{t}), \mathbf{m}_{\dot{X}}(\mathbf{t}), \mathbf{m}_{\ddot{X}}(\mathbf{t}))^T \tag{4}$$

and the covariance matrix is given by

$$\Sigma_{X\dot{X}\ddot{X}} = \begin{bmatrix} \Sigma_{XX} & \Sigma_{X\dot{X}} & \Sigma_{X\ddot{X}} \\ \Sigma_{\dot{X}X} & \Sigma_{\dot{X}\dot{X}} & \Sigma_{\dot{X}\ddot{X}} \\ \Sigma_{\ddot{X}X} & \Sigma_{\ddot{X}\dot{X}} & \Sigma_{\ddot{X}\ddot{X}} \end{bmatrix} \tag{5}$$

with

$$\Sigma_{XX} = \{\mathrm{Cov}[X_i(t_1),X_j(t_2)];\ 1 \le i,j \le m\}$$
$$\Sigma_{X\dot{X}} = \{\mathrm{Cov}[X_i(t_1),\dot{X}_j(t_2)];\ 1 \le i \le m,\ m+1 \le j \le nm\}$$
$$\Sigma_{X\ddot{X}} = \{\mathrm{Cov}[X_i(t_1),\ddot{X}_j(t_2)];\ 1 \le i \le m,\ nm+1 \le j \le nm(nm+1)/2\}$$
$$\Sigma_{\dot{X}X} = \{\mathrm{Cov}[\dot{X}_i(t_1),X_j(t_2)];\ m+1 \le i \le nm,\ 1 \le j \le m\}$$
$$\Sigma_{\dot{X}\dot{X}} = \{\mathrm{Cov}[\dot{X}_i(t_1),\dot{X}_j(t_2)];\ m+1 \le i \le nm,\ m+1 \le j \le nm\}$$
$$\Sigma_{\dot{X}\ddot{X}} = \{\mathrm{Cov}[\dot{X}_i(t_1),\ddot{X}_j(t_2)];m+1 \le i \le nm,nm+1 \le j \le nm(nm+1)/2\}$$
$$\Sigma_{\ddot{X}X} = \{\mathrm{Cov}[\ddot{X}_i(t_1),X_j(t_2)];\ nm+1 \le i \le nm(nm+1)/2,\ 1 \le j \le m\}$$
$$\Sigma_{\ddot{X}\dot{X}} = \{\mathrm{Cov}[\ddot{X}_i(t_1),\dot{X}_j(t_2)];nm+1 \le i \le nm(nm+1)/2,m+1 \le j \le nm\}$$
$$\Sigma_{\ddot{X}\ddot{X}} = \{\mathrm{Cov}[\ddot{X}_i(t_1),\ddot{X}_j(t_2)];\ nm+1 \le i \le nm(nm+1)/2,\ nm+1 \le j \le nm(nm+1)/2\} \tag{6}$$

Note that due to the symmetry of covariance matrices it suffices to determine the diagonal and the upper or lower triangle. Further, since the partial (mean square) derivatives of a quantity with respect to the parameters indexed by i and j and j and i are identical the vector $\ddot{X}$ can be reduced to a vector of $(n^2 + n)/2$ dimensions.

EXCURSIONS OF SCALAR–VALUED RANDOM FIELDS

The excursion set $\mathcal{A}$ for a scalar–valued random field is defined as

$$\mathcal{A} = \{\mathbf{t} \in \mathbf{B}\backslash\partial\mathbf{B};\ X(\mathbf{t}) \ge a(\mathbf{t})\} \tag{7}$$

where $\mathbf{B}$ is the considered domain, $\partial\mathbf{B}$ the boundary of $\mathbf{B}$ and $a(\mathbf{t})$ the threshold function. Alternatively, the excursion set $\mathcal{B}$ can be considered

$$\mathcal{B} = \{\mathbf{t} \in \mathbf{B}\backslash\partial\mathbf{B};\ Z(\mathbf{t}) \ge b(\mathbf{t})\} \tag{8}$$

where $Z(\mathbf{t})$ is a standardized scalar field defined by

$$Z(\mathbf{t}) = \left[\frac{X(\mathbf{t}) - m_X(\mathbf{t})}{\sigma_X(\mathbf{t})} \right] \tag{9}$$

$$\dot{Z}(\mathbf{t}) = \left\{ \frac{\dot{X}_i(\mathbf{t}) - m_{\dot{X}_i}(\mathbf{t})}{\sigma_{\dot{X}_i}(\mathbf{t})}, \ i=1,2,\dots n \right\}^T \tag{10}$$

$$\ddot{Z}(\mathbf{t}) = \left[\frac{\ddot{X}_{ij}(\mathbf{t}) - m_{\ddot{X}_{ij}}(\mathbf{t})}{\sigma_{\ddot{X}_{ij}}(\mathbf{t})}, \ i,j=1,2,\dots,n \right] \tag{11}$$

and

$$b(\mathbf{t}) = \left[\frac{a(\mathbf{t}) - \mu_X(\mathbf{t})}{\sigma_X(\mathbf{t})} \right] \tag{12}$$

$$\dot{b}(\mathbf{t}) = \left[\frac{\dot{a}_i(\mathbf{t}) - \mu_{\dot{X}_i}(\mathbf{t})}{\sigma_{\dot{X}_i}(\mathbf{t})}, \ i=1,2,\dots,n \right]^T \tag{13}$$

$$\ddot{b}(\mathbf{t}) = \left\{ \frac{\ddot{a}_{ij}(\mathbf{t}) - \mu_{\ddot{X}_{ij}}(\mathbf{t})}{\sigma_{\ddot{X}_{ij}}(\mathbf{t})}, \ i,j=1,2,\dots,n \right\} \tag{14}$$

For notational reasons the standardized random field $Z(\mathbf{t})$ is considered in the following.

The expected number of crossings of $Z(\mathbf{t})$ above $b(\mathbf{t})$ in $\mathbf{B}$ can be written as

$$E[N_Z(\mathbf{B})] = \int_{\mathbf{B}} \mu_Z(\tau)\, d\tau \tag{15}$$

where the mean rate of excursions $\mu_Z(\tau)$ is defined by

$$\mu_Z(\mathbf{t}) = \lim_{\Delta t_v \to 0} \frac{P_1(\mathbf{t}\ ;\ \Delta\mathbf{t})}{\Delta t_v} \tag{16}$$

where Δt_v is the volume of $\Delta\mathbf{t}$.

In order to access the probability of exactly one crossing within the volume element $\Delta\mathbf{t}$, $P_1(\mathbf{t};\Delta\mathbf{t})$, the excursion characteristic due to Bolotin C_B is considered. This excursion characteristic is based on the Euler characteristic of a set (see e.g. Morse & Chairn [12]). The Euler characteristic of the set $\mathcal{B}$, $\psi(\mathcal{B})$, is

$$\psi(\mathcal{B}) = \sum_{j=0}^{n} (-1)^j N_j + c \tag{17}$$

where N_j is the number of points in $\mathcal{B}$ where the Hessian matrix of the function $f(\mathbf{t})$ formed by the realizations of $Z(\mathbf{t}) - b(\mathbf{t}) = z(\mathbf{t}) - b(\mathbf{t})$ have j negative eigenvalues and where all first–order partial derivatives of $f(\mathbf{t})$ vanish. The usefulness of the Euler characteristic is based on the fact that the Euler characteristic of a set with an odd dimension is equal to 0 and for a set with an even dimension is equal to 2. By observing that the connectivity c of the set $\mathcal{B}$ is equal to 1 if the set is simply connected (assuming that $\mathcal{B}$ is simply connected involves assumptions on the regularity of the field and the threshold as well as assumptions on the level of the threshold) the Euler characteristic yields 1 if the set $\mathcal{B}$ is non–empty and zero otherwise. The expected number of excursions $N_Z(\mathbf{B})$ can then be written as

$$E[N_Z(\mathbf{B})] = \sum_{j=0}^{n} (-1)^{n-j}\, E[N_j(\mathbf{B})] \tag{18}$$

where the expected number of points of type j is

$$E[N_j(\mathbf{B})] = \int_{\mathbf{B}} \mu_j(\tau)\, d\tau \tag{19}$$

The conditional rate of points of type j is

$$\mu_j(\mathbf{t}) = \lim_{\Delta t_V \to 0} \frac{P_j(\mathbf{t}\ ;\ \Delta\mathbf{t})}{\Delta t_V} \tag{20}$$

The probability $P_j(\mathbf{t};\Delta\mathbf{t})$ of one point of type j within the volume element $\Delta\mathbf{t}$ can be estimated from

$$P_j(\mathbf{t};\Delta\mathbf{t}) = P\left[\begin{array}{l} \{Z(\mathbf{t}) > b(\mathbf{t}) - \Delta z\} \cap \\ \{\dot{Z}(\mathbf{t}) \in \Delta\dot{b}(\mathbf{t})\} \cap \\ \{\mathbf{H}(\mathbf{t}) \in \mathbb{H}_j\} \end{array}\right] \tag{21}$$

where $\mathbb{H}_j$ is the set of Hessian matrices of f($\mathbf{t}$) with exactly j negative eigenvalues.

After some manipulations $E[N_j(\mathbf{B})]$ can be written as

$$E[N_j(\mathbf{B})] = \int_{\mathbf{B}} \int_{b(\mathbf{t})}^{\infty} \int_{\ddot{z}\in\mathbb{H}_j} \varphi(z,\dot{\mathbf{b}}(\mathbf{t}),\ddot{z})\, |\det \mathbf{H}(\mathbf{t})|\, \mathbf{d}\ddot{z}\, dz\, dt \tag{22}$$

Because $|\det \mathbf{H}(\mathbf{t})|$ is equal to $(-1)^j \det H(\mathbf{t})$, the conditional expected number of outcrossings of $Z(\mathbf{t})$ is obtained as

$$\begin{aligned} E[N_X(\mathbf{B})] &= \sum_{j=0}^{n} (-1)^{n-j}\, E[N_j(\mathbf{B})] \\ &= (-1)^n \int_{\mathbf{B}} \int_{b(\mathbf{t})}^{\infty} \int_{\ddot{z}\in\mathbb{R}} \varphi(z,\dot{\mathbf{b}}(\mathbf{t}),\ddot{z})\, \det \mathbf{H}(\mathbf{t})\, \mathbf{d}\ddot{z}\, dz\, \mathbf{d}t \end{aligned} \tag{23}$$

or

$$E[N_X(B)] = \int_B \int_{b(t)}^{\infty} \int_{\ddot{z} \in \mathbb{R}^{n(n+1)/2}} \frac{1}{(2\pi)^k} \Psi(t,z) \exp(-p(z,\dot{b},\ddot{z};t))\, d\ddot{z}\, dz\, dt \quad (24)$$

with $k = (n(n+1)/2+n+1)/2$ and

$$\Psi(t,z) = \frac{(-1)^n \det \mathbf{H}(t)}{(\det \Sigma_{Z\dot{Z}\ddot{Z}})^{1/2}} \quad (25)$$

$$p(z,\ddot{z};t) = \frac{1}{2} (Z(t),\dot{b}(t),\ddot{Z}(t))^T \Sigma^{-1}_{Z\dot{Z}\ddot{Z}} (Z(t),\dot{b}(t),\ddot{Z}(t)) \quad (26)$$

LAPLACE'S ASYMPTOTIC INTEGRAL EXPANSION

The integral in equation (24) can be estimated using the asymptotic integral expansions due to Laplace. Consider an integral of the form

$$I(\lambda) = \int_D \Psi(\mathbf{y}) \exp[\lambda^2 p(\mathbf{y})]\, d\mathbf{y} \quad (27)$$

where $\Psi(\mathbf{y})$ and $p(\mathbf{y})$ are real at least twice differentiable functions, $\mathbf{D}$ is an integration domain with at least twice differentiable boundaries $\partial\mathbf{D}$, λ is a scalar and $\mathbf{y}$ is a n–dimensional vector. For large λ the integral can be solved analytically because only the local behavior of the functions $\Psi(\mathbf{y})$ and $p(\mathbf{y})$ and of the boundary $\partial\mathbf{D}$ in the point $\mathbf{y}^*$ where $p(\mathbf{y}^*)$ takes its maximum need to be taken into account. In particular, it suffices to expand these functions into Taylor series up to the first non–vanishing term.

In the present case $\mathbf{y}^*$ is a boundary point where $\Psi(\mathbf{y})$ and $\nabla p(\mathbf{y}) \neq 0$. $\mathbf{D}$ can be represented as $\mathbf{D} = \cap_{i=1}^{k} \mathbf{D}_i$ with $D_i = \{h_i(\mathbf{y}) \leq 0\}$ and the functions $h_i(\mathbf{y})$ are twice differentiable in a neighborhood of $\mathbf{y}^*$. Then the following solution has been derived by Breitung/Hohenbichler [13]

$$I(\lambda) = \int_D \Psi(\mathbf{y}) \exp[(\lambda^2 p(\mathbf{y})]\, d\mathbf{y}$$

$$\sim \Psi(\mathbf{y}^*) \exp[\lambda^2 p(\mathbf{y}^*)] \frac{(2\pi)^{\frac{n-k}{2}}}{\lambda^{n+k}} \frac{1}{\det \mathbf{A}} \left(\prod_{i=1}^{\delta} |\xi_i|^{-1}\right) \frac{1}{(|\det \mathbf{S}(\mathbf{y}^*)|)^{1/2}} \quad (28)$$

where $h_i(\mathbf{y}^*) = 0$ for $i = 1,2,..,\delta$, $(\delta \in \{1,2,..,k\})$, the gradients $\mathbf{a}_i = \nabla h_i(\mathbf{y}^*)$, $i = 1,2,..,\delta$ are linearly independent and $a_{ij} = 0$ for $i = 1,2,..,\delta$ and $j = \delta+1, \delta+2,..,n$. Further, there is

$$\mathbf{A} = (\mathbf{a}_1, \mathbf{a}_2, .., \mathbf{a}_\delta)$$

$$\nabla p(\mathbf{y}^*) = \sum_{i=1}^{\delta} \xi_i \mathbf{a}_i$$

with $\xi_i \leq 0$ and $\mathbf{S}(\mathbf{y}^*) = \{s_{ij};\ i,j = \delta+1,\delta+2,..,n\}$ with

$$s_{ij} = \frac{\partial^2 p(\mathbf{y}^*)}{\partial y_i \partial y_j} - \sum_{q=1}^{\delta} \xi_q \frac{\partial^2 h_q(\mathbf{y}^*)}{\partial y_i\, \partial y_j} \tag{29}$$

The aforementioned conditions can always be fulfilled by a suitable coordinate transformation. For $\delta = n$ there is det $\mathbf{S}(\mathbf{y}^*) = 1$.

SCALARIZATION SCHEME FOR VECTOR FIELDS

These results can be used for vector–valued Gaussian fields if an appropriate scalarization scheme is applied. As failure surfaces generally are non–linear a scalarization should be capable to represent the most significant characteristics of non–linearity. The idea of the scalarization scheme is to consider for fixed $\mathbf{t} = \mathbf{t}_0$ a random variable S associated with the failure surface such that when the vector–valued random field has a realization in the failure domain, the random field $S(\mathbf{t})$ has a realization exceeding a certain deterministic threshold $\|\mathbf{x}^*(\mathbf{t})\|$. This is achieved by assigning a proper expansion of the failure surface $\partial\Omega_f(\mathbf{t}_0)$ with respect to the components of $\mathbf{X}(\mathbf{t})$ to $S(\mathbf{t})$. For simplicity of notation it is assumed that the random field $\mathbf{X}(t)$ is standardized with zero mean and unit variance for $\mathbf{t} = \mathbf{t}_0$. This can always be assured by a proper transformation. Before proceeding it should be remembered that equation (24) for a vector–valued random field $\mathbf{X}(\mathbf{t})$ scalarized into the random field $R(\mathbf{t})$ can be written as

$$E[N_X(B)] = \int_B \int_{\Omega_f(\mathbf{t})} \int_{\ddot{\mathbf{r}} \in \mathbb{R}^{n(n+1)/2}} \frac{1}{(2\pi)^k} \Psi(\mathbf{t},\mathbf{x}) \exp(-\,p(\mathbf{x},\dot{\mathbf{b}},\ddot{\mathbf{r}};\mathbf{t}))\, d\ddot{\mathbf{r}}\, d\mathbf{x}\, d\mathbf{t} \tag{30}$$

with $k = (n(n+1)/2+n+m)/2$ and

$$\Psi(\mathbf{t},\mathbf{x}) = \frac{(-1)^n \ \det\ \mathbf{H}(\mathbf{t})}{(\det\ \Sigma_{\mathbf{X}\dot{\mathbf{R}}\ddot{\mathbf{R}}})^{1/2}} \tag{31}$$

$$p(\mathbf{x},\ddot{\mathbf{r}};\mathbf{t}) = \frac{1}{2}\,(\mathbf{X}(\mathbf{t}),\dot{\mathbf{b}}(\mathbf{t}),\ddot{\mathbf{R}}(\mathbf{t}))^T\, \Sigma^{-1}_{\mathbf{X}\dot{\mathbf{R}}\ddot{\mathbf{R}}}\, (\mathbf{X}(\mathbf{t}),\dot{\mathbf{b}}(\mathbf{t}),\ddot{\mathbf{R}}(\mathbf{t})) \tag{32}$$

with obvious notations.

Linearly bounded safe domain

The simplest case is when the failure surface $\partial\Omega_f(\mathbf{t})$ is given in terms of a linear combinations of the components of $\mathbf{X}(\mathbf{t})$, i.e. by

$$\partial\Omega_f(\mathbf{t}) = \{\|\mathbf{x}^*(\mathbf{t}_0)\| - \sum_{i=1}^{m} \gamma_i x_i = 0\} \tag{33}$$

where the γ_i's are constants. Then a scalarized Gaussian random field can be represented by

$$S(\mathbf{t}) = \sum_{i=1}^{m} \gamma_i \, X_i(\mathbf{t}) \tag{34}$$

with partial derivatives

$$\frac{\partial S(\mathbf{t})}{\partial t_i} = \sum_{j=1}^{m} \gamma_j \frac{\partial X_j(\mathbf{t})}{\partial t_i} \tag{35}$$

and

$$\frac{\partial^2 S(\mathbf{t})}{\partial t_i \partial t_j} = \sum_{k=1}^{m} \gamma_k \frac{\partial^2 X_k(\mathbf{t})}{\partial t_i \partial t_j} \tag{36}$$

It is seen that the joint density function of $S(\mathbf{t})$, $\dot{S}(\mathbf{t})$ and $\ddot{S}(\mathbf{t})$ is Gaussian and, hence, if the scalar field $S(\mathbf{t})$ and the threshold function $\|\mathbf{x}^*(\mathbf{t})\|$ are standardized corresponding to the operation in equations (9) to (14), then, equation (24) for the expected number of excursion characteristics applies directly. If the failure surface is not linear in the components of $\mathbf{X}(\mathbf{t})$ the linear scalarization scheme may be used as a first–order approximation.

<u>Non–linearly bounded safe domains</u>
For non–linearly bounded safe domains it is proposed to scalarize the vector–valued random field by a second order expansion of the failure surface thus keeping the argumentation for the asymptotic integral expansions as indicated before. It is assumed that the failure surface $\partial\Omega_f(\mathbf{t})$ is at least twice differentiable in $\mathbf{x}^*(\mathbf{t}_0)$. By performing a suitable transformation $\mathbf{y} = \mathbf{T}\,\mathbf{x}$ it is possible to rotate the failure surface such that all mixed derivatives vanish and that the expansion point $\mathbf{x}^*(\mathbf{t}_0)$ is positioned on the positive side of the axis of the m'th component of $\mathbf{y}$, i.e. y_m (see figure 1).

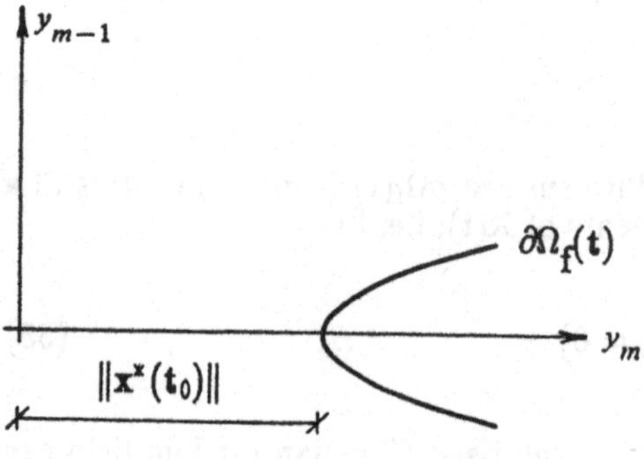

Figure 1. Illustration of the representation of the rotated failure surface.

The failure surface can then be approximated by

$$\partial\Omega_f(\mathbf{t}) = \{y_m - \|\mathbf{x}^*(\mathbf{t}_0)\| - \frac{1}{2}\sum_{i=1}^{m-1} \kappa_i\, y_i^2 = 0\} \tag{37}$$

where the κ_i are the principal curvatures of $\partial\Omega_f(\mathbf{t})$ in $\mathbf{x}^*(\mathbf{t}_0)$. The vector–valued random field $\mathbf{Y}(\mathbf{t})$ is now scalarized into $S(\mathbf{t})$ by

$$S(\mathbf{t}) = Y_m(\mathbf{t}) - \frac{1}{2}\sum_{i=1}^{m-1} \kappa_i\, Y_i(\mathbf{t})^2 \tag{38}$$

Excursions of $\mathbf{Y}(\mathbf{t})$ into $\Omega_f(\mathbf{t})$ are approximately equivalent to excursions of $S(\mathbf{t})$ above $\|\mathbf{x}^*(\mathbf{t}_0)\|$.

The only remaining problem is to assess the joint probability density function of $S(\mathbf{t})$, $\dot{S}(\mathbf{t})$ and $\ddot{S}(\mathbf{t})$ or, equivalently, for use in equation (32), $\mathbf{Y}(\mathbf{t})$, $\dot{S}(\mathbf{t})$ and $\ddot{S}(\mathbf{t})$. The components of $\dot{S}(\mathbf{t})$ and $\ddot{S}(\mathbf{t})$ can be written as

$$\frac{\partial S(\mathbf{t})}{\partial t_j} = \frac{\partial Y_m(\mathbf{t})}{\partial t_j} - \sum_{i=1}^{m-1} \kappa_i \frac{\partial Y_i(\mathbf{t})}{\partial t_j} Y_i(\mathbf{t}) \tag{39}$$

and

$$\frac{\partial^2 S(\mathbf{t})}{\partial t_j \partial t_k} = \frac{\partial^2 Y_m(\mathbf{t})}{\partial t_j \partial t_k} - \sum_{i=1}^{m-1} \kappa_i \left[\frac{\partial^2 Y_i(\mathbf{t})}{\partial t_j \partial t_k} Y_i(\mathbf{t}) + \frac{\partial Y_i(\mathbf{t})}{\partial t_j}\frac{\partial Y_i(\mathbf{t})}{\partial t_k} \right] \tag{40}$$

Inspection of equation (39) and (40) reveals that the components of $\dot{S}(\mathbf{t})$ are Gaussian conditional on the realization of the components of $\mathbf{Y}(\mathbf{t})$. In the same way the components of $\ddot{S}(\mathbf{t})$ are Gaussian conditional on the realizations of the components of $\mathbf{Y}(\mathbf{t})$ and $\dot{\mathbf{Y}}(\mathbf{t})$. The joint probability density function $f(\mathbf{y},\dot{\mathbf{s}},\ddot{\mathbf{s}};\mathbf{t})$ is hence readily obtained as

$$f(\mathbf{y},\dot{\mathbf{s}},\ddot{\mathbf{s}};\mathbf{t}) = \varphi_{\mathbf{X}}(\mathbf{y};\mathbf{t})\, \varphi_{\dot{\mathbf{S}}}(\dot{\mathbf{s}}\,|\,\mathbf{y};\mathbf{t})\, \varphi_{\ddot{\mathbf{S}}}(\ddot{\mathbf{s}}\,|\,\dot{\mathbf{s}},\mathbf{y};\mathbf{t}) \tag{41}$$

This is the function to be used in equation (30).

ILLUSTRATING EXAMPLE

For the illustrating example a vector field with two zero mean and unit variance, uncorrelated components in two dimensions is assumed. The state function is

$$g(z(t)) = \alpha(t) - \frac{(z_1 + 1)^2}{4^2} + \frac{z_2^2}{3^2}$$

with

$$\alpha(t) = 1 - \lambda(1-u)\, u^{\alpha}(1-v)\, v^{\beta}$$

Furthermore the covariance matrix of $(\mathbf{Z}, \dot{\mathbf{Z}}, \ddot{\mathbf{Z}})$ is given by

$$\mathbf{\Sigma}_{\mathbf{Z}\dot{\mathbf{Z}}\ddot{\mathbf{Z}}} = \begin{bmatrix} \mathbf{\Sigma}_{Z_1Z_2} & \mathbf{\Sigma}_{Z_1\dot{Z}_2} & \mathbf{\Sigma}_{Z_1\ddot{Z}_2} \\ \mathbf{\Sigma}_{\dot{Z}_1Z_2} & \mathbf{\Sigma}_{\dot{Z}_1\dot{Z}_2} & \mathbf{\Sigma}_{\dot{Z}_1\ddot{Z}_2} \\ \mathbf{\Sigma}_{\ddot{Z}_1Z_2} & \mathbf{\Sigma}_{\ddot{Z}_1\dot{Z}_2} & \mathbf{\Sigma}_{\ddot{Z}_1\ddot{Z}_2} \end{bmatrix} = \begin{bmatrix} 1 & 0 & 0 & 0 & 0 & 0 & 2 & 2 & 0 & 0 & 0 & 0 \\ 0 & 1 & 0 & 0 & 0 & 0 & 0 & 0 & 0 & 2 & 2 & 0 \\ 0 & 0 & 2 & 0 & 0 & 0 & 0 & 0 & 0 & 0 & 0 & 0 \\ 0 & 0 & 0 & 2 & 0 & 0 & 0 & 0 & 0 & 0 & 0 & 0 \\ 0 & 0 & 0 & 0 & 2 & 0 & 0 & 0 & 0 & 0 & 0 & 0 \\ 0 & 0 & 0 & 0 & 0 & 2 & 0 & 0 & 0 & 0 & 0 & 0 \\ 2 & 0 & 0 & 0 & 0 & 0 & 12 & 4 & 0 & 0 & 0 & 0 \\ 2 & 0 & 0 & 0 & 0 & 0 & 4 & 12 & 0 & 0 & 0 & 0 \\ 0 & 0 & 0 & 0 & 0 & 0 & 0 & 0 & 4 & 0 & 0 & 0 \\ 0 & 2 & 0 & 0 & 0 & 0 & 0 & 0 & 0 & 12 & 4 & 0 \\ 0 & 2 & 0 & 0 & 0 & 0 & 0 & 0 & 0 & 4 & 12 & 0 \\ 0 & 0 & 0 & 0 & 0 & 0 & 0 & 0 & 0 & 0 & 0 & 4 \end{bmatrix}$$

It is easy to see that also

$$\mathbf{m}_{\mathbf{Z}} = \mathbf{m}_{\dot{S}|\mathbf{Z}} = \mathbf{m}_{\ddot{S}|\mathbf{Z},\dot{S}} = 0$$

whereas we have

$$\mathbf{\Sigma}_{\mathbf{ZZ}} = \begin{bmatrix} 1 & 0 \\ 0 & 1 \end{bmatrix}$$

$$\mathbf{\Sigma}_{\dot{S}|\mathbf{Z}} = \begin{bmatrix} 2^2 + (\kappa_1 \; z_1)^2\, 2^2 & 0 \\ 0 & 2^2 + (\kappa_1 \; z_1)^2\, 2^2 \end{bmatrix}^{1/2}$$

$$\mathbf{\Sigma}_{\ddot{S}|\mathbf{Z},\dot{S}} = \begin{bmatrix} 12^2 + (\kappa_1 \; z_1)^2\, 12^2 & 4^2 + (\kappa_1 \; z_1)^2\, 4^2 & 0 \\ 4^2 + (\kappa_1 \; z_1)^2\, 4^2 & 12^2 + (\kappa_1 \; z_1)^2\, 12^2 & 0 \\ 0 & 0 & 4^2 + (\kappa_1 \; z_1)^2\, 4^2 \end{bmatrix}^{1/2}$$

For $\lambda = \alpha = \beta = 1$ the minimum point is found to be $(u^*, v^*)^T = (0.5, 0.5)^T$ where $(Z_1, Z_2)^T = (1.33, 2.43)^T$. The expected number of Bolotin's excursion characteristic can then be determined to be 0.0264.

CONCLUSIONS

Two suitable scalarization schemes for the failure surfaces are proposed. The linear scalarization scheme allows direct use of the results for scalar random fields based on either excursions by local maxima or by the more accurate excursion characteristic due to Bolotin. The linearization scheme can be proposed as a first order approximation also in non–linear cases. The use of a second order expansion of the failure surface in the local critical point requires considerable more computational effort. In both cases the first and second order derivative fields must exist. Especially the second requirement can restrict the applicability of the derived computation schemes.

REFERENCES

1. Belyayev, Y.K. Distribution of the Maximum of a Random Field and its Application to Reliability Problems, Eng. Cybernet., 2, pp. 269—276, 1970.

2. Hasofer, A.M. The Mean Number of Maxima Above High Levels in Gaussian Random Fields, Journal of Applied Probability, 13, pp. 377—379, 1976.

3. Vanmarcke, E.H., Random Fields — Analysis and Synthesis, MIT Press, Cambridge, 1983

4. Adler, R.J. and Hasofer, A.M. Level Crossings of Random Fields, Annals of Probability, 4,1, pp. 1—12, 1976

5. Adler, R.J. The Geometry of Random Fields, John Wiley & Sons, Chichester, New York, Brisbane, Toronto, 1981.

6. Bolotin, V.V. Warscheinlichkeitsmetoden zur Berechnung von Konstruktionen, VEB, Verlag für Bauwesen, Berlin, 1981.

7. Faber, M.H. Excursions of Gaussian Random Fields in Structural Reliability Theory. Ph.D.—thesis defended publicly at the University of Aalborg, February 1989. Institute of Building Technology and Structural Engineering. The University of Aalborg, Denmark, 1989.

8. Faber, M.H. and Rackwitz, R. Asymptotic Excursion Probabilities of Non—Homogeneous Gaussian Fields, (Ed. Thoft—Christensen, P.), Proceedings of the Second IFIP WG 7.5 Working Conference, London, England, Springer—Verlag 1988.

9. Faber, M.H., Plantec, J.—Y. and Rackwitz, R. Two Applications of Asymptotic Laplace Integrals in Structural Reliability, Proc. 5'th ICOSSAR'89, San Francisco, pp. 1263—1270, ASCE, New York 1990.

10. Faber, M.H. and Rackwitz, R. Asymptotic Mean Number of Excursions of Non—Homogeneous Gaussian Scalar Fields. Submitted to : Probabilistic Engineering Mechanics, 1989.

11. Bleistein, N. and Handelsman, R.A. Asymptotic Expansions of Integrals, Holt, Rinehart and Winston, New York, 1975.

12. Morse, M. and Cairns, S. Critical Point Theory in Global Analysis and Differential Topology, Academic Press, New York, 1969.

13. Breitung, K. and Hohenbichler, M. Asymptotic Approximations for Multivariate Integrals with an Application to Multivariate Probabilities, Journal of Multivariate Analysis, Vol. 30, No. 1, Academic Press, 1989.

Fitting-Adaptive Importance Sampling in Reliability Analysis

Y. Fujimoto, M. Iwata, Y. Zheng

Department of Naval Architecture and Ocean Engineering, Hiroshima University, Higashi-Hiroshima City 724, Japan

ABSTRACT

In this paper a fitting-adaptive approach to importance sampling is proposed, in which a fitting density function is introduced to calculate the failure probability instead of the importance sampling density function. The importance sampling density function is only used to generate sample points. The histogram is employed as the fitting adaptive density function for it is straightforward and fulfills the purpose. Moreover, if the fitting density function is not satisfactory, adaptive procedure is automatically performed. Several examples are presented to demonstrate the applicability of the proposed approach. It is shown that a significant reduction in variance of the failure probability can be achieved by use of this approach.

INTRODUCTION

In reliability analysis, the Monte Carlo Simulation(MCS) is often employed when the analytical solution is not attainable. However, the computational effort involved in MCS sometimes becomes exorbitant because of the enormous sample size, especially for those problems which take a lot of CPU time in each Monte Carlo run. In order to improve the efficiency and simultaneously keep the accuracy, a variety of variance reduction techniques have been proposed, among which the Importance Sampling(IS) technique is generally recognized as the most efficient[1,4,7-11]. Although many kinds of IS techniques have been read in the literature, they may basically be classified into three categories, viz. (1)IS using design points, (2)IS based on the information obtained from the direct MCS and (3)IS through adaptive procedures. The first one is the combination of IS and First-/Second-Order Reliability Methods[1,12], but this is only possible when the performance function, or the failure surface function, can be given explicitly. The second one needs the information of the direct MCS within the failure domain[7]. However, this is rather difficult for low failure probability problems. The third one refines the solution step by step by using the 'knowledge' just learnt from the problem under consideration[8,9]. This method generally gives good results if the number of random variables is not large[8].

To date there have appeared a great number of papers on the application of the IS technique to reliability analysis. However, most studies are concerned with the formation of the ideal importance sampling density function(i.s.d.f.), but little attention has been paid to the discrepancy between the i.s.d.f. and the one simulated by a finite number of sample points. In this paper, a fitting-adaptive approach is proposed to improve the accuracy and efficiency of the importance sampling technique. In this approach a fitting density function is used for calculating the failure probability so as to make full use of the samples generated from the selected i.s.d.f. Furthermore, the fitting density function can be improved by an adaptive procedure. The key point of the proposed approach is to exploit thoroughly the samples obtained, and hence achieve an improvement on the IS technique in solving problems that are very time-consuming.

As to the structural reliability analysis, the randomness of both the loads and the strength is taken into consideration. Because the COV values of strength parameters are usually small, the Neumann Expansion method is employed to speed up the structural analysis in every Monte Carlo run[5,6].

FITTING-ADAPTIVE APPROACH

As opposed to the conventional IS technique in which the i.s.d.f. is to be simulated by generating sufficient samples according to this very density function, in the fitting-adaptive approach a fitting function, named fitting-adaptive density function (f.a.d.f.), is introduced to properly fit the samples drawn from the selected i.s.d.f. This approach is described below after a brief presentation of the IS technique.

Let $\mathbf{X} = (x_1, x_2, ..., x_M)$ denote the basic random vector consisting of all the random variables considered and $f_X(\mathbf{X})$ denote the joint probability density function of $\mathbf{X}$, then the failure probability P_f is expressed as the multi-dimensional integral of $f_X(\mathbf{X})$, i.e.

$$\begin{aligned} P_f &= \int_{\Omega_f} f_X(\mathbf{X}) d\mathbf{X} \\ &= \int_{\Omega} I(\mathbf{X}) f_X(\mathbf{X}) d\mathbf{X} \end{aligned} \tag{1}$$

where Ω_f and Ω denote the failure domain and the entire domain, respectively. $I(\mathbf{X})$ is an indicator defined by the following expression.

$$I(\mathbf{X}) = \begin{cases} 1, & \text{if } \mathbf{X} \text{ in } \Omega_f \\ 0, & \text{otherwise} \end{cases} \tag{2}$$

The direct MCS of P_f can be formulated in terms of sample-mean[4], i.e.

$$P_{fM} = \frac{1}{N} \sum_{i=1}^{N} I(\mathbf{X}_i) \tag{3}$$

where N is the sample size, $\mathbf{X}_i (i = 1, 2, \ldots, N)$ is a sample taken from the density function $f_X(\mathbf{X})$.

Assume that $\mathbf{Y} = (y_1, y_2, \ldots, y_M) \in \Omega$ is another random vector, and is distributed with the joint probability density function $g_Y(\mathbf{Y})$, then, eq.(1) can be rewritten as

$$P_f = \int_{\Omega} \frac{I(\mathbf{X}) f_X(\mathbf{X})}{g_Y(\mathbf{X})} g_Y(\mathbf{X}) d\mathbf{X} \tag{4}$$

where $g_Y(\mathbf{Y})$ is the so-called importance sampling density function.

Similarly, the importance sampling MCS of eq.(4) can be obtained by means of sample-mean, i.e.

$$P_{fI} = \frac{1}{N}\sum_{i=1}^{N} I(\mathbf{Y}_i)\frac{f_X(\mathbf{Y}_i)}{g_Y(\mathbf{Y}_i)} \tag{5}$$

where $\mathbf{Y}_i$ is a sample generated according to $g_Y(\mathbf{Y})$.

The transformation of eq.(1) into eq.(4) only contains a mathematical manipulation of introducing a function $g_Y(\mathbf{Y})$, so eq.(4) will result in correct answer as long as the integrand is well-behaved. As for the simulation formula, eq.(5), the selection of i.s.d.f. $g_Y(\mathbf{Y})$ has a critical bearing on both the efficiency and the accuracy of the simulation. The variance of P_{fI} is proved[4] to be

$$Var(P_{fI}) = \frac{1}{N}\{\int_{\Omega} \frac{[I(\mathbf{Y})f_X(\mathbf{Y})]^2}{g_Y(\mathbf{Y})}d\mathbf{Y} - P_f^2\} \tag{6}$$

Therefore, the best choice of $g_Y(\mathbf{Y})$ is

$$g_Y(\mathbf{Y}) = I(\mathbf{Y})f_X(\mathbf{Y})/P_f \tag{7}$$

which can reduce the variance of P_{fI} to zero. Could an i.s.d.f. be chosen to meet eq.(7), only one sample would be enough to estimate the failure probability. However, eq.(7) contains P_f which is unknown. This renders the rigorous determination of eq.(7) useless.

The observation of eq.(3) and eq.(5) reveals that the i.s.d.f. must be simulated with adequate accuracy. $I(\mathbf{X}_i)$ in eq.(3) is a 'domain function', namely its value merely depends upon which domain $\mathbf{X}_i$ falls in. In one-dimensional case, P_{fM} can be interpreted as the area below the probability density curve over the failure domain. Whereas P_{fI} in eq.(5) should be compared, in a sense, to the moment of that area. Different locations within the same domain have different values of $f_X(\mathbf{Y}_i)/g_Y(\mathbf{Y}_i)$, providing eq.(7) is not strictly satisfied. It is, therefore, misleading to expect the importance sampling to yield reliable solution with an inadequately small number of samples.

When the sample size is small, the fidelity of the simulated density function is subject to many factors, such as the transforms adopted, the initial data(seed for generating uniform random numbers), the word-length of the available computer, and so forth. In the ordinary IS, in which a unique i.s.d.f. is used both for generating samples and for calculating the failure probability, errors due to these factors are unavoidable unless eq.(7) holds. It is thus crucial to construct a density function that is less affected by these factors. Pilot study showed that, if the generated sample points sufficiently constitute the i.s.d.f., the solution of an IS is almost not changed when the i.s.d.f. is only altered a little(i.e. a similar but different i.s.d.f. is utilized). On the contrary, if the generated sample points do not adequately constitute the i.s.d.f., the solution will be markedly affected. Therefore, an effective way to improve the solution is to introduce a fitting density function that best fits the obtained sample points, though it is to some extent different from the selected i.s.d.f. If the fitting density function is denoted by $h_Y(\mathbf{Y})$, the failure probability can be formulated as follows.

$$P_f = \int_{\Omega} \frac{I(\mathbf{X})f_X(\mathbf{X})}{h_Y(\mathbf{X})}h_Y(\mathbf{X})d\mathbf{X} \tag{8}$$

Then, the sample-mean of eq.(8) is of the form

$$P_{fH} = \frac{1}{N}\sum_{i=1}^{N} I(\mathbf{Y}_i)\frac{f_X(\mathbf{Y}_i)}{h_Y(\mathbf{Y}_i)} \tag{9}$$

What follows is the description of the procedure of the fitting-adaptive approach.

1. Choose an appropriate i.s.d.f., $g_Y(\mathbf{Y})$, for the random vector $\mathbf{Y}$ relevant to the basic random vector $\mathbf{X}$[1,4,11].
2. For $i = 1, 2, ..., N$, generate a sample $\mathbf{Y}_i$ according to $g_Y(\mathbf{Y})$, determine $I(\mathbf{Y}_i)$, and store $\mathbf{Y}_i$ and $I(\mathbf{Y}_i)$.
3. Fit the sample points $\mathbf{Y}_i (i = 1, 2, ..., N)$ with a proper f.a.d.f. $h_Y(\mathbf{Y})$.
4. If the f.a.d.f. is not satisfactory, then adapt it as described in the next chapter or reselect the i.s.d.f. $g_Y(\mathbf{Y})$, go to step 2 and proceed with the remaining samples. Otherwise, go to step 5.
5. Compute the failure probability by eq.(9).

It becomes clear at this point that, when the sample size is increased for adaptive operation, the samples already generated are still useful. This saves much computational time at the cost of additional storage for storing $\mathbf{Y}_i$ and $I(\mathbf{Y}_i)$.

The crux of the fitting-adaptive approach lies in the formation of the h-function. The simplest way is to fit the sample points with a function of the same type as the i.s.d.f. used. For example, if the i.s.d.f. of Y_i, $g_{Y_i}(Y_i)$, is chosen to be the normal distribution $N(\overline{Y_i}, \sigma^2_{Y_i})$, then the f.a.d.f. $h_{Y_i}(Y_i)$ can be taken as $N(\overline{Y_i}', (\sigma'_{Y_i})^2)$, with $\overline{Y_i}'$ and $(\sigma'_{Y_i})^2$ being the sample-mean and sample-variance. Unfortunately, this kind of f.a.d.f. has little, if any, improvement on importance sampling. In the present paper the histogram is employed as f.a.d.f. because it is straightforward and fulfills the purpose. Other kinds of f.a.d.f.'s may as well be used in further studies.

If well structured, a histogram can appropriately reflect the obtained samples, so that errors due to the insufficiency of samples may be removed to a certain degree. Furthermore, if an i.s.d.f. fails to produce enough failure samples, it has to be moved towards the failure domain to generate 'new' samples[11], and the moved i.s.d.f. may have different shape from the original one(Fig.1). Even in such a case the 'old' samples may still be used, because a histogram is capable of covering sample points which come from different distributions provided the density functions are not too far away. This is a distinctive feature of the fitting-adaptive approach.

HOW TO FORM THE HISTOGRAM

After the i.s.d.f. is determined, the goodness of an IS depends on the sample size. The solution gets improved as the sample size increases. On the other hand, if the sample size is also fixed, the goodness of an IS depends on how effectively the generated samples are made use of, i.e. how well the f.a.d.f. behaves. For simplicity, the discussion on the histogram is confined to uni-variate problems in this chapter.

Let W denote the width of each window, the histogram may be expressed by

$$h_Y(y) = \frac{1}{N}\sum_{i=1}^{N}\frac{1}{W}K(\frac{y_i - y_c}{W}) \tag{10}$$

in which y_i is a sample point, y_c is the coordinate of the mid-point of the window where point y belongs to, and $K(.)$ is defined as follows.

$$K(t) = \begin{cases} 1, & \text{if } -\frac{1}{2} \leq t < \frac{1}{2} \\ 0, & \text{otherwise} \end{cases} \tag{11}$$

Figure 2 illustrates the normal density curves and the associated histograms drawn to the same scale. The histograms are plotted with 30 windows within the interval $[-3\sigma, +3\sigma]$. In Fig.2(a) the histogram is formed with 400 sample points. It can be seen that the histogram turns out different from the corresponding normal density curve, and has an empty window at point y_2. At point y_1, the ratios

$$\frac{f_X(y_1)}{g_Y(y_1)} = \frac{a}{b} \quad \text{and} \quad \frac{f_X(y_1)}{h_Y(y_1)} = \frac{a}{c}$$

are quite different. Because a/c reflects the obtained samples more accurately than a/b does, the FA estimator P_{fH} has smaller variance than the IS estimator P_{fI}. The histogram shown in Fig.2(b) is formed using 800 sample points. It is closer to the normal density curve than that in Fig.2(a), but it still has an empty window at $y_4 (y_4 \neq y_2)$. In addition, it clearly deviates from the normal density curve at point y_3.

In order to get better results, the following two conditions should be taken into account in forming a histogram.

1. The histogram should narrowly cover all the sample points, that is, its lower and upper tails should, respectively, reach the minimum and maximum values of all sample points.

2. There must not be empty windows in the histogram in order not to lose information[11]. If empty windows emerge they must be eliminated either by increasing the sample size or by adjusting the window width.

NUMERICAL EXAMPLES

Several examples are presented in the following to demonstrate the applicability of the fitting-adaptive approach. Random numbers are generated by calling the computer(HITAC-M680H) library routines. In all the examples presented the IS and the FA are carried out under identical conditions so that comparison may be made.

Mathematical Problems

Example 1: Consider respectively the following three cases of failure mode, where Z follows $N(0,1)$ distribution.

Case 1: $f(Z) = 2 - Z$

Case 2: $f(Z) = 3 - Z$

Case 3: $f(Z) = 4 - Z$

For each case, the 'seed' for generating random numbers is initialized once for all, i.e. after each simulation the seed is saved to be used as the initial data in the next simulation. In this way the effect of initial data may be reflected. In order to examine the influence of different locations of the i.s.d.f., the same distribution $N(3,1)$ is used as the i.s.d.f. for all these three cases. Its expected value is situated at the design point of Case 2, in the failure domain of Case 1, and in the safety domain of Case 3.

Figures 3 through 5 show the failure probabilities of the three cases. In general, the fitting-adaptive approach behaves much better than the importance sampling technique. In Case 1 the failure probability is relatively high, so the direct MCS applies, exhibiting larger variance(Fig.3). In Case 2 the design point is utilized, so the IS is expected to yield good results. Still, the FA approach improves the accuracy of IS(Fig.4). Fig.5 highlights the applicability of the fitting-adaptive approach to low probability problems.

Example 2: This example is intended to examine the feasibility that samples from different distributions are used together to form the fitting adaptive density. The failure surface function is as follows.

$$f(X,Y,Z) = 2X + 3Y - 0.5Z$$

where X, Y and Z are independent normal variables, and

$$X \sim N(1,0.3^2),\ Y \sim N(0.5,0.2^2),\ Z \sim N(1.5,1^2)$$

First, an initial i.s.d.f. $g_1(x,y,z)$ is chosen and 300 samples are generated according to $g_1(.)$; then $g_1(.)$ is shifted towards the failure domain for a distance of $0.3\sigma_Z$ to form $g_2(.)$ and 400 samples are generated according $g_2(.)$; then $g_2(.)$ is shifted again for the same distance to form $g_3(.)$ and 500 samples are generated according $g_3(.)$. $g_3(.)$ arrives at the design point. The mean values $(\overline{X}, \overline{Y}, \overline{Z})$ with respect to $g_1(.)$, $g_2(.)$ and $g_3(.)$ are, respectively,

$$(0.49,0.16,2.32),\ (0.49,0.16,2.62) \text{ and } (0.49,0.16,2.92)$$

Both IS and FA are performed for three cases associated with $g_1(.)$, $g_2(.)$ and $g_3(.)$, respectively. The generated samples are all saved, and the FA is performed using the samples coming from more than one distributions. The results are listed in Table 1 from which the following can be observed: (1)For the first three cases, where single distribution is used to generate samples, FA behaves better than IS. (2)For the last three cases, where multiple distributions are used to generate samples, FA produces still better results than for the first three cases.

Table 1 Failure probability of Example 2

i.s.d.f.	P_{fI}	P_{fH}	i.s.d.f.	P_{fH}
g_1	2.414×10^{-3}	2.553×10^{-3}	$g_1\&g_2$	2.598×10^{-3}
g_2	2.465×10^{-3}	2.506×10^{-3}	$g_2\&g_3$	2.620×10^{-3}
g_3	2.680×10^{-3}	2.627×10^{-3}	$g_1\&g_2\&g_3$	2.612×10^{-3}

Exact failure probability: $P_f = 2.617 \times 10^{-3}$

Structural Problems

Structural analysis is performed in the same way as in Ref.[11]. For strength parameters, the Young's modulus, the yielding stress and the plastic modulus are treated as random variables because they are significant to the probability of failure[6]. Their COV values are all 0.05. The mean values of the Young's modulus and the yielding stress are $\overline{E} = 210\text{GPa}$ and $\overline{\sigma}_Y = 276\text{MPa}$, respectively. All load parameters are treated as random variables except for the topside weights of Example 2. COV values for random loads are all 0.30. Both load variables and strength variables are assumed normally distributed and mutually independent. In such a case, the i.s.d.f. can be reduced into an expression solely depending on one parameter V. V is defined as $V = (\overline{Y} - \overline{X})/\sigma_X$(Fig.6). It was concluded in Ref.[11] that IS can be applied to strength variables, load variables or both. In this paper, the IS is applied to load variables. The 'exact' solution in the figures refers to the result by IS with sample size N=500000.

The structural failure is defined as the occurrence of one or more plastic hinges due to the combined load effect of axial force, shearing force and bending moment[6].

Example 1: Figure 7 shows a portal frame structure to which two loads are applied. Numerical data are listed in Table 2. For this case, V=2.0.

Figure 8 shows the failure probability versus sample size. It can be seen that the FA approach exhibits noticeable superiority to the IS technique. FA yields very good result with hundreds of samples. Both FA and IS are feasible when thousands of samples are used.

Table 2 Numerical data of portal frame

Element end number	Cross sectional area (m^2)	Moment of inertia (m^4)	Mean value of plastic modulus (m^3)
1, 2	4.8×10^{-3}	3.58×10^{-5}	2.12×10^{-3}
3, 4	4.0×10^{-3}	4.77×10^{-5}	3.66×10^{-3}
5, 6	4.0×10^{-3}	4.77×10^{-5}	3.66×10^{-3}
7, 8	4.8×10^{-3}	3.58×10^{-5}	2.12×10^{-3}

$$\overline{L}_1 = 15\text{kN}, \qquad \overline{L}_2 = 25\text{kN}$$

Example 2: Figure 9 shows a simplified plane structure, which can be thought of as the residual strength of a platform after braces on the second floor and the third floor have been subjected to brittle failure[2]. The numerical data are given in Table 3. For this structure, V=2.5. The failure probability versus sample size is shown in Fig.10. It can be observed that FA improves IS markedly.

Table 3 Numerical data of plane frame

Element end number	Cross sectional area (m^2)	Moment of inertia (m^4)	Mean value of plastic modulus (m^3)
1,2; 3,4			
5,6; 7,8	5.37×10^{-2}	2.19×10^{-3}	9.19×10^{-3}
9,10; 11,12	3.73×10^{-2}	1.06×10^{-3}	5.32×10^{-3}
13,14	4.68×10^{-2}	1.66×10^{-3}	7.47×10^{-3}
15,16; 17,18	4.85×10^{-2}	1.78×10^{-3}	7.88×10^{-3}
19,20; 21,22	9.30×10^{-3}	9.30×10^{-6}	0.67×10^{-3}

$\overline{L}_1 = 85\text{kN}$, $\overline{L}_3 = 65\text{kN}$, $\overline{L}_5 = 45\text{kN}$, $L_7 = 2000\text{kN}$
$\overline{L}_2 = 75\text{kN}$, $\overline{L}_4 = 55\text{kN}$, $\overline{L}_6 = 35\text{kN}$, $L_8 = 2000\text{kN}$

CONCLUSIONS

In the proposed approach, a fitting-adaptive density function is introduced to calculate the failure probability instead of the importance sampling density function, while the importance sampling density function is only used for generating sample points. This approach is very effective when the sample size is small, it is therefore useful for time-consuming problems.

The fitting-adaptive density function allows the use of samples from different distributions. Therefore, after the importance sampling density function is shifted for adaptive operation, the generated samples may still be used to save computational time. When the design point is used, the proposed approach is also of value to variance reduction.

If the two conditions stated herein above are satisfied, the histogram gives very good results. Further studies on the fitting-adaptive density function would be beneficial. For example, the histogram with variable-width windows[3,7] may offer advantage.

REFERENCES

1. Schueller, G.I. and Stix, R. A Critical Appraisal of Methods to Determine Failure Probabilities, Structural Safety, 4(1987), pp.293-309.

2. Murotsu, Y., et al. Reliability Analysis of Frame Structures under Combined Load Effect, ICOSSAR'85 pp.I117-128.

3. Abramson, I.S. On Bandwidth Variation in Kernel Estimates - A Square Root Law, Ann. Statist., 10(1982), pp.1217-1223.

4. Rubinstein, R.Y. Simulation and the Monte Carlo Method, John Wiley & Sons, New York, 1981.

5. Yamazaki, F., et al. Neumann Expansion for Stochastic Finite Element Analysis, Stochastic Mechanics, Vol.1, pp.59-95, 1987.

6. Zheng, Y., Fujimoto, Y. and Iwata, M. Reliability Analysis of Frame Structure by Monte Carlo Simulation Using Neumann Expansion, Trans. of the West-Japan Society of Naval Architects, Vol.79, pp.103-112, 1990.

7. Ang, G.L., et al. Kernel Method in Importance Sampling Density Estimation, ICOSSAR'89, pp.1193-1200.

8. Karamchandani, A., et al. Adaptive Importance Sampling, ICOSSAR'89, pp.855-862.

9. Bucher, C.G. Adaptive Sampling - An Iterative Fast Monte Carlo Procedure, Structural Safety, 5(1988), pp.119-126.

10. Verma, D., Fu, G. and Moses, F. Efficient Structural System Reliability Assessment by Monte-Carlo Methods, ICOSSAR'89, pp.895-900.

11. Zheng, Y., Fujimoto, Y. and Iwata, M. On Reliability Assessment of Frame Structures Based on Monte Carlo Simulation, J. of The Society of Naval Architects of Japan, Vol.167, pp.199-204, 1990.

12. Madsen, H.O. First Order vs. Second Order Reliability Analysis of Series Structures, Structural Safety, 2(1985), pp.207-214.

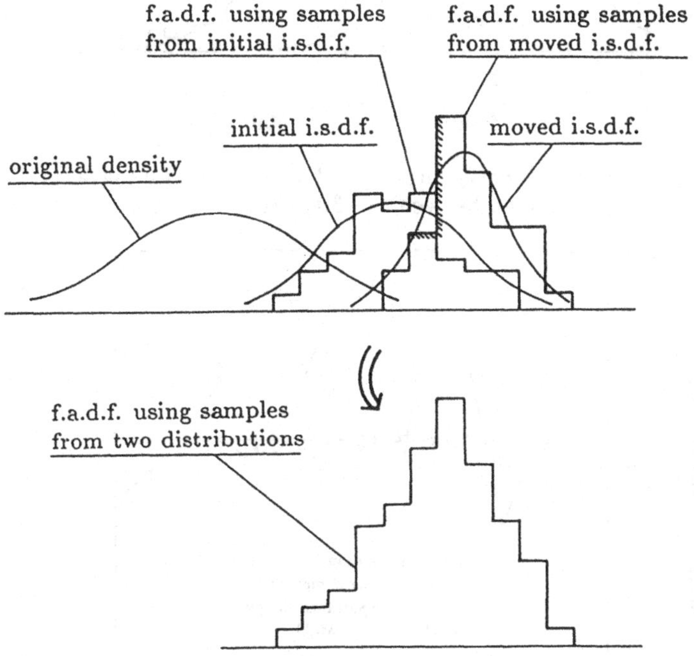

Fig.1 Schematic representation of moved i.s.d.f.

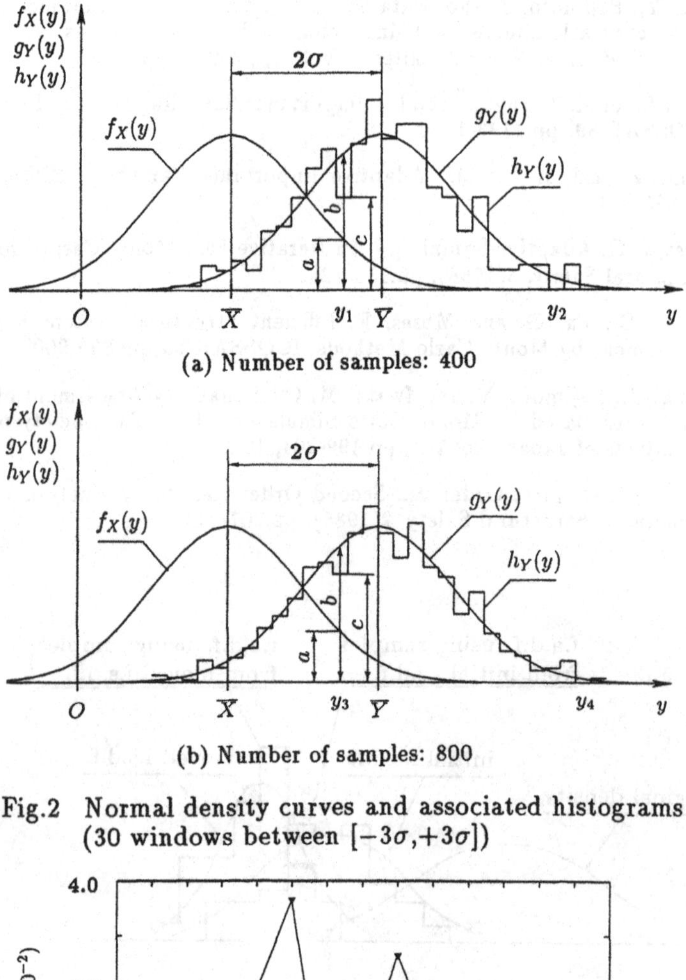

Fig.2 Normal density curves and associated histograms (30 windows between $[-3\sigma, +3\sigma]$)

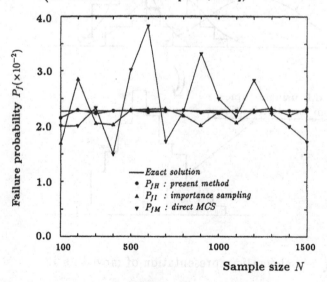

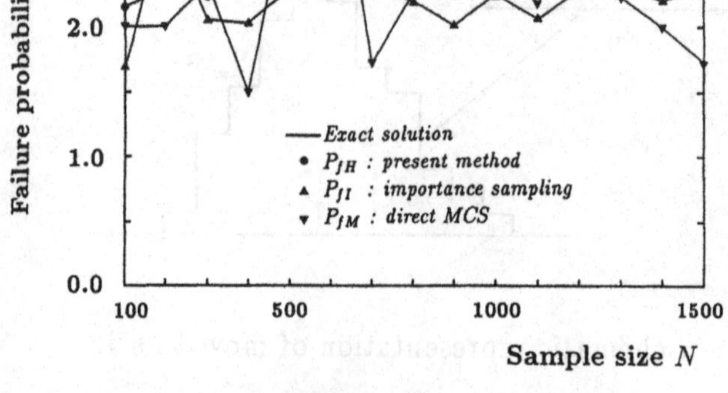

Fig.3 Failure probability of Case 1

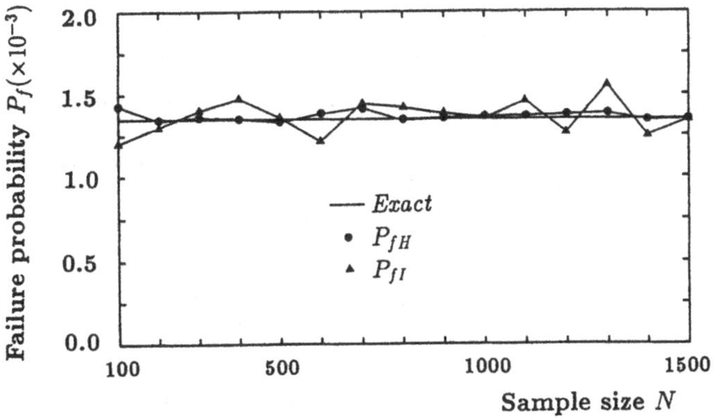

Fig.4 Failure probability of Case 2

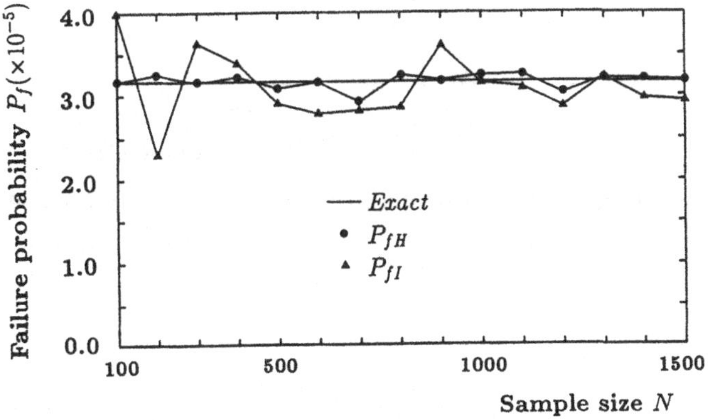

Fig.5 Failure probability of Case 3

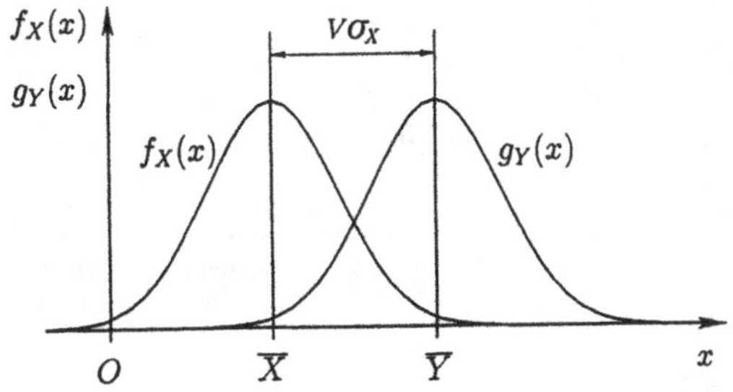

Fig.6 Schematic representation of parameter V

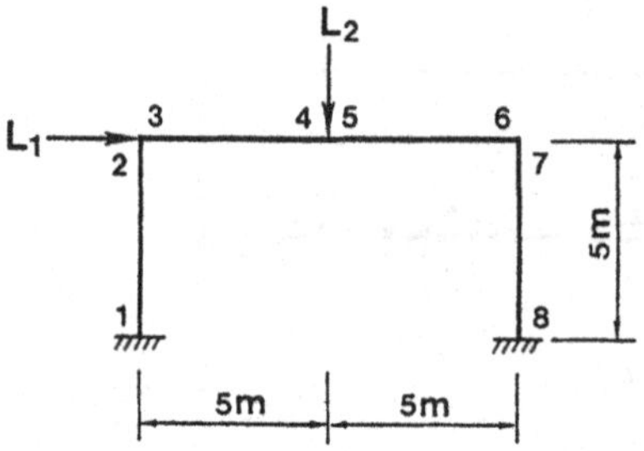

Fig.7 FE modeling of portal frame structure

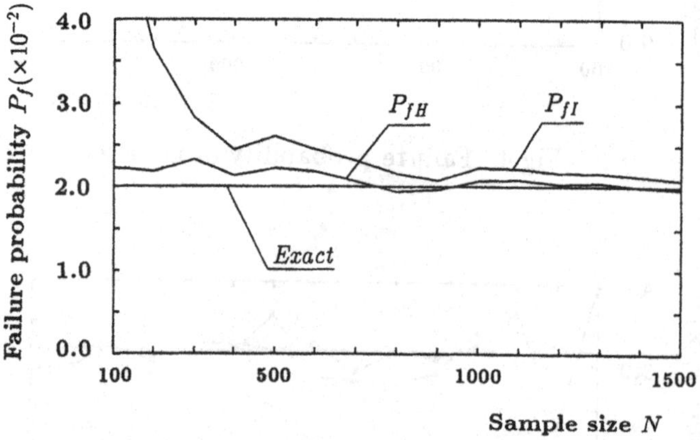

Fig.8 Failure probability of portal frame structure

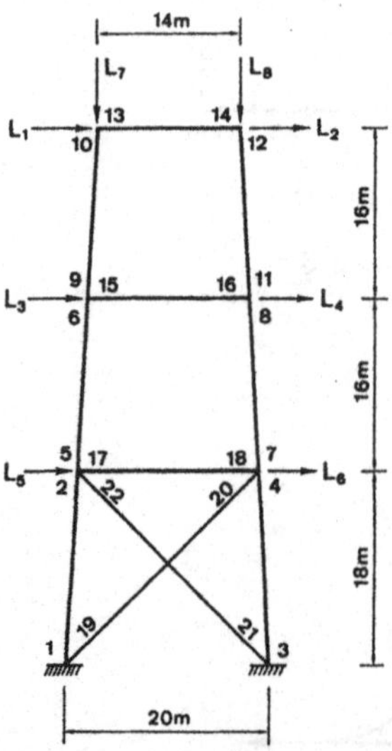

Fig.9 FE modeling of plane frame structure

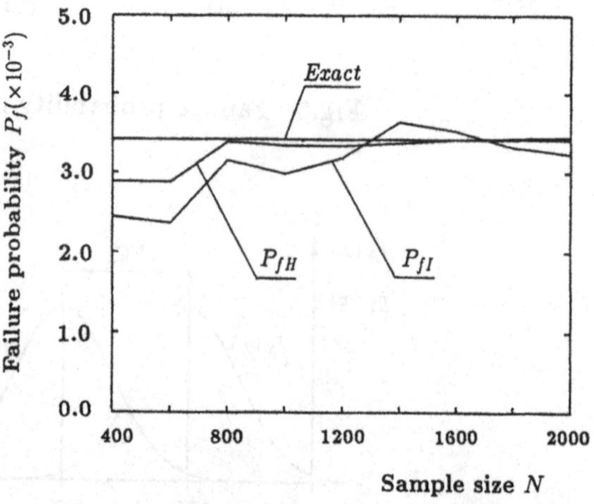

Fig.10 Failure probability of plane frame structure

Exceedance Probabilities Under Various Combinations of Rare Events

H. Katukura (*), M. Mizutani (**), S. Ogawa (*), T. Takada (*)

(*) Ohsaki Research Institute, Shimizu Corporation, 2-2-2 Uchisaiwaicho, Chiyoda-ku, Tokyo 100, Japan

(**) Tokyo Electric Power Services Co., Ltd., 1-3-1 Uchisaiwaicho, Chiyoda-ku, Tokyo 100, Japan

ABSTRACT

Several load combination methods are studied for simple rare events models by introducing multiple safety domains. The relationship between conventional load combination methods is summarized in terms of exceedance probability and occurrence number. Furthermore, approximations of exceedance probability including the upper and the lower bounds corresponding to special occurrence properties of elementary events such as 'always on' and nearly 'always off' are discussed. Several remarks on numerical issues concerning the calculation of the convolution integrals appearing in the load combination methods are presented as well.

INTRODUCTION

One of the most significant concerns for structures is the failure due to loading; the safety of structures. As hazardous loads to structures such as earthquake loads, wind loads, accident loads, etc. cannot fully anticipate its occurrence and strength, we apply a probabilistic approach to estimate the safety of structures; reliability analysis. Generally, structures are subjected to multiple elementary load events. Therefore, load combination methods play an important role in the reliability analysis of the structures. As to the load combination problems, Turkstra, Larrabee and Cornell, and Wen have developed the methods which are called Turkstra's rule, Point Crossing method, and Load Coincident method, respectively. Turkstra 1970 has discussed the basic rule of load combination which combines the maximum value of an event with the arbitrary-point-in-time(APIT) value of another event. This concept has been utilized in many parts of recent design codes, and based on Turkstra's rule with some modifications, we can evaluate probability characteristics of combined events. On the other hand, Larrabee and Cornell 1979 in the Point Crossing method, and Y.K.Wen 1977 in the Load Coincidence method have presented load combination using upcrossing rate functions. In this paper, we call the modified Turkstra's rule as MTR method , the Point Crossing method as PC-method and the Load Coincidence method as LC-method.

In general, a lifetime exceedance probability is prone to be adopted as a reliability index of a system. However, for the safety assessments of huge and complicated systems, the lifetime exceedance probability may not be the only one important index and other kinds of reliability indexes must be taken into consideration in order to estimate the overall reliability. Particularly, for decision making problems, various indexes should be examined in order to reply to all kind of human requirements. Therefore, it is very significant for reliability researches to realize the characteristics of the reliability indexes which will be obtained from various load combination methods. In this paper, various load combination methods of rare events are studied for simple event models by introducing multiple safety domains. Then, relationships among the various combination methods are explained in detail. Furthermore, the similarity of LC, PC, and MTR methods is pointed out. The upper and lower bounds of load combination effects are discussed as well by introducing special occurrence properties of elementary events such as always on and nearly always off.

LOAD COMBINATION METHODS

In this paper, in order to simplify the load combination problems, two independent events combination problem is selected as an object. The arbitrary- point-in-time(APIT) density functions of these events are assumed to have the form written as follows.

$$f_{si}(r)=p_i\delta(r)+q_if_{ri}(r)\,,\ f_{sj}(r)=p_j\delta(r)+q_jf_{rj}(r) \tag{1.1}$$

$$q_i=\lambda_i\mu_i\,,\ q_j=\lambda_j\mu_j \tag{1.2}$$

$$p_i=1-q_i\,,\ p_j=1-q_j \tag{1.3}$$

in which, q_i is the probability that the event i is on, λ_i is the average occurrence rate, μ_i is the average duration, and $f_{ri}(r)$ is the probability density function of the strength of event i given occurrence, p_i is the probability that the event i is off and $\delta(r)$ is Dirac's delta function.

Throughout this paper, the variables r_i, s_i, and r_{mi} are used for the conditional magnitude of event i given occurrence, APIT(unconditional) magnitude of event i, and lifetime maximum of event i, respectively. In addition, probability of failure is evaluated by means of the defining functions, U_{MTR}, U_{APIT}, U_{RIJ}, etc., which denote different safety domains.

Turkstra's Rule

Turkstra 1970 developed a load combination rule in which the lifetime maximum value of one event is combined with the arbitrary-point-in-time(APIT) value of the other events. This rule can be completely expressed if the density functions of the maximum and APIT values of each event are given. For a threshold r, the safety domain of a combination of two events, i and j, is expressed as follows.

$$U_{MTR}=U(r-(r_{mi}+s_j))U(r-(r_{mj}+s_i)) \tag{2}$$

where $U(r)$ denotes a step function; $U(r)=1$ for $0<r$, $U(r)=0$ for $r<0$, and $U(r)=0.5$ for $r=0$. The variables r_{mi} and r_{mj} indicate the maximum values during the period under consideration and s_i and s_j indicate the APIT values of events i and j. If, for the threshold level r, both $r_{mi}+s_j<r$ and $r_{mj}+s_i<r$ are satisfied, then $U_{MTR}=1$ holds. Therefore, $U_{MTR}=1$ indicates safety. Likewise, $U_{MTR}=0$ means failure.

Based on these characteristics, it may be possible to denote that U_{MTR} plays a role of the defining function representing the safety domain of the combined events.

From Eq.(2), we can determine the probability of failure P_{MTR} as follows.

$$1-P_{MTR}=\mathrm{E}[U_{MTR}] \tag{3}$$

where E[ ] denotes the expectation operation, taken with respect to the random variables. Introducing probability density functions, $f_{rmi}(r)$, etc., Eq.(3) is converted to the following equation.

$$1-P_{MTR}=(F_{rmi}(r)*f_{sj}(r))(F_{rmj}(r)*f_{si}(r)) \tag{4}$$

where the symbol $*$ denotes the convolution integral and $F_{rmi}(r)$ and $F_{rmj}(r)$ are probability distribution functions of the maximum values of events i and j, and $f_{si}(r)$ and $f_{sj}(r)$ are the density functions of APIT values of events i and j. Using the exceedance probability functions $G_{rmi}(r)$ and $G_{rmj}(r)$, Eq.(4) is also expressed as

$$1-P_{MTR}=(1-G_{rmi}(r)*f_{sj}(r))(1-G_{rmj}(r)*f_{si}(r)) \tag{5}$$

Eqs.(4) and (5) are the representation of the probability of failure based on Turkstra's rule. In the above derivation, the events $r_{mi}+s_j<r$ and $r_{mj}+s_i<r$ are assumed to be independent.

Upper and lower limit values of MTR

The upper and lower limit values of the MTR can be easily obtained by introducing the concept of safety domains. These are,

$$U_{MTRU}=U(r-(r_{mi}+r_j))U(r-(r_{mj}+r_i)) \tag{6.1}$$

$$U_{MTRL}=U(r-r_{mi})U(r-r_{mj}) \tag{6.2}$$

$$(\ U_{MTRU}\leqq U_{MTR}\leqq U_{MTRL}\ )$$

and the corresponding probabilities of failure are written as follows.

$$1-P_{MTRU}=(1-G_{rmi}(r)*f_{rj}(r))(1-G_{rmj}(r)*f_{ri}(r)) \tag{7.1}$$

$$1-P_{MTRL}=(1-G_{rmi}(r))(1-G_{rmj}(r)) \tag{7.2}$$

$$(\ P_{MTRL}\leqq P_{MTR}\leqq P_{MTRU}\ )$$

If $p_i=p_j=0$ is assumed, then Eq.(7.1) is derived from Eq.(5). Likewise, if $q_i=0$ or $q_j=0$ is assumed, then Eq.(7.2) is obtained from Eq.(5). The situation $p_i=p_j=0$ means that both events i and j always exist. On the contrary, $q_i=0$ or $q_j=0$ corresponds to the situation that these two events will never occur simultaneously and there is no need to consider the load combination effects with the other event.

Combination of APIT values

The combination effect of APIT values of events i and j can be obtained from the below safety domain.

$$U_{APIT}=U(r-(s_i+s_j)) \tag{8}$$

The corresponding probability of failure is written as

$$1-P_{APIT}=1-G_{si}(r)*f_{sj}(r)=1-G_{sj}(r)*f_{si}(r). \tag{9}$$

P_{APIT} is the unconditional probability, since the occurrence of events need not be assumed.

Conditional probability

Similarly the values for the conditional probability of the threshold level r being exceeded given the coincidence can be also determined . The following equations are the results.

$$U_{RIJ}=U(r-(r_i+r_j)) \tag{10}$$

$$1-P_{RIJ}=1-G_{ri}(r)*f_{rj}(r)=1-G_{rj}(r)*f_{ri}(r) \tag{11}$$

It will be shown that the conditional probability P_{RIJ} is included in the representation of many important probabilities such as P_{MTR}, P_{APIT}.

Other probabilities

We can introduce several other safety domains in the same manner. For example, similar to U_{APIT} and U_{RIJ},

$$U_{MIJ}=U(r-(r_{mi}+r_{mj})) \tag{12.1}$$

and similar to U_{MTRL},

$$U_{RL}=U(r-r_i)U(r-r_j) \tag{12.2}$$

$$U_{APITL}=U(r-s_i)U(r-s_j) \tag{12.3}$$

and to U_{MTR},

$$U_{MTRR}=U(r-(r_i+s_j))U(r-(r_j+s_i)) \tag{12.4}$$

are introduced. The corresponding probabilities are expressed as follows.

$$1-P_{MIJ}=1-G_{rmi}(r)*f_{rmj}(r) \tag{13.1}$$

$$1-P_{RL}=(1-G_{ri}(r))(1-G_{rj}(r)) \tag{13.2}$$

$$1-P_{APITL}=(1-G_{si}(r))(1-G_{sj}(r)) \tag{13.2}$$

$$1-P_{MTRR}=(1-G_{ri}(r)*f_{sj}(r))(1-G_{rj}(r)*f_{si}(r)) \tag{13.4}$$

These exceedance probabilities may also be important reliability indexes. For example, P_{MIJ} means the conditional exceedance probability when the lifetime maxima of events i and j occur simultaneously.

LC and PC methods

The concepts of LC and PC methods are essentially the same to each other. In these methods, load combinations in the occurrence number domain are considered. The upcrossing rate function $v(r)$ of the combined events is first determined in the occurrence number domain as follows.

$$v(r)=v_{ri}(r)*f_{sj}(r)+v_{rj}(r)*f_{si}(r) \tag{14}$$

where $v_{ri}(r)$ and $v_{rj}(r)$ are the upcrossing rate functions of events i and j and $f_{si}(r)$ and $f_{sj}(r)$ are the density functions of APIT values of events i and j. When $v(r)$ is given the probability of failure P_f can be evaluated as follows.

$$1-P_f=\exp[-v(r)T] \tag{15}$$

where T is the period of time under consideration.

For the LC method, Wen represented the expanded form instead of Eq.(13) (see Wen 1990). However, as Winterstein and Cornell 1984 pointed out, the expanded form can be derived from Eq.(14). Therefore, it may be said that there is no fundamental difference between PC and LC methods.

RELATIONS AMONG EVENT COMBINATION METHODS

The load combination methods explained in the above are summarized in Fig.1. In the figure, new operations are utilized: For example, $\{s_i+s_j\}$ means the event $r<(s_i+s_j)$ and $U[x_i+x_j]$ means the step function $U(r-(x_i+x_j))$. Then, $P\{x_i\}$ denotes the exceedance probability $P(r<x_i)$ and $N\{x_i\}$ is the corresponding occurrence number such that $r<x_i$. $\{s_i+s_j\}^c$ is the complement of $\{s_i+s_j\}$.

As indicated in Fig.1, the exceedance probabilities can be transformed into the occurrence number domain. The occurrence number of the union of two disjoint exceedance events are denoted by the sum of individual occurrence numbers. This may indicate that the occurrence number may be a measure of fields of sets composed of exceedance events. In Fig.1, load combination effects are discussed based on a set of the instances [APIT, a point in time when an elementary event takes place, a point in time time when the lifetime maximum event takes place], that is, $[s_i,r_i,r_{mi}]$ or $[s_j,r_j,r_{mj}]$. When we estimate the reliability of a system, we frequently imagine all of these situations and are interested in the corresponding reliability. Accordingly, it is possible to note that the information depicted in Fig.1 is important for the reliability assessments of systems. For almost every structural designs, however, the only combinations $r_{mi}+r_j$ and $r_{mj}+r_i$ are prone to be taken into account as a proper index of reliability as the conventional load combination methods such as MTR, PC, and LC methods.

Comparison of MTR, LC, and PC methods

For small exceedance probabilities, that is, for rare events, the following relations can be derived.

$$-\ln[1-P\{x_i\}]=N\{x_i\}\approx P\{x_i\} \tag{16.1}$$

$$-\ln[(1-P\{x_i\})(1-P\{x_j\})]=N\{x_i\}+N\{x_j\}\approx P\{x_i\}+P\{x_j\} \tag{16.2}$$

Eq.(16) indicates that the occurrence numbers are approximately equal to the exceedance probabilities. This deformation is done in order to compare the relations among load combination methods, especially to compare MTR method with LC and PC methods.

Using Eqs.(16.1) and (16.2), we can obtain the following relations.

$$N_{MTR}(r)\approx G_{rmi}(r)*f_{sj}(r)+G_{rmj}(r)*f_{si}(r) \tag{17.1}$$

$$N_{MTRU}(r)\approx G_{rmi}(r)*f_{rj}(r)+G_{rmj}(r)*f_{ri}(r) \tag{17.2}$$

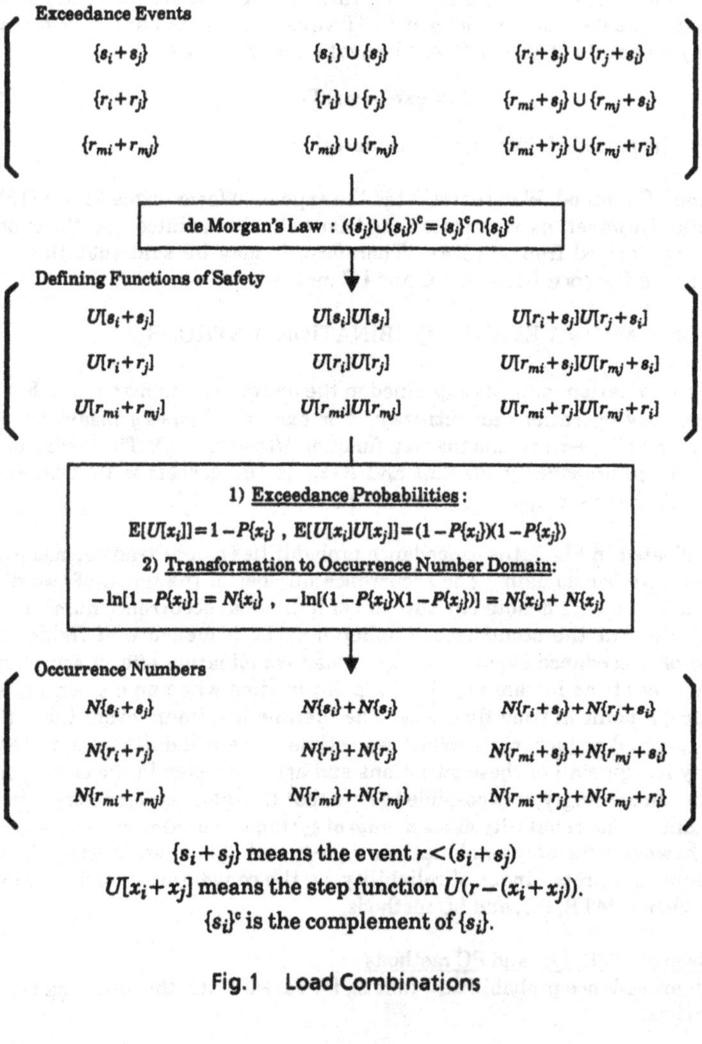

$\{s_i+s_j\}$ means the event $r<(s_i+s_j)$

$U[x_i+x_j]$ means the step function $U(r-(x_i+x_j))$.

$\{s_i\}^c$ is the complement of $\{s_i\}$.

Fig.1 Load Combinations

$$N_{MTRL}(r) \approx G_{rmi}(r) + G_{rmj}(r) \tag{17.3}$$

These are further deformed by using Eq.(1) and the approximate relation

$$G_{rmi}(r) \approx \lambda_i T\, G_{ri}(r) \tag{18}$$

The final results are

$$N_{MTR}(r) \approx p_j\lambda_i T G_{ri}(r) + q_j\lambda_i T P_{RIJ}(r) + p_i\lambda_j T G_{rj}(r) + q_i\lambda_j T P_{RIJ}(r) \tag{19.1}$$

$$N_{MTRU}(r) \approx \lambda_i T P_{RIJ}(r) + \lambda_j T P_{RIJ}(r) \tag{19.2}$$

$$N_{MTRL}(r) \approx \lambda_i T G_{ri}(r) + \lambda_j T G_{rj}(r) \tag{19.3}$$

where $P_{RIJ}(r)$ is the conditional probability determined by Eq.(11). From these equations, various information can be obtained. For example, by considering Eq.(14) and substituting the following relation to Eq.(14),

$$\nu_{ri}(r)T = \lambda_i T\, G_{ri}(r) \tag{20}$$

we notice the similarity of Eq.(19.1) and Eq.(14): From Eqs.(14), (20) and (1),

$$\nu(r)T = p_j\lambda_i TG_{ri}(r) + q_j\lambda_i TP_{RIJ}(r) + p_i\lambda_j TG_{rj}(r) + q_i\lambda_j TP_{RIJ}(r) \tag{21}$$

can be obtained. Namely, it is possible to denote that LC and PC methods are quite similar to MTR method, as is pointed out by Madsen et al. 1986.

From Eq.(7), the following significant relation is obtained as well.

$$\lambda_i TG_{ri}(r) + \lambda_j TG_{rj}(r) \leqq N_{MTR}(r) \leqq \lambda_i TP_{RIJ}(r) + \lambda_j TP_{RIJ}(r) \tag{22}$$

Eq.(22) presents that the occurrence number for MTR method is larger than the sum of occurrence numbers of each event and less than the occurrence number which is determined from the conditional probability that threshold level r is exceeded given the coincidence of events i and j.

Comparison of $N_{APIT}(r)$ and $N_{MTR}(r)$

Similarly, we can obtain the following occurrence number $N_{APIT}(r)$.

$$N_{APIT}(r) \approx p_i p_j U(-r) + p_j q_i G_{ri}(r) + p_i q_j G_{rj}(r) + q_i q_j P_{RIJ}(r) \tag{23}$$

For $r>0$, this expression of $N_{APIT}(r)$ is similar to the $N_{MTR}(r)$'s. To show the similarity, we deform Eq.(23) as follows.

$$N_{APIT}(r) \approx w_i p_j\lambda_i TG_{ri}(r) + w_{ij}q_j\lambda_i TP_{RIJ}(r) + w_j p_i\lambda_j TG_{rj}(r) + w_{ij}q_i\lambda_j TP_{RIJ}(r) \tag{24}$$

$$w_i = \mu_i/T\,,\; w_j = \mu_j/T\,,\; w_{ij} = w_i w_j/(w_i + w_j) \tag{25}$$

Eq.(24) indicates that $N_{APIT}(r)$ can be estimated from $N_{MTR}(r)$, when weighting constants w_i, w_j and w_{ij} are given. These weighting constants are generally very small since the load event durations μ_i and μ_j are far smaller than T which is the period of lifetime under consideration.

Comparison of occurrence numbers

By the same procedure, we can obtain individual occurrence number for each load combination in Fig. 1. In Table 1, we summarize the calculated results. From this table, we may able to grasp the load combination effect systematically. For example, from $N_{APITL}(r)$, $N_{RL}(r)$, and $N_{MTRL}(r)$, it is shown that if the magnitude of the occurrence number derived from r_i be 1, then the magnitude of the occurrence number from s_i becomes the order of q_i. Similarly, the magnitude of occurrence number from r_{mi} becomes the order of $\lambda_i T$. Furthermore, from $N_{MTRR}(r)$, $N_{MTR}(r)$, and $N_{MTRU}(r)$, it follows that $G_{ri}(r)$ is modified to $p_jG_{ri}(r) + q_jP_{RIJ}(r)$ by combining s_j and that $G_{ri}(r)$ is modified to $P_{RIJ}(r)$ by combining r_j. These may be basic knowledge when we discuss the load combination effects.

In Table 2, the occurrence numbers are summarized for the particular event cases for which $q_i \approx q_j \approx 1$ or $q_i \approx q_j \approx 0$ holds.

Table 1 Comparison of Occurrence Numbers

Name	Defining Function	Occurrence Number
$N_{APIT}(r)$	$U[s_i+s_j]$	$p_ip_jU(-r)+p_jq_iG_{ri}(r)+p_iq_jG_{rj}(r)+q_iq_jP_{RIJ}(r)$
$N_{RIJ}(r)$	$U[r_i+r_j]$	$P_{RIJ}(r)$
$N_{MIJ}(r)$	$U[r_{mi}+r_{mj}]$	$\lambda_iT\lambda_jTP_{RIJ}(r)$
$N_{APITL}(r)$	$U[s_i]U[s_j]$	$q_iG_{ri}(r) + q_jG_{rj}(r)$
$N_{RL}(r)$	$U[r_i]U[r_j]$	$G_{ri}(r) + G_{rj}(r)$
$N_{MTRL}(r)$	$U[r_{mi}]U[r_{mj}]$	$\lambda_iTG_{ri}(r) + \lambda_jTG_{rj}(r)$
$N_{MTRR}(r)$	$U[r_i+s_j]U[r_j+s_i]$	$p_jG_{ri}(r)+q_jP_{RIJ}(r) + p_iG_{rj}(r)+q_iP_{RIJ}(r)$
$N_{MTR}(r)$	$U[r_{mi}+s_j]U[r_{mj}+s_i]$	$p_j\lambda_iTG_{ri}(r)+q_j\lambda_iTP_{RIJ}(r) + p_i\lambda_jTG_{rj}(r)+q_i\lambda_jTP_{RIJ}(r)$
$N_{MTRU}(r)$	$U[r_{mi}+r_j]U[r_{mj}+r_i]$	$\lambda_iTP_{RIJ}(r) + \lambda_jTP_{RIJ}(r)$

Table 2 Comparison of Occurrence Numbers for Particular Events

Name	Defining Function	Occurrence Number	
		$p_i \sim p_j \sim 0$ and $q_i \sim q_j \sim 1$	$p_i \sim p_j \sim 1$ and $q_i \sim q_j \sim 0$
$N_{APIT}(r)$	$U[s_i+s_j]$	$P_{RIJ}(r)$	$U(-r)$
$N_{RIJ}(r)$	$U[r_i+r_j]$	$P_{RIJ}(r)$	$P_{RIJ}(r)$
$N_{MIJ}(r)$	$U[r_{mi}+r_{mj}]$	$\lambda_iT\lambda_jTP_{RIJ}(r)$	$\lambda_iT\lambda_jTP_{RIJ}(r)$
$N_{APITL}(r)$	$U[s_i]U[s_j]$	$G_{ri}(r) + G_{rj}(r)$	0
$N_{RL}(r)$	$U[r_i]U[r_j]$	$G_{ri}(r) + G_{rj}(r)$	$G_{ri}(r) + G_{rj}(r)$
$N_{MTRL}(r)$	$U[r_{mi}]U[r_{mj}]$	$\lambda_iTG_{ri}(r) + \lambda_jTG_{rj}(r)$	$\lambda_iTG_{ri}(r) + \lambda_jTG_{rj}(r)$
$N_{MTRR}(r)$	$U[r_i+s_j]U[r_j+s_i]$	$P_{RIJ}(r) + P_{RIJ}(r)$	$G_{ri}(r) + G_{rj}(r)$
$N_{MTR}(r)$	$U[r_{mi}+s_j]U[r_{mj}+s_i]$	$\lambda_iTP_{RIJ}(r) + \lambda_jTP_{RIJ}(r)$	$\lambda_iTG_{ri}(r) + \lambda_jTG_{rj}(r)$
$N_{MTRU}(r)$	$U[r_{mi}+r_j]U[r_{mj}+r_i]$	$\lambda_iTP_{RIJ}(r) + \lambda_jTP_{RIJ}(r)$	$\lambda_iTP_{RIJ}(r) + \lambda_jTP_{RIJ}(r)$

The relation $q_i \approx q_j \approx 1$ indicates that the events i and j are always on. In this case, occurrence numbers are essentially determined from the function $P_{RIJ}(r)$ which is the conditional probability of the threshold level r being exceeded given the coincidence. Then, the occurrence number $N_{MTR}(r)$ equals the upper limit value $N_{MTRU}(r)$. Similarly, the relation $q_i \approx q_j \approx 0$ indicates that the events i and j are almost always off. In this case, the occurrence number $N_{MTR}(r)$ equals the lower limit value $N_{MTRL}(r)$.

CONVOLUTION INTEGRALS IN LOAD COMBINATION METHODS

From the above discussion, it is evident that load combination methods include the following convolution integrals.

$$P_{RIJ}(r)=G_{ri}(r)*f_{rj}(r) \tag{26}$$

It becomes our next target to discuss how to evaluate this convolution integral numerically.

Fourier analysis of convolution integral

Before discussing the numerical problems, we will briefly summarize the mathematical relations concerning convolution integrals. The Fourier transform of Eq.(26) becomes

$$P_{RIJ}(\omega)=G_{ri}(\omega)f_{rj}(\omega) \tag{27}$$

where $G_{ri}(\omega)$ and $f_{rj}(\omega)$ are Fourier transforms of $G_{ri}(r)$ and $f_{rj}(r)$ and represented as follows.

$$G_{ri}(\omega)=\int_{-\infty}^{\infty}G_{ri}(r)e^{-i\omega r}\,dr \tag{28.1}$$

$$f_{rj}(\omega)=\int_{-\infty}^{\infty}f_{rj}(r)e^{-i\omega r}\,dr \tag{28.2}$$

Considering the characteristics of $G_{ri}(r)$, the Fourier transform $G_{ri}(\omega)$ will include the singularity nature. The singularity will often yield the data dependent characteristics of the FFT analyses. The singularity can be modified if the Fourier transform of the derivatives of Eq.(26) is introduced.

$$\int_{-\infty}^{\infty}d/dr\,P_{RIJ}(r)\,e^{-i\omega r}\,dr=i\omega\,G_{ri}(\omega)f_{rj}(\omega) \tag{29}$$

Since

$$d/dr\,G_{ri}(r)=d/dr\,(1-F_{ri}(r))=-f_{ri}(r) \tag{30}$$

we can obtain the following equation.

$$i\omega\,G_{ri}(\omega)f_{rj}(\omega)=-f_{ri}(\omega)f_{rj}(\omega) \tag{31}$$

In this case, the functions $f_{ri}(r)$ and $f_{rj}(r)$ tend to 0 approximately when r goes to $\pm\infty$. Therefore, we will be able to expect more correct results than those obtained from Eq.(27).

Numerical techniques to calculate convolution

It is frequently pointed out that the FFT analysis is effective to calculate the convolution integrals. However, FFT algorithm includes the summation procedures related to the Fourier integral. By the procedures, truncated errors are taken place often. This becomes a significant problem if extremely small probabilities are required to calculate.

The most simple way to solve this issue is to apply directly the numerical convolution technique to Eq.(26). In this case, the truncate errors similar to the ones associated with FFT analyses will not affect the results. Therefore, very small values of probability of failure may be correctly evaluated if direct integration method is used.

Load combination effects of rare events

The stochastic characteristics of rare events are sometimes modeled by the extreme value distribution functions. For example, the type Ⅱ extreme value distribution function is often adopted as a macro scale model of earthquakes. Furthermore, the occurrence rates of rare events subjected to structures are generally very small: By these reasons, there must be some different relations taken place among load combination methods. As the most prominent relation, the following equation can be derived for a sufficiently large threshold level r.

$$P_{RIJ}(r) \approx G_{ri}(r) + G_{rj}(r) \tag{32}$$

This approximate relation can easily derived by comparing the safety domains U_{RIJ} and U_{RL} and by extreme value characteristics of events.

Table 3 Comparison of Occurrence Numbers ($P_{RIJ}(r) \approx G_{ri}(r)+G_{rj}(r)$ is assumed)

Name	Defining Function	Occurrence Number ($P_{RIJ}(r) \approx G_{ri}(r)+G_{rj}(r)$)
$N_{APIT}(r)$	$U[s_i+s_j]$	$p_ip_jU(-r)+q_iG_{ri}(r)+q_jG_{rj}(r)$
$N_{RIJ}(r)$	$U[r_i+r_j]$	$G_{ri}(r)+G_{rj}(r)$
$N_{MIJ}(r)$	$U[r_{mi}+r_{mj}]$	$\lambda_jT\lambda_iTG_{ri}(r)+\lambda_iT\lambda_jTG_{rj}(r)$
$N_{APITL}(r)$	$U[s_i]U[s_j]$	$q_iG_{ri}(r) + q_jG_{rj}(r)$
$N_{RL}(r)$	$U[r_i]U[r_j]$	$G_{ri}(r) + G_{rj}(r)$
$N_{MTRL}(r)$	$U[r_{mi}]U[r_{mj}]$	$\lambda_iTG_{ri}(r) + \lambda_jTG_{rj}(r)$
$N_{MTRR}(r)$	$U[r_i+s_j]U[r_j+s_i]$	$(G_{ri}(r)+q_jG_{rj}(r)) + (G_{rj}(r)+q_iG_{ri}(r))$
$N_{MTR}(r)$	$U[r_{mi}+s_j]U[r_{mj}+s_i]$	$\lambda_iT(G_{ri}(r)+q_jG_{rj}(r)) + \lambda_jT(G_{rj}(r)+q_iG_{ri}(r))$
$N_{MTRU}(r)$	$U[r_{mi}+r_j]U[r_{mj}+r_i]$	$\lambda_iT(G_{ri}(r)+G_{rj}(r)) + \lambda_jT(G_{rj}(r)+G_{ri}(r))$

Here, we derive Eq.(32) by utilizing the Turkstra's rule.

$$U[r_i+r_j] \approx U[r_i+\bar{r}_j]U[r_j+\bar{r}_i] \tag{33}$$

Therefore, we can obtain

$$P_{RIJ}(r) \approx G_{ri}(r+\bar{r}_j)+G_{rj}(r+\bar{r}_i). \tag{34}$$

Considering large r, that is average values $\bar{r}_j$ and $\bar{r}_i$ are much smaller than r, and the extreme value characteristics of events,

$$G_{ri}(r+\bar{r}_j) \approx G_{ri}(r) \tag{35.1}$$

$$G_{rj}(r+\bar{r}_i) \approx G_{rj}(r) \tag{35.2}$$

can be assumed. Accordingly, Eq.(32) can be derived. Of course, Eq.(34) is more correct approximation than Eq.(32). It should be noted that the above approach using Turkstra's rule may become one of the interesting methods to evaluate the convolutional integral $P_{RIJ}(r)$.

If Eq.(32) is given, load combination effects can be summarized as shown in Table 3. For example, the occurrence number $N_{MTR}(r)$ is demonstrated as follows.

$$N_{MTR}(r) \approx \lambda_i T(G_{ri}(r) + q_j G_{rj}(r)) + \lambda_j T(G_{rj}(r) + q_i G_{ri}(r)) \tag{36}$$

The meaning of load combination is sufficiently explained even in Eq.(36), since the load combination effects is to modify $G_{ri}(r)$ to $G_{ri}(r) + q_j G_{rj}(r)$ and $G_{rj}(r)$ to $G_{rj}(r) + q_i G_{ri}(r)$ in the definition of $N_{MTRL}(r)$. The relations indicated in Table 3 will be effectively utilized in order to grasp the characteristic of load combination effects which is rather rough but sufficiently correct for so long as r is large.

Numerical example

In Fig.2, occurrence numbers are depicted for the case of $q_i = q_j = 0.1$. Though this is not rare case, the approximate occurrence number given by Eq.(36) is sufficiently explain the strict one. Fig.2 indicates that, for the combination of rare events, it is not necessary to consider not only the convolution integral but also the combination effects. That leads to the result that $N_{MTRL}(r)$ will give good estimation for occurrence number.

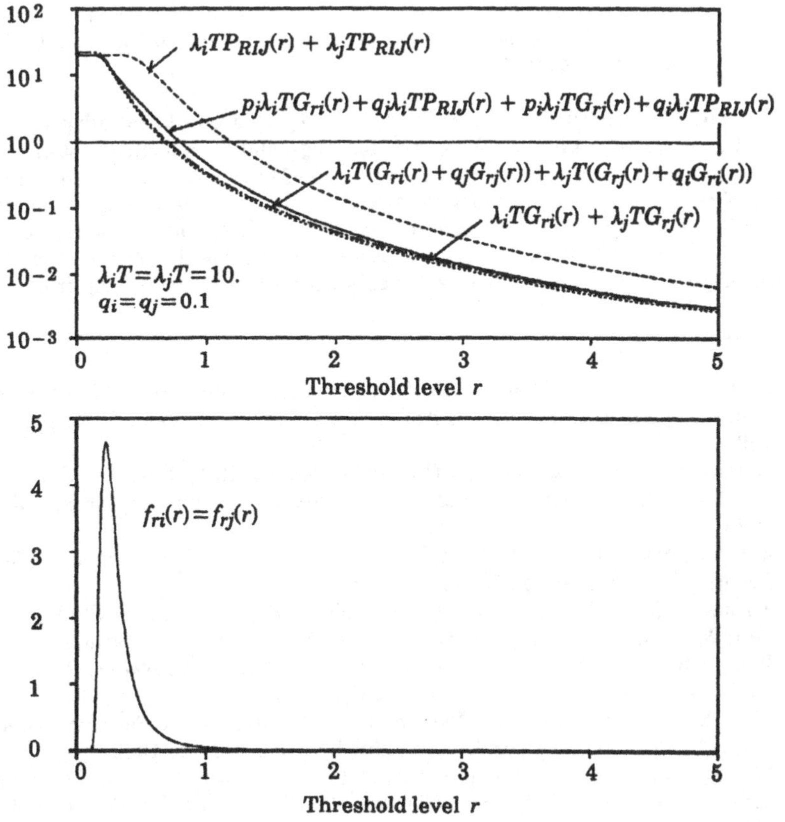

Fig. 2 Comparison of Occurrence Number $N_{MTR}(r)$

CONCLUSIONS

In this paper, various load combination methods are studied for simple rare event models introducing multiple safety domains. Especially, relationships among the event combination methods are summarized. Conclusions of this study are summarized as follows.

1) Various exceedance probabilities including conventional ones are explained in order to indicate a measure of the reliability of systems. These should cover not only structural design problems but also the almost every requirements of human being on reliability of systems.
2) A definite similarity of LC, PC, and MTR methods is shown in the occurrence number domain. Furthermore, the upper and lower bounds of load combination effects by MTR method can be explained by means of introducing special occurrence properties of elementary events such as always on and nearly always off.
3) The numerical techniques to calculate the convolution integrals are discussed. Particularly, the problem of using FFT techniques has been pointed out. In addition, for rare event cases, the approximation method to evaluate the convolution integrals is presented.
4) Several indexes considered significant in the reliability assessment corresponding multiple safety domains are summarized in Tables 1 to 3 by using the occurrence numbers.

In general, a lifetime exceedance probability is prone to be adopted as a reliability index of a system. However, for huge and complicated systems, the lifetime exceedance probability may be only one of the important indexes and all kinds of reliability indexes must be examined in order to evaluate the real reliability. Particularly, for decision making problems, various indexes should be given in order to reply to all kind of human requirement. The information obtained in this paper will be effectively applied to such decision making problems.

REFERENCES

1. Larrabee, R.D., and Cornell, C.A.," Combination of Various Load processes," *Journal of Structural Division*, ASCE, Vol.107, No.ST1, Jan., 1981, pp. 223-239.
2. Larrabee, R.D., and Cornell, C.A.,"Upcrossing Rate Solution for Load Combinations," *Journal of Structural Division*, ASCE, Vol.105, No.ST1, Jan., 1979, pp. 125-132.
3. Madsen, H.O., Krenk, S., and Lind, N.C., Method of Structural Safety, Prentice-Hall, Englewood Cliffs, New Jersey, 1986
4. Turkstra, C.J.," Theory of Structural Safety," Solid Mechanics Study, No.2, Solid Mechanics Division, University of Waterloo,Waterloo, Ontario, 1970.
5. Wen, Y.K., " Statistical Combination of Extreme Loads," *Journal of Structural Division*, ASCE, Vol.103, May 1977, pp. 1079-1093.
6. Wen, Y.K., Structural Load Modeling and Combination for Performance and Safety Evaluation, Elsevier Science Publishers, Amsterdam,1990.
7. Winterstein, S.R., and Cornell, C.A.," Load Combination and Clustering Effects," *Journal of Structural Division*, ASCE, Vol.110, No.11, November, 1984, pp. 2690-2707.

Skewness and Kurtosis of Safety Margin

S. Terada

Department of Architecture, Tokyo Metropolitan University 1-1, Minami-ohsawa, Hachiohji, Tokyo 192-03, Japan

INTRODUCTION

Reliability of a system will be evaluated through the safety margins those are expressed as the functions of basic random variables. Approximate failure probability for a specific mode is generally evaluated by the first two moments those are estimated by considering the distributions of constituent basic variables and correlation coefficients between them. They all are settled into a reliability index β.

This derives a method to calculate the high order moments of a safety margin that will be used to precisely evaluate the reliability. The method is a development of the idea that the variables may be decomposed to one common variables and the mutually independent ones [1]. Detailed examination on the idea yields the equations regarding to these variables at high order moments. Also, the correlation coefficients between two safety margins will precisely be estimated by these equations.

BASIC FORMULAS

Let a basic random variable $X_i (i = 1, 2)$ be an arbitrarily distributed standard one. Decompose X_i to an independent variable $\hat{D}_i$ and a perfectly correlated variable $\hat{C}_i$:

$$X_i = \hat{C}_i + \hat{D}_i \tag{1}$$

Rewrite the variables $\hat{C}_i$ and $\hat{D}_i$ by the standardized variables C and D_i :

$$X_i = (m_{2C_i})^{0.5} C + (m_{2D_i})^{0.5} D_i \tag{2}$$

in which, m_{2C_i} and m_{2D_i} are the unknown variances (second central moments) of $\hat{C}_i$ and $\hat{D}_i$, respectively. This means that we assume the correlation coefficients(ρ) as follows.

$$\rho(C, D_i) = \rho(D_1, D_2) = 0 \tag{3}$$

Examinations on the second central moment(m_2) and the joint second moments yield the relations on m_2's and ρ's in the following.

$$E(X_i^2) = m_{2X_i} = m_{2C_i} + m_{2D_i} = 1 \tag{4}$$

$$E(X_1X_2) = (m_{2C_1}m_{2C_2})^{0.5} = \rho(X_1, X_2) = \rho \tag{5}$$

$$(m_{2C_i})^{0.5} = \rho(X_i, C) \tag{6}$$

$$(m_{2D_i})^{0.5} = \rho(X_i, D_i) \tag{7}$$

where the symbol $E(\cdot)$ is the expectation.

Similarly, the third central moment(m_3) of X_i is expressed by,

$$E(X_i^3) = m_{3x_i} = (m_{2C_i})^{1.5}m_{3C} + (m_{2D_i})^{1.5}m_{3D_i} \tag{8}$$

in which, m_{3X_i}, m_{3C} and m_{3D_i} are third moments of X_i, C and D_i, respectively. With Equation 4,

$$m_{2C_i}[(m_{2C_i})^{0.5}m_{3C} - m_{3X_i}] + m_{2D_i}[(m_{2D_i})^{0.5}m_{3D_i} - m_{3X_i}] = 0 \tag{9}$$

Optimum solution to Equation 9 is obtained through Lagrange method as follows.

$$(m_{2C_i})^{0.5}m_{3C} = (m_{2D_i})^{0.5}m_{3D_i} = m_{3X_i} \tag{10}$$

From which, the following relations will be induced.

$$\left(\frac{m_{2C_2}}{m_{2C_1}}\right)^{0.5} = \frac{m_{3X_2}}{m_{3X_1}} = \alpha_3 \tag{11}$$

$$\rho m_{3C}^2 = m_{3X_1}m_{3X_2} = \alpha_3 m_{3X_1}^2 \tag{12}$$

and,

$$m_{2C_1} = \rho\frac{m_{3X_1}}{m_{3X_2}} = \rho\alpha_3^{-1} \tag{13}$$

If $m_{3X_1} = m_{3X_2}$, then

$$m_{2C_1} = m_{2C_2} = \rho(X_1, X_2) = \rho \tag{14}$$

and,

$$m_{2D_1} = m_{2D_2} = 1 - \rho \tag{15}$$

Similar calculations up to eighth moments yield the following general equations.

$$L_{kX} = m_{2C}^{0.5k}L_{kC} + m_{2D}^{0.5k}L_{kD} \tag{16}$$

where the notation L_k is the k-th cumulant and $\alpha_k = L_{kX_2}/L_{kX_1}$.

$$L_{kX} = m_{2C}^{0.5(k-2)}L_{kC} = m_{2D}^{0.5(k-2)}L_{kD} \tag{17}$$

$$\left(\frac{m_{2C_2}}{m_{2C_1}}\right)^{0.5(k-2)} = \frac{L_{kX_2}}{L_{kX_1}} = \alpha_k \tag{18}$$

Table 1. m_k

k	m_k
3	L_3
4	$L_4 + 3$
5	$L_5 + 10L_3$
6	$L_6 + 5(3m_4 + 2m_3^2 - 6) = L_6 + 5(3L_4 + 2L_3^2 + 3)$
7	$L_7 + 21(m_5 - 10m_3) - 35m_4m_3$
	$= L_7 + 7[3L_5 + 5L_3(L_4 + 3)]$
8	$L_8 + 7[4m_6 + 8m_3(m_5 - 10m_3) + 5m_4(m_4 - 12) + 90]$
	$= L_8 + 7[4L_6 + 8L_3(L_5 + 5L_3) + 5L_4(L_4 + 6) + 15]$

$$\rho^{k-2} L_{kC}^2 = L_{kX_1} L_{kX_2} = \alpha_k L_{kX_1}^2 \tag{19}$$

$$m_{2C_1}^{k-2} = \rho^{k-2} \frac{L_{kX_1}}{L_{kX_2}} = \rho^{k-2} \alpha_k^{-1} \tag{20}$$

in which, the contents of m's are given in Table 1.

If the variables are normally distributed, then the values of L_{kX}'s are 0. Moreover, observe that the method does not hold if ($\alpha_k < 0$ and k is even number).

JOINT MOMENTS

To facilitate the estimates on the high-order moments of a safety margin, the several joint moments of standardized variables X_1 and X_2 are calculated by using the preceding equations and they are summarized in Table 2.

SKEWNESS AND KURTOSIS

The Skewness(θ) and Kurtosis(γ) of a safety margin which is consisted of mutually correlated random variables, can be estimated through the equations in Table 1 and 2. Some simple examples will follow.

(Example 1) Consider a simple linear function

$$Z_1 = a_1 \hat{X}_1 \pm a_2 \hat{X}_2 \tag{21}$$

in which, a_i is a constant, $\hat{X}_i(\mu_i, \sigma_i)$ is an arbitrarily distributed random variable and the correlation coefficient between $\hat{X}_1$ and $\hat{X}_2$ is ρ. The variance of Z_1 is,

$$m_{2Z_1} = (a_1\sigma_1)^2 + (a_2\sigma_2)^2 \pm 2\rho a_1 a_2 \sigma_1 \sigma_2 \tag{22}$$

The third and fourth central moments are,

$$m_{3Z_1} = (a_1\sigma_1)^2(a_1\sigma_1 \pm 3\rho a_2\sigma_2) m_{3X_1} \pm (a_2\sigma_2)^2(a_2\sigma_2 \pm 3\rho a_1\sigma_1) m_{3X_2} \tag{23}$$

Table 2. $E(X_1^m X_2^n)$

Z	$E(Z)$
$X_1^{k-1}X_2$	ρm_{k1}
$X_1^2X_2^2$	$\rho[2\rho + \alpha_4^{\frac{1}{2}} L_{41}] + 1$
$X_1^3X_2^2$	$\rho[\alpha_5^{\frac{1}{3}} L_{51} + 3(\alpha_3 + 2\rho)L_{31}] + m_{31}$
$X_1^4X_2^2$	$\rho[\alpha_6^{\frac{1}{4}} L_{61} + 2(4\rho + 3\alpha_4^{\frac{1}{2}})L_{41} + 2(2\alpha_3 + 3\rho)L_{31}^2 + 12\rho] + m_{41}$
$X_1^3X_2^3$	$\rho[\alpha_6^{\frac{1}{2}} L_{61} + 3(1 + \alpha_4 + 3\rho\alpha_4^{\frac{1}{2}})L_{41} + 9\rho\alpha_3 L_{31}^2 + 3(3 + 2\rho^2)] + m_{31}m_{32}$
$X_1^5X_2^2$	$\rho[\alpha_7^{\frac{1}{5}} L_{71} + 10(\rho + \alpha_5^{\frac{1}{3}})L_{51} + 5(4\rho + 3\alpha_4^{\frac{1}{2}})L_{41}L_{31} + 5(3\alpha_3 + 16\rho)L_{31}] + m_{51}$
$X_1^4X_2^3$	$\rho[\alpha_7^{\frac{2}{5}} L_{71} + 3(1 + 2\alpha_5^{\frac{2}{3}} + 4\rho\alpha_5^{\frac{1}{3}})L_{51} + 2(2\alpha_4 + 15\rho\alpha_3)L_{41}L_{31} + 6\{5 + 6\rho(\alpha_3 + \rho)\}L_{31}] + m_{41}m_{32}$
$X_1^6X_2^2$	$\rho\begin{bmatrix}\alpha_8^{\frac{1}{6}} L_{81} + 2(15\rho + 13\alpha_5^{\frac{1}{3}})L_{51}L_{31} + 5(4\rho + 3\alpha_4^{\frac{1}{2}})L_{41}^2 + 3(4\rho + 5\alpha_6^{\frac{1}{4}})L_{61} \\ +30(7\rho + 2\alpha_3)L_{31}^2 + 15(10\rho + 3\alpha_4^{\frac{1}{2}})L_{41} + 90\rho\end{bmatrix} + m_{61}$
$X_1^5X_2^3$	$\rho\begin{bmatrix}\alpha_8^{\frac{1}{3}} L_{81} + (3 + 15\rho\alpha_6^{\frac{1}{4}} + 10\alpha_6^{\frac{1}{2}})L_{61} + 5\{11\rho\alpha_3 + 2(\alpha_5^{\frac{2}{3}} - \rho\alpha_5^{\frac{1}{3}})\}L_{51}L_{31} + 5(\alpha_4 + 6\rho\alpha_4^{\frac{1}{2}})L_{41}^2 \\ +15(3 + \alpha_4 + 4\rho^2 + 6\rho\alpha_4^{\frac{1}{2}})L_{41} + 30(1 + 3\rho^2 + 5\rho\alpha_3)L_{31}^2 + 15(3 + 4\rho^2)\end{bmatrix} + m_{51}m_{32}$
$X_1^4X_2^4$	$\rho\begin{bmatrix}\alpha_8^{\frac{1}{2}} L_{81} + 2\alpha_6^{\frac{1}{4}}\{3(1 + \alpha_6^{\frac{1}{2}}) + 8\rho\alpha_6^{\frac{1}{4}}\}L_{61} + 4\{\alpha_5 + \alpha_3 - \rho\alpha_5^{\frac{1}{3}}(\alpha_3 + \alpha_5^{\frac{1}{3}}) + 14\rho\alpha_3^{\frac{1}{2}}\alpha_5^{\frac{1}{3}}\}L_{51}L_{31} \\ +34\rho\alpha_4 L_{41}^2 + 3\{2(6 + 11\rho^2)\alpha_4^{\frac{1}{2}} + \rho(16 + \rho)(1 + \alpha_4)\}L_{41} \\ +4\{4(4 + 9\rho^2)\alpha_3 + 9\rho(1 + \alpha_3^2)\}L_{31}^2 + 24\rho(3 + \rho^2)\end{bmatrix} + m_{41}m_{42}$

Note ; When the correlation coefficient ρ is negative, the formulae should be transformed as follows.
(1) use absolute value $|\rho|$ in the formulae, if m or n is even number.
(2) $E(X_1^m X_2^n) = -E[(-X_1)^m X_2^n]$, if m and n is odd number, simultaneously.

and,

$$m_{4Z_1} = (a_1\sigma_1)^3(a_1\sigma_1 \pm 4\rho a_2\sigma_2)m_{4X_1} + (a_2\sigma_2)^3(a_2\sigma_2 \pm 4\rho a_1\sigma_1)m_{4X_2} + 6(a_1a_2\sigma_1\sigma_2)^2[1 + 2\rho^2 + \rho\alpha_4^{0.5}L_{4X_1}] \tag{24}$$

, respectively.

Following relations on θ and γ will be found if $m_{3X_1} = m_{3X_2} = \theta_X$, $m_{4X_1} = m_{4X_2} = \gamma_X$ and $a_1\sigma_1 = a_2\sigma_2$.

$$\hat{\theta} = \frac{\theta_Z}{\theta_X} = \frac{(1 \pm 1)(1 \pm 3\rho)}{[2(1 \pm \rho)]^{1.5}} \tag{25}$$

$$\hat{L} = \frac{L_{4Z}}{L_{4X}} = \frac{\gamma_Z - 3}{\gamma_X - 3} = \frac{1 + 3\rho \pm 4\rho}{2(1 \pm \rho)^2} \tag{26}$$

If the safety margin is expressed by the sum of X_1 and X_2, then $\hat{\theta}$ increases with ρ and the distribution of Z_1 is less skewed than that of X(that is : $\theta_{Z_1} < \theta_X$). Also, $\theta_Z = \theta_X$ and $L_{4Z} = L_{4X}$ if $\rho = 1$. If the safety margin is expressed by the difference of X_1 and X_2, then $\hat{\theta} \equiv 0$ and $\hat{L}$ increases with ρ. If the variables X_1 and X_2 are normally distributed, then $\theta_Z = L_{4Z} = 0$, regardless of ρ.

(Example 2) Let a nonlinear function Z_2 be

$$Z_2 = X_1X_2 \tag{27}$$

in which, X_i's are the arbitrarily distributed standard variables. The average value of Z_2 is ρ and the second to fourth central moments are,

$$m_{2Z_2} = 1 + \rho^2 + \rho\alpha_4^{\frac{1}{2}}L_{41} \tag{28}$$

$$m_{3Z_2} = \rho[\alpha_6^{\frac{1}{2}}L_{61} + 3(1 + \alpha_4 + 2\rho\alpha_4^{\frac{1}{2}})L_{41} + 9\rho\alpha_3L_{31}^2 + 2(3 + \rho^2)] + \alpha_3L_{31}^2 \tag{29}$$

and,

$$m_{4Z_2} = \rho\begin{bmatrix} \alpha_8^{\frac{1}{2}}L_{81} + 6\alpha_6^{\frac{1}{4}}(1 + \alpha_6^{\frac{2}{4}} + 2\rho\alpha_6^{\frac{1}{4}})L_{61} \\ +4[\alpha_5 + \alpha_3 - \rho\alpha_5^{\frac{1}{3}}(\alpha_3 + \alpha_5^{\frac{1}{3}}) + 14\rho\alpha_3^{\frac{1}{2}}\alpha_5^{\frac{1}{3}}]L_{51}L_{31} \\ +34\rho\alpha_4L_{41}^2 + 3[\rho(12 + \rho)(1 + \alpha_4) + 12(1 + \rho^2)\alpha_4^{\frac{1}{2}}]L_{41} \\ +12[5\alpha_3 + 3\rho(1 + \alpha_3^{\frac{2}{3}} + 3\rho\alpha_3)]L_{31}^2 + 3\rho(14 + 3\rho^2) \end{bmatrix} + m_{41}m_{42} \tag{30}$$

, respectively. Apply the absolute value of ρ in the formulae (28) and (30) when the correlation coefficient is negative($\rho < 0$).

If the variables X_i's are distributed normally, then the skewness(θ) and kurtosis(γ) are expressed by,

$$\theta_{Z_2} = \frac{2\rho(3 + \rho^2)}{(1 + \rho^2)^{1.5}} \tag{31}$$

$$L_{Z_2} = \gamma_{Z_2} - 3 = \frac{6}{(1+\rho^2)^2}[(1+\rho)^4 - 4\rho(1+\rho^2)] \tag{32}$$

or,

$$\gamma_{Z_2} = \frac{3(3+14\rho^2+3\rho^4)}{(1+\rho^2)^2} \tag{33}$$

Both θ_Z and L_Z increase with ρ.

CORRELATION COEFFICIENT BETWEEN SAFETY MARGINS

In general, the correlation coefficient between two safety margins which are expressed as the nonlinear functions of mutually correlated variables, is approximately estimated through the linearization of the functions. The formulas will help to improve the estimate of the correlation coefficient.

(Example 3) Evaluate the probable error by linear approximation in the calculation of the correlation coefficient between Z_1 and Z_2,

$$Z_1 = \hat{X}_1\hat{X}_2 = \mu^2(1+\delta X_1)(1+\delta X_2) \tag{34}$$

$$Z_2 = \hat{X}_1\hat{X}_3 = \mu^2(1+\delta X_1)(1+\delta X_3) \tag{35}$$

in which, δ is the coefficient of variance. Assume that the variables $\hat{X}_i$'s have the common parameters in μ, σ, θ and γ, and $\rho(X_i, X_j) = \rho$. Then, the first order estimation on $\rho(Z_1, Z_2)$ is expressed by

$$\rho(Z_1, Z_2) \simeq \frac{(1+3\rho)}{2(1+\rho)} \tag{36}$$

which is dependent to ρ. The following standard deviation(σ_Z) and the covariance between Z_1 and Z_2 (Cov.),

$$\sigma_Z^2 = \mu^4\delta^2[2(1+\rho) + 4\rho m_3\delta + (1+\rho^2+\rho L_4)\delta^2] \tag{37}$$

$$\mathrm{Cov}(Z_1Z_2) = \mu^4\delta^2[1+3\rho+4\rho m_3\delta + \rho(1+\rho+L_4)\delta^2] \tag{38}$$

yield a precise expression on $\rho(Z_1, Z_2)$:

$$\rho(Z_1, Z_2) = \frac{\mathrm{Cov}(Z_1Z_2)}{\sigma_{Z_1}\cdot\sigma_{Z_2}} = \frac{1+3\rho+4\rho m_3\delta+\rho(1+\rho+L_4)\delta^2}{2(1+\rho)+4\rho m_3\delta+(1+\rho^2+\rho L_4)\delta^2} \tag{39}$$

The value of $\rho(Z_1, Z_2)$ depends not only ρ but also δ, m_3 and m_4. Figure 1 shows the value of $\rho(Z_1, Z_2)$ in relation to ρ, for the variables X_i with normal, uniform, exponential and Type 1 extreme-largest value distributions. Observe that the error by the linear approximation increases with the increase of δ, m_3 and m_4.

RELIABILITY EVALUATION

As an example, consider a safety margin expressed by a simple performance function [2] :

$$Z = X_1X_2 - X_3 \tag{40}$$

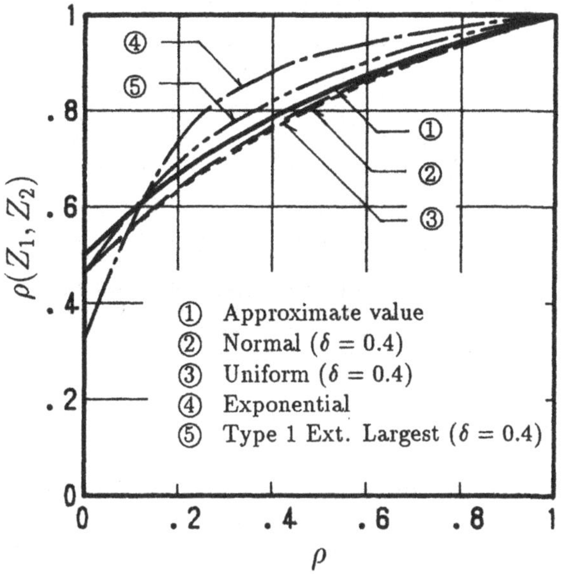

Figure 1. $\rho(Z_1, Z_2)$ in relation to ρ

Table 3 shows the assumed values of means and coefficients of variance. Four cases in Table 4 are chosen as the combination of the distribution of variable and the correlation coefficients between them. The first four moments are firstly calculated by the procedure as above. Then, the failure probabilities are calculated by a method that employs "Hermite polinomial expansion". The results are given in Table 5 in comparison to the estimates by normal tail approximation. The number in the table is the order of moment considered. It seems that the reasonable estimates are obtained by considering the first four moments as to the examples.

Table 3. μ and δ

X_i	μ	δ
X_1	4	0.125
X_2	5	0.05
X_3	10	0.2

Table 4. Assumed Condition

Case	Correlation Coefficient	Distribution
1	$\rho_{ij} = 0$	Normal
2	$\rho_{12} = 0.4$, $\rho_{13} = \rho_{23} = 0$	
3	$\rho_{ij} = 0$	X_1, X_2 ; Lognormal
4	$\rho_{12} = 0.4$, $\rho_{13} = \rho_{23} = 0$	X_3 ; Type 1 Ext. Largest

Table 5. Estimated Failure Probability

Case	$\theta \times 10^2$	γ	$P(F) \times 10^3$			
			2	3	4	(a)*
1	4.96	3.011	1.44	1.13	1.17	1.14
2	9.00	3.018	2.89	2.01	2.10	2.12
3	−8.15	3.393	1.44	1.65	3.12	3.02
4	0.40	3.304	2.98	2.96	4.27	3.94

* Values by equivalent normal distribution [2]

CONCLUDING REMARKS

A method is developed to calculate the skewness and kurtosis of a safety margin, in which the constituent random variables may mutually correlate. An idea, that the variables may be decomposed to a perfectly correlated common variable and the other independent variables, is employed to develop the formulas. The method is also applicable to the calculation of correlation coefficient between safety margins. As suggested in the paper, care should be taken on the limit of the mothod in application. Also, the method can be extended to the case of three or more basic rondom variables.

REFERENCES

1. Grigoriu, Mircea and Lind, Niels C. Probabilistic models for prototype testing. *J. Struc. Div.*, ASCE, 108(7), pp. 1511-1525, 1982.
2. Ang, A. H-S., and Tang, W. H. (Ed.). *Probability concepts in engineering planning and design.* Vol.2, John Wiley and Sons, Inc., New York, 1984.

A Curve-Fitting Method of Combined Distribution in Probabilistic Modeling of Random Variables

M. Kubo (*), H. Nakahara (*), Hide. Ishikawa (**), Hiro. Ishikawa (***)

(*) SOGO Engineering Inc., 3-5-9 Higashinakajima, Osaka 533, Japan

(**) Department of Information Systems, Kagawa College of Technology, 3202 Gunge-cho, Marugame-shi, Kagawa 763, Japan

(***) Department of Information Science, Kagawa University, 2-1 Saiwai-cho, Takamatsu-shi, Kagawa 760, Japan

ABSTRACT

The present paper proposes a practical curve-fitting method of combined distribution by using the statistics of measured data, which will be a useful technique in probabilistic modeling of random variables, especially needed, for example, in the analysis of load combination for structural design. Combined distributions are classified and defined into two types; a jointed type and a mixed type. The basic idea of fitting proposed in this study is to apply the least square method as well as the observed statistics of moments concurrently to the statistical inference of the distribution parameters. The methods of numerical calculation for both types of combined distributions are practically developed, in which the least square method with respect to two parameters is performed equivalently by numerical convergence in contour map of the function. Numerical examples are also given for both cases.

INTRODUCTION

Most of service loads subjected to practical structures would be of random nature with complicated temporal variation like wind load, seismic load and so forth. Hence, in order to assure the desired level of reliability for structural components under these random loads, it becomes of vital importance to clarify their statistical properties, which could be accomplished through proper probabilistic modeling of these random loads based upon their observations [1] ~[6]. To this end, it has been often the case to utilize simple probability density functions. However, a combined distribution is strongly recommended to describe a variability in, for instance, the annual maximum wind speed in a specific district [7] or the vehicle weight in the actual live load survey [8]. Unfortunately, however, neither the concrete curve-fitting method of combined distribution has been established, nor has the stable solution been obtained with the aid of the methods proposed to date [9].

In this respect, the present paper proposes a practical curve-fitting method of combined distribution by using the statistics of measured data, which will be a useful technique in probabilistic modeling of random variables, especially in the analysis of load combination for structural design. Combined distributions are classified and defined, according to the properties of the measured data, into two kinds of types; a jointed type and a mixed type. The basic idea of fitting proposed in this study is to apply the least square method as well as the empirical statistics of moments concurrently to the statistical inference of distribution parameters. The methods of numerical calculation for both types of combined distributions are practically developed, in which the least square method with respect to two parameters is performed equivalently by numerical convergence in contour map of the function. This method is believed to present a useful tool in various aspects of probabilistic approach in the field of structual engineering.

TYPES OF COMBINED DISTRIBUTION

When the probability density function $f_X(x)$ of the random variable X is composed of two different probability density functions $f_1(x)$ and $f_2(x)$, X is said to follow a combined distribution, where $f_1(x)$ and $f_2(x)$ are single probability density functions and are termed as original component functions for convenience' sake. Then, combined distributions are classified and defined into the following two kinds of types; a jointed type and a mixed type:

(1) Jointed type of combined distribution

$$f_X(x) = \left\{ \begin{array}{ll} q_1 f_1(x) & (-\infty < x < x_0) \\ q_2 f_2(x) & (x_0 \leq x < \infty) \end{array} \right\} \tag{1}$$

$$q_1 f_1(x_0) = q_2 f_2(x_0) \qquad (0 < q_1, q_2) \tag{2}$$

$$q_1 F_1(x_0) + q_2\{1 - F_2(x_0)\} = 1 \qquad (0 < q_1, q_2) \tag{3}$$

(2) Mixed type of combined distribution

$$f_X(x) = p_1 f_1(x) + p_2 f_2(x) \qquad (-\infty < x < \infty) \tag{4}$$

$$p_1 + p_2 = 1 \qquad (0 < p_1,\ p_2 < 1) \tag{5}$$

where $F_1(x)$ and $F_2(x)$ are the cumulative distribution functions corresponding to $f_1(x)$ and $f_2(x)$, respectively. q_1, q_2, p_1 and p_2 are constants to account for the degree of combination of the original component functions and are left unknown until the combined distribution is determined. In case of a jointed type of combined distribution, x_0 represents the boundary value to joint $f_1(x)$ to $f_2(x)$. It is basically desirable to obtain its value as the optimum solution with x_0 being an unknown variable. However, for the sake of simplicity, x_0 is given here as a prescribed constant. The distribution function of the combined distribution $F_X(x)$ is obtained from the following equation:

$$F_X(x) = \int_{-\infty}^{x} f_X(x)dx \qquad (-\infty < x < \infty) \tag{6}$$

The jointed type of combined distribution is exemplified in Fig.1 [10], where two kinds of probability distributions are partially jointed at a specific boundary of $x = x_0$. This type of combined distribution is well fitted to the case where measured data in the upper and lower region of the distribution are ruled by different factors as in, for instance, the annual maximum wind speed in a specific district. In general, the annual maximum speed is composed of those of typhoon and seasonal wind. The former lies in the upper region and the latter in the lower region. Although there is a mixture near the boundary, two kinds of probability distributions are jointed there in the probabilistic modeling of random variables.

On the other hand, a mixed type of combined distribution, as shown in Fig.2 [8], is composed of the mixture of two kinds of original component functions. This type of distribution is well fitted to the case where two kinds of observed data, which forms a bi-peak distribution, do not definitely show a large and small relationship. In this example, the upper region would correspond to loading and the lower to vacancy.

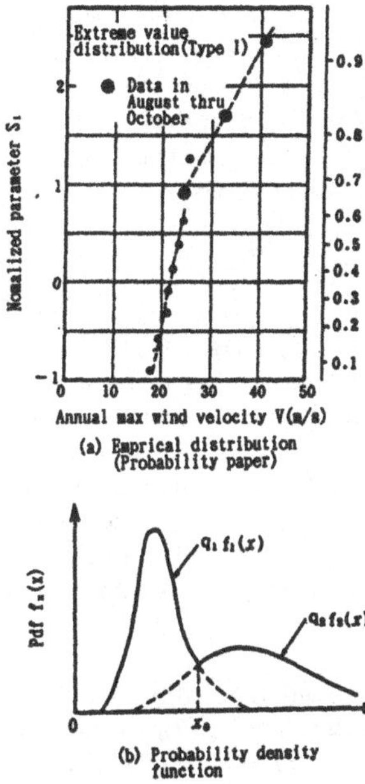

Fig.1 An example of a jointed type of combined distribution.

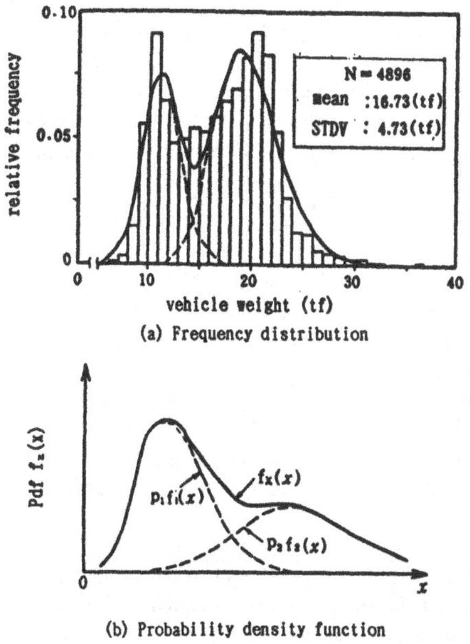

Fig.2 An example of a mixed type of combined distribution.

BASIC IDEA OF CURVE-FITTING METHOD

Usual method of curve-fitting

As for the statistical parameter estimatiom method, there are many ways such as the method by means of probability paper, the moment method, the maximum likelihood method and the least square method [11]. In general, any method can be sufficiently applicable to the practical case. Unfortunately, however, in the combined distribution, the concrete curve-fitting method has not yet been established to date. For more pertinent modeling, the statistical inference of parameters is desirable with the aid of analytical method.

Assumed that existing probability distributions are adopted as the original component functions, $f_1(x)$ and $f_2(x)$ usually contain two unknown parameters, and hence, six parameters are left unknown in curve-fitting of combined distribution of either type. The curve-fitting of combined distribution based upon measured data thus reduces to the determination of the aforementioned unknown parameters.

At this point, the evaluation method proposed by the authors [8],[9] is briefly summarized in what follows. The square error E in the estimation of the probability distribution function $F_X(x)$ based upon the measured data x_i $(i = 1, 2, \cdots, N)$ is given as

$$E = \frac{1}{N}\sum_{i=1}^{N}\{F_X(x_i) - \tilde{F}_X(x_i)\}^2 \tag{7}$$

where $\tilde{F}_X(x_i)$ represents the empirical probability distribution for the i-th order statistics x_i, which is given, by Gumbel [11], as

$$\tilde{F}_x(x_i) = i/(N+1) \qquad (i = 1, 2, \cdots, N) \tag{8}$$

As ststed above, considering a general case that $f_1(x)$ and $f_2(x)$ involve two parameters, A_1, B_1 and A_2, B_2, respectively, E in Eq.(7) becomes a function of these parameters as well as q_1, q_2 or p_1, p_2. Since the optimum parameters are determined by means of the least square method so as to minimize the estimated square error in Eq.(7), the unknown parameters in a mixed type of combined distribution can be obtained by solving the following simultaneous differential equations:

$$\frac{\partial E}{\partial p_1} = 0, \quad \frac{\partial E}{\partial A_1} = 0, \quad \frac{\partial E}{\partial B_1} = 0, \quad \frac{\partial E}{\partial A_2} = 0, \quad \frac{\partial E}{\partial B_2} = 0 \tag{9}$$

except p_2 which is calculated through Eq.(5). Generally speaking, Eq.(9) becomes very complicated non-linear simultaneous equations and hence, has to be solved by numerical convergence. Unfortunately, however, there has been no complete numerical computation method for non-linear simultaneous equations since the judgement of convergence is too difficult in most cases [12]. The desirable solutions can be hardly obtained by use of the Newton-Raphson method. Therefore, the statistical inference method of the distribution parameters is developed and proposed with the aid of the least square method as well as the statistics of moments observed from data.

Proposed method of curve-fitting

In the present study, instead of solving Eq.(9) directly, a statistical estimation method has been developed in the following fashion. Now, similarly to Eq.(9), assumed that unknown parameters are five such as p_1, A_1, B_1, A_2 and B_2. Let $\tilde{M}_k$ $(k = 1, 2, \cdots)$ be the k-th moment around the origin obtained from measured data, and $\tilde{M}_k$ can be easily calculated as follows:

$$\tilde{M}_k = \frac{1}{N}\sum_{i=1}^{N}(x_i)^k \qquad (k = 1, 2, \cdots) \tag{10}$$

On the other hand, assumed that M_k represents the k-th moment around the origin obtained from the probability density function $f_X(x)$ to be fitted, M_k is given as a function of p_1, A_1, B_1, A_2 and B_2, and hence expressed by M_k $(p_1, A_1, B_1, A_2, B_2)$ $(k = 1, 2, \cdots)$. Similarly, since the estimated square error E is also a function of these five unknown variables, this is expressed by E $(p_1, A_1, B_1, A_2, B_2)$.

At this point, let us establish two equations with respect to arbitrary two unknown variables by use of the least square method and three other equations by use of the moment method. Consequently, all unknown variables are determined to satisfy these five simultaneous equations. As a typical example, by applying the least square method on A_1 and A_2, the equations are given as follows:

$$\frac{\partial E(p_1, A_1, B_1, A_2, B_2)}{\partial A_1} = 0 \tag{11}$$

$$\frac{\partial E(p_1, A_1, B_1, A_2, B_2)}{\partial A_2} = 0 \tag{12}$$

$$\tilde{M}_1 = M_1(p_1, A_1, B_1, A_2, B_2) \tag{13}$$

$$\tilde{M}_2 = M_2(p_1, A_1, B_1, A_2, B_2) \tag{14}$$

$$\tilde{M}_3 = M_3(p_1, A_1, B_1, A_2, B_2) \tag{15}$$

where the moment relationships up to the third order are applied since in this case five unknown variables are assumed. In general, the moment relationships are utilized, in ascending order, up to the number of unknown variables less two. In case of a jointed type, Eqs.(2) and (3), Eqs.(11) and (12), and the necessary number of moment relationships are used.

The method to utilize Eqs.(11) ~ (15) needs to solve simultanious equations by numerical convergence with respect to only two or three unknown parameters. Hence, from the viewpoint of good convergence of solution, this method can provide a practical analysis. In odet to assure the convergence to the optimum solution, the solution to minimize the value of E by numerical convergence is obtained based upon sequential evalution of the value of E on the $A_1 - A_2$ plane, as shown in Fig.3, instead of directly solving Eqs.(11) and (12). In other words, this method has such an advantage as to obtain securely the optimum solution by sequential computation on the plane with respect to two unknown variables and as to establish comparatively simple equations associated with the remaining unknown variables.

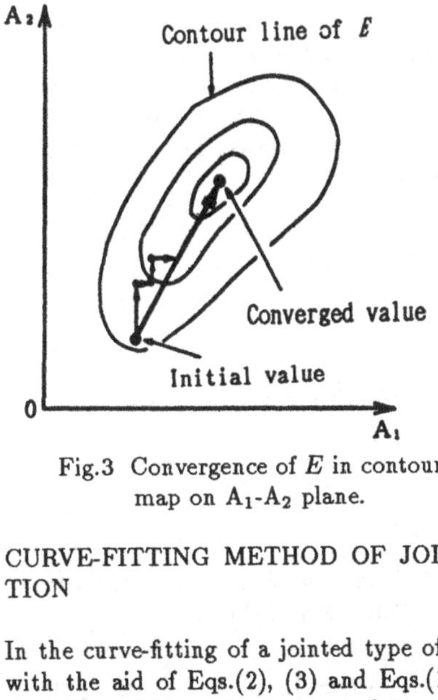

Fig.3 Convergence of E in contour map on A_1-A_2 plane.

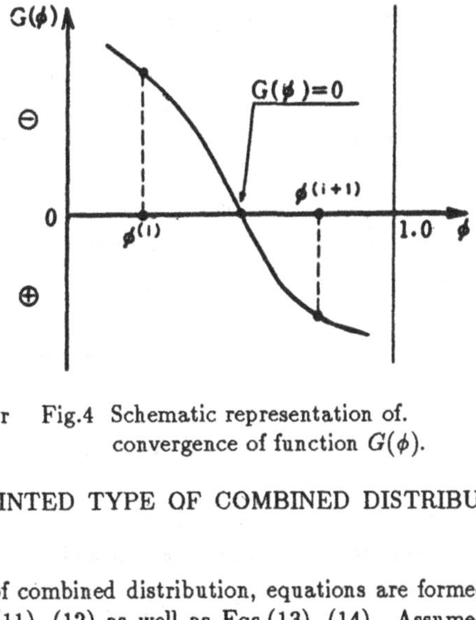

Fig.4 Schematic representation of. convergence of function $G(\phi)$.

CURVE-FITTING METHOD OF JOINTED TYPE OF COMBINED DISTRIBUTION

In the curve-fitting of a jointed type of combined distribution, equations are formed with the aid of Eqs.(2), (3) and Eqs.(11), (12) as well as Eqs.(13), (14). Assumed that the variable x_0 is given as a deterministic value determined, for example, through a probability paper, six parameters are left unknow such as q_1, q_2, A_1, B_1, A_2 and B_2. In applying the least square method, it is convenient to use such a combination of variables as A_1 and B_1 or A_2 and B_2, respectively. Hereinafter, the latter case is disussed, which result can be similarly applied to the former case. That is, the following equations are formed with recpect to A_2 and B_2:

$$\frac{\partial E}{\partial A_2} = 0, \quad \frac{\partial E}{\partial B_2} = 0 \tag{16}$$

Next, $\tilde{M}_k$'s $(k = 1, 2)$ are calculated by use of Eq.(10) based upon measured data. Moreover, by introducing M_k's $(k = 1, 2)$ in the righthand side of Eqs.(13) and (14) based upon the definition of Eq.(1), the following equations can be derived:

$$\tilde{M}_1 = q_1 \int_{-\infty}^{x_0} x f_1(x)dx + q_2 \int_{x_0}^{\infty} x f_2(x)dx \tag{17}$$

$$\tilde{M}_2 = q_1 \int_{-\infty}^{x_0} x^2 f_1(x)dx + q_2 \int_{x_0}^{\infty} x^2 f_2(x)dx \tag{18}$$

In order that Eqs.(17) and (18) are significant with respect to the unknown variables q_1 and q_2, the following relationship should hold:

$$\begin{vmatrix} \int_{-\infty}^{x_0} x f_1(x)dx & \int_{x_0}^{\infty} x f_2(x)dx \\ \int_{-\infty}^{x_0} x^2 f_1(x)dx & \int_{x_0}^{\infty} x^2 f_2(x)dx \end{vmatrix} \neq 0 \tag{19}$$

In the case that Eq.(19) does not hold even approximately, a careful attention must be paid since stable solutions of q_1 and q_2 are hard to obtain. In such a case, the curve-fitting of a jointed type of combined distribution by the definition of Eq.(1) is not suitable, and hence, the definition needs to be changed or the curve-fitting of a mixed type is more desirable.

The square estimation error E is evaluated on the A_2-B_2 plane with A_2 and B_2 being the parameters of the function $F_2(x)$. Eqs.(2), (3) and Eqs.(17), (18) are utilized to determine the remaining four unknown parameters after giving a set of suitable values to A_2 and B_2. In geneal, however, values of unknown parameters are difficult to derive in a closed form from these equations. Hence, they are solved by means of numerical convergence by introducing the following parameter:

$$\phi = q_1 F_1(x_0) \tag{20}$$

whose significant existing domain is defined from Eq.(3) as

$$0 \leq \phi \leq 1 \tag{21}$$

By putting $\phi^{(i)}$ as the i-th value in the iterative computation of ϕ, q_2 can be obtained through Eq.(3), which is termed as $q_2^{(i)}$.

$$q_2^{(i)} = \frac{1-\phi^{(i)}}{1-F_2(x_0)} \tag{22}$$

By substituting Eq.(22) into Eqs.(2), (17) and (18), the equations to give the i-th values of the remaining variables q_1, A_1 and B_1 are obtained as follows:

$$q_1^{(i)} f_1(x_0, A_1^{(i)}, B_1^{(i)}) = \frac{1-\phi^{(i)}}{1-F_2(x_0)} f_2(x_0) \tag{23}$$

$$q_1^{(i)} \int_{-\infty}^{x_0} x f_1(x_0, A_1^{(i)}, B_1^{(i)})dx = \tilde{M}_1 - \frac{1-\phi^{(i)}}{1-F_2(x_0)} \int_{x_0}^{\infty} x f_2(x)dx \tag{24}$$

$$q_1^{(i)} F_1(x_0, A_1^{(i)}, B_1^{(i)}) = \phi^{(i)} \tag{25}$$

where $A_1^{(i)}$ and $B_1^{(i)}$ are represented in the functions $f_1(x)$ and $F_1(x)$.

Next, by solving Eqs.(22) ~ (25), $q_1^{(i)}$, $q_2^{(i)}$, $A_1^{(i)}$ and $B_1^{(i)}$ are obtained. Then, these results as well as A_2 and B_2 are substituted into the following equation which is derived from Eq.(18):

$$G(\phi) = \tilde{M}_2 - q_1 \int_{-\infty}^{x_0} x^2 f_1(x)dx - q_2 \int_{x_0}^{\infty} x^2 f_2(x)dx \tag{26}$$

Consequently, the remaining four unknown variables can be obtained under certain prescribed values of A_2 and B_2 by getting $\phi^{(i)}$ which satisfies $G(\phi^{(i)}) = 0$ through numerical convergence based upon the change of sign in $G(\phi)$ as shown in Fig.4. The regula folsi method is considered practical in numerical convergence to $G(\phi) = 0$, althrough the simultaneous equations of Eqs.(23) ~ (25) need to be solved strictly according to the kinds of the original component functions. Trial calculations in this study show that the equations are comparatively easy to derive in case of ordinary distribution functions. Numerical convergence in $G(\phi)$ is comparatively good since ϕ is to be approximated to the ratio of the number of data smaller than x_0 to that of the total data, which could be understood from the definition of ϕ as in Eq.(20).

Further, with the aid of four variables thus obtained as well as A_2 and B_2, the square error in Eq.(7) can be estimated. By performing the above-stated process successively for various combinations of values of A_2 and B_2, the contour map of E can be numerically obtained on the $A_2 - B_2$ plane. The optimum solution of the unknown variables in the statistical inference can be obtained by determining the values of A_2 and B_2 so as to give the minimum value of E by numerical convergence and the corresponding values of q_1, q_2, A_1 and B_1. The result of numerical convergence on the $A_2 - B_2$ plane is equivalent to the satisfaction of the two equations in Eq.(16).

The convergence judgement of the solution is repeated until every variable satisfies each convergence condition. A typical example of the convergence condition on ϕ is shown as follows:

$$\left| \frac{\phi^{(i+1)} - \phi^{(i)}}{\phi^{(i+1)}} \right| \leq \epsilon \tag{27}$$

The iteration procedure is terminated when Eq.(27) is satisfied for a sufficiently small positive value of ϵ. Fig.5 represents the flow chart of the abovementioned procedure.

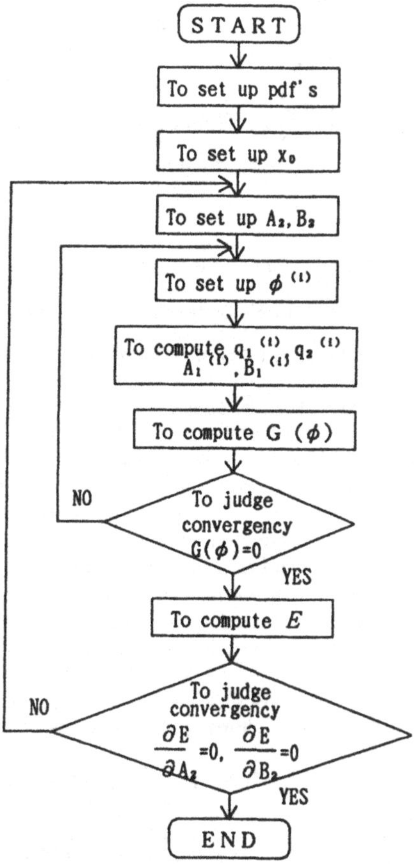

Fig.5 Iterative procedure for jointed type of combined distribution

An example of the curve-fitting of the jointed type of combined distribution is briefly illustrated in Fig.6. As the original component functions $f_1(x)$ and $f_2(x)$, two different extreme value distributions (Type I) were selected . With certain prescribed values of distribution parameters and x_0, the total number of $N = 400$ samples x_i's were first generated by Monte-Carlo simulation techniques, whose histogram is shown in Fig.6(a). The initial value of ϕ in Eq.(20) was chosen, based on these observations, as $\phi^{(i)} = 280/400 = 0.70$. According to the fitting procedure stated ealier, a jointed type of combined distribution was estimated and shown in Fig.6 by a solid line. A good agreement is observed, which implies the propriety of the present method.

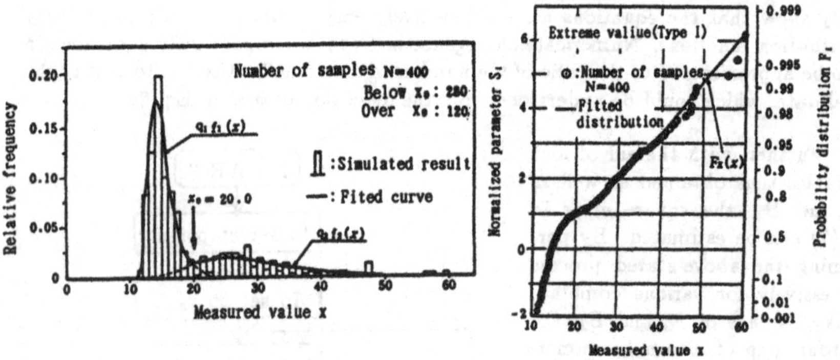

(a) Probability density function (b) Cumulative distribution function

Fig.6 An example of curve-fitting of jointed type of combined distribution

CURVE-FITTING METHOD OF MIXED TYPE OF COMBINED DISTRIBUTION

In case of different mean values

In the curve-fitting of a mixed type, Eq.(5), Eqs.(11) and (12), and Eqs.(13) ~ (15) are used. Now, suppose that the original component functions $f_1(x)$ and $f_2(x)$ have two unknown parameters, respectively, and that the unknown variables A_i and B_i ($i = 1, 2$) represent the expected value and the standard deviation of the respective distribution. The distribution parameters can be obtained with the aid of the moment method when the expected value and the standard deviation of the distribution are known. Therefore, in this case, the unknown parameters become six such as p_1, p_2, A_1, B_1, A_2 and B_2. It is convenient to minimize the square error estimate by use of A_1 and A_2 for the curve-fitting. The relationship becomes similar to Eqs.(11) and (12).

Next, with the aid of $\tilde{M}_k$'s ($k = 1, 2, 3$) in Eq.(10) and based upon M_k's derived by use of Eq.(4), the following equations can be established:

$$\begin{aligned}\tilde{M}_1 &= p_1 \int_{-\infty}^{\infty} x f_1(x)dx + p_2 \int_{-\infty}^{\infty} x f_2(x)dx \\ &= p_1 A_1 + p_2 A_2\end{aligned} \tag{28}$$

$$\begin{aligned}\tilde{M}_2 &= p_1 \int_{-\infty}^{\infty} x^2 f_1(x)dx + p_2 \int_{-\infty}^{\infty} x^2 f_2(x)dx \\ &= p_1(A_1^2 + B_1^2) + p_2(A_2^2 + B_2^2)\end{aligned} \tag{29}$$

$$\begin{aligned}\tilde{M}_3 &= p_1 \int_{-\infty}^{\infty} x^3 f_1(x)dx + p_2 \int_{-\infty}^{\infty} x^3 f_2(x)dx \\ &= p_1(\gamma_1 B_1^3 + 3A_1 B_1^2 + A_1^3) + p_2(\gamma_2 B_2^3 + 3A_2 B_2^2 + A_2^3)\end{aligned} \tag{30}$$

where γ_1 and γ_2 represent the skewness of the original component functions $f_1(x)$ and $f_2(x)$, respectively, and are given as a constant or a function of the distribution papameters. Table 1 represents the skewness of typical probability distributions [13].

Table 1. Skewness of typical probability distribution

Type of distribution	Skewness
Uniform	0
Normal	0
Log-Normal	$(B/A)^3+3(B/A)$
Exponential	2
Gamma	2(B/A)
Poisson	(B/A)
Notes	A:mean B:STDV

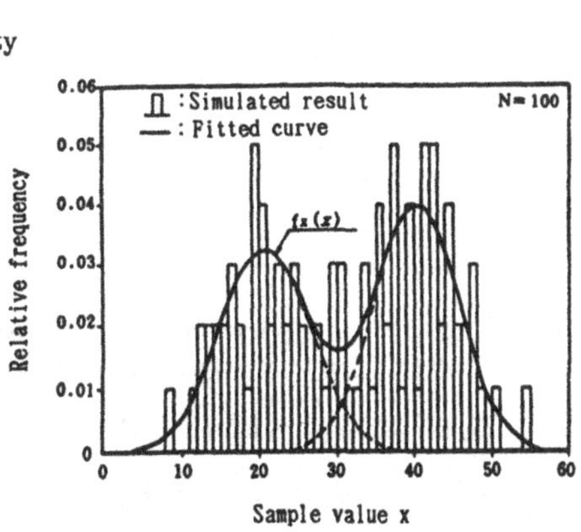

Fig.7 An example of curve-fitting of mixed type of combined distribution

According to the skewness of the original component distributions, the equations can be summarized in the following simpler form:

(a) In case of symmetric distribution ($\gamma_1 = \gamma_2 = 0$) Here, we discuss the simplest case where $\gamma_1 = \gamma_2 = 0$ under the condition that two expected values A_1 and A_2 are different. This corresponds to the case that the distribution is symmetric around the expected value. A typical example is the case to apply two normal distributions.

By substituting Eq.(5) into Eqs.(28) $\sim$ (30), p_1, B_1 and B_2 can be solved in terms of A_1 and A_2 where $A_1 < A_2$.

$$p_1 = \frac{A_2 - \tilde{M}_1}{A_2 - A_1} \tag{31}$$

$$B_1^2 = \frac{1}{3(\tilde{M}_1 - A_2)}\{\tilde{M}_3 - 3\tilde{M}_2 A_2 + \tilde{M}_1(2A_2^2 + 2A_1A_2 - A_1^2) - A_1A_2(2A_2 - A_1)\} \tag{32}$$

$$B_2^2 = \frac{1}{3(\tilde{M}_1 - A_1)}\{\tilde{M}_3 - 3\tilde{M}_2 A_1 + \tilde{M}_1(2A_1^2 + 2A_1A_2 - A_2^2) - A_1A_2(2A_1 - A_2)\} \tag{33}$$

Consequently, when A_1 and A_2 are arbitrarily given, the remaining four unknown parameters are computed with the aid of Eqs.(5) and (31) $\sim$ (33). As the result, the optimum values of the unknoown variables can be obtained by numerical convergence in contour map of E on the A_1-A_2 plane. However, under the condition that each unknown variable has a significant value, the significant domain of A_1 and A_2 is limited so as to satisfy the following relationships:

$$0 < p_1 < 1,\ A_1 < A_2,\ 0 \le B_1^2,\ 0 \le B_2^2 \tag{34}$$

An example of the curve-fitting in this case is illustrated in Fig.7. With the true parameter values of $p_1 = 0.40$, $p_2 = 0.60$, $A_1 = 20.0$, $B_1 = 5.0$, $A_2 = 40.0$ and $B_2 = 7.0$, one hundred samples ($N = 100$) were generated by means of Monte-Carlo Simulation techniques, to which a mixed type of combined distribution was fitted according to the proposed method. A good agreament is observed from Fig.7.

(b) In case of asymmetric distribution (γ_1 or $\gamma_2 \neq 0$) Here, we discuss that either γ_1 or γ_2 is not zero under the condition that the expected value is different each other. In this case, Eqs.(30) becomes more complicated compared with the above case (a), and

B_1 and B_2 in Eqs.(32) and (33) are more difficult to obtain in a closed form. As shown in Table 1, γ_1 and γ_2 in Eq.(30) become a function of unknown variables according to the kind of the original component function. Therefore, equations need to be solved by means of numerical computation, and hence, numerical convergence is effective by use of B_1 and B_2. For exsample, when B_1 is fixed to a certain value, B_2 is obtained from Eq.(29) as

$$B_2^2 = \frac{1}{1-p_1}\{\tilde{M}_2 - A_2^2 + p_1(A_1^2 + B_1^2 - A_2^2)\} \tag{35}$$

By substituting B_1 and B_2 into the following equation which is transformed from Eq.(30), it is solved so as to satisfy the relationship $h(B_1, B_2) = 0$.

$$h(B_1, B_2) = \tilde{M}_3 - p_1(\gamma_1 B_1^3 + 3A_1 B_1^2 + A_1^3) - p_2(\gamma_2 B_2^3 + 3A_2 B_2^2 + A_2^3) \tag{36}$$

where p_1 and p_2 are obtained through Eqs.(5) and (31). In general, since the function $h(B_1, B_2)$ becomes a comparatively monotonous curve, the equation $h(B_1, B_2) = 0$ can be easily solved, for instance, by the regula falsi method. If unknown variables are thus calculated, the subsequent procedure becomes similar to the case (a), although each unknown variable needs to satisfy the conditional equation of Eq.(34).

In case of the same mean values

In this case, it is convenient to minimize E on B_1 and B_2 under the condition that $A_1 = A_2 = A$. Equations corresponding to Eqs.(11) and (12) are provided as follows:

$$\frac{\partial E}{\partial B_1} = 0, \qquad \frac{\partial E}{\partial B_2} = 0 \tag{37}$$

The number of unknown variables are five such as p_1, p_2, A, B_1 and B_2, and the first and the second order moments are utilized. Therefore, Eqs.(5), (28), (29) and (37) are used. At first, $A = A_1 = A_2$ is obtained through Eq.(28), and this result as well as Eq.(5) are substituted into Eq.(29).

$$A = A_1 = A_2 = \tilde{M}_1 \tag{38}$$

$$\tilde{M}_2 = \tilde{M}_1^{\,2} + B_2^2 + p_1(B_1^2 - B_2^2) \tag{39}$$

(a) In case of different variances ($B_1^2 \neq B_2^2$) p_1 is obtained by use of Eq.(39).

$$p_1 = \frac{\tilde{M}_2 - \tilde{M}_1^2 - B_2^2}{B_1^2 - B_2^2} \qquad (B_1 \neq B_2^2) \tag{40}$$

Next, by drawing contour map of E on the $B_1 - B_2$ plane, the optimum solutions are obtained by numerical convergence.

(b) In case of the same variance ($B_1^2 = B_2^2$) This case makes sense as a mixed type of combined distribution only when different original component functions are utilized, though it is quite rare to apply this case to the probabilistic modeling in a practical sense. With the aid of Eqs.(28) and (29), the values of $A_1 = A_2$ and $B_1 = B_2$ are readily determined. In the case that $\gamma_1 \neq \gamma_2$, p_1 and p_2 are calculated with the aid of Eqs.(5) and (30). In the case that $\gamma_1 = \gamma_2$, the equation associated with the fourth order moment needs to be introduced. The peakedness of the distribution is concerned with the fourth order moment. However, in the probabilistic modeling utilized in the field of structural engineering, such a case is very rare and hence, it may be enough to refer to the above points. By the way, in this case, the curve-fitting can be performed by means of only the moment method without using the least square method.

NUMERICAL CONVERGENCE FOR OPTIMUM SOLUTIONS

The method of fitting proposed in the present study is to seek a set of unknown variables so as to produce the minimum square error estimate of E defined by Eq.(7). The numerical procedure for this purpose is termed as the "rolling method" whose details are briefly explained in what follows.

Let us suppose a curved surface, as an extention of Fig.3, by adding the third axis of E perpendicular to the $A_1 - A_2$ plane. This surface becomes concave with single hollow, the bottom point of which corresponds to the optimum solutions of A_1 and A_2. In case that the surfase has more than one hollow, it might be difficult to obtain the optimum solutions since plural sets of solutions will satisfy Eqs.(11) and (12). For example, when the number of measured data is insufficient and the fitting is unstable, this case might apply. In such a case, it is recommended to improve the measured data. Hence, in the following, only the case of the single minimum on E is considered

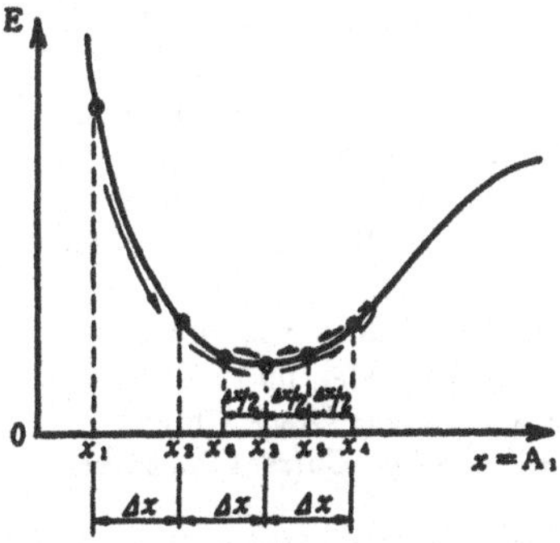

Fig.8 Principle of rolling method

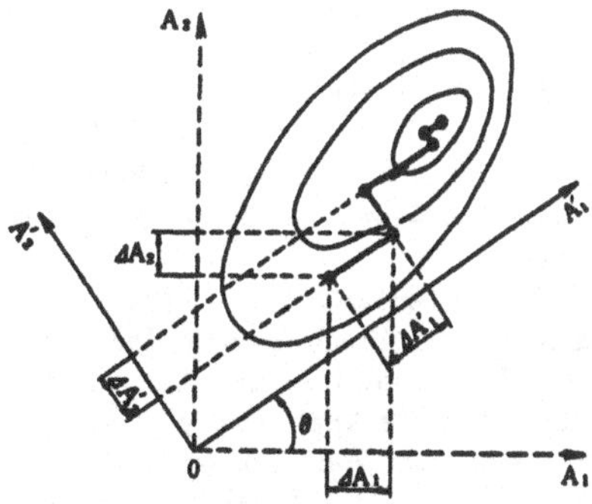

Fig.9 Rotation of axes and rolling distribution

The figure 8 shows the schmatic representation of the principle. Starting from the initial value x_1, the computation is repeated with an increment Δx toward the direction of decreasing E, and when the value of E becomes relatively large, the computation is repeated with half of the increment $\Delta x/2$ into the inverse direction until it reaches the minimum point for a certain fixed value of A_2. When the same procedure is repeated from that point with repect to A_2, the optimum solutions will be obtained finally. In order to assure efficiency of convergence, it is recomended to proceed this procedure according to the axes A_1' and A_2' of the surface near its principal axes by rotating the $A_1 - A_2$ plane with rotation angle θ, as shown in Fig.9. When an increment $\Delta A_1'$ is forwarded according to the axis A_1', the corresponding increments ΔA_1 and ΔA_2 on the $A_1 - A_2$ plane can be given as

$$\Delta A_1 = \Delta A_1' \cos\theta, \qquad \Delta A_2 = \Delta A_1' \sin\theta \tag{41}$$

Similarly, when the movement is made with an increment $\Delta A_2'$ according to A_2' axis,

$$\Delta A_1 = -\Delta A_2' \sin\theta, \qquad \Delta A_2 = \Delta A_2' \cos\theta \tag{42}$$

The convergence speed would be affected by the rotation angle θ adopted. Hence, in the practical fitting of the combined distribution, it is recommended first to draw an outline of the contour map on E, then to select suitable initial values in the vicinity of the bottom, and finally to choose a certain pertinent rotation angle.

CONCLUDING REMARKS

The present paper proposes practical methods of the curve-fitting of combined distributions to be utilized in the modeling of random variables in reliability-based structural design. The combined distribution is classified into two types; a jointed type and a mixed type. For each type, a pretinent numerical computation procedure is developed and numerical examples are provided to show the propriety of the proposed methods. Although discussions are limited to the case of two continuous component distributions, the procedure proposed in this study can be extended to other more complicated cases in a similar fashion.

REFERENCES

1. Takaoka, N. and Hoshiya, M., Application of Theories of Reliability and Probability to Problems in Civil Engineering, Journal of Japan Society of Civil Engineers, Vol.60, No.8, pp.61-69. 1975 (in Japanese).
2. Okamura, H. and Itagaki, H., Statistical Treatment of Strength, Baifukan, Tokyo, 1979 (in Japanese).
3. Fujino, Y., Itoh, M. and Endo, M., Evalution of Design Live Load for Road Bridges Based upon Simulation, Trans. JSCE, No.286, pp.1-13, 1979 (in Japanese).
4. Ishikawa, Hiro., A State-of-the-Art Survey of Reliability-Based Design of Structures, Journal of Japan Society of Steel Construction, Vol.18, No.192, pp.3-22, 1982 (in Japanese).
5. Tsurui, A. and Ishikawa, Hiro., Application of the Fokker-Planck Equation to a Stochastic Fatigue Crack Growth Model, Structural Safety, Vol.4, pp.15-22, 1986.
6. Ishikawa, Hiro., Tsurui, A. and Kimura, H., Stochastic Fatigue Crack Growth Model and Its Wide Applicability in Reliability-Based Design, Current Japanese Materials Research, Vol.2, pp.45-58, Elsevier, 1987.
7. Fujino, Y., Itoh, M. and Sakai, T., Study on Evaluation of Basic Design Wind Velocity Based on Annual Maximum Wind Velocity Record, Trans. JSCE, No.305, pp.23-34, 1981 (in Japanese).
8. Hanshin Expressway Public Corporation, HDL Committee Report, Part II, pp.88-105, 1984 (in Japanese).
9. Nakajima, M., Kubo, M. and Ishida, R., Application of Combined Distributions Based upon Observed Statistical Data, Preprint of Anaual Meeting of Kansai Branch, JSCE, No.I-97, 1985 (in Japanese).
10. Honshu-Shikoku Bridge Authority, Ten Years of Observation of Wind in Tarumi Observatory Tower, 1975 (in Japanese).
11. Kanda, T. and Fujita, N., Water Engineering: Probabilistic Method and Its Application, pp.18-23, Gihodo Press, Tokyo, 1982 (in Japanese).
12. Iri, M., Numerical Computation, pp.39-99, Asalura-Shoten, Tokyo, 1984 (in Japanese).
13. Ishikawa, H. and Kimura, H., Introduction to Statistics, pp.135-188, Kagawa Univ. Press, Takamatsu, 1980.

On Procedures to Calculate the Reliability of Structural Systems under Stochastic Loading

G.I. Schuëller, C.G. Bucher, H.J. Pradlwarter
Institute of Engineering Mechanics, University of Innsbruck, Innsbruck, Austria

ABSTRACT

Procedures to calculate the reliability of structural systems under both static and dynamic loading - by taking into account the statistical variations of loads and structural resistance respectively - are examined in view of their range of applicability as well as accuracy. Additional criteria are the computational efficiency as well as the degree of flexibility in mechanical modeling (e.g. by Finite Element procedures).

For structural systems under static loading the capabilities of failure mode analysis (e.g. branch and bound procedures), early developments in the response surface procedure as well a second order reliability analysis are briefly discussed.

In this paper it is shown that the response surface method (RSM) represents a unified approach which is applicable for both static and dynamic loading and which meets, most important, many requirements for practical application.

INTRODUCTION

The purpose of this paper is to examine methods to calculate the reliability of structural systems under both static and dynamic loading by taking into account the statistical variations of loads and structural resistance respectively. In this context the range of applicability as well as accuracy of the various methods are in the focus of interest. Additional criteria are the computational efficiency and the degree of flexibility in mechanical modeling (e.g. Finite Element procedures, etc.). Present approaches to analyze structural systems under stochastic loading either simplify the structural system or the procedure for stochastic analysis.

Current demands in structural analysis - particularly for practical application - require most realistic mechanical modeling irrespective of whether or not statistical uncertainties are taken into account. These requirements can certainly be met by applying simulation procedures. However, due to excessive computational efforts, classical Monte Carlo Simulation is somewhat limited in practical application, particularly for larger MDOF-systems. Hence so-called advanced simulation procedures utilizing various variance reduction techniques are applied quite successfully.

A procedure which may represent even complex systems with only few points - which in fact may be reused in combination with variance reduction simulation techniques - is the response surface method (RSM). Although the points on the response surface may be considered quite accurate as they reflect most sophisticated mechanical modeling, the entire response surface is the result of a fit procedure through these points and hence represents an approximation of the structural behavior, i.e. in terms of an interpolation. Procedures for quantifying the accuracy, however, are already available.

There is no disagreement among experts that stochastic methods are more likely to be accepted for application in practice if they reflect on one hand the state of the art of mechanical modeling and on the other hand if they are applicable to larger structural systems. The current paper is a contribution in this direction.

BRIEF DISCUSSION OF CURRENT METHODS

This brief review of current methods can certainly not be considered complete. However, it somehow reflects the state of the art of currently available procedures with respect to the capabilities within stochastic structural analysis.

In this context a finite element procedure was suggested [1] which utilizes second order reliability approximation (SORM). By this method larger systems can be handled, however, only with some limitations. Firstly, the method appears to be limited to those cases in which the gradients can be determined analytically. In other cases, i.e. where numerical differentiation is required the limit state function has to be sufficiently smooth. This is certainly true for systems with small nonlinearities. Consequently, this method covers seviceability rather than collapse problems.

The latter problem, however can be treated by the so-called branch and bound procedure as suggested e.g. in [2, 3, 4]. By applying this method the structure generally is modeled by idealized elasto-plastic system properties. Another major drawback of this procedure is the fact that one has to know beforehand where failure may occur; in other words, one has to identify the critical cross sections in advance. Moreover for larger structural models a very large number of failure mechanisms has to be analyzed which, of course, requires a considerable amount of computational effort.

Finally, some procedures are suggested which utilize the response surface method (RSM) in context with finite element procedures (see e.g. [5, 6, 7]). The major drawback of the methods at their early stage of development is the fact that the points on the limit state functions are not calculated directly, but may be obtained by extrapolation. This is a possible source of severe errors, particularly when dealing with highly nonlinear problems. Hence the results would require verification, which is generally not available.

It is shown in the following, however, that response surface procedures are indeed capable of representing realistically the limit state functions and hence reflecting the actual mechanical model as well as the failure condition as far as collapse both for static as well as dynamic loading conditions.

COSSAN - Computational Stochastic Structural Analysis

This section is intended to serve as a concept for the development of engineering software for practical application. Such a software requires combined efforts in both mechanical and stochastic modeling with respect to accuracy and efficiency. The following developments focus on one hand on the computational aspects of determining the failure probabilities of systems which are modeled by nonlinear Finite Elements. Certainly such computational procedures must be based on generally used discretization methods and should include various design critria such as serviceability, collapse, fatigue, etc. On the other hand, the computational efforts for the stochastic modeling must not exceed acceptable limits. Consequently, a COSSAN-software must consist of closely connected FE and probabilistic procedures.

It is well known that the design criterion of structures when taking the randomness of the systems parameters into account can be described by a so called limit state function $g(\underline{x})$ of the random variables $\underline{x}$, where the following definition applies:

$$g(\underline{x}) > 0 \quad \text{no failure}$$
$$g(\underline{x}) \leq 0 \quad \text{failure} \tag{1}$$

If time variance of the structural resistance, e.g. due to crack propagation, embrittlement, corrosion, etc., enters the problem the above equation writes:

$$g(\underline{x}, T) > 0 \quad \text{no failure}$$
$$g(\underline{x}, T) \leq 0 \quad \text{failure} \tag{2}$$

where T is the time range of observation, e.g. the design life of a particular structure.

In order to obtain information of the effect of the randomly distributed variables on the response of the system a sensitivity analysis may be performed. The procedure which is used here allows the simultaneous determination of the response surface. In other words the limit state function as defined by eq.(1) or (2) will be approximated by interpolation and hence by a function $\bar{g}(\underline{x})$ (cf Fig. 1).

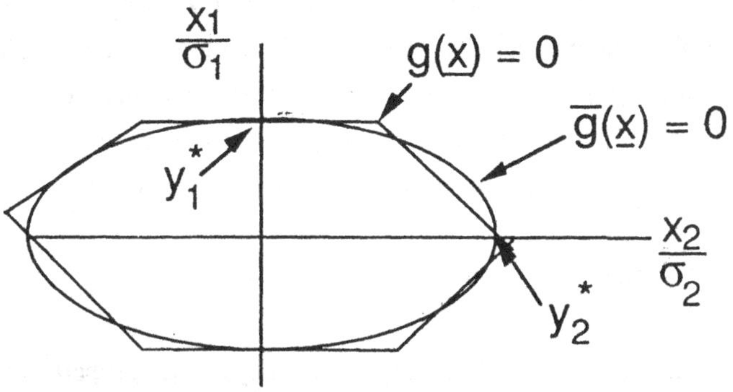

Fig. 1: Limit State Function and Response Surface

This response surface $\bar{g}(\underline{x})$ may be characterized most conveniently by

$$\bar{g}(\underline{x}) = a + \sum_{i=1}^{n} b_i x_i + \sum_{i=1}^{n} \sum_{j=1}^{n} c_{ij} x_i x_j \quad (3)$$

where the coefficients a, b and c are to be determined by structural analysis and n represents the number of basic random variables. As already mentioned above, this analysis may include all important mechanical effects, i.e. the most advanced structural analysis procedures can be utilized. At this stage it should be emphasized that the function $\bar{g}(\underline{x})$ represents just an approximation of the limit state, i.e. the region of transition from the survival to the failure state respectively and hence need not represent the actual system response. Naturally the approximations as described above require an error estimation which will be shown in detail in APPENDIX I.

Depending on the design criterion the respective failure probability must be determined in a final step. It can be calculated by evaluating the following integral:

$$p_f(T) = \int_{g(\underline{x},T)\leq 0} f_{\underline{x}}(\underline{x})\, d\underline{x} \tag{4}$$

where $f_{\underline{x}}(\underline{x})$ represents the joint probability density of the random variables $\underline{x}$ and $g(\underline{x},T) \leq 0$ defines the time variant failure domain (eq.(2)). For evaluating eq.(4) various procedures are available. Most advantageously both computationally efficient and accurate methods, such as advanced simulation procedures (see e.g. [8] and [9], etc.) are utilized. Needless to say that the replacement of the actual limit state surface $g(\underline{x})$ by the response surface $\bar{g}(\underline{x})$ results in an additional significant reduction of computational effort.

NUMERICAL EXAMPLE

A grid structure consisting of 22 elements with hysteretic properties subjected to a double sine-pulse load is analysed. A sketch of the geometry is given in Fig.2

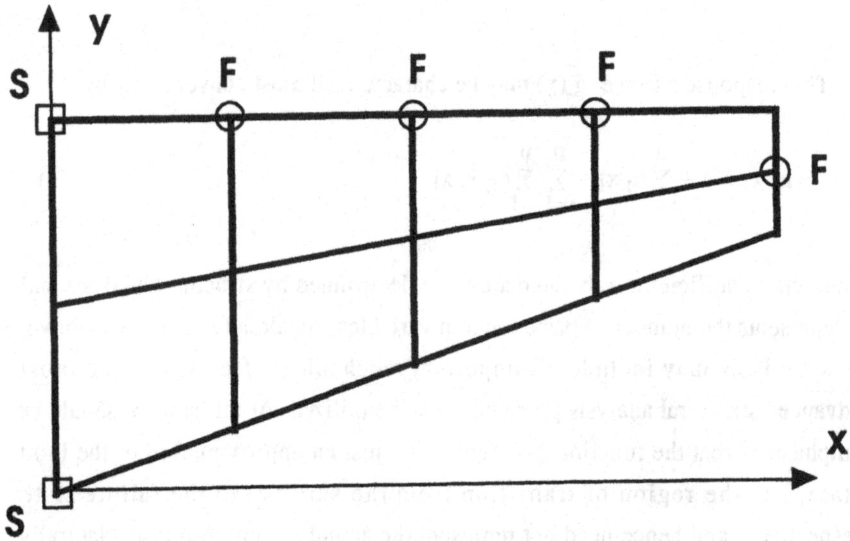

Fig. 2: Structural geometry for numerical example

Points S indicate supports where 3 DOF's are zeroed and points F indicate application of the external loads F(t) (cf. Fig. 3)

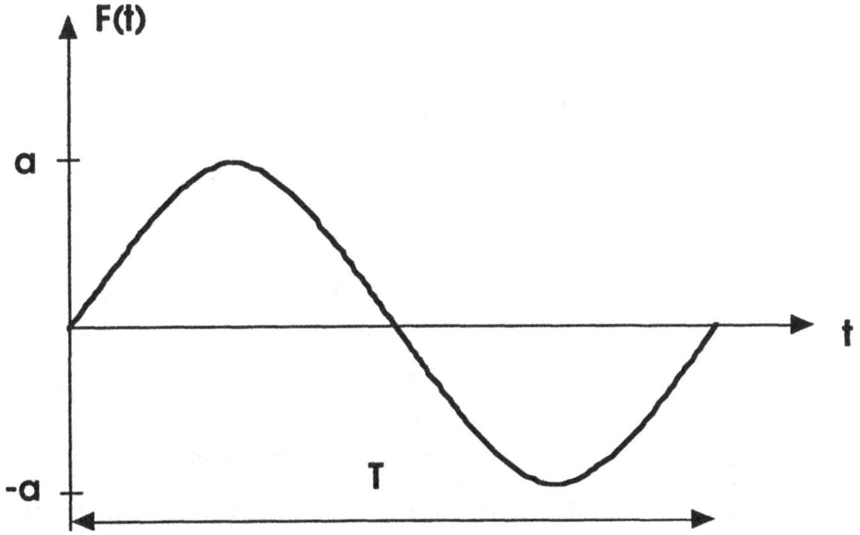

Fig. 3: Double sine pulse loading

These loads are described by

$$F(t) = a \sin \frac{2\pi t}{T} \qquad 0 \le t \le T$$
$$= 0 \qquad \text{elsewhere} \qquad (4)$$

Both a and T are assumed to be random variables. The grid elements are assumed to behave hysteretically, i.e. the moment curvature relation is given as follows

$$\frac{\partial M}{\partial \kappa} = EJ \cdot \begin{cases} 1 & \text{for} \quad \dot{M} > 0 \text{ and } M \leq M_y^+ \\ & \text{or} \quad \dot{M} < 0 \text{ and } M \geq M_y^- \\ \dfrac{M_p^+ - M}{M_p^+ - M_y^+} & \text{for} \quad \dot{M} > 0 \text{ and } M > M_y^+ \\ \dfrac{M - M_p^-}{M_y^- - M_p^-} & \text{for} \quad \dot{M} < 0 \text{ and } M < M_y^- \end{cases} \tag{5}$$

where M denotes the bending moment, κ the curvature and $\dot{M}$ the bending moment increment. The hysteretic characteristic of the cross section is defined by the four constants M_p^-, M_y^-, M_y^+ and M_p^+, where M_y denotes the yielding moment and M_p the ultimate plastic bending moment. The superscript "+" and "-" indicates these constants in the positive and negative direction of the bending moment.

The yield moment $M_y = M_y^+ = -M_y^-$ is assumed to be a random variable i.e. fully correlated in the entire structure. Statistical parameters characterizing the random variables are given in Table 1

Random Variable	Type of Distribution	Mean Value	Standard Deviation	Coeff. of Variation	Units
a	lognormal	10	2	0.2	kN
$f = \frac{1}{T}$	lognormal	0.23	0.02	0.087	s^{-1}
M_y	lognormal	1000	50	0.05	kNm

Tab. 1: Statistical parameters for random variables

The parameters a and f are assumed to be correlated with $\rho = 0.8$. M_y is assumed independent of a and f. The flexural rigidity $EJ = 2000\ kNm^2$ and mass per unit length $\mu = 1kg/m$ describe the linear properties of the structure. The structure is assumed to fail whenever the bending moment in a particular element exceeds the threshold value of $M_p^+ = -M_p^- = 1.5\ M_y^+$. Carrying out the sensitivity analysis as described in sect. 3 results in a response surface corresponding to eq.(3).

The failure probability is then determined by the adaptive sampling method [10] resulting in $p_f = 3 \cdot 10^{-6}$.

5. CONCLUSIONS

A procedure to utilize sophisticated mechanical modeling when taking into account statistical uncertainties for characterizing the load as well as the structural properties is suggested. In order to show the capabilities of the approach, a numerical problem which reflects these requirements has been solved. The successful solution of this problem immediately shows the advancement of the procedure when compared with methods currently available as discussed in section 2. The suggested approach which makes use of finite element procedures in context with the response surface method (RSM) as well as variance reduction simulation techniques is shown to be computationally efficient. Finally, most important, the method reveals the capabilities to be extended, i.e. applied to more complex systems.

REFERENCES

[1] Der Kiureghian, A., De Stefano, M.: "An Efficient Algorithm for Second-Order Reliability Analysis", Department of Civil Engineering, University of California at Berkeley, Berkeley, California, Report No. UCB/SEMM-90/20, October 1990.

[2] Murotsu, Y., Okada, H., Yonezawa, M., Grimmelt, M.J. and Taguchi, K.: "Automatic Generation of Stochastically Dominant Modes of Structural Failure in Frame Structure", *Bulletin of University of Osaka Prefecture*, Series A, Vol. 30, No. 2, 1981.

[3] Grimmelt, M.J., Schuëller, G.I., Murotsu, Y.: "On the Evaluation of Collapse Probabilities", Proc. of the 4th Engr. Mech. Division Specialty Conf. on "Recent Advances in Engr. Mechanics and their Impact on Civil Engr. Practice", EM-ASCE, Vol.II, W.F. Chen and A.D.M. Lewis (Ed.), American Society of Civil Engineers, New York, 1983, pp. 859-862.

[4] Thoft-Christensen, P., Murotsu, Y.: "Application of Structural Systems Reliability Theory", Springer Verlag, Berlin, Heidelberg, New York, Tokyo, 1986.

[5] Bucher, C.G., Bourgund, U.: "A Fast and Efficient Response Surface Approach for Structural Reliability Problems", *Structural Safety*, Vol. 7, No. 1, 1990, pp. 57-66.

[6] Faravelli, L.: "Response-Surface Approach for Reliability Analysis", *Journal of Engineering Mechanics*, Vol. 115, No. 12, December 1989, pp. 2763-2762.

[7] Schuëller, G.I., Bucher, C.G., Bourgund, U., Ouypornprasert, W.: "On Efficient Computational Schemes to Calculate Structural Failure Probabilities", *Probabilistic Engineering Mechanics*, Vol. 4, No. 1, March 1989, pp. 10-18.

[8] Schuëller G.I. and Stix, R. A Critical Appraisal of Methods to Determine Failure Probabilities. *Structural Safety* , Vol. 4, 1987, pp. 293-309.

[9] Bucher C.G., Nienstedt, J., Ouypornprasert, W. Adaptive Strategies in ISPUD V3.0 - A User's Manual. Report No. 25-89, Institute of Engineering Mechanics, University of Innsbruck, Austria, 1989.

[10] Bucher C.G. Adaptive Sampling - An Iterative Fast Monte-Carlo Procedure. *Structural Safety* , 1988, Vol. 5, No. 2, 1988, pp. 119-126.

[11] Ouypornprasert W., Bucher, C.G., Schuëller, G.I. On the Application of Conditional Integration in Structural Raliability. Structural Safety and Reliability (Eds. A.H-S. Ang, M. Shinozuka and G.I. Schuëller), Proc. ICOSSAR '89, 1990,Vol. III, pp. 1683-1689.

[12] Schuëller, G.I., Bucher, C.G.: "Computational Stochastic Structural Analysis - A Contribution to the Software Development for the Reliability Assessment of Structures Under Dynamic Loading", *PEM*, accepted for publication, to appear 1991.

APPENDIX I - Error Estimation

Since the response surface as defined by eq.(3) is based on interpolation procedures there are regions in the space of basic variables where errors, i.e. deviations, may occur when comparing to the exact limit state surface. As shown in Fig.A1 these errors may contribute both positive and negative values to the failure probability. Quite clearly, the major contributions to the error result from regions close to the limit state surface. Consequently, in order to check the accuracy of the response surface, a sampling scheme which produces samples close to $g(\underline{x}) = 0$ is most appropriate. Such a scheme, called conditional sampling, is described in [11]. For the present problem the conditional sampling procedure is carried out as follows:

(1) In order to obtain the optimum sampling densities $h_{\underline{Y}}(\underline{x})$ for the vector $\underline{x}$ and for the safety margin Z perform adaptive sampling [10] using the response surface.

(2) Identify the most important variable y_m^* in terms of sensitivity factors.

(3) Generate samples for $\underline{x}$ from the sampling density $h_{\underline{Y}}(\underline{x})$ and discard the m-th component y_m^*.

(4) Generate samples for the safety margin $Z = \bar{g}(\underline{x})$ near $Z = 0$ and calculate y_m^* from $\underline{x}$ and Z (cf. eq.(3)). Check whether $g(\underline{x})$ and $\bar{g}(\underline{x})$ have identical sign. If not, add to the error estimate (cf. Fig.A1).

This procedure ensures that the generated samples are close to the limit state surface thus yielding a reliable error estimate. Numerical examples showing the application of this error estimation procedure are given in [12].

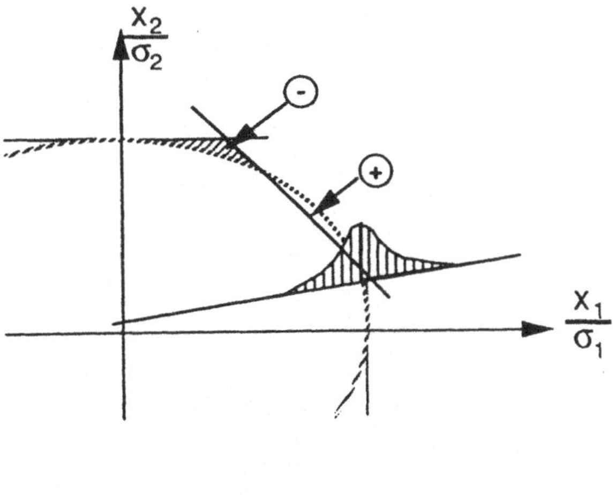

Fig. A1: Error Estimation for Response Surface Method

The Sensitivity Analysis of Stochastic Hysteretic Dynamic Systems

L. Socha, G. Zasucha

Institute of Transport, Silesian Technical University, ul.Krasińskiego 8, 40-019 Katowice, Poland

ABSTRACT

The purpose of this paper is to develop the idea of sensitivity analysis with respect to a parameter of stochastic dynamic systems presented by Socha [1]. This approach is applied for a simple nonlinear hysteretic model of practical sliding isolation system discussed by Constantinou and Papageorgiou [2]. The derived analytical solutions are compared with results of Monte Carlo simulations and the degree of accuracy of each solution is established.

INTRODUCTION

The problem of the random vibration of nonlinear systems in general and hysteretically degrading systems in particular has been often considered in recent years. It has been applied in the response analysis of structural or mechanical systems subjected to earthquakes, winds, acoustic, road and wave loadings. Several models of systems with various types of piecewise linear or smoothly varying hysteresis behavior have been discussed in the literature. Baber and Noori [3],[4], Davoodi and Noori [5], Wen [6], Roberts and Spanos [7] reviewed much of this work. To obtain characteristics of the response of hysteretic systems several approximate techniques have been applied, for instance, equivalent linearization, gaussian and non-gaussian techniques. The sensitivity analysis of these characteristics with respect to parameters of the model has been done in the literature indirectly, it means the characteristics were calculated and plotted for various values of parameters and sensitivity effects were established from diagrams. The direct sensitivity analysis for deterministic hysteretic models has been proposed by Ray et al. [8]. The objective of this paper is to develop the idea of direct sensitivity analysis of stochastic dynamic systems presented by

Socha [1] for a simple nonlinear hysteretic model. We use the model of practical sliding isolation system discussed by Constantinou and Papageorgiou [2]. The derived analytical solutions are compared with results of Monte Carlo simulations and the degree of accuracy of each solution is established.

FORMULATION OF THE PROBLEM

The nonlinear model to be studied herein is the Constantinou and Papageorgiou smooth hysteresis model of the practical sliding isolation system in which the coefficient of friction exhibits a strong dependency on the velocity of sliding. The equation of motion for the slip displacement x is:

$$\ddot{x} + 2\zeta_o\omega_o\dot{x} + \omega_o^2 x + F_f = -\ddot{x}_g \tag{1}$$

$$F_f = \mu(\dot{x})gz \tag{2}$$

$$\dot{z} = -\frac{1}{r}(\gamma|\dot{x}|z|z|^{n-1} + \beta\dot{x}|z|^n + a\dot{x}) \tag{3}$$

$$\mu(\dot{x}) = f_{max} - D_f\exp\{-\alpha|\dot{x}|\} \quad , \tag{4}$$

where ω_o is natural frequency of the system, ζ_o is the corresponding damping ratio, F_f is the mobilized frictional force, $\mu(\dot{x})$ is the coefficient of sliding friction, f_{max}, D_f and α are constant parameters which define the coefficient $\mu(\dot{x})$, $\ddot{x}_g$ is the ground acceleration, z is variable taking values in the interval [-1,1], r, γ, β, a and n are parameters which define the Bouc-Wen model of hysteresis; in this case a=1, n=3, $\beta+\gamma=1$. Applying equivalent linearization to system (1)-(4) Constantinou and Papageorgiou [2] obtained the following equations

$$\ddot{x} + C_{11}\dot{x} + \omega_o^2 x + K_{12}z = -\ddot{x}_g \tag{5}$$

$$\dot{z} = -\frac{1}{r}(C_{21}\dot{x} + K_{22}z) \tag{6}$$

in which $\ddot{x}_g$ represents evolutionary white noise. Coeffictients C_{11}, C_{21}, K_{12} and K_{22} take the following form :

$$C_{11} = 2\zeta\omega_o + \left(\frac{2}{\pi}\right)^{1/2} aD_f g\,\frac{E[\dot{x}z]}{\sigma_{\dot{x}}}\Phi(B) \tag{7}$$

$$K_{12} = g[f_{max} - D_f\exp(B^2)\mathrm{erfc}(B)] \tag{8}$$

$$B = \frac{\alpha\,\sigma_{\dot{x}}}{2^{1/2}} \tag{9}$$

$$\Phi(B) = 1-\pi^{1/2}B\exp(B^2)\mathrm{erfc}(B) \tag{10}$$

$$C_{21} = \gamma\left(\frac{2}{\pi}\right)^{1/2}\sigma_z^3\rho(3-\rho^2) + 2\beta\left(\frac{2}{\pi}\right)^{1/2}\sigma_z^3 - a \tag{11}$$

$$K_{22} = 3\gamma\left(\frac{2}{\pi}\right)^{1/2}\sigma_z^2\sigma_{\dot{x}}(1+\rho^2) + 6\beta\left(\frac{2}{\pi}\right)^{1/2}\sigma_{\dot{x}}\sigma_z^2 \tag{12}$$

$$\rho = \frac{E[\dot{x}z]}{\sigma_{\dot{x}}\sigma_z} \tag{13}$$

The first and second order moment equations for the system (5)-(6) have the closed form:

$$\frac{dm(t)}{dt} = A(p)m(t) \tag{14}$$

$$\frac{dS(t)}{dt} = A(p)S(t) + S(t)A^T(p) + Q \tag{15}$$

where $m=E[X]$, $S=E[XX^T]$, $X=[X_1,X_2,X_3]^T$, $X_1=x$, $X_2=\dot{x}$, $X_3=z$, $Q_{ij}=0$ except $Q_{22}=2\pi G_o$, the matrix $A(p)$ is given in the Appendix, p is vector of parameters, for instance: ω_o^2, ζ_o, a, α, β, γ. G_o is the constance power spectral density of two-sided white noise input and $I(t)$ is the intensity function. To obtain moment sensitivity equations, we differentiate moment equations (14)-(15) with respect, for instance, to parameter $p_j=a$.

$$\frac{dm_a(t)}{dt} = A_a(p)m(t) + A(p)m_a(t) \tag{16}$$

$$\frac{dS_a(t)}{dt} = A_a(p)S(t)+S(t)A_a^T(p)+A(t)S_a(t)+S_a(t)A^T(t) \tag{17}$$

where $m_a=\partial m/\partial a$, $S_a=\partial S/\partial a$, $A_a(p)=\partial A(p)/\partial a$, the matrix $A_a(p)$ is in the Appendix.
This approach we apply also to the parameter $p_j=\omega_o$.

NUMERICAL STUDIES

The accuracy of the derived approximate solutions is investigated by comparing to results of numerical simulation. Analytical sensitivities of characteristics of stationary responses obtained via linearization approach have been found by solving the system of nonlinear matrix equations (14)-(15)

and (16)-(17) for $t \to \infty$. Similar to Constantinou and Papageorgiou [2] numerical responses are obtained by creating a realization of stationary zero mean Gaussian process of psd G_o (two-sided) and integrating numerically equations (1) through (4) up to 20 sec. 200 realizations are taken into account. For sensitivity characteristics such as, for instance,

$$2E[X_1Y_1] = \frac{\partial E[X_1^2]}{\partial a}, \quad 2E[X_2Y_2] = \frac{\partial E[X_2^2]}{\partial a},$$

both approximate solutions obtained via linearization approach and simulations are calculated for the following parameters: $T_o=2\pi/\omega_o= 0.5$ sec, $\zeta_o=0.05$, $a=1.0$, $\alpha=0.236$ sec/cm, $\beta=\gamma=0.5$, $n=3$, $r=0.025$cm, $D_f=0.12$, $f_{max}=0.19$, $I(t)=1$.

Figures 1 and 2 compare approximate sensitivities of characteristics obtained by analitical method and simulations for $\frac{\partial E[X_1^2]}{\partial a}$, $\frac{\partial E[X_2^2]}{\partial a}$ and $\frac{\partial E[X_1^2]}{\partial \omega_o}$, $\frac{\partial E[X_2^2]}{\partial \omega_o}$ respectively.

Comparisons of results with Monte Carlo solutions indicate that the accuracy of presented solutions is good in considered range of system parameters and response level. However, it is not true in a wide range of system parameters.

The results of Figures 1 and 2 show that the sensitivities of the characteristics of stationary response for both parameters a and ω_o are decreasing functions with respect to normalized density.

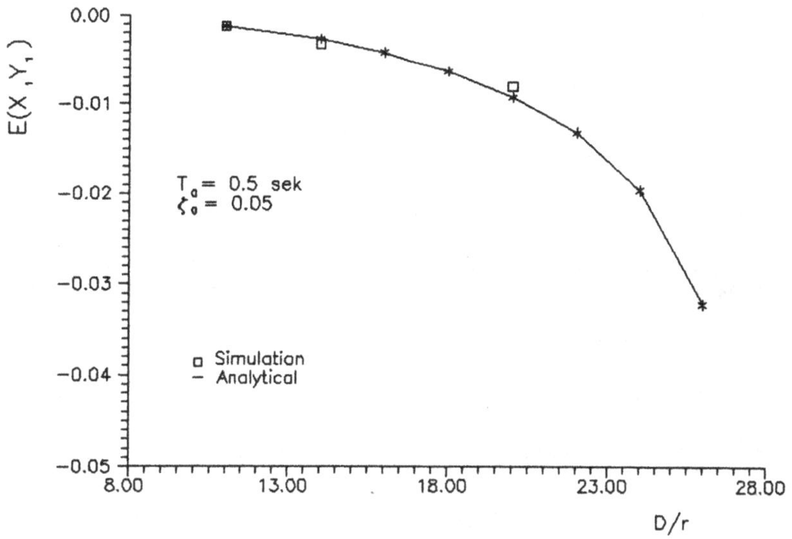

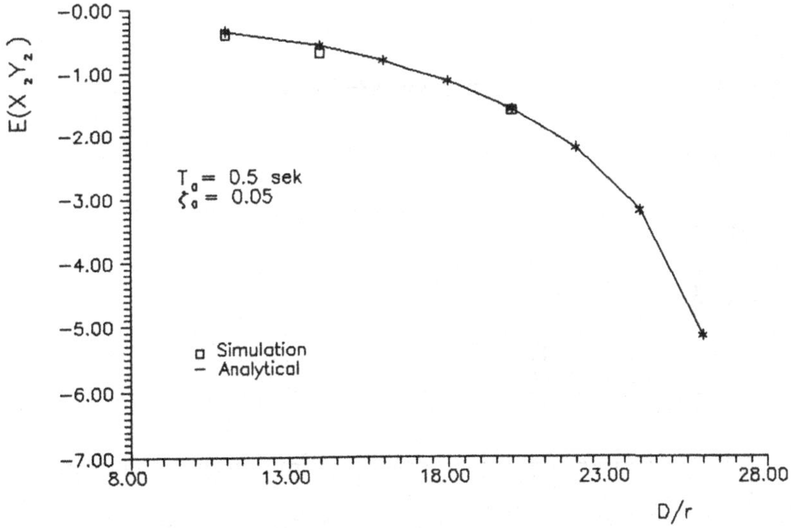

Fig. 1. Stationary sensitivity response with respect to parameter a of a sliding system with $T_o = 0.5$ sec.

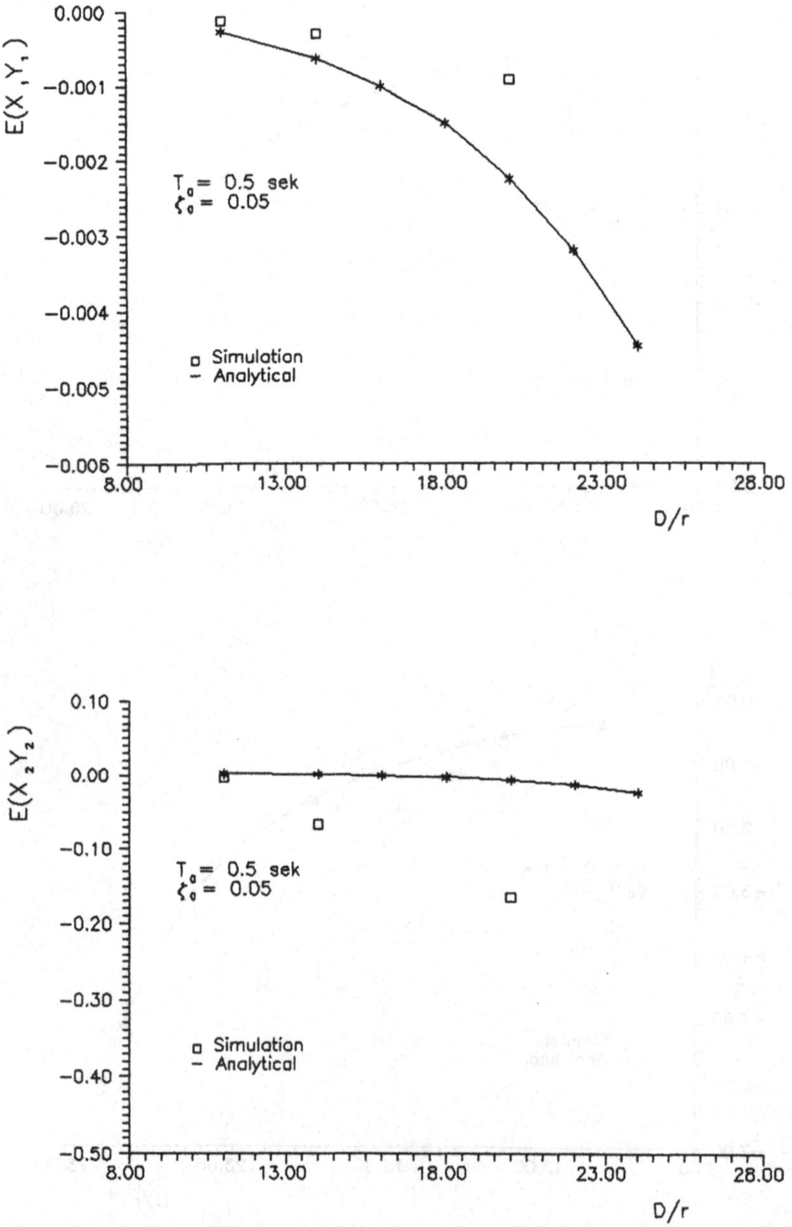

Fig. 2. Stationary sensitivity response with respect to parameter ω_o of a sliding system with $T_o = 0.5$ sec.

REFERENCES

1. Socha, L. The Sensitivity Analysis of Stochastic Non-linear Dynamical Systems, Journal of Sound and Vibrations, Vol.110, pp.271-288, 1986.
2. Constantinou, M.C. andPapageorgiou, A.S. Stochastic Response of Practical Sliding Isolation Systems, Probabilistic Engineering Mechanics, Vol.5, pp 27-34, 1990.
3. Baber, T. and Noori, M. Random Vibration of Degrading, Pinching Systems, ASCE, Journal of Engineering Mechanics Division, Vol.111, pp.1010-1027, 1985.
4. Baber, T. and Noori, M. Modeling General Hysteresis Behavior and Random Vibration Application, Trans ASME, Journal of Vibration, Acoustics, Stress, and Reliability in Design, Vol.108, pp.411-420, 1986.
5. Davoodi, H. and Noori,M. Extension of an ITO-Based General Approximation Technique for Random Vibration of a BBW General Hysteris Model, Part II: Non-Gaussian Analysis, Journal of Sound and Vibration, Vol.140, pp.319-339, 1990.
6. Wen, Y.K. Methods of Random Vibration for Inelastic Structures, Applied Mechanics Review, Vol.42, pp.39-52, 1989.
7. Roberts, I.B. and Spanos, P.D. Random Vibration and Statistical Linearization, John Willey & Sons, Chichester 1990.
8. Ray, A.,Pister, K.S. and Polak,E Sensitivity analysis for hysteretic dynamic systems: theory and applications, Computers methods in applied mechanics and engineering 14,pp. 179-208, 1978.

APPENDIX

Matrices A and A_p have the following form:

$$A = \begin{bmatrix} 0 & 1 & 0 \\ -\omega_o^2 & -C_{11} & -K_{12} \\ 0 & -C_{21}/r & -K_{22}/r \end{bmatrix}, \quad A_p = \begin{bmatrix} 0 & 0 & 0 \\ 0 & P_{11} & P_{12} \\ 0 & P_{21} & P_{22} \end{bmatrix}$$

where

$$P_{11} = -\frac{\partial C_{11}}{\partial a} = \left(\frac{2}{\pi}\right)^{1/2} \alpha D_f g \left(R_1 + \frac{E[X_2Y_2]}{(E[X_2^2])^{1/2}} \frac{\partial \Phi(B)}{\partial a} \right)$$

$$R_1 = \left(\frac{((E[X_3Y_2]+E[X_2Y_3])(E[X_2^2])^{1/2}}{E[X_2^2]} - \frac{E[X_2X_3]E[X_2Y_2]}{(E[X_2^2])^{3/2}} \right) \Phi(B)$$

$$P_{12} = -\frac{\partial K_{12}}{\partial a} = gD_f\left(2B\frac{\partial B}{\partial a}\exp(B^2)\mathrm{erfc}(B) + \exp(B^2)\frac{\partial\,\mathrm{erfc}(B)}{\partial a}\right)$$

$$P_{21} = -\frac{1}{r}\frac{\partial C_{21}}{\partial a} = -\frac{1}{r}\left(\gamma\left(\frac{2}{\pi}\right)^{1/2} R_2 + 6\beta\left(\frac{2}{\pi}\right)^{1/2} R_3 - 1\right)$$

$$R_2 = \left(3E[X_3Y_3](E[X_3])^{1/2}\rho(3-\rho^2) + (E[X_3^2])^{1/2}\frac{\partial\rho}{\partial a}((3-\rho^2)-2\rho^2\right)$$

$$R_3 = E[X_3Y_3](E[X_3^2])^{1/2}$$

$$P_{22} = -\frac{1}{r}\frac{\partial K_{22}}{\partial a} = -\frac{1}{r}\left(3\gamma\left(\frac{2}{\pi}\right)^{1/2} R_4 + 6\beta\left(\frac{2}{\pi}\right)^{1/2} R_6\right)$$

$$R_4 = \left(2E[X_3Y_3](E[X_3^2])^{1/2}(1+\rho^2) + E[X_3^2]R_5\right)$$

$$R_5 = \frac{E[X_2Y_2]}{(E[X_2^2])^{1/2}}(1+\rho^2) + (E[X_3^2])^{1/2}2\rho\frac{\partial\rho}{\partial a}$$

$$R_6 = (E[X_2^2])^{1/2}E[X_3]\frac{\partial\rho}{\partial a} + \rho R_7$$

$$R_7 = \frac{E[X_2Y_2]}{(E[X_2^2])^{1/2}}E[X_3] + 2(E[X_2^2])^{1/2}E[X_3Y_3]$$

$$\frac{\partial\rho}{\partial a} = \frac{(E[X_3Y_2]+E[X_2Y_3])(E[X_2^2]E[X_3^2])^{1/2}}{(E[X_2^2]E[X_3^2])} - R_8$$

$$R_8 = \frac{E[X_2X_3](E[X_2Y_2]E[X_3^2] + E[X_3Y_3]E[X_2^2])}{(E[X_2^2]E[X_3^2])^{3/2}}$$

$$\frac{\partial\Phi(B)}{\partial a} = -(\pi)^{1/2}\left(\frac{\partial B}{\partial a}\exp(B^2)\mathrm{erfc}(B) + BR_9\right)$$

$$R_9 = 2B\frac{\partial B}{\partial a}\exp(B^2)\mathrm{erfc}(B) + \exp(B^2)\frac{\partial\,\mathrm{erfc}(B)}{\partial a}$$

$$\frac{\partial \mathrm{erf}(B)}{\partial a} = (2\pi)^{-1/2}\exp(-B^2/2)\frac{\partial B}{\partial a}$$

$$\frac{\partial B}{\partial a} = \frac{\alpha E[X_2 Y_2]}{(2E[X_2^2])^{1/2}}$$

Solution of Random Eigenvalue Problem by Crossing Theory and Perturbation

M. Grigoriu

School of Civil & Environmental Engineering, Cornell University, Ithaca, New York 14853, U.S.A.

ABSTRACT

Two methods are developed for finding probabilistic characteristics of the eigenvalues and eigenvectors of a stochastic matrix. They are based on the mean zero-crossings rate of the characteristic polynomial of this matrix and a perturbation approach. The methods are applied to characterize probabilistically the natural frequencies of an uncertain dynamic system and to find the first two moments of the displacement of a simple oscillator with random damping and stiffness that is subject to white noise.

INTRODUCTION

Consider a real-valued symmetric matrix $\underline{A}$ with random components and dimension (n, n). The eigenvalues $\{\Lambda_i\}$, i = 1, ..., n, of $\underline{A}$ satisfy the characteristic equation

$$\det (\underline{A} - \Lambda \underline{I}) = 0 \tag{1}$$

in which $\underline{I}$ denotes the (n, n)-identity matrix. There are several techniques for finding probabilistic characteristics of the eigenvalues of $\underline{A}$. For example, the relationships between Λ_i and the components of $\underline{A}$ can be used to obtain the joint probability distribution of the eigenvalues of this matrix. The method can only be applied when the dimension n of $\underline{A}$ is small (Grigoriu [3,4]). Variational principles and other techniques can be used to develop simple approximations of and bounds on the eigenvalues of $\underline{A}$. Probabilistic descriptors based on the approximations and bounds are usually unsatisfactory (Grigoriu [3]). Perturbation and crossing theory for random processes were also applied to characterize the random variables $\{\Lambda_i\}$ (Bharucha-Reid [1], Boyce [2], Grigoriu [3,4]). However, available results focus on the second-moment characterization of the eigenvalues and eigenvectors of $\underline{A}$ in the perturbation method (Boyce [2]) and on the frequency of occurrence of zeros of polynomials with random coefficients in the crossing method (Bharucha-Reid [1]). Simulation method can also be applied to characterize the eigenvalues of $\underline{A}$. The method is inefficient because it requires to solve a complete eigenvalue problem for every realization of $\underline{A}$.

The objectives of the paper are to (1) find probabilistic characteristics of the eigenvalues of $\underline{A}$ and (2) demonstrate the use of these characteristics in the response analysis of linear uncertain systems subject to random dynamic excitations. The analysis is based on the crossing theory of random processes and the perturbation method.

CROSSING METHOD

Let

$$V(\lambda) = \det(\underline{A} - \lambda\,\underline{I})$$
$$= \lambda^n + B_1\,\lambda^{n-1} + \ldots + B_{n-1}\,\lambda + B_n \qquad (2)$$

the characteristic polynomial of $\underline{A}$, where $\underline{B}^T = \{B_1, \ldots, B_n\}$ are random coefficients. These coefficients depend on $\underline{A}$ and are equal to (Grigoriu [4])

$$B_1 = -T_1$$
$$B_2 = -\frac{1}{2}(B_1\,T_1 + T_2) \qquad (3)$$
$$\vdots$$
$$B_n = -\frac{1}{n}\left[B_{n-1}\,T_1 + B_{n-2}\,T_2 + \ldots + B_1\,T_{n-1} + T_n\right]$$

where $T_k = \mathrm{tr}\,(\underline{A}^k)$ is the trace of $\underline{A}^k$. The polynomial $V(\lambda)$ is a nonstationary differentiable process of argument λ. The eigenvalues of $\underline{A}$ coincide with the zero-crossings of $V(\lambda)$.

According to the Rice formula the mean rate at which $V(\lambda)$ crosses zero at λ, or the mean zero-crossing rate at λ, is

$$\nu(\lambda) = \int_{-\infty}^{\infty} |z|\; f_{V(\lambda),\,V'(\lambda)}(0, z)\, dz \qquad (4)$$

in which $f_{V(\lambda),\,V'(\lambda)}$ is the joint probability of $\{V(\lambda), V'(\lambda)\}$.

The exact calculation of the probability of $\{V(\lambda), V'(\lambda)\}$ is difficult because of the complex dependence of the vector of random coefficients $\underline{B}$ on the components of matrix $\underline{A}$. Even moments of $\underline{B}$ are difficult to obtain analytically for large matrices $\underline{A}$. However, the mean and covariance matrices of $\underline{B}$ can be determined by simulation efficiently from Eq. 3. The second-moment characterization of $\underline{B}$ describes fully the random process $V(\lambda)$ if it assumed that $\underline{B}$ is a Gaussian vector. In this case the mean zero crossing rate $\nu(\lambda)$ is

$$\nu_G(\lambda) = \sigma(\lambda)\ \chi\left[\frac{m(\lambda)}{\sigma(\lambda)}\right]\frac{1}{\sqrt{\gamma_{00}(\lambda,\lambda)}}\ \phi\left[-\frac{\mu(\lambda)}{\sqrt{\gamma_{00}(\lambda,\lambda)}}\right] \tag{5}$$

in which

$$m(\lambda) = \mu'(\lambda) - \frac{\gamma_{01}(\lambda,\lambda)}{\gamma_{00}(\lambda,\lambda)}\ \mu(\lambda)$$

$$\sigma(\lambda)^2 = \gamma_{11}(\lambda,\ \lambda) - \frac{\gamma_{01}(\lambda,\ \lambda)\ \gamma_{10}(\lambda,\ \lambda)}{\gamma_{00}(\lambda,\ \lambda)}$$

$$\chi(\alpha) = \alpha[2\ \Phi(\alpha) - 1] + 2\phi\ (\alpha)$$

$$\Phi(\alpha) = \int_{-\infty}^{\alpha} \phi(u)\ du \tag{6}$$

and $\phi(u) = (2\pi)^{-1/2} \exp(-\ 0.5\ u^2)$. The other functions in Eq. 6 are $\mu(\lambda) = E\ V(\lambda)$, $\mu'(\lambda) = d\mu(\lambda)/d\lambda$, $\gamma_{00}(\lambda,\ \rho) = \mathrm{Cov}\ (V(\lambda),\ V(\rho))$, $\gamma_{01}(\lambda,\ \rho) = \partial\gamma_{00}(\lambda,\ \rho)/\partial\rho$, $\gamma_{10}(\lambda,\ \rho) = \partial\ \gamma_{00}(\lambda,\ \rho)/\partial\lambda$, $\gamma_{11}(\lambda,\ \rho) = \partial^2\ \gamma_{00}(\lambda,\ \rho)/\partial\lambda\partial\rho$, and E is the expectation operator.

The mean zero-crossing rate of $V(\lambda)$ contains useful probabilistic information on the eigenvalues of $\underline{A}$. Suppose that $\underline{A}$ has distinct eigenvalues almost surely. Let $(\lambda,\ \lambda+\Delta\lambda)$ be a sufficiently small interval on the real line such at $\underline{A}$ can have at most one eigenvalue in it. Let p_λ be the probability of a "success", i.e., an eigenvalue of $\underline{A}$ in $(\lambda,\ \lambda+\Delta\lambda)$. This probability corresponds to a Bernoulli variable Z taking the values zero and one (success) with probabilities $1-p_\lambda$ and p_λ, respectively. The average value of Z is p_λ and represents the average number of eigenvalues in $(\lambda,\ \lambda+\Delta\lambda)$. Therefore,

$$p_\lambda = \int_{\lambda}^{\lambda+\Delta\lambda} \nu(\rho)\ d\rho \simeq \nu(\lambda)\ \Delta\lambda \tag{7}$$

for small increments $\Delta\lambda$.

<u>Example 1</u>. Consider a system with n identical masses m connected in series by n springs of random stiffnesses K_1, ..., K_n. The first spring of stiffness K_1 is attached to a fixed support. Let

$$\underline{A} = \begin{bmatrix} S_1+S_2 & -S_2 & & & & \\ -S_2 & S_2+S_3 & -S_2 & \underline{0} & & \\ & \cdot & \cdot & \cdot & \cdot & \\ \underline{0} & & \cdot & \cdot & \cdot & \\ & & & -S_{n-1} & S_{n-1}+S_n & -S_n \\ & & & & -S_n & S_n \end{bmatrix} \tag{8}$$

be a normalized stiffness matrix of the system where $S_i = K_i/m$, $i = 1, \ldots, n$. Figure 1 shows the approximate mean zero-crossing rate $\nu_G(\lambda)$ in Eq. 5 for $n = 4$ and independent Gaussian stiffnesses with means $E\ K_i$ equal to 30,000; 20,000; 10,000; and 10,000 lb/in for $i = 1, 2, 3$, and 4. The coefficient of variation of these variables is 0.15. The mass m is equal to 100 lb sec^2/in. The figure also shows an estimate of the exact mean zero-crossing rate $\nu(\lambda)$ obtained by simulation.

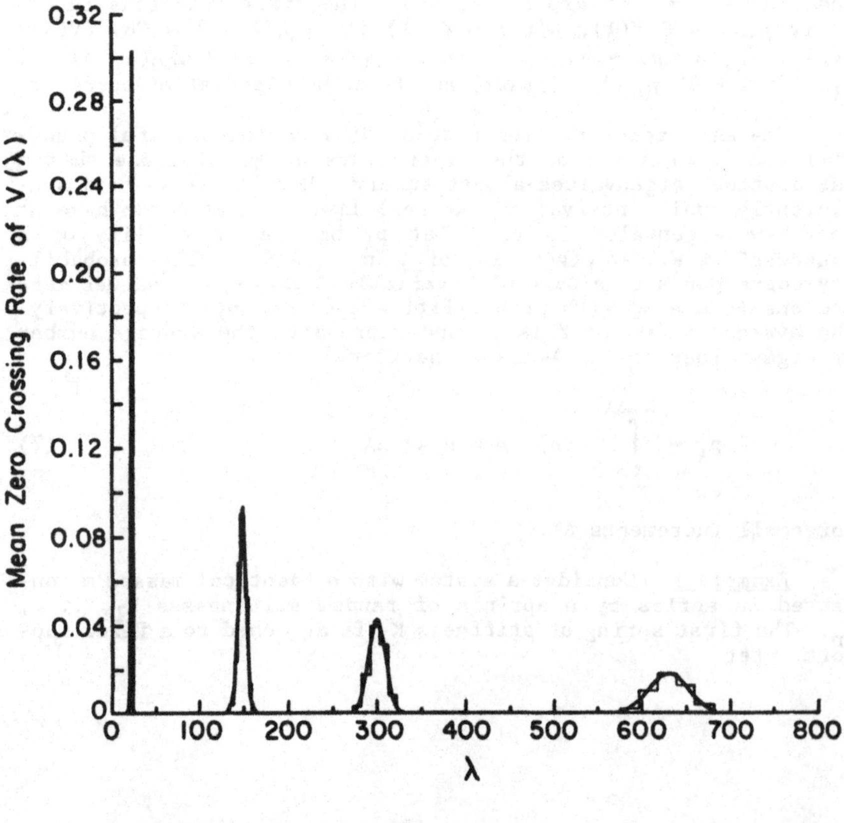

Fig. 1 Mean zero-crossing rates for the uncertain system in Example 1: Gaussian hypothesis and simulation solutions.

Results in the example suggest that (i) ν_G provides an accurate approximation of ν; (ii) the most likely location of the eigenvalue of $\underline{A}$ can be simply obtained from the maxima of ν_G and ν; and (iii) the uncertainty in the natural frequencies of the system increases rapidly with the frequency number. Small fluctuations in the system stiffness matrix have a limited effect on the first natural frequency but can significantly affect the frequencies of higher modes. This observation may be related to the common difficulty of measuring higher order natural frequencies.

PERTURBATION METHOD

Suppose that the uncertainty in $\underline{A}$ is small and let

$$\underline{A} = \underline{a} + \underline{R} \tag{9}$$

where $\underline{a} = E\,\underline{A}$ and $\underline{R} = \underline{A} - \underline{a}$ is the stochastic component of the matrix. The perturbation series solutions

$$\Lambda_i = \sum_{s=0}^{\infty} \Lambda_{i,s} \tag{10}$$

and

$$\underline{\Psi}_i = \sum_{s=0}^{\infty} \underline{\Psi}_{i,s} \tag{11}$$

can be used to approximate the eigenvalues Λ_i and eigenvectors $\underline{\Psi}_i$, i = 1, ..., n, of $\underline{A}$. It can be shown that the first order approximation of Λ_i and Ψ_i are (Boyce [2], Grigoriu [3,4])

$$\Lambda_i \simeq \lambda_{i,0} + \Lambda_{i,1} = \lambda_{i,0} + \underline{\psi}_{i,0}^T\, \underline{R}\, \underline{\psi}_{i,0} \tag{12}$$

and

$$\underline{\Psi}_i \simeq \underline{\psi}_{i,0} + \underline{\Psi}_{0,1} = \underline{\psi}_{i,0} + \sum_{\substack{p=1\\ p\neq i}}^{n} \frac{\underline{\psi}_{p,0}^T\, \underline{R}\, \underline{\psi}_{i,0}}{\lambda_{p,0} - \lambda_{i,0}}\, \underline{\psi}_{p,0} \tag{13}$$

in which $\lambda_{i,0} = \Lambda_{i,0}$ and $\underline{\psi}_{i,0} = \underline{\Psi}_{i,0}$, i = 1, ..., n, are the eigenvalues and eigenvectors of the deterministic component $\underline{a}$ of matrix $\underline{A}$. It is assumed that $\underline{a}$ has distinct eigenvalues.

The approximations in Eqs. 12 and 13 can be used to calculate the means and covariances of the eigenvalues and eigenvectors of $\underline{A}$. For example,

$$E\,\Lambda_i \simeq \lambda_{i,0} \tag{14}$$

and

$$\mathrm{Cov}\,(\Lambda_i, \Lambda_j) \simeq \sum_{p,q,s,t=1}^{n} \psi_{i,0}^{(p)}\, \psi_{i,0}^{(q)}\, \psi_{j,0}^{(s)}\, \psi_{j,0}^{(t)}\, \gamma_{pq,st} \tag{15}$$

in which $\underline{\psi}_{i,0}^{T} = \{\psi_{i,0}^{(1)}, \psi_{i,0}^{(2)}, \ldots, \psi_{i,0}^{(n)}\}$ and $\gamma_{pq,st} = E\, R_{pq}\, R_{st}$. These probabilistic characteristics can be used to describe approximately the response of linear dynamic systems with uncertain parameters.

Example 2. Consider a simple oscillator with perfectly known mass m but uncertain damping C and stiffness K that is at rest at the initial time t = 0. The system is subject to the excitation m W(t) where W(t) is a zero-mean stationary Gaussian white noise with one-sided power spectral density $G(\omega)$ of intensity G_0. The oscillator displacement X(t) satisfies the differential equation

$$m\,\ddot{X}(t) + C\,\dot{X}(t) + K\,X(t) = m\,W(t) \tag{16}$$

or, equivalently

$$\dot{\underline{X}}(t) = \underline{A}\,\underline{X}(t) + \underline{b}\,W(t) \tag{17}$$

where $\underline{X}(t)^T = \{X_1(t) = X(t),\ X_2(t) = \dot{X}(t)\}$, $\underline{b}^T = [0,\ 1]$ and $\underline{A} = \underline{a} + \underline{R}$ in which

$$\underline{a} = \begin{bmatrix} 0 & 1 \\ -\frac{k}{m} & -\frac{c}{m} \end{bmatrix}$$

$$\underline{R} = \begin{bmatrix} 0 & 0 \\ -\frac{U_1}{m} & -\frac{U_2}{m} \end{bmatrix} \tag{18}$$

The random variables in $\underline{R}$ are $U_1 = C - c$, $c = EC$, and $U_2 = K - k$, $k = EK$.

It can be shown that

$$\underline{X}(t) = \begin{bmatrix} \int_0^t g_1(t-s);\ \Lambda_1,\ \Lambda_2)\ W(s)\,ds \\ \int_0^t g_2\ (t-s);\ \Lambda_1,\ \Lambda_2)\ W(s)\,ds \end{bmatrix} \tag{19}$$

when $\underline{X}(0) = \underline{0}$ where

$$g_1(u;\ \Lambda_1,\ \Lambda_2) = \frac{1}{\Lambda_2-\Lambda_1}\left[-e^{\Lambda_1 u} + e^{\Lambda_2 u} \right]$$
$$g_2(u;\ \Lambda_1,\ \Lambda_2) = \frac{1}{\Lambda_2-\Lambda_1}\left[-\Lambda_1\, e^{\Lambda_1 u} + \Lambda_2\, e^{\Lambda_2 u} \right] \qquad (20)$$

To obtain approximate probabilistic characteristics of $\underline{X}(t)$, the random eigenvalues Λ_i, $i = 1, 2$, are replaced by their first order approximations obtained by the perturbation method and defined in Eq. 12. For example, the ith component of $\underline{X}(t)$, $i = 1, 2$, in Eq. 19 can be approximated to the first order by

$$X_i(t) \simeq \int_0^t [g_i(t-s) + g_{i,1}(t-s)(\lambda_{1,0} + \Lambda_{1,1}) + + g_{i,2}(t-s)(\lambda_{2,0} + \Lambda_{2,1})]\ W(s)\ ds \qquad (21)$$

in which $g_i(u) = g_i(u;\ \lambda_{1,0},\ \lambda_{2,0})$, $g_{i,1}(u) = \partial g(u;\ \alpha,\ \lambda_{2,0})/\partial\alpha$ for $\alpha = \lambda_{1,0}$; and $g_{i,2}(u) = \partial g(u;\ \lambda_{1,0},\ \alpha)/\partial\alpha$ for $\alpha = \lambda_{2,0}$. The approximate mean and covariance matrices of $\underline{X}(t)$ can be obtained from Eq. 21 and second-moment characteristics of the eigenvalues of $\underline{A}$ because $\underline{X}(t)$ in Eq. 21 is linear in the eigenvalues of this matrix. For example, the first two moments of component $i = 1, 2$, of $\underline{X}(t)$ are

$$EX_i(t) = \lambda_{1,0}\int_0^t g_{i,1}(t-s)\ W(s)\ ds + \lambda_{2,0}\int_0^t g_{i,2}(t-s)\ W(s)\ ds \qquad (22)$$

and

$$EX_i(t)^2 = \pi G_0 \left\{ \int_0^t g_i(t-u)^2\ du + E\ \Lambda_{1,1}^{\ 2}\int_0^t g_{i,1}(t-u)^2\ du + 2\ E\Lambda_{1,1}\ \Lambda_{1,2}\int_0^t g_{i,1}(t-u)\ g_{i,2}(t-u) + E\ \Lambda_{1,2}^{\ 2}\int_0^t g_{i,2}(t-u)^2\ du \right] \qquad (23)$$

Figure 2 shows the evolution of the variance $\mathrm{Var}\ X_1(t) = EX_1(t)^2 - (E\ X_1(t))^2$ of the displacement $X_1(t) = X(t)$ of the oscillator obtained by simulation and from Eqs. 22 and 23. Numerical results are for $\pi G_0 = 1$; $k/m = 1$, $c/m = 0.10$; $U_{i,2}$= independent Gaussian variables with mean zero and variance σ_i^2; $\sigma_1 = 0.2$; and $\sigma_2 = 0.03$. The approximation is satisfactory in this case and is based on the first order perturbation approximation.

CONCLUSION

Crossing theory of random processes and perturbation methods were applied to obtain probabilistic characteristics of the eigenvalues of stochastic matrices and of the response of linear uncertain systems subject to random dynamic excitations. The perturbation method can only be used for stochastic matrices with small uncertainties.

Two numerical examples were presented to demonstrate the use of the proposed methods and evaluate their accuracy. Results suggest that both the crossing theory and the perturbation methods are simple and sufficiently accurate.

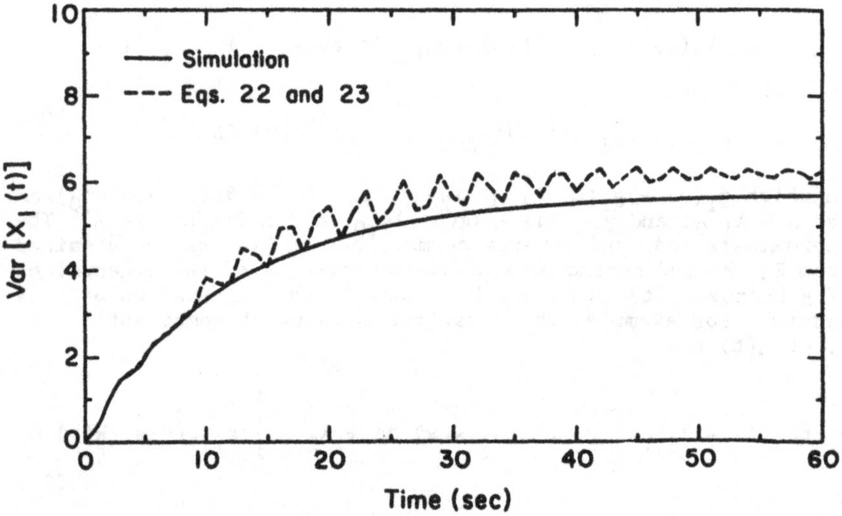

Fig. 2 Evolution of displacement variance for a simple oscillator with uncertain damping and stiffness: first order perturbation and simulation solutions.

REFERENCES

1. Bharucha-Reid, A.T. and Sambandham, M. Random Polynomials. Academic Press, Inc., New York, 1986.

2. Boyce, W.E. Random Eigenvalue Problems, Probabilistic Methods in Applied Mathematics, Vol. I, (Ed. Bharucha-Reid, A.T.), Academic Press, New York, pp. 2-73, 1968.

3. Grigoriu, M. Eigenvalue Problem for Uncertain Systems, pp. 283-284, Proceedings of the 2nd Pan American Congress of Applied Mechanics, Valparaiso, Chile, 1991.

4. Grigoriu, M. A Solution of Random Eigenvalue Problem by Crossing Theory, Report 91-2, School of Civil and Environmental Engineering, Cornell University, Ithaca, New York.

SORM Analysis Using Quasi-Newton Optimization

P. Geyskens (*), A. der Kiureghian (**),
G. De Roeck (*)
(*) Katholieke Universiteit te Leuven, Belgium
(**) University of California at Berkeley, U.S.A.

Abstract

In second order reliability analysis, an approximation of the failure probability is calculated using a parabolic approximation of the limit-state function at the design point. Principal curvatures of the limit-state surface at the design point are required in order to use closed formulas for the probability content of the approximate failure set. After a brief review of current procedures for obtaining the principal curvatures, an alternative algorithm for simultaneous determination of the design point and the principal curvatures using Quasi-Newton Optimization Methods is introduced. Example applications demonstrate the accuracy and efficiency of the method.

Introduction

Solving a component reliability problem involves evaluating the probability integral

$$p_f = \int_{g(\mathbf{x})\leq 0} f_{\mathbf{X}}(\mathbf{x})d\mathbf{x} \tag{1}$$

where $\mathbf{x} = [x_1, x_2, \ldots, x_n]^T$ is the set of basic random variables, describing the behavior of the component, $f_{\mathbf{X}}(\mathbf{x})$ is the joint probability density function of the random variables and $g(\mathbf{x})$ is the limit-state function. Component failure is conventionally defined by $g(\mathbf{x}) \leq 0$. Assuming that a full distribution model is chosen in accordance with the available data on the basic variables [1], it is possible to simplify the probability integration (1) by using a First or Second-Order Reliability Method (FORM or SORM).

In FORM analysis, the limit-state function is linearized at the design point $\mathbf{u}^*$, after transformation of the basic random variables to the standard-normal space $\mathbf{u} = \mathbf{T}(\mathbf{x})$. The design point $\mathbf{u}^*$ is defined as the point on the limit-state surface $G(\mathbf{u}) = g(\mathbf{T}^{-1}(\mathbf{u})) = 0$ that is closest to the origin of the standardized space [2]. It can be determined by solving the nonlinear programming problem [3]

$$\begin{cases} \text{minimize } h(\mathbf{u}) = \frac{1}{2}\mathbf{u}^T\mathbf{u} \\ \text{subject to } G(\mathbf{u}) = 0 \end{cases} \tag{2}$$

Liu and Der Kiureghian have investigated different algorithms for solving the above optimization problem [4].

Although FORM reveals to give sufficiently accurate approximations for most structural engineering problems, use of second-order approximations of the limit-state function $G(\mathbf{u})$ has been extensively investigated, resulting in a SORM approach to the probability integration. Interest in parabolic approximation is boosted by the availability of closed form expressions for their probability content in the standard normal space [5][6].

Attempts to fit a full quadratic to the limit-state function have proven to be too cumbersome [7]. In practical applications parabolic approximations of the limit-state function are used. Once the design point is located, the parabolic fitting problem can be tackled in different ways. If the Hessian of the constraint $G(\mathbf{u}) = 0$ can be constructed, principal curvatures can be determined by eigenvalue solution. Calculation of the Hessian proves to be a tedious job. Most often one has to resort to a numeric differentiation based on finite difference approximations. A second approach, described in a publication by Der Kiureghian et al. [8], consists in fitting the paraboloid by using sampling points on the limit-state function.

In a recent report [9], Der Kiureghian and De Stefano combine iteration to the design point and determination of the principal curvatures of the constraint $G(\mathbf{u}) = 0$, using information from the final steps of a gradient projection (or HL-RF).

Procedures for unconstrained optimization that simultaneously construct second-order information of the objective function, are known to be the most efficient general purpose algorithms for solving nonlinear programs. In this paper, a Quasi-Newton Optimization Method (QNOM) is introduced for solving the reliability problem, giving fast convergence to the design point, but also providing second-order information about the limit-state function that is needed for SORM analysis.

SORM-Analysis: review

Second Order Approximations

Assuming that $G(\mathbf{u}) = 0$ is twice differentiable in the neighborhood of $\mathbf{u}^*$, expanding in Taylor series around $\mathbf{u}^*$ and truncating after second-order terms, leads to the quadratic approximation

$$q(\mathbf{u}) = -\hat{\alpha}^*(\mathbf{u} - \mathbf{u}^*) + \frac{1}{2}(\mathbf{u} - \mathbf{u}^*)^T\mathbf{D}(\mathbf{u} - \mathbf{u}^*) = 0 \tag{3}$$

where $-\hat{\alpha}^* = \nabla G(\mathbf{u}^*)/\,|\nabla G(\mathbf{u}^*)|$ and $\mathbf{D} = \nabla^2 G(\mathbf{u}^*)/\,|\nabla G(\mathbf{u}^*)|$. In order to obtain a convenient formulation of the parabolic approximation, the standard space $\mathbf{u}$ is rotated to $\mathbf{u}'$, so that $\mathbf{u}^*$ is located on the nth axis of the transformed system. This results in the following formulation for the quadratic approximation

$$q(\mathbf{u}') = -(u'_n - \beta) + \frac{1}{2}\begin{bmatrix} \tilde{\mathbf{u}} \\ u'_n - \beta \end{bmatrix}^T \mathbf{A}_n \begin{bmatrix} \tilde{\mathbf{u}} \\ u'_n - \beta \end{bmatrix} = 0 \tag{4}$$

in which $\tilde{\mathbf{u}} = [u'_1, u'_2, \ldots, u'_{n-1}]^T$ and $\mathbf{A}_n = \mathbf{R}\mathbf{D}\mathbf{R}^T$, where $\mathbf{R}$ is an orthonormal matrix with $\hat{\alpha}^*$ as its nth row. Solving (4) for u_n and including only second-order terms in $\tilde{\mathbf{u}}$ results in a parabolic approximation of the limit-state function

$$u'_n = \beta + \frac{1}{2}\tilde{\mathbf{u}}^T \mathbf{A}_{n-1}\tilde{\mathbf{u}} \tag{5}$$

in which $\mathbf{A}_{n-1}$ is equal to $\mathbf{A}_n$ without the nth row and column. After rotation to the spectral coordinate system of $\mathbf{A}_{n-1}$, we obtain the principal-axes formulation of the paraboloid

$$u''_n = \beta + \frac{1}{2}\sum_{i=1}^{n-1} \kappa_i (u''_i)^2 \tag{6}$$

with κ_i equal to eigenvalues of $\mathbf{A}_{n-1}$ and the principal curvatures of the paraboloid.

In most engineering problems, the limit-state function is defined in load-effect space $g(\mathbf{v}(\mathbf{x}))$. An algorithmic mechanical transformation $\mathbf{S} : \mathbf{x} \mapsto \mathbf{v}$ (e.g. finite element method) defines the relation between basic variables (load/resistance) and decision variables (load-effects). A complex, nonlinear probability transformation [10][1] $\mathbf{T} : \mathbf{x} \mapsto \mathbf{u}$ governs the transition to the standard-normal space. Due to the implicit definition of the limit-state function $G(\mathbf{u}) = g(\mathbf{S}(\mathbf{T}^{-1}(\mathbf{u})))$, the Hessian (and gradient) of the constraint are not available in a closed form. Liu and Der Kiureghian [11] have reported on pseudo-analytical techniques for obtaining gradients in the Finite Element Reliability Method (FERM). In most cases one would have to resort to numerical differentiation for the calculation of the second-order sensitivity information. When computation of curvatures is problematic, the alternative point-fitting algorithm [8] can be used.

Iterative Curvature Method

This method, proposed by Der Kiureghian and De Stefano [9] allows for the second-order sensitivities to be determined in a very convenient way: curvature information of the surface $G(\mathbf{u})$ is gathered during the final steps of the optimization algorithm, used to solve (2). It can be proven [9] that for both the Gradient Projection method (GP) and the HL-RF algorithm [4] under certain conditions, the sequence of search directions $\{\mathbf{d}^{(k)}\}$ converges to the major principal axis of the limit-state function $G(\mathbf{u})$, provided the sequence of trial points $\{\mathbf{u}^{(k)}\}$ converges to the design point $\mathbf{u}^*$. Once the design point and the principal axis are determined, the principal curvature is computed using gradient information of the last two iteration steps

$$\kappa_i = \frac{\text{sign}[\hat{\alpha}^{(N-1)}(\mathbf{u}^{(N)} - \mathbf{u}^{(N-1)})]}{\mathbf{u}^{(N)} - \mathbf{u}^{(N-1)}} \arccos(\hat{\alpha}^{(N-1)}[\hat{\alpha}^{(N)}]^T) \tag{7}$$

with $\hat{\alpha}^{(k)}$ the negative unit gradient vector at iteration point (k). By restarting the procedure in a subspace orthogonal to the previously found principal axes, a descending sequence of the principal curvatures (and corresponding axes) is constructed. It is interesting to note the similarities between the Iterative Curvature Method and the general principle of Hessian upgrading, described in the following section.

Quasi-Newton Optimization

General Principle

Using Newton optimization for unconstrained nonlinear programming requires the knowledge of the Hessian of the objective at each iteration point. Quasi-Newton methods (QNOM) use approximations of the Hessian, thus avoiding evaluation of the second-order derivatives. Most (performing) algorithms use information gathered during the optimization process to reconstruct the (inverted) Hessian of the objective function.

The updating of the Hessian is based on the fact that for an objective function $h(\mathbf{x})$, $\mathbf{x} \in E^n$ with continuous second-order partial derivatives, we can write

$$\mathbf{q}^{(k)^T} = [\nabla h(\mathbf{x}^{(k+1)}) - \nabla h(\mathbf{x}^{(k)})]^T \approx \nabla^2 h(\mathbf{x}^{(k)})(\mathbf{x}^{(k+1)} - \mathbf{x}^{(k)}) \tag{8}$$

If a set of n linearly independent directions can be found then it is possible to determine the inverted Hessian $\mathbf{H} = \{\nabla^2 h\}^{-1}$ by solving $\mathbf{H}^{(k+1)}[\mathbf{q}_i^{(k)}]^T = (\mathbf{x}_i^{(k+1)} - \mathbf{x}_i^{(k)})$, for $i = 1(1)n$. For $i < n$ there are an infinity of solutions to this equation, allowing for different updating algorithms. In this paper an introduction to only the Davidon-Fletcher-Powell method will be given, since this updating method will be used to solve the reliability programming problem (2).

Davidon-Fletcher-Powell Method

This method, originally introduced by Davidon, but rigorously described by Fletcher and Powell [12], is considered as the most efficient general-purpose algorithm for unconstrained optimization [13]. It can be identified as a Conjugate Directions Method [14], thus allowing for eventual non-gradient alternatives [15][16]. Stewart [17] used the Davidon-Fletcher-Powell (DFP) algorithm for efficient selection of the step size, when approximating gradients by finite differences.

In the DFP-algorithm, one searches along a direction $\mathbf{d}^{(k)} = -\mathbf{H}^{(k)}\nabla h(\mathbf{x}^{(k)})$, with $\mathbf{x}^{(k)}$ the current coordinates of the optimum and $\mathbf{H}^{(k)}$ the approximation of the Hessian of the objective $h(\mathbf{x})$. After solving a linear optimization problem along this direction, a new trial point is found $\mathbf{x}^{(k+1)} = \mathbf{x}^{(k)} + \mathbf{s}^{(k)} = \mathbf{x}^{(k)} + \sigma^{(k)}\mathbf{d}^{(k)}$. The inverted Hessian $\mathbf{H}$ is updated using a rank-two correction

$$\mathbf{H}^{(k+1)} = \mathbf{H}^{(k)} + \mathbf{A}^{(k)} + \mathbf{B}^{(k)} \tag{9}$$

with

$$\mathbf{A}^{(k)} = \frac{\mathbf{s}^{(k)}\mathbf{s}^{(k)T}}{\mathbf{q}^{(k)}\mathbf{s}^{(k)}} \qquad \mathbf{B}^{(k)} = -\frac{\mathbf{H}^{(k)}\mathbf{q}^{(k)T}\mathbf{q}^{(k)}\mathbf{H}^{(k)}}{\mathbf{q}^{(k)}\mathbf{H}^{(k)}\mathbf{q}^{(k)T}} \tag{10}$$

Both corrections are determined by applying the DFP-method to a quadratic programming problem. The Hessian of a quadratic objective is a constant matrix $\nabla^2 h(\mathbf{x}) \equiv \mathbf{C}$. It can be shown [12] that in that case the directions $\{\mathbf{s}^{(k)}\}$ are $\mathbf{C}$ conjugate and that they are eigenvectors of $\mathbf{H}^{(k)}\mathbf{C}$. This implies

$$\mathbf{s}^{(m)T}\mathbf{C}\mathbf{s}^{(l)} = 0 \quad \text{for} \quad m < l < k \tag{11}$$

$$\mathbf{H}^{(k)}\mathbf{C}\mathbf{s}^{(m)} = \mathbf{s}^{(m)} \quad \text{for} \quad m < k \tag{12}$$

Both conditions guarantee (quadratic) convergence of the algorithm in n iteration steps. The same number of iterations is sufficient to construct the inverted Hessian $\mathbf{H}^{(n)} = \mathbf{C}^{-1}$, and this independently of the initial choice of $\mathbf{H}^{(0)}$. Moreover, Fletcher and Powell continue to prove that if $\mathbf{H}^{(0)}$ is chosen to be positive definite, all subsequent $\mathbf{H}^{(k)}$ will be positive definite. Since a nonlinear objective function behaves quadratically in the vicinity of a (local) minimum, the aforementioned principles can succesfully be extended to general optimization problems.

DFP for the Reliability Problem

Constrained optimization – Lagrange methods

Applying the DFP-method to the reliability problem requires extension of this quasi-newton optimization procedure to solve constrained, nonlinear programming problems. In this section the extension is restricted to the very specific nature of the reliability problem, i.e., a simple objective function $h(\mathbf{u}) = \frac{1}{2}\mathbf{u}^T\mathbf{u}$ with a single nonlinear equality constraint $G(\mathbf{u}) = 0$. Luenberger [18] suggests effective use of the Lagrangian equations for solving constrained optimization problems.

In order for a problem (2) to have a solution, the Lagrange first-order necessary conditions have to be met, meaning that $\mathbf{u}^*$ must be a solution of the nonlinear system of equations

$$\begin{aligned} \nabla_u L(\mathbf{u}, \lambda) &= 0 \\ G(\mathbf{u}) &= 0 \end{aligned} \tag{13}$$

with $L(\mathbf{u}, \lambda) = f(\mathbf{u}) + \lambda G(\mathbf{u})$ the Lagrangian of the programming problem and λ the Lagrange multiplier. Applying Newton's method to solve the system of equations, results in

$$\begin{bmatrix} \mathcal{L}^{(k)} & \nabla G^{(k)T} \\ \nabla G^{(k)} & 0 \end{bmatrix} \begin{bmatrix} \Delta \mathbf{u}^{(k)} \\ \Delta \lambda^{(k)} \end{bmatrix} = - \begin{bmatrix} \nabla_u L^{(k)} \\ G^{(k)} \end{bmatrix} \tag{14}$$

with $\mathcal{L}^{(k)} = \nabla_u^2 L(\mathbf{u}^{(k)}, \lambda)$ and superscript (k) meaning that all function values are evaluated at $\mathbf{u}^{(k)}$. The next iteration point is $\mathbf{u}^{(k+1)} = \mathbf{u}^{(k)} + \Delta\mathbf{u}^{(k)}$ with the associated Lagrange multiplier $\lambda^{(k+1)} = \lambda^{(k)} + \Delta\lambda^{(k)}$. When $\mathcal{L}^{(k)}$ is positive definite, explicit solution of (14) is possible and is given by

$$\Delta\lambda = \frac{[\, G - \nabla G \mathcal{L}^{-1} \nabla_u L \,]}{\nabla G \mathcal{L}^{-1} \nabla G^T} \tag{15}$$

$$\Delta\mathbf{u} = -\mathcal{L}^{-1}[\, \mathbf{I}_n - \frac{\nabla G^T \nabla G \mathcal{L}^{-1}}{\nabla G \mathcal{L}^{-1} \nabla G^T} \,] \nabla_u L - \frac{\mathcal{L}^{-1} \nabla G^T G}{\nabla G \mathcal{L}^{-1} \nabla G^T} \tag{16}$$

where the superscript (k) has been dropped for notational convenience.

DFP–Lagrange Method and SORM

From (14), it is seen that a quasi-Newton scheme can easily be implemented, approximating only the Hessian of the Lagrangian $\mathcal{L}^{(k)}$. Considering equations (15) and (16), it is obvious that DFP-updating is particularly well-suited since only the inverted Hessian enters the formulas. DFP also guarantees the positive definiteness of $\mathcal{H}^{(k)}$, approximation to $[\mathcal{L}^{(k)}]^{-1}$, so that explicit solution of (14) is always possible.

When finally converged, the DFP–Lagrange method provides the required information for SORM analysis

$$\mathcal{H}^{(N)} \rightarrow [\mathcal{L}^*]^{-1} = (\mathbf{I}_n + \lambda^* \nabla^2 G(\mathbf{u}^*))^{-1} \tag{17}$$

By solving (17), the $\nabla^2 G(\mathbf{u}^*)$ is easily obtained, thus providing all necessary information for the curvature-fitted SORM.

It is interesting to note that in the proposed procedure no cumbersome and inaccurate line-searches are performed. In finite element reliability, the high numerical cost for solving (17) and the associated eigenvalue solution is negligible in comparison to the time-consuming evaluation of the limit-state function and its gradient.

Numerical Example

The example used by Der Kiureghian and De Stefano [9] is employed here to evaluate the accuracy and efficiency of the proposed method. The limit-state function is defined to be

$$G(\mathbf{u}) = \beta + \frac{1}{2}\sum_{i=1}^{9} \kappa_i u_i^2 - u_{10} \tag{18}$$

Since the design point $\mathbf{u}^*$ and principal curvatures of the limit-state function are known, performance of the algorithm can easily be evaluated. Convergence to the design point is determined by the criterion

$$\max[\,|\,G(\mathbf{u}^{(k)})\,|, \max_i |\,u_i^{(k)} - \hat{\alpha}^{(k)}\mathbf{u}^{(k)}\alpha_i^{(k)}\,|\,] \leq tol \tag{19}$$

with $\hat{\alpha}$ the unit negative gradient of the limit-state function.

If accurate second-order derivative information is required, a convergence check on the Hessian should be included. A commonly applicable criterion is

$$\frac{\|\mathbf{H}^{(k+1)}\| - \|\mathbf{H}^{(k)}\|}{\|\mathbf{H}^{(k)}\|} \leq tol \tag{20}$$

Aside from the fact that this criterion puts an additional numerical load on the optimization software, it shows unreliable behavior due to instability in the Hessian updating after a certain number of iteration steps.

Knowing that for quadratic programming problem the DFP-method converges in n steps to a local minimum and to the Hessian (see previous section), and knowing that a function behaves quadratically in a close neighborhood of a local optimum, the use of an alternative stopping criterion is suggested. Once convergence to the failure point is achieved, perform n additional iteration steps to construct the Hessian of the limit-state function. Convergence to the failure point is evaluated using (19). Applying the DFP-Lagrange method with the aforementioned stopping criterion to (18), the following results are obtained:

	$\beta = 3$			$\beta = 6$		
$tol =$	10^{-4}	10^{-6}	10^{-8}	10^{-4}	10^{-6}	10^{-8}
κ_i	N=18	N=20	N=22	N=22	N=26	N=29
0.30	0.298	0.298	0.299	0.290	0.292	0.292
0.29	0.288	0.288	0.289	0.279	0.280	0.281
0.28	0.280	0.278	0.279	0.268	0.268	0.268
0.27	0.275	0.269	0.270	0.258	0.261	0.263
0.26	0.267	0.259	0.259	0.250	0.252	0.254
0.25	0.263	0.249	0.250	0.235	0.242	0.242
0.24	0.235	0.239	0.239	0.230	0.231	0.235
0.23	0.219	0.229	0.229	0.214	0.221	0.223
0.22	0.017	0.219	0.219	0.206	0.206	0.213

	$\beta = 3$			$\beta = 6$		
$tol =$	10^{-4}	10^{-6}	10^{-8}	10^{-4}	10^{-6}	10^{-8}
κ_i	N=18	N=20	N=22	N=22	N=26	N=29
0.30	0.292	0.298	0.298	0.298	0.298	0.299
0.25	0.237	0.249	0.249	0.248	0.249	0.250
0.20	0.199	0.199	0.200	0.200	0.200	0.200
0.15	0.149	0.149	0.150	0.149	0.149	0.150
0.10	0.102	0.102	0.102	0.099	0.100	0.100
0.05	0.049	0.049	0.049	0.050	0.050	0.050
0.00	-0.004	-0.004	-0.004	-0.012	-0.012	-0.012
-0.05	-0.051	-0.051	-0.051	-0.049	-0.049	-0.049
-0.10	-0.102	-0.102	-0.100	-0.100	-0.100	-0.100

These values have been calculated by starting the iteration procedure with $\mathbf{u}^{(0)} = [1, 1, \ldots, 1]^T$ and $\lambda^{(0)} = 1$. Note that for all practical purposes accurate curvatures are obtained in few iteration steps. The parameters of the problem were specifically chosen to represent ill-conditioned situations.

Sensitivity to the starting values $\mathbf{u}^{(0)}$ and $\lambda^{(0)}$ is within reasonable boundaries although influence on the curvature values is unmistakable. When choosing $\mathbf{u}^{(0)} = [0, 1, \ldots, 1]^T$, i.e. a point on one of the principal axes as the starting point, the algorithm converges to the correct failure point but the Hessian of the limit-state function is singular and the largest principal curvature is not found. This situation can be detected and the procedure automatically can be restarted using a perturbed trial point as starting point.

The accuracy and convergence rate of the Hessian upgrading obviously depends on the conditioning of the problem. Schemes for improving the convergence rate, such as scaling, partial updating, ..., are currently being compared and implemented.

Conclusion

Using Quasi-Newton optimization algorithms for solving the reliability optimization problem shows interesting features: the sequential approximation of the Hessian not only provides a means of accelerating convergence to the minimum, but can also (indirectly) be used for determining second-order derivative information about the limit-state function, thus immediately enabling SORM-analysis by Curvature Fitting.

References

[1] Der Kiureghian A. and Liu P-L. Structural reliability under incomplete probability information. *Journal of the Engineering Mechanics Division, ASCE*, 112(1):85–104, 1986.

[2] Madsen H.O, Krenk S., and Lind N.C. *Methods of Structural Safety.* Prentice-Hall, Englewood Cliffs, New Jersey, USA, 1986.

[3] Shinozuka M. Basic analysis of structural safety. *Journal of Structural Engineering, ASCE*, 109(3):0–0, 1983.

[4] Liu P-L. and Der Kiureghian A. Optimization algorithms for structural reliability. *Structural Safety*, 9(0):161–177, 1991.

[5] Breitung K. Asymptotic approximation for multinormal integrals. *Journal of Engineering Mechanics, ASCE*, 110(3):357–366, 1984.

[6] Tvedt L. Distribution of quadratic forms in normal space – application to structural reliability. *Journal of Engineering Mechanics, ASCE*, 116(6):1183–1197, 1990.

[7] Fiessler B., Neumann H-J., and Rackwitz R. (1979). Quadratic limit states in structural reliability. *Journal of the Engineering Mechanics Division, ASCE*, 105(4):0–0, 1979.

[8] Der Kiureghian A., Lin H-Z., and Hwang S-J. Second order reliability approximations. *Journal of Engineering Mechanics, ASCE*, 113(8):1208–1225, 1987.

[9] Der Kiureghian A. and De Stefano M. An efficient algortihm for second-order reliability analysis. Technical Report UCB/SEMM-90/20, Department of Civil Engineering, Universtity of California at Berkeley, Berkeley, California, USA, 1990.

[10] Hohenbichler M. and Rackwitz R. Non-normal dependent vectors in structural safety. *Journal of the Engineering Mechanics Division, ASCE*, 107(6):1227–1238, 1981.

[11] Liu P-L. and Der Kiureghian A. Finite-element reliability methods for geometrically nonlinear stochastic systems. Technical Report UCB/SEMM-89/05, Department of Civil Engineering, University of California at Berkeley, Berkeley, California, USA, 1989.

[12] Fletcher R. and Powell M.J.D. A rapidly convergent method for minimization. *Computer Journal*, 6(0):163–168, 1963.

[13] Rao S.S. *Optimization: Theory and Applications.* Wiley & Sons, New Delhi, India, second edition, 1984.

[14] Luenberger D.G. *Linear and Nonlinear Programming.* Addison-Wesley, Reading, Massachusetts, USA, second edition, 1984.

[15] Powell M.J.D. An efficient method for finding the minimum of a function of several variables without calculating derivatives. *Computer Journal*, 7(0):155–162, 1964.

[16] Brent R.P. *Algorithms for Minimization without Derivatives.* Prentice-Hall, Englewood Cliffs, New Jersey, USA, 1973.

[17] Stewart G.W. A modification of davidon's minimization method to accept difference approximation of derivatives. *Journal of the Association for Computing Machinary*, 14(1):72–83, 1967.

[18] Luenberger D.G. The conjugate residual method for constrained minimization problems. *SIAM Journal of Numerical Analysis*, 7(1):390–398, 1970.

Method of Stochastic Linearization Revised and Improved*

I. Elishakoff

Center for Applied Stochastics Research and Department of Mechanical Engineering, Florida Atlantic University, Boca Raton, FL 33431-0991, U.S.A.

ABSTRACT

There is a vast literature devoted to the stochastic linearization technique generated during almost three decades. Can anything new be said on this technique? The paper demonstrates that the answer to this question is affirmative. New versions of the stochastic linearization technique are described. The methods are based on the requirement associated with the mean-squares of the potential energies of the original nonlinear system and in their equivalent, linear counterparts. Two examples of the nonlinear systems are numerically evaluated. In one of the cases considered a new stochastic linearization criterion yields results coinciding with the exact solution.

INTRODUCTION

Stochastic linearization is a most versatile method for analysis of single and multidegree-of-freedom nonlinear systems, as well as of continuous structures. It has been suggested in now classical papers by Booton [1] and by Kazakov [2].

Consider the single-degree-of-freedom system governed by equation

$$m\ddot{X} + c\dot{X} + f(X) = Q(t) \tag{1}$$

where m is mass, c - damping, f (X) - nonlinear restoring force, Q(t)

---

* Dedicated to 60th Birth Anniversary of Professor M. Shinozuka.

stochastic excitation, X -stochastic displacement. Within the stochastic linearization technique, Eq. (1) is replaced by the following linear equation

$$m\ddot{X} + c\dot{X} + k_{eq}X = Q(t) \tag{2}$$

where k_{eq} is the spring constant chosen in the manner, to be equivalent to the nonlinear restoring force f(X).

The equivalence is understood [1,2] in the following sense: it is required that the mean-square deviation of f(X) and of k_{eq} X is minimal:

$$E\{[f(X) - k_{eq}X]^2\} = \min \tag{3}$$

or

$$\frac{\partial}{\partial k_{eq}} E\{[f(X) - k_{eq}X]\} = 0 \tag{4}$$

This results in the expression

$$k_{eq} = \frac{E[Xf(X)]}{E(X^2)} \tag{5}$$

In almost 30 years since this method was discovered, numerous studies were undertaken for various nonlinear systems. The accounts of these investigations were given by, e.g., Sinitsin [3] and Roberts and Spanos [4]. The first structural applications were given by Caughey [5]. In these investigations the equivalent spring constant k_{eq}, determined through Eq. (5) was applied.

Recently, new versions of stochastic linearization was suggested by Elishakoff and Zhang [8] and by Zhang, Elishakoff and Zhang [9]. In order to elucidate new versions of the stochastic linearization, let us consider the differential equation studied previously in the literature in detail [10,11]:

$$\ddot{X} + \beta\dot{X} + \epsilon X^3 = Q(t) \tag{6}$$

where β is the damping constant, α is the non-linear stiffness constant and Q(t) is a Gaussian white noise with

$$E[Q(t)] = 0\ , \quad E[Q(t)Q(t + \tau)] = 2d\beta\delta(\tau) \tag{7}$$

where $\delta(\tau)$ is Dirac's delta function, α and β are positive constants.

The exact stationary probability density function of the above system, obtained by the Fokker-Plank approach is

$$f_X(x) = C_1 \exp\left[-\frac{\epsilon}{4d}x^4\right] \tag{8}$$

where C_1 is the normalization constant. The exact mean-square displacement

$$\sigma_X^2 = E(X^2) = \int_{-\infty}^{\infty} x^2 f_X(x)\,dx \tag{9}$$

is found by using the formula

$$\int_0^{\infty} x^{r-1}\exp(-ax^h)\,dx = \frac{\Gamma(r/h)}{h a^{r/h}} \tag{10}$$

where $\Gamma(x)$ is the Gamma function. The exact mean-square displacement becomes

$$\sigma_X^2 = \left[\frac{4d}{\epsilon}\right]^{1/2}\frac{\Gamma(3/4)}{\Gamma(1/4)} = 0.6760\left[\frac{d}{\epsilon}\right]^{1/2} \tag{11}$$

Conventional equivalent linearization yields (Eq.5):

$$k_{eq}^{(1)} = \frac{E(\epsilon X^4)}{E(X^2)} \tag{12}$$

We use an assumption, that X is normally distributed. Hence Eq. 12 becomes

$$k_{eq}^{(1)} = 3\epsilon E(X^2) \tag{13}$$

The mean-square response of the system (2), with parameters $m \equiv 1$, $c=\beta$, $k_{eq} \equiv k_{eq}^{(1)}$

$$E(X^2) = \int_{-\infty}^{\infty} \frac{S_Q(\omega)\,d\omega}{(k_{eq} - \omega^2)^2 + \beta^2\omega^2} \tag{14}$$

equals

$$E(X^2) = \frac{\pi S}{k_{eq}^{(1)}\beta} \tag{15}$$

where $S = 2d\beta/2\pi = d\beta/\pi$ is the spectral density of the excitation. Mean-square displacements is

$$E(X^2) = \frac{d}{3\epsilon E(X^2)} \tag{16}$$

which results in

$$E(X^2) = 0.5774 \left[\frac{d}{\epsilon}\right]^{1/2} \tag{17}$$

According to the version due to Elishakoff and Zhang [8] we require that the original nonlinear system and its linear counterpart possess equal mean-square values of potential energies, i.e.

$$E\left[U^2(X)\right] = E\left[\frac{1}{2}k_{eq}^{(2)}X^2\right]^2 \tag{18}$$

In the system under consideration

$$U(X) = \frac{1}{4}\epsilon X^4 \tag{19}$$

Hence

$$k_{eq}^{(2)} = \frac{1}{2}\epsilon\sqrt{\frac{E(X^8)}{E(X^4)}} \tag{20}$$

Gaussian assumption for distribution of X yields

$$E(X^8) = 105E(X^2), \quad E(X^4) = 3E(X^2) \tag{21}$$

Therefore

$$k_{eq}^{(2)} = \frac{\sqrt{105}}{2\sqrt{3}}\,\epsilon E(X^2) \tag{22}$$

Eq. 15 becomes under new circumstances

$$E(X^2) = \frac{\pi S}{k_{eq}^{(2)}\beta} = \frac{2\sqrt{3}}{\sqrt{105}\,\epsilon E(X^2)} \tag{23}$$

yielding

$$E(X^2) = \left[\frac{12}{105}\right]^{1/4} \left[\frac{d}{\epsilon}\right]^{1/2} = 0.5814 \left[\frac{d}{\epsilon}\right]^{1/2} \tag{24}$$

This represents a slight improvement over the result furnished by the conventional stochastic linearization technique (Eq. 17).

In Ref. 9, Zhang, Elishakoff and Zhang presented an additional criterion for equivalence. It consists in the requirement that the mean-square deviation between the potential energies possessed by the original nonlinear system and its linear imitation be minimal, i.e. that

$$\frac{\partial}{\partial k_{eq}} E \left[U(X) - \frac{1}{2}k_{eq}^{(3)}X^2\right]^2 = 0 \tag{25}$$

This requirement yields

$$k_{eq}^{(3)} = \frac{1}{2}\epsilon\frac{E(X^6)}{E(X^4)} \tag{26}$$

Within Gaussian approximation

$$E(X^6) = 15\,[E(X^2)]^3 \tag{27}$$

$$k_{eq}^{(3)} = \frac{5}{2}\epsilon E(X^2) \tag{28}$$

Therefore, with Eq. 15 in mind and replacement $k_{eq}^{(1)} \rightarrow k_{eq}^{(3)}$ we obtain

$$E(X^2) = \frac{d}{2.5\epsilon E(X^2)} \tag{29}$$

yielding

$$\sigma_X^2 = \left[0.4\frac{d}{\epsilon}\right]^2 = 0.6325\left[\frac{d}{\epsilon}\right]^{1/2} \tag{30}$$

Comparison between the exact solution and the approximate solutions reveals that for the system under consideration the latter approximation is the best. Indeed, conventional stochastic linearization yields an error of 14.6%. In contrast the error associated with the minimum mean-square deviation of potential energies constitutes only 6.4%.

Consider now condition Eq. 6 but with linear restoring force added to yield a Duffing oscillator:

$$\ddot{X} + \beta\dot{X} + \alpha X + \epsilon X^3 = Q(t) \tag{31}$$

The mean-square displacement for the system with $\epsilon \equiv$ is obtained from Eq. 15 by formally replacing $k_{eq}^{(1)}$ by α

$$E(X^2)\,|_{\epsilon=0} = \frac{\pi S}{\alpha\beta} \equiv e_0^2 \tag{32}$$

The exact probability density of X is uniformly available in the literature and will allow a comparison with conventional and new stochastic linearization methods:

$$f_X(x) = C_2 \exp\left[-\frac{\pi S}{\alpha\beta}\left(\frac{1}{2}x^2 + \frac{1}{4}\frac{\epsilon}{\alpha}x^4\right)\right] \tag{33}$$

where C_2 is a normalization constant. In view of Eq. 32, Eq. 33 can be rewritten as

$$f_X(x) = C_2 \exp\left[-\frac{1}{e_0^2}\left(\frac{1}{2}x^2 + \frac{1}{4}\frac{\epsilon}{\alpha}x^4\right)\right] \tag{34}$$

We introduce new variables

$$x = p\tau, \quad p = \left(\frac{4\alpha e_0^2}{\epsilon}\right)^{1/4}, \quad y = \frac{1}{4e_0}\left(\frac{\alpha}{\epsilon}\right)^{1/2} \tag{35}$$

The normalization condition

$$C_2 \int_{-\infty}^{\infty} \exp\left[-\frac{1}{e_0^2}\left(\frac{1}{2}x^2 + \frac{1}{4}\frac{\epsilon}{\alpha}x^4\right)\right] dx = 1 \tag{36}$$

yields

$$C_2 = \frac{1}{pZ_1(y)} \tag{37}$$

where

$$Z_1(y) = 2 \int_0^{\infty} \exp(-\tau^4 - 4y^2\tau^2)d\tau \tag{38}$$

Mean square displacement

$$E(X^2) = \int_{-\infty}^{\infty} x^2 f_X(x)dx \tag{39}$$

reads

$$E(X^2) = \frac{p^2 Z_2(y)}{Z_1(y)} = 2e_0 \frac{Z_2(y)}{Z_1(y)} \left[\frac{\alpha}{\epsilon}\right]^{1/2} \tag{40}$$

where

$$Z_2(y) = 2 \int_0^{\infty} \tau^2 \exp(-\tau^4 - 4y^2\tau^2)dx \tag{41}$$

Functions $Z_1(y)$ and $Z_2(y)$ are defined and tabulated for certain values of y by Stratonovich [12]. Note that these functions can also be reduced to cylindrical functions of a fractional order [13-15]. Consider a particular case

$$e_0^2 = 0.54 \ , \quad \frac{\alpha}{\epsilon} = 1 \tag{42}$$

In these circumstances,

$$y = 0.34021 \ , \quad Z_1(y) = 1.26368 \ , \quad Z_2(y) = 0.26310 \tag{43}$$

The exact mean square value becomes

$$E(X^2) = 0.306 \tag{44}$$

Let us now contrast the performance of the conventional and new stochastic linearization techniques. Conventional stochastic linearization yields

$$k_{eq}^{(1)} = \alpha + 3\epsilon E(X^2) \tag{45}$$

Substitution into Eq. 15 yields

$$E(X^2) = \frac{\pi S}{\beta[\alpha + 3\epsilon E(X^2)]} \tag{46}$$

or, in view of Eq. 32

$$E(X^2) = \frac{e_0^2}{1 + 3(\epsilon/\alpha)E(X^2)} \tag{47}$$

resulting in a quadratic

$$3(\epsilon/\alpha)E^2(X^2) + E(X^2) - e_0^2 = 0 \tag{48}$$

For numerical values adopted in Eq. 44, we have, instead of Eq. 48

$$3E^2(X^2) + E(X^2) - 0.54 = 0 \tag{49}$$

with attendant mean-square value

$$E(X^2) = 0.289 \tag{50}$$

which constitutes a difference of 6.47% with the exact solution

Consider now the energy based stochastic linearization method. Eq. 25 yields

$$k_{eq}^{(3)} = \alpha + 2.5\epsilon E(X^2) \tag{51}$$

Substitution $k_{eq}^{(3)}$ instead of $k_{eq}^{(1)}$ in Eq. (15) yields

$$E(X^2) = \frac{\pi S}{\beta[\alpha + 2.5\epsilon E(X^2)]} \tag{52}$$

or in view of Eq. 32

$$E(X^2) = \frac{e_0^2}{1 + 2.5(\epsilon/\alpha)E(X^2)} \tag{53}$$

This results in a quadratic

$$2.5(\epsilon/\alpha)E^2(X^2) + E(X^2) - e_0^2 = 0 \tag{54}$$

For the values in Eq. 44, we get an equation

$$2.5E^2(X^2) + E(X^2) - 0.54 = 0 \tag{55}$$

with attendant mean-square value

$$E(X^2) = 0.306 \tag{56}$$

which coincides with the exact value given in Eq. 45. For value of e_0^2 in vicinity of .54 and ratio ϵ/α in vicinity of unity the relative error furnished by a new stochastic linearization technique may constitute about one percent, much smaller than the one produced by the conventional stochastic linearization technique.

The discovered coincidence of the stochastic linearization result with exact solution suggests that for specific set of parameters the new version of stochastic linearization constitutes "true" linearization, in terminology of Kozin [16].

CONCLUSION

Preliminary results for an example problem have been presented to illustrate the power of the new stochastic linearization techniques. It turns out that the energy based equivalence criteria are superior to the conventional stochastic linearization scheme in some ranges of parameters. It is demonstrated that for specific set of parameters one of the energy based stochastic linearization techniques yields results which coincide with exact solution, thus representing a "true" stochastic linearization. It appears that additional investigation is needed to explore the advantages of the new stochastic linearization criteria.

ACKNOWLEDGEMENT

Helpful discussions with and continous encouragement of Dr. S.C. Liu of the National Science Foundation, are greatly appreciated.

REFERENCES

1. Booton, R.C., The Analysis of Nonlinear Control Systems with Random Inputs, Proceedings Symposium on Nonlinear Circuit Analysis, Vol. 2, 1953.

2. Kazakov, I.E., Approximate Method of Statistical Investigation of Nonlinear Systems, Proceedings of V.V.I.A., Vol. 394, 1954 (in Russian).

3. Sinitsyn, I.N., Methods of Statistical Linearization (Survey), Automation and Remote Control, Vol. 35, 1974, pp. 765-776.

4. Roberts, J.B. and Spanos, P.D., Random Vibration and Statistical Linearization, Wiley, Chichester, 1990.

5. Caughey, T.K., Equivalent Linearization Techniques, Journal of Acoustical Society of America, Vol. 35 (11), pp. 1706-1711, 1963.

6. Popov, E.P. and Paltov, I.N., Approximate Methods of Investigation of Nonlinear Automatic Systems," Fizmatgiz" Publishers, Moscow, 1960 (in Russian).

7. Bolotin, V.V., Application of the Methods of Theory of Probability and the Theory of Reliability in Design of Structures, "Izdatelstvo Literatury po Stroitelstvu" Publishers, Moscow, 1961 (in Russian).

8. Elishakoff, I. and Zhang, X., An Appraisal of Different Stochastic Linearization Techniques, Journal of Sound and Vibration, 1991 (to appear).

9. Zhang, X.T. Elishakoff, I. and Zhang, R.Ch., A Stochastic Linearization Technique Based on Minimum Mean Square Deviation of Potential Energies, in "Stochastic Structural Dynamics-New Theoretical Developments" (Eds. Lin, Y.K. and Elishakoff, I.), Springer, Berlin, 1991, to appear.

10. Atalik T.S. and Utku, S., Stochastic Linearization of Multi-degree-of-freedom Non-linear Systems, Earthquake Engineering and Structural Dynamics, Vol. 4, pp. 411-420, 1976.

11. Roberts, J.B., Response of Nonlinear Mechanical Systems to Random Excitation. Part 2: Equivalent Linearization and Other Methods, The Shock and Vibration Digest, Vol. 13 (5), pp. 15-29, 1981.

12. Stratonovich, R.L., Selected Porblems of Theory of Fluctuations in Radiotechnics, "Sovietskoe Radio" Publishers, Moscow, 1961, pp. 368 and 545.

13. Bolotin, V.V., Random Vibrations of Elastic Systems, Martinus Nijhoff Publishers, The Hague, 1984, pp. 290-292.

14. Piszczck, K. and Niziol, J., Random Vibration of Mechanicsl Systems, Ellis Horwood Limited, Chichester, 1986, pp. 173-175.

15. Constantinou, M.C., Vibration Statistics of the Duffing Oscillator, Soil Dynamics and Earthquake Engineering, Vol. 4, 1985, pp. 221-223.

16. Kozin, F., The Method of Statistical Linearization for Non-Linear Stochastic Vibrations, in " Nonlinear Stochastic Dynamic Engineering Systems" (Eds. Ziegler, F. and Schuëller, G.I.), pp.45-56 Springer, Berlin, 1987.

Maximal Lyapunov Exponent for a Stochastically Perturbed Co-Dimension Two Bifurcation

N. Sri Namachchivaya, S. Talwar

Department of Aeronautical and Astronautical Engineering, University of Illinois at Urbana-Champaign, Urbana, Il 61801, U.S.A.

ABSTRACT

Almost-sure asymptotic stability of a 3 dimensional codimension two dynamical system under small intensity stochastic excitations is investigated. The method of stochastic averaging is used to derive a set of approximate Itô equations. These equations, along with their sample properties, are then examined to obtain the almost-sure stability conditions. The sample properties of the process are based on the boundary behavior of the associated scalar diffusion process of the amplitude Itô equations. The maximal Lyapunov exponent is calculated using the ergodic scalar diffusive process, which in turn yields the almost-sure stability conditions.

INTRODUCTION

An autonomous system undergoes a co-dimension two bifurcation when the linear part of the vector field is doubly degenerate. There are three cases of such degeneracies with two, three and four dimensional center manifolds where the linear part can have (1) two zero eigenvalues, (2) a pair of pure imaginary eigenvalues and a single zero eigenvalue and (3) two pairs of pure imaginary eigenvalues without 1:1 resonance, respectively. The aim of this analysis is to study the effects of random parametric excitations on such co-dimension two bifurcations. As a first step, one has to examine the linear stochastic system to determine the parameter values at which the system will exhibit bifurcations. It has been shown by Arnold and Kliemann [1] that for systems with stochastic parametric excitation the Lyapunov exponents are analogous to the real part of the eigenvalue and the maximal Lyaponov exponent yields the almost-sure stability boundaries in the system parameter space. To this end, this paper is concerned with the calculation of the Lyapunov exponents and thus the bifurcation points in the parameter space.

Even though a large amount of research pertaining to sufficient conditions is available, an exact stability boundary, which relies on a necessary and sufficient almost-sure stability condition was first obtained by Khasminskii [2]. A complete study of second order systems, taking into consideration all possible singularities that can exist in one dimensional diffusion, was done by Nishioka [3]. In most of the studies reported thus far, the necessary and sufficient conditions were obtained only for second order systems. However, to understand most physical phenomena, it is necessary to obtain results for multidegree-of-freedom systems. The level of mathematical difficulty encountered in this type of analysis, due to the associated multidimensional diffusion problem, has restricted major developments in this area.

The method presented herein has three important steps. As the first step approximate Itô equations in amplitudes and phase are obtained assuming the fluctuating coupling terms have zero mean with auto-correlation that goes to zero rapidly for large time. For the three dimensional case, the Itô equations represent a diffusion on a cylinder (r,θ,z). Due to the fact that the amplitude (r) process and the process on the real line (z) are decoupled from the phase (θ) equation, we can transform these two coordinates to a half circle. Thus, the second step is to examine the diffusion on this half circle. Making use of the boundary behavior of the singular points, the diffusion is shown to be ergodic. Finally, the maximal Lyapunov exponent is evaluated using the stationary density of the ergodic diffusive process. This provides the almost-sure asymptotic stability condition and thus the stability boundary in the system parameter space.

LYAPUNOV EXPONENTS

As mentioned previously co-dimension two bifurcation problems occur in one of three general forms. However, only the case associated with a simple zero and a pair of pure imaginary eigenvalues is considered. The deterministic version of this system is represented by the following equations:

$$\dot{r} = \mu_1 r + arz + \left(cr^3 + drz^2\right)$$

$$\dot{z} = \mu_2 + br^2 - z^2 + \left(er^2 z + fz^3\right)$$

In this paper, we investigate the almost-sure stability of the above bifurcation problem. The stochastic linear system corresponding to the above can be written as

$$\dot{x} = Ax + \eta f(t)Bx \quad , \qquad x \in \mathbf{R}^3 \tag{1}$$

where

$$A = \begin{bmatrix} \gamma_1 & \omega & 0 \\ -\omega & \gamma_1 & 0 \\ 0 & 0 & \gamma_2 \end{bmatrix} \quad , \qquad B = [b_{ij}] \; , \qquad i,j = 1, 2, 3$$

and B is assumed to be a symmetric matrix. The term f(t) is a real valued zero mean stochastic process with an auto-correlation that goes to zero rapidly for large times, i.e. it satisfies a strong mixing condition

$$\int_0^\infty \tau^m R(\tau)d\tau < \infty \quad , \qquad m = 0, 1, 2 \tag{2}$$

This condition implies that the power spectral density of the process and its first 2 derivatives are bounded at zero frequency and, hence, indicates that the auto-correlation of the process is decreasing as the time is increased. The method of stochastic averaging [4] can be used on models satisfying this assumption.

Introducing a scaling parameter ε into (1) and defining

$$\gamma_1 = -\varepsilon\delta_1 \qquad \gamma_2 = -\varepsilon\delta_2 \qquad \eta = \varepsilon^{1/2}\sigma$$

yields

$$\dot{x} = A_0 x - \varepsilon A_1 x + \varepsilon^{1/2}\sigma f(t)Bx$$

where

$$A_o = \begin{bmatrix} 0 & \omega & 0 \\ -\omega & 0 & 0 \\ 0 & 0 & 0 \end{bmatrix} \quad , \quad A_1 = \begin{bmatrix} \delta_1 & 0 & 0 \\ 0 & \delta_1 & 0 \\ 0 & 0 & \delta_2 \end{bmatrix}$$

In order to apply the method of stochastic averaging, the new variables $r(t)$, $\varphi(t)$ and $z(t)$ are introduced by means of the transformation

$$x_1 = r\sin\Phi \qquad x_2 = r\cos\Phi \qquad x_3 = z \qquad \Phi = \omega t + \varphi(t) \tag{3}$$

where r and z are regarded as the amplitude and the real axis respectively, and φ is the phase of the response processes on the cylinder. The use of this transformation yields equations in what is known as "standard form". These are

$$\dot{r} = -\varepsilon r\delta_1 + \varepsilon^{1/2}\sigma f(t)F_r \quad , \qquad \dot{\varphi} = \varepsilon^{1/2}\sigma f(t)F_\varphi$$

$$\dot{z} = -\varepsilon z\delta_2 + \varepsilon^{1/2}\sigma f(t)F_z \tag{4}$$

where

$$F_r = \frac{1}{2}r\left[J_+ + J_-\cos2\Phi + 2b_{12}\sin2\Phi\right] + z\left[b_{13}\sin\Phi + b_{23}\cos\Phi\right]$$

$$F_\varphi = \frac{1}{2}\left[2b_{12}\cos2\Phi - J_-\sin2\Phi\right] + \frac{z}{r}\left[b_{13}\cos\Phi - b_{23}\sin\Phi\right]$$

$$F_z = r\left[b_{31}\sin\Phi + b_{32}\cos\Phi\right] + zb_{33} \quad , \qquad J_\pm = b_{22} \pm b_{11}$$

The exact solution of (4) for an arbitrary random processes f(t) is not available. However, for processes satisfying the strong mixing condition (2), the solution converges weakly to a diffusive Markov process with the infinitesimal generator

$$L(\cdot) = m_i(x)\frac{\partial}{\partial x_i}(\cdot) + \frac{1}{2}[\sigma\sigma^T]_{ij}\frac{\partial^2(\cdot)}{\partial x_i \partial x_j} \quad , \qquad x = \{r, \varphi, z\} \tag{5}$$

Using the procedure for stochastic averaging outlined in Khasminskii [4], the drift and diffusion terms are found to be

$$m_r = -\delta_1 r + \frac{\sigma^2}{8} r J_+^2 S_{ff}(0) + \frac{3r\sigma^2}{16}\left(4b_{12}^2 + J_-^2\right)S_{ff}(2\omega)$$
$$+ \frac{\sigma^2}{4r}\left(b_{13}^2 + b_{23}^2\right)\left(r^2 + z^2\right)S_{ff}(\omega)$$

$$m_\varphi = -\frac{\sigma^2}{4}\left(b_{13}^2 + b_{23}^2\right)\Psi_{ff}(\omega) - \frac{\sigma^2}{8}\left(4b_{12}^2 + J_-^2\right)\Psi_{ff}(2\omega)$$

$$m_z = -\delta_2 z + \frac{\sigma^2 z}{2} b_{33}^2 S_{ff}(0) + \frac{\sigma^2 z}{2}\left(b_{13}^2 + b_{23}^2\right)S_{ff}(\omega) \tag{6}$$

$$\left[\sigma\sigma^T\right]_{rr} = \frac{r^2\sigma^2}{4} J_+^2 S_{ff}(0) + \frac{r^2\sigma^2}{8}\left(4b_{12}^2 + J_-^2\right)S_{ff}(2\omega)$$
$$+ \frac{z^2\sigma^2}{2}\left(b_{13}^2 + b_{23}^2\right)S_{ff}(\omega)$$

$$\left[\sigma\sigma^T\right]_{zz} = \sigma^2 z^2 b_{33}^2 S_{ff}(0) + \frac{\sigma^2 r^2}{2}\left(b_{13}^2 + b_{23}^2\right)S_{ff}(\omega)$$

$$\left[\sigma\sigma^T\right]_{\varphi\varphi} = \frac{\sigma^2 z^2}{2r^2}\left(b_{13}^2 + b_{23}^2\right)S_{ff}(\omega) + \frac{\sigma^2}{8}\left(4b_{12}^2 + J_-^2\right)S_{ff}(2\omega)$$

$$\left[\sigma\sigma^T\right]_{rz} = \frac{\sigma^2}{2} r z J_+ b_{33} S_{ff}(0) + \frac{\sigma^2}{2} r z\left(b_{13}^2 + b_{23}^2\right)S_{ff}(\omega)$$

$$\left[\sigma\sigma^T\right]_{r\varphi} = \left[\sigma\sigma^T\right]_{z\varphi} = 0$$

where

$$S_{ff}(\omega) = 2\int_0^\infty R(\tau)\cos(\omega\tau)d\tau \qquad \Psi_{ff}(\omega) = 2\int_0^\infty R(\tau)\sin(\omega\tau)d\tau$$

$$R(\tau) = E\left[f(t)f(t+\tau)\right]$$

$E[\cdot]$ denotes the expectation operator.

It is evident that the Itô equations for amplitude, r and z, are uncoupled from that of the phase φ. Using the logarithmic polar transformation (or Khasminskii transformation)

$$\xi = \frac{1}{2}\ln\left(r^2 + z^2\right) \quad , \qquad \theta = \tan^{-1}\left(\frac{z}{r}\right)$$

the necessary and sufficient condition for almost-sure stability can be obtained using the concepts of maximal Lyapunov exponents and multiplicative ergodic theory. The resulting Itô equations are

$$d\xi = Q(\theta)dt + \sum \mu^{\xi}_{ij}dW \quad , \qquad d\theta = L(\theta)dt + \sum \mu^{\theta}_{ij}dW \tag{7}$$

where

$$Q(\theta) = \lambda_1 \cos^2\theta + \lambda_2 \sin^2\theta + \psi^2(\theta)$$

$$L(\theta) = \frac{1}{2}(\lambda_2 - \lambda_1)\sin 2\theta + \frac{1}{8}\rho \sin 4\theta - \frac{1}{2}\beta_2 \tan\theta \tag{8}$$

$$\left[\mu^{\xi}\mu^{\xi^T}\right] = \beta_1 \cos^4\theta + \beta_2 \sin^2 2\theta + \frac{\sigma^2}{4}\left(J_+ \cos^2\theta + 2b_{33}\sin^2\theta\right)^2 S_{ff}(0)$$

$$\left[\mu^{\theta}\mu^{\theta^T}\right] = \psi^2(\theta) = \beta_2 + \frac{1}{4}\rho \sin^2 2\theta$$

$$\lambda_i = -\delta_i + \beta_i \qquad\qquad \beta_1 = \frac{\sigma^2}{8}\left(J_-^2 + 4b_{12}^2\right)S_{ff}(2\omega)$$

$$\beta_2 = \frac{\sigma^2}{2}\left(b_{13}^2 + b_{23}^2\right)S_{ff}(\omega) \qquad\qquad \rho = \frac{\sigma^2}{4}(J_+ - 2b_{33})^2 S_{ff}(0) + \beta_1 - 4\beta_2$$

From the definition of β_1 and β_2 it is obvious that they are always positive or identically zero. It is clear that the θ-process is itself a diffusion process on the unit half circle, $\theta \in [0,\pi]$. If $\psi(\theta)$ vanishes in this region, or if $L(\theta)$ goes to infinity, the process is singular. Examination of $L(\theta)$ indicates that a singularity is always present at $\theta=\pi/2$. Singularities can also when $\beta_2=0$ and $\theta=0,\pi/2$, or π simultaneously. β_2 is null if $S_{ff}(\omega) = 0$ or if $b_{13}=b_{23}=0$.

The diffusion behavior at the boundary points and any singularities are classified according to the scheme of Feller [5]. To this end, we introduce scale and speed measures defined, respectively, as

$$S(\theta) = \int^{\theta} s(x)dx \qquad\qquad M(\theta) = \int^{\theta} m(x)dx \tag{9}$$

where

$$s(x) = \mathrm{Exp}[-B(x)] \qquad\qquad m(x) = \frac{1}{\psi^2(x)s(x)}$$

$$B(x) = \int^{x} \frac{2L(\eta)}{\psi^2(\eta)} d\eta \tag{10}$$

s(x) and m(x) are called the scale and speed densities. Calculating B(x) yields

$$B(x) = \ln(|\cos x|) + \frac{1}{2}\ln\left(\beta_2 + \frac{1}{4}\rho\sin^2 2x\right) + \frac{2(\lambda_1 - \lambda_2)}{4\beta_2 + \rho}\int^{\cos 2x} \frac{du}{1 - \left(\frac{\rho}{4\beta_2 + \rho}\right)u^2}$$
$$+ \int^{\cos x} \frac{\rho u^3 - \rho u}{\beta_2 + \rho u^2 - \rho u^4} du \tag{11}$$

which is dependent on the sign of ρ. Consider first the case when $\rho=0$ which, in the presence of noise, is possible only for particular forms of B. For an arbitrary spectrum, this condition is equivalent to $B=\alpha I_{3x3}$ which results in no coupling between the z-process and the others. This may also represent a linear form that arises when one adds a fluctuating term to the unfolding parameter. It is worth noting that under this condition there is a reflective symmetry in the x_3 axis. However, for white noise the condition becomes

$$\frac{1}{4}(J_+ - 2b_{33})^2 + \frac{1}{8}\left(J_-^2 + 4b_{12}^2\right) - 2\left(b_{13}^2 + b_{23}^2\right) = 0 \tag{12}$$

which is satisfied by $B=\alpha I_{3x3}$ but can also be satisfied by less sparse B matrices.

Analytical solutions of (9) can be obtained for the case of $\rho=0$ and $\beta_2 \neq 0$. The scale and speed densities are found to be

$$s(\theta) = |\sec\theta| \mathrm{Exp}\left[\Gamma \sin^2\theta\right] \qquad m(x) = \frac{1}{\beta_2}|\cos\theta| \mathrm{Exp}\left[-\Gamma\sin^2\theta\right] \tag{13}$$

where

$$\Gamma = \frac{\lambda_1 - \lambda_2}{\beta_2}$$

The presence of a "singularity" in $s(\theta)$ at $\theta=\pi/2$ indicates a change in the diffusion process at this point and verifies the earlier prediction of a singularity at $\pi/2$. Thus, it is necessary to determine the boundary classifications of the set of points $\{0,\pi/2,\pi\}$. It is shown in Appendix A that $\theta=\pi/2$ is an entrance, and that both $\theta=0$ and $\theta=\pi$ are regular points, which can be further classified as purely reflecting [10]. The diffusion process thus evolves on the two quarter circles separately, $[0,\pi/2]$ and $[\pi/2,\pi]$, since once a particle enters one of the regions it can never pass the entrance point, $\theta=\pi/2$, to reach the other region. For a regular diffusion process, i.e. one which has no singularities within its domain, the diffusion is governed by the Fokker-Planck equation

$$\frac{dp}{d\tau}=\frac{1}{2}\frac{d^2}{d\theta^2}\left[\psi^2(\theta)p(\theta)\right]-\frac{d}{d\theta}\left[L(\theta)p(\theta)\right] \tag{14}$$

whose stationary solution is given as

$$p_{st}(\theta)=m(\theta)\left[c_1 S(\theta)+c_2\right] \tag{15}$$

where c_1 and c_2 are determined by normality and boundary conditions. The maximal Lyapunov exponent is then

$$\lambda=\int^{\theta} Q(\eta)p_{st}(\eta)\,d\eta \tag{16}$$

Examine the nonsingular diffusion on the quarter circle $\theta\in[0,\pi/2)$. The remaining derivations are valid on the quarter circle $\theta\in[0,\pi/2)$, but similar results can be obtained for the second region, $\theta\in(\pi/2,\pi]$. Due to the boundary conditions, $c_1=0$ and hence the stationary probability density is $p_{st}=c_2 m(\theta)$, where c_2 is determined from the normality condition

$$\int_0^{\pi/2} p_{st}(\eta)\,d\eta=1 \tag{17}$$

The probability density function for $\rho=0$ is found to be

$$p_{st}(\theta)=\frac{2\cos\theta}{\mathrm{erf}\left(\sqrt{\Gamma}\right)}\sqrt{\frac{\Gamma}{\pi}}\,\mathrm{Exp}\left[-\Gamma\sin^2\theta\right] \tag{18}$$

and the maximal Lyapunov exponent is

$$\lambda = \lambda_1 + \frac{\beta_2}{2} + \beta_2\sqrt{\frac{\Gamma}{\pi}}\frac{\mathrm{Exp}[-\Gamma]}{\mathrm{erf}(\sqrt{\Gamma})} \tag{19}$$

The maximal Lyapunov exponent can also be written using the confluent hypergeometric function [6] as

$$\lambda = \lambda_1 + \frac{\beta_2}{2} + \frac{\beta_2}{2M\left(1, \frac{3}{2}, \Gamma\right)} \tag{20}$$

It can be readily seen from (20) that the form of the Lyapunov exponent is correct by considering the case of zero stochastic disturbance. In this case, the maximal exponent should be the larger of δ_1 and δ_2 which is satisfied by (20). Figures 1 and 2 show representative forms of the probability density and maximal Lyapunov exponents for a given B matrix and specific values of the stochastic disturbance.

The cases for positive and negative ρ could not be completely solved analytically. The scale and speed measures are

(i) $\rho > 0$

$$s(\theta) = \frac{|\sec\theta|}{\left(\beta_2 + \frac{\rho}{4}\sin^2 2\theta\right)^{1/4}}\left[\frac{2\beta_2\sec^2\theta + \rho + \sqrt{\rho(\rho + 4\beta_2)}}{2\beta_2\sec^2\theta + \rho - \sqrt{\rho(\rho + 4\beta_2)}}\right]^{\left[\frac{\frac{\rho}{4} + \lambda_2 - \lambda_1}{\sqrt{\rho(\rho + 4\beta_2)}}\right]}$$

$$m(\theta) = \frac{|\cos\theta|}{\left(\beta_2 + \frac{\rho}{4}\sin^2 2\theta\right)^{3/4}}\left[\frac{2\beta_2\sec^2\theta + \rho - \sqrt{\rho(\rho + 4\beta_2)}}{2\beta_2\sec^2\theta + \rho + \sqrt{\rho(\rho + 4\beta_2)}}\right]^{\left[\frac{\frac{\rho}{4} + \lambda_2 - \lambda_1}{\sqrt{\rho(\rho + 4\beta_2)}}\right]} \tag{21a}$$

(ii) $\rho < 0$

$$s(\theta) = \frac{|\sec\theta|}{\left(\beta_2 + \frac{\rho}{4}\sin^2 2\theta\right)^{1/4}}\mathrm{Exp}\left[\frac{4(\lambda_1 - \lambda_2) - \rho}{2\sqrt{-\rho(\rho + 4\beta_2)}}\tan^{-1}\left(\frac{2\beta_2\sec^2\theta + \rho}{\sqrt{-\rho(\rho + 4\beta_2)}}\right)\right]$$

$$m(\theta) = \frac{|\cos\theta|}{\left(\beta_2 + \frac{\rho}{4}\sin^2 2\theta\right)^{3/4}}\mathrm{Exp}\left[\frac{4(\lambda_2 - \lambda_1) + \rho}{2\sqrt{-\rho(\rho + 4\beta_2)}}\tan^{-1}\left(\frac{2\beta_2\sec^2\theta + \rho}{\sqrt{-\rho(\rho + 4\beta_2)}}\right)\right] \tag{21b}$$

Note that both cases have the same singularity at $\theta=\pi/2$. The behavior of the process is equivalent irrespective of the sign of ρ. The maximal Lyapunov exponent for $\theta\in[0,\pi/2]$ can now be written as

$$\lambda = \frac{1}{2}(\lambda_1 + \lambda_2) + \frac{c_2}{2}(\lambda_1 - \lambda_2)\int_0^{\pi/2} \cos 2\theta \; m(\theta) d\theta + c_2 \int_0^{\pi/2} \frac{d\theta}{s(\theta)} \tag{22}$$

where c_2 is determined from the normality condition (17). The maximal Lyapunov exponent for $\theta\in[\pi/2,\pi]$ is determined similarily.

ACKNOWLEDGMENTS

This research was partially supported by the National Science Foundation through Grant MSS 90-57437 PYI.

APPENDIX A

The boundary behavior, as classified by Feller [5], is determined for the case of $\rho=0$. For both $\rho>0$ and $\rho<0$ the proof is similar and, therefore, omitted. Furthermore, since the region is symmetrical, consider only the first quadrant. On the region $\theta\in[0,\pi/2)$,the function s(x) is defined by

$$s(x) = \sec x \, \mathrm{Exp}[\Gamma \sin^2 x]$$

and can be bounded above and below by $K_1|\sec x| \le s(x) \le K_2|\sec x|$ where $K_1 \le 1$ and $K_2 \ge \mathrm{Exp}[\Gamma]$. Then, at the boundary points, we have

$$\int_0^{\theta} K_2 \sec x \, dx = \log\left[\tan\left(\frac{\pi}{4} + \frac{x}{2}\right)\right]_0^{\theta} < \infty$$

$$\int_{\theta}^{\pi/2} K_1 \sec x \, dx = \log\left[\tan\left(\frac{\pi}{4} + \frac{x}{2}\right)\right]_{\theta}^{\pi/2} = \infty \quad , \quad \theta \in \left(0, \frac{\pi}{2}\right)$$

Thus the behavor at the two boundary points differs. The speed measure is calculated to be

$$M[\theta_1,\theta_2] = K\left(\mathrm{erf}\left(\sqrt{\Gamma}\sin\theta_2\right) - \mathrm{erf}\left(\sqrt{\Gamma}\sin\theta_1\right)\right)$$

where K is a constant. The speed measure is finite for all $\theta \in [0,\pi/2]$. Thus the point $\theta=0$ is a regular point [5].

Consider now $M[\theta,\pi/2) < K$. Defining $N(\pi/2)$ as

$$N\left(\frac{\pi}{2}\right) = \int_{\theta}^{\pi/2} s(x)M[x,\pi/2)dx$$

we can say

$$N\left(\frac{\pi}{2}\right) < \int_{\theta}^{\pi/2} KK_2 \sec x dx = \infty$$

Thus, as $N(\pi/2)$ is finite for $\theta \in (0,\pi/2)$ the point $\theta=\pi/2$ is an entrance.

REFERENCES

[1] Arnold, L. and Kliemann, W., Probalistic Analysis and Related Topics, Vol. 3, ed. A.T. Bharucha-Reid, Academic Press, New York, Qualitative Theory of Stochastic Systems, 1983.

[2] Khasminskii, R.Z.,"Necessary and Sufficient Conditions for the Asymptotic Stability of Linear Stochastic Systems", Theory of Probability and Applications, Vol. 12(1), pp. 144-147, 1967.

[3] Nishioka, K., "On the Stability of Two-Dimensional Linear Stochastic Systems", Kodai Mathematics Seminar, Rep. 27, pp. 211-230, 1976.

[4] Khasminskii, R.Z., "A Limit Theorem for Solutions of Differential Equations With Random Right-Hand-Side", Theory of Probability and Applications, Vol. 11(3), pp. 390-406, 1966.

[5] Feller, W., "Diffusion Process in One Dimension", Trans. of Amer. Math Soc., Vol. 97, pp. 1-31, 1954.

[6] Abramowitz, M. and Stegun, I.A., Handbook of Mathematical Functions, Dover, New York, 1972.

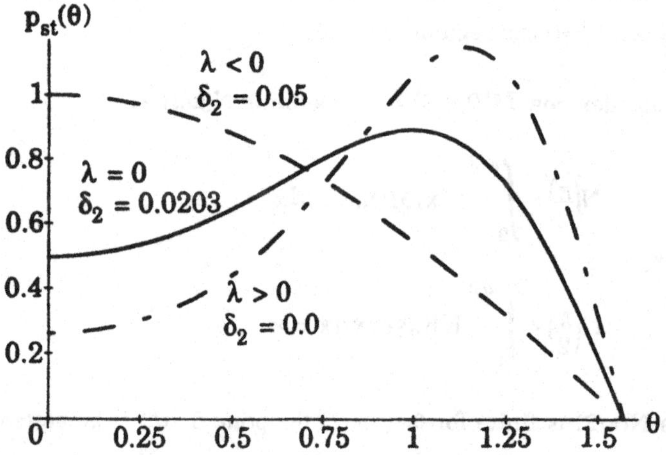

Figure 1: Stationary probability density function for ρ=0, white noise with magnitude 0.5, σ=1, δ_1=0.05 and the B matrix given below.

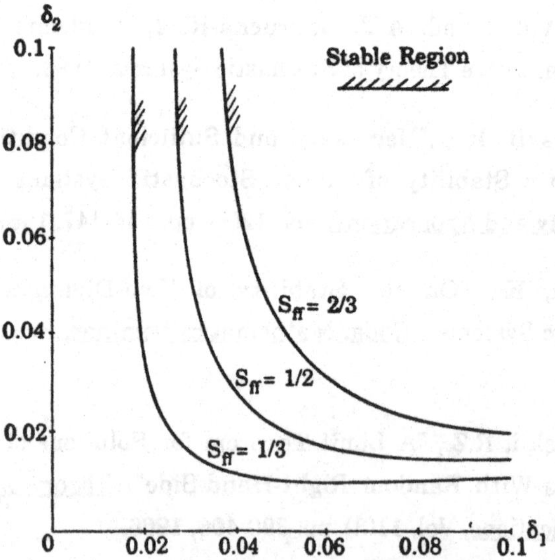

Figure 2: Almost-sure stability boundary for ρ=0, σ=1 and the B matrix given below.

$$B = \begin{bmatrix} 1 & \frac{3\sqrt{2}}{16} & \frac{3}{16} \\ \frac{3\sqrt{2}}{16} & 1 & \frac{3}{16} \\ \frac{3}{16} & \frac{3}{16} & 1 - \frac{3\sqrt{3}}{16} \end{bmatrix}$$

Updating of a Model and its Uncertainties Utilizing Dynamic Test Data

J.L. Beck, L.S. Katafygiotis

California Institute of Technology, Pasadena, CA 91125, U.S.A.

ABSTRACT

The problem of updating a structural model and its associated uncertainties by utilizing structural response data is addressed. Using a Bayesian probabilistic formulation, 6the updated "posterior" probability distribution of the uncertain parameters is obtained and it is found that for a large number of data points it is very peaked at some "optimal" values of the parameters. These optimal parameters can be obtained by minimizing a positive-definite measure-of-fit function. This paper focuses on the identifiability of the optimal parameters. The problem of finding the whole set of optimal models that have the same output at the observed degrees of freedom for a given input is resolved for the first time, by presenting an algorithm which methodically and efficiently searches the parameter space. Also, a simplified expression is given for the weighting coefficients associated with each optimal model which are involved in the probability distribution for the predicted response.

INTRODUCTION

The uncertainties encountered when modeling a structure with a model out of a specified class can be divided into uncertainties of the "model parameters", concerned with which model out of the specified class is the most appropriate to describe the system, and uncertainties of the parameters describing the "model error", which is the error arising because any mathematical model is only an approximation of the real dynamic behavior of the structure. For example, if we choose the class of linear dynamic models, then there are uncertainties in the values of the parameters, such as Young's modulus E, which should be chosen for the structure. Furthermore, for each model in the class, we know that the corresponding predicted response will not be identical to the actual structural response because of model error.

In order to properly describe the modeling uncertainties, a probabilistic formulation can be followed, but from the Bayesian point of view, that is, probability is viewed as a multi-valued logic for plausible reasoning. Note that for most of the applications of interest in this study, the common interpretation of probability as a relative frequency of occurences in the long run does not make sense. Therefore, in order to quantify the uncertainty associated with a parameter, a probability distribution will be assigned describing how plausible each value is for the parameter, on the basis of the given information.

During the preliminary design of a structure, when no records of structural response are available, the modeling uncertainties must be estimated subjectively, on the basis of any available information and experience dealing with similar structures [Katafygiotis and Beck 1991]. When records from dynamic testing or earthquake records of structural response become available, the information contained in these records can be used to update the initial estimates of the modeling uncertainties by applying Bayes' Theorem. The updated probability model can be used for response predictions, for improving the performance of a control system, or for health monitoring of the structure by detecting changes in its stiffness distribution.

For a large number of available data points, the posterior distribution of the uncertain parameters resulting from Bayes' Theorem is very peaked at some optimal values of the parameters, therefore making these values much more plausible than the other values. Beck [1990] showed that it is then asymptotically correct for response predictions to use only the models corresponding to these optimal parameters, appropriately weighted. This result is very important, since the high-dimensional integrations which are required to calculate the uncertainties in the predictive response are computationally prohibitive [Katafygiotis and Beck, 1991], but the asymptotic result implies that they can be replaced by a weighted sum over all optimal parameters, assuming their number is finite. However, the implementation of these asymptotic results requires that the problem of finding the set of all optimal parameters be solved.

It is usually the case in structural model updating that there is more than one optimal value of the parameters for given input and output data, so the concepts of model and system identifiability are introduced to provide a framework to handle this nonuniqueness in the optimal parameters. Also, a new algorithm is presented to methodically and efficiently search the high-dimensional model parameter space by following only a finite set of one-dimensional curves. This algorithm finds all other optimal model parameters corresponding to models with identical model response at the observed degrees of freedom, thus resolving the problem of model identifiability. This important problem of finding all optimal parameters corresponding to models which are "output-equivalent" is solved for the first time.

FORMULATION OF THE PROBLEM

Let $\mathcal{D}_N$ denote the set of observed data for a structural system, consisting of a sampled observed input history $\hat{Z}_{1,N} = \{\hat{\underline{z}}(n) \in \mathbb{R}^{N_I} : n = 1,2,\ldots,N\}$ and output history $\hat{X}_{1,N} = \{\hat{\underline{x}}(n) \in \mathbb{R}^{N_o} : n = 1,2,\ldots,N\}$. Usually, the measured output consists of the acceleration histories at certain degrees of freedom (dof), which are referred to as observed or measured dof. Let $\underline{z}(n) \in \mathbb{R}^{N_I}$ and $\underline{x}(n; Z_{1,N}) \in \mathbb{R}^{N_o}$ denote the vector of the corresponding *system* input and output at time $t_n = n\Delta t$, where Δt is the sampling interval for the data in $\mathcal{D}_N$. Assuming that for modern instrumentation the measurement noise is negligible compared with the model error, it follows that $\underline{x}(n) = \hat{\underline{x}}(n)$ and $\underline{z}(n) = \hat{\underline{z}}(n)$ for $n \leq N$. Assume that an N_d-degree of freedom theoretical model $\mathcal{M}$ has been chosen to describe the input-output behavior of the system, and let $\underline{a} \in \mathbb{R}^{N_a}$ be the vector of the uncertain model parameters. $\mathcal{M}$ provides a functional relationship between the *model* output vector $\underline{q}(n;\underline{a}) \in R^{N_o}$ at time $t_n = n\Delta t$ and the system input $Z_{1,n}$:

$$\underline{q}(n;\underline{a}) = \underline{q}(n;\underline{a}, Z_{1,n}, \mathcal{M}) \tag{1}$$

In the following, the dependence of $\underline{q}(n;\underline{a})$ on the input $Z_{1,n}$ and the theoretical model $\mathcal{M}$ as well as the dependence of $\underline{x}(n)$ on the input $Z_{1,n}$ will be suppressed in the notation.

The model error $e(\underline{n})$ is defined to be the difference between the system output and the model output, so:

$$\underline{x}(n) = \underline{q}(n;\underline{a}) + \underline{e}(n) \tag{2}$$

In order to account for the model error, a class of probability models $\mathcal{P}$ is chosen which prescribes a function h_M giving the probability density function of the sequence $E_{1,M} = \{\underline{e}(\underline{n}); n = 1, \ldots, M\}$. The class $\mathcal{P}$ selected here assumes that $E_{1,M}$ is a zero-mean stationary Gaussian white-noise sequence. In addition, it assumes that the variance of the model error is the same at all observed degrees of freedom, denoted by σ^2. Therefore, σ is the only model-error parameter required to specify a particular probability model out of the class $\mathcal{P}$ and h_M is:

$$\begin{aligned} p(E_{1,M}|\sigma, \mathcal{P}) &= h_M(E_{1,M};\sigma) \\ &= \frac{1}{(2\pi\sigma^2)^{\frac{MN_o}{2}}} \exp\left(-\frac{1}{2\sigma^2}\sum_{n=1}^{M}\sum_{i=1}^{N_o} e_i(n)^2\right) \end{aligned} \tag{3}$$

The selection of the classes $\mathcal{M}$ and $\mathcal{P}$ defines the class $\mathcal{M}_P$ parameterized by $\underline{\tilde{a}} = [\underline{a}^T, \sigma]^T$ prescribing the pdf:

$$\begin{aligned} p(X_{1,M}|\underline{\tilde{a}}, Z_{1,M}, \mathcal{M}_P) &= f_M(X_{1,M};\underline{\tilde{a}}, Z_{1,M}) \\ &= \frac{1}{(2\pi\sigma^2)^{\frac{MN_o}{2}}} \exp\left(-\frac{1}{2\sigma^2}\sum_{n=1}^{M}\sum_{i=1}^{N_o}(x_i(n) - q_i(n;\underline{a}))^2\right) \end{aligned} \tag{4}$$

In order to account for the uncertainty in the parameters $\underline{a}$ and σ, it is assumed that $\mathcal{M}_P$ also specifies a function $\pi_{\underline{a},\sigma}$ corresponding to their prior pdf:

$$p(\underline{a}, \sigma|\mathcal{M}_P) = \pi_{\underline{a},\sigma}(\underline{a}, \sigma) \equiv \pi_{\underline{\tilde{a}}}(\underline{\tilde{a}}) \tag{5}$$

The updated, or "posterior", joint pdf of $\underline{\tilde{a}}$ is given by Bayes' Theorem:

$$\begin{aligned} p(\underline{\tilde{a}}|\mathcal{D}_N, \mathcal{M}_P) &= \frac{p(\hat{X}_{1,N}|\underline{\tilde{a}}, \hat{Z}_{1,N}, \mathcal{M}_P)p(\underline{\tilde{a}}|\mathcal{M}_P)}{p(\hat{X}_{1,N}|\hat{Z}_{1,N}, \mathcal{M}_P)} \\ &= kp(\hat{X}_{1,N}|\underline{\tilde{a}}, \hat{Z}_{1,N}, \mathcal{M}_P)p(\underline{\tilde{a}}|\mathcal{M}_P) \\ &= kf_N(\hat{X}_{1,N};\underline{\tilde{a}}, \hat{Z}_{1,N})\pi_{\underline{\tilde{a}}}(\underline{\tilde{a}}) \end{aligned} \tag{6}$$

where

$$k^{-1} = p(\hat{X}_{1,N}|\hat{Z}_{1,N}, \mathcal{M}_P) = \int_{S(\underline{\tilde{a}})} p(\hat{X}_{1,N}|\underline{\tilde{a}}, \hat{Z}_{1,N}, \mathcal{M}_P)p(\underline{\tilde{a}}|\mathcal{M}_P)d\underline{\tilde{a}} \tag{7}$$

As can be seen from Equation (6), the effect of utilizing the available records to update the pdf of the model parameters $\underline{a}$ and the model-error parameter σ is contained in the term $kf_N(\hat{X}_{1,N};\tilde{\underline{a}},\hat{Z}_{1,N})$, where k serves as a normalizing constant. The effect of the function f_N on the updated distribution is much more drastic than that of the prior distribution, since it can be shown that f_N is very peaked at some optimal values of the parameters, making these values much more probable than the other ones.

Define the *optimal* parameters $\hat{\tilde{\underline{a}}} = [\hat{\underline{a}}^T, \hat{\sigma}]^T$ to be the values of the parameters $\tilde{\underline{a}}$ that globally maximize $f_N(\hat{X}_{1,N};\tilde{\underline{a}},\hat{Z}_{1,N})$. Maximizing $f_N(\hat{X}_{1,N};\tilde{\underline{a}},\hat{Z}_{1,N})$ with respect to $\tilde{\underline{a}}$ is equivalent to maximizing:

$$\ln f_N(\hat{X}_{1,N};\tilde{\underline{a}},\hat{Z}_{1,N}) = -c - NN_o\ln\sigma - \frac{1}{2\sigma^2}\sum_{n=1}^{N}\sum_{i=1}^{N_o}(\hat{x}_i(n) - q_i(n;\underline{a}))^2 \quad (8)$$

At $\tilde{\underline{a}} = \hat{\tilde{\underline{a}}}$ the following conditions hold:

$$\left.\frac{\partial \ln f_N(\hat{X}_{1,N};\tilde{\underline{a}},\hat{Z}_{1,N})}{\partial \tilde{\underline{a}}}\right|_{\tilde{\underline{a}}=\hat{\tilde{\underline{a}}}} = 0 \quad (9)$$

For fixed $\underline{a}$, maximizing $\ln f_N(\hat{X}_{1,N};\tilde{\underline{a}},\hat{Z}_{1,N})$ with respect to σ requires:

$$\hat{\sigma}(\underline{a})^2 = \frac{1}{NN_o}\sum_{n=1}^{N}\sum_{i=1}^{N_o}(\hat{x}_i(n) - q_i(n;\underline{a}))^2 \quad (10)$$

This shows how the most probable variance $\hat{\sigma}(\underline{a})^2$, for given $\underline{a}$, depends on the choice of the model parameters $\underline{a}$. Obviously, the condition for the overall most probable variance $\hat{\sigma}^2$ is given by (10) when $\underline{a} = \hat{\underline{a}}$. Substituting Equation (10) into Equation (8):

$$\ln f_N(\hat{X}_{1,N};\underline{a},\hat{\sigma}(\underline{a}),\hat{Z}_{1,N}) = -c - NN_o\ln\hat{\sigma}(\underline{a}) - \frac{NN_o}{2} \quad (11)$$

Thus $\hat{\underline{a}}$ is given by minimizing $\hat{\sigma}(\underline{a})$ or, equivalently, minimizing:

$$J(\underline{a}) = NN_o\hat{\sigma}(\underline{a})^2 = \sum_{n=1}^{N}\sum_{i=1}^{N_o}(\hat{x}_i(n) - q_i(n;\underline{a}))^2 \quad (12)$$

Since $J(\underline{a})$ might attain its minimum at more than one value $\hat{\underline{a}}$, the identifiability of the optimal model parameters $\hat{\underline{a}}$ must be resolved. On the other hand, the optimal model error parameter $\hat{\sigma}$ given by:

$$\hat{\sigma} = \left(\frac{1}{NN_o}\min_{\underline{a}\in S(\underline{a})} J(\underline{a})\right)^{\frac{1}{2}} \quad (13)$$

is uniquely determined and is therefore said to be globally identifiable. Here $S(\underline{a})$ denotes the space of permissible values of $\underline{a}$. It can be shown that:

$$\begin{aligned} \frac{p(\underline{a},\hat{\sigma}(\underline{a})|\mathcal{D}_N,\mathcal{M}_P)}{p(\hat{\underline{a}},\hat{\sigma}(\hat{\underline{a}})|\mathcal{D}_N,\mathcal{M}_P)} &= \frac{\pi_{\underline{a},\sigma}(\underline{a},\hat{\sigma}(\underline{a}))}{\pi_{\underline{a},\sigma}(\hat{\underline{a}},\hat{\sigma}(\hat{\underline{a}}))}\left(\frac{\hat{\sigma}(\hat{\underline{a}})}{\hat{\sigma}(\underline{a})}\right)^{NN_o} \\ &= \frac{\pi_{\underline{a},\sigma}(\underline{a},\hat{\sigma}(\underline{a}))}{\pi_{\underline{a},\sigma}(\hat{\underline{a}},\hat{\sigma}(\hat{\underline{a}}))}\left(\frac{J(\hat{\underline{a}})}{J(\underline{a})}\right)^{\frac{NN_o}{2}} \end{aligned} \tag{14}$$

The posterior pdf $p(\tilde{\underline{a}}|\mathcal{D}_N,\mathcal{M}_P)$ can be approximated locally, in the neighborhood $\mathcal{H}(\tilde{\underline{a}};\hat{\tilde{\underline{a}}})$ of an optimal parameter $\hat{\tilde{\underline{a}}}$, with a multi-dimensional Gaussian distribution with mean $\hat{\tilde{\underline{a}}}$ and an $(N_o+1)\times(N_o+1)$ covariance matrix $A_N^{-1}(\hat{\tilde{\underline{a}}})$ [Beck 1990]:

$$p(\tilde{\underline{a}}|\mathcal{D}_N,\mathcal{M}_P) \simeq p(\hat{\tilde{\underline{a}}}|\mathcal{D}_N,\mathcal{M}_P)\exp(-\frac{1}{2}[\tilde{\underline{a}}-\hat{\tilde{\underline{a}}}]^T A_N(\hat{\tilde{\underline{a}}})[\tilde{\underline{a}}-\hat{\tilde{\underline{a}}}]) \ ; \ \tilde{\underline{a}} \in \mathcal{H}(\tilde{\underline{a}};\hat{\tilde{\underline{a}}}) \tag{15}$$

where the elements $[A_N(\hat{\tilde{\underline{a}}})]_{ij}$ are given by:

$$\begin{aligned} [A_N(\hat{\tilde{\underline{a}}})]_{ij} &= -\left.\frac{\partial^2 \ln f_N(\hat{X}_{1,N};\tilde{\underline{a}},\hat{Z}_{1,N})}{\partial\tilde{a}_i\partial\tilde{a}_j}\right|_{\tilde{\underline{a}}=\hat{\tilde{\underline{a}}}} - \left.\frac{\partial^2 \ln \pi_{\tilde{\underline{a}}}(\tilde{\underline{a}})}{\partial\tilde{a}_i\partial\tilde{a}_j}\right|_{\tilde{\underline{a}}=\hat{\tilde{\underline{a}}}} \\ &\cong -\left.\frac{\partial^2 \ln f_N(\hat{X}_{1,N};\tilde{\underline{a}},\hat{Z}_{1,N})}{\partial\tilde{a}_i\partial\tilde{a}_j}\right|_{\tilde{\underline{a}}=\hat{\tilde{\underline{a}}}} \end{aligned} \tag{16}$$

The elements of A_N are $\mathcal{O}(N)$ and, therefore, for a large number N of available data points, which is usually the case with dynamic tests or earthquake records of structural response, the pdf $p(\tilde{\underline{a}}|\mathcal{D}_N,\mathcal{M}_P)$ becomes very peaked at the optimal parameters $\hat{\tilde{\underline{a}}}$; this result can also be concluded by viewing Equation (14). It has been shown [Beck 1990], that if the number of optimal parameters is finite, it is asymptotically correct for prediction purposes to use out of the class $\mathcal{M}_P$ only the probability models corresponding to the optimal parameters $\hat{\tilde{\underline{a}}}_k$, $k=1,\ldots,K$, since for large N, the updated predictive distribution is given by:

$$p(X_{1,M}|\mathcal{D}_N,Z_{N+1,M},\mathcal{M}_P) \simeq \sum_{k=1}^{K} w_k p(X_{1,M}|\hat{\tilde{\underline{a}}}_k,Z_{1,M},\mathcal{M}_P) \tag{17}$$

The predictive distribution for each of the optimal models is weighted proportionally to the volume of the posterior pdf $p(\tilde{\underline{a}}|\mathcal{D}_N,\mathcal{M}_P)$ under its Gaussian-shaped peak positioned at the corresponding optimal parameters. The mathematical expression for the weighting coefficient w_k corresponding to the kth vector of optimal parameters $\hat{\tilde{\underline{a}}}_k$ is [Beck 1990]:

$$w_k = \frac{w'_k}{\sum_{k=1}^{K} w'_k} \tag{18}$$

where

$$w'_k = \pi_{\tilde{\underline{a}}}(\hat{\tilde{\underline{a}}}_k)|A_N^{-1}(\hat{\tilde{\underline{a}}}_k)|^{1/2} \tag{19}$$

Notice that the prior pdf $\pi_{\tilde{\underline{a}}}(\tilde{\underline{a}})$ does not need to be specified over the whole domain $S(\tilde{\underline{a}})$. Instead, only the relative values for the optimal parameters $\hat{\tilde{\underline{a}}}_k$ need to be specified. The elements of $A_N(\hat{\tilde{\underline{a}}}_k)$ can be evaluated numerically through Equation (16). Numerical examples have shown that these calculations can be very sensitive to roundoff errors if the vector of the observed model parameters $\underline{a}$ does not consist of modal quantities exclusively. In addition, independent of the choice of $\underline{a}$, the matrix $A_N(\hat{\tilde{\underline{a}}}_k)$ is often ill-conditioned, which results in numerical errors when calculating $|A_N^{-1}(\hat{\tilde{\underline{a}}}_k)|$. Thus, in general, the weighting effect of $|A_N^{-1}(\hat{\tilde{\underline{a}}}_k)|^{\frac{1}{2}}$ cannot be estimated reliably by calculating it directly. A reliable alternative expression is presented later to overcome this difficulty.

MODEL AND SYSTEM IDENTIFIABILITY OF THE OPTIMAL PARAMETERS

Let $Q_{1,N}(\underline{a}; \hat{Z}_{1,N}) = \{\underline{q}(n; \underline{a}, \hat{Z}_{1,N}, \mathcal{M}) \in R^{N_o} : n = 1, 2, \ldots, N\}$ denote the *model* output history, which corresponds to the observed quantities, for the given input $\hat{Z}_{1,N}$ and for a model $M(\underline{a}) \in \mathcal{M}$, and let $S(Q_{1,N}; \hat{Z}_{1,N})$ denote the space formed by the range of $Q_{1,N}(\underline{a}; \hat{Z}_{1,N})$ as $\underline{a}$ ranges over $S(\underline{a})$. There is a natural mapping of the models in the class $\mathcal{M}$ onto $S(Q_{1,N}; \hat{Z}_{1,N})$, but it may happen that several models in $\mathcal{M}$ are "output-equivalent", that is, they get mapped into the same output under the specified input, making the inverse problem non-unique for that input and output. Investigating the uniqueness of this inverse problem constitutes the problem of *model* identifiability of the model parameters.

Let $S_{opt}(M(\hat{\underline{a}}); \hat{Z}_{1,N}) \subset \mathcal{M}$ denote the set of all optimal models which are output-equivalent to model $M(\hat{\underline{a}})$ under input $\hat{Z}_{1,N}$. Let $S_{opt}(\hat{\underline{a}}; \hat{Z}_{1,N}) \subset S(\underline{a})$ denote the set of all corresponding optimal model parameters. The following definitions are introduced:

M1. A parameter a_j of $\underline{a}$ is *globally M-identifiable* ("model identifiable") at $\hat{\underline{a}}$ for the input $\hat{Z}_{1,N}$ if $S_{opt}(\hat{\underline{a}}; \hat{Z}_{1,N})$ contains only one optimal parameter or, if not, then:

$$\hat{\underline{a}}^{(1)}, \hat{\underline{a}}^{(2)} \in S_{opt}(\hat{\underline{a}}; \hat{Z}_{1,N}) \Rightarrow \hat{a}_j^{(1)} = \hat{a}_j^{(2)} \tag{20}$$

Definition M1 implies that a_j is uniquely specified by $\hat{Z}_{1,N}$ and $Q_{1,N}(\hat{\underline{a}}; \hat{Z}_{1,N})$.

M2. A parameter a_j of $\underline{a}$ is *locally identifiable* at $\hat{\underline{a}}$ for the input $\hat{Z}_{1,N}$ if there exists a positive number ϵ_j such that:

$$\hat{\underline{a}}^{(1)}, \hat{\underline{a}}^{(2)} \in S_{opt}(\hat{\underline{a}}; \hat{Z}_{1,N}) \Rightarrow |\hat{a}_j^{(1)} - \hat{a}_j^{(2)}| > \epsilon_j \text{ or } \hat{a}_j^{(1)} = \hat{a}_j^{(2)} \tag{21}$$

Definition M2 implies that a_j is uniquely specified within a neighborhood of each of its possible values by $\hat{Z}_{1,N}$ and $Q_{1,N}(\hat{\underline{a}}; \hat{Z}_{1,N})$, and that if $S(\underline{a})$ is a closed-bounded parameter set, there are only a finite number of possible values for a_j under the given input and model output. Note that if a_j is globally M-identifiable at $\hat{\underline{a}}$, then it is also locally M-identifiable at $\hat{\underline{a}}$.

M3. A parameter a_j of $\underline{a}$ is M-identifiable at $\hat{\underline{a}}$ for the input $\hat{Z}_{1,N}$ if it is either locally or globally M-identifiable.

The above definitions can be extended as follows: The parameter vector $\underline{a}$, or a portion of it, is globally (locally) M-identifiable at $\hat{\underline{a}}$ if all its elements are globally (locally) M-identifiable at $\hat{\underline{a}}$. The parameter vector $\underline{a}$ is not M-identifiable at $\hat{\underline{a}}$ if at least one of its elements is not M-identifiable at $\hat{\underline{a}}$.

As discussed earlier, it is of particular importance to investigate the identifiability of the optimal parameter vector $\hat{\tilde{\underline{a}}}$ based on the input and output data $\mathcal{D}_N$ from a structural system. An *optimal model* $M_P(\hat{\tilde{\underline{a}}})$ for given data $\mathcal{D}_N$ is defined to be any model in $\mathcal{M}_P$ such that:

$$f_N(\hat{X}_{1,N};\hat{\tilde{\underline{a}}},\hat{Z}_{1,N}) = \max_{\tilde{\underline{a}}\in S(\tilde{\underline{a}})} f_N(\hat{X}_{1,N};\tilde{\underline{a}},\hat{Z}_{1,N}) \tag{22}$$

where the parameters $\hat{\tilde{\underline{a}}} = [\hat{\underline{a}}^T,\hat{\sigma}]^T$ are called optimal parameters. Let $S_{opt}(M_P(\hat{\tilde{\underline{a}}});\mathcal{D}_N) \subset \mathcal{M}_P$ denote the set of all optimal models in the class $\mathcal{M}_P$ and $S_{opt}(\hat{\tilde{\underline{a}}};\mathcal{D}_N) \subset S(\tilde{\underline{a}})$ denote the set of all corresponding optimal parameters. The following definitions are introduced:

S1. A parameter $\tilde{a}_j$ of $\tilde{\underline{a}}$ is *globally S-identifiable* ("system identifiable") at $\hat{\tilde{\underline{a}}}$ for the input and output data $\mathcal{D}_N$ if $S_{opt}(\hat{\tilde{\underline{a}}};\mathcal{D}_N)$ contains only one optimal parameter, or, if not, then:

$$\hat{\tilde{\underline{a}}}^{(1)},\hat{\tilde{\underline{a}}}^{(2)} \in S_{opt}(\hat{\tilde{\underline{a}}};\mathcal{D}_N) \Rightarrow \hat{\tilde{a}}_j^{(1)} = \hat{\tilde{a}}_j^{(2)} \tag{23}$$

S2. A parameter $\tilde{a}_j$ of $\tilde{\underline{a}}$ is *locally S-identifiable* at $\hat{\tilde{\underline{a}}}$ for the input and output data $\mathcal{D}_N$ if there exists a positive number ϵ_j such that:

$$\hat{\tilde{\underline{a}}}^{(1)},\hat{\tilde{\underline{a}}}^{(2)} \in S_{opt}(\hat{\tilde{\underline{a}}};\mathcal{D}_N) \Rightarrow |\hat{\tilde{a}}_j^{(1)} - \hat{\tilde{a}}_j^{(2)}| > \epsilon_j \text{ or } \hat{\tilde{a}}_j^{(1)} = \hat{\tilde{a}}_j^{(2)} \tag{24}$$

S3. A parameter $\tilde{a}_j$ of $\tilde{\underline{a}}$ is S-identifiable at $\hat{\tilde{\underline{a}}}$ for the input and output data $\mathcal{D}_N$ if it is either locally or globally S-identifiable.

Given an optimal model $M_P(\hat{\underline{a}},\hat{\sigma})$ in the class $\mathcal{M}_P$, all other models $M(\underline{a}^*)$ $\in S_{opt}(M(\hat{\underline{a}});\hat{Z}_{1,N})$, if any, having the same observed model output as $M(\hat{\underline{a}})$, correspond to an optimal model $M_P(\underline{a}^*,\hat{\sigma}) \in S_{opt}(M_P(\hat{\underline{a}},\hat{\sigma});\mathcal{D}_N)$ in the class $\mathcal{M}_P$. Another way of looking at this result is that if the parameter vector $\underline{a}$ is not globally M-identifiable at $\hat{\underline{a}}$, then $\tilde{\underline{a}}$ cannot be globally S-identifiable at $[\hat{\underline{a}}^T,\hat{\sigma}]^T$. Furthermore, the number of optimal probability models in $S_{opt}(M_P(\hat{\underline{a}},\hat{\sigma});\mathcal{D}_N)$ $\subset \mathcal{M}_P$ must be at least as large as the number of optimal models in $S_{opt}(M(\hat{\underline{a}});\hat{Z}_{1,N}) \subset \mathcal{M}$. However, there can be models in $S_{opt}(\mathcal{M}_P(\hat{\underline{a}},\hat{\sigma});\mathcal{D}_N)$ which do not have the same model response $Q_{1,N}(\hat{\underline{a}};\hat{Z}_{1,N})$ but still give an equally good fit to the data.

This paper addresses M-identifiability for linear MDOF (multi-degree-of-freedom) models and therefore makes an important first step in solving the very difficult problem of S-identifiability for such a class of models.

MDOF LINEAR STRUCTURAL MODELS

Consider the class $\mathcal{M}_{N_d}$ of N_d-degree of freedom linear structural models starting from rest, and subject to a base excitation:

$$M\ddot{\underline{q}} + C\dot{\underline{q}} + K\underline{q} = -M\underline{b}\ddot{z}(t) \quad ; \quad \underline{q}(0) = \dot{\underline{q}}(0) = 0 \tag{25}$$

The $N_d \times N_d$ matrices M, C, and K are the mass, the damping and the stiffness matrix, respectively. It is assumed that classically damped modes exist. The vector

$\underline{q} = [q_1, q_2, \ldots, q_{N_d}]^T$ consists of the generalized displacements relative to the base of each degree of freedom. The components of the vector $\underline{b} = [b_1, b_2, \ldots, b_{N_d}]^T$ are called pseudo-static influence coefficients, and they are known from the prescribed geometry of the structural model.

The response at the ith dof can be expressed as a superposition of the first $N_m \leq N_d$ modal contributions, where the higher modal contributions are neglected:

$$q_i(t) \cong \sum_{r=1}^{N_m} q_i^{(r)}(t) \quad ; \quad N_m \leq N_d \tag{26}$$

The equation of motion for the contribution of the rth mode to the response at the ith degree of freedom (dof) is:

$$\ddot{q}_i^{(r)} + 2\zeta_r \omega_r \dot{q}_i^{(r)} + \omega_r^2 q_i^{(r)} = -\beta_i^{(r)} \ddot{z}(t) \quad ; \quad q_i^{(r)}(0) = \dot{q}_i^{(r)}(0) = 0 \tag{27}$$

where ω_r is the rth modal frequency, ζ_r is the damping ratio of the rth mode, and $\beta_i^{(r)}$ is the effective participation factor of the rth mode at the ith dof [Beck 1978]. The elements of the mass matrix M are assumed to be determinstically known, since they can be estimated accurately enough from the structural drawings. The uncertainties concerning the damping of the structural model is accounted for by uncertainties of the modal damping ratios ζ_r. Finally, the uncertainty in the stiffness distribution is parameterized through a set of nondimensional positive parameters $\theta_i, i = 1, \ldots, N_\theta$, so that:

$$K = K_0 + \sum_{i=1}^{N_\theta} \theta_i K_i \quad ; \quad \theta_i > 0 \;,\; i = 1, \ldots, N_\theta \tag{28}$$

Each parameter θ_i scales the stiffness contribution K_i of a certain substructure to the total stiffness matrix and K_0 accounts for the stiffness contributions of those substructures with deterministic stiffnesses.

It follows from the above that the vector of uncertain model parameters is:

$$\underline{a} = [\theta_1, \theta_2, \ldots, \theta_{N_\theta}, \zeta_1, \zeta_2, \ldots, \zeta_{N_M}]^T \tag{29}$$

M-IDENTIFIABILITY OF THE MODEL PARAMETERS

Assume that a set of optimal parameters $\hat{\underline{a}}$, where $\underline{a}$ is given by Equation (29), has been found by globally minimizing $J(\underline{a})$, given by Equation (12). The issue here is the M-identifiability of the optimal parameters $\hat{\underline{\theta}} = [\hat{\theta}_1, \ldots, \hat{\theta}_{N_\theta}]^T$ and $\hat{\underline{\zeta}} = [\hat{\zeta}_1, \ldots, \hat{\zeta}_{N_M}]^T$

Let $\mathcal{L}^o$ and $\mathcal{L}^u$ denote the set of integers corresponding to the observed and unobserved degrees of freedom, respectively. The two sets are related as follows:

$$\mathcal{L}^u = \{1, 2, \ldots, N_d\} - \mathcal{L}^o \tag{30}$$

It has been shown [Beck 1978] that the parameters $\{\omega_r, \zeta_r, \beta_i^{(r)}, r = 1, 2, \ldots, N_d, i \in \mathcal{L}^o\}$ which comprise the elements of $\underline{a}$ are globally M-identifiable from the input and output if the following conditions are met: (a) the model has no repeated modes, that is, no two modes have the same modal frequencies and damping ratios, (b) there are no modes with a zero participation factor, and (c) no mode has a node at each coordinate at which the response is measured. Conditions (b) and (c) can be stated as follows: for each mode $r = 1, 2, \ldots, N_d$, there exists at least one $i \in \mathcal{L}^o$, such that $\beta_i^{(r)} \neq 0$. Notice that if this condition is not satisfied, that is, if $\beta_i^{(r)} = 0$ for each $i \in \mathcal{L}^o$, the rth mode will be missing from the output and hence ω_r and ζ_r will not be able to be determined from the input and output.

Utilizing the previous result, it follows that out of the elements of $\hat{\underline{a}}$, the vector $\hat{\underline{\zeta}}$ is globally M-identifiable and therefore, only the M-identifiability of the optimal stiffness parameters $\hat{\underline{\theta}}$ needs to be resolved. Udwadia et al [1978] showed that the problem of identifying the stiffness distribution of an N_d-story shear building from its base input and its response at the roof is a non-unique problem with an upper bound on the number of stiffness distributions corresponding to "output-equivalent" models of $N_d!$. However, the exact number of these "output-equivalent" solutions and an algorithm for obtaining these solutions was not given.

An algorithm is presented here to resolve the M-identifiability of the optimal parameters $\hat{\underline{\theta}}$.

ALGORITHM FOR M-IDENTIFIABILITY

Assume that there exists a finite number of "output-equivalent" models $M(\hat{\underline{\theta}}^{(k)}, \hat{\underline{\zeta}})$; $k = 1, 2, \ldots, K$. Let $S_{opt}(\hat{\underline{\theta}}; \hat{Z}_{1,N})$ denote the set of all corresponding optimal stiffness parameters $\hat{\underline{\theta}}^{(k)}$; $k = 1, 2, \ldots, K$. According to the earlier stated result by Beck [1978], all these stiffness parameters have the following corresponding modal quantities identical:

$$\omega_r(\hat{\underline{\theta}}^{(k)}) = \hat{\omega}_r \quad ; \quad r = 1, \ldots, N_m \quad , \quad k \in \{1, 2, \ldots, K\} \tag{31}$$

$$\beta_i^{(r)}(\hat{\underline{\theta}}^{(k)}) = \hat{\beta}_i^{(r)} \quad ; \quad r = 1, \ldots, N_m \quad , \quad i \in \mathcal{L}^o \quad , \quad k \in \{1, 2, \ldots, K\} \tag{32}$$

Let $\Theta_{\hat{\Omega}}$ denote the set of all parameters $\tilde{\underline{\theta}}^{(k)} \in S(\underline{\theta})$ with corresponding set of modal frequencies $\hat{\Omega} = \{\hat{\omega}_r, r = 1, \ldots, N_m\}$, that is:

$$\Theta_{\hat{\Omega}} = \{\tilde{\underline{\theta}} \in S(\underline{\theta}) : \omega_r(\tilde{\underline{\theta}}) = \hat{\omega}_r; r = 1, \ldots, N_m\} \tag{33}$$

It is obvious from the definition of $\Theta_{\hat{\Omega}}$ that it is a superset of $S_{opt}(\hat{\underline{\theta}}; \hat{Z}_{1,N})$:

$$S_{opt}(\hat{\underline{\theta}}; \hat{Z}_{1,N}) \subseteq \Theta_{\hat{\Omega}} \tag{34}$$

The methodology for finding the set $S_{opt}(\hat{\underline{\theta}}; \hat{Z}_{1,N})$ consists of two steps. First, the parameter space $S(\underline{\theta})$ is searched methodically, using a new proposed algorithm, to find all elements of $\Theta_{\hat{\Omega}}$. It can be shown that an upper bound on the elements of $\Theta_{\hat{\Omega}}$ is $N_d!$. After $\Theta_{\hat{\Omega}}$ has been found, the second step is taken, consisting of an elimination process, to determine which elements of $\Theta_{\hat{\Omega}}$ satisfy (32), belonging, therefore, in the desired set $S_{opt}(\hat{\underline{\theta}}; \hat{Z}_{1,N})$.

In this paper, we will assume that the number N_θ of uncertain stiffness parameters is equal to the number of contributing modes N_m. If $\Theta_{\hat{\omega}_i} = \{\underline{\theta} \in S(\underline{\theta}) : \omega_i(\underline{\theta}) = \hat{\omega}_i\}$ is the $(N_\theta - 1)$-dimensional hypersurface in the N_θ-dimensional space $S(\underline{\theta})$ where the ith modal frequency remains constant and equal to $\hat{\omega}_i$, then the required set $\Theta_{\hat{\Omega}}$ is given by:

$$\Theta_{\hat{\Omega}} = \cap_{i=1}^{N_\theta} \Theta_{\hat{\omega}_i} \tag{35}$$

Let $c_k(\underline{\theta};\underline{\theta}^*)$ denote a one-dimensional curve in the space $S(\underline{\theta})$, passing through a point $\underline{\theta}^*$, with the property that along this curve all of the first N_m modal frequencies remain fixed except for the kth modal frequency, which is allowed to vary, that is:

$$\begin{aligned} c_k(\underline{\theta};\underline{\theta}^*) = &\text{the largest connected subset of} \\ &\{\underline{\theta} \in S(\underline{\theta}) : \omega_r(\underline{\theta}) = \omega_r(\underline{\theta}^*); r = 1,\ldots,k-1,k+1,\ldots,N_m\} \\ &\text{containing } \underline{\theta}^* \end{aligned} \tag{36}$$

Following all the different curves $c_k(\underline{\theta};\underline{\tilde{\theta}}^{(1)})$, passing through the known optimal point $\underline{\tilde{\theta}}^{(1)} = \underline{\hat{\theta}}^{(1)}$, and monitoring when, if ever, the "released" frequency ω_k corresponding to each $\underline{\theta} \in c_k(\underline{\theta};\underline{\tilde{\theta}}^{(1)})$ becomes equal to $\hat{\omega}_k$, the "frequency-equivalent" set $\Theta_{\hat{\Omega}}$ is built up. The algorithm has a systematic way of following all possible curves $c_k(\underline{\theta};\underline{\tilde{\theta}}^{(l)}), k = 1,\ldots,K$, for all $\underline{\tilde{\theta}}^{(l)} \in \Theta_{\hat{\Omega}}$ that are found. Each curve $c_k(\underline{\theta};\underline{\tilde{\theta}}^{(l)})$ is followed solving a system of linear equations involving the gradient of the function $\underline{\omega}(\underline{\theta})$. The following simple analytical expression can be used to calculate $\frac{\partial\omega_r}{\partial\theta_i}$:

$$\frac{\partial\omega_r}{\partial\theta_i} = \frac{1}{2\omega_r}\underline{\phi}^{(r)^T} K_i \underline{\phi}^{(r)} \tag{37}$$

Figure 1 illustrates schematically the concepts of the proposed algorithm for the case of a uniform two-degree of freedom shear building with parameters θ_1 and θ_2 scaling the interstory stiffnesses of the first and second floor respectively.

It can be shown [Katafygiotis 1991] that the earlier Equation (19) can be replaced by the following:

$$w'_k = \pi_{\underline{\theta}}(\underline{\hat{\theta}}^{(k)})\mathcal{J}^{-1}(\underline{\hat{\theta}}^{(k)}) \tag{38}$$

where $\mathcal{J}(\underline{\hat{\theta}}^{(k)}) = |\nabla\underline{\omega}(\underline{\hat{\theta}}^{(k)})|$ is the Jacobian of the transformation $\underline{\theta} \rightarrow \underline{\omega}(\underline{\theta})$ calculated at $\underline{\theta} = \underline{\hat{\theta}}^{(k)}$. It is interesting to notice that the weighting coefficient w_k, given by the expressions (18) and (38), does not depend explicitly on the measured output. This is surprising at first, since the term $|A_N^{-1}(\underline{\hat{a}}_k)|^{\frac{1}{2}}$ in the earlier expression (19) for w'_k clearly depended on the measured output.

SOME NUMERICAL RESULTS

Consider all possible stiffness distribution solutions $\underline{\hat{\theta}}^{(k)}$ that are obtained for a linear planar shear building with a number of degrees of freedom N_d when the observed degree of freedom is the one corresponding to the roof. The mass distribution was assumed to be uniform and known. The number K of "output-equivalent" solutions found by our algorithm is much smaller than the upper bound of $(N_d!)$ derived by Udwadia et al [1978], and for the tested cases, where N_d ranged from two to ten dof, is given by $K = 2^{\mathrm{INT}(\frac{N_d}{2})}$.

Table 1 shows the eight "output-equivalent" solutions for a six-story $(N_d = 6)$ uniform shear building, when the observed degree of freedom is the one corresponding to the roof. Figure 2 shows the effective participation factors of the first three modes at the different floor levels corresponding to all the different optimal solutions $\hat{\underline{\theta}}^{(i)}, i = 1, \ldots, 8$, shown in Table 1. It can be seen that while all these different optimal solutions have exactly the same effective participation factors at the observed degree of freedom, their values at the lower degrees of freedom become increasingly scattered. It can be concluded that if predictions are to be made at the roof, then any of these optimal solutions is going to give the same results, while if the response at a lower degree of freedom is to be predicted, the predictions of all optimal models must be included, with their probabilities appropriately weighted through the coefficients w_k. The weighting factors w_k for each model are given in the last column of Table 1, based on (18) and (38), and under the assumption that the models are equally plausible *a priori*, so that the factors $\pi_{\underline{\theta}}(\hat{\underline{\theta}}^{(k)})$ can be omitted.

CONCLUSIONS

An algorithm to investigate model identifiability of the optimal model parameters in structural dynamics is presented. It is shown that choosing just a single model, as usually done by estimating the model parameters through optimization of the model and measured responses at certain degrees of freedom, can lead to unreliable response predictions at the unobserved degrees of freedom, when the model used is not globally identifiable.

REFERENCES

Beck, J.L., "Determining Models of Structures from Earthquake Records," EERL Report No, 78-01, California Institute of Technology, 1978.

Beck, J.L., "Statistical System Identification of Structures," *Structural Safety and Reliability*, ASCE, II, 1395-1402, 1990.

Katafygiotis, L.S., "Treatment of Model Uncertainties in Structural Dynamics," EERL Report No. 91-01, California Institure of Technology, 1991.

Katafygiotis, L.S. and Beck, J.L., "An Efficient Treatment of Model Uncertainties for the Dynamic Response of Structures," *Proceedings 1st International Conference on Computational Stochastic Mechanics*, Corfu, Greece, 1991.

Udwadia, F.E. and Shah, P.C., "Some Uniqueness Results Related to Building Structural Identification," *SIAM Journal of Applied Mathematics*, Vol. 24, No. 1, 1978.

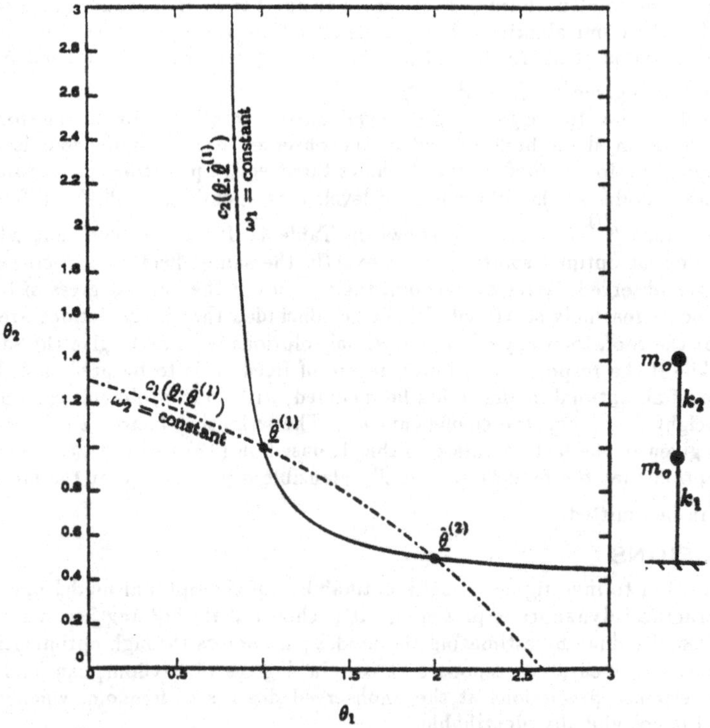

Figure 1. Schematic representation of the proposed alogrithm investigating the model indentifiability of the stiffness parameters θ for the case of a two-story uniform shear building

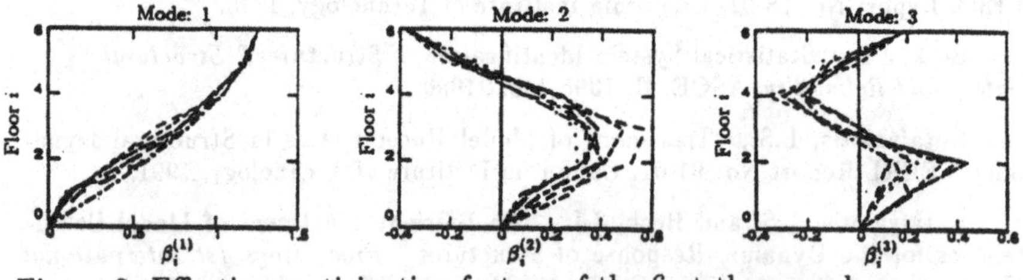

Figure 2. Effective participation factors of the first three modes corresponding to the " output-equivalent" stiffness parameters of Table 1.

No.	$\hat{\theta}_1$	$\hat{\theta}_2$	$\hat{\theta}_3$	$\hat{\theta}_4$	$\hat{\theta}_5$	$\hat{\theta}_6$	$w_k(\%)$
1	1.0000	1.0000	1.0000	1.0000	1.0000	1.0000	21.35
2	1.5848	0.6963	1.2875	0.7574	1.1766	0.7898	13.49
3	1.9970	0.7980	0.7095	1.3848	0.7113	0.8980	4.91
4	2.0000	1.0000	1.0000	0.5000	1.0000	1.0000	21.35
5	2.0932	1.0476	0.7240	0.7374	0.6705	1.2738	17.07
6	2.2911	0.6304	0.9321	1.1774	0.9515	0.6631	6.46
7	2.4913	0.8777	0.6514	1.1106	0.6672	0.9475	7.40
8	2.8252	0.6753	0.8826	0.9021	0.8753	0.7520	7.97

Table 1. "Output-equivalent" stiffness distribution for a six-story uniform shear building, when the only observed degree of freedom is the one corresponding to the roof.

SECTION 2: DAMAGE ANALYSIS

Identification of Structural Dynamic Systems by Sequential Prediction Error Method

C-B. Yun (*), C-G. Lee (*), H-N. Cho (**)
(*) Department of Civil Engineering, Korea Advanced Institute of Science and Technology, Seoul, Korea
(**) Department of Civil Engineering, Hanyang University, Seoul, Korea

ABSTRACT

The modal parameter estimations of linear multi-degree-of-freedom structural dynamic systems are carried out in time domain. For this purpose, the equation of motion is transformed into the auto-regressive and moving average model with auxiliary stochastic input (ARMAX) model. The parameters of the ARMAX model are estimated by using the sequential prediction error method. Then, the modal parameters of the system are obtained thereafter. Experimental results are given for a 3-story building model subject to ground exitations.

INTRODUCTION

In this paper, a method for the time domain identification of the modal parameters of a linear multi-degree-of-freedom structural dynamic system is studied. For the time domain identification, it is very important to obtain a reasonable model for the sampled data system, since the observation data are commonly measured at discrete time instances and contaminated by the measurement noises(Shinozuka et al.[1], Yun et al. [2]). The auto-regressive and moving average(ARMAX) model is used for this purpose.

Moreover, in actual experiments, the response of a system may be measured in terms of displacement, velocity or acceleration depending on sensor types. Hence, different discrete time models depending on the measured response components are derived. Then, the sequential prediction error method is applied to the model to estimate its parameters. The modal parameters are obtained thereafter. Experimental results are given for a 3-story building model. It has been found that the present method utilizing the ARMAX model and the sequential prediction error method yields good estimates of the modal parameters.

MATHEMATICAL MODEL

Experimental modal analysis

Consider a structural system that can be modeled as an n-degrees of freedom linear system. Then, the dynamics of the system can be described by the equation of motion as

$$M_0\ddot{\xi}(t) + C_0\dot{\xi}(t) + K_0\xi(t) = L_0u(t) \tag{1}$$

where $\xi(t)$, $\dot{\xi}(t)$ and $\ddot{\xi}(t)$ are n-dimensional response vectors; $u(t)$ is an m-dimensional input vector; M_0, C_0 and K_0 are mass, damping and stiffness matrices, respectively; L_0 is the $n \times m$ input coefficient matrix; and $(\ \cdot\ )$ denotes differentiation with respect to time.

In actual experiments, the responses can be measured in terms of acceleration, velocity or displacement depending on the sensor type. Therefore, it is necessary to give a different expression of the system by using the measured response component. Also , in general, the responses cannot be measured at all the nodes of the equation of motion. Hence, it is analytically convenient that the coupled equation of motion is transformed into a set of uncoupled equations by using the modal decomposition. After taking the Laplace transform, the transfer functions corresponding to the response components can be written as (Lee and Yun[3])

for $d(t) = \xi(t)$;

$$D(s) = \sum_{i=1}^{n} [\frac{\Psi_i\Psi_i^T}{s-\lambda_i} + \frac{\Psi_i^*\Psi_i^{*T}}{s-\lambda_i^*}]\, L_0U(s) \tag{2}$$

for $v(t) = \dot{\xi}(t)$;

$$V(s) = \sum_{i=1}^{n} [\frac{\lambda_i\Psi_i\Psi_i^T}{s-\lambda_i} + \frac{\lambda_i^*\Psi_i^*\Psi_i^{*T}}{s-\lambda_i^*}]\, L_0U(s) \tag{3}$$

for $a(t) = \ddot{\xi}(t)$;

$$A(s) = \sum_{i=1}^{n} [\frac{\lambda_i^2\Psi_i\Psi_i^T}{s-\lambda_i} + \frac{\lambda_i^{*2}\Psi_i^*\Psi_i^{*T}}{s-\lambda_i^*}]\, L_0U(s) + M_0^{-1}L_0U(s) \tag{4}$$

where $D(s), V(s), A(s)$ and $U(s)$ are the Laplace transforms of $d(t), v(t), a(t)$ and $u(t)$, respectively and Ψ_i is the complex mode shape which is purely imaginary when the damping matrix is proportional; and λ_i is the system pole which is related to the natural frequency and damping ratio associated with the mode shape; and asterisk (*) denotes the complex conjugate.

In this study, the response of the system is assumed to be measured in terms of acceleration. Under the assumption of zero-order hold reconstruction (Phillips and Nagle[4]), the discrete time equations corresponding to Eq. 4 can be simply obtained as

$$a(k) = \sum_{i=1}^{n} [\, \frac{\gamma_i \Psi_i \Psi_i^T}{z - \mu_i} + \frac{\gamma_i^* \Psi_i^* \Psi_i^{*T}}{z - \mu_i^*} \,] \, L_O u(k) + M_0^{-1} L_O u(k) \tag{5}$$

where z can be regarded as one time ahead operator for the present purpose and

$$\mu_i = \exp(\lambda_i \Delta t), \quad \gamma_i = (\mu_i - 1)\lambda_i \tag{5.a}$$

Reducing the fractions to a common denominator, Eq. 5 can be rewritten as

$$a(k) = \frac{Q(z)}{p(z)} u(k) + M_0^{-1} L_O u(k) \tag{6}$$

where

$$\begin{aligned} p(z) &= (z - \mu_1)(z - \mu_1^*)(z - \mu_2) \cdots (z - \mu_n^*) \\ &= z^{2n} + p_1 z^{2n-1} + p_2 z^{2n-2} + \cdots + p_{2n} \end{aligned} \tag{6.a}$$

$$\begin{aligned} Q(z) &= \sum_{i=1}^{n} [\, \frac{\gamma_i p(z) \Psi_i \Psi_i^T}{z - \mu_i} + \frac{\gamma_i^* p(z) \Psi_i^* \Psi_i^{*T}}{z - \mu_i^*} \,] \, L_O \\ &= Q_1 z^{2n-1} + Q_2 z^{2n-2} + \cdots + Q_{2n} \end{aligned} \tag{6.b}$$

It is apparent that the polynomial $p(z)$ and the polynomial matrix $Q(z)$ have real coefficients. It is noted from Eq. 6 that the coupled equation of motion is decomposed into a set of ARMA models for the individual output.

From the estimates of $p(z)$ and $Q(z)$, the modal parameters can be obtained as described below :
The discrete time system poles μ_i 's may be obtained from

$$p(z)|_{z=\mu_i} = 0 \tag{7}$$

Then, from Eq. 5.a,

$$\lambda_i = \frac{1}{\Delta t} \ln(\mu_i), \quad \gamma_i = (\mu_i - 1)\lambda_i \tag{8}$$

where the value of $\ln(\mu_i)$ are chosen as its principal values.

It is well known that λ_i is related to the natural frequency ω_i and damping ratio ζ_i by

$$\lambda_i = -\zeta_i\omega_i + j\omega_i\sqrt{1-\zeta_i^2} \tag{9}$$

where $j = \sqrt{-1}$. Hence, ω_i and ζ_i are given as

$$\omega_i = |\lambda_i|, \quad \zeta_i = -\frac{Re(\lambda_i)}{|\lambda_i|} \tag{10}$$

Finally, it can be easily shown from Eq. 6.b that the mode shape can be obtained as

$$\Psi_i\Psi_i^T L_0 = \left.\frac{(z-\mu_i)Q(z)}{\gamma_i p(z)}\right|_{z=\mu_i} \tag{11}$$

It is noted that for a multi-input case the resultant matrix from Eq. 11 consists of m-column vectors which are proportional to the i-th mode shape but with different proportionality constants, where m is the number of inputs.

Earthquake loading

When the input force is given by ground acceleration the input term can be written as

$$L_0 u(t) = -M_0\{1\}\ddot{u}_g(t) \tag{12}$$

where $\ddot{u}_g(t)$ is the base acceleration and $\{1\}$ is a column vector of which the elements consist of unities.

Hence, the corresponding discrete time equation for acceleration measurement can be written as

$$a(k) = \frac{Q(z)}{p(z)}\ddot{u}_g(k) - \{1\}\ddot{u}_g(k) \tag{13}$$

In Eq. 13, $a(k)$ denotes the floor acceleration relative to the base defined by

$$a(k) = a^t(k) - \{1\}\ddot{u}_g(k) \tag{14}$$

where $a^t(k)$ is the absolute acceleration. Therefore, Eq. 13 can be rewritten as

$$a^t(k) = \frac{Q(z)}{p(z)}\ddot{u}_g(k) \tag{15}$$

It is interesting to note from Eq. 15 that for a structure subjected to a ground acceleration, the ARMA model can be constructed in terms of the absolute acceleration responses, but without the concurrent input term.

Noise consideration (ARMAX Model)
Defining $y(k)$ and $u(k)$ as the measured acceleration at a node and the measured ground acceleration of Eq. 15, they can be written as

$$y(k) = a_j^t(k) + \eta(k) \tag{16}$$

$$u(k) = \ddot{u}_g(k) + \omega(k) \tag{17}$$

where $a_j^t(k)$ denotes the total acceleration at j th node and $\eta(k)$ and $\omega(k)$ are measurement noises which are generally assumed as white processes with finite variances.

Based on the Kalman's innovations theory (Kailath and Frost[5]), Eq. 15 can be asymtotically represented by using the measured data $y(k)$ and $u(k)$ and innovations sequences $e(k)$ as

$$p(z)y(k) = q(z)u(k) + c(z)e(k) \tag{18}$$

where $q(z)$ is a polynomial related to the measured point of $n \times 1$ polynomial matrix $Q(z)$ and has the following form :

$$q(z) = q_1 z^{2n-1} + q_2 z^{2n-2} + \cdots + q_{2n} \tag{18.a}$$

and $c(z)$ is a polynomial to describe the innovations process and can be written as

$$c(z) = z^{2n} + c_1 z^{2n-1} + \cdots + c_{2n} \tag{18.b}$$

SEQUENTIAL PREDICTION ERROR METHOD

Eq. 18 can be equivalently rewirtten as

$$y(k) = \phi^T(k-1) \cdot \theta + e(k) \tag{19}$$

where

$$\begin{aligned} \phi^T(k-1) \; = < \; & y(k-1), y(k-2), \cdots, y(k-2n), \\ & u(k-1), u(k-2), \cdots, u(k-2n), \\ & e(k-1), e(k-2), \cdots, e(k-2n) > \end{aligned} \tag{19.a}$$

$$\theta^T = < -p_1, -p_2, \cdots, -p_{2n}, q_1, q_2, \cdots, q_{2n}, c_1, c_2, \cdots, c_{2n} > \tag{19.b}$$

It may be impossible in the stochastic situation to obtain the value of θ for which the prediction error $e(k,\theta)$ is equal to zero at every instances of time. Instead, one can only estimates the value of θ which minimizes the prediction errors. The following criterion is generally taken as the measure of the quality of approximation :

$$V_N(\theta) = \frac{1}{N}\sum_{i=1}^{N} e^2(k,\theta) \tag{20}$$

where N is the number of data points. It is not possible to minimize the criterion with respect to θ by analytical means, since $e(k,\theta)$ is depentent on θ implicitly as well as explicitly. The numerical search procedure can be approximately implemented in the sequential form. The algorithm of the sequential prediction error method is summarized as follows (Soderstrom[6], Goodwin and Sin[7]) :

$$\hat{\theta}(k) = \hat{\theta}(k-1) + F(k-1)\phi_f(k-1)(y(k) - \hat{\phi}(k-1)\hat{\theta}(k-1)) \tag{21}$$

$$F^{-1}(k) = F^{-1}(k-1) + \phi_f(k-1)\phi_f^T(k-1) \tag{22}$$

$$\phi_f(k-1) = \hat{c}^{-1}(z,k-1)\hat{\phi}(k-1) \tag{23}$$

where

$$\begin{aligned} \hat{\phi}^T(k-1) \;&= <\; y(k-1), y(k-2), \cdots, y(k-2n), \\ &\quad\;\; u(k-1), u(k-2), \cdots, u(k-2n), \\ &\quad\;\; \hat{e}(k-1), \hat{e}(k-2), \cdots, \hat{e}(k-2n) \;> \end{aligned} \tag{24}$$

$$\hat{e}(k-1) = y(k-1) - \hat{\phi}(k-2)\hat{\theta}(k-1) \tag{25}$$

As shown in Eq. 23, the gradient vector $\phi_f(k-1)$ is obtained by passing the regression vector $\hat{\phi}(k-1)$, through a time varying filter $\hat{c}^{-1}(z,k-1)$ which is estimated at time $k-1$. In order to keep away from the divergence problem, the projection technique can be employed to make the filter remain in the stability region.

NUMERICAL EXAMPLES AND DISCUSSIONS

The present estimation method is applied to a three-story building model to evaluate the modal parameters. As shown in Fig. 1, the input to the model is given by the base acceleration. The accelerations are measured at the base and three floors. The measurement records of two different experiments are shown in Fig. 2 and Fig. 3. One experiment is for the case with a long duration input and the other is for an impulse. For the present purpose, the structural model is approximately regarded as a system with three degrees of freedom.

Estimations of the modal parameters are carried out for three different sets of experiments. In Table 1, the results are compared with those obtained by the structural analysis, in which the structure used in the experiment is idealized as a shear building model. The natural frequencies and the damping ratios are evaluated redundantly from the acceleration records at each floor. In this study, the natural frequencies and damping ratios are obtained by taking the average of the estimates. The estimated natural frequencies and the first mode shapes from three different experiments are found to be very consistent with each other. As for the second and third mode shapes, however, a little discrepancies can be observed between the results from different experiments. It has been found that the estimated values for the damping ratios are not quite consistent. It may be caused by the fact that the damping values are so small (i.e. less than 0.01) that the properties of damping cannot be observed well. Compared with the estimated natural frequencies, those calculated by the eigenvalue analysis on the shear building model are found to be slightly large. It may be caused by the fact that the fixity conditions at the ends of the vertical members might be idealized more rigidly in the shear building model than the realities of the actual model used in the experiment. The responses are re-evaluated based on the estimated parameters and compared with the measured records in Figs. 4 and 5. It can be seen that the formers are almost identical to the latters, except that the measurement noises are filtered out.

CONCLUSIONS

In this paper, methods for the estimation of the modal parameters of linear multi-degrees-of-freedom structural dynamic systems are presented. The following is the summary of this study:

(1) The auto-regressive and moving-average models with the auxiliary stochastic input (ARMAX model) are derived for three different types of the measured response components; i.e. displacement, velocity and acceleration.

(2) Estimated modal parameters of a building model from different sets of experimental data by the present method are found to be very consistent, particularly for the natural frequencies and the first two mode shapes. The estimated responses based on the identified parameters show good agreements with the measurements.

ACKNOWLEDGEMENTS

This research was jointly supported by the Korea Science and Engineering Foundation and National Science Foundation of United States. Those supports are gratefully acknowledged.

REFERENCES

1. Shinozuka, M., Yun, C.B. and Imai, H., Identification of Linear Structural Dynamic Systems, J. Eng. Mech., ASCE, Vol.108, pp.1372–1390, 1982.

2. Yun, C.B., Kim, W.J. and Ang, A.H–S., Damage Assessment of Bridge Structures by System Identification, Proc. of the 5th ICOSSAR, San Francisco, CA, 1989.

3. Lee, C.G. and Yun, C.B., Parameter Identification of Linear Structural Dynamic Systems, Accepted in J. of Computers and Structures, 1990.

4. Phillips, C.L. and Nagle, H.T., Digital Control System Analysis and Design, Prentice-Hall, Englewood Cliffs, N.J., 1984.

5. Kailath, T. and Frost, P.A., An Innovations Approach to Least-Squares Estimation. Part 2: Linear Smoothing in Additive White Noise, IEEE Trans. on Automatic Control, AC-13, pp.655-660, 1968.

6. Soderstrom, T., An On-line Algorithm for Approximate Maximum Likelihood Identification of Linear Dynamic Systems, Report 7308, Lund Ins. Tech., Dept. Auto. Control, 1973.

7. Goodwin, G.C. and Sin, K.S., Adaptive Filtering Prediction and Control, Prentice-Hall, Englewood Cliffs, N.J., 1984.

Table 1. Estimated Modal Parameters.

Modal properties \ Cases		Experiment 1		Experiment 2		Experiment 3		Structural Analysis	
ω_1 , ζ_1		22.6 ,	.0037	22.9 ,	.0076	22.8 ,	.0070	25.5 ,	-
ω_2 , ζ_2		63.9 ,	.0015	62.8 ,	.0013	62.8 ,	.0030	71.1 ,	-
ω_3 , ζ_3		94.4 ,	.0075	92.9 ,	.0032	92.8 ,	.0029	103.6 ,	-
ψ_1	3F	1.	(0)	1.	(0)	1.	(0)	1.	(0)
	2F	.809	(2)	.790	(-1)	.816	(-1)	.802	(0)
	1F	.440	(-1)	.444	(-1)	.447	(-1)	.445	(0)
ψ_2	3F	1.	(0)	1.	(0)	1.	(0)	1.	(0)
	2F	.307	(175)	.400	(175)	.556	(175)	.554	(180)
	1F	1.106	(177)	1.152	(178)	1.103	(176)	1.250	(180)
ψ_3	3F	1.	(0)	1.	(0)	1.	(0)	1.	(0)
	2F	3.307	(-177)	2.226	(-174)	1.824	(-168)	2.242	(180)
	1F	1.641	(-8)	1.481	(4)	1.402	(-6)	1.801	(0)
Input Types		Long duration		Impulsive		Impulsive			

Note : 1. The unit of ω_i is in rad/sec.

2. Values in parentheses denote the phase angles of the complex mode shapes in degrees.

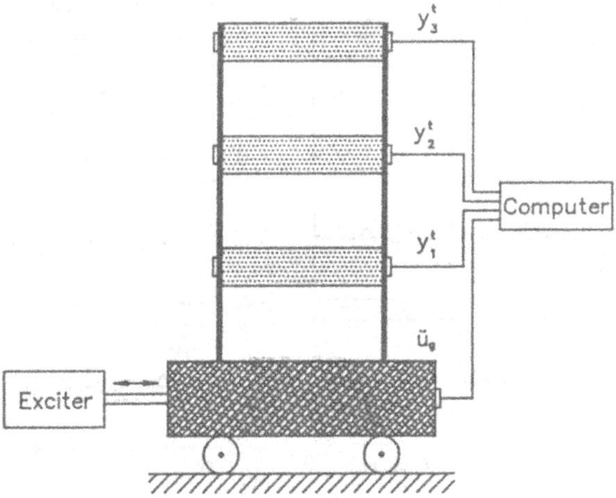

Fig. 1. Experimental Set-up of a Building Model.

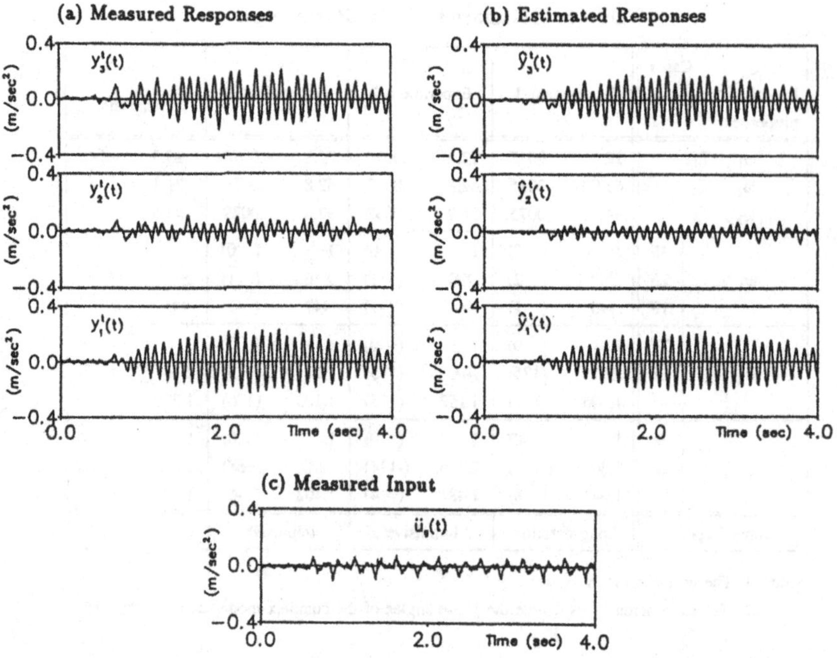

Fig. 2. Time Histories of Measured and Estimated Acceleration Responses(Exp. 1)

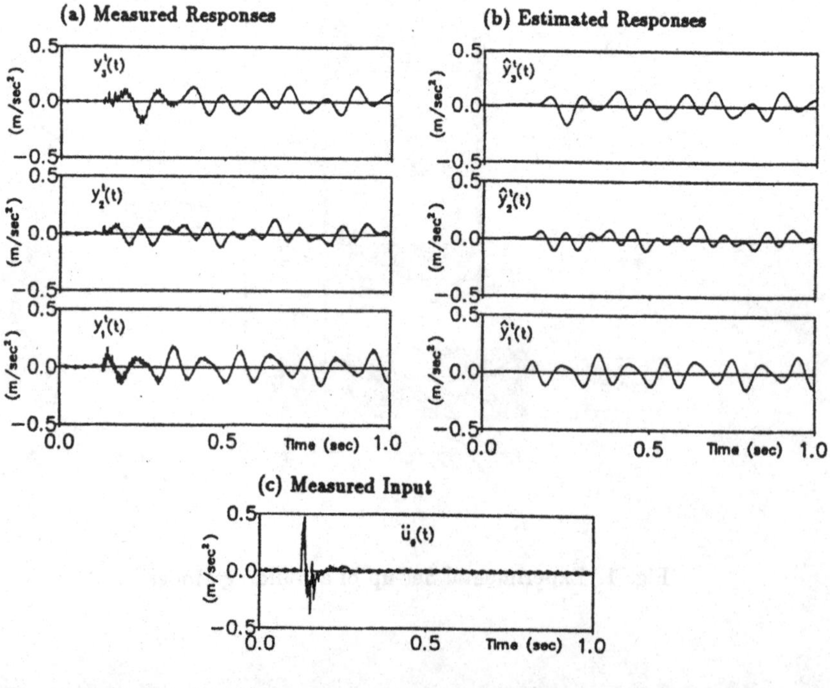

Fig. 3. Time Histories of Measured and Estimated Acceleration Responses(Exp. 2)

Computer and Test Aided Modelling and its Application to Identify Structural Damage

H.G. Natke (*), J.T.P. Yao (**)

(*) Curt-Risch-Institut für Dynamik, Schall-und Messtechnik, University of Hannover, Appelstrasse, D-300 Hannover 1, Germany

(**) Department of Civil Engineering, Texas A&M University, College Station, Texas 77843-3136, U.S.A.

ABSTRACT

In this paper, several classical methods for the identification of structural damage are critically reviewed. The test and computer aided modelling is then introduced and discussed. Finally, several research topics are selected and outlined.

INTRODUCTION

To date, several mathematical models have been used along with measurements from existing structures in safety evaluation studies. Although finite element (FE) methods are useful, it may produce models with uncertainties. In general, the user does not know the various degrees of uncertainty in such models. Computational mechanics are very useful in the formulation of mathematical models. However, fast algorithms for simulation and optimization may give additional uncertainty and/or fail due to divergent iteration. In spite of the progress in modelling to date, there is a need for further investigations with respect to the stability, consistency, efficiency, and robustness of numerical methods.

Consequently, test-supported modelling can be used to overcome some of these difficulties (Natke and Yao [1]). Nondestructive test data resulting from artificial or natural (e.g., microseismic) excitation are used for detection and localization of errors in the mathematical model and for subsequent correction. The result can be a validated mathematical model with known confidence and sufficiently small errors. This model is dependent on the discrete value of the lifetime of the structure due to the time of measurement. Therefore, the measured data contain the state of the structure including all structural damage up to this time and in consequence the mold will reflect it also. We can use such a model to detect and to classify significant deviations during the structural life resulting in model/software redundant (with insufficient measurements) monitoring and assessment of structure damage.

CRITICISMS OF CLASSICAL METHODS

Classical vibrational monitoring and damage assessment refer to those investigations using vibrational signals and signatures including model quantities of the structure (Braun [2], Collacott [3]). Data reduction gives some features which yield a decision vector. Decisions in practice are often made using thresholds (e.g., classification theory). Meanwhile, assessment needs must be performed by an expert with special knowledge of the structural behavior.

For very special structures such as pressurized water reactors (Stölben and Wehling [4]) and pipelines (Beller and Schneider [5]), these procedures seem to work well. However, accurate procedures for structures in general, even for such frame structures as platforms, can not be deduced easily.

TEST AND COMPUTER AIDED MODELLING (TACAM)

Nondestructive test data are taken for updating the spatially discretized mathematical model resulting from theoretical analysis. If the model structure is verified for the defined purpose, the problem is reduced to that of parameter estimation (Natke [6,7,8]). These procedures are well documented in the literature. They work well only for a relatively small number of degrees of freedom (on the order of 10^3) and a few parameters to be estimated (on the order of 10^2) due to memory limitations of the computer and the associated numerical difficulties. The latter are due to the ill-poseness of inverse problems with erroneous and incomplete data. They can be overcome by introducing modified operators as by inclusion of the prior mathematical model (Bayesian approach, see Natke [9,10]). Experimental design (Cottin [11], Ben-Haim and Natke [12]) is another way to obtain a stable and unique solution.

The usage of updated models for damage detection is twofold. At first the detection of detectable faults is done by comparing the models describing different states of the structure. Deviations between these models will indicate the faults and they are able to locate the faults within the model. The specific damage detection of the structure with the use of these models is illustrated in Figure 1. Studies of possible faults (theoretical or from experience) can be helpful in system modelling, which consists of an initial mathematical model and assumed faults. Using such a system model, we may calculate the system response. Meanwhile, the observed and measured system response can be compared with the calculated response with assumed fault. If the agreement is acceptable (i.e. $|\text{residuals}|<\epsilon$), the assumed faults are said to be identified. Otherwise, the system modelling is modified with different mathematical model and assumed faults. The process is repeated until the faults are identified.

The advantages of the total procedure are listed as follows:

(a) In each stage of the structure lifetime, a model with known covariances will be formulated.
(b) It allows for computational studies of sensitive quantities of the structure due to various assumed faults and thus leads to symptoms.
(c) It is a model/software redundant monitoring with a minimum set of measurement data.
(d) It enables damage detection.
(e) It additionally localizes the damage by, for example, submodelling.
(f) It adds to the knowledge base.

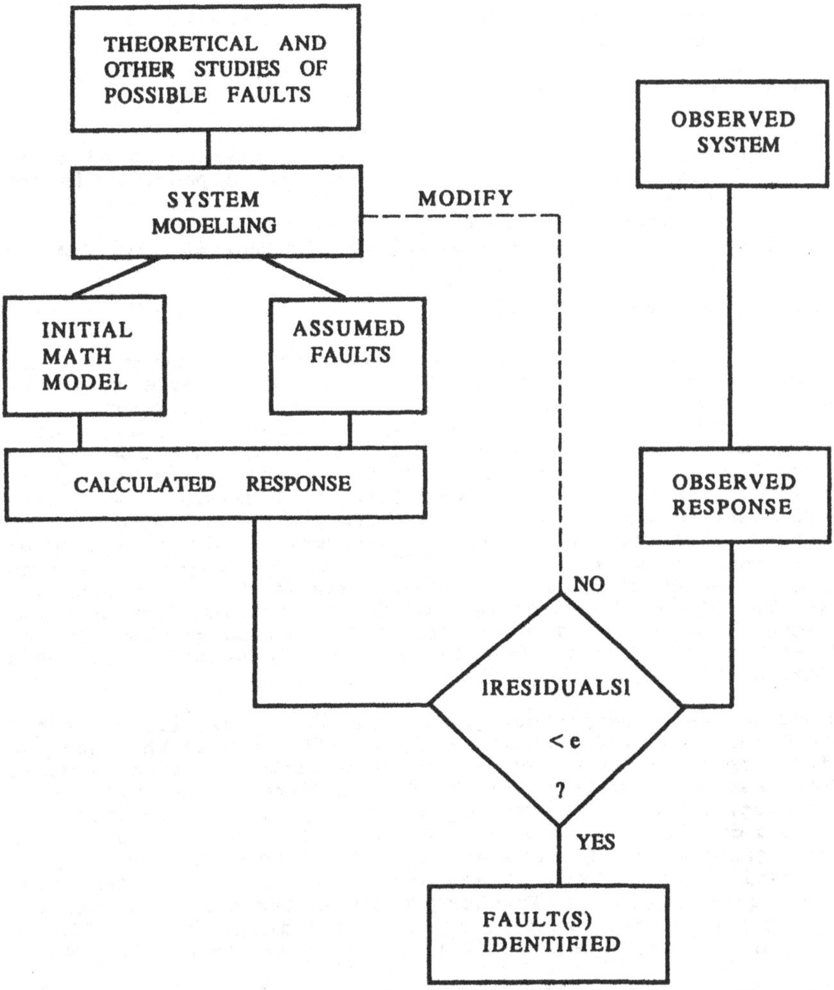

FIGURE 1 METHODOLOGY

(g) It helps to find the cause(s) of damage by analyzing the system with most results already available using this procedure.
(h) It can be automated with the use of neural network and/or fuzzy logic (e.g., Villarreal and Baffes [13]; Villarreal and Shelton [14]).

Several disadvantages are as follows:

(a) It is not a real-time procedure.
(b) At any given time, the model is assumed to be time-invariant. Therefore, the damage propagation is not modelled directly.
(c) Linear models are generally used.
(d) The model is predefined and assumed as sufficiently accurate.

SELECTED RESEARCH TOPICS

It seems that a pointwise (with respect to lifetime) time-invariant model meets the need for a model/software redundant methodology. Modifications to a linear model are concerned with the type of damping and the number of degrees of freedom of a band-limited model. Because the knowledge of damping is generally limited and its precise modelling is extremely difficult, equivalent viscous or structural damping seems to be sufficient (except for damping identification purposes). Damping ratios as damage indicators can be very sensitive symptoms. Relatively inaccurate estimates of damping ratio may be obtained when compared with eigenfrequency estimates. Detection and handling of additional degrees of freedom within a prespecified frequency interval is a problem which can be reduced to the determination of the effective number of degrees of freedom (Zamirowski [15]).

Among several other investigators (Bschorr et al. [16]), Stubbs and Osegueda [17,18] applied a first difference to the equation of motion of an undamaged structure to obtain changes in model stiffness as a measure of structural damage. Results of a numerical example of a simple beam and experimental verification using a cantilever beam with a reduced section were given. It is not clear whether this type of method will be effective for more realistic structures which are much more complicated than those in the given examples. Further procedures based on linear models are published, for example, in the proceedings edited by Natke and Yao (see Natke and Yao [1]) and by Hjelmstad et al. [19].

Linearized modelling of nonlinear systems will usually suppress certain characteristics of real system behavior which may be important for damage detection. If this nonlinearity must be modelled, then the identification of the nonlinear model structure is important in reducing the problem to that of parameter estimation.

In such cases, one must classify and investigate those faults which lead to nonlinear system behavior at first. Secondly, the effects of these nonlinearities must be studied in the direct problem to find symptoms or discriminants (or feature vectors). Then sensitivity analysis should be performed in order to estimate the parameters for solving the nonlinear inverse problem including:

i) global detection (by deviation of the linear

behavior through nonlinear characteristics) and residuals,

ii) modification distinguishability in the model (type of model structure),

iii) model structure identification,

iv) parameter identification, and

v) damage localization.

This part can be done using the finite element modelling along with approximation (instead of recalculation) by interpolation and/or extrapolation of available results for pregiven faults.

At present, model structure identification will be restricted to polynomials as class of models. According to the Stone-Weierstrass theorem, single-valued function defined in a finite interval is the limit of some uniformly convergent sequence of polynomials. This class of models will cover a wide range of systems. Structure identification here includes the estimation of the maximum power of the polynomial as well as its coefficients. The obvious advantage is that neither a priori choice of the structure nor a posteriori model validation is necessary.

A first step in this direction is already done by Natke and Zamirowski [20] for separable restoring and damping forces. This approach should be extended concerning coupled nonlinear element description (non-separable forces) and the application of regularization methods in order to make the problem convex and to include a priori information with the use of a penalty term.

OUTLOOK

Mathematical models are uncertain and need verification otherwise correction using measured values of the existing system. Such an updated mathematical model can be used for damage investigations. The measurements of the damaged system can serve for damage detection and location using the updated initial (previous) model. Additionally, it can be used for assessment and decision for further steps. The procedure described is a software redundant monitoring.

The usage of linear models are advantageous for many reasons. However, the damage process is usually highly nonlinear and cannot be described satisfactorily with any linear model. To make such a modelling process and its practical application more useful, it is desirable to develop automation. The use of neural network and/or fuzzy logic seems to be desirable and feasible at present.

In order to describe the damage process, the system time-invariance during the lifetime has to be studied. Furthermore, one has to investigate possible nonlinear behavior. Their description by models even with the help of system identification is unsolved. Data-bases and information systems here can be helpful tools.

REFERENCES

1. Natke, H. G. and Yao, J. T. P., "System Identification Approaches in Structural Safety Evaluation" in: Natke, H. G., Yao, J. T. P. (Eds.) Structural Safety Evaluation Based

on System Identification Approaches, View eg. Braunschweig, Wiesbaden, 460-473, 1988.

2. Braun, S. (Ed.), "Mechanical Signature Analysis-Theory and Applications," Academic Press, 1986.

3. Collacott, R. A., "Structural Integrity Monitoring," Chapman & Hall, London, New York, 1985.

4. Stölben, H. and Wehling, H.-J., "Vibration Monitoring of Kraftwerk Union Pressurized Water Reactors - Review, Present Status, and Further Development," Nuclear Technology, 80, 400-411, Mar. 1988.

5. Beller, M. and Schneider, U., "Pipeline Inspection with the Intelligent Ultrasonic Pig" (in German), Erdöl Erdgas Kohle, 106, 1, 47-51, 1990.

6. Natke, H. G. (Ed.), "CISM Courses and Lectures No. 272," Inter-nat. Centre for Mechanical Sciences, Identification of Vibrating Structures, Springer-Verlag, Wien, New York, 1982.

7. Natke, H. G., "Einführung in Theorie und Praxis der Zeitreihen- und Modalanalyse," Friedr. Vieweg & Sohn, Braunschweig/Wiesbaden, 2. Aufl., 1988.

8. Natke, H. G. (Ed.), "CISM Courses and Lectures No. 296," Inter-nat. Centre for Mechanical Sciences, Application of System Identification in Engineering, Springer-Verlag Wien, New York, 1988.

9. Natke, H. G., "On Regularization Methods Applied to the Error Localization of Mathematical Models," IMAC, 1991/1992.

10. Natke, H. G., "Error Localization within Spatially Finite-Dimensional Mathematical Models - A Review of Methods and the Application of Regularization Techniques" (to be published in Computational Mechanics), 1991.

11. Cottin, N., "On Optimum Experimental Design for Parametric Identification of Linear Elastomechanical Systems, in W. B. Kratzig, et al. (Eds.), Structural Dynamics, Vol. 1, A. A. Balkema, Rotterdam, Brookfield, 347-354, 1991.

12. Ben-Haim, Y. and Natke, H. G, "Diagnosis of Changes in Elastic Boundary Conditions in Beams by Adaptive Vibrational Testing" (to be published), 1991.

13. Villarreal, J. A. and Baffes, P., "Sunspot Prediction Using Neural Networks," Presented at the Third Annual Workshop on Space Operations Automation and Robotics, NASA Conference Publication 3059, L. B. Johnson Space Center, Houston, Texas, 1989.

14. Villarreal, J. A. and Shelton, R. O, "A Space-Time Neural Network (STNN)," Presented at the Second International Joint Conference on Neural Network and Fuzzy Logic, L. B. Johnson Space Center, Houston, Texas, 1990.

15. Zamirowski, M, "Some Tests for Determination of the Effective Number of Modes," (to appear in ZAMM 1991).

16. Bschorr, O. and Mittmann, J., "Qualitatskontrolle und Mustervergleich mittels differentieller Modalanalyse,", VDI-Zeitschrift 126, No. 9, 5.308-5.310, 1984.

17. Stubbs, N. and Osegueda, R., "Global Non-destructive Damage Evaluation in Solids," The International Journal of Analytical and Experimental Modal Analysis, 5(2), 67-79, April 1990.

18. Stubbs, N. and Osegueda, R., "Global Damage Detection in Solids - Experimental Verification," International Journal of Analytical and Experimental Modal Analysis, 5(2), 81-97, April 1990.

19. Hjelmstad, K. D., Wood, S. L., and Clark, S. J., "Parameter Estimation in Complex Linear Structures," Department of Civil Engineering, University of Illinois, Urbana, Illinois, Report UILU-ENG-90-2015, 1990.

20. Natke, H. G. and Zamirowski, M., "On Methods of Structure Identification for the Class of Polynomials within Mechanical Systems," ZAMM, Z. angew. Math. Mech. 70, 10, 415-420, 1990.

The Effect of Model Uncertainty on the Accuracy of Global Nondestructive Damage Detection in Structures

N. Stubbs, J.T. Kim, K.G. Topole

Department of Civil Engineering, Texas A&M University, College Station, TX 77843, U.S.A.

ABSTRACT

An investigation of the effect of model uncertainty on the accuracy of damage localization is presented. A theory of damage localization is summarized. The validity of the theory is demonstrated experimentally. To evaluate the effect of model uncertainty on localization accuracy, a numerical experiment is performed to evaluate the effect of model uncertainty on localization accuracy. It is shown that localization accuracy depends on the magnitude of the model uncertainty, the location of the uncertainty, the location of the damage, and the relative magnitude of the damage and the model uncertainty.

INTRODUCTION

This paper addresses the general problem of damage localization in structures via pre and post-damage changes in the model response of the structure. A solution to this class of problems is important for at least two reasons. Firstly, damage localization is the first step in the broader problem of damage assessment. Secondly, the conduct of a timely damage assessment could lead to desirable consequences such as saving of lives, a reduction of human suffering, protection of property, increased reliability, increased productivity of operations, and a reduction in maintenance costs and time.

During the past decade, a significant amount of research has been conducted in the area of damage detection via changes in the modal response of a structure. Research studies have related changes in eigenfrequencies to changes in beam properties (Gudmunson [4]), located defects in beam elements from changes in eigenfrequencies (e.g., Cawley and Adams

[2]), located cracks in joints (Dimarogonas and Chondos [3]), detected faults in mechanical structures (Ju et al [5]), attempted to monitor the integrity of offshore platforms (e.g., Nataraja [6]),; attempted to monitor the structural integrity of bridges (e.g., Biswas et al [1]), and investigated the feasibility of damage detection in space structures using changes in modal parameters (Stubbs, et al [11]).

Despite this modest combined research effort, many problems remain unsolved. A need remains to develop more robust theories of damage detection to simultaneously include changes in all modal parameters. A need also remains to circumvent the reality of being capable of measuring only a few modes, particularly in large civil engineering structures. In addition, a need remains to account for omnipresent experimental errors in measured modal parameters. Finally, even if the damage detection theories were perfect, many modal parameters were available, and experimental error were insignificant, the need still exists to account for the uncertainty associated with modelling the properties and boundary conditions of real structures.

The objective of this paper is to investigate the effect of model uncertainty on the accuracy of nondestructive damage localization in structures using a theory of damage detection proposed by Stubbs [8]. Here model uncertainty will mean a lack of knowledge regarding the geometry or material characterization of a structure. The level of accuracy of the damage localization scheme is related to the error magnitude between the true location of damage and the predicted location of damage. The investigation is presented in two parts. In the first part, we present the theory of the approach which includes a summary of the theory of damage localization and a validation of the theory using experimental results. In the second part, we examine the effect of model uncertainty on damage localization accuracy. To our knowledge, this relationship has not been studied previously. Furthermore, the findings of this work may lead to a clearer understanding of the impact of model uncertainty on damage localization.

THEORY OF DAMAGE LOCALIZATION

Theory of Approach

Suppose an undamaged structure with N degrees of freedom and B elements has mass, damping, and stiffness matrices given by M, C, and K, respectively. The free-vibration response of this structure will be characterized by the 4N modal quantities: damped natural frequencies (ω_{di}), natural frequencies (ω_{oi}), mode shape vectors (ϕ_i), and modal damping (ξ_i) (i=1,...N). Similarly, if the structure is damaged or modified, the spatial properties become M+ΔM, C+ΔC, and K+ΔK, while the modal properties become $\omega_{di}+\Delta\omega_{di}$, $\omega_{oi}+\Delta\omega_{oi}$, $\phi_i+\Delta\phi_i$, and $\xi_i+\Delta\xi_i$. If modal equations are written for both states of the structure and the equations are combined, the result is a system of N linear equations in B unknowns of the form:

$$Z_d + Z_m + Z_{damp} + Z_{modes} = (F_o + \Delta F)\alpha \tag{1}$$

In which, Z_d, Z_m, Z_{damp}, and Z_{modes} are NX1 matrices containing, respectively, fractional changes in the damped eigenvalues, fractional changes in the modal masses, fractional changes in the modal damping ratios, and fractional changes in the norm of the mode shapes; α is a BX1 matrix containing the fractional losses in stiffness of each element; F_o is a NXB sensitivity matrix relating changes in modal parameters to changes in element stiffness (assuming damage is small); and ΔF is the change in the sensitivity matrix due to changes in mode shapes resulting from the damage.

All items on the left hand side of Equation (1) can be obtained from pre and post-damage modal tests. The sensitivity matrix, F_o, can be obtained either from a dynamic finite element analysis of a model of the undamaged structure or from the stiffness matrix of the structure and measured undamaged modes. The matrix ΔF can be obtained only from measured pre and post-damage modes and a stiffness matrix of the structure. Solution strategies for Equation (1) are discussed elsewhere (See e.g., Stubbs and Osegueda [9], Stubbs and Osegueda [10]).

Validation of Theory

The localization model will be validated using data published in a recent paper by Silva and Gomes [7]. Those researchers performed an extensive set of dynamic analysis experiments on free-free uniform beams with the goal of providing objective data to validate proposed techniques for damage localization.

Here the results of 32 experiments reported by Silva and Gomes [7] are utilized: 16 experiments on undamaged beams and 16 experiments on damaged beams. The following experimental procedure was utilized. The first four bending frequencies were measured for each of 16 undamaged free-free beams. A cut, simulating a crack, was introduced into each beam and the same four bending frequencies were measured.

Referring to Equation (1), the experiments of Silva and Gomes [7] do not provide information on Z_m, Z_{damp}, Z_{modes} or ΔF. Therefore, the model to be used to locate damage is an approximation of Equation (1) and is given by:

$$Z_d = F_o \alpha \tag{2}$$

For analysis purposes we divide the 72 cm- beams into 24 elements of 3 cm each. Therefore, since four frequencies are measured, F_o is a 4X24 matrix. In other words, for this experiment, Equation (2) is a system of four equations with 24 unknowns. The 96 elements of the matrix F_o can be obtained from a single dynamic analysis of a finite element analysis of the free-free beam model (Stubbs [8]). Assuming that the sensitivities F_{ij} (i=1,4;j=1,24) have been determined, the location of damage, j=K, is found via the following strategy. Since the structure is damaged in a single location, Equation (2) may be rewritten (for i=1,...4)

$$Z_{di} = F_{iK} \alpha_K \tag{3}$$

Dividing any two of the equations we get

$$\frac{Z_{di}}{Z_{dj}} = \frac{F_{iK}}{F_{jK}} \tag{4}$$

i.e., the damage is located at the element for which the measured ratio of the fractional changes in frequency for two modes (i&j) equals the ratio of the theoretically obtained sensitivities for those modes and that location. This criterion was used as the basis for localizing the damage.

The results for the sixteen cases are summarized in Figure 1. All but two of the predictions are within 1.5 cm of the true location. Note also that the magnitude of the localization error is a non-obvious function of both the size and the location of the crack. In summary, the approximate formulation given by Equation (2) can predict the accuracy of location to within approximately 2 percent of the beam span (i.e., 1.5/72 x 100).

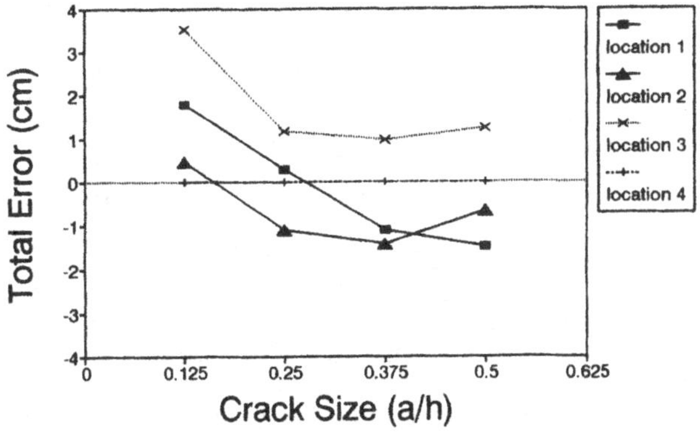

Fig. 1 Accuracy of Damage Localization Using Experimental Data [7].

EFFECT OF MODEL UNCERTAINTY ON LOCALIZATION ACCURACY

The Design of the Experiment

We wish to investigate the effect of our lack of knowledge of the structure in which damage is to be located on the accuracy of the damage localization prediction. That lack of knowledge is manifested by the uncertainty in the geometric and material properties of the structure or the nature of the boundary conditions. For analytical purposes we can simulate the lack of knowledge by perturbing geometric and/or material properties at locations of interest. In this paper we will restrict the investigation to those occurrences of uncertainty (or ignorance) in which the uncertainty can be modelled by perturbations in the structural stiffness at various locations. Thus the investigation will isolate the effects of model uncertainty in the forms of the magnitude of stiffness perturbation and the location of stiffness perturbation on the accuracy of damage localization using Equation (4).

An experiment to evaluate the effect of the location of model uncertainty on the accuracy of damage localization may be carried out as follows: (1) Select a test structure; (2) Select the magnitude of stiffness change representing model uncertainty to be introduced; (3) Introduce the stiffness change at a desired location in the structure; (4) Introduce simulated damage to the same structure; (5) Simulate the dynamic response of the structure with the simulated damage and the model uncertainty; (6) Simulate the response of the structure assuming no damage and no model uncertainty, and compute the sensitivity matrix; (7) Use the sensitivities and reference frequencies in Step 6 and the simulated response in Step 5 to locate the damage in the structure with model uncertainty (using Equation 4); (8) Correct for the effect of the error arising from the simplified mathematical model; (9) Determine the accuracy of the localization procedure; (10) Repeat Step 3 to Step 9 for other locations; (11) Relate the accuracy of damage localization to the location of model uncertainty.

A Similar experiment to evaluate the effect of the magnitude of model uncertainty on the accuracy of damage localization can be carried out. In this case the steps 1,2, and 10 are modified as follows: (1) Select the same test structure as above; (2) Select a single location of model uncertainty to be studied; (10) Repeat Step 3 to Step 9 for other magnitudes of damage.

Step 1 to Step 7 and Step 9 to Step 11 in the strategies outlined above are relatively straightforward; however, we must clarify Step 8. The objective of the experiments is to investigate the effect of model uncertainty. Therefore, the effect of model uncertainty must be extracted from the total response. The location predicted in Step 7 will contain errors arising from the mathematical model used to locate the damage (Eq. 4) as well as errors arising from model uncertainty. Before correction, the error in the predicted location may be given by

$$\epsilon_T = \epsilon_D + \epsilon_M \tag{5}$$

where ε_D is the error due the mathematical model and ε_M is the error due to the model uncertainty. Note that ε_T is associated with the structure with simulated damage and model uncertainty. The error, ε_D, is associated with the structure with the same simulated damage but without any model uncertainty. The error, ε_D, can be obtained by repeating Step 1 to Step 8 (in both experiments) and using the appropriately damaged structure with no model uncertainty. The error due solely to model uncertainty can then be obtained using Equation 5.

The Numerical Experiment

The structure selected for the numerical experiment was a free-free beam with the same geometric and material properties as the beam used by Silvas and Gomes [7]. This structure was selected for two reasons: (a) the previous experimental results and analysis could provide a useful basis of comparison; and (b) the structure contains well defined boundary conditions.

The effect of more complicated boundary conditions is being investigated in a parallel study. A schematic of the analytical test beam is shown in Figure 2. For this analysis the beam has again been subdivided into 24 finite elements each of which are three centimeters long.

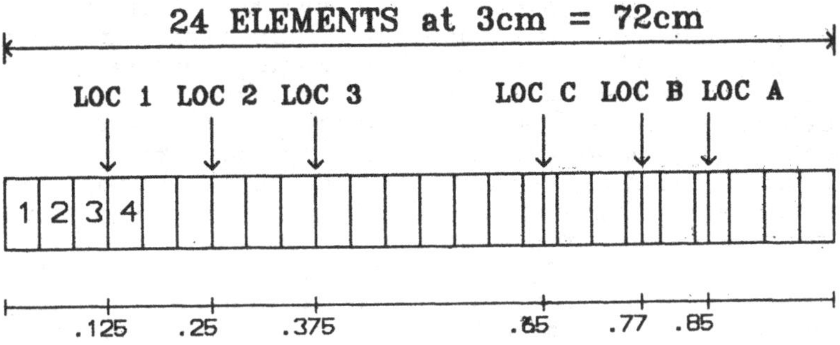

Fig. 2 Test Beam for Numerical Experiment

Model uncertainty was introduced by perturbing the stiffness of the beam by known amounts at the locations of interest. Uncertainty Locations A (Z/L = 0.854), B (Z/L = 0.771) and C (Z/L = 0.646) for this study corresponded to the nodes of modes 2, 1, and 3, respectively. The magnitudes of model uncertainty selected corresponded to fractional stiffness decreases of 0.025, 0.1, and 0.5.

The preselected locations of damage corresponded to the locations of the cracks in the experimental beam. These locations corresponded to the division line between two elements in Figure 2. Damage was introduced by uniformly decreasing the stiffness of the two adjacent elements. For each location, the simulated damage corresponded to a fractional loss in stiffness of one of the following values: 0.125, 0.25, 0.375 and 0.5.

Using the model in Figure 2, we introduced the appropriate damage and model uncertainties as outlined in steps 2-5 above. The sensitivity matrix for the system was computed for the structure with no model uncertainty and simulated damage. Step 7 to Step 11 was then performed for each experiment. Equation (4) was then used to locate the damage. The results are shown in Figures 3 to Figure 8.

Discussion of Results

The effect of the magnitude of the model uncertainty on the accuracy of damage localization is shown in Figure 3 to Figure 5. In these figures the uncertainty was restricted to Location B but the magnitude of the uncertainty assumed values of 0.025, 0.1, and 0.5. From Figure 3, in which the behavior of damage at Location 1 (Z/L = 0.125) is presented, we observe that as the magnitude of the model uncertainty increases, the localization error increases and, as the magnitude of the simulated damage increases, the localization error decreases. From Figure 4, in which the behavior of damage at Location 2 (Z/L = 0.25) is presented, we observe that, as the magnitude of the model uncertainty increases, the localization error increases but the magnitude of the inflicted damage appears to have little influence on the localization error. From Figure 5, in which the behavior of damage at Location 3 (Z/L = 0.375) is presented, we again observe that as the magnitude of the model uncertainty increases, the localization error increases and as the size of the inflicted damage increases, the localization error decreases. From these observations, we conclude that, for this model the accuracy of the prediction of damage location decreases with the magnitude of the model uncertainty. Furthermore, the localization accuracy, per se, depends upon the specific location of the simulated damage relative to the location of the model uncertainty. Finally, the relationship between the localization accuracy and the magnitude of the inflicted damage depends upon the location of the damage.

The effect of the location of the model uncertainty on the accuracy of localization is shown in Figure 6 to Figure 8. In these figures, the magnitude of the model uncertainty was held constant ($\alpha_u = 0.10$) while the location of the model uncertainty was varied from Location A to Location B to Location C. From Figure 6, in which the results for Location 1 are presented, we observe that the localization error depends on the location of the model uncertainty and strongly depends on the magnitude of the inflicted damage. From Figure 7, in which the results for damage at Location 2 are presented, we again observe that the localization error depends on the location of the model uncertainty, however, the dependence of the localization error on the simulated damage magnitude is quite weak. Finally, in Figure 8, in which the information for damage at Location 3 is presented, we again observe that the localization error depends upon the location of the model uncertainty while the localization error appears to depend slightly upon the magnitude of the simulated

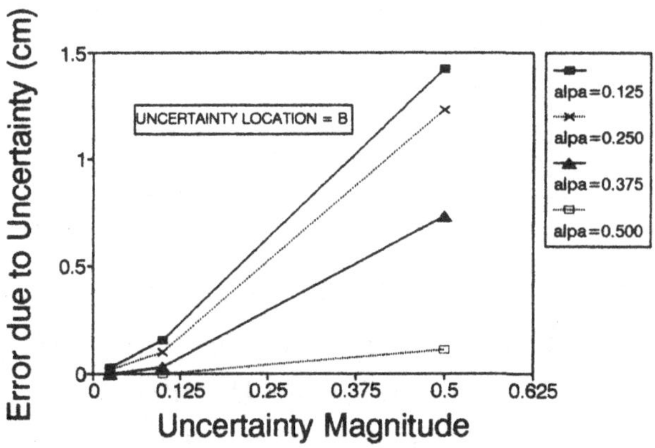

Fig. 3 Localization Accuracy Vs Uncertainty Magnitude (Loc. 1)

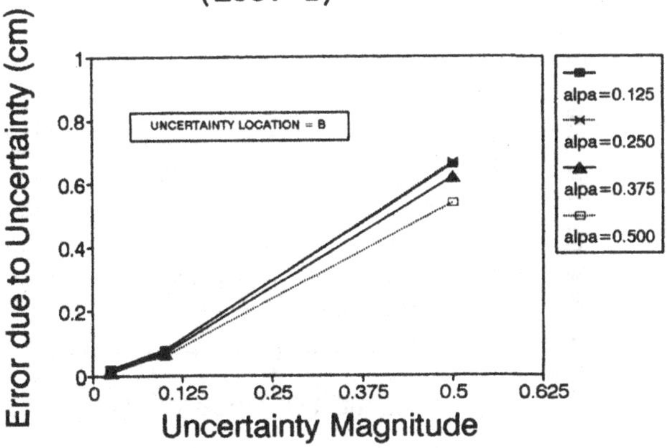

Fig. 4 Localization Accuracy Vs Uncertainty Magnitude (Loc. 2)

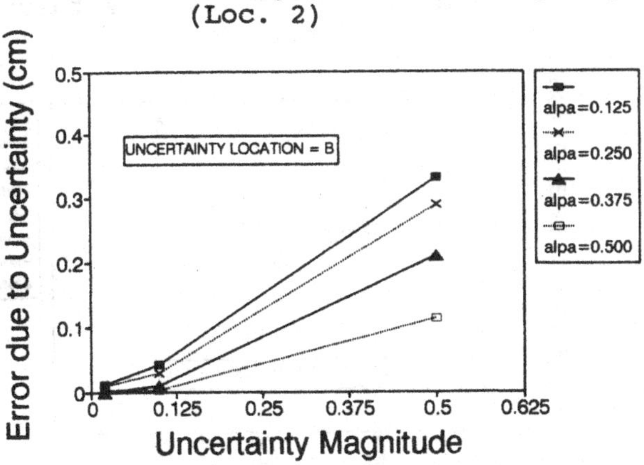

Fig. 5 Localization Accuracy Vs Uncertainty Magnitude (Loc. 3)

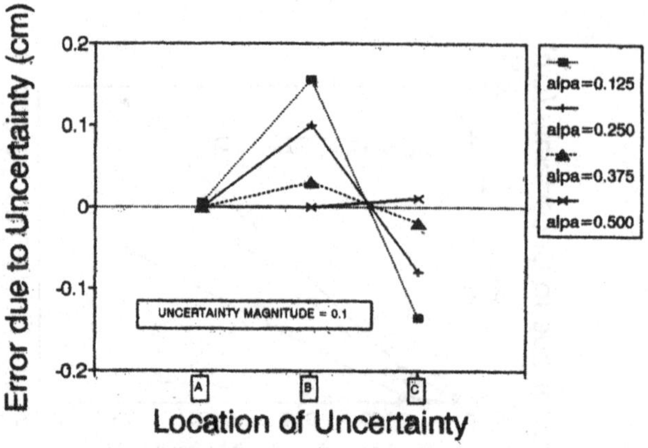

Fig. 6. Localization Accuracy Vs Uncertainty Location (Loc. 1)

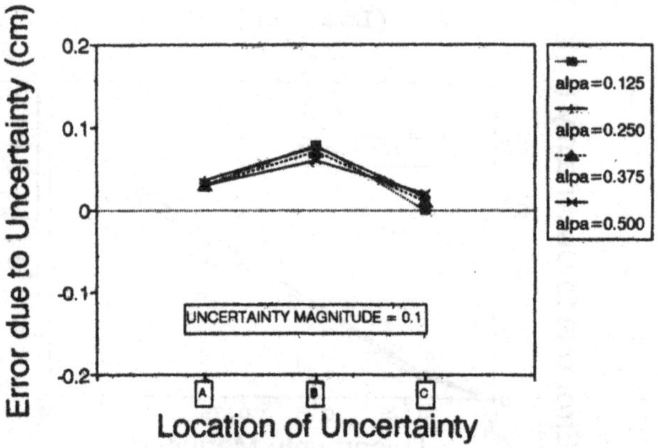

Fig. 7 Localization Accuracy Vs Uncertainty Location (Loc. 2)

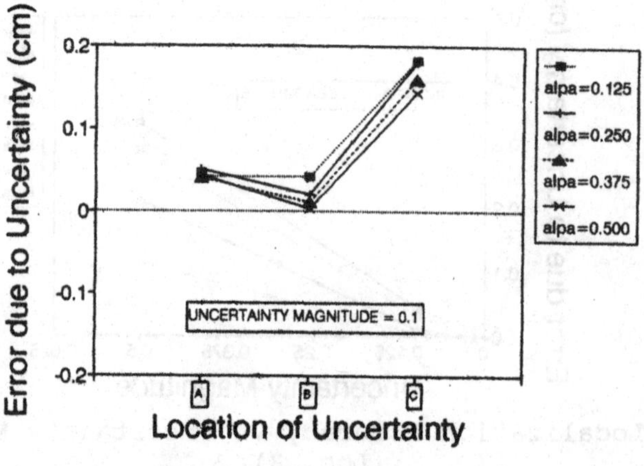

Fig. 8 Localization Accuracy Vs Uncertainty Location (Loc. 3)

damage. From these observations we conclude that the accuracy of damage localization depends upon the location of the model uncertainty. The localization accuracy also appears to be influenced by the relative magnitude of the damage relative to the magnitude of the model uncertainty. Finally, although these results indicate an influence of the location of model uncertainty on localization error, no clear relationship between accuracy and damage magnitude can be stated at this time.

CONCLUSIONS

On the basis of results from the numerical experiment, we arrived at the following conclusions regarding the damage behavior of the free-free beam: (1) The accuracy of damage localization decreases as the magnitude of model uncertainty increases; (2) For a constant magnitude of model uncertainty, the accuracy of damage localization depends upon the location of the damage; (3) The relationship between the accuracy of damage localization and the magnitude of the simulated damage depends upon the location of the damage; (4) The accuracy of damage localization depends upon the location of the model uncertainty; and (5) The accuracy of damage localization appears to be influenced by the magnitude of the inflicted damage relative to the magnitude of the model uncertainty.

REFERENCES

1. Biswas, M., Pandey, A.K. and Samman, M.M., Diagnostic Experimental Spectral/Modal Analysis of a Highway Bridge, The International Journal of Analytical and Experimental Modal Analysis, Vol. 5, No. 1, pp. 33-42, 1989.

2. Cawley, P., Adams, R.D. The Location of Defects in Structures from Measurements of Natural Frequencies, J. Strain Anal. Eng. Des., Vol 14, pp. 49-57, 1979.

3. Dimarogonas, A.D. and Chondros, T.G., Identification of Cracks in Welded Joints of Complex Structures, J. Sound and Vib., Vol. 69, No. 4, 1980.

4. Gudmundson, P. Eigenfrequency Changes of Structures Due to Cracks, Notches or Other Geometrical Changes. J. Mech. Phys. Solids, Vol. 30, pp. 339-353, 1982.

5. Ju, F.D., Wong, E.T. and Paez, T.L., Modal Method in Diagnosis of Fracture Damage in Simple

Structures, Productive Applications of Mechanical Vibrations, ASME Publications, 1982.

6. Nataraja, R. Structural Integrity Monitoring in Real Seas, Vol. 2, pp. 221-228, Proc. 15th Annu. Conf. on Offshore Tech., Houston, Offshore Technology Conference, 1983, Houston, Texas, 1983.

7. Montalvao E. Silva, J.M., and Araujo Gomes, A.J.M. Experimental Dynamic Analysis of Cracked Free-Free Beams, Experimental Mechanics, Vol. 20, pp. 20-25, 1990.

8. Stubbs, N. A General Theory of Non-Destructive Damage Detection in Structures in Structural Control (Ed. Leipholz H.H.E.), pp. 694-713, Proceedings of the Second International Symposium on Structural Control, Waterloo, Canada, 1985, Martinus Nijhoff, Dordrecht, Neteherlands, 1987.

9. Stubbs, N. and Osegueda, R., "Global Non-Destructive Damage Evaluation in Solids, The International Journal of Analytical and Experimental Modal Analysis, Vol. 5, No. 2, pp. 67-79, 1990.

10. Stubbs, N. and Osegueda, R., Global Damage Detection in Solids - Experimental Verification, The International Journal of Analytical and Experimental Modal Analysis, Vol. 5, No. 2, pp. 81-97, 1990.

11. Stubbs, N., Broome, T.H. and Osegueda, R., Nondestructive Construction Error Detection in Large Space Structures, AIAA Journal, Vol. 28, No. 111, pp. 146-152, 1990.

Evaluation of Maxium Softening as a Damage Indicator for Reinforced Concrete Under Seismic Excitation

S.R.K. Nielsen (*), A.S. Cakmak (**)

(*) Department of Building Technology and Structural Engineering, University of Aalborg, DK-9000 Aalborg, Denmark

(**) Department of Civil Engineering and Operations Research, Princeton University, Princeton, NJ 08544, U.S.A.

ABSTRACT

The properties of maximum softening of DiPasquale and Cakmak (1990) as a global damage indicator for reinforced concrete structures subjected to seismic loads have been evaluated. Especially the Markov property of the damage indicator is investigated, which is mandatory if statements on the post earthquake structural reliability is intended solely from knowledge of the latest recorded damage value. The transition probability density function (tpdf) of the damage indicator during sequential earthquakes has been determined by Monte-Carlo simulation for a planar 3 storey 2 bay reinforced concrete frame designed according to the Uniform Building Code (UBC) zone 4. The applied earthquake excitation model is the non-stationary single earthquake ARMA-model of Ellis and Cakmak (1987), which depends on the magnitude of the earthquake, the epicentral distance, the duration of strong shaking and a local soil quality parameter. The dynamic analysis of the structure is performed by the SARCF-II program of Chung et al. (1987). For each realization of the earthquake process a sample of the damage indicator during the earthquake is obtained by numerical integration of the structural equations of motion. From the observed data sample, the Markov assumption of the damage indicator sequence is justified for the mean value of the tpdf, whereas some deviation is observed for the variance and higher order statistical moments. In addition a regression analysis is performed for the increment of the damage indicator versus the sample peak acceleration and the sample peak frequency of the causing realization of the earthquake process, from which is concluded, that the maximum softening is completely uncorrelated to these quantities.

1. INTRODUCTION

Local damage manifested as stiffness and strength deterioration takes place in reinforced concrete structures during severe seismic excitation. A global damage indicator is a non-negative scalar functional of such local damage, which adequately characterizes the overall damage state and serviceability of the structure. Ideally a global damage indicator should have at least the following properties:

1. When recorded for any damaged structure, the damage indicator should attain the same value, irrespectively of the actual earthquake excitation leading to structural

failure.
2. The damage indicator should be a non-decreasing function of time.
3. The damage indicator should be observable by measurements.
4. Post earthquake reliability estimates for a partly damaged structure should be possible, solely from the latest recorded value of the damage indicator.

The second requirement implies, that a structure, which is hit by an earthquake, is always considered at least as damaged as before the seismic excitation. Further the requirement rules out the possibility of material reversibility in agreement with observations on reinforced concrete structures.

The fourth requirement implies that two initially identical structures, for which the same present value of the global damage indicator is recorded, should be considered to have the same structural reliability, when exposed to the same future earthquake process, independent of the preceeding different loading histories. Mathematically this means, that the values of the global damage indicator after each earthquake is assumed to form a Markov chain. If the local damage as measured by the stiffness and strength deterioration of all beams and columns are point-wise the same throughout the two structures, the probability distributions of the local damage will be identical after the next earthquake, irrespectively of the arbitrary and different preceeding loading histories. This establishes the Markov property for the finite or infinite dimensional vector process made up of all local damage. However, generally, the Markov property does not apply to the scalar global damage indicator processes. A sufficient condition for this is obtained, if the local damages have the same spatial distribution in all loading histories, so that the damage state can be described by a single scaling parameter. Then the local damage are identical in the two structures, if the global damage indicators are same, which establishes the Markov property for the global damage indicator. On the other hand any violations of the Markov property of the global damage indicator can be attributed to different spatial distributions of the local damage in different loading sequences.

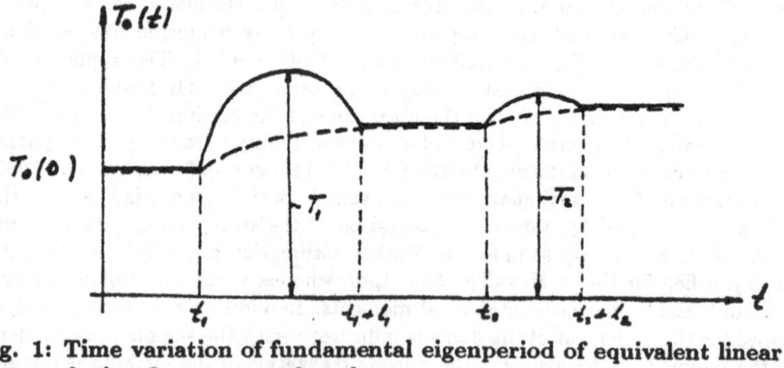

Fig. 1: Time variation of fundamental eigenperiod of equivalent linear system during first two earthquakes.

In this paper a study with reference to the above requirements will be performed for the maximum softening global damage indicator defined by DiPasquale and Cakmak (1990). For the i'th earthquake, maximum softening is defined as:

$$D_i = \max_{t\in[t_i,t_i+l_i]} \left(1 - \frac{T_0(0)}{T_0(t)}\right) \tag{1}$$

$T_0(0)$ is the linear fundamental eigenperiod of the initial undamaged structure, $T_0(t)$ is the eigenperiod of an equivalent linear system at time t, t_i is the time of initiation and l_i is the length of the i'th earthquake. Intuitively the maximum softening measures the combined stiffness degradation due to stiffness deterioration in the elastic range after previous plastic deformations, and the averaged effect of the loss of incremental stiffness at extreme excursions into the plastic range. The latter contribution is a measure of the maximum plastic strains observed during the earthquake, which is measured by the commonly used damage indicator for reinforced concrete structures called the ductility ratio. Fig. 1 shows the variation of $T_0(t)$ during two earthquakes initiated at times t_1 and t_2. The stippled curve indicates the variation of $T_0(t)$ due to the deterioration of the linear elastic stiffness. This curve is allways non-decreasing with time for reinforced concrete structures. The unbroken curve shows the increase of the fundamental eigenperiods due to loss of incremental stiffness during plastic deformations. The local maxima T_1 and T_2 of this latter curve determine the maximum softenings D_1 and D_2 during the first two earthquakes according to (1).

It was shown by DiPasquale and Cakmak (1989), that the maximum softening can be interpreted as an average of local stiffness degradations over the body of the structure with the square of the generalized strain tensor in the first eigenmode as weighting factor, in case the response of the structure is dominated by the first eigenmode. On the other hand the maximum softening concept can only be expected to work properly, when the structure behaves this way.

It was proven by Beck and Jennings (1980) that a linear structural system described in terms of the modal parameters can be identifiable. When the structural behavior is nonlinear, the system identifiable algorithm based on the linear models will yield equivalent linear estimates. Obviously (1) depends on the definition of the equivalent linear system. In the analysis of DiPasquale and Cakmak (1988) equivalent time-dependent single and two degree-of-freedom linear systems were determined, with parameters identified by a maximum likelihood procedure from corresponding segments of the ground acceleration and storey displacement signals. Rodriguez-Gomez (1990) determined $T_0(t)$ by a smoothing procedure over eigenperiods determined from the instantaneous incremental stiffness matrix of the system and obtained a high correlation to the method of DiPasquale and Cakmak. This suggests that the damage indicator is not very sensitive to the applied linearization procedure. In case of a stochastic analysis, $T_0(t)$ must be determined based on a statistical linearization of the equations of motion and the constitutive equations of the system.

Based on the analysis of empirical data DiPasquale and Cakmak (1987) demonstrated that the maximum softening does fulfill the first of the above requirements. Further the maximum softening can easily be measured by a frequency analysis of an accelerometer recording. The second requirement, however, need not to be fulfilled. If large plastic deformations are present in the first earthquake and only insignificant plastic deformations in the second earthquake, it may happen that $T_2 < T_1$, as shown in fig. 1, implying $D_2 < D_1$. In the following only significant earthquakes are considered, for which this event occurs with negligible probability. This assumption additionally rules out the possibility of elastic performance of the structure during any considered earthquake, resulting in an increasing sequence of global damage values.

The main advantage of the maximum softening is its robustness for structures, where it can be applied. Rodriguez-Gomez (1990) considered a class of plane framed concrete structures designed according to the UBC, which were exposed to the El Centro NS 1940 earthquake twice, with the time scale compressed with a factor of 2.5. The structures all survived the first eartquake, but some failed during the second earthquke. A comparison with other global damage indicators showed the appropriateness of the maximum softening in separating failed structures from those surviving the second earthquake.

The main purpose of the present paper is to investigate the Markov properties of the maximum softening as global damage indicator. Assuming the Markov assumption to be fulfilled, it is initially demonstrated how the probability of failure, the distribution of the first passage time to structural failure and other long term statistics can be evaluated, if only a stochastic model is available for arrival times of earthquakes of a certain magnitude, epicentral distance, duration of strong shaking. Next the Markov property of the maximum softening is investigated by Monte-Carlo simulations on a planar 3 storey 2 bay reinforced concrete frame designed according to UBC zone 4. Finally a regression analysis is performed for the increment of the damage indicator versus various characteristics of the input realization of the earthquake process such as the sample peak acceleration and the sample peak frequency.

The stochastic earthquake model used in the simulation studies is the variance and frequency stabilized single event earthquake ARMA (2,2)-model of Ellis and Cakmak (1987). Based on a regression analysis of 148 earthquake accelerograms recorded in California the coefficients of the ARMA-model as well as the parameters of the variance and frequency stabilizing functions were related to the earthquake magnitude, epicentral distance, duration of strong shaking and a local soil quality parameter. The model has been extended to deal with multiple event earthquakes, originating from several independent sources along the fault, as registrated in some earthquakes in Japan , Ellis and Cakmak (1988).

The dynamic analysis of the frame is performed by the SARCF-II program of Chung et al. (1988). Comparing the predictions of a number of available analysis programs with the results of shaking table experiments by Sozen and his associates (Healey and Sozen (1978), Cecen (1979)), Rodriguez-Gomez (1990) concluded, that the best agreement was obtained with this program.

2. PROBABILITY OF FAILURE BASED ON MAXIMUM SOFTENING

A certain structure s within a class of structures S, constructed at time t_0, and intended in use until the time T is considered. The structure is exposed to earthquakes for which a stochastic model need to be formulated. The model is most conveniently separated into a long term model, specifying the arrival times of significant earthquakes along with their magnitudes, epicentral distances, duration of strong shakings etc., and a short term stochastic model, specifying the actual acceleration signal at the ground surface at the site of the building, on condition that an earthquake with the indicated long term parameters occurs.

The arrival times $t_1 < t_2 < \cdots < t_{N(t)} \leq T$ of significant earthquakes is specified by a counting process $\{N(t), t \in (t_0, T]\}$. The counting process is assumed to be regular (the probability of one event in the interval Δt is then proportional to Δt, and the probability of more than one event in Δt is ignorable). No assumptions are made about the correlation of the arrival times. Hence $\{N(t), t \in (t_0, T]\}$ may be a Poisson process, a renewal process etc. Each earthquake is related with a mark $\mathbf{M}_i$, $i = 1, 2, \ldots$, which are all identically distributed as a stochastic vector $\mathbf{M}$, and which are all stochastically independent of the counting process $\{N(t), t \in (t_0, T]\}$. The components of $\mathbf{M}_i$ specify the intensity, duration of strong shaking, frequency contents, etc. of the i'th earthquake. In the following it is assumed, that $\mathbf{M} = [\mathbf{M}_0, \Gamma]$, $\mathbf{M}_0 = [M, R, \tau]$, where M is the magnitude of the earthquake, R is the epicentral distance, τ is the duration of strong shaking, and Γ is a local soil quality parameter. Γ specifies the local amplification of earthquake waves in soft sedimental sublayers. A low value of Γ specifies a relatively soft soil. All the subvectors $\mathbf{M}_{0,i}$ are assumed to be mutually independent and identically distributed as $\mathbf{M}_0$. The soil quality parameters Γ_i are equivalent stochastic variables, and are assumed to be stochastically independent of all parameter subvectors $\mathbf{M}_{0,i}$.The ground acceleration process of the i'th earthquake is denoted by $\{A_i(t), t \in [t_i, \infty)\}$. The following short term model will be applied for this quantity

$$A_i(t) = A_i(t, \{W_i(u), u \in [t_i, t]\}, \mathbf{M}_i) \tag{2}$$

$\{W_i(u), u \in (-\infty, \infty)\}$, $i = 1, 2, \ldots$ are assumed to be mutual independent unit intensity Gaussian white noise processes. It is assumed that $\{A_i(t), t \in [t_i, \infty)\}$ is obtained by a causal linear filtering of a restriction to the index set $[t_i, t]$ of these processes. The parameters of the corresponding linear functional are determined by the mark of the i'th earthquake. The development of the maximum softening is modelled by the damage indicator process $\{D(t), t \in [t_0, T]\}$, which is assumed to fulfill all the requirements specified in the introduction. Hence failure of the structure will take place, when the damage process for the first time exceeds an allowable level d_0. The failure probability in the interval $[t_0, T]$ of the structure s then becomes

$$P_f([t_0, T]; s) = P\big(\max_{u \in [t_0, T]} D(u) > d_0; s\big) = P\big(D(T) > d_0; s\big) = 1 - F_{D(T)}(d_0; s) \tag{3}$$

The first passage time to failure L has the probability distribution function

$$F_L(t; s) = P(L \le t; s) = P\big(D(t) > d_0; s\big) = 1 - F_{D(t)}(d_0; s) \tag{4}$$

$F_{D(t)}(\cdot; s)$ is the probability distribution of $D(t)$ for the structure s. The expected time to failure can then be calculated from

$$E[L; s] = \int_0^\infty \big(1 - F_L(t; s)\big)\, dt = \int_0^\infty F_{D(t)}(d_0; s)\, dt \tag{5}$$

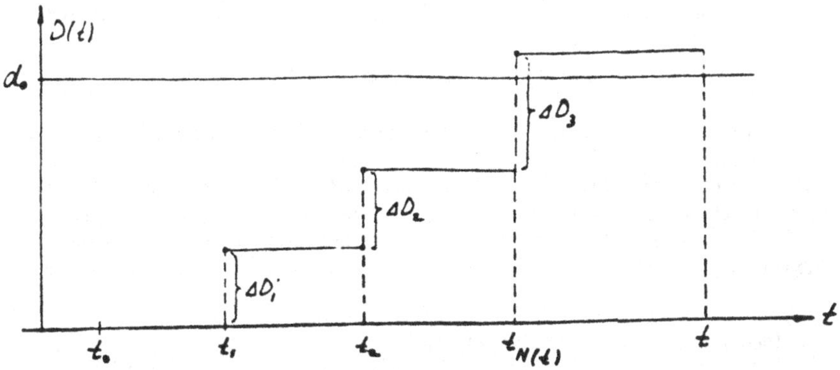

Fig. 2: Realization of global damage indicator process.

A realization of $\{D(t), t \in [t_0, T]\}$ is shown in fig. 2. Assuming the length of the earthquakes to be insignificant compared to the intervals between significant earthquakes, these realizations becomes stepfunctions. During the i'th earthquake a finite increment ΔD_i of damage takes place. $D(t)$ is the sum of all increments up to and including the time t, i.e.

$$D(t) = \sum_{i=1}^{N(t)} \Delta D_i \tag{6}$$

Clearly ΔD_i will be a functional of all previous acceleration processes up to and including the i'th , i.e. $\Delta D_i = \Delta D_i\left(\bigcup_{j=1}^{i}\{A_j(u), u \in [t_j, \infty)\}\right)$. This general formulation takes into consideration the times, at which the earthquakes are initiated, corresponding to an implicit dependency of ΔD_i on the counting process $\{N(t), t \in [t_0, T]\}$. This is important in cases of additional deterministic material deterioration due to chemical or mechanical exposures. Such aging processes are not considered in the present study, and the damage increment may consequently be considered independent of the initiation times of previous earthquakes, if only these are applied in the correct order. The fourth requirement in the introduction implies, that the information on the first preceeding $i-1$ earthquakes are inbedded in the accumulated damage $D(t_{i-1})$ caused by these. Consequently the following functional dependency is valid for the i'th damage increment of the structure s

$$\Delta D_i = \Delta D_i(\{A_i(u), u \in [0, \infty)\}, D(t_{i-1}); s) \tag{7}$$

(7) implies that two realizations of the long term earthquake process resulting in the same accumulated damage $D(t_{i-1})$ at time t_{i-1}, will result in the same distribution of ΔD_i, if the same acceleration stochastic process is next applied. Due to the mutual independence of the marks $\mathbf{M}_{0,i}$, the stochastic sequence $\{D_i,\ i = 1, 2, \ldots\}$, $D_i = D(t_i)$, on condition of the local soil quality parameter $\Gamma = \gamma$, then forms a Markov chain. Because the stochastic processes $\{A_i(u), u \in [0, \infty)\}$ on condition of $\Gamma = \gamma$ are identically distributed, the transition probability density function of the structure s on condition of $\Gamma = \gamma$ is stationary in the following sense

$$q_{D_i|D_{i-1}}(d_i \mid d_{i-1}, \gamma; s) = q_{D_2|D_1}(d_i \mid d_{i-1}, \gamma; s), \quad i = 2, 3, \ldots \tag{8}$$

Let $q_{D_2|D_1}(d_i \mid d_{i-1}, m, r, \tau, \gamma; s)$ be the transition probability density function on condition of $\mathbf{M}_0 = [m, r, \tau]$ and $\Gamma = \gamma$. $q_{D_2|D_1}(d_i \mid d_{i-1}, \gamma; s)$ is then obtained from the convolution integral

$$q_{D_2|D_1}(d_i \mid d_{i-1}, \gamma; s) = \int_0^\infty\int_0^\infty\int_0^\infty q_{D_2|D_1}(d_i \mid d_{i-1}, m, r, \tau, \gamma; s) f_{\mathbf{M}_0}(m, r, \tau)\, dm dr d\tau \tag{9}$$

$f_{\mathbf{M}_0}(m, r, \tau)$ is the joint probability density function of $\mathbf{M}_0 = [M, R, \tau]$.

If $q_{D_2|D_1}(d_i \mid d_{i-1}, \gamma; s)$ and the stochastic structure of the counting process $\{N(t), t \in (t_0, T]\}$ are known, long term statistics can easily be calculated. Let $R_1(d_0 \mid \gamma; s)$, $R_2(d_0 \mid \gamma; s), \ldots$ be the probability of surviving the first earthquake, the second earthquake etc. of the structure s on condition of $\Gamma = \gamma$, in case the failure boundary is d_0. Using the Markov property of the conditioned damage indicator, these are calculated in the following way

$$\left.\begin{aligned}
R_1(d_0 \mid \gamma; s) &= \int_0^{d_0} q_{D_2|D_1}(x_1 \mid 0, \gamma; s)\, dx_1 \\
R_2(d_0 \mid \gamma; s) &= \int_0^{d_0} q_{D_2|D_1}(x_1 \mid 0, \gamma; s)\, dx_1 \int_{x_1}^{d_0} q_{D_2|D_1}(x_2 \mid x_1, \gamma; s)\, dx_2 \\
&\vdots \\
R_i(d_0 \mid \gamma; s) &= \int_0^{d_0} q_{D_2|D_1}(x_1 \mid 0, \gamma; s)\, dx_1 \int_{x_1}^{d_0} q_{D_2|D_1}(x_2 \mid x_1, \gamma; s)\, dx_2 \cdots \\
&\int_{x_{i-1}}^{d_0} q_{D_2|D_1}(x_i \mid x_{i-1}, \gamma; s)\, dx_i \\
&\vdots
\end{aligned}\right\} \tag{10}$$

The unconditioned probability of surviving the first i earthquakes can next be calculated from

$$R_i(d_0;s)=\int_0^\infty R_i(d_0\mid\gamma;s)f_\Gamma(\gamma)\,d\gamma,\quad i=1,2,\cdots \tag{11}$$

$f_\Gamma(\cdot)$ is the probability density function of the local soil quality parameter. The probability of no failure in the interval $[t_0,T]$ of the structure s is given by $1-P_f([t_0,T];s)$. The failure probability can then be written

$$P_f([t_0,T];s)=1-\sum_{i=1}^{\infty}P(\{\text{no failure in interval}[t_0,T]\}\mid N(T)=i\ ;\ s)P(N(T)=i)$$

$$=1-\sum_{i=1}^{\infty}R_i(d_0;s)\ P(N(T)=i) \tag{12}$$

From (3) and (4) follows, that the accumulated damage $D(t)$ and the first passage time to failure has the distribution functions

$$F_{D(t)}(d;s)=\sum_{i=1}^{\infty}R_i(d;s)\ P(N(t)=i) \tag{13}$$

$$F_L(t;s)=1-\sum_{i=1}^{\infty}R_i(d_0;s)\ P(N(t)=i) \tag{14}$$

From (5) and (13) follows, that the expected time to failure can be written

$$E[L;s]=\sum_{i=1}^{\infty}R_i(d_0;s)\int_0^\infty P(N(t)=i)\ dt \tag{15}$$

3. SIMULATION RESULTS

The considered planar 3 storey 2 bay reinforced concrete frame has been designed according to the UBC specifications for earthquake zone 4. The geometrical and structural details of the frame is indicated on fig. 3. The linear fundamental eigenperiod of the undamaged structure is $T_0(t_0)$ =0.695 sec.

In the simulation study the following long term earthquake parameters are kept constant at the values r=2 km, τ=20 sec and γ=0.1. Notice, γ=0.1 indicates a relatively soft sedimental sublayer, and hence specifies a significant earthquake amplification. The condidered magnitudes of the earthquake is m=6.8 and m=8.8. For each combination of these parameters 160 independent single event horizontal earthquake realizations are generated by a variance and frequency stabilized ARMA (2,2)-model with parameters calibrated from earthquakes in Japan. Assuming the frame has accumulated the damage $D_{i-1}=d_{i-1}$ during the first $i-1$ earthquakes, these realizations are applied as the i'th earthquake. The maximum softening D_i during the i'th earthquake are next determined by a dynamic analysis of the frame with the SARCF-II program in combination with the method of moving averages of instantaneous eigenperiods by Rodriguez-Gomez (1990).The eigenvalue solver sometimes fails to converge when the structure temporarily is in the plastic range, and the incremental stiffness matrix hence becomes singular or ill-conditioned. Being a defect of the analysis program, rather than of the structure, it was decided to abandon the calculation of the maximum softening for these cases. Further, if the number of samples n obtained for the damage indicator in this way was less than 2/3 of the number of performed simulations equal to 160, the whole sample was abandoned on a suspicion of biased selections.

From the obtained data sample a histogram of the probability density function on condition of the damage sequense $D_1 = d_1, \ldots, D_{i-1} = d_{i-1}$, $q_{D_i|D_{i-1}\cdots D_1}(d_i \mid d_{i-1}, \ldots, d_1, m, r, \tau, \gamma; s)$, can be estimated, along with the conditional mean value and conditional variational coefficient, which will shortly be referred to as $E[D_i \mid d_{i-1}, \ldots, d_1]$ and $V[D_i \mid d_{i-1}, \ldots, d_1]$.

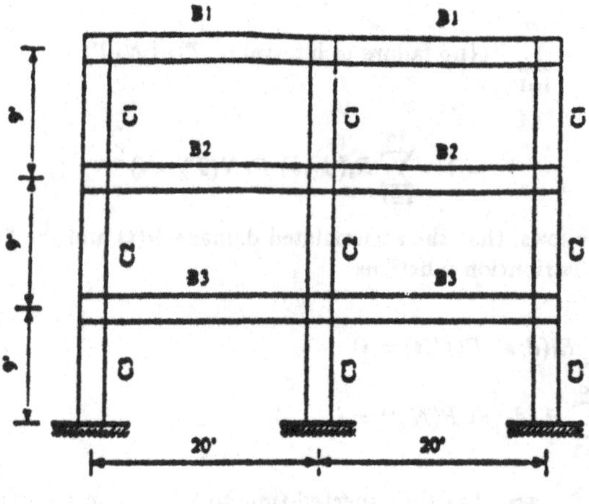

ELEVATION
Frame spacing 20'

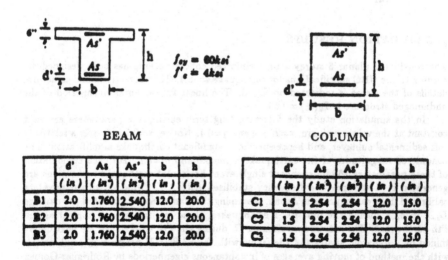

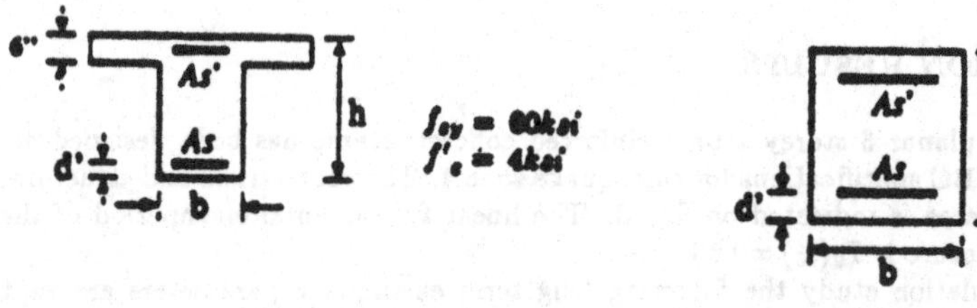

BEAM

COLUMN

	d' (in)	As (in²)	As' (in²)	b (in)	h (in)
B1	2.0	1.760	2.540	12.0	20.0
B2	2.0	1.760	2.540	12.0	20.0
B3	2.0	1.760	2.540	12.0	20.0

	d' (in)	As (in²)	As' (in²)	b (in)	h (in)
C1	1.5	2.54	2.54	12.0	15.0
C2	1.5	2.54	2.54	12.0	15.0
C3	1.5	2.54	2.54	12.0	15.0

Fig. 3: Details of 3 storey 2 bay reinforced concrete frame, UBC zone 4. (Rodriguez-Gomez (1990)).

Tables 1 and 2 show the mean value $E[D_2 \mid d_1]$ and the variational coefficient $V[D_2 \mid d_1]$ in the 2nd earthquake on condition of the damage d_1 in the 1st earthquake, when two earthquakes of respectively magnitudes $m_1 = m_2 = 8.8$ and $m_1 = m_2 = 6.8$ hit the structure. The values in row 1-3 of table 1 and row 1-4 of table 2 indicate the conditional statistical moments in the case d_1=0.0000, corresponding to the limit, where the structure remains in the elastic range during the 1st earthquake. These conditional expectations fulfill the relations

$$E[D_2 \mid 0] = E[D_1], \quad V[D_2 \mid 0] = V[D_1] \tag{16}$$

The right hand sides in (16) indicate the unconditional mean value and variational coefficient after the 1st earthquake of magnitudes $m_1 = 8.8$ and $m_1 = 6.8$ has hit the initial undamaged structure. The variability of the indicated estimates of $E[D_2 \mid 0]$ and $V[D_2 \mid 0]$ can be attributed entirely to statistical uncertainties due to the limited sample size. In the tables the remaining estimates of $E[D_2 \mid d_1]$ and $V[D_2 \mid d_1]$ are presented in groups of cases with almost equal values of d_1. These results seem to have a somewhat larger scatter than those of the case $d_1 = 0.0000$. If the Markov assumption of the global damage indicator was perfectly fulfilled, this should not be observed. This leads to the conclusion, that the Markov property most likely only will be approximately fulfilled. Physically this means that different spatial distributions of local damages after the 1st earthquake, even though they map into the same global damage indicator value, cause somewhat different probability distributions of the global damage indicator in the next earthquake.

Additionally, in rows 4,12,15,18 of table 1 and rows 5,8,11,14 of table 2 are shown the weighted averages of d_1, $E[D_2 \mid d_1]$ and $V[D_2 \mid d_1]$ with the number of samples n used as weight factor for each group of results with almost identical values of d_1.

Table 3 shows estimates of the mean value $E[D_3 \mid d_2, d_1]$ and the variational coefficient $V[D_3 \mid d_2, d_1]$ in the 3th earthquake on condition of the damage d_1 in the 1st earthquake and on condition of the damage d_2 in the 2nd earthquake, when three earthquakes of respectively magnitudes $m_1 = m_2 = 6.8$ and $m_3 = 8.8$ hit the structure. d_2 are fixed approximately at the value 0.31, whereas d_1 is varied in the interval $[0, d_2]$. These conditional statistical moments fulfill the relations

$$E[D_3 \mid d_2, 0] = E[D_2 \mid d_2], \quad V[D_3 \mid d_2, 0] = V[D_2 \mid d_2] \tag{17}$$

$$E[D_3 \mid d_2, d_2] = E[D_2 \mid d_2], \quad V[D_3 \mid d_2, d_2] = V[D_2 \mid d_2] \tag{18}$$

The left hand sides sides of (17) and (18) indicates the limit cases, where the structure remained elastic in respectively the 1st and the 2nd earthquake. The right hand sides of these equations are interpreted as the statistical moments of the damage in the 2nd earthquake with the magnitude $m_2 = 8.8$ on condition of an 1st earthquake of magnitude $m_1 = 6.8$, resulting in the damage $D_1 = d_2$. For these the averaged damage estimate in row 12 of table 1 has been applied in table 3. Because the 1st earthquake in table 1 is of magnitude $m_1 = 8.8$, this is actually an approximation, which relies on the Markov property of the global damage indicator. In table 3 weighted averages have also been calculated, in cases where several results are available with almost identical values of d_1 and d_2. If the Markov assumption of the global damage indicator is fulfilled, $E[D_3 \mid d_2, d_1]$ and $V[D_3 \mid d_2, d_1]$ are independent of d_1. Because of (17) and (18) it is necessary, that $E[D_3 \mid d_2, d_1]$ and $V[D_3 \mid d_2, d_1]$ attain either a maximum or a minimum as d_1 is varied through the interval $(0, d_2)$, unless the Markov assumption is valid. As seen from table 3, $E[D_3 \mid d_2, d_1]$ seems to fulfil the Markov property, whereas it is significant, that $V[D_3 \mid d_2, d_1]$ attains a maximum as a function of d_1 somewhere in the interval $(0, d_2)$. This is a further indication, that the Markov assumption for the maximum softening as a global damage indicator will only be approximately fulfilled. In fig. 4, 5, 6 estimated distributions of $q_{D_2|D_1}(d_2 \mid d_1, m, r, \tau, \gamma; s)$ for the data corresponding to the cases in row 2, 9, 16 of table 1 are shown assuming Beta distributed data. Then the transition probability distribution function is given by

$$q_{D_2|D_1}(d_2 \mid d_1, m, r, \tau, \gamma; s) = \frac{1}{1-d_1} \frac{\Gamma(\alpha+\beta)}{\Gamma(\alpha)\Gamma(\beta)} \left(\frac{d_2-d_1}{1-d_1}\right)^{\alpha-1} \left(\frac{1-d_2}{1-d_1}\right)^{\beta-1}, \quad d_2 \in [d_1, 1] \tag{19}$$

The parameters $\alpha = \alpha(d_1, m, r, \tau, \gamma; s)$ and $\beta = \beta(d_1, m, r, \tau, \gamma; s)$ are related to the conditional expectations by the following relations

$$E[D_2 \mid d_1, m, r, \tau, \gamma; s] = d_1 + (1-d_1)\,\frac{\alpha}{\alpha+\beta} \tag{20}$$

$$V[D_2 \mid d_1, m, r, \tau, \gamma; s] = (1-d_1)\,\frac{\sqrt{\alpha\beta}}{\sqrt{\alpha+\beta+1}\,(\alpha+\beta d_1)} \tag{21}$$

Fig 4, 5, 6 show the histograms of the data samples along with the fitted beta distribution. From these results it is concluded, that the tpdf can be modelled by a 2 parameter beta distribution. It should be emphasized, that the indicated examples are representative for all considered cases. In any case the tpdf turned out to be monomodal.

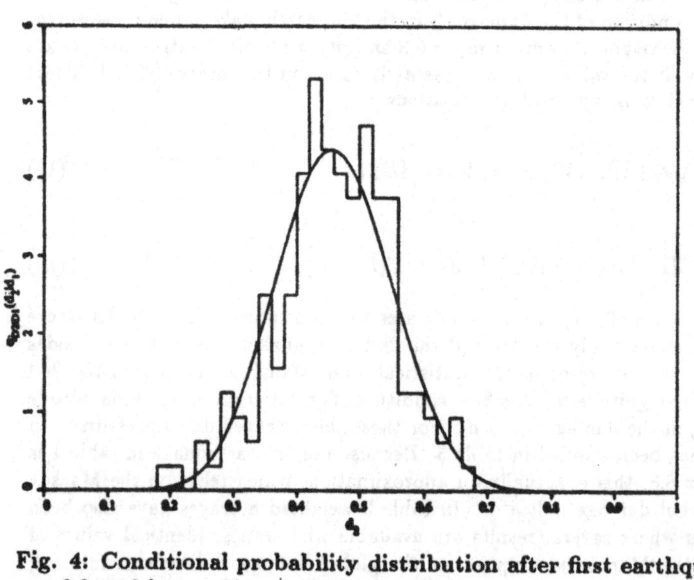

Fig. 4: Conditional probability distribution after first earthquake.
m_1=8.8, r=2 km, τ=20 s, γ=0.10.
$E[D_1 \mid 0]$=0.459, $V[D_1 \mid 0]$=0.194, n=160.
Histogram and fitted beta distribution, α=13.92, β=16.40.

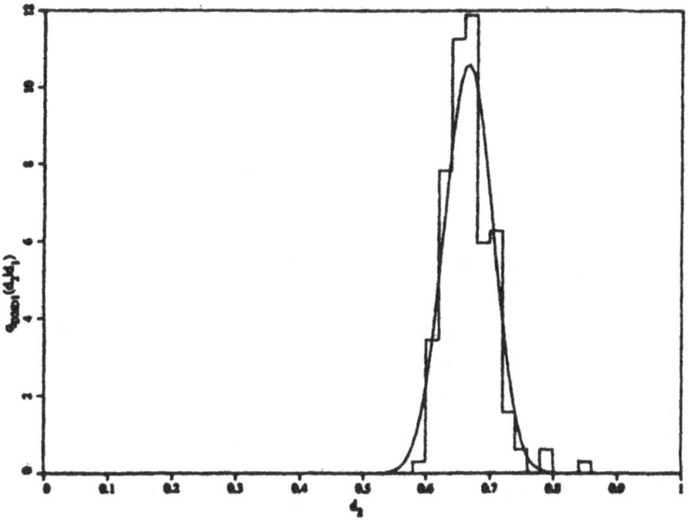

Fig. 5: Conditional probability distributions after second earthquake.
m_1=8.8, m_2=8.8, r=2 km, τ=20 s, γ =0.10.
d_1=0.3153, $E[D_2 \mid d_1]$ =0.666, $V[D_2 \mid d_1]$ =0.056, n=160.
Histogram and fitted beta distribution, α =42.62, β =40.59.

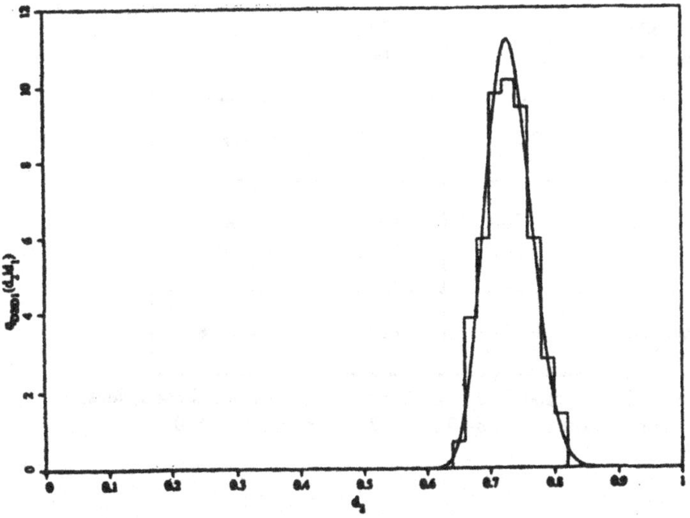

Fig. 6: Conditional probability distributions after second earthquake.
m_1=8.8, m_2=8.8, r=2 km, τ=20 s, γ =0.10.
d_1=0.5577, $E[D_2 \mid d_1]$ =0.730, $V[D_2 \mid d_1]$ =0.048, n=143.
Histogram and fitted beta distribution, α =14.37, β =22.52.

d_1	$E[D_2 \mid d_1]$	$V[D_2 \mid d_1]$	n
0.0000	0.453	0.192	147
0.0000	0.459	0.194	160
0.0000	0.466	0.184	160
0.0000	0.460	0.190	average
0.3121	0.683	0.059	160
0.3139	0.681	0.070	158
0.3142	0.645	0.058	157
0.3152	0.624	0.080	118
0.3153	0.666	0.056	160
0.3163	0.656	0.075	160
0.3174	0.678	0.076	160
0.3149	0.663	0.067	average
0.4332	0.727	0.071	160
0.4350	0.691	0.064	159
0.4341	0.709	0.068	average
0.5577	0.730	0.048	143
0.5583	0.795	0.036	160
0.5580	0.764	0.042	average

Table 1: Conditional mean and conditional variational coefficient. Earthquake sequence: 8.8+8.8. r =2 km, τ =20 s, γ = 0.1.

d_1	$E[D_2 \mid d_1]$	$V[D_2 \mid d_1]$	n
0.0000	0.237	0.393	160
0.0000	0.244	0.381	160
0.0000	0.247	0.375	160
0.0000	0.256	0.367	160
0.0000	0.246	0.379	average
0.0753	0.343	0.203	160
0.0756	0.348	0.197	160
0.0755	0.346	0.200	average
0.1195	0.391	0.136	159
0.1200	0.406	0.136	160
0.1198	0.399	0.136	average
0.1804	0.433	0.106	157
0.1808	0.509	0.079	159
0.1806	0.471	0.092	average

Table 2: Conditional mean and conditional variational coefficient. Earthquake sequence: 6.8+6.8. r =2 km, τ =20 s, $\gamma = 0.1$.

d_1	d_2	$E[D_3 \mid d_2, d_1]$	$V[D_3 \mid d_2, d_1]$	n
0.0000	0.3149	0.663	0.067	average
0.0753	0.3168	0.665	0.080	122
0.0756 0.0756 0.0756	0.3123 0.3148 0.3171	0.661 0.651 0.625	0.088 0.072 0.084	141 150 155
0.0756	0.3148	0.645	0.081	average
0.1195	0.3179	0.671	0.097	110
0.1200	0.3143	0.650	0.118	108
0.3149	0.3149	0.663	0.067	average

Table 3: Conditional mean and conditional variational coefficient. Earthquake sequence: 6.8+6.8+8.8. r=2 km, τ =20 s, $\gamma = 0.1$.

4. CORRELATION ANALYSIS

	$\hat{D}_1$	$\hat{f}_p$	$\hat{a}_{max}$	$\hat{\sigma}_a$	$\hat{S}_a(f_0)$	$\hat{S}_a(\hat{f}_p)$
$\hat{D}_1$	1.000	-0.142	0.060	0.391	-0.105	0.193
$\hat{f}_p$	-0.142	1.000	-0.152	-0.227	-0.180	-0.107
$\hat{a}_{max}$	0.060	-0.152	1.000	0.407	0.187	0.400
$\hat{\sigma}_a$	0.391	-0.227	0.407	1.000	0.189	0.830
$\hat{S}_a(f_0)$	-0.105	-0.180	0.187	0.189	1.000	0.202
$\hat{S}_a(\hat{f}_p)$	0.193	-0.107	0.400	0.830	0.202	1.000

Table 4: Correlation coefficient matrix.
$m = 8.8$, $r = 2$ km, $\tau = 20$ s, $\gamma = 0.1$, UBC zone 4, $n = 160$.

Table 4 shows an estimate based on 160 realizations from the same earthquake process of the correlation coeffient matrix of the vector made up of the global damage indicator in the first earthquakex $\hat{D}_1$, the sample peak frequency $\hat{f}_p$, the sample maximum acceleration $\hat{a}_{max}$, the sample standard deviation $\hat{\sigma}_a$, the sample value of the autospectral density of the acceleration process $\hat{S}_a(f_0)$ and $\hat{S}_s(\hat{f}_p)$, taken respectively at the fundamental eigenfrequency $f_0(t_0) = 1.439$ Hz and at the sample peak frequency. The average sample peak frequency of the process was estimated to $E[\hat{f}_p] = 0.754$ Hz. The peakfactor of the acceleration process was estimated to $E[\hat{a}_{max}]/E[\hat{\sigma}_a] = 4.15$, which is relatively high compared to an average number of zero up-crossings of $\approx 0.754 \cdot 20 = 15.8$. According to stochastic vibration theory any response quantity of a linear structure will be highly correlated to $\hat{S}_a(f_0)$. As seen from table 4 this is not the case for the damage increment $\hat{D}_1$. Further $\hat{D}_1$ seems to be completely uncorrelated to the sample peak frequency $\hat{f}_p$ and the sample maximum acceleration $\hat{a}_{max}$. However, there is a significant, but still not very large correlation, between $\hat{D}_1$ and the sample standard deviation $\hat{\sigma}_a$. This study shows that the earthquake generation model is a good one, in that no significant differences can be found in each realization, that effect the value of the damage indicator.

5. CONCLUSIONS

A reliability theory for reinforced concrete structures has been presented, based on damage accumulation with the maximum softening used as global damage indicator. The basic assumption of the approach is that the maximum softening during each significant earthquake forms a Markov chain with a stationary tpdf. Based on Monte-Carlo simulations this hypotesis has been justified for the mean value of the tpdf, whereas some violation is observed for the variance and also, possibly also, for higher order statistical moments.

Further the simulations have indicated that the tpdf in case of any significant major earthquake may be represented by a 2 parameter beta distribution. Because the lower order statistical moments of the tpdf in principle can be determined analytically based on a statistical linearization approach, this finding provides an analytical determination of the failure probability, if only the long term statistical properties of the earthquake excitation process is known.

Finally a regression analysis has shown that the maximum softening during a certain earthquake is completely uncorrelated to such measures of the corresponding earthquake realization as the peak acceleration and the peak frequency of the causing realization of the earthquake process.

6. ACKNOWLEDMENTS

The authors want to thank Prof. Erhan Cinlar of the Princeton University for suggesting the beta distribution as a possible tpdf in damage accumulation theories.

7. REFERENCES

(1) Beck, J.L., Jennings, P.C. Structural Identification Using Linear Models and Earthquake Records, Eng. Struc. Dyn., Vol.8, pp. 145-160, 1980.

(2) Cecen, H. Response of Ten Storey, Reinforced Concrete Models Frames to Simulated Earthquakes, Ph.D. Dissertation, University of Illinois at Urbana-Champaign,1979.

(3) Chung, Y.S.,Meyer, C. and Shinozuka, M. Seismic Damage Assessment of Reinforced Concrete Members, Technical Report NCEER-87-0022, National Center for Earthquake Engineering Research, State University of New York at Buffalo, 1987.

(4) Chung, Y.S., Shinozuka, M. and Meyer, C. SARCF User's Guide. Seismic Analysis of Reinforced Concrete Frames, Technical Report NCEER-88-0044, National Center for Earthquake Engineering Research, State University of New York at Buffalo,1988.

(5) DiPasquale, E. and Cakmak, A.S. Detection and Assessment of Structural Damage, Technical Report NCEER-87-0015, National Center for Earthquake Engineering Research, State University of New York at Buffalo,1987.

(6) DiPasquale, E. and Cakmak, A.S. Identification of the Serviceability Limit State and Detection of Seismic Damage, Technical Report NCEER-88-0022, National Center for Earthquake Engineering Research, State University of New York at Buffalo, 1988.

(7) DiPasquale, E. and Cakmak, A.S. On the Relation Between Local and Global Damage Indices, Technical Report NCEER-89-0034, National Center for Earthquake Engineering Research, State University of New York at Buffalo,1989.

(8) DiPasquale, E. and Cakmak, A.S. Detection of Seismic Structural Damage Using Parameter-Based Global Damage Indices, Probabilistic Engineering Mechanics, Vol.5, No.2, pp. 60-65, 1990.

(9) Ellis, G.W. and Cakmak, A.S. Modelling Earthquakes in Seismically Active Regions Using Parametric Time Series Methods, Technical Report NCEER-87-0014, National Center for Earthquake Engineering Research, State University of New York at Buffalo, 1987.

(10) Ellis, G.W. and Cakmak, A.S. Modelling Strong Ground Motion from Multiple Event Earthquakes, Technical Report NCEER-88-0042, National Center for Earthquake Engineering Research, State University of New York at Buffalo, 1988.

(11) Healey, T.J. and Sozen, M.A. Experimental Study of the Dynamic Response of a Ten Story Reinforced Concrete Frame with a Tall First Story, Report No. UILU-ENG-78-2012, SRS 450, University of Illinois at Urbana-Champaign, 1978.

(12) Rodriguez-Gomez, S. Evaluation of Seismic Damage Indices for Reinforced Concrete Structures, M.Sc.Eng. Dissertation. Department of Civil Engineering and Operations Research, Princeton University, Princeton, N.J., 1990.

(13) Uniform Building Code, Earthquake Regulations,1988.

SECTION 3: APPLIED RELIABILITY ANALYSIS

Empirical Formulae to Presume the Resistance Performance of High Speed Craft Models

T. Nagai (*), Y. Yoshida (**)
(*) Department of Mechanical Engineering, Kanagawa Institute of Technology, 1030, Shimoogino Atsugi City, Kanagawa Prefecture, 243-02, Japan
(**) 4th Division of First Research Center, Technical Research & Development Institute, Japan Defense Agency, 2-2-1, Nakameguro Meguro-ku, Tokyo 153, Japan

ABSTRACT

This paper follows the previous one published at PACON 90. Analyzing 78 towing test data in still water of high speed craft 2.5 m length models by so-called principal component analyses, a presumption of resistance performance including total resistance coefficient, change of trim and of draft is proposed by using eleven non-dimensional hull form parameters including length, width, height of chine at midship, initial trim and so on. Results by two kinds of empirical formulae are favorably compared with the test data. Five more new models were tested to reconfirm those formulae, and we obtained reasonable correspondencies. Both formulae, therefore, are adequately clarified as applicable, defining two kinds of applicable ranges inside such as either the tolerance ellipse containing 95% data or the Cramér's concentrated ellipse containing 75 % data.

INTRODUCTION

This paper describes the empirical formulae to presume the resistance performance of high speed crafts in still water at the initial design stage, following the previous paper. [1] The principal component analyses are applied to 78 towing test data of about 2.5 m length models in order to classify models, and obtain empirical formulae, which enable designers to presume the resistance performance approximately by using eleven nondimensional hull form parameters, such as total length, half width and height of chine at midship, initial trim, and so on.

From the correlation coefficients' matrix composed corresponding to each parameter, the principal component having the main characteristics is extracted as the form of eigenvectors, and named the first component. The second one follows next, and so on. By using the orthogonal relation between these principal components the orthogonal co-ordinates system is formed, on which all test models are plotted with proper classification. This means that each plotted model corresponds with a point defined by the set of the scores calculated by the inner product of eigenvector and deviation vector.

We get, therefore, (1) reasonable empirical formula utilizing the orthogonal which inheres in principal components and (2) simple formula including only few nondimensional hull form parameters effective to presume the resistance performance. Comparing both results of the formulae with the towing test data, good correspondencies are obtained from low to high speed range. Another 5 models are used for towing tests to reconfirm those formulae. Their results coincide with the test data reasonably. Considering that the formulae should be used within the ellipse including many experimental data, the tolerance ellipse including 95 % data and the Cramér's concentrated ellipse including 75 % are recommended as the usable range.

We propose a new technique by applying the principal component analyses to a series of towing test data through the classification and arrangement process, under the computer aided mechanics.

NOMENCLATURE

V : model speed.

∇ : initial displacement volume.

g : gravity accerelation.

ρ : fluid dencity.

F_∇ : Froude number, $F_\nabla = V / \sqrt{\nabla^{1/3} \cdot g}$.

R : total resistance.

R' : total resistance coefficient, $R' = R / \{(1/2)\rho V^2 \nabla^{2/3}\}$.

$\Delta\theta$: trim. Change in angle from the initial trim θ. The unit is measured by minutes, and the positive sign should be taken for rising bow against the water line. (see Fig. 1)

ΔD : nondimensional change in draft divided by $\nabla^{1/3}$ and measured at the cross point of keel line and Ord.10 line at the stern. The positive sign should be taken for rising stern. (see Fig. 1)

X_i (i = 2,3, …… ,12) except X_{11} : nondimensional hull form parameters divided by $\nabla^{1/3}$. (see Fig. 1)

X_2 : height of chine at Ord.10.

X_3 : height of chine at Ord.5.

X_4 : height of chine at Ord.2.

X_5 : height of keel at Ord.2.

X_6 : height of keel at Ord.0.

X_7 : distance between Ord.0 and Ord.10 (total length).

X_8 : half distance between chines at Ord.5 (half width).

X_9 : half distance between chines at Ord.2 (half width).

X_{10} : curvature of transom.

X_{11} : initial trim. Angle between water line and keel line from Ord.5 to 10 at rest. $X_{11} = \theta$.

X_{12} : half distance between chines at Ord.10 (half width).

N : total number of models, N (= 78).

$\overline{X}_i$: mean value of X_i divided by N.

σ_i : standard deviation of X_i.

A : correlation coefficients' matrix of the hull form parameters by N sets of X_i (i = 2,3, …… ,12).

$\beta_{i\nu}$ (ν = 2,3, …… ,12) : ν's element of β_i.

Z_i (i = 1,2,3, …… ,11) : principal component score.

$\bar{Z}_i$: mean value of Z_i divided by N.

b_i (i = 1,2,…,5) : set of coefficients in the principal component empirical formula Equation (2). b_1 is the constant.

d_i (i = 1,2,…,5) : set of coefficients in the simple empirical formula Equation (3). d_1 is the constant.

f_i (i = 1,2,…,7) : set of coefficients in the empirical formula with four hull form parameters and their quadratic terms Equation (4). f_1 is the constant.

{ΔZ} : deviation vector, which means { $Z_1 - \bar{Z}_1$, $Z_2 - \bar{Z}_2$, $Z_3 - \bar{Z}_3$, $Z_4 - \bar{Z}_4$ }.

{ΔX} : deviation vector of hull form parameters. The elements are such as $X_2 - \bar{X}_2$, $X_3 - \bar{X}_3$, etc.. $X_7^2 - \overline{X_7^2}$, etc. are added in case Equation (4).

$\{\Delta X\}^T$: transposed vector of {ΔX}.

B : variance and covariance matrix made from N sets of 78 X_i (i = 2,3, …… ,12)

B_Z : variance and covariance matrix made from N sets of 78 (Z_1, Z_2, Z_3, Z_4).

B_4 : variance and covariance matrix made from N sets of 78 (X_3, X_7, X_8, X_{11}).

B_6 : variance and covariance matrix made from N sets of 78 (X_3, X_7, X_8, X_{11}, X_7^2, X_{11}^2).

1. Classification of Test Models

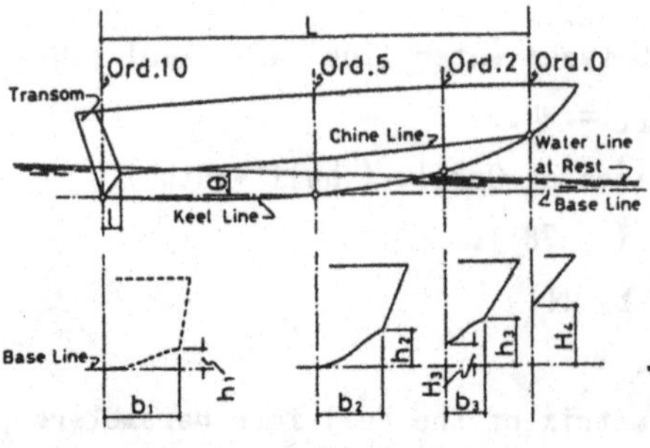

Fig.1 Principal Dimension

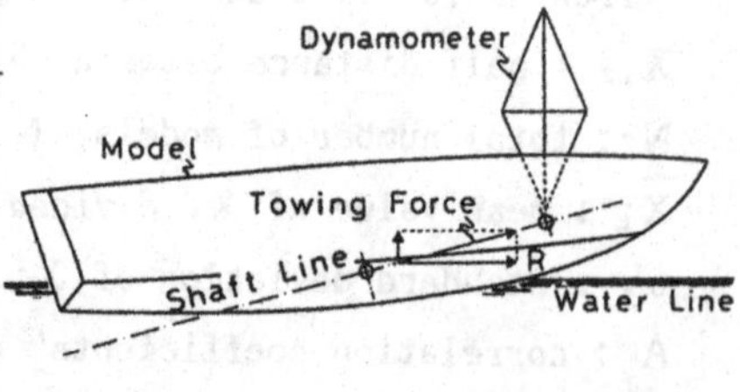

Fig.2 Towing Test

Thirteen models of about 2.5 m length with wave-shaped bottom were used for towing test changing 6 kinds of displacements and center of gravities for each model, which mean total 78 models. (see Fig. 2) We apply the principal component analyses to classify 78 models. Expressing the complex correlation existing among eleven hull form parameters by correlation coefficient matrix A, the principal component scores Z_i are obtained by using the eigenvectors β_1, β_2, β_3, and β_4 corresponding to A's eigenvalues $\lambda_1 > \lambda_2 > \lambda_3 > \lambda_4$.

$$Z_i = \sum_{\nu=2}^{12} \beta_{i\nu} \frac{X_\nu - \overline{X}_\nu}{\sigma_\nu} \qquad i = 1,2,3,4 \qquad (1)$$

Hence, scores Z_i can be calculated as the inner product of the eigenvector β and deviation vector of hull form parameters normalized by each standard deviation $(X - \overline{X})/\sigma$.

A principal component which reveals the main part of A's characteristics is extracted as the first principal component Z_1, and followed by the second one Z_2 and so on. Each principal component including the corresponding eigenvector is orthogonal. Taking the origin at the mean point of Z_i ($i = 1,2,3,\cdots$), the rectangular co-ordinates system is composed. Therefore, all test models are plotted like ellipse as shown in Fig. 3, in which 13 kinds of clusters were obtained. [2] One cluster is composed of six kinds of models, where displacement and center of gravity are changed for the same kind of model. Similar hull forms are expressed near cluster, but non-similar ones far clus-

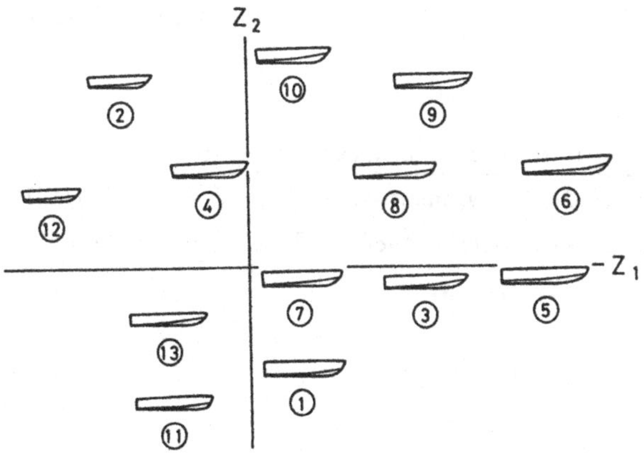

Fig. 3 Classification of Models

ter.

We first try to get rigorous empirical formula using the orthogonal which inheres among principal components.

2. Principal Component Analyses and Empirical Formulae

Four major principal components Z_i are successively selected. Their corresponding eigenvalues become greater than 1: $\lambda_1 > \lambda_2 > \lambda_3 > \lambda_4 > 1$. As the first principal component Z_1 has the longest part of data distribution on its axis, this component reveals most predominantly A's characteristics and eigenvalue λ_1 becomes the largest. For Z_1 main hull form parameters are height of chine at Ord.5 X_3 of 0.87 in correlation coefficient, height of chine at Ord. 2 X_4 of 0.93, height of keel at Ord.0 X_6 of 0.96, and half width at Ord.5 X_8 of 0.63. The second principal component axis Z_2 is formed perpendicularly to Z_1 axis, and the data distribution length follows next to Z_1's. The eigenvalue λ_2 is less than λ_1: $\lambda_2 < \lambda_1$. Next to Z_1, Z_2 reveals A's major characteristics. For Z_2 main hull form parameters are distance between Ord.0 and Ord.10 (total length) X_7 of -0.67 in correlation coefficient, curvature of transom X_{10} of 0.89, and half width at Ord.10 X_{12} of 0.80. Keeping perpendicular to Z_2, we have for Z_3 half width at Ord.5 X_8 of -0.65 in correlation coefficient, and half width at Ord.2 X_9 of -0.78. Similarly Z_4's main hull form parameter is initial trim X_{11} of -0.65.

2.1 Formula composed of principal component scores

For total N sets of principal component scores Z_i (i = 1,2,3,4) and R', applying the least square method [3] we have following principal component empirical formula.

$$R' = b_1 + b_2 Z_1 + b_3 Z_2 + b_4 Z_3 + b_5 Z_4 \qquad (2)$$

Substituting $\Delta\theta$ and ΔD for R', we have the similar formula as Equation (2). Then, we will discuss simple empirical formula including only 4 nondimensional hull form parameters, which are extracted from Z_i's 11 terms.

2. 2 Simple empirical formula

Eleven nondimensional hull form parameters were used to obtain Equation (2). Now, neglecting orthogonal requirement, we try to get a more simple one composed of only four hull form parameters, which have much greater connection with Z_i ($i = 1,2,3,4$), such as height of chine at Ord.5 X_3, total length X_7, half distance between chines at Ord.5 X_8, and initial trim X_{11}. The deadrise angle at midship, which is one of the most interesting terms for designers, is taken into consideration by X_3 together with X_8. Using these terms, R' will be expressed as follows.

$$R' = d_1 + d_2X_3 + d_3X_7 + d_4X_8 + d_5X_{11} \tag{3}$$

$\Delta\theta$ and ΔD are also expressed by the similar formula as Equation (3). In Equation (3), if R' decreases as X_7 increases, the sign of X_7 should be minus. After checking these signs based on correlation coefficients of hull form parameters X_i against R', they were confirmed fine.

Adding the quadratic terms of the total length X_7 and of the initial trim X_{11}, we have from Equation (3) for R'

$$R' = f_1 + f_2X_3 + f_3X_7 + f_4X_8 + f_5X_{11} + f_6X_7^2 + f_7X_{11}^2 \tag{4}$$

$\Delta\theta$ and ΔD will also be expressed by the similar formula as Equation (4).

3. Comparison of Formulae (2), (3) and (4) with Test Data

Test results obtained for following hull form parameters:

X_2 = 0.1738434, X_3 = 0.2764890, X_4 = 0.4040412,
X_5 = 0.0000000, X_6 = 0.5829162, X_7 = 6.6822780,

$X_8 = 0.7972644$, $X_9 = 0.5814067$, $X_{10} = -0.0241519$,
$X_{11} = 68.75$, $X_{12} = 0.6090808$

were compared with the results by Equation (2), (3) and (4) for

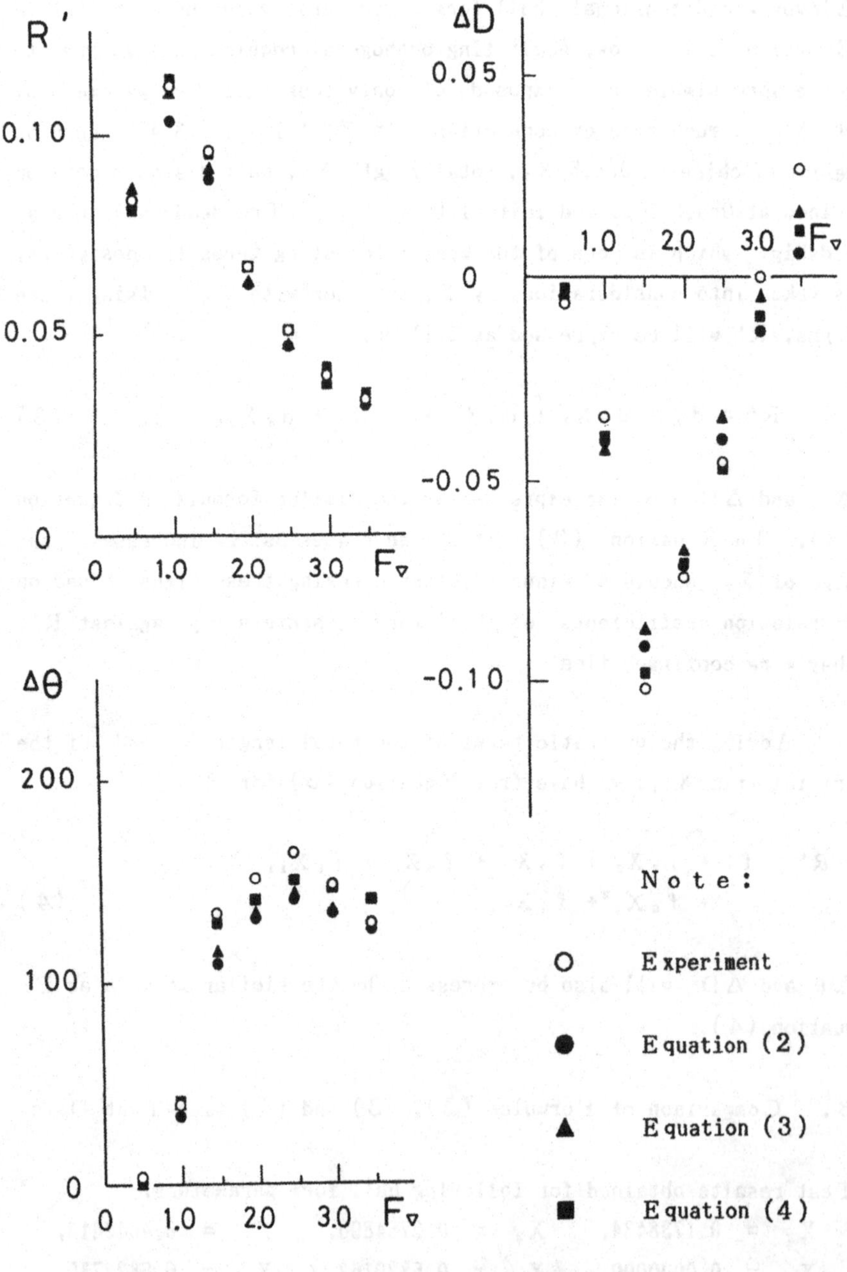

Fig.4 Resistance Performance of Model

the R', $\Delta\theta$ and ΔD, showing good coincidencies from low to high range of F_∇ as shown in Fig.4.

The usable ranges of these formulae Equation (2), (3) and (4) follow next.

4. Usable Ranges of Empirical Formulae

Empirical formulae are recommended to use within the range, where as many models as possible exist. We propose here tolerance ellipse and Cramér's concentrated ellipse.

4.1 Tolerance ellipse

As the origin of the principal component co-ordinates system is determind by keeping $X_i = \overline{X}_i$ (i = 2,3, ……, 12), test models are distributed as shown in Fig.3. The distance between the point a model exist and the origin is called Maharanobis's distance M.

$$M = \{\Delta X\}^T B^{-1} \{\Delta X\} \tag{5}$$

For normal distribution of models' points the usable range is limited within 95 %, which is called the tolerance ellipse.

4.2 Cramér's concentrated ellipse

For non-normal distribution of models' points the usable range is limited within 75 %, which is inside than the case of normal distribution and called " Cramér's concentrated ellipse ". [4] Hence, the range of Cramér's concentrated ellipse is smaller than that of the tolerance ellipse, but the Cramér's enables us to determine the limit of range even for any form of arbitrary data's distribution. The limit of range M equals to the numbers of principal components, and of hull form parameters plus 2. This means for Cramér's concentrated ellipse that the limit M of Equation (5) equals to 13, because of 11 X_i's (i = 2,3, …… 12) + 2.

$$\{\Delta X\}^T B^{-1} \{\Delta X\} \leq 13 \tag{6}$$

Satisfying Equation (6), the following requirements are added for Equation (2), (3) and (4).

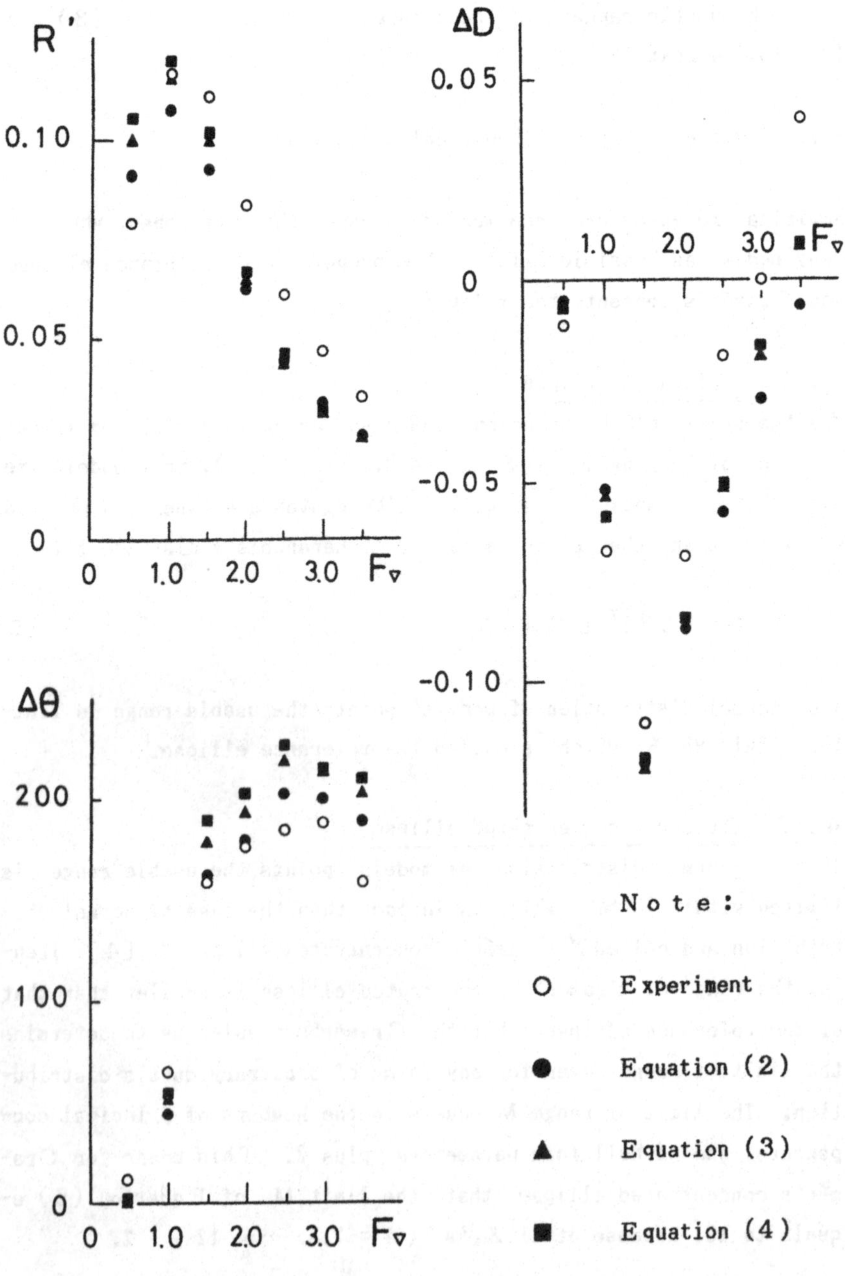

Fig.5 Resistance Performance of a New Model

For the principal component empirical formula Equation (2), the limit M equals to 6, because of 4 Z_i's (i = 1,2,3,4) + 2, $\{\Delta Z\}^T B_Z^{-1} \{\Delta Z\} \leqq 6$. For the simple empirical formula Equation (3), the limit M equals to 6, because of 4 X_i's (i = 3,7,8,11)+ 2, $\{\Delta X\}^T B_4^{-1} \{\Delta X\} \leqq 6$. And for the quadratic empirical formula Equation (4), the limit M equals to 8, as the same way, $\{\Delta X\}^T B_6^{-1} \{\Delta X\} \leqq 8$.

Five more new models were used to confirm proposed formulae.

5. Confirmation Test

For hull form parameters of a new model:

X_2 = 0.0340648, X_3 = 0.2094872, X_4 = 0.3550159,
X_5 = 0.0231734, X_6 = 0.5712233, X_7 = 5.7933400,
X_8 = 0.6639168, X_9 = 0.5081918, X_{10} = 0.0000000,
X_{11} = 0.0000000, X_{12} = 0.6574282,

comparison of Equation (2),(3) and (4) with the experimental data concerning R', $\Delta\theta$ and ΔD is shown in Fig.5, satisfying reasonable coincidencies. Equation (2) shows good correspondence for $\Delta\theta$ and smaller magnitude in R' than test data. Equation (3) shows good coincidencies around R's hump and for ΔD within high speed range. Equation (4) shows the same tendencies for R' and ΔD also. Anyway the most practical formula considered is Equation (3) of the least terms.

CONCLUSIONS

At the initial design stage, in order to presume the resistance performance of high speed crafts having wave-shaped bottom hull in still water, the principal component analyses technique is applied in order to obtain empirical formulae by using 78 towing test data. Two kinds of empirical formulae obtained are (1) principal component empirical formula including major four principal components selected by the analyses, and (2) simple formula including several nondimensional hull form parameters. Comparing these formulae with towing test data for the resistance performance, they show good results.

For the usable range where many test data exist the tollerance ellipse containing 95 % of the data and Cramér's concentrated ellipse containing 75 % are recommended.

Five more new models were used for towing test to confirm the proposed formulae, and reasonable coincidencies were obtained.

REFERENCES

Paper in Conference Proceedings

1. Yoshida, Y. and Nagai, T. Numerical Determination of the Lift Distribution of High Speed Craft, pp.235-242, Vol. I, Proceedings of the Fourth Pacific Congress on Marine Science and Technology (PACON 90), Tokyo, Japan, Jul.16-20, 1990.

Paper in journals

2. Yoshida, Y. and Nagai, T. Principal Component Analysis and Factor Analysis referring to Model Test Results in Still Water of High Speed Craft, Jour. Soc. Naval Arch. of Japan, Vol.140, pp.58-66, 1976

3. Nagai, T., Tanaka, H. and Yoshida, Y. Statistical Analyses of Model Test Results in Still Water of High Speed Craft, Jour. Soc. Naval Arch. of Japan, Vol.137, pp.48-57, 1975.

Book

4. Cramér, H. Mathematical Methods of Statistics, Princeton University Press, Princeton, 1946.

Computational Experience with Damage-Tolerant Optimization of Structural Systems

D.M. Frangopol (*), M. Klisinski (**), M. Iizuka (*)
(*) Department of Civil Engineering, University of Colorado at Boulder, Boulder, CO 80309-0428, U.S.A.
(**) Department of Structural Mechanics, Chalmers University of Technology, S-412 96 Göteborg, Sweden

ABSTRACT

A complete design methodology for structural systems must address the problem of damage tolerance. To incorporate damage tolerance goals directly, designs based on consideration of system safety with respect to damaged states have to be considered. In this paper, some of the computational experience gained by the authors with damage-tolerant optimization of structural systems is presented.

INTRODUCTION

The deteriorating structural health of the infrastructure of public works and recent failures in highway bridges and offshore platforms indicate the desirability for damage tolerability in structural systems [2,3,4,7,9]. Design based on conventional limit state requirements cannot always prevent occurrence of damage during the lifetime of a structure. To guard against catastrophic failure due to damaged states that may occur during the lifetime of a structure, one needs to anticipate these conditions and account for them in the initial design process [11,14]. According to Arora et al. [1] a structure is called damage-tolerant if it continues to perform its basic function even after it sustains a specified level of damage. The ultimate goal in design of damage-tolerant structures is optimization [8,10]. This may be defined as the establishment of the best damage-tolerant structure satisfying a prescribed objective and a given set of constraints including both intact and damaged states. In this paper, which is based on previous work of the authors [6], techniques for damage-tolerant optimization of structural systems are presented along with a numerical example. Additionally, some of the computational experience with damage-tolerant structural optimization at the University of Colorado, Boulder, is reported.

DAMAGE-TOLERANT OPTIMIZATION FORMULATIONS

Both deterministic [6] and reliability-based [8] damage-tolerant optimization formulations are presented in this study.

Deterministic formulations

A general model for the multiobjective deterministic damage-tolerant optimization formulation problem is to find the design variable vector $\mathbf{x}$

$$\mathbf{x} \in \Omega \tag{1}$$

which minimizes

$$\mathbf{f}(\mathbf{x}) \tag{2}$$

where $\mathbf{f}$ is a vector objective function given by

$$\mathbf{f}(\mathbf{x}) = [f_1(\mathbf{x}), f_2(\mathbf{x}), ..., f_i(\mathbf{x}), ..., f_m(\mathbf{x})]^t \tag{3}$$

and $f_i(\mathbf{x})$ is the ith objective. The feasible set Ω belongs to an Euclidean space with n dimensions (i.e., $\mathbf{x} \in R^n$) determined by a set of equality (i.e., $\mathbf{g}(\mathbf{x}) = \mathbf{0}$) and inequality (i.e., $\mathbf{h}(\mathbf{x}) \leq \mathbf{0}$) conditions.

If only weight, intact capacity, residual capacity, and nodal displacements have to be optimized, the objective function (3) may be written as

$$\mathbf{f}(\mathbf{x}) = [W, -C_{ULT}, -C_{RESD}, \mathbf{\Delta}]^t \tag{4}$$

where W is the weight of the structure, C_{ULT} and C_{RESD} are the actual ultimate and residual load carrying capacities of the structural system, respectively, and $\mathbf{\Delta} = [\Delta_1, \Delta_2, ..., \Delta_{m-3}]^t$ is the vector of nodal displacements. The objective C_{RESD} addresses the damage-tolerability requirement over the lifetime of the structure. If all the structural members consist of the same material, the minimum weight design and the minimum volume design are identical. Consequently, the objective function (4) may be expressed as follows:

$$\mathbf{f}(\mathbf{x}) = [V, -C_{ULT}, -C_{RESD}, \mathbf{\Delta}]^t \tag{5}$$

where V is the total material volume of the structure. For truss systems, the design variable vector $\mathbf{x}$ consists of member areas

$$\mathbf{x} = [A_1, A_2, ..., A_n]^t \tag{6}$$

where A_i is the cross-sectional area of group i of members. By imposing bounds for cross-sectional areas, the feasible set Ω of the deterministic damage-tolerant multicriterion formulation is defined by

$$\Omega = \{\mathbf{x} \in R^n \ : \ \underline{A_i} \leq A_i \leq \bar{A}_i, \ \ i = 1, 2, ..., n\} \tag{7}$$

where $\underline{A}_i$ and $\bar{A}_i$ are the lower and upper bounds of the design variable A_i, respectively.

The above multiobjective problem may be converted to a series of single objective minimizations using the ϵ-constraint method. One of these single-objective optimizations consists in finding the design variable vector $\mathbf{x}$ (see Eq. (6)) such that the total volume of the structure is minimized

$$\min \quad V \tag{8}$$

and the following constraints are satisfied

$$C_{ULT} \geq C^0_{ULT} \tag{9}$$

$$C_{RESD} \geq C^0_{RESD} \tag{10}$$

$$\Delta_j^{max} \leq \Delta_j^0 \qquad ,j = 1, 2, ..., m-3 \tag{11}$$

$$\underline{A}_i \leq A_i \leq \bar{A}_i \qquad ,i = 1, 2, ..., n \tag{12}$$

where the value of the required residual capacity C^0_{RESD} varies to cover the entire solution set. Let us notice that in this study we are mainly interested in the influence of the residual capacity requirement (10) on the optimization results. For this reason, the serviceability requirements (11) are not explicitly considered in the optimization formulation.

In this study damage conditions are generated by accidents in which a member is completely removed from the system. All accidents are considered possible. Consequently, any one member may be removed from the structural system. The removal of a member creates a new structure which may have some capacity under a load proportional to the nominal load. The most critical is the damaged structure with the lowest capacity. This capacity is defined as the residual capacity of the original structural system, C_{RESD}, as follows

$$C_{RESD} = \min(C_1, C_2, ..., C_i, ..., C_n) \tag{13}$$

where C_i is the capacity of the structure having member i completely removed. For statically determinate structures it is obvious that the presence of all members is essential (i.e., $C_{RESD} = 0$). On the other hand, for statically indeterminate structures, removal of any one and only one member will not necessarily constitute collapse (i.e., $C_{RESD} \neq 0$). However, to obtain residual capacity, adequate structural configurations must be chosen for these indeterminate systems.

Reliability-based formulation

The general model for the multiobjective reliability-based damage-tolerant optimization formulation is similar to that used for the deterministic formulation previously presented (see, Eqs. (1) to (3)). If only weight (i.e., volume), system probabilities with respect to collapse, damaged states, and excessive deformations have to be optimized simultaneously, the objective function (3) may be written as follows:

$$\mathbf{f}(\mathbf{x}) = [V, P_{f(int)}, P_{f(dmg)}, P_{f(def)}]^t \tag{14}$$

where V=total material volume of the structure, $P_{f(int)}$= probability of system collapse initiated from intact state, $P_{f(dmg)}$= probability of system collapse initiated from damaged states (i.e., with any one component removed), and $P_{f(def)}$= probability of excessive elastic deformation of the intact structure. The overall system residual reliability (equal to one minus the probability of system collapse initiated from damaged states, $1 - P_{f(dmg)}$) is used as a damage tolerability objective.

SOLVING DAMAGE-TOLERANT OPTIMIZATION PROBLEMS

Both deterministic and reliability-based structural system optimization problems represented by minimizing the vector objective functions (5) and (14), respectively, were solved by using the ϵ-constraint method [12] along with the Rosenbrock's optimization algorithm [13]. The flowchart of the deterministic optimization program using, as an intermediate step of the ϵ-constraint method, the total material volume of the structure as the objective to be minimized is shown in Fig. 1 [6]. For solving the multiobjective reliability-based damage-tolerant optimization problem whose vector objective is defined in (14), Fu and Frangopol [8] presented a three-step solution strategy which has been implemented in a computer code.

NUMERICAL EXAMPLE

Fig. 2 defines the geometry and loading of a 10-bar truss example under four proportional loads Q. Geometrical, mechanical and loading characteristics are assumed to be deterministic. The modulus of elasticity is $E = 29,000ksi$. As shown in Fig. 2 the member cross-sectional areas A_i are assembled into five groups, taking into account the symmetry of the structure. The material is assumed to be elastic perfectly plastic with yielding stresses of $-10ksi$ and $20ksi$ in compression and tension, respectively. The minimum and maximum cross-sectional areas are assumed to be $\underline{A}_i = 0.1in^2$ and $\bar{A}_i = \infty$, respectively. The initial cross-sectional areas are $1in^2$ except for the diagonal elements whose areas are $0.25in^2$.

Let us first concentrate on the behavior of the initial structure which has the volume $V_{intact} = 1176in^3$. Table 1 shows the stress distribution for intact and all (i.e., ten) possible damage scenarios. The capacity of the intact structure is $C_{ULT} = 1.664kips$ and its residual capacity is (see Eq. (13)) $C_{RESD} = 0.332kips$. Therefore, the residual strength factor is $R = C_{RESD}/C_{ULT} = 0.332/1.664 = 0.20$ (see also Table 2(a)). The

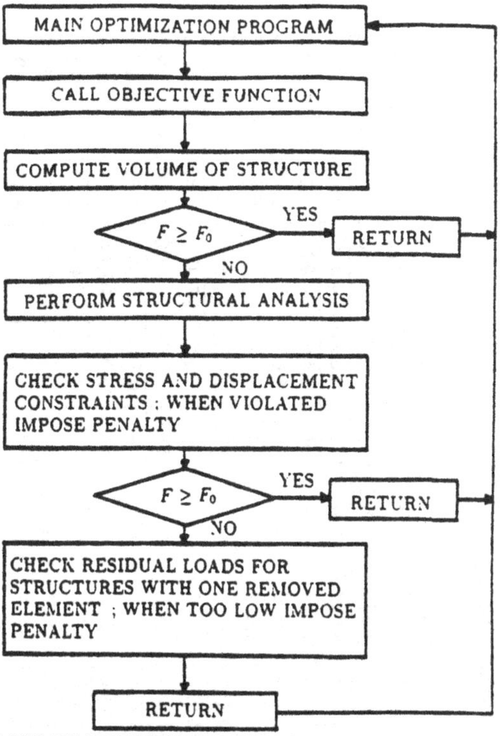

F - VALUE OF OBJECTIVE FUNCTION
F_0 - REFERENCE VALUE OF OBJECTIVE FUNCTION

Figure 1. Computer Program Flowchart.

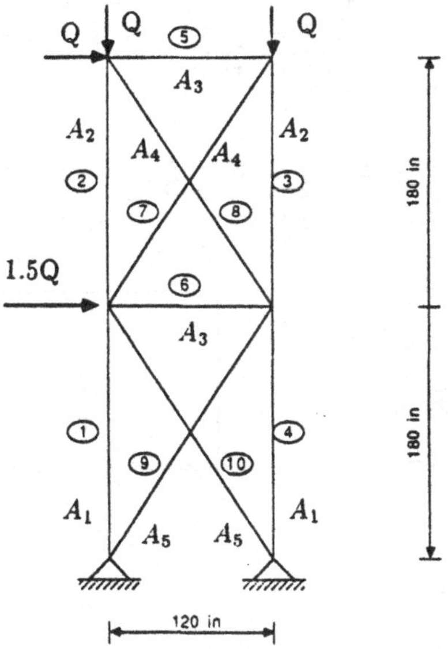

Figure 2. Ten-Bar Truss : Geometry and Loading.

Table 1. Ten-Bar Intact and Damaged Trusses: Ultimate Capacities and Associated Stresses.

$L = 120$ in ; $H = 2 \times 180$ in $= 360$ in ; $E = 29{,}000$ ksi $A_1 = A_2 = A_3 = 1.0\ in^2$; $A_4 = A_5 = 0.25\ in^2$; $V_{INTACT} = 1176\ in^3$ DUCTILE BEHAVIOR (-10 ksi; 20 ksi)											
ELEMENT REMOVED	LOAD LEVEL (kips)	STRESSES IN ELEMENTS (ksi)									
		1	2	3	4	5	6	7	8	9	10
NONE	1.031	2.596	-0.101	-1.647	-4.367	-0.411	-0.572	2.962	-4.474	8.596	**-10.00**
NONE	**1.664**	2.912	-0.139	-2.635	-8.320	-0.648	-1.757	4.669	-7.330	**20.00**	**-10.00**
1	**0.978**	—	-0.051	-1.518	-6.603	-0.360	-2.154	2.596	-4.456	**20.00**	2.351
2	1.032	2.595	—	-1.548	-4.371	-0.344	-0.506	2.481	-4.961	8.610	**-10.00**
2	**1.664**	2.912	—	-2.496	-8.320	-0.555	-1.664	4.000	-8.000	**20.00**	**-10.00**
3	**0.832**	2.069	1.248	—	-3.547	0.555	0.409	-4.000	**-10.00**	7.051	-7.948
4	**0.332**	2.242	-0.058	-0.556	—	-0.149	0.736	1.075	-1.319	-4.000	**-10.00**
5	1.037	2.598	0.519	-1.037	-4.402	—	-0.169	0.000	-7.478	8.697	**-10.00**
5	**1.386**	2.773	0.693	-1.386	-6.582	—	-0.692	0.000	**-10.00**	15.00	**-10.00**
6	0.854	2.507	0.268	-1.013	-3.257	-0.106	—	0.763	-5.396	5.396	**-10.00**
6	**1.109**	2.635	0.970	-0.694	-4.851	0.277	—	-2.000	**-10.00**	10.00	**-10.00**
7	1.037	2.598	0.519	-1.037	-4.402	0.000	-0.169	—	-7.478	8.697	**-10.00**
7	**1.386**	2.773	0.693	-1.386	-6.582	0.000	-0.692	—	**-10.00**	15.00	**-10.00**
8	1.023	2.591	-1.023	-2.557	-4.314	-1.023	-1.171	7.377	—	8.445	**-10.00**
8	**1.664**	2.912	-1.664	-4.160	-8.320	-1.664	-2.773	12.00	—	**20.00**	**-10.00**
9	**0.554**	2.355	-0.071	-0.902	-1.385	-0.232	0.322	1.674	-2.321	—	**-10.00**
10	**1.109**	0.555	-0.068	-1.731	-6.931	-0.415	-2.078	2.992	-5.005	**20.00**	—

Note : Ultimate capacities of intact and damaged structures and stresses at the ductile failure of the critical elements are indicated in bold characters.

Table 2. Ten-Bar Truss : (a) Initial Structure; (b) Optimized Structure under Ultimate Requirements; (c) Optimized Structure under Ultimate and Damage-Tolerability Requirements.

L = 120 in ; H = 2 x 180 in = 360 in ; E = 29,000 ksi DUCTILE BEHAVIOR (-10 ksi, 20 ksi)										
AREAS (in^2)					VOLUME (in^3)	ULTIMATE CAP. (kips)	RESIDUAL CAP. (kips)	RESIDUAL FACTOR	MAX. EL. DISPL. (in) (Q = 1 kip)	
A_1	A_2	A_3	A_4	A_5	V	C_{ULT}	C_{RESD}	$R_2 = C_{RESD}/C_{ULT}$	Δ_X^{max}	Δ_Y^{max}
(a) INITIAL STRUCTURE										
1.000	1.000	1.000	0.250	0.250	1176	1.66	0.332	0.200	0.272	0.036
(b) OPTIMIZED STRUCTURE UNDER ULTIMATE CAPACITY REQUIREMENT $V \rightarrow \min$ $C_{ULT} \geq 1.66$ kips $0.1 in^2 \leq A_i \leq \infty$; i=1,2,3,4,5										
0.830	0.287	0.191	0.155	0.249	623	**1.66**	0.332	0.200	0.341	0.063
(c) OPTIMIZED STRUCTURES UNDER BOTH ULTIMATE AND RESIDUAL CAPACITIES REQUIREMENTS $V \rightarrow \min$ $C_{ULT} \geq 1.66$ kips $C_{RESD} \geq Q_{RESD}$ $0.1 in^2 \leq A_i \leq \infty$; i=1,2,3,4,5										
0.726	0.289	0.124	0.151	0.374	623	**1.66**	**0.498**	0.300	0.320	0.066
0.655	0.248	0.129	0.201	0.499	659	**1.66**	**0.664**	0.400	0.297	0.072
0.560	0.208	0.158	0.249	0.623	692	**1.66**	**0.830**	0.500	0.302	0.082
0.672	0.219	0.190	0.299	0.748	830	1.99	**0.996**	0.500	0.252	0.069
0.781	0.290	0.221	0.319	0.873	969	2.32	**1.162**	0.500	0.216	0.059
0.896	0.332	0.253	0.399	0.998	1107	2.65	**1.328**	0.500	0.189	0.052

Note : Active constraints are indicated in bold characters.

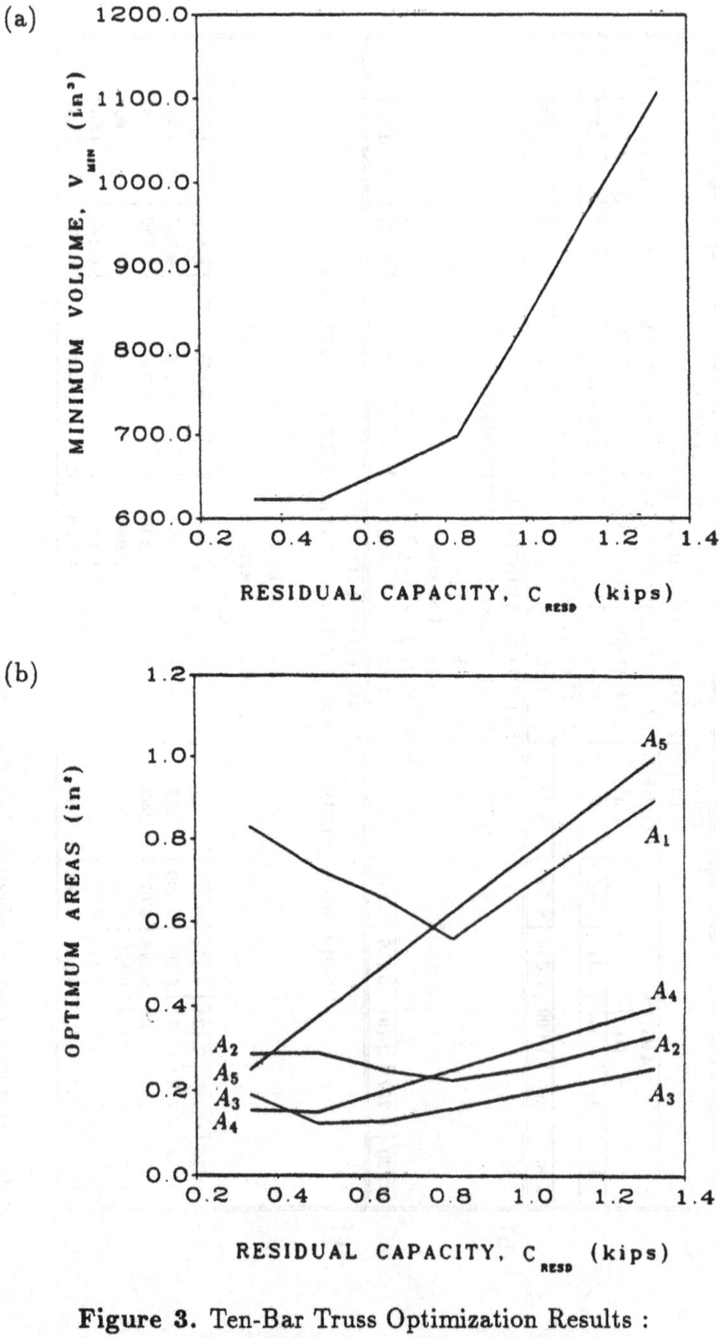

Figure 3. Ten-Bar Truss Optimization Results :
(a) Optimum Volume versus Residual Capacity;
(b) Optimum Areas versus Residual Capacity.

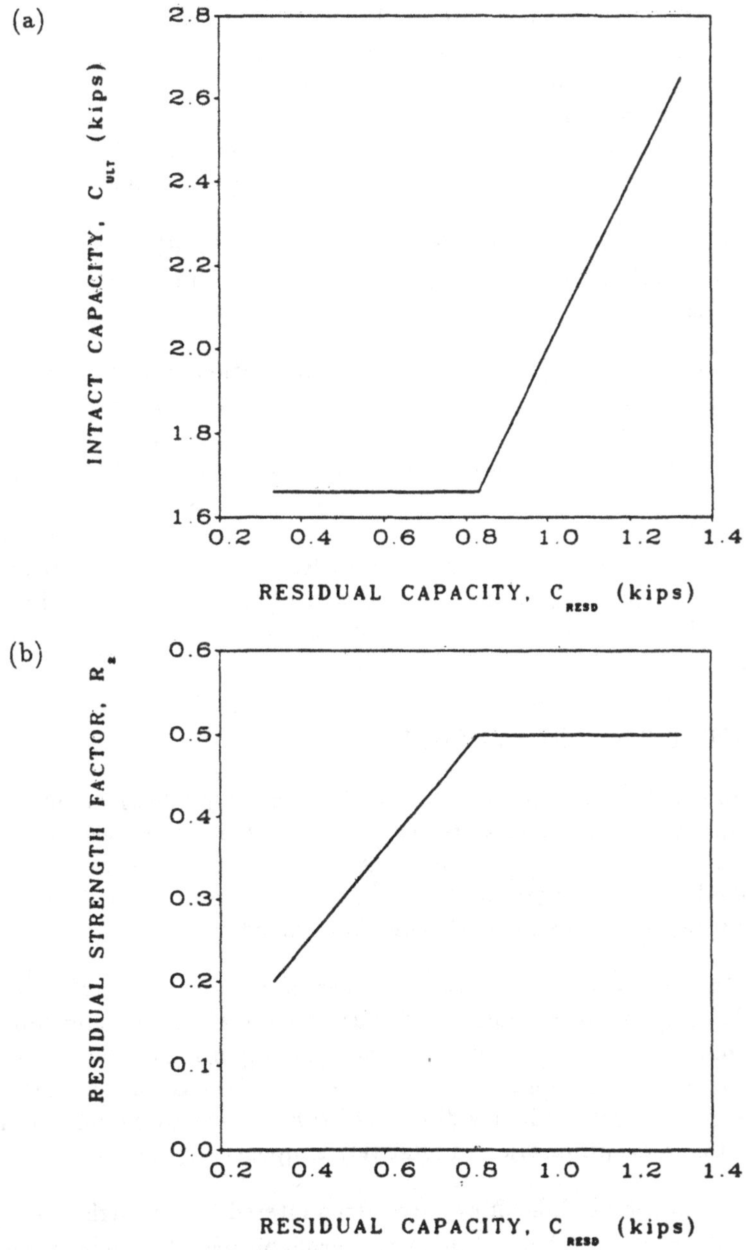

Figure 4. Ten-Bar Truss Optimization Results :
(a) Intact Capacity versus Residual Capacity;
(b) Residual Strength Factor versus Residual Capacity.

classical optimization of the truss in Fig. 2 is based on minimum volume design under ultimate capacity requirement only, $C_{ULT} \geq 1.664 kips$ (see Table 2(b)). The resulting optimized structure has a volume of $623 in^3$ (i.e., 47% less material). The residual capacity and residual strength factor of the optimized and initial trusses are identical. However, considering both ultimate and residual capacity requirements (i.e., $C_{ULT} \geq 1.664 kips$ and $C_{RESD} \geq Q_{RESD}$) in the optimization process (see Table 2(c)) it is possible by using 41% less material (i.e., $V_{opt} = 692 in^3$) to assure the same capacity of the optimum structure (i.e., $C_{ULT} = 1.664 kips$) and increase the residual capacity 2.5 times from $0.332 kips$ to $0.830 kips$, so that the residual strength factor of the damage-tolerant optimum structure R is equal to 0.5.

The results of the damage-tolerant optimization of the ten-bar truss in Fig. 2, are shown in Figs. 3 and 4. Fig. 4(b) and the last four rows of Table 2 show that the residual strength factor cannot exceed the value of 0.50. This is because after reaching a certain residual capacity level (i.e., $C_{RESD} = 0.830 kips$) the increase in the residual capacity also affects the intact ultimate capacity (i.e., both C_{RESD} and C_{ULT} increase proportionally so that R is constant). Also, the maximum horizontal and vertical displacements, Δ_X^{max} and Δ_Y^{max}, of the initial and optimized trusses under the load $Q = 1 kip$ are shown in the last two columns of Table 2.

COMPUTATIONAL EXPERIENCE

Owing to space limitations, this section contains only a brief summary of the computational experience gained by the authors in solving damage-tolerant optimization problems by using the ϵ-constraint method [12] and the Rosenbrock's optimization algorithm [13] for both deterministic [6] and nondeterministic [8,10] steel truss and framed systems.

Damage-tolerant optimization of structural systems involves considerable computational effort. Therefore, the efficiency of algorithms is a very important factor. Several methods and algorithms can be used to treat damage-tolerant optimization problems. The ϵ-constraint method and the Rosenbrock's minimization algorithm appear to be suitable for damage-tolerant optimization of relatively simple structures.

Fundamental difficulties were encountered when high residual strength factors R or low ratios of the probabilities of system collapse initiated from damaged and intact states $P_{f(dmg)}/P_{f(int)}$ were imposed. In deterministic design it was found that for a given structure and material behavior there is always a maximum value of the residual strength factor above which there is no feasible design.

In order to check deterministically or probabilistically residual capacity requirements for structural systems we should, theoretically, perform analyses for all possible damage conditions. These analyses have to be done by removing each member of a structural system and verifying whether (a) the deterministic residual system capacity is higher than the imposed one, or (b) the residual system probability of failure is less than the imposed one. In many practical studies it is possible to identify critical members and, therefore, restrict the number of examined structures. Such an approach was recently proposed by Iizuka [10].

CONCLUDING REMARKS

1. To achieve a prescribed level of invulnerability of structural systems during their expected lifetime , formulations and techniques for damage tolerant optimum design have been proposed and tested on simple steel structures.
2. Computational experience with damage-tolerant optimization techniques has been reported.
3. Further research is needed on damage-tolerant optimization of structural systems in connection with (a) realistic damage scenarios, (b) nonlinear and dynamic behavior, (c) reanalysis and failure mode identification techniques, and (d) reliability, generality and efficiency of optimization algorithms.

ACKNOWLEDGEMENTS

This work was supported by the National Science Foundation under Grants No. MSM-8618108 and MSM-9013017, Dr. Ken P. Chong, Program Director. This support is gratefully acknowledged.

REFERENCES

1. Arora, J. S., Haskell, D. F. and Govil, A. K. Optimal Design of Large Structures for Damage Tolerance, AIAA Journal, Vol. 118, No. 5, pp. 563-570, 1980.

2. De, R. S. Offshore structural system reliability : Wave-load modeling, system behavior, and analysis, Ph. D. Thesis, Department of Civil Engineering, Stanford University, Stanford, California, 1990.

3. Feng, Y. S. and Moses, F. Optimum Design, Redundancy and Reliability Analysis of Structural Systems. Computers and Structures, Vol. 24, No. 2, pp. 239-251, 1986.

4. Frangopol, D. M., Iizuka, M. and Yoshida, K. Redundancy measures for design and evaluation of structural systems, ASME Paper No. OMAE 91-1242, 1991 (in press).

5. Frangopol, D. M. and Klisinski, M. Material Behavior and Optimum Design of Structural Systems, Journal of Structural Engineering, ASCE, Vol. 115, No. 5, pp. 1054-1075, 1989.

6. Frangopol, D. M., Klisinski, M. and Iizuka, M. Optimization of Damage-Tolerant Structural Systems, Proceedings of the International Conference on Computational Structures Technology , Edinburgh, Scotland, August 20-22, 1991 (submitted).

7. Frangopol, D. M. and Nakib, R. Effects of Damage and Redundancy on the Safety of Existing Bridges, Transportation Research Record No. 1290, Vol. 1, TRB, NRC, Washington. D.C., pp. 9-15, 1991.

8. Fu, G. and Frangopol, D. M. Reliability-Based Vector Optimization of Structural Systems, Journal of Structural Engineering, ASCE, Vol. 116, No. 8, pp. 2143-2161, 1990.

9. Fu, G., Yingwei, L. and Moses, F. Management of structural system reliability, in Reliability and Optimization of Structural Systems '90 (Eds. Der Kiureghian, A. and Thoft-Christensen, P.), Lecture Notes in Engineering, Vol. 61, Springer-Verlag, Berlin, pp. 113-128, 1991.

10. Iizuka, M. Time invariant and time variant reliability analysis and optimization of structural systems , Ph. D. Thesis, Department of Civil Engineering, University of Colorado, Boulder, 1991.

11. Nguyen, D. T. and Arora, J. S. Fail-Safe Optimal Design of Complex Structures with Substructures, Journal of Mechanical Design, Vol. 104, pp. 861-868, 1982.

12. Osyczka, A. Multiobjective Optimization in Engineering. Ellis Horwood Ltd., Chicester, England, 1984.

13. Rosenbrock, H. H. An Automatic Method for Finding the Greatest or Least Value of Function, Computer Journal, Vol. 3, pp. 175-184, 1960.

14. Sun, P. F., Arora, J. S. and Haug, E. J. Fail-Safe Optimal Design of Structures, Engineering Optimization, Vol. 2, pp. 43-53, 1976.

On the Effect of Stochastic Imperfections on the Buckling Strength of Certain Structures

G.V. Palassopoulos

Department of Applied Mechanics, Military Academy of Greece, 41 Blessa Street, Papagos, Athens 15669, Greece

ABSTRACT

The present paper addresses the problem of the response variability of structures, which experience bifurcation buckling. The buckling strength of such structures may be very sensitive to the small structural imperfections, which are practically inevitable in all real structures. A new general method is presented, which is particularly suitable for the treatment of the stochastic nature of the structural imperfections. The new method is exemplified with two well known buckling problems. The first problem is the buckling of a column on a linear elastic foundation. The shape imperfections of the column are treated as a weakly stationary random process with a pre-specified auto-correlation function. The second problem is the buckling of a thin cylindrical shell under axial compression. The shape imperfections of the shell are treated as a broad-band random Gaussian process with an arbitrarily specified power spectral density function. In both cases, representative numerical results are presented for the purpose of improving our understanding of the response variability of these structures.

1. INTRODUCTION

The response variability of engineering structures is caused by both the variability of the applied loads and the variability of the material and the geometric properties of the structure itself. The determination of the response variability of practical engineering structures is a very interesting problem, with direct application to the reliability-based design of these structures, and has been the focus of increased interest among engineers and researchers recently [1, 4, 11].

To the author's knowledge, the research efforts up to date have dealt with the more general cases, in which the variability of the loads and/or the material and geometric properties of a structure has only a mild effect on its response. However, there are certain exceptional cases, in which the structural response may be affected radically, both quantitatively and qualitatively, by the variability of the loads and/or the material and geometric properties. One such prominent and well known case, which will be the focus of the present paper, is the bifurcation buckling of the structures. Example structures, which experience bifurcation buckling and may be affected radically by the variability of the applied loads and/or the material and geometric properties, include most thin shells, many beams with thin walled sections and certain types of columns, trusses, frames and arches [3].

Starting point for understanding the buckling phenomenon in these structural cases is Koiter's Theory [5]. Koiter proved that, asymptotically close to the bifurcation buckling load of a structure, its displacement field is completely dominated by its classical buckling mode, as this mode is determined from classical (linearized) buckling analysis. This fact effectively reduces the analysis of the structure, asymptotically close to its bifurcation buckling load, from an infinite dimensional continuum problem to a one-dimensional problem in terms of its classical buckling mode. It can then be proved that a real structure, whose loads and/or material and geometric properties deviate inevitably from those of the assumed idealized structure, may experience limit point buckling (instead of bifurcation buckling) at a load much lower than the bifurcation buckling load of the idealized structure. It can be further proved that this decrease of the buckling load of a real structure depends predominantly on small imperfections of its geometry in the shape of the classical buckling mode. Such small shape imperfections are practically inevitable in all real structures and are stochastic in nature.

Koiter's Theory has been the basis of very extensive investigations up to date on the sensitivity of the buckling load of various structures to small shape imperfections. However, it is the opinion of the present author that there has been a misunderstanding of the range of applicability of Koiter's analysis in some of these investigations. Namely, all the analysis in Koiter's Theory is based on the dominance of the classical buckling mode, which is valid only asymptotically close to the bifurcation buckling load of the idealized perfect structure. For engineering purposes, one may further assume that this dominance is also approximately valid in some small finite neighborhood of the bifurcation buckling load. But, in the great majority of structures with medium and high imperfection sensitivity, the buckling load of the real structure is far below the bifurcation buckling load (up to 70% below in the notorious case of the axially compressed thin cylindrical shell). For such structures, the dominance of the classical buckling mode cannot be accepted, even as a crude approximation, and many more shape imperfection modes, as well as other types of imperfections, need to be taken into account for a realistic evaluation of their response variability. Thus, Koiter's Theory is fundamentally inadequate for the cases of medium and high imperfection sensitivity. This inadequacy of Koiter's Theory will be clearly demonstrated numerically in the two buckling problems of the present paper.

It is for this reason that the present paper addresses the problem of the response variability of structures with bifurcation buckling by means of a new general method. Two major departures from the traditional approach are combined in this new method. The first major departure, which can be traced back to an idea introduced in 1973 by Shinozuka and the present author [6, 7, 10], is related to the range of the analysis. Traditionally, the analytical relations cover both the pre-buckling and the post-buckling range and are inherently nonlinear. In the new method, attention is focused only on the pre-buckling range, which is carefully isolated by means of a regular perturbation expansion and is effectively linearized. The second major departure from the traditional approach was introduced in two recent papers by the present author [8, 9] and is related to the control variable. Traditionally, the inception of buckling is determined in terms of the smallest critical load for given structural imperfections [3, 5, 13]. In the new method, the inception of buckling is determined in terms of the smallest critical magnitude of the imperfections at pre-specified values of the applied loads.

In what follows, the new general method will be presented first. Next, the new general method will be applied to two well known buckling problems, both because these problems are interesting in their own right and in order to exemplify the new method. The first problem is the buckling of a column on a linear elastic foundation. The shape imperfections of the column will be treated as a weakly stationary random process with a pre-specified auto-correlation function. The second problem is the buckling of a thin cylindrical shell under axial compression. The shape imperfections of the shell will be treated as a broad-band random Gaussian process with an arbitrarily specified power spectral density function. In both cases, representative numerical results will be presented for the purpose of improving our understanding of the response variability of these structures.

2. OUTLINE OF THE NEW GENERAL METHOD

2.1. Basic Assumptions

As in Koiter's Theory [5], the fundamental assumption of the method is that the structure, together with the applied loads, constitutes a conservative elastic system with a single valued total potential energy. For engineering applications, the great majority of structures satisfies this assumption.

Two additional assumptions will be introduced in the present paper for purposes of simplicity of the presentation. The first additional assumption is that the continuum problem has been effectively discretized and all spatial configurations of the structure are adequately described by a finite number M of generalized coordinates q_i, i=1...M, which are measured incrementally from the pre-buckling equilibrium state of the idealized perfect structure. In terms of these generalized coordinates, the imperfections of the structure will be represented as the product of a normalized imperfection pattern p_i, i=1...M, and a unique magnitude parameter ε. Although this assumption is not necessary theoretically [8], it illuminates important relations between the data of the problem and the resulting formulas and corresponds to a necessary intermediate step in most practical analyses (e.g., finite element method).

The second additional assumption is that all the loads on the structure can be uniquely specified as the product of a unit load system and a single load parameter φ. Again, this assumption is not necessary from a theoretical point of view and it would be quite interesting to investigate the behavior of the structure under multiple load parameters. However, this assumption is implicit in most practical analyses of engineering structures and significantly simplifies the presentation of the method.

2.2. Potential Energy and Equilibrium Equations

Starting point of the method is the total potential energy V of the structure in terms of the generalized coordinates q_i, i=1...M, as it is assumed in Koiter's Theory [5]:

$$V=\varepsilon a_{ij}q_ip_j + b_{ij}q_iq_j + c_{ijk}q_iq_jq_k + d_{ijkl}q_iq_jq_kq_l + ... \qquad (1)$$

In this expression and in what follows, the usual Cartesian tensor summation convention from 1 to M is implied over all repeated subscripts in the terms of a product, unless stated otherwise.

The coefficients a_{ij}, b_{ij}, c_{ijk}, d_{ijkl}, ... in Equation 1, which are functions of the applied loads and the material and geometric properties of the structure, can always be selected, so as to be invariant with respect of any permutation of their indices, i.e.

$$a_{ij} = a_{ji} \quad b_{ij} = b_{ji} \qquad i,j = 1...M \tag{2a}$$

$$c_{ijk} = c_{ikj} = c_{jik} = c_{jki} = c_{kij} = c_{kji} \qquad i,j,k = 1...M \tag{2b}$$

$$d_{ijkl} = d_{ijlk} = d_{ikjl} = d_{iklj} = ... \qquad i,j,k,l = 1...M \tag{2c}$$

In this form, the coefficients b_{ij} are the components of the well-known geometric stiffness matrix of the idealized perfect structure. The coefficients a_{ij} represent the primary contribution of the structural imperfections to the potential energy expression, while the coefficients c_{ijk} and the d_{ijkl} represent higher order terms, necessary for the investigation of the effect of imperfections on the structure.

An alternate starting point of the analysis is the non-linear equilibrium equations of the structure:

$$\varepsilon a_{ij}p_j + 2b_{ij}q_j + 3c_{ijk}q_jq_k + 4d_{ijkl}q_jq_kq_l + ... = 0 \qquad i = 1...M \tag{3}$$

which result directly from Equation 1 by setting the first variation δV of the potential energy V equal to zero.

In both cases, the neutral equilibrium equations have the form:

$$(b_{ij} + 3c_{ijk}q_k + 6d_{ijkl}q_kq_l + ...)\ (\delta q_j) = 0 \qquad i = 1...M \tag{4}$$

The buckling load of the structure is determined from Equations 4, as the smallest value of the load parameter φ, for which these equations have a non-trivial solution, besides the trivial solution $\delta q_j=0$, j=1...M.

<u>2.3. Regular Perturbation Expansion and Eigenvalue Problem</u>

It is an experimental fact that small structural imperfections have only a mild effect on the structural response in the pre-buckling equilibrium range of a structure. Based on this fact, it is postulated that, in the pre-buckling equilibrium range of a structure (and only in this range), the generalized coordinates q_i, i=1...M, may be expanded around the pre-buckling equilibrium state q_i=0, i=1...M, of the corresponding idealized perfect structure in terms of the imperfection magnitude parameter ε in the following regular perturbation expansion:

$$q_i = \varepsilon q_{i1} + \varepsilon^2 q_{i2} + \varepsilon^3 q_{i3} + ... \qquad i=1...M \tag{5}$$

As is usual in perturbation expansions, it is anticipated that, for sufficiently small values of ε, only the first few terms of the expansion will be necessary for a satisfactory approximation. Substituting the generalized coordinates from Equation 5 into Equation 4, the neutral equilibrium equations result in the following form:

$$(b_{ij} + \varepsilon\,\beta_{ij1} + \varepsilon^2\beta_{ij2} + ...)\ (\delta q_j) = 0 \qquad i=1...M \tag{6}$$

where:

$$\beta_{ij1} = 3c_{ijk}q_{k1} \qquad i,j = 1...M \tag{7a}$$

$$\beta_{ij2} = 3c_{ijk}q_{k2} + 6d_{ijkl}q_{k1}q_{l1} \qquad i,j = 1...M \tag{7b}$$

$$...$$

It can be immediately recognized that Equation 6 represents a generalized eigenvalue problem in terms of ε. This problem can be directly solved, if the inverse (with a minus sign) $[r_{ij}]$ of the geometric stiffness matrix $[b_{ij}]$ is introduced, i.e.:

$$r_{ij}b_{jk} = -\ \delta_{ik} \qquad i,k = 1...M \tag{8}$$

where δ_{ik} is the Kronecker delta. For values of φ below the bifurcation buckling load φ_{cl} of the idealized perfect structure, the geometric stiffness matrix $[b_{ij}]$ is always positive definite and, therefore, the inverse matrix $[r_{ij}]$ always exists.

Pre-multiplying Equations 6 with r_{mi}, m=1...M, and adding from i=1 to M (i.e. pre-multiplying with $[r_{ij}]$ in matrix form), the following generalized eigenvalue problem in ε results:

$$(\varepsilon\ \gamma_{ij1} + \varepsilon^2\gamma_{ij2} + ...)\ (\delta q_j) = (\delta q_i) \qquad i = 1...M \tag{9}$$

where:

$$\gamma_{ij1} = 3r_{im}c_{mjk}q_{k1} \qquad i,j = 1...M \tag{10a}$$

$$\gamma_{ij2} = 3r_{im}c_{mjk}q_{k2} + 6r_{im}d_{mjkl}q_{k1}q_{l1} \qquad i,j = 1...M \tag{10b}$$

$$...$$

Only the absolutely smallest (dominant) eigenvalue $\varepsilon=\varepsilon_{cr}$ of Equation 9 is of practical significance. It will be denoted as critical imperfection magnitude ε_{cr} and will completely signify the inception of buckling at any given value φ of the load parameter. This eigenvalue can be easily determined numerically by methods described in [2] and [8].

2.4. Pre-Buckling Equilibrium State

In order to specify completely the eigenvalue problem in Equation 9, the pre-buckling equilibrium state q_{i1}, q_{i2},..., i=1...M, which appears in Equations 10, must be determined. This can be accomplished by the direct substitution of the q_i, i=1...M, from Equation 5 into the non-linear equilibrium Equations 3. Equating with zero the coefficients of similar powers of ε, the following successive algebraic systems of equations in q_{j1}, q_{j2},..., j=1...M, result:

$$2b_{ij}q_{j1} + a_{ij}p_j = 0 \qquad i = 1...M \tag{11a}$$

$$2b_{ij}q_{j2} + 3c_{ijk}q_{j1}q_{k1} = 0 \qquad i = 1...M \tag{11b}$$

$$...$$

Pre-multiplying Equations 11 with r_{mi}, m=1...M, and adding from i=1 to M (i.e. pre-multiplying with $[r_{ij}]$ in matrix form), the following solution to Equations 11 is formally obtained:

$$q_{i1} = \xi_{im}p_m \tag{12a}$$

$$q_{i2} = \xi_{imn}p_mp_n \tag{12b}$$

$$...$$

where:

$$\xi_{im} = \frac{1}{2}\, r_{ij}a_{jm} \qquad i,m = 1...M \tag{13a}$$

$$\xi_{imn} = \frac{3}{2}r_{ij}c_{jkl}\xi_{km}\xi_{ln} \qquad i,m,n = 1...M \tag{13b}$$

$$...$$

Substituting q_{i1}, q_{i2}, ..., i=1...M, from Equations 12 and 13 into Equations 10, the coefficients γ in Equation 9 result in the form:

$$\gamma_{ij1} = \tau_{ijm}p_m \qquad i,j = 1...M \tag{14a}$$

$$\gamma_{ij2} = \tau_{ijmn}p_mp_n \qquad i,j = 1...M \tag{14b}$$

$$...$$

where:

$$\tau_{ijm} = 3r_{iv}c_{vjk}\xi_{km} \qquad i,j,m = 1...M \tag{15a}$$

$$\tau_{ijmn} = 3r_{iv}(c_{vjk}\xi_{kmn} + 2d_{vjkl}\xi_{km}\xi_{ln}) \qquad i,j,m,n = 1...M \tag{15b}$$

$$...$$

Equations 14, together with 8, 13 and 15, completely specify all the coefficients γ of the generalized eigenvalue problem in Equation 9. It is worth noting that the imperfection components appear only in Equations 14, with the constant coefficients τ as weighting factors. Therefore, the present formulation is particularly suitable for the determination of the relative significance of the various imperfection components and the treatment of the stochastic nature of the structural imperfections.

2.5. First Order Approximations

The more general case of imperfection sensitivity is the case, in which the third order coefficients c_{ijk} in Equation 1 do not all vanish identically. In this case, the first order approximation to Equation 9 is the following ordinary eigenvalue problem:

$$\tau_{ijm}p_m(\delta q_j) = \frac{1}{\varepsilon_{cr}}(\delta q_i) \quad i = 1...M \tag{16a}$$

It is obvious that only the dominant eigenvalue of Equation 16 has a physical meaning and need be computed. This case corresponds the asymmetric bifurcation buckling in Koiter's Theory [5].

Due to certain symmetry conditions in many practical engineering problems, it may happen that all the third order coefficients c_{ijk} in Equation 1 vanish identically. In this case, the second order terms in ε must be taken into account for a first order approximation. Then Equation 9 is reduced to the following ordinary eigenvalue problem:

$$\tau_{ijmn}p_m p_n(\delta q_j) = \frac{1}{\varepsilon_{cr}^2}(\delta q_i) \quad i = 1...M \tag{16b}$$

Again, it is obvious that only the dominant positive eigenvalue of Equation 17 has a physical meaning and need be computed. This case corresponds to the unstable symmetric bifurcation buckling in Koiter's Theory [5].

2.6. Orthonormal Displacement Modes

Formulas 15, together with 8 and 13, for the constant weighting factors τ can be significantly simplified, if the linear space of the classical buckling modes is employed for the discretization of the problem. In this case, both the stability matrix $[b_{ij}]$ and its inverse $[-r_{ij}]$ are diagonal. Then, Equation 15a takes the form:

$$\tau_{ijm} = \frac{3c_{ijm}a_{mm}}{2b_{ii}b_{mm}} \quad i,j,m = 1...M \tag{17a}$$

while Equation 15b takes the form:

$$\tau_{ijmn} = -\frac{3d_{ijmn}a_{mm}a_{nn}}{2b_{ii}b_{mm}b_{nn}} \quad i,j,m,n = 1...M \tag{17b}$$

where no summation is implied for repeated indices in both cases.

2.7. Discussion of the General Method and Comparison With Koiter's Analysis

It is the opinion of the present author that the major advantage of the new general method is its conceptual simplicity. A simple regular perturbation expansion in the imperfection magnitude (see Equation 5) leads directly to a well posed eigenvalue problem in the same variable (see Equation 9). In order to appreciate the usefulness of this result, all that is required is a simple change in the control variable. Instead of considering the load parameter φ as the independent variable, φ is treated as a constant and the imperfection magnitude parameter ε is treated as the independent variable.

Within this framework, the new general method has certain distinct advantages:

1. All the imperfection components can be taken into account with no restrictions (see Equations 14). Therefore, it is particularly suitable for the realistic simulation of structural imperfections in actual structures. It is noted in this respect

that Koiter's Theory is restricted to imperfection components only in the shape of the classical buckling mode.

2. The weighting factors τ provide a direct indication on the relative significance of the various imperfection components for the buckling strength of the structure (see Equations 14). This capability is important in reducing the size of the resulting matrix problem. In the two buckling problems of the present paper, this capability will be also employed in order to show the inadequacy of Koiter's analysis in cases of structures with medium and high imperfection sensitivity.

3. It can be identified theoretically as a systematic minimization of the total potential energy of the (imperfect) structure. This was done by the present author in [8].

4. It provides a systematic capability for higher order approximations and, therefore, can give a clear indication of the accuracy achieved with the first order approximation (compare Equation 9 with 16a and 16b).

5. It is equally well applicable to both simple and compound buckling (two or more coincident classical buckling modes). It is noted in this respect that the application of Koiter's Theory in cases of compound buckling presents significant analytical difficulties.

6. It provides a conceptually simple framework for the treatment of the interaction among different types of structural imperfections, as well as among multiple load parameters. This potentiality of the method has not been tested yet.

7. It can be readily extended conceptually to the cases of non-conservative systems and secondary bifurcations, although it can no longer be justified as a systematic minimization of the total potential energy of the structure in these cases. Neither this potentiality of the method has been tested yet.

8. It is numerically simple and efficient. All that is required is the inversion of the geometric stiffness matrix of the structure at the load under consideration and the computation of the dominant eigenvalue of the resulting square matrix (see Equations 16a and 16b). There exist very efficient algorithms for this computation.

9. The higher the imperfection sensitivity of the structure is, the more accurate are the numerical results of the method. This rather unusual fact, which is opposite from what is valid in Koiter's Theory, is due to the much faster convergence of the perturbation expansion in Equation 5, as the imperfection sensitivity of the structure increases.

3. BUCKLING PROBLEM I: COLUMN ON A LINEAR ELASTIC FOUNDATION

3.1. Problem Statement and Analysis

The first problem is the buckling of a simply supported column on a linear elastic foundation. This is a rather simple problem, which has proved very interesting as the testing ground of the various theories on the buckling of shells. The mathematical analysis of the problem has been presented in all due detail by the same author elsewhere [8, 9].

In non-dimensional form, the significant quantities of the problem are the axial coordinate x, which varies from 0 to π, the foundation stiffness f, the lateral deflection u(x), the deviation ε.w(x) of the axis of the column from the straight line (imperfection) and the applied load φ. The pre-buckling equilibrium state of the column with no imperfections, i.e. of the idealized perfect structure, is u(x) = 0. The displacement u(x) and the imperfection w(x) are discretized around this state in the form of sine Fourier series. Then, it is a simple algebraic task to obtain the total potential energy V of the structure in the following discretized form:

$$V = \varepsilon a_{ii}q_i p_i + b_{ii}q_i q_i + d_{ijkl}q_i q_j q_k q_l + \dots \quad (18)$$

where a_{ii}, b_{ii} and d_{ijkl} are known constant coefficients. It is noted that the displacement modes employed in Equation 18 are orthogonal. For a first order approximation, the problem is reduced to the numerical solution of the eigenvalue problem 16b for its dominant eigenvalue, where the weighting factors τ are computed from Equation 17b.

3.2. Relative Mode Significance

Figure 1 is representative of the numerical results obtained on the relative significance of the various imperfection components. One mode deterministic imperfection patterns in the form sin(ix) are assumed. For each one-mode imperfection pattern, the critical imperfection magnitude $\varepsilon_{cr}(i)$ is computed and the relative mode significance $\theta(i)$ is determined as the ratio $(\varepsilon_{cr}(i_o)/\varepsilon_{cr}(i))$, where i_o is the mode number of the classical buckling mode. Three values of the applied load φ are employed, equal to 90%, 80% and 70% of the bifurcation buckling load φ_{cl} respectively, and the relative significance $\theta(i)$ is plotted versus the mode number i. It is clearly observed that the dominance of the classical buckling mode fades away quickly, as the buckling load decreases (i.e. the imperfection sensitivity of the structure increases), and that Koiter's analysis is fundamentally inadequate for the present problem.

3.3. Monte-Carlo Simulation

The question of the representation of the structural imperfections is a difficult problem, which has not been settled yet. In the present buckling problem, the structural imperfections are represented as a weakly stationary random process with a pre-specified auto-correlation function. Based on arguments presented in [8], the following auto-correlation function is chosen:

$$K(x_1, x_1) = \exp[-\mu|\, x_1 - x_2 \,|] \cos[2\omega(x_1 - x_2)] \quad (19)$$

where μ and ω are suitable spectral parameters [12].

Having specified the auto-correlation function, the Monte-Carlo simulation proceeds in 4 steps. The first step is the determination of the covariance matrix $[h_{ij}]$, i,j=1...M, of the imperfection components, i.e. of the terms:

$$h_{ij} = E\left[(p_i - E[p_i])(p_j - E[p_j])\right] \quad (20)$$

where the symbol E[.] denotes the ensemble average. This determination can be done easily analytically from the expansion of w(x). The second step is the decomposition of the covariance matrix $[h_{ij}]$ into an upper triangular matrix $[g_{mi}]$ and its transpose, i.e.

$$h_{ij} = \sum_{i=1}^{M} g_{mi}\, g_{mj} \quad (21)$$

This decomposition can always be performed numerically, because $[h_{ij}]$ is always positive definite. The third step is the simulation of the imperfection components by means of the relation:

$$p_i = E[p_i] + \sum_{i=1}^{M} (g_{ij}\, z_j) \quad (22)$$

where all the components z_j, j=1...M, are statistically independent and normally distributed N(0,1). It can be easily proved that the covariance matrix of p_i from Equation 22 is given by Equation 21. The fourth step is the computation of ε_{cr}

for any particular realization of the imperfection components p_i, i=1...M. If the resulting ε_{cr} are arranged in ascending order, the design value ε_R of the critical imperfection magnitude ε_{cr} at any desired reliability level R can be immediately determined.

Due to the space limitations of the present paper, it is not possible to include a detailed account of the available numerical results. The presentation will be limited to one figure (Fig. 2), which is representative of effect of the foundation parameter f on ε_R. It is noted in this figure that increasing the nondimensional foundation parameter f does not result in a corresponding increase of the buckling strength of the structure, as anticipated from classical stability theory, due to the simultaneous increase of the imperfection sensitivity of the structure. This numerical result is not intuitively anticipated and can be tested experimentally.

4. BUCKLING PROBLEM II: AXIALLY COMPRESSED THIN CYLINDRICAL SHELL

4.1. Problem Statement and Analysis

The second problem is the buckling of a thin cylindrical shell under axial compression. This is an old and rather notorious problem in Solid Mechanics, because the buckling load is very sensitive to small shape imperfections [13]. For example, shape imperfections less than 1/10 of the shell thickness may result in a 70% reduction of the buckling load.

In non-dimensional form, the significant quantities of the problem are the axial and circumferential coordinates x and y, which vary from 0 to λ_x and from 0 to λ_y respectively, the lateral deflection u(x,y), the deviation ε.w(x,y) from the perfect cylindrical shape (imperfection), the applied compressive load φ, the stress function f(x,y) and the material constant ρ. The pre-buckling equilibrium state of the shell with no imperfections, i.e. of the idealized perfect structure, is u(x,y) = 0 and $f(x,y) = -(\varphi/\rho)y^2$. The displacement u(x,y), the stress function f(x,y) and the imperfection w(x,y) are discretized around this state in the form of complex Fourier series. Employing the well-known von-Karman-Donell equations, it is a simple algebraic task to obtain the non- linear equilibrium equations of the shell in the form:

$$\varepsilon a_{ii}p_i + 2b_{ii}q_i + 3c_{ijk}q_jq_k + ... = 0 \tag{23}$$

where a_{ii}, b_{ii} and c_{ijk} are known constant coefficients. It is again noted that the displacement modes employed in Equations 23 are orthogonal in the region $0 \le x \le \lambda_x$ and $0 \le y \le \lambda_y$. For a first order approximation, the problem is reduced to the numerical solution of the eigenvalue problem in Equation 16a for its dominant eigenvalue, where the weighting factors τ are computed from Equation 17a.

4.2. Relative Mode Significance

Figure 3 is representative of numerical results obtained on the relative significance of the various imperfection components. The results are normalized with respect to the classical buckling mode and the classical buckling load. As in the first buckling problem, it is again clearly observed that the dominance of the classical buckling mode fades away quickly, as the buckling load decreases (i.e. as the imperfection sensitivity of the structure increases), and that Koiter's analysis is fundamentally inadequate also for the present problem.

4.3. Monte-Carlo Simulation

In view of the very large number of significant displacement modes in the present problem and the practically complete lack of measurements of imperfections in actual shells, the imperfection pattern of the cylindrical shell is represented as a broad-band random Gaussian process with an arbitrarily specified power spectral density function. More specifically, the imperfection components p_j, j=1..M, of the shell are simulated by the formula:

$$p_j = \frac{1}{\sqrt{M}} g_j \sum_{m=1}^{M} \left(z_{jm} \cos\omega_{jm} + z'_{jm}\sin\omega_{jm}\right) \tag{24}$$

where z_{jm} and z'_{jm} are statistically independent random variables, normally distributed with mean 0 and variance 1, ω_{jm} are deterministic angles selected so that the Gaussian process is appropriately "filtered", and g_j are deterministic weight factors. Since the random variables are statistically independent, it can be easily verified that:

$$E[p_j] = 0 \quad \text{and} \quad E[p_j^2] = g_j^2 \tag{25}$$

where the symbol E[.] denotes the ensemble average. Therefore, the g_j^2, j=1..M, can be understood as representing the power spectral density of the imperfection pattern.

As in the first buckling problem, the space limitations of the present paper prevent the detailed presentation of the available numerical results. The presentation will be limited to one figure (Fig. 4), which is representative of effect of the assumed form of the power spectral density on the critical imperfection magnitude ε_{cr}. Two extreme cases of power spectral density are considered. Case 1 corresponds to a uniform power spectral density, while case 2 corresponds to an exponentially decreasing one. Both the average value and the 95% confidence intervals of the buckling load φ are plotted versus ε_{cr}. The significant effect on ε_{cr} of the assumed power spectral density is clearly observed.

5. CONCLUSIONS

A new general method has been presented for the determination of the response variability of structures, which experience bifurcation buckling. The method has been applied to two well-known buckling problems. From the presentation of the method and the numerical results of the two buckling problems, it can be concluded that this method:

1. Is conceptually simple.

2. Can account for all the imperfection components of the structure with no restrictions.

3. Provides direct indication on the relative significance of the various components for the buckling strength of the structure.

4. Is equally well applicable to both simple and compound buckling.

5. Is numerically simple and efficient.

Furthermore, the numerical results of the two buckling problems demonstrate both the inadequacy of Koiter's analysis for such problems and the importance of the structural imperfections in the reliability-based design of imperfection sensitive engineering structures. Effects of imperfections, which completely counterbalance intuitively anticipated design improvements, have been observed.

6. REFERENCES

1. Bucher, C. G., and Shinozuka, M. (1986). "Structural Response Variability II," J. Engrg. Mech., ASCE, 114(12), 2035-2054.

2. Gourlay, A. R., and Watson, G. A. (1973). Computational methods for matrix eigenproblems. Wiley and Sons, New York, N.Y.

3. Hutchinson, J. W., and Koiter, W. T. (1970). "Postbuckling theory." Appl. Mech. Rev., 23, 1353-1366.

4. Kardara, A., Bucher, C.G., and Shinozuka, M. (1989). "Structural Response Variability III." J. Engrg. Mech., ASCE, 115(8), 1726-1747.

5. Koiter, W. T. (1963). "Elastic stability and post-buckling behavior." Non-linear Problems, R. E. Langer, ed., Univ. of Wisconsin Press, Madison, Wisc.

6. Palassopoulos, G. V. (1973). "On the buckling of axially compressed thin cylindrical shells." J. Struct. Mech., 2(3), 177-193.

7. Palassopoulos, G. V. (1980). "A probabilistic approach to the buckling of thin cylindrical shells with general random imperfections: Solution of the corresponding deterministic problem." Theory of Shells, W. T. Koiter and G. K. Mikhailov, eds., North-Holland, Amsterdam, The Netherlands, 417-443.

8. Palassopoulos, G. V. (1989). "Optimization of imperfection sensitive structures." J. Engrg. Mech., ASCE, 115(8), 1663-1682.

9. Palassopoulos, G. V. (1989). "Reliability-based design of imperfection sensitive structures." J. Engrg. Mech., ASCE, 117(6), 1220-1240.

10. Palassopoulos, G. V., and Shinozuka, M. (1973). "On the elastic stability of thin shells." J. Struct. Mech., 1(4), 439-449.

11. Shinozuka, M. (1986). "Structural Response Variability." J. Engrg. Mech., ASCE, 113(6), 825-842.

12. Yaglom, A. M. (1962). Stationary random functions. Dover, New York, N.Y.

13. Yamaki, N. (1984). Elastic stability of circular cylindrical shells. North-Holland, New York, N.Y.

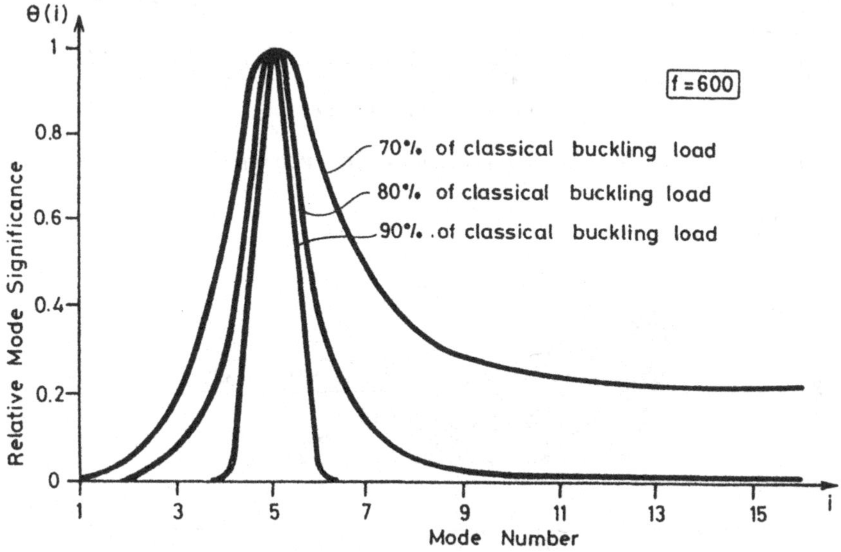

Figure 1. Relative Mode Significance in Problem I

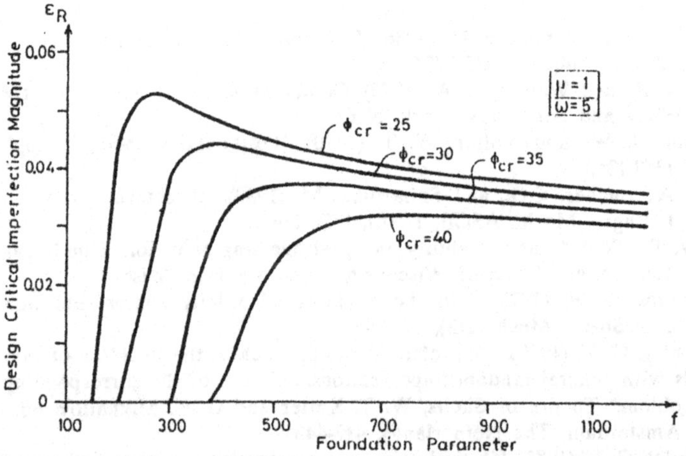

Figure 2. Effect of Foundation Stiffness in Problem I.

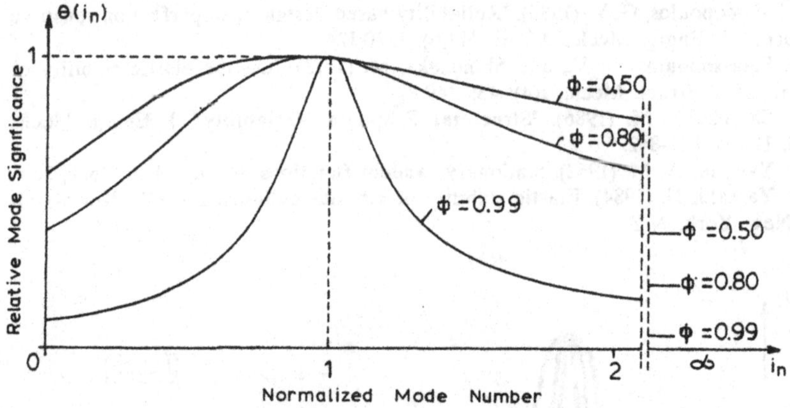

Figure 3. Relative Mode Significance in Problem II.

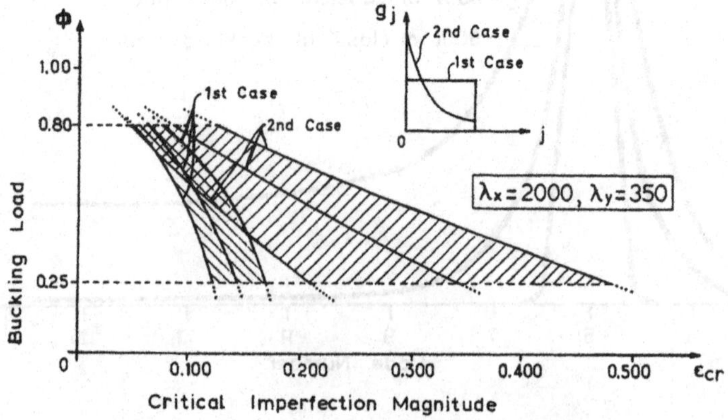

Figure 4. Mean Values and Confidence Intervals in Problem II.

Chaos, Stochasticity and Stability of a Nonlinear Oscillator with Control Part I: Analysis

C.Y. Yang (*), K. Hackl (**), A. H-D. Cheng (*)
(*) Civil Engineering Department, and
(**) Mathematics Department, University of Delaware, Newark, DE 19716, U.S.A.

ABSTRACT

In this first part on analysis of the title problem, we consider a structural model with a nonlinear soft spring of the Duffing type subjected to a deterministic harmonic force. A linear feedback control system with time delay is provided to suppress structural oscillations. The nonlinear control structural problem is formulated in a three-dimensional first order system with perturbation on the harmonic force, structural damping and control force. Based on the exact solution of the unperturbed simple but strongly nonlinear two-dimensional system the global stability of the degenerated system is examined. The global chaotic behavior of the perturbed three dimensional full system is then analyzed using the well known Melnikov function. An analytic criterion for chaos is found. The key engineering problem of control efficiency for linear and nonlinear systems is evaluated through an amplitude reduction ratio for both regular and chaotic oscillations. All digital simulated solutions are presented in part II of our study.

INTRODUCTION

Research in the area of control of civil engineering structures has a history of about twenty years. In 1972 Yao [1] presented the pioneer paper on concept of structural control. Since then increasing research activities have been reported in three International Symposiums on Structural Control, 1979 [2], 1985 [3] and 1990 [4], respectively. Text books on Control of Structures have been published by Leipholz and Abdel-Rohman [5], Meirovitch [6] and Soong [7]. The important area of nonlinear structural control has been investigated only recently by Masri, Bekey and Caughey in 1982 [8], Reinhorn and Manolis in 1987 [9] and Yang, Long and Wong in 1988 [10].

In 1990 we reported our initial work on chaotic dynamics

in nonlinear structural control [11], following our previous studies on an inelastic structural system with hysteresis and degradation [12]. In this and the companion paper [13], we investigate the long time behavior of chaos, stability, stochasticity and control of a simple but fundamental nonlinear elastic structural model subjected to a deterministic harmonic excitation. The important transient behavior is under investigation but not reported here. This paper covers the analytic part. The digital simulated solutions are given by the companion paper [13]. The analytic part here includes, in sequence, the problem formulation, linear solution as a reference to the nonlinear behavior, the simple exact nonlinear solution, theoretical criterion of chaos, stochasticity, stability and chaos, nonlinear control stability and efficiency, a concluding summary of the analysis here and finally an introduction to the digital simulations in the companion paper [13].

FORMULATION

Consider a single degree of freedom structural model with a nonlinear soft spring, a linear damper and a linear feedback control system. The governing equations of the controlled structural system subjected to a deterministic harmonic excitation are given by

$$m\ddot{x} + c\dot{x} + kx - \beta k^3 x^3/W^2 = f(t) - z \tag{1}$$

$$\dot{z} + \alpha\omega_n z = \alpha\gamma k\omega_n x \tag{2}$$

where

m = mass	c = damping coefficient
k = linear spring constant	W = gravity force of model
β = nonlinear force parameter	
$f(t) = f_o \cos\omega t$	ω = forcing frequency
z = control force	α = control speed parameter
ω_n = linear natural frequency	γ = control gain parameter

It is clear from equations (1) and (2) that the nonlinearity of the system arises from the soft spring restoring force $\beta k^3 x^3/W^2$ of the Duffing type. The linear feedback control force z is provided by a mechanical control device external to the structural model and is governed by the structural displacement signal x received from a sensor device. The control device is governed by a control speed parameter α and a control gain γ as shown in equation (2). Nonlinear control systems similar to the above have been studied by Ueda, Doumoto and Nobumoto in 1979 [14], Holmes and Moon in 1983 [15] and Holmes in 1985 [16]. Chaos and stability of the unconrolled Duffing oscillator with soft spring have been studied by Huberman and Crutchfield in 1979 [17], Szemplinska-Stupnicka in 1988 [18] and Nayfeh and Sanchez in 1989 [19].

The normalized equations in a convenient first order form with a deterministic harmonic force are given by

$$\dot{u} = v \tag{3}$$

$$\dot{v} = -u + u^3 + F\cos\Omega\tau - \delta v - \gamma w \tag{4}$$

$$\dot{w} = \alpha u - \alpha w \tag{5}$$

where

$$u = \frac{k\sqrt{\beta}}{W}\, x \qquad v = \frac{k\sqrt{\beta}}{W}\, \dot{x}$$

$$w = \frac{\sqrt{\beta}}{\gamma W}\, z \qquad F = \frac{\sqrt{\beta}}{W}\, f_o$$

$$\Omega = \omega/\omega_n \qquad \delta = \frac{c}{m\omega_n} = 2\xi$$

$$\tau = \omega_n t$$

Equations (3), (4) and (5) represent a three dimensional non-linear non-autonomous system. As a background reference, the degenerated linear system is investigated first.

LINEAR CONTROL BEHAVIOR

Consider the degenerated linear control system

$$\dot{u} = v \tag{6}$$

$$\dot{v} = -u + e^{i\Omega\tau} - \delta v - \gamma w \tag{7}$$

$$\dot{w} = \alpha u - \alpha w \tag{8}$$

where for convenience the complex harmonic excitation $e^{i\Omega\tau}$ is used in equation (7) in place of $F\cos\Omega\tau$ in equation (4). The complex frequency response function $U(\Omega)$ for u can be easily found. The amplitude ρ_c of $U(\Omega)$ is defined and given by

$$U(\Omega) = \rho_c e^{i\theta} \tag{9}$$

$$\rho_c = \frac{\alpha^2+\Omega^2}{\sqrt{[(1-\Omega^2)(\alpha^2+\Omega^2) + \gamma\alpha^2]^2 + [\delta\Omega(\alpha^2+\Omega^2)-\gamma\alpha\Omega]^2}} \tag{10}$$

The subscript c in ρ_c indicates controlled amplitude. For the uncontrolled system with $\gamma = 0$, the amplitude ρ of $U(\Omega)$ at linear resonant frequency ($\Omega = 1.0$) is

$$\rho = \frac{1}{\delta} \tag{11}$$

The amplitude reduction ratio, ρ_C/ρ, which serves as a measure of structural control efficiency is simply

$$\frac{\rho_C}{\rho} = \sqrt{\frac{1+\alpha^2}{\alpha^2+[\alpha(\gamma/\delta)-1]^2}} \tag{12}$$

When the structural control is instantaneous, the control speed $\alpha \to \infty$, and the control efficiency is governed by

$$\frac{\rho_C}{\rho} = \sqrt{\frac{1}{1+(\gamma/\delta)^2}}$$

depending on a single ratio γ/δ. As the control speed, $\alpha \to 0$ so that no control is effective in finite time, $\rho_C/\rho \to 1.0$ with no amplitude reduction. For the general case, the variation of amplitude reduction at resonant frequency is plotted as a function of the two parameters α and γ/δ in Fig. 1 according to equation (12).

The analysis for the simple linear control system shows that it is always stable and uniquely predictable, and the structural control is generally effective in suppressing resonant vibration.

EXACT NONLINEAR SOLUTION

As a first step towards the solution of the full nonlinear system, equations (3), (4) and (5), a perturbation parameter ϵ is introduced in the last three terms of equation (4) to form the full system with perturbation.

$$\dot{u} = v \tag{13}$$

$$\dot{v} = -u + u^3 + \epsilon\ (F\cos\Omega\tau - \delta v - \gamma w) \tag{14}$$

$$\dot{w} = \alpha u - \alpha w \tag{15}$$

For $\epsilon \to 0$, the perturbed full system is reduced to a two dimensional autonomous nonlinear system with u and v.

$$\dot{u} = v \tag{16}$$

$$\dot{v} = -u + u^3 \tag{17}$$

This two dimensional system is Hamiltonian or simply energy conservative since there is no damping involved. The exact solution is easily obtained as

$$\frac{u^2}{2} - \frac{u^4}{4} + \frac{v^2}{2} = E \tag{18}$$

where E represents the total energy of the system, consisting of the potential energy and the kinetic energy. Depending on the initial conditions corresponding to the initial energy level E, equation (18) represents trajectories or orbits in the u-v phase plane. Note that the potential energy V as shown in Fig. 2

$$V = \frac{u^2}{2} - \frac{u^4}{4} \tag{19}$$

indicates one stable equilibrium point at the origin u = 0 and two unstable equilibrium points at u = ±1 in the V versus u plot of this well known one potential well problem. These equilibrium points may be obtained by equating the left hand side of equation (16) and (17) to zero and solving for u and v. They are (u,v) = (0,0) and (±1,0).

The local stability of the three equilibrium points can be determined by solving the eigenvalue problem of the Jacobian matrix

$$[J] = \begin{bmatrix} 0 & 1 \\ (-1+3u^2) & 0 \end{bmatrix} \tag{20}$$

with eigenvalues λ and eigenvectors {u} governed by

$$[J]\{u\} = \lambda\{u\} \tag{21}$$

where the Jacobian matrix is evaluated at the equilibrium point. For the equilibrium point at the origin (0,0) of the phase plane u-v, the eigenvalues are $\lambda_{1,2} = \pm i$, indicating that the origin is a center. For the equilibrium points (±1,0), the eigenvalues are $\lambda_{1,2} = \pm\sqrt{2}$ with corresponding eigenvectors $(1, \pm\sqrt{2})$. Thus these two points (±1,0) are unstable saddles with intersecting and diverging trajectory for $\lambda_1 = +\sqrt{2}$ and eigenvector $(1,+\sqrt{2})$ and converging trajectory for $\lambda_2 = -\sqrt{2}$ and eigenvector $(1,-\sqrt{2})$. These particular trajectories separate stable and unstable regions in the u-v phase plane and are called separatrix as shown in Fig. 3 together with an elliptic orbit around the center at the origin.

The above exact nonlinear solution (18) for the degenerated two dimensional autonomous system equations (16) and (17), corresponding to the initial energy level of E = 1/4, defines the separatrix. It is called a heteroclinic orbit, since the orbit connects two distinct saddle points (±1,0). The orbit moving away from the saddle point at (-1,0) is called an unstable manifold W^u and that moving towards the point is a stable manifold W^s. These two manifolds, W^u and W^s, intersect at the saddle point.

When the perturbation parameter ϵ is introduced, the heteroclinic orbit will break away from the closed orbit of Fig. 3. This breaking away due to perturbation is seen to have generated by the external forcing, the structural damping and the structural control, given by the last three terms in equation (14). To investigate the effect of perturbation on the heteroclinic orbit we must consider the full three dimensional system, equations (13), (14) and (15), using the explicit unperturbed exact solutions,

$$u_o = \pm \tanh \frac{\tau}{\sqrt{2}} \tag{22}$$

$$v_o = \mp \frac{1}{2} \operatorname{sech}^2 \frac{\tau}{\sqrt{2}} \tag{23}$$

$$w_o = \pm \alpha e^{-\alpha\tau}\left[c + \int_0^{\tau} e^{\alpha s}\tanh \frac{s}{\sqrt{2}}\, ds\right] \tag{24}$$

corresponding to the heteroclinic orbits with E = 1/4 in Fig. 3. The integration constant C is so chosen as to ensure that the quantity in square bracket of (24) vanishes as $\tau \to -\infty$. Using the unperturbed heteroclinic orbit as a base, the perturbed and broken stable and unstable manifolds W^s and W^u move around and may intersect with each other an infinite number of times. Such intersections lead to topological deformation of trajectories in the phase space known as the horseshoe map which guarantees the extreme sensitivity to initial conditions, or simply chaos. This is the concept of the method of Melnikov [20]. It was applied by Guckenheimer and Holmes [21] Ariaratnam, Xie and Vrscay [22], for the proof of global existence of chaos. See also Moon [23,24] for more engineering concepts and application of Melnikov's method. This method is used in the following.

THEORETICAL CRITERION OF CHAOS BY MELNIKOV

We apply Melnikov's analytic criterion to identify chaos in the full nonlinear control system with perturbation, equation (13), (14) and (15). The basic idea is to establish the condition for the infinitely many intersections of stable and unstable manifolds W^s and W^u around the base heteroclinic orbit, which guarantees the existence of chaos. This condition is satisfied by setting to zero the distance along the normal to the base heteroclinic orbit between the two surrounding manifolds, and is given by the following equation

$$M(\phi) = \int_{-\infty}^{\infty} v_o[F\cos(\Omega\tau+\phi) - \delta v_o - \gamma w_o]d\tau = 0 \tag{25}$$

where $M(\phi)$ is known as the Melnikov function, ϕ is a phase angle and v_o and w_o are the unperturbed solution (23) and (24). The integrand represents the scalar product of the perturbed vector $(u,v) = (0, F\cos\Omega\tau - \delta v_o - \gamma w_o)$ in equation (13) and (14) and the normal to the unperturbed vector $(u,v) = (v_o, -u_o + u_o^3)$.

Substituting v_o and w_o from equations (23) and (24) into equation (25) and carrying out the first two integrals, we obtain the condition

$$M(\phi) = \frac{2\pi\Omega F}{\sqrt{2}\sinh\frac{\pi\Omega}{\sqrt{2}}}\cos\phi - \frac{2\sqrt{2}}{3}\delta + \frac{1}{\sqrt{2}}\alpha K(\alpha)\gamma = 0 \qquad (26)$$

$$K(\alpha) = 4\int_0^1 s^{\frac{\alpha}{\sqrt{2}}-1}\left[\frac{1}{2}\frac{s+1}{s-1} - \frac{s}{(s-1)^2}\ln s\right]ds \qquad (27)$$

Some mathematics on integration is given in the Appendix. Since $-1 \leq \cos\phi \leq +1$, the criterion for the existence of chaos, equation (26), becomes

$$-F \leq \frac{\sinh\frac{\pi\Omega}{\sqrt{2}}}{\pi\Omega}\left(\frac{2}{3}\delta - \frac{\alpha K(\alpha)}{2}\gamma\right) \leq +F \qquad (28)$$

The above criterion for chaos involves five system parameters: the forcing amplitude F, frequency Ω, structural damping δ, structural control gain γ and the control speed α.

It is important to note that in the unforced case with F = 0, this criterion predicts that no finite chaotic region exists in the parameter space of Ω, δ and α. This simply means that no chaotic vibration exists.

For the simple limiting case of an uncontrolled structure ($\gamma = 0$) with forcing at linear resonant frequency ($\Omega = 1.0$), the criterion reduces to

$$-1.034\,F \leq \delta \leq 1.034\,F \qquad (29)$$

which defines a wedge region of chaos in the δ-F plane bounded by two lines in Fig. 4.

For the primary problem of controlled structures, the criterion for chaos depends on the integral $K(\alpha)$ of equation (27). This integral can not be evaluated analytically because of the irrational number $\sqrt{2}$ in the term $s^{\alpha/\sqrt{2}-1}$, but it is bounded and is evaluated numerically as a function of the control speed α. See Appendix for the limiting analysis and the numerically computed curve of $K(\alpha)$ versus α. For a given α (=1.0) and a structural damping δ (=0.5), the criterion for chaos at linear resonant frequency ($\Omega = 1.0$) is defined by a wedge region in the γ-F plane by two lines in Fig. 5. The criterion of chaos is

$$\left(-0.689\,F - \frac{2\delta}{3}\right) \leq \frac{-\alpha K(\alpha)\gamma}{2} \leq \left(0.689\,F - \frac{2\delta}{3}\right) \qquad (30)$$

CHAOS, STOCHASTICITY AND STABILITY

For the nonlinear control system with perturbation, theoretical analysis based on Melnikov's method in the previous section gives the criterion for the existence of chaotic vibration by equation (28). After the existence of chaotic vibration is determined, it is then necessary to investigate, in the initial value space of u(o), v(o) and w(o), the region of extreme sensitivity of the structural vibration to negligible changes of the initial values, leading to the fractal boundaries in the basin of attraction.

For the structural control problem, the engineering significance rules out the need for any consideration of the initial control force w(o) and thus the problem is reduced to the simplified two dimensional (u(o), v(o)) basin of attraction. The existence of chaotic vibration based on Melnikov's criterion together with a digital simulation of the fractal boundaries in the basin of attraction will enable us to identify the chaotic vibrations of a nonlinear structural control system.

When the vibration of the nonlinear structural control system is found to be chaotic, the vibration wave form (or time series) is erratic and unpredictable (or indefinite), the power spectral density function contains both the periodic spikes and the broad banded spectrum and the Liaponov exponents contain real positive values. See Moon, 1986 [24]. In spite of all such irregularities, chaotic vibrations are found to be orderly in at least two major aspects. First, it is a bounded vibration and the periodically plotted trajectories in the phase space, the so-called Poincaré map, take a geometric form known as the strange attractor. Secondly, the stochastic characterization of the deterministically unpredictable vibration has been found to be predictable. See Yang, Cheng and Roy [12]. Such stochastic predictability has a very important engineering meaning. In fact, engineers have used the stochastic predictability in structural designs for many years since the early fifties in the established discipline of random vibration (See Yang 1986 [25]). The difference between random and chaotic vibration is simply the input excitation. For random vibration, the excitation is assumed to be random whereas for chaotic vibration, it is assumed to be deterministic. Chaotic vibration in a way serves as a middle link between the classical deterministic vibration with definite excitation and the random vibration with uncertain but stochastically definite excitation.

The next inter-related phenomenon is the instability in nonlinear structural control responses. By the engineering concept of instability in vibration, the response grows out of bounds under the deterministic harmonic excitation. In such unstable vibrations there is no need for further investigation. The system simply fails or out of satisfactory operation. However, when the response is stable and is bounded, it may or may not be chaotic depending on whether the vibration response is extremely sensitive to initial conditions or not. This is equivalent to deterministically predictable vibration or unpredictable chaotic vibration.

These engineering considerations lead to the digital simulation in two major groups. The first group concentrates on the stability (bounded vibration) studies of the controlled nonlinear structural system. Regions of stability are found in the initial value plane (u(o), v(o)). The second group concentrates on the fractal geometry in the basin of attraction. Regions of chaotic and regular vibrations are found in the initial value plane, located within the stable regions identified by the first group of simulations.

CONTROL STABILITY AND EFFICIENCY

For linear controlled structures the vibration under steady state harmonic excitation is shown earlier in the paper to be always stable and regular. This means that the amplitude is always finite and bounded and is uniquely predictable. The control efficiency is investigated through a simple amplitude reduction ratio. For nonlinear controlled structures, the problem is much more complicated and interesting. First of all, the controlled structural vibration under the specified simple harmonic forcing must be limited to within clearly defined stability boundaries in the parameter space. Then and only then an investigation of control efficiency is meaningful.

As it is introduced in the linear case, the simple amplitude reduction ratio ρ_c/ρ can again be employed in the nonlinear control problem. When the structural vibration is regular in both the controlled and the uncontrolled cases, the control efficiency $(1 - \rho_c/\rho)$ is a deterministic quantity. When the vibration of either the controlled structure or the uncontrolled structure is chaotic, the control efficiency must be stochastic. All the three parametric studies; 1. basin of stability, 2. basin of attraction and 3. control efficiency are carried out by digital simulations in our companion paper [13].

SUMMARY ON ANALYSIS

In this first part of our study on chaos, stochasticity, stability and control of nonlinear structural vibrations, we formulate the nonlinear problem in a convenient system of three first order equations, with perturbation. For the unperturbed degenerate problem, the system is reduced to a nonlinear two dimensional autonomous system of the classical Duffing type. This system is Hamiltonian with no damping and no structural control. Exact solution trajectories and explicit solutions are used as a basis for the full three dimensional perturbed system with harmonic forcing, structural damping and structural control. This study was limited to long time vibration behavior as a first step to the more important transcient studies to follow.

For the full and perturbed system the existence of chaotic vibration is investigated by the Melnikov's analytic method and regions of stable but chaotic vibrations are found in terms of five system parameters. For controlled structures with stable vibration, whether regular or chaotic, the structural control efficiency is studied through a simple measure of amplitude reduction ratio, which is deterministic in the regular vibration region and stochastic in the chaotic vibration region. Melnikov's criterion of chaotic vibration, basin of stability, basin of attraction and the control efficiency are all investigated through the companion paper [13]. This study shows the occurrence of unstable and/or chaotic vibration on one hand and the control efficiency on the other. Thus the satisfactory application of structural control concept in nonlinear systems depends on the understanding of the complex dynamics and the careful evaluation of the system design.

ACKNOWLEDGMENT

K. Hackl is supported in part by a Feodor-Lynen Fellowship of the A.v. Humboldt Foundation.

APPENDIX A: REFERENCES

1. Yao, James T. P. Concept of Structural Control, Proceedings of ASCE, Structural Div., Vol. 98, No. ST7, July, 1972.

2. Leipholz, H. H. (Ed.) Structural Control, Proceedings of the International IUTAM Symposium on Structural Control, University of Waterloo, Ontario, Canada, June 4-7, 1979, North Holland Publishing Company.

3. Leipholz, H. H. (Ed.) Second International Symposium on Structural Control, University of Waterloo, Ontario, Canada, July 15-17, 1985.

4. Chong, K., Liu, S. C. and Li, J. C. (Ed.) Intelligent Structures, Elsevier Applied Science, 1990.

5. Leipholz, H. H. and Abdel-Rothman Control of Structures, Martinus Nijhoff Publishers, 1986.

6. Meirovitch, Leonard Dynamics and Control, John Wiley and Sons, 1985.

7. Soong, T. T. Active Structural Control, Longman Scientific & Technical Publishing Co., 1990.

8. Masri, S. F., Bekey, G. A. and Caughey, T. K. On-Line Control of Nonlinear Flexible Structures, Journal of Applied Mechanics, Vol. 49, Dec., 1982, pp. 877-884.

9. Reinhorn, A. M. and Manolis, G. D. Active Control of Inelastic Structures, Journal of Engineering Mechanics, ASCE, March, 1987, Vol. 113(3), pp. 315-333.

10. Yang, J. N., Long, F. X. and Wong, D. Optional Control of Nonlinear Structures, Journal of Applied Mechanics, ASME, Vol. 55, pp. 931-938, Dec., 1988.

11. Yang, C. Y. and Cheng, A. H-D Chaotic Dynamics and Structural Control, Proceedings of a Workshop on Structural Control, University Southern California, Oct., 1990.

12. Yang, C. Y., Cheng, A. H-D. and Roy, V. Chaotic and Stochastic Dynamics for a Nonlinear Structure System with Hysteresis and Degradation, Second International Conference on Stochastic Structural Dynamics, Boca Raton, FL, May, 1990.

13. Cheng, A. H-D, Hackl, K. and Yang, C. Y., Chaos, Stochasticity, Instability of a Nonlinear Oscillator with Control, Part II: Simulations, Proceedings of the First International Conference on Computational Stochastic Mechanics, Corfu, Greece, Sept., 1991.

14. Ueda, Y., Doumoto, H. and Nobumoto, K. An Example of Random Oscillations in Three-Order Self Restoring System, Proc. Electric and Electronic Communications Joint Meeting, Kansai District, Japan, Oct., 1978.

15. Holmes, P. J. and Moon, F. C. Strange Attractors and Chaos in Nonlinear Mechanics, Journal Applied Mechanics, ASME, Dec., 1983, Vol. 50/1021-1032.

16. Holmes, P. J. Dynamics of a Nonlinear Oscillator with Feedback Control: Local Analysis, Journal of Dynamics, Measurement and Control, ASME, June, 1985, Vol. 107/159-165.

17. Huberman, B. A. and Crutchfield, Chastic States of Anharmonic Systems in Periodic Fields, Physical Review Letters, Vol. 43, No. 23, 1743-1747, Dec., 1979.

18. Szemplinska-Stupnicka, W. Bifurcations of Harmonic Solution Leading to Chaotic Motion in the Softening Type Duffing's Oscillator, International Journal of Nonlinear Mechanics, Vol. 23, No. 4, 257-277, 1988.

19. Nayfeh, A. H. and Sanchez, N. E. Bifurcations in A Forced Softening Duffing Oscillator, International Journal of Nonlinear Mechanics, Vol. 24, No. 6,483-497, 1989.

20. Melnikov, V. K. On the Stability of the Center for Time Periodic Perturbation, Trans. Moscow Math. Soc. 12, 1-57, 1963.

21. Guckenheimer, J. and Holmes, P. J. Nonlinear Oscillations, Dynamical Systems and Bifurcations of Vector Field, Springer Verlag, New York, 1983.

22. Ariaratnam, S. T., Xie, W. C. and Viscay, E. R. Chaotic Motion under Parametric Excitation, Dynamics and STability of Systems, Vol. 4, No. 2, 111-130, 1989.

23. Moon, F. C. Experiments on Chaotic Motions of a Forced Nonlinear Oscillator: Strange Attractors, Journal Applied Mechanics, ASME, Vol. 47, 1980, pp. 638-644.

24. Moon, F. C. Chaotic Vibrations, Wiley Interscience, 1987.

25. Yang, C. Y. Random Vibration of Structures, Wiley Interscience Pub Co., 1986.

APPENDIX B: MATHEMATICS SUPPLEMENTS

In Eq. (25) the first two integrals can be carried out analytically and the third integral designated as K(α) can be reduced to a simple form given by Eq. (27). Some analysis of the three integrals are given as follows.

$$\frac{F}{\sqrt{2}}\int_{-\infty}^{\infty}\text{sech}^2\frac{\tau}{\sqrt{2}}\cos(\Omega\tau+\phi)d\tau = \frac{F}{\sqrt{2}}\cos\phi\int_{-\infty}^{\infty}\text{sech}^2\frac{\tau}{\sqrt{2}}\cos\Omega\tau\, d\tau$$

$$+\frac{F}{\sqrt{2}}\sin\phi\int_{-\infty}^{\infty}\text{sech}^2\frac{\tau}{\sqrt{2}}\sin\Omega\tau\, d\tau$$

$$= \frac{F}{\sqrt{2}}\cos\phi\cdot 2\pi\Omega\,\text{csch}\frac{\pi\Omega}{\sqrt{2}} \tag{B·1}$$

$$-\frac{\delta}{2}\int_{-\infty}^{\infty}\text{sech}^4\frac{\tau}{\sqrt{2}}d\tau = -8\sqrt{2}\,\delta\int_{0}^{\infty}\frac{u^3}{(1+u^2)^4}du = -\frac{2\sqrt{2}}{3}\delta \tag{B·2}$$

$$K(\alpha) = \int_{-\infty}^{\infty}e^{-\alpha\tau}\,\text{sech}^2\frac{\tau}{\sqrt{2}}\left[c+\int_{0}^{\tau}e^{\alpha s}\tanh\frac{s}{\sqrt{2}}ds\right]d\tau$$

$$c = \int_{-\infty}^{-0} e^{\alpha s} \tanh \frac{s}{\sqrt{2}} \, ds$$

$$K(\alpha) = \int_{-\infty}^{\infty} e^{-\alpha\tau} \operatorname{sech}^2 \frac{\tau}{\sqrt{2}} \int_{-\infty}^{\tau} e^{\alpha s} \tanh \frac{s}{\sqrt{2}} \, ds d\tau$$

$$= 8 \int_0^{\infty}\int_0^{V} \frac{1}{(1+v^2)^2} \cdot \frac{u^2-1}{u^2+1} \cdot \left(\frac{u}{v}\right)^{\sqrt{2}\alpha-1} dudv$$

$$= 2 \int_0^1 x^{\frac{\alpha}{\sqrt{2}} - 1} \left[\frac{x^2+1-2x\ell nx}{(x-1)^2}\right] dx \tag{B·3}$$

Boundedness of the integral $K(\alpha)$ above and in Eq. (27) in the text is shown in the following.

$$\lim_{x\to 1} \frac{x^2-1-2x\ell nx}{(x-1)^2} = \lim_{x\to 1} \frac{2x-2\ell nx-2}{2(x-1)} = 0 \tag{B·4}$$

$$\left|\frac{x^2-1-2x\ell nx}{(x-1)^2}\right| \le C$$

$$\therefore \left|K(\alpha)\right| \le 2C \int_0^1 x^{\frac{\alpha}{\sqrt{2}} - 1} dx = \frac{2\sqrt{2}C}{\alpha} \text{ for } \alpha > 0 \tag{B·5}$$

Since the term $\alpha K(\alpha)$ is used in the Melnikov's function, we determine its limits as $\alpha \to 0$ and $\alpha \to \infty$

$$\lim_{\alpha\to 0} \alpha K(\alpha) = - \overline{2}\sqrt{2} \tag{B·6}$$

$$\lim_{\alpha\to\infty} \alpha K(\alpha) = 0 \tag{B·7}$$

The function is evaluated using numerical quadrature rules and presented as $-\alpha K(\alpha)$ versus α in Figure 6.

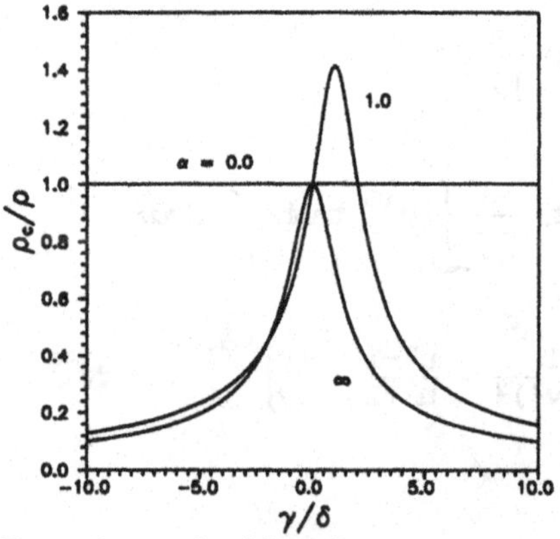

Figure 1: Effect of α and γ/δ on linear structural control.

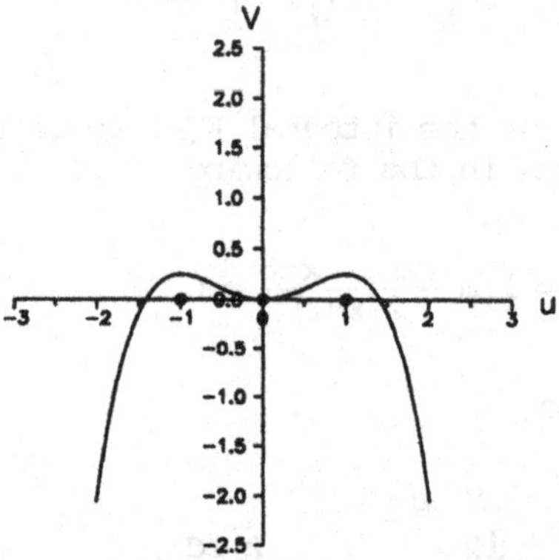

Figure 2: Potential energy V versus u with three equilibrium points.

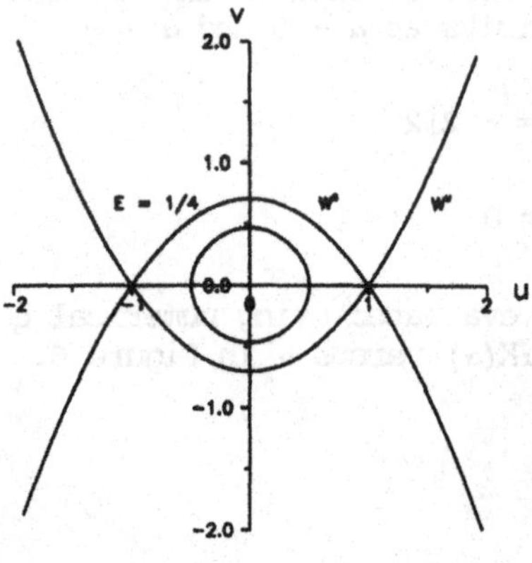

Figure 3: Trajectories of Hamiltonian system.

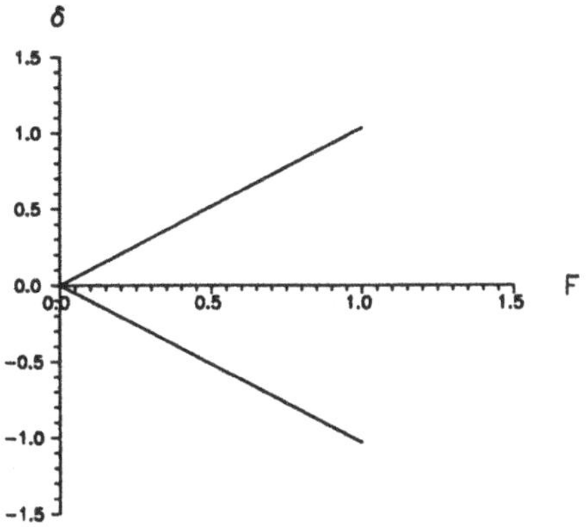

Figure 4: Criterion for chaos, F versus δ.

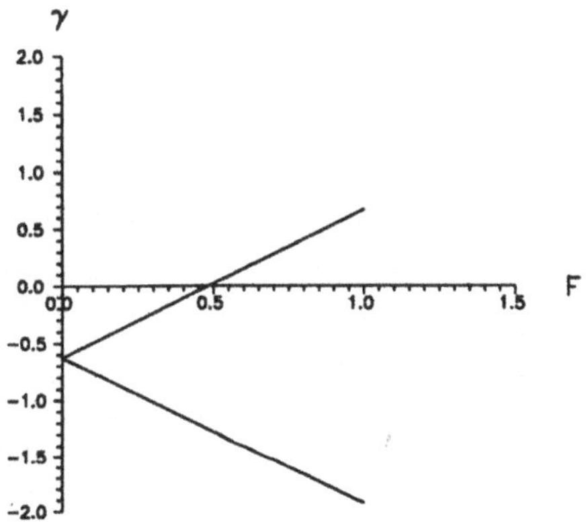

Figure 5: Criterion for chaos, F versus γ.

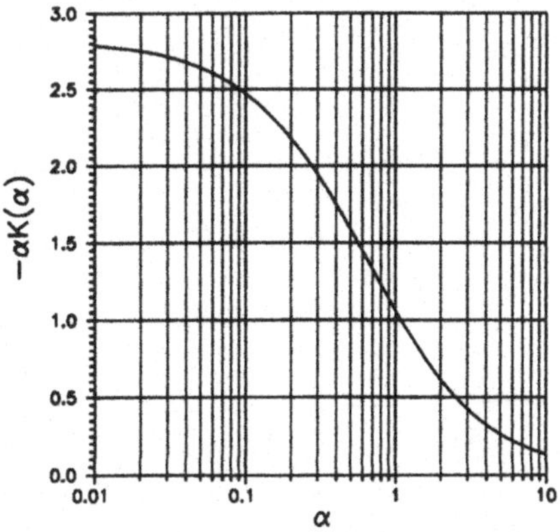

Figure 6: Plot of $-\alpha K(\alpha)$ versus α.

Chaos, Stochasticity, and Stability of a Nonlinear Oscillator with Control Part II: Simulation

A. H-D. Cheng (*), K. Hackl (**), C.Y. Yang (*)
(*) Department of Civil Engineering,and (**) Department of Mathematical Science, University of Delaware, Newark, Delaware 19716, U.S.A.

Abstract

Computer simulations are carried out to examine the cubic nonlinear soft spring structural control problem and to investigate the theoretical asymptotic analysis reported in Part I. Basins of stability, limit cycle solutions, evidence of chaotic vibrations, Poincaré maps, and control efficiency, relating to the forcing and material parameters, are investigated.

Introduction

For a nonlinear, controlled structure with a cubic soft spring of the Duffing type subjected to an external harmonic forcing function, the structural response can be either unstable, stable and regular, or stable but chaotic. An asymptotic global nonlinear analysis based on the Melnikov criterion has been presented in the first part of the paper.[1] In this second part, computer simulated results via a Runge-Kutta numerical algorithm are reported. In particular, we shall demonstrate the trajectories of bounded and unbounded solutions, chaotic and regular behaviors based on the criterion of sensitivity to initial conditions, Poincaré map for chaotic solutions (strange attractor), the basin of attraction for stability, effect of control on the region of stability, the investigation of the Melnikov criterion for the existence of chaos derived in Part I, the structure control efficiency under chaotic and non-chaotic conditions, among others.

Limit Cycles

For easy reference, the governing equations presented in Part I are repeated herein

$$\dot{u} = v \tag{1a}$$

$$\dot{v} = -u + u^3 + F\cos\Omega\tau + \delta v - \gamma w \tag{1b}$$

$$\dot{w} = \alpha(u - w) \tag{1c}$$

where u, v and w are respectively the normalized displacement, velocity and control force, F is related to the forcing magnitude, Ω the forcing frequency, τ the time, δ the damping coefficient, γ the control gain parameter, and α a control speed parameter, all of them in dimensionless forms. See Part I for their complete definitions.

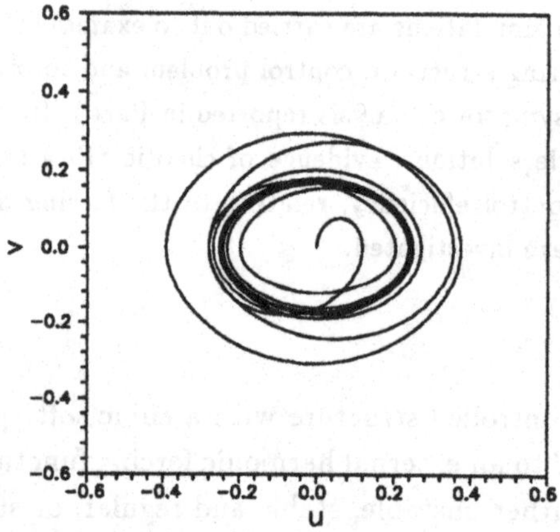

Figure 1: Attraction to limit cycle.

To fix ideas, we select a group of parameters within the realistic range of a controlled structure: $F = 0.1$, $\Omega = 0.7$, $\delta = 0.05$, $\gamma = -0.1$, and $\alpha = 1.0$. If we start the solution from an initial condition for (u, v, w) as $(0, 0, 0)$ and present the solution as a u–v phase plane trajectory (Figure 1), we find after a decaying transient the trajectory being attracted to a *limit cycle* solution with the maximum displacement amplitude about 0.245. On the other hand, if the solution begins with the initial condition $(-0.3, 0.3, 0.0)$, we find a different limit cycle solution with the maximum amplitude about 0.819. Those two limit cycles are plotted without the transient as Figure 2.

To find the *basin of attraction*, namely the collection of initial conditions which are attracted to the same limit cycle solution, we perform a sweep

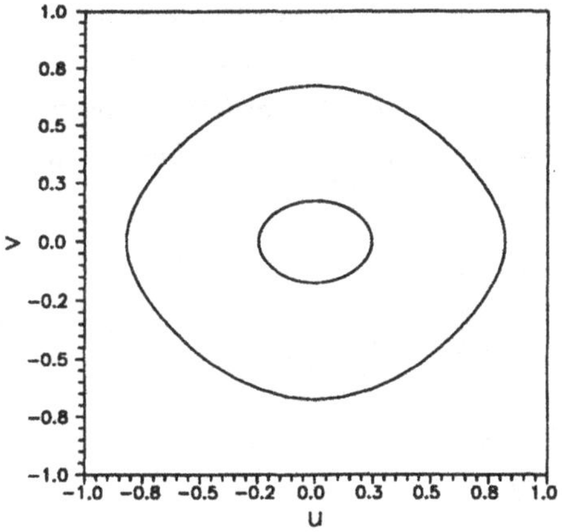

Figure 2: Two coexisting limit cycle solutions.

of the field by testing a large number of initial conditions located within $-2 \leq u \leq 2$, $-2 \leq v \leq 2$, and $w = 0$. We choose not to vary w because of the practical consideration that it is unlikely for a structure to undergo a non-zero initial control force. The result is illustrated as Figure 3 with u as the horizontal and v the vertical axis. In the figure the dark region represents initial conditions whose trajectory is attracted to the outer limit cycle, while the grey area contains those drawn to the inner limit cycle. The un-shaded region corresponds to points which become unbounded with time. The geometry of the map is of *fractal* nature.

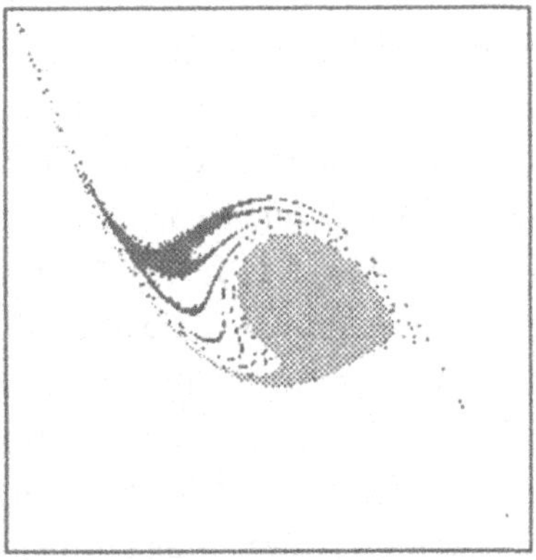

Figure 3: Basin of attraction of initial conditions.

The co-existence of different limit cycle solutions at a given frequency is known as the *jump phenomenon* in classical nonlinear vibration literature.[2] For the current system with the set of parameters given above (except for the frequency, which is scanned) we record all the possible limiting cycle amplitudes and present them as Figure 4. The 'jump' takes place around $\Omega = 0.65$ and 0.76.

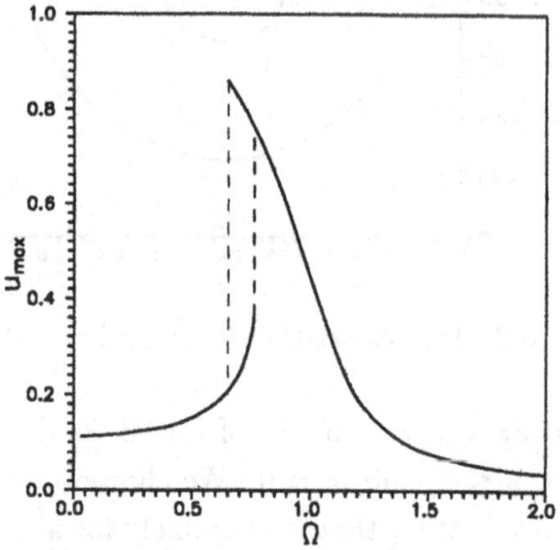

Figure 4: The jump phenomenon.

Chaotic Solution

For the cubic soft spring system without control force, which is equivalent to setting the parameter γ to zero in the current system, chaos has been reported by Huberman and Crutchfield[3] and confirmed by others.[4,5] Chaos in the soft spring system is actually very hard to spot, and when found, its behavior is quite mild as compared to the hard spring system.[6,7] As an illustration, we adopt a case reported in Szemplińska-Stupnicka,[4] which, when converted for the current normalization, is characterized by the following parameters: $\delta = 0.4$, $\Omega = 0.5255$ and $F = 0.23$. By performing a scan for initial conditions for basins of attraction, the result is presented as Figure 5, which again represents a region of $-2 \leq u \leq 2$ and $-2 \leq v \leq 2$. The basin leading to chaotic solutions is marked in black. The lightly shaded region correspond to the limit cycle solution (there is only one limit cycle). The white area is again the unbounded solution. We notice that chaos is limited in extent and is very close to the unbounded domain. The chaotic basin appears to be fractal.

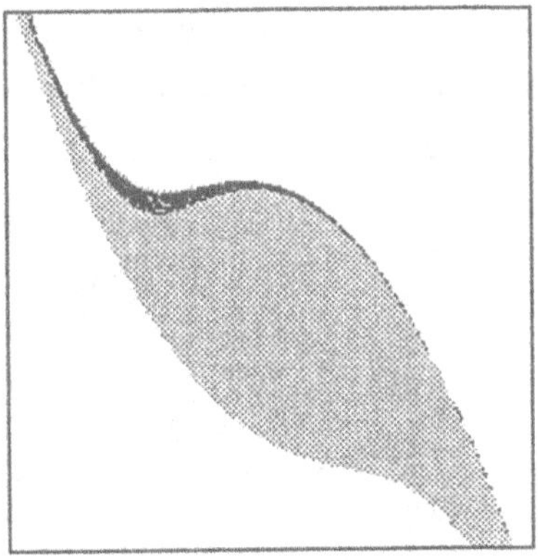

Figure 5: Basin of attraction for chaotic solution.

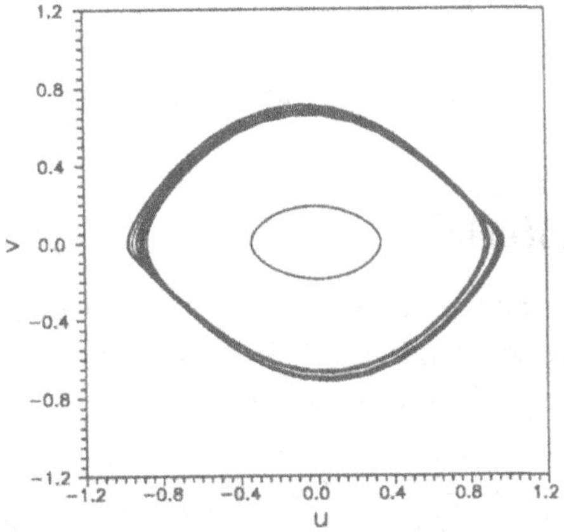

Figure 6: Trajectories of a chaotic and a limit cycle solution.

Figure 6 depicts the trajectory of a chaotic solution with initial condition $(-0.3, 0.7, 0.0)$ and the limit cycle of the non-chaotic solution. We have cut off the initial transient trajectory for the clarity of the picture. We notice that the chaotic solution has a much larger amplitude than the non-chaotic ones. The scattering of the chaotic trajectory seems to be relatively narrow, and not as dramatic as that for the hard spring system reported elsewhere.[7]

Figure 7 shows the *Poincaré map*, namely the locus of the phase plane trajectory sampled stroboscopically at multiples of the forcing period, to reveal the geometry of the *strange attractor*. We have simultaneously presented in the same diagram attractors sampled at eight different phase angles—from 0 to 7 times $T/8$, where T is the forcing period. These attractors appears to be non-fractal under the current graphical scale, but magnification reveals that they are indeed fractal.

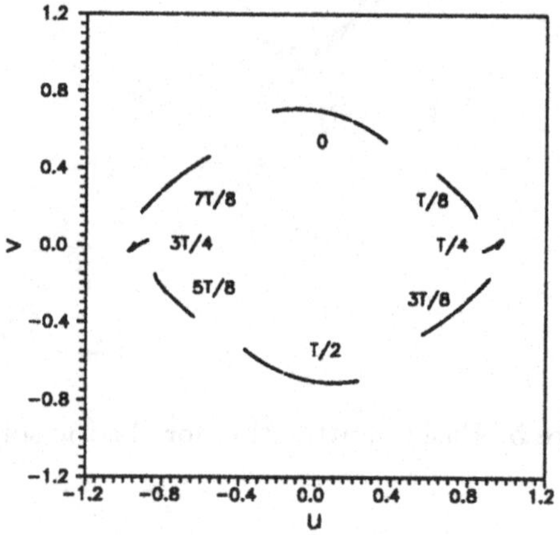

Figure 7: Strange attractors.

Melnikov Criterion

It is of interest to examine the Melnikov criterion derived in the companion paper. According to equation (28) in that paper, which is written in a different form below, chaos can exist if

$$\frac{2}{\alpha K(\alpha)}\left[\frac{2}{3}\delta + \frac{\pi F\Omega}{\sinh(\pi\Omega/\sqrt{2})}\right] \leq \gamma \leq \frac{2}{\alpha K(\alpha)}\left[\frac{2}{3}\delta - \frac{\pi F\Omega}{\sinh(\pi\Omega/\sqrt{2})}\right] \quad (2)$$

With parameters used in the preceding example, $\delta = 0.4$, $\Omega = 0.5255$ and $F = 0.23$, and choosing $\alpha = 1$, equation (2) is evaluated as $-0.997 \leq \gamma \leq -0.00945$. It appears that chaos should not exist for the previous case in which $\gamma = 0$. However, bearing in mind that the Melnikov criterion requires that $\delta, F, \gamma \ll 1$, which is not satisfied in the current set of parameters, we expect some deviation from the criterion. After some testing, we are surprised to find that chaos disappears very rapidly. It seems to exist only within the bounds of $-0.003 < \gamma < 0.001$. This means that with the introduction of the linear structural control system, chaotic oscillation can indeed be eliminated.

To gain further insight, we run several simulations with the same initial condition (0.15, 0.65, 0.0). We fix the parameters as those given above but vary γ for these values: -0.004, -0.002, 0.0009 and 0.0011. The trajectories are presented as Figures 8 to 11. We observe that at $\gamma = -0.004$ the solution quickly converges to a limit cycle (Figure 8). At $\gamma = -0.002$ it crosses into the chaotic region and the trajectory covers a relatively narrow band (Figure 9). The chaotic behavior becomes more violent and the band wider as γ is increased to 0.0009 (Figure 10). Further increase of γ causes another crossing. At $\gamma = 0.0011$ the permanent large amplitude motion disappears. After some transient, the trajectory eventually falls into the attraction of the inner limit cycle hence settles down to a non-chaotic solution. The above phenomenon indeed confirms that there exist some bounds for γ such that the solution is chaotic within, and non-chaotic without. The bounds however do not match those derived in Part I of the paper. The causes for the discrepancy is currently being investigated.

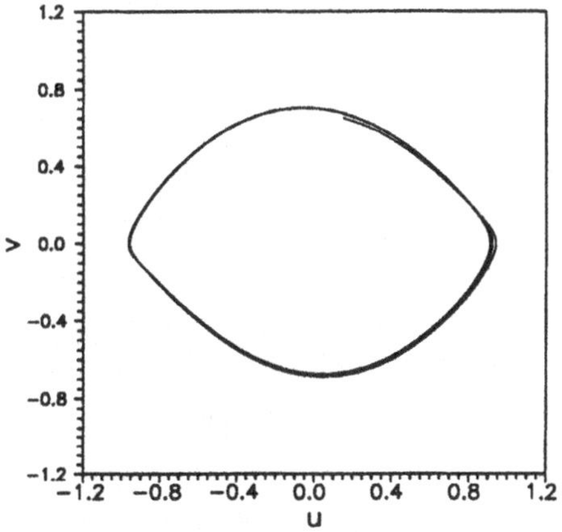

Figure 8: Trajectory for $\gamma = -0.004$—attraction to outer limit cycle.

Basin of Stability

Under a harmonic forcing function, the vibration of the linear controlled structural system is found to be always stable in the sense of boundedness and uniquely predictable for all times in the parametric space. In the nonlinear case, however, vibrations can become unstable and grow out of bounds within specified time. This instability phenomenon is especially critical for the soft than for the hard spring system due to the reduction of spring constant with deformation amplitude.

From practical considerations, it is of interest to seek for the *basin of stability* which defines the theoretical bound for the acceptable vibration

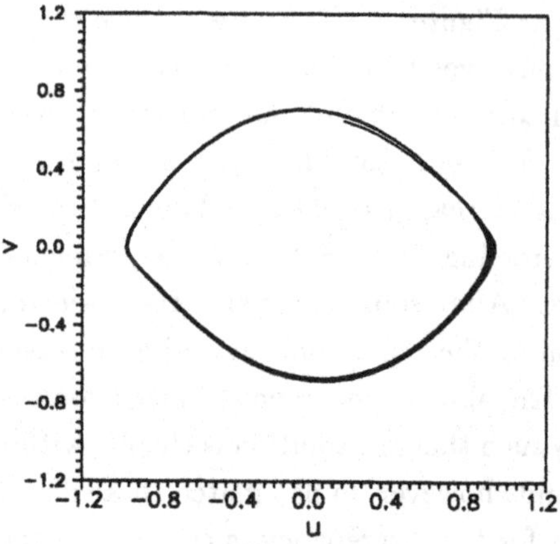

Figure 9: Trajectory for $\gamma = -0.002$—chaotic.

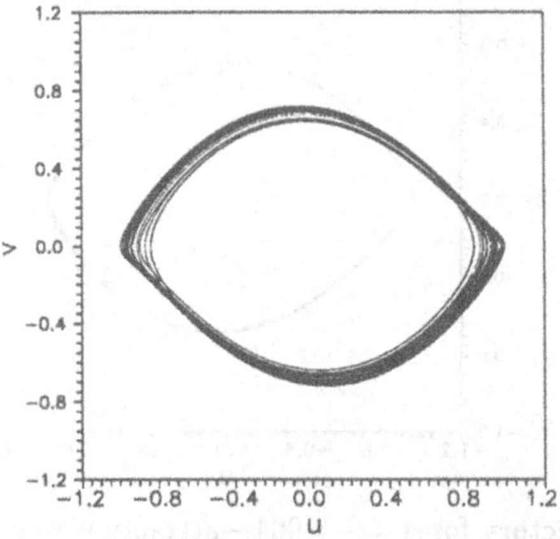

Figure 10: Trajectory for $\gamma = 0.0009$—chaotic.

range of the structure against unbounded, catastrophic failure. To understand the fundamental shape and the evolution of the basin of stability, we begin with the autonomous case, namely a system without forcing, damping and control force ($F = \delta = \gamma = 0$). Figure 12 demonstrate the initial condition basin in the range of $-2 \leq u, v \leq 2$. The white region is the collection of initial conditions that lead to bounded solution (opposite to the previous convention). The dark area corresponds to unbounded solutions. We have shaded the dark region according to their rate of divergence. We observe two lightly shaded 'arms' in which the rate of divergence is slower than their neighboring region.

Next we turn on the forcing and damping, but not the control. For

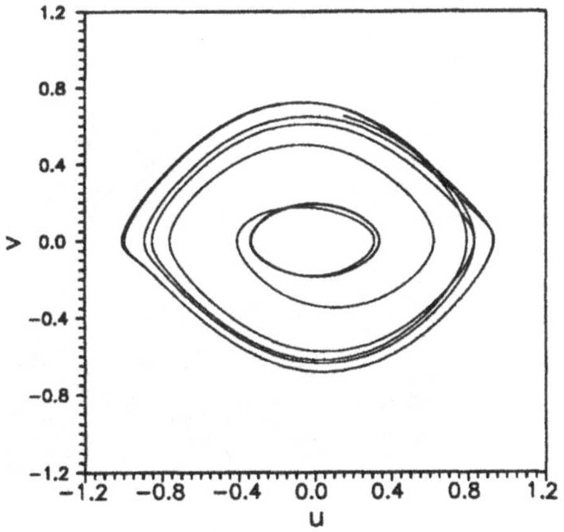

Figure 11: Trajectory for $\gamma = 0.0011$—attraction to inner limit cycle.

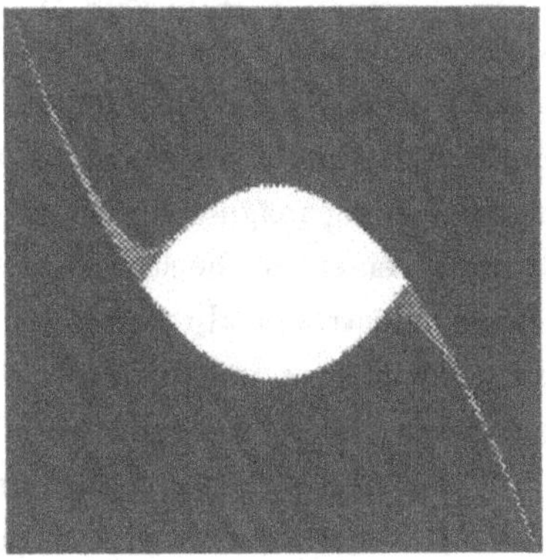

Figure 12: Basin of stability for the autonomous case.

$F = 0.1$, $\Omega = 1.0$, $\delta = 0.05$ and $\gamma = 0.0$, the basin of stability is presented as Figure 13. We observe that the stability region opens up and propagate along the arms. This kind of behavior has been discussed by Guckenheimer and Holms.[8] We also notice that there is only one limit cycle (see Figure 4 with $\Omega = 1.0$) and the basin is connected (except near the fractal region). This is to contrast with the disconnected, two-limit-cycle case shown as the shaded regions in Figure 3.

To examine the effect of structural control, we perform a series of simulation with $F = 0.1$, $\Omega = 1.0$, $\delta = 0.05$, $\alpha = 1$ and various γ values. Some selected results are shown as Figure 14A to F to demonstrate the evolution of the basin. We notice that the size of the stability basin seems

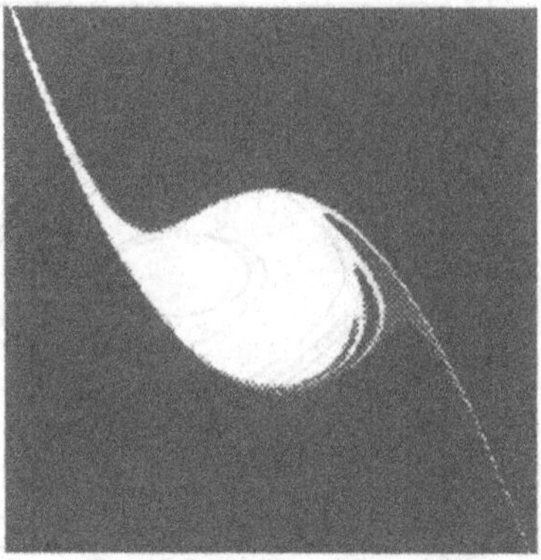

Figure 13: Basin of stability for the uncontrolled case.

to slightly improve as γ decreases from zero to the negative range, and rapidly deteriorate as γ moves into the positive range. When γ is pushed to larger values at both ends, the stability region disappears, i.e. there is no bounded solution in existence. This set of simulation seems to suggest that under the current forcing magnitude the implementation of control force provides little improvement to the structure stability, while too much of it can actually cause catastrophical results.

Next, we increase the forcing magnitude to $F = 0.3$, while keeping $\Omega = 1.0$, $\delta = 0.05$ and $\alpha = 1$. In this case the solution is fairly unstable without control $\gamma = 0$ (Figure 15A). It is in fact unstable at the zero start initial condition $(0,0,0)$. When control with negative gain is introduced, the stability first significantly improves ($\gamma = -0.4$, Figure 15B), and then degrades ($\gamma = -0.8$, Figure 15C).

Finally, we investigate the control speed effect by realizing that as $\alpha \to \infty$ the control reaction is instantaneous, and as the control speed $\alpha \to 0$ the system becomes uncontrolled. With the set of parameters, $F = 0.3$, $\Omega = 1.0$, $\delta = 0.05$ and $\gamma = -0.4$, the stability basins for $\alpha = 0.3$, 0.5, 1.0 and 5.0 are respectively given as Figures 16A, 16B, 15B, and 16C. We observe that the stability is very sensitive to the control speed.

Structure Control Efficiency

Another issue of interest to structure control is the amplitude modulation. By selecting the following parameters, $F = 0.1$, $\delta = 0.05$ and $\Omega = 1.0$, and

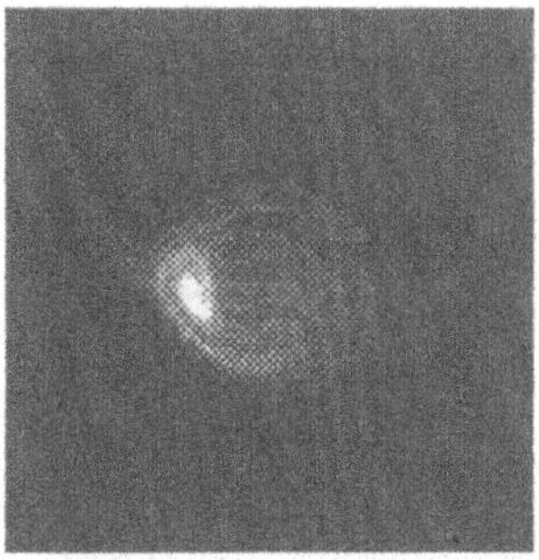

Figure 14A: Basin of stability at weak forcing, $\gamma = 0.2$.

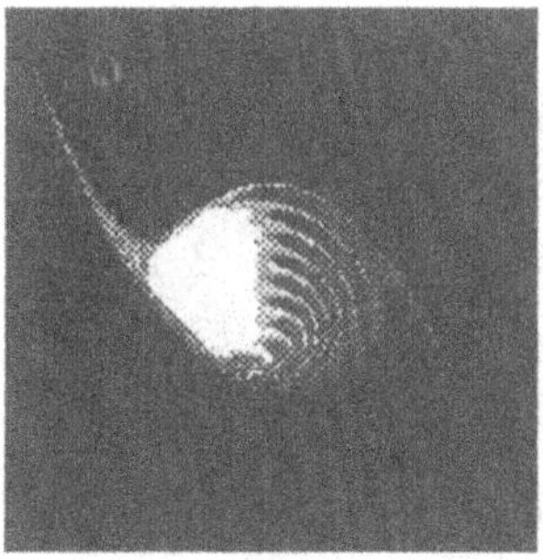

Figure 14B: $\gamma = 0.1$.

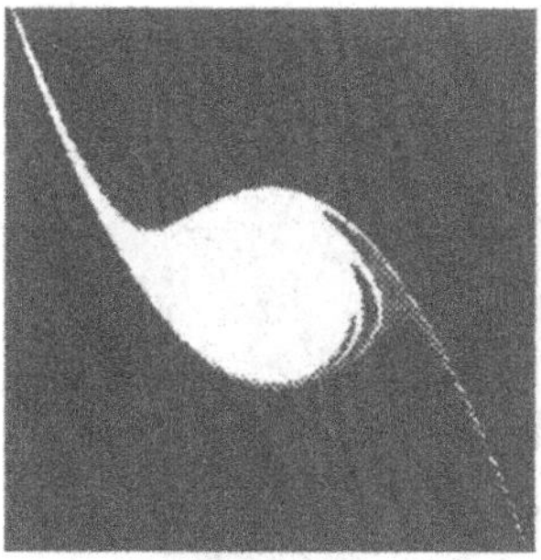

Figure 14C: $\gamma = 0.0$.

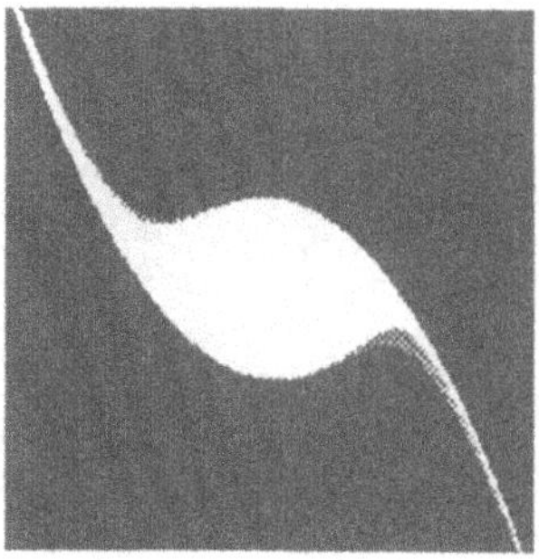

Figure 14D: $\gamma = -0.2$.

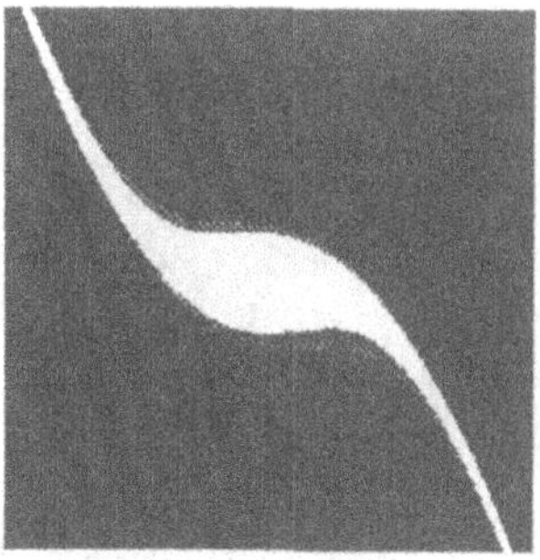

Figure 14E: $\gamma = -0.8$.

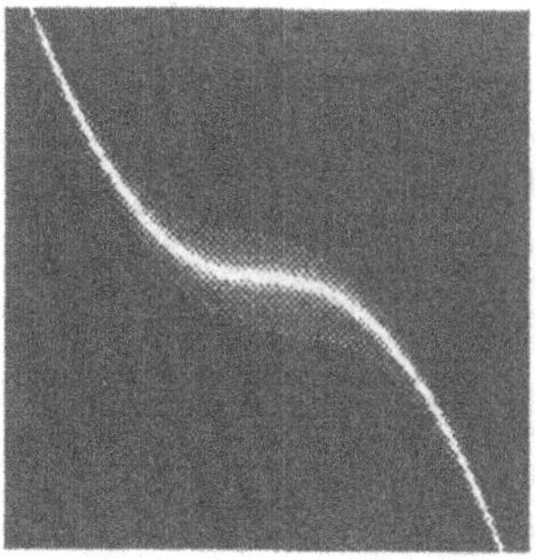

Figure 14F: $\gamma = -1.0$.

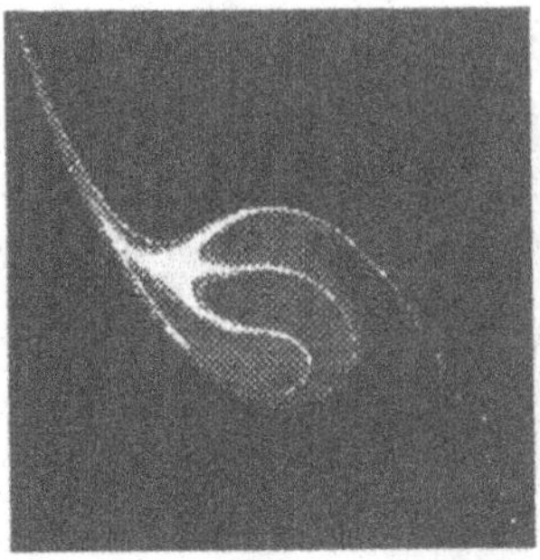

Figure 15A: Basin of stability at strong forcing, $\gamma = 0.0$.

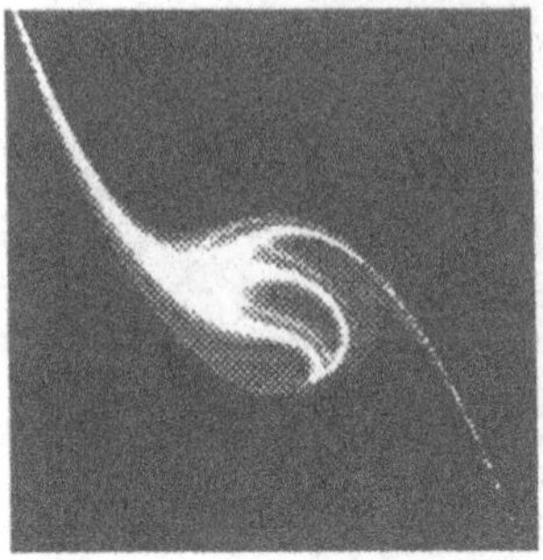

Figure 16A: Basin of stability, $\alpha = 0.3$.

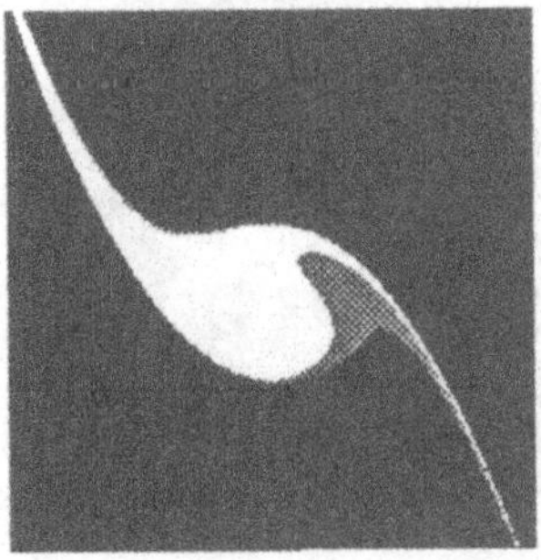

Figure 15B: $\gamma = -0.4$.

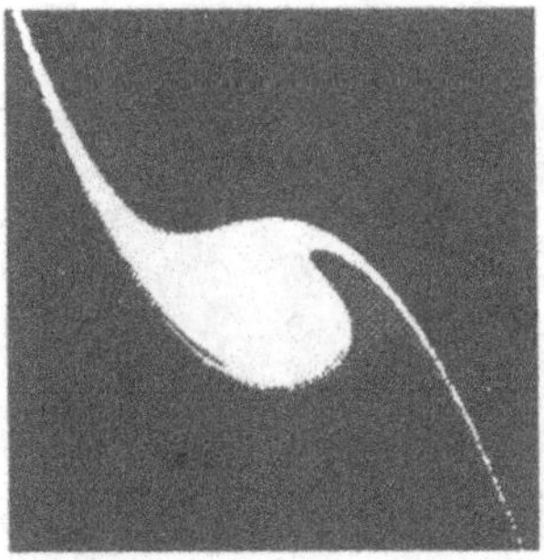

Figure 16B: $\alpha = 0.5$.

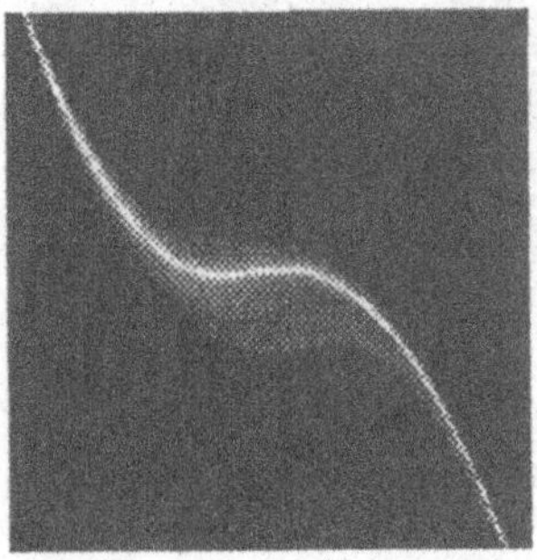

Figure 15C: $\gamma = -0.8$.

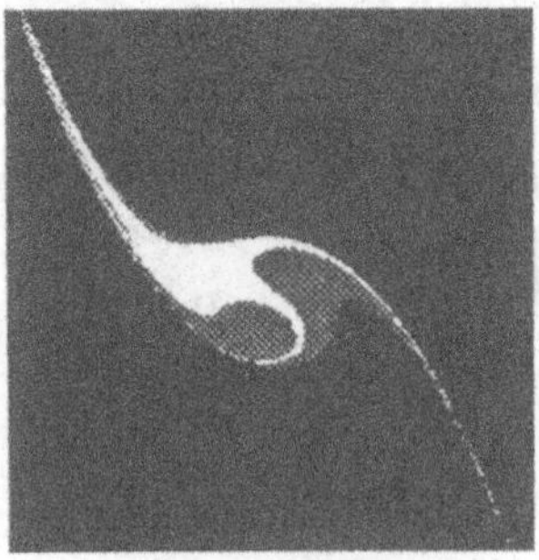

Figure 16C: $\alpha = 5.0$.

the initial condition $(0,0,0)$, the maximum displacement amplitude of the limit cycle can be recorded as a function of α and γ. The result is presented as Figure 17 for a few α values. In particular, $\alpha = 0$ implies no control. The amplitude is independent of γ. The amplitude ρ_c for other α values is hence normalized by that of the no control case ρ. The curves for $\alpha = 1.0$ and 10.0 are truncated because they become unbounded beyond those γ values. Figure 17 generally shows that negative γ tends to damp out the vibration amplitude. However, at low γ values, the amplitude reduction is achieved at the cost of loss of stability region.

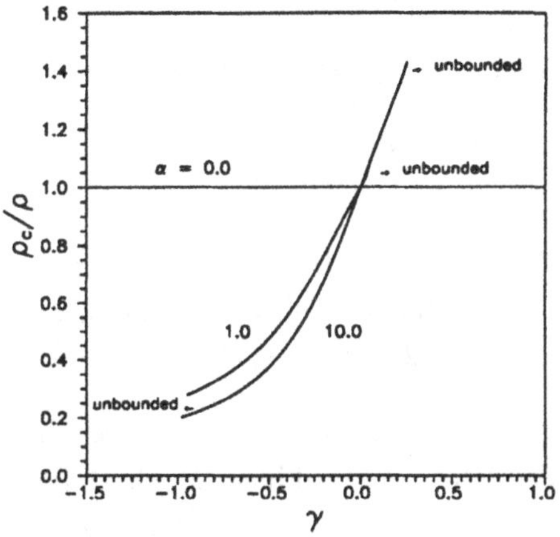

Figure 17: Amplitude modulation due to control at linear resonant frequence.

Summary and Conclusions

Through our preliminary numerical simulation, various aspects of the effectiveness of structure control are examined. Although chaos is present in the cubic soft spring controlled system, its occurrence is rare. Furthermore, the chaotic oscillation can be eliminated by introducing the linear control system. It is rather the basin of stability for bounded solution that becomes the most critical issue. The basin of stability of the oscillation is sensitively dependent on the parameters of the nonlinear structure model and the linear control system. We have examined the sensitivity of the basin to various parameters. We have supporting evidence that when applied properly, the introduction of the control system helps the structure in resisting externally applied force. The impact these connotations have on enginnering design will be further investigated.

Acknowledgment

K. Hackl is supported in part by a Feodor-Lynen Fellowship of the A.v. Humboldt Foundation.

References

[1] Yang, C.Y., Hackl, K. and Cheng, A.H-D., "Chaos, stochasticity, and stability of a nonlinear oscillator with control. Part I: Analysis," in this Proceeding.

[2] Stoker, J.J., *Nonlinear Vibrations,* Interscience, New York, 1950.

[3] Huberman, B.A. and Crutchfield, J.P., "Chaotic states of anharmonic systems in periodic fields," Phys. Rev. Let., **43**, 1743-1747, 1979.

[4] Szemplińska-Stupnicka, W., "Bifurcations of harmonic solution leading to chaotic motion in the softening type Duffing's oscillator," Int. J. Non-Linear Mech., **23**, 257-277, 1988.

[5] Nayfeh, A.H. and Sanchez, N.E., "Bifurcations in a forced softening Duffing oscillator," Int. J. Non-Linear Mech., **24**, 483-497, 1989.

[6] Ueda, Y., "Randomly transitional phenomena in the system governed by Duffing's equation", J. Statistical Phys., **20**, 181-196, 1979.

[7] Yang, C.Y., Cheng , A. H-D. and Roy, R.V., "Chaotic and stochastic dynamics for a nonlinear structural system with hysteresis and degradation," to appear, J. Probabilistic Eng. Mech, 1991.

[8] Guckenheimer, J. and Holmes, P., *Nonlinear oscillations, dynamical systems, and bifurcations of vector fields,* Springer-Verlag, 1983.

SECTION 4: THEORETICAL RANDOM VIBRATIONS

Stochastic Response of Nonlinear Multi-Degrees-of-Freedom Structures Subjected to Combined Random Loads

T. Mochio

Nagasaki R&D Center, Mitsubishi Heavy Industries Ltd., 5-717-1 Fukahori-machi Nagasaki, Japan

ABSTRACT

A method to estimate the stochastic response of hysteretic structures under combined load is investigated. The response statistical moments of structures, with hysteretic characteristics approximated by Wen's mathematical model, are obtained by using the Fokker-Planck equation method with a cumulant truncation technique. This paper deals with the formulation for the multi-degrees-of-freedom system, and demonstrates that resulting variances of the response for a three-degrees-of-freedom structure, as a numerical example, agree well with a Monte Carlo simulation.

INTRODUCTION

In those structures which are subject to random dynamic loads such as earthquake, gust of wind, wave, etc., there are the possibilities of the structures receiving loads with very severe, great input level in future. The structures frequently show elasto-plastic behavior represented by hysteretic characteristics in the occurrence of such loads. In the evaluation of reliability, all loads which are likely to occur in future should be considered for the load level which is essentially uncertainty, and the loads should naturally include loads with the level at which the structures show nonelastic response. Therefore, probabilistic treatment based on the nonlinear random vibration theory is finally required in the dynamic reliability analysis. Further, only any one of the above various dynamic loads does not always occur during the service life of the structure, but also several loads generally occur, and yet some loads may be applied at the same time.

Under these circumstances, this paper describes a method to obtain the response statistical values of hysteretic structures subject to some random dynamic loads simultaneously, for the purpose of the eventual reliability analysis of hysteretic structures taking into consideration the coincidence effect of dynamic loads.

STOCHASTIC RESPONSE OF NONLINEAR STRUCTURE

There are various methods of solving random nonlinear vibration, but here, the Fokker-Planck method is utilized. This is because this method necessarily includes the information of probability density required for fatigue evaluation and others, for estimating the statistical values of dynamic system by stochastic differential equations, and this fact is very convenient when it is attempted to evaluate response values and eventual dynamic reliability under the unified conception.

Equations of motion for nonlinear structure

Using a multi-degree-of-freedom hysteretic model as shown in Fig. 1 as an object of the analysis, a combined load problem where loads by base excitation and general external force effect simultaneously is dealt with. Equations of motion generally for Fig. 1 are as follows:

$$\left.\begin{aligned}
&m_1\ddot{y}_1+C_1\dot{y}_1-C_2(\dot{y}_2-\dot{y}_1)+Q_1(y_1)-Q_2(y_2-y_1)\\
&\quad=-m_1\ddot{x}_g+F_1\\
&\vdots\\
&m_k\ddot{y}_k+C_k(\dot{y}_k-\dot{y}_{k-1})-C_{k+1}(\dot{y}_{k+1}-\dot{y}_k)+Q_k(y_k-y_{k-1})\\
&\quad-Q_{k+1}(y_{k+1}-y_k)=-m_k\ddot{x}_g+F_i\\
&\vdots\\
&m_N\ddot{y}_N+C_N(\dot{y}_N-\dot{y}_{N-1})+Q_N(y_N-y_{N-1})=-m_N\ddot{x}_g+F_M
\end{aligned}\right\}\quad(1)$$

where

m_k : Concentrated mass at point k

C_k : Damping coefficient between points k and (k−1)

$Q_k(y_k-y_{k-1})$: Nonlinear restoring force between points k and (k−1)

F_i : General external force of i-th kind

x_g : Ground displacement

x_k : Absolute displacement at point k

y_k : Relative displacement against ground displacement at point k ($=x_k-x_g$)

As a nonlinear restoring force model required in equation (1), the following ones approximated by Y.K. Wen's model[1] are employed.

$$Q(u_i,\ t) = \alpha k u_i+(1-\alpha)kZ \quad (2)$$

$$\dot{Z} = A\dot{u}_i-\beta|\dot{u}_i||Z|^{n-1}Z-\gamma\dot{u}_i|Z|^n \quad (3)$$

where u_i is relative displacement between mass points.

Substituting equations (2) and (3) for (1), the final equations of motion are obtained as follows:

$$\left.\begin{aligned}
&m_1\ddot{y}_1+C_1\dot{y}_1-C_2(\dot{y}_2-\dot{y}_1)+\alpha_1k_1y_1-\alpha_2k_2(y_2-y_1)\\
&\quad+(1-\alpha_1)k_1Z_1-(1-\alpha_2)k_2Z_2=-m_1\ddot{x}_g+F_1\\
&\qquad\vdots\\
&m_k\ddot{y}_k+C_k(\dot{y}_k-\dot{y}_{k-1})-C_{k+1}(\dot{y}_{k+1}-\dot{y}_k)+\alpha_kk_k(y_k-y_{k-1})\\
&\quad-\alpha_{k+1}k_{k+1}(y_{k+1}-y_k)+(1-\alpha_k)k_kZ_k-(1-\alpha_{k+1})k_{k+1}Z_{k+1}\\
&\quad=-m_k\ddot{x}_g+F_i\\
&\qquad\vdots\\
&m_N\ddot{y}_N+C_N(\dot{y}_N-\dot{y}_{N-1})+\alpha_Nk_N(y_N-y_{N-1})+(1-\alpha_N)k_NZ_N=-m_N\ddot{x}_g+F_M\\
&\dot{Z}_1=A_1\dot{y}_1-\beta_1|\dot{y}_1||Z_1|^{n_1-1}Z_1-\gamma_1\dot{y}_1|Z_1|^{n_1}\\
&\qquad\vdots\\
&\dot{Z}_k=A_k(\dot{y}_k-\dot{y}_{k-1})-\beta_k|\dot{y}_k-\dot{y}_{k-1}||Z_k|^{n_k-1}Z_k-\gamma_k(\dot{y}_k-\dot{y}_{k-1})|Z_k|^{n_k}\\
&\qquad\vdots\\
&\dot{Z}_N=A_N(\dot{y}_N-\dot{y}_{N-1})-\beta_N|\dot{y}_N-\dot{y}_{N-1}||Z_N|^{n_N-1}Z_N-\gamma_N(\dot{y}_N-\dot{y}_{N-1})|Z_N|^{n_N}
\end{aligned}\right\}\qquad(4)$$

Estimation of the response covariance using the Fokker-Planck method

Using the Fokker-Planck method, the response covariance is estimated by equation (4). First, when the equation (4) is expressed by the state variables with (5),

$$\left.\begin{aligned}
&y_1 = X_1,\quad y_2 = X_2,\quad \cdots\cdots\cdots,\ y_N = X_N\\
&\dot{y}_1 = X_{N+1},\quad \dot{y}_2 = X_{N+2},\quad \cdots\cdots\cdots,\ \dot{y}_N = X_{2N}\\
&Z_1 = X_{2N+1},\quad Z_2 = X_{2N+2},\quad \cdots\cdots\cdots,\ Z_N = X_{3N}
\end{aligned}\right\}\qquad(5)$$

the equation (6) can be obtained.

$$\left.\begin{aligned}
&\dot{X}_1 = X_{N+1}\\
&\quad\vdots\\
&\dot{X}_N = X_{2N}\\
&\dot{X}_{N+1} = -2\xi_1\omega_1X_{N+1}+2\rho_2\xi_2\omega_2(X_{N+2}-X_{N+1})-\alpha_1\omega_1{}^2X_1\\
&\qquad+\alpha_2\rho_2\omega_2{}^2(X_2-X_1)-(1-\alpha_1)\omega_1{}^2X_{2N+1}\\
&\qquad+(1-\alpha_2)\rho_2\omega_2{}^2X_{2N+2}-\ddot{x}_g+\frac{F_1}{m_1}\\
&\qquad\vdots
\end{aligned}\right\}\qquad(6)$$

$$
\left.
\begin{aligned}
\dot{X}_{2N} &= -2\xi_N\omega_N(X_{2N}-X_{2N-1})-\alpha_N\omega_N^2(X_N-X_{N-1})\\
&\qquad -(1-\alpha_N)\omega_N^2X_{3N}-\ddot{x}_g+\frac{F_M}{m_N}\\
\dot{X}_{2N+1} &= A_1X_{N+1}-\beta_1|X_{N+1}||X_{2N+1}|^{n_1-1}X_{2N+1}\\
&\qquad -\gamma_1X_{N+1}\,|X_{2N+1}|^{n_1}\\
&\qquad \vdots\\
\dot{X}_{3N} &= A_N(X_{2N}-X_{2N-1})-\beta_N|X_{2N}-X_{2N-1}||X_{3N}|^{n_N-1}X_{3N}\\
&\qquad -\gamma_N(X_{2N}-X_{2N-1})|X_{3N}|^{n_N-1}
\end{aligned}
\right\}
$$

where

$$k_i/m_i = \omega_i^2$$

$$C_i/m_i = 2\xi_i\omega_i$$

$$m_{i+1}/m_i = \rho_{i+1}$$

Then, if $\ddot{x}_g$ and F_i in equation (6) are considered to be random process, the equations (6) become probability differential equations with state random variables of (X_1, ..., X_{3N}).

Assuming that $\ddot{x}_g$ and $F_1 \sim F_M$ in equation (6) are independently stationary random processes, they are presumed as follows:

$$\ddot{x}_g = w_g(t) \tag{7}$$

$$F_i(t)/m_k = b_i w_i(t) \quad (i=1, \cdots\cdots, M) \tag{8}$$

where b_i is a fixed constant to stipulate the load pattern acting on a structure, and $w_g(t)$ is to be Gaussian white noise with the following characteristics.

$$
\left.
\begin{aligned}
&E[w_g(t)] = 0\\
&E[w_g(t)w_g(t+\tau)] = 2\pi S_{og}\delta(\tau)
\end{aligned}
\right\} \tag{9}
$$

where
$E[\cdot]$: Ensemble average
S_{og} : Power spectral intensity of $w_g(t)$
$\delta(\tau)$: Delta function

Also $w_1(t)$ through $w_M(t)$ being independently random process, it is presumed that each power spectral intensity is equal Gaussian white noise as follows:

$$
\left.
\begin{aligned}
&E[w_i(t)] = 0\\
&E[w_i(t)w_i(t+\tau)] = 2\pi S_{of}\delta(\tau) \quad (i=1, \cdots\cdots, M)
\end{aligned}
\right\} \tag{10}
$$

Since $\ddot{x}_g$ and F_i $(i=1,\ldots, M)$ in equation (6) are independent Gaussian white noise, state random variables, $X=(X_1,\ldots, X_{3N})^T$, may be considered to be Markov vector component, and its transient probability density will satisfy the Fokker-Planck equation. If the initial condition with a probability of 1 is known, the relational equations which $P_{\{x\}}(X^T)$, probability density function, for the system should satisfy, will be obtained automatically. Derived hereunder are the concretely relational equations for the case where most standard nonlinear parameter is $n_k=1$.

$$
\begin{aligned}
\frac{\partial P}{\partial t} = & -\sum_{k=1}^{N} X_{N+k}\frac{\partial P}{\partial X_k} - \sum_{k=1}^{N}\Big[-2\xi_k\omega_k P - 2\xi_k\omega_k X_{N+k}\frac{\partial P}{\partial X_{N+k}} \\
& + 2\xi_k\omega_k X_{N+k-1}\frac{\partial P}{\partial X_{N+k}} + 2\rho_{k+1}\xi_{k+1}\omega_{k+1}X_{N+K+1}\frac{\partial P}{\partial X_{N+k}} \\
& - 2\rho_{k+1}\xi_{k+1}\omega_{k+1}P - 2\rho_{k+1}\xi_{k+1}\omega_{k+1}X_{N+k}\frac{\partial P}{\partial X_{N+k}} \\
& - \alpha_k\omega_k{}^2(X_k - X_{k-1})\frac{\partial P}{\partial X_{N+k}} + \alpha_{k+1}\rho_{k+1}\omega_{k+1}{}^2(X_{k+1} - X_k)\frac{\partial P}{\partial X_{N+k}} \\
& - (1-\alpha_k)\omega_k{}^2 X_{2N+k}\frac{\partial P}{\partial X_{N+k}} \\
& - (1-\alpha_{k+1})\rho_{k+1}\omega_{k+1}{}^2 X_{2N+k+1}\frac{\partial P}{\partial X_{N+k}}\Big] \\
& - \sum_{k=1}^{N}\Big[A_k(X_{N+k} - X_{N+k-1})\frac{\partial P}{\partial X_{2N+k}} \\
& - \beta_k|X_{N+k} - X_{N+k-1}|\frac{\partial}{\partial X_{2N+k}}(X_{2N+k}P) \\
& - \gamma_k(X_{N+k} - X_{N+k-1})\frac{\partial}{\partial X_{2N+k}}(|X_{2N+k}|P)\Big] \\
& + \pi\sum_{k=1}^{N}\sum_{\ell=1}^{N}(S_{og} + S_{of}b_k{}^2)\frac{\partial^2 P}{\partial X_{N+k}\partial X_{N+\ell}}
\end{aligned}
\tag{11}
$$

Further, the terms of $|X_{N+k}-X_{N+k-1}|_1$, $|X_{2N+k}|$ in equation (11) are approximated as follows:

$$|X_{N+k}-X_{N+k-1}| \approx a_{1k}(X_{N+k}-X_{N+k-1})^2 \tag{12}$$

$$|X_{2N+k}| \approx a_{2k}X_{2N+k}^2 \tag{13}$$

where a_{1k} and a_{2k} are so determined that the statistical mean square errors by approximation become minimum. Thus a_{1k} and a_{2k} are obtained so as to satisfy the equations (14) and (15),

$$\frac{\partial}{\partial a_{1k}}E[\{|X_{N+k}-X_{N+k-1}|-a_{1k}(X_{N+k}-X_{N+k-1})^2\}^2] = 0 \tag{14}$$

$$\frac{\partial}{\partial a_{2k}}E[(|X_{2N+k}|-a_{2k}X_{2N+k}^2)^2] = 0 \tag{15}$$

and then presuming that X_{N+k} , X_{N+k-1} and X_{2N+k} are approximately in accordance with the normal distribution for a mean value of 0, the followings are obtained.

$$a_{1k} = \frac{4}{3\sqrt{2\pi}} \frac{1}{\sqrt{\sigma^2_{X_{N+k}} + \sigma^2_{X_{N+k-1}} - 2\kappa_{X_{N+k}X_{N+k-1}}}} \tag{16}$$

$$a_{2k} = \frac{4}{3\sqrt{2\pi}} \frac{1}{\sigma_{X_{2N+k}}} \tag{17}$$

where σ^2 and κ show variance and covariance respectively.

Employing equation (11) for which both of equations (12) and (13) are substituted, the equations for up to second moment relating to $X_1 \ldots\ldots, X_{3N}$ (total number of equations$_{3N}H_2 = \frac{3N(3N+1)}{2}$) are derived as follows:

$$\frac{\partial}{\partial t}E[X_k^2] = 2E[X_kX_{N+k}] \tag{18}$$

$$\frac{\partial}{\partial t}E[X_kX_\ell] = E[X_\ell X_{N+k}] + E[X_kX_{N+\ell}] \tag{19}$$

$$\begin{aligned}\frac{\partial}{\partial t}E[X_{N+k}^2] = &-4\xi_k\omega_kE[X_{N+k}^2] + 4\xi_k\omega_kE[X_{N+k-1}X_{N+k}] \\ &+4\rho_{k+1}\xi_{k+1}\omega_{k+1}E[X_{N+k}X_{N+k+1}] - 4\rho_{k+1}\xi_{k+1}\omega_{k+1}E[X_{N+k}^2] \\ &-2\alpha_k\omega_k^2E[X_kX_{N+k}] + 2\alpha_k\omega_k^2E[X_{k-1}X_{N+k}] \\ &+2\alpha_{k+1}\rho_{k+1}\omega_{k+1}^2E[X_{k+1}X_{N+k}] - 2\alpha_{k+1}\rho_{k+1}\omega_{k+1}^2E[X_kX_{N+k}] \\ &-2(1-\alpha_k)\omega_k^2E[X_{N+k}X_{2N+k}] \\ &+2(1-\alpha_{k+1})\rho_{k+1}\omega_{k+1}^2E[X_{N+k}X_{2N+k+1}] + 2\pi(S_{og} + S_{of}b_k^2)\end{aligned} \tag{20}$$

$$\begin{aligned}\frac{\partial}{\partial t}E[X_{N+k}X_{N+\ell}] = &-2\xi_k\omega_kE[X_{N+k}X_{N+\ell}] - 2\xi_\ell\omega_\ell E[X_{N+k}X_{N+\ell}] \\ &+2\xi_k\omega_kE[X_{N+k-1}X_{N+\ell}] + 2\xi_\ell\omega_\ell E[X_{N+\ell-1}X_{N+k}] \\ &+2\rho_{k+1}\xi_{k+1}\omega_{k+1}E[X_{N+k+1}X_{N+\ell}] + 2\rho_{\ell+1}\xi_{\ell+1}\omega_{\ell+1}E[X_{N+k}X_{N+\ell+1}] \\ &-2\rho_{k+1}\xi_{k+1}\omega_{k+1}E[X_{N+k}X_{N+\ell}] - 2\rho_{\ell+1}\xi_{\ell+1}\omega_{\ell+1}E[X_{N+k}X_{N+\ell}] \\ &-\alpha_k\omega_k^2E[X_kX_{N+\ell}] + \alpha_k\omega_k^2E[X_{k-1}X_{N+\ell}] - \alpha_\ell\omega_\ell^2E[X_\ell X_{N+k}] \\ &+\alpha_\ell\omega_\ell^2E[X_{\ell-1}X_{N+k}] + \alpha_{k+1}\rho_{k+1}\omega_{k+1}^2E[X_{k+1}X_{N+\ell}] \\ &-\alpha_{k+1}\rho_{k+1}\omega_{k+1}^2E[X_kX_{N+\ell}] + \alpha_{\ell+1}\rho_{\ell+1}\omega_{\ell+1}^2E[X_{\ell+1}X_{N+k}] \\ &-\alpha_{\ell+1}\rho_{\ell+1}\omega_{\ell+1}^2E[X_\ell X_{N+k}] - (1-\alpha_k)\omega_k^2E[X_{N+\ell}X_{2N+k}] \\ &-(1-\alpha_\ell)\omega_\ell^2E[X_{N+k}X_{2N+\ell}] + (1-\alpha_{k+1})\rho_{k+1}\omega_{k+1}^2E[X_{N+\ell}X_{2N+k+1}] \\ &+(1-\alpha_{\ell+1})\rho_{\ell+1}\omega_{\ell+1}^2E[X_{N+k}X_{2N+\ell+1}] + 2\pi S_{og}\end{aligned} \tag{21}$$

$$\frac{\partial}{\partial t}E[X_{2N+k}^2]=2A_kE[X_{N+k}X_{2N+k}]-2A_kE[X_{N+k-1}X_{2N+k}]$$
$$-2\beta_k a_{1k}E[X_{N+k}^2X_{2N+k}^2]+4\beta_k a_{1k}E[X_{N+k-1}X_{N+k}X_{2N+k}^2]$$
$$-2\beta_k a_{1k}E[X_{N+k-1}^2X_{2N+k}^2]-2\gamma_k a_{2k}E[X_{N+k}X_{2N+k}^3]$$
$$+2\gamma_k a_{2k}E[X_{N+k-1}X_{2N+k}^3] \qquad (22)$$

$$\frac{\partial}{\partial t}E[X_{2N+k}X_{2N+\ell}]=A_kE[X_{N+k}X_{2N+\ell}]-A_kE[X_{N+k-1}X_{2N+\ell}]$$
$$+A_\ell E[X_{N+\ell}X_{2N+k}]-A_\ell E[X_{N+\ell-1}X_{2N+k}]-\beta_k a_{1k}E[X_{N+k}^2X_{2N+k}X_{2N+\ell}]$$
$$+2\beta_k a_{1k}E[X_{N+k-1}X_{N+k}X_{2N+k}X_{2N+\ell}]-\beta_k a_{1k}E[X_{N+k-1}^2X_{2N+k}X_{2N+\ell}]$$
$$-\beta_\ell a_{1\ell}E[X_{N+\ell}^2X_{2N+k}X_{2N+\ell}]+2\beta_\ell a_{1\ell}E[X_{N+\ell-1}X_{N+\ell}X_{2N+k}X_{2N+\ell}]$$
$$-\beta_\ell a_{1\ell}E[X_{N+\ell-1}^2X_{2N+k}X_{2N+\ell}]-\gamma_k a_{2k}E[X_{N+k}X_{2N+k}^2X_{2N+\ell}]$$
$$+\gamma_k a_{2k}E[X_{N+k-1}X_{2N+k}^2X_{2N+\ell}]-\gamma_\ell a_{2\ell}E[X_{N+\ell}X_{2N+k}X_{2N+\ell}^2]$$
$$+\gamma_\ell a_{2\ell}E[X_{N+\ell-1}X_{2N+k}X_{2N+\ell}^2] \qquad (23)$$

$$\frac{\partial}{\partial t}E[X_kX_{N+\ell}]=E[X_{N+k}X_{N+\ell}]-2\xi_\ell\omega_\ell E[X_kX_{N+\ell}]+2\xi_\ell\omega_\ell E[X_kX_{N+\ell-1}]$$
$$+2\rho_{\ell+1}\xi_{\ell+1}\omega_{\ell+1}E[X_kX_{N+\ell+1}]-2\rho_{\ell+1}\xi_{\ell+1}\omega_{\ell+1}E[X_kX_{N+\ell}]$$
$$-\alpha_\ell\omega_\ell^2E[X_kX_\ell]+\alpha_\ell\omega_\ell^2E[X_kX_{\ell-1}]+\alpha_{\ell+1}\rho_{\ell+1}\omega_{\ell+1}^2E[X_kX_{\ell+1}]$$
$$-\alpha_{\ell+1}\rho_{\ell+1}\omega_{\ell+1}^2E[X_kX_\ell]-(1-\alpha_\ell)\omega_\ell^2E[X_kX_{2N+\ell}]$$
$$+(1-\alpha_{\ell+1})\rho_{\ell+1}\omega_{\ell+1}^2E[X_kX_{2N+\ell+1}] \qquad (24)$$

$$\frac{\partial}{\partial t}E[X_{N+k}X_{2N+\ell}]=-2\xi_k\omega_kE[X_{N+k}X_{2N+\ell}]+2\xi_k\omega_kE[X_{N+k-1}X_{2N+\ell}]$$
$$+2\rho_{k+1}\xi_{k+1}\omega_{k+1}E[X_{N+k+1}X_{2N+\ell}]-2\rho_{k+1}\xi_{k+1}\omega_{k+1}E[X_{N+k}X_{2N+\ell}]$$
$$-\alpha_k\omega_k^2E[X_kX_{2N+\ell}]+\alpha_k\omega_k^2E[X_{k-1}X_{2N+\ell}]$$
$$+\alpha_{k+1}\rho_{k+1}\omega_{k+1}^2E[X_{k+1}X_{2N+\ell}]-\alpha_{k+1}\rho_{k+1}\omega_{k+1}^2E[X_kX_{2N+\ell}]$$
$$-(1-\alpha_k)\omega_k^2E[X_{2N+k}X_{2N+\ell}]+(1-\alpha_{k+1})\rho_{k+1}\omega_{k+1}^2E[X_{2N+\ell}X_{2N+k+1}]$$
$$+A_\ell E[X_{N+k}X_{N+\ell}]-A_\ell E[X_{N+k}X_{N+\ell-1}]-\beta_\ell a_{1\ell}E[X_{N+k}X_{N+\ell}^2X_{2N+\ell}]$$
$$+2\beta_\ell a_{1\ell}E[X_{N+k}X_{N+\ell}X_{N+\ell-1}X_{2N+\ell}]$$
$$-\beta_\ell a_{1\ell}E[X_{N+k}X_{N+\ell-1}^2X_{2N+\ell}]-\gamma_\ell a_{2\ell}E[X_{N+k}X_{N+\ell}X_{2N+\ell}^2]$$
$$+\gamma_\ell a_{2\ell}E[X_{N+k}X_{N+\ell-1}X_{2N+\ell}^2] \qquad (25)$$

$$\frac{\partial}{\partial t}E[X_kX_{2N+\ell}]=E[X_{N+k}X_{2N+\ell}]+A_\ell E[X_kX_{N+\ell}]-A_\ell E[X_kX_{N+\ell-1}]$$
$$-\beta_\ell a_{1\ell}E[X_kX^2_{N+\ell}X_{2N+\ell}]+2\beta_\ell a_{1\ell}E[X_kX_{N+\ell}X_{N+\ell-1}X_{2N+\ell}]$$
$$-\beta_\ell a_{2\ell}E[X_kX^2_{N+\ell-1}X_{2N+\ell}]-\gamma_\ell a_{2\ell}E[X_kX_{N+\ell}X^2_{2N+\ell}]$$
$$+\gamma_\ell a_{2\ell}E[X_kX_{N+\ell-1}X^2_{2N+\ell}] \tag{26}$$

The equations (18) through (26) are the moment equations to be obtained, however these do not come to the closed form equations since higher-order moment is included. Accordingly for these higher-order moments as generated, approximations by cumulant truncation technique proposed by Wilcox and Bellman[2] are made.

$$E[X_iX_jX_kX_\ell] = E[X_iX_j]E[X_kX_\ell]+E[X_iX_k]E[X_jX_\ell]+E[X_iX_\ell]E[X_jX_k] \tag{27}$$

Substituting the equations (16) and (17) as well as (27) for the equations (18) through (26), the closed form moment equations are obtained. Further by limiting to obtain stationary response only in this paper, the moment equations for the subject correspond to the cases where the left sides for the equations (18) through (26) are made to be zero, therefore the stationary response moment values can be obtained by numerical calculation.

In reliability evaluation it is necessary to specify limit state, i.e. when viewing from the reliability or safety standpoint, there are two cases:

a) Maximum displacement of a certain mass point from the ground is a problem in terms of maintainability and serviceability, or
b) Relative displacement between two adjoining mass points is a problem from viewpoint of structural strength.

As for a) above, there is no problem because the state variable is taken with y, a displacement from the ground, however as for b), the variances for relative displacement u_k and relative velocity $\dot{u}_k$ are further required. Considering that $u_k=y_k-y_{k-1}$ and the right sides in equations (18) and (19) are equal to zero in case of stationary response, those values can be obtained easily as follows:

$$E[u_k^2] = E[y_k^2]+E[y_{k-1}^2]-2E[y_{k-1}y_k] \tag{28}$$

$$E[\dot{u}_k^2] = E[\dot{y}_k^2]+E[\dot{y}_{k-1}^2]-2E[\dot{y}_{k-1}\dot{y}_k] \tag{29}$$

$$E[u_k\dot{u}_k] = 0 \tag{30}$$

NUMERICAL CALCULATION

Numerical calculation was made for the case of three mass points and three acting general external forces in the model shown in Fig. 1. Constants used in the calculation for the hysteretic model and those indicating the dynamic characteristics of the structure are as follows:

$$A_1 = A_2 = A_3 = 1.0$$

$$\alpha_1 = \alpha_2 = \alpha_3 = 0.05$$

$$\beta_1 = \beta_2 = \beta_3 = 0.75$$

$$\gamma_1 = \gamma_2 = \gamma_3 = 0.47$$

$$\omega_1 = \omega_2 = \omega_3 = 70.0 \text{ (rad/s)}$$

$$\xi_1 = \xi_2 = \xi_3 = 0.01$$

$$\rho_2 = \rho_3 = 1.0$$

Considering that the terms relating to time can be neglected in case of stationary random process, the response moments can be obtained by solving the equations (18) through (26) which are the general nonlinear simultaneous equations. Figs. 2 and 3 show some of the response moments obtained as functions of S_{0g} ($=S_{0f}$). In case of multi-degrees-of-freedom system, limit states defined by response displacements and limit states defined by structural integrity are considered for reliability evaluation. To correspond to the above, response moment relating to the relative displacement y_k of the ground is shown in Fig. 2, while the one relating to the story displacement u_k in Fig. 3.

Numerical experiment results are plotted by employing Monte Carlo method using 200 artificial input waves together with the theoretical analysis in Figs. 2 and 3 for comparison, both agree quite well, and thus it is confirmed that the analytical procedure is appropriate.

CONCLUSIONS

The response statistical moments of a multi-degrees-of-freedom structure, with hysteretic characteristics approximated by Y.K. Wen's mathematical model, are obtained by using the Fokker-Planck equation method. Since the obtained moment equations of nonlinear systems by F-P method are not closed form, the cumulant truncation technique is used to get closed formed equations.

For the response moments obtained by applying the cumulant truncation technique to the three-degrees-of-freedom structure, the Monte Carlo simulation is also performed. The comparison of both results shows good agreement and describes that the analytical theory is valid.

REFERENCES

1. Wen, Y.K. Method for Random Vibration of Hysteretic Systems, Journal of Engineering Mechanics, Vol.102, No.2, pp.249-263, 1976.

2. Wilcox, R.M. and Bellman, R. Truncation and Preservation of Moment Properties for Fokker-Planck Moment Equations, Journal of Mathematical Analysis and Applications, Vol.32, pp.532-542, 1970.

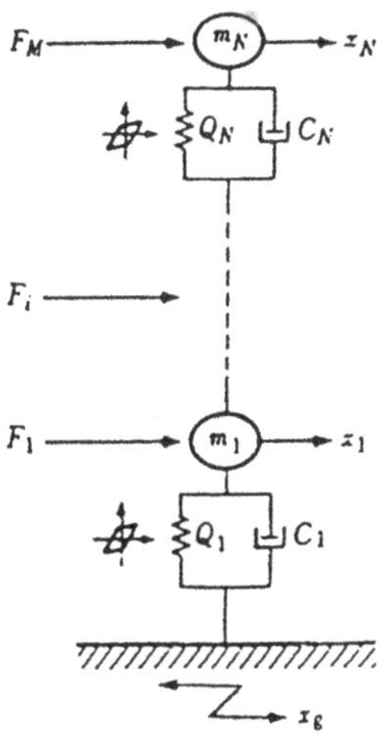

Fig.1 Mathematical model of nonlinear multi-degrees-of-freedom system

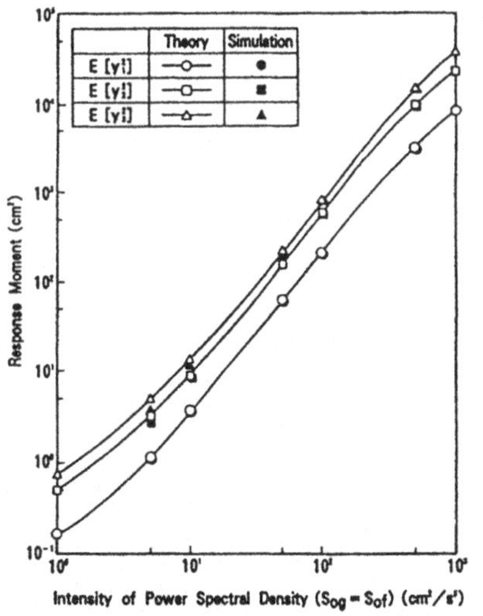

Fig.2 Relationship between power spectral intensity and response moment(in case of relative displacement y_k)

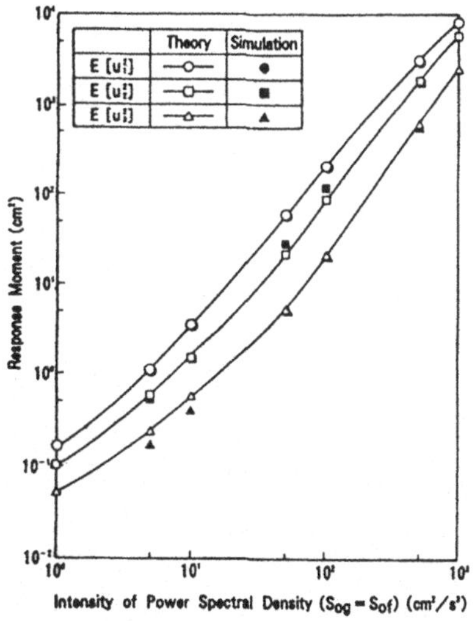

Fig.3 Relationship between power spectral intensity and response moment(in case of story displacement u_k)

Fig. 1 [illegible]

Fig. 2 Relationship between power spectral intensity and response amount (in case of relative displacement x)

Fig. 3 Relationship between power spectral intensity and response amount (in case of story displacement [illegible])

Moment Equations Approach to Nonstationary Responses of a Nonlinear System Subjected to Nonwhite Random Excitation

K. Kimura, M. Sakata

Department of Mechanical Engineering Science, Tokyo Institute of Technology, Ookayama, Meguro-ku, Tokyo 152, Japan

ABSTRACT

A technique for obtaining nonstationary responses of a nonlinear system subjected to nonwhite excitation is investigated. The moment equations method through use of the augmented system consisting of a main system and a shaping filter is discussed, in which the influence of the filter dynamics can be eliminated by substituting the stationary values of the moments with respect to the filter outputs into the inhomogeneous terms. The method of the direct moment equations, which are derived immediately from the main system and whose inhomogeneous terms are determined by using a recurrence algorithm, is reviewed. Characteristics of both moment equations method are compared.

INTRODUCTION

The study of the dynamic behavior of a system under random excitation is of considerable importance in safety design and reliability assessment of structures. Actual random inputs such as earthquakes, blast and sea waves are considered to be nonwhite random excitation with finite correlation time. The authors have developed [1][2] an approximate method, consisting of modification of the equivalent linearization technique and the use of the moment equations of the equivalent linear system, to calculate the nonstationary response of a nonlinear system subjected to nonwhite excitation with an arbitrary correlation function and have applied it to response analysis for various systems [3-5].

It has been pointed out (Bryson and Johansen [6], Åström [7], Gelb [8]) that a stochastic process possessing a rational spectrum may be represented as the output of some white input shaping filter and consequently analysis to nonwhite excitation would be reduced to that of a white excitation problem by combining an additional filter with the original system. However, this state-space augmentation technique is not necessarily effective in the case of nonwhite excitation with a non-rational spectrum since one is not always successful in finding a corresponding shaping filter. Moreover, in employing this technique for calculating the nonstationary response of a system to nonwhite excitation, one has to attend to the transient response stage caused by the dynamics of an additional shaping filter. It has been suggested (Bryson and Johansen [6], Deyst [9]) that suitable selection of the initial condition is necessary to solve the moment equations derived from the augmented system.

In this paper, characteristics of both moment equations methods, i.e., one is the direct approach and the other is the augmented system approach, are investigated.

PROBLEM STATEMENT

Consider a single degree-of-freedom nonlinear system described by the equation of motion

$$\ddot{x} + 2\zeta\dot{x} + x + g(x,\dot{x}) = f(\tau) \tag{1}$$

where overdots denote differentiation with respect to the dimensionless time τ, x is the displacement of the system, ζ is the damping ratio, $g(x,\dot{x})$ is the nonlinear function with respect to displacement and velocity and $f(\tau)$ is the random forcing function. It is assumed that f is expressed as an amplitude modulated nonwhite random excitation (Shinozuka and Henry [10], Bolotin [11], Barnosky and Maurer [12]) in the form

$$f(\tau) = e(\tau)n(\tau) \tag{2}$$

$$E[n(\tau)] = 0 \ , \quad E[n(\tau)n(\tau+v)] = R(v) \tag{3}$$

where $e(\tau)$ is a deterministic envelope function and $n(\tau)$ is a Gaussian stationary nonwhite noise with zero mean and correlation function $R(v)$. The system is assumed to be weakly nonlinear and initially at rest.

SHAPING FILTER AND AUGMENTED SYSTEM

When a Gaussian stationary nonwhite noise $n(\tau)$ possesses a rational spectrum, $n(\tau)$ can be expressed as an output process of a shaping filter with zero mean white noise input $w(\tau)$, as shown Figure 1. According to the representation theorem of the Gaussian process (Åström [7]), it is well known that the stationary response given by

$$n_{st}(\tau) = \int_{-\infty}^{\infty} h_f(\tau-u)w(u)du \tag{4}$$

satisfies the property of equation (3), white the transient response given by

$$n(\tau) = \int_{0}^{\tau} h_f(\tau-u)w(u)du \tag{5}$$

dose not satisfy it since its correlation function cannot be expressed in terms of the time interval. Here $h_f(\tau)$ in equations (4) and (5) is the impulse response function of the filter. Input-output relationship of the main system under consideration is represented as a diagram in Figure 2. The state vector of the system is generally non-Markovian. Input-output relationship of the augmented system, which is constructed by replacing $n(\tau)$ in equation (2) with the shaping filter, is represented as a diagram in Figure 3. The augmented system consists of the main system, the envelop function $e(\tau)$ and the shaping filter. The input of the augmented system is now a white noise process so that the state vector of the system turns out to be Markovian. It is noted that the influence of the filter dynamics has to be eliminated when using the augmented system approach for the nonstationary response analysis.

RESPONSE ANALYSIS (USE OF AUGMENTED SYSTEM)

Figure 1 Shaping filter

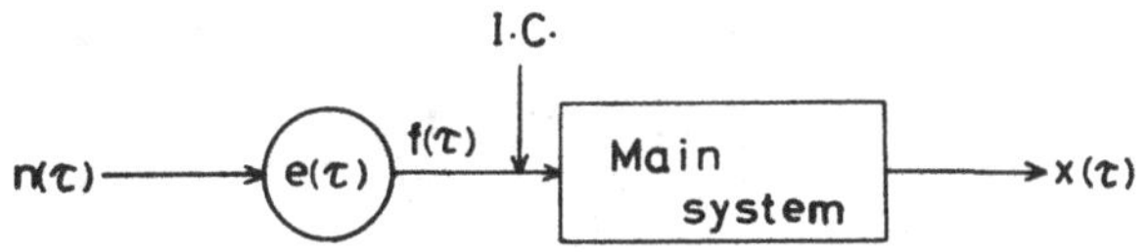

Figure 2 Input-output relationship of the main system

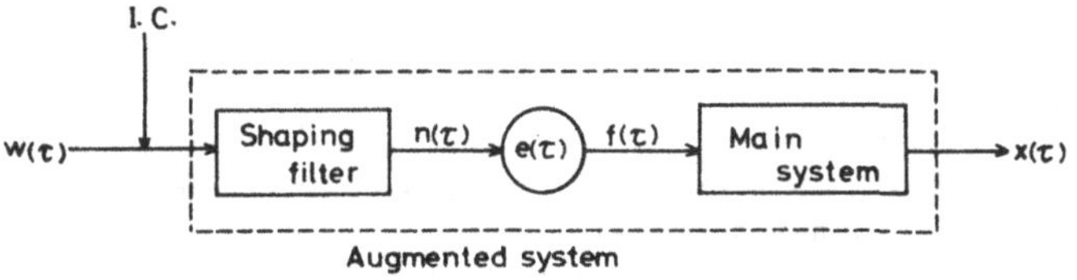

Figure 3 Input-output relationship of the augmented system

Equation of motion for the augmented system

The nonwhite noise $n(\tau)$ shown in Figure 1 can be expressed as the stationary output process of the following linear filter

$$n(\tau) = \boldsymbol{\alpha}^T \mathbf{z}\,, \quad \dot{\mathbf{z}} = \mathbf{B}\mathbf{z} + \boldsymbol{\beta} w(\tau) \tag{6}$$

where, $\mathbf{z}$ is an m-dimensional state vector, $\boldsymbol{\alpha}$ and $\boldsymbol{\beta}$ are m-dimensional coefficient vectors, $\mathbf{B}$ is an $(m \times m)$ coefficient matrix, ^{T} denotes the operation of transpose and $w(\tau)$ is a one-dimensional white noise process with zero mean and intensity I_w, i.e., $E[w(\tau)] = 0$ and $E[w(\tau)w(\tau + v)] = I_w \delta(v)$. By using the state vector $\mathbf{y} = (y_1, y_2)^T = (x, \dot{x})^T$, the augmented system which combines equation (1) with equation (6) is expressed as the stochastic differential equation (Arnold [13])

$$d\begin{pmatrix} \mathbf{y} \\ \mathbf{z} \end{pmatrix} = \left\{ \begin{pmatrix} \mathbf{A} & \mathbf{C} \\ \mathbf{0} & \mathbf{B} \end{pmatrix} \begin{pmatrix} \mathbf{y} \\ \mathbf{z} \end{pmatrix} - \begin{pmatrix} \mathbf{g} \\ \mathbf{0} \end{pmatrix} \right\} d\tau + \begin{pmatrix} \mathbf{0} \\ \boldsymbol{\beta} \end{pmatrix} dW_\tau \tag{7}$$

where $\mathbf{A}$ and $\mathbf{C}$ are (2×2) and $(2 \times m)$ coefficient matrices, respectively. $\mathbf{g}$ is a 2-dimensional nonlinear function

$$\mathbf{A} = \begin{pmatrix} 0 & 1 \\ -1 & -2\zeta \end{pmatrix}, \quad \mathbf{C} = \mathbf{C}(\tau) = \begin{pmatrix} 0 \cdots 0 \\ e(\tau)\boldsymbol{\alpha}^T \end{pmatrix}, \quad \mathbf{g} = \begin{pmatrix} 0 \\ g \end{pmatrix} \tag{8}$$

and W_τ is a one-dimensional Wiener process with properties, $E[dW_\tau] = 0$ and $E[(dW_\tau)^2] = I_w d\tau$. $\mathbf{C}$ is a function of time because of inclusion of the envelope function, $e(\tau)$.

Moment equations of the augmented system

The moment equations of the system described by equation (7) are expressed as

$$\frac{d}{d\tau}\begin{pmatrix} \mathbf{P}_{yy} & \mathbf{P}_{yz} \\ \mathbf{P}_{zy} & \mathbf{P}_{zz} \end{pmatrix} =$$
$$\begin{pmatrix} \mathbf{AP}_{yy}+\mathbf{P}_{yy}\mathbf{A}^T+\mathbf{CP}_{zy}+\mathbf{P}_{yz}\mathbf{C}^T & \mathbf{AP}_{yz}+\mathbf{P}_{yz}\mathbf{B}^T+\mathbf{CP}_{zz} \\ \mathbf{P}_{zy}\mathbf{A}^T+\mathbf{BP}_{zy}+\mathbf{P}_{zz}\mathbf{C}^T & \mathbf{BP}_{zz}+\mathbf{P}_{zz}\mathbf{B}^T \end{pmatrix}$$
$$-\begin{pmatrix} \mathbf{P}_{gy}+\mathbf{P}_{yg} & \mathbf{P}_{gz} \\ \mathbf{P}_{zg} & \mathbf{0} \end{pmatrix}+\begin{pmatrix} \mathbf{0} & \mathbf{0} \\ \mathbf{0} & \boldsymbol{\beta}\boldsymbol{\beta}^T \end{pmatrix} I_w \tag{9}$$

where

$$\mathbf{P}_{yy} = E[\mathbf{y}\mathbf{y}^T]\ ,\quad \mathbf{P}_{yz} = E[\mathbf{y}\mathbf{z}^T] = \mathbf{P}_{zy}^T\ ,\quad \mathbf{P}_{zz} = E[\mathbf{z}\mathbf{z}^T] \tag{10}$$

$$\mathbf{P}_{gy} = E[\mathbf{g}\mathbf{y}^T] = \mathbf{P}_{yg}^T\ ,\quad \mathbf{P}_{gz} = E[\mathbf{g}\mathbf{z}^T] = \mathbf{P}_{zg}^T \tag{11}$$

$\mathbf{P}_{yy}$, $\mathbf{P}_{yz}$ and $\mathbf{P}_{zz}$ are respectively (2×2),$(2\times m)$ and $(m\times m)$ second moment matrices. $\mathbf{P}_{gy}$ and $\mathbf{P}_{gz}$ are respectively (2×2) and $(2\times m)$ higher order moment matrices since g is a nonlinear function.

It is noted that the independent variables of the moment equations (9) can be classified into three types of moments. These are (1) $\mathbf{P}_{yy}$, moments with respect to the output process $\mathbf{y}$ of the main system, (2) $\mathbf{P}_{yz}$, coupled moments between the output processes $\mathbf{y}$ and $\mathbf{z}$ of the main system and filter, respectively and (3) $\mathbf{P}_{zz}$, moments with respect to the output process $\mathbf{z}$ of the filter. Thus equations (9) can be written alternatively in the form

$$\begin{aligned} \dot{\mathbf{P}}_{yy} &= v_1[\mathbf{P}_{yy}(\tau), \mathbf{P}_{yz}(\tau), \mathbf{P}_{gy}(\tau)] \\ \dot{\mathbf{P}}_{yz} &= v_2[\mathbf{P}_{yz}(\tau), \mathbf{P}_{gz}(\tau), \mathbf{P}_{zz}(\tau)] \\ \dot{\mathbf{P}}_{zz} &= v_3[\mathbf{P}_{zz}(\tau)] \end{aligned} \tag{12}$$

where, v_1, v_2 and v_3 are functions of their arguments.

Under the Gaussian assumption for the response the higher order moments can be expressed in terms of the second moments of the response (Lin [14]) and consequently $\mathbf{P}_{gy}$ and $\mathbf{P}_{gz}$ are written by

$$\mathbf{P}_{gy}(\tau) = w_1[\mathbf{P}_{yy}(\tau)]\ ,\quad \mathbf{P}_{gz}(\tau) = w_2[\mathbf{P}_{yy}(\tau), \mathbf{P}_{yz}(\tau)] \tag{13}$$

where, w_1 and w_2 are generally nonlinear function of their arguments. Substituting equation (13) into equation (12), the solvable moment equations are obtained as follows

$$\begin{aligned} \dot{\mathbf{P}}_{yy} &= v_1\{\mathbf{P}_{yy}(\tau), \mathbf{P}_{yz}(\tau), w_1[\mathbf{P}_{yy}(\tau)]\} \\ \dot{\mathbf{P}}_{yz} &= v_2\{\mathbf{P}_{yz}(\tau), w_2[\mathbf{P}_{yy}(\tau), \mathbf{P}_{yz}(\tau)], \mathbf{P}_{zz}(\tau)\} \\ \dot{\mathbf{P}}_{zz} &= v_3[\mathbf{P}_{zz}(\tau)] \end{aligned} \tag{14}$$

It is found that the equation with respect to $\mathbf{P}_{zz}$ is itself of the closed form, while $\mathbf{P}_{yy}$ is dependent upon $\mathbf{P}_{yz}$ and also $\mathbf{P}_{yz}$ upon $\mathbf{P}_{zz}$.

Elimination of filter dynamics

Substituting the stationary value, $\mathbf{P}_{zz}(\infty) = \mathbf{P}_{zz}(\tau)|_{\tau\to\infty}$, of the filter output into equation (14) leads to the following form of moment equations

$$\begin{aligned} \dot{\mathbf{P}}_{yy} &= v_1\{\mathbf{P}_{yy}(\tau),\, \mathbf{P}_{yz}(\tau),\, w_1[\mathbf{P}_{yy}(\tau)]\} \\ \dot{\mathbf{P}}_{yz} &= v_2\{\mathbf{P}_{yz}(\tau),\, w_2[\mathbf{P}_{yy}(\tau),\, \mathbf{P}_{yz}(\tau)]\ ;\ \mathbf{P}_{zz}(\infty)\} \end{aligned} \tag{15}$$

where $\mathbf{P}_{zz}(\infty)$ can be readily determined in advance by solving the algebraic equation, $v_3 = 0$, which is derived by equaling the corresponding time derivative in equation (14) to zero. In equation (15) the main system moments $\mathbf{P}_{yy}(\tau)$ and the coupled moments $\mathbf{P}_{yz}(\tau)$ are the independent variables and $\mathbf{P}_{zz}(\infty)$ is the known inhomogeneous term so that the dimension of the equations is reduced in comparison with that of equations (9). It is emphasize that the influence of the transient response of the filter output is eliminated in equation (15)

RESPONSE ANALYSIS (DIRECT APPROACH)

An approximate analytical method [1][2] is reviewed, in which the solvable moment equations are derived directly from the equation of motion (1) for the main system shown Figure 2 and the nonstationary responses are calculated by solving these equations.

Equivalent linear system

In order to obtain an approximate solution of equation (1), the following equivalent linearized equation is considered

$$\ddot{x} + 2\zeta_e\dot{x} + k_e x = f(\tau) \tag{16}$$

where ζ_e and k_e are equivalent linear coefficients and are in general expressed as functions of the response (Lin [14], Atalik and Utku [15]). Under the assumption of Gaussian responses the equivalent linear coefficients can be expressed in terms of the second moments of the responses. By using the state vector $\mathbf{y}$, equation (16) is written as

$$\dot{\mathbf{y}} = \mathbf{A}_e\mathbf{y} + \mathbf{f} \tag{17}$$

where $\mathbf{A}_e$ is a (2×2) equivalent linear coefficient matrix and $\mathbf{f}$ is a 2-dimensional forcing function vector.

$$\mathbf{A}_e = \begin{pmatrix} 0 & 1 \\ -k_e & -2\zeta_e \end{pmatrix} ,\quad \mathbf{f} = \begin{pmatrix} 0 \\ f \end{pmatrix} \tag{18}$$

Moment equations

The moment equations derived from the linearized equation (17) are expressed as

$$\dot{\mathbf{P}}_{yy} = \mathbf{A}_e\mathbf{P}_{yy} + \mathbf{P}_{yy}\mathbf{A}_e^T + \mathbf{P}_{yf} + \mathbf{P}_{yf}^T \tag{19}$$

where the inhomogeneous term, $\mathbf{P}_{yf}$, is a (2×2) cross-correlation matrix between the response and the forcing function.

$$\mathbf{P}_{yf} = E[\mathbf{y}\mathbf{f}^T] = \mathbf{P}_{fy}^T \tag{20}$$

The relationship between the independent variables of the moment equations is expressed as

$$\dot{\mathbf{P}}_{yy} = v_0[\mathbf{P}_{yy}(\tau), \mathbf{P}_{yf}(\tau)] \tag{21}$$

where v_0 is the function of its arguments.

Recurrence algorithm for determination of $E[\mathbf{y}\mathbf{f}^T]$

One takes the equispaced time sequence τ_0, τ_1, τ_2, $\cdots$ and assumes that the equivalent coefficients take on the constant values during each interval

$$\zeta_{ei} = \zeta_e(\tau) = \zeta_e(\tau_{i-1}) \;, \quad k_{ei} = k_e(\tau) = k_e(\tau_{i-1}) \;, \quad \tau \in [\tau_{i-1}, \tau_i) \tag{22}$$

$E[\mathbf{y}\mathbf{f}^T]$ at an arbitrary reference time $\tau \in [\tau_{n-1}, \tau_n)$ can be determined by the following recurrence algorithm.

$$E[\mathbf{y}(\tau_i)\mathbf{f}^T(\tau)] = \int_{\tau_{i-1}}^{\tau_i} \mathbf{H}_{ei}(\tau_i - v)\, E[\mathbf{f}(v)\mathbf{f}^T(\tau)]dv + \mathbf{H}_{ei}(\Delta\tau)\, E[\mathbf{y}(\tau_{i-1})\mathbf{f}^T(\tau)]$$

$$(i = 1, 2, \cdots, n-1) \tag{23}$$

$$E[\mathbf{y}(\tau)\mathbf{f}^T(\tau)] = \int_{\tau_{n-1}}^{\tau} \mathbf{H}_{en}(\tau - v)\, E[\mathbf{f}(v)\mathbf{f}^T(\tau)]dv + \mathbf{H}_{en}(\tau - \tau_{n-1})\, E[\mathbf{y}(\tau_{n-1})\mathbf{f}^T(\tau)] \tag{24}$$

where $\mathbf{H}_{ei}(\tau)$ is the impulse response function matrix during the time interval $[\tau_{i-1}, \tau_i)$. It is noted that since the present recurrence algorithm includes the reference time τ as the fixed parameter, for each step of the Runge-kutta-Gill

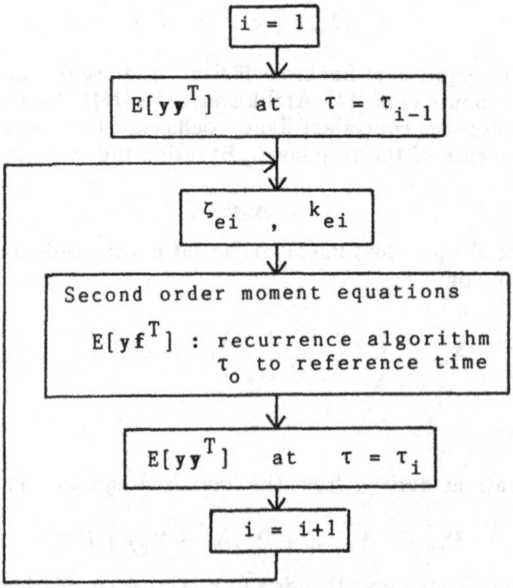

Figure 4 Flow chart for response computation (direct moment equations method)

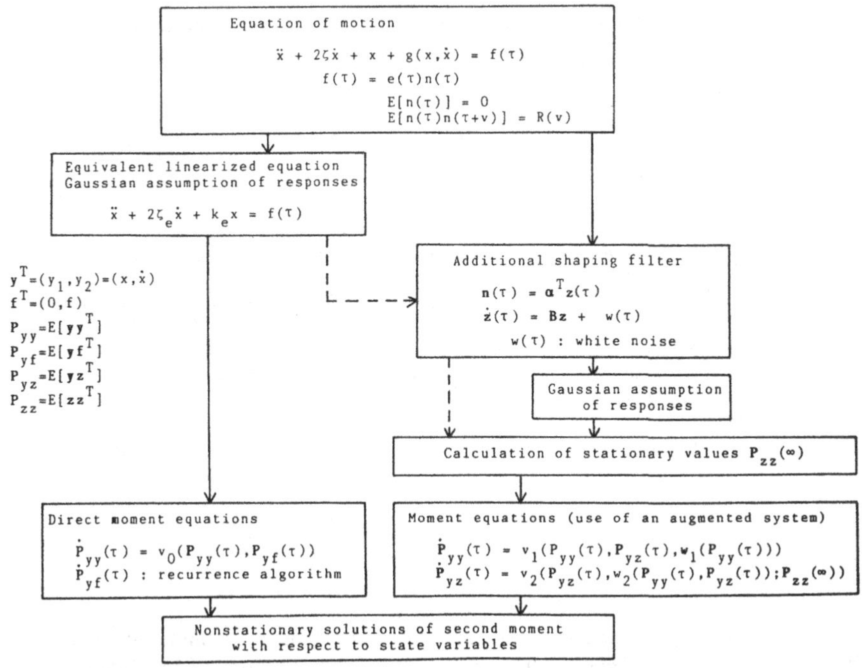

Figure 5 Flow chart for response analysis by moment equations methods : direct approach and augmented system approach

scheme the recurrence procedure has to be repeated over the whole interval from the initial time to the reference time. This procedure seems to be the key to the response analysis for the nonwhite excitation. The flow chart for response computation by using the direct moment equations method is given in Figure 4.

COMPARISON OF DIRECT AND AUGMENTED SYSTEM APPROACHES

Flow chart for response analysis by moment equations methods, i.e., one is the direct approach and the other is the augmented system approach, is shown in Figure 5, where the associated equations are also provided. In the augmented system approach, the same moment equations are also obtained by the equivalent linear system with an additional shaping filter as shown by the broken line in Figure 5. Advantages and disadvantages of both moment equations methods are given in Table 1.

The direct moment equations method can be applied to the general case in which the nonwhite excitation possesses an arbitrary correlation function or spectrum, although a recurrence algorithm is needed in determining the inhomogeneous terms. The dimension of the equations is small, for instance, it is ${}_{2n}H_2$ (H : repeated combination) for a n degree-of-freedom system.

Once the augmented system is constructed, the equation of motion is described by the stochastic differential equation and the Markov property of the

Table 1 Characteristics of moment equations methods : direct approach and augmented system approach

	DIRECT MOMENT EQUATIONS METHOD	MOMENT EQUATIONS METHOD (USE OF AN AUGMENTED SYSTEM)
ADVANTAGES	Can be applied to noise with an arbitrary autocorrelation function	Equation of motion of an augmented system can be described by a stochastic differential equation
	Dimension of moment equations is small	Impulse response functions are not needed
		Can be applied to systems with parametric excitation
DISADVANTAGES	Requires a recurrence algolritm for determining cross-correlations of responses and forcing functions	Is limited to use only for noise with a rational power spectrum
	Impulse response functions are needed	Requires an additional shaping filter corresponding to given noise
		Increases dimension of moment equations

state vector is used for response analysis. The dimension of the moment equations becomes large, for instance, it is ${}_{2n}H_2 + 2nm$ for a n degree-of-freedom system with an additional m-dimensional shaping filter. It is noted that the recurrence algorithm in the direct approach might be equivalent to the substitution procedure of the stationary values of the moments with respect to the filter outputs in the augmented system approach. The method using an augmented system can be effectively employed for the nonwhite noise with a rational spectrum. Thus, the direct moment equations method has to be employed in the case of nonwhite excitation with a non-rational spectrum.

ILLUSTRATIVE EXAMPLES

A Duffing system is considered as a typical example of nonlinear system

$$g(x, \dot{x}) = g(y_1, y_2) = \epsilon y_1^3 \tag{25}$$

where ϵ is a nonlinearity parameter. The random forcing function in the form of equation (2) has been assumed in the preceding analysis. A unit step function or an exponential function is used as an envelope function : these are, respectively

$$e(\tau) = U(\tau) \tag{26}$$

$$e(\tau) = (\mathrm{e}^{-a\tau} - \mathrm{e}^{-b\tau}) / \max(\mathrm{e}^{-a\tau} - \mathrm{e}^{-b\tau}) \tag{27}$$

Equation (27) has been employed to simulate the nonstationariness of the earthquake excitation (Shinozuka and Sato [16]). Noise with an exponentially decaying harmonic correlation function in the form of

$$R(v) = R_0 \mathrm{e}^{-\alpha|v|} \cos pv \tag{28}$$

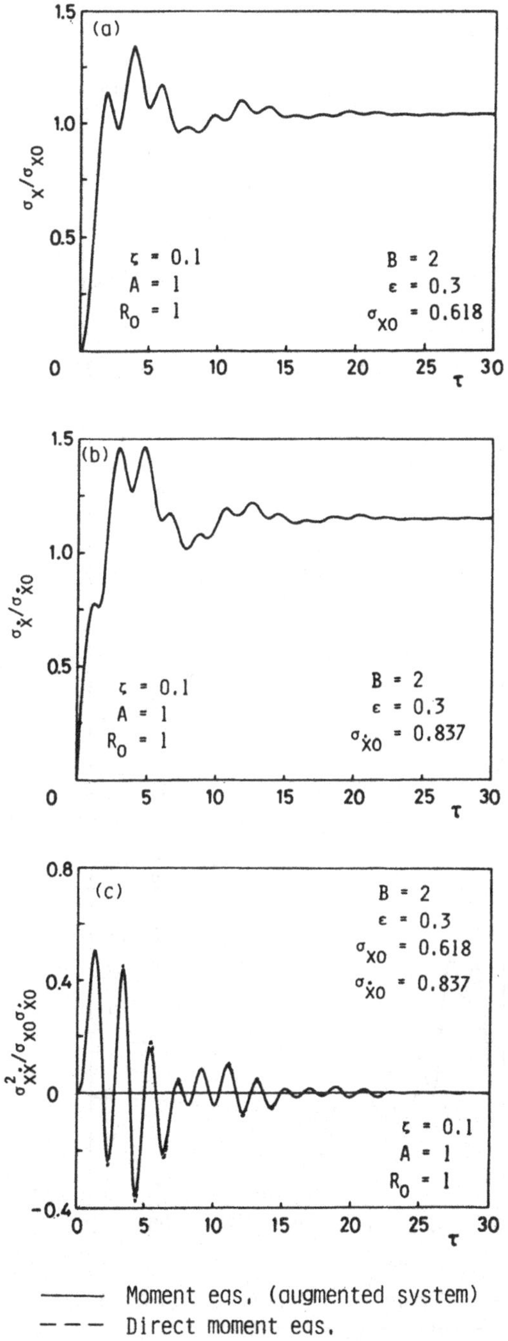

Figure 6 Transient rms and cross-correlation responses for displacement and velocity (unit step envelope function)

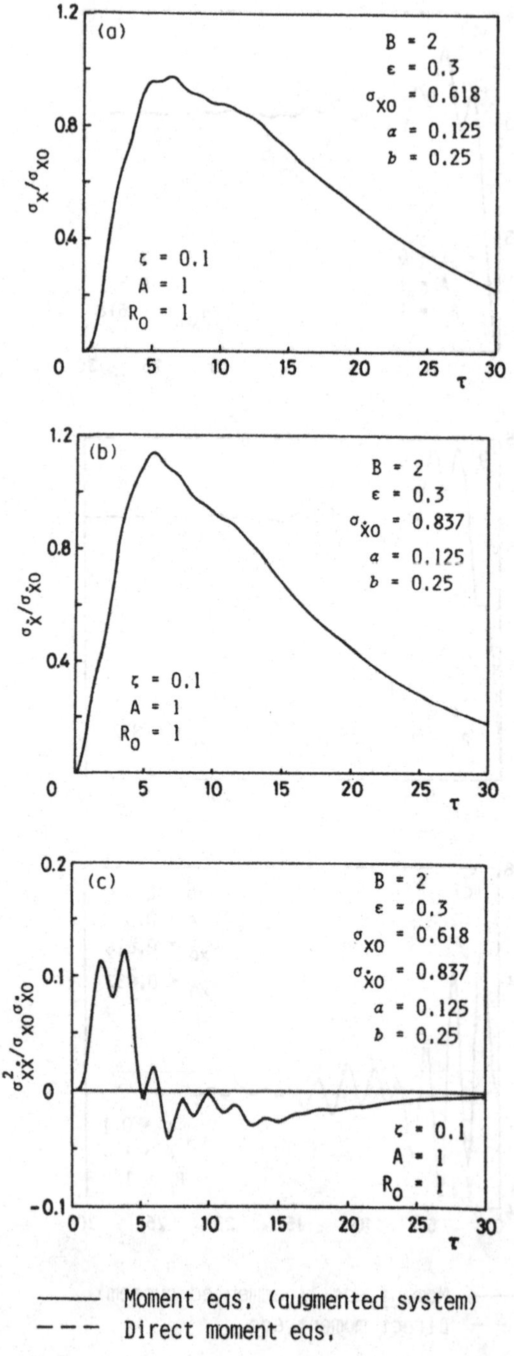

Figure 7 Transient rms and cross-correlation responses for displacement and velocity (exponential envelope function)

is considered. The typical form of equation (6) describing the shaping filter, which can generate this nonwhite noise, is expressed as

$$\begin{aligned} n(\tau) &= z_1(\tau) \\ \dot{z}_1 &= z_2 + w \\ \dot{z}_2 &= -\beta^2 z_1 - 2\alpha z_2 + (\beta - 2\alpha)w \end{aligned} \tag{29}$$

where $\beta^2 = \alpha^2 + \rho^2$ and the intensity of white noise is $2R_0\alpha$. It is convenient to express the exponential decay constant α and the dominant frequency ρ by the following ratios A and B, respectively

$$A = \alpha/\zeta \ , \quad B = \rho/\sqrt{1-\zeta^2} \tag{30}$$

Dimension of a system of the moment equations (15) through use of the augmented system and the direct moment equations (21) are, respectively, 7 and 3.

Figure 6 shows the transient rms and cross-correlation responses for displacement and velocity to the unit step envelope function. Figure 7 shows those to the exponential envelope function. The responses are normalized by the stationary values, σ_{X0} and $\sigma_{\dot{X}0}$, of the responses of the linear system ($\epsilon = 0$). The solid and broken lines represent the results obtained by the moment equations method through use of the augmented system and the direct moment equations approach, respectively. The validity of the direct approach has been demonstrated [1] by comparing with the digital simulation results. It is natural that an almost perfect agreement is found between both results, although the subtle difference is seen in Figure 6(c).

CONCLUSIONS

A technique for obtaining nonstationary responses of a nonlinear system subjected to amplitude modulated nonwhite random excitation is investigated. The moment equations method through use of the augmented system, which consists of a main system and a shaping filter, is presented, in which the stationary values of the moments with respect to the filter outputs are substituted into the inhomogeneous terms in order to eliminate the influence of the filter dynamics. This method can be effectively employed for response analysis when the excitation process has a rational spectrum. The direct moment equations approach is reviewed, in which the equations are derived immediately from the main system and their inhomogeneous terms, the cross-correlations of the state variables and the forcing function, are determined by using a recurrence algorithm. This approach can be applied to the general case in which the nonwhite random excitation possesses an arbitrary correlation function or spectrum.

REFERENCES

1. Sakata, M. and Kimura, K. Calculation of the Nonstationary Mean Square Response of a Nonlinear System Subjected to Nonwhite Excitation, J. Sound and Vibration, Vol.73, pp.333-343, 1980.

2. Kimura, K. and Sakata, M. Nonstationary Responses of a Nonsymmtric Nonlinear System Subjected to a Wide Class of Random Excitation, J. Sound and Vibration, Vol.76, pp.261-272, 1981.

3. Kimura, K., Yagasaki, K. and Sakata, M. Nonstationary Responses of a System With Bilinear Hysteresis Subjected to Nonwhite Random Excitation, J. Sound and Vibration, Vol.91, pp.181-194, 1983.

4. Sakata, M., Kimura, K. and Utsumi, M. Non-Stationary Random Vibration of an Elastic Circular Cylindrical Liquid Storage Tank to a Simulated Earthquake Excitation, Trans. JSME, Vol.50, C, No.458, pp.2004-2012, 1984.

5. Kimura, K. and Sakata, M. Nonstationary Response Analysis of a Nonsymmetric Nonlinear Multi-Degree-of-Freedom System to Nonwhite Random Excitation, JSME Int. J., Series III, Vol.31, No.4, pp.690-697, 1988.

6. Bryson, A.E. Jr. and Johansen, D.E. Linear Filtering for Time-Varying Systems Using Measurements Containing Color Noise, IEEE Transactions on Automatic Control, Vol.AC-10, pp.4-10, 1965.

7. Åström, K.J. Introduction to Stochastic Control Theory, Academic Press, 1970.

8. Gelb, A. (Ed.). Applied Optimal Estimation, Cambridge, Massachusetts, MIT Press, 1974.

9. Deyst, J.J. Derivation of the Optimum Continuous Linear Estimation for Systems With Correlated Measurement Noise, AIAA J., Vol.7, pp.2116-2119, 1969.

10. Shinozuka, M. and Henry, L.R. Random Vibration of a Beam-Column, Proc. ASCE, J. Eng. Mech. Div., Vol.91, No.EM5, pp.123-143, 1969.

11. Bolotin, V.V. Statistical Methods in Structural Mechanics: Holden-Day, 1969.

12. Barnosky, R.L. and Maurer, L.R. Mean Square Response of Simple Mechanical System to Nonstationary Random Excitation, Trans. ASME, J. Appl. Mech., Vol.36, pp.221-227, 1969.

13. Arnold, L. Stochastic Differential Equations : Theory and Applications, John Wiley, New York, 1974.

14. Lin, Y.K. Probabilistic Theory of Structural Dynamics, McGraw-Hill, New York, 1967.

15. Atalik, T.S. and Utku, T. Stochastic Linearization of Multi-Degree-of-Freedom Nonlinear Systems, Earthquake Eng. Struct. Dyn., Vol.4, pp.411-420, 1976.

16. Shinozuka, M. and Sato, Y. Simulation of Nonstationary Random Process, Proc. ASCE, J. Eng. Mech. Div., Vol.93, No.EM1, pp.11-40, 1967

The Path Integral Solution Technique Applied to the Random Vibration of Hysteretic Systems

A. Naess, J.M. Johnsen

Department of Civil Engineering, Norwegian Institute of Technology, N-7034 Trondheim, Norway

ABSTRACT

In the paper the path integral solution technique will be described. It is shown how this solution technique can be exploited to provide estimates of the response statistics of nonlinear single-degree-of-freedom dynamic systems excited by white noise or even filtered white noise. The solution technique is singularly well suited to deal with nonlinear systems as there are apparently few limitations on the kind of nonlinearities that can be accomodated. In this paper emphasis is given to the random vibration of hysteretic systems.

INTRODUCTION

In recent years there has been a significant progress in the development of numerical methods for providing estimates of the response statistics of nonlinear dynamic systems subjected to random excitation. A fairly good picture of the situation up to 1986 is obtained from the reviews given by Roberts [1,2] and Spencer [3].

Markov diffusion processes have played a prominent role in this development work. An important feature of these processes is that their transition probability densities (TPD) and probability density functions (PDF) satisfy a partial differential equation, viz. the Fokker-Planck-Kolmogorov (FPK) equation. Most of the previous research concerned with the application of Markov processes in nonlinear stochastic dynamics has been directed towards a numerical solution of the FPK equation, or the associated Pontryagin-Witt (PW) equations for the statistical moments.

An alternative approach to providing the solution of the FPK equation in this context is indicated in a rather inconspicuous note published by Kapitaniak [4]. In that paper the amplitude probability density function of a nonlinear Duffing oscillator excited by filtered white noise was obtained by using the path integral solution technique [5]. The path integral solution is a formal solution of the FPK equation which exploits the fact that the associated Markov vector process locally looks like a Brownian motion. This makes it possible to express the TPD of the Markov process in closed form for a short time step. The Markov character can then be invoked to build a global (in time) solution from local (in time) TPDs. Kapitaniak's work was based on the development of numerical methods in theoretical physics for calculating the path integral solution [6]. Since the appearance of [4], we have been engaged in work aimed at further developing the numerical methods described in [6], but it is only during the last couple of years that active research on these methods has been carried out on our part.

The main efforts in the early stages of our development work were invested in establishing efficient numerical interpolation and smoothing procedures. This was done to be able to fully exploit what we discovered to be one of the most important characteristics of the path integral solution, viz. the possibility to obtain accurate solutions down to very low probability levels, which is important for extreme value prediction. Another important feature of the path integral solution is the numerical stability that can be attained in the solution procedure. These properties are intimately related to the way the path integral solution is established, viz. by following the evolution in time of the statistics of the system. By this, the numerical calculation of the statistical distribution of the state space vector is closely connected to the physics of the dynamic model. This is in contrast to a direct numerical solution of the FPK equation, where such a connection is only implicit. Recent progress in the direct numerical solution of the FPK equation is reported by Langtangen [7] and Enneking et al. [8].

The accuracy of the solution procedure we have developed, has been documented previously for specific two- and tree-dimensional problems [9-11]. Our experience indicates that any two-dimensional problem can be solved routinely with high accuracy requiring a few minutes CPU time on a work station (DEC 3100). However, the solution of three-dimensional problems requires more care in the sense that computer capacity starts to become an issue of importance. In such cases the CPU time easily runs into many hours. This point will be further discussed below for the specific problems considered here.

Recently, Sun and Hsu [12] published a paper where also in effect the path integral solution technique, although they gave it another name, has been used to study the response statistics of nonlinear dynamic systems. Their work was limited to two-dimensional problems only. A particular feature of their approach, was the assumption that the short-time TPD was (nondegenerate) Gaussian. In general, this makes it necessary to solve a moment equation. The solution procedure presented here avoids this additional complication.

In the present paper the applicability of the numerical path integral solution procedure we have developed, will be demonstrated by application to the random vibration of hysteretic systems. As commonly done, the model adopted for describing the hysteretic behaviour is the Baber-Bouc-Wen (BBW) model [13-15].

THE PATH INTEGRAL SOLUTION

To bring out the basic ideas behind the path integral solution technique in a simple manner, it is expedient to explain the method for a scalar Markov process. Hence, let $X(t)$ denote a scalar Markov process with a TPD $p(x,t|x',t')$ satisfying the FPK equation

$$\frac{\partial p(x,t|x',t')}{\partial t} = -\frac{\partial}{\partial x}\{\ \alpha(x,t)\ p(x,t|x',t')\ \} + \frac{1}{2}\frac{\partial^2}{\partial x^2}\{\ \beta(x,t)\ p(x,t|x',t')\ \} \tag{1}$$

The functions $\alpha(x,t)$ and $\beta(x,t) \geq 0$ are determined by the coefficients in the SDE of the dynamic model, see e.g. ref. [16].

The key to the solution technique we have applied, lies in the observation that an approximate solution to equation (1) can always be obtained if the time-step $t - t'$ is small enough and the fact that every TPD can be expressed in terms of a product of short-time TPD's. Specifically, it can be shown that up to corrections of order τ^2, the following relation obtains [5]

$$p(x,t+\tau|x',t) = \frac{1}{\sqrt{2\pi\beta(x',t)\tau}}\exp\{-\frac{(x-x'-\alpha(x',t)\tau)^2}{2\beta(x',t)\tau}\} \tag{2}$$

Also, by dividing the time interval (t',t) into N small time intervals of length $\tau = (t-t')/N$, it is obtained that ($t_j = t' + j\tau$, $t = t_N$, $x = x^{(N)}$, $t' = t_0$, $x' = x^{(0)}$)

$$p(x,t|x',t') = \underbrace{\int\cdots\cdots\int}_{(N-1)\text{-}fold} \prod_{j=1}^{N} p(x^{(j)},t_j|x^{(j-1)},t_{j-1})dx^{(1)}\dots dx^{(N-1)} \tag{3}$$

Hence, by combining equations (2) and (3), a formal (approximate) solution to the FPK-equation can be written down. Equation (3), which is often referred to as a path integral solution, is the gist of our numerical solution procedure. It is realized that a numerical solution according to this method, implies that one follows the evolution in time of the probability density function (PDF) of the Markov process $X(t)$ after the start condition $X(t') = x'$.

The numerical implementation of the solution technique involves the following two elements: The stepping-stone for the time evolution of the PDF is given by the relation

$$p(x^{(j)},t_j|x',t') = \int_{-\infty}^{\infty} p(x^{(j)},t_j|x^{(j-1)},t_{j-1})\, p(x^{(j-1)},t_{j-1}|x',t')dx^{(j-1)} \tag{4}$$

In the numerical solution, the discretization of the state space makes it appropriate to employ an interpolation procedure to increase the numerical efficiency. We have found that interpolation by B-splines [17] offers the desired flexibility and accuracy. So, at each time step $t_{j-1} \rightarrow t_j$, $p(x^{(j-1)},t_{j-1}|x',t')$ is represented as a B-spline series

$$p(x^{(j-1)},t_{j-1}|x',t') = \sum_{k=1}^{n} \Gamma_{j-1,k}\, B_k(x^{(j-1)}) \tag{5}$$

where n = number of grid points in state space, $\{B_k(\cdot)\}_{k=1}^{n}$ is a basis of B-splines [17] and $\Gamma_{j-1,k}$ are the interpolation coefficients.

HYSTERESIS MODEL

The dynamic system to be studied is a SDOF nonlinear oscillator. Its equation of motion is the following

$$\ddot{X} + 2\xi\omega_o\dot{X} + aX + bZ = \sqrt{2\pi G_o}\, N(t) \tag{6}$$

Here $\sqrt{2\pi G_o}\, N(t)$ is the normalized forcing function, $N(t)$ is assumed to be a stationary, zero-mean Gaussian white noise, satisfying $E[N(t)\,N(t+\tau)] = \delta(\tau)$ where $\delta(\cdot)$ denotes Dirac's delta function. ξ, ω_o, a and b are constant parameters. $Z = Z(t)$ is a hysteretic restoring 'force' modelled by the following constitutive equation [14]

$$\eta\dot{Z} + \gamma|\dot{X}|Z|Z|^{n-1} + \beta\dot{X}|Z|^{n} - A\dot{X} = 0, \tag{7}$$

where β, γ, η and A are constants. The parameter n is an integer controlling the shape of the transition from elastic to plastic response [14].

Introducing the state space $y = (y_1,y_2,y_3)^T$, where $y_1 = x$, $y_2 = \dot{x}$, $y_3 = z$, equations (6) and (7) can be expressed as a set of three ordinary first-order differential equations, viz.

$$\dot{Y}_1 = m_1(Y) = Y_2 \tag{8}$$

$$\dot{Y}_2 = m_2(Y) = -aY_1 - 2\xi\omega_o Y_2 - bY_3 + \sqrt{2\pi G_o}\, N(t) \tag{9}$$

$$\dot{Y}_3 = m_3(Y) = \eta^{-1}\{AY_2 - \gamma|Y_2|Y_3|Y_3|^{n-1} - \beta Y_2|Y_3|^n\} \tag{10}$$

Written in this manner, $Y(t) = (Y_1(t), Y_2(t), Y_3(t))^T$ becomes a Markov vector process whose TPD $p = p(y,t|y',t')$ satisfies the FPK equation

$$\frac{\partial p}{\partial t} = -\sum_{j=1}^{3} \frac{\partial(m_j p)}{\partial y_j} + \pi G_o \frac{\partial^2 p}{\partial y_2^2} \tag{11}$$

Proceeding as in Risken [5], it can be shown that the TPD pertaining to the particular dynamic model considered here assumes the form ($\tau \to 0$)

$$\begin{aligned} p(y,t+\tau|y',t) = \; & \delta(y_1 - y_1' - m_1(y')\tau) \\ & \cdot \tilde{p}(y_2, t+\tau|y_2',t) \\ & \cdot \delta(y_3 - y_3' - m_3(y')\tau) \end{aligned} \tag{12}$$

where $\tilde{p}(y_2,t+\tau|y_2',t)$ is given by the relation

$$\tilde{p}(y_2,t+\tau|y_2',t) = \frac{1}{2\pi\sqrt{G_o\tau}} \exp\left\{-\frac{(y_2 - y_2' - m_2(y')\tau)^2}{4\pi G_o \tau}\right\} \tag{13}$$

By combining equations (12) and (13) and applying the solution technique described in the previous section, the TPD $p(y,t|y',t')$ for large $t-t'$ can be calculated. By this, the time evolution of the system when it starts from e.g. rest, can be studied. The stationary PDF is obtained in the limit as $t - t' \to \infty$.

NUMERICAL EXAMPLES

Two particular examples of hysteretic systems taken from the literature [18,19], will constitute the basis for the numerical studies presented here. However, the numerical results that are given in this section, should be considered as preliminary. The main reason for this, is that for both examples studied, the results we obtain by the path integral solution technique deviate from those of the original sources. Of course, there may be several causes for this. If our results are in error, the two most likely causes for this seem to be intrinsic difficulties connected with the numerical calculation of the path integral solution of the specific examples considered, or to bugs in the computer program. The last factor does not appear to be involved. This conclusion is based on the fact that the methodology and the computer program have been tested on a variety of two-dimensional (2D) and three-dimensional (3D) problems which have analytical solutions, and in all cases complete agreement has been found [9-11]. Also, for the hysteretic systems studied here, the velocity and hysteretic 'force' are to a large extent decoupled from the displacement variable. This makes it possible to investigate the response statistics of these two variables by solving a 2D problem. And as indicated, the accuracy of 2D solutions appears to be very good. Then comes the observation, as will be shown below, that there is no practical discrepancy between the PDFs of the velocity and the hysteretic 'force' obtained from either the 2D or the 3D solution.

Concerning intrinsic difficulties with the numerical solution procedure itself, this is a somewhat harder problem to resolve on the basis of the obtained results. A closer scrutiny of this aspect is clearly necessary. The particular difficulty that might arise is connected with the possibility of having a positive probability assigned to the boundary values of the hysteretic 'force' variable. However, no trace of such boundary effects were observed in the results.

Example 1

The first example is taken from [18]. In this case the equation of motion is given by equations (6) and (7) with the following set of parameters: $\xi = 0.1$, $\omega_0 = 1.0$ rad/sec. $a = \alpha\omega_0^2$, $b = (1-\alpha)\omega_0^2$, where α is the post-yield to pre-yield stiffness ratio. Here $\alpha = 1/21$. Further, $\eta = 1.0$, $A = 1.0$, $\beta = \gamma = 0.5$ and $n = 1$. The numerical calculations have been carried out for two specific values of the (two-sided) power spectral density of the white noise used, viz. $G_0 = 0.1$ m^2/sec.^{3} and $G_0 = 0.6$ m^2/sec.^{3}.

The results of the numerical calculations regarding the stationary PDFs for the case $G_0 = 0.1$ are given in Figures 1 - 3. The associated standard deviations are calculated to be $\sigma_1 = 1.62$ m, $\sigma_2 = 0.693$ m/sec. and $\sigma_3 = 0.444$ m. The corresponding values given in [18] are $\sigma_1 \approx 2.5$ m, $\sigma_2 \approx 0.8$ m/sec. and $\sigma_3 \approx 0.55$ m. Due to this discrepancy and the fact that the term bY_3 to a certain extent dominates the term aY_1 in equation (9), it was decided, in accordance with the introductory discussion of this section, to investigate the velocity, Y_2, and hysteretic 'force', Y_3, separately by replacing equation (9) by the 'decoupled' equation

$$\dot{Y}_2 = m_2(Y) = -2\xi\omega_o Y_2 - bY_3 + \sqrt{2\pi G_o}\; N(t) \tag{14}$$

By this, the problem of calculating estimates of the stationary PDFs of Y_2 and Y_3 is reduced to a 2D problem, which is much easier to handle computationally than a full 3D calculation. For the present problem the full 3D computation takes some hours, the exact time requirement being dependent on the time step and the grid size in state space. On the other hand, a 2D computation takes typically a few minutes. Please note that so far we have made no serious effort to develop numerical schemes with the specific aim of speeding up the 3D computations.

The full 3D computations were carried out with a time step $\tau = 2.5\times10^{-2}$ sec. and a grid size $N_1 \times N_2 \times N_3 = 63\times23\times41$, where N_j denotes the number of grid points on the y_j-axis. Two separate 2D computations were performed, one with the same time step and grid size as for the 3D analysis, the other with a time step of $\tau = 2.5\times10^{-3}$ sec.

and a grid size of 81×131. The two 2D computations were performed as a check on the influence of time step and grid size, but no such influence was found. The results of the 2D calculations are also given in Figures 2 and 3, and it is seen that the agreement between the 2D and 3D computations is quite satisfactory. The observed deviations can be explained by the effect of the coupling that is neglected in the 2D computations. Also, no indication of possible boundary effects in the PDF of the hysteretic 'force' were detected in the computations. It is therefore tempting to conclude that with the given input data the calculations per se are correct. It may also be mentioned that computations were carried out with different initial conditions, but as expected, this only affected the time required to reach stationarity.

Example 2

This example, which is adopted from [19], deals with an hysteretic oscillator representing a sliding base isolation system with sticking and slipping modelled by the use of the theory of viscoplasticity. The equation of motion assumes the form given by equations (6) and (7) with the following set of parameters: $\xi = 0.05$, $\omega_0 = \pi$ rad/sec. ($T_0 = 2$ sec.), $a = \omega_0^2, b = \mu g$, $\mu = 0.15$, $g = 386$ in/sec.^{2} $\eta = 0.01$ inches (0.25 mm) represents the elastic displacement prior to sliding (yield displacement). $A = 1$, $\beta = 0.1$, $\gamma = 0.9$ and $n = 3$. The numerical calculations have been carried out for two specific values of the (two-sided) power spectral density level G_0 of the white noise excitation, viz. $G_0 = 5$ in^2/sec.^{3} and $G_0 = 25$ in^2/sec.^{3}.

The marginal stationary PDFs obtained from the path integral solution for the case $G_0 = 5$ are shown in Figures 4 - 6. Calculating the standard deviations from these PDFs result in $\sigma_1/\eta = 7$ and $\sigma_2/\omega_0\eta = 22$, where $\sigma_j = \sqrt{Var[Y_j]}$, $j = 1,2$. The corresponding values given in [19] are $\sigma_1/\eta \approx 13$ and $\sigma_2/\omega_0\eta \approx 20$. The value of σ_3 is not given. Since for the present example the term bY_3 completely dominates the term aY_1 in equation (9), a similar check of our caculations as in the previous example can be carried out. In this particular example, there is one feature that complicates quite considerably the calculation of the complete 3D path integral solution, viz. the presence of two different time scales. While the PDFs of the velocity and hysteretic 'force' reach stationarity after starting from rest rather quickly (less than 1 sec.), the PDF of the displacement response, on the other hand, needs more than 10 sec. to approach stationarity [19]. Taking proper account of both these time scales leads to very long computer runs (of the order of 1 day). The short time scale requires a small time step τ, with the consequence that the long time scale inflates the length of the computer run to reach stationary conditions. The full 3D computation was carried out with the time step $\tau = 10^{-3}$ sec. and grid size 63×23×41. In contrast to this, the 2D solution was obtained in less than 10 minutes with a time step $\tau = 5\times10^{-5}$ sec. and grid size 81×131. The results of the 2D computation have also been given in Figures 5 and 6, and it is seen that from a practical point of view there is complete agreement between the two calculations. Again, no indication of boundary effects on the PDF of the hysteretic 'force' was seen. It therefore seems justified to conclude that with the given input data, the calculations per se are correct.

Similarly as for Example 1, for the second case study with $G_0 = 25$, the complete 3D calculation was carried only as far as was necessary to ensure that the PDFs of velocity and hysteretic 'force' had reached their stationary form. The computations were performed with $\tau = 2.5\times10^{-2}$ sec. and grid size 23×41. The corresponding PDFs have been plotted in Figures 5 and 6. The results from calculating the same PDFs by using the decoupled 2D approximation with $\tau = 2.5\times10^{-3}$ sec. and grid size 81×131 are also given in Figures 5 and 6, and again it is seen that the agreement between the two sets of PDFs is good. However, the calculated standard deviation of $\sigma_2/\omega_0 Y = 32$ is much less than the value of $\sigma_2/\omega_0 Y \approx 60$ reported in [19].

CONCLUSIONS

A general method for obtaining the response statistics of nonlinear dynamic systems excited by white noise or filtered white noise has been described. Application of the method leads to numerical estimates of the joint PDF of the state variables. Its practical usefulness has been studied by applying it to the random vibration of SDOF hysteretic oscillators excited by stationary white noise.

The specific examples investigated in the paper have been taken from the published literature [18,19], where results for the second-order statistical moments have been given. It has been shown that our computations lead to results that deviate somewhat from the corresponding data given in the sources. The causes for these discrepancies have been difficult to find. Control computations for a part of the full problem have been carried out for all case studies, and these corroborate fully our initial computations. In the control computations a program was used that has previously been tested against a range of analytical solutions, and in all cases tested complete agreement was found. It is pointed out that one possible source of error could be related to boundary effects in the PDF of the hysteretic variable. However, such effects could not be seen from the obtained results for the examples presented.

The method is still under development, and there is definitely room for improvements. Nevertheless, until a closer study of the hysteretic systems considered becomes available, the results presented here to some extent support the (preliminary) conclusion that our computations per se are correct.

REFERENCES

1. Roberts, J. B. Response of Nonlinear Mechanical Systems to Random Excitations. Part 1: Markov Metods, Shock and Vibration Digest, Vol. 13, No. 5, pp.17-28, 1981.

2. Roberts, J. B. First-Passage Probabilities for Randomly Excited Systems: Diffusion Methods, Probabilistic Engineering Mechanics, Vol. 1, No. 2, pp. 66-81, 1986.

3. Spencer, Jr., B. F. Reliability of Randomly Excited Hysteretic Structures, Lecture Notes in Engineering, Springer Verlag Berlin, Heidelberg 1986.

4. Kapitaniak, T. Stochastic Response with Bifurcations to Non-Linear Duffing's Oscillator, Journal of Sound and Vibration, Vol. 102, No. 3, pp. 440-441, 1985.

5. Risken, H. The Fokker Planck Equation, Second Edition, Springer Verlag Berlin, Heidelberg 1989.

6. Wehner, M. F. and Wolfer, W. G. Numerical evaluation of path-integral solutions to Fokker-planck equations, Physical Review A, Vol. 27, No. 5, pp. 2663-2670, 1983.

7. Langtangen, H. P. Estimation of Reliability of Dynamic Systems by Numerical Solution of Fokker-Planck and Backward Kolmogorov Equations, Proc. Scandinavian Forum for Stochastic Mechanics, Lund Institute of Technology, Lund, Sweden, August 1990.

8. Enneking, T. J., Spencer Jr., B. F. and Kinnmark, I. P. E. Stationary Two-State Variable Problems in Stochastic Mechanics, Journal of Engineering Mechanics, Vol. 116, No. 2, pp. 343-358, 1991.

9. Naess, A. and Johnsen, J. M. Direct Numerical Simulation of the Response Statistics of Nonlinear Dynamic Systems, Proc. Scandinavian Forum for Stochastic Mechanics, Lund Institute of Technology, Lund, Sweden, August 1990.

10. Naess, A. and Johnsen, J. M. Response Statistics of Nonlinear Dynamic Systems by Path Integration, Proc. IUTAM Symposium on Nonlinear Stochastic Mechanics, Torino, Italy, July 1991.

11. Johnsen, J. M. and Naess, A. Time Variant Wave Drift Damping and its Effect on the Response Statistics of Moored Offshore Structures, Proc. 1st Int Conference on Offshore and Polar Engineering (ISOPE-91), Edinburgh, August 1991.

12. Sun, J. Q. and Hsu, C. S. The Generalized Cell Mapping Method in Nonlinear Random Vibration Based Upon Short-Time Gaussian Approximation, Journal of Applied Mechanics, Vol. 57, pp. 1018-1025, 1990.

13. Bouc, R. Forced Vibration of Mechanical Systems with Hysteresis, Proc. 4th

Conference on Nonlinear Oscillation, Prague, Czechoslovakia, 1967.

14. Wen, Y. K. Method for Random Vibration of Hysteretic Systems, Journal of the Enginering Mechanics Division, ASCE, Vol. 102, No. EM2, pp. 249-263, 1976.

15. Baber, T. T. and Wen, Y. K. Stochastic Equivalent Linearization for Hysteretic Degrading Multistory Structures, Civil Engineering Studies, Structural Research Series No. 471, University of Illinois, Urbana, Illinois, December 1979.

16. Wong, E. and Hajek, B. Stochastic Processes in Engineering Systems, Springer Verlag, New York, 1985.

17. de Boor, C. A Practical Guide to Splines, Springer Verlag, New York, 1978.

18. Noori, M., Davoodi, H. and Saffar, A. An Itô-based General Approximation Method for Random Vibration of Hysteretic Systems. Part I: Gaussian Analysis, Journal of Sound and Vibration, Vol. 127, No. 2, pp. 331-342, 1988.

19. Papageorgiou, A. S. and Constantinou, M. C. Response of sliding structures with restoring force to stochastic excitation, Probabilistic Engineering Mechanics, Vol. 5, No. 1, pp. 19-26, 1990.

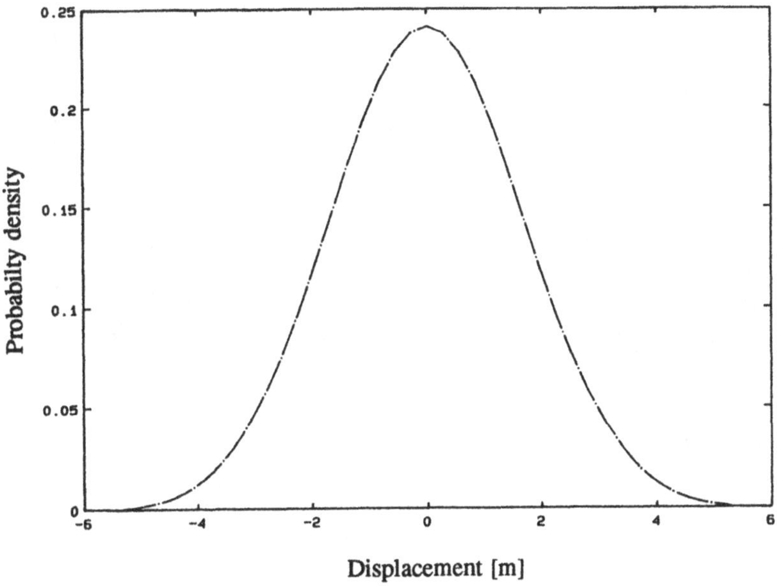

Figure 1 Calculated PDF of displacement response for example 1.

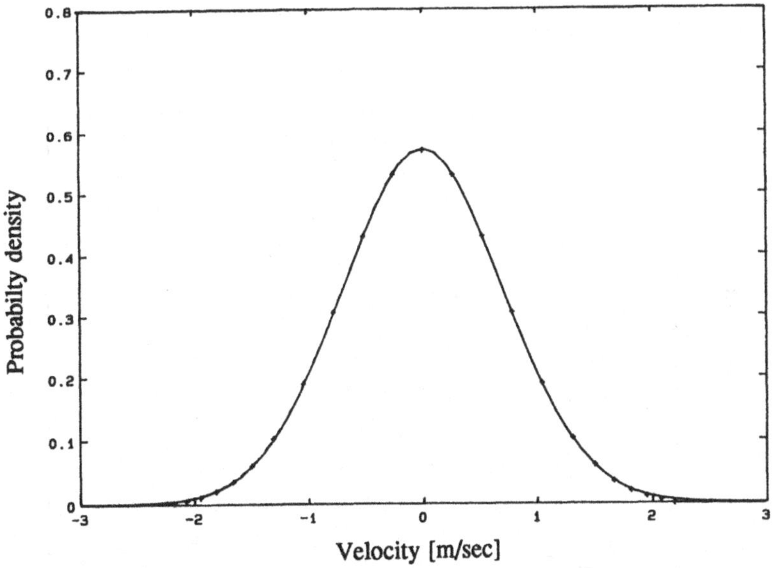

Figure 2 Calculated PDFs of the velocity response for example 1.
+++ 3D calculation
——— 2D calculation

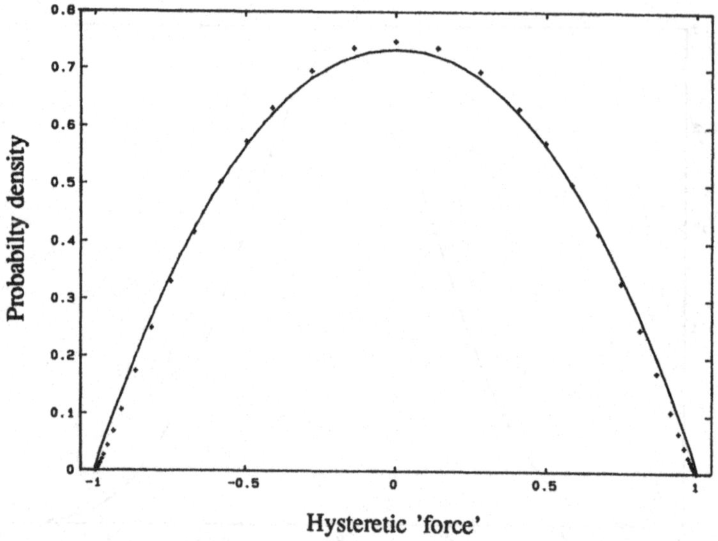

Figure 3 Calculated PDFs of the hysteretic 'force' for example 1.
+ + + 3D calculation.
——— 2D calculation.

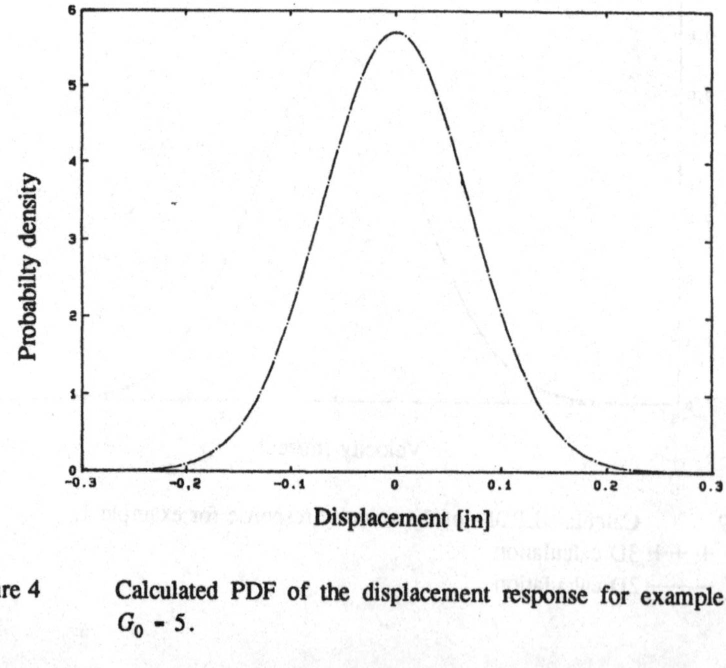

Figure 4 Calculated PDF of the displacement response for example 2 with $G_0 = 5$.

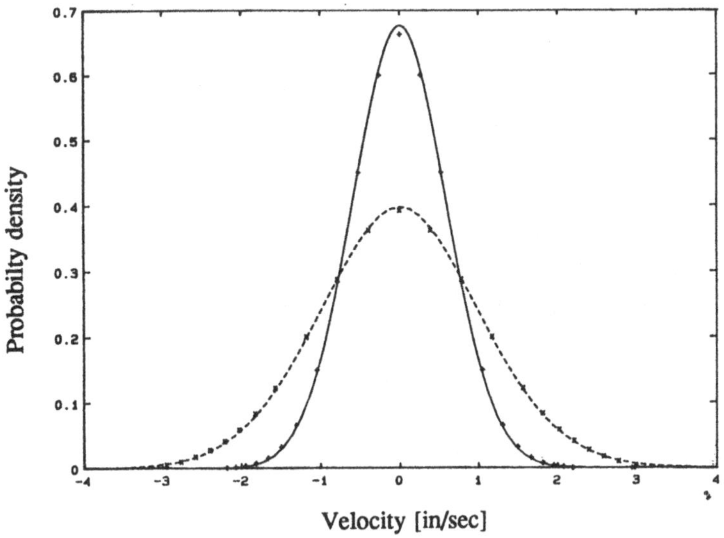

Figure 5 Calculated PDFs of the velocity response for example 2.
+ + + 3D calculation, G_0 - 5
——— 2D calculation, G_0 - 5
× × × 3D calculation, G_0 - 25
– – – 2D calculation, G_0 - 25

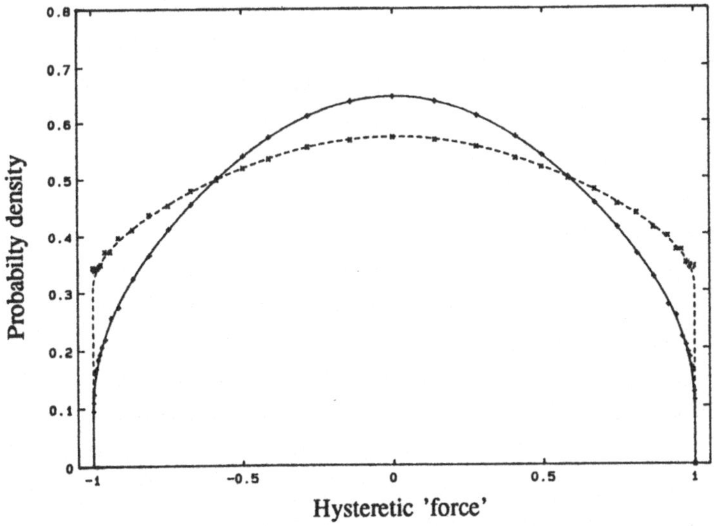

Figure 6 Calculated PDFs of the hysteretic 'force' for example 2.
+ + + 3D calculation, G_0 - 5
——— 2D calculation, G_0 - 5
× × × 3D calculation, G_0 - 25
– – – 2D calculation, G_0 - 25

Non-Stationary Probabilistic Response of Linear Systems Under Non-Gaussian Input

M. Di Paola, G. Muscolino

Dipartimento di Ingegneria Strutturale e Geotecnica, Università di Palermo, Viale delle Scienze, I-90128 Palermo, Italy

ABSTRACT

The probabilistic characterization of the response of linear systems subjected to non-normal input requires the evaluation of higher order moments than two. In order to obtain the equations governing these moments, in this paper the extension of the Ito's differential rule for linear systems excited by non-normal delta correlated processes is presented. As an application the case of the delta correlated compound Poisson input process is treated.

INTRODUCTION

In many cases of engineering interest it has become quite common to use stochastic processes to model loading such as results from earthquakes, turbulent winds or ocean waves. In these circumstances the structural response needs to be adequately evaluated in a probabilistic sense by means of the cumulants or the moments of any order of the response (see e.g. Stratonovich [1], Lin [2]). In particular, for linear systems excited by normal input, the response process is normal too and the moments or the cumulants up to the second order fully characterize the probability density function of both input and output processes. Fortunately, many practical problems involve processes which are approximately normal and the effect of the non-normality is often considered to be negligible, and this explains the popularity of the second order analysis.

For linear or non-linear systems driven by normal delta-correlated processes, the problem of finding the probabilistic characterization of the response can be faced by means of Ito's stochastic calculus (Ito [5]) which in connection with Ito's rule (Ito [6], Jazwinski [7], Arnold [8]) allows the complete characterization of the response process. It can be obtained, for example, by means of the moment equation approach (Stratonovich [1], Lin [2], Ibrahim et al. [9]). The method consist first in converting the equation of motion into an Ito type stochastic differential equation and then applying Ito's rule, in evaluating the differential equations governing the response moments of any order. Unfortunately this powerful approach fails when the input is non-normal.

On the other hand, in the simpler case of linear systems driven by non-normal input, the response is non-normal too and the probabilistic characterization of the response requires higher order moments. Recently Lutes [3] and Lutes et al. [4] investigated the case of linear systems excited by non-normal delta-correlated or filtered processes by means of two methods making it possible to evaluate the cumulants up to the fourth order of the response. These methods, however, do not give a general tool for obtaining higher order moments of the response and they do not seem to be easily extendible to non-linear systems.

The particular class of linear or non-linear dynamical systems under non-normal delta-correlated trains of Poisson-distributed impulses has been treated by Roberts [10], who gives a generalization of the Fokker-Planck-Kolmogorov equation governing the joint probability density function of the non-diffusive Markov vector process representing the response to random impulses. In this framework Iwankiewicz et al [11] have investigate the dynamic response of non-linear systems of a duffing oscillator driven by a Poisson-distributed train of impulses. Recently Di Paola and Falsone [12] treated the general case of non-normal delta correlated external or parametric input processes appropriately extending the Ito rule to take the input features into account.

In this paper the basic aspect of the case of linear systems excited by non-normal delta correlated processes is treated. Moreover, in the numerical example the case of systems driven by non-normal compound Poisson filtered processes is presented.

PRELIMINARY CONCEPTS, SDOF SYSTEM

In this section the case of a single oscillator is treated. Let the equation of motion be given in the form

$$\dot{Z} = A(t)\,Z + G(t)\,W(t) \tag{1}$$

where A(t) and G(t) are deterministic time dependent functions and W (t) represent the excitation. We assume that W(t) is a zero mean white noise process (see Stratonivich [1], Grigoriu [13]), delta correlated up to s-th order, that means

$$\begin{aligned} &k_r(W\,;t_1,t_2,\ldots,t_r) = q^{(r)}\,\delta(t_1-t_2)\,\delta(t_1-t_3)\ldots\delta(t_1-t_r)\,; \quad r=2,3,\ldots,s \\ &k_{s+p}(W\,;t_1,t_2,\ldots,t_{s+p}) = 0\,; \quad p=1,2,\ldots \end{aligned} \tag{2}$$

in which $\delta(\cdot)$ is the Dirac's delta, $q^{(r)}$ is an intensity coefficient, $k_r(\cdot)$ is the r-th cumulant of W. The latter is related to the log-characteristic function by means of the following relationship:

$$k_r\left[W\,;t_1,t_2,\ldots,t_r\right] = k_r\left[W(t_1),W(t_2),\ldots,W(t_r)\right] =$$

$$\left[\frac{1}{(-i)^r}\,\frac{\partial^r}{\partial\vartheta_1\,\partial\vartheta_2\ldots\partial\vartheta_r}\,\ln M_W(\underset{\sim}{\vartheta}_r,\underset{\sim}{t}_r)\right]_{\underset{\sim}{\vartheta}_r=\underset{\sim}{0}} \tag{3}$$

in which $\underset{\sim}{\vartheta}_r$ is a vector of real parameters collecting $\vartheta_1,\vartheta_2,\ldots,\vartheta_r$, $\underset{\sim}{t}_r$ is the vector collecting $t_1,t_2,\ldots,t_r$, and $M_W(\underset{\sim}{\vartheta}_r,\underset{\sim}{t}_r)$ is the characteristic function related to the probability density function by means of the Fourier transform operator that is:

$$M_W(\underset{\sim}{\vartheta}_r,\underset{\sim}{t}_r) = \int_{-\infty}^{\infty}\!\!\ldots\int_{-\infty}^{\infty}\exp(-i\,\underset{\sim}{\vartheta}_r^T\,\underset{\sim}{w})\,p_W(\underset{\sim}{w}\,;\underset{\sim}{t}_r)\,dw_1\,dw_2\ldots dw_r \tag{4}$$

where the apex T means transpose and $p_{\underset{\sim}{W}}(\underset{\sim}{w}; \underset{\sim}{t}_r)$ is the multidimensional probability density function of the random variables $W(t_1), W(t_2), \ldots, W(t_r)$ collected in the vector $\underset{\sim}{W}$. A delta correlated process can be interpreted as a formal derivatives of a process L(t), that is:

$$W(t)\, dt = dL(t) \tag{5}$$

in this way the r-th cumulant of dL is given in the form:

$$k_r [dL(t_1), dL(t_2), \ldots, dL(t_r)] = q^{(r)} \delta(t_1 - t_2)\, \delta(t_1 - t_3) \ldots \delta(t_1 - t_r)\, dt_1\, dt_2 \ldots dt_r \tag{6}$$

letting $t_1 = t_2 = t_3 = \ldots = t_r = t$ we obtain

$$\begin{aligned} &k_1 [dL(t)] = 0\,; \quad k_r [dL(t)] = q^{(r)}\, dt\,; \qquad r = 2, 3, \ldots, s \\ &k_{s+p} [dL(t)] = 0\,; \quad p = 1, 2, \ldots \end{aligned} \tag{7}$$

Taking into account the relationships between moments and cumulants, it can be easily seen that, neglecting higher order infinites than dt, the moments of dL(t) coincide with the cumulants, that is

$$m_r [dL] = E[(dL)^r] = k_r [dL] \tag{8}$$

in which $E[\cdot]$ means stochastic average. If $s = 2$ then the process $W(t)$ is a normal white noise, otherwise the process is a delta correlated one.

Equation (1) can be rewritten in the form:

$$dZ = A(t)\, Z\, dt + G(t)\, dL \tag{9}$$

If the probabilistic characterization of the response Z is required then the moment equation approach can be applied. In this case it is necessary to write the differential equations governing the moments of the response. In order to obtain these equations it is necessary to extend Ito's differential rule to take into account the non normality of the input. For this purpose, we consider an arbitrary scalar real valued function $\phi(Z, t)$, s-times differentiable with respect to Z and continuously differentiable on t. Then an increment $\Delta\phi$ of ϕ can be written in the following form (Di Paola and Falsone [12]):

$$\Delta\phi = \frac{\partial \phi}{\partial t} dt + \sum_{r=1}^{s} \frac{1}{r!} \frac{\partial^r \phi}{\partial Z^r} (dZ)^r \tag{10}$$

Taking into account equation (9) and neglecting infinitesimal of higher order than dt, the following relationship hold:

$$(dZ)^k = (G\, dL)^k\,; \qquad k = 2, 3, \ldots, s \tag{11}$$

equation (10) becomes:

$$\Delta\phi = \frac{\partial \phi}{\partial t} dt + \frac{\partial \phi}{\partial Z} (A\, Z\, dt + G\, dL) + \frac{1}{2} \frac{\partial^2 \phi}{\partial Z^2} G^2 (dL)^2 + \sum_{r=3}^{s} \frac{1}{r!} \frac{\partial^r \phi}{\partial Z^r} G^r (dL)^r \tag{12}$$

Equation (12) is the extension of the Ito differential rule for linear system to non-normal delta correlated input. The first three terms in equation (12) are present in the classical Ito rule, while the last one represent the effect of the non-normality of the input.

Particularizing the function ϕ in order to obtain the moments of order k of the response, we let

$$\phi(Z, t) = Z^k \tag{13}$$

inserting equation (13) in equation (12), with the aid of equations (7) and (8), making the stochastic average and dividing for dt, we obtain:

$$\dot{m}_k[Z] = A_k(t)\, m_k[Z] + \sum_{r=2}^{k} \frac{G_{rk}(t)}{r!} q^{(r)} \tag{14}$$

where

$$A_k(t) = k\,A(t)\,; \quad G_{rk}(t) = k\,(k-1)\,....\,(k-r+1)\, m_{k-r}\, G^r(t) \tag{15}$$

In equation (14) use has been made of the fact that the stochastic average of the product of any function $f(Z)$ for $(dL)^r$ obeys to the following relationship, in virtue of Ito's interpretation:

$$E[f(Z)\,(dL)^r] = E[f(Z)]\,E[(dL)^r] = E[f(Z)]\,q^{(r)}\,dt\,; \qquad r = 2, 3, \ldots, s$$

$$E[f(Z)\,dL] = 0\,; \quad E[f(Z)\,(dL)^{s+p}] = 0\,; \quad p = 1, 2, ... \tag{16}$$

The particularization of equation (14) for $k = 1, 2, 3, 4$ gives the differential equations governing the first four moments of the response in the form:

$$\dot{m}_1[Z] = A\,m_1[Z] \tag{17a}$$

$$\dot{m}_2[Z] = 2\,A\,m_2[Z] + G^2 q^{(2)} \tag{17b}$$

$$\dot{m}_3[Z] = 3\,A\,m_3[Z] + 3\,m_1[Z]\,G^2 q^{(2)} + G^3 q^{(3)} \tag{17c}$$

$$\dot{m}_4[Z] = 4\,A\,m_4[Z] + 6\,m_2[Z]\,G^2 q^{(2)} + 2\,m_1[Z]\,G^3 q^{(3)} + G^4 q^{(4)} \tag{17d}$$

From these equations, one can observe that the differential equations governing the first two moments exactly coincide with those obtained for a normal white noise input, while the effect of the non-normality of the input makes contributions starting from the moments of third order. In particular it is important to note that in the equation governing the moment of order r the intensity coefficients up to the r-th order appear. So, if the input is delta correlated up to the s-th order, only the moments of the response up to the s-th order fully characterize the response process from a probabilistic point of view.

EXTENSION TO MULTIDEGREE-OF-FREEDOM SYSTEMS

The extension of the previous concepts to muldidegree-of-freedom systems become straigthforward using the Kronecker algebra. Readers unfamiliar with this algebra are referred to Feller [14], Ma [15].

Let the n-dof system be cast in the form:

$$d\underset{\sim}{Z} = \underset{\sim}{A}(t)\,\underset{\sim}{Z}\,dt + \underset{\sim}{G}(t)\,d\underset{\sim}{L} \tag{18}$$

where $\underset{\sim}{Z}(t)$ is an n-vector of state variables, $\underset{\sim}{A}(t)$ and $\underset{\sim}{G}(t)$ are deterministic time dependent matrices of order $n \times n$ and $n \times m$ respectively and $d\underset{\sim}{L}$ is an m-vector of delta zero mean non-normal correlated processes that is:

$$E[d\tilde{L}] = 0; \quad E[(d\tilde{L})^{[r]}] = \tilde{q}^{(r)}\,dt; \qquad r = 2, 3, \dots, s$$
$$E[(d\tilde{L})^{[s+p]}] = 0; \qquad p = 1, 2, \dots \tag{19}$$

where the exponent into square brackets means Kronecker power:

$$E[(d\tilde{L})^{[r]}] = E[d\tilde{L} \otimes d\tilde{L} \otimes \dots \otimes d\tilde{L}] \tag{20}$$

$\tilde{q}^{(r)}$ is an m^r vector of strength of the vector processes $d\tilde{L}$.

The non-classical Ito rule given in equation (10) to the multidimensional case is given in the form:

$$\Delta \tilde{\phi}(\tilde{Z}, t) = \frac{\partial \tilde{\phi}(\tilde{Z}, t)}{\partial t}\,dt + \sum_{r=1}^{s} \frac{1}{r!}\left[\tilde{\nabla}_Z^{[r]^T} \otimes \tilde{\phi}(\tilde{Z}, t)\right](d\tilde{Z})^{[r]} \tag{21}$$

where $\tilde{\phi}(\tilde{Z}, t)$ is an l-vector of real valued function of $\tilde{Z}$ and t and the del-vector $\tilde{\nabla}_Z$ is the differential vector given as:

$$\tilde{\nabla}_Z^T = \left[\partial/\partial Z_1 \;\; \partial/\partial Z_2 \dots \partial/\partial Z_n\right] \tag{22}$$

Notice that, according to Ito's rule, the choice of $\tilde{\phi}(\tilde{Z}, t)$ depends mainly on the type of statistical function to be evaluated. If the moments of response $\tilde{Z}$ of order k are required $\tilde{\phi}(\tilde{Z}, t)$ may be replaced by $\tilde{Z}^{[k]}$, in this way $E[\tilde{Z}^{[k]}] = \tilde{m}_k$ (of order n^k) contains all possible moments of order k of $\tilde{Z}$.

Once this position is attained by taking the stochastic average of both sides of equation (21), we obtain the differential equation of the moments of order k in the form:

$$\dot{\tilde{m}}_k = \tilde{A}_k(t)\,\tilde{m}_k + \sum_{r=2}^{k} \frac{1}{r!}\tilde{G}_{rk}(t)\,\tilde{q}^{(r)}; \qquad \forall k \le s \tag{23}$$

where $\tilde{A}_k$ and $\tilde{G}_{rk}$ are defined as follows:

$$\tilde{A}_k(t) = \tilde{A}(t) \otimes \tilde{I}_n^{[k-1]} + \tilde{I}_n \otimes \tilde{A}(t) \otimes \tilde{I}_n^{[k-2]} + \dots + \tilde{I}_n^{[k-1]} \otimes \tilde{A}(t) =$$
$$\tilde{A}(t) \otimes \tilde{I}_n^{[k-1]} + \tilde{I}_n \otimes \tilde{A}_{k-1}(t) \tag{24}$$

$$\tilde{G}_{rk}(t) = \tilde{Q}_k(\tilde{Q}_{k-1} \otimes \tilde{I}_n) \dots (\tilde{Q}_{k-r+1} \otimes \tilde{I}_n^{[r-1]})(\tilde{m}_{k-r} \otimes \tilde{I}_n^{[r]})\,\tilde{G}^{[r]}(t) \tag{25}$$

in which $\tilde{I}_n$ is the identity matrix of order $n \times n$, and $\tilde{Q}_j$ is defined in the form:

$$\tilde{Q}_j = \sum_{s=0}^{j=1} \tilde{E}_{n^{j-s},\,n^s}; \qquad j = k, k-1, \dots, k-r+1 \tag{26}$$

where $\underset{\sim}{E}_{q,p}$ denotes a permutation matrix of order (qp) x (pq) consisting of q x p array of p x q dimensional elementary submatrix $\underset{\sim}{E}^{ij}$. The q x p dimensional elementary submatrix $\underset{\sim}{E}^{ij}$ has one in the (i, j)-th position and zero in all other positions.

The numerical solution of equation (23) is very hard because the number of state variables increases with the power law, so in order to solve these equations the eigenproperties of $\underset{\sim}{A}_k$ (order n^k x n^k) have to be evaluated. However, in the case in which $\underset{\sim}{A}$ is constant, as has been demonstrated in (Di Paola and Muscolino [16]) the eigenproperties of $\underset{\sim}{A}_k$ are strictly related to the eigenproperties of $\underset{\sim}{A}$ in a very simple way. Let $\underset{\sim}{\Phi}_k$ be the (complex) modal matrix of $\underset{\sim}{A}_k$ and $\underset{\sim}{\Phi}$ the modal matrix of $\underset{\sim}{A}$, then

$$\underset{\sim}{\Phi}_k = \underset{\sim}{\Phi}^{[k]} \tag{27}$$

Moreover, if $\underset{\sim}{\Lambda}$ is the diagonal matrix whose elements are the (complex) eigenvalues of $\underset{\sim}{A}$, that is

$$\underset{\sim}{A}\,\underset{\sim}{\Phi} = \underset{\sim}{\Phi}\,\underset{\sim}{\Lambda} \tag{28}$$

then the diagonal matrix $\underset{\sim}{\Lambda}_k$ containing the eigenvalues of $\underset{\sim}{A}_k$ is related to $\underset{\sim}{\Lambda}$ by means of the following relationship

$$\underset{\sim}{\Lambda}_k = \underset{\sim}{\Lambda}_{k-1} \otimes \underset{\sim}{I}_n + \underset{\sim}{I}_n^{k-1} \otimes \underset{\sim}{\Lambda} \tag{29}$$

Moreover, denoting as $\underset{\sim}{\Theta}$ (t) the transition matrix, the solution of

$$\begin{aligned} \dot{\underset{\sim}{\Theta}}(t) &= \underset{\sim}{A}\,\underset{\sim}{\Theta}(t) \\ \underset{\sim}{\Theta}(0) &= \underset{\sim}{I}_n \end{aligned} \tag{30}$$

where $\underset{\sim}{I}_n$ is the identity matrix, the correspondent transition matrix of $\underset{\sim}{A}_k$ is related to $\underset{\sim}{\Theta}$ by means of the following relationship

$$\underset{\sim}{\Theta}_k(t) = \underset{\sim}{\Theta}^{[k]}(t) \tag{31}$$

It follows that the solution of equation (23) in the case in which $\underset{\sim}{A}$ is a constant matrix and $\underset{\sim}{G}_{rk}$ (t) is a smooth function is given as

$$\underset{\sim}{m}_k(t_p + \Delta t) = \underset{\sim}{\Theta}^{[k]}(\Delta t)\,\underset{\sim}{m}_k(t_p) + (\underset{\sim}{I}_n^{[k]} - \underset{\sim}{\Theta}^{[k]}(\Delta t)) \sum_{r=2}^{k} \frac{1}{r!}\, \underset{\sim}{G}_{rk}(t_p)\, \underset{\sim}{q}^{(r)} \tag{32}$$

where Δt is a small temporal step. According to the initial conditions $m_k(t_0)$ equation (32) provides a step by step solution technique to find the equation of moments of every order.

EXAMPLES

Results in equations (14) and (23) are applied to obtain probabilistic descriptions of the response for linear time-independent systems subjected to non Gaussian input.

Example 1. Let Z_2 be a scalar real valued process satisfying the differential equation

$$\begin{aligned} dZ_2 &= -B\,Z_2\,dt + \alpha\,dL \\ Z_2(0) &= 0 \end{aligned} \tag{33}$$

where B and α are constant and $dL/dt = W(t)$ is a stationary delta correlated input constituted as a train of Dirca's delta impulses $\delta(t - t_j)$ occurring at random times t_j, given in the form

$$W = \sum_{j=1}^{N(t)} P_j \delta(t - t_j) \tag{34}$$

we assume that the occurrences are Poisson-distributed, hence N (t) is a Poisson counting process giving the number of impulses in the time interval (0, t). The Poisson process N (t) is completely characterized by the average arrival rate $\nu(t)$ that is:

$$E[N(dt)] = \nu(t)\, dt \tag{35}$$

where N (dt) denotes the number of impulses occurring in the time interval $[t, t + dt]$. P_j is a family of independent and identically distributed random variables. Al least the process N (t) and the sequence P_j are assumed to be independent. Under these conditions the stochastic process W (t) is the so-called compound Poisson process.

The corresponding process dL, related to the compound-Poisson process W (t), is characterized by the following moments:

$$E[(dL)^k] = \nu(t)\, E[P^k]\, dt \;; \qquad k = 1, 2, 3, \ldots.. \tag{36}$$

so that the compound Poisson process is delta correlated up to infinite order. Departure of the excitation process from normality depends on the value of the mean arrival rate $\nu(t)$. If $\nu(t) \to \infty$ and $\nu(t) E[P^2]$ is kept constant, then because $E[P^k] \to 0\,(dt^k)$ then the compound-Poisson process becomes a normal one. Without loss of generality the random variable P is assumed to be symmetric zero mean distributed and $\nu(t) = \nu$ is kept as a constant. The analysis of the response moment is carried out up to the four-th order.

Because the initial condition is assumed to be zero, the mean value and the odd moments of the response are always zero. The equations of moments up to four-th order, according to equation (14), are then given as:

$$\dot{E}[Z_2^2] = -2BE[Z_2^2] + \alpha^2 q^{(2)} \tag{37a}$$

$$\dot{E}[Z_2^4] = -4BE[Z_2^4] + 6E[Z_2^2]\,\alpha^2 q^{(2)} + \alpha^4 q^{(4)} \tag{37b}$$

where

$$q^{(2)} = \nu E[P^2]\,; \qquad q^{(4)} = \nu E[P^4] \tag{38}$$

The second moment coincides with the second cumulant because $E[Z_2] = 0$, and the four-th cumulant equation is simply given in the form:

$$\dot{k}_4[Z_2] = -4B k_4[Z_2] + \alpha^4 q^{(4)} \tag{39}$$

from this equation we can observe that the four-th order cumulant is only related to the four-th order moment of the excitation. It can be easily seen that analogous results can be obtained for higher order cumulants so the equation of 2p-th order cumulant is given in the form:

$$\dot{k}_{2p}[Z_2] = -2pB k_{2p}[Z_2] + \alpha^{2p} q^{(2p)} \tag{40}$$

The transient response for these cumulants is given as:

$$\frac{k_{2p}[Z_2]}{q^{(2p)}} = \frac{\alpha^{2p}}{2\,pB}\,(1 - \exp(-2\,p\,B\,t)) \tag{41}$$

Such an example the transient excess coefficient $\gamma_e[Z_2] = k_4[Z_2]/k_2[Z_2]^2$ is given in the form:

$$\gamma_e[Z_2] = \frac{q^{(4)}}{q^{(2)^2}}\,B\,\frac{1 - \exp(-4\,B\,t)}{(1 - \exp(-2\,B\,t))^2} \tag{42}$$

<u>Example 2</u>. Let Z_2 be the response of the stochastic differential equation (33), representing the input on another differential equation. The complessive system will be given in the form:

$$\begin{aligned} d\,Z_1 &= -A\,Z_1\,dt + Z_2\,dt \\ d\,Z_2 &= -B\,Z_2\,dt + \alpha\,dL \end{aligned} \tag{43}$$

where $dL/dt = W(t)$ is the coupond-Poisson process described in the Example 1. The analysis has been carried out up to four-th order. The equation of second order (obtained by equation (23) particularized for $k = 2$) as:

$$\dot{E}[Z_1^2] = -2\,A\,E[Z_1^2] + 2\,E[Z_1\,Z_2] \tag{44a}$$

$$\dot{E}[Z_1\,Z_2] = -(A + B)\,E[Z_1\,Z_2] + E[Z_2^2] \tag{44b}$$

$$\dot{E}[Z_2^2] = -2\,B\,E[Z_2^2] + \alpha^2 q^{(2)} \tag{44c}$$

The equation of four-th order moment (obtained by equation (23) particularized for $k = 4$) is:

$$\dot{E}[Z_1^4] = -4\,A\,E[Z_1^4] + 4\,E[Z_1^3\,Z_2] \tag{45a}$$

$$\dot{E}[Z_1^3\,Z_2] = -(3\,A + B)\,E[Z_1^3\,Z_2] + 3\,E[Z_1^2\,Z_2^2] \tag{45b}$$

$$\dot{E}[Z_1^2\,Z_2^2] = -2\,(A + B)\,E[Z_1^2\,Z_2^2] + 2\,E[Z_1\,Z_2^3] + E[Z_1^2]\,\alpha^2 q^{(2)} \tag{45c}$$

$$\dot{E}[Z_1\,Z_2^3] = -(3\,B + A)\,E[Z_1\,Z_2^3] + E[Z_2^4] + 3\,E[Z_1\,Z_2]\,\alpha^2 q^{(2)} \tag{45d}$$

$$\dot{E}[Z_2^4] = -4\,B\,E[Z_2^4] + 6\,E[Z_2^2]\,\alpha^2 q^{(2)} + \alpha^4 q^{(4)} \tag{45e}$$

The stationary solution, obtained by setting $\dot{E}[Z_j\,Z_k] = 0,\ \forall j, k = 1, 2$ and $\dot{E}[Z_j\,Z_k\,Z_r\,Z_s] = 0,\ \forall j, k, r, s = 1, 2$ gives for Z_1 and Z_2

$$E[Z_1^2] = \frac{\alpha^2 q^{(2)}}{2\,A\,B\,(A + B)}; \quad E[Z_2^2] = \frac{\alpha^2 q^{(2)}}{2\,B} \tag{46}$$

$$E[Z_2^4] = 3\,E[Z_2^2]^2 + \frac{\alpha^4 q^{(4)}}{4\,B}; \quad E[Z_1^4] = 3\,E[Z_1^2]^2 + \frac{3}{4}\,\frac{\alpha^4 q^{(4)}}{AB\,(A + B)\,(A + 3B)\,(3A + B)} \tag{47}$$

it follows that the four-th cumulant of Z_1 and Z_2 are given in the form:

$$k_4[Z_2] = \frac{\alpha^4 q^{(4)}}{4B}; \quad k_4[Z_1] = \frac{3}{4}\,\frac{q^{(4)}}{A\,B\,(A+B)\,(A+3\,B)\,(3\,A+B)} \tag{48}$$

from these results one can observe that even in the case of filtered processes the fourth cumulant only depends on the fourth moment of the excitation. Moreover, the moments of a fixed order do not depend on the higher order terms.

CONCLUSIONS

In this paper the problem of the probabilistic characterization of the response of linear systems driven by non-normal delta correlated processes has been examined. The probabilistic description of the non-normal response has been treated by means of Ito's differential rule, here extended in order to take into account the non-normality of the input.

The equations of the moments of a given order k of the response have been obtained, showing that they involve moments of equal or lesser order than k of both input and output processes. This implies that, if the input is delta correlated up to the s-th order, then the moments up to the s-th order of the response fully characterize the response processes. Moreover, in the linear case, the equations of the cumulant of the response of a fixed order k involve only the cumulants of order k of the input. These properties remain unchanged in the case of filtered processes.

REFERENCES

1. Stratonovich, R.L., Topics in the theory of random noise, Part 1, Gordon and Breach, New York, 1963.

2. Lin, Y.K., Probabilistic theory of structural dynamics, McGraw-Hill, New York, 1967.

3. Lutes, L.D., Cumulants of stochastic response for linear systems, Journal of Engineering Mechanics Division, ASCE, Vol. 112, pp. 1062-1075, 1986.

4. Lutes, L.D. and Hu, S.L.J., Non-normal stochastic response of linear systems, Journal of Enginering Mechanics Division, ASCE, Vol. 112, pp. 127-141, 1986.

5. Itô, K., On a formula concerning stochastic differential, Nagoya Mathematical Journal, Vol. 3, pp. 55-65, 1951.

6. Itô, K., Lectures on stochastic processes, Tata Fundamental Research, Bombay, India, 1961.

7. Jazwinski, A.H., Stochastic processes and filtering theory, Academic Press, New York, N.Y., 1973.

8. Arnold, L., Stochastic differential equation: theory and applications, John Wiley & Sons, New York, N.Y., 1975.

9. Ibrahim, R.A., Somclarajan, A., Heo, H., Stochastic response of nonlinear dynamic systems based on non-Gaussian closure, Journal of Applied Mechanics, Vol. 52, pp. 965-970, 1987.

10. Roberts, J.B., System response to random impulses, Journal of Sound and Vibration, Vol. 24, pp. 23-24, 1972.

11. Iwankiewicz, R., Nelsen, S.R.K. and Thaft-Christensen, P., Dynamic response of non-linear systems to Poisson-distributed pulse trains: Markv approach, Structural Safety, Vol. 8, pp. 223-238, 1990.

12. Di Paola, M. and Falsone, G., Stochastic dynamic systems driven by non-normal delta-correlated processes, to appear.

13. Grigoriu, M., White noise processes, Journal of Engineering Mechanics Division, ASCE, Vol. 113, pp. 757-765, 1987.

14. Feller, M., Special matrices and their applications in numerical mathematics, Martinus Nijthoff Publishers, Dordrecht, The Netherlands, 1986.

15. Ma, F., Extension of second order moment analysis to vector valued and matrix-valued functions, Int. Non-Linear Mech., Vol. 22, pp. 251-260, 1987.

16. Di Paola, M. and Muscolino, G., Differential moment equations of FE modelled structures with geometrical nonlinearities, Int. Journ. Non-Linear Mech., Vol. 25, pp. 363-373, 1990.

Stochastic Linearization for the Response of MDOF Systems Subjected to External and Parametric Gaussian Excitations

G. Falsone

Dipartimento di Ingegneria Strutturale e Geotecnica, DISEG, Università di Palermo, Viale delle Scienze, 90128 Palermo, Italy

ABSTRACT

The stochastic linearization approach is examined for the most general case of non zero-mean response of non-linear MDOF systems subjected to parametric and external Gaussian white excitations. It is shown that, for these systems too, stochastic linearization and Gaussian closure are two equivalent approaches if the former is applied to the coefficients of the Itô differential rule. Moreover, an extension of the Atalik-Utku approach to non zero-mean response systems allows to obtain simple formulations for the linearized drift coefficients. Some applications show the good accuracy of the method.

INTRODUCTION

The stochastic linearization (SL) method is one of the most used approaches in solving non-linear structural and mechanical systems subjected to random excitations.

The SL can be considered as an extension of the equivalent linearization method originated by Krylov and Bogoliubov[1] for the treatment of non-linear systems under deterministic excitations. Caughey was one of the first researchers to extend this method to solve the problems of randomly excited non-linear systems[2]. The basic idea of the SL approach is to replace the original non-linear system by a linear one in such a way that the difference between the two systems is minimized in some statistical sense. In this way, the parameters which appear in this linearized system involve unknown statistics of the response which are evaluated approximating the response to a Gaussian process, in accordance with the equivalent linearization technique. This basic idea has been developed by many researchers, obtaining various versions of this method[3−6].

The use of the Gaussian approximation for the response suggests that SL is very close to the Gaussian closure (GC) method[7,8], which consists in solving the first two order moments differential equations of the response, obtained by means of the Itô calculus[9,10], evaluating the statistics that appear in these equations under the assumption that the response is a Gaussian process. In the case of systems excited by purely external loads, many researchers have pointed out that, if properly applied, the two approaches lead to the same results[11,12].

When parametric loads are present too, the applications of the SL are rare and based on different approaches. For example, recently Chang and Young applied this approach, linearizing the equation of motion of the system[13], while Wu linearized the Itô equation of the non-linear system considered[14]. Moreover Wu showed that, if properly considered, GC and SL, applied to a generic SDOF non-linear system subjected to parametric type of loads, are equivalent; for MDOF systems he limited himself to showing the equivalence of results for some simple examples. Furthermore the presence of some terms of the linearized systems reported by Wu are not fully justified.

In this paper is first shown that, when parametric excitations are present, the only way to obtain the same results applying either the GC or the SL is to apply the latter on the coefficients of the Itô diffrential rule. The particularization of the approach here outlined, valid for MDOF non-linear systems, to the case on SDOF systems coincides with the approach proposed by Wu; in this way the terms obtained by Wu are considered in a new light.

Moreover the expression of the linearized coefficients is ulteriorly simplified by adopting an extension of the Atalik-Utku approach[4] which takes into account the fact that the response can be a non-zero mean process. This can happen either when the excitations have non-zero means or when the non-linear terms in the motion equations are symmetric functions. The extension above mentioned allows to obtain directly the expression of the linearized drift coefficients without separing the response process as the sum of its mean and a zero-mean process, as made by other researchers[3].

At last, the proposed procedure is applied to the study of a non-linear oscillator, excited by parametric and external white noises, showing its great efficiency.

ITÔ'S CALCULUS

In this section some well-known concepts of the Itô calculus are briefly summarized for clarity's sake.

The equation of motion of a MDOF non-linear structural system subjected to parametric and external excitations can be written in the following canonical form:

$$\dot{\mathbf{z}} = \mathbf{g}(\mathbf{z}, t) + \mathbf{G}(\mathbf{z}, t)\mathbf{f}(t) \tag{1}$$

where the dot means time derivative, $\mathbf{z}$ is the $2n-$state variables vector (n being the freedom degrees of the system), $\mathbf{g}(\mathbf{z}, t)$ is the $2n-$vector of internal loads non-linearly depending on $\mathbf{z}$, $\mathbf{G}(\mathbf{z}, t)$ is a $2n \times m$ matrix whose dependence on $\mathbf{z}$ is connected with the presence of parametric loads, and, lastly, $\mathbf{f}(t)$ is the $m-$vector of the loads exciting the system. If the matrix $\mathbf{G}(\mathbf{z}, t)$ depends only on t, then the system is said to be excited only by external loads. If $\mathbf{f}(t)$ is a vector of Gaussian white noise processes, then the response $\mathbf{z}$ is a vector of Markov processes, and the $i-$th equation of the system (1) can be written in the following Itô standard form[10]:

$$dz_i = m_i(\mathbf{z}, t)dt + G_{ij}(\mathbf{z}, t)dw_j \tag{2}$$

in which the summation convention of the tensorial calculus has been adopted; in this equation $dw_i(t) = f_i(t)dt$ is the i–th element of the increment vector $d\mathbf{w}(t)$, $\mathbf{w}(t)$ being the so-called Wiener processes m–vector; these increments have the following characteristics:

$$E\left[dw_i(t)\right] = 0; \qquad E\left[dw_i(t)dw_j(t)\right] = Q_{ij}(t)dt \tag{3}$$

in which $E[\ ]$ means stochastic average of $[\ ]$ and $Q_{ij}(t)$ is the generic element of the strength matrix $\mathbf{Q}(t)$ of the white noises $\mathbf{f}(t)$ connected with $\mathbf{w}(t)$, that is:

$$E\left[f_i(t_1)f_j(t_2)\right] = Q_{ij}(t_1)\delta(t_2 - t_1) \tag{4}$$

$\delta(\ )$ being the Dirac delta function. If $\mathbf{Q}(t)$ does not depend on t, then the white noise processes are stationary. In equation (2) $\mathbf{m}(\mathbf{z}, t)$ is the drift coefficient vector which takes into account the Wong-Zakai (WZ) vector correction terms[15] $\mathbf{p}(\mathbf{z}, t)$, that is :

$$\mathbf{m}(\mathbf{z}, t) = \mathbf{g}(\mathbf{z}, t) + \mathbf{p}(\mathbf{z}, t) \tag{5}$$

where the i–th entry of the vector $\mathbf{p}(\mathbf{z}, t)$, $p_i(\mathbf{z}, t)$ is given as[10]:

$$p_i(\mathbf{z}, t) = \frac{1}{2}G_{iu,j}(\mathbf{z}, t)G_{jv}(\mathbf{z}, t)Q_{uv}(t) \tag{6}$$

in which $G_{iu,j}(\mathbf{z}, t) = \partial G_{iu,j}(\mathbf{z}, t)/\partial z_j$. It is easy to note that, in the case of purely external excitations, the vector $\mathbf{p}(\mathbf{z}, t)$ is zero, $\mathbf{G}$ being independent of $\mathbf{z}$.

The fundamental tool to write the differential equation governing any function $\phi(\mathbf{z}, t)$ of the response process $\mathbf{z}$, continuously differentiable on t and twice differentiable with respect to the components of $\mathbf{z}$, is the Itô differential rule, which can be written as follows[10]:

$$d\phi(\mathbf{z}, t) = \frac{\partial\phi(\mathbf{z}, t)}{\partial t}dt + \phi_{,i}(\mathbf{z}, t)dz_i + \frac{1}{2}\phi_{,ij}(\mathbf{z}, t)G_{iu}(\mathbf{z}, t)G_{jv}(\mathbf{z}, t)Q_{uv}(t)dt \tag{7}$$

where $\phi_{,i}(\mathbf{z}, t) = \partial\phi(\mathbf{z}, t)/\partial z_i$ and $\phi_{,ij}(\mathbf{z}, t) = \partial\phi(\mathbf{z}, t)/\partial z_i \partial z_j$. Particularizing $\phi(\mathbf{z}, t)$ in an opportune way, it is possible to write the differential equation of the desired characteristic of the random response $\mathbf{z}$.

STOCHASTIC LINEARIZATION APPROACH

Following the SL approach, it is necessary to approximate the original non-linear equations by appropriate linear ones in such a way that the difference between the two systems of equations is minimized in some statistical sense.

In the case of purely external excitations, it has been showed that this type of linearization can be applied indifferently on the equations of motion (1) or on the Itô equations (2)[16]. When parametric excitations are present too, the applications of SL are rare and they are based on different procedures that lead to different results. In this case, the results obtained by applying the stochastic linearization on the equations of motions[13] or on the corresponding Itô equations are different, due to the presence of the WZ correction terms. Moreover, while in the case of purely external excitations, the SL approach always gives the same results of the GC approach[3], in the case of parametric loads, using both the above mentioned SL procedures, the results differ from those obtained by means of the GC method. Recently Falsone showed that, when parametric loads are present, the only way to obtain the same results of the GC approach is applying the SL on the coefficients of the Itô differential rule[16]. Following this procedurethe Itô differential rule given in equation (7) must be replaced by the following one, in which the coefficients are linearized:

$$\begin{aligned} d\phi(\mathbf{z},t) = \frac{\partial\phi(\mathbf{z},t)}{\partial t}dt + \phi_{,i}(\mathbf{z},t)\left[\left(A_{ij}(t)z_j + a_i(t)\right)dt + B_{ij}(t)dw_j(t)\right] \\ + \phi_{,ij}(\mathbf{z},t)\bar{B}_{ijuv}Q_{uv}(t)dt \end{aligned} \tag{8}$$

Comparing this equation with equation (7), the following three errors can be defined:

$$e_i^{(1)} = A_{ij}(t)z_j + a_i(t) - m_i(\mathbf{z},t) \tag{9}$$

$$e_{ij}^{(2)} = B_{ij}(t) - G_{ij}(\mathbf{z},t) \tag{10}$$

$$e_{ijuv}^{(3)} = \bar{B}_{ijuv}(t) - G_{iu}(\mathbf{z},t)G_{jv}(\mathbf{z},t) \tag{11}$$

Minimizing the mean square error $E\left[e_i^{(1)}e_i^{(1)}\right]$ with respect to the element of $\mathbf{A(t)}$ and $\mathbf{a(t)}$ one respectively obtains:

$$\frac{\partial}{\partial A_{kl}}E\left[e_i^{(1)}e_i^{(1)}\right] = 0 \Rightarrow A_{kj}(t)E\left[z_jz_l\right] + a_k(t)E\left[z_l\right] - E\left[m_k(\mathbf{z},t)z_l\right] = 0 \tag{12}$$

$$\frac{\partial}{\partial a_k}E\left[e_i^{(1)}e_i^{(1)}\right] = 0 \quad \Rightarrow \quad a_k(t) + A_{kj}E\left[z_j\right] - E\left[m_k(\mathbf{z},t)\right] \tag{13}$$

Minimizing the mean square error $E\left[e_{ij}^{(2)}e_{ij}^{(2)}\right]$ with respect to the elements of $\mathbf{B}$, one obtains the following relationship:

$$\frac{\partial}{\partial B_{ku}}E\left[e_{ij}^{(2)}e_{ij}^{(2)}\right] = 0 \quad \Rightarrow \quad B_{ku}(t) - E\left[G_{ku}(\mathbf{z},t)\right] = 0 \tag{14}$$

Lastly, minimizing the mean square error $E\left[e_{ijuv}^{(3)}e_{ijuv}^{(3)}\right]$ with respect to the elements of $\bar{\mathbf{B}}$, one obtains:

$$\frac{\partial}{\partial \bar{B}_{klwq}}E\left[e_{ijuv}^{(3)}e_{ijuv}^{(3)}\right] = 0 \quad \Rightarrow \quad \bar{B}_{klwq}(t) - E\left[G_{kw}(\mathbf{z},t)G_{lq}(\mathbf{z},t)\right] \tag{15}$$

After some algebra, equations (12),(13),(14) and (15) give the following optimal linearized coefficients:

$$A_{kj}(t) = \left(E\left[m_k(\mathbf{z},t)z_l\right] - E\left[m_k(\mathbf{z},t)\right]E\left[z_l\right]\right)\Xi_{jl} \tag{16}$$

$$a_k(t) = E[m_k(\mathbf{z},t)] - A_{kj}(t)E[z_j] \tag{17}$$

$$B_{ku}(t) = E[G_{ku}(\mathbf{z}t)] \tag{18}$$

$$\bar{B}_{klwq}(t) = E[G_{kw}(\mathbf{z},t)]\,E[G_{lq}(\mathbf{z},t)] \tag{19}$$

where Ξ_{jl} is the generic element of the inverse matrix of the covariance matrix of $\mathbf{z}$, $\boldsymbol{\Sigma}$ which is given by:

$$\boldsymbol{\Sigma} = E\left[\mathbf{z}\mathbf{z}^T\right] - E[\mathbf{z}]\,E\left[\mathbf{z}^T\right] \tag{20}$$

It is to be noted that the linearized coefficients $\bar{B}_{klwq}(t)$ cannot be obtained by multipling the coefficients $B_{ku}(t)$; indeed, from equations (18) and (19), it is easy to show that:

$$\begin{aligned} &B_{kw}(t)B_{lq}(t) = E[G_{kw}(\mathbf{z},t)]\,E[G_{lq}(\mathbf{z},t)] \neq \\ &E[g_{kw}(\mathbf{z},t)G_{lq}(\mathbf{z},t)] = \bar{B}_{klwq}(t) \end{aligned} \tag{21}$$

Moreover, it can be shown that the relationship (19) represents the generalization of the relationship given by Wu[14] for the SL of a SDOF system subjected to a parametric load, altough he claims to apply the SL to the Itô equation.

As it will be seen in the numerical examples, the approach here outlined gives more accurate results with respect to the other SL approaches and, as it will be seen in the next section, it gives just the same results of the GC approach.

Summarizing we can affirm that the best SL approach for MDOF systems subjected to parametric loads consists in considering the following Itô linearized equation:

$$dz_i = (A_{ij}(t)z_j + a_i(t))\,dt + B_{ij}(t)dw_j(t) \tag{22}$$

where $A_{ij}(t)$, $a_i(t)$ and $B_{ij}(t)$ are given in equations (16),(17) and (18); and the following Itô differential rule with linearized coefficients:

$$d\phi(\mathbf{z},t) = \frac{\partial\phi(\mathbf{z},t)}{\partial t}dt + \phi_{,i}(\mathbf{z},t)dz_i + \frac{1}{2}\phi_{,ij}(\mathbf{z},t)\bar{B}_{ijuv}(t)Q_{uv}(t)dt \tag{23}$$

where $\bar{B}_{ijuv}(t)$ is given in equation (19).

COMPARISON BETWEEN SL AND GC

When the system is linearized, since the response of a linear system to a Gaussian input is a Gaussian process too, the moments of the first and second order fully characterize the response. The differential equations governing these moments can be obtained by setting $\phi(\mathbf{z},t) = z_k$ and $\phi(\mathbf{z},t) = z_k z_l$ in equation (23), applying the stochastic average and dividing by dt; in this way we obtain:

$$\dot{E}[z_k] = A_{kj}(t)E[z_j] + a_k(t) \tag{24}$$

$$\dot{E}[z_k z_l] = A_{kj}(t)E[z_j z_l] + A_{lj}(t)E[z_k z_j] + a_k(t)E[z_l] + a_l(t)E[z_k] + \frac{1}{2}\left(\bar{B}_{kluv}(t)Q_{uv}(t) + \bar{B}_{lkuv}(t)Q_{uv}(t)\right) \tag{25}$$

Replacing equations (16),(17) and (18) into equations (24) and (25), the latter ones can be rewritten as follows:

$$\dot{E}[z_k] = E[m_k(\mathbf{z},t)] \tag{26}$$

$$\dot{E}[z_k z_l] = E[m_k(\mathbf{z},t)z_l] + E[m_l(\mathbf{z},t)z_k] + E[G_{ku}(\mathbf{z},t)G_{lv}(\mathbf{z},t)]\,Q_{uv}(t) \tag{27}$$

The averages which appear at the right hand side of these equations can be evaluated under the assumption that the response $\mathbf{z}$ is a Gaussian process.

Now we will show that, applying the GC approach, just the same differential equations governing the first and the second moments of the response are obtained. This fact ensures that the two approaches are equivalent. Indeed, following the GC approach these equations can be obtained by setting $\phi(\mathbf{z},t) = z_k$ and $\phi(\mathbf{z},t) = z_k z_l$ in the Itô diffrential rule defined in equation (7), applying the stochastic average and dividing by dt; in this way we obtain:

$$\dot{E}[z_k] = E[m_k(\mathbf{z},t)] \tag{28}$$

$$\dot{E}[z_k z_l] = E[m_k(\mathbf{z},t)z_l] + E[m_l(\mathbf{z},t)z_k] + E[G_{ku}(\mathbf{z},t)G_{lv}(\mathbf{z},t)]\,Q_{uv}(t) \tag{29}$$

which exactly coincide with the equations obtained by applying the SL approach on the coefficient of the Itô differential rule. While, it can be showed that, using other SL approaches the equations governing the first two orders moments are different and they lead to less accurate approximated results, as it will be showed in the numerical example.

EXTENSION OF ATALIK-UTKU APPROACH

In this section the expression of the linearized elements $A_{ij}(t)$ are ulteriorly simplified by using an extension of Atalik-Utku approach that takes into account the fact that, in general, the response $\mathbf{z}$ is a non zero-mean process. This extension can be obtained by considering the quantity $E[\psi_{k,j}(\mathbf{z},t)] = E[\partial\psi_k(\mathbf{z},t)/\partial z_j]$, $\psi(\mathbf{z},t)$ being any vector function of $\mathbf{z}$ and t, which, by using the definition of stochastic average, can be written as follows

$$E[\psi_{k,j}(\mathbf{z},t)] = \int_{-\infty}^{\infty} \cdots (2n) \int_{-\infty}^{\infty} \psi_{k,j}(\mathbf{z},t)p_{\mathbf{z}}(\mathbf{z})dz_1 dz_2 \cdots dz_{2n} \tag{30}$$

$p_{\mathbf{z}}(\mathbf{z})$ being the probability density function of the process $\mathbf{z}$, which, under the assumption that $\mathbf{z}$ is a Gaussian process, is characterized by the following expression:

$$p_{\mathbf{z}}(\mathbf{z}) = \left[(2\pi)^{2n} det(\Sigma)\right]^{-\frac{1}{2}} \exp\left[-\frac{1}{2}(z_i - E[z_i])(z_j - E[z_j])\Xi_{ij}\right] \tag{31}$$

where $det(\boldsymbol{\Sigma})$ is the determinant of the matrix $\boldsymbol{\Sigma}$ and Ξ_{ij} is the general element of the inverse of the matrix $\boldsymbol{\Sigma}$. If $\boldsymbol{\psi}(\mathbf{z},t)$ is such a vector function that $|\psi_i(\mathbf{z},t)| < c\exp\left(\sum_{j=1}^{2n} z_j^{\alpha}\right)$, with $\alpha < 2$, for some arbitrary c and any $\mathbf{z}$, then, integrating by parts the right hand side of equation (30), we obtain:

$$E\left[\psi_{k,j}(\mathbf{z},t)\right] = r_k - \int_{-\infty}^{\infty}\cdots(2n)\int_{-\infty}^{\infty}\psi_k(\mathbf{z},tp_{\mathbf{z},j}(\mathbf{z})dz_1dz_2\cdots dz_{2n} \quad (32)$$

where r_k is the $k-$th entry of the vector $\mathbf{r} = [\boldsymbol{\psi}(\mathbf{z},t)p_{\mathbf{z}}(\mathbf{z})]_{-\infty}^{\infty}$, which is a zero-vector for the property of the vector function $\boldsymbol{\psi}(\mathbf{z},t)$ above mentioned[17]. Taking into account the equation (31), the derivative $p_{\mathbf{z},j}(\mathbf{z})$ assumes the form:

$$p_{\mathbf{z},j}(\mathbf{z}) = -\left(z_l - E\left[z_l\right]\right)\Xi_{lj}p_{\mathbf{z}}(\mathbf{z}) \quad (33)$$

In virtue of equation (33), equation (32) becomes:

$$E\left[\psi_{k,j}(\mathbf{z},t)\right] = \Xi_{lj} \int_{-\infty}^{\infty}\cdots(2n)\int_{-\infty}^{\infty}\left(\psi_k(\mathbf{z},t)z_l - \psi_k(\mathbf{z},t)E\left[z_l\right]\right)p_{\mathbf{z}}(\mathbf{z})dz_1dz_2\cdots dz_{2n} \quad (34)$$

from which the following relationship can be obtained:

$$E\left[\psi_k(\mathbf{z},t)z_l\right] = E\left[\psi_{k,j}(\mathbf{z},t)\right]\Sigma_{lj} + E\left[\psi_k(\mathbf{z},\mathbf{t})\right]E\left[z_l\right] \quad (35)$$

or in matricial form:

$$E\left[\mathbf{z}\boldsymbol{\psi}^T(\mathbf{z},t)\right] = \boldsymbol{\Sigma}\left(\nabla_{\mathbf{z}}\boldsymbol{\psi}^T(\mathbf{z},t)\right) + E[\mathbf{z}]E[\boldsymbol{\psi}(\mathbf{z},t)] \quad (36)$$

where $\nabla_{\mathbf{z}}$ is a differentila vector operator whose $i-$th entry is the partial derivative with respect to the $i-$th element of the vector $\mathbf{z}$. Equation (35), or equivalently equation (36), which are valid for all the vector functions which satisfy the above mentioned conditions, can be considered as an extension of the Atalik-Utku theorem which takes into account that $\mathbf{z}$ is not a zero-mean process.

Setting $\boldsymbol{\psi}(\mathbf{z},t) = \mathbf{m}(\mathbf{z},t)$ in equation (35) and replacing this equation in the expressions of $A_{kj}(t)$ and $a_k(t)$ given in equations (16) and (17), we obtain:

$$a_k(t) = E\left[m_k(\mathbf{z},t)\right] - E\left[m_{k,j}(\mathbf{z},t)\right]E\left[z_j\right] \quad (37)$$

$$A_{kj}(t) = E\left[m_{k,j}(\mathbf{z},t)\right] \quad (38)$$

It is important to note that, particularizing these equations to the case of systems excited by purely external loads, they coincide with those obtained by Roberts and Spanos[3]. However, it is to be noted that these equations are obtained here directly differentiating the vector function $\mathbf{m}(\mathbf{z},t)$, without setting the response as the summation of its mean and a zero-mean process. This fact allows to overcome the confusion about the differentiation of $\mathbf{m}(\mathbf{z}, t)$.

NUMERICAL EXAMPLE

In this section a numerical example is reported in order to evidence the great accuracy of the proposed procedure with respect to the SL based on the linearization of the motion equations. The system considered is a SDOF non-linear system excited by external and parametric Gaussian stationary zero-mean white noises, whose motion equation is:

$$\ddot{x} + (\alpha + \beta x F_1(t))\,\dot{x} + \left(\gamma + \delta x + \epsilon x^2 + \eta F_2(t)\right) x = \nu F_3(t) \tag{39}$$

where $\alpha, \beta, \delta, \gamma$ and ν are constant coefficients and the forces $F_i(t)$ (with $i = 1, 2, 3$) have the following correlation functions:

$$E\left[F_i(t_1)F_j(t_2)\right] = \begin{cases} 0 & \text{for } i \neq j \\ S_i\delta(t_2 - t_1) & \text{for } i = j \end{cases} \tag{x40}$$

The presence of a quadratic non-linear term in equation (38) makes the response a non zero-mean process, even if the excitations are zero-mean processes.

Following the procedure proposed in this paper, it is necessary to write the Itô equations corresponding to equation (39). Letting $x = x_1$ and $\dot{x} = x_2$, these equation are written in the form:

$$\begin{aligned} dx_1 &= x_2 dt + p_1 \\ dx_2 &= \left(-\alpha x_2 - \gamma x_1 - \delta x_1^2 - \epsilon x_1^3\right) dt - \beta x_1 x_2 dw_1 - \eta x_1 dw_2 \\ &\quad + \nu dw_3 + p_2 dt \end{aligned} \tag{41}$$

where p_1 and p_2 are the components of the WZ correction terms-vector, which are given by:

$$p_1 = 0; \quad p_2 = \frac{1}{2}\beta^2 x_1^2 x_2 S_1 \tag{42}$$

Applying the SL on the coefficients of the Itô differential rule, we obtain the following components of the matrix $\mathbf{A}(t)$ and of the vector $\mathbf{a}(t)$:

$$\begin{aligned} &A_{11}(t) = 0; \quad A_{12}(t) = 1; \quad A_{21}(t) = -\gamma - 2\delta E[x_1] - 3\epsilon E[x_1^2] \\ &+ \beta^2 E[x_1 x_2] S_1; \quad A_{22}(t) = -\alpha + \frac{1}{2}\beta^2 E[x_1^2] S_1; \\ &a_1(t) = 0; \quad a_2(t) = \delta\left(2E[x_1]^2 - E[x_1^2]\right) + \epsilon\left(3E[x_1^2]E[x_1] - E[x_1^3]\right) \\ &- \beta^2 E[x_1 x_2] E[x_1] + \frac{1}{2}\beta^2 S_1 \left(E[x_1^2 x_2] - E[x_1^2] x_2\right) \end{aligned} \tag{43}$$

Applying the linearized Itô differential rule the equations of the moments of the first and second order can be obtained in the form:

$$\begin{aligned} \dot{E}[x_1] &= E[x_2] \\ \dot{E}[x_2] &= -\alpha E[x_2] - \gamma E[x_1] - \delta E[x_1^2] - \epsilon E[x_1^3] + \frac{1}{2}\beta^2 S_1 E[x_1^2 x_2] \end{aligned} \tag{44}$$

$$
\begin{aligned}
\dot{E}[x_1^2] &= 2E[x_1x_2] \\
\dot{E}[x_1x_2] &= E[x_2^2] - \alpha E[x_1x_2] - \gamma E[x_1^2] - \delta E[x_1^3] - \epsilon E[x_1^4] \\
&\quad + \frac{1}{2}\beta^2 S1 E[x_1^3 x_2] \\
\dot{E}[x_2^2] &= -2\alpha E[x_2^2] - 2\gamma E[x_1x_2] - 2\delta E[x_1^2x_2] - 2\epsilon E[x_1^3x_2] \\
&\quad + 2\beta^2 S_1 E[x_1^2x_2^2] + \eta^2 E[x_1^2] S_2 + \nu^2 S_3
\end{aligned}
\tag{45}
$$

The moments of the third and the fourth which appear in these equations can be evaluated in terms of the moments of the first two orders by means of well-known relationships valid for Gaussian variables.

On the contrary, if the SL is applied on the motion equation, the corresponding moments equations are:

$$
\begin{aligned}
\dot{E}[x_1] &= E[x_2] \\
\dot{E}[x_2] &= -\alpha E[x_2] - \gamma E[x_1] - \delta E[x_1^2] - \epsilon E[x_1^3] + \frac{1}{2}\beta^2 S_1 E[x_1]E[x_1x_2]
\end{aligned}
\tag{46}
$$

$$
\begin{aligned}
\dot{E}[x_1^2] &= 2E[x_1x_2] \\
\dot{E}[x_1x_2] &= E[x_1^2] - \alpha E[x_1x_2] - \gamma E[x_1^2] - \delta E[x_1^3] - \epsilon E[x_1^4] \\
&\quad + \frac{1}{2}\beta^2 S_1 \left(E[x_1]E[x_2]E[x_1^2] + 2E[x_1]^2 E[x_1x_2] - 2E[x_1]^3 E[x_2]\right) \\
\dot{E}[x_2^2] &= -2\alpha E[x_2^2] - 2\gamma E[x_1x_2] - 2\delta E[x_1^2x_2] - 2\epsilon E[x_1^3x_2] \\
&\quad + \beta^2 \left(E[x_1]E[x_2]E[x_1x_2] + 2E[x_1]^2 E[x_2^2] - 2E[x_1]^2 E[x_2]^2\right) \\
&\quad + \beta^2 S_1 E[x_1x_2]^2 + \eta^2 E[x_1]^2 S_2 + \nu^2 S_3
\end{aligned}
\tag{47}
$$

It is worth to note that the main differences between the two groups of moments equations are connected with the terms which are multiplied for the intensity S_1, that is the terms which are connected with the WZ correction terms. These two systems of non-linear differential equations have been solved by using the Runge-Kutta method with $\Delta t = 0.01 sec$. In Fig.1 and in Fig.2 the covariances of the displacement x and of the velocity $\dot{x}$ are respectively depicted for the particular case in which the parameters which appear in the equations have the following values:

$$
\alpha = 0.1; \quad \beta = 1.0; \quad \gamma = 5.0; \quad \delta = \epsilon = \eta = 0.1; \quad \nu = 0.4 \tag{48}
$$

The results obtained have been compared with those obtained by means of a Monte-Carlo simulation. From these figures it is easy to notice the great accuracy of the proposed method with respect to the other SL approach; the latter falls when the system is characterized by a strong parametric excitation on the velocity; this is well evidenced in Fig.3 and in Fig.4 where the covariances of the displacement and of the velocity are depicted for $\beta = 0.5$. Indeed, in this case, in which the parametric excitation on the velocity is less strong than in the previous case, the two response are very close. This fact is confirmed by the graphics in Fig.5, where the stationary covariances of the displacement and of the velocity varying the parameter β and using the two SL procedures are reported. These graphics confirm that when strong parametric load are applied in a non-linear system, the most appropriate linearization technique is the one based on the linearization of the Itô differential rule.

CONCLUSIONS

The SL applied to MDOF non-linear systems, characterizing by a non zero-mean response and subjected to parametric and external excitations has been examined. It has been showed that if one wants to obtain the same results between applying the SL or the GC, even when parametric excitations are present, it is necessary to apply the SL to the coefficient of the Itô differential rule. Applications have also showed that this procedure is the most accurate among the SL approaches, expecially when strong parametric excitations on velocity are present. At last an extension of the Atalik-Utku approach is presented in order to evalute the linearized drift coefficients simply by differentiation of the original non-linear ones, even for non-zero mean response processes.

REFERENCES

1. N. Krilov and N. Bogoliubov, *Introduction to nonlinear mechanics*, Princeton University Press, Princeton, 1943.
2. T.K. Caughey, Equivalent linearization technique, *Journal of Acoustical Society of America*, **35**(11), 1963, pp.1706-1711.
3. J.B. Roberts and P.T.D. Spanos, *Random vibration and statistical linearization*, John Wiley and Sons, New York, N.Y., 1990.
4. T.S. Atalik and S. Utku, Stochastic linearization of multi-degree-of-freedom nonlinear systems, *Earthquake Engineering and Structural Dynamics*, **4**, 1976, pp.411-470.
5. N.D. Iwan and J.-M. Yang, Statistical linearization for nonlinear structures, *Journal of Engineering Mechanical Division*, **97**, 1971, pp.1609,1623.
6. Y.K. Wen, Equivalent linearization for hysteretic systems under random excitations, *Journal of Applied Mechanics*, **47**, 1980, pp.150-154.
7. R.A. Ibrahim, *Parametric random vibration*, Research Studies Press, Letchworth, England, 1985.
8. W.F. Wu and Y.K. Lin, Cumulant-neglect closure for nonlinear oscillators under random parametric and external excitations, *International Journal of Non-Linear Mechanics*, **19**, 1983, pp.349-362.
9. K. Itô, On a formula concerning stochastic differential, *Nagoya Mathematical Journal*, **3**,1951, pp.55-65.
10. A.H. Jazwinski, *Stochastic processes and filtering theory*, Academic Press, New York, N.Y., 1973.
11. R.A. Ibrahim, A. Soundararajan and H. Heo, Stochastic response of nonlinear dynamic systems based on a non-Gaussian closure, *Journal of Applied Mechanics*, **52**(12), 1985, pp.965-970.
12. S.H. Crandall, Heuristic and equivalent linearization techniques for random vibrations of nonlinear oscillators, *Proceedings of Eight International Conference of Nonlinear Oscillations*, Academica, Prague, 1978, pp.211-226.
13. R.I. Chang and G.E. Young, Methods and Gaussian criterion for statistical linearization and stochastic parametrically and externally excited nonlinear systems, *Journal of Applied Mechanics*, **56**, 1989, pp.179-186.

14. W.F. Wu, Comparison of Gaussian closure technique and equivalent linearization method, *Probabilistic Engineering Mechanics*, **2**(1), 1987, pp.2-8.
15. E. Wong and M. Zakai, On the relation between ordinary and stochastic differential equations, *International Journal of Engineering Science*, **3**, 1965, pp.213-229.
16. G. Falsone, Stochastic linearization of MDOF systems under parametric excitations, to appear on *International Journal of Non-Linear Mechanics*.
17. A. Papoulis, *Probability, Random Variables and Stochastic Processes*, McGraw-Hill, New York, 1965.

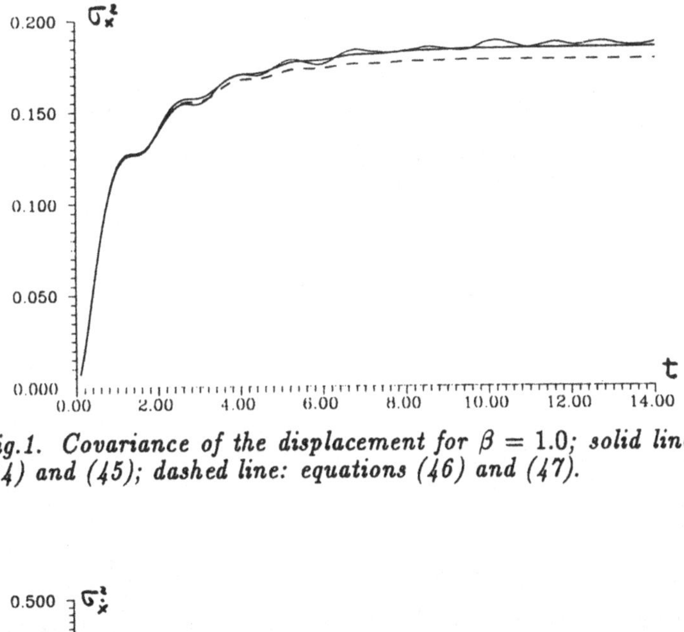

Fig.1. Covariance of the displacement for $\beta = 1.0$; solid line: equations (44) and (45); dashed line: equations (46) and (47).

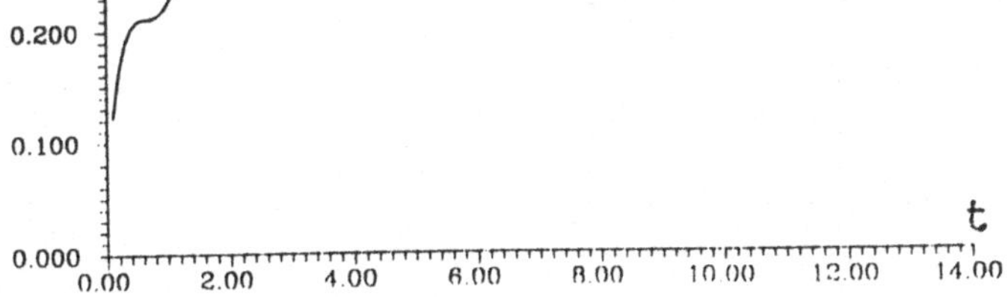

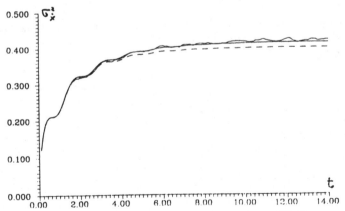

Fig.2. Covariance of the velocity for $\beta = 1.0$; solid line: equations (44) and (45); dashed line: equations (46) and (47).

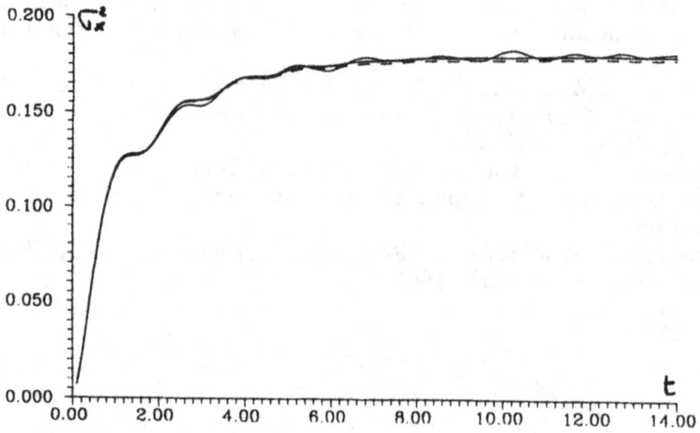

Fig.3. Covariance of the displacement for $\beta = 0.5$; solid line: equations (44) and (45); dashed line: equations (46) and (47).

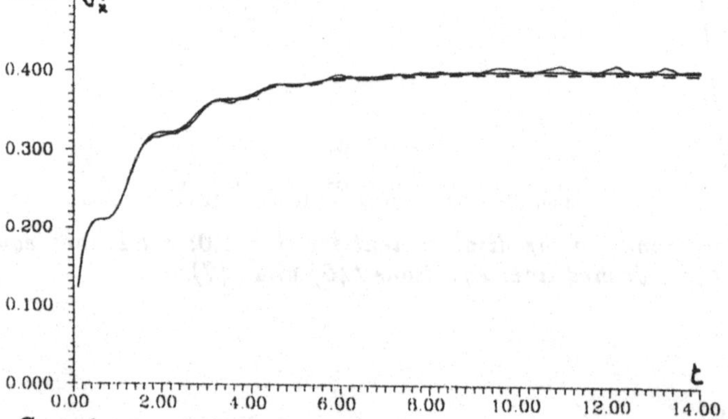

Fig.4. Covariance of the velocity for $\beta = 0.5$; solid line: equations (44) and (45); dashed line: equations (46) and (47).

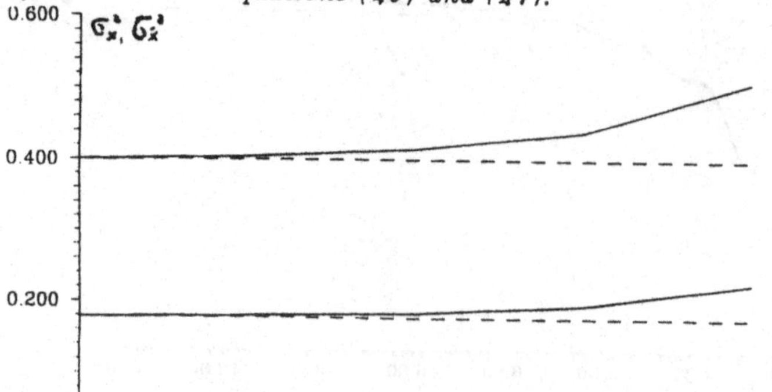

Fig.5. Stationary covariances of displacement and velocity varying β; solid line: equations (44) and (45); dashed line: equations (46) and (47).

Equivalent Linearization of a Newly Introduced General Hysteretic Model

M. Noori (*), M. Saffar (*), G. Ghantous (*),
A. Guran (**), H. Davoodi (***), T.T. Baber (****)
(*) Mechanical Engineering Department, Worcester Polytechnic Institute, Worcester, MA 01609, U.S.A.
(**) Department of Mechanical Engineering, University of Toronto, Ontario, M5S 1A4, Canada
(***) Department of Mechanical Engineering, University of Puerto Rico, Mayaguez, PR 00708, Puerto Rico
(****) Civil Engineering Department, University of Virginia, Ch'ville, VA 22901, U.S.A.

ABSTRACT

A new general hysteresis model which is developed based on the mathematical modelling technique of Baber-Noori[3] is presented. The capabilities of the model in generating numerous hysteretic models such as Wen-Baber, Pinching hysteresis and bilinear are discussed. In addition, the ability of the model in simulating unloading stiffness degradation, which no other available model can predict, is shown. The method of Equivalent Linearization is used to predict the first and second order statistics for a SDOF system having this hysteretic restoring force and under different levels of input excitation. It is shown that the linearized response statistics are in good agreement with Monte Carlo simulation results.

INTRODUCTION

Most structures under severe loading, such as strong seismic excitations, may respond inelastically. This means, not only may the restoring force behavior of such structures become highly nonlinear, it also depends on the time history of the response. This history-dependence phenomenon is referred to as hysteresis, study of which has attracted considerable attention in the field of earthquake engineering.

Under a dynamic excitation, the restoring force may deteriorate in strength, stiffness or both when the oscillation progresses for an extended period of time. The exact nature of the system degradation is a function of the materials and configuration which varies with the type of system.

Degradation may lead to progressive weakening and collapse of structures. To predict such highly nonlinear response, mathematical models are needed that can predict the energy absorption, hysteretic response, and the resulting system evolution through degradation. Constructing such models requires, in general, the deployment of systematic modelling procedures.

Several mathematical models have been proposed to describe the hysteretic behavior of structures excited beyond the elastic range. These range from the simple elastoplastic model to general hysteresis models [3,4,8-12,13,20-22].

To facilitate the analysis of nonlinear structural response, e.g. under random excitation, mathematically traceable hysteresis models are important. Rate-type models for hysteresis elements appear to be quite useful in this regard. Wen[23,24], Baber and Wen[5,6], Baber and Noori[3,4], Park et al[16], and Noori et. al[13,14] developed several models for general hysteresis including pinching behavior. However, none of the existing hysteresis models, is capable of reproducing unloading stiffness deterioration. Nevertheless, this degradation is inherent and is always observed in pinching hysteretic behavior, e.g. in reinforced concrete structural members.

In this paper, a new general hysteresis model is presented which has the ability of reproducing the complicated unloading stiffness degradation. Development of this model is based on the general modelling technique proposed earlier by Baber and Noori[3]. In addition, random vibration analysis of a SDOF system having this type of nonlinear restoring force is considered and Gaussian response statistics are obtained. Method of equivalent linearization as proposed by Atalik and Utku[1], Spanas[18], and Roberts and Spanas[17] is utilized to generate up to second order response statistics under a Gaussian white noise excitation.

MODELING OF A GENERAL HYSTERETIC RESTORING FORCE.

Consider a single-degree-of-freedom (SDOF) oscillator with a response time history-dependent, hysteretic restoring force z(t), under some general excitation f(t). The motion u(t) of such a system is given by

$$\ddot{u} + 2\xi\omega_0\dot{u} + \alpha\omega_0^2 u + (1-\alpha)z(t) = f(t)/m \tag{1}$$

In equation (1), two displacement-dependent restoring force components exist, the hysteretic restoring force z, and a cycle independent "postyield" portion given by $\alpha\omega^2 u$ term. Typically α maybe taken as the ratio of postyield to preyield stiffness. A general rate-type hysteresis model is described by the differential equation

$$\dot{z} = g(z;\ \dot{u};\ z(\tau),\ \tau \leq t; \ldots\ldots) \tag{2}$$

For example, Wen[23, 24] and Baber and Wen[5, 6] extended Bouc's[7] smooth system rate model to the form

$$g = (1/\eta)\{\ A\dot{u} - \nu[\beta|\dot{u}||z|^{n-1}z + \gamma\dot{u}|z|^{n}]\} \tag{3}$$

where A, ß, γ and n are shape parameters determining the type of hysteresis shape, and ν and η are additional response history-dependent parameters which, along with A allow system deterioration to be accomodated. For convenienvce, A, ν and η were taken as linearly varying functions of element total energy dissipation, $\epsilon(t)$.

Utilizing this model, it was found that satisfactory random vibration response estimates could be obtained by the method of equivalent linearization. Subsequently, Sues et al.,[19] modified the deterioration rate control to depend upon maximum displacement, and also obtained satisfactory random vibration response estimates.

Although the model of equations (1) and (3) allows stiffness, strength or combined deterioration it does not allow pinching deterioration to be incorporated. Baber and Noori[3,4] and Noori and Baber[13,14] developed several models for pinching hysteresis and they introduced two mathematical techniques for developing general hysteresis models. Models developed based on these two approaches are the extended and generalized forms of Baber-Wen[5,6] and to the best of the authors' knowledge are the only rate-type models capable of reproducing combined stiffness, strength and pinching degradation.

Extensive discussion on these models, including their nonzero mean random vibration analyses has been presented elsewhere [3,4,13,14]. It is noteworthy to mention these nonzero mean analyses along with the work of Baber[2] are the only nonzero mean random vibration of hysteretic systems which has appeared in random vibration literature.

Although the aforementioned general hysteresis models [3,4,13,14] are capable of reproducing stiffness, strength and pinching deterioration, they are all based on the assumption that hysteresis pinching is essentially a slipping phenomenon occuring when z changes signs. Additional pinching however, may also occur, attributable to general unloading stage stiffness decrease. Pinching behavior of this type may be observed in reinforced concrete elements during unloading, to cite an example. Such degradation may be described mathematically with the aid of dz/du versus z considerations. The nature of the expected behavior is illustrated in the z versus u plot of Figure 1, with the corresponding qualitative dz/du versus z curve shown in Figure 2. Modelling, in this case, includes a decrease in the unloading stiffness as well as the pinching associated with load reversal slip. One hysteretic model, as proposed in this paper, which incoroporates the desired effects is given by

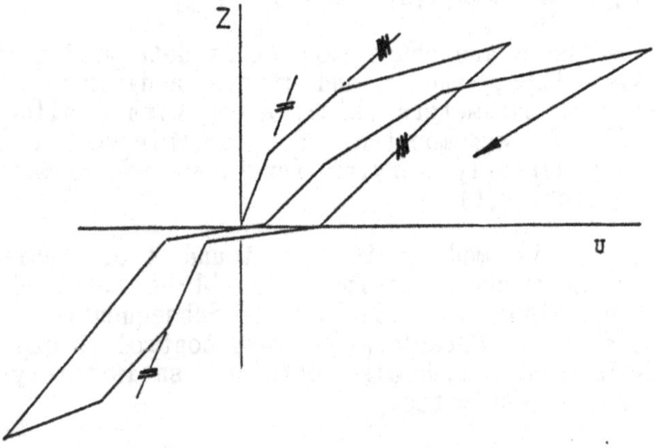

Figure 1 - Typical loop-pinching behavior for a reinforced concrete element.

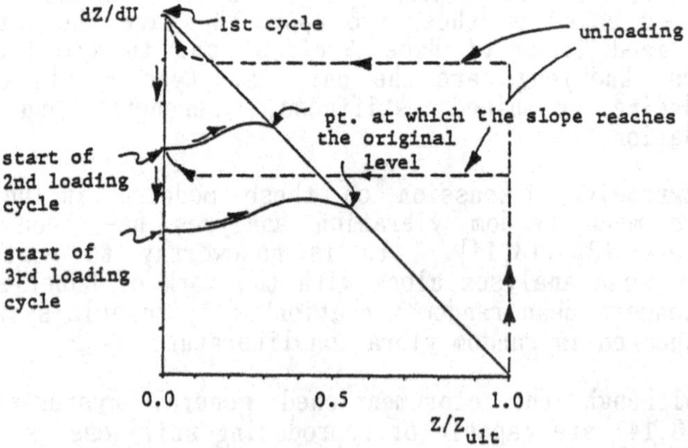

Figure 2 - Corresponding dz/du variation for the loop-pinching.

$$\dot{z} = \left[1-\zeta_1 \exp\left(-\frac{z^2}{\zeta_2^2}\right)\right] . \frac{A_1}{4} \Bigg[\Bigg[\left[\frac{1}{2} + \frac{1}{\pi} \mathrm{Arctan}\left(\frac{z - z_{cp}}{z_{ch}}\right)\right] . \left[1 - \left(\frac{z}{z_u}\right)^n\right]$$

$$+ \left[\frac{1}{2} - \frac{1}{\pi} \mathrm{Arctan}\left(\frac{z - z_{cp}}{z_{ch}}\right)\right] . \alpha_1 \Bigg] (\dot{u} + |\dot{u}|) \qquad (4)$$

$$+ \Bigg[\left[\frac{1}{2} + \frac{1}{\pi} \mathrm{Arctan}\left(\frac{z - z_{cn}}{z_{ch}}\right)\right] . \alpha_1$$

$$+ \left[\frac{1}{2} - \frac{1}{\pi} \mathrm{Arctan}\left(\frac{z - z_{cn}}{z_{ch}}\right)\right] . \left(1 + \frac{z}{z_u} \left|\frac{z}{z_u}\right|^{n-1} \right) \Bigg] . (\dot{u} - |\dot{u}|) \Bigg]$$

where $(1-\zeta_1)\ \exp(-z^2/\zeta^2_2)$ is the pinching multiplier function (PMF) used in Single Element Pinching model (SEP) proposed earlier [3] by Noori and Baber[14]. The rather lengthy expression which follows describes loading and unloading regions. For $u > 0$, transition from loading to unloading occurs at z_{cp}, the length of the transition region being controlled by the magnitude of z_{ch}, and the function $1/\pi$ Arctan $[(z - z_{cp})/z_{ch}]$. For $u < 0$, the transition is controlled by z_{cn}, z_{ch}, and the equation $1/\pi$ Arctan$[(z - z_{cn})/z_{ch}]$. α_1 is the ratio of unloading-to-loading stiffness, and z_u is the value of z at which final yielding occurs. Two sample plots of this model hysteresis, obtained by subjecting a SDOF system having this type of restoring force to a sinusodial loading, are shown in Figures 3 and 4. In these plots $z_{cp} = 0.2$, $z_{cn} = -0.2$, $z_u = 1$, $A_1 = 2$, $z_{ch} = 0.04$, $\omega_0 = 1$, $\zeta_{10} = 0.9$, $\lambda = 0.2$, $\delta\xi = 0.1$, $\xi = 0.25$ and $p = 1$. For Figure 3, $\alpha_1=0.3$ and for Figure 4, $\alpha_1=0.4$. This model demonstrates the value in using the modeling technique for developing general hysteretic behavior proposed by Baber and Noori[3].

RANDOM VIBRATION ANALYSIS USING USR MODEL

Herein, random vibration application is presented for zero mean excitation of this newly proposed unloading stiffness reduction (USR) model. The response statistics for the SDOF oscillator described by this model cannot presently be obtained in closed form. The response statistics are derived numerically and by the method of equivalent linearization[18].

STOCHASTIC EQUIVALENT LINEARIZATION

The nonlinear model containing this hysteretic restoring force model is described by equations (1) and (4), where in

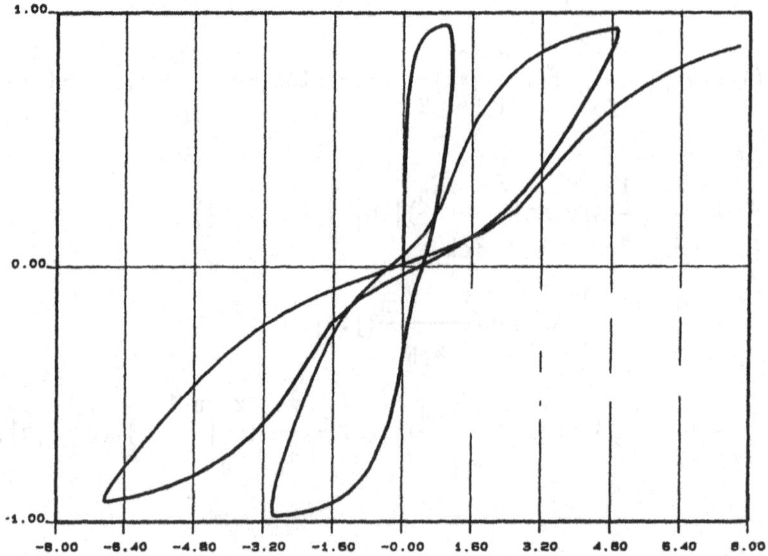

Figure 3 - Behavior of the new hysteresis model under a cyclic load (severe unloading stiffness deterioration).

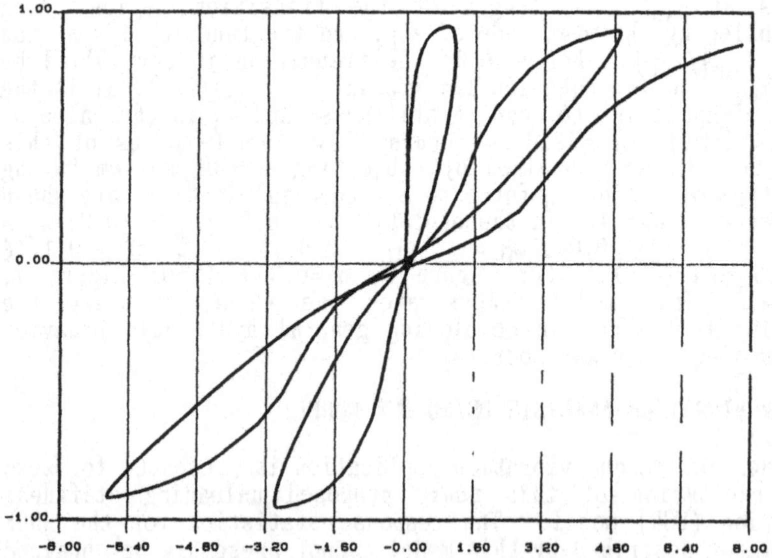

Figure 4 - Behavior of the new hysteresis model under a cyclic load (moderate unloading stiffness deterioration).

equation (4) parameters ζ_1 and ζ_2 are given by the following functions

$$\zeta_1 = \zeta_{10} \, [\, 1 - \exp(-p \cdot \epsilon) \,] \tag{5}$$

$$\zeta_2 = (\xi_0 + \delta\xi)\,(\lambda + \zeta_1) \tag{6}$$

In these two equations ϵ is the total energy dissipated by the hysteretic element, the pinching development rate of ζ_1 is controlled by p, the maximum value of ζ_1 is ζ_{10}, while λ, ξ_0 and $\delta\xi$ control the evolution of ζ_2. The governing equations (1) and (4) - (6) can be rewritten as

$$\dot{y}_1 = y_2 \tag{7-a}$$

$$\dot{y}_2 = -\alpha\omega^2 y_1 - 2\zeta\omega_0 y_2 - (1-\alpha)\omega^2{}_0 y_3 + a(t) \tag{7-b}$$

$$\dot{y}_3 = g(\bullet) \tag{7-c}$$

where $g(\bullet)$ is the function given by equation (4) describing the hysteretic restoring force.

Equations (7) can be rewritten in the matrix form

$$\dot{\mathbf{y}} + h(\mathbf{y}) = \mathbf{f} \tag{8}$$

With the zero mean response assumption, equation (7-c) may be replaced by the linearized form

$$\dot{y}_3 = C_e y_2 + K_e y_3 \tag{9}$$

where the coefficients C_e and K_e are defined by

$$\begin{aligned} C_e &= E\,[\, \partial g(\bullet)/\partial y_2 \,] \\ K_e &= E\,[\, \partial g(\bullet)/\partial y_3 \,] \end{aligned} \tag{10}$$

Under the assumption that y_2 and y_3 are jointly Gaussian, and given the response statistics σ_2, σ_3 and ρ_{23}, equation (10) can be evaluated, however, not in closed form. The expected values included in these evaluations are reduced to a single Gauss-Hermite quadrature and are evaluated numerically. Detailed derivations of 45 expected values involved are not provided in this paper due to space limitations. Equation (9) along with (7-a) and (7-b) now form a linearized set of simultaneous stochastic differential equations, of the form similar to equation (8). From this equation the covariance equation

$$\dot{\mathbf{s}} + \mathbf{G}\mathbf{s} + \mathbf{s}\mathbf{G}^T = \mathbf{B} \tag{11}$$

for the zero time-lag covariance matrix $\mathbf{s} = E[yy^T]$ can be derived. Procedure for the evaluation of **G** matrix and response statistics σ_1, σ_2, ρ_{23} at each time step is similar to the approach previously reported[3,13].

NUMERICAL STUDIES

The numerical studies which are reported in this paper illustrate the relative validity of the approximate random vibration analysis of USR model by equivalent linearization, assuming zero mean response. In these studies, only the parameters ζ_1 and ζ_2 were degraded. Nonstationary response statisitcs were obtained under stationary white-noise input beginning at zero initial conditions. Filtering and modulation of the input noise excitation can be easily incoroporated into the model.

The approximate response of a SDOF system model with USR hysteresis restoring force model was obtained using several values of input power spectral density. In the studies for

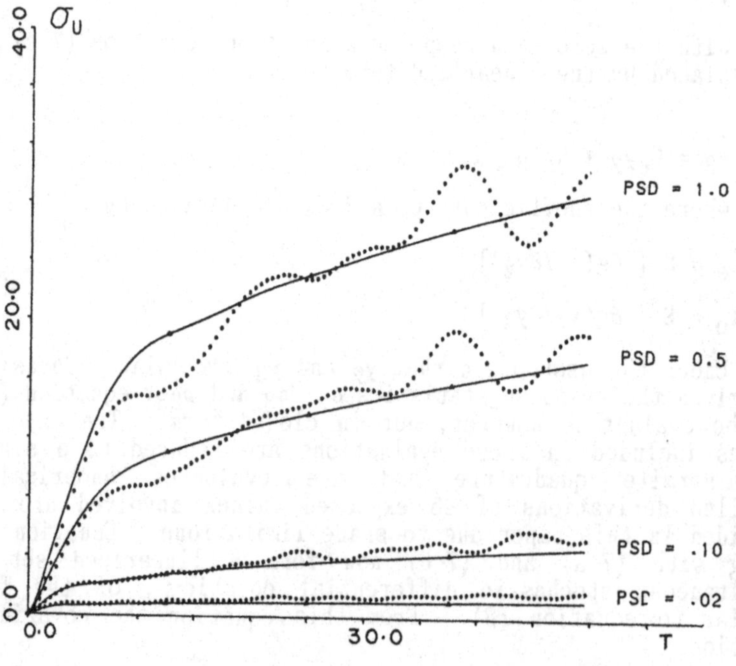

Figure 5 - RMS displacement response of the model under Gaussian white noise excitation.

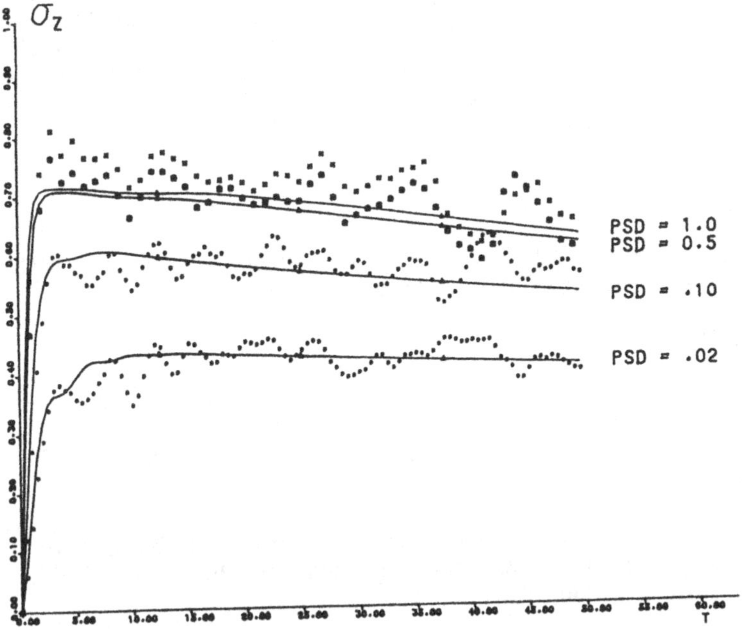

Figure 6 - RMS restoring force response of the model under Gaussian white noise excitation.

computation of zero time-lag covariance matrix, "high" pinching with parameter values of ζ_{10} = 0.85 and λ = 0.24 were considered. A system viscous damping ratio of 3% was chosen and parameter values of $\alpha_1 = 0.3$, $z_u = 1$, $z_{cp} = 0.2$, $z_{cn} = -0.2$, $z_{ch} = 0.02$ and $A_1 = 2.6$ ad $\alpha = 0.46$, $\omega_0 = 1r/s$, $\xi_0 = 0.2$, and $\delta\xi = 0.05$ were used. Simulated response estimates were obtained via 500 samples of Monte Carlo simulations. Figures 5 and 6 show the rms response displacement and restoring force predictions respectively for various levels of input excitation. As these figures show, the rms response prediction for the displacement, σ_u, is in good agreement with the simulation at all levels of excitation. For σ_z, the prediction underestimates the response at high excitation level, however, the trend of the response is well estimated.

CONCLUSION

A new general hysteresis model to predict very complicated deterioraton behavior of structural elements has been presented. This model is capable of reproducing various types of hysteretic behaviors including pinching and unloading stiffness degradation. Behavior of a SDOF system of this model under cyclic and random loading was studied. The obtained results indicate the versatility of the technique proposed earlier by Baber and Noori[3] for developing various types of general hysteresis. The model has a form that can be suitable for approximate solution by equivalent linearization technique. Nevertheless, the rather lengthy form of the governing equation should be simplified to allow for nonzero mean or non-Gaussian response analysis.

ACKNOWLEDGEMENTS

Authors would like to thank Mr. Pushpan and Ms. Elham Falsafi for their help in preparing the manuscript.

REFERENCES

1. Atalik, T. S. and Utku, S. Stochastic Linearization of Multidegree-of-Freedom Nonlinear System, Earthquake Engineering and Structural Dynamics, Vol. 4, pp. 411-420, 1976.

2. Baber, T. T. Nonzero Mean Random Vibration of Hysteretic Systems, Journal of Engineering Mechanics Division, ASCE, Vol. 110(7), 1036-1049, 1984.

3. Baber, T.T. and Noori, M.N. Modeling General Hysteresis Behavior and Random Vibration Application, ASME Journal of Vibration, Acoustics, Stress and Reliability in Design, Vol. 108, pp. 411-420, 1986.

4. Baber, T.T. and Noori, M.N. Random Vibration of Degrading Pinching Systems, ASCE Journal of Engineering Mechanics, Vol. III, No. 8, pp. 1010-1026, 1985.

5. Baber, T. T., and Wen, Y. K., Random Vibration of Hysteretic Degrading Systems, Journal of Engineering Mechanics Division, ASCE, Vol. 107, No. EM6, pp. 1069, 1981.

6. Baber, T. T. and Wen, Y. K. Stochastic Response of Multistory Yielding Frames. Earthquake Engineering and Structural Dynamics, 10, 403-416, 1982.

7. Bouc, R. Forced Vibration of Mechanical Systems with Hysteresis, Abstract Proceedings, 4th Conference on Nonlinear Oscillation, Prague, Czechoslovakia, 1967.

8. Casciati, F. Nonlinear Stochastics Dynamics of Large Structural Systems by Equivalent Linearization, Proceedings of the ICASP5, the Fifth International Conference on Application of Statistics and Probabiltity in Soil and Structural Engineering, University of British Columbia, Vancouver, 1987.

9. Cifuentes, A.O. System Identification of Hysteretic Systems, Ph. D. Dissertation, California Institute of Technology, Pasadena, 1984.

10. Clough, W. and Johnston, S.B. Effects of Stiffness Degradation on Earthquake Ductility Requirements Proceedings of the Second Earthquake Engineering Symposium, Tokyo, 227-232, October 1966.

11. Iwan, W.D. A Distributed-Element Model for Hysteresis and its Steady-State Dynamic Response, Journal of Applied Mechanics, ASME, Vol. 33, No. 4, 893-900, 1966.

12. Jayakumar, P. Modeling and Identification in Structural Dynamics, Ph. D. Dissertation, California Institute of Technology, Pasadena, 1987.

13. Noori, M.N., Choi, J.D. and Davoodi, H. Zero and Nonzero Mean Random Vibration Analysis of A New General Hysteresis Model, J. of Probabilistic Engineering Mechanics, Vol. 1, No. 4, pp. 192-201, 1986.

14. Noori, M. N., and Baber, T. T. Random Vibration of Degrading Systems With General Hysteresis, Technical Report on Research Grant No. CEE81-12584, submitted to the National Science Foundation, 1984.

15. Ozdemir, H. Nonlinear Transient Dynamic Analysis of Yielding Structures, PhD Dissertation, University of California, Berkeley, California, 1976.

16. Park, Y.J., Wen, Y.K., and Ang, A.H-S Random Vibration of Hysteretic Systems Under Bi-directional Ground Motions, Earthquake Engineering and Structural Dynamics, Vol. 14, pp. 543- 557, 1986.

17. Roberts, J. B. and Spanos P. D. Random Vibration and Statistical Linearization, John Wiley and Sons, NY, 1990

18. Spanos, P. D. Stochastic Linearization in Structural Dynamics, Applied Mechanics Review, 34(1), 1981.

19. Sues, R. H., Wen, Y. K., and Ang. A. H-S. Stochastic Seismic Performance Evaluating Building, Civil Engineering Studies Structural Research Series, No. 506, University of Illinois, 1981.

20. Suzuki, Y. and Minai, R. A Method of Seismic Response Analysis of Hysteretic Structures Based on Stochastic Differential Equations, Proceedings of the Eighth World Conference in Earthquake Engineering, Vol. 4, 459-466, 1984.

21. Takeda, T., Sozen, M.A. and Nielsen, N.N., Reinforced Concrete Response to Simulated Earthquakes, Journal of the Structural Division, Transactions of the ASCE, Vol. 96, 2557-2573, 1970.

22. Wen, Y.K. Stochastic Response and Damage Analysis of Inelastic Structures, Probabilistic Engineering Mechanics, Vol. 1, No. 1, 49-57, 1986.

23. Wen, Y. K. Equivalent Linearization for Hysteretic Systems Under Random Excitations, Journal of Applid Mechanics, Transactions ASME, 47(1), 150-154, 1980.

24. Wen, Y-K. Method for Random Vibration of Hysteretric Systems. Journal of Engineering Mechanics Division, ASCE. 102, EM4, 249-264, 1976.

A Method for Solving the Diffusion Equation with Random Coefficients

L. Carlomusto (*), A. Pianese (*), L.M. de Socio (**)
(*) Facoltà di Ingegneria, Università di Cassino, Italy
(**) Dipartimento di Meccanica e Aeronautica, Università "La Sapienza", Roma, Italy

Abstract

This papers introduces a procedure for solving a non-linear system of partial differential equations of the parabolic type with random coefficients. An adaptive method transforms the system into a set of ordinary differential equations which is solved after a suitable expansion of the probability density of the unknown random field in terms of orthogonal polynomials.

1. INTRODUCTION AND BASIC EQUATIONS

Several methods have been proposed for the solution of partial differential equations with random coefficients. Yet, a sufficiently general answer to the many questions arising when one deals with this kind of problem is still to come, in spite of the large amount of work already done (see, for references, [1], [2]). One of the reasons for the lack of a complete systematic treatment, apart from the intrinsic difficulties, is the noticeable variety of mathematical forms assumed by the models of physical situations related to these problems. Physics is the ground where the simulation of complex phenomena introduces, by necessity, simplified descriptions of real facts which correspond to PDEs with random coefficients.

Engineers' and applied mathematicians' need for sufficiently fast and accurate procedures to be implemented when dealing with these equations is, in many cases, satisfied by the adoption of *ad hoc* methods, the origin of which can be found in the solution algorithms for deterministic problems.

In this paper we show how systems of non-linear PPDs of the parabol i c type with random coefficients can be solved by a relatively simple method which is based on the Differential Quadrature Method [8]. For a simpler approach, the method will be introduced at the same time with an application. A particular attention situation will be considered in order to give the basic guidelines of the procedure. The application is, by itself, of relevance in engineering problems like those associated with energy storage systems in porous media, transpiration cooling of materials in a hot environment and so on.

Consider an infinite porous slab bounded by two parallel walls at a dimensionless distance equal to unity. One of the boundaries is kept adiabatic whereas the other one is subjected to radiative/convective conditions. At the time zero two initial temperatures profiles $T_{f_0}(x)$ and $T_{s_0}(x)$ are assigned along the cross coordinate x to the two phases which are present in the slab, *i.e.*: to the fluid and to the solid matrix.

The problem is to determine the evolution with the time t of the temperature distributions T_f and T_s in the fluid and in the solid phase, all physical quantities being made dimensionless with respect to proper reference units.

The governing equation is the energy conservation equation written for each one of the two media, subjected to the already indicated initial and boundary conditions [3]. In particular the differential problem can be given the following form:

$$\frac{\partial T_f}{\partial t} = A(T_s - T_f) - U\frac{\partial T_f}{\partial x} + B\frac{\partial^2 T_f}{\partial x^2} \tag{1}$$

$$\frac{\partial T_s}{\partial t} = L(T_f - T_s) + H\frac{\partial^2 T_s}{\partial x^2} \tag{2}$$

$$\frac{\partial T_f}{\partial t} = M(T_f - T_e) \quad \text{for } t > 0 \text{ at } x = 0 \tag{3}$$

$$\frac{\partial T_s}{\partial t} = N(T_s - T_e) + R(T_s^4 - T_e^4) \quad \text{for } t > 0 \text{ at } x = 0 \tag{4}$$

$$\frac{\partial T_f}{\partial x} = 0 \quad \text{for } t \geq 0 \text{ at } x = 1 \tag{5}$$

$$\frac{\partial T_s}{\partial x} = 0 \quad \text{for } t \geq 0 \text{ at } x = 0 \tag{6}$$

$$T_f(x,0) = T_{f_0}(x) \tag{7}$$

$$T_s(x,0) = T_{s_0}(x) \tag{8}$$

The characteristic number U, which appears in the convective term of equation (1), is indicative of a possible cross flow, whereas T_e is the external temperature which has been assumed to be the same either in the convective heat exchanges at $x = 0$ wall (eq.s 3,4) and in the radiative transfer (eq. 4). The dimensionless parameters A, B, L, H, M, N, R can in general, be functions of the temperature. This fact and condition (4) make the differential problem nonlinear. In addition, the thermophysical properties of the system, which appear in the parameters, may be random with assigned distributions of the probability density. Therefore both T_f and T_s must be considered as random processes.

In the following, Bellman's Differential Quadrature method, when applied to the system of PDEs, yields a finite set of ODEs which can be solved by means of one of the standard procedures for dealing with the initial value problems. A similar method was originally applied to an hyperbolic problem in [4], whereas the extension of Bellman's ideas to the random heat equation was first proposed in [5] as an adaptive method. Furthermore, this work follows a related study [3] where an analogous problem was solved in the case where the parameters were deterministically assigned but the initial conditions were given a probabilistic form.

2. ANALYSIS

For the sake of simplicity, it will be assumed here that B, L, H, M, N, R, U, T_e are all constant and that the characteristic parameters A and L alone follow the law

$$A = A_0[1 + \mu_k(\omega)] \ ; \ L = L_0[1 + \mu_k(\omega)] \tag{9}$$

where the zero mean random variable $\mu(\omega)$ has the $s + 1$ discrete realizations

$$\mu_k = 2\mu_s(k - s/2) \ / \ s \tag{10}$$

associated to a binomial distribution P_μ. The quantity ω is taken in the space of elementary events Ω. It is obvious that any other realistic distribution can be assumed instead of eq.s (9), depending on the particular physical situation and on the available experimental data.

The domains of the independent time (t) and space (x) variables are $I = (0, t*]$ and $D = [0, 1]$. The temperatures are thus stochastic processes

$$T_f(\omega, x, t) \ : \ \Omega \cdot I \cdot D \longrightarrow R \qquad ; \qquad T_s(\omega, x, t) \ : \ \Omega \cdot I \cdot D \longrightarrow R$$

with the related probability densities $P_f(T_f; x, t)$ and $P_s(T_s; x, t)$. In this problem the reference dimensional units were chosen in such a way so that $0 \leq T_f \leq 1; 0 \leq T_s \leq 1$.

Under appropriate hypotheses on the existence and uniqueness of the solutions $T_f(\omega, x, t)$ and $T_s(\omega, x, t)$ [6], the adaptive method can be applied and the probability densities $P_f(T_f; x, t)$ and $P_s(T_s; x, t)$ can be evaluated in order to determine the statistics of the two stochastic processes. To this end, after discretizing the space interval, the original set of PDEs is reduced to a finite system of ODEs by assuming that each unknown variable is given the expression

$$T(\omega, x, t) \approx \sum_{j=1}^{N} p_j T_j(\omega, t) \tag{11}$$

where j is a generic node and N is the total number of equispaced nodes in D, such that $j = 1$ corresponds to $x = 0$ and $j = N$ to $x = 1$. The p_j's are Lagrange-type polynomials, $T_j = T(\omega; x_j, t)$ and $p_i(x_j) = \delta_{ij}$ [3, 4, 5].

The boundary conditions have been, at this point, immediately implemented, by writing relations (3-6) in the proper end nodes. The initial conditions for the system of ODEs are also soon obtained from the imposed $T_{f_0}(x)$ and $T_{s_0}(x)$. The determination of the unknown stochastic field is then obtained by numerically solving the initial value problem corresponding to the finite set of ODEs, for a sufficient number of realizations of parameter A in the interval $D_\mu = [-\mu_s, \mu_s]$. Tis step can be made by one of the many available standard methods.

In order to evaluate the probability densities of the two random processes, as in [4], one considers the expansion

$$P_\alpha(T_\alpha; x, t) = w(T_\alpha) \cdot \sum_{n=0}^{\infty} c_n(x,t) f_n(T_\alpha) \tag{12}$$

where α stays either for f or for s, w is a weight function associated to the orthogonal polynomials f_n's and c_n's are the expansion coefficients. These coefficients are to be evaluated by imposing the orthonormality conditions. In the considered domain D, a proper choice leads to assuming Jacobi's polynomials $G_n(r, v, T_\alpha)$ as orthogonal polynomials, and, therefore, for them the weight function $\omega(T_\alpha)$ will be a β-distribution. Recall that one has

$$f_m(y) = \sum_{q=0}^{m} a_{m_q} y^q$$

and

$$c_m = \sum_{q=0}^{m} a_{m_q} E\{y^q\}$$

As a consequence one has, in each node and for each process

$$P_\alpha(T_{\alpha j}, t) = \omega(T_{\alpha j}) \cdot \left[1 + \sum_{m=1}^{\infty} G_m(r, v, T_{\alpha j}) \cdot \sum_{q=0}^{m} a_{mq} E(T_{\alpha j}^q) \right] \tag{13}$$

where the a_{mq}'s are constants and $E(T_{\alpha j}^q)$ is the q-order moment of $T_{\alpha j} = T_\alpha$ $(x = x_j)$

$$E(T_{\alpha j}^q = \int_{D_\mu} T_{\alpha j}^q(t;\mu) \cdot P_\alpha(\mu) d\alpha = \frac{1}{2^s} \sum_{k=0}^{s} \binom{s}{k} T_{\alpha j}^q(t;\mu_k) \tag{14}$$

which is a function of time.

3. RESULTS AND CONCLUSIONS

Some numerical calculations were carried out in order to prove the feasibility of the method and the numerical convergence characteristic of the procedure.

The parameters A_0, B, L_0, H, M, N, R were all taken equal to one, whereas $U = 0.1$ and $T_e = 0.1$. In (10), $s = 10$ and the number of nodes N was 11. Jacobi polynomials were for $r = 7, v = 4$, so that the β-distribution for the weight function was

$$\omega = (1 - T_\alpha)^3 \, T_\alpha^3$$

and the expansion (13) was arrested for $m = 5$.

As for the initial conditions, the ones reported in the graphs of Fig. 1.a were adopted.

The calculated probability densities P_f and P_s, in two sample nodes ($x = 0.5, x = 0.8$) are shown in Fig. 2 at the selected dimensionless time $t = 1, 4, 10$. However, as one would have expected, starting from the initial time, $t = 0$, when both P_f and P_s tend to infinity at $T_{f_0}(x = 0.5)$, $T_{f_0}(x = 0.8)$ and $T_{s_0}(x = 0.5)$, $T_{s_0}(x = 0.8)$, afterwards the probability densities spread around maximum values which initially decrease with time up to about $t = 10$ and subsequently tend again to infinity as $t \longrightarrow \infty$, while their domains of non-zero values (ΔT_f and ΔT_s) decrease. This is proved in Figs. 3.a and 3.b where the behaviour of ΔT_f and ΔT_s versus t is shown.

This is easily understood if one considers that $T_f(\omega; x, t \longrightarrow \infty) = 0.1$ and $T_s(\omega; x, t \longrightarrow \infty) = 0.1$, as shown in Fig. 1.b and can be seen by simply inspecting Eqs. (1) to (6).

The errors involved in the calculations derive from the discretization connected with the adopted algorithm for solving the initial value problem (in this case a 4-th order Runge-Kutta procedure) and from arresting the expansion in terms of orthonormal polynomials of the probability densities. The first two sources of error can be discussed following the guidelines given in [7]. As far as the error connected with the truncation of the expansion (13) is concerned, an indication about the convergence of the calculations is given by the values of the higher order moments of the variables. In this regard, Tables 1.A and 1.B show the evolutions of $E(T^q_{sj})$ and $E(T^q_{fj})$ for $t = 1$ and $t = 10$, respectively. It is evident that the results are quite good.

As a conclusion one can say that a simple fast and accurate enough approach to the solution of non-linear systems of parabolic, partial differential equations with random coefficients has been proposed, which can prove to be very suitable when dealing with the mathematical models of physics and engineering.

ACKNOWLEDGEMENTS

This work has been partially supported by C.N.R. - CT-87.02710.07, and C.N.R. - G.N.F.M..

REFERENCES

1. Adomian G., *Stochastic Systems*, Academic Press, New York, 1983.
2. Soong T.T., *Random Differential Equations in Science and Engineering*, Academic Press, New York, 1973.
3. Carlomusto L., Pianese A. and de Socio L.M., "Heat Transfer for the Two Phases of a Porous Storage Matrix", in *Energy Storage Systems* (Kilkis B. and Kakac S. eds.), pp. 721-728, Kluwer Ac. Publ., The Netherlands, 1989.

q \ x_j	$j=2$	$j=4$	$j=6$	$j=8$	$j=10$
0	1	1	1	1	1
1	$1.01\cdot10^{-1}$	$2.27\cdot10^{-1}$	$3.38\cdot10^{-1}$	$5.21\cdot10^{-1}$	$5.75\cdot10^{-1}$
2	$1.03\cdot10^{-2}$	$5.17\cdot10^{-2}$	$1.50\cdot10^{-1}$	$2.72\cdot10^{-1}$	$3.30\cdot10^{-1}$
3	$1.04\cdot10^{-3}$	$1.18\cdot10^{-2}$	$5.82\cdot10^{-2}$	$1.42\cdot10^{-1}$	$1.90\cdot10^{-1}$
4	$1.06\cdot10^{-4}$	$2.67\cdot10^{-3}$	$2.26\cdot10^{-2}$	$7.37\cdot10^{-2}$	$1.09\cdot10^{-1}$
5	$1.07\cdot10^{-5}$	$6.08\cdot10^{-4}$	$8.75\cdot10^{-3}$	$3.84\cdot10^{-2}$	$6.27\cdot10^{-2}$
6	$1.09\cdot10^{-6}$	$1.38\cdot10^{-4}$	$3.39\cdot10^{-3}$	$2.00\cdot10^{-2}$	$3.60\cdot10^{-2}$
7	$1.11\cdot10^{-7}$	$3.14\cdot10^{-5}$	$1.31\cdot10^{-3}$	$1.04\cdot10^{-2}$	$2.07\cdot10^{-2}$
8	$1.12\cdot10^{-8}$	$7.14\cdot10^{-6}$	$5.10\cdot10^{-4}$	$5.44\cdot10^{-3}$	$1.19\cdot10^{-2}$
9	$1.14\cdot10^{-9}$	$1.62\cdot10^{-6}$	$1.99\cdot10^{-4}$	$2.83\cdot10^{-3}$	$6.83\cdot10^{-3}$
10	$1.16\cdot10^{-10}$	$3.69\cdot10^{-7}$	$7.66\cdot10^{-5}$	$1.48\cdot10^{-3}$	$3.93\cdot10^{-3}$

Table 1.A: Values of $E(T^q_{x_j})$, for $t = 1$.

q \ x_j	$j=2$	$j=4$	$j=6$	$j=8$	$j=10$
0	1	1	1	1	1
1	$4.50\cdot10^{-2}$	$6.25\cdot10^{-2}$	$1.00\cdot10^{-1}$	$1.51\cdot10^{-1}$	$1.81\cdot10^{-1}$
2	$2.04\cdot10^{-3}$	$3.93\cdot10^{-3}$	$1.01\cdot10^{-2}$	$2.29\cdot10^{-2}$	$3.26\cdot10^{-2}$
3	$9.34\cdot10^{-5}$	$2.48\cdot10^{-4}$	$1.02\cdot10^{-3}$	$3.47\cdot10^{-3}$	$5.90\cdot10^{-3}$
4	$4.29\cdot10^{-6}$	$1.58\cdot10^{-5}$	$1.03\cdot10^{-4}$	$5.26\cdot10^{-4}$	$1.07\cdot10^{-3}$
5	$1.99\cdot10^{-7}$	$1.00\cdot10^{-6}$	$1.04\cdot10^{-5}$	$7.99\cdot10^{-5}$	$1.93\cdot10^{-4}$
6	$9.26\cdot10^{-9}$	$6.44\cdot10^{-8}$	$1.06\cdot10^{-6}$	$1.21\cdot10^{-5}$	$3.50\cdot10^{-5}$
7	$4.34\cdot10^{-10}$	$4.14\cdot10^{-9}$	$1.07\cdot10^{-7}$	$1.85\cdot10^{-6}$	$6.35\cdot10^{-6}$
8	$2.05\cdot10^{-11}$	$2.68\cdot10^{-10}$	$1.10\cdot10^{-8}$	$2.82\cdot10^{-7}$	$1.15\cdot10^{-6}$
9	$9.72\cdot10^{-13}$	$1.74\cdot10^{-11}$	$1.12\cdot10^{-9}$	$4.30\cdot10^{-8}$	$2.09\cdot10^{-7}$
10	$4.64\cdot10^{-14}$	$1.14\cdot10^{-12}$	$1.15\cdot10^{-10}$	$6.57\cdot10^{-9}$	$3.81\cdot10^{-8}$

Table 1.B: Values of $E(T^q_{x_j})$, for $t = 10$.

4. Bonzani I., Riganti R., "On the Probability Density of Random Fields in Continuum Mechanics", Stochastic Analysis Appl., 1, pp. 1-18, 1989.

5. Bellomo N., de Socio L.M. and Monaco R., "Random Heat Equation: Solutions by the Stochastic Adaptive Interpolation Method", Comp. Math. Appl., 16, pp. 759-766, 1988.

6. Bellomo N., Flandoli F., "Stochastic Partial Differential Equations in Continuum Physics: on the Foundations of the Stochastic Adaptive Interpolation Method for Ito's Type Equations", Math. Comp. Simul., 2nd special issue on *Stochastic Systems Modelling*, (to appear).

7. Carlomusto L., de Socio L.M., Gaffuri G., and Pianese A., "An Efficient Method for Solving the Boundary Layer Equations", Proc. 5th Int. Conf. on Boundary and Interior Layers - computational and Asymptotic Methods, BAIL V, pp. 103-108, 1988, Shanghai, China.

8. Bellman R., Adomian G., *Partial Differential Equations*, Reidel, New York, 1985.

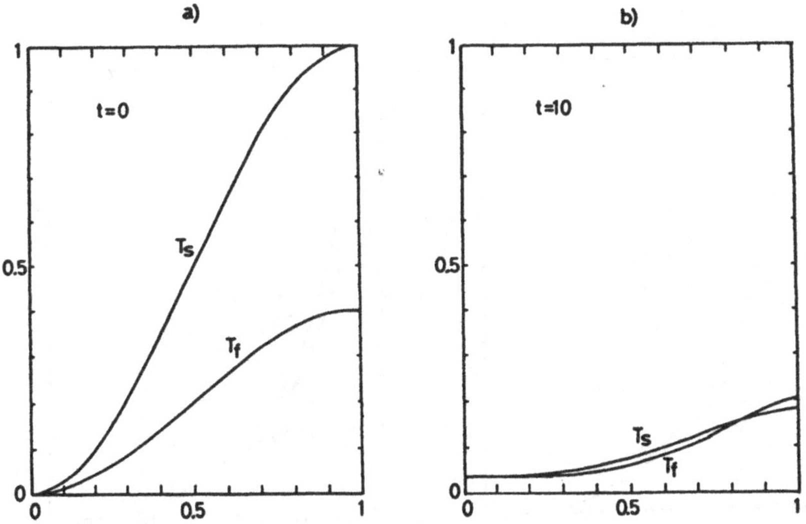

Fig. 1 - Temperature distributions in the two phases at $t = 0$ (a) and $t = 10$ (b).

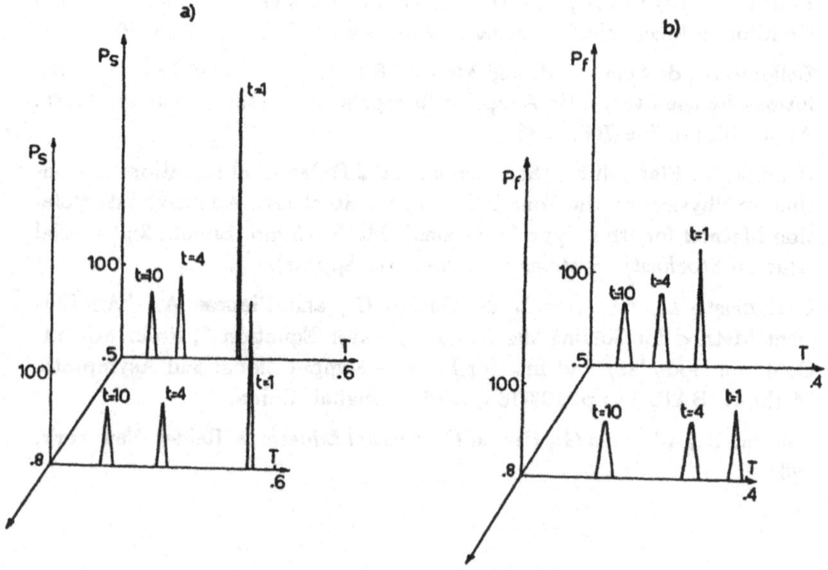

Fig. 2 - Probability density distributions of the temperature in the solid (a) and in the fluid (b) at selected times.

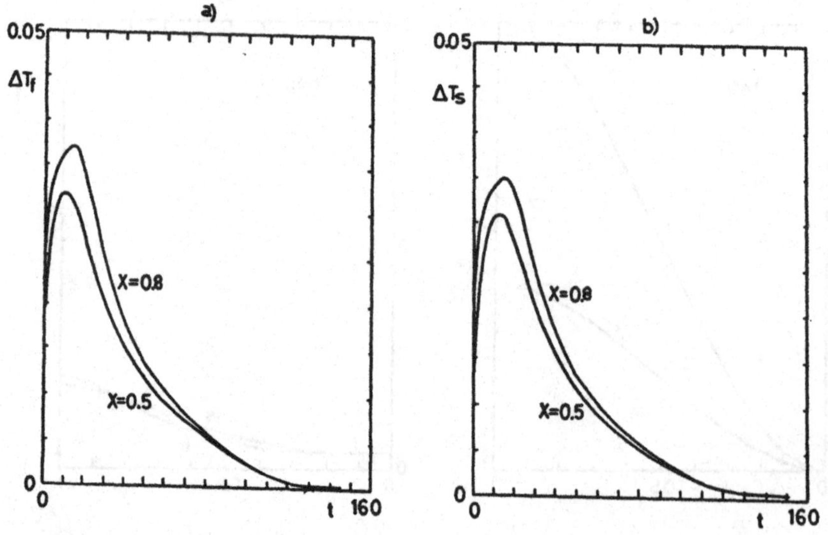

Fig. 3 - Domains of non-zero temperature values in the fluid (a) and in the solid (b) vs. time in two indicative locations.

Response Statistics of Nonlinear Systems Under Variations of Excitation Bandwidth

P.K. Koliopulos (*), S.R. Bishop (*),
G.D. Stefanou (**)
(*) University College London, Gower Street, London, WC1E 6BT, U.K.
(**) University of Patras, Greece

ABSTRACT

The response of a hardening Duffing oscillator to coloured noise excitation is considered. It is shown that for certain combinations of excitation intensity and bandwidth the system realises multi-valued response states. Theoretical predictions of bounds in the parameter space which define regions of multiple statistical moments are supported by numerical simulation results. The effect of the occurrence and persistence of multi-level responses on the probability distribution of the peaks is also considered.

INTRODUCTION

It is well known that under a periodic input, nonlinear equations of motion exhibit many interesting phenomena, including the folding over of the resonance response curve or sub-harmonic motions both of which introduce a multiplicity of solutions. Systems driven by narrow-band random excitation often display similar behaviour to their deterministic counterparts but are less well understood. The statistical significance of the response jumping between competing attractors is reflected in non-unique "temporal" statistical moments and possible changes of the shape of the distribution of peaks, leading to clear consequences for fatigue damage accumulation.

The effective use of the powerful theory of Markov processes for the analysis of such systems is precluded, without limiting approximations, through the inherent assumption of a broad-band input. Alternative approaches are based on the implementation of "non-local" Quasi-Static methods [1] or the "local" methods such as Stochastic Linearisation, Multiple time scaling, Stochastic averaging. These techniques have been used in the past to produce estimates of response statistical moments with variable success [2-8]. Relatively few are concerned with the problem of estimating probability densities [9-11]. What is lacking is a parametric study which will assess the relative merits of both "local" and "non-local" methods in predicting the occurrence and persistence of multiple moments and the effects on further response statistics.

In this study a methodology is developed for the class of problems which involve a hardening Duffing oscillator subjected to filtered white noise. The shaping filter considered is of order 2. The set of stochastic differential equations considered are:

$$\ddot{x} + 2\alpha\dot{x} + \omega_o^2(x + \gamma x^3) = F(t) \tag{1}$$

where

$$\ddot{F} + 2\xi_f\omega_f\dot{F} + \omega_f^2 F = W(t) \tag{2}$$

The main objectives of the methodology is to enable the designer to analyse a nonlinear structure in random environment and assess whether multi-valued response statistics are likely to occur and in such a case how to estimate them. Furthermore, the effect of these jumps on the statistical distribution of the peaks of the response is of great interest and a suggested mixture technique to predict the shape of the distribution is tested.

PREDICTION OF THE REGION OF MULTIPLE RESPONSE STATES

This section is concerned with the development of theoretical methods for the construction of bounds in the parameter space which will define the regions of possible multiple states of the response. In particular the Equivalent Linearisation and the Quasi-Static methods are implemented and their performance assessed via extensive numerical simulation results.

Equivalent Linearisation Approach (Local)

Assumptions - Limitations: In the classical version of this method the response of the system is assumed to be near-Gaussian. This approximation may be considered to be reasonable in the cases where the excitation is wide-band and the true probability density of the response has only one peak (indicating no jumps). In fact simulation studies have confirmed that predictions of the second order response statistics of a variety of nonlinear oscillators to wide-band noise compare well with numerical results [12]. The accuracy of this approach for narrow-band excitation where domains of multiple solutions may occur is clearly doubtful. A number of digital experiments confirm that in certain cases the probability density function is significantly different from the Gaussian, and the theoretical predictions of response variance do not correlate well with experiments [5,6]. To answer these questions and establish the regimes of falsity of the above analysis extensive simulation studies are supplied in the end of this section.

Achievements: The original nonlinear system is replaced by an "equivalent" linear system. For a Duffing oscillator of the form of equation (1) the corresponding "equivalent" system can be written as:

$$\ddot{x} + 2\alpha\dot{x} + \omega_l^2 \cdot x = F(t) \tag{3}$$

where ω_l is defined so as to minimise the square error of such an approximation, i.e.

$$\omega_l^2 = \omega_o^2(1 - \gamma\langle x^4\rangle/\langle x^2\rangle) \tag{4}$$

At this point one has to make an assumption for the relationship between the fourth statistical moment of the response with its variance. If the assumption is that the response is nearly Gaussian then $\langle x^4\rangle \approx 3\langle x^2\rangle^2 = 3\sigma_x^4$ and the corresponding natural frequency of the equivalent linear system is given as

$$\omega_l^2 \approx \omega_o^2(1 + 3\gamma\sigma_x^2) \tag{5}$$

If the assumption is that a very narrow-band input will cause an almost sinusoidal response then the relationship between second and fourth moment has to be modified to $\langle x^4\rangle \approx 1.5\langle x^2\rangle^2 = 1.5\sigma_x^4$ while the corresponding natural frequency of the system now becomes

$$\omega_l^2 \approx \omega_o^2(1 + 3\hat{\gamma}\sigma_x^2) \tag{6}$$

with $\hat{\gamma} = \gamma/2$. This indicates that the rest of the analysis can be carried out under the Gaussian assumption and the final results can be of use even if the response is of a sinusoidal type. In such a case a modified value for the parameter of the nonlinearity has to used; the modification consisting of a scaling of the parameter value by the ratio of the assumed kurtosis to the Gaussian one. In what follows, the equation (5) will be used unless otherwise stated.

Employing the conventional input-output relations for the extended system of the shaping filter (2) followed by the linearised oscillator (3, 5) an expression for the variance of the response is found as follows:

$$r = 1 + \frac{\{(3\eta/\delta)\cdot[\delta r^2 + \varepsilon\nu^2 + 4\varepsilon\delta(\varepsilon+\delta)]\}}{\{r(r-\nu^2)^2 + 4r(\varepsilon+\delta)\cdot(\varepsilon r + \delta\nu^2)\}} \tag{7}$$

where

$$\nu = \omega_f/\omega_o, \quad \delta = \alpha/\omega_o, \quad \varepsilon = \nu\cdot\xi_f, \quad r = 1 + 3\cdot\gamma\cdot\sigma_x^2, \quad \eta = \gamma\cdot\sigma_F^2/\omega_o^4$$

The above expression is almost identical with that obtained by Dimentberg [1]; however there are some errors in the final formula found in the above reference. The correct equation should read:

$$r = 1 + \frac{\{12\cdot(\varepsilon+\delta)\cdot(1+4\varepsilon\delta)^{-1}\cdot\lambda^2\cdot[\delta r + \varepsilon\nu^2 + 4\varepsilon\delta(\varepsilon+\delta)]\}}{\{r(r-\nu^2)^2 + 4r(\varepsilon+\delta)\cdot(\varepsilon r + \delta\nu^2)\}} \tag{8}$$

The new parameter λ is related to the variance of the response of an original linear system with an input exactly tuned to resonance (i.e. $\gamma = 0, \nu = 1$) and can be expressed in terms of the scaled variance η, as:

$$\lambda^2 = \eta\cdot\frac{1+4\varepsilon\delta}{4\delta(\delta+\varepsilon)} \tag{9}$$

It is clear that λ is a combination of input intensity and input bandwidth (proportional to ε) and hence equation (8) is not suitable for parametric studies of constant input intensity and variable bandwidth. In such cases equation (7) is preferred.

The advantage of the above analysis is that a computer search of the real and positive roots of equations (7) or (8) can be implemented to define regions in the parameter space of multiple solutions. A numerical search of multiple real roots of equation (7) over a wide range of parameter values, using the Laguerre's method [13], produced the regions in the parameter space where multiple response amplitudes are possible. These results are presented in figure 1 as functions of the central excitation frequency ν, the excitation bandwidth parameter ε and scaled excitation variance η (eta).

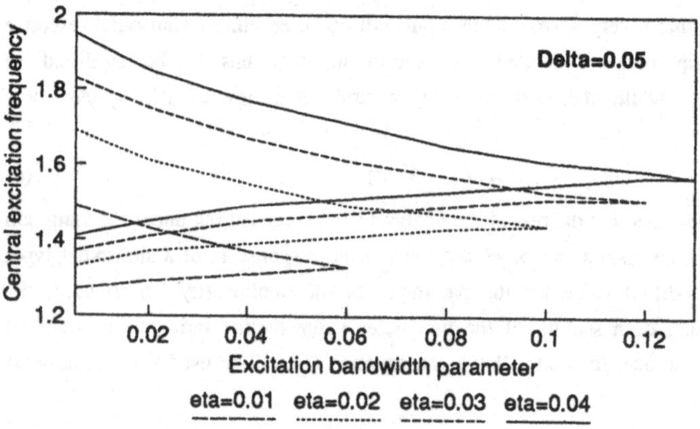

Fig.1 Regions of multiple response states as predicted via the Equivalent Linearisation technique.

The figure indicates that the domains of non-unique solutions are drastically reduced with the increase of the parameter ε which is proportional to the bandwidth of the excitation. The same trend has been observed with the increase of the system's damping, proportional to the parameter δ. If the forcing characteristics (intensity and bandwidth) are kept constant then the degree of nonlinearity determines whether or not the Statistical Linearisation method leads to non-unique solutions for the response variance σ_x^2. It is also interesting to note that for a given system and forcing bandwidth the increase of the excitation intensity results in a shift of the region towards higher excitation frequencies. This is expected since the increase of hardening nonlinearity results to an increase of the natural frequency of the system. Finally it is worth noting a remarkable observation made by Dimentberg [1], that the boundaries corresponding to the maximum input frequencies ν_{max} for possible non-unique response statistics follow a "single universal" curve in the (λ, ν) plane, defined by the equation $\nu_{max} = 1 + \lambda/2$. Along this curve the highest positive root of equation (8) corresponds to the case where the excitation frequency is equal to the "effective" natural frequency of the nonlinear system defined in equation (4). This law is still valid but not manifested in the (ε, ν) plane. In our notation the law can be expressed as follows:

$$\nu_{max} = 1 + 1/4 \cdot \sqrt{\eta \frac{(1+4\varepsilon\delta)}{\delta(\delta+\varepsilon)}} \tag{10}$$

Quasi-Static Approach (Non-local)

Assumptions - Limitations: Both excitation and response are narrow-band so that both can be represented as sinusoids with slowly varying amplitude and phase. In addition the excitation process is considered to be narrow-band with respect to the system's bandwidth, so that the forcing amplitude and phase are approximated as constant random variables within the cycles of oscillation. For example, in the case of a Duffing oscillator of the form of equation (1) subjected to a force described in (2), the Quasi-Static approach implies that ξ_f is very small, $\alpha \ll \omega_o$ and $\xi_f \omega_f \ll \alpha$. Under these assumptions the input force $F(t)$ and the resulting response $x(t)$ may be represented as:

$$x(t)=X_o(t)\cos(\omega_f t+\phi(t)) \quad , \quad F(t)=F_o(t)\cos(\omega_f t+\psi(t)) \tag{11}$$

where $X_o(t), F_o(t)$ and $\phi(t), \psi(t)$ all have slowly varying amplitudes and phase.

Achievements: The well-known closed form relation between the amplitude of the response and that of the excitation, obtained via classical deterministic analysis (e.g. Harmonic-Balance method or Perturbation techniques), can be used as the basis of a complete statistical analysis. For the case under consideration this relation reads:

$$\left[(\omega_o^2-\omega_f^2)X_o+\frac{3}{4}\gamma\omega_o^2X_o^3\right]^2+4\alpha^2X_o^2\omega_f^2=F_o^2 \tag{12}$$

This expression can now be used to express various statistical characteristics of $X_o(t)$ in terms of those of $F_o(t)$. In particular, critical excitation amplitudes defining the range of possible jumps of the response are easily obtained. In terms of the parameters introduced in equation (7) and defining two new variables, $y(t)$ and $\theta(t)$, the above relation reads:

$$y^3+\frac{8}{3}(1-\nu^2)y^2+\frac{16}{9}[(1-\nu^2)^2+4\delta^2\nu^2]y=\frac{32}{9}\theta \tag{13}$$

where $y(t)=\gamma X_o(t)^2$ and $\theta(t)=F_o(t)^2\gamma/2\omega_o^4$. Note that in both equations (12) and (13) the explicit dependence on time is neglected due to the slowly varying assumption. For the special case of small damping (i.e $\delta \ll |\nu-1|$) the threshold values which define the domain of possible jumps, are:

$$\theta_{min}=(8/3)\delta^2\nu^2(\nu^2-1) \qquad \theta_{max}=(8/81)(\nu^2-1) \tag{14}$$

It is important to note that equation (13) is a memoryless nonlinear transformation of the random variable θ and hence it cannot be treated in the same fashion as the deterministic equation (7). One simplistic way to proceed the analysis further is to replace θ with its mean value. Since the input process $F(t)$ models the output of a linear shaping filter driven by white noise follows the Gaussian distribution and its amplitude F_o is Rayleigh distributed

$$p(F_o)=\frac{F_o}{\sigma_F^2}\cdot\exp\left[-\frac{1}{2}(F_o/\sigma_F)^2\right] \quad , \quad \langle F_o^2\rangle=2\sigma_F^2 \tag{15}$$

Substituting this result into the expression of θ and recalling the definition of η in equation (7), it is clear that $\langle\theta\rangle=\eta$. A computer search can once again be implemented to solve the inverse problem rather than the one addressed in equation (14), that is of finding the range of excitation frequencies for which the given forcing amplitude lies within the "jump" region. For the special case when equation (14) is valid the following analytic expression holds:

$$\nu_{min}^2 = 1 + (81\eta/8) \quad , \quad \nu_{max}^2 = 0.5 + \sqrt{0.25 + (3\eta/8\delta)} \tag{16}$$

The above analysis however, suffers from two drawbacks; i) It is based on the rather "crude" replacement of the forcing term by its mean value and ii) due to the inherent assumptions it is independent of the excitation bandwidth. Despite these simplifications however, the implementation of this technique produces some very useful bounds which can be supplemented by those obtained via the Equivalent Linearisation method, as reported in [14].

Another approach suggested by Robin Langley [15] is to treat equation (13) in a purely probabilistic fashion. For each pair of parameters (δ, ν), the corresponding threshold values of $(\theta_{min}, \theta_{max})$ or equivalently (F_{min}, F_{max}) are found. Then one can compute the probability that the random excitation amplitude F_o will lie between those bounds, i.e.

$$P(F_{min} \leq F_o \leq F_{max}) = \int_{F_{min}}^{F_{max}} F_o dF_o = \exp[-0.5(F_{min}/\sigma_F)^2] - \exp[-0.5(F_{max}/\sigma_F)^2] \tag{17}$$

This information however is of limited use and one could go further and calculate the persistence of such an excursion into the specified range. The mean upcrossing rate of the envelope of a Gaussian narrow-band process is given by

$$\nu_F^+ = \frac{q}{\sqrt{2\pi}} \cdot \left(\frac{\sigma_{\dot{F}}}{\sigma_F}\right) \cdot \left(\frac{F}{\sigma_F}\right) \cdot \exp[-0.5(F/\sigma_F)^2] \tag{18}$$

where q is the Vanmarcke's bandwidth parameter defined in terms of the moments of the one-sided power spectrum $G_F(\omega)$ of the input process as

$$q = \sqrt{1 - \frac{m_1^2}{m_0 m_2}} \quad , \quad m_j = \int_0^\infty \omega^j G_F(\omega)\, d\omega \tag{19}$$

The expected duration of an excursion is given by:

$$\langle T_c \rangle = \frac{P(F_{min} \leq F \leq F_{max})}{\nu_{F_{min}}^+ + \nu_{F_{max}}^+}$$

Finally setting the average period of excitation equal to $2\pi(\sigma_{\dot{F}}/\sigma_F)$, the average number of cycles between the bounds (F_{min}, F_{max}) is evaluated

$$\langle N_c \rangle = \left(\frac{1}{\sqrt{2\pi} q}\right) \cdot \left\{ \frac{\exp(-0.5F_{min}^2/\sigma_F^2) - \exp(-0.5F_{max}^2/\sigma_F^2)}{(F_{min}/\sigma_F) \cdot \exp(-0.5F_{min}^2/\sigma_F^2) + (F_{max}/\sigma_F) \cdot \exp(-0.5F_{max}^2/\sigma_F^2)} \right\} \tag{20}$$

This final relation can be easily recast in terms of the forcing parameters (θ, η) as follows:

$$\langle N_c \rangle = \left(\frac{1}{\sqrt{2\pi} q}\right) \cdot \left\{ \frac{\exp(-\theta_{min}/\eta) - \exp(-\theta_{max}/\eta)}{\sqrt{(2\theta_{min}/\eta)} \cdot \exp(-\theta_{min}/\eta) + \sqrt{(2\theta_{max}/\eta)} \cdot \exp(-\theta_{max}/\eta)} \right\} \tag{21}$$

This new approach does not suffer from the drawbacks mentioned in the simplistic analysis allowing for a more complete probabilistic treatment and perhaps more importantly it takes into account the bandwidth of the excitation via the Vanmarcke's parameter q. For the input process modelled according to equation (2) the bandwidth parameter is related to the damping ratio of the filter as

$$q^2 \approx 4\frac{\xi_f}{\pi}\cdot(1-1.1\xi_f) \quad , \quad \xi_f \leq 0.2 \tag{22}$$

Despite the above statements however, one has to remember that it is still in the framework of Quasi-Static analysis and thus valid only if the severe inherent assumptions are met. Additionally there is a subjective element in the choice of a critical number of cycles in assessing whether multiple states are likely or not. This later problem has proved to be a major one since the predicted number of cycles varied considerably for different choices of input intensity and bandwidth. It is also clear that as the central excitation frequency increases the number of forcing cycles in the multiple state regime will increase but the duration may not be long enough for the system to respond. To overcome this difficulty a comparison is needed between the number of forcing cycles and the number of response cycles during the same duration. This calibration requires the knowledge of the effective natural frequency of the system and to this end the results of the Equivalent Linearisation technique are employed.

Simulation studies.

A number of simulation studies have been performed for the system given by equations (1) and (2). The parameters of the oscillator are set as $\alpha=0.05, \omega_o^2=10, \gamma=0.2$ corresponding to the same nonlinear model as the one studied in [16]. Four different values of scaled variance η for the forcing process are considered (0.01-0.04) and for each variance level five values of the bandwidth parameter ε were used (0.005, 0.02, 0.04, 0.06, 0.08). These target input characteristics were obtained by varying accordingly the parameters of the shaping filter given in equation (2) and scaling the white noise intensity. The equation of motion of the Duffing oscillator was integrated via a 4th order Runge-Kutta routine with a time step equal to 1/40 of the dominant excitation period. The total length of each generated record was equal to 2000 forcing cycles from which only the last 1000 were kept in order to ensure stationarity. A total of 20 such records were generated in order to form the ensemble in which the statistical analysis was performed. Typical extracts of simulated response histories for constant input intensity (η) and a variation of central excitation frequency (ν) and bandwidth parameter (ε) are shown in figures (2-5). For clarity the maximum recorded response per forcing cycle is plotted. It is clear that under sufficient narrow banded input process the system mimics the deterministic behaviour. There is a fundamental difference however that in the random case the system realises both possible states during the same record independently of the initial conditions, in accordance with observations reported in [7]. In the cases where multiple states exist and are persistent an estimation of the global response statistical moments is rather misleading and "temporal" statistics are of major importance for the reliability of the system. On the other hand the effect of an increased bandwidth is to make the two states indistinguishable and hence global statistics are more meaningful.

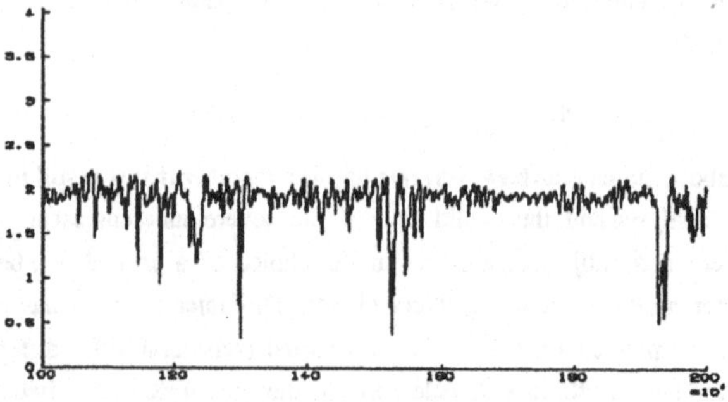

Fig.2 Response maxima $\eta=0.01$, $\varepsilon=0.005$, $\nu=1.25$

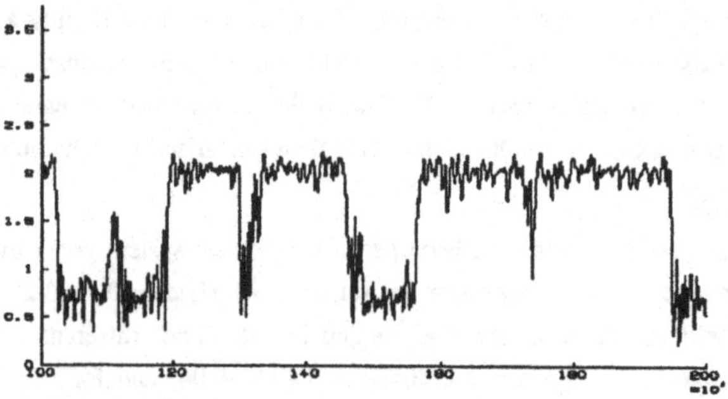

Fig.3 Response maxima $\eta=0.01$, $\varepsilon=0.005$, $\nu=1.28$

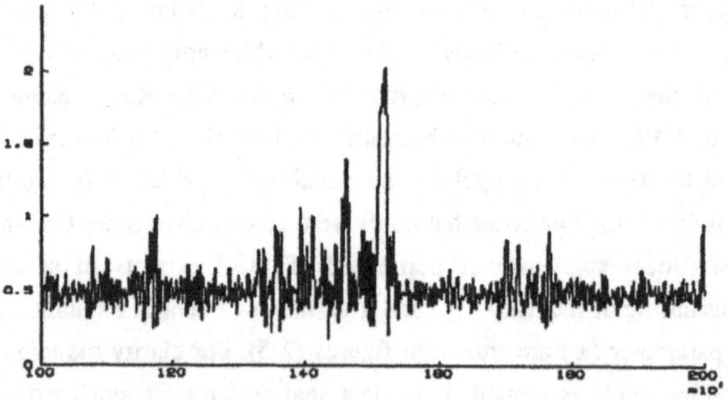

Fig.4 Response maxima $\eta=0.01$, $\varepsilon=0.005$, $\nu=1.35$

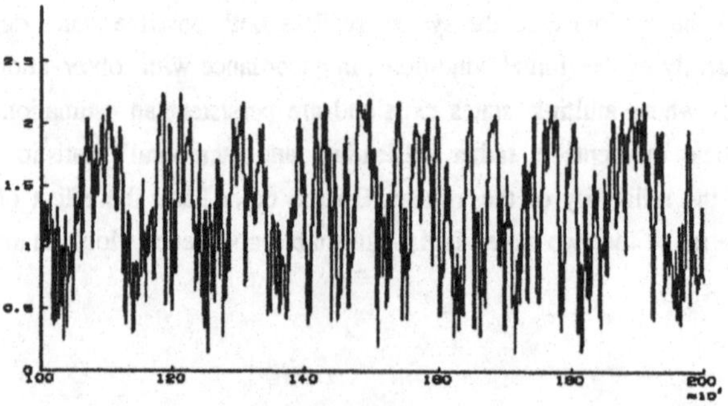

Fig.5 Response maxima $\eta=0.01$, $\varepsilon=0.04$, $\nu=1.28$

A comparison between the performance of the Equivalent Linearisation and the Quasi-Static techniques in predicting regions of multiple response moments with results obtained via simulation for two values of input intensity ($\eta=0.01, 0.03$) are shown in figures 6 and 7. The pair of lines labelled (E-L) are obtained via the Equivalent Linearisation method (equation 7) while those labelled (Q-S), are obtained via the Quasi-Static technique (equation 21). The middle dotted line labelled as (Q-S max) corresponds to the frequency at which equation 21 predicts the maximum expected number of forcing cycles during which the excitation remains within the range of critical values of amplitude; hence this frequency may be regarded as the one most inclined to result in multiple response states.

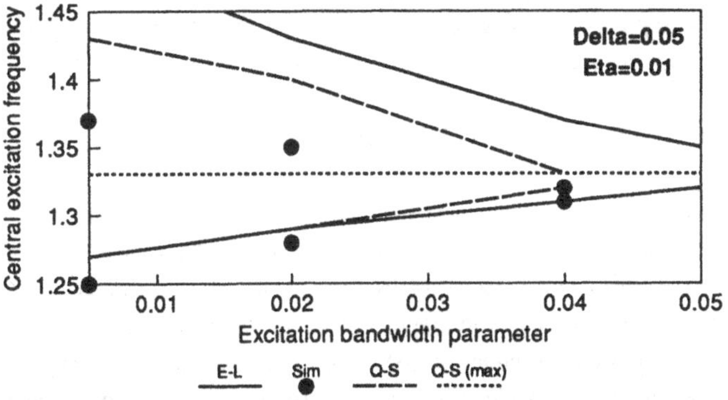

Fig.6 Regions of multiple solutions as predicted via Equivalent Linearisation (E-L), Quasi-Static (Q-S) and Simulation techniques.

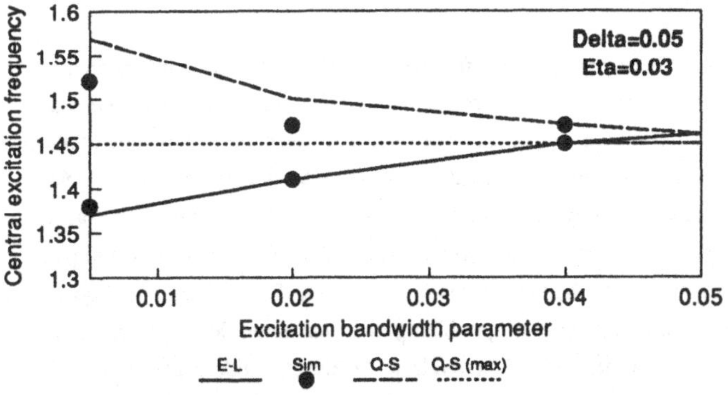

Fig.7 Regions of multiple solutions as predicted via Equivalent Linearisation (E-L), Quasi-Static (Q-S) and Simulation techniques.

The above graphs reveal the following:

i) The Equivalent Linearisation produces a very accurate lower bound of this region but a conservative upper bound. This tendency is related to an over-estimate of the effective natural frequency of the system and becomes stronger as the excitation intensity increases. On the other hand, the lower bound corresponds to frequencies which are related to the square root

of the effective frequency and hence the effect an error is significantly reduced.
ii) The Quasi-Static approach predicts a closer pair of bounds, provided that it is calibrated with the lower bound found via the Equivalent Linearisation technique. This calibration determines the minimum required number of cycles of the forcing process to lie between the threshold amplitude values in order that the system has sufficient time to experience multiple response levels.
iii) The effect of the increase of the excitation bandwidth is to reduce the region of possible multiple solution and eventually this region disappears.
iv) The frequency in which the Quasi-Static approach predicts the maximum probability of occurance of jump phenomena, is independent of input bandwidth. It is also interesting to note that this value is very close to the single value in which the bounds obtained via Equivalent Linearisation are reduced as the bandwidth increases (see figure 1).

PREDICTION OF THE TEMPORAL RESPONSE VARIANCE

Having established the regions of interest the next step is to try to estimate the multi-valued temporal response variance. Several publications have reported that the accuracy of the Equivalent Linearisation technique in the case of narrow-band random noise is variable [5,6]. In these works however no assessment on the regions in which the method is valid was made. The same applies in reports concerning the performance of the Quasi-Static method [16]. In the present study the results obtained from the simulation runs are compared with those obtained via the Equivalent Linearisation technique (numerical search of the roots of equation 8) and the simple version of Quasi-Static approach (numerical search of equation 13 by replacing θ with its mean value η).

Typical results of this comparison are for the case of $\eta=0.04$ are shown in figures 8-10. The shaded part corresponds to the region of possible jumps as predicted via the procedure explained in the previous section. It is seen that when the bandwidth is narrow enough the maximum recorded temporal variances compare well with the results obtained via the Quasi-Static method. Even in these cases however, it seems that more accurate estimates for the lower values are obtained via the Equivalent Linearisation technique. As the input bandwidth increases and the two states are less clearly distinguished, the Equivalent Linearisation becomes more accurate in predicting the maximum level. Finally, as the bandwidth increases even further, the input tends to become a wide-band noise and the response statistics are expected eventually to approach the unique solutions obtained by the Fokker-Planck formulation.

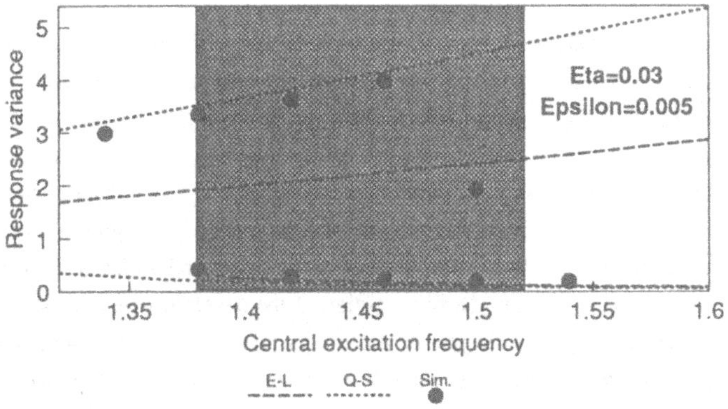

Fig.8 Response variance as predicted via Equivalent Linearisation (E-L), Quasi-Static (Q-S) and Simulation techniques.

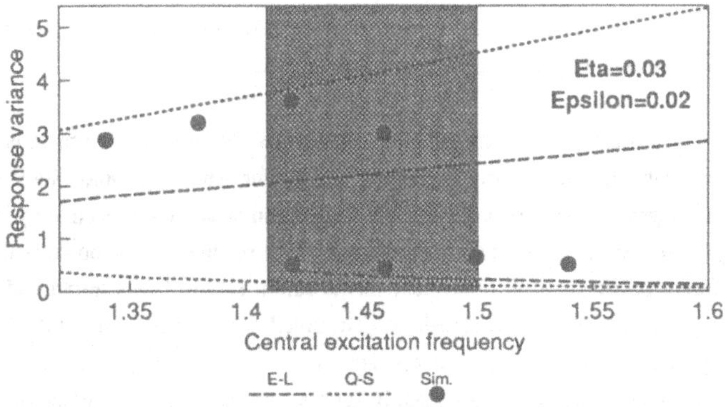

Fig.9 Response variance as predicted via Equivalent Linearisation (E-L), Quasi-Static (Q-S) and Simulation techniques.

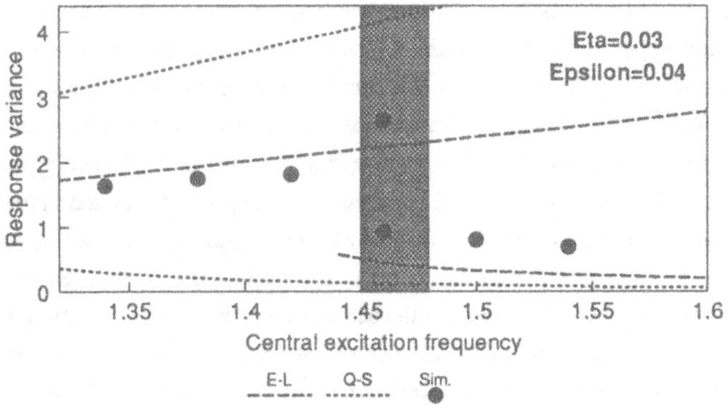

Fig.10 Response variance as predicted via Equivalent Linearisation (E-L), Quasi-Static (Q-S) and Simulation techniques.

PROBABILITY FUNCTION OF RESPONSE MAXIMA

The evaluation of the response statistical moments is not sufficient information for the determination of the reliability of a structure. Certain types of failure (e.g. fatigue damage accumulation) depend on the statistical behaviour of the peaks of the response process. Thus the effect of multiple response levels to the probability function of response maxima is of major importance. Among the reported works on this subject [9-11], only the most recent one [10], appears to have considered this problem for the class of excitation models similar to those adopted in the present work. The authors of the paper under discussion derived an approximate form of the probability density function of the envelope of the response via the stochastic averaging technique. The suggested probability density function shows two maxima in the cases where the response is multi-valued. After some algebra the suggested form can be expressed in our notation as follows:

$$p(X_o)=C\cdot X_o\cdot \exp\left\{\frac{-2\nu^2\delta\gamma}{(\varepsilon+\delta)\eta}X_o^2\cdot\left[(\varepsilon+\delta)^2+\frac{(\nu^2-1)^2}{4\nu^2}-\frac{3(\nu^2-1)\gamma X_o^2}{16\nu^2}+\frac{9\gamma^2X_o^4}{192\nu^2}\right]\right\} \tag{23}$$

where C is the scaling constant. A numerical integration of this relation results in the cumulative density function (cdf) for the envelope X_o.

An alternative method which incorporates the results obtained in the previous sections, is to first decide whether the system is operating in a regime where the Quasi-Static approach is valid or in a regime where the Equivalent Linearisation yields more accurate predictions. If for example, the excitation bandwidth is relatively high so that there is no clear distinction between different states then the Equivalent Linearisation should be preferred and it would predict that the response maxima are Rayleigh distributed. On the other hand if the bandwidth is sufficiently small so that the system operates in the multiple response region then the Quasi-Static formulation should be adopted. In such a case, the amplitude coincides with the envelope or maxima; hence for the rest of this section we will concentrate on the response amplitude.

Equation (13) is a memoryless transformation between the input random variable θ and the response variable y which is related to the response random amplitude X_o. For for the range of parameters where equation (13) is multi-valued, the standard probability transformation techniques should be supplemented with additional information of the relative stay time in the upper and lower response state. This relative stay time can be interpreted as the ratio of the probabilities that the system will be attracted to the upper or lower state provided that it operates within the multi-valued regime. This ratio can be simply evaluated by calculating the relative contribution of the two states to the total expected duration of an excursion given immediately after relation (19). Having estimated this ratio the cdf of the envelope process can be constructed as a mixture of two types of cdf. A numerical search of the roots of equation (13) in the region of multiple solutions yields the two sets of values of solutions; one corresponding to the lower solutions of equation (13) and the other to the upper. Each set is related to the same range of input values; hence the probability that the response stays within the lower set is equal to the probability that the input lies within the two threshold values of the multi-valued region, scaled by the ratio defined above. It is clear that the final result exhibits a step-function shape.

To test the performance of the above techniques the construction of the cdf of the response maxima for the two cases corresponding to figures 3 and 5 was attempted and the comparison between the theoretical and simulation results is shown in figure 11. The accuracy of the technique based on equation (23), labelled as "Davies", was very poor in the case of a very narrow input. Hence in the plot only the curve corresponding to the wide-band case is included. The Rayleigh approximation was adopted for $\varepsilon=0.04$ with the value of response variance predicted by the Equivalent Linearisation technique obtaining a satisfactory performance. Finally for the $\varepsilon=0.005$ case the Quasi-Static approach was implemented. The graph reveals that the Quasi-Static approach produces very accurate results within the region its validity (e.g. very small bandwidth).

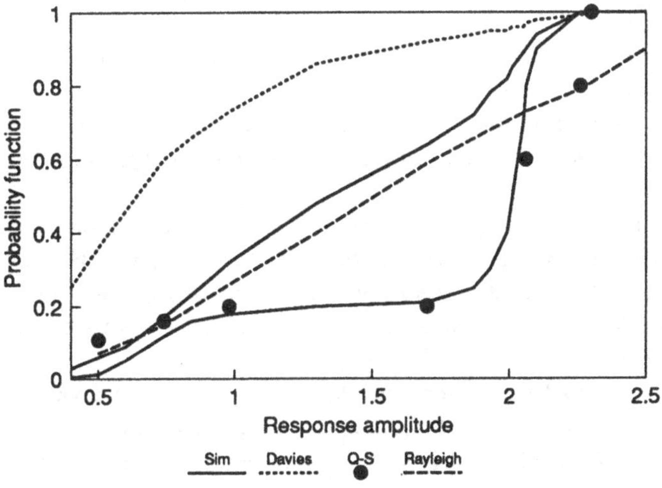

Fig.11 Cumulative probability function of the response maxima.
Input parameters: $\eta=0.01, \nu=1.28, \varepsilon=0.005, 0.04$

CONCLUSIONS

The present study examines the effect of a variation of the input bandwidth on the occurrence and persistence of multiple response moments. It is shown that for a given excitation intensity, bounds defining the region (in the parameter space) of possible multiple response states can be predicted utilising both local and non-local techniques. Furthermore the associated effects on the probability distribution of response maxima are also considered and a theoretical prediction based on a mixture technique is presented. A comparison with simulation results confirmed the accuracy of the proposed methodology.

REFERENCES

[1]. Dimentberg, M.F. Statistical Dynamics of Nonlinear and Time-Varying Systems, Research Studies Press Ltd., Somerset, England, 1988.

[2]. Davies, H.G. And Nandlall, D. Phase Plane for Narrow Band Random Excitation of a Duffing Oscillator, Journal of Sound and Vibration, Vol.104, No.2, pp.277-283, 1986.

[3]. Davies, H.G. And Rajan, S. Random Superharmonic Response of a Duffing Oscillator, Journal of Sound and Vibration, Vol.111, No.1, pp.61-70, 1986.

[4]. Fan, F.G. And Ahmadi, D. On Loss of Accuracy and Non-Uniqueness of Solutions Generated by Equivalent Linearization and Cumulant-Neglect Methods, Journal of Sound and Vibration, Vol.137, No.3, pp.385-401, 1990.

[5]. Iyengar, R.N. Stochastic Response and Stability of the Duffing Oscillator under Narrowband Excitation, Journal of Sound and Vibration, Vol.126, No.2, pp.255-263, 1988.

[6]. Langley, R.S. An Investigation of Multiple Solutions Yielded by the Equivalent Linearisation Method, Journal of Sound and Vibration, Vol.127, No.2, pp.271-281, 1988.

[7]. Rajan, S. And Davies, H.G. Multiple Time Scaling of the Response of a Duffing Oscillator to Narrow-Band Random Excitation, Journal of Sound and Vibration, Vol.123, No.3, pp.497-506, 1988.

[8]. Richard, K. And Anand, G.V. Non-Linear Resonance in Strings under Narrow Band Random Excitation, Part I: Planar Response and Stability, Journal of Sound and Vibration, Vol.86, No.1, pp.85-98, 1983.

[9]. Davies, H.G. The Response and Distribution of Maxima of a Non-Linear Oscillator with Band Limited Excitation, Journal of Sound and Vibration, Vol.90, No.3, pp.333-339, 1983.

[10]. Davies, H.G. And Liu, Q. The Response Envelope Probability Density Function of a Duffing Oscillator with Random Narrow- Band Excitation , Journal of Sound and Vibration, Vol.139, No.1, pp.1-8, 1990.

[11]. Roberts, J.B. Energy Methods for Non-Linear Systems with Non-White Excitation, Random Vibrations and Reliability, pp.285-294, Akademie-Verlag, Berlin, 1983.

[12]. Roberts, J.B. and Spanos, P.D. Random Vibration and Statistical Linearisation, John Wiley & Sons, Chichester, England, 1990.

[13]. Press, W.H., Flannery, B.P., Teukolsky, S.A. and Vetterling, W.T. Numerical Recipes: The Art of Scientific Computing (Fortran Version), Cambridge University Press, New York, 1989.

[14]. Koliopulos, P.K. And Bishop, S.R. Sensitivity of Multiple Responses in Nonlinear Systems Subjected to Random Excitation, accepted for presentation and proceedings of International Conference on Computational Engineering Science ICES-91, Melbourne, Australia, August 11-16, 1991.

[15]. Langley, R.S. Private communication.

[16]. Fang, T. And Dowell, E.H. Numerical Simulations of Jump Phenomena in Stable Duffing Systems, International Journal of Nonlinear Mechanics, Vol.22, No.3, pp.267-274, 1987.

Stochastic Equivalent Linearization of Nonlinear Complex Structures

F. Casciati (*), L. Faravelli (**), P. Venini (*)
(*) Department of Structural Mechanics, University of Pavia, I-27100 Pavia, Italy
(**) Institute of Energetics, University of Perugia, I-06100 Perugia, Italy

ABSTRACT

This paper studies the dynamical behaviour of hysteretic frames under stochastic excitation. They are discretized into elastic elements interconnected at potential plastic hinges where the whole inelastic deformation concentrates. The Bouc-Wen endochronic model is adopted to describe the hysteretic behaviour of the critical sections. The stochastic input is assumed to be stationary. A stochastic equivalent linearization of the equations of motion is introduced.

The iterative solution method requires, at each step, the study of a linear system of first order differential equations. This is done via complex modal analysis. The variation of the eigenproperties of the structures is studied and the mechanical behaviour is interpreted in terms of these eigenproperties.

INTRODUCTION

The dynamics of 2-D lumped mass frames, excitated at the base by forces of stochastic nature, is analyzed.

The connections between consecutive masses are idealized as linear-elastic elements. Rotations which concentrate at potential plastic hinges (critical sections) , localized at the end of each elastic beam-element, idealize the inelastic behaviour of the system under strong earthquake inputs. The adoption of the Bouc-Wen endochronic model[1][2]makes it possible to write the constitutive law at each critical section by expressing the generalized force as the linear combination of the generalized deformation and of an auxiliary variable. The latter one is related to its derivative and to the derivative of the deformation by a strongly nonlinear differential relationship (Bouc-Wen model). Stochastic equivalent linearization [3] substitutes this relationship with a linearized form of it [4][5].

A computer code [6][7] helps in assembling the structural matrices governing the dynamics of the frame. When the local constitutive equations are coupled with the global equations of motion, complex modal analysis can be used to find the solution. A pair of complex and conjugate eigenvalues, associated with each vibrating mass, and one real eigenvalue, related to the constitutive law in every plastic hinge, are found. The variation of the real eigenvalues, as the iteration proceeds, gives information about the sequential activation of the different plastic hinges, which turns into a variation of the linearization coefficients.

For stationary excitation the estimation of the linearization coefficients is managed after the evaluation of the stationary covariance matrix of the response. This matrix can be obtained via a classical time domain approach, but here a frequency domain approach is discussed. The true linearization coefficients are found at the end of an iterative process. The procedure generalizes the one of Refs [8] and [9] in view of exploiting the advantages of mode neglecting [9] in the analysis of complex structural systems [10][11][12].

A single story frame involving six potential plastic hinges is numerically studied. A discussion of the frequency domain solution is presented.

GOVERNING RELATIONS

This section discusses the evaluation of the stationary covariance matrix of the response via a frequency domain approach. For the determination of the motion equations the reader is referred to Refs. [6] and [7]. Nevertheless, without entering details, let

$$\mathbf{m}\ddot{\mathbf{u}} + \mathbf{g}(\mathbf{u}, \dot{\mathbf{u}}) = \mathbf{f}(t) \tag{1}$$

be the equation of motion of a generic multi degree of freedom system characterized by linear inertial forces: $\mathbf{m}$ is the mass matrix and $\mathbf{f}(t)$ is the vector of external excitations. The restoring force $\mathbf{g}$ is linearly related to the velocity while its dependency on the displacements can be written by separating the contribution of the elastic deformation from that of the inelastic one. The structural relation is written

$$\mathbf{V} = \mathbf{E}_1\mathbf{u} + \mathbf{E}_2\mathbf{q} \tag{2}$$

and the vectorial restoring force is

$$\mathbf{g} = \mathbf{E}_3\mathbf{V} + \mathbf{c}\dot{\mathbf{u}} \tag{3}$$

In equations (2) and (3) $\mathbf{E}_1, \mathbf{E}_2$ and $\mathbf{E}_3$ are structural matrices, $\mathbf{c}$ is the damping matrix, $\mathbf{u}$ is the vector of elastic generalized displacements and $\mathbf{q}$ is the vector of the inelastic ones.

The Bouc-Wen endochronic model for the local constitutive laws is then considered. This leads to additional relations involving $\mathbf{V}$ and $\mathbf{q}$:

$$\mathbf{V} = \mathbf{A}\mathbf{z} + \mathbf{B}\mathbf{q} \tag{4}$$

$\mathbf{z}$ being a vector of auxiliary variables related to $\mathbf{q}$ by means of the linearized equation

$$\dot{\mathbf{z}} = \mathbf{C}\mathbf{q} + \mathbf{H}\mathbf{z} \tag{5}$$

In equation (5) $\mathbf{C}$ and $\mathbf{H}$ are matrices of linearization coefficients. Substituting equations (2) and (3) into equation (1), one obtains

$$\mathbf{m}\ddot{\mathbf{u}} + \mathbf{c}\dot{\mathbf{u}} + \mathbf{E}_3\mathbf{E}_1\mathbf{u} + \mathbf{E}_3\mathbf{E}_2\mathbf{q} = \mathbf{f}(t) \tag{6}$$

It can be rewritten in the equivalent form

$$\dot{\mathbf{y}_1} - \mathbf{y}_2 = \mathbf{0} \tag{7}$$

$$\dot{\mathbf{y}_2} + \mathbf{m}^{-1}\mathbf{c}\mathbf{y}_2 + \mathbf{m}^{-1}\mathbf{E}_3\mathbf{E}_1\mathbf{y}_1 + \mathbf{m}^{-1}\mathbf{E}_3\mathbf{E}_2\mathbf{y}_3 = \mathbf{m}^{-1}\mathbf{f}(t) \tag{8}$$

In equations (7) and (8) the notation $\mathbf{y}_1 = \mathbf{u}, \mathbf{y}_2 = \dot{\mathbf{u}}$ and $\mathbf{y}_3 = \mathbf{q}$ has been used.

Solving equations (2) and (4) for $\mathbf{z}$ and substituting the resulting expression into equation (5) one obtains

$$\dot{\mathbf{q}} = \mathbf{D}_1\mathbf{u} + \mathbf{D}_2\dot{\mathbf{u}} + \mathbf{D}_3\mathbf{q} \tag{9}$$

or equivalently

$$\dot{\mathbf{y}_3} = \mathbf{D}_1\mathbf{y}_1 + \mathbf{D}_2\mathbf{y}_2 + \mathbf{D}_3\mathbf{y}_3 \tag{10}$$

Equations (7), (8) and (10) all together give rise to the following linear system of differential equations:

$$\dot{\mathbf{d}} + \mathbf{L}\mathbf{d} = \mathbf{w} \tag{11}$$

where $\mathbf{d}$, $\mathbf{w}$ and $\mathbf{L}$ represent assembled vectors and matrix respectively. Further details about the forcing vector $\mathbf{w}$ are reported in Ref. [7].

Application to framed structures

Equation (11) is somewhat abstract but its specification to the case of framed structures follows straightforward as detailed in Ref. [7]. In this case the structural relation (2) can be specified as

$$\mathbf{M} = \mathbf{K}_1\mathbf{u} + \mathbf{K}_2\boldsymbol{\theta} \tag{12}$$

that is to say the vector of generalized forces $\mathbf{V}$ is represented by the vector of bending moments $\mathbf{M}$ at the critical sections and the inelastic generalized deformations $\mathbf{q}$ are nothing but plastic rotations $\boldsymbol{\theta}$ which take place at the end of each beam element.

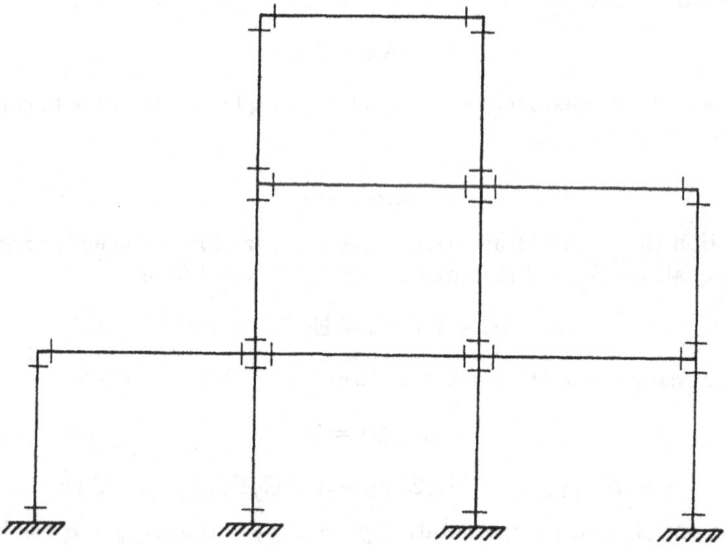

Figure 1: Generic multi-story frame and potential plastic hinges. One mass is lumped at each story: N=3; N_p=30

The equation of motion (6) can also be specified to the case of frames with the help of any general purpose computer code which automatically accounts for the influence of plastic rotations in the vibration of the lumped masses.

From now on, a N degree of freedom lumped mass frame will be considered, as the one shown in figure 1, where the location of the potential plastic hinges is sketched too. Let N_p be the number of potential plastic hinges for the whole structure; (N_p is 30 for the case of figure 1). The order N_t of the matrix **L** in equation (11) thus results

$$N_t = 2N + N_p \tag{13}$$

In more detail, system (11) is made by N equations of motion of the vibrating masses, which have been splitted in $2N$ first order differential equations, and by N_p constitutive laws, one for each section which is a potential location of plastic rotations.

SOLVING PROCEDURE

Equation (11) is solved via a complex modal analysis as reported in Refs. [8] and [9]. If $G(\omega)$ is the power spectral density function of the ground acceleration, then the stationary cross covariance between two response quantities is shown to be

$$E[y(l)y(m)] = \int_{-\infty}^{\infty} G(\omega)e^{i\omega(t_1-t_2)} \sum_{j=1}^{N_t}\sum_{k=1}^{N_t} \frac{q_j(l)}{i\omega + p_j}\frac{q_k(m)}{-i\omega + p_k}d\omega \tag{14}$$

for any time interval $t_1 - t_2$.

In equation (14), l and m indicate the l-th and the m-th component of **d** respectively, p_j is the jth eigenvalue of the matrix **L** and $q_j(l)$ is the product of the j-th participation factor with the m-th component of the j-th eigenvector of **L**.

The expected nature of the N_t complex eigenvalues p_j follows from a physical examination of the system under study. Matrix **L** provides N_p real eigenvalues, say

$$p_j = \nu_j \tag{15}$$

one for each potential plastic hinge and N pairs of complex and conjugate eigenvalues, say

$$p_j = \omega_j\beta_j \pm i\omega_j\sqrt{1-\beta_j^2} \tag{16}$$

giving N equivalent frequencies ω_j and N equivalent damping ratios β_j of the vibrating system.

The term with the summation over j, in the right hand side of equation (14), can now be specified as

$$\sum_{j=1}^{N_t} \frac{q_j(l)}{i\omega + p_j} = \sum_{j=1}^{N_p} \frac{q_{1j}(l)}{i\omega + \nu_j} + 2\sum_{J=1}^{N} \frac{R_J(l) + ia_J(l)\omega}{\omega_J^2 - \omega^2 + 2i\beta_J\omega_J\omega} \tag{17}$$

In equation (17) $q_{1j}(l)$ is the real $q_j(l)$ term associated with the real eigenvalue ν_j; $R_J(l) = \omega_J[a_J(l)\beta_J + b_J(l)\sqrt{1-\beta_J^2}]$ and $a_J(l), b_J(l)$ are the real and imaginary part of the complex term $q_J(l)$ associated with the complex eigenvalue p_j. The summation over k is easily written by substituting in equation (17) j, J and l with k, K and m respectively, and by taking the conjugate of the derived expression.

If one compares equation (17) with the similar form of Ref. [9], where one summation index was required, it is self evident that a further summation has been introduced. The reason is that the formulation of Ref. [9] provided an equal number of real eigenvalues and of pairs of complex ones because the number of vibrating masses was taken equal to the number of hysteretic elements. In the case now under study, on the contrary, each mass is related to several potential plastic hinges leading to an asymmetry between complex and real terms.

Substituting equation (17) in (14) and integrating, the following final expression is found for the stationary cross covariance:

$$E[y(l)y(m)] = T_1 + T_2 + T_3 \tag{18}$$

with the following meaning for the terms in the right hand side

$$T_1 = \sum_{j=1}^{N_p} q_{1j}(l)q_{1j}(m)I_{1j} + \sum_{j=1}^{N_p}\sum_{k=1}^{N_p} q_{1j}(l)q_{1k}(m)[A_{jk}I_{1j} + B_{jk}I_{1k}] \tag{19}$$

$$T_2 = 2\sum_{j=1}^{N}[A'_jI_{1j} + B'_jI_{2j} + C'_jI_{3j}] +$$

$$2q_j(l)\sum_{j=N+1}^{N_p}\sum_{k=1}^{N}[A''_{jk}I_{1j} + B''_{jk}I_{2k} + C''_{jk}I_{3k}] +$$

$$2q_k(m)\sum_{k=N+1}^{N_p}\sum_{j=1}^{N}[A''_{kj}I_{1k} + B''_{kj}I_{2j} + C''_{kj}I_{3j}] \tag{20}$$

$$T_3 = 4\sum_{j=1}^{N}[R_j(l)R_j(m)I_{2j} + a_j(l)a_j(m)I_{3j}] +$$

$$4\sum_{j=1}^{N}\sum_{k=1}^{N}[A^*_{jk}I_{2j} + B^*_{jk}I_{3j} + C^*_{jk}I_{2k} + D^*_{jk}I_{3k}] \tag{21}$$

In the last three equations every double summation term has to be calculated for $j \neq k$. Each group of coefficients A_{jk} B_{jk}, A'_j B'_j C'_j, A''_{jk} B''_{jk} C''_{jk} and A^*_{jk} B^*_{jk} C^*_{jk} D^*_{jk} is calculated via a partial fraction technique as reported in Ref. [9]. The terms I_{1j}, I_{2j}, I_{3j} are the following frequency integrals :

$$I_{1j} = \int_{-\infty}^{\infty} \frac{G(\omega)}{\omega^2 + \nu_j^2} d\omega \tag{22}$$

$$I_{2j} = \int_{-\infty}^{\infty} G(\omega)|F_j|^2 d\omega \tag{23}$$

$$I_{3j} = \int_{-\infty}^{\infty} G(\omega)|F_j|^2\omega^2 d\omega \tag{24}$$

$F_j(\omega)$ being the frequency response function of an oscillator governed by the classical second order linear differential equation.

Formula (18) is useful in calculating the elastic and the plastic contribution to the whole story drift. The T_3 term, in fact, represents the interaction of complex eigenvalues with themselves that is to say the effect of elastic equivalent frequencies and damping ratios on the story displacement. T_1 and T_2 represent the interaction of real eigenvalues with themselves and the cross term interaction respectively. Since the iterative equivalent linearization method starts with linearization coefficients close to the elastic values, one obtains a sort of elastic response at the first step and this is mainly provided by T_3. With the proceeding of the iteration, however, real eigenvalues grow up so that the other two terms become larger, giving the effect of the activation of the plastic hinges on the total displacement.

Figure 2: Single story frame and six potential plastic hinges

NUMERICAL EXAMPLE

A single story frame forced at the base by an idealized filtered white noise of two-sided power spectral density function of Kanai-Tajimi type (G_0=200/π sqi/s^3; ω_g=15.6 rad/s; ξ_g=0.69) is numerically studied. The physical and geometrical properties of the system are given in Ref. [5]. The sequence of the results as the iteration proceeds are given in Table 1 in terms of variances of the horizontal displacement of the girder and of the corresponding velocity. In the same table the evolution of the complex-conjugate eigenvalues is illustrated.

The presence of six potential plastic hinges is considered (Figure 2). Nevertheless this analysis exploits the symmetry of the structure and, hence, the number of plastic hinges is halved. The main assumption about the nature of complex eigenvalues of the system is numerically confirmed. Moreover one can easily associate each hinge with a real eigenvalue and localize sections where plastic rotations are in progress, simply verifying the growth of real eigenvalues as the iteration proceeds.

Table 2 describes the evolution of the elastic and inelastic properties of the system, as the iteration proceeds. The variation of real eigenvalues and of linearization coefficients is presented for some steps of calculus. The different behaviour of real eigenvalues as the iteration proceeds is worth nothing. Less meaningful eigenvalues and correspondent inactivate plastic hinges may be disregarded in the computation of the covariance matrix of the response after few steps when one has surely localized the sections which rotate and the sections which do not.

The results perfectly coincide with the ones reached in Ref.[5] by a time domain approach. They can also be achieved by working on the matrix **L** built over six plastic hinges. (In this case the corresponding real eigenvalues at step 7 are ν_1=0.0235 , ν_2=0.0236 , ν_3=0.0376 , ν_4=0.289, ν_5=0.4337 , ν_6=0.645; equivalent frequence and damping ratio are respectively ω=18.83, β=0.149)

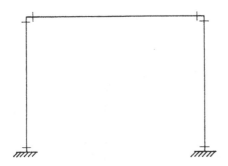

Table 1: Variances of the horizontal displacement (in inch) and the velocity (in inch/s) of the girder of the frame in Figure 2 as the solving iteration procedure proceeds. Evolution of the complex conjugate eigenvalue.

ITERATION	Var [u]	Var [$\dot{u}$]	ω	β
0	0.0703	21.474	19.09	0.129
1	0.0725	20.238	18.97	0.137
2	0.0740	19.534	18.90	0.142
3	0.0749	19.130	18.87	0.145
4	0.0755	18.897	18.85	0.147
5	0.0758	18.763	18.84	0.148
6	0.0760	18.685	18.83	0.149
7	0.0761	18.640	18.83	0.149

CONCLUSIONS

A frequency domain approach is presented in order to calculate the seismic response of multi story frames subjected to inputs of stochastic nature. Plasticity effects are idealized with plastic hinges localized at the end of each beam and column.

The Bouc-Wen endochronic model helps in writing the constitutive laws at each critical section and the consequent nonlinearity is approached via an equivalent linearization technique.

A good agreement is numerically shown between the presented method and the classical time domain approach. As already observed in Ref.[9] the possibility of neglecting higher modes is one of the main advantages of this frequency domain method: higher eigenfrequencies of the system in that case, the contribution of inactivate plastic hinges in this case can be disregarded via a similar algorithm.

ACKNOWLEDGEMENT

This research has been supported by grants from both the Italian Ministry of University and Research (MURST) and the National Research Council (CNR)

Table 2: Evolution of the linearization coefficients and of the real eigenvalues as the iteration proceeds for the example of figure 2.(CB=column bottom; CT=column top; G=girder)

ITERATION	C	H	loc.	ν_1	ν_2	ν_3
0	0.999	-0.02685	BC			
0	0.999	-0.02685	TC	0.0004	0.0004	0.0060
0	0.999	-0.05375	G			
1	0.937	-1.099	BC			
1	0.963	-0.655	TC	0.0105	0.0152	0.1784
1	0.928	-1.274	G			
2	0.905	-1.723	BC			
2	0.945	-0.986	TC	0.0165	0.0246	0.2863
2	0.891	-2.010	G			
3	0.889	-2.087	BC			
3	0.936	-1.158	TC	0.0199	0.0303	0.3516
3	0.871	-2.458	G			
4	0.880	-2.299	BC			
4	0.932	-1.247	TC	0.0217	0.0337	0.3903
4	0.860	-2.731	G			
5	0.875	-2.422	BC			
5	0.930	-1.292	TC	0.0227	0.0357	0.4129
5	0.854	-2.896	G			
6	0.872	-2.492	BC			
6	0.929	-1.314	TC	0.0232	0.0369	0.4261
6	0.851	-2.996	G			
7	0.871	-2.533	BC			
7	0.928	-1.325	TC	0.0235	0.0376	0.4337
7	0.849	-3.056	G			

References

[1] Bouc R., Forced Vibrations of a Mechanical System with Hysteresis, *Proc.4th Conf. Nonlinear Oscillations, Prague, Czechoslovakia*, (1967),

[2] Wen Y.K., Equivalent Linearization for Hysteretic Systems under Random Excitation, *J. App. Mech., Vol.47, pp.150-154*, (1980),

[3] Roberts J.B. and Spanos P.D., Random Vibration and Statistical Linearization, *John Wiley and Sons Ltd.*, (1990),

[4] Casciati F., Non Linear Stochastic Dynamics of Large Structural Systems by Equivalent Linearization, *Proc ICASP5, Vancouver, Vol 2, pp.1165-1172*, (1987),

[5] Casciati F. and Faravelli L., Fragility Analysis of Complex Structural Systems, *Research Studies Press Ltd., Taunton*, (1991)

[6] Casciati F. and Faravelli L., Methods of nonlinear Stochastic Dynamics for the Assessment of Structural Fragility, *Nucl. Eng. Des., Vol.90, pp.341-356*, (1985)

[7] Casciati F., Approximate Methods in Non-Linear Stochastic Dynamics, *in Schueller G. and Shinozuka M. (eds.), Stochastic Methods in Structural Dynamics* (1987),

[8] Singh M.P., Maldonado G., Heller R. and Faravelli L., Modal Analysis of Nonlinear Hysteretic Structures for Seismic Motions, *in Ziegler F.and Schueller G. (eds.), Nonlinear Structural Dynamics in Engineering Systems, Springer-Verlag, Berlin, pp.443-454*, (1988),

[9] Casciati F., Faravelli L. and Venini P., Neglecting Higher Complex Modes in Nonlinear Stochastic Vibration, *Proc. 4th Int. Conf. on Recent Advances in Str. Dyn.*, (1991),

[10] Casciati F. and Faravelli L., Stochastic Equivalent Linearization for 3-D Frames, *J. Eng. Mech., ASCE, Vol. 114(10), pp.1760-71*, (1988),

[11] Casciati F. and Faravelli L., Hysteretic 3-D Frames under Stochastic Excitation, *Res Mechanica, Vol.26, pp.193-213*, (1989),

[12] Casciati F. and Faravelli L., Non-Linear Stochastic Finite-Element Analysis of Continua, *Proc. ICOSSAR, Vol 2, pp. 1105-1112* (1989)

Numerical Solution of the Fokker-Planck Equation for First Passage Probabilities

B.F. Spencer, Jr.(*), L.A. Bergman (**)
Department of Civil Engineering, University of Notre Dame, Notre Dame, Indiana 46556-0767, U.S.A.
(**) Department of Aeronautical and Astronautical Engineering, University of Illinois at Urbana-Champaign, Urbana, Illinois 61801-0236, U.S.A.

ABSTRACT

An alternate method of accurately computing first passage probabilities for discrete linear and nonlinear systems of two dimensions subjected to additive white noise excitation is presented. The transient Fokker-Planck equation is solved on the safe domain defined by *D* barriers using the finite element method. Open boundary conditions are prescribed, and their physical basis is reviewed. A linear system and a Van der Pol system are examined. In both cases, no discernible difference was detected between solutions of the forward equation for distribution of first passage time and well-established first passage solutions of the backward Kolmogorov equation.

INTRODUCTION

An important problem in stochastic dynamics is the determination of the probability that a system, subjected to random disturbances, will malfunction during a specified period of time. When failure occurs upon the first excursion of the response process out of the prescribed safe region, the failure process is generally referred to as a first passage problem. This class of problems has been, and continues to be, one of considerable interest due to its relationship to the more general system reliability problem.

While the first passage problem has eluded analytical solution for all but the simplest scalar systems (*cf.* Darling and Siegert [6]), a vast body of approximation theory with which to address the problem has evolved over the past forty years. The interested reader can refer to Roberts [7] and Bergman [1] for an extensive account of the literature in this area.

Of particular relevance to the present work are those methods that utilize Markov process theory in conjunction with computational procedures to effect approximate solutions of the first passage problem for systems of two or more dimensions. Among these include the random walk method of Toland [12] and Toland and Yang [13], the cellular diffusion method of Chandiramani [3] and Crandall, Chandiramani and Cook [4], the finite element solution to the backward Kolmogorov equation of Bergman and Spencer [2] and Spencer [9], and the generalized cell mapping algorithms of Sun and Hsu [10], [11].

In all of these, the response of the dynamical system is assumed to be Markovian, leading to a differential or integral equation governing the evolution of the transition probability density function in phase space. The rate at which probability mass is lost through boundaries in the phase space defines the failure process for the system.

Crandall [5] described the first passage problem in terms of the Fokker-Planck equation and conjectured that the problem is well-posed with open boundaries. Yang and Shinozuka [14] proposed that the backward Kolmogorov equation, rather than the forward Kolmogorov (Fokker-Planck) equation, be solved directly for the reliability function, subject to boundary conditions imposed on physical grounds. Spanos [8], employing stochastic averaging, utilized the Fokker-Planck equation to analytically obtain the reliability function associated with the first passage of the amplitude process of a single degree of freedom oscillator. Sri Namachchivaya [10] examined boundaries associated with one dimensional first passage problems and formalized their classification in terms of scale and speed measures.

In this paper, a boundary value problem is formulated for first passage probabilities in terms of the Fokker-Planck equation. Solution by the finite element method is presented, and several examples are given. The formulation is shown to be computationally tractable for two dimensional systems.

FORMULATION

An appropriate stochastic differential equation describing a time-independent discrete dynamical system subject to additive white noise is given by

$$\frac{d\mathbf{x}}{dt} = \mathbf{g}(\mathbf{x}) + \mathbf{H}\mathbf{w}(t) \tag{1}$$

Here, $\mathbf{x}$ and $\mathbf{g} \in R^n$, $\mathbf{w}(t) \in R^m$ is a vector white noise, and $\mathbf{H}$ is an $n \times m$ dimensional matrix of constant coefficients. The moments of the white noise are

$$E[\mathbf{w}(t)] = 0 \tag{2}$$

$$E[\mathbf{w}(t)\mathbf{w}^T(t+t')] = 2\mathbf{D}\delta(t') \tag{3}$$

where $\delta(\cdot)$ is the Dirac delta function and $\mathbf{D}$ is an $m \times m$ matrix with constant coefficients.

It is well known that Eqs. (1)-(3) define a vector Markov process, the behavior of which is completely determined by the transition probability density function $p(\mathbf{x}, t | \mathbf{x}_0)$, where $\mathbf{x}_0 = \mathbf{x}(0)$. The density satisfies both the forward Kolmogorov (Fokker-Planck) equation

$$\frac{\partial p}{\partial t} = \mathcal{L}_{\mathbf{x}} p \quad \text{on } \Omega \tag{4}$$

and backward Kolmogorov equation

$$\frac{\partial p}{\partial t} = \mathcal{L}^*_{\mathbf{x}_0} p \quad \text{on } \Omega \tag{5}$$

subject to the initial condition

$$p(\mathbf{x}, 0 | \mathbf{x}_0) = \delta(\mathbf{x} - \mathbf{x}_0) \tag{6}$$

The operators $\mathcal{L}_{\mathbf{x}}$ and $\mathcal{L}^*_{\mathbf{x}_0}$ are, respectively,

$$\mathcal{L}_{\mathbf{x}} u = \sum_{i=1}^{n} \sum_{j=1}^{m} \sum_{k=1}^{m} \sum_{l=1}^{n} D_{jk} \frac{\partial}{\partial x_l} \left[H_{ij} \frac{\partial}{\partial x_l} (H_{lk} u) \right] - \sum_{l=1}^{n} \frac{\partial}{\partial x_l} (g_l u) \tag{7}$$

$$\mathcal{L}^*_{\mathbf{x}_0} v = \sum_{i=1}^{n} \sum_{j=1}^{m} \sum_{k=1}^{m} \sum_{l=1}^{n} D_{jk} H_{ij} \frac{\partial}{\partial x_{0_l}} (H_{lk} \frac{\partial v}{\partial x_{0_l}}) + \sum_{l=1}^{n} g_l \frac{\partial v}{\partial x_{0_l}} \tag{8}$$

and are formal adjoints. Here D_{ij} is the $(i\text{-}j)^{th}$ element of $\mathbf{D}$.

If the domain of interest Ω extends infinitely in all dimensions, boundary conditions need not be prescribed for either Eq. (4) or (5). However, when the domain is finite in one or more dimensions (*e.g.* as occurs in the first passage problem), boundary conditions are required and are often imposed based upon physical consideration. For example, when $n = 2$ and $m = 1$, it is well known that the first passage problem is given by the well-posed initial boundary value problem on Ω, the safe domain, based upon the backward equation,

$$\frac{\partial R}{\partial t} = D \frac{\partial^2 R}{\partial \dot{x}_0} - g_1(x_0, \dot{x}_0) \frac{\partial R}{\partial x_0} - g_2(x_0, \dot{x}_0) \frac{\partial R}{\partial \dot{x}_0} \tag{9}$$

where the reliability function

$$R = R(t | x_0, \dot{x}_0; \Omega) = \int_{\Omega} p(\mathbf{x}, t | \mathbf{x}_0; \Omega) \, d\mathbf{x} \tag{10}$$

is obtained by integrating the transition probability density function, now conditional on being in the safe domain Ω and not previously have left and returned, with respect to the forward variables, **x**.

The initial condition for Eq. (9) is given by

$$R\,(0|\,x_0, \dot{x}_0) \;=\; 1 \qquad \forall\; x_0, \dot{x}_0 \in \Omega \tag{11}$$

and the boundary conditions depicted in Fig. 1 by

$$R\,(0|\,B, \dot{x}_0) \;=\; 0, \qquad \dot{x}_0 > 0 \tag{12}$$

$$R\,(0|\,-B, \dot{x}_0) \;=\; 0, \qquad \dot{x}_0 < 0 \tag{13}$$

$$R\,(0|\,x_0, \dot{x}_0) \rightarrow 0, \qquad |\dot{x}_0| \rightarrow \infty \tag{14}$$

Here, the displacement boundaries are symmetrically disposed at $\pm B$, which is not a general requirement. As can be seen, the absence of the second derivative with respect to x_0 in Eq. (9) implies that there are only partial boundaries in the initial displacement direction. The boundaries in the initial velocity direction are natural, or placed at plus and minus infinity. The physical significance of these boundaries can be seen by considering several possible initial states of the oscillator. Assume that the initial displacement of the oscillator is at $B - \epsilon$ and the initial velocity is positive. For an infinitesimal ϵ any positive velocity will immediately cause the oscillator to collide with the absorbing barrier at $+B$, implying zero reliability. Thus a boundary is required at $+B$ for all positive initial velocities. Again, consider the initial displacement $B - \epsilon$, but let the initial velocity be negative. Regardless of the excitation, the oscillator will not collide

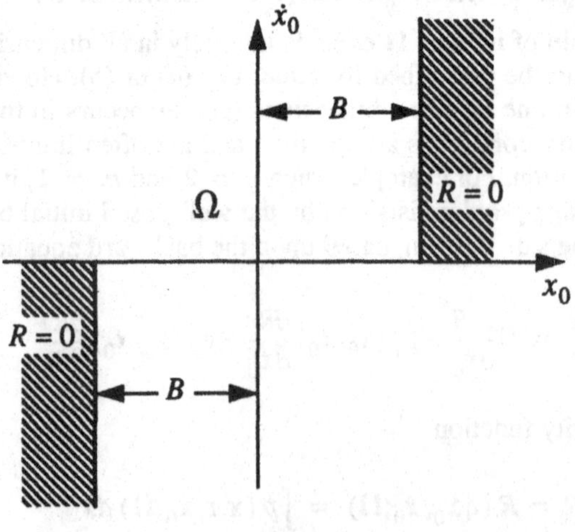

Figure 1. Phase space safe domain for D-barriers in the initial boundary value problem for the reliability function *R*.

with the barrier for some finite time. Rather, for any ϵ, the trajectory will be into the "safe" domain. Thus, the reliability is not necessarily zero, and a boundary is not needed at $+B$ when the initial velocity is negative. A similar argument can be made for boundaries in the other half of the phase plane due to symmetry of the problem about the origin.

An alternate formulation, based upon the Fokker-Planck equation, is also possible. Again, when $n = 2$ and $m = 1$, an initial boundary value problem for the transition probability density of the oscillator is given by

$$\frac{\partial p}{\partial t} = D\frac{\partial^2 p}{\partial \dot{x}^2} - \frac{\partial}{\partial x}[g_1(x,\dot{x})p] - \frac{\partial}{\partial \dot{x}}[g_2(x,\dot{x})p] \tag{15}$$

where $p = p(x,\dot{x},t|x_0,x_0;\Omega)$ is the probability of being between x and $x+dx$, and $\dot{x}$ and $\dot{x}+d\dot{x}$ at time t having started at x_0, $\dot{x}_0$ and not having previously left and re-entered Ω. The initial condition is given by

$$p(x,\dot{x},t|x_0,x_0;\Omega) = \delta(x-x_0)\,\delta(\dot{x}-\dot{x}_0) \tag{16}$$

and the boundary conditions, depicted in Fig. 2, by

$$p(B,\dot{x},t|x_0,x_0;\Omega) = 0, \quad \dot{x}<0 \tag{17}$$

$$p(-B,\dot{x},t|x_0,x_0;\Omega) = 0, \quad \dot{x}>0 \tag{18}$$

$$p(B,\dot{x},t|x_0,x_0;\Omega) \rightarrow 0, \quad |\dot{x}| \rightarrow 0 \tag{19}$$

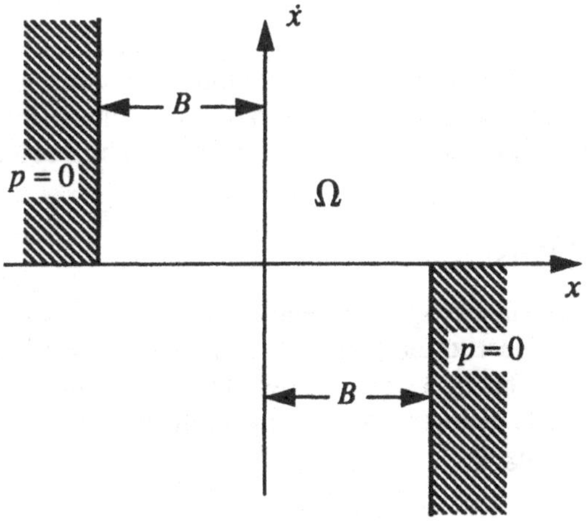

Figure 2. Phase space safe domain for D-barriers for the first passage problem employing the Fokker Planck

Again, the displacement boundaries are symmetrically disposed at $\pm B$, and the absence of a second derivative with respect to x in Eq. (15) implies partial boundaries. Now, however, the physical interpretation of the boundary conditions changes.

Consider an element of probability mass at a displacement B with a negative velocity. Since sample function trajectories are clockwise spirals, the boundary $x = B$ can only be reached from outside the safe domain, Ω. However, the transition probability density in the first passage problem is conditioned upon the trajectory not having previously been outside Ω; therefore it must be zero at $x = B$ for all negative velocities. In contrast, when the velocity is positive, the trajectories have originated in the safe domain, and thus the probability density is not necessarily zero. A similar argument applies in the other half of the phase plane, again due to symmetry about the origin (*cf.* Fig. 2).

When random initial conditions are prescribed, for example as a symmetric binormal density given by

$$f_{\mathbf{X}_0}(\mathbf{x}_0) = f_{\mathbf{X}_0}(x_{0_1}, x_{0_2}) = \phi\left(\frac{x_{0_1} - \mu_{x_{01}}}{\sigma_0}\right)\phi\left(\frac{x_{0_2} - \mu_{x_{02}}}{\sigma_0}\right) \tag{20}$$

where $\phi(\cdot)$ is the standard normal density function, and $\boldsymbol{\mu}_{x_0}$ is the expected value of the initial condition vector $\mathbf{X}_0$, the solution of Eq. (7) is a joint probability density function given by

$$f_{\mathbf{X}}(x_1, x_2, t;\Omega) = \int_{-\infty}^{\infty}\int_{-\infty}^{\infty} p(\mathbf{x}, t|\mathbf{x}_0;\Omega) f_{\mathbf{X}_0}(\mathbf{x}_0)\, d\mathbf{x}_0 \tag{21}$$

which, as $t \to \infty$, approaches zero as all of the probability mass diffuses out of the safe domain.

The solution of Eq. (15) in terms of $f_{\mathbf{X}}$, for n greater than one and for arbitrary $\mathbf{g}(\mathbf{x})$, generally defies analytical solution. Restricting n to two, or perhaps three, the finite element method provides a convenient and efficient method to obtain approximate solutions to a wide range of problems.

DISCRETIZATION

Consider the Fokker-Planck equation with $n = 2$, $m = 1$ and $\mathbf{H} = [0\ 1]^T$ (*i.e.* Eq. (15)) and subject to the boundary and initial conditions of Eq. (16)-(19). We seek to solve this by a straightforward Bubnov-Galerkin finite element method. Within a single element, the joint probability density given by Eq. (21) is interpolated according to

$$f(\mathbf{x}, t) = \sum_{r=1}^{n_p} N_r(\mathbf{x}) f_r^e(t) \tag{22}$$

where the $N_r(\mathbf{x})$, $r = 1, \ldots, n_p$, are the element trial functions of class C^0, n_p is the number of nodes in a single element, and the f_r^e are the nodal values of the density. This expression is substituted into the weak form of Eq. (15) on the element domain, Ω_e, leading to expressions for the element "stiffness" and "mass" matrices

$$k_{rs} = \int_{\Omega_e} \{D\frac{\partial N_r}{\partial x_2}\frac{\partial N_s}{\partial x_2} + N_r\left(g_1\frac{\partial N_s}{\partial x_1} + g_2\frac{\partial N_s}{\partial x_2}\right) + N_r N_s\left(\frac{\partial g_1}{\partial x_1} + \frac{\partial g_2}{\partial x_2}\right)\}\, d\Omega \quad (23)$$

and

$$m_{rs} = -\int_{\Omega_e} N_r(\mathbf{x})\, N_s(\mathbf{x})\, dx_1 dx_2 \quad (24)$$

Transformation of each element matrix into global coordinates and summation over the number of elements in the mesh, N_e, leads to the global evolution equation

$$\mathbf{M}\dot{\mathbf{f}} + \mathbf{K}\mathbf{f} = 0 \quad (25)$$

subject to the initial condition

$$\mathbf{f}(0) = \mathbf{f}_{\mathbf{X}_0} \quad (26)$$

where $\mathbf{f}_{\mathbf{X}_0}$ is a vector of global dimension representing the initial condition given in Eq. (20).

Equation (25) is further discretized in time using the well known Crank-Nicholson method [9]. A rectangular mesh comprised of bilinear four node quadrilateral elements is generated over the computation domain Ω. Displacement boundary conditions are applied in accordance with Eqs. (16)-(17). The limiting value of velocity is set to a constant C large enough to ensure that negligible probability mass is inadvertently lost.

THE LINEAR OSCILLATOR

The efficacy of integrating the forward equation to obtain first passage probabilities is best demonstrated for the linear oscillator, where ample numerical solutions of the backward equation and Monte Carlo simulations exist for comparison.

Here,

$$\mathbf{g}(\mathbf{x}) = \begin{bmatrix} x_2 \\ -2\zeta\omega_0 x_2 - \omega_0^2 x_1 \end{bmatrix} \quad (27)$$

and

$$\mathbf{H} = \begin{bmatrix} 0 \\ 1 \end{bmatrix} \tag{28}$$

Oscillator parameters are given by $\zeta = 0.08$ and $\omega_0 = 1$; excitation by $D = 1$; and initial conditions by $\mathbf{x}_0 = [0\,0]$ and $\sigma_0 = 1/3$. A barrier level, B, corresponding to two stationary standard deviations of the response, or $B = 5$, is examined. A nonuniform mesh is employed, allowing for additional refinement in the vicinity of $x_2 = 0$, in order to accurately model the singular behavior near the boundaries.

The solution of the Fokker-Planck equation is presented in Fig. 3 in terms of the reliability function (*cf.* Eq. (10)). Comparison is given to a solution of the backward Kolmogorov equation which has been shown to be indiscernible from extensive Monte Carlo simulation [2], [9]. The density function for first passage time is depicted in Fig. 4 where we see the characteristic oscillations expected in the solution. As can be seen, the forward and backward solutions are identical.

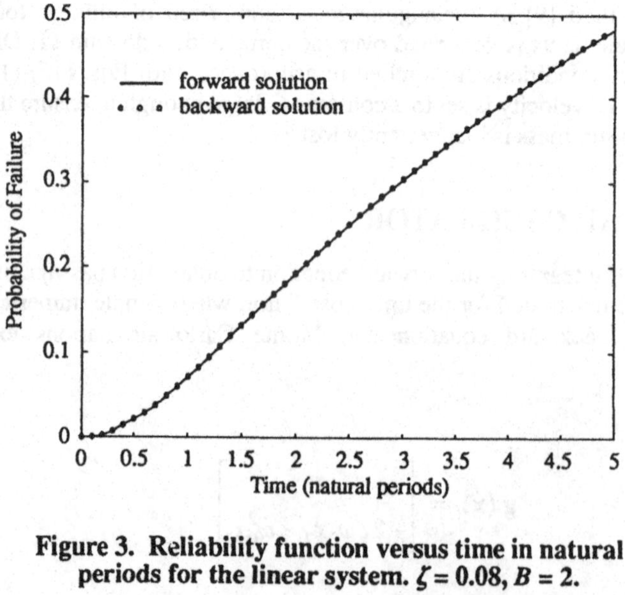

Figure 3. Reliability function versus time in natural periods for the linear system. $\zeta = 0.08$, $B = 2$.

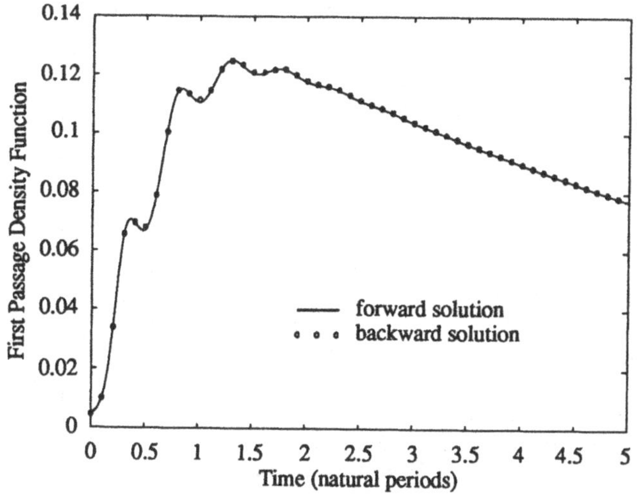

Figure 4. First passage density function versus time in natural periods for the linear system. $\zeta = 0.08$, $B = 2$.

THE VAN DER POL OSCILLATOR

Here,

$$\mathbf{g}(\boldsymbol{x}) = \begin{bmatrix} x_2 \\ -2\zeta\omega_0 x_2(\epsilon x_1^2 - 1) - \omega_0^2 x_1 \end{bmatrix} \tag{29}$$

and

$$\mathbf{H} = \begin{bmatrix} 0 \\ 1 \end{bmatrix} \tag{30}$$

The oscillator exhibits limit cycle behavior which provides an additional test of the method's robustness. The oscillator parameters are given by $\zeta = 0.5$, $\omega_0 = 1$, and $\epsilon = 1$. Excitation and initial conditions are as in the linear example. The barriers are placed at $B = 3$. The oscillator parameters chosen correspond to a system exhibiting limit cycle behavior and a stationary probability density function [3].

Comparison of the results from the forward and backward equations is given in Fig. 5, where the reliability function is plotted versus time. The associated density function is presented in Fig. 6. Once again, the two solution procedures produce identical results.

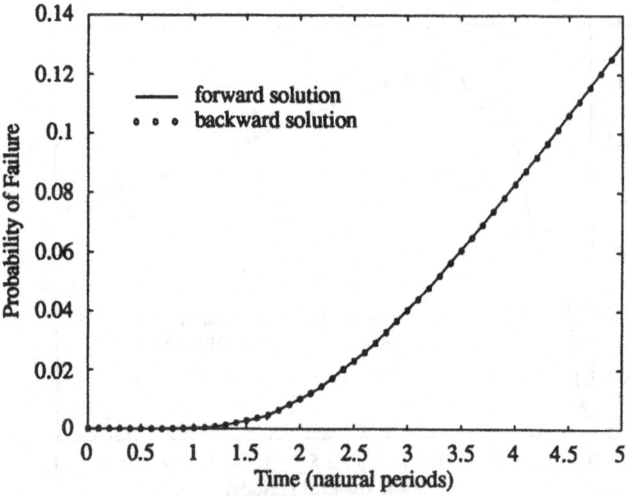

Figure 5. First passage density function versus time in natural periods for the linear system. $\zeta = 0.05$, $B = 3$.

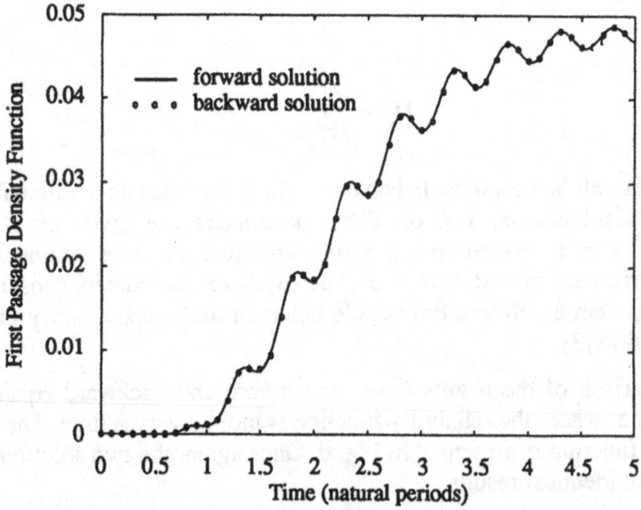

Figure 6. First passage density function versus time in natural periods for the Van der Pol system. $\zeta = 0.05$, $B = 3$.

CONCLUSIONS

An alternate method of accurately computing first passage probabilities for discrete linear and nonlinear systems of two dimensions subjected to additive white noise excitation has been given. The transient Fokker-Planck equation is solved on the safe domain using the finite element method. Open boundary conditions are prescribed, and their physical basis is reviewed. A linear system and a Van der Pol system are examined. In both cases, no discernible difference was detected between solutions of the backward and forward equations.

Integration of the backward equation is desirable from the standpoint that the solution for all admissible initial conditions is obtained in a single analysis, while integration of the forward equation requires prescription of a single set of initial conditions or, alternately, of a distribution of initial conditions. On the other hand, integration of the forward equation is straightforward and directly yields the response probability density as a function of time. The approach uses a standard Bubnov-Galerkin formulation, and allowable bound width appears to be limited only by the precision of the computer. This is in contrast to the more complex Petrov-Galerkin (upwind) formulation and dense mesh requirements that place severe restraints on solution of the backward equation (*cf.* [1], [2]).

Clearly, though, both approaches provide the ability to effectively solve the first passage problem for a broad range of nonlinear systems, with accuracy not readily obtainable with Monte Carlo simulation or other available methods.

REFERENCES

1 Bergman, L.A. Numerical Solutions of the First Passage Problem in Stochastic Structural Dynamics, in *Computational Mechanics of Probabilistic and Reliability Analysis*, ElmePress International, Lausanne, Switzerland, pp. 479-508, 1989.

2 Bergman, L.A. and Spencer, B.F., Jr. Solution of the First Passage Problem for Simple Linear and Nonlinear Oscillators by the Finite Element Method, Department of Theoretical and Applied Mechanics, University of Illinois at Urbana-Champaign, *T & AM Report No. 461*, 1983.

3 Bergman, L.A. and Spencer, B.F., Jr. Robust Numerical Solution of the Transient Fokker-Planck Equation for Nonlinear Dynamical Systems, to appear in the *Proceedings of the IUTAM Symposium on Nonlinear Stochastic Mechanics*, Torino, Italy, July 1-5, 1991.

4 Chandiramani, K.L. First-Passage Probabilities of a Linear Oscillator, Ph.D. Thesis, Department of Mechanical Engineering, Massachusetts Institute of Technology, Cambridge, Massachusetts, USA, 1964.

5 Crandall, S.H., Chandiramani, K.L. and Cook, R.G. Some First-Passage Problems in Random Vibration, *Journal of Applied Mechanics, ASME*, Vol. 33, pp. 532-538, 1966.

6 Crandall, S.H. First Crossing Probabilities of the Linear Oscillator, *Journal of Sound and Vibration*, Vol. 12, No. 3, pp. 285-299, 1970.

7 Darling, D.A. and Siegert, A.J.F. The First Passage Problem for a Continuous Markov Process, *Annals of Mathematical Statistics*, Vol. 24, pp. 624-639, 1953.

8 Roberts, J.B. First-Passage Probabilities for Randomly Excited Systems: Diffusion Methods, *Probabilistic Engineering Mechanics*, Vol. 1, No. 2, pp. 66-81, 1986.

9 Spanos, P-T.D. and Solomos, G.P. Barrier Crossing due to Transient Excitation, *Journal of Engineering Mechanics, ASCE*, Vol. 110, No. 1, pp. 20-36, 1984.

10 Spencer, B.F., Jr. *On the Reliability of Nonlinear Hysteretic Structures Subjected to Broadband Random Excitation*, Lecture Notes in Engineering (series editors: C.A. Brebbia and S.A. Orszag), Vol. 21, Springer-Verlag, 1986.

11 Sri Namachchivaya, N., "Instability Theorem Based on the Nature of the Boundary Behavior for One Dimensional Diffusion," *SM Archives*, V. 14, Nos. 3&4, pp. 131-142, 1989.

12 Sun, J.-Q. and Hsu, C.S. First-Passage Time Probability of Non-Linear Stochastic Systems by Generalized Cell Mapping Method, *Journal of Sound and Vibration*, Vol. 124, 233-248, 1988.

13 Sun, J.-Q. and Hsu, C.S. The Generalized Cell Mapping Method in Nonlinear Random Vibration Based Upon Short-Time Gaussian Approximation, *ASME Journal of Applied Mechanics*, Vol. 57, pp. 1018-1025, 1990.

14 Toland, R.H. Random Walk Approach to First-Passage and other Random Vibration Problems, Ph.D. Thesis, University of Delaware at Newark, Delaware, 1969.

15 Toland, R.H. and Yang, C.Y. Random Walk Model for First Passage Probability, *Journal of Engineering Mechanics Division, ASCE*, Vol. 97, No. EM3, pp. 791-807, 1971.

16 Yang, J-N. and Shinozuka, M. First Passage Time Problem, *The Acoustical Society of America*, Vol. 47, No. 2, pp. 393-394, 1970.

Approximate Analysis of Nonstationary Random Vibrations of MDOF Systems

K. Papadimitriou (*), J.L. Beck (**)
(*) Department of Civil Engineering, Texas A&M University, College Station, TX 77843, U.S.A.
(**) Department of Civil Engineering, Cal Tech, Pasadena, CA 91125, U.S.A.

ABSTRACT

This paper considers the nonstationary response of classically-damped MDOF linear systems to a general class of nonstationary non-white excitations. A new approximate method is developed to considerably simplify the equations for the covariance of the response. The method of analysis provides the general conditions involving the system and excitation characteristics for which the approximations are reliable. The derived simplified equations reveal the similarities between the nonstationary and stationary response, and they provide valuable insight into the nonstationary response characteristics. The performance of the new method is demonstrated by numerical examples.

INTRODUCTION

The problem of determining the covariance response of a linear MDOF (multi-degree-of-frredom) system subjected to nonstationary non-white excitation has been well-formulated. The existing time domain and frequency domain methods of analysis, however, involve complex mathematical operations that provide little physical insight into the characteristics of the nonstationary response and their dependence on the system and excitation parameters. In the case of lightly-damped structures and wide-banded excitations, approximate methods (Spanos [6], Igusa [2]) have been proposed to considerably simplify the analysis of SDOF systems. For excitations with nonstationarities in both the amplitude and the frequency content encountered in earthquake engineering problems, the existing approximate methods often result in significant errors (Papadimitriou [4]). The reason is that in the range of system parameters and excitation characteristics of practical interest, the assumptions for the validity of the approximations are violated. Bucher [1] has extended the approximations to treat MDOF systems by utilizing properties of the stationary response to simplify the nonstationary analysis. However, Bucher's approximate technique, which neglects important dynamical effect, results in unreliable response estimates.

In this paper, a new method of analysis is developed to considerably improve the existing approximations for the covariance of the nonstationary response of linear systems for a broader range of system and excitation parameters. The basic idea is to approximate the dynamics of the corresponding Liapunov differential matrix equation for the mean-square response by the dynamics of a much simpler set of equations, neglecting secondary dynamical effects. The approximations cover any range of system parameters and they are not restricted to the case of lightly-damped systems as in Spanos [6] and Igusa [2]. Similarities and important differences with Bucher's approximation are discussed in the MDOF case. The resulting simple equations provide insight into the characteristics of the response and its dependence on the system and excitation characteristics.

FORMULATION OF THE MEAN-SQUARE RESPONSE

Consider the excitation of an N-degree-of-freedom, viscous damped, linear system having classical modes and initially at rest. The response at some point of the system can be expressed in terms of modal contributions as $x(t) = \sum_{i=1}^{N} p_i x_i(t)$ where p_i is the effective participation factor for mode i at the point of interest, and $x_i(t)$ is the response of the i-th normal mode satisfying the modal equation

$$\ddot{x}_i(t) + 2\zeta_i\omega_i\dot{x}_i(t) + \omega_i{}^2 x_i(t) = G(t) \tag{1}$$

In this equation the forcing functions $G(t)$ could be the base excitation, and ω_i and ζ_i ($\zeta_i < 1$) are the i-th modal frequency and damping, respectively. The mean-square response $R(t) = E[x^2(t)]$ can be written as $R(t) = \sum_{i=1}^{n}\sum_{j=1}^{n} p_i p_j R_{ij}(t)$ where $R_{ij}(t) = E[x_i(t)x_j(t)]$ is the covariance response of two modes i and j. If $\underline{x}_i(t) = (x_i(t), \dot{x}_i(t))^T$ is the state vector of the modal equation (1), then the covariance matrix $Q^{(ij)}(t) = E[\underline{x}_i(t)\underline{x}_j^T(t)]$ of a two-mode system satisfies the matrix differential equation

$$\dot{Q}^{(ij)}(t) = A_i Q^{(ij)}(t) + Q^{(ij)}(t)A_j^T + L_i(t) + L_j^T(t) \tag{2}$$

with initial condition $Q^{(ij)}(0) = 0$, where

$$L_i(t) = \int_0^t \Phi_i(\tau)\begin{pmatrix} 0 & 0 \\ 0 & r_e(t, t-\tau) \end{pmatrix} d\tau, \quad A_i = \begin{pmatrix} 0 & 1 \\ -\omega_i^2 & -2\zeta_i\omega_i \end{pmatrix}, \tag{3}$$

$\Phi_i(t)$ is the principal solution matrix of the state-space form of equation (1) satisfying $\dot{\Phi}_i(t) = A_i\Phi_i(t)$ with $\Phi_i(0) = I$, and $r_e(t, t-\tau) = E[G(t)G(t-\tau)]$ is the autocovariance function of the excitation process. The equation for $R_{ij}(t)$ is next formulated and approximations are introduced to simplify the original exact equations.

For $i = j$, the modal covariance matrix $Q^{(ii)}$ is symmetric. Eliminating $Q_{12}^{(ii)}(t) = Q_{21}^{(ii)}(t)$ and $Q_{22}^{(ii)}(t)$ from the system of equations (2), a third-order differential equation in terms of the mean-square modal displacement $Q_{11}^{(ii)}(t) \equiv R_{ii}(t)$ is obtained. The characteristic equation of the third-order differential equation has one real negative root $\rho_1 = -2\zeta_i\omega_i$ and two complex roots $\rho_2 = -2\zeta_i\omega_i + i2\omega_i\sqrt{1-\zeta_i^2}$ and $\rho_3 = -2\zeta_i\omega_i - i2\omega_i\sqrt{1-\zeta_i^2}$. Therefore, the third-order differential equation for $R_{ii}(t)$ can be split into the first-order differential equation

$$\dot{R}_{ii}(t) + 2\zeta_i\omega_i R_{ii}(t) = 2r_{ii}(t) \tag{4}$$

associated with the real negative root, and the second-order differential equation

$$\ddot{r}_{ii}(t) + 4\zeta_i\omega_i\dot{r}_{ii}(t) + 4\omega_i^2 r_{ii}(t) = g_i(t) \tag{5}$$

for the term $r_{ii}(t)$, which is associated with the two complex roots. The forcing term $g_i(t)$ has the form

$$g_i(t) \equiv g(t;\omega_i,\zeta_i) = 2L_{i,22}(t) + 4\zeta_i\omega_i L_{i,12}(t) + \dot{L}_{i,12}(t) \tag{6}$$

and it simplifies to $g_i(t) = f^2(t)$ for a modulated white-noise excitation with modulation function $f(t)$.

For $i \neq j$, the components of $Q^{(ij)}$ for the two-mode system satisfy a four-dimensional system of first-order differential equations. Eliminating $Q_{12}^{(ij)}$, $Q_{21}^{(ij)}$, and $Q_{22}^{(ij)}$ from this system, a fourth-order equation for $Q_{11}^{(ij)}(t) \equiv R_{ij}(t)$ is obtained. Treating for simplicity the case $\zeta_i = \zeta_j = \zeta$, the characteristic polynomial of the fourth-order equation has the roots $\rho_{1,2} = -2\zeta\omega \pm 2i\omega\sqrt{1-\zeta^2}$ and $\rho_{3,4} = -2\zeta\omega \pm 2i\omega\lambda\sqrt{1-\zeta^2}$ where $\omega = \frac{\omega_i+\omega_j}{2}$ is the average frequency and $\lambda = \frac{\omega_i-\omega_j}{\omega_i+\omega_j}$ is the fractional difference of the modal frequencies of the two-mode system. The fourth-order differential equation for $R_{ij}(t)$ can be split into two second-order differential equations as follows

$$\ddot{R}_{ij} + 4\zeta\omega\dot{R}_{ij} + 4\omega^2[\zeta^2 + \lambda^2(1-\zeta^2)]R_{ij} = 2r_{ij}(t) \tag{7}$$

$$\ddot{r}_{ij}(t) + 4\zeta\omega\dot{r}_{ij}(t) + 4\omega^2 r_{ij}(t) = \dot{F}_{ij}(t) + 2\zeta\omega F_{ij}(t) \tag{8}$$

where

$$2F_{ij}(t) = g_i(t) + g_j(t) - 2\zeta\omega\lambda(L_{i,12}(t) - L_{j,12}(t)) \tag{9}$$

with $g_k(t) \equiv g(t;\omega_k,\zeta)$ and $L_{k,12}(t) \equiv L_{k,12}(t;\omega_k,\zeta)$, $k=i,j$.

An analysis follows which shows that under specific conditions between the system and excitation characteristics: a) the integral in equations (6) and (9) can be adequately replaced in most cases of practical interest by approximate simplified algebraic expressions without significant loss of accuracy, and b) the dynamics of the first-order differential equation (4) or the second-order differential equation (7) provide all the essential characteristics of the response while the dynamics of the second-order differential equations (6) and (8) have only secondary effects on the mean-square response. The approximate method of analysis is first demonstrated for the mean-square modal response $Q^{(ii)}(t)$ and then it is extended to simplify the equation for the covariance response $Q^{(ij)}(t)$.

APPROXIMATE ANALYTICAL EXPRESSIONS FOR $g(t;\omega,\zeta)$

Let $T_c(t)$ be the length of the time interval of non-zero correlation of the stochastic process $G(t)$ at time t, that is, $r_e(t,t-\tau) \neq 0$ for $\tau \in [0,T_c(t)]$ and $r_e(t,t-\tau) \approx 0$ for $\tau > T_c(t)$. In most physical processes encountered in mechanics problems, the autocovariance function at time t reduces from its peak value to a negligible small amount in a relatively short time period $T_c(t)$. The more broadband the process is, the less the correlation length $T_c(t)$. Therefore, the integration of (3) in the interval $[T_c(t),t]$ can be neglected. Finally, because of the decay of $r_e(t,t-\tau)$ to insignificantly small values at $T_c(t)$, the definite integrals in (3) can be approximated by indefinite integrals with their value computed at time $\tau = 0$.

Next, consider a general group of nonstationary stochastic excitation with autocovariance function having the form, or approximated by the form,

$$r_e(t,t-\tau) = q_e(t)\, r(\underline{\theta}_e(t),\tau) \qquad \text{with} \qquad r(\underline{\theta}_e(t),0) = 1 \tag{10}$$

where $q_e(t)$ is the variance of the stochastic excitation process at time t, and $r(\bullet,\tau)$ is the autocorrelation function of the process at time t with $\underline{\theta}_e(t)$ accounting for the time variation of the frequency content of the excitation process. Subclasses of such stochastic processes with time-invariant $\underline{\theta}_e$ have been extensively used in the past to model environmental loads and especially earthquake loads. These subclasses include white noise, modulated white noise, and the modulated non-white broadband processes known as modulated filtered white noise and filtered modulated white noise, with rational (usually second-order) time-invariant transfer function for the filter. Substituting the general expression (10) into equation (6), the forcing term $g(t;\omega_i,\zeta_i)$ takes the form

$$g(t;\omega_i,\zeta_i) = q_e(t)\, R_e(\omega_i,\zeta_i;\underline{\theta}_e(t))\{\, 1 + \epsilon_R(\omega_i,\zeta_i;\underline{\theta}_e(t))\} \tag{11}$$

where

$$R_e\left(\omega_i,\zeta_i;\underline{\theta}_e(t)\right) = 2\, I_c\left(\omega_i,\zeta_i;\underline{\theta}_e(t)\right) + 2\zeta_i\omega_i\, I_s\left(\omega_i,\zeta_i;\underline{\theta}_e(t)\right) \tag{12}$$

$$I_c\left(\omega_i,\zeta_i;\underline{\theta}_e(t)\right) = \frac{1}{\omega_d}\left[\int \exp\left[-\zeta_i\omega_i\tau\right] r\left(\underline{\theta}_e(t),\tau\right)\cos\left[\omega_{i,d}\tau\right] d\tau\right]_{\tau=0} \tag{13}$$

$$I_s\left(\omega_i,\zeta_i;\underline{\theta}_e(t)\right) = \frac{1}{\omega_d}\left[\int \exp\left[-\zeta_i\omega_i\tau\right] r\left(\underline{\theta}_e(t),\tau\right)\sin\left[\omega_{i,d}\tau\right] d\tau\right]_{\tau=0} \tag{14}$$

$$\epsilon_R\left(\omega_i,\zeta_i;\underline{\theta}_e(t)\right) = \frac{\dot{q}_e(t)}{2\omega_0 q_e(t)}\frac{2\omega_i I_s(t)}{R_e(t)} + \frac{\dot{I}_s(t)}{R_e(t)} \tag{15}$$

and $\omega_{i,d} = \omega_i\sqrt{1-\zeta_i^2}$. The functional $r(\bullet,\bullet)$ is a general functional operator of $\underline{\theta}_e(t)$ and τ. For specific functions $r(\underline{\theta}_e(t),\bullet)$ of τ, such as polynomials, exponentials, trigonometric functions or a combination of these, the integration in (13) and (14) can be carried out analytically without requiring the exact time variation of the excitation model parameters $\underline{\theta}_e(t)$. In such cases, the functionals $R_e\left(\omega_i,\zeta_i;\underline{\theta}_e(t)\right)$ and $\epsilon_R\left(\omega_i,\zeta_i;\underline{\theta}_e(t)\right)$ depend only on the form of the functional operator $r_e(\bullet,\bullet)$, that is, the general structure of the excitation model. For processes with slowly-varying intensity and frequency content, i.e., slowly-varying $q_e(t)$ and $\underline{\theta}_e(t)$, usually the case in most applications, the forcing term $g_i(t)$ is a slowly-varying function of time as well. Furthermore, the term $\epsilon_R(t)$ is small ($\epsilon_R(t) << 1$) and its contribution to $g_i(t)$ can be neglected. As ζ approaches zero, it can be shown (Papadimitriou [4]) that the form of $q_e(t)R_e\left(\omega_i,\zeta_i;\theta_e(t)\right)$ approaches the evolutionary power spectral density $S(\omega_i,t)$ of the excitation process.

APPROXIMATION FOR THE MEAN-SQUARE RESPONSE

An efficient way to derive the approximations for the mean-square modal response and justify their validity under slowly-varying forcing term $g_i(t)$ is to use the two-timing method to solve the second-order differential equation (5). The slow time is governed by t and the fast time $\tau = \omega_i t$ is governed by the reciprocal of the "high" angular frequency ω_i. Doing so, a series expansion for $r_{ii}(t)$ is derived (Papadimitriou [4]) in the form:

$$r_{ii}(t) = \frac{g_i(t)}{(2\omega_i)^2}\left\{1 + \sum_{n=1}^{\infty} c_n\, \frac{g_i^{(n)}(t)}{(2\omega_i)^n g_i(t)} + \frac{g_i^{(n)}(0)}{(2\omega_i)^n g_i(t)}\,\varphi\left(\rho_n,\phi_n\right)\right\} \tag{16}$$

where $\varphi(\rho,\phi) = \rho\exp\left(-2\zeta_i\omega_i t\right)\cos\left(2\omega_{i,d}t+\phi\right)$. For example, the values of c_n, ρ_n and ϕ_n for the first and second term in the expansion are: $c_1 = -2\zeta$, $c_2 = -(1-2\zeta^2)$, $\rho_1 = \rho_2 = 1/\sqrt{1-\zeta^2}$, $\sin\phi_1 = -(1-2\zeta^2)$ and $\sin\phi_2 = -\zeta(3-4\zeta^2)$. The solution for $r_{ii}(t)$ consists of a non-oscillatory term plus an exponentially-decaying oscillatory term with period of oscillations $\pi/\omega_{i,d}$. For *heavily-damped* modes, that is for large ζ_i, the oscillatory terms decay quickly to zero. For *lightly-damped* oscillators, that is, $\zeta_i << 1$, the oscillatory terms persist for several cycles of oscillations. The magnitude of the oscillatory terms in $r_{ii}(t)$ depends on $g_i^{(n)}(0)/(2\omega_i)^n$, $n = 1, 2, \ldots$.

The proposed approximations are a) to neglect the small oscillatory terms in the series expansion, an assumption that preserves the essential features of the response, and b) to treat $g_i(t)$ as a slowly-varying function and to retain only the dominant term $g_i(t)/(4\omega_i^2)$ in the expansion. Substituting into equation (4), the mean-square displacement of the response can be obtained by solving the much simpler first-order differential equation

$$\dot{R}_{ii}(t) + 2\zeta_i\omega_i R_{ii}(t) = \frac{g_i(t)}{2\omega_i^2} \tag{17}$$

For sufficiently slowly-varying correlation structure and sufficiently small modal damping ζ_i, $g_i(t) \approx S(\omega_i, t)$ and therefore equation (17) is the same as the one obtained by Spanos [6] for linear SDOF systems. The conditions for neglecting the higher-order terms are directly determined by the series expansion, and they are mathematically stated as

$$\frac{g_i^{(n)}(t)}{(2\omega_i)^n g_i(t)} << 1, \qquad n = 1, 2, \ldots \tag{18}$$

These conditions specify how slow the forcing term $g_i(t)$ should vary with time so that the dominant solution is an adequate approximation, and they will be referred to as the "slowly-varying" conditions for $g_i(t)$. For example, the condition for $n = 1$ is roughly that the fractional change of $g_i(t)$ over a cycle of oscillation is much less than 4π. Conditions (18) are violated only at time intervals where $g_i(t)$ has sufficiently small values. An advantage of the series expansion (16) is that even if the slowly-varying conditions are violated, the dominant term can be corrected to any degree of accuracy by including the next higher-order terms in the expansion.

Using the series expansion (16) to solve equation (8), and retaining only the dominant term, assuming that $\dot{F}_{ij}(t)+2\zeta\omega F_{ij}(t)$ is slowly-varying, the covariance of the i and j modes can be obtained approximately by solving the second-order differential equation

$$\ddot{R}_{ij} + 4\zeta\omega\dot{R}_{ij} + 4(\zeta\omega)^2[1+\rho^2]R_{ij} = \frac{\dot{F}_{ij}(t) + 2\zeta\omega F_{ij}(t)}{2\omega^2} \tag{19}$$

where $\rho = \frac{\lambda}{\zeta}\sqrt{1-\zeta^2}$. Conditions (18) with ω_i and $g_i(t)$ replaced by ω and $\dot{F}_{ij}(t) + 2\zeta\omega F_{ij}(t)$ respectively, determine the conditions for the approximation to be valid. The terms in expression $F_{ij}(t)$ can be approximated according to the analysis developed previously for $g(t;\omega,\zeta)$.

The formulation for the covariance of two modes i and j is further simplified for the following four cases.

Case 1: For closely spaced modal frequencies ω_i and ω_j such that the variation of the terms in equation (9) with respect to ω is approximately linear between ω_i and ω_j , the function $F_{ij}(t)$ can be approximated by

$$F_{ij}(t) = g(t;\omega,\zeta) + \zeta\omega\lambda \left.\frac{\partial L_{12}}{\partial \omega}\right|_{\omega} (2\omega) \tag{20}$$

where only first-order partial derivatives with respect to ω have been retained.

Case 2: For sufficiently small modal damping, i.e., lightly-damped systems, the term $F_{ij}(t)$ takes the form

$$2F_{ij}(t) = (S(\omega_i, t) + S(\omega_j, t)) \tag{21}$$

which can be further approximated by $S(\omega, t)$ if the conditions of case 1 apply.

Case 3: For closely spaced modal frequencies ω_i and ω_j such that $\lambda^2 << \zeta^2$, the second-order differential equation (19) simplifies to the first-order differential equation

$$\dot{R}_{ij} + 2\zeta\omega R_{ij} = \frac{F_{ij}}{2\omega^2} \tag{22}$$

Case 4: For an envelope modulated white-noise input with modulation function $f(t)$, $F_{ij}(t) = g_i(t) = f^2(t)$.

Qualitative differences exist between the second-order differential equation (19) and the first-order differential equation derived by Bucher [1] using a quasi-stationary approximation. Both formulations would yield the same results only if condition $\lambda^2 << \zeta^2$ applies.

Another advantage of the present formulation is that it simplifies considerably the relationship between the mean-square velocity or absolute acceleration and the mean-square displacement. The absolute modal acceleration is defined by $a_i(t) = -2\zeta\omega_i\dot{x}_i(t) - \omega_i^2 x_i(t)$. Let $Q_a^{(ij)} = E[a_i(t)a_j(t)]$ be the covariance of the absolute acceleration of two modes i and j. Assuming for simplicity that the two modal frequencies are closely spaced, that is, λ is sufficiently small ($\lambda << 1$), it can be shown that the velocity covariance $Q_{22}^{(ij)}$ and the absolute acceleration covariance $Q_a^{(ij)}$ of two modes i and j are approximately related to the displacement covariance $Q_{11}^{(ij)} \equiv R_{ij}$ by the simple algebraic expressions

$$Q_{22}^{(ij)}(t) = \omega^2 Q_{11}^{(ij)}(t) - L_{12}(t;\omega,\zeta) \tag{23}$$

and

$$Q_a^{(ij)}(t) = \omega^4 \left\{1 + 4\zeta^2\right\} Q_{11}^{(ij)}(t) - 4\zeta^2\omega_0^2 L_{12}(t;\omega,\zeta) \tag{24}$$

respectively. In the stationary case, these equations are exact and independent of the time t. Therefore, the nonstationary mean-square velocity and absolute acceleration can be obtained in terms of the mean-square displacement by the stationary algebraic relations, after replacing the time-invariant quantities in these relations with the time-varying ones corresponding to time t.

NUMERICAL RESULTS

An envelope modulated white-noise excitation is first used and a nondimensional analysis is employed to provide valuable information about the accuracy of the method. An example of practical interest is the Gamma modulation (Saragoni and Hart [5]) given by $f(t) = f_m e(\tau)$ where $e(\tau) = \tau^\beta e^{\beta(1-\tau)}$, $\tau = t/t_m$, f_m is the maximum intensity, t_m is the time the modulation achieves its maximum, and β is a nondimensional measure of the duration of the excitation. The most informative way to present the results is to define the following nondimensional quantities: the nondimensional time $\tau = t/t_m$, the number of cycles of oscillations $\eta = t_m/T$ with frequency $\omega = 2\pi/T$ needed to reach the time at which the maximum of the modulation occurs, and the normalized response quantity

$$q_{ij}(\tau) = \frac{R_{ij}(t_m\tau)}{R^s_{ij}}, \qquad \text{where} \qquad R^s_{ij} = \frac{f_m^2}{4\zeta\omega_0^3(1+\rho^2)}, \tag{25}$$

which compare the nonstationary covariance response of the two modes i and j to the equivalent stationary response R^s_{ij} obtained for a constant modulation with power f_m^2. Substituting in equation (17), the nondimensional response q_{ii} satisfies

$$q'_{ii}(\tau) + 4\pi(\zeta\eta)q_{ii}(\tau) = 4\pi(\zeta\eta)\tau^{2\beta}\, e^{2\beta(1-\tau)} \tag{26}$$

Using equation (19), the nondimensional response q_{ij} satisfies

$$q''_{ij}(\tau) + 2(4\pi\zeta\eta)q'_{ij}(\tau) + \Omega^2 q_{ij}(\tau) = \Omega^2\left[e(\tau) + \frac{e'(\tau)}{(4\pi\zeta\eta)}\right] \tag{27}$$

where $\Omega = (4\pi\zeta\eta)\sqrt{1+\rho^2}$. The dominant characteristics of the approximate modal response q_{ii} and the correlation response q_{ij} of two modes i and j depend on the system parameters ρ (for $i \neq j$) and the product $\zeta\eta$, and on the excitation parameter β. In contrast, the exact responses for q_{ii} and q_{ij} obtained by using the original matrix equation (2) depend explicitly on ζ and η and not on the product $\zeta\eta$ only.

Comparisons between the approximate and the exact solutions are shown in Figure 1 for $q_{ii}(\tau)$, and in Figure 2 for $q_{ij}(\tau)$ with $\rho = 1.0$ and

5.0. The input modulation function and Bucher's approximation are also plotted in Figure 2. The effects of the input characteristics are studied by considering two values of the modulation parameter β: $\beta = 0.5$ corresponding to long duration excitations and $\beta = 4.0$ corresponding to short duration excitations. Several sets of curves are presented corresponding to different values for the product $\zeta\eta$. The value of $\eta = 1$ for calculating the exact response in the figures is purposely chosen to almost violate the slowly-varying conditions (18) upon which the approximate theory is based. A very good agreement between the exact and the approximate formulation is shown in the Figures 1 and 2. Based on the slowly-varying conditions (18), the higher the value of η, the smaller the discrepancies. In fact, for $\eta = 2$ and greater, the approximate responses are found to be almost identical to the exact ones. For values of $\eta < 1$, the approximations preserve the essential features of the response by neglecting the high frequency oscillatory terms. Bucher's approximation is inaccurate and in several cases even fails to predict the qualitative features of the response (e.g., Figure 2(c) and (d)).

Additional insight into the nonstationary response characteristics and the importance of the modal correlation on the response can be gained by noting that at any response time $t = t_m\tau$,

$$R_{ij}(t) = \frac{1}{1+\rho^2}\frac{q_{ij}(\tau)}{q_{ii}(\tau)}R_{ii}(t) \tag{28}$$

For stationary response, the contribution of the modal correlation to the response depends only on the parameter ρ of the two-mode system because the ratio $\frac{q_{ij}(\tau)}{q_{ii}(\tau)} = 1$. The higher the value of ρ, the less significant the modal correlation. For the nonstationary case, however, the contribution of the modal correlation on the response also depends on the ratio $\frac{q_{ij}(\tau)}{q_{ii}(\tau)}$. When the damped oscillatory frequency $\Omega_d = 4\pi(\zeta\eta)\rho$ of the second-order differential equation (27) is close to the dominant frequency of the modulation, $q_{ij}(t)$ $(i \neq j)$ can take values much greater than one. This is depicted in Figure 2(c,d). Since q_{ii} is always bounded by one, the ratio $\frac{q_{ij}(\tau)}{q_{ii}(\tau)}$ takes even greater values, especially at times τ corresponding to small $q_{ii}(\tau)$. Therefore, the nonstationary characteristics of the excitation can be such that the modal correlation can not be neglected in the nonstationary response even for well separated modes or more strictly speaking, higher values of ρ.

The accuracy of the approximations for non-white excitation is demonstrated by using the ground motion model (Papadimitriou and Beck [3])

$$\ddot{G}(t) + 2\alpha_e(t)\dot{G}(t) + \omega_e^2(t)G(t) = f(t)w(t) \tag{29}$$

where $w(t)$ is a white noise process. For sufficiently broadband processes, analytical algebraic expressions can been derived for $g_i(t)$ in the form (11)

(Papadimitriou [4]) with $\underline{\theta}_e(t) = (\omega_e(t), \alpha_e(t))$. Fitting the model to the Orion Blvd. ground accelerogram (Papadimitriou and Beck [3]) recorded during the 1971 San Fernando earthquake, the following variations are obtained: $q_e(t) = q_m(t/t_m)^{2\beta} \exp(2\beta(1 - t/t_m))$ ($\sqrt{q_m} = 79$, $\beta = 2.4$ and $t_m = 9.8$), $\omega_e(t) = \omega_e(\infty) + (\omega(0) - \omega(\infty) \exp(-\lambda t)$ ($\omega_e(0) = 7.4$Hz, $\omega_e(t_m) = 3.1$Hz and $\omega_e(\infty) = 1.1$Hz) and $\alpha_e(t) = 0.71$Hz. This recording shows a significant time variation of the model characteristics and therefore it is supposed to be a representative case for testing the validity of the technique when applied to strong ground motion stochastic processes. An efficient but still time-consuming method for evaluating the exact mean-square modal response is to rewrite the second-order differential equations (1) and (29) as a four-dimensional first-order vector equation and numerically integrate the corresponding Liapunov matrix equation for the mean-square response. This results in a 10-dimensional system of first-order differential equations for obtaining the exact modal mean-square response. Its numerical solution provides the basis for assessing the approximate results. Similarly, the exact covariance response $R_{ij}(t)$ $(i \neq j)$ is obtained by solving a 21-dimensional system. Using the approximations, the computing time is reduced significantly by a factor of 20 or higher depending on the applications. The discrepancies between the exact and the approximate solution for R_{ii}, shown in Figure 3(a) and 3(b), are small. Using equation (11) instead of the evolutionary power spectal density used by Spanos [6], we improve the accuracy for a broader range of parameters, without significantly increasing the computational time as shown in Figure 3(b). For higher values of ω smaller differences were found between the exact and the approximate response, which is consistent with the slowly-varying conditions (18) of the mathematical analysis. The same order of accuracy between the exact and the approximate solution were also found for $R_{ij}(\tau)$.

CONCLUSIONS

A new method was developed to approximately simplify the formulation for the covariance response of MDOF systems for a broad range of system and excitation characteristics. The method is very efficient and reductions in computer time ranging from one to two orders of magnitude are achieved, depending on the application. The approximate solutions are in very good agreement with the exact ones for modal periods shorter than the time scale of the modulation of the excitation. For modal periods of the order of the modulation, however, the formulation reliably predicts the qualitative features of the response. Existing approximations based on the quasi-stationary analysis [Bucher [1]) of MDOF systems were found to be unreliable. The formulation has been applied successfully to the nonstationary analysis (Papadimitriou [4]) of nonlinear SDOF systems modeled by equivalent linear systems. The new approximate analysis when applied to seismic analysis of structures can provide a better insight into the response characteristics and their dependence on both the system and excitation characteristics.

REFERENCES

1. Bucher, C.G. Approximate Nonstationary Random Vibration Analysis for MDOF Systems, Journal of Applied Mechanics, ASME, Vol. 55, pp. 197-200, 1988.

2. Igusa, T. Characteristics of Response to Nonstationary White Noise: Theory, J. of Engrg. Mech. Div., ASCE, Vol. 115(9), pp. 1904-1934, 1989.

3. Papadimitriou, K. and Beck J.L. Nonstationary Stochastic Characterization of Strong-Motion Accelerograms, Proceedings of the Fourth U.S. Nat. Conf. on Earthquake Engineering, Palm Springs, USA, 1990.

4. Papadimitriou, K. Stochastic Characterization of Strong Ground Motion and Applications to Structural Response, Ph.D Thesis, California Institute of Technology, Report No. EERL 90-03, 1990.

5. Saragoni, G.R. and Hart, G.C. Simulation of Artificial Earthquakes, Earthquake Engineering and Structural Dynamics, Vol. 2, 249-267, 1974.

6. Spanos, P.T.D. Probability of Response to Evolutionary Process, J. of Engrg. Mech. Div., ASCE, Vol. 106, No. EM2, pp. 213-224, 1980.

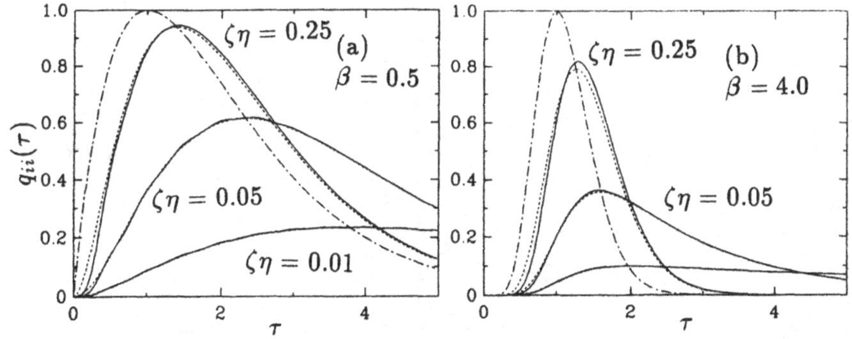

Figure 1. Comparison between the exact (———) and proposed (············) solution for the mean-square modal response. Input $e(\tau)$ (-·-·-·-·-). $\eta = 1$.

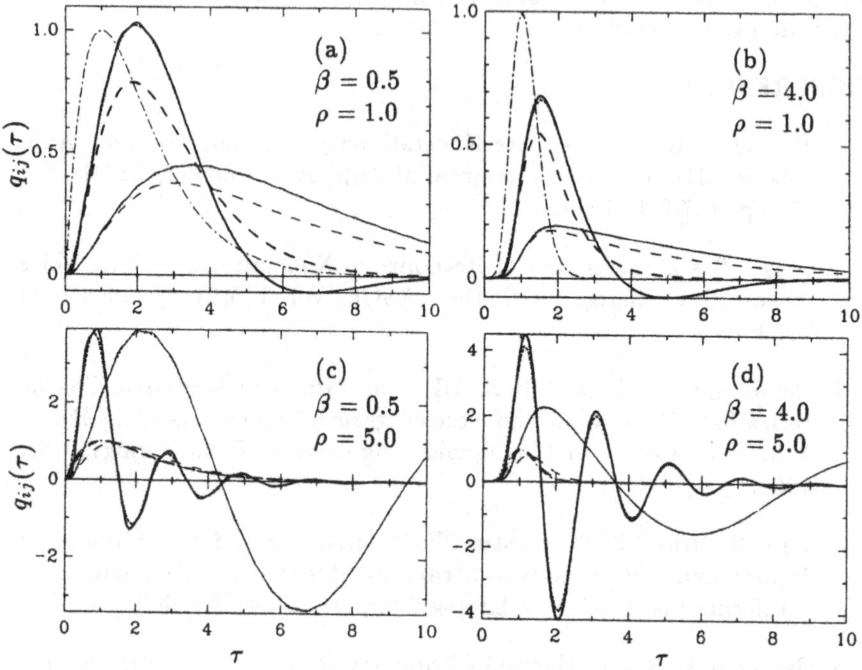

Figure 2. Comparison between the exact (———), proposed (···········) and Bucher's (- - - - -) solution for the covariance response of two modes i and j. $\zeta\eta$ = 0.01 (thin lines) and 0.05 (thick lines). Input $e(\tau)$ (-·-·-·-). $\eta = 1$.

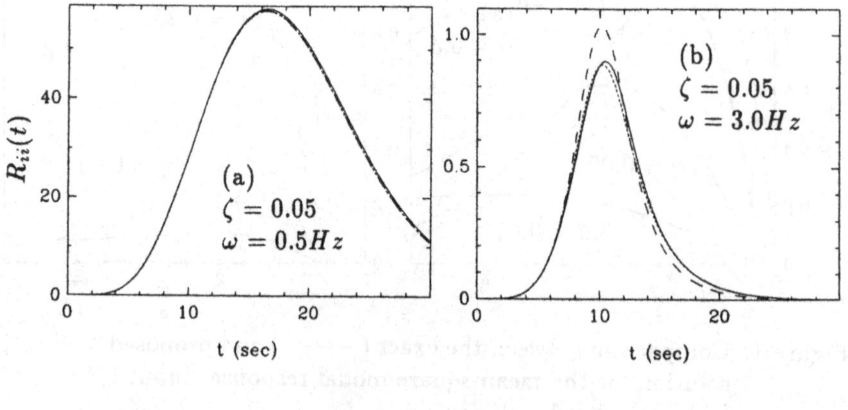

Figure 3. Comparison between the exact (———), proposed (···········) and Spanos's (- - - -) solution for the mean-square modal response to non-white excitation.

SECTION 5: STOCHASTIC FINITE ELEMENT CONCEPT

First Order Reliability Analysis Using Stochastic Finite Element Methods

S. Reh, F. Böhm, A. Brückner-Foit,
H. Riesch-Oppermann
Institute for Materials Research (IMF II), Kernforschungszentrum Karlsruhe, P.O. Box 3640, W-7500 Karlsruhe 1, Germany

ABSTRACT

Stochastic finite element methods are suitable for the reliability analysis of components with stochastic material, loading and geometry parameters, which are described by correlated and non-gaussian random variables and random fields. The first order reliablility method (FORM) can be considered for calculating the failure probability. The design point is usually evaluated within FORM by means of an optimization scheme, which requires the gradient of the structural response with respect to the random problem parameters. The large number of random variables of the probabilistic model requires the usage of the adjoint method to obtain this gradient information. The complexity of the mechanical model necessitates the use of commercial finite element codes. In the present contribution the finite element equations of the real and the adjoint problem are derived. Then, these equations are solved with the commercial finite element code ABAQUS. For this purpose, "adjoint elements" are defined for linear-elastic isotropic and anisotropic material behaviour. As an illustration, the proposed method is applied to a model turbine blade.

INTRODUCTION

Stochastic finite element methods generally consist of two parts, namely the mechanical model describing the deterministic structural behaviour and the probabilistic model evaluating the stochastic properties. For both parts, there are well known and established methods.

The mechanical model is represented by a finite element model, which in turn is realized using a commercial finite element package. Commercial FEM packages are favoured in order to take full advantage of the capabilities of these packages such as the possibility to handle components with complex geometry and sophisticated material models and the ability of the input and output software to process the information in a user friendly form.

Reliability methods are considered to describe the probabilistic model. Reliability methods of first and second order (FORM and SORM) are well established procedures to approximate the failure probability of a component with arbitrary stochastic properties. Compared with simulation methods the computational effort is relatively small. It is well known that the accuracy of these methods is increasing with decreasing failure probability [1], which makes them even more attractive for applications where the safety requirements are very high.

The intention of the combination of commercial FEM codes with reliability methods is therefore twofold. First, the mechanical model can be described as correctly as possible and second the probability information is obtained with good accuracy and reasonable computational effort. The combination introduced in this paper can be used for any FEM code provided that the element level can be accessed at least via an interface. If this assumption is fulfilled, "adjoint elements" can be implemented. It will be shown that "adjoint elements" are a computationally efficient method to calculate the gradient of the structural response with respect to the stochastic variables involved.

The paper is divided into six major sections. The first gives an introduction to reliability methods, the second describes the basics of the adjoint method, which leads to the definiton of "adjoint elements". The third section deals with some aspects of the implementation of "adjoint elements" and the fourth is concerned with the description of anisotropic material behaviour. The fifth section contains an application example which is a turbine blade with all material parameters assumed to be correlated non-normal stochastic variables or fields. A preview of further investigations followed by a conclusion will close this paper.

RELIABILITY METHODS

The parameters of the mechanical model such as material properties, geometry and external loads show in general scatter in their values. This leads to considering them as random variables or random fields. Random fields are usually described by a set of correlated random variables [2,3]. Let the entire vector of random variables including those derived from random fields be denoted as $\underline{b}$, where the underscore is used to characterize a vector. Its entries b_i are called basic variables, $i = 1,\dots,n$ and n is the number of basic variables.

For linear elastic problems the displacement field $\underline{u}$ is calculated by the fundamental equation of the FEM

$$\underline{\underline{K}}(\underline{b})\cdot\underline{u} = \underline{f}(\underline{b}) \quad , \tag{1}$$

where $\underline{\underline{K}}(\underline{b})$ is the global stiffness matrix with the double underscore as the notation for a matrix and $\underline{f}(\underline{b})$ is the nodal vector of external loads. It can be seen that the displacement field is implicitly depending on the basic variables due to the explicit dependence of $\underline{\underline{K}}$ and $\underline{f}$. The calculation of the failure probability requires the definition of a measure that is capable to describe the failure of the structure. This measure is the performance function $g(\underline{b}, \underline{u})$, which follows the convention:

$$\begin{aligned} g(\underline{b}, \underline{u}) &\le 0 : \text{component has failed (fail set)} \\ g(\underline{b}, \underline{u}) &> 0 : \text{component is still operable (safe set).} \end{aligned} \tag{2}$$

The boundary $g(\underline{b}, \underline{u}) = 0$ is called the limit state surface. The failure probability is then expressed by the multidimensional integral of the joint probability density function $f_{\underline{b}}(\underline{b})$ over the failure domain $g(\underline{b}, \underline{u}) < 0$:

$$P_f = \int\limits_{g(\underline{b}, \underline{u}) \le 0} f_{\underline{b}}(\underline{b})\cdot db_1\cdot\ldots\cdot db_n \quad . \tag{3}$$

Due to the fact that the performance function is not given explicitly but in an algorithmic form this integral cannot be solved directly. Reliability methods treat the problem of solving this integral by approximating the integration limit, i.e. the limit state surface, in the space of the independent standard normal variables, which are denoted here as $\underline{r}$. First order reliability methods (FORM) approximate the limit state surface by a hyper-plane in the so-called design point $\underline{r}^*$, which is the point on the limit state surface, with minimal distance to the origin of the $\underline{r}$-space.

For this purpose the correlated and usually non-normally distributed basic variables have to be transformed into standard normal uncorrelated variables. Then FORM approximates the integral in equation (3) by

$$P_f \simeq \Phi(-\beta) \quad , \tag{5}$$

where Φ is the cumulative distribution function of the standard normal distribution and β is called the Hasofer-Lind reliability index and is the distance of the design point from the origin, i.e. the length of the vector $\underline{r}^*$. Details concerning FORM are given for example in [1,2].

Thus, reliability methods replace the problem of solving the multidimensional integral in equation (3) by the task of finding the design point. Due to the geometrical interpretation of the design point in the $\underline{r}$-space this is a constrained optimization problem, which is written as follows:

$$\begin{array}{ll} \text{minimize} & d^2 = \underline{r}^T \cdot \underline{r} \ , \\ \text{constraint} & g(\underline{b}, \underline{u}) = 0 \ . \end{array} \tag{6}$$

It is beyond the scope of this paper to discuss the different optimization schemes available to solve this kind of problem. But it should be emphasized that most of the optimization algorithms need the gradient of the limit state surface with respect to the basic variables to obtain fast convergence. There are three possibilities to calculate the gradient, namely the method of finite differences, the direct differentiation method and the adjoint method.

Without going into the details of these methods it should be mentioned that if n is the number of basic variables and m is the number of different possible failure modes, i.e. the number of performance functions, finite differencing needs $2 \cdot n$ finite element computations, while the direct method needs $n+1$ and the adjoint method requires $m+1$ FEM calculations [5].

The number of failure modes m is usually one or at least much smaller than n especially when random fields have to be taken into account. Therefore, preference will be given to the adjoint method, which is described in the next section.

ADJOINT METHOD

The adjoint method is based on the solution of the equation of the adjoint problem, which reads

$$\underline{\underline{K}}(\underline{b}) \cdot \underline{\lambda} = \left(\left. \frac{\partial g(\underline{b}, \underline{u})}{\partial \underline{u}} \right|_{\underline{b}} \right)^T \ . \tag{7}$$

Here, $\underline{\lambda}$ is an auxiliary vector. Due to the analogous form of equations (1) and (7) $\underline{\lambda}$ shall be called the "adjoint displacement vector" and the right side in equation (7) is called the "adjoint force vector". It can be shown [4] that equation (7) is a direct consequence of the Lagrange function solving the optimization problem (6) by introducing $\underline{\lambda}$ as a Lagrange multiplier.

It should be noted that even for non-linear problems the adjoint problem is always linear in $\underline{\lambda}$. Once equation (7) is solved the gradient of the performance function with respect to the basic variables is

$$\frac{dg(\underline{b}, \underline{u})}{d\underline{b}} = \left.\frac{\partial g(\underline{b}, \underline{u})}{\partial \underline{b}}\right|_{\underline{u}} + \underline{\lambda}^T \frac{\partial \underline{f}(\underline{b})}{\partial \underline{b}} - \underline{\lambda}^T \frac{\partial \underline{\underline{K}}(\underline{b})}{\partial \underline{b}} \cdot \underline{u} \ . \tag{8}$$

This equation contains $\underline{\lambda}$ as solution of the adjoint problem and $\underline{u}$ as solution of the real problem as well. Hence, the realization of the adjoint method consists of the following steps:

1. Solve the "real problem" in equation (1) in order to obtain $\underline{u}$.
2. Choose a performance function, which describes the failure mode according to conditions (2).
3. Calculate the "adjoint force vector" of equation (7).
4. Solve the "adjoint problem" (7) in order to obtain $\underline{\lambda}$.
5. Calculate the partial derivatives of g, $\underline{f}$ and $\underline{\underline{K}}$ with respect to the basic variables that are required in equation (8).
6. Compute the gradient according to equation (8).

IMPLEMENTATION OF ADJOINT ELEMENTS

The implementation of "adjoint elements" is based on the requirement that the FEM code, in which these elements have to be implemented, has to allow the access of the element level directly or via an interface.

The FEM package used in our case is ABAQUS. It provides an interface to define elements of arbitrary properties via FORTRAN coding, which are called USER-elements. For a structure built up with USER-elements ABAQUS needs the element stiffness matrix and the nodal forces for each element individually.

This feature allows the solution of the adjoint problem and the calculation of the gradient of the limit state surface with respect to the basic variables strictly on the element level. Thus, "adjoint elements" are those, which solve the equation

$$\underline{\underline{K}}_{el}(\underline{b})\cdot\underline{\lambda} = \left(\left.\frac{\partial g(\underline{b}, \underline{u})}{\partial \underline{u}_{el}}\right|_{\underline{b}} \right)^T , \tag{9}$$

where the quantities indexed with *el* refer to a single element of the structure. In other words, the element stiffness matrix is calculated in the usual manner but instead of the real forces the "adjoint forces" are returned to the FEM code. Exploiting the fact that the assembly of the finite elements can be represented as the sum of elemental quantities, as shown in [5], equation (8) transforms to

$$\frac{dg(\underline{b}, \underline{u})}{d\underline{b}} = \left.\frac{\partial g(\underline{b}, \underline{u})}{\partial \underline{b}}\right|_{\underline{u}} + \sum_{el=1}^{n_{el}} \left(\underline{\lambda}_{el}^{T} \frac{\partial \underline{f}_{el}(\underline{b})}{\partial \underline{b}} - \underline{\lambda}_{el}^{T} \frac{\partial \underline{\underline{K}}_{el}(\underline{b})}{\partial \underline{b}} \cdot \underline{u}_{el} \right) , \quad (10)$$

where n_{el} denotes the number of elements of the discretized component. Bearing in mind that ABAQUS processes one element after the other as mentioned above this equation can be implemented within the coding of the "adjoint element" instead of being a part of post processing. Hence, the required gradient information can be calculated within one FEM computation run.

It should be mentioned that for a random variable that relates only to a substructure of the component the summation in equation (10) yields a contribution only for the elements, which are part of the substructure.

Choice of the performance function

Most applications of linear elastic and static analysis are dealing with stresses or displacements. Thus, without claiming completeness, three performance functions have been chosen.

1. Equivalent stress according to the v. Mises criterion:

$$g(\underline{b}, \underline{u}) = \sigma_{crit} - \left.\sigma_{eq}(\underline{b}, \underline{u})\right|_{Mises}$$

2. Equivalent stress according to the Tresca criterion:

$$g(\underline{b}, \underline{u}) = \sigma_{crit} - \left.\sigma_{eq}(\underline{b}, \underline{u})\right|_{Tresca}$$

3. Displacement criterion:

$$g(\underline{b}, \underline{u}) = \Delta l_{max} - \underline{u}$$

It has to be taken into account that these criteria relate to the most critical point of the structure and in addition the displacement criterion relates to the most crucial direction.

Required derivatives

The derivatives in equation (10) can be calculated analytically, which reduces the computation time. Suffice it to say that equation (10) usually simplifies if the random variable b_i is a parameter of the material or of the external load. In the first case the derivative of the external force vector vanishes and in the second case the derivative of the stiffness matrix does not appear.

DESCRIPTION OF LINEAR ELASTIC MATERIAL MODELS

Hooke's law is usually written in matrix form. Using Cartesian coordinates for the general three-dimensional situation, it reads

$$\underline{\sigma} = \underline{D}(\underline{b}, \underline{u}) \cdot \underline{\varepsilon} \ , \quad (11)$$

where

$$\underline{\sigma} = (\sigma_x, \sigma_y, \sigma_z, \tau_{yz}, \tau_{zx}, \tau_{xy})^T , \qquad \underline{\varepsilon} = (\varepsilon_x, \varepsilon_y, \varepsilon_z, \varepsilon_{yz}, \varepsilon_{zx}, \varepsilon_{xy})^T$$

and $\underline{\underline{D}}$ is the symmetric material matrix.

Isotropic materials

Isotropic materials have only two independent material parameters, Young's modulus E and Poisson's ratio ν. Using these parameters, the entries of the material matrix may be taken from the literature, e.g. [6].

Transversely isotropic materials

Transverse isotropy is a special case of anisotropy, which describes the behaviour of directionally solidified materials. These materials show isotropic behaviour in every plane normal to the growth direction of the grains, which are therefore called the planes of isotropy. All directions in these planes are elastically equivalent.

Following the notation used in [7], transversely isotropic materials are characterized by five independent constants. The parameters E_1, ν_1 and G_1 are associated with the orientation normal to the plane of isotropy, i.e. the growth direction. E_2 and ν_2 describe the behaviour within the plane of isotropy. If the z axis of the Cartesian coordinate system is normal to the plane of isotropy, then it can be deduced from [8] that the material matrix has the entries:

$$
\begin{aligned}
D_{11} &= \frac{E_2\left(1 - p\cdot\nu_1^2\right)}{(1+\nu_2)\left(1-\nu_2-2p\nu_1^2\right)} = D_{22} , \\
D_{12} &= \frac{E_2\left(\nu_2 + p\cdot\nu_1^2\right)}{(1+\nu_2)\left(1-\nu_2-2p\nu_1^2\right)} = D_{21} , \\
D_{13} &= \frac{E_2\cdot\nu_1}{\left(1-\nu_2-2p\nu_1^2\right)} = D_{31} = D_{23} = D_{32} , \\
D_{33} &= \frac{E_1\left(1-\nu_2\right)}{\left(1-\nu_2-2p\nu_1^2\right)} \\
D_{44} &= G_1 = D_{55} , \\
D_{66} &= \frac{E_1}{2\left(1+\nu_2\right)} ,
\end{aligned}
\tag{12}
$$

with $p = E_2/E_1$ and the entries not shown are zero.

APPLICATION EXAMPLE

As application example a simple model of a guide vane was chosen. The material is a directionally solidified nickel-based superalloy. The cooling conditions during the solidification process lead to a relatively small grain size at the thin parts of the blade. Thus, the grain structure at the edges can be regarded as quasi-homogenous and isotropic. In contrast, the thicker inner part shows an orientation of the grains along the growth direction, which is the radial axis of the blade.

Mechanical model

The turbine blade was discretized using 10 element stratas in the radial direction with 15 isoparametric elements with quadratic shape functions in each layer. The FEM mesh discretization is shown in Figure 1.

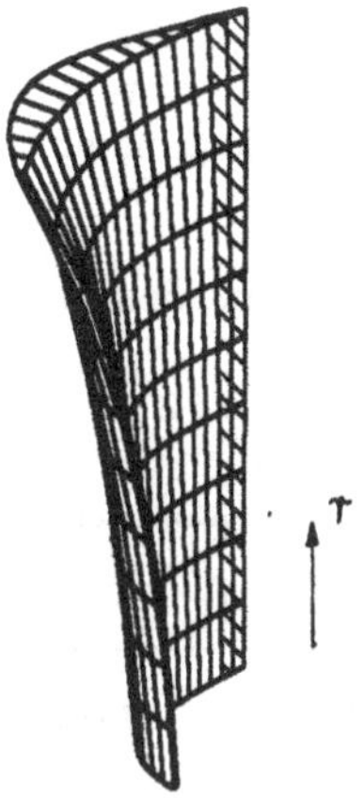

Figure 1. Turbine blade FEM mesh

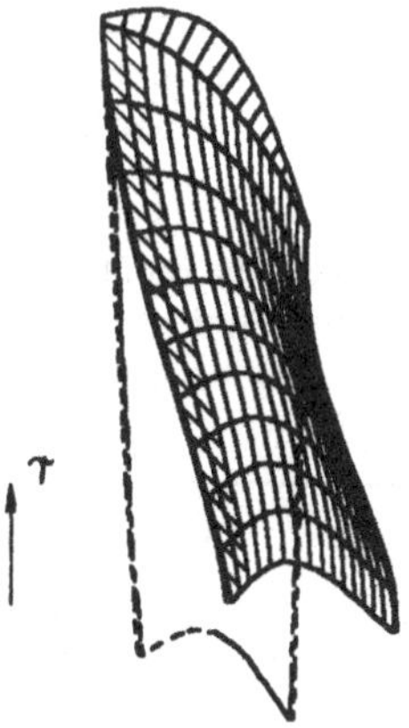

Figure 2. Deflected turbine blade

The guide vane is kept fixed at the outer radial face. The difference of the pressure at the inner side of the blade and the outer side is modelled with equivalent nodal forces. The deflection of the guide vane resulting from this loading is shown in Figure 2.

The isotropic parts of the blade, i.e. the edges, are assumed to cover two elements of each element layer respectively. The isotropic rear edge is shaded in Figure 1 and the leading edge is shaded in Figure 2. The residual inner part covering 110 elements is treated as transversely isotropic.

Probabilistic model

Two different probabilistic approaches to the problem have been chosen one with and one without random fields. In both cases the isotropic constants of the leading edge E_{lead} and ν_{lead}, of the rear edge E_{rear} and ν_{rear} and the anisotropic parameters of the inner part G_1, ν_1 and ν_2 are random variables.

In Example 1 the anisotropic Young's moduli E_1 and E_2 are treated as random variables as well. Hence, for this example the total number of basic variables is 9.

In Example 2 the anisotropic Young's moduli E_1 and E_2 are treated as random fields. According to [2,3] a stochastic mesh has to be chosen in order to discretize random fields. For this example the stochastic mesh is identical to the mechanical mesh. Then, each element of the inner part of the guide vane is associated with one random variable. Thus, because of the two random fields in the inner part, the total number of basic variables is 227.

In both examples the gradient is calculated at the expectation point. According to [9] the expected values are given by:

$$E_{lead} = E_{rear} = 191.06\,\text{GPa}$$
$$\nu_{lead} = \nu_{rear} = 0.3328$$

$$E_1 = 125.1\,\text{GPa}$$
$$\nu_1 = 0.391$$
$$G_1 = 118.6\,\text{GPa}$$
$$E_2 = 166\,\text{GPa}$$
$$\nu_2 = 0.143$$

Results

The method of finite differences is used to verify the results obtained with the adjoint method. In addition, a comparison of these two methods is used to illustrate the efficiency of the adjoint method especially for problems with a large number of basic variables.

For Example 1 the results of the adjoint method (AM) and finite differencing (FD) are compared in Table 1.

Table 1: Results Example 1

Perf.fc.	*v. Mises*		*Tresca*		*Displacement*	
method	*AM*	*FD*	*AM*	*FD*	*AM*	*FD*
dg/dE_{lead}	$4.64298\cdot10^{-6}$	$4.6335\cdot10^{-6}$	$4.68320\cdot10^{-6}$	$4.6712\cdot10^{-6}$	$2.30994\cdot10^{-7}$	$2.31005\cdot10^{-7}$
dg/dv_{lead}	-0.41088	-0.4176	-0.41504	-0.4203	$7.38896\cdot10^{-4}$	$7.3865\cdot10^{-4}$
dg/dE_{rear}	$-1.92066\cdot10^{-5}$	$-1.9268\cdot10^{-5}$	$-1.94519\cdot10^{-5}$	$-1.9502\cdot10^{-5}$	$4.08569\cdot10^{-8}$	$4.0875\cdot10^{-8}$
dg/dv_{rear}	$5.26453\cdot10^{-2}$	$5.1215\cdot10^{-2}$	$1.31695\cdot10^{-1}$	$1.29315\cdot10^{-1}$	$-1.38911\cdot10^{-4}$	$-1.389\cdot10^{-4}$
dg/dE_1	$4.96069\cdot10^{-6}$	$4.9135\cdot10^{-6}$	$5.39227\cdot10^{-6}$	$5.3358\cdot10^{-6}$	$3.70702\cdot10^{-7}$	$3.70675\cdot10^{-7}$
dg/dv_1	-0.70260	-0.69755	-0.99949	-0.99714	2.25038	2.25039
dg/dG_1	$3.95145\cdot10^{-6}$	$3.966\cdot10^{-6}$	$4.20003\cdot10^{-6}$	$4.2191\cdot10^{-6}$	$5.81849\cdot10^{-8}$	$5.8195\cdot10^{-8}$
dg/dE_2	$1.02005\cdot10^{-5}$	$1.0343\cdot10^{-5}$	$0.993371\cdot10^{-5}$	$1.00797\cdot10^{-5}$	$1.738877\cdot10^{-7}$	$1.73895\cdot10^{-7}$
dg/dv_2	-1.43794	-1.4406	-1.55314	-1.5565	$-1.89555\cdot10^{-2}$	$-1.89554\cdot10^{-2}$
CPU time	772 sec.	1690 sec.	765 sec.	1656 sec.	775 sec.	1674 sec.

Table 1 demonstrates that the values of the gradient computed with the adjoint method are in good agreement with those obtained by finite differencing. As expected, the computational effort for finite differencing is much higher than for the adjoint method.

Due to the large number of basic variables in the second example only the required CPU time for both methods are listed in Table 2. For the method of finite differencing only arbitrarilly chosen entries of the gradient have been calculated. Therefore, the CPU time for finite differencing is obtained by 2×227 times the CPU time of one FEM computation. Especially Example 2 clearly demonstrates the computational efficiency of the adjoint method.

Table 2: CPU time in seconds for Example 2

performance function	*Adjoint method*	*Finite differences*
v. Mises	*771*	*43357*
Tresca	*770*	*43298*
Displacement	*770*	*42253*

As expected, a comparison of the computational effort of both examples shows that the amount of CPU time for the adjoint method is nearly independent of the number of basic variables, while for the method of finite differencing it increases linearly with the number of basic variables.

PREVIEW OF FURTHER INVESTIGATIONS

As mentioned in the section "Reliability Methods" the search for the design point consists of two major parts, the transformation of the basic variables into the space of non-correlated standard normal variables and the optimization scheme.

The implementation of the transformation has to take into account that random variables may be correlated among each other. Then, as a consequence the entries of the gradient become correlated too. There are several possiblilities to realize this transformation [2]. Due to its advantageous properties preference will be given to the Nataf-model in future works.

As demonstrated in this paper with the use of the adjoint method one possibility to save computational effort is already considered. The implemention of a fast converging optimization scheme provides the potential for additional savings of CPU time [2]. This implementation will be the subject of further investigations as well.

In the application example is assumed that the axis prependicular to the plane of isotropy is coaxial with the radial axis of the guide vane. As demonstrated in [10], this is a rough idealization, since in reality there is an angle between these two axes and it scatters in a relatively wide range, due to the manufacturing process. Therefore, this angle has to be treated as a random quantity as well.

CONCLUSIONS

In the present contribution a method is introduced to compute the gradient of the structural response with respect to the stochastic variables by means of "adjoint elements". As demonstrated, the formulation the adjoint method strictly on the element level allows the computation of the gradient with commercial FEM packages within one calculation run.

The application examples illustrate the efficiency of the adjoint method especially for cases with a large number of stochastic variables, which are always obtained for random fields. Thus, "adjoint elements" are highly suitable to reduce the computational effort of a stochastic finite element analysis of arbitrary structures.

REFERENCES

1 P. Bjerager, Probability Computation Methods in Structural and Mechanical Reliability, Computational Mechanics of Probabilistic and Reliability Analysis, Edited by Wing Kam Liu and Ted Belytschko, Elmepress International,Lausanne (1989) 47 - 67.

2 A. Der Kiureghian, P.-L. Liu, Finite Element Reliability Methods for Geometrical Nonlinear Stochastic Structures, Report No. UCB/SEMM-89/05, Department of Civil Engineering, University of California, Berkeley, 1 (1989).

3 E. Vanmarcke, Random Fields: Analysis and Synthesis, The MIT Press, Cambridge, Massachusetts, (1983).

4 G. H. Besterfield, W. K. Liu, M. A. Lawrence, T. B. Belytschko, Brittle Fracture Reliability by Probabilistic Finite Elements, Journal of Engineering Mechanics, 116, 3 (1990) 642 - 659.

5 E. J. Haug, K. K. Choi, V. Komkov, Design Sensitivity Analysis of Structural Systems, Academic Press Inc. (London) Ltd., Mathematics in Science and Engineering, 177 (1986).

6 O. C. Zienkiewicz, The Finite Element Method, McGraw-Hill Book Company Limited, Maidenhead·Berkshire, (1977).

7 O. C. Zienkiewicz, Y. K. Cheung, K. G. Stagg, Stresses in Anisotropic Media with Particular Reference to Problems of Rock Mechanics, The Journal of Strain Analysis, 1, 2 (1966) 172 - 182.

8 S. G. Lekhnitskii, Theory of Elasticity of an Anisotropic Body, Mir Publishers, Moscow, (1963).

9 C. H. Wells, The Elastic Constants of a Directionally Solidified Nickel-Base Superalloy, ASM Transactions Quarterly, 60, 2 (1967) 270 - 271.

10 J. H. Laflen, Analysis of an Idealized Directionally Solidified FCC Material, Journal of Engineering Materials and Technology, 105, 10 (1983) 307 - 312.

Analysis of Two-Dimensional Stochastic Systems by the Weighted Integral Method

G. Deodatis, W. Wall, M. Shinozuka
Department of Civil Engineering and Operations Research, Princeton University, Princeton, NJ 08544, U.S.A.

ABSTRACT

In this paper, the weighted integral method is extended to calculate the response variability of two-dimensional stochastic systems in conjunction with the stochastic finite element method. Plane stress and plane strain problems are examined using the constant stress/strain triangular element. The stochastic stiffness matrix of the structure is calculated in terms of integrals of the stochastic field describing the random material property over the area of each element. These integrals are random variables called weighted integrals. The covariance matrix of these weighted integrals is calculated numerically using Gaussian quadrature formulas. As a consequence, the method is considerably more accurate than the conventional stochastic finite element method that uses the midpoint method to reduce the stochastic field involved in the problem to a series of random variables. Then, a Taylor series expansion is used to calculate the response variability of the response displacements and stresses. Finally, a numerical example involving a stochastic circular plate is examined.

INTRODUCTION

A stochastic system is defined herein as any structural system that possesses uncertainties in its material properties and/or geometry. Then, the analysis of the response variability of stochastic systems consists of evaluating the probabilistic characteristics of the response of such systems subjected to deterministic or random loads.

A small number of analytic solutions to such problems is available, mainly for simple linearly elastic structures under static loads[1–3]. The majority of research work in this area, however, has focused on developing various stochastic finite element methods (SFEM) in order to obtain the solutions numerically. The most widely-used SFEM approach is based on mean-centered perturbation techniques[4–11]. As far as the reliability analysis of stochastic systems is concerned, work has been done using either first- or second-order reliability methods[12–15]. A more recently developed SFEM approach is based on the Neumann expansion of the inverse of the stiffness matrix of the system[16-17]. In this regard, Monte Carlo simulation-based SFEM are also used not only for validating the perturbation and other approximate methods, but also in conjunction with the Neumann expansion technique as an integral part of the solution method. Finally, the weighted integral method was introduced in conjunction with the stochastic finite element method to calculate the response variability and the reliability of stochastic systems[18-23]. The weighted integral method calculates the stochastic stiffness matrix as a function of random variables obtained as integrals of the stochastic field describing the random material properties multiplied by a deterministic function. These random variables are called weighted integrals. As a consequence, a finite element mesh practically identical to the one that would be used in a deterministic analysis can be used for any value of the correlation distance of the stochastic field involved in the problem.

The present work extends the weighted integral method to calculate the response variability of two-dimensional stochastic systems.

STOCHASTIC ELEMENT STIFFNESS MATRIX

Consider the constant stress/strain triangular element (CST element) shown in Fig. 1 with six degrees of freedom. Assume that the elastic modulus of the element varies randomly over its area according to the following form:

$$E^{(e)}(x,y) = E_0^{(e)}\left[1 + f^{(e)}(x,y)\right] \tag{1}$$

where $E_0^{(e)}$ = mean value of the elastic modulus and $f^{(e)}(x,y)$ = two-dimensional, uni-variate, zero-mean, homogeneous stochastic field. To avoid the possibility of obtaining non-positive values of the elastic modulus, the stochastic field $f^{(e)}(x,y)$ is assumed to be bounded as follows:

$$-1+\theta \leq f^{(e)}(x,y) \leq 1-\theta \qquad \{x,y\} \in \left[A^{(e)}\right] \tag{2}$$

where θ must satisfy the condition $0 < \theta < 1$ and $A^{(e)}$ denotes the area of the element. Using the standard finite element analysis methodology based on the principle of stationary potential energy (e.g. Segerlind[24]), the element nodal displacement vector is denoted by:

$$\left\{U^{(e)}\right\} = \lfloor u_{2i-1} \quad u_{2i} \quad u_{2j-1} \quad u_{2j} \quad u_{2k-1} \quad u_{2k} \rfloor^T \tag{3}$$

and the stress-strain relationship for element (e) is given by:

$$\{\sigma^{(e)}\} = \left[D^{(e)}\right]\{\epsilon^{(e)}\} \tag{4}$$

where $\{\sigma^{(e)}\}$ and $\{\epsilon^{(e)}\}$ are the stress and strain vectors for element (e), respectively. For plane stress, matrix $\left[D^{(e)}\right]$ is given by:

$$\left[D^{(e)}\right] = \frac{E^{(e)}(x,y)}{1-\mu^2}\begin{bmatrix} 1 & \mu & 0 \\ \mu & 1 & 0 \\ 0 & 0 & \frac{(1-\mu)}{2} \end{bmatrix} \tag{5}$$

and for plane strain:

$$\left[D^{(e)}\right] = \frac{E^{(e)}(x,y)}{(1+\mu)(1-2\mu)}\begin{bmatrix} (1-\mu) & \mu & 0 \\ \mu & (1-\mu) & 0 \\ 0 & 0 & \frac{1-2\mu}{2} \end{bmatrix} \tag{6}$$

where μ denotes Poisson's ratio. Then, the stochastic element stiffness matrix $\left[K^{(e)}\right]$ can be calculated from:

$$\left[K^{(e)}\right] = \int_{V^{(e)}} \left[B^{(e)}\right]^T \left[D^{(e)}\right] \left[B^{(e)}\right] dV^{(e)} \tag{7}$$

where $\left[B^{(e)}\right]$ is the gradient matrix of element (e) and $V^{(e)}$ denotes the volume of element (e). Since the element thickness $t^{(e)}$ and matrix $\left[B^{(e)}\right]$ are both constant within each CST element, Eq. 7 can be reduced to:

$$\left[K^{(e)}\right] = t^{(e)} \left[B^{(e)}\right]^T \int_{A^{(e)}} \left[D^{(e)}\right] dA^{(e)} \left[B^{(e)}\right] \tag{8}$$

Finally, carrying out the integration indicated in Eq. 8, the stochastic element stiffness matrix is calculated as:

$$\left[K^{(e)}\right] = \left[K_0^{(e)}\right] + X_0^{(e)} \cdot \left[\Delta K_0^{(e)}\right] \tag{9}$$

where $\left[K_0^{(e)}\right]$ and $\left[\Delta K_0^{(e)}\right]$ are deterministic matrices and $X_0^{(e)}$ is a random variable defined as:

$$X_0^{(e)} = \int_{A^{(e)}} f^{(e)}(x,y)\, dA^{(e)} \tag{10}$$

The random variable $X_0^{(e)}$ defined in Eq. 10 is called a weighted integral. Note that the stochastic element stiffness matrix (Eq. 9) is a linear function of the weighted integral $X_0^{(e)}$. Note also that matrix $\left[K_0^{(e)}\right]$ is the mean value of matrix $\left[K^{(e)}\right]$ since random variable $X_0^{(e)}$ has mean value equal to zero, as can be easily seen from Eq. 10. Equivalently, it can be said that matrices $\left[K_0^{(e)}\right]$ and $X_0^{(e)}\left[\Delta K_0^{(e)}\right]$ are the deterministic and stochastic parts of the element stiffness matrix $\left[K^{(e)}\right]$, respectively.

STOCHASTIC GLOBAL STIFFNESS MATRIX

Using the standard finite element analysis methodology, the global stiffness matrix $[K]$ is assembled as follows:

$$[K] = \sum_{e=1}^{N_e} \left[K^{(e)}\right] \tag{11}$$

where N_e is the total number of finite elements and $\left[K^{(e)}\right]$ is the extended version of the element stiffness matrix. Finally, after introducing the appropriate boundary conditions, the equations of equilibrium are given by:

$$[K]\{U\} = \{P\} \tag{12}$$

where $\{U\}$ is the global nodal displacement vector and $\{P\}$ is the deterministic global force vector.

ANALYSIS OF RESPONSE VARIABILITY OF NODAL DISPLACEMENTS

The global stiffness matrix $[K]$ appearing in Eq. 12 involves weighted integrals X_{e0} which are random variables of the form:

$$X_{e0} = \int_{A^{(e)}} f^{(e)}(x,y)dA^{(e)} \tag{13}$$

where $e = 1, 2, \ldots, N_e$. As already mentioned, the mean values of random variables X_{e0}, denoted by $\bar{X}_{e0}$, are all equal to zero:

$$\bar{X}_{e0} = 0 \ ; \ e = 1, 2, \ldots, N_e \tag{14}$$

Another obvious conclusion drawn from Eq. 12 is that the global nodal displacement vector $\{U\}$ will also be a function of random variables X_{e0} :

$$\{U\} = \{U\} \ (X_{e0} \ ; \ e = 1, 2, \ldots, N_e) \tag{15}$$

Denoting now by $\{U_0\}$ and $[K_0]$ the values of the global nodal displacement vector and global stiffness matrix evaluated at $\bar{X}_{e0}$; $e = 1, 2, \ldots, N_e$, respectively, the following relation holds between $\{U_0\}$ and $[K_0]$:

$$\{U_0\} = [K_0]^{-1}\{P\} \tag{16}$$

Consider now the first-order approximation of the Taylor expansion of function $\{U\}$ around the mean values of random variables X_{e0} ; $e = 1, 2, \ldots, N_e$:

$$\{U\} \cong \{U_0\} + \sum_{e=1}^{N_e} (X_{e0} - \bar{X}_{e0}) \left[\frac{\partial\{U\}}{\partial X_{e0}}\right]_E \tag{17}$$

where the symbol $[\bullet]_E$ denotes evaluation at the mean values of random variables X_{e0} ; $e = 1, 2, \ldots, N_e$. The term $\left[\frac{\partial\{U\}}{\partial X_{e0}}\right]_E$ appearing in Eq. 17 can be calculated by partially differentiating Eq. 12 with respect to X_{e0} and then evaluating the result at $\bar{X}_{e0}$; $e = 1, 2, \ldots, N_e$ to obtain:

$$\left[\frac{\partial\{U\}}{\partial X_{e0}}\right]_E = -[K_0]^{-1}\left[\frac{\partial[K]}{\partial X_{e0}}\right]_E \{U_0\} \tag{18}$$

Substituting now Eq. 18 into Eq. 17, the following result is obtained:

$$\{U\} \cong \{U_0\} - \sum_{e=1}^{N_e} [K_0]^{-1}\left[\frac{\partial[K]}{\partial X_{e0}}\right]_E \{U_0\} \cdot X_{e0} \tag{19}$$

Hence, the first-order approximation of the mean value and the co-variance matrix of $\{U\}$ are now easily evaluated as:

$$\mathcal{E}[\{U\}] = \{U_0\} \tag{20}$$

$$\mathrm{Cov}[\{U\},\{U\}] = \mathcal{E}[(\{U\} - \{U_0\})(\{U\} - \{U_0\})^T] =$$

$$= \sum_{e_1=1}^{N_e}\sum_{e_2=1}^{N_e} [K_0]^{-1}\left[\frac{\partial[K]}{\partial X_{e_1 0}}\right]_E \{U_0\}\{U_0\}^T \left[\frac{\partial[K]}{\partial X_{e_2 0}}\right]_E^T ([K_0]^{-1})^T \cdot$$

$$\cdot\mathcal{E}\{X_{e_1 0}X_{e_2 0}\} \tag{21}$$

Using Eq. 13, the expected value shown in Eq. 21 can be written as:

$$\mathcal{E}\{X_{e_1 0}X_{e_2 0}\} = \int_{A^{(e_1)}}\int_{A^{(e_2)}} \mathcal{E}\left[f^{(e_1)}(x_1,y_1)f^{(e_2)}(x_2,y_2)\right] dA^{(e_1)}dA^{(e_2)} \tag{22}$$

Considering now that all finite elements are characterized by the same stochastic field $f(x,y)$, the expected value appearing in the integrand in Eq. 22 is calculated in the following way (see Fig. 2):

$$\mathcal{E}\left[f^{(e_1)}(x_1,y_1)f^{(e_2)}(x_2,y_2)\right] = R_{ff}(x_2 - x_1, y_2 - y_1) = R_{ff}(\xi,\eta) \tag{23}$$

where R_{ff} is the autocorrelation function of stochastic field $f(x,y)$. The double integral shown in Eq. 22 will be calculated numerically in the section "Numerical Integration of Equation 22" that follows.

Using now Eqs. 9 and 11, it is straightforward to show that the only remaining term to be calculated in Eq. 21 is given by:

$$\left[\frac{\partial[K]}{\partial X_{e0}}\right]_E = \left[\Delta K_0^{(e)}\right] \tag{24}$$

Substituting then Eqs. 22, 23 and 24 into Eq. 21, the following expression is obtained for the covariance matrix of $\{U\}$:

$$\text{Cov}[\{U\},\{U\}] =$$

$$= \sum_{e_1=1}^{N_e} \sum_{e_2=1}^{N_e} [K_0]^{-1} \left[\Delta K_0^{(e_1)}\right] \{U_0\}\{U_0\}^T \left[\Delta K_0^{(e_2)}\right]^T ([K_0]^{-1})^T \cdot$$

$$\cdot \int_{A^{(e_1)}} \int_{A^{(e_2)}} R_{ff}(\xi,\eta) dA^{(e_1)} dA^{(e_2)} \qquad (25)$$

Finally, the variance vector of $\{U\}$ consisting of the diagonal elements of the covariance matrix of $\{U\}$ is calculated as:

$$\text{Var}[\{U\}] =$$

$$= \sum_{e_1=1}^{N_e} \sum_{e_2=1}^{N_e} \text{diag}\left([K_0]^{-1} \left[\Delta K_0^{(e_1)}\right] \{U_0\}\right) [K_0]^{-1} \left[\Delta K_0^{(e_2)}\right] \{U_0\} \cdot$$

$$\cdot \int_{A^{(e_1)}} \int_{A^{(e_2)}} R_{ff}(\xi,\eta) dA^{(e_1)} dA^{(e_2)} \qquad (26)$$

where diag($\bullet$) represents a diagonal matrix whose diagonal components consist of the vector within the parenthesis.

NUMERICAL INTEGRATION OF EQUATION 22

As already mentioned, the integrals appearing in Eqs. 25 and 26 will be computed numerically. The most appropriate formulas for the numerical integration of a function over a triangular area are the Gaussian quadrature formulas. In this case, the area integral over a triangular region can be written as:

$$\int_\Omega f d\Omega = c \sum_{l=1}^{n_i} f_l \cdot j_l \cdot W_l \qquad (27)$$

where $c = 1/2$, n_i = number of integration points, j_l = determinant of Jacobien at integration point l, f_l = value of function f at integration point l and W_l = weight for integration point l.

The integration points are usually defined in triangular coordinates, e.g. Cowper[26], and therefore have to be transformed to global coordinates in order to calculate $R_{ff}(\xi,\eta)$ appearing in Eqs. 25 and 26. Taking into account that the CST element is an isoparametric one and that the triangular coordinates shown in Fig. 1 are identical to the shape functions used, it is easy to calculate the Jacobien of the transformation. The determinant of the Jacobien is found to be identical for all integration points and therefore can be taken out of the summation shown in Eq. 27

The 12-point formula was eventually chosen, after comparing it to results obtained from an analytic integration of a special case.

VARIABILITY RESPONSE FUNCTION

In order to get insight into the underlying mechanisms controlling the response variability of a stochastic system, the "variability response function" is of great importance. The concept of the "variability response function" of a stochastic system introduced in Deodatis and Shinozuka[18] was extended to trusses and frames analyzed by the finite element method in Deodatis[19]. It is a very useful tool to calculate bounds on response variability of stochastic systems. In the following, it is briefly shown how to extend this concept to two-dimensional structures.

The spectral decomposition of a homogeneous stochastic field $f(x,y)$, using the two-dimensional version of the Wiener-Khintchine transform pair, is given by:

$$R_{ff}(\xi,\eta) = \int_{-\infty}^{\infty}\int_{-\infty}^{\infty} S_{ff}(\kappa_1,\kappa_2)\cdot \exp[i(\xi\kappa_1+\eta\kappa_2)]d\kappa_1 d\kappa_2 \tag{28}$$

where ξ and η denote separation distances and κ_1 and κ_2 wave numbers. Using now Eq. 28, Eq. 22 can be rewritten in the following form:

$$\mathcal{E}\{X_{e_1 0}X_{e_2 0}\} = \int_{-\infty}^{\infty}\int_{-\infty}^{\infty} S_{ff}(\kappa_1,\kappa_2)\cdot$$

$$\cdot\int_{A^{(e_1)}}\int_{A^{(e_2)}} \exp[i(\xi\kappa_1+\eta\kappa_2)]dA^{(e_1)}dA^{(e_2)}d\kappa_1 d\kappa_2 \tag{29}$$

Substituting now Eq. 29 into Eq. 26, the variance vector of $\{U\}$ can be written in the following form:

$$\mathrm{Var}[\{U\}] = \int_{-\infty}^{\infty}\int_{-\infty}^{\infty} S_{ff}(\kappa_1,\kappa_2)\{V(\kappa_1,\kappa_2)\}d\kappa_1 d\kappa_2 \tag{30}$$

where:

$$\{V(\kappa_1,\kappa_2)\} = \sum_{e_1=1}^{N_e}\sum_{e_2=1}^{N_e} \mathrm{diag}\left([K_0]^{-1}\left[\Delta K_0^{(e_1)}\right]\{U_0\}\right)\cdot$$

$$\cdot[K_0]^{-1}\left[\Delta K_0^{(e_2)}\right]\{U_0\}\int_{A^{(e_1)}}\int_{A^{(e_2)}} \exp[i(\xi\kappa_1+\eta\kappa_2)]dA^{(e_1)}dA^{(e_2)} \tag{31}$$

Note that $\{V(\kappa_1,\kappa_2)\}$ is interpreted as the first-order approximation of the variability response function as defined in Deodatis and Shinozuka[18]. Finally, it should be mentioned that a similar analysis was performed for the internal stresses.

NUMERICAL EXAMPLE

The weighted integral method developed in this paper is now applied to the problem of a stochastic circular plate. Fig. 3 shows one quarter of the circular plate subjected to a deterministic uniform load along its curved edge. The spatial variation of the elastic modulus is described by Eq. 1 and the autocorrelation function characterizing stochastic field $f(x,y)$ is chosen to be:

$$R_{ff}(\xi,\eta) = \sigma_{ff}^2 \exp\left\{-\frac{|\xi|+|\eta|}{b}\right\} \tag{32}$$

where the coefficient of variation of the elastic modulus σ_{ff} is set equal to $\sigma_{ff} = 0.10$.

The coefficient of variation (COV) of the response displacement and stress (at locations indicated in Fig. 3) are plotted in Fig. 4 as functions of the non-dimensional quantity $\frac{b}{R}$ (R being the radius of the circular plate). Note that both COV approach zero as $\frac{b}{R} \to 0$ and that the COV of the displacement approaches σ_{ff} while the COV of the stress approaches zero as $\frac{b}{R} \to \infty$. Both behaviours were expected since the stochastic field $f(x,y)$ degenerates to a random variable as $\frac{b}{R} \to \infty$ while it becomes a finite power white noise as $\frac{b}{R} \to 0$.

CONCLUSIONS AND FUTURE WORK

This work extended the weighted integral method to account for two-dimensional stochastic systems (plane stress and plane strain problems). A numerical example was examined involving a stochastic circular plate. In future work, rectangular finite elements will be considered alone and in combination with triangular ones, as well as plate bending problems.

ACKNOWLEDGMENTS

This work was supported by the National Science Foundation under Grant No. CES-8813923 with Dr. S. -C. Liu as Program Director. The second author wishes to acknowledge the support of the Austrian Bundeswirtschaftskammer and of the Austrian-American Educational Commission.

REFERENCES

1. Shinozuka, M. (1987). "Structural Response Variability," *Journal of Engineering Mechanics*, ASCE, Vol. 113, No. 6, pp. 825-842.
2. Bucher, C. G. and Shinozuka, M. (1988). "Structural Response Variability II," *Journal of Engineering Mechanics*, ASCE, Vol. 114, No. 12, pp. 2035-2054.

3. Kardara, A., Bucher, C. G. and Shinozuka, M. (1989). "Structural Response Variability III," *Journal of Engineering Mechanics,* ASCE, Vol. 115, No. 8, pp. 1726-1747.

4. Cambou, B. (1975). "Applications of First-Order Uncertainty Analysis in the Finite Elements Method in Linear Elasticity," *Proceedings of the 2nd International Conference on Applications of Statistics and Probability in Soil and Structural Engineering,* Aachen, West Germany, pp. 67-87.

5. Baecher, G.B. and Ingra, T.S. (1981). "Stochastic FEM in Settlement Predictions," *Journal of the Geotechnical Engineering Division,* ASCE, Vol. 107, No. GT4, pp. 449-463.

6. Handa, K. and Andersson, K. (1981). "Application of Finite Element Methods in the Statistical Analysis of Structures," *Proceedings of the 3rd International Conference on Structural Safety and Reliability,* Trondheim, Norway, pp. 409-417.

7. Hisada, T. and Nakagiri, S. (1981). "Stochastic Finite Element Method Developed for Structural Safety and Reliability," *Proceedings of the 3rd International Conference on Structural Safety and Reliability,* Trondheim, Norway, pp. 395-408.

8. Hisada, T. and Nakagiri, S. (1985). "Role of Stochastic Finite Element Method in Structural Safety and Reliability," *Proceedings of the 4th International Conference on Structural Safety and Reliability,* Kobe, Japan, pp. I.385-I.394.

9. Liu, W. K., Belytschko, T. and Mani, A. (1986). "Probabilistic Finite Elements for Nonlinear Structural Dynamics," *Journal of Computer Methods in Applied Mechanics and Engineering,* Vol. 56, pp. 61-81.

10. Liu, W. K., Mani, A. and Belytschko, T. (1987). "Finite Element Methods in Probabilistic Mechanics," *Journal of Probabilistic Engineering Mechanics,* Vol. 2, No. 4, pp. 201-213.

11. Spanos, P.D. and Ghanem, R. (1989). "Stochastic Finite Element Expansion for Random Media," *Journal of Engineering Mechanics,* ASCE, Vol. 115, No. 5, pp. 1035-1053.

12. Der Kiureghian, A. and Liu, P-L. (1986). "Structural Reliability Under Incomplete Probability Information," *Journal of Engineering Mechanics,* ASCE, Vol. 112, No. 1, pp. 85-104.

13. Liu, P-L. and Der Kiureghian, A. (1989). "Finite Element Reliability Methods for Geometrically Nonlinear Stochastic Structures," Report No. UCB/SEMM-89/05, Department of Civil Engineering, University of California at Berkeley, Berkeley, California.

14. Ghanem, R. and Spanos, P.D. (1991). "Stochastic Finite Elements: A Spectral Approach," Springer-Verlag, New York.

15. Lawrence, M., Liu, W.K., Besterfield, G. and Belytschko, T. (1990). "Fatigue Crack-Growth Reliability," *Journal of Engineering Mechan-*

ics, ASCE, Vol. 116, No. 3, pp. 698-708.

16. Shinozuka, M. and Deodatis, G. (1988). "Response Variability of Stochastic Finite Element Systems," *Journal of Engineering Mechanics*, ASCE, Vol. 114, No. 3, pp. 499-519.

17. Yamazaki, F., Shinozuka, M. and Dasgupta, G. (1988). "Neumann Expansion for Stochastic Finite Element Analysis," *Journal of Engineering Mechanics*, ASCE, Vol. 114, No. 8, pp. 1335-1354.

18. Deodatis, G. and Shinozuka, M. (1989). "Bounds on Response Variability of Stochastic Systems," *Journal of Engineering Mechanics*, ASCE, Vol. 115, No. 11, pp. 2543-2563.

19. Deodatis, G. (1990). "Bounds on Response Variability of Stochastic Finite Element Systems," *Journal of Engineering Mechanics*, ASCE, Vol. 116, No. 3, pp. 565-585.

20. Deodatis, G. (1990). "Bounds on Response Variability of Stochastic Finite Element Systems: Effect of Statistical Dependence," *Probabilistic Engineering Mechanics Journal*, Vol. 5, No. 2, pp. 88-98.

21. Takada, T. (1990). "Weighted Integral Method in Stochastic Finite Element Analysis," *Journal of Probabilistic Engineering Mechanics*, Vol. 5, No. 3, pp. 146-156.

22. Deodatis, G. (1991). "The Weighted Integral Method. I: Stochastic Stiffness Matrix," *Journal of Engineering Mechanics*, ASCE, Vol. 117, No. 8.

23. Deodatis, G. and Shinozuka, M. (1991). "The Weighted Integral Method. II: Response Variability and Reliability," *Journal of Engineering Mechanics*, ASCE, Vol. 117, No. 8.

24. Segerlind, L. J. (1984). "Applied Finite Element Analysis," John Wiley & Sons.

25. Cowper, G.R. (1973). "Gaussian Quadrature Formulas for Triangles," *International Journal for Numerical Methods in Engineering*, Vol. 7, pp. 405-408.

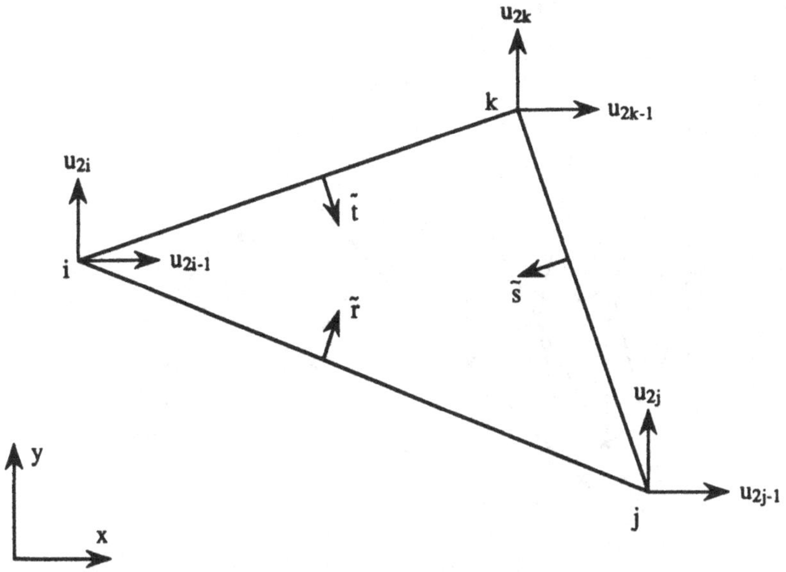

Figure 1: Nodal displacements and natural coordinates for CST element used.

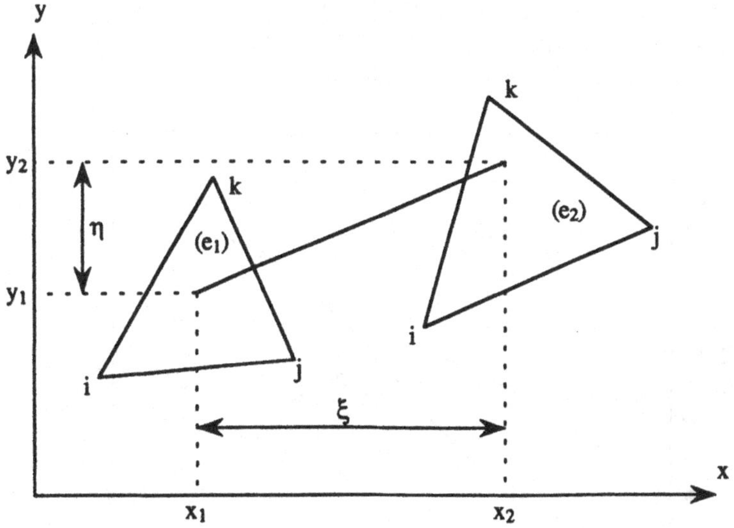

Figure 2: Configuration of relative positions of finite elements (e_1) and (e_2).

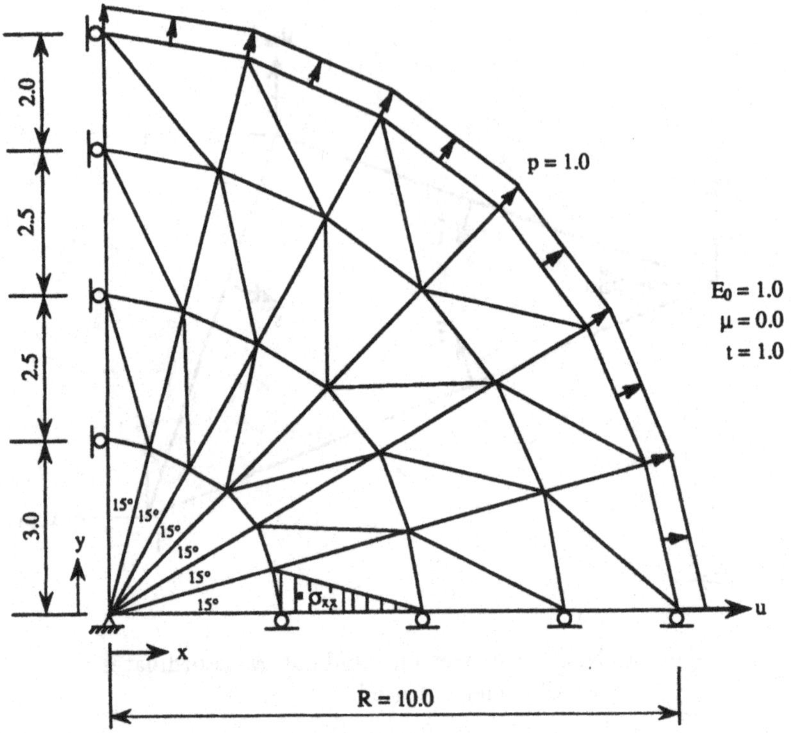

Figure 3: Finite element mesh used with 42 elements and 58 degrees of freedom. Identification of locations where response displacement u and stress σ_{xx} are calculated.

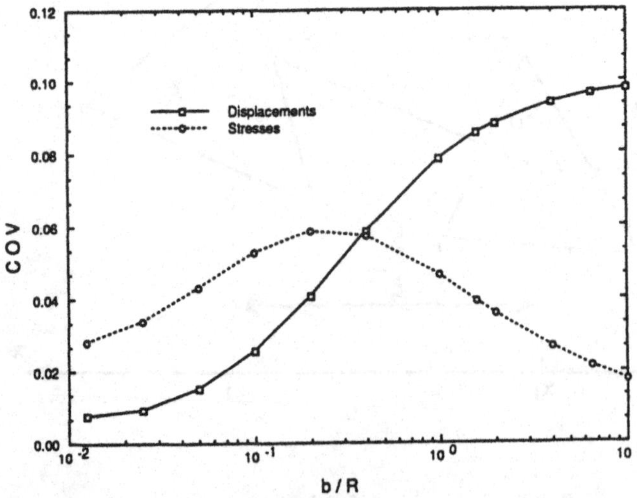

Figure 4: Coefficient of variation of response displacement and stress as a function of dimensionless parameter b/R.

Non Conservatively Loaded Stochastic Columns

S.A. Ramu (*), T.S. Sankar (**), R. Ganesan (**)
(*) Concordia University, Montreal, Canada
(**) Indian Institute of Science, Bangalore, India

ABSTRACT

The finite element method for analysis of nonconservatively loaded stochastic columns is developed. The governing equations for a column whose material properties are having stochastic fluctuations are derived. The resulting non self adjoint random eigenvalue problem is examined to derive the stochastic characteristics of its eigen solution. The free vibration problems of stochastic Beck's column and stochastic Leipholz columns whose Young's modulus and mass density are distributed stochastically are considered. Numerical results are presented.

INTRODUCTION

Slender structural elements employed in aerospace technology are frequently subjected to complex loading environments. In particular, when the loading is nonconservative in nature a dynamic stability investigation of the element is called for. This in turn requires the examination of the precise nature of the natural frequencies of vibration of the nonconservatively loaded structure. However, this task is complicated by the fact that in such important structures, several uncertainties in structural parameters like length, moment of inertia, cross-section area, material parameters like elasticity modulus, mass density etc. and the loading parameters, are always present and unavoidable. These uncertainties could at best be described as stochastic fluctuations over their mean values. A detailed investigation is necessary to assign the probability limits of various characteristics of the system, such as level excursions, peak statistics, envelope statistics etc. of eigenvalues of the system. The usage of modern engineering materials, which are characterised by their inherent, unavoidable uncertainties, does necessitate such an analysis to ascertain the safety measures.

Extensive literature exists on the analysis of deterministic structures under nonconservative loads. A finite element solution of non conservatively loaded deterministic columns is developed in [1]. However, literature survey indicates that stochastic structures, subjected to random non-conservative loading are not analysed yet. Only columns subjected to random loadings and having random material properties, resulting in a conservative system were analysed by many authors [2-5] using Euler's criterion i.e. statical method. Ariaratnam [6] considered the stability of a deterministic column subjected to a random loading in time.

The stochastic finite element method in the field of structural analysis where finite element methods are used in a probabilistic setting has got extensive development recently with different approaches [7]-[11]. Nakagiri et al [12] used a version of stochastic finite element method to analyse the eigenvalue problem of laminated composite plate considering stochastic stiffness variation. Further applications of stochastic finite element method for self-adjoint problems can be seen in [13].

In the present work, a different version of the stochastic finite element method developed in [13] is extended and used for solving the free vibration problem of non self-adjoint type arising when the material property of the non conservative systems such as Young's modulus, mass density are random processes in space.

In this paper, Beck's column and Leipholz column whose Young's modulus and mass per unit length are having a stochastic variation are analysed. The stochastic variation of Young's modulus and mass per unit length are considered to be spatially distributed one demensional, univariate stochastic fields. The standard perturbation analysis is employed.

DESCRIPTION OF THE PROBLEM

Consider a free-free column as shown in Fig.1 which is subjected to two end loads P at x=0 inclined at angle of $\alpha_0\phi_0$ to the undeformed axis of the column, where ϕ_0 is the angle between the tangent to the deformed axis of the column and the undeformed axis of the column at x=0 and $(P+Q)$ at x=1, inclined at an angle of $\alpha_1\phi_1$, where ϕ_1 is the angle between the tangent to the deformed axis and the undeformed axis of the column at x=1. Also, a distributed follower load of uniform intensity p/unit length inclined at the angle $\alpha_D\phi$ with respect to the undeformed axis of the column is considered, where ϕ is the angle made by the tangent to the deformed axis at any arbitrary point with the undeformed axis of the column. Here x is measured from the end from which the distributed load is directed towards and α_0, α_1 and α_D define the degrees of the non-conservativeness of the respective forces.

STOCHASTIC CHARACTERIZATION OF THE SYSTEM PARAMETERS

The Young's modulus and mass density are distributed randomly, along the undeformed axis of the column. The fluctuations over their mean values are assumed to constitute independent one dimensional, univariate, homogeneous, real, spatial stochastic fields. The Young's modulus and mass density are thus given by

$$E(x)=\bar{E}[1+a(x)], \quad m(x)=\bar{m}[1+b(x)] \tag{1,2}$$

where $\bar{E}$, $\bar{m}$ are the mean values of the Young's modulus and mass density respectively. $a(x)$ and $b(x)$ are independent, one dimensional, univariate, homogeneous, real, spatial stochastic fields. The processes are characterized by their respective autocorrelation functions R_{aa} and R_{bb} (or their equivalent power spectral density functions S_{aa} and S_{bb}) and scale of fluctuations Θ_E and Θ_m. The variances are σ_E^2 and σ_m^2 respectively.

GENERAL CONSISTENT FINITE ELEMENT FORMULATION

The column is discretized into n elements. Let the i-th element be considered. The nodal degrees of freedom are taken to be w_1^i, Θ_1^i, w_2^i, and Θ_2^i for the i-th element. w_1^i, w_2^i are the transverse deflections of the two ends of the element and Θ_1^i and Θ_2^i are the rotation of the tangents to the deformed axis of the element with respect to undeformed axis. The transverse displacement $w(x,t)$ at any arbitrary point of this element is given by

$$w(x,t)=N^i q^i, \quad N^i=\{N_1\ N_2\ N_3\ N_4\}^i, \text{ and } q_1^T=\{w_1, \Theta_1, w_2, \Theta_2\}^i \tag{3,4,5}$$

N^i is a row vector of interpolation polynomials (cubic Hermitian polynomials) which are the same as that of a deterministic structure. The components of q^T are function of t, the time.

If T^i is the kinetic energy stored in the i-th element,

$$T^i=\frac{1}{2}\int_0^{\ell_i} m(x)A_i\left(\frac{\partial w}{\partial t}\right)^2 dx=\frac{1}{2}q_t^{i^T}\left[\int_0^{\ell_i} N^{i^T}\bar{m}(1+b(x))A_i N^i dx\right]q_t^i=\frac{1}{2}q_t^{i^T}M^i q_t^i \tag{6}$$

where $M^i=\bar{M}^i+M^i(\Omega)$, and the differentiation w.r.t. time is indicated by the subscript t. The elements of $\bar{M}^i$ are given by

$$\bar{m}_{ij}=\int_0^{\ell_i}\bar{m}AN_iN_j dx, \quad m_{ij}(\Omega)=\bar{m}A\int_0^{\ell_i} b(x)N_iN_j dx \text{ and } m_{ij}=\bar{m}_{ij}+m_{ij}(\Omega) \tag{7,8}$$

If U^i is the strain energy stored in the i-th element

$$U^i=\frac{1}{2}\int_0^{\ell_i}\bar{E}(1+a(x))I_i\left(\frac{\partial^2 w}{\partial x^2}\right)^2 dx=\frac{1}{2}q_i^T\left[\int_0^{\ell_i} N''^{i^T}\bar{E}(1+a(x))I_i N''^i dx\right]q_i=\frac{1}{2}q^{i^T}K^i q^i$$

Here primes indicate differentiation w.r.t. x, and

$$K^i=\int_0^{\ell_i} N''^{i^T}\bar{E}(1+a(x))I_i N''^i dx=\bar{K}^i+K^i(\Omega) \tag{9}$$

The coefficients of the stiffness matrix are thus given by

$$k_{ij}=\int_0^{\ell_i} N''_i\bar{E}(1+a(x))I_i N''_j dx$$

$$=\bar{E}I_i\int_0^{\ell_i} N''_i N''_j dx+\bar{E}I_i\int_0^{\ell_i} (a(x))N''_i N''_j dx=\bar{k}_{ij}+k_{ij}(\Omega) \tag{10}$$

Because of the presence of the axial component of the distributed follower load, the i-th element is subjected to a uniformly varying axial compression increasing from F_1^i to F_2^i. (Ref. Fig.1). The axial compression at any arbitrary section of the element is given by $(F_1^i+px)=F_1^i+(Q/\ell)x)$, where x is measured +ve in the increasing direction of the axial compression, i.e. from end 1 to end 2. Therefore, the work done by the axial compression in the element W_c^i is given by

$$W_c^i=\frac{1}{2}\int_0^{\ell_i} (F_1^i+Qx/\ell)\left(\frac{\partial w}{\partial x}\right)^2 dx=\frac{1}{2}F_1^i q^{i^T} k_{GC}^i q^i+\frac{1}{2}Qq^{i^T} k_{GD}^i q^i \tag{11}$$

where

$$k_{GC}^i=\int_0^{\ell_i} N'^{i^T} N'^i dx,\quad k_{GD}^i=\int_0^{\ell_i} N'^{i^T} (L_2\ell_i/\ell)N'^i dx \tag{12,13}$$

where L_1 and L_2 are the natural coordinates.

k_{GC}^i (which is actually the geometric stiffness matrix used in the stability analysis when the compressive force is constant) and k_{GD}^i (which is the geometric stiffness matrix which accounts for the linearly varying compression starting from zero at one end) are deterministic coefficients.

The lateral components of the follower forces are dependent on the slopes of the tangents to the deformed axis of the column at the points of application of these forces and hence there is no unique scalar work function corresponding to work done by these forces. Therefore, the virtual work done by these forces is appended as an external term in the usual form of Hamilton's principle.

The virtual work done by the non-conservative components of the forces is worked out considering the virtual displacements to be the variations of the actual displacements. The virtual work done by the distributed follower force in the element i, δW_{DNC}^i is given by

$$\delta W_{DNC}^i=-\int_0^{\ell_i} p\sin(\alpha_D\phi_0)\delta w dx=\int_0^{\ell_i} p\alpha_D \frac{\partial w}{\partial x}\cdot\delta w dx$$

$$=p\alpha_D \delta q^{1^T}\left[\int_0^{\ell_1} N^{1^T} N'^1 dx\right] q^1 = [\alpha_D Q/\ell]\delta q^{1^T} k^1_{DNC} q^1 \tag{14}$$

where $k^1_{DNC}=\int_0^{\ell_1} N^{1^T} N'^1 dx$. The virtual work done by the concentrated follower force at the ends may be taken together as δW_{CNC} equal to

$$-p\sin(\alpha_0\phi_0)\delta w_0-(P+Q)\sin(\alpha_1\phi_1)\delta w_1=P\alpha_0\delta w_0\left(\frac{\partial w}{\partial x}\right)_0-(P+Q)\alpha_1\delta w_1\left(\frac{\partial w}{\partial x}\right)_1 \tag{15}$$

Now, the total kinetic energy of the entire structure is

$$T=\sum_{l=1}^{n} T^l=\left(\sum_{l=1}^{n}\frac{1}{2}q_t^{l^T} M^l q_t^l\right)=\sum_{l=1}^{n}\frac{1}{2}q_t^{l^T}(\bar{M}^l+M^l(\Omega))q_t^l \tag{16}$$

The total strain energy is given by

$$U=\sum_{l=1}^{n} U^l=\left(\sum_{l=1}^{n}\frac{1}{2}q_t^{l^T} K^l q_t^l\right)=\sum_{l=1}^{n}\frac{1}{2}q_t^{l^T}(\bar{K}^l+K^l(\Omega))q_t^l \tag{17}$$

Total work done by the conservative part of the forces, W_c is given by

$$W_c=\sum_{l=1}^{n} W_c^l=\sum_{l=1}^{n}\frac{1}{2}F_1^l q^{l^T} K^l_{GC} q^l+\sum_{l=1}^{n}\frac{1}{2}Q q^{l^T} K^l_{GD} q^l \tag{18}$$

The total virtual work done by the non-conservative components of the forces,

$$\delta W^l_{DNC}+\delta W_{CNC}=\sum_{l=1}^{n}\{(\alpha_D Q/\ell)\delta q^{l^T} K^l_{DNC} q^l+P\alpha_0\delta W_0\left(\frac{\partial w}{\partial x}\right)_0-(P+Q)\alpha_1\delta W_1\left(\frac{\partial w}{\partial x}\right)_1 \tag{19}$$

The classical Hamilton's principle, if modified for non-conservative systems becomes

$$\int_{t_1}^{t_2}[\delta(T-U+W_c)+\delta W_{NC}]dt=0 \tag{20}$$

Substituting for T, U, W_c and δW_{NC}, we get

$$\int_{t_1}^{t_2}\delta\left[\sum_{l=1}^{n}\frac{1}{2}q_t^{l^T}(\bar{M}^l+M^l(\Omega))q_t^l\right]dt-\int_{t_1}^{t_2}\delta\left[\sum_{l=1}^{n}\frac{1}{2}q_t^{l^T}(\bar{K}^l+K^l(\Omega))q_t^l\right.$$

$$+\sum_{l=1}^{n}F_1^l\frac{1}{2}q^{l^T}K^l_{GC}q^l+\sum_{l=1}^{n}Q\frac{1}{2}q^{l^T}K^l_{GD}q^l+$$

$$\int_{t_1}^{t_2}\sum_{l=1}^{n}(\alpha_D Q/\ell)\delta q^{l^T}K^l_{DNC}q^l+P\alpha_0\delta W_0\left(\frac{\partial w}{\partial x}\right)_0-(P+Q)\alpha_1\delta W_1\left.\left(\frac{\partial w}{\partial x}\right)_1\right]dt=0 \tag{21}$$

The summations are made in the sense of finite element assemblage, taking the global displacement vector to be q. In addition, w_0 is taken as q_1, $(\partial w/\partial x)_0$ as q_2, w_1 as q_{2n+1} and $(\partial w/\partial x)_1$ as q_{2n+2}. Now, if contemporaneous variations of q are taken, while integrating the first term by parts, the δW_{NC} is given as follows:

$$\delta W_{NC}=\int_{t_1}^{t_2}[-\delta q^T M q_{tt}-\delta q^T Kq+\delta q^T pK_{GC}q+(\delta q^T QK^*_{GC}q+\delta q^T QK_{GD}q)$$

$$+\delta q^T(\alpha_D Q/\ell)K_{NDC}q+\delta q^T\alpha_0 PK_{CNC1}q-q^T\alpha_1 PK_{CNC2}q-q^T\alpha_1 QK_{CNC2}q]dt=0 \qquad (22)$$

F^i_1 in the equation (18) is put in the form: $F^i_1=P+\beta^i Q$, where $\beta^i<1.0$ corresponds to the fraction of the total distributed load acting at the trailing node of the element i and $K^{i*}_{GC}=\beta^i K^i_{GC}$.

Matrices K_{CNC1} and K_{CNC2} contain non-zero elements only in one location each namely (1, 2) and ($2n$+1, $2n$+2); and therefore, these matrices need not be assembled but the corresponding terms in K_{GC} and K^*_{GC} may be modified by the addition of the non-zero terms in K_{CNC1} and K_{CNC2} appropriately, after multiplying them by the respective factors shown in equation (22).

Substituting $q=\bar{q}^* e^{st}$ in equation (22) and considering the arbitrariness of the variation of q, we obtain the equation for a non self adjoint random eigenvalue problem:

$$\{-s^2(\bar{M}+M(\Omega))+(-\bar{K}-K(\Omega))+PK^*_{GP}+QK^*_{GQ}\}\bar{q}^*=0 \qquad (23)$$

where $K^*_{GP}=(K_{GC}+\alpha_0 K_{CNC1}-\alpha_1 K_{CNC2})$ and $K^*_{GQ}=K^*_{GC}+K_{GD}+\alpha_D K_{DNC}-\alpha_1 K_{CNC2}$ (24,25).

Equations governing the particular cases for a cantilever column with a concentrated follower force at the free end (Beck column) and a uniformly distributed follower force along its length (Leipholz column) can be derived from the above general equation. The onset of instability by divergence or by flutter is indicated by the appearance of s^2 values which are real, +ve and complex respectively. The calculation of the eigenvalue statistics will give the statistical information about the instability loads of the nonconservatively loaded columns.

STOCHASTIC CHARACTERISATION OF INFLUENCE COEFFICIENTS

Since $\langle a(x)\rangle$ and $\langle b(x)\rangle$ are zero the means and variances of the coefficients of mass and stiffness matrices of equations (8) and (10) can be written as follows:

$$\langle k_{ij}\rangle=\langle\bar{k}_{ij}+k_{ij}(\Omega)\rangle=\bar{E}I\int_0^{\ell_e} N''_i(x)\cdot N''_j(x)dx=\bar{k}_{ij} \qquad (26)$$

Variance $(k_{ij})=\mathrm{var}(k_{ij})=\mathrm{Var}(\bar{k}_{ij}+k_{ij}(\Omega))=\mathrm{var}(k_{ij}(\Omega))$

$$=\bar{E}^2I^2\int_0^{\ell_e}\int_0^{\ell_e} R_{aa}(\xi_1-\xi_2)\cdot N''_i(\xi_1)\cdot N''_j(\xi_1)\cdot N''_i(\xi_2)\cdot N''_j(\xi_2)d\xi_1 d\xi_2 \qquad (27)$$

Mean of $m_{ij}=\bar{m}_{ij}$ and $\mathrm{Var}(m_{ij})=\mathrm{Var}(m_{ij}(\Omega))=$

$$\bar{m}^2A^2\int_0^{\ell_e}\int_0^{\ell_e} R_{bb}(\xi_1-\xi_2)N_i(\xi_1)\cdot N_j(\xi_1)\cdot N_i(\xi_2)\cdot N_j(\xi_2)d\xi_1 d\xi_2 \qquad (28)$$

As can be seen the evaluation of the variances using the above expressions will be tedious. Also, the full correlation function expression is needed for its evaluation. Such an adequate information is seldom available and the experimental data seldom allows one to distinguish among competing analytical models for the correlation function. So, a simplified treatment is adopted. The stochastic processes $a(x)$, $b(x)$ are assumed to be characterised by three parameters in each case: the means, which are zero, the standard deviations σ_E, σ_m and the scale of fluctuations Θ_E, Θ_m. The second order properties are covered by either the autocorrelation function or power spectral densities or by the variance functions [11]. So, the fluctuating parts of stiffness and mass elements are evaluated using the local averages over elements of material property fluctuations. For the i-th finite element, the local averages are

$$E_i=\frac{1}{\ell_i}\int_0^{\ell_i} a(x)dx;\quad m_i=\frac{1}{\ell_i}\int_0^{\ell_i} b(x)dx \qquad (29)$$

where ℓ_i=length of the i-th element. The properties of local averages are then:

$$<E_i>=\frac{1}{\ell_i}\int_0^{\ell_i} <a(x)>dx=0 \text{ and } <m_i>=\frac{1}{\ell_i}\int_0^{\ell_i} <b(x)>dx=0$$

The variance of local averages are

$$\mathrm{Var}(E_i)=\frac{1}{\ell_i^2}\int_0^{\ell_i}\int_0^{\ell_i} R_{aa}(\xi_1-\xi_2)d\xi_1 d\xi_2 \text{ and } \mathrm{Var}(m_i)=\frac{1}{\ell_i^2}\int_0^{\ell_i}\int_0^{\ell_i} R_{bb}(\xi_1-\xi_2)d\xi_1 d\xi_2$$

In terms of variance functions, which characterize the dependence of variance of local average on the size of the element,the variances can be written as,

$$\mathrm{Var}(E_i)=\sigma_E^2\gamma_E(\ell_i)=\sigma_E^2\cdot\frac{\Theta_E}{\ell_i} \text{ and } \mathrm{Var}(m_i)=\sigma_m^2\gamma_m(\ell_i)=\sigma_m^2\cdot\frac{\Theta_m}{\ell_i};\quad \ell_i\gg\Theta_E,\ \Theta_M$$

where $\gamma_E(\ell_i)$, $\gamma_m(\ell_i)$ are the variance functions of spatial averages of E and m respectively and Θ_E, Θ_m are the scale of fluctuations of E and m respectively. The covariance functions of stiffness coefficients can be calculated for any two coefficients $k_{ij}(\Omega)$ and

$k_{rs}(\Omega)$, where i, j and r, s are the nodal point indices for those two elements through the use of spatial averages as in [11]:

$$\{\bar{E}^2 I^2 \int_0^{\ell} N_i''(x) N_j''(x) dx \int_0^{\ell} N_r''(x) N_s''(x) dx\} \frac{\sigma_E^2}{2} [\ell_0^2 \gamma_E(\ell_0) - \ell_1^2 \gamma_E(\ell_1) + \ell_2^2 \gamma_E(\ell_2) - \ell_3^2 \gamma_E(\ell_3)]$$

where ℓ_0, ℓ_1, ℓ_2, ℓ_3 are as shown in Fig. 2.

The above equation can be written in terms of the correlation functions ρ_a also. Similar expressions can be written for $\langle m_{ij}(\Omega) \cdot m_{rs}(\Omega) \rangle$. The covariance matrix between the different coefficients of stiffness and mass matrices can be generated using these equations.

EIGENSOLUTION STATISTICS

The eigenvalue problem is written as

$$[K^{tot}]\{x\} = \lambda [M]\{x\} = [\bar{K}^{tot} + K^{tot}(\Omega)]\{x\} = \lambda [\bar{M} + M(\Omega)]\{x\} \tag{30}$$

where $[K^{tot}] = [K] + P[K_{GC} + \alpha_0 [K_{CNC1}]]$ for Beck's column

and $[K^{tot}] = [K] + Q[K^*_{GC} + \alpha_D [K_{DNC}]]$ for Leipholz column.

Moreover, since only the structural parameters are stochastic

$$K^{tot}(\Omega) = [K(\Omega)]$$

The perturbations of eigenvalues λ_i can be shown to be [2],

$$d\lambda_i = \sum_{r=1}^{n} \sum_{s=1}^{n} \frac{\partial \lambda_i}{\partial k_{rs}^{tot}} dk_{rs}^{tot} + \sum_{r=1}^{n} \sum_{s=1}^{n} \frac{\partial \lambda_i}{\partial m_{rs}} dm_{rs} \tag{31}$$

Since the K^{tot} and M elements are regular functions of the influence coefficients, if we seek the derivatives of each eigenvalue about averaged eigenvalue, we get,

$$\frac{\partial \lambda_i}{\partial k_{rs}^{tot}} = y_{ri} x_{si} / y_i^T M x_i \quad \text{and} \quad \frac{\partial \lambda_i}{\partial m_{rs}} = -\lambda_i (y_{ri} x_{si} / y_i^T M x_i)$$

In the above, y_i is the left eigenvector and x_i is the right eigenvector, so that

$$([K^{tot}] - \lambda_i [M])\{x\} = 0 \quad \text{and} \quad \{y_i\}^T ([K^{tot}] - \lambda_i [M]) = 0$$

If the elements of random matrices are resulting from continuous stochastic processes describing material property fluctuations, for sufficiently small perturbations, we can write,

$$\lambda_i=\bar{\lambda}_i+\sum_{i=1}^{n}\sum_{j=1}^{n}\frac{\partial\lambda_i}{\partial k_{ij}^{tot}}(k_{ij}^{tot}(\Omega))+\sum_{i=1}^{n}\sum_{j=1}^{n}\frac{\partial\lambda_i}{\partial m_{ij}}(m_{ij}(\Omega)) \tag{32}$$

As $\langle k_{ij}^{tot}(\Omega)\rangle=\langle m_{ij}(\Omega)\rangle=0$, the mean values are $\langle\lambda_i\rangle=\bar{\lambda}_i$.

The covariance between any two eigenvalue is given by,

$$\langle(\lambda_i-\bar{\lambda}_i)(\lambda_j-\bar{\lambda}_j)\rangle=\langle\{\sum_{i=1}^{n}\sum_{j=1}^{n}\frac{\partial\lambda_i}{\partial k_{ij}^{tot}}(k_{ij}^{tot}(\Omega))+\sum_{i=1}^{n}\sum_{j=1}^{n}\frac{\partial\lambda_i}{\partial m_{ij}}(m_{ij}(\Omega))\}\cdot$$

$$\{\sum_{r=1}^{n}\sum_{s=1}^{n}\frac{\partial\lambda_j}{\partial k_{rs}^{tot}}(k_{rs}^{tot}(\Omega))+\sum_{r=1}^{n}\sum_{s=1}^{n}\frac{\partial\lambda_j}{\partial m_{rs}}(m_{rs}(\Omega))\}\rangle$$

The variance is therefore given by,

$$\sigma_i^2=\mathrm{var}(\lambda_i)=\sum_{i=1}^{n}\sum_{j=1}^{n}\sum_{r=1}^{n}\sum_{s=1}^{n}\frac{\partial\lambda_i}{\partial k_{ij}^{tot}}\frac{\partial\lambda_j}{\partial k_{rs}^{tot}}\langle k_{ij}^{tot}(\Omega)\ k_{rs}^{tot}(\Omega)\rangle+$$

$$+\sum_{i=1}^{n}\sum_{j=1}^{n}\sum_{r=1}^{n}\sum_{s=1}^{n}\frac{\partial\lambda_i}{\partial m_{ij}}\frac{\partial\lambda_j}{\partial m_{rs}}\langle m_{ij}(\Omega)\ m_{rs}(\Omega)\rangle \tag{33}$$

Since the stiffness and mass matrices are derived using stochastic finite element method to be consisting of a determinstic component and zero mean fluctuating components, the mean values of any random stiffness or mass matrix element is given by the corresponding components of deterministic parts of those random matrices.

Since $\mathrm{Var}(k_{ij}^{tot})=\mathrm{Var}(k_{ij}(\Omega))$, the covariance between two stiffness elements and between two mass matrix elements are given by the expressions previously derived using spatial averages. Hence, the covariance matrix between stiffness and mass elements can be found.

As a result, the mean values of eigenvalues becomes the eigenvalues got by solving the unperturbed eigenproblem, $[\bar{K}^{tot}]\{\bar{x}_i\}=\bar{\lambda}_i[\bar{M}]\{\bar{x}_i\}$, where $[\bar{K}^{tot}]$, $[\bar{M}]$ are formed using the deterministic component of k_{ij}^{tot} and m_{ij}. Then the variances of eigenvalues are given by

$$\sigma_i^2=\mathrm{var}(\lambda_i)=\sum_{i=1}^{n}\sum_{j=1}^{n}\sum_{r=1}^{n}\sum_{s=1}^{n}\frac{\partial\lambda_i}{\partial k_{ij}^{tot}}\frac{\partial\lambda_i}{\partial k_{rs}^{tot}}\mathrm{COV}(k_{ij}^{tot}(\Omega)\ k_{rs}^{tot}(\Omega))$$

$$+\sum_{i=1}^{n}\sum_{j=1}^{n}\sum_{r=1}^{n}\sum_{s=1}^{n}\frac{\partial\lambda_i}{\partial m_{ij}}\frac{\partial\lambda_j}{\partial m_{rs}}\mathrm{COV}(m_{ij}(\Omega)\ m_{rs}(\Omega)) \tag{34}$$

and the cc variance between any two eigenvalues is given by

$$\mathrm{COV}(\lambda_i,\lambda_j)=\sum_{i=1}^{n}\sum_{j=1}^{n}\sum_{r=1}^{n}\sum_{s=1}^{n}\frac{\partial\lambda_i}{\partial k_{ij}^{tot}}\frac{\partial\lambda_j}{\partial k_{rs}^{tot}}\mathrm{COV}(k_{ij}^{tot}(\Omega)\ k_{rs}^{tot}(\Omega))$$

$$+\sum_{i=1}^{n}\sum_{j=1}^{n}\sum_{r=1}^{n}\sum_{s=1}^{n}\frac{\partial\lambda_i}{\partial m_{ij}}\frac{\partial\lambda_j}{\partial m_{rs}}\mathrm{COV}(m_{ij}(\Omega)\ m_{rs}(\Omega)) \tag{35}$$

Similar expressions can be written to evaluate the covariance or eigenvector elements. The covariance matrix of the eigenvalues and eigenvectors can be constructed.

EXAMPLE

The Beck's column as shown in Fig.3 is taken as an example. The elastic modulus distribution is considered to be stochastic. The material property values are: $\bar{E}=2.1\times10^{5}N/mm^{2}$, $m=7.83\times10^{-9}N\text{-}sec/mm^{4}$, the length of the beam=7.35m, cross sectional area=4329mm^{2}, moment of inertia=2.672x10$^{7}mm^{4}$, and The tip axial load $P=0.25P_{fl}$, where P_{fl} is the flutter load of determinstic Beck's column.

The stochastic process $a(x)$ representing the fluctuating components of elastic modulus is represented by the exponential type correlation function relationship. This is the first-order autoregressive or Markov process representation. The correlation function is given by $\rho(\xi)=e^{-|\xi|/c}$, c=constant, ξ=sepration distance; the variance function is given by $\gamma(U)=2\left(\frac{c}{U}\right)^{2}\left(\frac{U}{c}-1+e^{-U/c}\right)$. The covariance between stiffness elements and mass elements is zero and between stiffness elements and geometric stiffness elements as we know, are also zero. The variances of the fundamental eigenvalue as well as the covariances between the eigenvalues for different values of input variance are listed in Tables 1 and 2.

CONCLUSIONS

The stochastic finite element method is developed to solve the general random nonself-adjoint eigenvalue problems like free-vibration of stochastic Beck's column and stochastic Leipholz column. The virtual work formulation is adopted. The first order perturbation method is used. The concepts of adjoint system are used for nonself-adjoint system perturbations. In the numerical example employing the most commonly observed correlation structure the superiority of the method developed is demonstrated.

REFERENCES

1. Venuraju, M.T. and Ramu, S.A. Influence of an Additional Discrete Elastic Support on the Stability of a Leipholz Column. Computers and Structures, Vol.36, No.4, pp.769-775, 1990.

2. Collins, D. and Thompson, W.T. The Eigenvalue Problem for Structural Systems with Statistical Properties. AIAA Journal, Vol.7, pp.642-648, 1969.
3. Shinozuka, M. and Astill, C.A. Random Eigenvalue Problems in Structural Analysis. AIAA Journal, Vol.10, pp.456-462, 1972.
4. Hoshiya, M. and Shah,H.C. Free Vibration of a Stochastic Beam-Column. Journal of Engineering Mechanics, ASCE, Vol.97, pp.1239-1255, 1971.
5. Augusti, G., Borri, A. and Casciati, F. Structural Design under Random Uncertainties: Economical Return and 'Intangible' Quan tities. In ICOSSAR-81, pp.483-494, Proceedings of 3rd International Conference on Structural Safety and Reliability, Elsevier, Amsterdam, 1981.
6. Ariaratnam, S.T. Dynamic Stability of a Column under Random Loading. Proceedings of International Congress Held at Northwestern University, Evanston, Illinois, October 18-20, 1965, Pergamon Press, Oxford, UK.
7. Contreras, H. The Stochastic Finite Element Method. Comput. Struc., Vol.12, pp.341-348, 1980.
8. Liu, W.K., Belytschko,T. and Mani, A. Random Field Finite Elements. Int. J. Numer. Methods Eng., Vol.23, pp.1831-1845, 1986.
9. Yamazaki, F., Shinozuka, M. and Dasgupta, G. Neumann Expansion for Stochastic Finite Element Analysis. J. Eng. Mech., Vol.114, pp.1335-1354, 1985.
10. Shinozuka, M. and Deodatis, G. Response Variability of Stochastic Finite Element Systems. ASCE J. Eng. Mech., Vol.114, pp.399-419, 1988.
11. Vanmarcke, E. and Grigoriu, M. Stochastic Finite Element Analysis of Simple Beams. J. Eng. Mech., Vo.109, pp.1203-1214, 1983.
12. Nakagiri, S., Takabatake, H. and Tani, S. Uncertain Eigenvalue Analysis of Composite Laminated Plates by the Stochastic FEM. Trans. ASME, J. of Eng. for Industry, Vol.109, pp.9-12, 1987.
13. Ramu, S.A. and Ganesan, R. Analysis of Random Beam-Column Subjected to Random Loading Using Stochastic FEM. Journal of Computers and Structures, (in press)

Table 1. Variances of Vibration Frequencies of Beck's Column with Uncorrelated E Values for P/P_{fl}=0.25

Input variance	$\mathrm{var}\omega_1^2$	$\mathrm{var}\omega_2^2$	$\mathrm{var}\omega_3^2$	$\mathrm{var}\omega_4^2$	$\mathrm{var}\omega_5^2$	$\mathrm{var}\omega_6^2$
0.01	23.5708	0.4218	0.0772	4.0229	100.3446	24.7048
0.02	47.1417	0.8436	0.1545	8.0457	200.6892	49.4097
0.03	70.7125	1.2654	0.2317	12.0686	301.0337	74.1145
0.04	94.2833	1.6872	0.3089	16.0914	401.3783	98.8195
0.05	117.8541	2.1090	0.3862	20.1143	501.7229	123.5242
0.06	141.4250	2.5308	0.4634	24.1372	602.0675	148.2291
0.07	164.9958	2.9526	0.5406	28.1600	702.4120	172.9340
0.08	188.5666	3.3744	0.6179	32.1829	802.7566	197.3888
0.09	212.1375	3.7962	0.6951	36.2057	903.1012	222.3436
0.10	235.7083	4.2180	0.7723	40.2286	1003.4458	247.0485

Table 2. Covariance Matrix of Eigenvalues for P/P_{fl}=0.25 and Input Variance=0.01

23.5708	-48.1849	9.6245	1.2205	-2.8253	-12.3817
	0.4218	-19.1673	-2.1016	5.7389	24.4282
		0.0772	0.4812	-1.2195	-2.9141
Symmetric			4.0229	-0.1422	-0.4426
				100.3446	0.8862
					24.7048

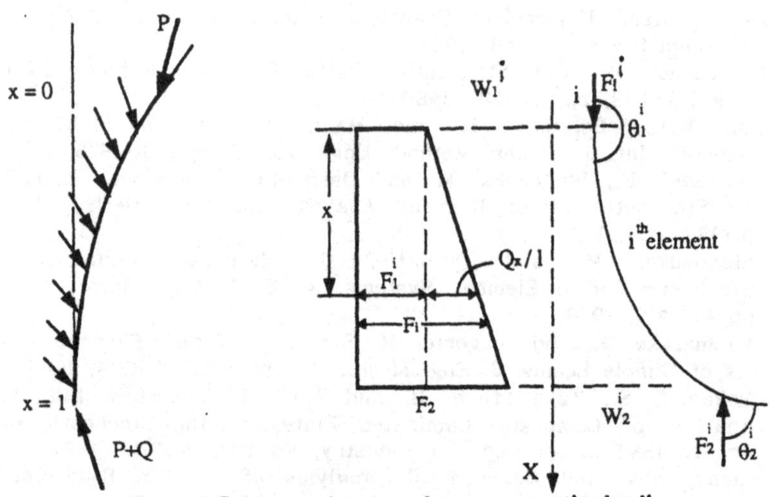

Figure:1 Column under general non-conservative loading

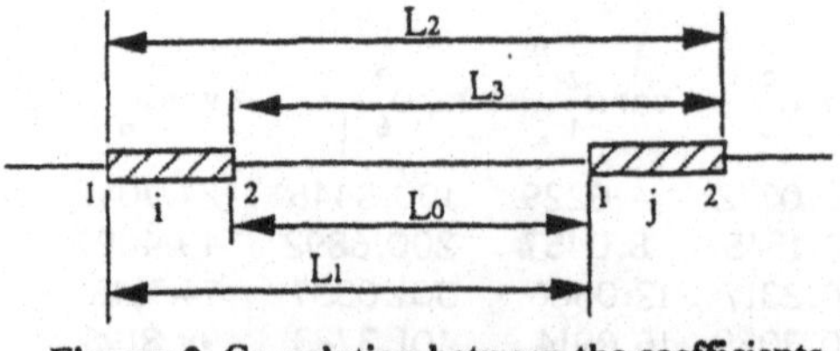

Figure: 2 Correlation between the coefficients corresponding to two arbitrarily located finite elements i and j

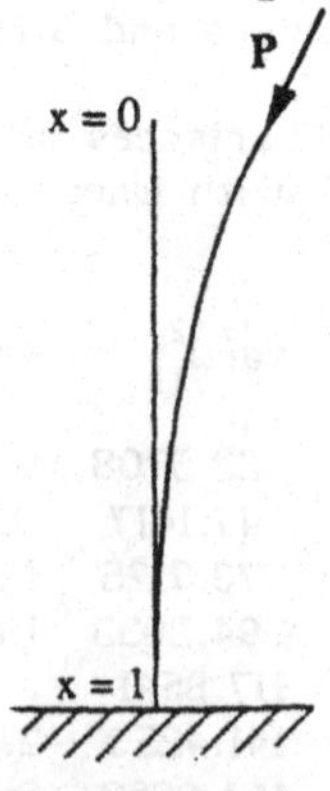

Figure:3 Beck's column Example

SECTION 6: FATIGUE/FRACTURE

Inspection Strategy for Deteriorating Structures Based on Cost Minimization Approach

Y. Fujimoto (*), M. Mizutani (**), S. Swilem (*), M. Asaka (**)
(*) Department of Naval Architecture and Ocean Engineering, Hiroshima University, Higashi-Hiroshima 724, Japan
(**) Tokyo Electric Power Services Co., Ltd., 1-3-1 Uchisaiwaicho, Chiyoda-ku, Tokyo 100, Japan

ABSTRACT

A sequential cost minimization method and its consistent formulation are presented for the inspection planning problem of deteriorating structures. The method aims to find an optimal inspection strategy so that the total cost expected in the period between the present inspection and the next be minimum. The optimization variables are the inspection methods at the present inspection and the interval to the next inspection for the structure. The optimization is repeatedly carried out at every inspection. The applicability of the proposed method is discussed through the numerical analyses of a hypothetical structural member set as well as a hypothetical structure.

INTRODUCTION

The total cost minimization is the optimal criterion of decision making for design and maintenance of structures[1][2]. In inspection planning problems, optimization can be achieved by the appropriate selections of inspection interval, inspection method, repair quality and so on.

In this study, a sequential cost minimization method and its consistent formulation are presented for deteriorating structures. The method aims to find an optimal inspection strategy so that the total expected cost in an inspection interval, the period between the present inspection and the next, be minimum. In this method the optimization parameters are the inspection methods for each member set and the interval to the next inspection for the structure. The optimization is repeatedly carried out at every inspection. The optimal inspection method is selected from the following five methods: 1) no inspection, 2) visual inspection, 3) mechanical inspection, 4) visual and conditional mechanical inspection, and 5) sampling mechanical inspection. The cost evaluation equation corresponding to each inspection method is developed, where the following cost items are included: Inspection cost, repair cost, risk of member failure and catastrophic failure, and the loss due to service suspension by scheduled and accidental system downs.

Numerical analyses are carried out for a hypothetical structural member set as well as for a hypothetical structure which includes a variety of member sets. Fatigue crack initiation, propagation and continuous failure are treated as deterioration damages of the members. Markov Chain Model is employed to describe the entire probabilistic feature of the fatigue process.

SEQUENTIAL COST MINIMIZATION METHOD

The interval of inspections is usually determined for the structures, such as once a year or once every two years, in line with the prefixed schedule based on the synthetic consideration of service plan, economical trend and loss due to the service suspension during maintenance. When a damage is detected or a failure occurs, the member is repaired generally with such quality that the similar damage will never take place again in the member after that. On such conditions the optimization is mainly achieved by the appropriate selections of inspection method and inspection interval.

Total expected cost for structure at the present inspection and in the interval until the next inspection

The structure includes a variety of member sets which consist of different numbers of members (see Figure 1). In this study the following assumptions are made for the formulation of the problem. All the structural members in each set have the same strength property and are subjected to the same loading condition. Each member has a possibility of failure due to the deterioration damage. If any member fails, the service of the structure is suspended urgently and the failed member is to be repaired. With certain probability, the member failure might result in a catastrophic failure. Inspections are periodically carried out during the service life to find damages. And the detected damages are perfectly repaired.

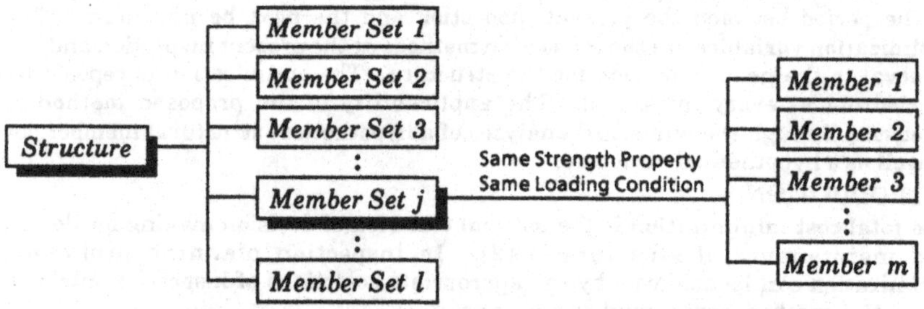

Figure 1 Hierarchy of structure, member sets and members

At a certain inspection time, the total expected cost for the structure in the succeeding inspection interval are classified into two groups: 1) costs necessary in the present inspection, and 2) risks (expected costs) during the service period until next inspection:

1)
- *C*(*inspection*): Inspection cost for each member set.
- *C*(*repair*): Expected repair cost of the damages detected in the inspection.
- *C*(*scheduled system down*) Loss due to the service suspension caused by the scheduled system down for the inspection.

2)
- *C*(*member failure*): Expected loss due to a member failure. This includes repair cost of failed members, loss due to the service suspension caused by member failure, and other economical losses accompanied with the accidental system down.
- *C*(*catastrophic failure*): risk against a catastrophic failure which may occur starting from member failure with a certain probability.

Among above cost items, only *C(scheduled system down)* is defined on the whole structure and all the other costs are imposed on each member set.

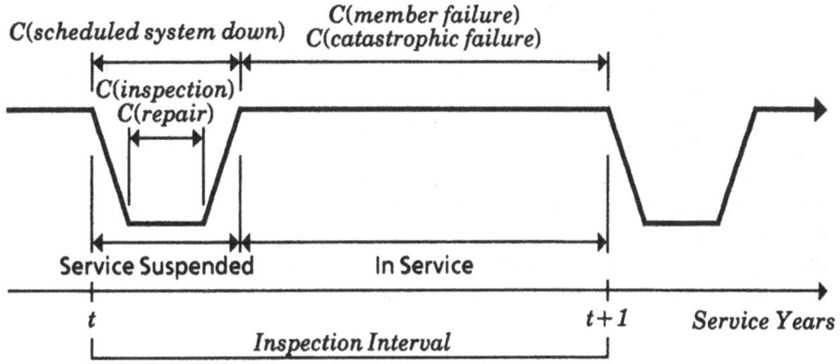

Figure 2 Costs Required for Structure In an Inspection Interval

The total expected cost for the whole structure in an inspection interval, from time t to time $t+1$, denoted by $CT(t, t+1)$, can be written as:

$$CT(t, t+1) = \sum_{j=Set\,1}^{Set\,l} C_j(t, t+1) + C_{SWS} \tag{1}$$

where,

$$C_j(t, t+1) = C(inspection) + C(repair) + C(member\ failure) + C(catastrofic\ failure) \tag{2}$$

$$C_{SWS} = C(scheduled\ system\ down) \tag{3}$$

In the above equations, $C_j(t, t+1)$ is the expected cost during the inspection interval $(t, t+1)$ for member set j.

Selection of optimal inspection method for a member set

If it is assumed that C_{SWS} in eq.(1) is independent on the applied inspection methods and repair qualities, then the minimization of $CT(t, t+1)$ for the structure is achieved when the inspection method for each member set is selected so as to minimize $C_j(t, t+1)$, ($j = 1$ to l). The selection is carried out at every inspection time from the following five methods: No inspection (*NO*), visual inspection (*VI*) method, mechanical (precise) inspection (*MI*) method, visual and conditional mechanical inspection (*V&M*) method, and sampling mechanical inspection (*SM*) method.

In *V&M* method, visual inspection is carried out for all the members at first. If no defect is found then the inspection is terminated. Else, if defects are found at least in one member, then mechanical inspection is carried out for all the members. In *SM* method, first mechanical inspection is carried out for limited number of sample members. If no defect is found among the sample members, then the inspection is terminated. Else, if defects are found at least in one member, then mechanical inspection is applied for all the rest members.

<u>Selection of suitable inspection interval for a structure</u>

The selection of suitable inspection interval for the structure is to be achieved by comparison of total expected costs evaluated for possible interval plans applying the proposed method. For simplicity, let us assume a case of the choice from two inspection interval plans, once a year and once every two years.

1) First, set the inspection interval for one year (t, $t+1$) and select the optimal inspection methods for all the member sets of the structure following the procedure mentioned in the previous section.
2) Then, evaluate the total expected cost for the structure following eq. (1) in the period between time t and time $t+1$: $CT(t, t+1)$.
3) Repeat 1) and 2) for the next inspection interval ($t+1$, $t+2$) to evaluate $CT(t+1, t+2)$.
4) Sum up these two costs for the structure and obtain the total expected cost in the period from the inspection time t to time $t+2$:

$$CT(t,t+2) = \sum_{j=Set\,1}^{Set\,l} C_j(t,t+1) + \sum_{j=Set\,1}^{Set\,l} C_j(t+1,t+2) + 2C_{SWS} \tag{4}$$

5) Secondly, change the inspection interval to two years (t, $t+2$), and carry out the selection of the optimal inspection method for each member set.
6) And evaluate the total expected cost during two years adding a system down cost for the present inspection at time t. The total cost for this case is given by the following equation.

$$CT(t,t+2) = \sum_{j=Set\,1}^{Set\,l} C_j(t,t+2) + C_{SWS} \tag{5}$$

7) The choice of the optimal inspection interval for the structure is to be done comparing these two costs evaluated by above two equations.

COST EVALUATION EQUATIONS

In order to estimate $C_j(t, t+1)$ corresponding to each member set, appearing in eqs. (1), (2), (4) and (5), the following assumptions were made.

1) The detection of defects in visual or mechanical inspection is probabilistic.
2) The repair of the detected damage or failed member is treated by the perfect repair model, that is, repair is carried out perfectly, such that damage and failure will never take place in the same member after the repair throughout remaining service life. Then the repair cost is to be a function of service time as the repair quality is determined considering the period of remaining service life, as well as a function of defect size.
3) Costs due to the service suspension caused by accidental system down are to be taken into consideration. Member failure does not necessarily mean a collapse of the structure. However, the service of the structure is suspended urgently and the failed member is to be repaired. This accidental system down requires considerably larger cost than that of the scheduled (predetermined) system down.
4) A member failure may result in a catastrophic failure with a certain transition probability. When the catastrophic failure occurs, the cost is due not only to the loss of the structure but also to the losses received from different portions of society such as owners, client, insurance, related industries and so on.

The expected cost for a member set in an inspection interval $(t, t+1)$ is evaluated by the following equations for the five inspection methods.

No inspection (*NO*)

$$C_j(t, t+1 \mid NO) = G \times P_{F1} \times C_F \tag{6}$$

Visual inspection (*VI*) method

$$C_j(t, t+1 \mid VI) = G \times \{C_{VI} + P_{DV} \times (C_{ZMI} + C_{RD}) + (1 - P_{DV}) \times P_{F2} \times C_F\} \tag{7}$$

Mechanical inspection (*MI*) method

$$C_j(t, t+1 \mid MI) = G \times \{C_{MI} + P_{DM} \times C_{RD} + (1 - P_{DM}) \times P_{F3} \times C_F\} \tag{8}$$

Visual and conditional mechanical inspection (*V&M*) method

$$\begin{aligned} C_j(t, t+1 \mid V\&M) = G \times [C_{VI} + P_{DV} \times (C_{ZMI} + C_{RD}) \\ + (1 - P_{DV}) \times [(1 - P_{DV})^{G-1} \times P_{F2} \times C_F \\ + \{1 - (1 - P_{DV})^{G-1}\} \times \{C_{MI} + P_{DM} \times C_{RD} + (1 - P_{DM}) \times P_{F3} \times C_F\}]] \end{aligned} \tag{9}$$

Sampling mechanical inspection (*SM*) method

$$\begin{aligned} C_j(t, t+1 \mid SM) = \alpha \times G \times \{C_{MI} + P_{DM} \times C_{RD} + (1 - P_{DM}) \times P_{F3} \times C_F\} \\ + (1 - \alpha) \times G \times [\{1 - (1 - P_{DM})^{\alpha G}\} \times \{C_{MI} + P_{DM} \times C_{RD} \\ + (1 - P_{DM}) \times P_{F3} \times C_F\} + (1 - P_{DM})^{\alpha G} \times P_{F1} \times C_F] \end{aligned} \tag{10}$$

In the above equations,

$$C_F = (1 - P_{FC}) \times (C_{SWA} + C_{RF}) + P_{FC} \times C_{CF} \tag{11}$$

$$G = m \times P_{SV} \tag{12}$$

m : Number of members in the member set.

α : Sampling rate of members in *SM* method.

P_{F1}, P_{F2}, P_{F3} : Occurrence probabilities of member failure in succeeding inspection interval on the conditions that *NO*, *VI* and *MI* methods are applied at the present inspection, respectively.

P_{DV}, P_{DM} : Probabilities of detecting a defect by *VI* and *MI* methods, respectively.

P_{FC} : Probability that a member failure develops into a catastrophic failure of the structure.

P_{SV} : Probability that a member has not experienced repair and failure until the present inspection.

C_{VI}, C_{MI} : Visual and mechanical inspection costs of a member, respectively.

C_{ZMI} : C_{ZMI} equals C_{MI} when the detected damage needs to be inspected mechanically for sizing to determine the repair method, otherwise C_{ZMI} is zero.

C_{RD} : Repair cost of a damaged member detected by visual or mechanical inspection.

C_{RF} : Repair cost of a failed member.

C_{CF} : Risk of a catastrophic failure.

C_{SWA} : Loss due to the service suspension caused by accidental system down.

COST ITEMS AND PROBABILITIES IN COST EVALUATION EQUATIONS

Cost items and their contents

In this study the structural members which have possibilities of fatigue failure are chosen as the object of the analysis. The fatigue deterioration includes both crack initiation and crack propagation processes and also the succeeding failure.

Figure 3 shows the content of each cost item assumed for the numerical analysis in the next chapter, in which C_{RD} and C_{RF} are the functions of the service time by aforementioned reason. Further, repair cost is treated as a function of crack size. The rate of price inflation is not considered in this study.

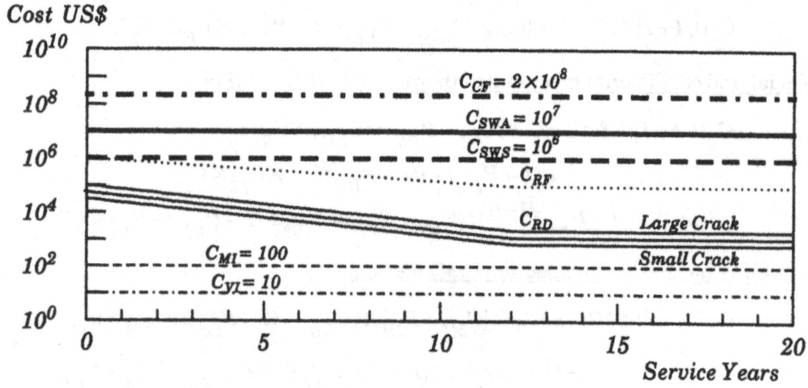

Figure 3 Cost Items and Their Contents

Probability estimation using Markov Chain Model

The probabilities appearing in the cost evaluation equations are calculated by the Markov Chain Model (MCM). Basically, in the simplest Bogdanoff and Kozin stationary MCM[3], an initial state vector $\mathbf{A}(0)$ and a duty cycle independent basic transition matrix $\mathbf{P}$ are sufficient to describe the entire probabilistic feature of fatigue process. In this study fatigue crack initiation and propagation processes are incorporated into the transition matrix of a single MCM[4]. The probability of member failure is also considered in the absorbing term $a_k(t)$ in the state vector $\mathbf{A}(t)$. One of the merits of applying MCM to the fatigue reliability analysis is that the procedure of periodic or non-periodic in-service inspections as well as that of pre-service inspection can be incorporated into the calculation easily. Furthermore, the existence probability of initial defects and the distribution of defect size can also be considered in the initial state vector of MCM.

Let $\mathbf{A}_{BI}(t)$ be the state vector just before the inspection at time t.

$$\mathbf{A}_{BI}(t)=\{a_1(t), a_2(t), a_3(t), \ldots, a_{k-1}(t), a_k(t)\} \tag{13}$$

The state vector right after the inspection can be given by the below equation for the perfect repair model.

$$\mathbf{A}_{AI}(t)=\{a'_1(t), a'_2(t), a'_3(t), \ldots, a'_{k-1}(t), a_k(t)\} \tag{14}$$

in which

$$a'_i(t)=a_i(t)\times\{1-D(i)\}, \quad i=1,2,3,\ldots,k-1 \tag{15}$$

In the above equation, $D(i)$ is the detection probability of the crack whose size is classified to the i-th state.

The probability P_{DV} as well as P_{DM} are calculated by:

$$P_{DV}=\sum_{i=1}^{k-1}\{a_i(t)\times D_{VI}(i)/P_{SV}\} \tag{16}$$

$$P_{DM}=\sum_{i=1}^{k-1}\{a_i(t)\times D_{MI}(i)/P_{SV}\} \tag{17}$$

where

$$P_{SV}=\sum_{i=1}^{k-1}a_i(t) \tag{18}$$

The failure probabilities P_{F1}, P_{F2} and P_{F3} are calculated following below procedure: When no inspection is carried out in the present time, the failure probability in the succeeding inspection interval is obtained from the equation:

$$P_{F1}=\{a_k(t+1)-a_k(t)\}/P_{SV} \tag{19}$$

in which $a_k(t+1)$ and $a_k(t)$ are the absorbing terms of the state vector at the time $t+1$ and time t, respectively. When visual inspection is applied, first we calculate the state vector right after the inspection, $\mathbf{A}_{AI}(t\,|VI)$, from eq. (14). Then calculate the state vector at the next inspection with h times state transitions.

$$\mathbf{A}_{BI}(t+1)=\mathbf{A}_{AI}(t\,|\,VI)\times\mathbf{P}^h \tag{20}$$

The failure probability P_{F2} is obtained by the below equation.

$$P_{F2}=\{a_k(t+1)-a_k(t)\}/P_{SV} \tag{21}$$

in which $a_k(t+1)$ is the absorbing term of $\mathbf{A}_{BI}(t+1)$ in eq.(20). Similarly, if we apply mechanical inspection and calculate $\mathbf{A}_{AI}(t\,|MI)$, the failure probability P_{F3} can be obtained.

<u>Markov state vector right after *V&M* and *SM* methods</u>
If *V&M* method is selected as the cost minimum, then the Markov state vector right after the inspection is expressed by the combination of $\mathbf{A}_{AI}(t\,|VI)$ and $\mathbf{A}_{AI}(t\,|\,MI)$:

$$\mathbf{A}_{AI}(t\,|\,V\&M)=(1-P_{DV})^G\times\mathbf{A}_{AI}(t\,|\,VI)+\{1-(1-P_{DV})^G\}\times\mathbf{A}_{AI}(t\,|\,MI) \tag{22}$$

If *SM* method is selected, the state vector right after the inspection is expressed as following equation.

$$\begin{aligned}\mathbf{A}_{AI}(t\,|\,SM)=\alpha\times\mathbf{A}_{AI}(t\,|\,MI)+(1-\alpha)\times[(1-P_{DM})^{\alpha G}\times\mathbf{A}_{BI}(t)\\+\{1-(1-P_{DM})^{\alpha G}\}\times\mathbf{A}_{AI}(t\,|\,MI)]\end{aligned} \tag{23}$$

NUMERICAL ANALYSES

In order to examine the applicability of this sequential cost minimization method, numerical analyses are carried out for a hypothetical structural member set and a hypothetical structure with fatigue deterioration. For the member set, the problem of prefixed interval inspections is treated, where the proposed method is used to find the optimal inspection method at each inspection time. For the problem of structure, not only the inspection method but also the inspection interval is chosen as the optimization variables.

Analysis of a structural member set

The assumed structural member set consists of *10* members which have statistically identical strength properties and are put under the same loading condition. The surface fatigue crack initiated from the weld toe of a plate with *20 mm* thickness is treated as the deterioration damage. The distributions of fatigue crack initiation and propagation lives follow two parameter Weibll distributions with shape parameter, $\gamma = 2.5$ and *4.0*, respectively. The mean crack growth curve is described by Paris' equation with stress intensity factor range ΔK calculated by linear elastic fracture mechanics. Member failure was defined when the crack depth reaches the plate thickness. Table 1 shows the fatigue property and the initial defect condition of the member set for five cases analyzed.

Table 1 Fatigue Property and Initial Defect Condition of Member Sets for Five Cases Analyzed

	Mean of Crack Initiation Life $\bar{N}_c$ (years)	Mean of Crack Propagation Life $\bar{N}_p$ (years)	Prob. of Initial Defects P_{ID}
Case 1	20	15	0.00
Case 2	30	15	0.00
Case 3	50	15	0.00
Case 4	30	15	0.01
Case 5	30	15	0.10

Periodic inspections are to be carried out for the member set once a year based on the predetermined schedule. Figure 4 shows the assumed inspection capability of the visual and mechanical inspections in terms of the crack depth. All the probabilities except P_{FC} used in the cost evaluation equations are calculated by the MCM. For the probability P_{FC}, *0.005* was assumed. The assumed value of each cost item is shown in Figure 3. The scheduled system down cost, C_{SWS}, is not included in this calculation, because the inspection interval is prefixed for the whole structure regardless of the inspection method for each member set.

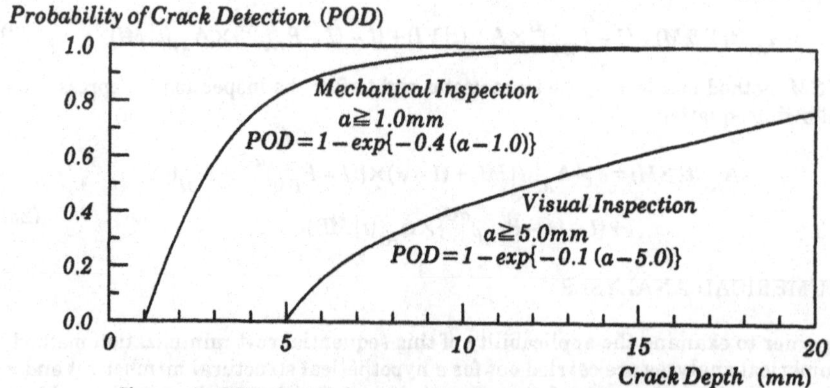

Figure 4 POD Curves for Visual and Mechanical Inspection

Table 2 shows the inspection methods selected at each inspection during the period of *20 years'* service. From the table it is clear that the first inspection timing becomes later when the crack initiation life of the member is longer. Visual and mechanical inspections are alternately applied after the first inspection. It is obvious that the first inspection timing becomes earlier when P_{ID} increases.

Table 2 Selected Inspection Methods for Member Sets

	Selected Method in Each Year																			
	1	2	3	4	5	6	7	8	9	10	11	12	13	14	15	16	17	18	19	20
Case 1	-	-	-	-	-	-	V	V	M	V	M	V	M	M	M	V	M	M	M	M
Case 2	-	-	-	-	-	-	-	-	V	V	M	V	M	V	M	V	M	M	V	M
Case 3	-	-	-	-	-	-	-	-	-	-	V	V	V	M	V	V	M	V	M	V
Case 4	-	-	-	V	V	M	V	M	V	V	M	V	M	V	M	V	M	M	V	M
Case 5	-	-	V	V	M	M	M	V	M	V	M	V	M	V	M	V	M	M	V	M

- : No Inspection V : Visual Inspection M : Mechanical Inspection

Figure 5 shows an example of the cumulative failure probability, P_F, of a member during service for the selected inspection methods. The other three curves in the figure are the P_F's calculated under the respective conditions that *NO* method, *VI* method and *MI* method are continually applied at every inspection.

In the above examples only the *NO*, *VI* and *MI* methods were selected as optimal, and *V&M* and *SM* methods were never selected (α, the sampring rate of *SM*, is 10% in numerical analyses). The reason for this is: When *V&M* method is applied, the estimated cost always lies between the two costs estimated by *VI* and *MI* methods. That is, either *VI* or *MI* method gives smaller cost than *V&M* method. When *SM* method is applied, the estimated cost always lies between those by *NO*

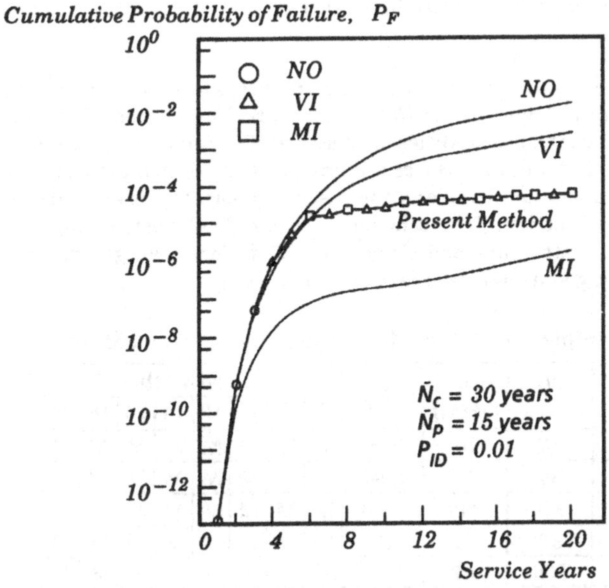

Figure 5 Cumulative Probability of Failure for Member Set Case 4

and *MI* methods. That is, either *NO* or *MI* method gives smaller cost than *SM* method. This can be proved from the structure of the cost evaluation equations.

The profit of *V&M* and *SM* methods

In this section, *V&M* and *SM* methods is examined applying to a member set, for which estimation of fatigue property includes an error. Assume a member set including *100* statistically identical members. Both crack initiation and crack propagation lives of the members are misestimated to the longer side as shown in Table 3. It is also assumed that the other probabilities and costs are accurately estimated and they have the same values as in the above section.

Table 3 Truth and Misestimated Fatigue Lives of the Member Set

	Mean of Crack Initiation Life $\bar{N}_c$ (years)	Mean of Crack Propagation Life $\bar{N}_p$ (years)
Truth	*50*	*15*
Wrong Estimation	*100*	*25*

Figure 6 compares the selected inspection methods and the evaluated $C_j(t, t+1)$ with the correct fatigue property estimation and those with wrong estimation. From the figure it is obvious that the expected costs become very large when estimation error exists in the fatigue property. The solid and dashed lines drawn in this figure express $C_j(t, t+1)$'s calculated under the conditions that *V&M* and *SM* methods are continually applied from the first inspection, respectively. These two lines lie considerably below the $C_j(t, t+1)$ of the wrong estimation case at the latter stage of the service.

Figure 7 compares the cumulative costs for the above four cases. As is seen from the figure, *V&M* and *SM* methods are profitable when the fatigue property of member includes an estimation error. The same tendency was observed for the case that the fatigue life was misestimated to the shorter side.

Analysis of a structure

The proposed method is applied to an assumed structure which consists of five member sets. The number of members is *100* for sets A, B, C and D and *500* for set E. Also sets C and D have the probabilities of existence of initial defects with *0.10* and *0.01*, respectively. The values of the cost items, the distribution properties of crack initiation and propagation lives and the inspection capabilities are all the same as in the previous section. The scheduled system down cost, C_{SWS}, was considered for the whole structure and $C_{SWS}=10^6$ *USdollars* was given. The selected inspection timing and qualities are shown in Table 4.

Table 4 Selected Inspection Methods for the Structure (C_{SWS} = US10^6)

Member set	# of Members	$\bar{N}_c$ (years)	$\bar{N}_p$ (years)	P_{ID}	Selected Methods 4	8	10	12	14	16	18	20
A	*100*	*20*	*10*	*0.00*	M	M	M	M	M	M	M	M
B	*100*	*30*	*15*	*0.00*	-	V	M	M	M	M	M	M
C	*100*	*50*	*20*	*0.10*	M	M	M	M	M	M	M	V
D	*100*	*50*	*15*	*0.01*	M	M	M	M	V	M	M	M
E	*500*	*100*	*15*	*0.00*	-	-	-	-	-	V	M	V

- : No Inspection V : Visual Inspection M : Mechanical Inspection

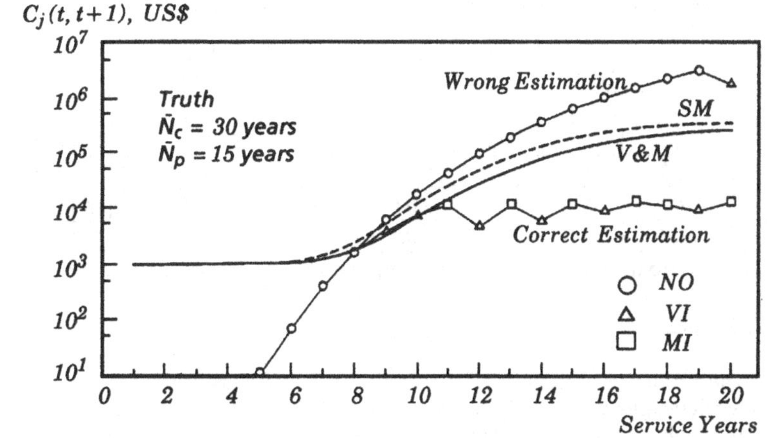

Figure 6 Effect of Estimation Error of Fatigue Life on Inspection Strategy

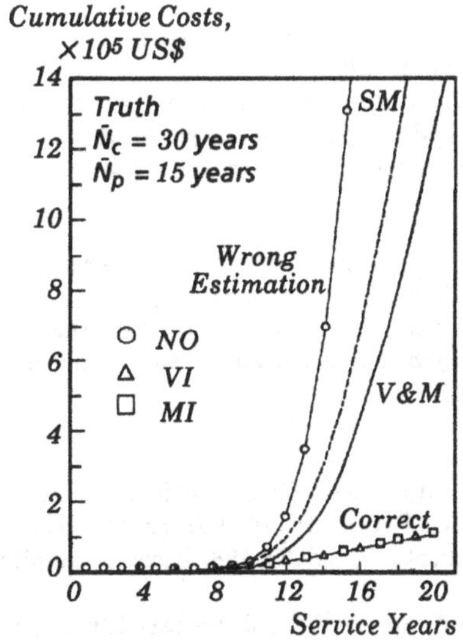

Figure 7 Cumulative Costs for the Member Set

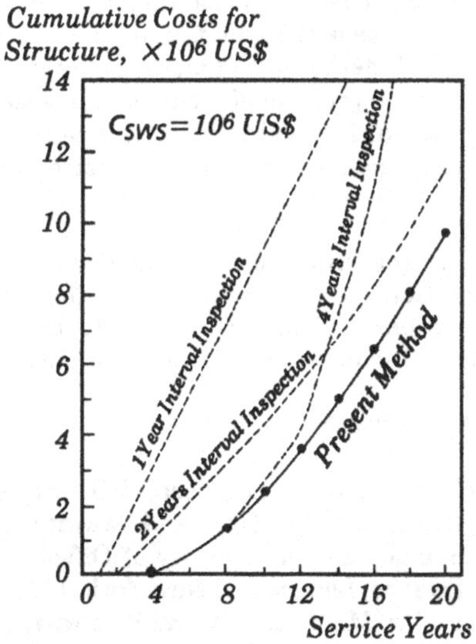

Figure 8 Cumulative Costs for the Structure by Present Method

Figure 8 shows the superiority of present method. The three dashed curves express the cumulative costs required for the structure when the prefixed inspection intervals, one year, two years or four years, are applied. In these cases optimal inspection methods are selected to each member set at every inspection time. The solid curve with black circles is the cumulative cost calculated by the present method and this cost is always the minimum throughout the service time. When the C_{SWS} is reduced to 3×10^5 *USdollars* for the same structure, the selected inspection timing and qualities become as given in Table 5.

Table 5 Selected Inspection Methods for the Structure (C_{SWS} = US\$3 × 10^5)

Member set	# of Members	$\bar{N}_c$ (years)	$\bar{N}_p$ (years)	P_{ID}	Selected Methods									
					4	8	10	12	14	15	16	18	19	20
A	*100*	*20*	*10*	*0.00*	M	M	M	M	M	M	M	M	M	M
B	*100*	*30*	*15*	*0.00*	-	V	M	M	M	V	M	M	M	M
C	*100*	*50*	*20*	*0.10*	M	M	M	M	M	V	M	V	V	M
D	*100*	*50*	*15*	*0.01*	M	M	M	M	V	V	M	M	V	M
E	*500*	*100*	*15*	*0.00*	-	-	-	-	-	V	V	V	V	M

- : No Inspection V : Visual Inspection M : Mechanical Inspection

CONCLUSIONS

Through the numerical analyses of a hypothetical structural member set and a structure with fatigue deterioration, the following findings can be summarized.

1) For the problem of prefixed interval inspections, the first inspection timing and the optimal inspection method can be sequentially selected at each inspection so as to minimize the total expected cost until the next inspection.
2) For the statistically determined problem, neither *V&M* nor *SM* method is selected as the optimal inspection method.
3) In the case that the fatigue life of the members contains an estimation error, *V&M* method or *SM* method may become profitable.
4) For the problem of structure for which the inspection interval is not fixed, the optimal interval to the next inspection and the optimal inspection method for each member set can be selected simultaneously so as to minimize the total expected cost.

In the inspections of actual structures, it is rather often the case that initial inspection plan is updated with the additional information obtained through inspections and maintenances during the service[5]. It is expected that the proposed sequential cost minimization approach becomes useful in such situations.

REFERENCES

[1.] Madsen, H.O., Sorensen, J.D., Probability-Based Optimization of Fatigue Design, Inspection and Maintenance, Proceedings 4th International Symposium on Integrity of Offshore Structure, pp.421-438, Elsevier Applied Science, London and New York, 1990.

[2.] Fujita, M., Schall, G. and Rackwitz, R., Adaptive Reliability-Based Inspection Strategies for Structures Subject to Fatigue, Proceedings of 5th International Conference on Structural Safety and Reliability, Vol.2, pp.1619-1626, ASCE Publications, San Francisco, 1989.

[3.] Bogdanoff, J.L. and Kozin, F., Probabilistic Models of Cumulative Damage, John Wiley & Sons, Inc., New York, 1985.

[4.] Fujimoto, Y., Ideguchi, A. and Iwata, M., Reliability Assessment for Deteriorating Structure by Markov Chain Model, Journal of Society of Naval Architects of Japan, Vol. 166, pp.303-314, 1989 (in Japanese).

[5.] Fujimoto,Y., Itagaki, H., Itoh, S., Asada, H. and Shinozuka, M., Bayesian Reliability Analysis of Structures with Multiple Components, Proceedings of 5th International Conference on Structural Safety and Reliability, Vol.3, pp.2143-2146, ASCE Publications, San Francisco, 1989.

Probabilistic Approach to Cumulative Damage of Car Components

G. Righetti, A. Barbagelata, D. Diamantidis

D'Appolonia S.p.A., Via Siena 20, 16146 Genova, Italy

ABSTRACT

The present contribution summarizes results from a study regarding the probabilistic approach to cumulative damage of mechanical components. Bases of a predictive system for fatigue damage assessment of car components are discussed. In such a context a representative case is considered to illustrate the feasibility of the project and to demonstrate the advantages of the developed methodology.

INTRODUCTION

Fatigue is a major failure mode in mechanical components of automobiles which respond dynamically to random loading. In fact in motor vehicles various mechanical parts are subjected to fatigue stresses of variable amplitude during their service lives. An important aspect of the development of fatigue resistant car components is the ability to predict component fatigue life and thus to anticipate possible problems. This is often a difficult task due to the complications involved in input data and analysis procedures.

In addition one should bear in mind that most parameters influencing the fatigue life of mechanical components are random. Material properties determined on the basis of test specimens, whose microstructure, geometry and surface characteristics are representative of those of the component, show considerable scatter. The loading time history of the component is basically a random process. Moreover there are uncertainties due to limitations inherent in the analytical models. All these uncertainties contribute to a probability that the components do not perform as intended. It appears therefore necessary and beneficial that a reliability based approach will be adopted to significantly increase the possibility of a reliable prediction of the performance of such components.

All the aforementioned aspects indicate that fatigue life prediction combines problems from different domains such as experimental analysis, deterministic and probabilistic modelling. A complete approach for a fatigue life prediction shall include the following steps:

- modelling of cyclic stress-strain material behaviour;
- experimental investigation of fatigue parameters;
- statistical description of loading, resistance, and model uncertainty parameters;
- formulation of damage accumulation model;
- reliability analysis to compute failure probability of components;
- interpretation of results and assessment of lifetime.

METHODOLOGY

A model based on crack initiation which utilizes the fatigue properties of the steel obtained from smooth laboratory test specimens [1] is combined with a first-order reliability approach [2] for the prediction of fatigue life of a steering knuckle. It is assumed in this case that the fatigue behavior is equivalent to that of a smooth specimen (made of steel having similar mechanical properties, metallurgical structure and surface conditions) subjected to a stress history corresponding to the stress acting in the critical zone of the component. This method utilizes the most recent knowledge acquired on mathematical modelling of the behavior of materials subjected to random time histories. The following sections describe the necessary steps for the development of this approach.

Cyclic stress-strain analysis

The cyclic stress-strain curve of a material provides the necessary relation for the cyclic plasticity analysis. The cyclic and monotonic stress-strain relationships demonstrate that cyclic loading can significantly affect the deformation behavior of a material.

The relation between strain amplitude, $\Delta\epsilon/2$, and stress amplitude, $\Delta\sigma/2$, is given by:

$$\frac{\Delta\epsilon}{2} = \frac{\Delta\sigma}{2E} + \left(\frac{\Delta\sigma}{2K'}\right)^{1/n'} \tag{1}$$

where:

E = modulus of elasticity;

n', K' = material parameters.

The material parameters n' and K' representing the plastic range of the stress-strain relationship are calibrated from test results on representative samples. In the numerical analysis, the curve of Equation (1) is discretized into an appropriate number of segments [3].

Damage analysis

The assessment of damage for the various events in a complex sequence given by the strain time history is combined with the stress-strain analysis. At the end of each hysteresis cycle, damage is determined using constant amplitude strain-life curves [4]. The strain-life curve is defined by the Manson-Coffin relationship, which relates the strain amplitude $\Delta\epsilon/2$ to the reversals to failure $2N_f$ as follows:

$$\frac{\Delta\epsilon}{2} = \frac{\sigma'_f}{E} (2N_f)^b + \epsilon'_f (2N_f)^c \tag{2}$$

where σ'_f, E, b, ϵ'_f and c are material properties obtained from experimental results on smooth specimen by applying regression analysis techniques.

Assuming a linear accumulation of damage, the total damage D_T for an endurance L results as:

$$D_T(L) = \sum_{i=1}^{n(L)} D_i = \sum_{i=1}^{n(L)} \frac{1}{N_{f_i}} \tag{3}$$

where:

n(L) = total number of cycles for a given endurance L (in kilometers);

D_i = damage related to i-th cycle.

The damage parameter utilized in this study , given by the number of cycles N_{f_i} leading to failure, is proposed by Smith, Watson and Topper [5] , who adopt the Manson-Coffin Equation (2) and include also the effect of mean strain of each cycle (through $\sigma_{max,i}$):

$$\sigma_{max,i} \frac{\Delta\epsilon_i}{2} E = \sigma'_f{}^2 (2N_{f_i})^{2b} + \sigma'_{f_i} E (2N_{f_i})^{b+c} \quad (4)$$

where:

$\sigma_{max,i}$ = maximum stress in the i-th cycle;
$\Delta\epsilon_i$ = strain variation in the i-th cycle.

The other parameters are defined in the Manson-Coffin relationship of Equation (2). Equation (4) is then solved iteratively for N_{f_i} to determine the damage associated with each cycle defined by the couple of values DE_i, $\sigma_{max,i}$.

<u>Probabilistic approach to fatigue damage</u>
Since the main parameters entering fatigue damage analysis are in nature random, a reliability based approach has been implemented to account for all types of uncertainties involved in the analyses. First-order reliability methods (FORM) have been used since they have been proven as very efficient and sufficiently accurate.A detailed description of the methods can be found in [6].

An analysis procedure has been developed herein to account for randomness in the input parameters, such as load time history uncertainties, model uncertainties, material properties. Thereby an efficient iterative procedure for the linearization of the limit state surface has been implemented.

The limit state function is defined in our case by the following relationship:

$$g(L) = \Delta - D_T(L) \quad (5)$$

where:

Δ = accumulated fatigue damage in case of failure;

$D_T(L)$ = cumulative actual fatigue damage for a given endurance L.

The resistance parameter Δ can be considered a random variable while $D_T(L)$ is a function of several stochastic variables as described above.

On the basis of the limit state function of Equation (5), the cumulative distribution of the fatigue life can be computed. This is realized by specifying levels of endurance for which the analyses shall be performed. Input to the program is the statistical description of the influencing variables, i.e. type of statistical distribution, mean value and coefficient of variation (c.o.v.). Output is the probability of failure, i.e. the probability of not exceeding the defined endurance level and also the sensitivity of the basic variables.

ANALYSIS AND COMPARISON WITH TEST RESULTS

Based on the methodology described and on the corresponding software, selected cases were analyzed in order to validate the developed tools and to demonstrate their applicability to practical cases. Figure 1 illustrates the applied procedure.

The component considered for the analyses is a steering knuckle used in the front suspensions of light commercial vehicles. The most highly stressed zone is the radius connecting the cylindrical nose and the flange, where the points of attachment to the steering system are located.

The steering knuckle is manufactured from 42 CrMo4 steel; the required strength after quenching and tempering is in the range of 880 - 1030 MPa. The load history was taken from recordings on a test track with corrugations.

Table 1 summarizes the input values for the basic variables used in the analyses. For all random variables lognormal distributions have been fitted since they take only positive values and they are associated to rather small variabilities (c.o.v.'s less than 0.4).

The results of the performed analyses may be presented as the cumulative distribution of endurance $F_L(L)$ corresponding to the fatigue life distribution:

$$F_L(L) = P\,[\Delta - D_T(L) \leq 0] \tag{6}$$

Figure 2 illustrates the fatigue life distribution for three representative values for the c.o.v. of the stress concentration factor K_t, i.e. 0.1, 0.25 and 0.40. For a c.o.v. of 0.40 the fatigue life distribution has, as expected, the largest scatter. However, also in the other cases the scatter is considerable, thus demonstrating that a probabilistic approach to fatigue life prediction is more suitable than classical deterministic procedures. In addition the frequently assumed two-parameter Weibull distribution of the fatigue life cannot model in all cases this large scatter.

Figure 3 illustrates the fatigue life distribution for a range of values related to the c.o.v. of the uncertainty inherent in the recorded time history. The differences between the various curves are small indicating that the influence of the basic variable X_{th} is not as large as the influence of K_t.

The results have been compared to experimental values derived from five replicate bench tests on steering knuckles built up with materials identical to those used for the fatigue analysis. Thereby the original time history magnified by a factor of 1.9 has been applied. Probabilistic prediction overestimates the component fatigue life; in fact the mean value of fatigue life obtained through experiments is of 43000 kilometers corresponding to a probability level P of 0.3 in the theoretically derived curve (see Figure 2) instead of P ≃ 0.5 as one would expect. This discrepancy is attributed to an underestimate of the stress concentration factor uncertainty.

However, it is believed that the c.o.v. of 0.40, considered in some analyses, is already representing a large value and that a better agreement between test results and theoretical results can be achieved by introducing a bias for the uncertainty in the stress concentration factor. Figure 4 illustrates that such a bias value is of the order of 1.10 - 1.20. It is noted here that more experimental results are needed to better verify the fatigue life of the investigated components.

CONCLUSIONS

A probabilistic procedure for the prediction of the fatigue life of car components has been developed and has been applied to a case study. By applying such procedure the fatigue life distribution of the components can be computed in a rational and systematic way. Thereby all sources of uncertainty are accounted for. The following conclusions may be drawn from the case study:

- the fatigue life of car components is associated to considerable scatter and thus cannot be simply modeled by a two-parameter Weibull distribution as it is usually done;

- the largest sources of uncertainty in fatigue life prediction are related to loading parameters such as stress concentration factor and applied time history;

- the probabilistic approach helps to understand better and to verify experimental results; in fact a good agreement between test results and theoretical predictions can be found when appropriate bias and c.o.v. for the stress concentration factor are determined.

Table 1
Input Values for Basic Variables

SYMBOL	DESCRIPTION	DISTRIB. TYPE	MEAN VALUE	C.O.V.
n'	Material parameters	LN	0.1008	0.063
E	Modulus of elasticity	LN	207907 MPa	0.05
$X_{stw}^{(3)}$	Model uncertainty	LN	0.1	$0.03^{(1)}$ $0.05^{(2)}$
Δ	Damage factor	LN	1.0	0.1
X_{th}	Uncertainty in time history	LN	1.0	0.05-0.25
K_t	Stress concentration factor	LN	1.62	0.10-0.40

Legend:

LN = Lognormal

Notes:

(1) Valid for $2N_{f_i} = 50000$

(2) Valid for $2N_{f_i} = 1000000$

(3) The other parameters entering the Smith-Topper-Watson equation are deterministic, i.e:

$\sigma_f' = 1164$ MPa

$\epsilon_f' = 1.66$

$b = -0.068$

$c = -0.73$

REFERENCES

[1] Blarasin, A. and Farsetti P. A Procedure for the Rational Choice of Microalloyed Steels for Automative Hot-Forged Components Subjected to Fatigue Loads, International Journal of Fatigue 11, No. 1, pp. 13-18, 1989.

[2] Hohenbichler, M. and Rackwitz R. Non-Normal Dependent Vectors in Structural Safety, Journal of Engineering Mechanics Division, ASCE, Vol. 109, pp. 1227-1228, 1981.

[3] Neuber, H. Theory of Stress Concentration for Shear Strained Prismatical Bodies with Arbitrary Non-Linear Stress-Strain Law, Journal of Applied Mechanics, 1965.

[4] Landgraf, R. W., F. D. Richards and La Posute N. R. Fatigue Life Predictions for a Notched Member Under Complex Load Histories, Automotive Engineering Congress and Exposition, Detroit, MI, 1975.

[5] Smith, R. M., P. Watson and Topper T. H. A Stress Strain Function of Fatigue Metals, Journal of Materials JMLSA, Vol. 5, pp. 767-778, 1970.

[6] Madsen, H. O., S. Krenk and Lind N. C. Methods of Structural Safety, Prentice-Hall, New Jersey, 1986.

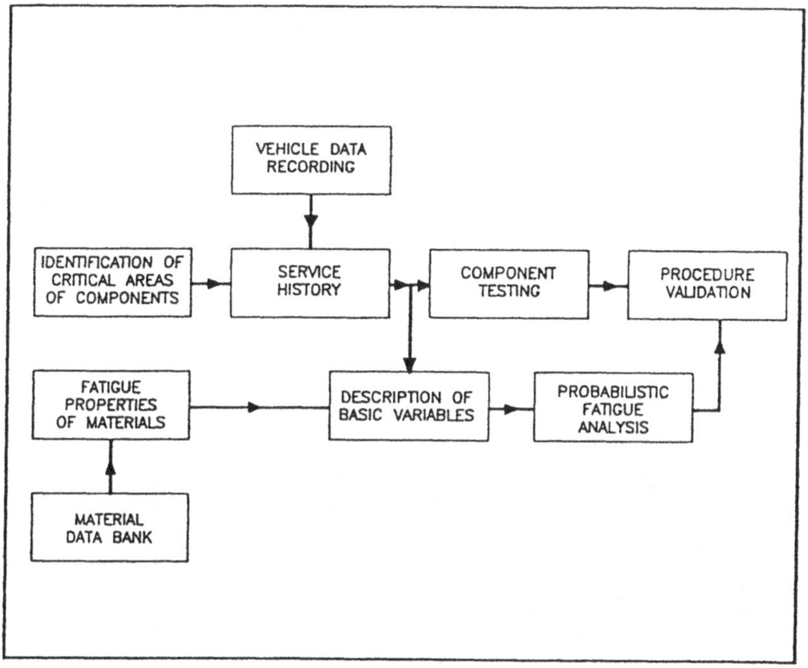

Figure 1: Flow chart for the fatigue life prediction procedure

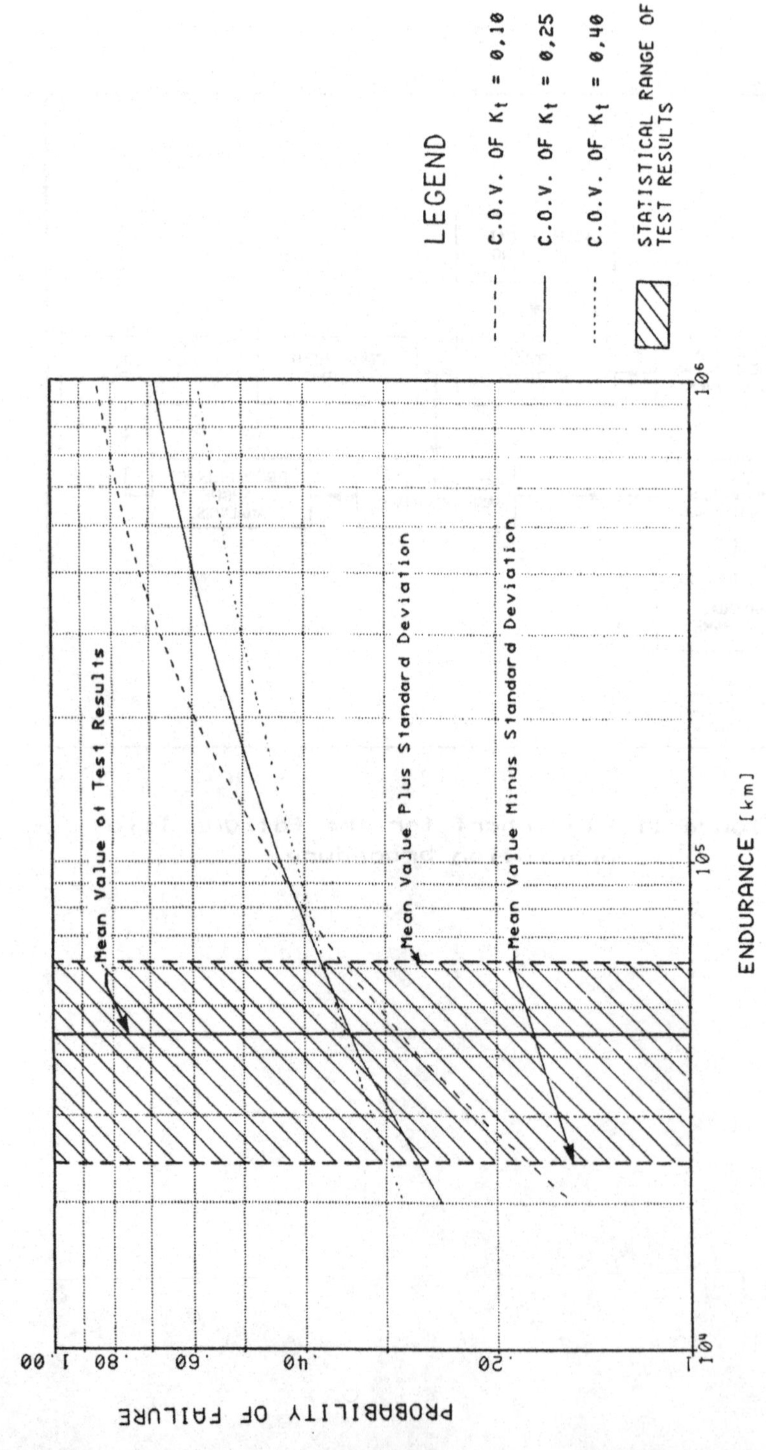

Figure 2: Fatigue life distribution for different variabilities of the stress concentration factor K_t

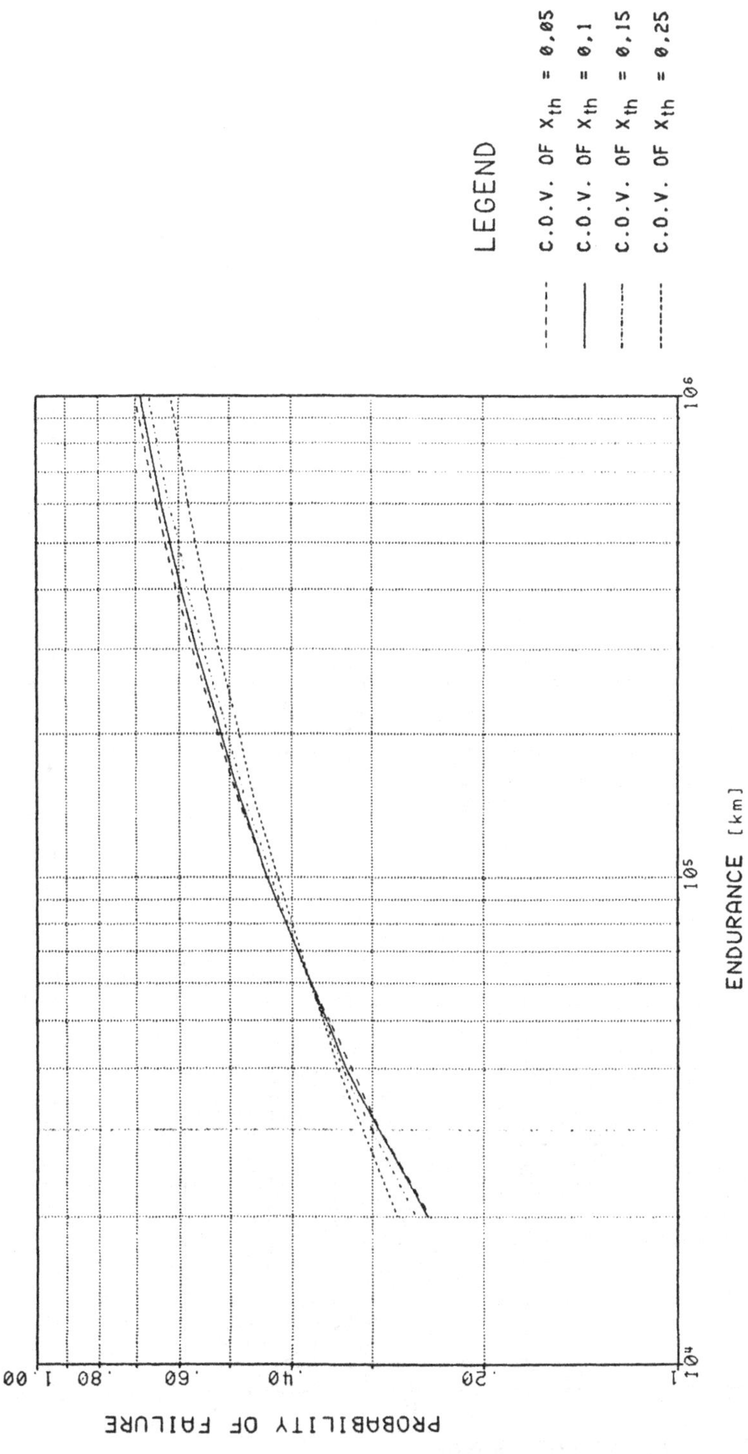

Figure 3: Fatigue life distribution for different uncertainties inherent in the applied time history

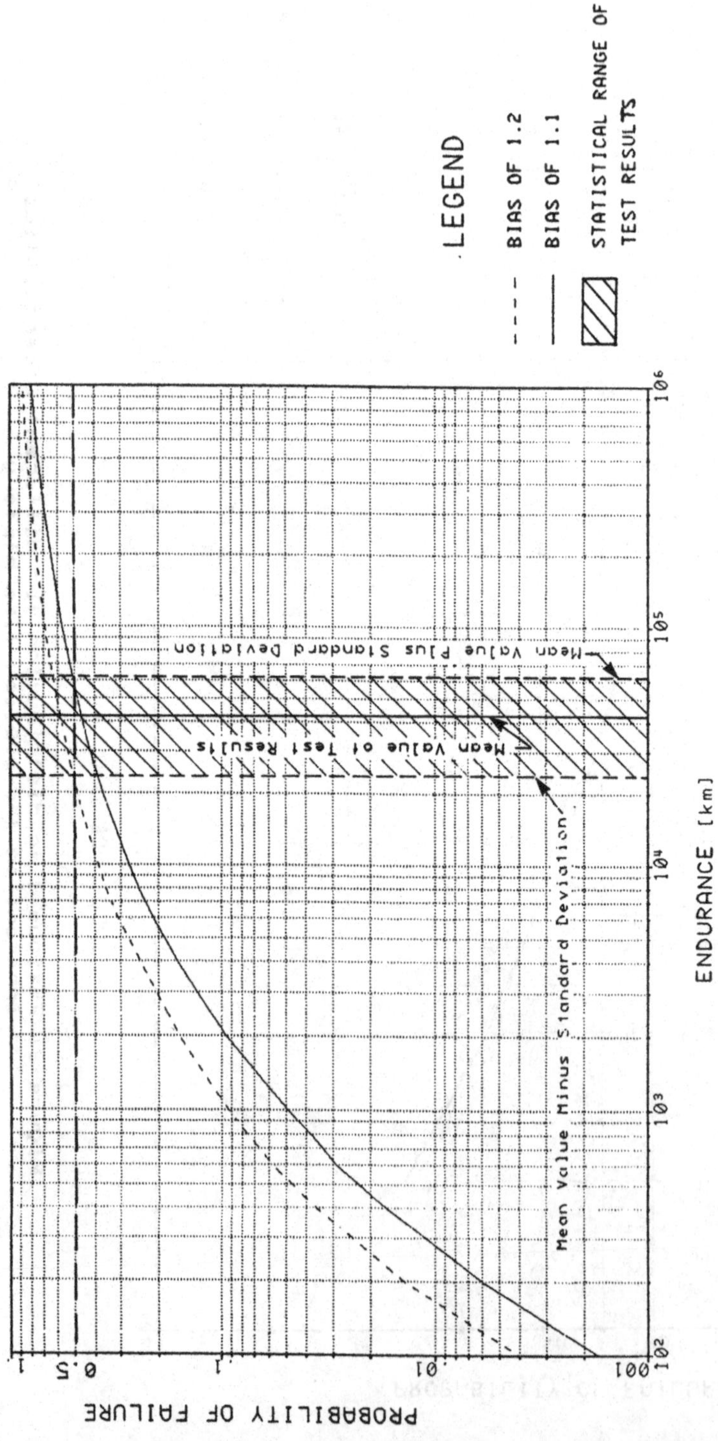

Figure 4: Influence of stress concentration factor bias on fatigue life distribution

Stochastic Two-Dimensional Fatigue Crack Growth Analysis Including the Effect of Crack Coalescence

P.F. Hansen (*), E.H. Cramer (**), H.O. Madsen (***)

(*) The Engineering Academy of Denmark, Building 373, DK-2800 Lyngby, Denmark

(**) Department of Naval Architecture and Offshore Engineering, U.C. Berkeley, Berkeley, CA 94720, U.S.A.

(***) Det norske Veritas, Danmark A/S, Nyhavn 16, DK-1051 Copenhagen K, Denmark

ABSTRACT

Two-dimensional fatigue crack growth from undercuts in longitudinal welds is investigated. The initial crack defects are assumed Poisson distributed with independent identically distributed initial shape parameters along a weld seam with identical material parameters. All the initial defects are exposed to the same stochastic loading.

The individual cracks are described as semi-elliptical surface cracks, and the Paris and Erdogan equation for fatigue crack growth is applied. An empirical equation for the stress intensity factor, given by Newman and Raju, for semi-elliptical surface cracks in a plate loaded in combined bending and tension, modified for the influence of interaction from other cracks, is used together with a magnification factor accounting for the additional stiffness due to the presence of the weld. The crack growth equation (two coupled first-order differential equations) are solved using an ordinary differential equation solver.

A weld seam with specified crack defect intensity is analysed, where the effect of crack coalescence is included. When cracks coalesce, a new crack configuration is defined, having a crack length equal to the sum of the coalesced cracks. A safety margin for exceedance of a critical crack depth is formulated.

The influence of the initial defect intensity on the estimated fatigue failure probability and on the joint density function for the crack geometry of the deepest crack at design life is investigated.

1. INTRODUCTION

Flaws are inherent in most metallic structures due to notches, welding defects etc. From these flaws, cracks may initiate and propagate under time varying load, and possibly grow to a critical size causing failure. Since fatigue crack growth in most steel structures under time varying loading can hardly be prevented in the long run, it is important to be able to predict the propagation of cracks in order to control the reliability.

The fatigue crack growth mechanism will not be reviewed here; instead refer to e.g. Madsen [1] or Broek [2]. It is only stated, that crack growth is the geometrical consequence of slip and blunting at the crack tip. As these events are governed by the local stresses at the crack tip, these stresses must be described as a function of the applied stress and the crack geometry. Such a relationship can be established using theory of elasticity on the cracked element.

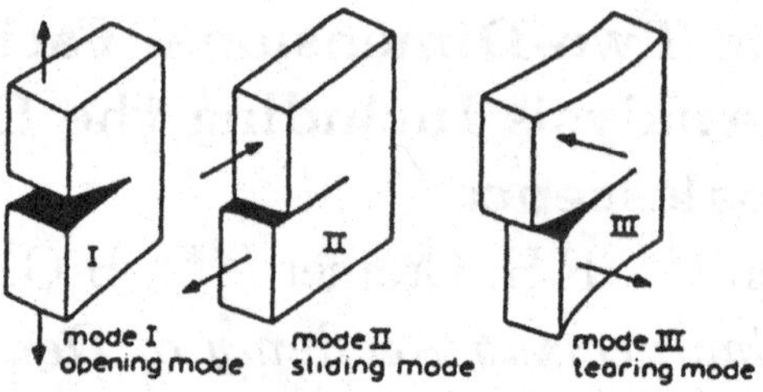

Figure 1: The three modes of loading. Broek [2]

In order to describe the problem, the stress field at the crack tip is classified as one of three different basic types, each associated with a local mode of deformation as shown in Fig. 1. The three different modes are referred to as

- *I* opening mode or tension mode
- *II* in-plane shear mode
- *III* out-of-plane shear mode or tearing mode

Only mode *I* is considered relevant for this analysis.

The stress field solution for mode *I* can be expressed as:

$$\sigma_{ij}(r,\theta) = \frac{K_I}{\sqrt{2\pi r}} f_{ij}(\theta) \tag{1.1}$$

where σ_{ij} are the stresses at a distance r from the crack tip (Fig. 2) and at an angle θ from the crack plane; - f_{ij} are dimensionless functions of θ; - K is the stress intensity factor.

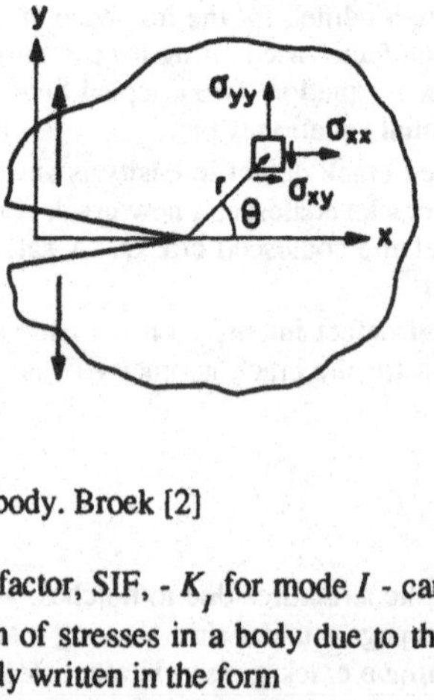

Figure 2: Crack in an arbitrary body. Broek [2]

Physically, the stress intensity factor, SIF, - K_I for mode *I* - can be interpreted as a parameter which reflects the redistribution of stresses in a body due to the introduction of a crack. The stress-intensity factor is normally written in the form

$$K = Y\sigma\sqrt{\pi a} \tag{1.2}$$

where the geometry function Y accounts for the effect of all the boundaries, σ is a reference stress at the analysed point and a is the crack depth.

The conditions governing the fatigue crack growth are the geometry of the structure, the crack initiation site, the material characteristics, the environmental conditions and the loading. These conditions are of random nature, consequently, analysis and design based on probabilistic methods is most relevant. The design criterion is therefore generally a limiting value for the probability that the crack depth exceeds the critical value during the design lifetime.

Experiments [3] show that the influence of the welding procedure on the initial crack depth and intensity of initial crack defects, is very important. The different initial cracks may in the early crack growth phase coalesce to longer and more shallow crack configurations. Consequently, it may be important to include the effect of coalescence in the probabilistic fatigue crack growth model, in order to properly describe the fatigue crack growth propagation.

2. FATIGUE CRACK GROWTH MODEL

In order to predict the fatigue crack growth of a single surface crack, it is assumed that the crack growth per stress cycle at any point along the crack front can be predicted from the Paris and Erdogan equation. This equation states that, at a specific point along the crack front, the increment in crack size $dr(\phi)$ - see Fig. 3 - during a load cycle dN, is related to the range of the SIF $\Delta K_r(\phi)$ for that specific load cycle through:

$$\frac{dr(\phi)}{dN} = C_r(\phi)(\Delta K_r(\phi))^m \;, \qquad \Delta K_r(\phi)>0 \tag{2.1}$$

where $C_r(\phi)$ and m are material parameters for that specific point along the crack front. The differential equation, Eq. (2.1), must be satisfied at all points along each of the crack fronts.

To simplify the problem it is assumed that the fatigue crack initially has a semi-elliptical shape, and that the shape remains semi-elliptical as the crack propagate. This implies that the crack depth (a), and the crack length $(2c)$, are sufficient parameters for the description of the crack front, see Fig. 3. When cracks coalesce, it is assumed that a new semi-elliptical crack is immediately formed.

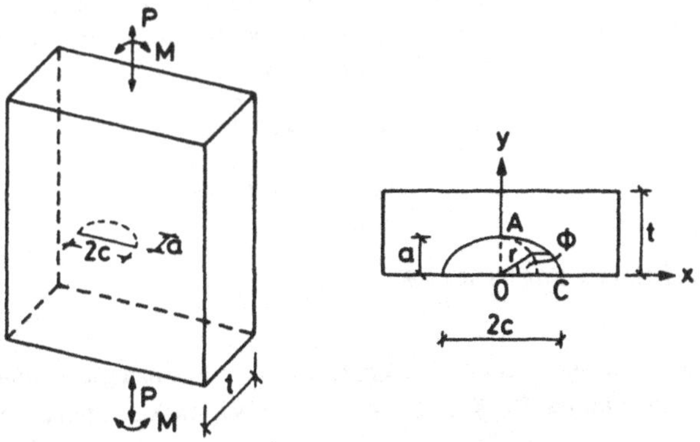

Figure 3: Semi-elliptical surface crack in a plate under tension and bending fatigue loads.

As a consequence of the assumption of a semi-elliptical crack, the general differential equation, Eq. (2.1), is replaced by a pair of coupled differential equations for each crack, Madsen [4], Shang-Xian [5]:

$$\frac{da}{dN} = C_A(\Delta K_A)^m \quad ; \quad a(N_0) = a_0 \tag{2.2}$$

$$\frac{dc}{dN} = C_C(\Delta K_C)^m \quad ; \quad c(N_0) = c_0 \tag{2.3}$$

The subscripts A and C refer to the deepest point and an end point of the crack at the surface, respectively. Due to the general triaxial stress field, experimental results and SIFs, as calculated by e.g. Newman and Raju [6], differ slightly. To account for this, the material parameters C_A and C_C may be taken as different values. It is proposed, Shang-Xian [5], Newman and Raju [6], that C_A and C_C are related by the formula:

$$C_A = 1.1^m C_C$$

The material property m depends mainly on the fatigue crack propagation, Shang-Xian [6]. It is therefore reasonably assumed to be independent of the crack size, both in the depth and surface directions.

Normally the failure criterion refers to a critical value of the crack depth a or the crack length c. Therefore the equations are conveniently rewritten as:

$$\frac{dc}{da} = \frac{C_C}{C_A}\left[\frac{\Delta K_C}{\Delta K_A}\right]^m \quad ; \quad c(a_0) = c_0 \tag{2.4}$$

$$\frac{dN}{da} = \frac{1}{C_A(\Delta K_A)^m} \quad ; \quad N(a_0) = N_0 \tag{2.5}$$

Alternatively, Eq. (2.4) and (2.5) may be written in terms of c as the independent parameter.

The general expression for the stress-intensity factor (SIF) is $K = Y\sigma\sqrt{\pi a}$, where the geometry factor Y accounts for the effect of all the boundaries, i.e. the relevant dimensions of the structure (width, thickness, crack front curvature etc.). The individual effects of these boundaries can be found in handbooks, and their composite effect is obtained by multiplying all the individual effects. As an example of compounding, Broek [7] lets the various boundary effects be due to back free surface (BFS), front free surface (FFS), width (w) and crack front curvature (CFC). By including the effect of the additional stiffness due to the presence of the weld and the presence of more than one crack, the SIF is written as:

$$K = Y_{BFS} Y_{FFS} Y_w Y_{CFC} Y_{mult} Y_{weld} \sigma\sqrt{\pi a} = Y\sigma\sqrt{\pi a}$$

$$Y = Y_{BFS} Y_{FFS} Y_w Y_{CFC} Y_{mult} Y_{weld}$$

3. STRESS INTENSITY FACTOR (SIF)

3.1 Stress intensity factor equation for a single surface crack in a plane plate

An empirical equation for the SIF $K_I(\phi)$ for a single surface crack in a finite plate subjected to tension and bending is given by Newman and Raju [6]. The equation has been fitted from finite-element results for two different types of loads applied to the surface cracked plate: remote uniform tension and remote bending. The remote uniform tension stress is designated as σ_t and the remote outer-fiber bending stress as σ_b. The SIF equation for combined tension and bending loads is:

$$K_I(\phi) = (\sigma_t + H\sigma_b)\sqrt{\pi\frac{a}{Q}}\,F\left(\frac{a}{t}, \frac{a}{c}, \frac{c}{b}, \phi\right) \tag{3.1}$$

for $0 < \frac{a}{c} \le 1.0$, $0 \le \frac{a}{t} < 1.0$, $\frac{c}{b} < 0.5$ and $0 \le \phi \le \pi$.

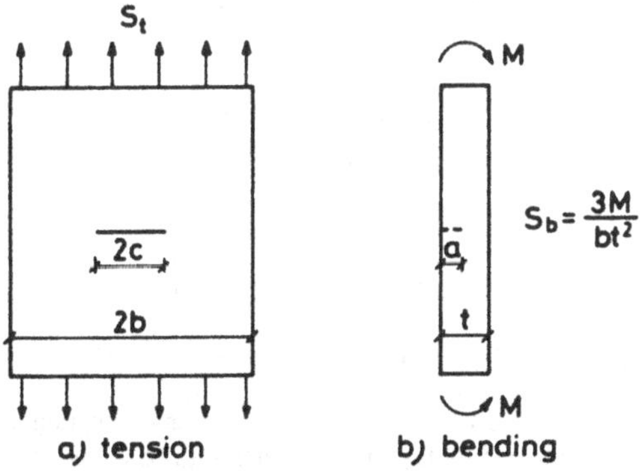

Figure 4: Geometry of a surface-cracked plate to tension or bending loads.

ϕ is the angle that defines the position of the point under consideration, a is the crack depth (semi-minor axis), c is half the crack length (semi-major axis), t is the thickness of the plate, and b is the half width of the panel, see Fig. 4. Q is the shape factor, and F and H define the boundary-correction factors. The Newman-Raju formula, Eq. (3.1), has passed several independent published investigations as the best available parametric formula.

With these functions it is seen, that the solution of the differential equation , Eq. (2.4), depends only on the exponent m, the geometry, the ratio between bending and membrane stress and the initial condition a_0 and c_0, but is independent of the loading magnitude.

A useful approximation for the shape factor Q is given by:

$$Q = 1.0{+}1.464(a/c)^{1.65} \quad ; \quad a/c \leq 1 \tag{3.2}$$

The functions F and H are defined in such a way that the boundary-correction factor for tension is equal to F and the boundary-correction factor for bending is equal to the product of H and F. The following expression for F is recommended by Newman and Raju [6]:

$$F = [M_1{+}M_2(a/t)^2{+}M_3(a/t)^4]\, f(\phi)\, g(\phi)\, f_w \tag{3.3}$$

where

$$M_1 = 1.13{-}0.09(a/c)$$

$$M_2 = -0.54+\frac{0.89}{0.2+(a/c)}$$

$$M_3 = 0.5+14(1-a/c)^{24}-\frac{1}{0.65+a/c}$$

$$g(\phi) = 1+[0.1+0.35(a/t)^2](1-\sin\phi)^2$$

The function $f(\phi)$, an angular function from the embedded elliptical crack solution, is:

$$f(\phi) = [\,(a/c)^2\cos^2\phi+\sin^2\phi]^{1/4}$$

The function f_w, a finite-width correction, is:

$$f_w = [sec(\frac{\pi c}{2b}\sqrt{a/t}\,)]^{1/2}; \quad c/b \leq 0.5$$

The function H taking bending into account, has the form:

$$H = H_1+(H_2-H_1)\sin^p\phi$$

where

$$p = 0.2+(a/c)+0.6(a/t)$$

$$H_1 = 1-0.34(a/t)-0.11(a/c)(a/t)$$

$$H_2 = 1+G_1(a/t)+G_2(a/t)^2$$

$$G_1 = -1.22-0.12(a/c)$$

$$G_2 = 0.55-1.05(a/c)^{0.75}+0.47(a/c)^{1.5}$$

Newman and Raju have investigated several combinations of parameters for which $a/t \leq 0.8$, and in all cases the SIF obtained by Eq. (3.1) was found to be within ± 5% of the finite-element result.

3.2 Correction in stress intensity factor due to crack interaction

For multiple cracks along a weld, the SIF for a single crack is influenced by the presence of other cracks. The expression for the SIF, described in Chapter 3.1, then has to be modified to account for this influence.

At present no analytical solutions covers fully the case of stress intensity factors in welds with multiple cracks at arbitrary locations along a line. Tada, Paris and Irwin [8] suggest several different stress intensity factors for two and more equally spaced through-thickness cracks in an infinite plate, but give no solution for multiple surface cracks or arbitrary distributed cracks.

In the present study, the increase in the SIF due to the presence of multiple cracks, is modelled as the relative increase in the SIF for a through-thickness crack due to the influence of other cracks. This relative increase in SIF is then applied as a multiplication factor on the Newman and Raju SIF in the length direction for the two dimensional semi-elliptical crack:

$$K_{plate} = K_c \left[1 + \sqrt{a/t}\,(\bar{Y}_{mult} - 1) \right] \tag{3.4}$$

where K_{plate} is the SIF for multiple surface cracks in a finite plate, $K_c = K_I(\phi{=}0)$ - the Newman and Raju SIF in the c direction, $\bar{Y}_{mult}$ is the relative increase of the SIF for through-thickness cracks due to presence of other cracks and $\sqrt{a/t}$ accounts the degree of influence from the presence of other cracks as a function of the relative crack depth. No multiplication factor is applied in the depth direction.

Tada, Paris and Irwin [8] give two applicable procedures: Westergaard's method for an infinity number of equally spaced, identically sized through-thickness cracks and Muskhelishvili's method for two identically sized through-thickness cracks.

The relative increase in the SIF according to Westergaard's and Muskhelishvili's method, Fig. 5, is:

$$\bar{Y}_{mult,West.} = \sqrt{\frac{w\,\tan(\pi c/w)}{\pi c}} \tag{3.5}$$

$$\bar{Y}_{mult,Muskh.} = \frac{E(k)/K(k) - (a/b)^2}{a/b\,\sqrt{1-(a/b)^2}} \tag{3.6}$$

where

$$K(k) = \int_0^{\pi/2} \frac{1}{\sqrt{1-k^2\sin^2\lambda}}\,d\lambda$$

$$E(k) = \int_0^{\pi/2} \sqrt{1-k^2\sin^2\lambda}\, d\lambda$$

$a = w/2 - c$, $b = w/2 + c$ and $k = \sqrt{1-(a/b)^2}$.

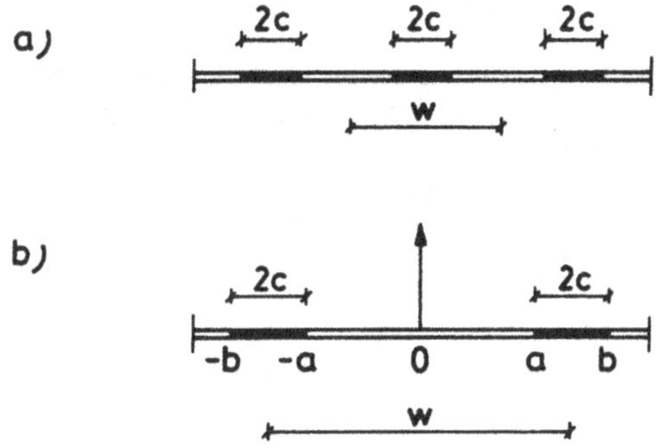

Figure 5: Location of defects. Westergaard's method[a)] and Muskhelishvili's method[b)].

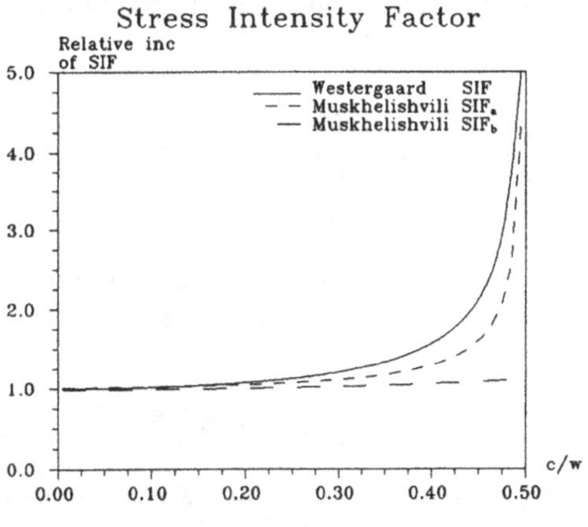

Figure 6: Relative increase in SIF due to the presence of more than one crack.

The relative influence on the SIF for the two different approaches is shown in Fig. 6. The influence of a neighbouring crack is negligible when c/w is less than 0.25, leading to the conclusion that the relative increase in the SIF for multiple cracks primarily is governed by the nearest crack. Consequently, applying the relative increase in SIF due to Westergaard is judged to be too conservative. Instead, a symmetric formulation based on Muskhelishvili's method is applied.

3.3 Stress intensity factor due to presence of weld

For a surface crack resulting from a weld undercut in a welded plate, the SIF is taken as

$$K_{welded\,plate} = K_{plate}\,M_k \tag{3.7}$$

where M_k is a magnification factor accounting for the additional stiffness due to the presence of the weld. A magnification factor based on FEM calculations by Smith and Hurworth [9] on weld defects in butt welded joints is applied:

$$M_k(a) = 1.0 + 1.24e^{-22.1a/t} + 3.17e^{-357a/t} \tag{3.8}$$

The last exponent, $-357a/t$, is so sensitive that the results may appear rather arbitrary for small a/t. For a/t relevant in fatigue reliability analysis, this last term is quite small, and a possible error in the exponent has a limited effect.

The same magnification factor is used independent of the bending/membrane stress ratio.

4. THRESHOLD ON STRESS INTENSITY FACTOR

The crack growth equation, Eq. (2.1), was formulated without a threshold value. Two formulations including a threshold are:

$$\frac{dr(\phi)}{dN} = C_r(\phi)(\Delta K_r(\phi))^m\,; \qquad \Delta K_r(\phi) > \Delta K_{thr} \tag{4.1}$$

$$\frac{dr(\phi)}{dN} = C_r(\phi)[(\Delta K_r(\phi))^m - (\Delta K_{thr})^m]\,; \qquad \Delta K_r(\phi) > \Delta K_{thr} \tag{4.2}$$

The first formulation is the most conservative while the second is proposed in the Fatigue Handbook [10]. The first approach is used in the following.

The SIF $K_I(\pi/2)$ is:

$$\begin{aligned} K_I\left(\frac{\pi}{2}\right) &= (\sigma_t+\sigma_b)\frac{1+H\sigma_b/\sigma_t}{1+\sigma_b/\sigma_t}\sqrt{\frac{\pi a}{Q}}\,F\left(\frac{a}{c},\frac{a}{t},0,\frac{\pi}{2}\right) \\ &= \sigma_s\sqrt{\pi a}\,Y_a(a/t,a/c) \end{aligned} \tag{4.3}$$

where the stress at the surface is $\sigma_s = \sigma_b + \sigma_t$. The bending/membrane ratio σ_b/σ_t is assumed constant and may be determined from a FEM analysis of the actual detail. With $\Delta K_{thr} = 0$, Eq. (2.2) is:

$$\frac{da}{dN} = C_a(\sqrt{\pi a})^m Y_a(a/t,a/c)^m(\Delta\sigma_s)^m\,; \qquad S > 0 \tag{4.4}$$

The long term stress range distribution for $\Delta\sigma_s = S$ is of Weibull type with distribution scale parameters A and shape parameter B. The cumulative distribution function for the stress range is:

$$F_S(s) = 1-\exp(-(s/A)^B)\;, \quad S > 0 \tag{4.5}$$

The factor S^m varies from cycle to cycle, and the expected value is used in the differential equation which becomes:

$$\frac{da}{dN} = C_a(\sqrt{\pi a})^m Y_a(a/t,a/c)^m A^m \Gamma\left(1+\frac{m}{B}\right) \tag{4.6}$$

where $\Gamma(\,)$ is the Gamma function.

Based on the same procedure Eq. (4.1) and (4.2) become for $\phi = \frac{\pi}{2}$, respectively:

$$\frac{da}{dN} = C_a(\sqrt{\pi a})^m Y_a(a/t, a/c)^m A^m \Gamma\left[1+\frac{m}{B}; \left[\frac{\Delta K_{thr}}{AY_a(a/t, a/c)\sqrt{\pi a}}\right]^B\right] \tag{4.7}$$

$$\frac{da}{dN} = C_a\left[(\sqrt{\pi a})^m Y_a(a/t, a/c)^m A^m \Gamma\left[1+\frac{m}{B}; \left[\frac{\Delta K_{thr}}{AY_a(a/t, a/c)\sqrt{\pi a}}\right]^B\right]\right.$$
$$\left. -(\Delta K_{thr})^m \exp\left[-\left[\frac{\Delta K_{thr}}{AY_a(a/t, a/c)\sqrt{\pi a}}\right]^B\right]\right] \tag{4.8}$$

where $\Gamma(\, ; \,)$ is the incomplete Gamma function.

The effect of using two different threshold models, Eq. (4.1) and (4.2), have been studied in Hansen and Madsen [11].

5 SOLUTION STRATEGY

Since the SIFs $K_A = K_I(\pi/2)$ and $K_C = K_I(0)$ depend on the crack size in a very complicated manner, it is not possible to obtain an analytical closed form solution to the coupled differential equations, Eq. (2.2) and (2.3). Numerical integration techniques therefore have to be applied in the solution procedure. In the present study the integration routine LSODA, provided in the NALIB library, is used. LSODA solves a number of coupled ordinary first order differential equations using the Livermore strategy. A complete description of LSODA is given in [12] and the routine has been extensively tested for many different problems.

For a given set of the Poisson distributed locations of the initial defects along the weld seam, the solution of the individual cracks are performed in a successive manner using Eq. (2.2) and (2.3). All cracks are solved simultaneously for a specified time increment, and the crack growth formulations are used in order to evaluate the crack growth of the individual cracks during this time increment. After each of these time increments it is checked whether or not coalescence of any of the cracks along the weld seam has occurred. The specified time increment in each iteration is calculated using a second order Taylor expansion for the crack growth of two neighbouring cracks. The estimated lowest number of cycles resulting in coalescence is applied as the time increment for the next evaluation of crack growth of the individual cracks. The number of cycles estimated from the Taylor expansion always yields a lower bound on the estimated number of cycles to coalescence.

Each of the cracks are assigned to an individual set of initial conditions, e.g. initial crack depth a_0 and aspect ratio a_0/c_0. Although no crack initiation period is included in the present analysis, the formulation used gives the possibility of assigning an individual crack initiation period to each of the cracks.

When two cracks coalesce, it is assumed that a new semi-elliptical crack configuration is immediately created. The new semi-elliptical crack is defined as a crack having a crack length equal to the sum of the lengths of the two coalescing cracks and a crack depth defined from equality in the crack-size-area before and after coalescence.

The solution strategy has been implemented as a limit state function in the reliability analysis program PROBAN [13, 14].

6. SAFETY MARGIN

The failure criterion is defined as the crack depth exceeding a critical crack depth a_C, equal to the plate thickness t, during the design life period of the structure having n_L load cycles. This failure criterion can be defined through a safety margin M given by:

$$M = a_C - a(n_L) \tag{6.1}$$

where failure takes place when $M \leq 0$. The failure probability P_F is:

$$P_F = P(M \leq 0) \tag{6.2}$$

7. NUMERICAL EXAMPLE

In order to show the effect of crack coalescence, a two meter long weld seam is analysed, Fig. 7. The initial defects are assumed Poisson distributed with defect intensity ν along the weld seam, where the initial crack depths and aspect ratios are independent and identically distributed. It is assumed, that all the cracks over the weld seam are exposed to the same stochastic loading process and are having identical material parameters.

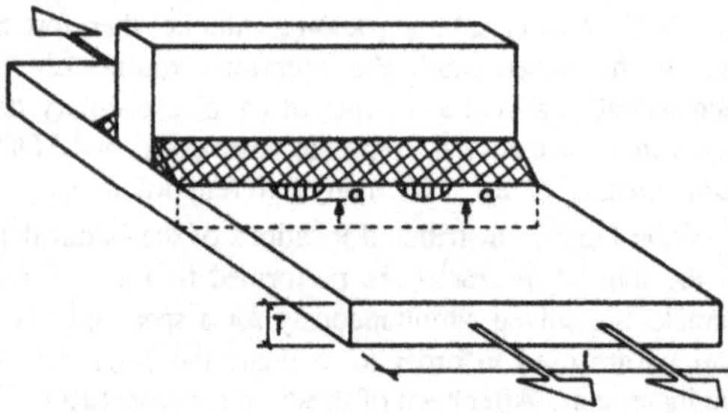

Figure 7: Weld seam investigated in the analysis.

The numerical data applied in the analysis are suggested based on literature review, see Table 1.

In the probabilistic study of the effect of crack coalescence, two different approaches are considered, a 'Hot-Spot' approach and a 'Series System' approach. The 'Hot-Spot' approach directly yields the change in the reliability level of a specified defect solely due to the effect of coalescence. The 'Series System' approach yields the change in the reliability level of the total weld seam, partly due to the effect of coalescence on the reliability level of the separate defects along the weld, and partly due to the reduction of the number of defects along the weld seam due to coalescence of cracks.

For the 'Hot-Spot' approach, an initial defect is present at the center of the weld, having additionally Poisson distributed defects on both sides. Failure is defined as through-thickness growth of the investigated crack at the 'Hot-Spot'. The calculated failure probability of the 'Hot-Spot' crack for a specified defect intensity, is found through successive full-distribution-reliability-method evaluations (FORM) [15], using PROBAN [13, 14] with simulated outcomes of the Poisson distributed defect locations. The failure probability is approximated by:

$$P(M \leq 0 \mid \nu) \approx \sum_{i=1}^{n_C} P(M \leq 0 \mid \nu, C_i) \frac{1}{n_C} \tag{7.1}$$

TABLE 1. Suggested distributions for basic variables.

Variable	Distribution	Comment
Initial crack depth a_0	Exponential $E[a_0] = 0.11$ mm	Bokalrud and Karlsen [16]
Initial aspect ratio a/c	Lognormal $E[a/c] = 0.3$; $COV = 0.50$	selected
Critical crack depth a_C	fixed $a_C = 20.0$ mm	Thickness of plate
Number of cycles n_L	fixed $n_L = 2.0 \cdot 10^7$	selected
Length of weld seam l	fixed $l = 2.0$ m	selected
Defect configuration	Poisson defect intensity ν m^{-1}	selected
Crack growth parameter m	fixed $m = 3.1$	DnV [17]
Crack growth parameter C_a	Lognormal $E[ln(C_a)]=-29.8$ $D[ln(C_a)]=0.55$	DnV [17]
Threshold ΔK_{thr}	fixed $\Delta K_{thr} = 3.0\ MPa\sqrt{m} = 95 Nmm^{-3/2}$	Fatigue Handbook[10]
Model uncertainty SIF Y_1	Lognormal $E[Y_1] = 1.0$; $COV = 0.10$	selected
Weibull parameter A	Normal $E[A] = 7.0$; $COV = 0.12$	selected
Weibull parameter B	fixed $B = 0.7$	selected
Bend./memb. ratio σ_b/σ_t	fixed $\sigma_b/\sigma_t = 4.0$	selected

where C_i is a simulated outcome of defect locations along the weld and n_C is the number of simulated events. In Fig. 8a, the mean value of the estimated reliability index, including the 90% confidence interval, is shown after n_C=20 simulations for different defect intensities. Defect intensity ν=0, represents the limiting case of only one crack present. For increasing defect intensities, the results directly yield the change in the estimated reliability index solely due to the effect of crack coalescence.

In the 'Series System' approach, failure is defined as through-thickness crack growth of any of the Poisson distributed defects along the weld seam. For this approach it is not possible to apply FORM, due to the form of the limit state function. The results were therefore obtained through simultaneous Monte Carlo simulations of defect locations and the stochastic parameters involved in the fatigue crack growth analysis. In Fig. 8b, the estimated reliability index for the 'Series System' of the two meter weld seam is presented with 90% confidence interval after 20.000 simulations. The results are compared to the reliability index of the weld neglecting the effect of crack coalescence, estimated by:

$$P_{F\,weld}(\mathbf{x} \mid \nu) = \sum_{n=0}^{\infty} P_{F\,sys}(\mathbf{x} \mid n)\, P_N(n \mid \nu) = \sum_{n=0}^{\infty} P_{F\,sys}(\mathbf{x} \mid n)\, \frac{(\nu l)^n}{n!} e^{-\nu l} \tag{7.2}$$

where the system reliability is equal to [18]:

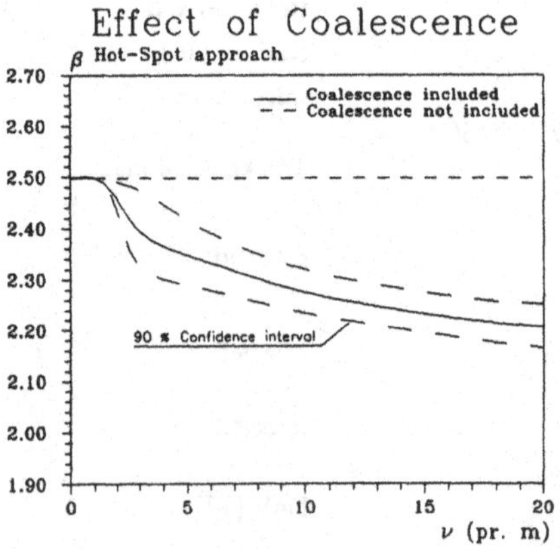

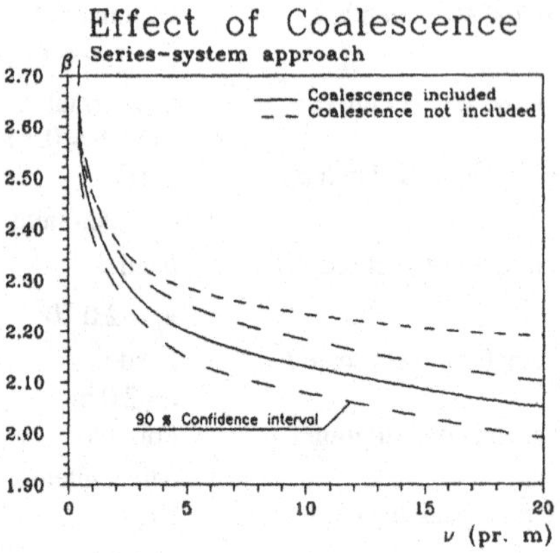

Figure 8: Estimated reliability index for the 'Hot-Spot' approach[a)] and the 'Series System' approach[b)]

$$P_{F_{sys}}(\mathbf{x} \mid n) = 1 - \int_{-\infty}^{\infty} \phi(t)\,[\Phi(\frac{\beta_e - \sqrt{\rho}t}{\sqrt{1-\rho}})]^n\, dt \tag{7.3}$$

ν is the defect intensity, β_e is the reliability index of a single defect. ρ is the common correlation coefficient between any pair of safety margins in the standard normal space:

$$\rho = \rho[M_i, M_j] = \alpha^T{}_i\, \alpha_j \quad ; \qquad i \neq j \tag{7.4}$$

where α_i is the unit normal vector to the failure surface at design point for component i, in the transformed standard normal space. In the present analysis, the different safety margins are highly correlated, $\rho = 0.96$, due to the large importance of the common loading and material parameters, and the less influence of the independent initial crack sizes and aspect ratios.

Coalescence of two separate cracks leads to a crack with a lower aspect ratio, and as shown in Fig. 8a, further leads to a more rapid fatigue crack growth in the depth direction, and consequently a decrease in the reliability index for the crack. From Fig. 8b, it is seen that the neglection of coalescence has less influence on the estimated reliability for the 'Series System' approach than for the 'Hot-Spot' approach. This is due to the fact that coalescence leads to a significant decrease in the total number of defects for higher defect intensities, and consequently a reduction of the series system effect.

Figure 9 shows the propagation of the aspect ratio for the 'Hot-Spot' as a function of the relative crack depth for defect intensities $\nu = 3,5,10$ and 20. The aspect ratios are obtained at the design point, and plotted for all of the 20 simulations of the defect configurations. The results indicate a quite large variations of the aspect ratios due to the effect of the coalescence of two or more defects in the initial phase of the crack growth. The aspect ratios are, however, approaching towards a stable relationship for larger crack depths.

Figure 10 shows the estimated joint density function of the crack geometry at design life, $f_{AC}(a,c)$, for different initial defect intensities. It is seen that an increase in the defect intensity leads to shallower crack configurations.

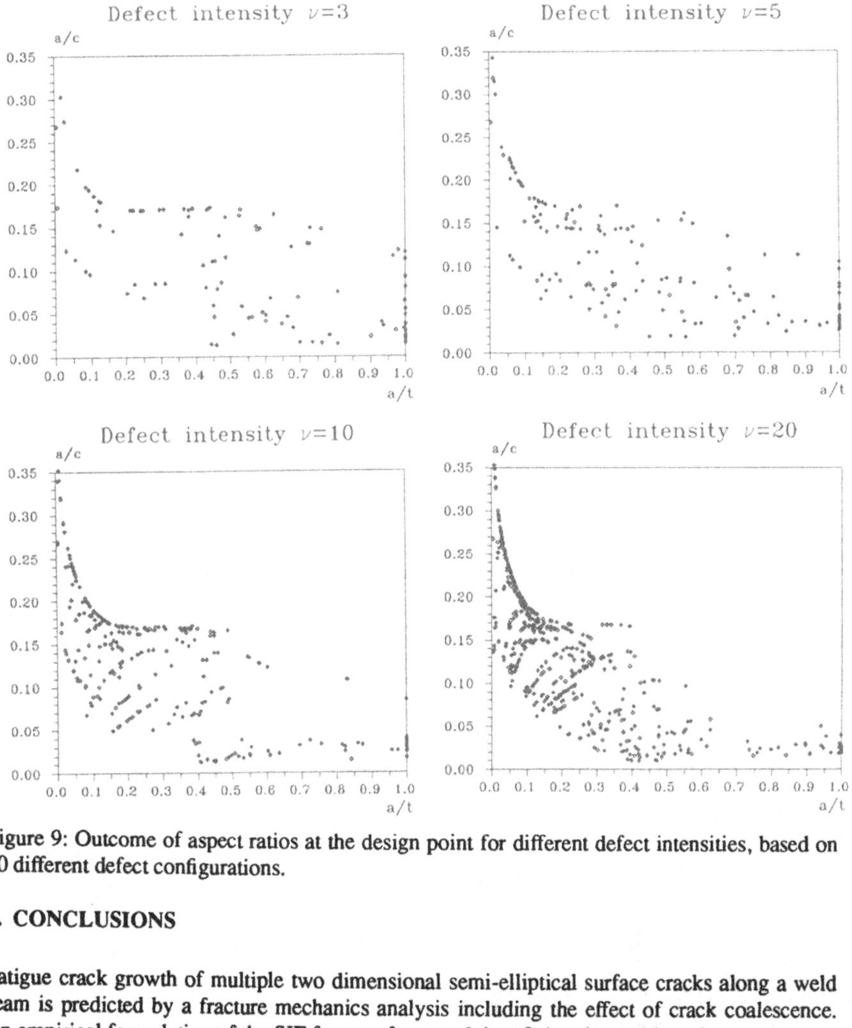

Figure 9: Outcome of aspect ratios at the design point for different defect intensities, based on 20 different defect configurations.

8. CONCLUSIONS

Fatigue crack growth of multiple two dimensional semi-elliptical surface cracks along a weld seam is predicted by a fracture mechanics analysis including the effect of crack coalescence. An empirical formulation of the SIF for a surface crack in a finite plate subjected to tension and bending is applied. A modification factor accounting for the presence of other cracks and the additional stiffness due to the presence of the weld is included. Two formulations of including a threshold value on the SIF are described.

A safety margin for crack growth through the thickness is defined, and the corresponding failure probability as a function of service time is computed by full-distribution-reliability method and ordinary Monte Carlo simulation technique.

It is seen that the effect of coalescense of two separate cracks leads to a more rapid fatigue crack growth in the depth direction, and consequently a lower reliability index for the defect. The neglection of coalescence has less influence on the estimated reliability for the total weld seam than for a specified 'Hot-Spot'. This is due to the fact that coalescence of cracks leads to a smaller number of cracks over the specified weld length, which reduces the series system effect.

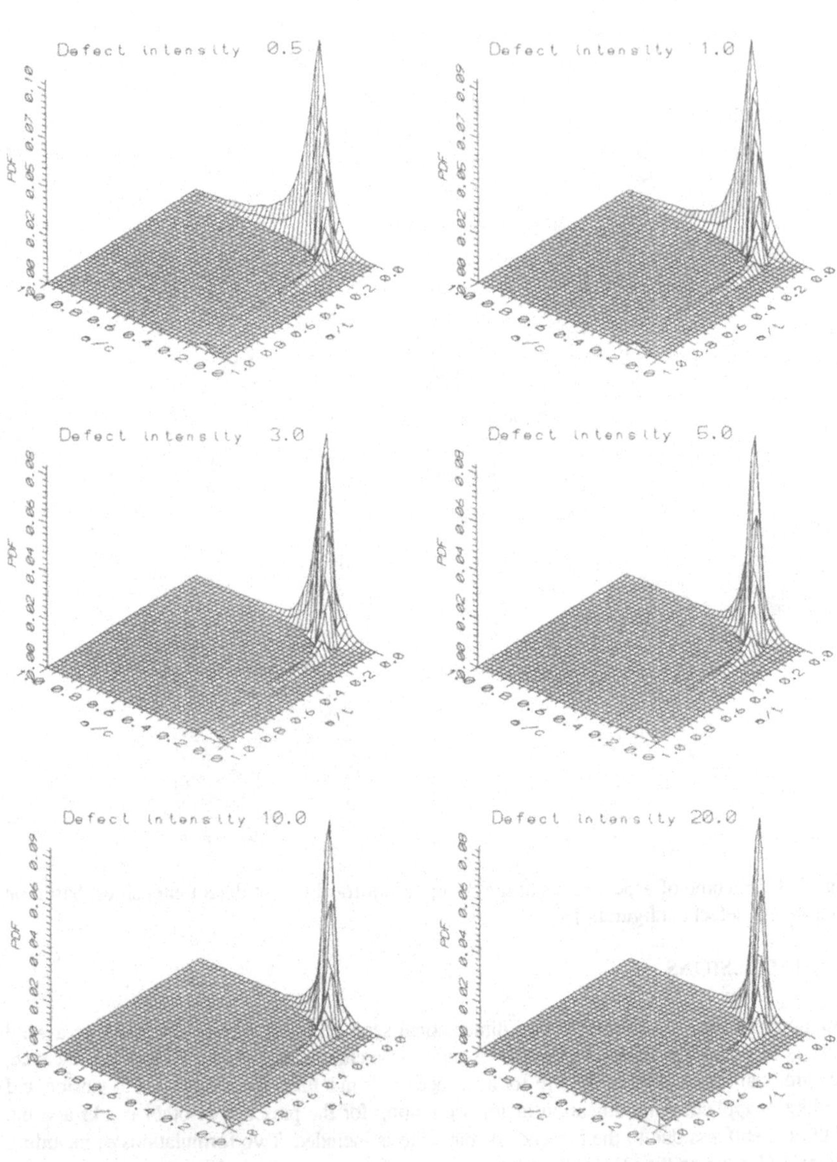

Figure 10: Estimated joint density function $f_{AC}(a,c)$ at design life for different defect intensities.

The analysis indicates that the effect of coalescence results in a significant decrease in the reliability index for higher defect intensities.

It is concluded that it is of importance to include the effect of coalescence in the fatigue reliability analysis. Further research has to be performed in order to obtain a better formulation of the SIF.

The present formulation is very computer time consuming. A calibration of a simpler probabilistic model to the one presented would therefore be desirable for practical applications.

9. ACKNOWLEDGEMENT

The work of Mr. Espen H. Cramer has been supported by the Danish Research Academy.

10. REFERENCES

[1] H.O.Madsen: 'Deterministic and probabilistic models for damage accumulation due to time varying loading', Dialog, Danish Engineering Academy, No 5, 1982.

[2] D.Broek: 'Elementary engineering fracture mechanics', Martinus Nijhoff Publishers, Dordrecht, 1986.

[3] T.Lassen: 'The Effect of the Welding Process on the Fatigue Crack Growth in Welded Joints', Agder Regional College of Engineering, Grimstad, Norway, 1990 (manuscript to be published in Welding Journal).

[4] H.O. Madsen: 'Stochastic Modelling of Fatigue Crack Growth', Lecture notes for ISPRA Course: Structural Reliability, November 13-17, 1989, Lisboa.

[5] Wu Shang-Xian: 'Shape Change of Surface Crack during Fatigue Growth', Engineering Fracture Mechanics, Vol 22, No 5, pp. 897-913, 1985.

[6] J.Newman and I.Raju: 'An empirical stress-intensity factor for the surface crack', Engineering Fracture Mechanics, Vol 22, No 6, pp. 185-192, 1981.

[7] D.Broek: 'The Practical Use of Fracture Mechanics', Kluwer Academic Publishers, The Netherlands, 1988.

[8] H. Tada, P.C. Paris and G.R. Irwin: 'The Stress Analysis of Cracks Handbook', Del Research Corporation, St. Louis, Missour.

[9] I.J.Smith and S.J.Hurworth: 'Probabilistic Fracture Mechanics Evaluation of Fatigue from Welds Defects in Butt Welded Joints' Fitness for Purpose Validation of Welded Constructions, London 17-19 November 1981

[10] Fatigue Handbook, Offshore Steel Structures, Ed by A.Almar Næss, Tapir, 1985.

[11] P.Friis Hansen and H.O.Madsen: 'Stochastic Fatigue Crack Growth Analysis Program', Danish Engineering Academy, 1990.

[12] NALIB: fortran subroutines library.

[13] PROBAN-2: 'Users manual', A.S Veritas Research Report No. 89-2024.

[14] PROBAN-2: 'Example manual', A.S Veritas Research Report No. 89-2025.

[15] H.O.Madsen, S.Krenk and N.C.Lind: 'Methods of Structural Safety', Prentice-Hall Inc., Englewood Clifs, New Jersey, 1986.

[16] T.Bokalrud and A.Karlsen: 'Probabilistic Fracture Mechanic Evaluation of Fatigue from Weld Defects in Butt Welded Joints', Fitness for purpose Validation of Welded Construction, London 17-19 November, 1981.

[17] Det norske Veritas (DnV): 'Rules for the Design, Construction and Inspection of Offshore Structures, Appendix C Steel Structures, Reprint with corrections, 1982.

[18] C.W. Dunnett and M. Sobel: ' Approximations to the Probability Integral and Certain Percentage Points of Multivariate Analogue of Students' t-Distribution', Biometrica, Vol.42, 1955, pp.258-260.

Stochastic Fatigue Damage in Welded Cruciforms

S. Sarkani (*), D.P. Kihl (**), J.E. Beach (**)
(*) School of Engineering and Applied Science, The George Washington University, Washington, DC, U.S.A.
(**) David Taylor Research Center, Bethesda, MD U.S.A.

ABSTRACT

Twenty high strength welded steel cruciforms have been tested under narrowband loadings to determine the effect of the root-mean-square (rms) of the loading process on the rate of fatigue damage accumulation. Comparison of experimental results with Rayleigh approximation technique, indicates that for loadings with relatively high rms values (compared to the specimen's yield strength) the analytical predictions are about twice as long as the experimental fatigue lives. For loadings with intermediate rms values, the analytical predictions are generally in good agreement with experimental results. For loadings with low rms values, the analytical predictions can be twice as long as the experimental results depending on the interpretation of constant amplitude test results in the high cycle region.

INTRODUCTION

The problem of variable amplitude fatigue life prediction of structural members is not new, but many issues still remain unclear. Perhaps the main reasons are the complexity of welded members (in terms of geometry, residual stresses, initial flaw size, etc.) and the variety of service loadings that such members are subjected to. In practice, because of these difficulties, fatigue life under service loads is normally predicted on the basis of constant amplitude laboratory test results. Such constant amplitude tests are usually run at two or more stress levels and the results are plotted as the so-called S/N curve. The variable amplitude service life is then predicted by somehow identifying the variable amplitude cycles and using a damage model, usually the Palmgren-Miner (P-M) hypothesis (Palmgren 1924, Miner 1945), to find the damage due to each cycle. The cumulative total of all cyclic damages is then used to predict service life. There has been a widely held belief that this procedure results in conservative fatigue life predictions for welded members.

However, recently some experimental studies have concluded that the use of the P-M damage model may be non-conservative for welded members subjected to some class of variable amplitude loading. Specifically, the study by Sarkani and Lutes (1988) revealed that for some welded members subjected to narrowband loadings, the experimental fatigue life was only half as long as the predicted theoretical life. This is somewhat surprising since narrowband loadings are closer to constant amplitude loadings than other variable amplitude loadings and therefore one would expect better agreement between experimental and analytical results for such loadings. The loadings used by Sarkani and Lutes had very high root-mean-square (rms) values and there was some concern as to whether this had any influence on the results.

In order to better understand the fatigue damage accumulation under narrowband variable amplitude loadings and also to determine whether the P-M damage model can give consistent predictions for such loadings, an experimental study was undertaken. Five sets of variable amplitude loadings with various rms values were used. The highest rms value, 22.5 ksi (155 MPa), was one-fourth of the nominal yield stress of the specimens, which is consistent with the loadings used in the study by Sarkani and Lutes (1988). The lowest rms value, 5 ksi (34 MPa), was two-ninths of the highest rms value. All the loadings were narrowband, where the typical problem of cycle identification is not encountered. In order to compare experimental results with analytical predictions, constant amplitude tests were also performed to establish the S/N curve for these cruciforms. The analytical fatigue life predictions were obtained using the constant amplitude test results and the P-M damage model to calculate the damage in a variable amplitude cycle and the Rayleigh approximation technique (Miles, 1954) to identify the stress amplitude in the individual cycles. Comparison of the experimental and analytical results indicate that the P-M damage model is generally non-conservative for loadings having high rms levels, and produces good estimates for loadings with intermediate rms values. For loadings with low rms values, depending on how the constant amplitude results are interpreted, the P-M model gives predictions that can be twice as long as the experimental results.

EXPERIMENTAL SETUP

The basic configuration and dimensions of the test specimens were as shown in Figure 1. These cruciforms were made from A-710 Grade A, Class 3 steel plate, as rolled 7/16 inches (1.1 cm) thick. The yield stress and ultimate stress of the steel plate were 92.5 ksi (638 MPa) and 99.0 ksi (683 MPa), respectively, in the direction transverse to the roll direction. The horizontal stems were attached to the mid-section of the vertical leg by full penetration GMAW fillet welds. The cruciforms were loaded axially in the vertical direction with loads applied to the ends of the vertical legs by means of hydraulic grips. Tests were run in closed loop servo-hydraulic test machines. All specimens were tested in load control mode.

Due to the presence of stress concentrations and residual stresses at the weld toe, the fatigue cracks typically began at the toe of the welds at the center of the specimen and grew out toward the edges. Failure was defined to be when the flexibility of the specimen exceeded two times that at the initiation of the test. When the flexibility reached this value, cracks were growing so rapidly that total fracture of the specimen was imminent.

For the variable amplitude tests, a sequence of 10,000 random amplitudes were generated on a computer. These random amplitudes were used as peaks and valleys of the variable amplitude loading. An analog waveform synthesizer was used to produce an analog voltage having these peaks and valleys connected by smooth haversine curves. This loading block was repeated as many times as necessary to fail the specimen.

TEST LOADING

The variable amplitude tests in this study were designed such that the rate of fatigue damage accumulation under narrowband loadings could be investigated. Narrowband Gaussian loadings are somewhat different than other stochastic loadings in that the extrema are very nearly Rayleigh distributed and neighboring extrema are highly correlated. Therefore it is possible to identify a narrowband Gaussian loading by a single bandwidth parameter, namely, the rms value. It must also be noted that since narrowband loadings are closer to constant amplitude than most other stochastic loadings it is essential to first understand the effectiveness of the P-M damage model and the widely used cycle identification techniques for such loadings before any attempt is made to use these models for more complicated loadings. Therefore experiments with narrowband loadings were performed. In order to minimize other effects, all the loadings were identical in terms of distribution and correlation of extrema, except that the loading rms was varied from one set of tests to another.

The sequence of random extrema in this study were efficiently generated by a first-order autoregressive technique (Sarkani 1990). This technique directly generates Rayleigh distributed random variables, which were then used as the time history extrema. The even numbered extrema were used as peaks and the odd numbered ones were multiplied by -1 and used as valleys. Each neighboring peak and valley was then joined by a smooth haversine curve. Processes of this type have the same number of peaks as they do zero up-crossings i.e., they have unit irregularity factor (ratio of zero crossings rate to extrema rate). The set of extrema that were used in this study had adjoining (preceding or following) valley correlation values, ρ, of -0.95. Each set of experiments consisted of testing four specimens to failure under identical loadings.

For each loading, these extrema were multiplied by a constant multiplier so that the loading process would have the desired rms value. The five variable amplitude loadings had rms values of 22.5, 15.0, 10.0 ,7.5 and 5.0 ksi (155, 103, 69, 52, and 34 MPa), respectively.

The actual stress at any particular location within the specimen under a given load is not precisely known. Nonetheless, it is important to use some nominal stress in order to gain a sense of the appropriate magnitude for the testing loads. The nominal stress used here was calculated by simply dividing the load by the cross sectional area of the vertical leg; i.e., nominally 1.75 in^2 (11.29 cm^2). The maximum load during each variable amplitude test was chosen to be four times the loading rms value. This resulted in the maximum nominal stress under the loading with the largest rms to be approximately equal to the nominal yield stress of the base metal. When the actual stochastic simulation produced extrema that were larger than four times the RMS value, then these extrema were clipped to the chosen level. Such clipping occurred no more frequently than about one in each 3,000 extrema.

The simulated extrema were carefully examined to assure that they did have the desired distribution and correlation. In fact each of the first ten moments of the unclipped extrema were within 0.3% of the theoretical moment values. After clipping, the first nine moments were within 4% of the theoretical moment values and the tenth moment was within 6%.

VARIABLE AMPLITUDE TEST RESULTS

As discussed previously, five sets of variable amplitude tests were performed. Each set consisted of testing four specimens to failure under identical loading. This resulted in running a total of twenty specimens under variable amplitude loadings. Results of all the variable amplitude fatigue tests are shown in Table 1. Also presented in Table 1 are geometric means for each set of variable amplitude loadings. As expected the geometric mean of fatigue lives increases as the loading rms decreases.

It is now useful to compare the experimental results against analytical predictions. In order to properly analyze the results of variable amplitude tests and to come up with analytical predictions, it is necessary to know the constant amplitude S/N curve for these cruciforms shown in Table 2. Such constant amplitude tests have been performed and Kihl et. al (1991) give an analysis and explanation of those results. Only the results of that analysis will be presented here. In particular it was concluded that a single line S/N curve described by

$$N=10^{10.714}S^{-4.020} \qquad \text{Equation (1)}$$

(where N is the cycles to failure, and S is the applied stress amplitude in ksi) and a bi-linear S/N curve described by

$$N=10^{10.267}S^{-3.744} \qquad for\ S>11.8\ ksi$$

$$=10^{17.402}S^{-10.408} \qquad for\ S<11.8\ ksi \qquad \text{Equation (2)}$$

are the most appropriate S/N curves to be used for these cruciforms.

Since the loadings used in this study are fairly narrowband, the only analytical method used is the Rayleigh approximation. In this technique, the number of cycles is taken to be the number of zero up-crossings and the average damage per cycle is calculated by taking each stress amplitude to be a Rayleigh distributed amplitude for the loading process. The P-M damage hypothesis is used to justify simply adding these incremental damage values to obtain the total damage. Since all the variable amplitude loadings in this study have identical rate of zero up-crossings, the damage calculated from one set of tests to the other would only vary as a function of the rms value.

The Rayleigh approximation results in the following expression for the average number of cycles to failure

$$N_{ave} = \frac{K\, 2^{-m/2}\, \sigma_x^{-m}}{\Gamma(m/2+1)} \qquad Equation\ (3)$$

in which K and m are the slope and itercept of the S/N curve in a Log-Log plot. σ_x is the rms of the loading process and $\Gamma(\bullet)$ denotes the Gamma function.

$$\Gamma(a) = \int_0^{\infty} t^{a-1} e^{-t} dt \qquad Equation\ (4)$$

Equation (3), must be slightly modified before it can be used to predict the theoretical fatigue life of the cruciforms. The modification results from the fact that the variable amplitude loadings were truncated at four times the rms level. Modification of Equation (3), results in the following expression

$$N_{ave} = \frac{K\, 2^{-m/2} \sigma_x^{-m}}{\Gamma(m/2+1) - \Gamma(m/2+1, 8)} \qquad Equation\ (5)$$

in which $\Gamma(a,b)$ denotes the complementary incomplete Gamma function (Abramowitz and Stegun, 1965).

$$\Gamma(a,b) = \int_b^{\infty} t^{a-1} e^{-t} dt \qquad Equation\ (6)$$

Note that in order to use the bi-linear S/N curve of Equation (2) to predict the variable amplitude fatigue lives, Equation (5) must be further modified as follows:

$$N_{ave} = \frac{1}{\dfrac{\gamma(m_2/2+1, S_0^2/2\sigma_x^2)}{K_2 2^{-m_2/2} \sigma_x^{-m_2}} + \dfrac{\Gamma(m_1/2+1, S_0^2/2\sigma_x^2) - \Gamma(m_1/2+1, 8)}{K_1 2^{-m_1/2} \sigma_x^{-m_1}}} \qquad Equation\ (7)$$

in which the subscripts 1 and 2 on K and m represent parameters from the two portions of the bi-linear S/N curve, and $\gamma(a,b)$ denotes the incomplete Gamma function (Abramowitz and Stegun, 1965). The variable S_0 in

$$\gamma(a,b) = \int_0^{b} t^{a-1} e^{-t} dt \qquad Equation\ (8)$$

Equation (7) denotes the stress amplitude at which the two portions of the bi-linear S/N curve intersect.

Table 3 presents the variable amplitude fatigue life predictions based on the modified Rayleigh approximation method using the S/N curve of Equation (1). Also shown in Table 3 are the geometric mean of the experimental fatigue lives for variable amplitude loadings and the theoretical fatigue life predictions using the bi-linear S/N curve of Equation (2).

Examination of the results in Table 3 indicates that for the loadings with the highest rms value (22.5 ksi, 155 MPa) the geometric mean of experimental results is only 57% to 60% of the analytical predictions depending on the choice of the S/N curve. These results are consistent with the earlier results obtained by Sarkani and Lutes (1988) that used similar loadings but different specimen geometry. For loadings with intermediate rms values(15.0 and 10.0 ksi, 103 and 69 MPa, respectively), the experimental results are in generally good agreement with analytical predictions using either S/N curves. For the loading with the 7.5 ksi (52 MPa) rms, the geometric mean of the experimental results is about 63% to 74% of the analytical predictions depending on the choice of the S/N curve.

Finally, for the loadings with the lowest rms value the geometric mean of experimental results is only 41% to 47% of the analytical prediction using the bi-linear and single line S/N curves respectively. This is somewhat disturbing since loadings that have low rms levels can be typical of those encountered in engineering design applications.

It must also be noted that for the loading with the 5 ksi (34 MPa) rms value, 86% of the cycles have amplitudes lower than 10 ksi (69 MPa). In other words, 86% of the cycles have amplitudes below the lowest level of constant amplitude test data. It is therefore important to investigate what fraction of the total fatigue damage is caused by these cycles.

For Rayleigh distributed amplitudes, the fraction of total damage caused by cycles having stress amplitudes between "0" and "s" (using the Rayleigh approximation procedure and the P-M damage model) can be obtained (Equation 9) in which $f_S(s)$ is the probability density function of the Rayleigh distributed amplitudes. An expression for the intensity of damage caused by a Rayleigh distributed amplitude "s" can be obtained by differentiating Equation (9)

$$D(s) = \frac{\int_0^s s^m f_S(s)\,ds}{\int_0^\infty s^m f_S(s)\,ds} \qquad \textit{Equation (9)}$$

with respect to "s".

$$D'(s) = \frac{\left(\frac{s}{\sigma_x}\right)^{m+1} e^{-\frac{1}{2}\left(\frac{s}{\sigma_x}\right)^2}}{2^{m/2}\Gamma(m/2+1)} \qquad \textit{Equation (10)}$$

Note that both Equations (9) and (10) were derived assuming a linear S/N curve on a Log-Log paper (as opposed to bi-linear or other forms). Using Equation (10) it can easily be shown (using a single line S/N curve with m=4.02) that about 30% of the damage comes from cycles having stress amplitudes below two times the rms value. Therefore, for the loadings with the 5 ksi (34 MPa) rms value, about 30% of the damage is presumed to have been caused by the 86% of the cycles having stress amplitudes below the lowest stress constant amplitude test data (near where the "endurance limit" appears to be located). This finding raises some questions about the validity of the variable amplitude fatigue life predictions for the loading with the 5 ksi (34 MPa) rms value. Furthermore, results observed by Keating and Fisher (1989), indicate that under variable amplitude loadings, the constant amplitude "endurance limit" vanishes as long as at least one of the variable amplitude loading cycles has an amplitude that is larger than the constant amplitude "endurance limit".

In order to check this scenario, a single line S/N curve was obtained using the four highest stress level test data. This in effect ignores the apparent existence of the "endurance limit" seen at low constant amplitude stress levels. Using this S/N curve (Log(K)=10.267, m=3.744) variable amplitude fatigue life predictions were obtained such that the geometric mean of the experimental results are 68%, 82%, 118%, 104% and 58% of the analytical predictions for the 5, 7.5, 10, 15, and 22.5 ksi (34, 52, 69, 103, and 155 MPa) rms loadings, respectively. Considering an S/N curve (Log(K)=9.737, m=3.349) based on only the constant amplitude data from stress levels 15, 30, and 45 ksi (103, 207, 310 MPa), where the constant amplitude S/N curve exhibits the least amount of scatter, results in even slightly better agreement with experimental results. In this case, the geometric mean of the experimental results are 90%, 93%, 120%, 89%, and 42% of the analytical predictions for the 5, 7.5,

10, 15, and 22.5 ksi (34, 52, 69, 103, and 155 MPa) rms loadings, respectively.

Even though the variable amplitude fatigue life predictions using these S/N curves are closer to the experimental results than the previous predictions, it is not clear as to which one of the four S/N curves is the most appropriate one to use in practice. Perhaps if variable amplitude test data for rms values less than 5 ksi (34 MPa) were available one could obtain stronger evidence regarding the accuracy of the various S/N curves for variable amplitude fatigue life predictions. For instance, for loadings with 3.5 ksi (24 MPa) rms values, the four S/N curves (original linear, bi-linear and two additional linear curves) ignoring any "endurance limit" would predict fatigue lives of approximately 42, 58, 26 and 17 million cycles, respectively. It seems likely that experimental results for this rms value would be fairly valuable in resolving the uncertainty regarding the choice of the appropriate S/N curve for variable amplitude fatigue life predictions. Indeed the authors plan to run experiments with 3.5 ksi (24 MPa) rms values, however, the results will not be available for at least another year or so.

The other experimental data which did not match analytical predictions were those at the highest rms level. For all S/N curve representations considered, analytical estimates for the 22.5 ksi (155 MPa) rms loading are all about the same, being non-conservative by about a factor of two. Observing this type of behavior with loadings having many stress cycles near yield is not too surprising and, as indicated earlier, is consistent with results from other tests conducted by Sarkani and Lutes (1988).

CONCLUSIONS

Examination of the experimental results indicates that as the loading rms decreases the fatigue life increases. Comparison of experimental results with analytical predictions reveals that for loadings with the highest rms value, the analytical predictions are generally greater than the experimental results. For loadings with intermediate rms values the analytical predictions are generally in good agreement with experimental results. For loadings with the lowest rms value the analytic predictions were found to vary from being non-conservative by nearly a factor of two to being in good agreement with experimental results depending on how the constant amplitude test results are interpreted, particularly in the high cycle region, for fatigue damage calculations. Better agreement between variable amplitude fatigue life predictions and experimental results for the lowest rms loading was found using S/N curves that were obtained by ignoring the constant amplitude data around the "endurance limit" of the specimens.

REFERENCES

1. Abramowitz, M., and Stegun, I.A. (editors) (1965)."Handbook of Mathematical Functions," Dover Publications, Inc., New York, NY.

2. Keating, P.B., and Fisher, J.W. (1989). "High-Cycle, Long-Life Fatigue Behavior of Welded Steel Details." Journal of Construction Steel Research, **12**, pp. 253-259.

3. Kihl, D.P., Sarkani, S. and Beach J.E. (1991). "Interim Fatigue Test Results of HSLA-80 Steel Narrowband Loadings," David Taylor Research Center Report, 1991.

4. Miles, J.W. (1954). "On Structural Fatigue Under Random Loading." Journal of Aeronautical Science, pp. 753-762.

5. Miner, M.A. (1945). "Cumulative Damage in Fatigue," Journal of Applied Mechanics, Vol. 12, pp. A159-A164.

6. Palmgren, A. (1924). "Die Lebensdauer Von Kugallagern, "Ver. Deut. Ingr., Vol. 68, pp. 339-341.

7. Sarkani, S., and Lutes, L.D. (1988). "An Experimental Investigation of Fatigue of Welded Joints Under Pseudo-Narrowband Loadings." Journal of Structural Engineering, ASCE, Vol. 114(8), pp. 1901-1916.

8. Sarkani, S. (1990). "Feasibility of Auto-Regressive Simulation Model for Fatigue Studies." Journal of Structural Engineering, ASCE, Vol. 116(9), pp. 2481-2495.

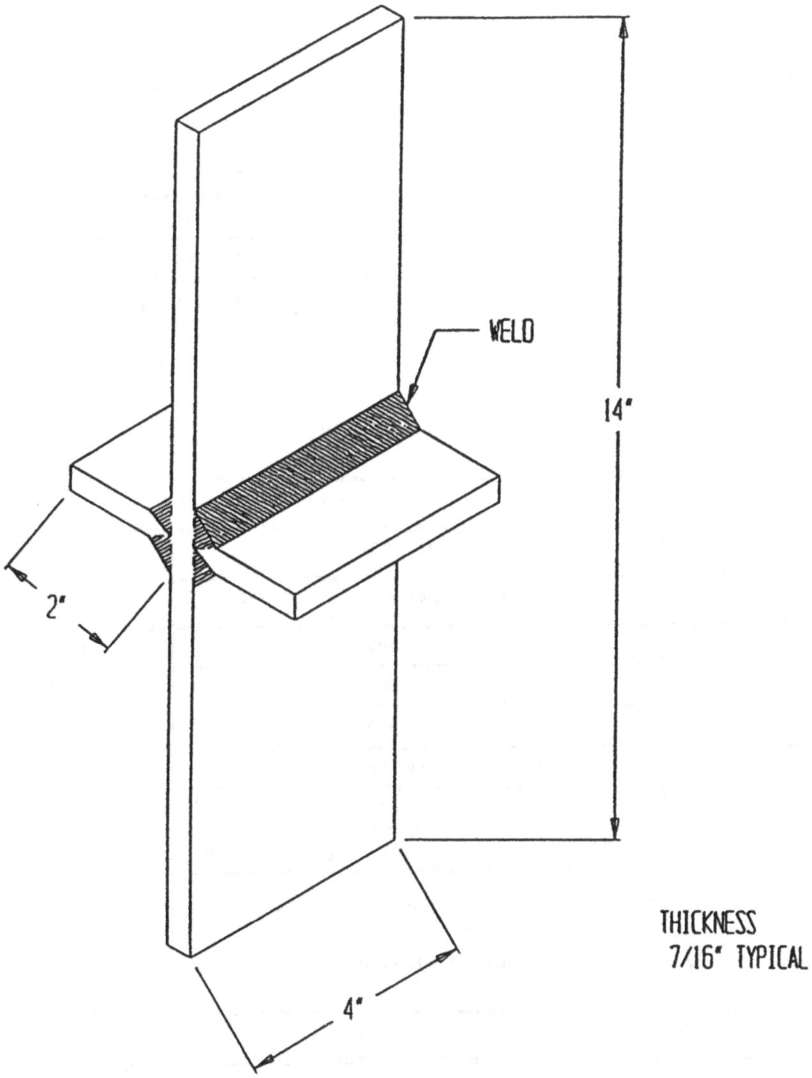

Figure 1: Cruciform Fatigue Test Specimen

Table 1 - Variable Amplitude Test Results

RMS STRESS LEVEL (KSI)	CYCLES TO FAILURE				
	INDIVIDUAL EXPERIMENTS				GEOMETRIC MEAN
	EXP #1	EXP #2	EXP #3	EXP #4	
5.0	2,685,000	5,496,200	7,863,200	4,240,000	4,709,700
7.5	1,504,200	1,111,300	1,178,100	1,216,300	1,244,100
10.0	488,000	686,700	901,700	463,000	611,600
15.0	93,600	112,600	128,000	141,200	117,500
22.5	13,200	13,000	14,300	17,300	14,300

Table 2 - Constant Amplitude Fatigue Test Results

CONSTANT AMPLITUDE STRESS LEVEL (KSI)	CYCLES TO FAILURE			
	INDIVIDUAL EXPERIMENTS			GEOMETRIC MEAN
	EXP #1	EXP #2	EXP #3	
10.0	12,634,300	2,903,700	26,195,300+	9,868,300
12.0	775,600	3,732,900	1,118,600	1,479,500
15.0	572,000	779,500	515,000	612,300
30.0	66,500	61,900	70,600	66,300
45.0	14,500	14,800	16,300	15,200
70.0	1,010	2,020	2,680	1,760

plus sign (+) indicates the test was suspended, no failure occurred

Table 3 - Comparison of Analytical and Experimental Fatigue Lives

RMS STRESS LEVEL (KSI)	EXPERIMENTAL AND PREDICTED CYCLES TO FAILURE		
	EXPERIMENTAL RESULTS	SINGLE LINE S/N CURVE	BI-LINEAR S/N CURVE
5.0	4,709,700	10,013,200	11,445,400
7.5	1,244,100	1,961,900	1,671,800
10.0	611,600	617,200	530,600
15.0	117,500	120,900	113,800
22.5	14,300	23,700	24,900

Statistical Evaluation of the Distribution of Crack Propagation Fatigue Life by Simulating the Crack Growth Process

T. Sasaki (*), S. Sakai (**), H. Okamura (**)

(*) Graduate School, Department of Mechanical Engineering, and

(**) Department of Mechanical Engineering, University of Tokyo, 7-3-1, Hongo, Bunkyo-ku, Tokyo 113, Japan

ABSTRACT

It is widely recognized that the fatigue crack propagation is fundamentally a random process which can be predicted only in terms of probability. The primary source of statistical variation of fatigue crack propagation is material inhomogeneity. To explain its effects, the authors proposed in the previous paper a new stochastic model which treats the material's resistance against fatigue crack growth as a spatial stochastic process along the path of the crack. This paper investigates the influence of the parameter variation on the results using Monte-Carlo simulations for the proposed model in the case of the constant load amplitude. It is shown that the statistical properties of random crack propagation resistance has great influence on the distribution of the crack propagation fatigue life. The results are also compared with well-known experimental data sets and satisfactory agreements are obtained.

KEYWORDS

Fatigue, Crack Propagation, Structural Reliability, Statistical Treatment, Spectral Analysis, Distribution of Crack Propagation Fatigue Life, Random Crack Propagation Resistance

INTRODUCTION

The structural reliability analysis for crack growth fatigue life is often needed in a reliability based design of fatigue critical structures. To obtain a detailed geometry of the left side base of the fatigue life distribution function, which is especially important in the reliability analysis, a lot of fatigue crack growth tests are required. It is time-consuming and laborious to conduct such fatigue crack growth tests and more practical method which needs only a few crack growth tests is desired.

The distribution of crack propagation fatigue life can be obtained through the crack growth simulations if the fatigue crack growth is essentially a stochastic process. The evaluation of the stochastic nature of crack growth process requires only a few fatigue tests, and is effectively used in combined with simulations for the evaluation of the distribution of crack growth fatigue life [1]. In view of this situation, the authors proposed a new stochastic model and the parameters of the model were estimated through the constant ΔK fatigue crack growth tests [2]. In this paper, the numerical simulations of the crack growth process are conducted in order to confirm the adequacy of the proposed model, and the influence of the parameter variation on the results is investigated. The results are also compared with well-known experimental data sets and satisfactory agreements are obtained.

STOCHASTIC MODEL OF FATIGUE CRACK GROWTH

In order to consider the random nature of fatigue crack growth, the non-dimensional spatial stochastic process Z named crack growth resistance is introduced into the power crack growth law and randomized as follows:

$$\frac{da}{dN} = \frac{1}{Z} C \left(\frac{\Delta K}{K_0} \right)^m , \tag{1}$$

where a = crack size, N = number of cycles, ΔK = the stress intensity factor range, C and m are constants. K_0 is a unit quantity with the same dimension as ΔK. Z is supposed to be ergodic and both the ensemble mean of Z, $E[Z]$ and the spatial mean of Z, $\overline{Z}$ are normalized as

$$E[Z] = \overline{Z} = 1. \tag{2}$$

The spectral analysis of Z with the results of constant ΔK tests for SUS304 and Cr–Mo–V casting steel shows following statistical properties of Z.

(1) The distribution function of Z is approximated by the normal distribution.
(2) Z can be approximated by the mixture of a first-order Markov process and a white noise process.
(3) The statistical properties of Z is not greatly affected by the value of ΔK within the region of stage 2 fatigue crack growth.

The crack growth process can be simulated, if the random process Z satisfies these conditions.

METHOD FOR SIMULATION

The sample function of the stational Gaussian stochastic process Y with zero mean can be simulated as:

$$Y = 2 \sum_{k=1}^{N} \sqrt{S_Y(\omega_k) \Delta\omega} \cos(\omega_k a + \theta_k), \tag{3}$$

$$\Delta\omega = (\omega_U - \omega_L)/N, \tag{4}$$

$$\omega_k = \omega_L + (k - \frac{1}{2}) \Delta\omega, \tag{5}$$

where S_Y = two–sided power spectral density function, θ_k = random variable uniformly distributed between 0 and 2π, ω_U = the upper limit of angular frequency, ω_L = the lower limit of angular frequency [3].

The two-sided power spectral density function of a first order Markov process M is formulated as

$$P_M(f) = \frac{2\alpha_M \beta_M}{\alpha_M^2 + 4\pi^2 f^2}, \tag{6}$$

and that of a white noise process W is formulated as

$$P_W(f) = \beta_W, \tag{7}$$

where f = spatial frequency. Once the parameters, α_M, β_M and β_W are determined, Z can be simulated using eqn (3).

CONDITIONS OF SIMULATION

Specimen geometry used in the simulation is ASTM standard compact tension type, with a width $W = 50$ mm and a thickness $B = 12$ mm. Constant load amplitude $\Delta P =$ 5000 N is supposed to be loaded for this specimen. The simulations are performed from initial crack length $a_i = 15$ mm to the failure crack length $a_f = 35$ mm. The statistical parameters of crack growth resistance Z are $\alpha_M = 3 \times 10\ \mathrm{m}^{-1}$, $\beta_M = 2 \times 10^{-2}$, $\beta_W = 1 \times 10^{-6}$, that have been determined for Cr–Mo–V casting steel in the previous paper. The constants C and m are also determined just as 2.9×10^{-13} m/cycle and 4 respectively. In this simulations, the number of load cycles required to grow the crack to given lengths is recorded spaced at equal intervals, $\Delta a = 0.2$ mm.

The stress intensity factor K for CT specimen is calculated using the following equation proposed by Srawly [4].

$$K = \frac{P(2+\xi)}{B\sqrt{W}(1-\xi)^{1.5}}(0.886 + 4.64\xi - 13.32\xi^2 + 14.72\xi^3 - 5.6\xi^4), \tag{8}$$

where $\xi = a/W$.

RESULTS OF SIMULATION

Using the method shown above, crack growth simulations are conducted for 80 specimens. Then the distribution of crack growth rate and the crack growth fatigue life are investigated using these results.

Distribution of crack growth rate

Fig. 1 and Fig. 2 show the crack growth curves for 80 replicate simulations and the crack growth rate which is calculated using a secant method, respectively. These figures are similar to those obtained by Virkler *et al.* [5]. As the several researchers has been mentioned, it appears to be necessary to account for inhomogeneous crack propagation properties at two levels: (1) the mean behavior of each individual test and (2) the irregularity within the test. Because of the former variation, if the power crack growth law without Z defined by:

$$\frac{da}{dN} = C_s \left(\frac{\Delta K}{K_0}\right)^{m_s}, \tag{9}$$

is applied to each specimen, the best fit of C_s and m_s would be scattered. We have to pay attention to the fact that C_s and m_s in eqn (9) are random variables, on the other

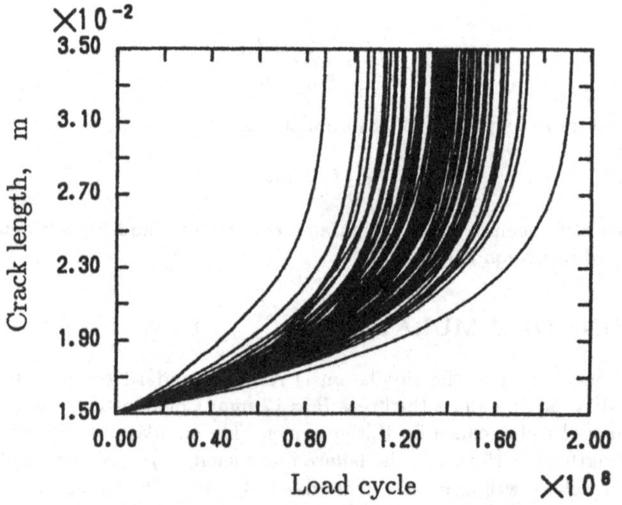

Fig.1 Simulated sample functions of crack propagation time history.

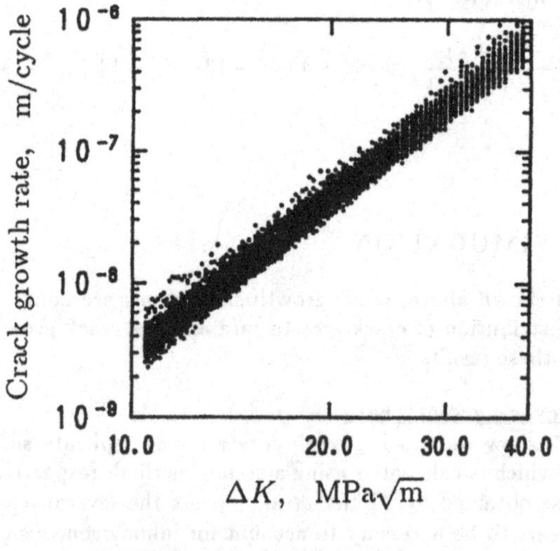

Fig.2 Simulated crack growth rate.

hand C and m in eqn (1) are constants. They are the different quantities and should be made a distinction.

The inter-specimen variation of C_s and m_s were investigated in detail by Tanaka *et al.* [6]. They showed that (1) C_s and m_s may be regarded to be fitted to the log-normal distribution and the normal distribution respectively and (2) C_s and m_s exhibit

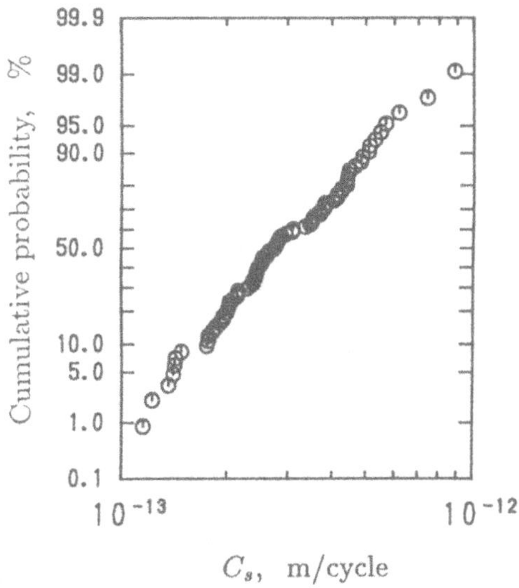

Fig.3 Distribution of C_s (log-normal probability paper).

a correlation of the form

$$\log C_s = a + bm_s, \tag{10}$$

where a and b are the constants. In order to confirm whether the proposed model can reproduce this properties, eqn (9) is applied to the results of simulations with a least square approximation. Then the distribution of C_s and m_s are plotted on a log-normal probability paper and a normal probability paper respectively as shown in Fig. 3 and Fig. 4. It can be confirmed that the log-normal distribution and the normal distribution provide good fits for C_s and m_s respectively, which is same as the experimental results.

Recently, Cortie *et al.* [7] has revealed that the C_s - m_s correlation does not possess any fundamental significance. But for the comparison with other works [6], the relation between C_s and m_s calculated from the results of simulation are plotted shown as Fig. 5. It can be confirmed that simulated C_s and m_s exhibit a good correlation like the form eqn (10), too.

Distribution function of crack growth fatigue life

The total fatigue life N_f which is normally obtained by a fatigue test is the summation of the crack initiation life N_i and the crack propagation life N_p. The crack initiation life and the crack growth life are defined as the load cycles required to reach to the fatigue crack initiation and the load cycles between the crack initiation and the final failure, respectively. Though the distribution of total fatigue life has been investigated by many researchers, the distribution of crack growth fatigue life is not fully investigated so far. The crack growth fatigue life will be equivalent to the total fatigue life if the load stress is relatively high, because the ratio of the crack growth life to the total fatigue life increases in proportion to the load stress. There have been two different opinions concerning the

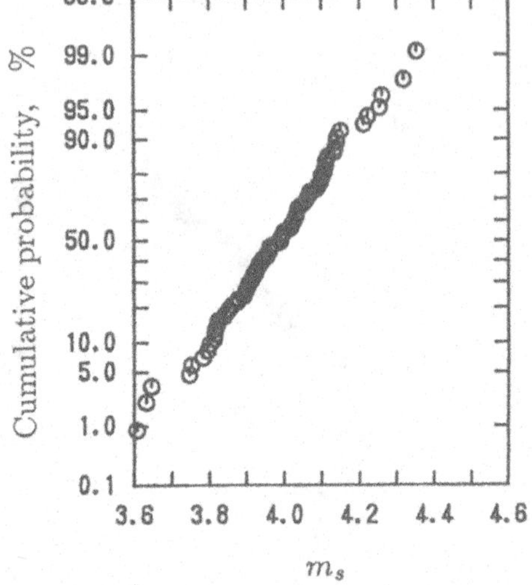

Fig.4 Distribution of m_s (normal probability paper).

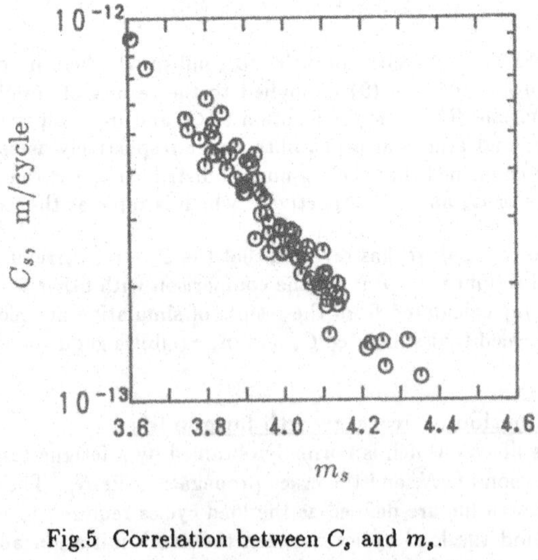

Fig.5 Correlation between C_s and m_s.

distribution function for total fatigue life under such a high load stress condition: one is the log-normal distribution, the other is the two parameter Weibull distribution. To date, the solution has not been given yet concerning which distribution provides a better

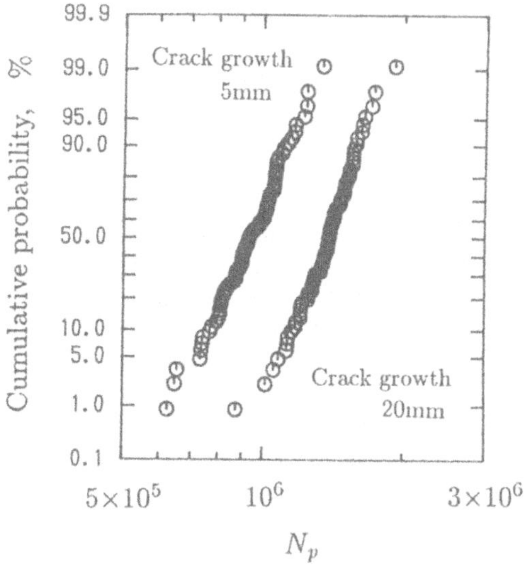

Fig.6 Distribution of simulated crack growth life (log-normal probability paper).

fit.

Then, the distribution of crack growth fatigue life is examined on both a log-normal probability paper and a Weibull probability paper as Fig. 6 and Fig. 7. The results are shown for the cases of crack growth length of 5 mm and 20 mm in the figures. These two figures do not show any distinct difference concerning the adaptability to those distribution functions.

Statistical properties of crack growth fatigue life

In this section, two statistical parameters η_1 and η_2 are investigated in order to evaluate the statistical properties of crack growth fatigue life. η_1 is the coefficient of variation defined as

$$\eta_1 = \frac{\sigma[N_p]}{E[N_p]}, \quad (11)$$

on the other hand, η_2 is defined as

$$\eta_2 = \frac{Var[N_p]}{E[N_p]}, \quad (12)$$

where $Var[N_p]$ = the variation of N_p, $\sigma[N_p]$ = the standard deviation of N_p and $E[N_p]$ = the mean of N_p. η_2 was considered by Shimada *et al.* [8]. They showed that η_2 would be constant for all crack length if the fatigue crack growth can be treated as a discrete Markov process.

Shimada *et al.* [8] conducted constant load amplitude fatigue crack growth tests for a high tensile strength steel APFH60 and showed the relation between η_1, η_2 and crack

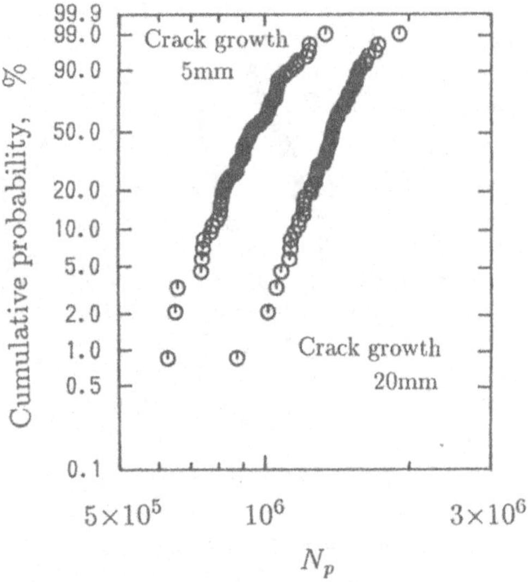

Fig.7 Distribution of simulated crack growth life (Weibull probability paper).

length. Ichikawa *et al.* [9] reported the same relation for 2024–T3 aluminum alloy. These experimental results are shown in Fig. 8 and Fig. 9. The figures show that the results obtained by these researchers are quite different for both η_1 and η_2. It seems that the magnitude of α_M in eqn (3) provides this difference. In order to confirm this conjecture, the crack growth simulation were performed for three cases of α_M: 3m^{-1} $3\times10\text{m}^{-1}$, $3\times10^2\text{m}^{-1}$. The conditions of simulation were the same as the previous simulation except for α_M. Fig. 10 and Fig. 11 shows the effect of α_M on η_1 and η_2. It becomes clear that α_M gives serious effect on the relation between η_2 and crack length.

In order to check the adequacy of the proposed model, the simulations were performed to fit η_1, η_2 vs. crack length relation for the experimental results. At the beginning, the simulation was started for arbitrary α_M, β_M and β_W. Then, α_M, β_M and β_W were modified repeatedly until the satisfactory results were obtained. The values determined were α_M=$2\times10^3\text{m}^{-1}$, β_M=7×10^{-2}, β_W=5×10^{-6} for the experiments by Shimada *et al.* and α_M=$1\times10^2\text{m}^{-1}$, β_M=2×10^{-2} β_W=3×10^{-5} for the experiments by Ichikawa *et al.* Simulated η_1, η_2 vs. crack length relations using these parameters are also shown in Fig. 8 and Fig. 9. It is obvious that the proposed model provides a good fit for the experimental results. This adaptability shows the adequacy of the proposed model.

CONCLUSIONS

In order to confirm the adequacy of the previously proposed stochastic crack growth model, numerical crack growth simulations were performed under constant load amplitude condition. The conclusions developed from this study can be summarized as follows:

(1) When the power crack growth low without considering random crack propagation resistance was applied to the results of simulations, the log-normal and the normal

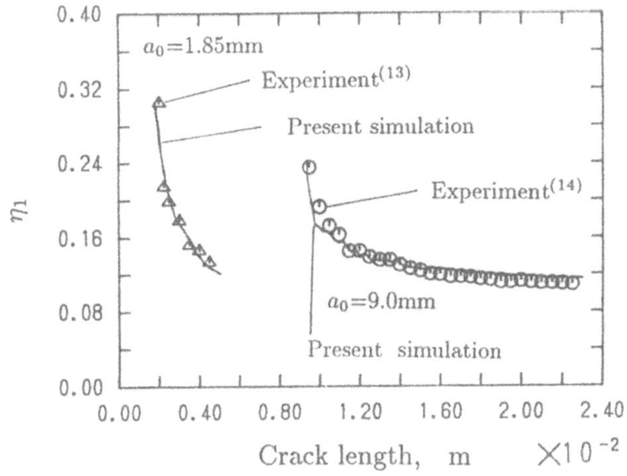

Fig.8 η_1 vs. crack length relation.

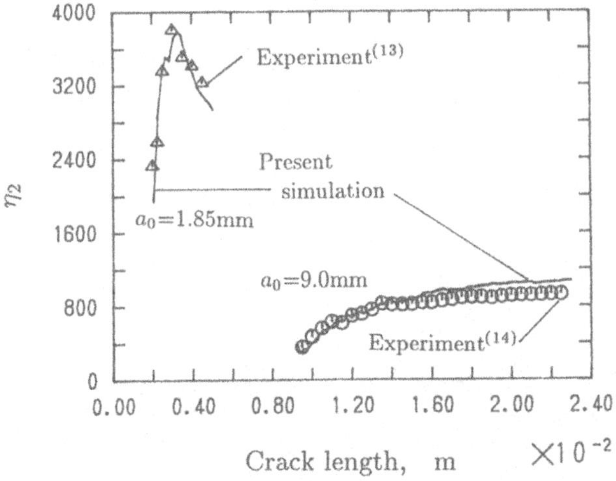

Fig.9 η_2 vs. crack length relation.

distribution provided a good fit for the inter-specimen variability of C_s and m_s respectively, and the linear relationship was obtained between $\log C_s$ and m_s;

(2) Either, the log-normal and the two parameters Weibull distribution provided a good fit for the distribution of simulated crack growth fatigue life N_p;

(3) Both the coefficient of variation of N_p, η_1 vs. crack length and the variation divided by the mean of N_p, η_2 were seriously affected by the statistical properties of random crack propagation resistance, but the simulated η_1 and η_2 vs. crack length

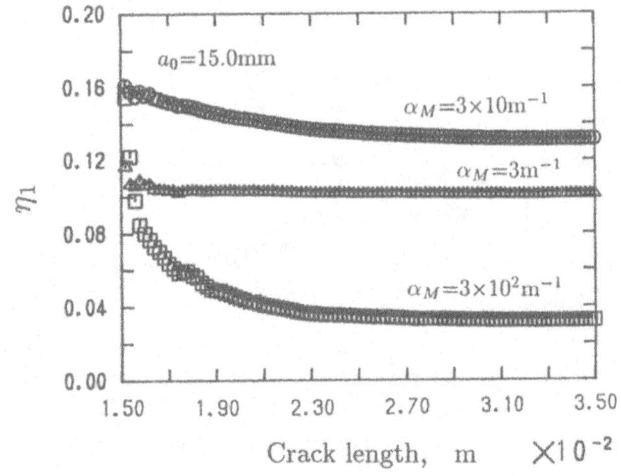

Fig.10 Effect of α_M on η_1.

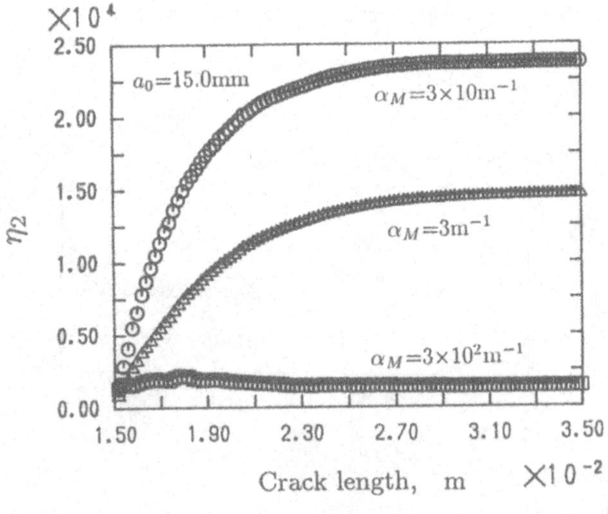

Fig.11 Effect of α_M on η_2.

relations gave a good fit for the experimental data if the parameters of random crack propagation resistance were properly determined.

These results seems to indicate the adequacy of the proposed model.

REFERENCES

[1] Itagaki, H., Ishizuka, T. and Huang, P., Reliability Assessment by Simulation of Fatigue Crack Growth, *Journal of the Society of Naval Architects of Japan*, Vol.165, pp. 253–264, 1989. (in Japanes)

[2] Sasaki, T., Sakai, S. and Okamura, H., Estimating the statistical properties of Fatigue Crack Growth Using Spectral Analysis Technique, to appear in *Proceeding of the 6th Int. Conf. of Mechanical Behavior of Material*, Kyoto, Japan, 1991. Pergamon, Oxford, 1991.

[3] Shinozuka, M., Simulation of Multivariate and Multidimensional Random Process, *The Journal of the Acoustical Soc. of America*, Vol.49, pp. 357–368, 1971.

[4] Srawly, J. E., Wide Range Stress Intensity Factor Expression for ASTM E-399 Standard Fracture Toughness Specimens, *Int. Journal of Fracture*, Vol.12, pp. 475–476, 1976.

[5] Virkler, D. A., Hillberry, B. M. and Goel, P. K., The Statistical Nature of Fatigue Crack Propagation, *ASME Journal of Engineering Materials and Technology*, Vol.101, pp. 148–153, 1979.

[6] Tanaka, S., Ichikawa, M. and Akita, S., Variability of m and C in the Fatigue Crack Propagation Law, *Int. Journal of Fracture*, Vol.17, pp. R121-125, 1981.

[7] Cortie, M. B. and Garrett, G. G., On the correlation between the C and m in the Paris Equation for Fatigue Crack Propagation, *Engineering Fracture Mechanics*, Vol.30, pp. 49–58, 1988.

[8] Shimada, S., Nakagawa, T. and Tokuno, H., Reliability Analysis of Fatigue Crack Propagation Life by Markov Chain, *Journal of the Society of Materials Science, Japan*, Vol.33, pp. 475–481, 1984. (in Japanese)

[9] Ichikawa, M. and Nakamura, T., Methods for Randomization of Parameters in the Fatigue Crack Propagation Law $da/dN = C(\Delta K)^m$, *Journal of the Society of Materials Science, Japan*, Vol.34, pp. 321–326, 1985. (in Japanese)

Filter Technique for Stochastic Crack Growth

F. Casciati (*), P. Colombi (*), L. Faravelli (**)
Department of Structural Mechanics, University of Pavia, I-27100 Pavia, Italy
(**) Institute of Energetics, University of Perugia, I-06100 Perugia, Italy

ABSTRACT

A fracture mechanics approach based, on Markov processes and filter techniques, is coupled with response surface methodology to model the randomness which affects the crack growth. The fatigue lifetime distribution is evaluated by a technique based on Hermite moments. These Hermite moments are calculated by classical Itô calculus. The fracture mechanics parameters (the stress intensity factor and the J integral) are evaluated by means of the line spring method. The results of the structural analysis are summarized by introducing, for given values of the geometrical parameters, a polynomial form for the relationship between the crack length and the root mean square of the stress intensity factor. The effects on the fatigue crack size distribution of uncertainties on the coefficients of this polynomial is estimated by a response surface approach.

INTRODUCTION

Fatigue failure is reached when a continuously growing crack exceeds a specified crack length. This limit length can be defined as the crack depth that will cause leakage, the one that will give the out-of-service conditions or the event which iniates catastrofic failure. The problem is characterized by uncertainties in the determination of the external load, the initial crack size and the material parameters. As a consequence the actual strength capacity of the structure cannot be estimated without a probabilistic analysis.

A considerable effort has been done recently for reaching a satisfactory probabilistic modelling of fatigue crack growth. Some authors, e.g. Lin and Yang[17], Sobczyk[21], Spencer and Tang[23], Solomos [22] and Ditlevesen [10], solve the crack growth equation by classical stochastic variable separation. Other authors make use of level II reliability methods to solve the crack growth equation in its non separable form, e.g. Dolinsky [11] and Madsen [18]. All these studies make use of elastic fracture mechanics theory.

In the present paper a stochastic model (Casciati et al. [4] [5] [6]) is used, together with a response surface technique (Faravelli[14] [15]), to model the uncertainties which affect the crack growth. The randomness of the fatigue strength, of the initial crack length and the one of the quantities which govern the stress intensity factor calculation are regarded as data.

The randomness in the fatigue strength is directly introduced, as usual, in the deterministic crack growth model. In this way the resulting stochastic differential equation is separable in a stochastic sense (see Section 2 for further details). The lifetime distribution is evaluated by a technique based on Hermite moments which are calculated via Itô calculus. A finite elements analysis is required for calculating the stress intensity factor for any crack length. A polynomial interpolation of the results of the structural analysis is introduced. The randomness of the corresponding interpolation coefficients and of the initial crack length is propagated by a response surface approach (see Casciati and Faravelli [8]). Its effect on the survival probability after a given number of duty cycles is estimated.

The application to one of the fatigue crack growth problems which characterize a pressure water reactor illustrates the main aspects of the approach.

STOCHASTIC MODEL FOR THE PROBABILITY DISTRIBUTION OF THE LIFETIME PROBABILITY

For design purposes the quantity of practical interest is the probability that the crack size does not exceeds a limit value a_L during a given time period. Let $\underline{N}(a_L)$ (underlined symbols denote random quantities) be the random number of duty cycles when the crack size $\underline{a}(t)$ reaches a specific value a_L. Since the event $\underline{N}(a_L) \geq N$ coincides with the event $\underline{a}(N) < a_L$, the complement to one of the probability distribution function $P_{\underline{N}}(N)$ of $\underline{N}(a_L)$ coincides with the probability distribution that no failure occurs for a given N. Then, the quantity of interest becomes the crack size probability distribution.

The Paris-Erdogan equation is one of the most commonly used crack growth model. It is written as :

$$\frac{da}{dN} = C(\Delta K)^m \tag{1}$$

$$a(0) = a_0 \tag{2}$$

where ΔK is the stress intensity factor range and C and m are material parameters. In a general situation $\underline{C}$,$\underline{m}$ and $\underline{a}_0$ are random variables and $\Delta\underline{K}$ is a random function of the crack length and of the geometric quantities $\{\underline{X}\}$ which influence the calculation of the stress intensity factor. Using classical stochastic variable separation to solve Eq. (1), one has to put on one side of the equation the input random quantities and on the other side the output random quantity. Eq.(1) is then re-written as:

$$\int_{\underline{a}_0}^{\underline{a}} \frac{da}{\Delta K^m} = C \int_{N_0}^{N} dN \tag{3}$$

Nevertheless, in this way on the left side of Eq.(3) one has an input random function, $\Delta\underline{K}$, which depends on the output random variable $\underline{a}(N)$. Then, Eq.(1) is not separable in a stochastic sense: information on the distribution of the output random variable $\underline{a}(N)$ cannot be obtained by taking the expectation of both sides of Eq.(3).

In this paper this inconvenient is avoided in two step. The randomness of the initial crack size a_0 and of the stress intensity factor ΔK. will be considered directly in the second step and propagated by a response surface technique [8]. Presently they are regarded as deterministic. A stochastic model based on a continuous Markov approach, was proposed (Casciati et al.[4] [5] [6]) in order to take into account the randomness of the fatigue strength (material parameters C and m). It assumes that the mean value of the crack growth rate can be obtained by substituting in Eq. (1) ΔK with the root mean square of the stress intensity factor $s_{\underline{K}}$. In this way one takes into account the randomness of the external loading by an "equivalent" ($s_{\underline{K}}$) constant quantity which characterizes , in the average, the random applied stress. Following Lin and Yang [17] the fluctuation around the mean value was modelled in Casciati et al. [4] by multiplying the right hand side of Eq.(1) by a stochastic process $\underline{x}(N)$. A filter, driven by a white noise, was used to define the process $\underline{x}(N)$.

The resulting stochastic differential equation for the crack growth is:

$$d\underline{a}(N) = Cs_{\underline{K}}{}^{m}[\mu_{\underline{x}} + \rho cos\underline{\phi}(N)]dN \tag{4}$$

$$d\underline{\phi}(N) = \sigma d\underline{w}(N); \tag{5}$$

$$\underline{\phi}(0) = 0; \underline{a}(0) = a_0 \tag{6}$$

where $\underline{w}(t)$ is a Wiener process , ρ and σ are model parameters and $\mu_{\underline{x}}$ is the mean value of process $\underline{x}(N)$. Eq.(1) is now separable in a stochastic sense; the lifetime distribution for a given N can be evaluated by a technique based on Hermite moments (Winterstein and Ness [24]). These Hermite moments are evaluated by Itô calculus (Arnold [1]).

The root mean square of the stress intensity factor is available in literature only for simple structural situation. In the general case, $s_{\underline{K}}$ is given, for a given values X of the random geometrical parameters $\{\underline{X}\}$, by the following relation:

$$s_{\underline{K}} = s_{\underline{S}} Y(a, \{X\})\sqrt{a} \tag{7}$$

where $s_{\underline{S}}$ is the root mean square of the external load. The evaluation of s_K for a complex structure requires finite element analysis. In preliminary step this analysis make use of the mean value of $\{\underline{X}\}$. The corresponding root mean square of the stress intensity factor is calculated for several cracks size under this condition. The results are then used to fit a suitable polynomial interpolation:

$$s_{\underline{K}}(a) = c_0 + c_1 a + c_2 a^2 \tag{8}$$

Of course the actual nature of the coefficients c_0, c_1 and c_2 is random due to the randomness of the quantities $\{\underline{X}\}$ which form the input of the finite element model. An accurate analysis would require to estimate the randomness of these coefficients. This could be done, for instance, by introducing an appropriate response surface technique as discussed in (Casciati and Faravelli [8]).

In this paper,one assumes that the coefficients c_0 c_1 and c_2 are know in a probabilistic sense. In particular, one assumes they are known functions of three independent random variables. A response surface technique is used to propagate this uncertainty in the evaluation process of the lifetime probability. In this scheme (see Casciati and Faravelli [8]) any response variable y of a mechanical system is expressed as a given function of independent random variables $\{q\}$:

$$y = f[\{q\}, \{\theta\}] + \epsilon \tag{9}$$

In Eq. (9) f is a function of known form, $\{\theta\}$ is a vector of unknown parameters and ϵ is a random error. The model parameters $\{\theta\}$ and ϵ are estimated on the basis of the results of appropriately planned numerical experiments (see Casciati and Faravelli [8]).

A second order polynomial for the function f is selected in this paper as in Faravelli [15]. Consider several realizations $\{q_1\}, \ldots, \{q_n\}$ of the vector $\{q\}$, according to experimental design theory (central composite design), and compute the corresponding response variables $y_1, \ldots, y_n$ for each of them. A regression scheme is used to evaluate the vector $\{\theta\}$ and the variance of the zero mean error term (see Casciati and Faravelli [8]). The calculation of the cumulative distribution function $P_{\underline{y}}$ requires then an approximate procedure. Level-2 reliability methods have been shown to be convenient (see Faravelli [14]) for this purpose.

In summary the proposed method for fatigue crack growth analisis consists of the following steps :

1. the stress intensity factor is estimated for some crack size values by a suitable finite element code. The random geometrical variables which form the input for that code are introduced by their mean value;
2. the results are used to fit a suitable polynomial interpolation to the relationship between the root mean square of the stress intensity factor range and the crack size;
3. the coefficients of the polynomial interpolation c_0, c_1 and c_2 and of the initial crack length a_0 are regarded as random variables;
4. a set of experiments is planned, according to experiment design theory, in the space of the independent random variables where the polynomial coefficients c_0, c_1 and c_2 and the initial crack length a_0 have been mapped;
5. the lifetime probability is evaluated, for given crack length a_L and given number of duty cycles by the method based on Hermite moments illustrated in Casciati et al. [7];
6. the resulting lifetime probability is modelled by an appropriate response surface;
7. the cumulative distribution function of the lifetime probability is estimated by a level-2 probabilistic method (Augusti et al. [2], Faravelli [13] and Casciati and Faravelli [8]).

A CASE STUDY

Mechanical model

The example presented herein regards a 1/5 scale pressure water reactor (PWR) built at the Joint Research Center of Ispra in the framework of the Reactor Safety Program. The main goal of the project is the evaluation of the degree of deterioration of the structural integrity and the estimation of the residual lifetime of the pressurized component. Two type of cracks should be studied:

- fabrication cracks;
- crack nucleated during service;

In the present work, however, only fabrication cracks are considered. The welded connection between the nozzle and the safe end of the PWR vessel is analysed. The mean values of the basic data regarding the geometry of the 1/5 scale PWR vessel, are given in Fig.1. These data are used in the evaluation of the stress intensity factor by the

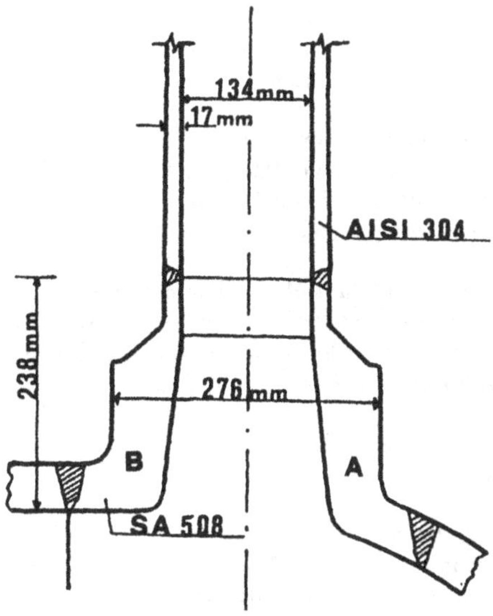

Fig. 1. Nozzle to safe-end connection of the 1:5 scale PWR vessel (A circumferential section; B longitudinal section).

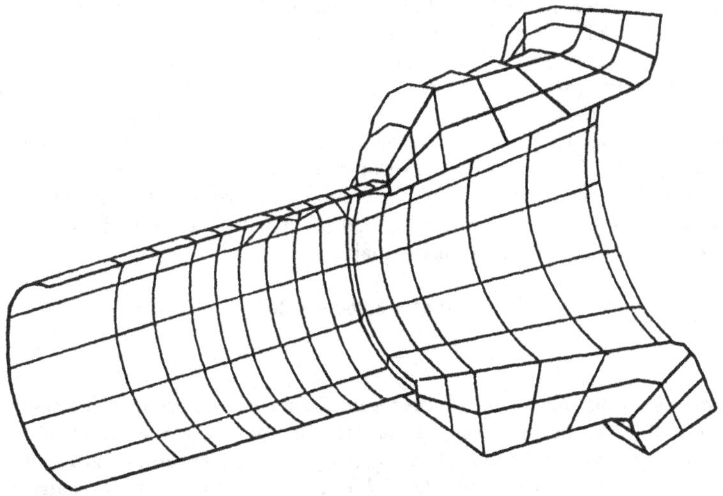

Fig. 2. The finite element mesh used for the structural analysis.

computer code ABAQUS (Hibbit et al. [16]).

The geometry of the crack under investigation is planar and its shape is semielliptical. The length of the longer radius b is 34 mm, ; the following values are considered for the other radius a : 1 mm - 3 mm - 7 mm - 10 mm.

The crack grows, initially, along the thickness but, since it has reached the leak point, it propagates upon the surface until the critical dimension is achieved and catastrophic failure takes place. In this paper attention is focussed on the first stage, i.e. crack propagation occurs along the thickness s , and a_L is lower than s.

The vessel and the safe end are built of two different materials : SA 508 (the nozzle and the weld between the nozzle and the safe end) and AISI 304 (the safe end). The material behaviour is elastoplastic with isotropic hardening. The external loads consist of an internal pressure of 3 Mpa and a concentraded load P at the free end of the safe-end. This latter excitation is not deterministic and is modelled by a Gaussian process with zero mean and a standard deviation of 42 KN. It represents the vibration of the structural pipe due to cyclic external events as earthquakes.

The finite element computer code ABAQUS offers special facilities for two- and three-dimensional fracture mechanics problems. The evaluation of the stress intensity factor is possible for any crack front, for any loading and for elastic and elastoplastic material behaviour. For three dimensional situations the value of the stress intensity factor is estimated by the "virtual crack extension method" (Parks [19]). In the presence of semielliptical surface cracks in plate or shell structures, the "line spring elements" (Parks [20]) can be used to evaluate the stress intensity factor. The latter technique provides an inexpensive estimation of the fracture mechanics parameters with accuracy satisfactory for technical purposes.

Modelling the safe end with 76 isoparametric, 8 nodes shell elements and the nozzle with 160 isoparametric, 20 nodes brick elements (Colombi[9]) , the line spring element methodology was used for the problem of Fig. 1. The mesh (841 nodes) was created with the computer code PATRAN and the relevant plot is given in Fig 2. The connection between the nozzle (three dimensional elements) and the safe-end (shell element) was realizated by the MPC (MultiPoints Constraint) facility of ABAQUS. A detail of the mesh around the crack tip is shown in Fig.3. The crack growth was modelled using 6 line spring elements in order to obtain a good approximation of the stress intensity factor around the crack front. The root mean square of the stress intensity factor was calculated for the values of the crack length reported in Section 3.1, according to Eq. (7). In this phase the mean value of the random geometrical input quantities is used. The results are interpolated by the second order polynomial:

$$s_{\underline{K}} = -.15a^2 + 3.86a - 1.308 \tag{10}$$

In the work of (Evans and Bache[12]) the result of tests on a compact tension specimen were made available. From these one evaluates the fatigue strength parameters C and m and the values of the filter parameters μ_x, ρ and σ. The experimental results are reported in Fig. 4 as $log(da/dN)$ versus $log(\Delta K)$. The best-fit straight line is also given in fig. 4. The value of the parameters which were derived in Casciati et al. [7] are:

$$C = 3.67 \times 10^{-12};$$

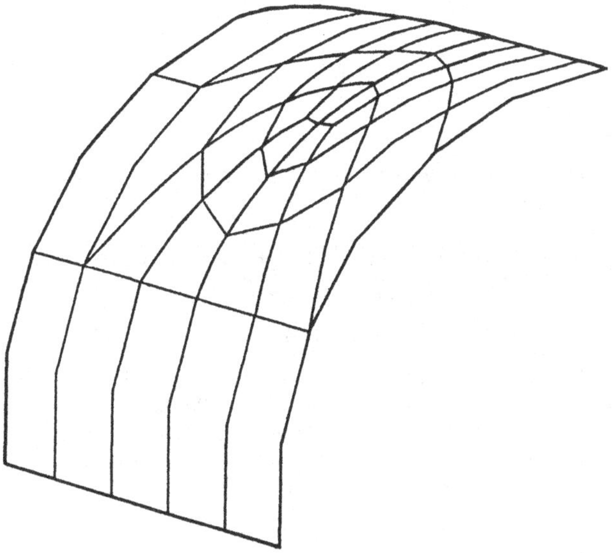

Fig. 3. Detail of the finite element mesh around the crack tip.

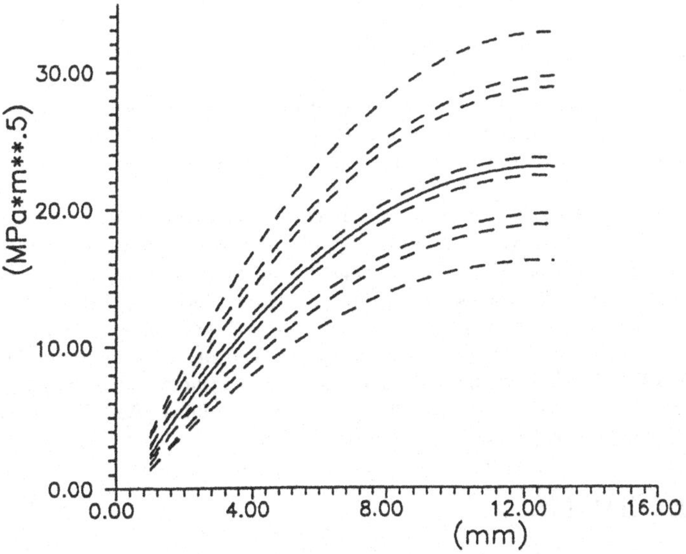

Fig. 4. Root mean square of the stress intensity factor vs. crack size for the values of the interpolation coefficients obtained from experimental design theory (the solid line shows the polinomial form obtained for the mean values of the interpolation coefficients).

$$m = 3.071;$$
$$\mu_{\underline{x}} = 1.121;$$
$$\rho = 1.17;$$
$$\sigma = 4.5;$$

Stochastic model

The resulting fatigue crack growth rate for the SA 508 is given by the following relation (see Eq. (4),(5) and (6)):

$$d\underline{a}(N) = 3.67 \times 10^{-12} s_{\underline{K}}^{3.071}[1.21 + 1.17cos\underline{\phi}(N)]dN \tag{11}$$

$$d\underline{\phi}(N) = 4.5d\underline{w}(N) \tag{12}$$

$$\underline{\phi}(0) = 0; \underline{a}(0) = a_0 \tag{13}$$

where $s_{\underline{K}}$ is given by Eq. (7). It is assumed that the interpolation coefficient $\underline{c}_0$ is a Gaussian variable with mean value equal to -1.308 and coefficient of variation -0.3, while the second coefficient $\underline{c}_1$ is modelled as a lognormal random variable with median 3.86 and coefficient of variation 0.2. Let $\underline{a}_v$ be the value of the crack length at which the first derivative of the function $s_{\underline{K}}(a)$ is equal to zero. One has:

$$\underline{a}_v = \frac{-\underline{c}_1}{2 \times \underline{c}_0} \tag{14}$$

It is assumed that the random variable $\underline{a}_v$ has a mean value equal to 12.86 and a dispersion around the mean value given by a Gaussian random factor $\underline{\Omega}$ with mean value one and standard deviation .1. Then the third coefficient $\underline{c}_2$ is given by:

$$\underline{c}_2 = \frac{-\underline{c}_1}{2 \times 12.86 \times \underline{\Omega}} \tag{15}$$

Realization of the interpolation coefficient and of the initial crack length was selected according to experimental design theory. The initial crack length is idealized to be a Gaussian random variable with mean value equal to 3 mm and standard deviation 1 mm.

Using experimental design theory, 20 experimental points were obtained in the space of the random variables $\underline{c}_0$, $\underline{c}_1$, $\underline{a}_0$, $\underline{\Omega}$ being regarded as a secondary variable, due to its narrow variability. For each of the planned experiments Eq.(8) is written. The relevant plots are given in Fig. 5. In this figure the solid line is the plot of Eq.(10). The lifetime probability was computed for the planned experiments and for a value of the limit crack length a_L equal to 8 mm. The probability distribution of the lifetime probability was evaluated by a level-2 reliability method. The corresponding probability density is then derived. Plots of the probability density of the lifetime probability after 360000, 380000, 400000, 420000 and 440000 duty cicles are given in Fig. 6.

CONCLUSIONS

A stochastic model is coupled with a response surface technique to study the uncertainties which affect crack growth. Finite element calculations (for instance by the computer code ABAQUS) permit one to evaluate the stress intensity factor. The mean values of the quantities which affect the estimation of the fracture mechanics parameters are used. The relationship between the crack length and the root mean square of the stress intensity factor is summarized into a polynomial interpolation of the structural results. The randomness of the fatigue strength capacity is modelled by Markov processes and

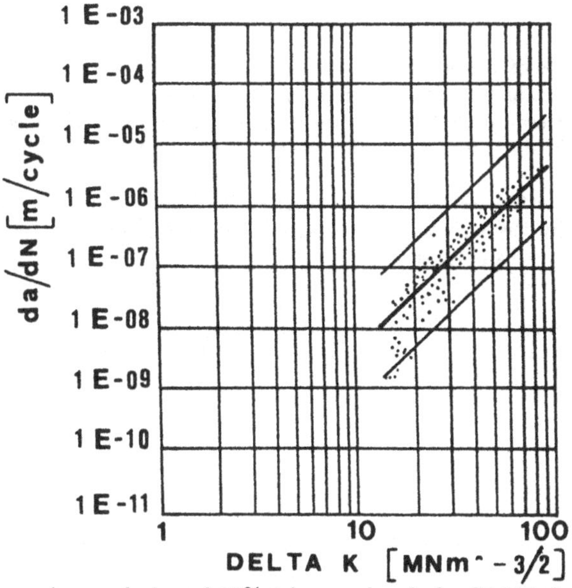

Fig. 5. Regression analysis and 95% tolerance bands for SA508 cl. B steel crack growth tests (from Ref.[12]).

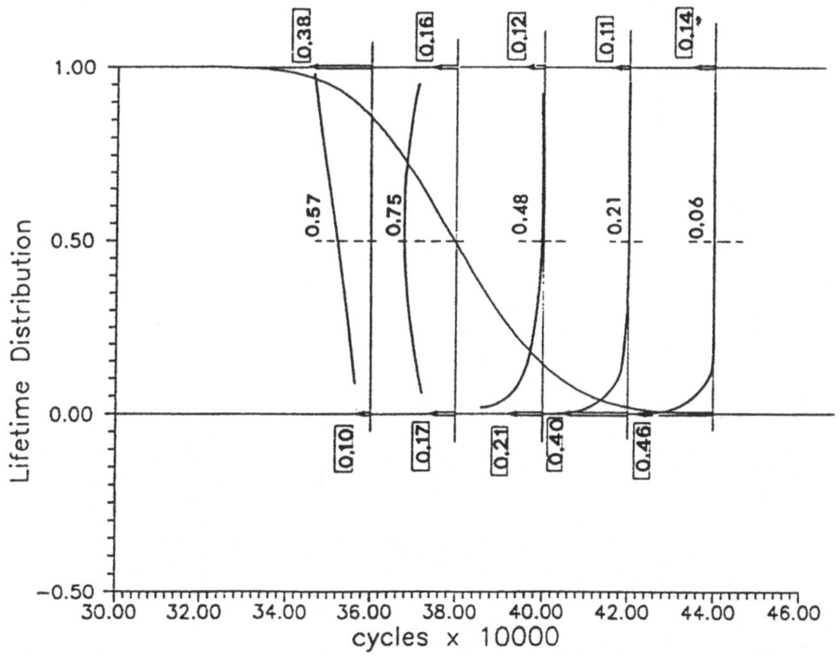

Fig. 6. Probability density function of the lifetime probability for $a_L = 8$ mm and some values of N, namely N equal to 360000, 380000, 400000, 420000 and 440000 duty cycles.

filter techniques. The lifetime probability is evaluated by a method which makes use of Hermite moments.

The randomness of the quantities which form the input of the finite element model and the randomness of the initial crack length is propagated by a response surface approach. At the present stage the statistics of the interpolation coefficients are regarded as data. The probability density function of the lifetime probability is then calculated.

Nevertheless, further research is still required for a more accurate modelling of the results achieved by the structural analysis. A response surface scheme seems to be appropriate also for this purpose.

Acknowledgement

The research has been supported by funds from both the Italian Mininstry of University and Research (MURST) and the National Research Council (CNR)

References

[1] L. Arnold, "Stochastic Differential Equations: Theory and Applications", John Wiley and Sons, Inc., New York, 1974.

[2] Augusti G., Baratta A., Casciati F. , "Probabilistic Methods in Structural Engineering", Chapman & Hall, London, 1984.

[3] Broek D. "Elementary Engineering Fracture Mechanics", Martinus Nijhoff Publishers, 1982.

[4] Casciati F., Colombi P., Faravelli L. "Filter Technique for Stochastic Crack Growth", Proceedings of the VIII European Congress on Fracture, Torino (Italy), pp. 1435-1440, 1990.

[5] Casciati F., Colombi P., Faravelli L. , "Un Modello Stocastico di Propagazione di Cricca per Materiali Metallici" (in Italian), Proceedings of X Congresso Nazionale Aimeta, Pisa (Italy), pp.123-128, 1990.

[6] Casciati F., Colombi P., Faravelli L., " Stochastic crack growth and Reliability Analysis", 10th International Conferences on Offshore Mechanics and Artic Engineering, Stavanger Forum, Stavanger, (Norway), 1991.

[7] Casciati F., Colombi P. and Faravelli L. "Stochastic Crack Growth by Filter Technique", Proceedings of the Sixth International Conference on Applications of Statistics and Probability in Civil Engineering, Mexico City, (Mexico), 1991.

[8] Casciati F., Faravelli L. , " Fragility Analysis of Complex Structural System", Research Studies Press Ltd., Taunton, 1991.

[9] Colombi P. , "Propagazione di Cricche in Materiali Metallici: Aspetti Meccanici e Stocastici" (in Italian), Master Thesis , Department of Structural Mechanics, Univiversity of Pavia, 1989.

[10] Ditlevsen O. "Random Fatigue Crack Growth - a First Fassage Problem", Eng. Fracture Mech., Vol. 23(2), pp.467-477, 1986.

[11] Dolinski K. "Stochastic Loading and Material Inhomogeneity in Fatigue Crack Propagation", Eng. Fracture Mech., Vol. 25 (5/6), pp.809-818, 1986.

[12] Evans W. J., Bache M. R., "Crack Formation Lives and Crack Propagation Behaviour of Pressure Vessel Steel" (confidential material).

[13] Faravelli L. , "Sicurezza Strutturale" (in Italian), Pitagora Editrice, Bologna, 1988.

[14] Faravelli L. , "Response Surface Approach for Reliability Analysis", J. of Eng. Mech., Vol. 115 (12), pp. 2763-2780, 1989.

[15] Faravelli L. "Finite Element Analysis of Stochastic Nonlinear Continua", (Ed. Liu W. K. and Belytschko T.), Computational Mech. of Probability and Reliab. Analysis, Elmepress, Lausanne, pp. 263-280, 1989.

[16] Hibbit, Karlsson & Sorensen Inc., "ABAQUS Manual: Vol.1, User's Manual; Vol. 2, Theory Manual; Vol 3., Example Problems Manual; Providence, RI, Usa, 1982.

[17] Lin Y.K. and Yang J.N. "A Stochastic Theory of Fatigue Crack Propagation", AIAA Journal, Vol. 23(1), pp. 117-124, 1985.

[18] Madsen H.O. "Probabilistic and Deterministic Models for Predicting Damage Accumulation due to Time Varyng Loading", DIALOG 5-82, Danish Engineering Academy, Lingby, Denmark, 1983.

[19] Parks D.M., "The Virtual Crack Extension Method for Nonlinear Material Behaviour", Comp. Meth. in Appl. Mech. Eng., Vol. 2, 1977.

[20] Parks D.M. , "The Inelastic Line Spring: Estimates of Elastic-Plastic Fracture Mechanics Parameters for Surface-cracked Plates and Shells", Jour. Pressure Vessel Tech., Vol.103, 1981.

[21] Sobczyk K. "Modelling of Random Fatigue Crack Growth", Eng. Fracture Mech., Vol. 24(4), pp. 609-623, 1986.

[22] Solomos G.P. "First-Passage Solutions in Fatigue Crack Propagation" , Prob. Eng. Mech., Vol. 4(1), pp. 33-39, 1989.

[23] Spencer B. F. and Tang J., "Markov Process Model for Fatigue Crack Growth" , Journal of Eng. Mech., vol. 114, (12), pp. 2134-2157, 1988.

[24] Winterstein S.R. and Ness O.B. (1989) "Moment Analysis of Non-linear Random Vibration", (Ed. Liu W. K. and Belytschko T.) Computational Mech. of Probability and Reliab. Analysis", Elmepress, Lausanne, pp. 451-478, 1989.

SECTION 7: MONTE CARLO SIMULATIONS

Probabilistic Seismic Response Analysis of a Steel Frame Structure Using Monte Carlo Simulation

H. Seya (*), H.H.M. Hwang (**), M. Shinozuka (***)

(*) Technical Research Laboratory, Takenaka Corp., Tokyo, Japan

(**) Center for Earthquake Research and Information, Memphis State University, Memphis, TN 38152, U.S.A.

(***) National Center for Earthquake Engineering Research, State University of New York, Buffalo, NY 14214, U.S.A.

ABSTRACT

In this study we present a probabilistic seismic analysis of a steel frame structure to generate fragility curves and to investigate the correlation between the response modification factor and the system ductility factor. As an example, we use a five-story hypothetical steel frame structure located in New York City and designed according to the provisions specified in the ANSI standard A58.1-1982 and the AISC specification. We quantify uncertainties in seismic and structure parameters that define the earthquake-structure system by choosing several representative values for each parameter. Then, by using the Latin hypercube sampling technique, we combine these representative values to establish 18 samples analysis. The nonlinear seismic analysis is performed by using the DRAIN-2D computer program.

For fragility analysis, we define five limit states representing nonstructural damage to collapse of structure and then establish the probabilistic structural capacity corresponding to each limit state. From the probabilistic structural response and capacity, the limit state probabilities at various levels of the peak ground accelerations can be computed and used to construct the fragility curves. For the response modification factor R, we perform nonlinear and corresponding linear analyses of the 18 samples subject to earthquakes with various levels of peak ground acceleration. The correlation between the R factor to the system ductility ratio is presented. Note that this correlation is based on one steel frame structure. Additional studies are needed to establish appropriate R factors for the design of steel structures.

INTRODUCTION

The potential damage caused by a large earthquake is of great concern to government officials, professionals, and the general public. However, we can neither predict the occurrence of an earthquake nor precisely estimate its amplitude, frequency contents, and duration. The computed structural responses may deviate from the actual values because of the idealization employed in structural modeling. Furthermore, the structural capacity cannot be accurately determined since basic parameters such as material strength always exhibit statistical variation. Thus, a probabilistic approach, which can incorporate uncertainty in the analysis, is appropriate for evaluating seismic responses of structures.

Fragility data are essential for seismic risk assessment studies to evaluate potential earthquake-induced loss of life and property damage, to estimate economic and societal risk, and to develop an effective emergency response plan. On the other hand, for the design of new structures, the nonlinear structural response can be obtained approximately by using the response modification factor. Thus, the fragility data and the response modification factor for a structure are very useful information. therefore, we present a probabilistic analysis to evaluate the seismic fragility curves and the response modification factor for a steel frame structure.

DESIGN OF STEEL FRAME STRUCTURE

The structure used in this study is a hypothetical five-story office building located in New York City. A typical plan and section are shown in Fig.1. Vertical and lateral loads are resisted by steel moment-resisting frames. The beams and columns are made of standard W-shapes of ASTM A36 steel (fy = 36ksi). The floor systems consist of 3-1/4 inch lightweight concrete topping on 2-inch 20-gauge metal decking designed for composite action. The floors are supported by spandrel beams 8 feet 4 inch on center. The roof system, consisting of roofing materials, insulation, and metal decking, is also supported by spandrel beams 8 feet 4 inch on center. Exterior walls are made of lightweight formed metal and rigid insulation in a panel system.

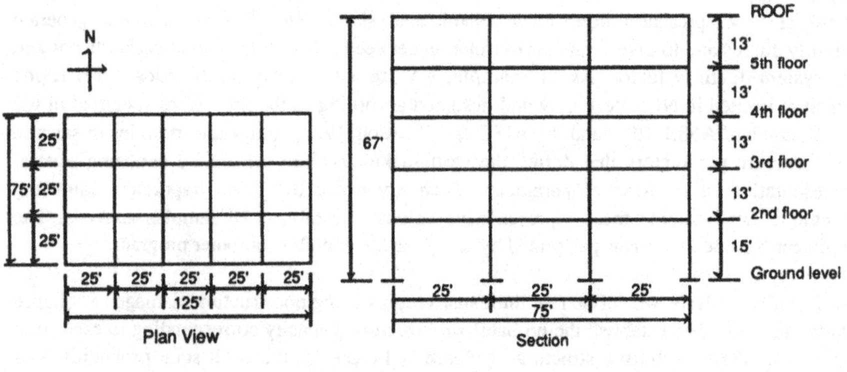

Fig 1. Plan and section of steel frame building

In this study, only a typical interior frame in the N-S direction is designed according to the ANSI standard A58.1-1982[1] and the AISC specification[2]. Although wind load may exceed seismic load depending on the choice of factors for wind load computation, only seismic load is considered in accordance with the purpose of this study. Furthermore, ANSI A58.1-1982 requires the design for horizontal torsional moment due to eccentricity of the building plus an accidental eccentricity caused by 5% of the dimension of the building perpendicular to the direction of applied load. Since this building is assumed to be symmetric and the effect of accidental torsion on the resultant forces is less than 3%, torsion is not included in the design. Lastly, the foundation design is assumed to be adequate and therefore is not addressed in our study. Vertical loads are assumed as follows:

Dead Loads	
Roof system (roofing, insulation, roof deck, spandrel beams)	67 psf
Floor system (concrete topping, metal decking, spandrel beams)	74 psf
Exterior walls	15 psf
Live Loads	
2nd through 5th floors	50 psf
Roof	25 psf

These dead and live loads acting on the frame are analyzed according to conventional procedure.

Earthquake Load

The design base shear V specified in the ANSI standard A58-1982 is

$$V = ZIKCSW \tag{1}$$

where Z is the zone factor. For seismic zone 2, Z is equal to 3/8. I is the importance factor, which is equal to 1 for an office building. K is the building system factor. For an ordinary steel frame structure, K is equal to 1. S is the soil factor. The building is assumed to be located at a site of S_2 soil conditions; thus, S is equal to 1.2. W is total dead load of the building and calculated as 4086 kips. C is determined by

$$C = \frac{1}{15\sqrt{T}} \tag{2}$$

in which T is the fundamental period of the building.

$$T = C_T h_n{}^{3/4} \tag{3}$$

where $C_T = 0.035$ for steel frame structures. h_n is the total height of the building above the base. For h_n of 67 ft, T is calculated as 0.82 second and C as 0.07. Thus, the design base shear V calculated by using equation (1) is 135.4 kips. The member forces caused by the design earthquake can be determined from this base shear.

The members (beams and columns) of the structure are designed according to the AISC specification considering the following two load combinations:

Case 1. Dead Load + Live Load
Case 2. Dead Load + Live Load + Earthquake

Note that the allowable stress can be increased one-third for case 2, which includes seismic load. Through an iteration process, the final sizes of beams and columns are determined and shown in Tables 1 and 2, respectively.

The story drift Δ_x can be computed by the following equation:

$$\Delta_x = \Delta_e / K < 0.005h \tag{4}$$

where h is the height of a story; Δ_e is the story drift calculated from the design lateral forces. From the calculations, the maximum drift, Δ_x = 0.0256 inch, occurs at the second story. Since the allowable story drift of the second story is 0.78 inch, the building satisfies the code drift requirements.

Table 1. Beam size

Floor	Beam size	Area (in^2)	I (in^4)	Sx (in^3)
2	W21x44	13	843	81.6
3	W18x46	13.5	712	78.8
4	W18x46	13.5	712	78.8
5	W18x46	13.5	712	78.8
roof	W18x35	10.3	510	57.6

Table 2. Column size

Story	Column size	Area (in^2)	I (in^4)
1	W14x90	26.5	999
2	W14x68	20.0	723
3	W14x61	17.9	640
4	W14x48	14.1	485
5	W14x48	14.1	485

BI-LINEAR HYSTERESIS

A bi-linear hysteretic relationship with the strain hardening ratio of 0.05 is used for all the members. The yield strength of a column is affected by the magnitude of axial force of the column. Nominal values of bending moment M_n and axial force P_n are calculated for each column and then the interaction surface for each column is determined by using the equations in Fig.2 as specified in the AISC LRFD Manual[3]. The values of interaction surface for all the column are summarized in Table 3.

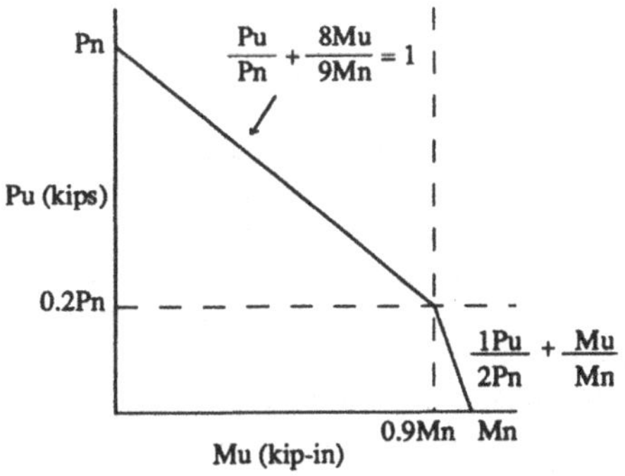

Fig 2. Interaction surface for column under combined bending and axial force

Table 3. Column interaction surface values

Story	Column size	M_n (k-in)	P_n (kips)	0.9 M_n (k-in)	0.2 P_n (kips)
1	W14x90	5652	773	5057	155
2	W14x68	4147	501	3732	100
3	W14x61	3667	447	3300	89.4
4	W14x48	2827	259	2544	51.8
5	W14x48	2827	259	2544	51.8

Since beams are not subject to axial force, the yield surfaces simply consist of positive and negative limits for bending moments. Although the floor slabs are continuously supported by the beams and act together in bending, the effective span of the beam is very small. For this reason, the nominal yield strength of the beam is determined from the nominal strengths of the beam without the contribution from slabs. The nominal strengths of all the the beams are summarized in Table 4.

Table 4. Nominal yield strength of beams

Floor	Beam size	M_n (kip-in)
2	W21x44	±3440
3	W18x46	±3267
4	W18x46	±3267
5	W18x46	±3267
roof	W18x35	±2400

UNCERTAINTY IN STRUCTURE

The response of a structure to earthquake loading is affected by energy-dissipation characteristics of the structure. The input seismic energy may be dissipated through the structural viscous damping and hysteretic mechanism. However, the selection of a representative viscous damping value for dynamic analysis requires considerable judgment. The hysteretic behavior (nonlinearity) of a structure is mainly affected by the member yielding strength, which also cannot be determined precisely. Thus, in this study, the structural viscous damping and member yield strength are treated as random variables. Uncertainty in the critical damping ratio is represented by three values: 0.01, 0.02, and 0.03. On the other hand, uncertainty in the yield strength of each member is represented by a log-normal distribution. For yield moment and axial force, the median values are 1.07 and 1.08, respectively, times the nominal values listed in Tables 3 and 4. The coefficients of variation for bending moment and axial force are 0.13 and 0.15, respectively[4]. Given these statistics, 18 yield strength values with equal probability can be generated for each of the 35 members. By using Latin hypercube sampling technique[5], a total of 18 structural models can be made from the combination of three representative values of the critical damping ratio and 18 values of the member yield strength.

UNCERTAINTY IN GROUND MOTION

The Kanai-Tajimi (K-T) power spectrum is used in this study to simulate earthquake acceleration time histories. The K-T power spectrum is a function of three parameters S_0, ω_g and ζ_g:

$$S_g(\omega) = S_0 \frac{1 + 4\zeta_g^2\left(\frac{\omega}{\omega_g}\right)^2}{\left[1 - \left(\frac{\omega}{\omega_g}\right)^2\right]^2 + 4\zeta_g^2\left(\frac{\omega}{\omega_g}\right)^2} \tag{5}$$

where

$S_g(\omega)$: one-sided earthquake power spectrum

S_0 : intensity of the spectrum

ω_g : dominant ground frequency

ζ_g : critical damping parameter

In this study, ω_g and ζ_g are considered as random variables. Hwang and Jaw[6] evaluated the variation of ω_g and ζ_g corresponding to different soil conditions. For sites of soft soil, ω_g is estimated to be in the range of 2.4π to 3.5π (rad/sec), whereas for rock site, ω_g ranges from 8π to 10π. For stiff soil, ω_g falls somewhere between the values estimated for rock and soft soils. The soil conditions in the New York metropolitan area range from soft soil to rock. Three values of ω_g, 2.4π, 5π, and 8π, are selected as the representative values for soft soil, stiff soil, and rock, respectively. It is estimated that the mean value of ζ_g is about 0.6 for rock and stiff soil, and 0.85 for soft soil. The parameter S_0 in the K-T power spectrum is related to peak ground acceleration (PGA). If the PGA value is specified, then S_0 is not a random variable. The parameters of the K-T power spectra for three soil conditions are summarized in Table 5. The total duration of earthquake acceleration for three soil conditions is also listed in Table 5.

Given a power spectrum, the earthquake time histories can be generated by using the method suggested by Shinozuka[7]. This method requires random phase angles and an envelope function. From the power spectrum for each soil condition, six time histories are generated. Thus, we have 18 earthquake time histories for three soil conditions. Note that 18 different sets of random phase angles are used to generate these nonstationary ground accelerations. By using the Latin hypercube sampling technique, these earthquake time histories are then matched to the structural models so that 18 samples of the earthquake-structural system are made for nonlinear seismic response analyses.

Table 5. Ground motion parameters

Soil Type	ω_g (rad/sec)	ζ_g	Total duration (sec)
Rock	8π	0.60	15
Stiff Soil	5π	0.60	20
Soft Soil	2.4π	0.85	25

NONLINEAR SEISMIC ANALYSIS

In this study, nonlinear seismic analysis is performed by using the computer program DRAIN-2D[8]. The structure is modeled as a frame with beams and columns, permitting column extensions, but ignoring beam extensions. Masses of the structure calculated based on dead load are associated with only the horizontal displacements at the floor level, i.e., the vibration in the vertical direction is not considered. In addition, the P-Δ effect is not taken into account.

The i-th story ductility ratio μ_i is defined as the ratio of the maximum absolute interstory displacement $U_{max,i}$ to the yielding displacement $U_{y,i}$.

$$\mu_i = \frac{U_{max,i}}{U_{y,i}} \tag{6}$$

According to Akiyama[9], the yielding displacement $U_{y,i}$ of i-th story can be defined as

$$U_{y,i} = \frac{h_i}{150} \tag{7}$$

where h_i is the height of i-th story.

As an illustration, Fig.3 shows the results of four dynamic analyses of the 18th sample of the structure. A black dot at the ends of a member represents a plastic hinge. When $\mu_m = 1.35$, all beams except those on the top floor have yielded, while less than half of the columns are yielded. Almost all columns have yielded at $\mu_m = 4.92$, that is, the structure has reached the collapse limit state.

Figures 4 through 6 show the story ductility ratios of all the samples according to soil conditions, rock, stiff soil, and soft soil, respectively. Because the fundamental frequency of the building 2.43π is close to that for soft soil ($\omega_g = 2.4\pi$), the story ductility ratios in Fig.6 are larger than those in Figs.4 and 5.

LIMIT STATES

In order to generate the fragility curve for a structure, the limit states of a structure must be defined. A limit state generally represents a state of undesirable structural behavior, for example, instability or structural collapse. Both structural and nonstructural damages in the event of an earthquake can be attributed to excessive interstory displacement; thus, the interstory displacement is a good measure of damage. In this study, the limit state is defined in terms of the system ductility ratio μ_m that is the largest value among all story ductility ratios.

$$\mu_m = \max(\mu_i) \tag{8}$$

Five limit states representing nonstructural damage, slight structural damage, moderate structural damage, severe structural damage, and collapse are established to categorize the degree of damage incurred in buildings during an earthquake. For each limit state, the probabilistic capacity is assumed to be lognormally distributed. The median capacity $\tilde{\mu}_R$ and the logarithmic standard deviation β_R are based on engineering judgment and shown in Table 6.

Table 6. Structural damage categories

Limit State	$\tilde{\mu}_R$	β_R
1. Nonstructural damage	1.0	0.3
2. Slight structural damage	1.5	0.3
3. Moderate structural damage	2.0	0.3
4. Severe structural damage	3.0	0.3
5. Collapse	4.0	0.3

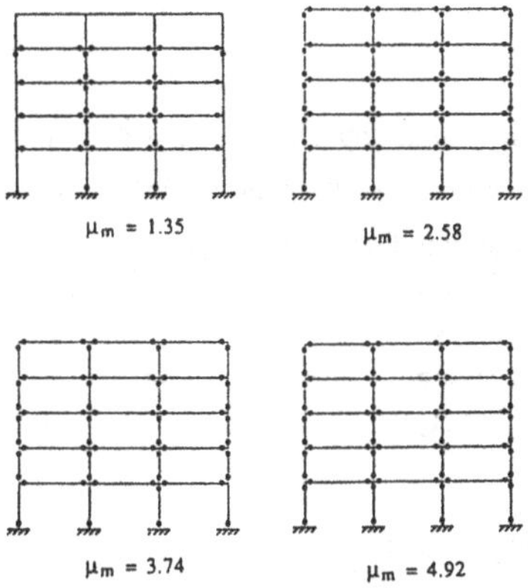

Fig 3. Nonlinear dynamic analysis - progression of failure for increasing values of μ_m

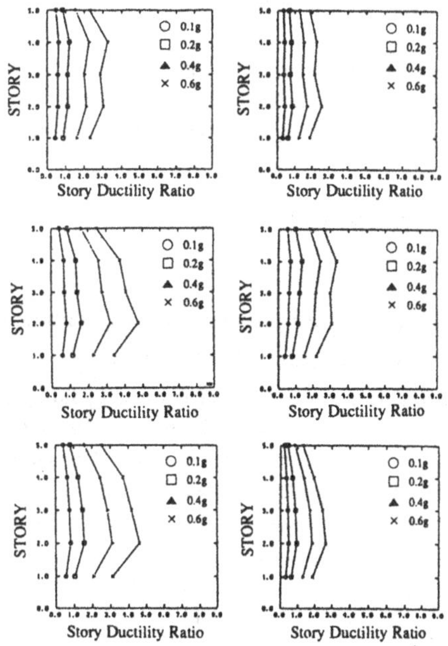

Fig 4. Story ductility ratio in six models with rock soils

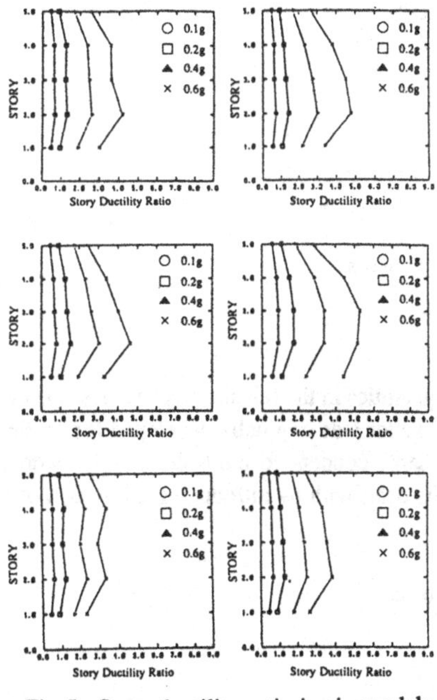

Fig 5. Story ductility ratio in six models with stiff soils

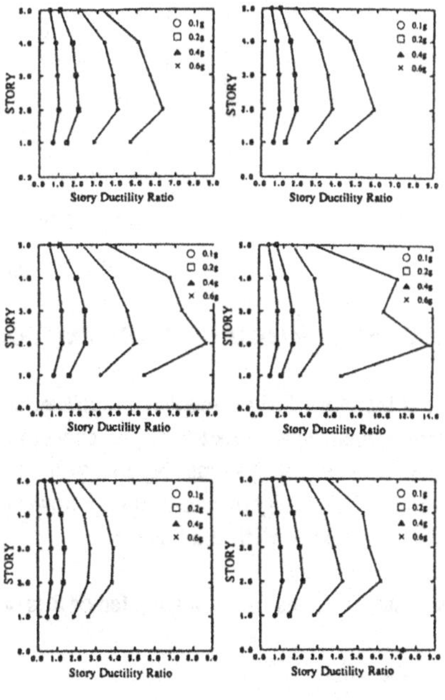

Fig 6. Story ductility ratio in six models with soft soils

FRAGILITY CURVES

For a given PGA level, the structural fragility with respect to a particular limit state is defined as the conditional probability that the structural response S exceeds the structural capacity R. The conditional limit state probability P_f can be determined as:

$$P_f = P_r\left(R<S\right) = \int_0^{\infty} \left[1 - F_S(r)\right] f_R(r)\,dr \tag{9}$$

where $f_R(\cdot)$ is the probability density function of structural capacity R and is determined according to the specified limit state. $F_s(\cdot)$ is the cumulative probability distribution of structural response S and is determined by fitting a lognormal distribution[10] to the computed response data. For example, 18 response data in terms of the system ductility ratio are obtained for PGA = 0.4 g. These data can be fitted by a lognormal distribution with the median value of 2.988 and the logarithmic standard deviation of 0.31. The fragility curve for a limit state can be constructed by evaluating P_f at different levels of PGA: 0.1, 0.2, 0.4 and 0.6 g. The fragility curves are plotted in Fig.7.

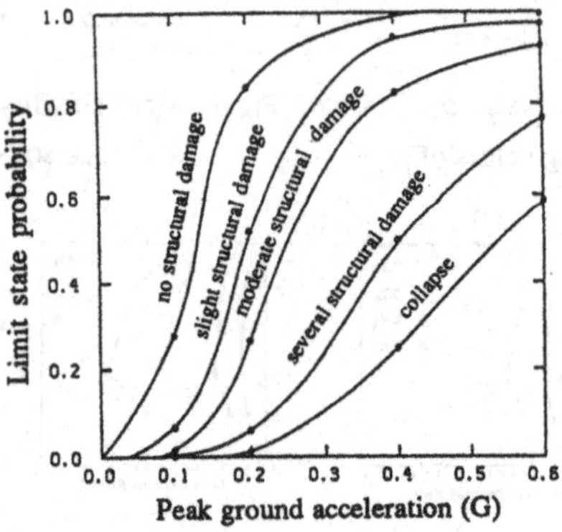

Fig 7. Fragility curves

RESPONSE MODIFICATION FACTOR

One of the more design-oriented and risk-related quantities in the seismic response issue is the response modification factor R. The concept of R factor was originally developed for single-degree-of-freedom systems. Application of the same concept to multi-degree-of-freedom systems is not straightforward. It can, however, be dealt with by introducing the notion of a system response modification factor.

The system response modification factor R is defined as

$$R = \frac{V_l}{V_n} \tag{10}$$

where V_n is the absolute maximum base shear obtained from a nonlinear time history analysis, while V_1 is the corresponding value obtained from a linear time history analysis using the same earthquake accelogram. The response modification factor is influenced by many parameters describing the earthquake-structure system. In this study, the R factor is considered as a function of μ_m.

$$R = f(\mu_m) \tag{11}$$

Figure 8 shows the correlation of the R factor and the system ductility ratio μ_m for all 18 samples subject to earthquakes with various PGA levels. The correlation proposed by Newmark and Hall[11] is also shown in the figure. Most of the R factor simulated in this study are smaller than the values proposed by Newmark and Hall. The difference may be caused by the multi-degree-of-freedom system used in this study. Note that these data are from a five-story steel frame structure. More data from other structures are needed to derive the relationship between R and μ_m.

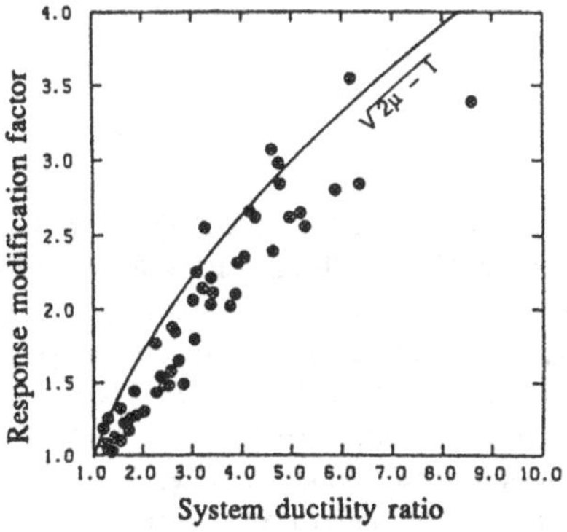

Fig 8. Response modification factor vs. system ductility ratio

SUMMARY AND CONCLUSIONS

In this paper we present a probabilistic analysis of a steel frame structure subject to earthquake ground motions. Uncertainties in seismic and structure parameters that define the earthquake-structural model are quantified by several representative values. Then, by using the Latin hypercube sampling technique, these representative values are combined to establish the samples of the earthquake-structure system. From this study, it shows that the Latin hypercube sampling technique can be used effectively to reduce the sample sizes. Nonlinear seismic analysis is performed by using the DRAIN-2D computer program. Thus, nonlinear seismic responses are explicitly taken into consideration.

For fragility analysis, five limit states representing nonstructural damage to collapse of structure are define and then the probabilistic structural capacity corresponding to each limit state can be established. On the basic of the probabilistic structural response and capacity, the limit state probabilities at various levels of peak ground accelerations can be computed and used to construct the fragility curves. These fragility curves can be used in probabilistic risk assessment of structures.

From the nonlinear and corresponding linear analyses, the response modification factor R for each sample of the earthquake-structure system can be computed. The correlation between the R factor and the system ductility ratio is presented. Further study of the R factor for steel structures is recommended to develop appropriate R factor for the design of structures.

ACKNOWLEDGMENTS

This paper is based on research sponsored by the National Center for Earthquake Engineering Research (NCEER) under contract number NCEER-88-1001 and 88-1003 (NSF Grant No. ECE086-07591). Any options, findings, and conclusions expressed in the paper are those of the authors and do not reflect the views of NCEER or NSF of the United States.

REFERENCES

1. American National Standard Institute. *Minimum Design Loads for Buildings and Other Structures*, ANSI A58.1-1982, New York, 1982.
2. American Institute of Steel Construction, Inc. *Manual of Steel Construction*, Eighth edition, Chicago, Illinois, 1986.
3. American Institute of Steel Construction, Inc. *Manual of Steel Construction, Load and Resistance Factor Design*, First edition, Chicago, Illinois, 1986.
4. Ellingwood, B. and Hwang, H.Probabilistic Descriptions of Resistance of Safety-Related Structures in Nuclear Plants, *Nuclear Engineering and Design*, 1985, 88, 169-178.
5. Iman, R.L and Conover, W.j. Small Sample Sensitivity Analysis Techniques for Computer Models With an Application to Risk Assessment, *Communications in Statics*, 1980, A9: 17, 1749-1842.
6. Hwang, H. and Jaw, J-W. *Statistical Evaluation of Response Modification Factors for Reinforced Concrete Structures*, Technical Report NCEER-89-0002, National Center for Earthquake Engineering Research, SUNY, Buffalo, New York, February 1989.
7. Shinozuka, M. Digital Simulation of Random Process in Engineering Mechanics with the Aid of FFT Technique, *Stochastic Problems in Mechanics*, S. T. Ariaratnam and H. E. E. Liepholz, eds., University of Waterloo Press, Waterloo, Ontario, Canada, 1974.
8. Kanaan, A. E. and Powell, G. H. *General Purpose Computer Program for Inelastic Dynamic Response of Plane Structures*, Technical Report 73-6, Earthquake Engineering Research Center, University of California, Berkeley, CA, April 1973.
9. Akiyama, H. *Earthquake-Resistant Limit-State Design for Buildings*, Tokyo University Press, 1985.
10. Hwang, H. and Jaw, J.-W. Probabilistic Damage Analysis of Structures, *Journal of Structural Engineering*, 1990, 116: 7.
11. Newmark, N. M. and Hall, W. J. Procedures and Criteria for Earthquake Resistant Design, in Building Practice for Disaster Mitigation, Building Science Series 46, National Bureau of Standards, Washington, DC, 1973, 209-236.

Galerkin Method to Analyze Systems with Stochastic Flexural Rigidity

T. Takada

Ohsaki Research Institute, Shimizu Corporation, 2-2-2 Uchisaiwai-cho, Chiyoda-ku, Tokyo 100, Japan

ABSTRACT

A new stochastic analysis method based on the Bubnov-Galerkin approximation is proposed herein for estimating the response variability of systems with spatially varying flexural rigidity. Such a flexural rigidity is idealized as a multi-dimensional, statistically homogeneous, continuous Gaussian stochastic field. Two kinds of techniques for evaluating the response statistics are utilized: a first-order perturbation technique and the Monte Carlo simulation technique. Numerical examples are presented in this paper.

INTRODUCTION

Stochastic responses of systems with spatially varying material properties, in general, become statistically non-homogeneous, non-Gaussian stochastic fields particularly when the material properties are idealized as stochastic fields[1]. Furthermore, it is extremely difficult to find an exact solution in most cases. Thus, many approximate treatments have been introduced.

Regarding the approximation, the present author would like to claim that even with linear elastic problems, most of stochastic analyses have not yet methodologically reached the present level already achieved by current deterministic analyses. Most analytical stochastic methods [2,3] simply utilize the deterministic solution and the perturbation technique, which must be verified by the Monte Carlo simulation method. Regarding most stochastic finite element methods (SFEMs) [4-6], discretization of the involved stochastic field can remarkably facilitate the analyses so that deterministic computer codes can conveniently be used. However, they sometimes require finer discretization that subsequently means an increase in computational cost. In other words, the conventional SFEMs are highly dependent upon the discretization for realizing the involved stochastic field. Therefore, regardless of whether the analytical method or the discretized method is used, considerable care is essential although they will obviously produce some sort of answer.

In this juncture, the analytical method such as is proposed in the literature[7,8] is quite useful for providing valuable insight into establishing a new approach[9,10] as well as into verifying the solution from any other approaches. Although this method is applicable to relatively simple systems consisting of trusses or beams, it is not applicable to complex systems with higher static indeterminacy or higher dimensionality. Needed, therefore, is a method that can treat such complex stochastic systems without using spatial discretization of the original stochastic field and that makes it feasible to perform even the Monte Carlo simulation technique. This is the major motivation behind the present study and this paper.

For these reasons, a stochastic analysis method will be proposed in this paper, based on the stochastic field theory and the Bubnov-Galerkin method which is one of the approximations even for deterministic problems. This method can treat the continuous stochastic field and is classified between the analytical and the discrete methods. Bending problems in which flexural rigidity of either beams or plates has spatial variation are systematically and conveniently formulated by virtue of the Galerkin approximation. It will then be shown that one of two kinds of approximations, either the first-order perturbation technique or the Monte Carlo simulation technique, can effectively be utilized when the response variability is evaluated. To exemplify the validity of the proposed method, numerical examples will be presented along with the conventional SFEM.

FORMULATION OF BENDING PROBLEMS BY GALERKIN METHOD

Stochastic Expression of Flexural Rigidity

Treated here will be such problems in which the flexural rigidity of either one-dimensional beams or two-dimensional plates spatially varies. Here, it is assumed that such flexural rigidity denoted by $K(\mathbf{x})$ is expressed as

$$K(\mathbf{x}) = K_0 (1 + f(\mathbf{x})), \tag{1}$$

in which K_0 is an expected rigidity $\langle K(\mathbf{x}) \rangle$ of $K(\mathbf{x})$, and $f(\mathbf{x})$ is a fluctuating part which is assumed to constitute a multi-dimensional, homogeneous, Gaussian stochastic field. Without lack of generality, $f(\mathbf{x})$ has zero expectation and the auto-correlation function such that

$$\langle f(\mathbf{x}) \rangle = 0, \text{and} \tag{2}$$

$$\langle f(\mathbf{x} + \xi)\, f(\mathbf{x}) \rangle = R_{ff}(\xi). \tag{3}$$

In Eq. (1), note that $K(\mathbf{x})$ represents EI for an elastic straight beam case, $Eh^3/12\,(1-\nu^2)$ for an elastic isotropic plate case with E being Young's modulus, I a sectional moment of inertia, ν a Poisson ratio, and h a plate thickness.

Derivation of Solution based on Galerkin Method

The problem to be solved here is finding an approximate solution $w(\mathbf{x})$ to the true deflection that can satisfy the following two equations, i.e., the fundamental equilibrium equation defined in a domain V and the boundary conditions at the boundaries S:

$$L(w) - p = 0 \quad \mathbf{x} \in V, \tag{4}$$

$$S(w) - q = 0 \quad \mathbf{x} \in S, \tag{5}$$

in which $L(\,)$ is a stochastic differential operator which is linear with respect both to the deflection and to the stochastic flexural rigidity $K(\mathbf{x})$, while $S(\,)$ is assumed to be a deterministic operator. p and q are deterministic quantities that imply respectively a distributed load and boundary conditions. Considering the beam case, $L(\ )$ is

$$L(\ ) = \frac{d^2}{dx^2}\left(EI\frac{d^2}{dx^2}\right), \tag{6}$$

while $L(\ )$ of the plate case can be found as

$$L(\ ) = \left\{\frac{\partial^2}{\partial x^2} \quad \frac{\partial^2}{\partial y^2} \quad 2\frac{\partial^2}{\partial x \partial y}\right\} K(x,y) \begin{bmatrix} 1 & \nu & 0 \\ \nu & 1 & 0 \\ 0 & 0 & (1-\nu) \end{bmatrix} \begin{Bmatrix} \frac{\partial^2}{\partial x^2} \\ \frac{\partial^2}{\partial y^2} \\ 2\frac{\partial^2}{\partial x \partial y} \end{Bmatrix}, \tag{7}$$

in which it should be noted that the Poisson ratio ν is considered to be deterministic in this paper to simplify the problems.

Since such a solution satisfying both Eqs. (4) and (5) is not always obtainable even in deterministic problems, many approximate procedures are proposed especially for deterministic plates, such as energy methods.

Returning to the stochastic problems, Eq. (4) is, in general, not solvable since the stochastic differential operator $L(\ )$ involves the stochastic field $f(\mathbf{x})$ characterized only in a stochastic manner. As an exception to this, Shinozuka[7] and Bucher et al.[8], whose work was based on the theory of differential equations, solved beam problems in which they made the same Gaussian assumption for the flexural flexibility $1/K(x)$ rather than the rigidity $K(x)$, as is seen in Eq. (1).

The Galerkin method begins with selecting trial functions which satisfy the boundary conditions prespecified by Eq. (5). For the stochastic problems, however, it is quite difficult to establish a set of stochastic trial functions compatible with the deterministic boundary conditions. In this paper, these trial functions are given deterministically. Therefore, the Galerkin solution to be derived in the following is approximate both in a deterministic and in a stochastic sense.

Using the deterministic trial functions $\phi_n(\mathbf{x})$, the approximated deflection $w(\mathbf{x})$ can be expressed in terms of the linear summation

$$w(\mathbf{x}) = \sum_{n=1}^{N} a_n \phi_n(\mathbf{x}) = \mathbf{\Phi}^t(\mathbf{x})\,\mathbf{a} \tag{8}$$

with a trial function vector $\mathbf{\Phi}(\mathbf{x})$, each component of which must be linearly independent and complete but not necessarily orthogonal:

$$\mathbf{\Phi}^t(\mathbf{x}) = \{\phi_1(\mathbf{x})\ \phi_2(\mathbf{x})\ \ldots\ \phi_N(\mathbf{x})\}, \tag{9}$$

where $\mathbf{a}$ is an undetermined coefficient vector which turns out to be stochastic.

The approximation of the deflection given in Eq. (8) obviously states that the true deflection as a non-homogeneous, non-Gaussian, continuous stochastic field is approximated by the stochastic function series that consists of the spatially continuous deterministic functions with random coefficients. This formulation may be an implicit discretization in which the non-homogeneous, continuous stochastic field constituted by the true deflection is decomposed into finite, non-homogeneous, continuous stochastic function spaces.

Using Eq. (8), the residual between the approximated and the true solutions is then orthogonalized to the trial functions in a domain V:

$$\int_V \{L(w) - p\} \Phi \, dx = 0. \tag{10}$$

Substituting Eq. (8) into the above and taking into account that $L(\,)$ is a linear operator with respect to w, the above equation turns out to be an algebraic equation of the undetermined vector **a**:

$$\int_V \Phi \mathbf{L}\left(\Phi^t\right) dx \; \mathbf{a} = \int_V p \, \Phi \, dx \tag{11}$$

with

$$\mathbf{L}\left(\Phi^t\right) = \{L(\phi_1) \; L(\phi_2) \; \ldots \; L(\phi_N)\}. \tag{12}$$

Using the following naming convention, Eq. (11) can be written in the simple form

$$\mathbf{K}\mathbf{a} = \mathbf{P}, \text{ where} \tag{13}$$

$$\mathbf{K} = \int_V \Phi \mathbf{L}\left(\Phi^t\right) dx, \text{ and} \tag{14}$$

$$\mathbf{P} = \int_V p \, \Phi \, dx. \tag{15}$$

The i-j component of the matrix **K** is

$$k_{ij} = \int_V L(\phi_i) \, \phi_j dx. \tag{16}$$

For the beam with the deterministic length l, various boundary conditions can be considered. If the boundary condition is homogeneous, k_{ij} become the simple form

$$k_{ij} = K_0 \int_0^l (1 + f(x)) \, \phi_i'' \phi_j'' dx, \tag{17}$$

in which the double prime means the second derivative with respect to the spatial coordinate x.

For a rectangular isotropic plate with $2\,l_x$ and $2\,l_y$ being respectively lengths in both directions, k_{ij} are found as

$$k_{ij} = K_0 \int_{-l_y}^{l_y} \int_{-l_x}^{l_x} (1 + f) \left[\left(\phi_i'' + \nu \phi_i^{\circ\circ} \right) \phi_j'' + 2\,(1 - \nu)\, \phi_i'^{\circ} \phi_j'^{\circ} + \left(\phi_i^{\circ\circ} + \nu \phi_i'' \right) \phi_j^{\circ\circ} \right] dx dy, \tag{18}$$

where $' = \partial/\partial x$, $^\circ = \partial/\partial y$. From the above two equations, it can be observed that k_{ij} can be divided into two parts: a deterministic and a stochastic part. The expressions of Eqs. (17) and (18) are quite similar to "a local integral or weghted integral" which the author has recently proposed[9-11]. It is interesting to note that even if ϕ_i are orthogonal each other, $\mathbf{K}$ does not become diagonal due to the presence of $f(\mathbf{x})$.

Equation (13) yields the solution for $\mathbf{a}$:

$$\mathbf{a} = \mathbf{K}^{-1}\mathbf{P}. \tag{19}$$

Finally, the approximate deflection can be evaluated as in

$$w(\mathbf{x}) = \Phi^t(\mathbf{x})\,\mathbf{a} = \Phi^t(\mathbf{x})\,\mathbf{K}^{-1}\mathbf{P}. \tag{20}$$

From Eq. (14), the components of the matrix $\mathbf{K}$ become Gaussian random variables since the stochastic differential operator $\mathbf{L}$ is linear with respect to $f(\mathbf{x})$. Note, however, that the vector $\mathbf{a}$ is no longer Gaussian since it requires the matrix inversion of the Gaussian $\mathbf{K}$. Now, the expectation and auto-correlation function of the deflection can formally be written as:

$$\langle w(\mathbf{x})\rangle = \Phi^t(\mathbf{x})\,\langle \mathbf{a}\rangle, \tag{21}$$

$$R_{ww}(\mathbf{x},\mathbf{y}) = \langle w(\mathbf{x})\, w(\mathbf{y})\rangle = \Phi^t(\mathbf{x})\,\langle \mathbf{a}\mathbf{a}^t\rangle \Phi(\mathbf{y}). \tag{22}$$

Next, assuming that the deflection can be obtained, the moment force denoted by $\sigma(\mathbf{x})$ is estimated employing the differentiation

$$\sigma(\mathbf{x}) = K(\mathbf{x})\, M(w(\mathbf{x})), \tag{23}$$

in which $M(\;)$ is a linear, deterministic differential operator with respect to the deflection. For the beam case, it is expressed as

$$M(\;) = -\frac{d^2}{dx^2}. \tag{24}$$

For the plate case, it becomes the vector operator

$$M(\;) = \begin{Bmatrix} -\left(\frac{\partial^2}{\partial x^2} + \nu\frac{\partial^2}{\partial y^2}\right) \\ -\left(\frac{\partial^2}{\partial y^2} + \nu\frac{\partial^2}{\partial x^2}\right) \\ (1-\nu)\,\frac{\partial^2}{\partial x \partial y} \end{Bmatrix} \tag{25}$$

where the first, second and third rows are associated with the moment forces M_{xx}, M_{yy} and M_{xy}, respectively.

Similarly to Eqs. (21) and (22), the expectation and the auto-correlation function of the moment force can be obtained:

$$\langle \sigma(\mathbf{x})\rangle = \langle K(\mathbf{x})\, M(w(\mathbf{x}))\rangle = \{\mathbf{M}(\Phi(\mathbf{x}))\}^t\,\langle K(\mathbf{x})\,\mathbf{a}\rangle, \tag{26}$$

$$\begin{aligned} \mathbf{R}_{\sigma\sigma}(\mathbf{x},\mathbf{y}) &= \langle K(\mathbf{x})\, M(w(\mathbf{x}))\, M^t(w(\mathbf{y}))\, K(\mathbf{y})\rangle \\ &= \mathbf{M}^t(\Phi(\mathbf{x}))\,\langle K(\mathbf{x})\,\mathbf{a}\mathbf{a}^t K(\mathbf{y})\rangle \mathbf{M}(\Phi(\mathbf{y}))\,' \end{aligned} \tag{27}$$

with

$$\mathbf{M}(\;) = \{M(\;)\; M(\;) \;...\; M(\;)\}^t . \tag{28}$$

First-order Approximation

As seen in Eqs. (21), (22), (26) and (27), it is impossible to rigorously evaluate the statistical moments of the undetermined random vector **a** since this vector is not Gaussian. To overcome this difficulty, the first-order approximation can be efficiently used as has been done in most stochastic analyses. Employing the Taylor expansion of Eq. (19) with respect to the random variables k_{ij} at their expected values and taking up to the first term, Eq. (19) turns out to be

$$\mathbf{a} \approx \mathbf{a}^0 + \sum_i^N \sum_j^N \mathbf{a}_{ij}^I \Delta k_{ij} \tag{29}$$

with

$$\Delta k_{ij} = k_{ij} - \langle k_{ij} \rangle, \tag{30}$$

where $\mathbf{a}^0$ and $\mathbf{a}_{ij}^I$ are

$$\mathbf{a}^0 = \langle \mathbf{K} \rangle^{-1} \mathbf{P}, \tag{31}$$

$$\mathbf{a}_{ij}^I = \frac{\partial \mathbf{a}}{\partial k_{ij}} \Big|_{k_{ij}=\langle k_{ij} \rangle} = -\langle \mathbf{K} \rangle^{-1} \frac{\partial \mathbf{K}}{\partial k_{ij}} \Big| \langle \mathbf{K} \rangle^{-1} \mathbf{P}. \tag{32}$$

Using Eq. (26), expectations appearing in Eqs. (21) and (22) are approximated as

$$\langle \mathbf{a} \rangle \approx \mathbf{a}^0, \tag{33}$$

$$\langle \mathbf{a}\mathbf{a}^t \rangle \approx \mathbf{a}^0 (\mathbf{a}^0)^t + \sum_i^N \sum_j^N \sum_k^N \sum_l^N \mathbf{a}_{ij}^I (\mathbf{a}_{kl}^I)^t \langle \Delta k_{ij} \Delta k_{kl} \rangle. \tag{34}$$

Similarly, the expectations appearing in Eqs. (26) and (27) become

$$\langle K(\mathbf{x})\, \mathbf{a} \rangle \approx K_0 \mathbf{a}^0, \tag{35}$$

$$\begin{aligned} \langle K(\mathbf{x})\, \mathbf{a}\mathbf{a}^t K(\mathbf{y}) \rangle \approx K_0^2 \Big[& \mathbf{a}^0 (\mathbf{a}^0)^t + \langle f(\mathbf{x})\, f(\mathbf{y}) \rangle \mathbf{a}^0 (\mathbf{a}^0)^t \\ & + \sum_i^N \sum_j^N \mathbf{a}^0 (\mathbf{a}_{ij}^I)^t \{ \langle f(\mathbf{x})\, \Delta k_{ij} \rangle + \langle f(\mathbf{y})\, \Delta k_{ij} \rangle \} \\ & + \sum_i^N \sum_j^N \sum_k^N \sum_l^N \mathbf{a}_{ij}^I (\mathbf{a}_{kl}^I)^t \langle \Delta k_{ij} \Delta k_{kl} \rangle \Big]. \end{aligned} \tag{36}$$

As seen in the above, the response statistics can finally be expressed in terms of the characteristics of the stochastic field $f(\mathbf{x})$. Using Eqs. (17) and (18), $\langle f(\mathbf{x})\, \Delta k_{ij} \rangle$ and $\langle \Delta k_{ij} \Delta k_{kl} \rangle$ in the above can be evaluated as in:

$$\langle f(\mathbf{x})\, \Delta k_{ij} \rangle = \int_V \Psi_{ij}(\mathbf{r})\, R_{ff}(\mathbf{x},\mathbf{r})\, d\mathbf{r}. \tag{37}$$

$$\langle \Delta k_{ij} \Delta k_{kl} \rangle = \int_V \int_V \Psi_{ij}(\mathbf{r}_1)\, \Psi_{kl}(\mathbf{r}_2)\, R_{ff}(\mathbf{r}_1,\mathbf{r}_2)\, d\mathbf{r}_1 d\mathbf{r}_2 \tag{38}$$

with

$$\Psi_{ij} = \begin{cases} \phi_i''\phi_j'' & \text{for beams} \\ \left(\phi_i'' + \nu\phi_i^{\circ\circ}\right)\phi_j'' + 2\,(1-\nu)\,\phi_i'^{\circ}\phi_j'^{\circ} + \left(\phi_i^{\circ\circ} + \nu\phi_i''\right)\phi_j^{\circ\circ} & \text{for plates.} \end{cases}$$

Monte Carlo Solution

The Monte Carlo simulation technique is an effective alternative when the first-order approximation is not appropriate primarily due to the strong nonlinear relationship between the original stochastic field and the response field. A digital generation technique of multi-dimensional, Gaussian stochastic field $f(\mathbf{x})$ is adopted firstly[12]. The samples of the random variables k_{ij} are secondly realized through the numerical integration of Eq. (16). Using these samples, Eq. (19) is solved and the response variability of the deflection and the bending moment are subsequently evaluated. In the simulation method, only the sample size should be carefully determined although no approximations are involved as far as the evaluation of the statistical response is concerned.

When one wants to evaluate only the variability of the deflection, only the term $\langle \mathbf{a}\mathbf{a}^t \rangle$ is needed, as seen in Eqs. (21) and (22). It is of great interest to note that it is not necessary to generate the original stochastic field $f(\mathbf{x})$ in multi-dimensional space, rather only the samples of the random vector $\mathbf{a}$ are needed. The samples of $\mathbf{a}$ can be realized, based on the covariance matrix of k_{ij}[8]. This confirms that the problem of the stochastic field is transformed into that of the finite random variables by means of the decomposition of the deflection field. This leads to tremendous savings in numerical effort in carrying out the Monte Carlo simulation.

NUMERICAL EXAMPLES AND DISCUSSIONS

Both End-fixed Stochastic Beam

A both end-fixed beam is analyzed. The uniformly distributed unit load is statically and deterministically acting along the beam axis. Unit mean flexural rigidity ($K_0 = 1$) is assumed. The following trial functions $\phi_n(x)$ are intuitively selected:

$$\phi_n(x) = x\,(l-x)\,\sin\frac{n\pi}{l}x \quad : (n = 1, 2, \ldots N)\;, \tag{39}$$

where l is the beam length. The boundary conditions at both ends can be easily found to be satisfied by the trial functions. These trial functions take into account that anti-symmetric deflection modes with respect to the midspan may exist since the flexural rigidity spatially varies despite the symmetry of the loading pattern and the boundary conditions.

The auto-correlation function $R_{ff}(\xi)$ is now taken as

$$R_{ff}(\xi) = \sigma_f^2 e^{-(\xi/b)^2}, \tag{40}$$

where σ_f is a standard deviation associated with $f(x)$ and b is "a correlation distance" that implies how fast the correlation decays along the beam axis. σ_f is set equal to 0.1. The b/l value is adopted in two ways: $b/l = 1.0$ and $b/l = 0.1$. The former case represents the more smoothly fluctuating stochastic field.

Adopting the first-order perturbation technique described in the previous section, the correlation terms, i.e., $\langle f(x)\,\Delta k_{ij}\rangle$ and $\langle \Delta k_{ij} \Delta k_{kl}\rangle$, must be evaluated. These integrations in Eqs. (37) and (38) are carried out numerically since it is difficult to evaluate these terms in an explicit form. The number of expansion N is changed to 2, 4 and 6 in order to see the solution convergence.

Figures 1 and 2 show the spatial distribution of the standard deviation of the deflection and the bending moment along with the results from the conventional first-order perturbation-based SFEM[5]. Here, 50 sub-elements for $b/l = 0.1$ and 10 sub-elements for $b/l = 1.0$ are used in the conventional SFEM. The convergence of the standard deviation of the deflection can be observed to be excellent while that of the bending moment is slower as N increases for $b/l = 0.1$.

From these results, the following statement can be made. Only a few trial functions are needed to produce accurate results in the proposed method, while the conventional SFEM requires a large number of finite elements. However, it should be noted that this agreement is meaningful only from the first-order approximation perspective, and the appropriateness of this approximation can be examined only through the Monte Carlo simulation technique.

Four Edge-clamped Stochastic Plate

A square plate with a stochastic flexural rigidity is now analyzed. The plate is subjected to a deterministic uniform unit load. Like the beam example, the trial functions in the two-dimensional space are selected as

$$\phi_n(x, y) = \phi_{n_x n_y}(x, y) = \left(l^2 - x^2\right)\left(l^2 - y^2\right) \sin\frac{n_x \pi}{2l}(x + l) \sin\frac{n_y \pi}{2l}(y + l) \quad (41)$$

Here, $2l$ is a plate length in both directions. The origin is selected at the center of the plate. The numerical analysis is done under the conditions $K_0 = 1$ and $\nu = 0.3$. The two-dimensional auto-correlation function $R_{ff}(\xi, \eta)$ is now taken as

$$R_{ff}(\xi, \eta) = \sigma_f^2 e^{-(|\xi|+|\eta|)/b}, \quad (42)$$

where σ_f is a standard deviation of $f(x, y)$ and is set equal to 0.1, and b is a two-dimensional correlation distance and is assumed to be $2l$. Since the above correlation function is separable into two directions, numerical integrations of $\langle f(x, y)\, \Delta k_{ij}\rangle$ and $\langle \Delta k_{ij} \Delta k_{kl}\rangle$ are not cumbersome jobs.

The first-order perturbation technique is utilized again in the proposed method. Similar to the previous example, the conventional SFEM[5] is also implemented by using a 10 × 10 mesh division. Here, a non-conforming rectangular bending plate element is adopted since it can yield good results in deterministic problems. The mesh division was determined not only from the solution accuracy in a deterministic sense, but also from the characteristics of the stochastic field $f(x, y)$.

Table 1 lists the response statistics at specific locations from both methods. The mean response from both methods can be compared with the deterministic exact solution[13] since the two methods adopt the first-order approximation. There is some slight difference in magnitude between the results from both methods. These differences in the standard deviation may come from those of the mean values.

Here, in order to see the solution convergence in the proposed method, the following sum of the square error to the specified case (N_0) is introduced:

$$\epsilon_{A_N}^2 = \frac{\int_V \left(\sqrt{Var\left[A_N(x,y)\right]} - \sqrt{Var\left[A_{N_0}(x,y)\right]}\right)^2 dV}{\int_V Var\left[A_{N_0}(x,y)\right] dV}, \quad (43)$$

where $Var[\ ]$ denotes the variance of the argument. $A_N(x,y)$ represents the response, e.g. w, M_{xx}, M_{yy} and M_{xy}, when the total number of superposition N is used. It can be observed from the table that the sum of the square error gradually approaches zero as N increases although the mean and variance values at the specific locations do not converge well. This behavior of the response at specific locations can be considered to result from the selected trial functions, which include anti-symmetric mode shapes having a zero value at the center of the plate. Therefore, the solution convergence cannot be improved even if N increases from 1 to 4 and from 9 to 16.

Figure 3 compares the results from both methods regarding the spatial distributions of the standard deviation of the response. As is evident, the proposed method agrees well with the overall tendency of the results from the conventional SFEM.

CONCLUSIONS

A new analysis method for systems with spatially fluctuating flexural rigidity was proposed. This method, based on the Galerkin approximation in which the trial functions are assumed to be deterministic, suggests that a new stochastic finite element method can be developed. Through numerical examples, the results from this method were confirmed to be close to those from the conventional stochastic finite element method. Finally, by virtue of the Galerkin approximation, this method is expected to be applicable to dynamic and/or nonlinear stochastic problems.

ACKNOWLEDGMENT

Some of the ideas presented herein arose during 1987, at which time the author stayed at Columbia University as a visiting scholar of Professor M. Shinozuka. The author deeply acknowledges his useful advice.

REFERENCES

[1] Takada, T. Response Variability and Reliability of Beams with Stochastic Mechanical Property, Journal of Structural and Construction Engineering, No. 409, pp. 67-74, 1990 (in Japanese)

[2] Bolotin, V. V. Statistical Method and Reliability Problems in Structural Design, Translated by S. Kobayashi, et al., Baifu-kan Publishing Company, 1981 (in Japanese)

[3] Baker, R., Zeitoun, D. G. and Uzan, J. Analysis of a Beam on Random Elastic Support, Soils and Foundations, Japanese Society of Soil Mechanics and Foundation Engineering, Vol. 29, No. 2, pp. 24-36, 1989

[4] Baecher, G. B. and Ingra, T. S. Stochastic FEM in Settlement Predictions, Journal of GT, ASCE, Vol. 107, No. 4, pp. 449-463, 1981

[5] Nakagiri, S. and Hisada, T. Introduction of Stochastic Finite Element Method, Baifu-kan Publishing Company, 1986 (in Japanese)

[6] Liu, W. K., and Belytschko, T. and Mani, A. Random Field Finite Elements, Int. Journal for Numerical Methods in Engineering, Vol. 23, pp. 1831-1845, 1986

[7] Shinozuka, M. Structural Response Variability, Journal of EM, ASCE, Vol. 113, No. 6, pp. 825-842, 1987

[8] Bucher, C. G. and Shinozuka, M. Structural Response Variability II, Journal of EM, ASCE, Vol. 114, No. 12, pp. 2035-2054, 1988

[9] Takada, T. Weighted Integral Method in Stochastic Finite Element Analysis, Journal of Probabilistic Engineering Mechanics, Vol. 5, No.3, pp.146-156, 1990

[10] Takada, T. Weighted Integral Method in Multi-dimensional Stochastic Finite Element Analysis, Journal of Probabilistic Engineering Mechanics, Vol. 5, No.4, pp.157-166, 1990

[11] Takada, T. and Shinozuka, M. Local Integration Method in Stochastic Finite Element Analysis, Proc. of the 5th ICOSSAR, Vol. II, pp. 1073-1080, San Fransisco, 1989

[12] Shinozuka, M. Stochastic Fields and Their Digital Simulation, Stochastic Methods in Structural Dynamics, Martinus Nijhoff Publishers, 1987

[13] Timoshenko, S. P. and Krieger, S. W. Theory of Plates and Shells, 2nd edition, McGraw-Hill, pp.197-205, 1959

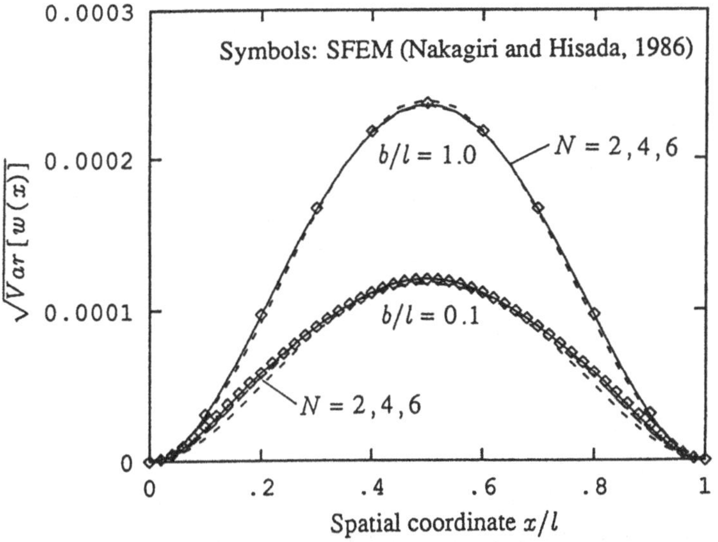

Fig. 1 Spatial distribution of standard deviation of deflection

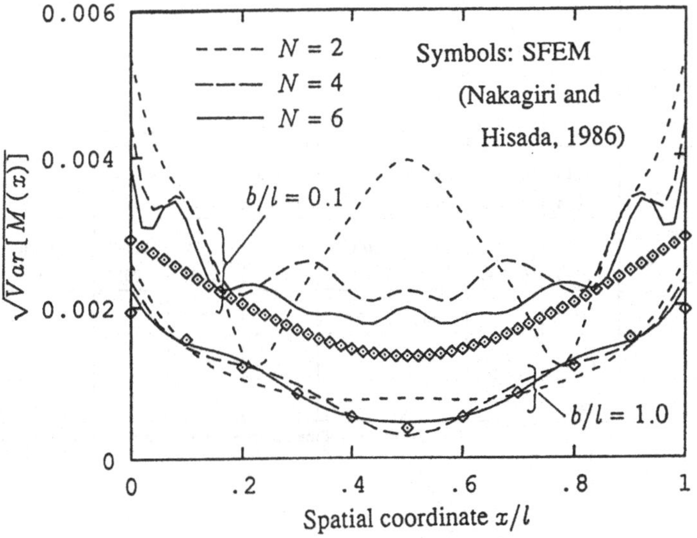

Fig. 2 Spatial distribution of standard deviation of bending moment

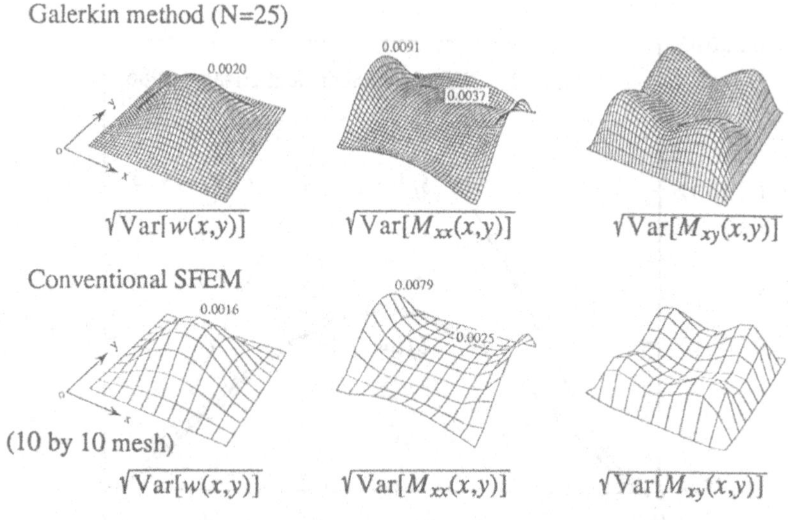

Fig. 3 Spatial distribution of standard deviation of various response

	Galerkin method					Conventional SFEM		Deterministic Exact Solution
	N=1	N=4	N=9	N=16	N=25	4×4	10×10	
Center deflection $\langle w(0,0)\rangle$	0.0243	0.0243	0.0229	0.0229	0.0232	0.0225	0.0207	0.0202
$\sqrt{Var[w(0,0)]}$	0.00213	0.00213	0.00201	0.00201	0.00203	0.00175	0.00158	–
Center moment $\langle M_{xx}(0,0)\rangle$	0.1410	0.1410	0.0997	0.0997	0.1143	0.1107	0.0946	0.0924
$\sqrt{Var[M_{xx}(0,0)]}$	0.0063	0.0063	0.0036	0.0036	0.0037	0.0032	0.0025	–
Edge moment $\langle M_{xx}(-1,0)\rangle$	-0.153	-0.153	-0.176	-0.176	-0.190	-0.192	-0.203	-0.2052
$\sqrt{Var[M_{xx}(-1,0)]}$	0.0116	0.0098	0.0097	0.0090	0.0091	0.0080	0.0079	–
Sum of Square Error ε_w (%)	6.2	4.9	1.0	0.8	–	–	–	–
Sum of Square Error ε_{Mxx} (%)	41.2	22.0	13.2	5.8	–	–	–	–
Sum of Square Error ε_{Mxy} (%)	34.2	12.0	4.6	4.2	–	–	–	–

ε are the sum of the square error to the specified case ($N_0=25$).

Table 1 Comparison of solution results

Monte Carlo Simulation of Beams on Winkler Foundation

P. de Simone, A. Ghersi, R. Mauro
Naples University, Istituto Tecnica, Fondazioni, Via Claudio 21, 80125 Napoli, Italy

ABSTRACT

Modulus of subgrade reaction and displacement of a beam foundation are considered respectively as input and output random functions, connected by a non-linear fourth order operator.
Relevant statistics of the output random functions are obtained through a Monte Carlo simulation, based on a Direct Boundary Element formulation of the problem, non-linear term being handled with an iterative technique.
An analysis is eventually performed on the output statistics, with special emphasis on covariance functions, its dependence on stiffness and loading conditions being put forward.

INTRODUCTION

Structural design of beam foundations is usually based on Winkler hypothesis of soil reaction simply proportional to the local displacement of the beam through some coefficient k, the modulus of subgrade reaction (see Hetényi[6], Terzaghi[9]). In current design practice k is assumed to be constant along the beam axis, despite its intrinsic variable character, which is a direct consequence of the intimate heterogeneous nature of soils.
In order to take in some way account of the variability of k, some Authors suggest to use two different values of k, namely the minimum and the maximum one, then performing two separate k-constant analyses; as a matter of fact this approach can give unsafe results as actual behaviour can lie beyond the range so obtained.
In a previous paper the authors[4] have adopted a probabilistic approach, considering k as a random function varying according to a lognormal law. Using a Monte Carlo simulation in conjunction with a DBEM algorithm they have analyzed a beam subjected to concentrated loading at the ends, comparing actual results with the ones of the aforementioned k-constant approach. Different probabilistic approaches have been followed by other Authors to handle similar problems (e.g. see Baratta[2]).
In this paper the previous Monte Carlo simulation by the authors[4] is extended to comprise more general loading conditions such as internal concentrated loading. A generalization to this situation of the previous DBEM algorithm obtained by the authors in a companion paper[5], which is particularly well suited to use in this Monte Carlo simulation, is adopted. An analysis of stationarity properties of output random functions is finally performed.

MODULUS OF SUBGRADE REACTION AND BEAM DISPLACEMENT AS RANDOM FUNCTIONS

Natural soils invariably exhibit marked heterogeneous behaviour, their properties being always characterized by some dispersion even when measurements are effectuated at close points. As a consequence, analyses based on the hypothesis of homogeneous medium provide results which can be distant from actual behaviour. On the other hand, owing to the small amount of measurements of soil properties usually made in geotechnical engineering, an approach based on a heterogeneous medium completely defined from a deterministic point of view is not feasible. An obvious way to overcome this situation is to perform a probabilistic analysis of the problem, where relevant soil properties are considered as random functions.

Such an approach can be actually followed to simulate variation of the modulus of subgrade reaction along the axis of a beam on a Winkler foundation; as a matter of fact, this property depends on the distribution within the ground of soil modulus, and can thus be considered as a random function itself.

From this viewpoint, the well known governing equation

$$EI\frac{d^4w}{dy^4}+kbw=q \tag{1}$$

where EI is the flexural rigidity, b the breadth, q the distributed load and y the abscissae along the axis of the beam, can be considered a differential transformation (for any given loading condition) of the input random function $k=\tilde{k}(y)$ into some output function w, which is then a random function $w=\tilde{w}(y)$ itself.

The equation can then be written

$$EI\frac{d^4\tilde{w}(y)}{dy^4}+\tilde{k}(y)b\tilde{w}(y)=q \tag{2}$$

and the differential operator of equation (2) can be considered as a filter acting on the input random function $k=\tilde{k}(y)$ to give the output one $w=\tilde{w}(y)$. The higher or lower degree of transformation inducted on the output random function by such a filtering process depends on the characteristics of the operator itself. On the other hand the kind of analysis too is strongly dependent on the type of operator; in case of linear differential operators relevant statistics of the output function can be analytically inferred from those of the input functions (e.g. see Papoulis[7]), while in case of non-linear operators the use of indirect techniques such as Monte Carlo simulation is more feasible. Owing to the non-linearity of the Winkler operator filter of equation (2), Monte Carlo simulation is the approach followed in this paper. Each random function $\tilde{F}(y)$, the input as well as the output ones, may be sampled both by the set of M realizations $F_m(y)$ $(1\le m\le M)$ and by the parametered family of random variables $\tilde{F}(y_n)$ at N selected sections $(1\le n\le N)$. An analysis has been done on the mean value, standard deviation and autocorrelation functions with reference to the cases analyzed, in order to explore the relation between characteristics of input and output random functions.

DIRECT BOUNDARY ELEMENT FORMULATION

In the Monte Carlo simulation of a beam on a Winkler medium, the modulus of subgrade reaction is a random function. The realization $w_m(y)$ of the output random function $\tilde{w}(y)$ corresponding to the generic sampling $k_m(y)$ of $\tilde{k}(y)$ is obtained solving equation (2), rewritten with reference to the m-th simulation

$$EI\frac{d^4w_m(y)}{dy^4}+k_m(y)bw_m(y)=q \tag{3}$$

together with given boundary conditions; owing to the arbitrary variability of k, an algorithm is then required for the solution of equation (3) which is versatile enough to be used in conjunction with any distribution of k.

In a previous paper the authors[4] have shown that an algorithm possessing such requirements can be obtained using a DBEM approach, in the case of beams subjected to concentrated loading at the ends. In a companion paper the authors[5] have extended the aforementioned algorithm to the case of beams on elastoplastic Winkler medium, subjected to internal concentrated loadings. By means of a systematic use of the Dirac delta distribution δ and its first generalized derivative δ' (see e.g. Schwartz[8]), the general case too can receive a very compact DBEM treatment, very similar to the one of the previous case and with final governing equations characterized by the same degree of complexity. This is precisely the algorithm used in this paper as a basis for the Monte Carlo simulation. In this section the main points of the algorithm will be outlined; for more details the readers are referred to the companion paper[5].

Introducing the non-dimensional quantities

$$L=l\bar{\lambda}$$

$$Y=y\bar{\lambda}$$

$$W=w\bar{\lambda}$$

$$K=\frac{k}{\bar{k}}$$

$$R=KW$$

where l is the length of the beam, $\bar{k}$ the mean value of $k_m(y)$ along the beam and $\bar{\lambda}$ the related characteristic according to Hetényi

$$\bar{\lambda}=\sqrt[4]{\frac{\bar{k}b}{4EI}}$$

the differential equation governing the problem of a beam on a k-variable Winkler medium subjected to non-dimensional distributed load $Q=q\bar{\lambda}/\bar{k}b$ and concentrated forces $F_i=f_i\bar{\lambda}^2/\bar{k}b$ and moments $M_j=m_j\bar{\lambda}^3/\bar{k}b$ respectively at i and j internal cross sections, can be written in the form

$$\frac{1}{4}\frac{d^4W}{dY^4}+R=Q+F_i\delta_i+M_j\delta_j' \tag{4}$$

Following the general pattern of the Direct Boundary Element Method (e.g. see Brebbia, Dominguez[3]), equation (4) is multiplied by a function W^* and then integrated by parts four times. In order to make use of a simple Green function already available in literature, the problem of a beam on a k-constant Winkler medium has been considered as the adjoint problem; with such a choice for W^* the Somigliana-type expression for W finally obtained assumes the form of an integral equation, namely

$$W(X) = [-S^{*}(Y,X)\,W(Y) + M^{*}(Y,X)\,T(Y) - T^{*}(Y,X)\,M(Y) + W^{*}(Y,X)\,S(Y)]_0^L +$$
$$+ \int_L W^{*}(Y,X)\,Q(Y)\,dY - \int_L (K-1)\,W^{*}(Y,X)\,W(Y)\,dY +$$
$$+ F_i\,W^{*}(Y_i,X) - M_j\,T^{*}(Y_j,X) \qquad (5)$$

T, M, S being respectively non dimensional rotation, bending moment and shear along the beam, and T^{*}, M^{*}, S^{*} similar quantities for the adjoint problem.
A similar identity for rotation (the Somigliana-type relation for the rotation) can be obtained by deriving relation (5) with respect to X, getting

$$T(X) = [W^{*}(Y,X)\,W(Y) - S^{*}(Y,X)\,T(Y) - 4\,M^{*}(Y,X)\,M(Y) - T^{*}(Y,X)\,S(Y)]_0^L$$
$$- \int_L T^{*}(Y,X)\,Q(Y)\,dY + \int_L (K-1)\,T^{*}(Y,X)\,W(Y)\,dY$$
$$- F_i\,T^{*}(Y_i,X) - 4M_j\,M^{*}(Y_j,X) \qquad (6)$$

As usual, collocation of equations (5), (6) at the ends of the beam, where forces and moments (or more generally two quantities among shear, moment, rotation and displacement) are known, gives a system of four (integral) equations.
The structure of equations (5), (6) suggests to solve the system by means of an iterative technique, considering the integral on the right side as a known quantity, then adjusting it successively until convergence is reached. As a matter of fact, the convergence of this algorithm is generally very fast, as has been verified by the authors in a large number of cases both in the present paper and in the previous one[4]; as a consequence it appears suitable for application in the present Monte Carlo simulation.

STATISTICAL ANALYSIS AND NUMERICAL RESULTS

In the present analysis the beam has been subdivided into fifty portions by means of equally spaced sections (N=51). The number of realizations M considered ranges from 100 to 1000, values which seem quite adequate to the kind of problem examined (see e.g. Augusti et al.[1]).
The input random function $\tilde{K}(Y)$ has been assigned through the N independent random variables $\tilde{K}(Y_n)$, which have been obtained by means of a generation of random numbers satisfying a lognormal law of probability density function and having mean value equal to one and the same variance at every section; the input random function is then strictly stationary (see e.g. Papoulis[7]).
A statistical analysis for each one $\tilde{F}(Y)$ of the output r.f. has been performed evaluating the mean value function $\mu_F(Y)$, the variance function $\sigma_F^2(Y)$ and the autocovariance function $\beta_F(Y_{n_1}, Y_{n_2})$ together with the normalized autocovariance (or autocorrelation) function $\rho_F(Y_{n_1}, Y_{n_2})$, evaluated using respectively the estimators

$$\bar{\mu}_F(Y) = \frac{1}{M}\sum_{m=1}^{M} F_m(Y)$$

$$\bar{\sigma}_F^2(Y) = \left[\frac{1}{M}\sum_{m=1}^{M} F_m^2(Y) - \bar{\mu}_F^2(Y)\right]\frac{M}{M-1}$$

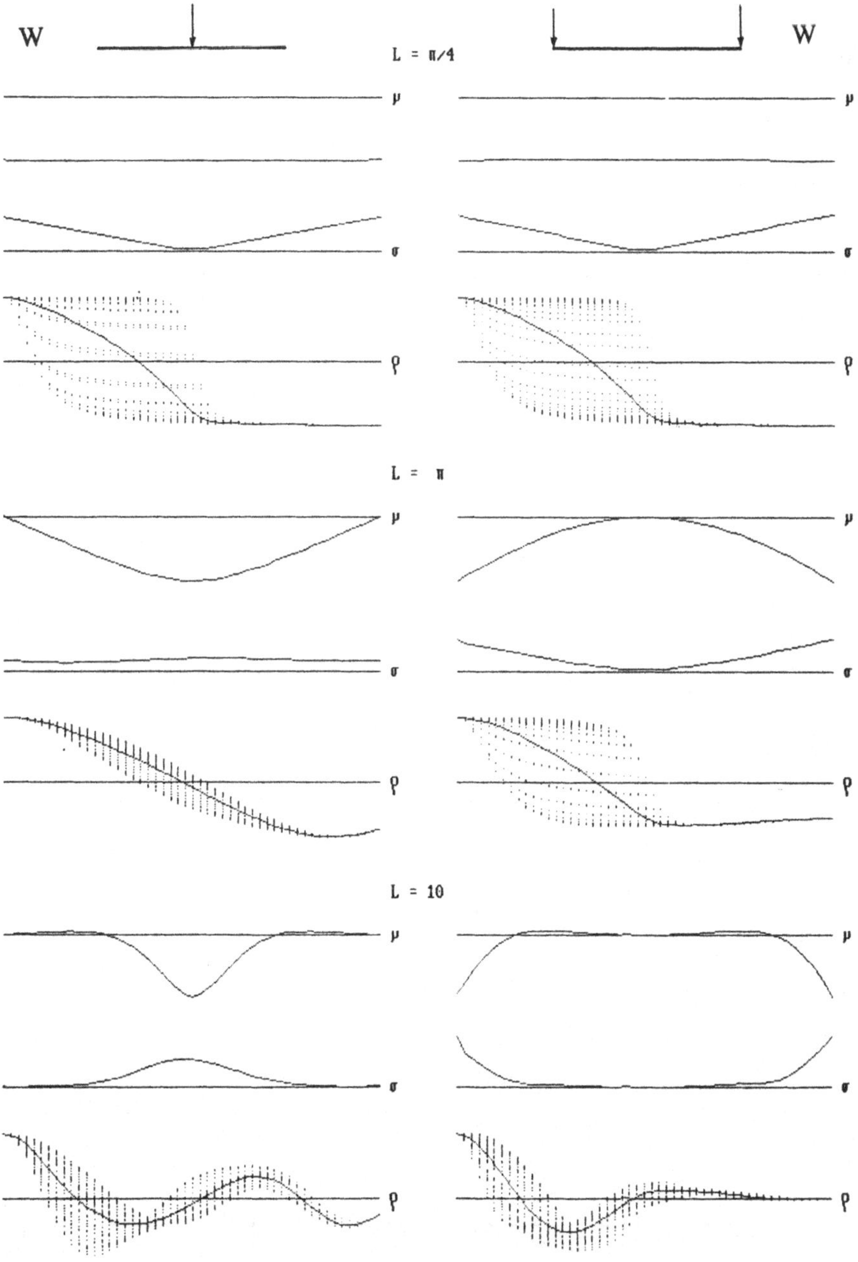

fig. 1 - Mean value μ_W, standard deviation σ_W and normalized autocovariance ρ_W for displacement, for beams subjected to one or two concentrated forces

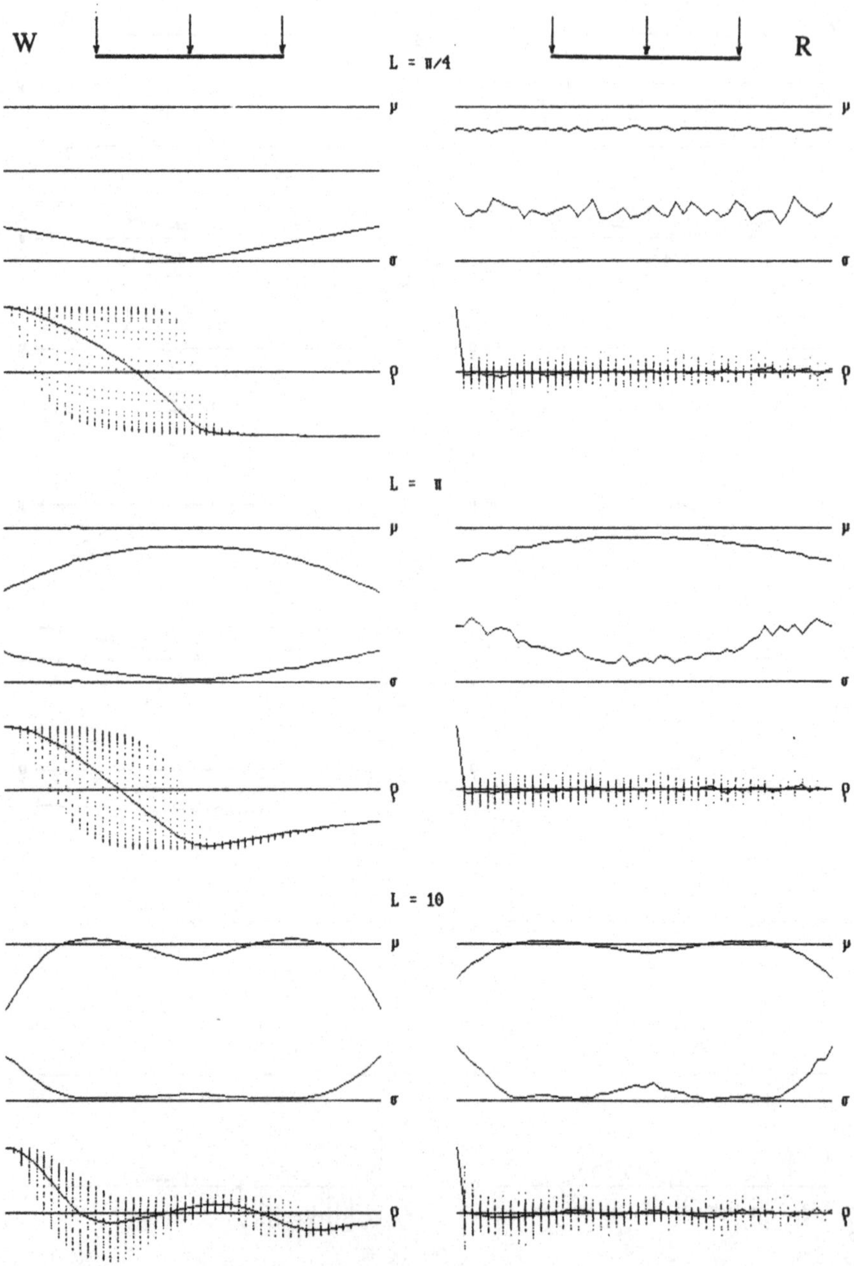

fig. 2 - Mean value, standard deviation and normalized autocovariance for displacement and soil reaction, for beams subjected to three concentrated forces

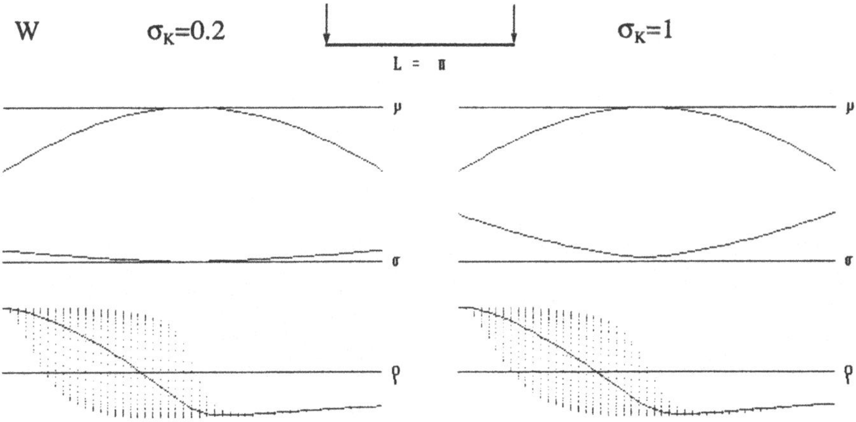

fig. 3 - Effect of standard deviation σ_K of input random function on numerical results

$$\overline{\beta}_F\left(Y_{n_1},Y_{n_2}\right)=\left[\frac{1}{M}\sum_{m=1}^{M}F_m\left(Y_{n_1}\right)F_m\left(Y_{n_2}\right)-\overline{\mu}_F\left(Y_{n_1}\right)\overline{\mu}_F\left(Y_{n_2}\right)\right]\frac{M}{M-1}$$

$$\overline{\rho}_F\left(Y_{n_1},Y_{n_2}\right)=\frac{\overline{\beta}_F(Y_{n_1},Y_{n_2})}{\overline{\sigma}_F(Y_{n_1})\,\overline{\sigma}_F(Y_{n_2})}$$

The analysis has been performed with reference to three load conditions (vertical forces symmetrically applied to 1, 2 and 3 sections) and to three values of the relative stiffness ($L=\pi/4$, π, 10). Some numerical results are synthesized in the diagrams of figs. 1 to 4, where estimated mean value $\mu(Y)$, standard deviation $\sigma(Y)$ and normalized autocovariance $\rho(\Delta Y)$ are shown. In plots of the normalized autocovariance functions each dot refers to a couple of sections n_1 and n_2, while solid lines indicate their mean values.

In the cases of stiff beams ($L=\pi/4$) the mean value of displacement μ_W is substantially constant while the standard deviation σ_W increases quite linearly moving from the mid-point of the beam. Single values of ρ_W greatly differ from their mean value when $\Delta Y < L/2$, but most of them are concentrated near the limiting values +1 and -1, the greater ΔY the more the -1 ones, positive values completely disappearing when $\Delta Y > L/2$. The trend of the normalized autocovariance function shows a negative linear correlation between sections farther then $L/2$, i.e. when $W_m(Y_{n_1}) > \mu_W(Y_{n_1})$ it must be $W_m(Y_{n_2}) < \mu_W(Y_{n_2})$ if $|Y_{n_1} - Y_{n_2}| > L/2$. The three functions μ, σ and ρ concordantly show that the difference between each realization and the mean value is mainly due to a rotation α of the beam around its mid-point. The random function $\tilde{W}(Y)$ may then be described with a good accuracy by a simple random function given by the product of a random variable $\tilde{\alpha}$ times a linear function, namely $W_m(Y) = (Y - L/2)\alpha_m$.

In the opposite cases of very flexible beams ($L=10$) the mean value function μ_W looks very similar to the wavy displacement diagram of an infinitely-long beam on a k-constant soil. The standard deviation function σ_W varies similarly to μ_W, the ratio between the values of the two functions being nearly constant. The range of normalized autocovariance function is narrower and the previous tendency to concentration disappeared; mean value of ρ_W varies in sign tending to zero as ΔY tends to L. Such sign variation corresponds to the waving of the μ_W diagram, suggesting a positive linear correlation between sections with the same sign of displacement, while the tendency

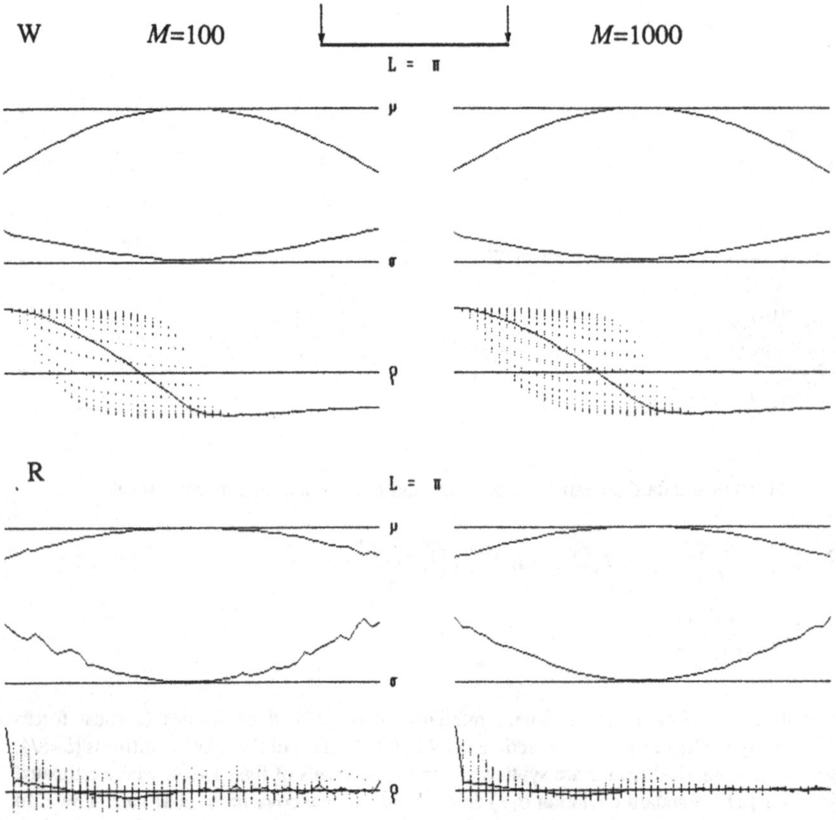

fig. 4 - Effect of number of realizations M on numerical results

to zero as ΔY increases corresponds to the absence of a linear correlation, which is in good agreement with the physically expected statistical independence of far sections. Accounting also for the substantial constancy of the ratio σ/μ, it can be concluded that random variability of k in this case has a mainly local effect with a nearly proportional increase or decrease of displacements. In the intermediate cases of medium-flexible beams ($L=\pi$) the aforementioned effects of rigid rotation (σ_W greater at the ends) and proportional increase (σ_W greater where μ_W is greater) are in some way superposed, loading conditions considerably affecting structural behaviour. As a matter of fact, statistical response for loading conditions where these effects are concordant (beam with forces at the ends) is similar to the one of stiff beams. Otherwise, when loading effects are opposed (beams with a force in the mid-section), standard deviation becomes nearly constant and the scattering of normalized autocovariance values is reduced, showing a substantial stationarity of the response.

Previous numerical results all refer to a value of the standard deviation of the input random function σ_K=0.5. In order to show the effects of σ_K on the output random functions, analyses with different values have been performed. In fig.3 some numerical results are given, referred to σ_K=0.2 and 1; from this picture it can be seen that σ_K influences only the standard deviation of the output functions, which varies nearly proportionally, while mean value and normalized autocovariance remain unaltered.

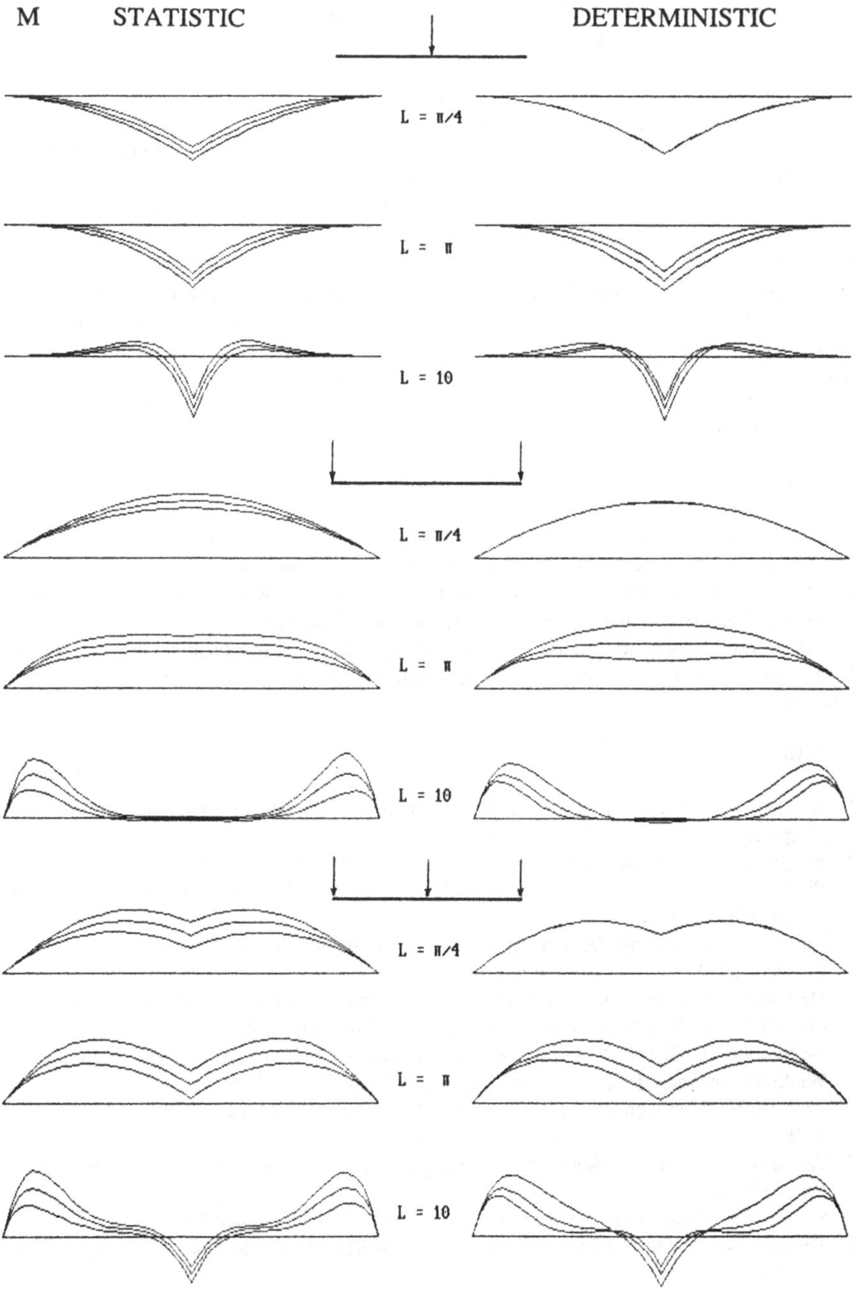

fig. 5 - Comparison between statistic and *k*-constant deterministic approaches

An analysis has then been performed in order to evaluate the effects of the number of realizations M on the numerical results. In fig.4 some of these comparisons are given, which show that a number of realizations M=100 can be sufficient for a correct evaluation of the statistical response of the beam. A larger number, M=1000, has no significant effect on the analysis of $\bar{W}(Y)$, and causes only slight variations in the $\bar{R}(Y)$ diagrams (fig.4)

A comparison has finally been made between numerical results of the statistic approach followed in the present work and those of the deterministic approach which takes into account the variability of k performing two different k-constant analyses, using a minimum and a maximum value of the modulus of subgrade reaction. Statistical results are shown in fig.5, where mean value μ_M of bending moment is plotted together with a range of minimum and maximum moment respectively exceeding the mean value by $\pm 3\sigma_M$, which according to the Chebyshev inequality contains actual bending moment with a probability of at least 89%. The minimum and maximum values of k used in the deterministic approach are selected so as to hold, with the same probability, the input random values of k. As already obtained by the authors in a previous paper[4] for a beam without internal loads, the statistical approach reveals analogous increases in bending moment whatever the stiffness of the beam, while the k-constant approach is very sensitive to it, under-estimating the moment variation in case of stiff beams.

CONCLUSIONS

An analysis has been performed of beams on Winkler foundation where modulus of subgrade reaction and soil displacement both are considered as random functions.

Statistics of output random functions have been obtained using a Monte Carlo simulation based on a Direct Boundary Element approach of the mechanical problem.

Numerical results show how statistical behaviour depends on stiffness and loading conditions. From a comparison with a simplified k-constant analysis, some conclusions are drawn for design practice.

REFERENCES

1. Augusti, G., Baratta, A., Casciati, F. Probabilistic Methods in Structural Engineering. Chapman & Hall Editors, London, 1984.
2. Baratta, A. Probabilistic treatment of structural problems with unilateral constraints. In G. Del Piero & F. Maceri (eds.), Unilateral Problems in Structural Analysis, pp. 21-32, Springer Verlag, Udine, 1985.
3. Brebbia, C.A., Dominguez, J. Boundary Elements, an Introductory Course. Computational Mechanics Publications, Southampton, 1989.
4. De Simone, P., Ghersi, A., Mauro, R. Statistical approach to beams on Winkler foundation. Proc. X Europ. Conf. Soil Mech. Found. Eng., Firenze, Italy, 1991a.
5. De Simone, P., Ghersi, A., Mauro, R. DBEM Analysis of beams on modified elastoplastic Winkler foundation. Proc. 13th BEM Int. Conf., Tulsa, USA, 1991b.
6. Hetényi, M. Beams on Elastic Foundation. The University of Michigan Press, Ann Arbor, 1946.
7. Papoulis, A. Probability, Random Variables, and Stochastic Processes. McGraw-Hill, New York, 1965.
8. Schwartz, L. Mathematics for the Physical Sciences. Hermann, Paris, 1966.
9. Terzaghi, K. Evaluation of coefficients of subgrade reaction. Géotechnique, V, pp. 297-326.

Simulation of Multi-Variate Stationary and Nonstationary Random Processes: A Recent Development

A. Kareem (*), Y. Li (**)

(*) Department of Civil Engineering, University of Notre Dame, Notre Dame, IN 46556-0767, U.S.A.

(**) Shell Development Company, Houston, TX 77001-0481, U.S.A.

ABSTRACT

This paper present computationally efficient procedures to simulate multivariate stationary random processes for any desired time duration, and nonstationary processes consistent with prescribed evolutionary spectrum. The simulation of multi-variate stationary random processes is based on the simple convenience of conventional FFT-based simulation schemes, but without the drawback of large computer memory requirement that has in the part precluded the generation of long-duration time series utilizing the FFT-based approaches. Central to this technique is a simulation of a large number of time series segments by utilizing the FFT algorithm, which is followed by their synthesis by means of a digital filter to obtain the desired length of simulated time series. The effectiveness of this methodology is demonstrated by means of examples concerning the simulation of a multi-variate random wind field and the spatial variation of wave kinematics in a random sea. The simulation of multi-variate nonstationary random processes is also based on FFT algorithm. The utilization of FFT has been made possible by a stochastic decomposition technique. The decomposed spectral matrix is expanded into a weighted summation of basic functions and time-dependent weights which are simulated by FFT. The effectiveness of the proposed technique is demonstrated by means of examples of the past earthquake events. This approach is computationally very efficient in simulating a large set of nonstationary processes, such as ground motion, that are needed for ensemble averaging of the response time histories.

INTRODUCTION

A number of papers describing the simulation of stationary random processes have been published in recent years with reference to applications in engineering mechanics. The simulation methods can be categorized into two classes: 1) Methods based on the summation of trigonometric functions (wave superposition); and 2) Methods based on the convolution of white noise with a kernel function or integration of a differential equation driven by white noise (linear filtering). These techniques vary in their applicability, complexity, computer memory and computing time (e.g., Borgman, 1969, and Shinozuka, 1970). Recently, auto-regressive (AR), moving-average (MA), and their combination, auto-regressive and moving-averages (ARMA) models have been utilized in a wide range of applications, e.g., earthquakes, waves, and wind (e.g. Li and Kareem, 1990; Kareem and Li, 1988; Naganuma et al., 1987; Spanos and Mignolet, 1986; Spanos & Mignolet, 1987; Samaras et al., 1985; and Reed and Scanlan, 1984). A hybrid combination of the two preceding basic approaches may be found in Li and Kareem, 1989a and Hasofar, 1989. In a series of papers, Shinozuka and his associates have extended the application of these approaches to a variety of stochastic systems with multi-variate and multi-dimensional attributes (e.g., Shinozuka, 1970; Shinozuka, 1971; Shinozuka, 1972; Shinozuka and Jan, 1971; Samaras et al., 1985; Naganuma et al., 1987; and Shinozuka and Deodatis, 1988).

Traditionally, the method based on the summation of trignometric series with random phase angles has been most popular, perhaps due to its simplicity. The summation process renders this approach computationally inefficient. In this context, it has been noted that the summation of trignometric terms may be carried out by utilizing a fast Fourier transform (FFT). The use of FFT algorithm, though, improves the computational efficiency drastically, but not without the expense of increased demand on computer memory. The limitation on the length of time series often imposed by available computer memory is eliminated by utilizing parametric time series models, such as an ARMA model. Unlike the FFT-based techniques, this approach does not demand a large computer memory; rather, only a limited amount of information, e.g.; coefficient matrices are stored and long-duration time series may be simulated through recursive relationships. However, the matching of an ARMA model for a certain application may not be very straightforward, e.g., simulation of ocean wave profile consistent with a target spectrum characterized by a sharp spike (e.g., Spanos and Mignolet, 1986 and Kareem and Li, 1988). Furthermore, in the parametric modelling approach, the model order selection process is also rather empirical. Therefore, despite the significant computational advantage of the parametric time series models over the FFT-based techniques, difficulty remains concerning a straightforward application of these models to a wide range of random processes by those not experienced in this field.

The preceding studies have been primarily focused on the simulation of stationary processes. The summation of trignometric series approach is applicable to both stationary and nonstationary processes. As noted earlier that the summation of the trignometric series may be carried out by utilizing the FFT algorithm which dramatically reduces the computational effort. However, in contrast with the direct summation of the trignometic series, the FFT approach has been limited primarily to the simulation of stationary processes. Most of the parametric time series modelling has focused primarily on the simulation of stationary processes, a limited effort has been devoted to the simulation of nonstationary processes (e.g., Kozin and Nakajima, 1980; Gersch and Kitagawa, 1985; Cakmak et al., 1985; Hoshiya et al., 1984; and Deodatis and Shinozuka, 1988). The simulation of ground motion records has also been accomplished by filtered white noise and filtered poisson models (Shinozuka and Deodatis, 1988).

This paper presents procedures based on FFT algorithm to simulate long-duration multi-variate correlated time processes efficiently without paying a premium for computer memory and to simulate efficiently multi-variate nonstationary processes. First the method for the simulation of stationary process is presented that is followed by the procedure for nonstationary processes.

STATIONARY RANDOM PROCESSES

A univariate Gaussian process $y(t)$ with a given power spectral density function $G(t)$ can be expressed by

$$y(t) = \int_{\infty}^{-\infty} \frac{1}{\sqrt{2}} D(f) \exp(j2\pi ft)\, dZ(f) \tag{1}$$

in which $D(f) = \sqrt{G(f)}$ and $D(f) = D^*(f)$, $dZ(f)$ is an orthogonal increment equal to $\varepsilon(t)\sqrt{df}$, $\varepsilon(f)$ is a zero mean, unit variance Gaussian white noise process, and * denotes complex conjugation. The preceding equation can be recast in a discrete form for simulation purposes

$$\hat{y}(m\Delta t) = \sum_{n=-N+1}^{N} \frac{1}{\sqrt{2}} D(n\Delta f) \exp\left(\frac{j\pi nm}{N}\right) \varepsilon_n \sqrt{\Delta f} \tag{2}$$

in which $D(n\Delta f) = \sqrt{G(n\Delta f)}$, ε_n is a zero mean, unit variance Gaussian white noise process and $\Delta f = \frac{1}{T}$, where T is the total duration of the process to be simulated. The preceding equation can be evaluated efficiently by means of a FFT algorithm. This approach has been utilized for the simulation of random processes (e.g., Shinozuka, 1971; Wittig and Sinha, 1975; Kareem, 1978; Kareem and Dalton, 1981; Shinozuka and Deodatis, 1988; and Shinozuka et al., 1989). The following derivation leads to a new approach for the simulation of long-duration time series without a high demand on computer memory.

We begin by assuming that Eq. 1 can be approximated by

$$y(t) = \sum_{s=-N_s}^{N_s} A_s(t) \frac{1}{\sqrt{2}} D(f_s) \exp(j2\pi f_s t) \tag{3}$$

where $A_s(t)$ is a modulating function which is introduced to account for both amplitude and phase modulations. $A_s(t)$, for $s = 0, 1, \ldots, N_s$, is a statistically independent Gaussian process which satisfies

$$A_s(t) = A^*_{-s}(t)$$

$$E[A_s(t) A_r^*(t)] = \delta_{sr}\Delta f, \text{ and}$$

$$S_{A_s}(f) = S_{A_r}(f) \tag{4}$$

in which * represent complex conjugation and $S_{A_s}(f)$ describes a double-sided spectral density.

Recasting Eq. 3 in terms of discrete time interval yields

$$y(\mu\Delta t) = \sum_{s=-N+1}^{N_s} A_s(\mu\Delta t)\frac{1}{\sqrt{2}}D(s\Delta f)\, exp\,(j\pi\frac{s\mu}{N_s}) \quad (5)$$

in which $N_s\Delta f = \frac{1}{2\Delta t}$, and $\mu = 0, 1, 2, \ldots, \alpha$.

The preceding expression may be referred to as the Modulated Inverse Discrete Fourier Transformation. The stationarity, ergodicity, and Gaussianity of $y(\mu\Delta t)$ can be assured, provided $A_s(\mu\Delta t)$ is a stationary, ergotic, and Gaussian process. The spectral density function of the time history simulated in this manner depends on the spectral density function of $A_s(\mu\Delta t)$. There are a number of ways to numerically evaluate the preceding equation. In this study, the FFT algorithm is utilized.

Let $\mu\Delta t = n\Delta T + \alpha\Delta t$, where $\alpha = 0, 1, 2, \ldots, N_\alpha - 1$, and $N_\alpha = \Delta T/\Delta t$ and $1/(\Delta f \bullet \Delta T)$ are integers ≥ 1. (Choice of ΔT or N_α is addressed elsewhere, Li and Kareem, 1989a). Substituting these expressions in Eq. 5 results in the following equation

$$y(n\Delta T + \alpha\Delta T) = \sum_{s=-N_s+1}^{N_s} A_s(n\Delta T + \alpha\Delta T)\frac{1}{\sqrt{2}}D(s\Delta f)\, exp\left(j\pi\frac{s(nN_\alpha + \alpha)}{N_s}\right) \quad (6)$$

$A_s(n\Delta T + \alpha\Delta t)$ may be expressed in the following form

$$A_s(n\Delta T + \alpha\Delta t) = \sum_{m=N_r^-}^{N_r^+} C_{m\alpha}\varepsilon_s(n+m) \quad (7)$$

where $C_{m\alpha}$ is a double-subscripted convolution coefficient (Li and Kareem, 1989), and $\varepsilon_s(n)$ is a Gaussian white noise such that $E[\varepsilon_s(n) \bullet \varepsilon_r(m)] = \delta_{sr}\delta_{mn}$. Generally, the double-subscripted convolution results in a nonstationary process. However, if $C_{m\alpha}$ is constructed such that both real and imaginary parts of the modulating function satisfy the following condition:

$$S_{A_s}(f) = 0 \qquad \text{for} \quad 1/(2\Delta T) > |f|. \quad (8)$$

It can be shown that $A_s(t)$ satisfies the stationarity, ergodicity, and Gaussianity (Li and Kareem, 1989). Therefore, Eq. 7 can be expressed in the following form,

$$y(n\Delta T + \alpha\Delta t) = \sum_{m=N_r^-}^{N_r^+} C_{m\alpha}y_{n+m}(n'N_\alpha + \alpha) \quad (9)$$

where n' is the integer remainder of $nN_\alpha/(2N_s)$, and $y_n(n'N_\alpha + \alpha)$ for $\alpha = 0, 1, 2, \ldots, N_\alpha - 1$ represents a time series that can be evaluated by the FFT algorithm

$$y_n(\beta) = \sum_{s=-N_s+1}^{N_s} \varepsilon_s(n) \frac{1}{\sqrt{2}} D(f_s) \, exp\,(j\pi \frac{\beta s}{N_s}) \tag{10}$$

where $\beta = n'N_\alpha + \alpha$.

At this stage, the simulation procedure can be summarized in the following steps: i) generate at set of complex random numbers of $\varepsilon_s(n+N_r)$, $s = 0, 1, 2, \ldots, N_s - 1$; ii) simulate $y_n(\beta)$ for $\beta = 0, 1, 2, \ldots, 2N_s - 1$ (Eq. 10) by FFT; iii) generate $y(n\Delta T + \alpha\Delta t)$ by the nonrecursive model (Eq. 9). The remaining concern is in regard to the selection of $A_s(t)$. In this study, two forms of $A_s(t)$ are used: the uniform modulating function and the linear modulating function. For the sake of brevity the linear modulating function is presented here, further details are available in Li and Kareem, 1989c.

Now the univariate case is expanded for the simulation of M multi-correlated random processes with prescribed spectral characteristics, e.g., $G(f)$, the cross-spectral density matrix (M, M). This matrix is factorized by a Cholesky or Schur decomposition into $D(f)$

$$G(f) = D(f) D^*(f) \tag{11}$$

where $D(f)$ is $(M,\wedge)$ matrix, in which $\wedge$ is the order of decomposition and * represents conjugation. The decomposition order, $\wedge$, is generally less than or equal to M. The aforementioned concept of decomposition has been utilized by Li and Kareem (1989b) for the dynamic analysis of random linear and quadratic systems. Further details are available in Li and Kareem (1989c).

Central to this technique is the decomposition of a set of correlated random processes into component random processes such that the relationship between any two subprocesses $y_{i/\mu}(t)$ and $y_{j/\lambda}(t)$ is either fully coherent for $\mu = \lambda$ or noncoherent if $\mu \neq \lambda$. In the frequency domain, the component processes are expressed in terms of decomposed spectrum $D_{i/\mu}(f)$. Typical elements of the spectral matrix of the parent process $Y(t)$ are given by

$$G_i(f) = \sum_{\mu=1}^{A} D_{i/\mu}(f) D^*_{i/\mu}(f)$$

$$G_{ij}(f) = \sum_{\mu=1}^{\Delta} D_{i/\mu}(f) D^*_{j/\mu}(f) . \tag{12}$$

It can be shown that the linear transformation that relates sub-processes also describes the corresponding decomposed spectra.

$$D_{i/\mu}(f) = H_{ij}(f) D_{j/\mu}(f) \tag{13}$$

in which $H_{ij}(f)$ represents the transfer function relating the ith and j[th] parent processes.

Once the target process is decomposed into the component subprocesses which are either fully coherent or noncoherent, the simulation of correlated processes can be accomplished by (Li and Kareem, 1989c).

$$y_i(n\Delta T + \alpha\Delta t) = \sum_{m=-1}^{2} C_{m\alpha} y_{i_n}(\beta) \tag{14}$$

$$y_{i_n}(\beta) = \sum_{\mu=1}^{\Delta} \left\{ \sum_{s=-N_s+1}^{N_s} \varepsilon_{s/\mu}(n) \frac{1}{\sqrt{2}} D_{i/\mu}(s\Delta f) \right\} exp\left(\frac{j\pi\beta s}{N_s}\right) \tag{15}$$

where n, ΔT, and β have been defined previously (Eq. 6, 7, and 9).

This procedure is utilized to simulate a half a million seconds of correlated wind velocity fluctuations at five levels on a building. Typical spectral characteristics of the wind field were used for this example (e.g., Kareem, 1987). The target and estimated power spectral density functions of wind fluctuations are presented in Figure 1. A near-perfect agreement is noted. This procedure is utilized to simulate long time histories of different sea states and associated wave particle kinematics. Details are provided elsewhere (Li and Kareem, 1989c).

NONSTATIONARY RANDOM PROCESSES

The present simulation scheme is based on a stochastic decomposition technique described briefly in the section on stationary processes. The evolutionary power spectra density representation of nonstationary processes is (Priestly, 1967) given by

$$G(f, t) = |A(f, t)|^2 K(f) . \tag{16}$$

The evolutionary spectral matrix $G(f, t)$ may be decomposed into time-dependent decomposed spectra

$$G(f, t) = D(f, t) D^*(f, t) . \tag{17}$$

A sub-process can be expressed in terms of the decomposed spectrum

$$y_{i/\mu}(t) = \int_0^{\infty} \sqrt{2} D_{i/\mu}(f, t)\, exp(j2\pi ft)\, dZ_\mu(f) . \tag{18}$$

Recall that each parent process is a summation of decomposed sub-processes, therefore, the time sample function for each parent process is given by

$$y_i(t) = \sqrt{2}\sum_{\mu=1}^{\Lambda}\int_0^{\infty} D_{i/\mu}(f,t)\, exp\,(j2\pi ft)\, dZ_{\mu}(f)\,. \tag{19}$$

A nonstationary process may be simulated by evaluating Eq. (19). In the following section, the FFT technique is implemented in the simulation scheme to enhance the computational efficiency.

Simulation by FFT Technique

The decomposed spectrum may be expressed as a product of frequency- and time-dependent functions

$$D_{i/\mu}(f,t) \approx \sum_{r=1}^{N_r} A^{(r)}(t)\,\Phi_{i/\mu}^{(r)}(f) \tag{20}$$

or in a matrix form,

$$D(f,t) \approx \sum_{r=1}^{N_r} A^{(r)}(t)\,\Phi^{(r)}(f)\,. \tag{21}$$

In the preceding equations, $D_{i/\mu}(f,t)$ can be viewed as a weighted summation of $\Phi_{i/\mu}^{(r)}(f)$. The definition of the sum of decomposed spectra suggests that subprocesses exist that correspond to the decomposed spectrum $\Phi_{i/\mu}^{(r)}(f)$

$$\phi_{i/\mu}^{(r)}(t) = \int_0^{\infty}\sqrt{2}\Phi_{i/\mu}^{(r)}(f)\, exp\,(j2\pi ft)\, dZ_{\mu}. \tag{22}$$

These subprocesses $\Phi_{i/\mu}^{(r)}(f)$ and $\Phi_{j/\lambda}^{(p)}(f)$ are fully coherent if $\mu = \lambda$. The sample time function of $y_{i/\mu}(t)$ is expressed by

$$y_{i/\mu}(t) = \sum_{r=1}^{N_r} A^{(r)}(t)\,\phi_{i/\mu}^{(r)}(t) \tag{23}$$

and consequently the time history of the target process is given by

$$y_i(t) = \sum_{r=1}^{N_r}\left(A^{(r)}(t)\sum_{\mu=1}^{\Lambda}\phi_{i/\mu}^{(r)}(t)\right). \tag{24}$$

The previous equation permits the simulation of a nonstationary vector process. Due to the nonstationary nature of the process, which is often characterized by a short duration time history, the frequency resolution may not be high (i.e., $\Delta f = \dfrac{1}{N_t\Delta t}$; N_t = total number of points to be simulated and Δt = time increment). Therefore, the frequency resolution may be improved by deriving $D(f, t)$ based on averaging the spectral matrix over a frequency interval Δf.

The procedure for simulating a multi-variate random process is summarized here. First, transpose the spectral matrix into a decomposed spectral matrix. Second, determine $A^{(r)}(n\Delta t)$ and $\Phi^{(r)}_{i/\mu}(m\Delta f)$ utilizing the decomposed spectral matrix. A matching procedure based on an error minimization procedure may be used to express the decomposed evolutionary spectrum in terms of time and frequency dependent functions according to Eq. 24. Details can be found in Li and Kareem (1989d). Third, simulate a complex white noise vector $\varepsilon_\mu(m)$ with $\mu = 1, 2, \ldots, \Lambda$, and $m = 0, 1, 2, \ldots, N_{r/2}$, such that the real and imaginary parts of $\varepsilon_\mu(m)$ are independent, zero mean Gaussian white noise proceses. Then by invoking the FFT algorithm

$$\phi_i^{(r)}(n\Delta t) = \sum_{m=0}^{N_t/2} \sqrt{2\Delta f}\left[\sum_{\mu=1}^{\Lambda} \Phi^{(r)}_{i/\mu}(f)\,\varepsilon_\mu(m)\right] exp\,(j2\pi\frac{nm}{N_t}) . \tag{25}$$

In this manner, the FFT algorithm is utilized (NxN_r) times to simulate a vector process $(Nx1)$. Finally, the ith element of the desired vector process is given by

$$y_i(n\Delta t) = \sum_{r=1}^{N_r} A^{(r)}(n\Delta t)\,\phi_i^{(r)}(n\Delta t) . \tag{26}$$

An example is presented to illustrate the effectiveness of the methodology presented here. The example relates to the simulation of the E-W component of the December 30, 1934 El Centro Earthquake (Liu, 1970). The single-point-evolutionary spectrum was analytically described by Deodatis and Shinozuka (1988)

$$G_{ii}(f, t) = \{\frac{exp(-at) - exp(-bt)}{max[exp(-at) - exp(-bt)]}\sqrt{\hat{K}_1(f)}$$

$$+ exp\left[-\frac{(t-m)^2}{2\sigma^2}\right]\sqrt{\hat{K}_2(f)}\ \}^2 \tag{27}$$

where, $a = 0.025$ and $b = 0.05$ and $m = 5.0$ sec., $\sigma = 1.0$ sec., $\hat{K}_1(f)$ and $\hat{K}_2(f)$ satisfy the Kanai-Tajimi spectrum given below

$$\hat{K}(f) = 4\pi S_0 \frac{f_g^4 + 4\zeta_g^2 f_g^2 f^2}{(f^2 - f_g^2)^2 + 4\zeta_g^2 f_g^2 f^2} \tag{28}$$

in which the parameters are $S_0 = 0.1$ cm^2 sec^{-3}, $f_g = 15/2\pi$ Hz, and $\zeta_g = 0.25$. In case of

$\hat{K}_2(f), f_g = \frac{30}{2\pi} hz$. To illustrate the multi-variate feature we are using an extension based on the empirical relationship for the cross-spectral density function given by Harichandran and Vanmarcke (1986),

$$G_{ij}(f) = G_{ii}(f)\,\rho(v_{ij}, f)\,exp(-j2\pi f v_{ij}/V_{ij}) \tag{29}$$

where v_{ij} denotes the distance between the ith and j[th] locations, V_{ij} denotes the apparent wave propagation velocity along the direction between the ith and j[th] locations, and $\rho(v_{ij}, f)$ is the coherence function, given by

$$\rho(v_{ij}, f) = a\,exp\left[-\frac{2v_{ij}}{\alpha\theta(f)}(1-a+\alpha a)\right]$$

$$+ (1-a)\,exp\left[-\frac{2v_{ij}}{\theta(f)}(1-a+\alpha a)\right] \tag{30}$$

and

$$\theta(f) = k\left[1+\left(\frac{f}{f_0}\right)^b\right]^{-1/2} \tag{31}$$

in which the parameters are: $a = 0.736$, $\alpha = 0.147$, $k = 5210$, and $f_0 = 1.09$.

The correlations obtained from the simulated E-W component of the December, 1934 El Centro Earthquake are given in Fig. (2). The comparison between the estimated and target correlations is excellent. N_r is equal to seven for this case. In this example, the time-and frequency-dependent components could be separated conveniently as a consequence of the functional form of the corresponding target spectral descriptions. There is a large class of evolutionary spectral descriptions that may not be amenable to a convenient separation into a time- and a frequency dependent function. A typical example is the evolutionary spectrum of the 1964 Niigata Earthquake. The ground motion time history has a peculiar feature that after 7 seconds the frequency contents of the signal change and it exhibits a dominant low frequency component believed to be due to the liquefaction of the ground (Deodatis and Shinozuka, 1988). The present approach helps to simulate ground motion that conforms with the special features of Niigata Earthquake. Complete details regarding the simulation and other examples are available in Li and Kareem 1989d.

CONCLUDING REMARKS

Computationally efficient, convenient and robust schemes are presented for simulating multi-variate stationary and nonstationary random processes. Both approaches are based on FFT algorithm and utilize the concept of stochastic decomposition. The stationary random processes involve simulation of a large number of time-series segments by utilizing FFT algorithm which are subsequently synthesized by means of a digital filter to provide the desired length of simulated processes. Long duration time series of the wind velocity field wave surface profiles and associated wave kinematics are simulated to demonstrate the effectiveness of this approach. The simulated data show excellent agreement with the target spectral characteristics. The proposed technique has immediate applications to the simulation of real time processes, e.g., driving the wave maker in an ocean system modelling basin or exciting an array of aerofoils in a gust-simulation tunnel. The effectiveness of the nonstationary technique is demonstrated by means of examples utilizing analytical description of ground motion of actual earthquakes. The simulated records exhibit an excellent agreement with the prescribed probabilistic characteristics, e.g., spectral density and correlation functions of different time lags. Applications are immediate in the time domain analysis of large-scale spatial structures spanning over large spatial areas to seismic excitation, e.g., pipelines, dams, and long-span bridges.

ACKNOWLEDGEMENTS

The support for this research was provided in part by the National Science Foundation Grant BCS90-96274 (BCS83-52223) and several industrial sponsors. Their support is gratefully appreciated.

REFERENCES

Borgman, L.E. (1969), "Ocean Wave Simulation for Engineering Design," J. Wtrwy. Harb. Div., ASCE, Vol. 95, No. 4.

Cakmak, A.S., Sherit, R.I., and Ellis, G. (1985), "Modelling Earthquake Ground Motions in California Using Parametric Time Series Methods," Soil Dynamics and Earthquake Engineering, 4(3), pp. 124-131.

Deodatis, G., and Shinozuka, M. (1988). "Auto-Regressive Model for Nonstationary Stochastic Processes," Journal of Engineering Mechanics, Vol. 114, No. 12, pp. 1995-2012.

Gersch, W., and Kitagawa, G. (1985), "A Time Varying AR Coefficient Model for Modelling and Simulating Earthquake Ground Motion," Earthquake Engineering and Structural Dynamics, 13(2), pp. 124-131.

Harichandran, R.S., and Vanmarcke, E. (1986). "Stochastic Variation of Earthquake Ground Motion in Space and Time," J. Engrg. Mech., ASCE, Vol. 112(2).

Hasofar, A.M. (1989). "Continuous Simulation of Gaussian Processes with Given Spectrum," Proceedings of the 5th International Conference on Structural Safety and Reliability, San Francisco, August 7-11, 1989, ASCE, NY.

Hoshiya, M., Ishii, K., and Nagata, S. (1984). "Recursive Covariance of Structural Response," J. Engineering Mechanics, ASCE, Vol. 110, No. 12, pp. 1743-1755.

Kareem, A. (1978). "Wind Excited Motion of Buildings," Ph.D. Thesis, Department of Civil Engineering, Colorado State University, Fort Collins, Colorado.

Kareem, A., and Dalton, C. (1982). "Dynamic Effects of Wind on Tension Leg Platforms," OTC 4229, Proceedings of Offshore Technology Conference, Houston, Texas.

Kareem, A. (1987). "Wind Effects on Structures: A Probabilistic Viewpoint," Probabilistic Engineering Mechanics, Vol., No. 4.

Kareem, A., and Li, Yousun (1988). "Stochastic Response of a Tension Leg Platform to Wind and Wave Loads," Department of Civil Engineering, Technical Report No. UHCE 88-18.

Kozin, R., and Nakajima, F. (1980), "The Order Determination Problem for Linear Time-Varying AR Models," IEEE Trans. Auto Control, AC-25(2), pp. 250-251.

Li, Yousun, and Kareem, A. (1990). "Modelling of ARMA Systems in Wind Engineering," Journal of Wind Engineering and Industrial Aerodynamics, Vol. 36.

Li, Yousun, and Kareem, A. (1989a). "Simulation an Multivariate Stationary Random Processes: A Combination of Digital Filtering and DFT," Department of Civil Engineering, University of Houston, Technical Report No. UHCE-89-11.

Li, Yousun, and Kareem, A. (1989b). "On Stochastic Decomposition and Its Applications in Probabilistic Dynamics," Proceedings of the 5th International Conference on Structural Safety and Reliability (ICOSSAR), August, 1989, San Francisco.

Li, Y., and Kareem, A.K. (1989c). "On Stochastic Decomposition and its Application in Probabilistic Dynamics," Proceedings of the 5th International Conference on Structural Safety and Reliability (ICOSSAR '89), San Francisco.

Li, Yousun, and Kareem, A. (1989d). "Simulation of Multi-Variate Nonstationary Random Processes by FFT," Department of Civil Engineering Tech. Report No. UHCE89-9.

Liu, S.C. (1970), "Evolutionary power spectral density of strong motion earthquakes," Bull. Seism. Soc. Am. 60(3), pp. 891-900.

Naganuma, T., Deodatis, G., and Shinozuka, M. (1987). "ARMA Model for Two-Dimensional Processes," J. Engrg., Mech. Div., ASCE, Vol. 113, No. 2.

Priestly, M.B. (1967), "Power Spectral Analysis of Non-stationary Random Processes," J. Sound Vib., 6:86-97.

Reed, D.A., and Scanlan, R.H. (1984). "Autoregressive Representation of Longitudinal, Lateral, and Vertical Turbulence Spectra," Journal of Wind Engineering and Industrial Aerodynamics, Vol. 17.

Samaras, E., Shinozuka, M., and Tsuri, A. (1985). "ARMA Representation of Random Vector Processes," Journal of Engineering Mechanics, ASCE, Vol. 111, No. 3.

Shinozuka, M. (1970). "Simulation of Multivariate and Multidimensional Random Processes," Journal of the Acoustical Society of America, Vol. 49, No. 1 (Part 2).

Shinozuka, M. (1971). "Simulation of Multivariate and Multidimensional Random Processes," The Journal of the Acoustical Society of America, Vol. 49, pp. 357-367.

Shinozuka, M. (1972). "Monte Carlo Solution of Structural Dynamics," Computer and Structures, Vol. 2.

Shinozuka, M. and Jan, C-M. (1972). "Digital Simulation of Random Processes and Its Applications," Journal of Sound and Vibration, Vol. 25, No. 1.

Shinozuka, M., and Deodatis, G. (1988). "Stochastic Process Models for Earthquake Ground Motion," Probabilistic Engineering Mechanics, Vol. 3, No. 3., pp. 114-123.

Shinozuka, M., Yun, C-B., and Sey, H. (1989), "Stochastic Methods in Wind Engineering," Proceedings of the Sixth U.S. National Conference on Wind Engineering, Houston, TX, March 8-10 (A. Kareem: Editor).

Spanos, P-T.D., and Mignolet, M.P. (1986). "Z-Transform Modelling of P-M Wave Spectrum," Journal of Engineering Mechanics, Vol. 112, No. 8.

Spanos, P-T.D., and Mignolet, M.P. (1987). "Recursive Simulation of Stationary Multi-Variate Random Processes, Parts I and II," Journal of Applied Mechanics, Vol. 54.

Wittig, L.E. and Sinha, A.K. (1975). "Simulation of Multicorrelated Random Processes Using the FFT Algorithm," J. Acoustical Society of America, Vol. 58, No. 3.

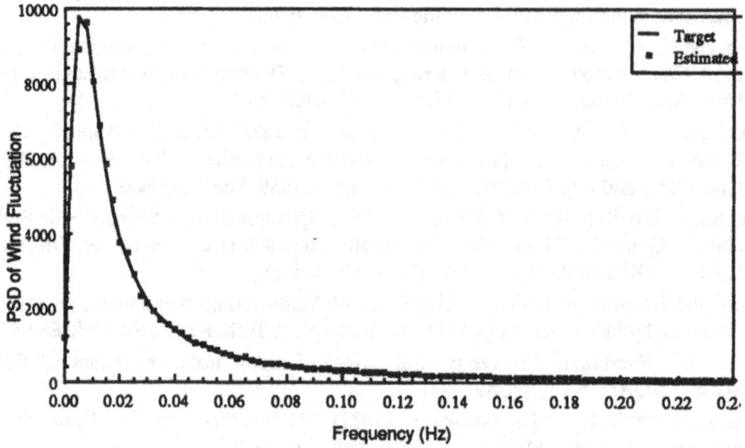

Fig. 1 Target and estimated PSD of wind velocity fluctuation.

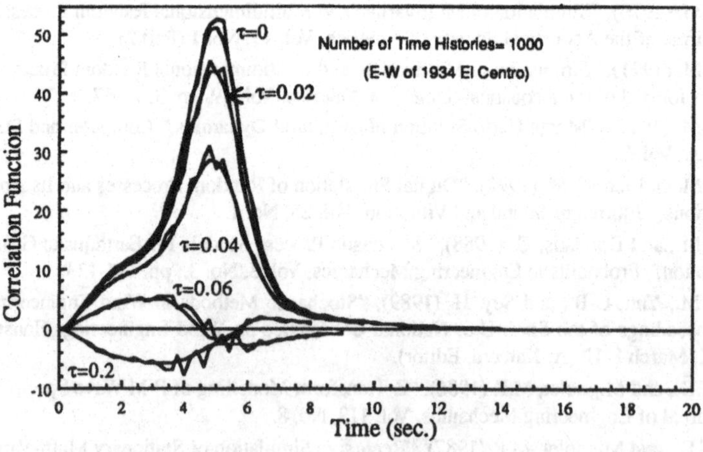

Fig. 2 Correlation function of ground motion (smooth solid line represents target correlation)

Optimality in the Estimation of a MA System from a Long AR Model

M.P. Mignolet (*), P.D. Spanos (**)
(*) Department of Mechanical and Aerospace Engineering, Arizona State University, Tempe, Arizona 85287-6106, U.S.A.
(**) L.B. Ryon Chair in Engineering, Brown School of Engineering, Rice University, Houston, Texas 77251, U.S.A.

ABSTRACT

The determination of moving average (MA) models from a prior autoregressive (AR) approximation of a specified (target) spectral matrix is addressed. First, an existing technique based on a direct modeling of the target expression is revisited. In this regard, the influence of the order of the prior AR approximation, and the number of its harmonics used in the determination of the MA model, is described. Further, a simple selection technique of these parameters is presented that leads to an optimum MA approximation. Next, the relationship between a method based on the Cholesky factorization of the covariance matrix and the present technique is investigated to derive additional insight into its convergence properties. Finally, an alternative modeling technique based on an AR representation of the inverse of the target spectral matrix is presented.

INTRODUCTION

Digital spectral analysis has been recognized over a period of years as an extremely useful tool in structural dynamics studies. In particular, it has led to the development of fast and reliable computational algorithms for both the estimation of structural parameters from a vibration record and the generation of time histories of forces associated with natural phenomena such as earthquake ground motions, turbulence, etc. These numerical techniques have demonstrated the usefulness of the class of autoregressive moving average (ARMA) discrete systems which possess a rational transfer function in terms of z transform. It has often been found, in the context of both parameter estimation and simulation, that the autoregressive (AR) part of the model, the denominator of the transfer function, is much easier to obtain than its moving average (MA) counterpart, the numerator of the transfer function. This situation is consistent with the determination of purely AR and MA models. Specifically, the AR approximation can be derived by solving a linear system of equation when relying on the linear prediction theory, while the computation of a MA model is inherently a nonlinear problem which involves the factorization of the specified (target) spectral matrix.

Various numerical techniques have been suggested to accomplish this task. Among these procedures are the methods of alternating projection, e.g. Wiener and Masani [1], the direct numerical solution of the nonlinear equations, e.g. Jezek and Kucera [2], the Cholesky decomposition of the infinite covariance matrix, e.g. Youla and Kazanjian [3], and others based on Riccati matrix equations, e.g.Tuel [4], Anderson et al. [5]. A different technique requiring only the solution of two systems of linear equations, has been suggested first by Durbin [6] in the context of parameter estimation and by Mignolet and Spanos [7] from the standpoint of simulation. This two-step procedure involves first the computation of a high order AR approximation and then the determination of the MA system from this prior AR model. The present article aims to analyze the influence of various parameters on the reliability of the final MA approximation. This investigation focuses in particular on the development of a selection strategy of these parameters that leads to an optimum matching between the target and the MA spectral matrices and on the convergence properties of the corresponding MA modeling technique. For completeness, the definition and some basic properties of MA and AR processes are briefly reviewed.

THE MOVING AVERAGE (MA) PROCESS

A n-variate causal moving average, or simply moving average, process $\underline{Y}'$ of order m can be expressed as

$$\underline{Y}'_r = \sum_{l=0}^{m} B_l' \, \underline{W}_{r-l} \tag{1}$$

where B_l' are real nxn matrices. The symbol $\underline{W}$ denotes an n-variate bandlimited, in the interval $[-\omega_b, \omega_b]$, white noise process. The autocorrelation of $\underline{W}$ is defined as

$$E\,[\,\underline{W}_i\,\underline{W}_j^{\dagger}\,] = 2\,\omega_b\,I_n\,\delta_{ij}, \tag{2}$$

where E [.] and $[.]^{\dagger}$ are the operators of mathematical expectation and transposition, respectively. The symbols I_n and δ_{ij} denote the nxn identity matrix and the Kronecker delta. The sampling period T and the cut-off frequency ω_b are related through the Nyquist relation

$$T = \frac{\pi}{\omega_b}. \tag{3}$$

The process $\underline{Y}'$ can be considered as the output to white noise input of a multi-degree-of-freedom dynamic system whose transfer function matrix is

$$H_{MA}(z) = N(z) = \sum_{l=0}^{m} B_l'\, z^{-l}. \tag{4}$$

Consequently, its spectral matrix can be expressed in the form

$$S_{\underline{Y}'\underline{Y}'}(\omega) = H_{MA}^{*}(e^{j\omega T})\, H_{MA}^{\dagger}(e^{j\omega T}) \tag{5}$$

where the symbol * denotes complex conjugation. The coefficients B_k' of the MA model, equation (1), are determined by imposing a matching requirement between the autocorrelations of the MA process, defined by the relation

$$R_{\underline{Y}'\underline{Y}'}(k) = E\left[\underline{Y}_r'\,\underline{Y}_{r+k}'^{\dagger}\right] = \int_{-\omega_b}^{\omega_b} S_{\underline{Y}'\underline{Y}'}(\omega)\, e^{jk\omega T}\, d\omega \tag{6}$$

and its target counterpart $R_{\underline{Y}\underline{Y}}(k)$, for $k=0,\ldots,m$. This condition can be shown to lead to a set of equations which are quadratic in the unknown matrices B_k' and require the use of an appropriate numerical algorithm.

THE AUTOREGRESSIVE (AR) PROCESS

A n-variate autoregressive (AR) process $\underline{\hat{Y}}$ of order $\hat{m}$ is a discrete random vector process whose r^{th} sample can be computed from the $\hat{m}$ previous ones in the following manner

$$\underline{\hat{Y}}_r = -\sum_{k=1}^{\hat{m}} \hat{A}_k\, \underline{\hat{Y}}_{r-k} + \hat{B}_0\, \underline{W}_r. \tag{7}$$

The symbols $\hat{A}_k$ and $\hat{B}_0$ denote real nxn matrices and $\underline{W}$ is defined by equation (2). The process defined by equation (7) can be considered as the response vector to white noise excitation of a discrete system whose transfer function matrix is

$$H_{AR}(z) = \hat{D}^{-1}(z)\,\hat{B}_0 \tag{8}$$

where

$$\hat{D}(z) = I_n + \sum_{k=1}^{\hat{m}} \hat{A}_k\, z^{-k}. \tag{9}$$

Given an arbitrary Hermitian positive definite matrix $S(\omega)$, the AR modeling problem involves the determination of the coefficients $\hat{A}_k$ and $\hat{B}_0$ so that the product $H_{AR}^{*}(e^{j\omega T})H_{AR}^{\dagger}(e^{j\omega T})$ is close in some sense to $S(\omega)$. It has been shown, e.g. Hannan [8], that this step can be achieved by selecting the AR parameters to satisfy the following equations (Yule-Walker equations)

$$R^{\dagger}(l) + \sum_{k=1}^{\hat{m}} \hat{A}_k\, R(k-l) = 0 \qquad \text{for } l=1,\ldots,\hat{m} \tag{10}$$

where $R(k)$ designates the Fourier coefficients of $S(\omega)$ which are defined similarly to equation (6). The parameter $\hat{B}_0$ is obtained by equating the means of the matrix $S(\omega)$ and of its AR approximation. That is,

$$\hat{B}_0\,\hat{B}_0^{\dagger} = \frac{1}{2\omega_b}\left\{ R(0) + \sum_{k=1}^{\hat{m}} \hat{A}_k\, R(k) \right\}. \tag{11}$$

It should be noted that the AR approximation $H_{AR}^{*}(e^{j\omega T})H_{AR}^{\dagger}(e^{j\omega T})$ becomes an exact representation of the matrix $S(\omega)$ as the system order $\hat{m}$ tends to infinity. That is,

$$H_{AR}^{*}(e^{j\omega T})\, H_{AR}^{\dagger}(e^{j\omega T}) \overset{\hat{m}\to\infty}{\longrightarrow} S(\omega). \tag{12}$$

THE TWO-STEP LINEAR MA MODELING TECHNIQUE

The convergence property described by equation (12) provides the basis for the two-step linear MA modeling technique, e.g. Durbin [6], Mignolet and Spanos [7]. Specifically, perform first an AR($\hat{m}$) modeling of the target spectral matrix $S_{\underline{YY}}(\omega)$ using the Yule-Walker equations, equations (10) and (11). Upon selecting an appropriate autoregressive order $\hat{m}$, the AR spectrum $H_{AR}^{*}(e^{j\omega T})\, H_{AR}^{\dagger}(e^{j\omega T})$ represents a good approximation of the target matrix $S_{\underline{YY}}(\omega)$ or equivalently

$$H_{AR}(e^{j\omega T}) = \hat{D}^{-1}(e^{j\omega T})\,\hat{B}_0 \approx H_{MA}(e^{j\omega T}). \tag{13}$$

This relation implies that the matrices B_k' and the AR coefficients $\hat{A}_k$ are approximately related through the following convolution equations

$$\sum_{l=\max(0,k-\hat{m})}^{\min(k,m)} \hat{A}_{k-l}\, B_l' \approx \hat{B}_0\, \delta_{k0} \quad \text{for } k=0,1,\ldots,m+\hat{m} \tag{14}$$

where the notation $\hat{A}_0 = I_n$ has been introduced for simplicity.

In the second step, a least squares solution of this overdetermined system of equations for the MA coefficients B_k' is sought. That is, select these matrices to minimize the deconvolution error

$$\varepsilon_{dec} = \sum_{k=0}^{\mu} \left| \sum_{l=\max(0,k-\hat{m})}^{\min(k,m)} \hat{B}_0^{-1}\, \hat{A}_{k-l}\, B_l' - I_n\, \delta_{k0} \right|^{p}, \tag{15}$$

for a given value of $\mu \geq m$. The symbol $|U|^2$ signifies the Euclidean norm of an arbitrary matrix of elements $[U]_{ij}$. That is,

$$|U|^2 = \sum_{i=1}^{n} \sum_{j=1}^{n} [U]_{ij}\, [U]_{ij}^{*}. \tag{16}$$

It can be shown that the minimum is attained when these matrices satisfy the linear system of equations

$$L_\mu^\dagger\, L_\mu\, B = L_\mu^\dagger\, \bar{I}_\mu \tag{17}$$

where L_μ, B and I_μ are $(\mu+1)n \times (m+1)n$, $(m+1)n \times n$ and $(\mu+1)n \times n$ matrices the block elements of which are, respectively,

$$\left\{L_\mu\right\}_{kl} = \hat{B}_0^{-1}\, \hat{A}_{k-l} \quad \textit{for } l \leq k \leq \min(l+\hat{m}, \mu) \text{ and } 0 \leq l \leq m \tag{18a}$$

$$\left\{L_\mu\right\}_{kl} = 0 \qquad \textit{otherwise} \tag{18b}$$

$$\left\{B\right\}_{k} = B_k' \qquad \textit{for } 0 \leq k \leq m \tag{19}$$

and

$$\left\{\bar{I}_\mu\right\}_{k} = I_n\, \delta_{k0} \qquad \textit{for } 0 \leq k \leq \mu. \tag{20}$$

The determination of the solution of the linear system of equations, equation (17), yields the MA coefficients B_k'.

OPTIMAL SELECTION OF THE PARAMETER μ

The selection of an appropriate value of the algorithmic parameter μ represents a delicate problem since this integer does not appear explicitly in either the prior AR approximation or the final MA model. A first approach is to select μ large enough to maintain an accurate modeling of the original target spectrum through the two-step approximation procedure. However, this strategy ignores the fact that the target expression is in fact completely specified by only $m+1$ values, the autocorrelation $R_{\underline{YY}}(k)$, $k=0,...,m$. Further, it has been noted by the authors that the low index AR coefficients $\hat{A}_k$ converge faster to their asymptotic values than their large k counterparts which are involved in the modeling procedure when μ is large. Thus, such a selection of this parameter could lead to the introduction of modeling errors in the numerical scheme and, eventually, to a poor MA approximation.

To resolve this issue, it is suggested to select the parameter μ to yield a minimum value of the following modeling error

$$\varepsilon_{mod} = \frac{1}{2\omega_b} \int_{-\omega_b}^{\omega_b} \left| H_{MA}^{*}(e^{j\omega T})\, H_{MA}^{\dagger}(e^{j\omega T}) - S_{\underline{YY}}(\omega) \right|^2 d\omega \tag{21}$$

which measures the closeness between the target spectrum and its MA approximation over the entire frequency domain of interest. Clearly, the reliability of the MA coefficients computed by the proposed two-step scheme is limited by the accuracy of the prior AR representation of the target spectrum. Consequently, it will be assumed in the following that the order $\hat{m}$ is large enough so that the AR spectrum $H_{AR}^{*}(e^{j\omega T})\, H_{AR}^{\dagger}(e^{j\omega T})$ closely matches the target expression $S_{\underline{YY}}(\omega)$. Under this assumption, this matrix can be replaced in equation

(21) by its AR approximation without unduly affecting the value of ε_{mod}.

The error ε_{mod} will first be evaluated in the case $\mu = m$. The corresponding matrix L_m is square and block lower triangular with its diagonal terms all equal to the nonsingular matrix $\hat{B}_0^{-1}$. Thus, L_m is invertible and equation (17) reduces to the following system of linear equations

$$L_m \; B = \tilde{I}_m . \tag{22}$$

The special form of the matrix L_m implies that the coefficients B_k' can be computed simply, recursively, from equation (22). Further, it is readily shown that these MA parameters satisfy exactly equation (14) for $k = 0,...,m$. Finally, it should be noted that the symmetry of these equations with respect to the MA and AR coefficients leads to a simple expression for the inverse L_m^{-1}. Specifically, introduce the $(m+1)n \times (m+1)n$ Toeplitz matrix M whose block elements are the MA parameters, that is

$$\left\{ M \right\}_{kl} = B'_{k-l} \qquad for\; 0 \le l \le k \le m \tag{23a}$$

and

$$\left\{ M \right\}_{kl} = 0 \qquad otherwise. \tag{23b}$$

Then, it is readily shown that

$$\left\{ L_m \; M \right\}_{kl} = \sum_{i=max(0,k-l-\bar{m})}^{min(k-l,m)} \hat{B}_0^{-1} \; \hat{A}_{(k-l)-i} \; B_i' = I_n \; \delta_{kl} \tag{24}$$

where the last equality follows from equation (14). The above relations can be written in the form

$$L_m \; M = M \; L_m = I_{(m+1)n} \tag{25}$$

so that

$$L_m^{-1} = M. \tag{26}$$

The computation of the value of the error ε_{mod} corresponding to the selection $\mu = m$ requires first the determination of $H_{MA}(e^{j\omega T})$. This transfer function can be evaluated in terms of its autoregressive counterparts by introducing the AR input-output crosscorrelations defined as follows, see Mignolet and Spanos [9],

$$R_{\underline{\hat{Y}W}}(l) = E\left[\underline{\hat{Y}}_r \; \underline{W}_{r+l} \right] = 0 \quad \text{for } l > 0 \tag{27a}$$

$$R_{\underline{\hat{Y}W}}(l) = 2\; \omega_b \; \hat{B}_0 \quad \text{for } l = 0 \tag{27b}$$

and

$$R_{\underline{\hat{Y}W}}(l) = - \sum_{k=1}^{min(m,-l)} \hat{A}_k \; R_{\underline{\hat{Y}W}}(k+l) \quad \text{for } l < 0. \tag{27c}$$

Combining equations (14) and (27c), it is readily shown that the MA coefficients B_k' satisfy the equations

$$\sum_{l=min(0,k-\bar{m})}^{k} \hat{A}_{k-l} \left[B_l' - \frac{1}{2\omega_b} R_{\underline{\hat{Y}W}}(-l) \right] = 0 \qquad for\; k=0,...,m \tag{28}$$

when $\mu = m$. The nonsingularity of the matrix $\hat{A}_0 = I_n$ implies that this system of equations admits only the trivial solution

$$B_k' = \frac{1}{2\omega_b} R_{\underline{f}\underline{W}}(-k) \qquad for\ k=0,...,m. \tag{29}$$

Thus, the MA parameters can be directly computed from the AR input-output crosscorrelations. The above relations can also be expressed in the form

$$B_k' = \left[\frac{1}{2\omega_b} R_{\underline{f}\underline{W}}(k) \right] \left[T_m(k) \right] \qquad for\ all\ k \tag{30}$$

where the time limiting matrix operator $T_m(k)$ is defined as

$$T_m(k) = I_n \qquad for\ 0 \le k \le m \tag{31a}$$

and

$$T_m(k) = 0 \qquad otherwise. \tag{31b}$$

According to equation (30), the set of MA parameters B_k' can be written as the product of two sequences, $\frac{1}{2\omega_b} R_{\underline{f}\underline{W}}(k)$ and $T_m(k)$. Thus, relying on the properties of the discrete Fourier transform, see Burrus and Parks [10], it can be shown that the MA transfer function can be written in the form

$$H_{MA}(e^{j\omega T}) = \left[\frac{1}{2\omega_b} \sum_{k=0}^{\infty} R_{\underline{f}\underline{W}}(-k)\, e^{-jk\omega T} \right] * \left[\sum_{k=0}^{\infty} T_m(k)\, e^{-jk\omega T} \right] \tag{32}$$

where the symbol * denotes the operation of Fourier convolution defined by the equation

$$\Psi(\omega) * \Phi(\omega) = \frac{1}{2\omega_b} \int_{-\omega_b}^{\omega_b} \Psi(\omega-\alpha)\, \Phi(\alpha)\, d\alpha. \tag{33}$$

It is readily shown, see Mignolet [11], that the first term in the right-hand-side of equation (31) simply reduces to $\hat{D}^{-1}(e^{j\omega T})\, \hat{B}_0$. Further, the second series can be directly evaluated as

$$\sum_{k=0}^{\infty} T_m(k)\, e^{-jk\omega T} = \frac{e^{-j\frac{m}{2}\omega T} \sin\left[\left(\frac{m+1}{2}\right)\omega T\right]}{\sin\left(\frac{\omega T}{2}\right)} I_n = e^{-j\frac{m}{2}\omega T}\, sinc_m(\omega T)\, I_n . \tag{34}$$

so that the MA transfer function corresponding to the selection $\mu = m$ can be expressed in the form

$$H_{MA}(e^{j\omega T}) = \left[\hat{D}^{-1}(e^{j\omega T})\, \hat{B}_0 \right] * \left[e^{-j\frac{m}{2}\omega T}\, sinc_m(\omega T)\, I_n \right]. \tag{35}$$

A similar relation can be derived in the case $\mu > m$. In this case, the number of equations involved, $\mu+1$, is greater than the number of unknowns, $m+1$, and the matrix L_μ is rectangular. Partitioning this matrix in the form

$$L_\mu^\dagger = \left[L_m^\dagger \quad F^\dagger \right], \tag{36}$$

it is readily shown that

$$L_\mu^\dagger\, L_\mu = L_m^\dagger\, L_m + F^\dagger\, F. \tag{37}$$

The right-hand-side of equation (17) can similarly be expressed in terms of L_m and F. Specifically, using the definition of the matrix $\bar{I}_\mu$, equation (20), it is found that

$$L_\mu^\dagger\, \bar{I}_\mu = L_m^\dagger\, \bar{I}_m \tag{38}$$

and finally,

$$B = L_m^{-1} \left[I_{(m+1)n} + L_m^{-t} F^t F L_m^{-1} \right]^{-1} \bar{I}_m. \tag{39}$$

Clearly, both the MA parameters, stacked in B, and the MA spectral matrix depend on μ through the matrix F. To derive a relation similar to equation (35) which is valid for $\mu > m$, introduce the $(m+1)n \times n$ matrix $E(\omega)$ whose k $n \times n$ block element is

$$\left\{ E(\omega) \right\}_k = e^{-jk\omega T} I_n \qquad for\ 0 \le k \le m. \tag{40}$$

Then, it is readily shown that the transfer function of the MA system can be written in the form

$$H_{MA}(e^{j\omega T}) = E^t(\omega)\, B = E^t(\omega)\, L_m^{-1} \left[I_{(m+1)n} + L_m^{-t} F^t F L_m^{-1} \right]^{-1} \bar{I}_m. \tag{41}$$

Further, note that the block elements of $E^t(\omega)\, L_m^{-1}$ can be expressed in terms of the AR input-output crosscorrelations according to equations (23), (26) and (29). Specifically, it is found that

$$\left\{ E^t(\omega)\, L_m^{-1} \right\}_k = \frac{1}{2\omega_b} \sum_{l=k}^{m} R_{\underline{YW}}(k-l)\, e^{-jl\omega T} = \frac{e^{-jk\omega T}}{2\omega_b} \sum_{l=0}^{m-k} R_{\underline{YW}}(-l)\, e^{-jl\omega T}. \tag{42}$$

Finally, proceeding as in equations (30) and (32), it is seen that

$$\frac{1}{2\omega_b} \sum_{l=0}^{m-k} R_{\underline{YW}}(-l)\, e^{-jl\omega T} = \sum_{l=0}^{\infty} \left[\frac{1}{2\omega_b} R_{\underline{YW}}(-l) \right] \left[T_{m-k}(l) \right] e^{-jl\omega T}$$

$$= \left[\hat{D}^{-1}(e^{j\omega T})\, \hat{B}_0 \right] * \left[e^{-j\left(\frac{m-k}{2}\right)\omega T} sinc_{m-k}(\omega T)\, I_n \right]. \tag{43}$$

so that the MA transfer function can be written in the form

$$H_{MA}(e^{j\omega T}) = \sum_{k=0}^{m} \left\{ \left[\hat{D}^{-1}(e^{j\omega T})\, \hat{B}_0 \right] * \left[e^{-j\left(\frac{m-k}{2}\right)\omega T} sinc_{m-k}(\omega T)\, I_n \right] \right\}$$

$$\times \left\{ \left[I_{(m+1)n} + L_m^{-t} F^t F L_m^{-1} \right]^{-1} \right\}_{k0} e^{-jk\omega T}. \tag{44}$$

for $\mu > m$. The above relation is also valid for $\mu = m$ since for this selection $F = 0$ and only the $k = 0$ term contributes to the summation so that equation (44) reduces to equation (35). If a larger value of the parameter μ is selected, F not vanish and the corresponding matrix $\left[I_{(m+1)n} + L_m^{-t} F^t F L_m^{-1} \right]^{-1}$ is, in general, full. Thus, all values of k contribute to the above sum and the MA transfer function involves terms of the form

$$\left[\hat{D}^{-1}(e^{j\omega T})\, \hat{B}_0 \right] * \left[e^{-j\left(\frac{m-k}{2}\right)\omega T} sinc_{m-k}(\omega T)\, I_n \right].$$

The above expression can be recognized from equation (43) as a Fourier series approximation of $\hat{D}^{-1}(e^{j\omega T})\hat{B}_0$ limited to $m-k$ harmonics. Then, it is readily shown that the matching between the above terms and $\hat{D}^{-1}(e^{j\omega T})\, \hat{B}_0$ will deteriorate as k increases. Thus, the choice $\mu > m$ is expected to yield a larger modeling error, ε_{mod}, than does the selection $\mu = m$.

Next, consider the result of an increase of the value of the deconvolution parameter from μ to $\bar{\mu}=\mu+\Delta\mu$. The corresponding matrix $\bar{F}$ can be partitioned as follows

$$\bar{F}^{t} = \left[F^{t} \quad \Delta F^{t} \right] \tag{45}$$

where ΔF has dimensions $\Delta\mu n \times (m+1)n$. Then, it is readily shown that

$$L_m^{-t}\, \bar{F}^{t}\, \bar{F}\, L_m^{-1} = \left[L_m^{-t}\, F^{t}\, F\, L_m^{-1} \right] + \left[L_m^{-t}\, \Delta F^{t}\, \Delta F\, L_m^{-1} \right]. \tag{46}$$

Further, note that the matrix $L_m^{-t}\, \Delta F^{t}\, \Delta F\, L_m^{-1}$ is positive semidefinite so that the eigenvalues of $L_m^{-t}\, \bar{F}^{t}\, \bar{F}\, L_m^{-1}$ are greater than the corresponding values of $L_m^{-t}\, F^{t}\, F\, L_m^{-1}$. Thus, the contribution of the spurious terms

$$\left[\hat{D}^{-1}(e^{j\omega T})\, \hat{B}_0 \right] * \left[e^{-j\left(\frac{m-k}{2}\right)\omega T} \; sinc_{m-k}(\omega T)\, I_n \right] \quad k>0$$

to the MA transfer function $H_{MA}(e^{j\omega T})$, increases with μ. It is then concluded that the modeling error, ε_{mod}, is a monotonically increasing function of this parameter so that its optimum value is $\mu=m$.

NUMERICAL EXAMPLE

The somewhat surprising result derived in the previous section will now be illustrated by considering the following target spectrum

$$S_{yy}(\omega) = \left[1 + e^{-j\omega} \right]^{*} \left[1 + e^{-j\omega} \right] = 2\,(1 + \cos\omega) \quad \omega \in [-\pi, \pi]. \tag{47}$$

The corresponding autocorrelation function is readily computed as

$$R_{yy}(k) = 4\,\pi \quad \text{for } k=0 \tag{48a}$$

$$R_{yy}(k) = 2\,\pi \quad \text{for } k=\pm 1 \tag{48b}$$

and

$$R_{yy}(k) = 0 \quad \textit{otherwise}. \tag{48c}$$

Clearly, the target spectrum $S_{yy}(\omega)$ can be written in the form of equations (4) and (5) with $m=1$ and

$$B'_0 = B'_1 = 1. \tag{49}$$

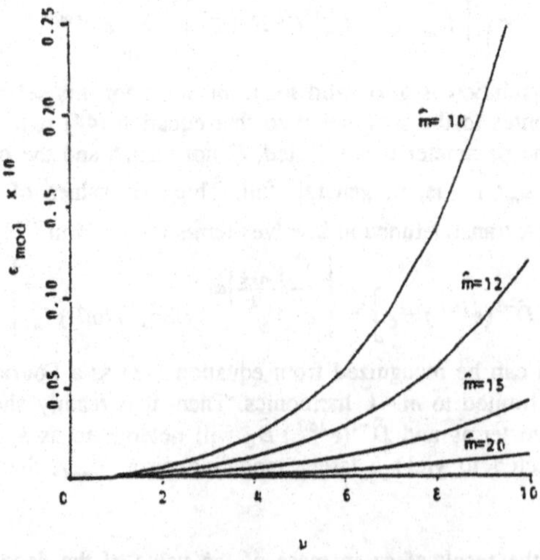

Fig. 1 Variations of the modeling error, ε_{mod}, as a function of μ for various values of $\hat{m}$.

The order $\hat{m}$ of the prior AR approximation has been selected much larger than the corresponding value of the true MA representation, $m=1$, so that the AR spectrum represents an accurate representation of $S_{yy}(\omega)$, equation (47). The variations of the modeling error, ε_{mod}, with the deconvolution parameter μ are displayed in Figure 1 for $\hat{m}=10, 12, 15$ and 20. As proved in the previous section, the quality of the MA approximation sharply deteriorates as the number of equations involved in the minimization procedure increases so that the selection $\mu=m$ yields the MA representation which is optimum in the sense of equation (21).

The variations of the modeling error ε_{mod} with the order $\hat{m}$ of the prior AR approximation should also be noted. Specifically, it is found that ε_{mod} decreases monotonically with increasing $\hat{m}$ independently of the selected value of the parameter μ. This behavior is expected; as the AR system order increases, the first AR parameters converge toward their asymptotic values, the set of equations (14) becomes compatible with a finite MA representation and a more accurate MA model is obtained.

ASYMPTOTIC EQUIVALENCE OF ALGORITHMS

To gain additional insight into the optimality of the selection $\mu=m$, the asymptotic behavior, $\hat{m}\to\infty$, of the solution of equation (22) and its relation to the Cholesky decomposition method, see Youla and Kazanjian [3], will be investigated.

Clearly, as $\hat{m}$, the AR system order, tends to infinity, $\hat{D}^{-1}(e^{j\omega T})\,\hat{B}_0$ approaches the exact MA(m) representation of the target spectral matrix

$$S_{\underline{YY}}(\omega) = \frac{1}{2\omega_b} \sum_{k=-m}^{m} R_{\underline{YY}}(k)\, e^{-jk\omega T}. \tag{50}$$

Thus, the bandlimiting postmultiplication by $T_m(k)$ in equation (30) becomes unnecessary, $H_{MA}(e^{j\omega T})$ tends to $\hat{D}^{-1}(e^{j\omega T})\,\hat{B}_0$, and ε_{mod} vanishes. The parameters of the AR($\hat{m}$) model used by the deconvolution procedure are obtained by solving the Yule-Walker equations (10) and (11) which can be written in the following compact form

$$A^{(\hat{m})}\, R^{(\hat{m})} = \hat{B}_0\, \hat{B}_0^{\dagger}\, \tilde{I}_{\hat{m}}^{\dagger} \tag{51}$$

where

$$\left\{ A^{(\hat{m})} \right\}_k = \hat{A}_k \qquad 0 \le k \le \hat{m}. \tag{52}$$

Further, the symbol $R^{(\hat{m})}$ denotes the covariance matrix of the target process defined by the relation

$$\left\{ R^{(\hat{m})} \right\}_{kl} = R_{\underline{YY}}(k-l) \qquad 0 \le k, l \le \hat{m}. \tag{53}$$

The positive semidefiniteness of $R^{(\hat{m})}$ implies the existence of a Cholesky decomposition of the form

$$R^{(\hat{m})} = \left[\hat{L}^{(\hat{m})} \right]^{\dagger} \hat{L}^{(\hat{m})} \tag{54}$$

where $\hat{L}^{(\hat{m})}$ designates a block lower triangular matrix. Combining this relation and equation (51), it is found that the AR coefficients satisfy the equation

$$A^{(\hat{m})} \left[\hat{L}^{(\hat{m})} \right]^{\dagger} = \hat{B}_0\, \hat{B}_0^{\dagger}\, \tilde{I}_{\hat{m}}^{\dagger} \left[\hat{L}^{(\hat{m})} \right]^{-1}. \tag{55}$$

Further, the special form of the matrices $\tilde{I}_{\hat{m}}$ and $\hat{L}^{(\hat{m})}$ implies that

$$\bar{I}_{\hat{m}}^{\dagger}\left[\mathcal{L}^{(\hat{m})}\right]^{-1}=\bar{I}_{\hat{m}}^{\dagger}\left\{\left[\mathcal{L}^{(\hat{m})}\right]^{-1}\right\}_{00}=\bar{I}_{\hat{m}}^{\dagger}\left\{\mathcal{L}^{(\hat{m})}\right\}_{00}^{-1}=\left\{\mathcal{L}^{(\hat{m})}\right\}_{00}^{-1}\bar{I}_{\hat{m}}^{\dagger} \tag{56}$$

so that equation (55) can be written in the form

$$A^{(\hat{m})}\left[\hat{\mathcal{L}}^{(\hat{m})}\right]^{\dagger}=\hat{B}_0\hat{B}_0^{\dagger}\left\{\mathcal{L}^{(\hat{m})}\right\}_{00}^{-1}\bar{I}_{\hat{m}}^{\dagger}. \tag{57}$$

Next, introduce the following block partition of $\hat{\mathcal{L}}^{(\hat{m})}$

$$\mathcal{L}^{(\hat{m})}=\begin{bmatrix} L_1^{(\hat{m})} & 0 \\ L_2^{(\hat{m})} & L_3^{(\hat{m})} \end{bmatrix} \tag{58}$$

where $L_1^{(\hat{m})}$ and $L_3^{(\hat{m})}$ are block lower triangular matrices of dimensions $(m+1)n \times (m+1)n$ and $(\hat{m}-m)n \times (\hat{m}-m)n$, respectively. The symbol $L_2^{(\hat{m})}$ designates a full $(\hat{m}-m)n \times (m+1)n$ matrix. A similar partitioning of $A^{(\hat{m})}$ yields the following representation

$$A^{(\hat{m})}=\left[A_1^{(\hat{m})} \quad A_2^{(\hat{m})}\right] \tag{59}$$

where $A_1^{(\hat{m})}$ and $A_2^{(\hat{m})}$ are $n \times (m+1)n$ and $n \times (\hat{m}-m)n$ matrices, respectively. Combining equations (57)-(59), it is readily found that the AR parameters satisfy the following relations

$$A_1^{(\hat{m})}\left[L_1^{(\hat{m})}\right]^{\dagger}=\hat{B}_0\hat{B}_0^{\dagger}\left\{L^{(\hat{m})}\right\}_{00}^{-1}\bar{I}_m^{\dagger} \tag{60}$$

and

$$A_1^{(\hat{m})}\left[L_2^{(\hat{m})}\right]^{\dagger}+A_2^{(\hat{m})}\left[L_3^{(\hat{m})}\right]^{\dagger}=0. \tag{61}$$

Next, note that the computation of the MA coefficients according to the two-step procedure requires only, when $\mu=m$, the prior determination of the AR parameters $\hat{A}_k$ for $k=1,\ldots,m$. Thus, in this case, it is only necessary to consider the system of equations (60). In this context, it has been shown, see Youla and Kazanjian [3], that $L_1^{(\hat{m})}$ converges, as $\hat{m}\rightarrow\infty$, to the block Toeplitz matrix $L_1^{(\infty)}$ whose block elements are

$$\left\{L_1^{(\infty)}\right\}_{kl}=B_{k-l}^{\dagger} \qquad 0\leq l\leq k\leq m+l \tag{62}$$

and

$$\left\{L_1^{(\infty)}\right\}_{kl}=0 \qquad \textit{otherwise.} \tag{63}$$

where the matrices B_k represent the parameters of the exact MA decomposition of the target spectrum $S_{\underline{YY}}(\omega)$. Using the above asymptotic relation, it can be seen that the system of equation (60) reduces to the following conditions

$$\sum_{l=0}^{k}\hat{A}_l\,B_{k-l}=\hat{B}_0\hat{B}_0^{\dagger}\,B_0^{-\dagger}\,\delta_{k0} \qquad \textit{for } k=0,1,\ldots,m. \tag{64}$$

In particular, for $k=0$, it is found that

$$B_0=\hat{B}_0\hat{B}_0^{\dagger}\,B_0^{-\dagger} \tag{65}$$

or

$$\hat{B}_0\hat{B}_0^{\dagger}=B_0B_0^{\dagger}. \tag{66}$$

It has been recognized in previous investigations of the AR modeling technique, see Mignolet and Spanos [9], that the value of the optimum AR parameter $\hat{B}_0$ is not unique, only the product $\hat{B}_0 \hat{B}_0^{\dagger}$ is determined and given by equation (11). The same result holds for B_0, see Youla and Kazanjian [3]; the Cholesky decomposition approach to the spectral factorization problem yields a specific value to $B_0 B_0^{\dagger}$ only. Assuming that the same form has been selected for both $\hat{B}_0$ and B_0, equation (66) implies that

$$\hat{B}_0 = B_0 \tag{67}$$

so that equation (64) can be written in the form

$$\sum_{l=0}^{k} \hat{A}_l \, B_{k-l} = \hat{B}_0 \, \delta_{k0} \qquad \textit{for } k = 0, 1, \dots, m \tag{68}$$

Comparing these relations with equation (14) and noting, as in equation (29), that this system of equations admits a unique solution, it is seen that

$$B_k{}' = B_k \qquad 0 \le k \le m. \tag{69}$$

Consequently, the two-step AR to MA modeling based on the selection $\mu = m$ is asymptotically, $\hat{m} \to \infty$, equivalent to a Cholesky decomposition of the target covariance matrix.

LINEAR MA MODELING BY SPECTRAL MATRIX INVERSION

The two-step AR to MA modeling technique investigated in the previous sections yields only approximate values of the MA parameters because of the finiteness of the AR order $\hat{m}$. Reliable and accurate AR approximations can easily be derived for most target spectral matrices. However, in some instances, see Mignolet and Spanos [12] for an extensive discussion, this step may either represent a delicate task or require a very large value of the order $\hat{m}$.

In such cases, an alternative scheme might be desirable. Such an approach can be devised by considering the inverse of the target spectral matrix. Indeed, if $S_{\underline{YY}}(\omega)$ can be exactly represented in the form of equations (4) and (5), then its inverse can be seen as the spectrum of an autoregressive process. Specifically, it is found that

$$S_{\underline{YY}}^{-1}(\omega) = \left[H_{MA}^{*}(e^{j\omega T}) \, H_{MA}^{\dagger}(e^{j\omega T}) \right]^{-1} = H_{MA}^{-\dagger}(e^{j\omega T}) \, H_{MA}^{-*}(e^{j\omega T}) \tag{70}$$

Proceeding with an AR modeling of $S_{\underline{YY}}^{-1}(\omega)$ yields a denominator polynomial $\hat{D}(z)$, a matrix $\hat{B}_0$ and the corresponding autoregressive spectrum

$$S_{AR}(\omega) = \hat{D}^{-*}(e^{j\omega T}) \, \hat{B}_0 \hat{B}_0^{\dagger} \, \hat{D}^{-\dagger}(e^{j\omega T}). \tag{71}$$

Upon comparing equations (70) and (71), it is found that a bonafide MA transfer function is simply

$$H_{MA}(e^{j\omega T}) = e^{-jm\omega T} \left[\hat{D}^{*\dagger}(e^{j\omega T}) \, \hat{B}_0^{-\dagger} \right] \tag{72}$$

or in terms of z transforms

$$H_{MA}(z) = z^{-m} \left[\hat{D}^{\dagger}\left(\frac{1}{z}\right) \hat{B}_0^{-\dagger} \right]. \tag{73}$$

This procedure is clearly conditioned by the availability of the inverse spectral matrix and of the corresponding autocorrelations. The determination of these quantities can represent an excessive computational burden if only the autocorrelations of the target process are available. On the contrary, this technique is especially expedient when the inverse of the spectral matrix can be derived analytically.

CONCLUDING REMARKS

In this paper, the applicability of autoregressive modeling to the derivation of a moving average representation of a target process has been investigated. An existing two-step procedure has been considered first in which a prior AR approximation of the target spectral matrix is used to produce a reliable MA model. It was found that the optimum value of the algorithmic parameter μ simply equals the order of the MA approximation. Further, it was recognized that the corresponding MA coefficients can be efficiently evaluated by a simple recurrence relation, equation (14). A numerical example was presented that confirmed the theoretical developments and demonstrated the importance of the reliability of the prior AR approximation. In this context, it was determined that the order $\hat{m}$ should be selected as large as possible and that in the limit $\hat{m} \rightarrow \infty$, the present procedure is equivalent to an existing scheme based on the Cholesky decomposition of an infinite-dimensional covariance matrix. Finally, an alternative approach has been described that yields an exact MA representation and which relies on the knowledge of the autocorrelations associated with the inverse of the target spectral matrix.

REFERENCES

1. Wiener N. and Masani, P. The Prediction Theory of Multivariate Stochastic Processes, I. The Regularity Condition, Acta Mathematica, Vol. 98, pp. 111-150, 1957.
2. Jezek, J. and Kucera, V. Efficient Algorithm for Matrix Spectral Factorization, Automatica, Vol. 21, pp. 663-669, 1985.
3. Youla, D.C. and Kazanjian, N.N. Bauer-Type Factorization of Positive Matrices and the Theory of Matrix Polynomials Orthogonal on the Unit Circle, IEEE Transactions on Circuits and Systems, Vol. CAS-25, pp. 57-69, 1978.
4. Tuel, W.G. Computer Algorithm for Spectral Factorization of Rational Matrices, IBM Journal of Research and Development, pp. 163-170, 1968.
5. Anderson, B.D.O., Hitz, K.L. and Diem, N.D. Recursive Algorithm for Spectral Factorization, IEEE Transactions on Circuits and Systems, Vol. CAS-21, pp. 742-750, 1974.
6. Durbin, J. Efficient Estimation of Parameters in Moving-Average Models, Biometrika, Vol. 46, pp. 306-316, 1959.
7. Mignolet, M.P. and Spanos, P.D. A Direct Determination of ARMA Algorithms for the Simulation of Stationary Random Processes International Journal of Non-Linear Mechanics, Vol. 25, pp. 555-568, 1990.
8. Hannan, E.J. Multiple Time Series, John Wiley and Sons, New York, 1970.
9. Mignolet, M.P. and Spanos, P.D. Recursive Simulation of Stationary Random Processes - Part I, Journal of Applied Mechanics, Vol. 54, pp. 674-680, 1987.
10. Burrus, C.S. and Parks, T.W. DFT/FFT and Convolution Algorithms, John Wiley and Sons, New York, 1984.
11. Mignolet, M.P. ARMA Simulation of Multivariate and Multidimensional Random Processes, Ph.D. Dissertation, Rice University, Houston, Texas, 1987.
12. Mignolet, P.D. and Spanos, P.D. Autoregressive Spectral Modeling: Difficulties and Remedies, International Journal of Non-Linear Mechanics, to appear 1991.

Random Fields. Digital Simulation and Applications in Structural Mechanics

E. Bielewicz, J. Górski, H. Walukiewicz

Department of Structural Mechanics, Technical University of Gdańsk, 80-952, Gdańsk, ul. Majakowskiego 11/12, Poland

ABSTRACT

A new method of simulation of nonhomogeneous scalar discrete random fields is proposed. These fields are simulated by assuming the sequential basis of several points located at nodes of the regular nets. Moreover, the global envelopes of the fields are taken into account which is needed in simulating arbitrary boundary conditions. The local covariance matrices play the main role in the method. The assumed group invariance of the matrices leads to an useful classification of the random fields. Generated realizations of geometrical imperfections are the input for numerical solutions of compressed elastic-plastic thin-walled structures.

INTRODUCTION

The computational study of random fields is a complex subject and it is necessary to restrict one's interest to specific areas. In stochastic structural mechanics loads acting on structures, geometrical imperfections and material characteristics of these structures should be treated as discretized spatial and/or temporal random fields. Description of these fields by the set of realizations (sample functions) is very clear and near to the engineering point of view and leads to the Monte Carlo simulation techniques for solving non-linear problems. The first step for numerical evaluation of the response of stochastic models of structures is generation of realizations of a dicretized random field. These functions, after the averaging, should reproduce the prescribed mean value vector and the global covariance matrix of the field.

Simulation of random fields is being developed by research workers at some leading universities (at Princeton University by Shinozuka and his co-workers [1], [2], [3], at Technion-Israel Institute of Technology by Adler [4]). These methods concern mainly homogeneous fields encountered in such domains as acoustics, turbulence, ocean engineering (Spanos [11]), tribology an so on. However, for many technological, environmental and physical problems nonhomogeneous fields are typical (Wilde [5], Sobczyk [6]).

Is seems that presented here a new method of simulation worked out for nonhomogeneous discrete fields will be especially useful in solving boundary value problems of two-dimensional and three-dimensional structures.

FORMULATION OF PROBLEM

The purpose of the main part of this work is to develop a new method of generating realizations of 2-D nonhomogeneous discrete fields described by local conditional covariance matrices and global envelopes of the fields. Assuming a group of symmetry of a regular net it is possible to obtain the special form of the local covariance matrix, with the induced symmetry.

The second part of the work is devoted to a numerical study of the effects of random geometrical imperfections on the critical points and the postcritical shapes of compressed elastic-plastic structures (a stiffened plate and a cruciform column). There is, as yet, no general theory available that relates the behaviour of imperfect elastic-plastic structures to the critical behaviour of the corresponding perfect structures. However, it is well known that elastic structures which have asymmetric or symmetric unstable bifurcation points show the sensitivity to initial imperfection (e.g. Waszczyszyn et al. [7]). Moreover, the imperfection sensitivity is typical in structures designed on the basis of the optimization theory (Budiansky and Hutchinson [8]). Therefore we have chosen the geometrical parameters of the models with care.

We did not applied here the complete Monte Carlo approach, in order to get the statistics of the response. The general methodology requires the large computational effort. Instead, considering only several realizations of the imperfections fields, we try to answer the question: is there any effect of the assumed type of imperfections on the limit loads and the deformed shapes of the structures?

In order to decrease the number of realizations an accumulation procedure has been proposed by Bielewicz, Górski and Skowronek [9]. The problem of reduction of a field into a small sample size with associated nonuniform probabilities will be presented elsewhere.

RANDOM FIELDS: DESCRIPTION AND CLASSIFICATION

The second order, real-valued random fields are determined once we specify the following two functions - the expected value:

$$\bar{\mathcal{X}}(r) = E\left(\mathcal{X}(r,\omega)\right) \tag{1}$$

and the correlation tensor:

$$\mathbb{K}(r_1,r_2) = E\left(\left(\mathcal{X}(r_1,\omega)-\bar{\mathcal{X}}(r_1)\right)\otimes\left(\mathcal{X}(r_2,\omega)-\bar{\mathcal{X}}(r_2)\right)\right), \tag{2}$$

where E is the expectation operator, ω denotes elementary events and sign $\otimes$ denotes tensor product; $r \in \mathbb{R}^3$ is the position vector.

The classification of the vector random fields can be based on the theory of groups of symmetry of tensors of the second order (Bielewicz and Walukiewicz [10]). The group of symmetry of the vector random field may be defined as a subgroup of the full orthogonal group O(3):

$$G_{\mathcal{X}}=\left\{Q \subset O(3): \forall_{r_1,r_2\in \mathbb{R}^3} Q^T\mathbb{K}(r_1,r_2)Q = \mathbb{K}(Qr_1,Qr_2)\right\}, \tag{3}$$

where all the operations are the contractions of the tensor product. For a discrete scalar random field the role of the correlation function plays the covariance matrix $\mathbb{K}(n\times n)$ and the definition reads

$$G = \left\{ Q \subset O(3): \forall_n Q^T\mathbb{K}\, Q = \mathbb{K} \right\}. \tag{4}$$

There are 5 types of plane nets which are invariable in translations and 10 types of point groups which are connected with rotations and reflections in the plane (in $\mathbb{R}^3$ there are 14 nets and 32 groups respectively). Choosing a definite group of symmetry we obtain the number of independent parameters and the form of the local covariance matrix. The method leading to these results is based on relation (4) and on idempotent and nilpotent operators. If G = SO(3) (rotations) we obtain the most simple form of the covariance matrix:

$\mathbb{K} = C\,\mathbb{I}, \ C > 0,$ $\qquad \mathbb{I}$ - the unit matrix nxn (5)

i.e. the discretized white noise. This follows from Schur's lemma: the only matrices which commute with all the matrices of an irreducible representation of a group are the scalar multiples of the unit matrix.

SIMULATION OF DISCRETE SCALAR 2-D RANDOM FIELD

We assume an envelope of the field in the whole domain which is defined as the set of intervals of random variable values for all the nodes of a regular net: $\{(a_i, b_i)\}$, $i=1,2, \ldots MxN$; $b_i > a_i$, where a_i is a lower bound, b_i is an upper bound and MxN is the total number of the nodes. We assume the expected value and the correlation function of the field.

The generation process (program POLE) consists of the following steps:

1. Determination of a local covariance matrix $\mathbb{K}$ (positive definite, dimension 9x9 or alternatively 5x5) on the basis of the assumed correlation function of a random field.
2. Determination of the conditional local covariance matrix $\mathbb{K}_W$. If we part $\mathbb{K}$ into

   $\mathbb{K} = \begin{bmatrix} \mathbb{K}_{11} & \mathbb{K}_{12} \\ \mathbb{K}_{21} & \mathbb{K}_{22} \end{bmatrix}$, where $\mathbb{K}_{11}$ is an mxm matrix, then

   the conditional covariance matrix is given by:

   $\mathbb{K}_W = \mathbb{K}_{11} - \mathbb{K}_{12}\mathbb{K}_{22}^{-1}\,\mathbb{K}_{21}.$
3. Generation of the m-dimensional vector w belonging to the interval (a_k, b_k): $w_k = (b_k - a_k)x_k + a_k$, $k=1,2,\ldots,m$; $1 \le m \le 8$ or $1 \le m \le 4$, where x_k is a random variable uniformly distributed in the interval $(0,1)$.
4. Generation of the random variable x_{m+1} from the interval $(0,1)$ and defining value $F = \Phi\, x_{m+1}$, where $\Phi = \left((2\pi)^m \det\mathbb{K}_W\right)^{-1/2}$ bounds the conditional density function from above.
5. Calculation of the conditional density function $f(w)$ under the assumption of the cut- off normal distribution.
6. Checking the condition $F \le f(w)$ (the von Neumann elimination). If this condition holds, the vector w is accepted and the next scheme is considered; if not, the calculation returns to point 3.

The standard deviation of the random variable at any node i can be connected with the envelope by the formula:

$$\sigma_i = \frac{b_i - a_i}{2s} , \tag{6}$$

where s is an integer ≥ 3.
The nine-point scheme (or alternatively the five-point "cross-scheme") (Fig.1a) is placed at the net nodes such that a sequence of the schemes covers all the nodes. After considering the whole sequence of the schemes (Fig.1b) one realization for the discrete random field is completed. Repeating this procedure NR-times we obtain the set of realizations with equal probabilities.

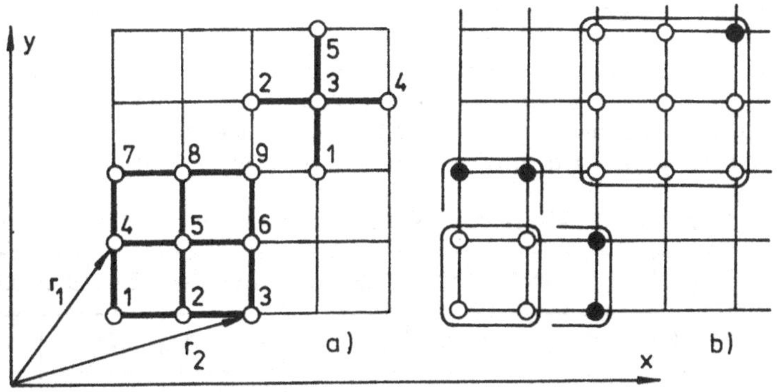

Fig.1 a) The nine-point and the five-point schemes,
b) Typical and border schemes in the nine-point version (o known, • generated values).

The estimate $\tilde{\mathbb{K}}$ of the global covariance matrix $\mathbb{K}$ is given by:

$$\tilde{\mathbb{K}} = \frac{1}{NR - 1} \sum_{j=1}^{NR} (\mathbf{w}_j - \tilde{\mathbf{w}})(\mathbf{w}_j - \tilde{\mathbf{w}})^T, \quad \text{where} \quad \tilde{\mathbf{w}} = \frac{1}{NR} \sum_{j=1}^{NR} \mathbf{w}_j \tag{7}$$

and NR denotes the number of realizations.

The global error GE with respect to the target covariance matrix is defined by:

$$GE = \frac{|\,\|\mathbb{K}\| - \|\tilde{\mathbb{K}}\|\,|}{\|\mathbb{K}\|} 100\% , \tag{8}$$

where $\|\mathbb{K}\| = \sqrt{tr(\mathbb{K}^2)}$ is the Euclidean norm. The analogous formula defines the global error for the variances.

Numerical examples

Several types of the correlation functions were considered in the testing examples. Here we present only two of them.

The Wiener process is a generalization of the one-dimensional Brown motion. The covariance function of the Wiener process on the plane is given by formula:

$$K_v(r_1,r_2) = \min(r_{1x},r_{2x})\min(r_{1y},r_{2y}),\quad E[W(r)]=0,\quad r\in\mathbb{R}^2_+.$$

In the second example the covariance function of a homogeneous process on the plane is considered (Yamazaki and Shinozuka [3]):

$$K_s(r_1-r_2) = \exp\left(-0.25\left((r_{1x}-r_{2x})^2+(r_{1y}-r_{2y})^2\right)\right),$$
$$E[S(r)]=0.$$

These two cases were generated for 11x11 nodes on the square regular net with the use of the nine-point scheme. The errors of our calculations are shown in Table 1. It should be noted that the applied procedure gives the full agreement between the estimates and the target mean value vectors.

Table 1. Errors with respect to number of realizations

ERRORS OF $\tilde{\mathbb{K}}$ MATRIX	WIENER PROCESS			HOMOGENEOUS PROCESS		
	NR=500	2000	10000	NR=500	2000	10000
GE of $\tilde{\mathbb{K}}$	6.4	3.0	2.0	12.2	9.5	8.6
GE of variances	4.1	3.4	1.8	1.3	0.6	0.1

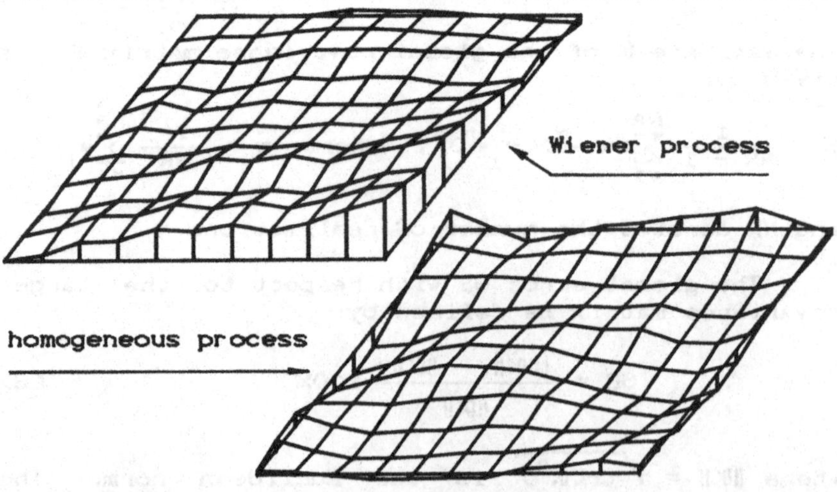

Fig. 2 Some realizations of the Wiener and homogeneous processes.

Our experience shows that the five-point "cross-scheme" gives acceptable results only for the diagonal elements of the global covariance matrix.

Generated realizations of the two fields are shown in Fig.2. The target covariance matrix and the simulated covariance matrix for the Wiener process are presented in Fig.3 and the corresponding matrices for the homogeneous process are presented in Fig 4.

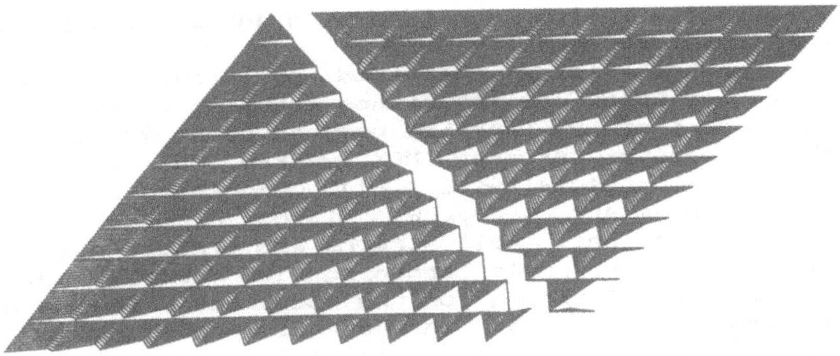

Fig.3 Wiener field: comparison of the target matrix (upper triangle) with the simulated matrix (lower triangle).

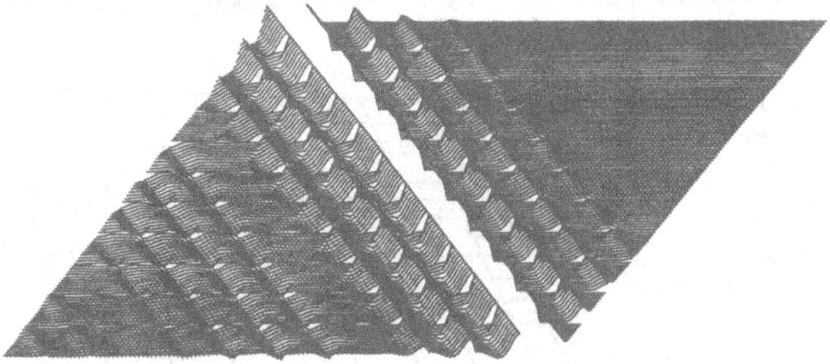

Fig.4 Homogeneous field: comparison of the target matrix (upper triangle) with the simulated matrix (lower triangle).

APPLICATIONS IN STRUCTURAL MECHANICS

The main purpose of this section is to apply some of the generated random fields in numerical solutions of some static elastic-plastic problems of structural mechanics. These problems are governed by the equation:

$$\mathbb{N}_{\omega}\big(w(r,\omega)\big) = \mathbb{F}(r,\omega) \ , \qquad (9)$$

where $\mathbb{N}_{\omega}$ is a nonlinear differential random operator, $w(r,\omega)$ is the displacement random vector and $\mathbb{F}(r,\omega)$ is the load random vector.

As basic forms of geometrical imperfections of the structures two discretized random fields have been assumed:

- white noise fields described by diagonal covariance matrices, see equation (5),
- degenerated fields described by single random variables (rank one covariance matrices), see Fig.5

In this part of the work the use has been made of the IBM-PC computer program BOX worked out at the Department (Chróścielewski [12]). General assumptions of the program are as follows: large displacements, moderate rotations, small strains and elastic-plastic material with isotropic hardening. The limit points and postcritical shapes for the models of structures are determined in an incremental procedure with iterations within each displacement increment Δ.

Numerical example: stiffened plate
An FEM model of the uniformly compressed elastic-plastic structure is considered: an eccentrically stiffened by a rib, simply supported rectangular plate (Fig.5).

$E = 2.1\times10^5$ MPa
model without hardening:
$\sigma_y = 250.0$ MPa
$\nu = 0.3$
$t = 0.01$ m
$h = 0.1047$ m

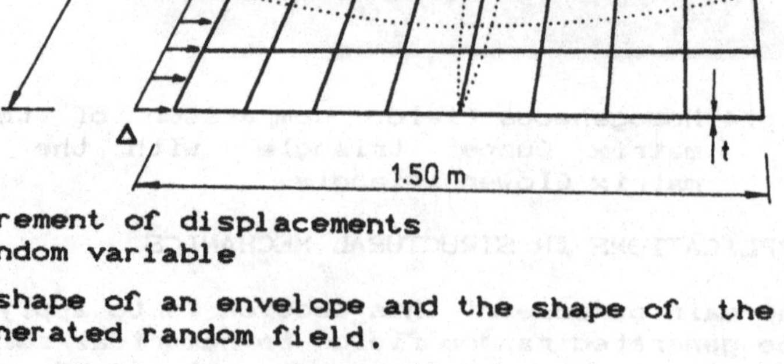

Δ - an increment of displacements
w - the random variable

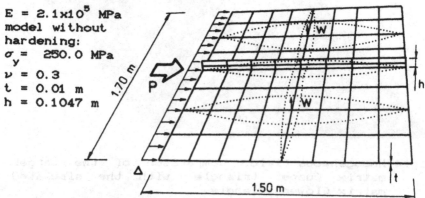

Fig.5 The shape of an envelope and the shape of the degenerated random field.

One type of in-plane boundary conditions is taken into account: all four edges of the plate remain straight throughout the loading history. The

parameters of the plate were chosen by numerical experiments in such a way that the global and the local buckling loads are close together.
First, the height of the rib was determined by a numerical experiment. For h=10.46 cm and the plate without imperfections we got the limit load P_{max} = 2.56 MN and the half-wave sinusoidal symmetric deformed shape. For h=10.48 cm and the plate without imperfections we got the limit load P_{max} = 2.63 MN and the three and half-wave sinusoidal asymmetric deformed shape.
Therefore the intermediate h=10.47 cm has been chosen for further study. In Table 2 the results for only six realizations are presented.

Table 2. Limit loads P_{max}

TYPE OF IMPERFECTIONS			
DEGENERATED FIELD		WHITE NOISE FIELD	
extremum of imperfection w	P_{max} [MN]	extremum of imperfection w	P_{max} [MN]
-0.06 cm	2.76	±0.60 cm	2.66
+0.06 cm	2.40	±0.60 cm	2.58
+0.60 cm	2.03	±0.60 cm	2.64

The postcritical shape corresponding to the degenerated field is presented in Fig.6 and this corresponding to the white noise field is presented in Fig.7.

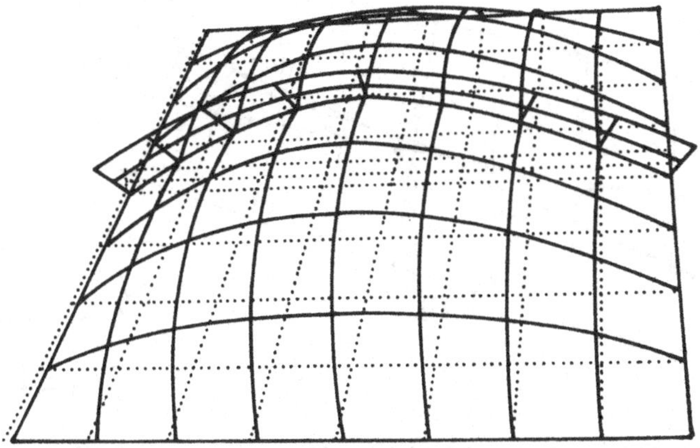

Fig.6 Postcritical shape (P = 2.03 MN) of the imperfect plate.

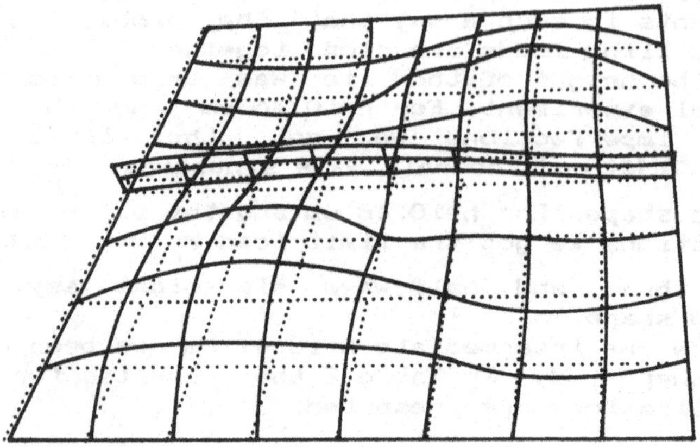

Fig. 7. Postcritical shape (P = 2.64 MN) of the imperfect plate.

Numerical example: cruciform column

The FEM model of a column with a cruciform cross-sections and with the same material characteristic as the stiffened plate is considered. One end of the column is simply supported, lateral displacements, twist and warp are prevented at the support. The second end is free, only warping is prevented in the process of central compressing. At the free end geometrical imperfections in the form of four-dimensional random variable $w=(w_1\ w_2\ w_3\ w_4)$ are assumed. Two types of the covariance matrix were considered: a degenerated case described by one random variable and the white noise case. Some results are shown in Fig. 8 and Fig. 9.

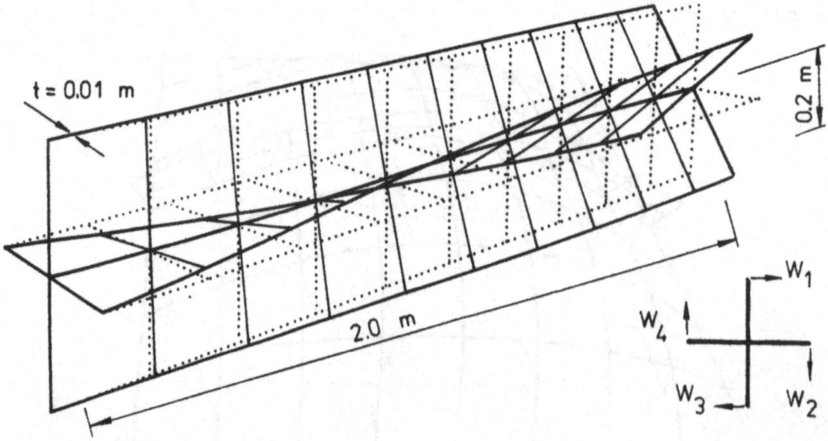

Fig. 8 Postcritical shape of the column with degenerated imperfections: $w=(0.1\ 0.1\ 0.1\ \ 0.1)$ [cm], $P_{max} = 1.67$ MN.

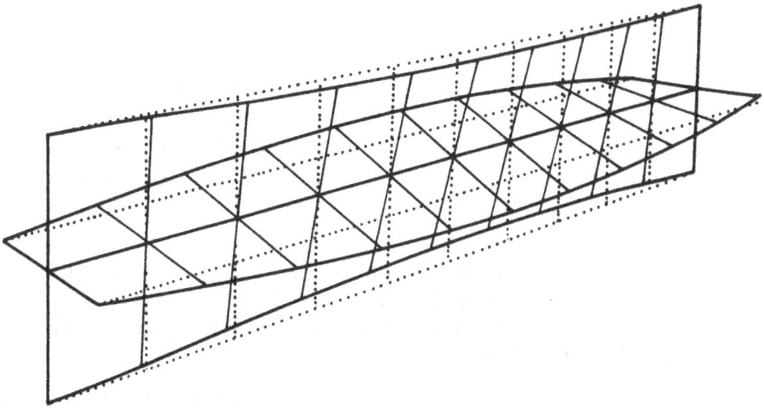

Fig.9 Postcritical shape of the column with degenerated imperfections: w=(0.1 0.1 -0.1 -0.1) [cm], P_{max}= 1.67 MN.

We note that a realization w=(-0.1209 -0.3445 -0.4138 0.0911) [cm] from the white noise random vector gives P_{max}= 1.60 MN and the shape analogous with that presented in Fig.8.

CONCLUSIONS

A new simulation method for two-dimensional nonhomogeneous random fields was examined in the main part of this study. The numerical results of simulation for the Wiener field are extremely good and those for the homogeneous field are relatively good. This observation suggests, in our opinion, that one should not look for some universal methods of simulation but rather should concentrate efforts on the methods suitable for some classes (in the group theoretical setting) of the correlation tensors.

The application of some generated random fields in the elastic-plastic models of structures indicates, that the deformed shapes and the limit loads are sensitive to the assumed type of geometrical imperfections.

Acknowledgement
We express our gratitude to Professor M.Shinozuka for sending us the three volumes [1] of papers by him and his associates.

REFERENCES

1. Shinozuka, M. (Ed.). Stochastic mechanics, Vol.I, Columbia University, 1987, Vol.II, Columbia University, 1987, Vol.III Princeton University, 1988.
2. Naganuma, T., Deodatis, G. and Shinozuka, M. An ARMA model for two-dimensional processes, ASCE Journal of Engineering Mechanics, Vol.113, No.2, pp.234-251, 1987.
3. Yamazaki, F. and Shinozuka, M. Simulation of stochastic fields by statistical preconditioning, ASCE Journal of Engineering Mechanics, Vol.116, No.2, pp.268-287, 1990.
4. Adler, R.J. The geometry of random fields, Wiley & Sons, 1981.
5. Wilde, P. Discretization of random fields in engineering calculations, PWN, Warsaw, 1981 (in Polish).
6. Sobczyk, K. Stochastic wave propagation, Elsevier Amsterdam, 1985
7. Waszczyszyn, Z., Cichoń, Cz. and Radwańska, M. Finite element method in stability of structures Arkady, Warsaw, 1990 (in Polish).
8. Budiansky, B. and Hutchinson, J.W. Buckling: progress and challenge in: Trends in Solid Mechanics, Delft University Press, 1979.
9. Bielewicz, E., Górski, J. and Skowronek, M. Generation of two-dimensional discrete random fields, Scientific Papers, Technical University of Gdańsk, No.408, Civil Eng. XLIV, pp.7-15, 1987, (in Polish).
10. Bielewicz, E. and Walukiewicz, H. On classification of random fields in mechanics, Proceedings of the 2-nd Conference KILiW PAN, Gdańsk, pp.15-22, 1985, (in Polish).
11. Spanos, P.T.D. ARMA algorithms for ocean wave modeling, Journal of Energy Resources Technology, Trans. of the ASME, Vol. 105, No. 3, pp.300-309, 1983.
12. Chróścielewski, J. Numerical nonlinear analysis of stiffened plates by FEM method, Ph.D. thesis, Technical University, Gdańsk, 1983, (in Polish).
13. Bielewicz, E., Skowronek, M. and Walukiewicz H. Random fields and their applications in nonlinear structural problems, Fachbereich A/RU/BT, Bauingenieurwesen, Universität Kaiserslautern, Arb.II, pp.81-94, 1990.

SECTION 8: EARTHQUAKE ENGINEERING APPLICATIONS

Stochastic Estimation of Orientation Error in Buried Seismometers

F. Yamazaki, L. Lu

Institute of Industrial Science, University of Tokyo, 7-22-1 Roppongi, Minato-ku, Tokyo 106, Japan

ABSTRACT

Array observation is an efficient tool to investigate various characteristics of earthquake ground motion. However, seismometers used in arrays may involve unexpected error in their orientation. Methods of orientation error estimation were developed in the three-dimentional space and effects of orientation error were demonstrated. The maximum cross-correlation method and the maximum coherence method were recommended because of their accuracy. The earthquake ground motions recorded in the Chiba array and other two arrays were used in numerical examples. Non-trivial orientation errors were detected for all these arrays. The cross-correlation coefficients and the coherence functions between two points increased significantly by correcting the estimated orientation error.

INTRODUCTION

A number of seismograph arrays have been installed at many places in the world in order to evaluate various engineering and/or seismological characteristics of earthquake ground motions. Recorded ground motions are often used for wave propagation analysis and soil amplification analysis. In these analyses, the correlation of earthquake ground motions between different locations or depths are of major concern. Thus the array seismometers must have correct orientation, otherwise such discussions cannot be made correctly.

It may not be a difficult task to place seismometers on the ground surface to preassigned orientation. However it is not so easy to set seismometers correctly in boreholes. Actually, orientation errors were reported for several arrays (Katayama and Sato, 1982; Real and Cramer, 1990). In these cases, orientation errors were detected based on orbits of two horizontal ground motion components, and they were estimated by maximizing the cross-correlation coefficients between the two components. This means orientation error is considered to be a rotation angle about the vertical axis.

However, in general, there are three components in earthquake ground motion. Hence as the orientation error for three-component seismometers, three rotation angles exist about three spatial coordinates. The two additional angles are the rotations about the two horizontal axes, which correspond to the tilt of the seismometers in their boreholes. To estimate these three angles accurately for buried seismometers, two methods using recorded ground motions are proposed in this paper. The maximum cross-correlation method compares the three pairs of the cross-correlation functions in time domain, while the maximum coherence method employs the three pairs of the coherence functions in frequency domain. These techniques are described and their examples are provided hereafter for three seismometer array systems.

METHODS FOR ORIENTATION ERROR ESTIMATION

Possible orientation error of a three-component seismograph can be fully expressed by three independent angles, α, β and γ as shown in Figure 1. Angle α describes the rotation of the casing about the vertical axis. Angle β defines the tilt of the casing in the borehole measured in the south-north and up-down plane, and angle γ provides the tilt measured in the east-west and up-down plane. Suppose ground motions at a point are represented by a three-component vector:

$$\mathbf{z}(t)=[z_1(t),\ z_2(t),\ z_3(t)]^T \tag{1}$$

where $z_1(t)$, $z_2(t)$ and $z_3(t)$ are the three components in the preassigned directions, i.e. the north-south (NS), east-west (EW) and up-down (UD) directions. These three components may be acceleration, velocity or displacement. Also these can be filtered waves for a selected frequency range in which high spatial coherence is observed. Denote the ground motion recorded by an instrument having orientation error α, β and γ as

$$\mathbf{y}(t)=[y_1(t),\ y_2(t),\ y_3(t)]^T \tag{2}$$

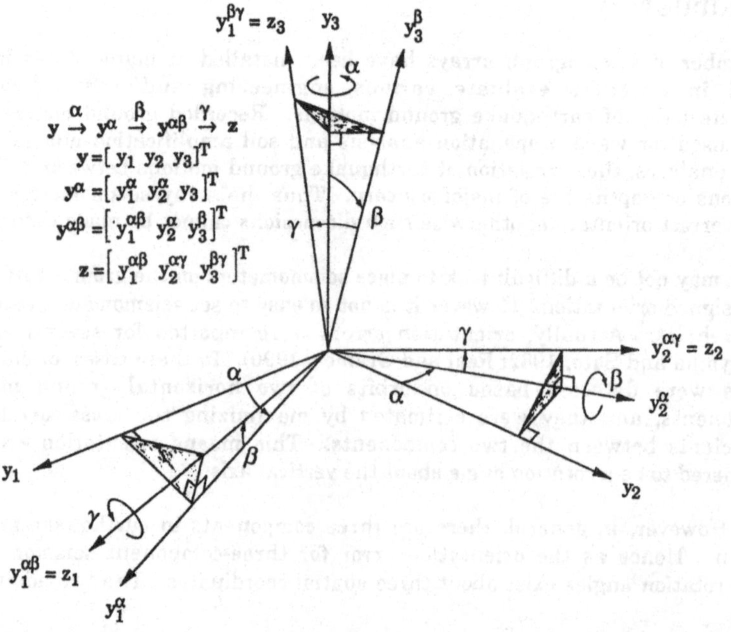

Figure 1 Rotation Angles in Three-dimensional Space

According to the coordinate transformation shown in Figure 1, the ground motion in the preassigned directions can be obtained by

$$\mathbf{z}(t) = \mathbf{T}(\alpha, \beta, \gamma)\, \mathbf{y}(t) \tag{3}$$

where

$$\mathbf{T} = \begin{bmatrix} \cos\alpha\cos\beta & \sin\alpha\cos\beta & \sin\beta \\ -\sin\alpha\cos\gamma - \cos\alpha\sin\beta\sin\gamma & \cos\alpha\cos\gamma - \sin\alpha\sin\beta\sin\gamma & \cos\beta\sin\gamma \\ \sin\alpha\sin\gamma - \cos\alpha\sin\beta\cos\gamma & -\cos\alpha\sin\gamma - \sin\alpha\sin\beta\cos\gamma & \cos\beta\cos\gamma \end{bmatrix} \tag{4}$$

The angles α, β and γ in matrix $\mathbf{T}$ are the orientation error to be estimated.

Maximum Cross-Correlation Method

Ground motions excited by an earthquake are not identical for different points even within a small area due mainly to the variation of local soil conditions. But they are correlated. These correlated seismic waves can be used for the estimation of the orientation error of array seismographs. Suppose $\mathbf{x}(t)$ is a ground motion vector recorded by a seismograph whose instrument orientation is correct (or at least known):

$$\mathbf{x}(t) = [x_1(t), x_2(t), x_3(t)]^T \tag{5}$$

and $\mathbf{y}(t)$ is a ground motion vector whose instrument orientation is dubious. The orientation error involved in $\mathbf{y}(t)$ may be estimated by maximizing the cross-correlation between $\mathbf{x}(t)$ and $\mathbf{z}(t)$. Since there are three components in these vectors, a sum of the three cross-covariance functions is considered:

$$S(\alpha,\beta,\gamma,\tau) = \sum_{i=1}^{3} R_{x_i z_i}(\tau) = \sum_{i=1}^{3} E[\,(x_i(t) - \bar{x}_i)\,(z_i(t+\tau) - \bar{z}_i)\,] \tag{6}$$

where the super bar indicates the mean value and E[ ] means the ensemble average. By substituting Equation (3) into Equation (6), we obtain

$$S(\alpha,\beta,\gamma,\tau) = \sum_{i=1}^{3}\sum_{j=1}^{3} t_{ij}\, R_{x_i y_j}(\tau) \tag{7}$$

in which t_{ij} is the element of the transformation matrix $\mathbf{T}$ and $R_{x_i y_j}(t)$ is defined by

$$R_{x_i y_j}(\tau) = E\left[\,(x_i(t) - \bar{x}_i)\,(y_j(t+\tau) - \bar{y}_j)\,\right] \tag{8}$$

where τ is the time lag between $\mathbf{x}(t)$ and $\mathbf{z}(t)$. In the actual calculation, however, the temporal average is taken instead by assuming the stationarity and ergodicity. Then Equation (8) yields

$$R_{x_i y_j}(\tau) = \frac{1}{n}\sum_{k=1}^{n} x_i(t_k)\, y_j(t_k+\tau) - \bar{x}_i\,\bar{y}_j \tag{9}$$

where

$$\bar{x}_i = \frac{1}{n}\sum_{k=1}^{n} x_i(t_k); \qquad \bar{y}_j = \frac{1}{n}\sum_{k=1}^{n} y_j(t_k+\tau) \tag{10}$$

in which n is the number of discretized time steps.

Since S in Equation (6) is a function of variables α, β, γ and τ, the estimation of orientation error becomes a multi-variable maximization problem. The revised quasi-Newton method is employed by converting the original problem to a minimization problem by multiplying both sides of the Equation (7) by -1. According to the method, for a given real function $f(\mathbf{x})$ with n variables and an initial vector $\mathbf{x}_0$, the vector $\mathbf{x}^*$ which gives the minimum of $f(\mathbf{x})$ and its function value $f(\mathbf{x}^*)$ can be obtained by an iterative process for a given allowable error. While time lag τ has to be treated discretely, variables α, β and γ can be dealt with as continuous variables. It is known that the high frequency contents of ground motion are less coherent than the low frequency contents. Thus, the low frequency contents will give a better accuracy for the method.

Maximum Coherence Method

The cross-correlation (or cross-covariance) function gives a correlation of ground motions in time domain, while the coherence function provides that in frequency domain. In this method, the coherence function is employed instead of the cross-correlation function for the detection of the orientation error. The quantity to be maximized for the estimation of the orientation error takes the following form:

$$C(\alpha,\beta,\gamma) = \sum_{i=1}^{3} \int_{f1}^{f2} coh^2_{x_i z_i}(f)\, df \tag{11}$$

where f1 and f2 indicate the range of frequency to be considered. The coherence function between x_i and z_i is defined by

$$coh^2_{x_i z_i}(f) = \frac{|S_{x_i z_i}(f)|^2}{S_{x_i x_i}(f)\, S_{z_i z_i}(f)} \tag{12}$$

in which, $S_{x_i x_i}(f)$, $S_{z_i z_i}(f)$ and $S_{x_i z_i}(f)$ are the power spectra of $x_i(t)$ and $z_i(t)$ and the cross spectrum between $x_i(t)$ and $z_i(t)$, respectively. This is also a multi-variable maximization problem. Because of the smoothing procedure used in the calculation of power and cross spectra, it is difficult to deal with this problem analytically. Hence, the problem will be solved numerically.

By giving an arbitrary combination of α, β and γ in matrix **T**, a value of C can be obtained. If we assume m_α cases for α, m_β cases for β and m_γ cases for γ, the total combination for α, β and γ becomes $m_\alpha \times m_\beta \times m_\gamma$. This is obviously not an efficient way to find the maximum value of C. Hence, at first, by setting $\beta = \gamma = 0$, α_1 is determined which maximizes C. Secondly, by using the α_1 thus determined, the first approximations for β_1 and γ_1 are obtained. Finally, by using these α_1, β_1 and γ_1 as initial values, a narrow range is searched to find the most appropriate error angles.

ORIENTATION ERROR ESTIMATION FOR THE CHIBA ARRAY

The Chiba Array

A dense seismometer array has been in operation since 1982 in the Chiba Experiment Station of the Institute of Industrial Science, the University of

Tokyo. In this array, 44 piezoelectric type accelerometers are placed very densely in a total of 15 boreholes shown in Figure 2. In each borehole, accelerometers are buried at different depths (Table 1). Orientation error was found for most of the seismometers soon after the start of the observation and 11 of them buried 1 m below the ground surface were excavated and re-installed (Sato and Katayama, 1983). However for the other seismometers, it was difficult to correct their orientations since they are buried deep in the ground. Hence the orientation error about the vertical axis α was estimated and it was corrected for the records.

Recently the Chiba array database was developed comprising strong ground motions from 27 major events (Katayama et al., 1990). At this time, the three-dimentional orientation error estimation was conducted and the results were utilized to the database.

<u>Estimation of Orientation Error</u>
First of all, in order to compare the results obtained by the two methods, the strong ground motion records of the 1987 Chibaken-Toho-Oki earthquake were used to estimate the orientation error for the Chiba array instruments. The

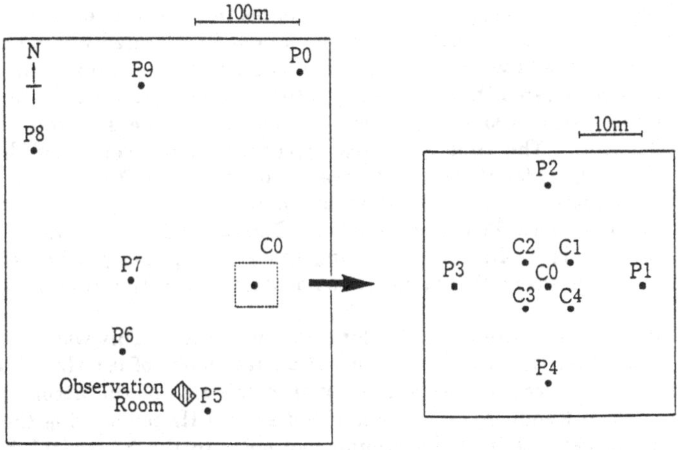

Figure 2 Layout of Boreholes in Chiba array

Table 1 Location of Borehole Accelerometers in Chiba Array

Depth (m)	Borehole														
	C0	C1	C2	C3	C4	P1	P2	P3	P4	P5	P6	P7	P8	P9	P0
1	○	○	○	○	○	○	○	○	○	○	○	○	○	○	○
5	○	○	○	○	○										
10	○	○	○	○	○	○	○	○	○	○	○				
20	○					○	○	○	○	○	○	○	○	○	○
40	○									○					

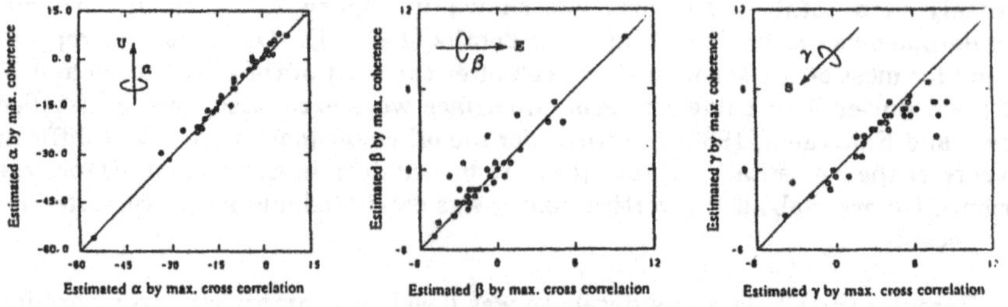

Figure 3 Comparison of the Orientation Error Estimated by the Two Suggested Methods

results estimated by the two methods are shown in Figure 3 under the same conditions for time span (T = 0 - 16 s) and frequency range (f = 0.1 - 1.5 Hz). The two methods give close values for α, β and γ. Hence we will use the maximum cross-correlation method hereafter.

It is possible to estimate the orientation error from the records of a single event. However, the average of estimated angles from several events should be more reliable due to the existence of randomness in recorded ground motions. Therefore seven events were selected which show relatively strong intensity at the site. It is considered that the stronger the ground motion is, the higher the signal-to-noise ratio becomes. Basic information on the selected events is listed in Table 2. The time histories recorded at the center of the array (accelerometer C001) for these seven events are shown in Figure 4, where the time spans adopted in the calculation are marked. The beginning and main shaking part were used for the calculation. The beginning part, where vertical motion is strong, is useful for estimating the angles β and γ, while the main part, where horizontal motion is dominant, is useful for estimating the angle α.

Ground motions were filtered before the calculation. This was realized by using a band-pass filter with the lower cut-off frequency of 0.1 Hz, which was determined by the frequency-response characteristics of the accelerometers. The upper cut-off frequency was determined as 1.5 Hz considering the spatial correlation characteristics of the ground motions. In the horizontal direction, the maximum separation distance between the accelerometers is about 250 m from the center accelerometer C001. Within this distance, ground motions are

Table 2 Information on Earthquake Records

No.	IEQK	JMA M	d (km)	Δ (km)	Azimuth (deg.)	Max. Acce. at C001 (cm/s/s) EW	NS	UD
1	8510	4.8	64	16	126.1	27.4	29.6	12.6
2	8519	6.1	78	28	9.0	59.2	82.2	23.5
3	8601	6.1	44	125	44.5	15.4	14.3	5.2
4	8602	6.5	73	105	147.7	54.0	40.7	21.5
5	8722	6.7	58	45	128.1	213.6	327.1	124.8
6	8806	5.2	48	38	133.3	54.9	97.8	19.8
7	8816	6.0	96	42	276.3	48.4	59.8	15.2

M=magnitude; d=focal depth; Δ =epicentral distance.
Azimuth: clockwise from north.

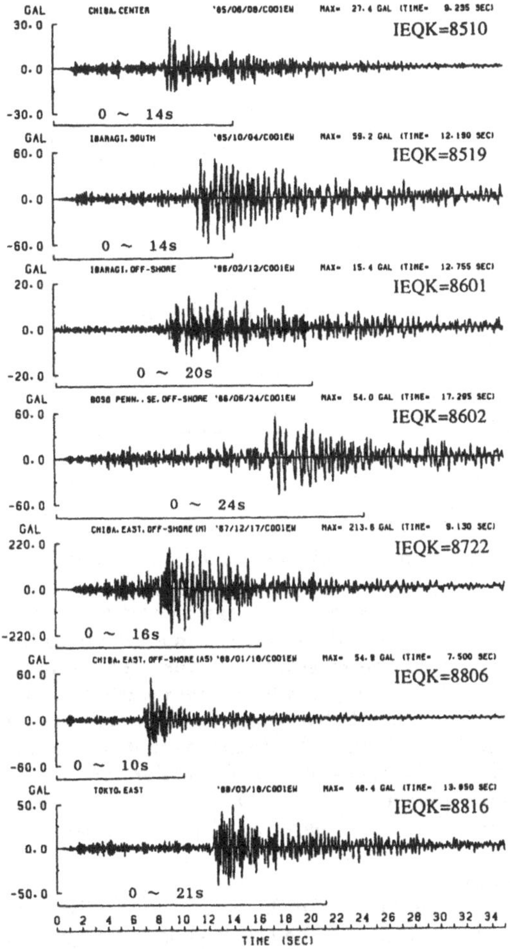

Figure 4 Time Span of the Acceleration Time Histories Used for the Orientation Error Estimation

found to be strongly coherent for the frequency contents under around 1.5 Hz. In the vertical direction, the deepest accelerometers are located 40 m below the ground surface. The fundamental period for the 40 m soil column is about 2.5 Hz. This means that the ground motion recorded at the ground surface has positive correlation with that recorded at GL - 40 m for the frequency contents lower than 2.5 Hz.

In the calculation, variables α, β and γ were dealt with as continuous variables, while time lag τ was treated discretely with a step of 0.005 s in the range of -0.25 to 0.25 s. Since accelerometer C001 was repositioned after the visual inspection, its orientation was considered to be correct. Also because this accelerometer is located at the center of the array, accelerometer C001 was taken as a reference. Hence, the estimated angles in the following part are the deviations from the directions of C001. Estimation was conducted by using ground motion records of a total of 45 accelerometers for each of the 7 selected events. 44 of them are the borehole accelerometers and one of them located on the ground floor of the observation building. The scatters of the estimated angles for the 7 events are plotted in Figure 5 showing an accelerometer having the maximum error α.

It is found that angle α is generally larger than angles β and γ. This had been expected because the tilt of the casing (β and γ) is restricted due to the limited margin between the casing and the wall of borehole. The standard deviations from the 7 events are less than 1 - 2 degrees in most cases for α, β and γ. The deviation of α is, in most cases, less than those of β and γ. This indicates that angle α can be better estimated than β and γ. This may be due to the fact that the intensity of vertical motion is usually less than that of the

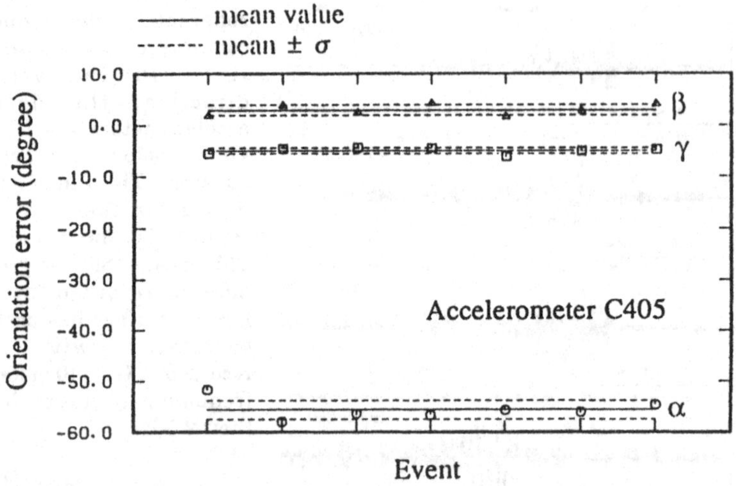

Figure 5 Mean Values and Standard Deviations of the Estimated Orientation Error from 7 Events for Accelerometer C405

horizontal motion. The standard deviations are of the same order both for the small and large orientation errors. Therefore the computation error seems to be independent of the absolute values of the orientation error. The standard deviations of the error are smaller for the accelerometers with the short separation distance from the reference C001 than those for the accelerometers with the longer separation distance from C001. This indicates that the accuracy of the proposed method is better when it is used for an accelerometer which locates close to the reference accelerometer. This is only natural because the coherence of ground motion decreases as the separation distance increases.

Examination of Estimated Results

The results obtained were carefully examined. In the case of the Chiba array, the orientation of the accelerometer installed in the observation building can be visually inspected. The accelerometer in the building was set along the axes of the building. The Y-direction of the accelerometer was measured to be S54°E. The orientation of the building accelerometer was estimated with respect to the reference accelerometer C001, which is about 100 m distant from the building. The estimated Y-direction of the accelerometer is S53.7°E. These two angles are very close to each other.

The actual directions of 11 accelerometers installed 1 m below the ground surface were visually inspected through excavation (Sato and Katayama, 1983). The estimated orientations for these excavated accelerometers were compared with the results of the visual inspection (Table 3). Although there exist some differences, the overall agreement for α, β and γ was good by considering that the estimated angle may include some error due to the random noise in the ground motions and that the visual measurement is also not very accurate. From these comparisons, the accuracy of the estimated orientation error was found to be satisfactory.

Table 3 Comparison of Measured and Estimated Orientation Error

seismo-meter	α (degree)		β (degree)		γ (degree)	
	Meas.	Est.	Meas.	Est.	Meas.	Est.
C001*	0	0	0	0.0	0	-3.0
C101	-13	-15.4	-2	-4.6	0	1.7
C201	-12	-10.6	0	0.2	-2	-3.2
C301*	0	-1.0	0	1.2	0	-0.2
C401*	0	-1.7	0	-2.2	0	0.2
P101	7	8.2	0	-0.5	0	-1.1
P201	-7	-8.9	0	0.0	0	2.1
P301	13	13.9	0	-2.5	0	-3.3
P401	-20	-20.6	-3	-1.3	0	1.1
P501*	0	2.0	0	2.4	0	2.8
P601*	0	-2.6	0	-1.5	0	-0.8

* : Re-orientated; Meas.=measured; Est.=estimated

Correction of Orientation Error

The ground motions recorded by the Chiba array were corrected by Equation (3) according to the mean values of angles α, β and γ estimated from the 7 events. The time histories, cross-correlation functions and coherence functions of the ground motions before correction were compared with those of the ground motions after correction.

Figure 6 shows the comparison of the filtered waveforms recorded by accelerometers C001 and C405 which are separated in a distance of 5 m and 4 m in horizontal and vertical directions, respectively. The estimated orientation differences are -55.6, 2.8 and -5.0 degrees for α, β and γ, respectively. It is found that the waveforms in the frequency range of 0.1 to 1.5 Hz, which was used in the estimation, become very close after the correction, although they show clear difference before the correction.

The changes of the maximum cross-correlation coefficient after the orientation error correction are shown in Figure 7 for 14 accelerometers. It is found that the cross-correlation coefficients between the records of the reference accelerometer C001 and those of the 14 accelerometers become higher after the correction of the instrument orientation error, with only one exception for the

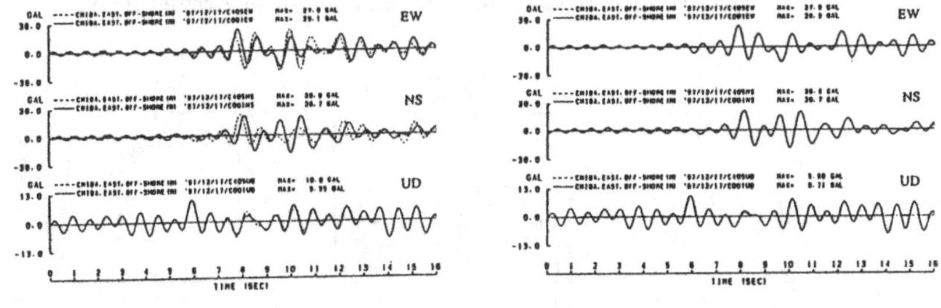

(a) before correction (b) after correction

Figure 6 Comparison of Filtered Waveforms Before and After Orientation Error Correction for C001 and C405

UD component between C001 and P220 which shows slight decrease of the cross-correlation coefficient.

Finally, the coherence functions from the original records were compared with those from the corrected records as shown in Figure 8. Considerable change is observed in the low frequency range. The coherence functions between C001 and C405 is very low for the two horizontal components before

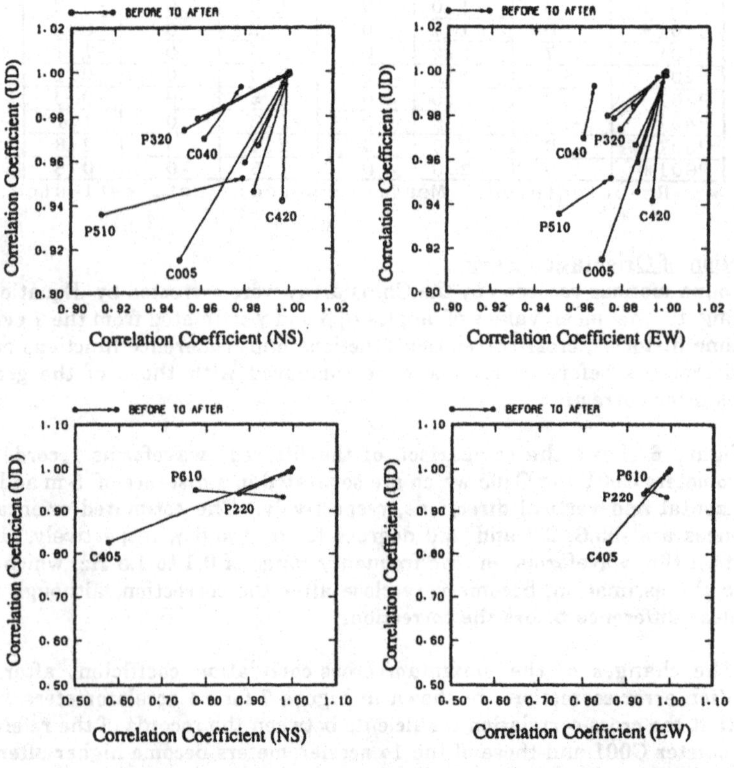

Figure 7 Change of the Maximum Cross-correlation Coefficient Between the Records of C001 and 14 Other Accelerometers

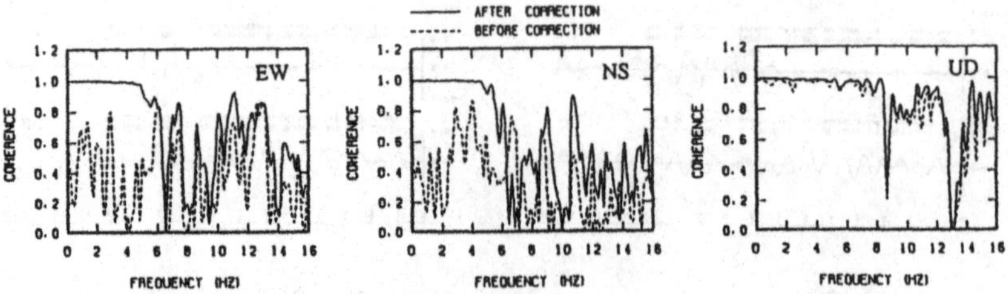

Figure 8 Comparison of the Coherence Function Before and After the Orientation Error Correction

correction because α is as large as about 55 degrees. From these comparisons, the instrument orientation error is found to have significant influence on the time history, cross-correlation function and coherence function. Therefore, it can be concluded that the correction of the orientation error is necessary when evaluating the characteristics of ground motions from array observations.

ORIENTATION ERROR ESTIMATION FOR OTHER ARRAYS

The orientation error was detected for the Chiba array accelerometers and the ground motions recorded by the array were corrected. It must be pointed out that this kind of orientation error is not unique to the Chiba array and that the similar orientation error may exist in other arrays. In order to demonstrate this fact, the proposed method was applied to two other arrays. One is called L array and the other is called Y array in the following.

Orientation Error in L Array

The L array is a dense three-dimensional array, aimed primarily at the observation of soil-structure interaction. In this array, three lines of accelerometers extend on the ground surface starting from points close to a concrete structure model to free field. For three accelerometers at the tip of the three lines, which are in the distance of about 50 m from the structure model, orientation error was estimated. The ground motion records from two events were employed with a frequency filter of 0.1 - 2.0 Hz. By using the maximum cross-correlation method a large orientation error was found for one of the accelerometers. The estimated angles were 30.3, 1.2 and 0.4 degrees with respect to α, β and γ for one event and 32.6, 1.2 and 1.3 degrees for the other event. The problem in this array indicates that the orientation error may occur even for the accelerometer installed on the ground surface, and that orientation error should first be inspected before using array records.

Orientation Error in Y Array

The Y array is an one-dimensional vertical array which is composed of three accelerometers located at 1, 18 and 89 m from the ground surface. Ground motion records from three events were used to estimate the orientation error. By taking the ground motions recorded at the ground surface as the reference, the orientation error of the other two accelerometers were estimated. In the calculation, the frequency range from 0.1 to 0.6 Hz was utilized. In the case of this array, an orientation error of about 10 degrees (α) was found for the two downhole accelerometers. According to the average of the errors from the three events, the ground motion records of the two downhole accelerometers were rotated. The coherence functions from the original ground motion records and the rotated ones were computed for the two sets of the accelerometer on the ground surface and the two downhole accelerometers. Again, the increase of coherence functions was clearly observed by correcting the orientation error.

CONCLUSIONS

The orientation error in array seismometers is a serious problem when the array records are used for wave propagation analysis or spatial variation analysis. Two methods were proposed to estimate the orientation error in three-dimensional space, one by maximizing the cross-correlation function and the other by maximizing the coherence function. The methods were first employed to the Chiba array accelerometers. The results were carefully inspected by

estimating the orientation of the accelerometer in the observation building, whose orientation is known, and by comparing the estimated orientation errors with those directly inspected by excavation. These comparisons show a satisfactory accuracy for both the proposed methods. The estimation was conducted for seven events and the deviations of the estimated orientation errors were found to be small. Applying the coordinate transform to the recorded motions, corrected motions were obtained. The cross-correlation coefficients and the coherence function between the motions increased significantly by this transformation. The method was further applied to two other arrays and the orientation error was found for both of them. Hence it is suggested that before conducting detailed studies using array records, orientation error estimation should be performed.

REFERENCES

Katayama, T. and Sato, N. (1982), Ground Strain Measurement by a very densely located seismometer array, Proc., 6th Japan earthquake Engineering Symposium-1982, pp. 241 - 248.

Katayama, T., Yamazaki, F., Nagata, S., Lu, L., and Turker, T. (1990), A Strong Motion Database for the Chiba Seismometer Array and its Engineering Analysis, Earthquake Engineering and Structural Dynamics, Vol. 19, No. 8, pp. 1089-1106.

Real, C. R. and Cramer, C. H. (1990), "Turkey Flat, USA, Site Effects Test Area: 'Blind' Test of Weak-Motion Soil Response Prediction," Proc., 4th U.S. National Conference on Earthquake Engineering, Vol. 1, pp. 535 - 544.

Sato, N. and Katayama, T. (1983), Estimation of Orientation for Buried Accelerometers, Proc., 17th JSCE Earthquake Engineering Symposium, pp. 115 - 118 (in Japanese).

Strains in Long Dams due to Spatially Correlated Random Ground Motions

O. Ramadan, M. Novak

Faculty of Engineering Science, The University of Western Ontario, London, N64 5B9, Canada

ABSTRACT

Horizontal response of long dams to spatially correlated random ground motion is theoretically investigated considering dam-foundation rock-reservoir interaction. The ground motions are generated using Shinozuka's approach and dam bending and twisting along its longitudinal axis are evaluated.

INTRODUCTION

The basic step in the analysis and design of structures to withstand earthquakes is an adequate representation of the seismic ground motion. Any improvement in this step should enhance the design of earthquake-resisting structures and provide a more realistic response prediction. Considerable efforts were made to improve the understanding and representation of seismic ground motion. One important feature of ground motion, relevant for the analysis of extended structures such as tunnels, pipelines, large dams and multispan bridges, is its spatial variability. This means that for large structures, different foundation parts may be subjected to ground motions with different amplitudes and phases and additional stresses may be produced due to this differential movement. These stresses may remain completely unforeseen if spatial variability is not accounted for as is usually done in current practice.

This paper outlines a procedure that makes it possible to account for spatial correlation of ground motion and foundation-structure interaction in the seismic response analysis. Using an example of lateral response of a concrete gravity dam, the effects of spatial correlation of seismic ground motions as well as structure-foundation interaction are investigated. The analysis is carried out in the frequency domain to account for the frequency dependence of the dynamic stiffness matrices. The dam-water interaction is accounted for using the reservoir dynamic stiffness parameters statically equivalent to the integral of dynamic water pressure over the contact area. The dynamic water pressure is obtained by solving the familiar Helmholtz equation with the appropriate boundary conditions.

Concrete gravity dams are usually built in short pieces (monoliths) with expansion joints to reduce tensile stresses due to shrinkage and seasonal effects. These joints are capable of transmitting shear forces between the dam parts. However, their capability for transmitting bending or torsional moments is dependent on their detailing. The presence of these joints affects the lateral response of the dam but this effect is not considered in this paper. This may be adequate for some dams. For example, the Old Aswan Dam in Egypt which is 2142 m long, and the Willow Creek Dam in Oregon, U.S.A., with a crest length of 543 m, were constructed without expansion joints. This paper complements an earlier study by Novak and Suen [4] which used a different technique and was limited to the vertical response, and a more recent report by the authors [5].

MATHEMATICAL MODEL

The substructuring method is used for the analysis of the dam-foundation-reservoir system. In this method, the dam structural stiffness matrix is calculated using the standard structural analysis techniques and the foundation stiffness matrix is obtained using the well established solutions for the halfspace. The dam material is homogeneous, isotropic, and linearly elastic with frequency independent (hysteretic) material damping and the foundation rock is modelled as a homogeneous, isotropic and linearly viscoelastic halfspace with frequency independent material damping. The total stiffness matrix is formed by superimposing the two matrices and the reservoir effect. Kinematic interaction is not considered.

The Dam

The dam is assumed to be built of concrete and to have a large aspect ratio (length/width), and is modelled by N beam elements accounting for shear deformation and featuring (N+1) nodes (Figure 1). Only horizontal components of ground motion perpendicular to the dam axis, u_g, are considered. This results in bending of the dam in the horizontal plane (Figure 2) and its twisting. The joint rotations in the horizontal plane are eliminated through static condensation leaving the model with 2(N+1) degrees of freedom. The masses, m, associated with the horizontal response and the polar mass moments of inertia, I, associated with torsional vibration are both lumped at the element nodes.

The Foundation

The effect of foundation flexibility is incorporated in the analysis through the complex, frequency dependent stiffness matrix established for a sequence of rectangular tributary areas indicated in Figure 1. The foundation stiffness matrix associated with

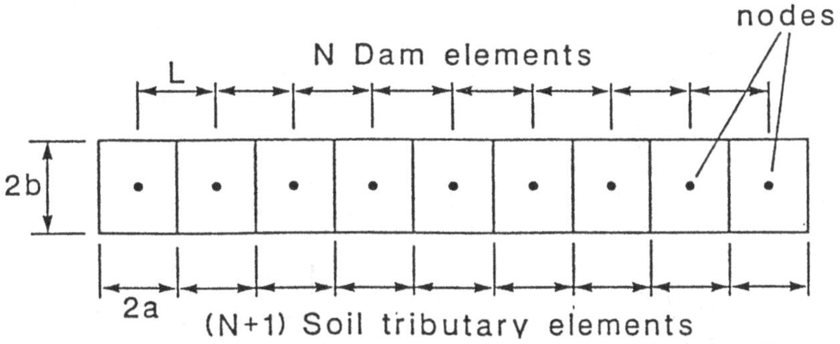

Figure 1 Schematic of dam elements and soil tributary elements

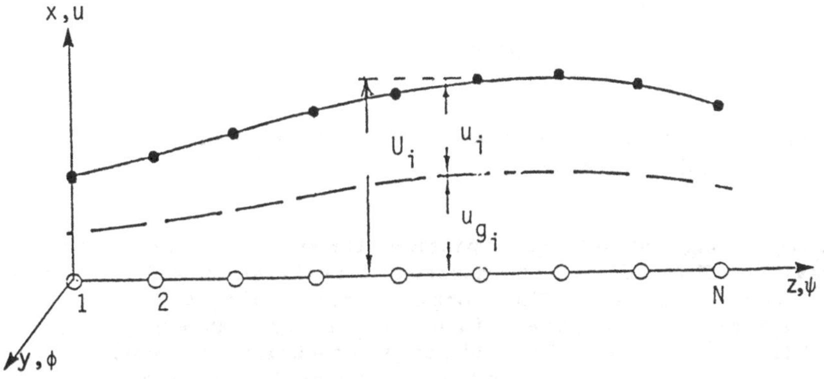

Figure 2 Dam displacements

horizontal translations is obtained by inversion of the full dynamic flexibility matrix $[f_{ij}]$ to account for the through-soil coupling of the tributary areas. The element f_{ij} is defined as the displacement at the centre of the soil element i due to unit harmonic load distributed over the soil element j. For the loaded element j, f_{jj} is evaluated at a point 2a/3 and 2b/3 away from node j, where 2a and 2b are the dimensions of the tributary area. This position was found to give displacements very close to those under a rigid base. The flexibility, or compliance, coefficients, f_{ij}, were calculated following Gaul's solution [3]. Figure 3a shows the variation of the horizontal compliance, f_{ij}, normalized by 1/Ga for square bases 2a x 2a and the separation normalized by a; G is the shear modulus of the foundation rock, V_s is the shear wave velocity of the foundation rock, and $A_o = \omega a/V_s$ where ω is circular frequency. While both the real and imaginary parts oscillate with separation, the absolute value decays monotonically and its rate of decay decreases with separation.

The part of the foundation stiffness matrix stemming from horizontal translations is first calculated at the dam base and then is transferred to the nodal points of the dam elements placed

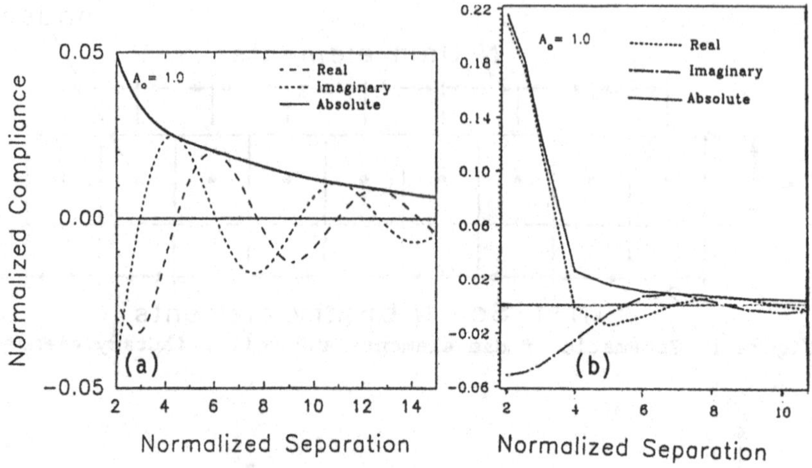

Figure 3 Foundation compliance: (a) - horizontal translation, (b) - rocking

on the longitudinal axis passing through the dam centre of gravity. This couples the horizontal translation and rocking of the dam elements. The rocking soil stiffness matrix was investigated approximately in a similar way. An example of the rocking compliance is plotted versus dimensionless separation in Figure 3b. The results show that through-soil coupling in rocking is insignificant and can be neglected. This suggestion is also supported by a more rigorous solution [7]. Therefore, the rocking foundation stiffness matrix features only diagonal terms. These were evaluated in a simpler way using the well known solution for rocking of a strip footing.

To approximately account for dam fixity at both ends, the foundation tributary elements, Figure 1, are assumed to exceed by one element the actual dam length. The foundation stiffness associated with dam bending in the horizontal plane is interpreted as the rocking stiffness on the plane of the dam cross-section. Two limiting cases may be considered: pure hinge when the bending stiffness is ignored and the dam ends may rotate freely; and total fixation. In the numerical example presented here, the dam is assumed to be hinged at both ends.

The Reservoir

An accurate representation of the dam-water interaction would require a three-dimensional finite element discretization of the fluid domain. Such idealization would allow one to represent any reservoir shape but would make the solution computationally impractical since the impounded water usually extends to great distances in the upstream direction. Therefore, in this study, the reservoir is assumed to be infinitely long with constant cross-section. In calculating the dynamic water pressure on the upstream face of the dam, the fluid domain is regarded, in the horizontal plane, as several subchannels of infinite length. Under these assumptions, and a few others, the dynamic water pressure is obtained by solving the Helmholtz equation in two dimensions for unit ground motion and an arbitrary shape of dam lateral displacement along its depth, as in Ref. 1. By assuming a linear variation of the dam displacement with height, the dynamic pressure associated with rocking is evaluated. The forces and moments about the dam axis resulting from these dynamic water pressures are statically equivalent to the frequency-dependent reservoir stiffness parameters associated with dam horizontal translation and rocking in the vertical plane.

REPRESENTATION OF INCOHERENT GROUND MOTION

While seismic response of structures can be conveniently analyzed in terms of random vibration as shown in [4], the deterministic approach is more common and can be very illustrative. For the latter approach, time histories of seismic ground motions are needed. These can be actual motions recorded in seismic events or digitally simulated motions generated to fit specified spectra. In this study, horizontal ground motions are generated using the technique due to Shinozuka and Deodatis [6]. For a homogeneous, stationary and spatially two-dimensional seismic ground motion with zero mean that results from Rayleigh waves and matches a prescribed power spectral density function, this technique generates the ground motion as

$$u_g(z,x,t) = \sqrt{2} \sum_{k_z=1}^{N_z} \sum_{k_x=1}^{N_x} [2\, S(ck_z,ck_x)\Delta k_z \Delta k_x]^{1/2}$$
$$\times \{\cos(ck_z z + ck_x x + g(ck_z,ck_x)t + \phi^{(1)}_{k_z,k_x})$$
$$+ \cos(ck_z z - ck_x x + g(ck_z,ck_x)t + \phi^{(2)}_{k_z,k_x})\} \qquad (1)$$

In Eq. 1, $ck_i = k_i\Delta k_i$, $\Delta k_i = k_{iu}/N_i$ with $i = x,z$, and k_{iu} = the upper limit of the wave number in the i-direction; $\phi^{(1)}_{k_z,k_x}, \phi^{(2)}_{k_z,k_x}$ are two independent random phase angles uniformly distributed between 0 and 2π. Finally, $S(ck_z,ck_x)$ is the prescribed power spectral density function with the circular frequency, ω, replaced by

$$\omega = \alpha\, g(ck_z,ck_x) = \alpha\sqrt{(ck_z)^2 + (ck_x)^2} \qquad (2)$$

where α is the wave phase velocity of the Rayleigh wave. The power spectral density function used in the numerical example is the modified Kanai-Tajimi spectrum [2], with $\zeta_f = \zeta_g$ and $\omega_f = 0.1\omega_g$, where ω_g, ω_f are resonance frequencies, and ζ_g, ζ_f are damping ratios. Figure 4 shows the seismic motions generated using this technique over an area 10 km x 10 km at different time instants. (These motions were generated using $\omega_g = 15.71\ s^{-1}$, $\zeta_g = 0.6$, $\alpha = 1400$ m/s, $k_{zu} = 0.08\ m^{-1}$, $k_{xu} = 0.00332\ m^{-1}$ and $N_1 = N_2 = 100$.) To account for nonstationarity, an envelope function, shown in Figure 5, is superimposed on the ground motion generated.

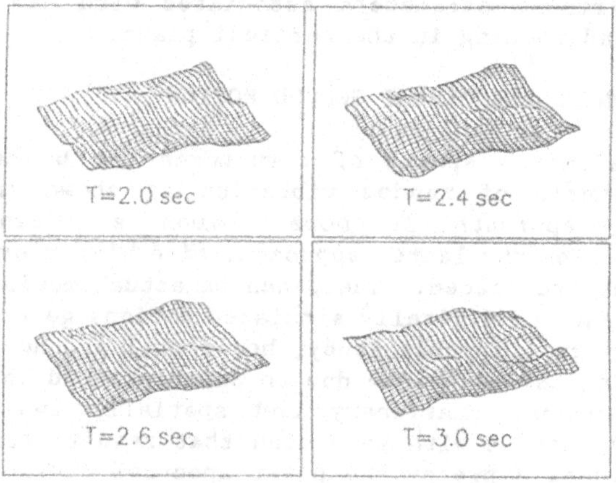

Figure 4 Simulated ground motion at four time instants over a 10 km x 10 km area

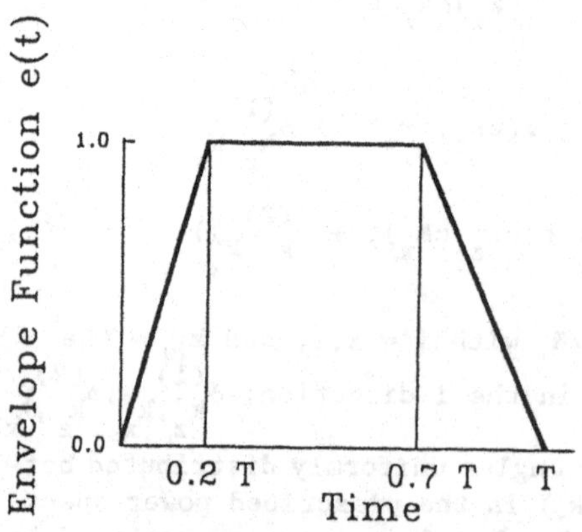

Figure 5 Envelope function

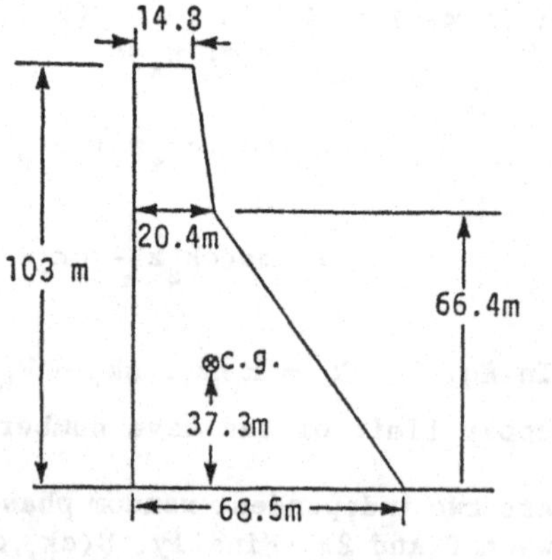

Figure 6 Dam cross-section

Governing Equations of Motion and Their Solution

Denote the horizontal ground motion at node i, u_{g_i}, the relative motion between the ground and the dam, u_i, the absolute horizontal motion of the dam, $U_i = u_i + u_{g_i}$ (Figure 2), and the rocking angle at node i, ψ_i. The governing equations in terms of absolute motion, preferred here because of subsequent internal forces calculations, becomes, after partitioning:

$$\begin{bmatrix} [\,m\,] & [0] \\ [0] & [\,I\,] \end{bmatrix} \begin{Bmatrix} \{\ddot{U}\} \\ \{\ddot{\psi}\} \end{Bmatrix} + \begin{bmatrix} [K_{uu}] & [K_{u\psi}] \\ [K_{\psi u}] & [K_{\psi\psi}] \end{bmatrix} \begin{Bmatrix} \{U\} \\ \{\psi\} \end{Bmatrix}$$

$$= \begin{bmatrix} [k_{uu}] & [k_{u\psi}] \\ [k_{\psi u}] & [k_{\psi\psi}] \end{bmatrix} \begin{Bmatrix} \{u_g\} \\ \{0\} \end{Bmatrix} \tag{3}$$

In Equation 3, $[K_{ij}]$ is the total dynamic stiffness matrix of the dam-foundation-reservoir system, $[k_{ij}]$ is the dynamic stiffness matrix of the foundation alone, and [m], [I] are the diagonal matrices of the lumped masses and mass polar moments of inertia, respectively. More details on this formulation can be found in [5]. As the foundation stiffness parameters are frequency dependent, Equation 3 is solved using the Fast Fourier Transform (FFT) and the procedure known as the complex response analysis.

NUMERICAL EXAMPLE

For the numerical application, an 853 m concrete gravity dam with a constant cross-section similar to that of the Koyna Dam in western India is chosen (Figure 6). The dam is 103 m high and has specific mass, Young's modulus and Poisson's ratio of 2300 kg/m^3, 30000 MPa and 0.2, respectively. The foundation is basalt with shear wave velocity, specific mass and Poisson's ratio of 1218 m/s, 2400 kg/m^3 and 0.3, respectively. The hysteretic material damping is assumed to be 0.02 and 0.05 for the dam and the soil, respectively. The dam is divided into ten elements and is assumed to lie on the Z-axis with its left end at the origin.

The damped natural frequencies and vibration modes of the dam are examined first by solving Equation 3 for the right hand side equal to zero. Due to frequency dependence of the stiffness parameters, the eigenvalue problem is a non-linear one. It is solved here by iteration and the complex eigenvalue analysis. The velocity of sound in water and the mass density of water are c = 1440 m/s and ρ_ω = 1000 kg/m^3, respectively. The first three damped natural frequencies were 18.54, 19.10 and 19.98 rad/s when the reservoir is empty and 14.26, 14.39 and 14.88 when the water depth, H, is 95 m. These frequencies are less than the fundamental frequency of the impounded water $\omega_f = \pi c/2H$ = 23.81 rad/s. Therefore, the effect of the water is characterized only by an added mass term which decreases both the natural frequencies and the apparent damping ratios.

The ground motions were simulated at the eleven nodes for a 10 second period with time steps 0.02 sec and assuming the parameters used for the shapes shown in Figure 4. The simulated motions are shown in Figure 7. Their standard deviations range from 5.58 to 5.87 mm and their peak values from (2.9 to 3.6) x standard deviation (Figure 8). The maximum ground acceleration is 0.15 g. Figure 9 shows the time history and Fourier amplitudes of the ground motion at the dam midpoint. The absolute values of Fourier amplitudes do not vary from node to node while the corresponding random phases do. The time history keeps the same features at all nodes with slight differences in the maximum values (17 to 20 mm) and their instants of occurrence.

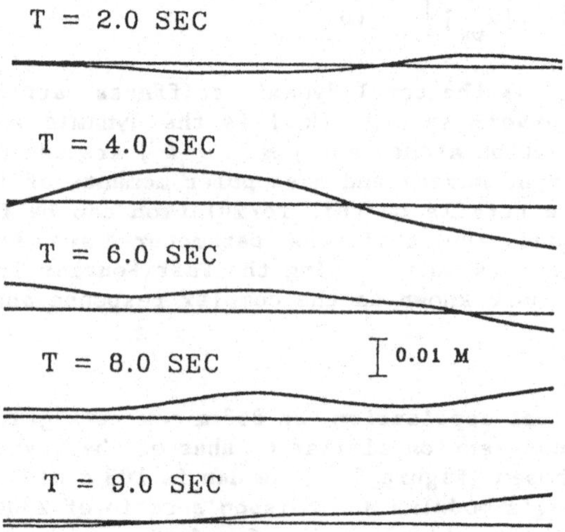

Figure 7 Simulated ground motions

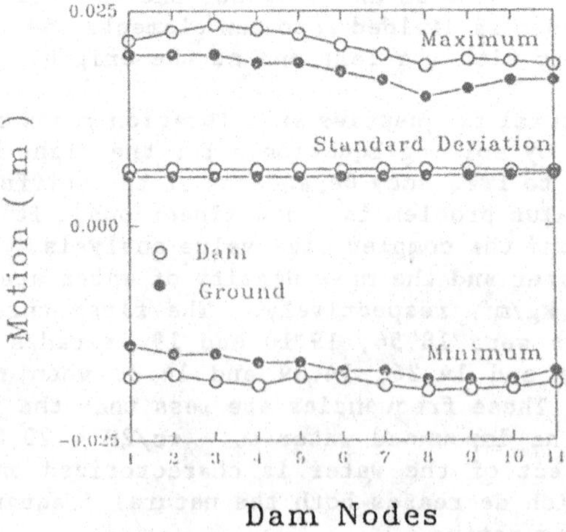

Figure 8 Variation of dam and ground motions

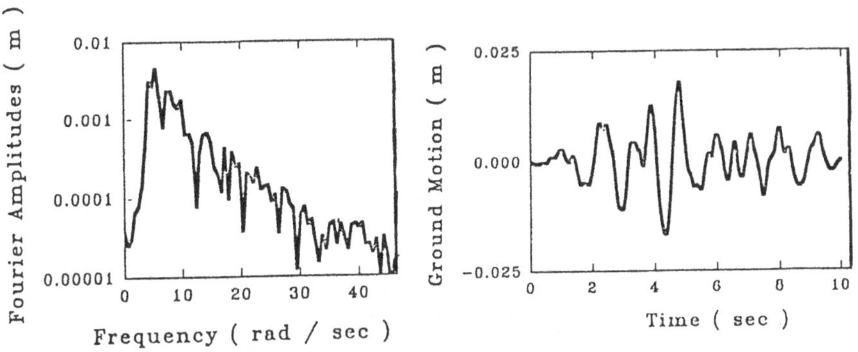

Figure 9 Ground motion at Node 6

To describe the degree of spatial correlation of ground motions, an apparent wave length can be estimated using three different approaches. By counting the number of wave lengths in the direction of the dam axis, the wave length is estimated as 3333 m. Comparing the variation of the correlation coefficient of the ground motion versus separation with that of a travelling wave, the wave length appears to be about 2700 m. Finally, by using the wave number corresponding to the maximum spectral amplitude, the wave length is estimated as 3000 m. All this suggests an apparent wave length of 3000 m which is about 3.5 times the dam length.

<u>Dam Response Analysis</u> (Empty Reservoir)

The dam response was calculated for the simulated spatially correlated motions (SC) and, for comparison, the fully correlated motions (FC). For the latter case, the ground motions at all eleven nodes were assumed to be identical to the motion at node 6 (the midpoint). Figure 10 shows the dam horizontal response for both FC and SC motions. While the dam responds almost as a rigid body under FC motions, it bends significantly under the SC ones. Figure 8 shows that the standard deviation of the dam motion does not vary much along the dam (6.21 to 6.49 mm) with the peak factor ranging between 3.0 and 3.62. The peak factor under the FC motion is almost invariant (3.21 to 3.27). The differences between the dam and the ground motion visible in Figure 8 are due to foundation-structure interaction. The amplification of the dam motion due to foundation-structure interaction is shown in Figure 11 where the ratio of the Fourier amplitudes of the dam motion to those of the ground motion is plotted. This ratio varies considerably between the dam nodes for the SC ground motions. Figure 11 shows that higher modes are present in the response to the SC ground motions. Distributions of dam bending and torsional moments are shown in Figures 12 and 13, respectively. While the maximum bending and combined shear stresses are 0.26 and 0.09 MPa under the FC ground motion, they are 1.24 and 0.36 MPa under the

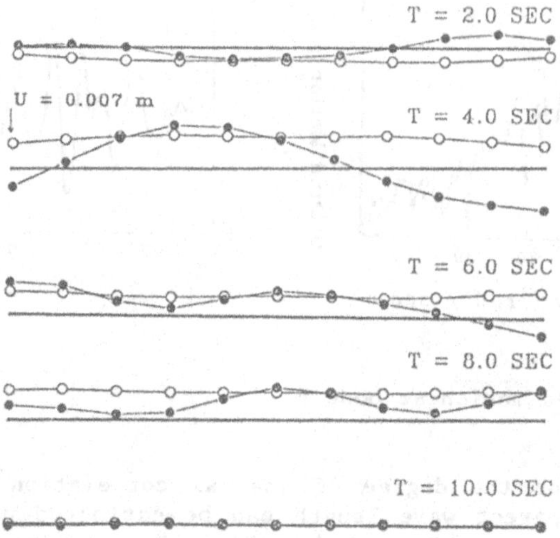

o FULLY CORRELATED • SPATIALLY CORRELATED

Figure 10 Dam horizontal response

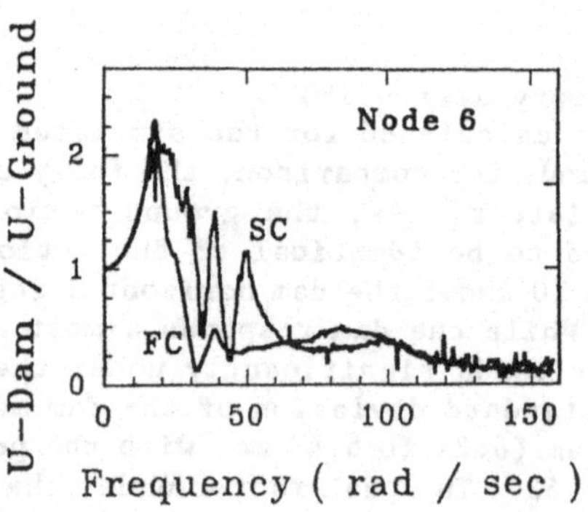

Figure 11 Dam response/ground motion

SC ones, respectively. The stresses for the SC case are comparable to the strength of the concrete used in this dam which are 7.59 and 2.41 MPa in compression and tension, respectively. It is important to emphasize that these stresses act on the dam vertical planes and are completely overlooked when the dam is analyzed using the standard slice or one monolith only.

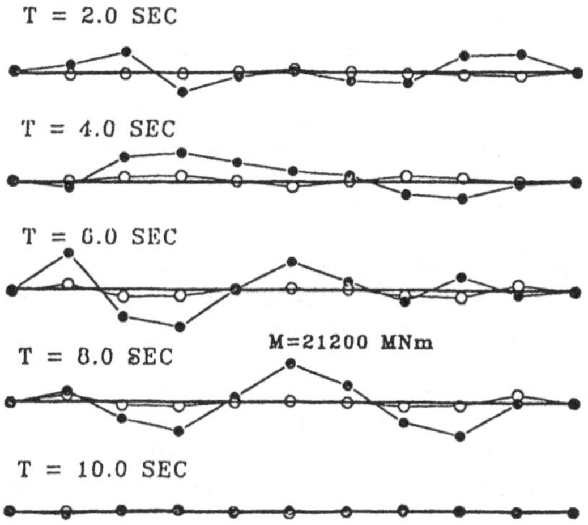

○ FULLY CORRELATED • SPATIALLY CORRELATED

Figure 12 Dam bending moments

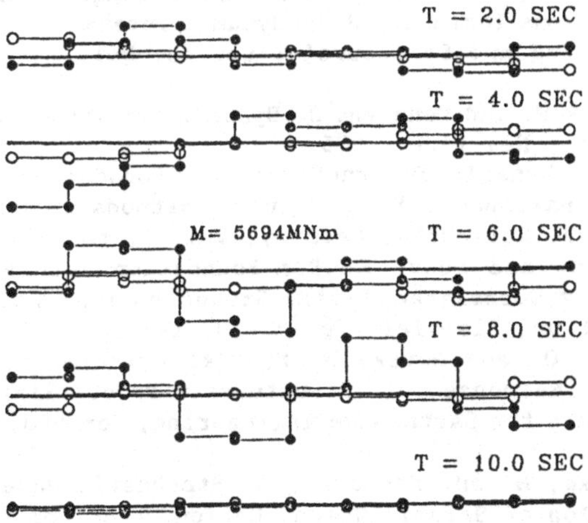

○ FULLY CORRELATED • SPATIALLY CORRELATED

Figure 13 Dam torsional moments

CONCLUSIONS

An approach for the response analysis of foundation-structure systems to spatially correlated seismic ground motions is presented. The following observations emerge:

(1) The dam response is strongly affected by the incoherence of ground motions. The dam responds almost as a rigid body to fully correlated motions but bends and twists significantly under incoherent motions.

(2) The internal forces and the corresponding dynamic stresses caused by the spatial variability of seismic ground motions are not insignificant even under moderate lack of coherence.

(3) The stresses due to ground motion incoherence remain completely unforeseen in the current design methods; they should be considered in the design of large structures such as tunnels, pipelines, large dams, and long bridges.

ACKNOWLEDGEMENT

The research reported here was supported by a grant from the Natural Sciences and Engineering Research Council of Canada.

REFERENCES

1. Chakrabarti, P. and Chopra, A. Earthquake Analysis of Gravity Dams Including Hydrodynamic Interaction, Earthquake Engineering and Structural Dynamics, Vol. 2, pp. 143-160, 1973.
2. Clough, R.W. and Penzien, J. Dynamics of Structures, McGraw-Hill Inc., New York, 1975.
3. Gaul, L. Dynamic Interaction of a Foundation With Visco-elastic Halfspace, Proc. Dynamic Methods in Soil and Rock Mechanics, Karlsruhe, Germany, Vol. 1, pp. 167-183, 1977.
4. Novak, M. and Suen, E. Dam-Foundation Interaction Under Spatially Correlated Random Ground Motion, Soil-Structure Interaction, Elsevier, pp. 25-33, 1987.
5. Ramadan, O. and Novak, M. Spatial Correlation Effects on Seismic Response of Structures, Proc. Sixth Canadian Conference for Earthquake Engineering, Toronto, June 12-14, 1991.
6. Shinozuka, M. and Deodatis, G. Stochastic Wave Models for Simulation of Seismic Ground Motion, Proc. of the Workshop on Spatial Variation of Earthquake Ground Motion, Dunwalke, Princeton University, November 7-9, 1988.
7. Triantafyllidis, Th. and Prange, B. Dynamic Subsoil-Coupling Between Rigid, Circular Foundations on the Halfspace, Journal of Soil Dynamics and Earthquake Engineering, Vol. 8, No. 1, pp. 9-21, 1989.

Seismic Reliability Assessment of Redundant Structural Systems

M. Ciampoli

Department of Structural and Geotechnical Engineering, University of Rome "La Sapienza", Rome, Italy

ABSTRACT

In a previous study, a method for the reliability assessment of structural systems made up of brittle elements and responding dynamically to actions modeled by means of multivariate continuous stochastic processes, was set up: the method is based on a Markov chain description of the evolution of the various stages of damage to the system. In order to evaluate its effectiveness, two numerical examples of application are discussed in this paper, one dealing with a space-frame steel structure supporting a hydraulic piping system, and the other with a simplified model of a truss steel jacket offshore platform. Both systems are subjected to seismic action, modeled as a multivariate gaussian stochastic process. The numerical examples demonstrate the importance of taking the progressive failure of the system into account in evaluating its reliability; and they also clarify that the proposed algorithm, although approximate, can be useful to gaining insights into the fundamental aspects of the problem, for example, making it possible to identify the more critical elements, that is, those whose failure defines the most likely sequences. This result can be assumed as a decision parameter in the design process.

KEYWORDS: System reliability, earthquake engineering, brittle elements, Markov processes.

INTRODUCTION

One of the most important areas of research in structural engineering is the development of practical methods for the reliability assessment of real redundant systems, characterized by stochastic properties (as concerns the element dimensions, the strengths of materials, and the structural response), and subjected to actions that can be modeled, in the more general case, as a vector of random fields or processes.

Currently, safety check on complex and realistic (i. e. not highly idealized) structures must rely on Monte Carlo simulation: in fact, the assessment of an analytical procedure is extremely involved, and it has been achieved only in particular cases [1], by adopting an idealized model of the structural response with respect to its possibilities of failure, and by making some simplifying assumption on the nature of the stochastic loadings or, more specifically, of the effects of the actions. Furthermore, the reliability assessment becomes more involved in time-dependent problems, i.e. in the case of systems subjected to stochastic processes inducing dynamic response, as, for example, a seismic action.

Moreover, and with reference to the evolution in time, two classes of (functional and structural) systems should be distinguished: the first is the class of *non-evolutive* systems, for which the safe and failure domains are invariant and hence uniquely defined; the second is the class of *evolutive* systems, for which failure can arrive through sequential degradation. The former is, for example, the class of series systems and of parallel systems, if the element behaviour can be represented as perfectly ductile; the latter is the class of redundant systems with brittle (or semi-brittle) elements, for which the classical concept of a failure domain is not applicable, failure being of the progressive type. The safe domain is then evolving in time, because the system is progressively deprived of some of its elements, as they fail and disappear: therefore, the state of the structure depends on the loading path, and on the redistribution of internal forces that results as the system state changes.

In previous studies [2] [3], a method of reliability evaluation of general series-parallel systems, responding dynamically to actions modeled by means of multivariate continuous stochastic processes, was set up and applied to the seismic safety evaluation of structural systems.

The reliability of these systems under the actions of multiple randomly varying loadings has been treated as a problem of outcrossings of the action effects vector process from the safe domain.

The proposed procedure is applicable to real redundant (functional or structural) systems, whose members can be assumed to behave linearly up to their collapse, this last occurring instantaneously and with complete loss of function. Redundancy implies that the system may reach failure through different paths, each of them being identified by an ordered sequence of element failures; fail-safe behaviour implies that some structural modifications occur, according to the sequential failures, and then that it is necessary to account for all the modes characterized by these sequential failures of the elements, that is, for the real evolutive nature of the failure process, and not only for the mode that corresponds to the simultaneous failure.

During the evolution toward collapse, the system state is represented by means of a homogeneous Markov process, continuous in time and discrete in space: therefore, the state transition rates coincide with the mean rates of outcrossing of the action effects vector process from the evolutive failure domains. These domains correspond to the effective transient configurations of the system, progressively deprived of some of its elements, according to its functional logic.

In this paper, after a brief introduction to the criteria governing the analytical procedure, two examples of applications to structural systems subjected to seismic actions are presented in detail. The first one is the case of a space-frame steel structure, supporting a hydraulic piping system; the second one is a partial model of a truss steel jacket platform, already proposed in [4], but analyzed there in the static case only. In the latter example, the stationary response is also updated, according to the sequential failures of the elements.

The numerical examples demonstrate the importance of the effect of load path, as well as of the progressive failure of the system to reliability evaluation; and also, they clarify that the proposed algorithm, although approximate, is useful to gaining insights into the fundamental aspects of the problem, as, for example, it makes it possible to identify the more critical system elements, that is, those whose failure defines the most likely sequences. This result can be assumed as a decision parameter in the design process.

SOME REMARKS ON THE ANALYTICAL PROCEDURE

As mentioned above, a redundant system, if progressively deprived of its elements, may reach failure through different paths corresponding to the alternative sequences of element collapses.

The possibility of a progressive type of collapse invalidates the notion of a fixed safe domain; therefore, in order to evaluate the system reliability it is necessary to consider all the possible states of the system, and to identify all the paths through the states that the system can follow before reaching failure.

The reliability assessment can be carried out according to a procedure consisting of the following steps:

a) the identification of the *system structure*, that is of the fault tree that reproduces the system functional logic (or, in the simplest case, the way the elements are connected in), and of the corresponding *"minimal path to failure"* or *"minimal out set"* representations;
b) the identification of the *system states*, each of them being characterized by ordered sequences of component failures;
c) the analysis of the state changes occurring in the system in its evolution towards collapse,

i.e. of the possible *transitions* from the various states to other states of higher order, characterized by a larger number of collapsed elements;

d) the evaluation of the *transition rates*;

e) the calculation of the *failure probabilities*, total or partial, these last being associated with each state of intermediate damage.

According to the procedure set forth in [2] and [3], the functional or structural system is represented by means of its "minimal path to failure", that is as an assemblage in parallel of a certain number of subsystems, each one made of serially connected elements. The formal definition of the safe D and failure F domains is then:

$$D = \cup_i (\cap_j D_{ij}) \qquad F = \cap_i (\cup_j F_{ij})$$

where: D_{ij} $(F_{ij} = D_{ij}^c)$ is the safe (failure) domain of the jth component of the ith subsystem.

The kth state of the system is identified by the particular set of collapsed, or by the complementary set of still active subsystems. The former is defined as the set of element collapses that corresponds to just that class of failed subsystems.

According to the hypothesized fail-safe behaviour, the total number of collapsed states is just $(2^n - 1)$, n being the number of subsystems connected in parallel. The initial zero state is that in which all the elements are safe; the last nth state corresponds to all failed subsystems.

The states are ordered by increasing level of degradation, since transitions are irreversible; in fact, during the application of dynamic actions, no repair is possible, and therefore the system can evolve only towards more degraded states.

The analysis proceeds defining all the possible transitions, corresponding to the collapse of predetermined sets of subsystems. If these sets are null, the corresponding transition is impossible: this is the case of a set of subsystems whose elements belong to other subsystems that remain still active in that transition. Only one rule must be observed in the generation of the system state changes. As the same element may belong to several subsystems, failure of a given subsystem (that is a series system by itself) requires the presence within it of at least one element not belonging to other unfailed subsystems. In fact, in a given state, the elements belonging to still active subsystems have obviously not yet failed.

The assumptions are that the system evolution can be described by a Markov process, one stationary and continuous in time and discrete in states, and that, during each transition, no other elements can fail; therefore, the transition rates coincide with the mean in-crossing rates of the action effects vector process $\{X\}$ into the appropriate evolving convex failure domains of the reduced systems whose failure produces the transition.

For ease of computation, only a first order Markov process is considered; this condition, exactly satisfied if the action can be represented as a white-noise process, is assumed here to make some algorithms of the advanced FOSM methods applicable. Therefore, the transition probabilities do not depend on previous history, but on starting and arrival times only, and the mean in-crossing rates can be evaluated by means of advanced FOSM methods, according to the well-known Rice formula and the procedure proposed by Ditlevsen [5].

The transition probabilities are derived by solving the associated Kolmogorov equation, the transition rate matrix being the ordered set of the rates of transition ν_{kj} from state k to state j. Denoting by $[P(\tau)]$ the matrix containing the probabilities of transition to a given vectorial state i (i = 1, ..., n) from a previous state j (j = 0, 1, ... n-1) in the time interval τ, the Kolmogorov equation is written as:

$$[P'(\tau)] = [P'(0)]\,[P(\tau)] \tag{1}$$

with primes indicating differentiation with respect to time.

The solution of eq. 1 can be cast in the form:

$$[P(\tau)] = [Y]\; diag\,[exp\,(\lambda_i\,\tau)]\;[Y]^{-1} \tag{2}$$

where: $[Y]$ is the matrix of the eigenvectors of $[P'(0)]$, and $diag\,[exp\,(\lambda_i \tau)]$ is a diagonal matrix formed by the eigenvalues of $[P'(0)]$.

Given the stationarity of the process, and the approximation: $P_{kj}\,(t,\, t+\Delta t) \cong \nu_{kj}\cdot\Delta t$, the matrix $[P'(0)]$ actually contains the transition rates for all the states (obviously, the terms $P_{kk}'(0)$ results equal to: $P_{kk}'(0) = -\,\Sigma_j\,\nu_{kj}$).

The last condition holds good under the assumption that the crossing process of the boundary of the evolutive failure domain is poissonian, which tends to become exact as the values of the associated probabilities becomes smaller, and as the crossing events becomes rarer and less correlated. For further details on the analytical procedure, refer to [2] and [3].

EXAMPLES OF APPLICATION

The first system analyzed is represented in fig. 1. It consists of a space-frame steel structure, supporting a hydraulic mechanical system, and subjected to a synchronous seismic excitation at its base.

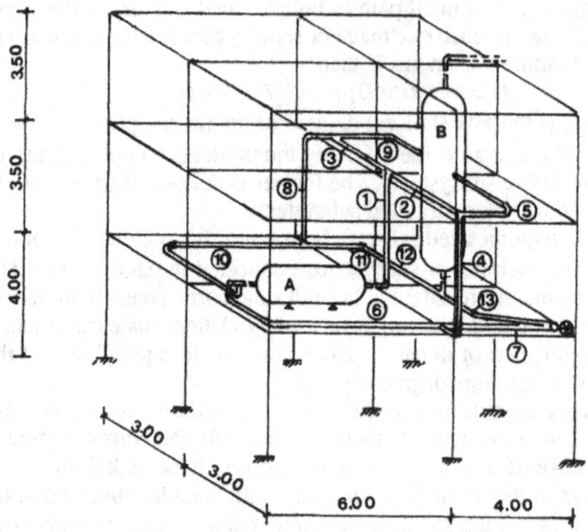

Fig. 1 - *Scheme of the analyzed frame-piping system*

The seismic motion is described by means of two independent processes, acting along two orthogonal directions and having the same spectral power density function. This latter corresponds to a selected standard response spectrum, i.e. that proposed by Eurocode no. 8 for intermediate soils. The ground motion intensity is assumed equal to 0.15 g.

The piping system is designed to circulate a fluid from vessel A to vessel B; the eight pipe elements performing this function are denoted by the numbers: 1-2-3-5-8-9-11-12. The piping system fault-tree is shown in fig. 2.

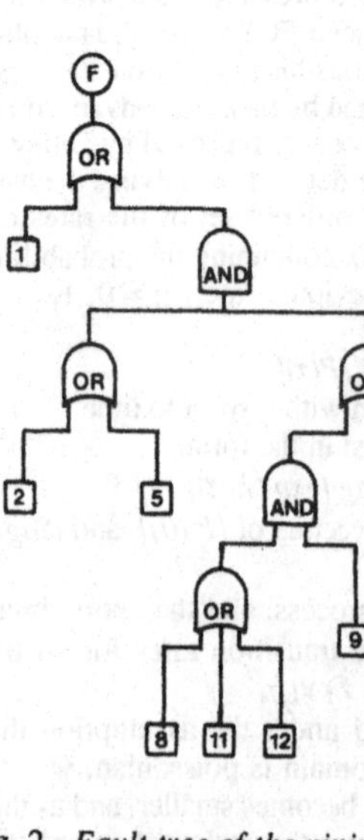

Fig. 2 - *Fault tree of the piping system*

Failure of the piping elements is assumed to occur when their respective flexural capacities are exceeded. For ease of computation, only the most stressed section of each pipe, labeled with the proper number, have been considered in the safety evaluation: only for pipe 1, which is serially connected to the other components, two different sections have been considered, here denoted by 1a and 1b.

The covariance matrix of the response process has been obtained using the first nine natural modes of vibration of the coupled structure-piping system. Decoupling is not possible because both masses and stiffnesses of the two systems are of comparable importance. In Table 1, the response correlation matrix, modified to include the standard deviation in the diagonal terms (instead of unity), is represented. It is evident that the action effects are strongly correlated.

Table 1 - *Response correlation matrix*

Elements	*1*	*1a*	*2*	*3*	*5*	*8*	*9*	*11*	*12*
1	2.30	.834	.785	.954	.789	.409	.434	.885	.408
1a		1.32	.619	.940	.857	.537	.388	.906	.482
2			.858	.765	.840	.679	.686	.663	.740
3				.861	.849	.462	.421	.914	.733
5					1.01	.821	.681	.847	.464
8						.605	.891	.595	.892
9		Symm.					1.10	.580	.877
11								.956	.600
12									.822

Three different cross sections diameters have been used, they having assigned to the elements so as to obtain an approximately uniform reliability level (as illustrated in Table 2).

The total number of system states is eight; in fact, according to the "minimal path to failure" representation, the system is formed of three subsystems, connected in parallel. These subsystems, denoted by S_i, are composed, respectively, of the elements:

$$S_1 = \{1, 1a, 2, 5\}$$
$$S_2 = \{1, 1a, 3, 9\}$$
$$S_3 = \{1, 1a, 3, 8, 11, 12\}$$

Table 2 - *Ultimate bending moments of the elements*

Elements	*1*	*1a*	*2*	*3*	*5*	*8*	*9*	*11*	*12*
Moments (tm)	15.0	15.0	7.02	7.02	7.02	7.02	7.02	7.02	4.53

The probability of system failure is found to be:

$$P_f = 0.044$$

It appears useful to note that, if only simultaneous failure of the parallel subsystems is considered, the corresponding failure probability is:

$$P_f = 0.0039$$

that is, it takes on a value eleven times smaller.

From this first example, the need then appears evident to consider all the possible paths to failure of a redundant system with brittle elements; in fact, while the simultaneous failure of all the components connected in parallel is certainly one possible path, it is not unique and, in some cases, it could also not be the most important, in probabilistic terms.

The second example deals with a partial model of the steel jacket offshore platform, represented in Fig. 3.

The plane truss structure, depicted in Fig. 4, is 3 times statically indeterminate and consists of 15 rotationally-symmetric tubular bars. The geometric variables, defined in Table 3, are assumed to be deterministic.

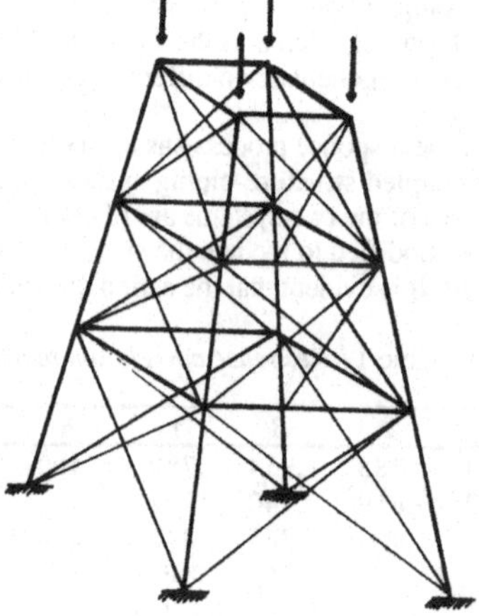

Fig. 3 - *Scheme of the 3D truss tower*

Table 3 - *Dimensions and mean tension yield forces of bars (as in [4]). The ratio between the diameter d and the wall thickness t is assumed to be: d/t = 60 for all bars.*

Bar	*d(m)*	*Area (m^2)*	*$E[N_i^+]$ (MN)*	*$E[N_i^-] / E[N_i^+]$*
1-2	2.5	0.324	103.78	0.50
3-4	2.5	0.324	103.78	0.68
5-6	2.5	0.324	103.78	0.85
7-8	1.5	0.116	37.07	0.50
9-10	1.2	0.074	23.73	0.50
11-12	0.9	0.042	13.35	0.50
13	2.0	0.210	67.20	0.54
14	1.5	0.116	37.07	0.87
15	1.0	0.053	16.80	0.96

The structure is subjected to four vertical dead loads at the top, each of magnitude P_1, to the horizontal wave loading, of magnitude proportional (by the factor γ) to P_2, and to a seismic action, applied at the base and modeled by means of a scalar gaussian process.

As in the first case, the spectral power density corresponds to the standard response spectrum proposed by Eurocode No. 8 for intermediate soils, and the ground motion intensity is assumed equal to 0.15 g.

The covariance matrix of the response process has been obtained using the first nine natural modes of vibration of the system.

In the analysis, no eccentricities are taken into account, so that only normal forces are assumed to be acting in the bars.

As in [4], the basic strength variables under compression and under tension: N^-, N^+; as well as the loading ones: $P_{1,2}$; are assumed to be normally distributed.

There is no correlation between strength and loading variable, and the constants in the wave load modelling are: $(\gamma_1; \gamma_2; \gamma_3) = (1.000; 0.667; 0.127)$.

It is assumed that system failure corresponds to the failure of any one of the vertical bars, numbered from 1 to 6, or of any one of the three pairs of diagonals. This assumption greatly simplifies the computation, because only the failure of diagonal bars (one for each level, as a maximum) defines the intermediate damage states of the system, whose number is therefore equal to 26.

The "minimal path to failure" representation corresponds to the connection in parallel of the eight subsystems:

$S_1 = \{1, 2, 3, 4, 5, 6, 7, 9, 11\}$ $S_2 = \{1, 2, 3, 4, 5, 6, 7, 9, 12\}$
$S_3 = \{1, 2, 3, 4, 5, 6, 7, 10, 11\}$ $S_4 = \{1, 2, 3, 4, 5, 6, 7, 10, 12\}$
$S_5 = \{1, 2, 3, 4, 5, 6, 8, 9, 11\}$ $S_6 = \{1, 2, 3, 4, 5, 6, 8, 9, 12\}$
$S_7 = \{1, 2, 3, 4, 5, 6, 8, 10, 11\}$ $S_8 = \{1, 2, 3, 4, 5, 6, 8, 10, 12\}$

while for the examined system the minimal cut sets are: $C_1 = \{1\}$; $C_2 = \{2\}$; $C_3 = \{3\}$; $C_4 = \{4\}$; $C_5 = \{5\}$; $C_6 = \{6\}$; $C_7 = \{7,8\}$; $C_8 = \{9,10\}$; $C_9 = \{11,12\}$.

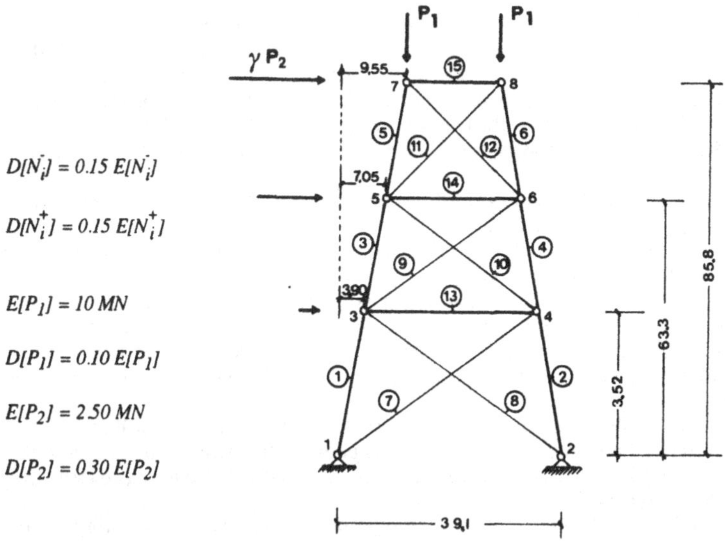

Fig. 4 - *Scheme of the partial model of the truss tower*

Table 4 sets forth the correlation matrix of the response, referred to the axial actions in the bars and, as usual, modified to include the standard deviations in the diagonal terms (instead of unity). It is evident that, in this case too, the action effects are strongly correlated.

Table 4 - *Response correlation matrix*

Elements	*1*	*2*	*3*	*4*	*5*	*6*	*7*	*8*	*9*	*10*	*11*	*12*
1	2.18	.975	.856	.798	.781	.723	.845	.753	.624	.587	.436	.421
2		1.84	.765	.912	.694	.713	.917	.863	.603	.758	.410	.446
3			.898	.684	.811	.653	.415	.583	.834	.752	.408	.391
4				.912	.761	.913	.434	.482	.575	.928	.425	.414
5					1.23	.973	.394	.412	.583	.728	.864	.912
6						1.38	.382	.409	.724	.758	.936	.952
7							1.46	.967	.875	.912	.784	.813
8								1.36	.924	.982	.872	.809
9		Symm.							1.21	.957	.924	.889
10										1.18	.898	.907
11											.995	.946
12												1.22

The probability of failure of the system is found to be:

$$P_f = 0.056$$

In an alternative analysis, the correlation matrix of the action effects was progressively updated, during the system evolution towards collapse. The schemes corresponding to the intermediate damage states are derived considering the brittle failure of (as a maximum) only one diagonal bar at each level.

The resulting system failure probability is:

$$P_f = 0.174$$

that is is more than two times the value previously determined.

CONCLUSIONS

Two examples of application of an analytical procedure for reliability evaluation of redundant structural systems with brittle elements, subjected to seismic action, have been illustrated.

From these examples, it may be observed that, for this class of systems, it is essential to consider all the modes of possible system collapse. The different sequences of element collapses exert a major influence on the realistic value of the failure probability, which can be seriously underestimated, if the analysis considers only the simultaneous failure of the parallel components.

One of the shortcomings of the proposed procedure is that it requires the examination of all the possible states of the given system, and the evaluation of all the rates of transition from each state to all those of higher order: therefore, the analysis becomes exponentially cumbersome with the increase of redundancy.

Some simplifications are however possible, if the system has some dominant failure modes, such that several terms of the matrix $[P'(O)]$ can be disregarded, or if the maximum value of the rate of transition to a state is less than a given threshold value. This last can be set equal to a convenient fraction of the rates found for the states with the same number of failed subsystems.

Obviously, consideration of the dominant failure modes only will give a lower bound for the actual failure probability. This limit is partially counteracted by the assumption of brittle behaviour, as concerns the element response, which usually leads to a conservative estimate of the real failure probability.

It is well known that the applied procedure is approximate, requiring drastic simplifications in the modelling of element behaviour. Nevertheless, while constituting a small improvement to the commonly used algorithms, it should also provide some useful information on the dynamic behaviour of complex systems, identifying the more critical elements, and some mutual-interacting sequences of element collapses. It could then be used as a design tool, perhaps even in the preliminary stage.

REFERENCES

1. Wen, Y.K. and Chen, H-C. System Reliability under Time Varying Loads: I - System Reliability under Time Varying Loads: II. Journal of Engineering Mechanics, ASCE, Vol. 115(4), pp. 808-823, pp. 824-839, 1989.
2. Ciampoli, M. Reliability Assessment of Systems Subjected to Stochastic Actions - Applications to the Analysis of the Seismic Safety of Structural Systems. Ph.D Thesis (in italian), Rome, 1987.
3. Ciampoli, M., Giannini, R., Nuti, C. and Pinto, P.E. Evolutive Failure of Systems Subjected to Continuous Stochastic Processes, in: Proceedings of the 5th Int. Conf. on Applications of Statistics and Probability in Soil and Structural Engineering (Ed. Lind, N.C.), pp. 829-836, University of British Columbia, Vancouver B.C., Canada, 1987.
4. Bjerager, P., Karamachndrani, A. and Cornell, C.A. Failure Tree Analysis in Structural System Reliability, in: Proceedings of the 5th Int. Conf. on Applications of Statistics and Probability in Soil and Structural Engineering (Ed. Lind, N.C.), pp. 985-996, University of British Columbia, Vancouver B.C., Canada, 1987.
5. Ditlevsen, O. Gaussian Outcrossing from Safe Convex Polyhedrons, Jnl. of Mech. Eng. Div., ASCE, Vol. 109(1), pp. 127-148, 1983.

Seismic Ground Strain Simulations for Pipeline Response Analysis

A. Zerva

Department of Civil and Architectural Engineering, Drexel University, Philadelphia, PA 19104, U.S.A.

ABSTRACT

The analysis and design of buried pipelines to resist seismic ground motions relies on estimates of the seismic ground strains. It is generally assumed that estimates of the maximum axial strain along the pipeline can be determined from the ratio of the maximum velocity recorded at the site over the velocity of the apparent propagation of the motions. The validity of this assumption is examined through strain and velocity simulations from models of spatially incoherent and propagating seismic ground motions in near-source regions. It is shown that strain estimates obtained through this assumption considerably underestimate the actual strains of the earthquake. It is also shown, that the consideration of the apparent propagation of the motions as the basic source for spatial variability produces only a fraction (approximately 50%) of the seismic strains induced by spatially incoherent and propagating motions.

INTRODUCTION

The analysis and design of buried pipelines to resist seismic motions differs fundamentally from that of conventional structures. One of the basic differences is that, because of their extended length, pipelines undergo differential motions during earthquakes. Furthermore, it has been postulated, both from experimental (recorded) data and from analytical evaluations, that inertia effects for buried pipelines may be neglected, that their movement during earthquakes is similar to that of the ground and that the maximum strain along a straight buried pipeline coincides with the maximum ground strain (Ariman and Muleski [1]; O'Rourke, Castro and Hossain [2]). Simplified guidelines for the design of pipelines to resist seismic motions determine the ground strain through the assumption of a plane, sinusoidal wave that propagates along the structure's axis with constant velocity without interference from other waves and without changing shape (Newmark and Rosenblueth [3]; ASCE Committee on Seismic Analysis [4]). Based on the above considerations, the maximum axial strain along the pipeline is obtained from the expression:

$$\varepsilon_{max} = v_{max}/c \tag{1}$$

in which, ε indicates strain, v_{max} is the maximum ground velocity of the recorded motion at the site, and c is the apparent propagation velocity of the motions on the ground surface. The description of the seismic motion by a sinusoidal wave, however, may not reproduce the actual ground strains, particularly in near-source regions, where the arrivals of body waves from an extended source and/or from scatterers may significantly influence the shape of the seismic motions at different locations on the ground surface- these effects are termed spatial incoherence in this study. The spatial variability of the motions, that results from both the incoherence and the apparent propagation, may induce significantly higher ground strains than the ones obtained through the assumption of unchanging, but propagating motions.

Current stochastic descriptions of the seismic ground motions take into consideration both the spatial incoherence and the apparent propagation of the motions. In the stochastic approach, the seismic ground motion is, generally, described by a space-time random field with its cross spectral density defined as:

$$S(\xi,\omega) = S(\omega)\ \rho(\xi,\omega)\ \rho_{wp}(\xi,c,\omega) \tag{2}$$

$S(\omega)$ is the power spectral density of the motions, ω being the frequency in rad/sec. The frequency dependent spatial correlation usually consists of two terms: The first term $\rho(\xi,\omega)$, with ξ indicating the separation distance between two stations on the ground surface, describes the incoherence of the waves (change in shape of the waves), and is given by an exponential function decaying with frequency and separation distance. The second term $\rho_{wp}(\xi,c,\omega)$ describes the propagation of the seismic motions; generally, it is assumed that the motions propagate with the same apparent propagation velocity c. The similarity between the two approaches (equations (1) and (2)) is that both assume that the entire motion propagates with the same apparent velocity c. The difference, however, between the stochastic approach (equation (2)) and the simplified design approximation (equation (1)) is that the incoherence of the seismic waves, as well as the contributions of the various types of waves during the strong shaking part of the motions, are taken into account in the stochastic formulation but not in equation (1).

This analysis examines the validity of the current estimates for the seismic ground strains (equation (1)) to be used in the seismic response of lifelines, through a comparison of strain ($\varepsilon(t)$) and strain estimate ($v(t)/c$) simulations from spatially variable seismic ground motion models. For this purpose, the stochastic description of the seismic motions (equation (2)) is implemented. It is noted that the analysis concentrates in near-source regions, for which the spatial incoherence of the motions becomes a significant factor. For regions where well-defined surface waves develop, that can appropriately be described by a sinusoidal wave, the validity of equation (1) is obvious. From the description of the random field through its power spectral density and its spatial correlation (equation (2)), the frequency wavenumber spectrum for strains and velocities is evaluated. The spectral representation method (Shinozuka [5], [6]; Zerva [7]) is used to obtain simulations for velocity and strain time histories. The correspondence between equation (1) and the actual strains obtained from the stochastic models is analyzed. The analysis may provide insight into the characteristics of ground strains to be used in the seismic response analysis of pipelines.

THE SPECTRAL REPRESENTATION METHOD

The simulated strain and velocity time histories are obtained through the spectral representation method. Shinozuka [5], [6] expanded Yang's [8] approach in simulating envelopes of random processes through the Fast Fourier Transform algorithm (FFT) to time history simulations from the frequency-wavenumber (F-K) spectra of random fields, provided that there exist an upper cut-off frequency ω_u and an upper cut-off wavenumber κ_u, above which the values of the F-K spectrum can be considered negligible for practical applications. For the simulated field, however, to be ergodic, it had to be assumed that the value of the F-K spectrum at the origin was zero. Zerva [7] introduced an improvement of the approach, that releases this condition and allows better convergence of the stochastic characteristics of the simulations to those of the random field. According to this methodology, simulations of a random field described by its F-K spectrum $S(\kappa,\omega)$ are obtained as follows:

Let the discrete wavenumber and frequency be given by:

$$\begin{aligned} \kappa_j &= (j+1/2)\Delta\kappa; \qquad j=0,\ldots\ldots,J-1 \\ \omega_n &= (n+1/2)\Delta\omega; \qquad n=0,\ldots\ldots,N-1 \end{aligned} \tag{3}$$

with J and N being powers of 2, and $\Delta\kappa$ and $\Delta\omega$ being the wavenumber and frequency step, respectively. The simulations of the random field are obtained according to the following equation:

$$\begin{aligned} f_{rs} = &\mathrm{Re}\Big[e^{i\pi r/M} e^{i\pi s/L}\Big\{\sum_{j=0}^{M-1}\sum_{n=0}^{L-1}\{[4S(\kappa_j,\omega_n)\,\Delta\kappa\Delta\omega]^{1/2} e^{i\phi_{jn}^{(1)}}\} e^{i2\pi rj/M} e^{i2\pi sk/L}\Big\}\Big] \\ &+\mathrm{Re}\Big[e^{i\pi r/M} e^{-i\pi s/L}\Big\{\sum_{j=0}^{M-1}\sum_{n=0}^{L-1}\{[4S(\kappa_j,-\omega_n)\,\Delta\kappa\Delta\omega]^{1/2} e^{i\phi_{jn}^{(2)}}\} e^{i2\pi rj/M} e^{-i2\pi sk/L}\Big\}\Big] \end{aligned} \tag{4}$$

in which,

$$f_{rs} = f(x_r, t_s)$$

$$x_r = r\Delta x; \qquad \Delta x = 2\pi/(M\Delta\kappa); \qquad r=0,.......,M-1$$
$$t_s = s\Delta t; \qquad \Delta t = 2\pi/(L\Delta\omega); \qquad s=0,.......,L-1 \qquad (5)$$

and $\phi_{jn}^{(1)}$ and $\phi_{jn}^{(2)}$ are two sets of independent random phase angles uniformly distributed between 0 and 2π. The FFT technique is then applied to the two expressions in the bold brackets in equation (4). M and L in equations (4) and (5) are powers of 2, and are given by:

$$M \geq 2J; \qquad \text{since } \Delta x \leq \pi/(J\Delta\kappa)$$
$$L \geq 2N; \qquad \text{since } \Delta t \leq \pi/(N\Delta\omega) \qquad (6)$$

so that aliasing effects are avoided. The following characteristics are inherent in the simulations: i) They are asymptotically Gaussian as J,N->∞ due to the central limit theorem; ii) they are periodic with period $T_o = 4\pi/\Delta\omega$ and wavelength $L_o = 4\pi/\Delta\kappa$; iii) they are ergodic, at least up to second moment, over an infinite time and distance domain or over the period and wavelength of the simulation; and iv) due to the orthogonality of the cosine functions in equation (4), with J,N->∞, and with the consideration of the cut-off wavenumber κ_u and frequency ω_u, the ensemble mean, covariance function and frequency-wavenumber spectrum of the simulations are identical to those of the random field.

SEISMIC GROUND MOTION MODEL

The power spectral density of displacements is assumed to be the same at all locations on the ground surface and is given by the Clough-Penzien spectrum (Clough and Penzien [9]), that allows finite values for the velocity and displacement spectra as ω ->0:

$$S(\omega) = S_o\,(\omega_g^4 + 4\zeta_g^2\omega_g^2\omega^2) / ((\omega_g^2 - \omega^2)^2 + 4\zeta_g^2\omega_g^2\omega^2) \,*$$
$$*\, 1/((\omega_f^2 - \omega^2)^2 + 4\zeta_f^2\omega_f^2\omega^2) \qquad (7)$$

The characteristics of the two filters assume the values : ω_g=15.46rad/sec, ω_f=1.636rad/sec, ζ_g=0.623, ζ_f=0.619, and S_o is assumed to be equal to 1cm^2/sec^3 in this study. The spatial incoherence model (second term in equation (2)) used in the simulations is of the form (Luco and Wong [10]):

$$\rho(\xi,\omega) = \exp\,[-a^2\omega^2\xi^2] \qquad (8)$$

in which, the incoherence parameter a assumes the values of $2{*}10^{-4}$sec/m and $1{*}10^{-3}$sec/m in this study. The apparent propagation term in equation (2) is of the form:

$$\rho_{wp}(\xi,c,\omega)=\exp[-i\omega(\mathbf{c}\cdot\boldsymbol{\xi})/c^2] \quad (9)$$

in which, $(\mathbf{c}\cdot\boldsymbol{\xi})$ indicates the dot product of the velocity vector $\mathbf{c}$ and the separation distance vector $\boldsymbol{\xi}$; the dot product for this one dimensional case specifies the direction of the propagation (along the positive or the negative x direction). In the absence of the apparent propagation term (equations (2) and (9)), which corresponds to $c\to\infty$, and with the spatial incoherence as defined in equation (8), the random field becomes quadrant symmetric (Vanmarcke [11]), and the simulated motions are superpositions of standing waves (Zerva [7]). The first value of the incoherence parameter examined herein ($a=2*10^{-4}$sec/m) renders high correlation at low frequencies, and results in almost identical displacement simulations over the range of distances of interest for lifeline analyses (Figure 1a). On the other hand, when $a=1*10^{-3}$sec/m, the simulated motions exhibit spatial variability (Figure 1b). When c assumes finite values, the seismic motions propagate with the apparent velocity c along the x-axis (Zerva [7]). The values for the apparent propagation velocity considered herein are $c\to\infty$ and c=1500m/sec. These values are realistic representations of the apparent propagation velocity, since the strong shaking part of the seismic motion generally propagates with the apparent velocity of the body waves at the bedrock-surface layer interface, and the angle of incidence of such waves could be as high as 90° ($c\to\infty$).

Based on equations (2), (7), (8) and (9), the F-K spectrum for the ground strains becomes:

$$S_{\varepsilon\varepsilon}(\kappa,\omega) = \kappa^2\, S(\omega)\exp[-(\kappa+\omega/c)^2/4a^2\omega^2]/[2\sqrt{(\pi a^2\omega^2)}\,] \quad (10)$$

and that for velocities:

$$S_{vv}(\kappa,\omega) = \omega^2\, S(\omega)\exp[-(\kappa+\omega/c)^2/4a^2\omega^2]/[2\sqrt{(\pi a^2\omega^2)}\,] \quad (11)$$

STRAIN AND VELOCITY SIMULATIONS

The simulated time histories of the seismic ground strains at an arbitrary location on the ground surface (x=500m) for incoherence parameter $a=2*10^{-4}$sec/m and apparent propagation velocities of $c\to\infty$ and c=1500m/sec are shown in Figure 2. For the simulations, the upper cut-off wavenumber was κ_u=0.06283rad/m, the upper cut-off frequency ω_u=62.83rad/sec, and J*N=128*128. When $c\to\infty$, the seismic ground strains in Figure 2 are induced by the incoherence effects only. It has to be noted, that the seismic ground strains resulting from incoherence effects only cannot be depicted by equation (1). Since the apparent propagation velocity tends to infinity, equation (1) would yield zero strain estimates. When the

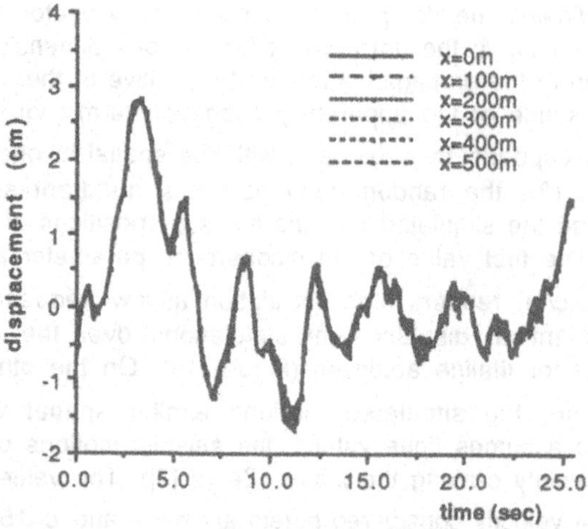

a) Displacement time histories for a=2*10^{-4}sec/m

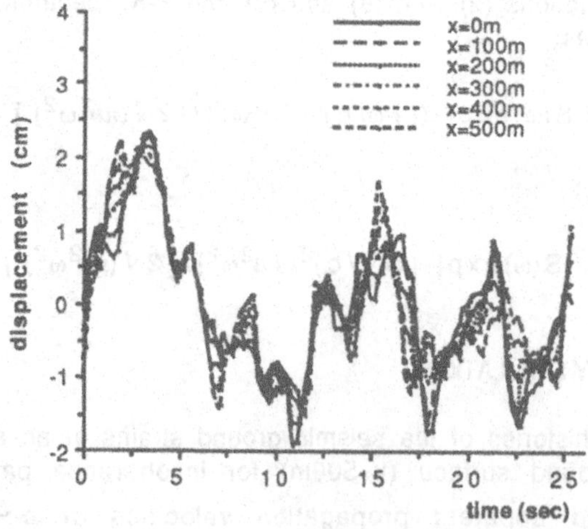

b) Displacement time histories for a=1*10^{-3}sec/m

Figure 1 Displacement time history simulations for the spatial incoherence model with c->∞

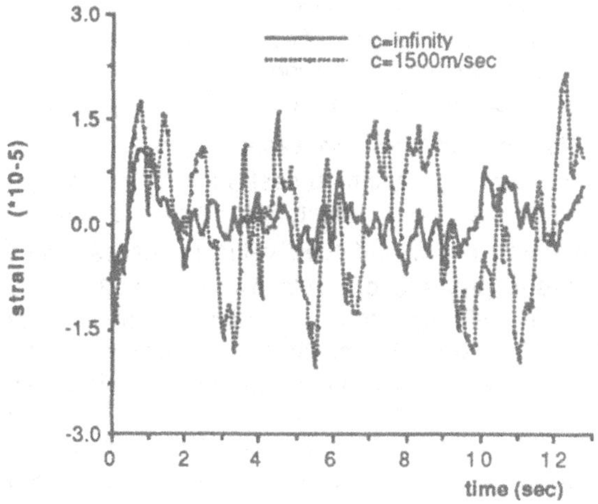

Figure 2 Seismic ground strain simulations at x=500m for c->∞ and c=1500m/sec; incoherence parameter a=2*10^{-4}sec/m

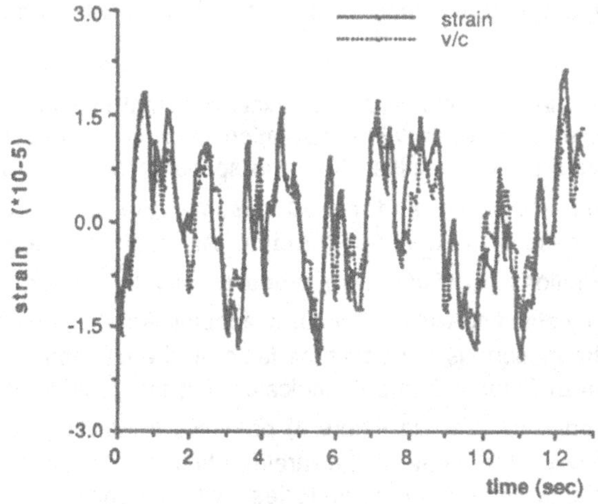

Figure 3 Strain and strain estimate simulations at x=500m for incoherence parameter a=2*10^{-4}sec/m and apparent propagation velocity c=1500m/sec

apparent propagation velocity assumes finite values (c=1500m/sec in Figure 2), the seismic ground strains are superpositions of both the incoherence and the apparent propagation effects; in this case, the seismic strains are significantly higher than the ones induced by incoherence effects only. This observation may be explained as follows: The displacement time histories for $a=2*10^{-4}$sec/m and $c\to\infty$ are essentially fully coherent; i.e., the displacement time histories remain almost identical in shape over distances of 500m (Figure 1a). As a result, the seismic strains assume relatively small values when $c\to\infty$, and their value increases significantly with decreasing apparent propagation velocity. Figure 3 presents a comparison between the seismic strains resulting from an apparent propagation of c=1500m/sec with the estimates of strains as they are currently used for design; i.e., velocity time histories are simulated from equations (4) and (11) and then divided by the value of the apparent propagation velocity. In order to obtain correspondence between the strain and velocity time histories, the same frequency and wavenumber steps ($\Delta\omega$ and $\Delta\kappa$) and the same seeds for the generation of the random phase angles are used in the simulations. Figure 3 suggests that the maximum of the simulated strains is only slightly higher than that of the maximum strain estimate v_{max}/c, thus confirming equation (1). Figure 3 also indicates that there is good agreement between the two time histories. This suggests that the dominant wave components propagate with the apparent propagation velocity; i.e., the dominant wavenumbers in equations (4), (10) and (11) are related to the frequency through $\kappa\approx|\omega|/c$, and, because of the relatively high coherence, wavenumbers different than $|\omega|/c$ have only a small contribution to the simulations.

It is, however, the rule rather than the exception, that seismic ground motions exhibit loss of coherence with separation distance. Presented in Figure 4 are the ground strains resulting from the spatial incoherence model with $a=1*10^{-3}$sec/m and $c\to\infty$ and c=1500m/sec; the upper cut-off wavenumber in this case was $\kappa_u=0.09425$rad/m. Although the apparent propagation of the motions (c=1500m/sec) produces slightly higher strains than those produced by standing waves ($c\to\infty$), it appears from Figure 4 that the incoherence of the motions is the dominant factor in the estimation of the strains. A comparison of Figures 2 and 4 indicates that the consideration of incoherence effects only (for $c\to\infty$ in Figure 4) produces a maximum ground strain that is almost twice the value of the strain obtained from unchanging but propagating displacement time histories with apparent velocity c=1500m/sec (Figure 2). This observation suggests that the spatial incoherence effects may dominate over the apparent propagation effects, and should, therefore, not be neglected in the seismic response analysis of pipelines.

In order to analyze the validity of equation (1) for the seismic ground motion model that involves both significant spatial incoherence and apparent propagation of the motions, the strain time history simulation is compared

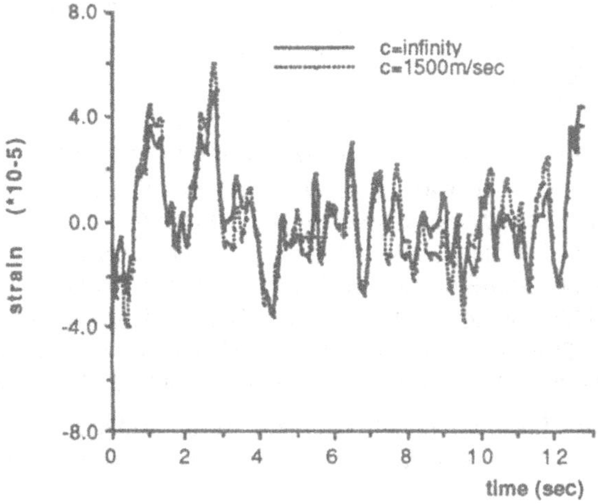

Figure 4 Seismic ground strain simulations at x=500m for c->∞ and c=1500m/sec; incoherence parameter $a=1*10^{-3}$sec/m

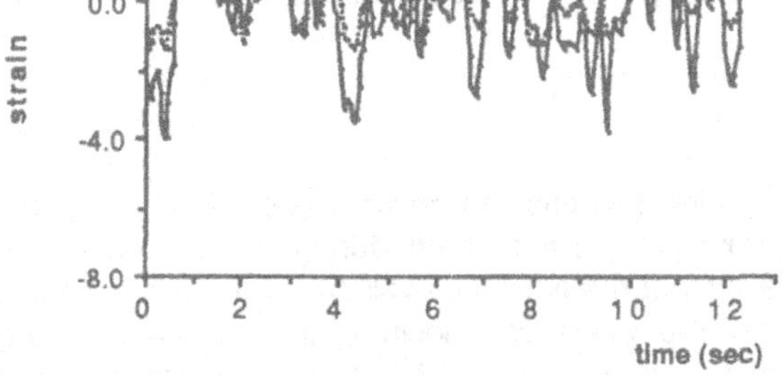

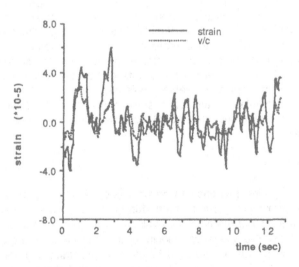

Figure 5 Strain and strain estimate simulations at x=500m for incoherence parameter $a=1*10^{-3}$sec/m and apparent propagation velocity c=1500m/sec

with the strain estimate (v(t)/c) for c=1500m/sec (Figure 5). The comparison reveals that the strain estimate v_{max}/c produces only half the value of the actual maximum strain, suggesting that strain estimates as they are currently used for design produce unconservative results in regions where incoherence effects are important. A match between the actual strain and the strain estimate (v_{max}/c) for $a=1*10^{-3}$sec/m was obtained when the apparent propagation velocity was reduced to the low value of 500m/sec (Figures 6 and 7); it is noted that this value of the propagation velocity falls more in the range of surface wave rather than body wave propagation velocities. The maximum value of the ground strain for c=500m/sec becomes significantly higher than the one obtained from the consideration of spatial incoherence only (Figure 6). Although the actual maximum strain is still higher than the strain estimate v_{max}/c in this case (Figure 7), the agreement between the two time histories (strain $\varepsilon(t)$ and strain estimate v(t)/c) in Figure 7 is reasonable, and resembles the comparison of strain and strain estimate in Figure 3.

It appears from Figures 4 through 7, that there exists a critical value for the apparent propagation velocity, below which the strain estimate (v_{max}/c) approaches from below the actual value of the ground strain. If the apparent propagation velocity, however, is higher than the critical value, then the strain estimate v_{max}/c may considerably underestimate the actual strains. For motions that exhibit loss of coherence, the critical propagation velocity is low, lower than realistic apparent propagation velocities in near-source regions. In this case, the strain estimate obtained through the ratio of the maximum velocity of the recorded motions over realistic values for the apparent propagation velocity produces unconservative results for lifeline seismic response analysis. It is interesting to note that a similar observation with regard to the critical propagation velocity was made by O'Rourke, Castro and Centola [12], who obtained estimates of seismic ground strains based on recorded displacement time histories at 22 stations during the 1971 San Fernando earthquake.

CONCLUSION

In establishing the pipeline response to seismic excitations, it is generally assumed that the maximum ground strain during an earthquake can be estimated through the ratio of the maximum velocity of the recorded motion at the site over the apparent propagation velocity of the motions on the ground surface. This assumption was investigated herein through strain and velocity simulations from models of spatially variable seismic ground motions. It was shown, that for displacement time histories that are essentially unchanging as they propagate along the ground surface with constant velocity, this assumption may produce realistic estimates for the ground strain. However, when the displacement time histories exhibit a significant loss of coherence, the spatial incoherence effects dominate over the propagation effects for realistic values of the apparent propagation velocity, thus rendering the

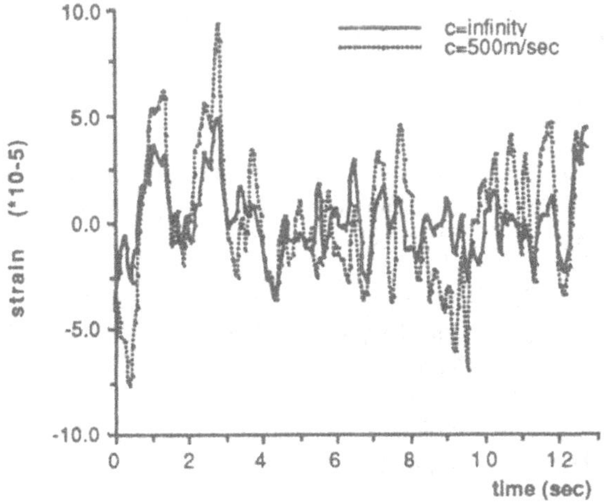

Figure 6 Seismic ground strain simulations at x=500m for c->∞ and c=500m/sec; incoherence parameter a=1*10^{-3}sec/m

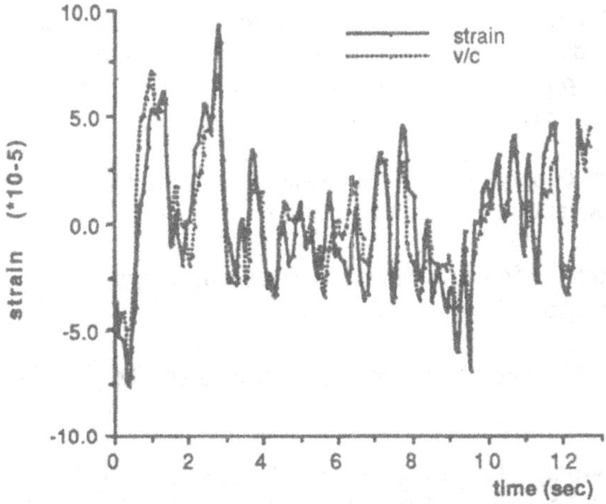

Figure 7 Strain and strain estimate simulations at x=500m for incoherence parameter a=1*10^{-3}sec/m and apparent propagation velocity c=500m/sec

assumption unconservative. In this case, alternative approaches for the estimation of the seismic ground strains, as, e.g., through simulations from seismic ground motion models, should be implemented.

REFERENCES

1. Ariman, T. and Muleski, G.E. A Review of the Response of Buried Pipelines under Seismic Excitations, Earthquake Engineering and Structural Dynamics, Vol.9, pp. 133-151, 1981.

2. O'Rourke, M.J., Castro, G. and Hossain, I. Horizontal Soil Strain due to Seismic Waves, Journal of the Engineering Mechanics Division, ASCE, Vol.110, pp. 1173-1187, 1984.

3. Newmark, N. M. and Rosenblueth, E. Fundamentals of Earthquake Engineering, Prentice-Hall, Englewood Cliffs, 1971.

4. Committee on Seismic Analysis, Seismic Response of Buried Pipes and Structural Components, ASCE, New York, 1983.

5. Shinozuka, M. Digital Simulation of Random Processes in Engineering Mechanics with the Aid of FFT Technique, Stochastic Problems in Mechanics, (Ed. Ariaratnam, S.T. and Leipholtz, H.H.E.), pp. 277-286 University of Waterloo Press, 1974.

6. Shinozuka, M. Stochastic Fields and their Digital Simulation, Chapter 3, Stochastic Methods in Structural Dynamics, (Ed. Schuëller, G.I. and Shinozuka, M.), pp. 93-133, Martinus Nijhoff Publishers, Dordrecht, The Netherlands, 1987.

7. Zerva, A. Seismic Ground Motion Simulations from a Class of Spatial Variability Models, submitted for publication.

8. Yang, J.N. Simulation of Random Envelope Processes, Journal of Sound and Vibration, Vol.21, pp. 73-85, 1972.

9. Clough, R.W. and Penzien, J. Dynamics of Structures, Mcgraw-Hill, New York, 1975.

10. Luco, J.E. and Wong, H.L. Response of a Rigid Foundation to a Spatially Random Ground Motion, Earthquake Engineering and Structural Dynamics, Vol.14, pp. 891-908, 1986.

11. Vanmarcke, E. Random Fields, MIT Press, Cambridge, Massachusetts, 1983.

12. O'Rourke, M.J., Castro, G. and Centola, N. Effects of Seismic Wave Propagation upon Buried Pipelines, Earthquake Engineering and Structural Dynamics, Vol.8, pp. 455-467, 1980.

Parametric Study of the Stochastic Seismic Response of Simple Nonlinear Structures

J.P. Conte (*), K.S. Pister (**), S.A. Mahin (**)
Department of Civil Engineering, Rice University, Houston, Texas 77251, U.S.A.
(**) Department of Civil Engineering, University of California, Berkeley, California 94720, U.S.A.

ABSTRACT

In the present study, time-dependent autoregressive - moving average (ARMA) models are used for stochastic modeling of earthquake ground motions. In a first stage, the earthquake models are checked against the real seismic data from which they have been identified. In a second stage, these models are used to simulate the stochastic seismic response of idealized linear and nonlinear structures. By varying the earthquake model parameters and the structural model parameters, a sensitivity study of the stochastic seismic response of these structural models is performed. A brief description of the developed methodology and some significant results obtained are presented in the sequel.

INTRODUCTION

When designing new structures or assessing existing structures, the earthquake engineer sees his task very much complicated by the uncertain nature of the potential earthquake excitation. The uncertain earthquake loading implies that the earthquake engineer should design a new structure or evaluate an existing structure in terms of the probability of survival (i.e., reliability) with respect to various limit states characterizing the structure. A structural reliability calculation requires two major components: (1) a quantitative model describing the uncertain loading; and (2) input-output relationships relating the response uncertainty to the loading uncertainty. The first requirement can be satisfied by considering either a large collection of real earthquake records or a random process-type earthquake model representative of real earthquake records. Unfortunately, at present, earthquake data are still scarce and it is impossible to find a large, homogeneous collection of records. In their heterogeneity, the seismic records available correspond to different source-site pairs and geophysical parameters which leads to a difficult interpretation of the statistical results on structural response. This "heterogeneity" problem does not exist if a random process-type model is used. Although stochastic earthquake models have been widely accepted in the engineering profession, many of them are not able to capture well enough the most salient features of actual seismic records.

Due to economic considerations, behavior of structures beyond the elastic range during severe earthquakes is not only tolerated but relied upon in order to decrease and dissipate the earthquake input energy into the structure. Also it is known from both experimental and analytical observations that inelastic behavior of structures may be very sensitive to the details of the earthquake ground motion time history. As a corollary, a strong need to better understand the cause-and-effect relationships existing between earthquake ground motion uncertainty, structural properties, and structural response uncertainty prevails. The research briefly described here attempts to shed some light on this complicated problem by considering a realistic earthquake model and idealized linear and nonlinear structures.

STOCHASTIC EARTHQUAKE GROUND MOTION MODEL

In this study, time-dependent or "dynamic" ARMA models are used to simulate real earthquake accelerograms. These models are explicitly formulated in discrete time. They are of equal generality with linear, continuous-time models (governed by stochastic differential equations), however, they have a number of significant advantages for digital analysis and simulation purposes (Conte [2]). The general time-dependent ARMA model of order (p,q), abbreviated ARMA(p,q), is represented by the following stochastic linear difference equation:

$$a_k - \phi_{1,k} a_{k-1} - \cdots - \phi_{p,k} a_{k-p} = e_k - \theta_{1,k} e_{k-1} - \cdots - \theta_{q,k} e_{k-q} \quad (1)$$

In this equation, $\{a_k = a(k\Delta t)\}$ represents the discrete earthquake ground acceleration process, Δt the sampling time interval, $\{e_k\}$ the driving uncorrelated Gaussian white noise with variance $\sigma^2_{e,k}$, $(\phi_{i,k}, i=1,p)$ and $(\theta_{i,k}, i=1,q)$ the time-dependent autoregressive and moving average coefficients, respectively. The stochastic model represented in equation (1) is able to reproduce the nonstationarity in both amplitude and frequency content characterizing real earthquake records. The nonstationarity in amplitude is modeled by the variance envelope $\sigma^2_{e,k}$ of the driving noise $\{e_k\}$, whereas the nonstationarity in frequency content is represented by the time-dependent ARMA filter coefficients. Both types of nonstationarity are extremely important in realistic prediction of nonlinear seismic response of structures, although the nonstationarity in frequency content has often been disregarded in the past.

From a real earthquake ground acceleration record called target accelerogram, a time-dependent ARMA model is estimated using an iterative Kalman filtering procedure (Conte [2,3]). This procedure requires a state-space formulation of the

ARMA model and updates in an optimum way the system parameters at each time step using the observation at this time step. The Kalman filter results are compared with the results produced by a more traditional nonstationary estimation technique based on the concept of a stationary moving time window. For a given time window, the latter technique assumes data stationarity and uses a standard maximum likelihood algorithm from time series analysis (Box and Jenkins [1]).

As an example, a time-dependent ARMA(2,1) model has been estimated from the well-known north-south component of the 1940 Imperial Valley earthquake recorded at El Centro and shown in figure 1(a). Figures 1 (b), (c), and (d) display the time histories of the parameter estimates $\hat{\phi}_1(t)$, $\hat{\phi}_2(t)$, and $\hat{\theta}_1(t)$, respectively, obtained using both Kalman filtering and moving time window techniques. The results obtained using these two different methods show the same trends except within the time segment between 8 and 12 seconds. The underlying physical system corresponding to the ARMA(2,1) model is a linear viscously damped single-degree-of-freedom (SDOF) oscillator with input support displacement $X(t)$ applied separately to the spring and the dashpot in proportions C_s and C_d, respectively (Conte [2], Chang [4]). The equation of motion of this equivalent SDOF system can be written as

$$\ddot{Z}(t) + 2\xi_g\omega_g\dot{Z}(t) + \omega_g^2 Z(t) = C_s\omega_g^2 X(t) + 2C_d\xi_g\omega_g\dot{X}(t) \tag{2}$$

where $Z(t)$ is the SDOF absolute displacement (measured with respect to a fixed reference). The equivalent SDOF system is characterized by the physical parameters ω_g (natural circular frequency), ξ_g (ratio of critical damping) and the ratio C_s/C_d. It can be shown that if the input acceleration process $\ddot{X}(t)$ is a continuous white noise process, then the continuous, stationary output process $a(t) = \ddot{Z}(t)$ is discretely coincident with the discrete output process of a stationary ARMA(2,1) model (Chang [4], Conte [2]). It is observed that the well-known Kanai-Tajimi stochastic earthquake model is discretely coincident with a subset of the discrete ARMA(2,1) model for which $C_s/C_d = 1$. The physical parameters ω_g and ξ_g indicate the predominant frequency and the frequency bandwidth of the ground shaking. By extension, a nonstationary (time-dependent) ARMA(2,1) model can be interpreted as a nonstationary output process of the aforementioned SDOF system with time-varying parameters $\omega(t)$, $\xi_g(t)$ and $C_s(t)/C_d(t)$. This interpretation remains valid as long as the trends of the physical parameters vary slowly in time compared to the natural period of the system $T_g = 1/F_g = 2\pi/\omega_g$. The interpretation of the ARMA parameters (e.g., ϕ_1, ϕ_2, θ_1) in terms of the physical parameters (e.g., ω_g, ξ_g, C_s/C_d) offers a potential link between seismology and ARMA modeling of earthquake ground motion time histories.

In the case of the El Centro 1940 record, the time histories of $\hat{F}_g$[Hz] and $\hat{\xi}_g$[–] obtained from $\hat{\phi}_1(t)$, $\hat{\phi}_2(t)$ and $\hat{\theta}_1(t)$ are shown in figures 1 (e) and (f), respectively. When observed in parallel with figure 1 (a), these two figures show that the important changes of frequency content of the target accelerogram are captured by the ARMA model. In figure 2, the corresponding ARMA(2,1) estimate of the normalized evolutionary power spectral density of El Centro 1940 is displayed. It shows the normalized frequency distribution of the earthquake energy as a function of time.

For ground motion simulation purposes, the estimation or analysis procedure is simply reversed. The sequence of operations performed is: (1) computer generation of a stationary, Gaussian, discrete white noise; (2) time modulation using $\hat{\sigma}_e(t)$; (3) ARMA filtering using $(\hat{\phi}_i(t), i=1,p)$ and $(\hat{\theta}_i(t), i=1,q)$; and (4) baseline correction using a high-pass Butterworth filter to eliminate the low frequency error introduced by the model. Figures 3 (a) and (b) show artificial ground motion samples corresponding to the target El Centro record.

STRUCTURAL MODELS AND RESPONSE CHARACTERIZATION

In order to better understand and gain insight into the inelastic seismic behavior of structures, comprehensive nonlinear mathematical models of real structures are needed. In general, the restoring forces of a real structure, modeled as a multi-degree-of-freedom system, depend on the past deformation history of all its structural components in a complex way. In this study, the approach has been to use, as a first step, idealized nonlinear SDOF representations of structures. The use of a SDOF idealization serves several purposes:

(1) to identify the effects and relative importance of the various factors influencing the stochastic seismic response of these simple systems;

(2) to identify general patterns and trends in stochastic inelastic seismic response behavior of structures;

(3) to provide foundation and guidance for the study of more complex structures.

Four simplified piecewise linear restoring force models are considered for this investigation, namely: (1) the bilinear nondegrading inelastic model, (2) Clough's stiffness degrading model, (3) the slip model, and (4) the bilinear elastic model. These restoring force models are characterized by the same bilinear skeleton curve, and the corresponding nonlinear dynamic SDOF systems are uniquely defined by the following structural parameters: (1) the initial natural period T_0 (or initial stiffness k_0), (2) the initial damping ratio $\xi_0 = c/2\sqrt{k_0 m}$ where c is the constant damping coefficient of the system and m the mass of the system, (3) the strength coefficient η expressing the yield strength of the system as a fraction of its own weight, and (4) the strain-hardening (or -softening) ratio α defined as the ratio of the post-yield stiffness to the initial stiffness.

The integration of the equation of motion of these nonlinear SDOF systems is performed analytically, taking advantage of the piecewise linearity of the force-deformation relationships and assuming piecewise linear ground acceleration time histories (Conte [2]). The nonlinear response time histories are then characterized by three types of non-dimensional (normalized) response parameters:

(1) the response parameters derived directly from the force-deformation response such as maximum, cumulative, and residual displacement ductility factors (μ_d, μ_{acc}, and μ_{res}), number and relative amplitudes of yield excursions;
(2) the maximum or cumulative normalized energy responses: kinetic energy (E_K), energy dissipated through viscous damping (E_D), recoverable elastic strain energy (E_E), and hysteretic energy dissipated (E_H);
(3) the peak rate of energy responses or maximum power responses ($P_{K,max}$, $P_{D,max}$, $P_{E,max}$, and $P_{H,max}$).
Some of these response parameters have been used extensively in analytical and experimental research as damage indices correlated to seismic damage imparted to real structures.

PARAMETRIC STUDY OF THE STOCHASTIC SEISMIC RESPONSE

Using the nonstationary stochastic earthquake model and the idealized nonlinear SDOF structural models described previously, a comprehensive parametric study has been conducted. In this study, the parameters of the general problem under consideration, i. e., the target earthquake record, the order of the fitted earthquake model, the structural model (type of structural behavior), and the structural parameters, are varied one by one to investigate their relative influence on the stochastic seismic response characterized in terms of response parameters which are random variables. The flow chart represented in figure 4 summarizes the various options of the parametric study undertaken. Two target (actual) earthquake records, very different in nature, have been used thus far. The first record (SMART 1 Array in Taiwan, Event 39, N-S comp., January 1986) is satisfactorily fitted by an ARMA(4,2) model (Conte [3]), whereas the second record (El Centro 1940, N-S comp.) is well fitted by an ARMA(2,1) model (figures 1, 2, 3). For each earthquake model, a population of 100 artificial earthquake ground acceleration time histories is generated. For a given set of structural parameters and for each structural model, the corresponding population of response time histories is obtained by deterministic integration of the equation of motion. The populations of response parameters, computed from the response time histories, are analyzed statistically. The response parameter statistics are presented in the form of histograms (figure 5), plots on probability papers to perform the tests of goodness-of-fit, probabilistic inelastic response spectra (PIRS, figures 6 and 8), and coefficient-of-variation (c.o.v.) versus initial natural period T_0 curves (figure 7).

The comprehensive probabilistic parametric study undertaken offers a large data base concerning the seismic response behavior of simple structural models. Numerous results were found, some of which confirm, in a probabilistic framework, previous results obtained from traditional deterministic methods. The most important results are:

(1) The type of probability distribution of the various response parameters considered appears to be insensitive to (i) the target accelerogram, (ii) the ARMA

model order, (iii) the type of hysteretic model, and (iv) the structural parameters.

(2) The statistical distributions of the inelastic response parameters considered are well approximated by lognormal distributions with the exception of the number of yield excursions and yield reversals which are better approximated by normal distributions. A plastic deformation range is defined as PDR = $\Delta u_p / U_y$ where Δu_p is an increment of plastic deformation and U_y is the deformation at first yield of the system. The lognormal distribution of the PDR's identified using the stochastic approach of this study (figure 5) confirms the result of another statistical study on the distribution of the PDR's based on a sample of real earthquake accelerograms (Lashkari-Irvani [5]). This result is particularly important for cumulative damage analysis (e.g., low cycle fatigue).

(3) The judgement on the relative destructiveness of two target earthquake motions with respect to a given structural model can differ depending on whether it is based on a comparison of the corresponding deterministic inelastic response spectra or on the corresponding probabilistic inelastic spectra. The comparison of the PIRS's provides a more reliable source of information.

(4) In general, the shape of the PIRS curves changes from one target earthquake to another, indicating that the dependence of inelastic response parameters on structural period varies across earthquakes. However, this question should be re-examined in the light of more than two target earthquakes. On the other hand, the general trends of the c.o.v. vs. period curves of the various response parameters are maintained across the two target earthquakes, but this should also be verified using a larger sample of target earthquakes.

(5) For a given probability level and among the different restoring force models investigated, the maximum normalized earthquake input energy and rate of earthquake input energy are lowest for the bilinear inelastic model indicating that the "fatter" the hysteresis loop, the greater the reduction in seismic input energy (figure 6). The c.o.v. of the seismic input energy is also lowest for the bilinear inelastic model (figure 7).

(6) For a fixed confidence level and with other structural parameters fixed, all inelastic response parameters considered in this study except the ratio $E_H/(E_D+E_H)$ are sensitive or extremely sensitive to structural period, to a small amount (few percent) of strain-hardening or strain-softening (P-Δ effect), or to the strength level in the short period range (figure 8). These are general results valid for all force-deformation models examined. They agree with the results of similar studies undertaken in a deterministic framework.

(7) The parameters of the probability distributions characterizing the various response parameters are significantly influenced by the spectral nonstationarity of the earthquake model.

CONCLUSIONS

The present paper describes briefly a methodology developed to investigate the complex interactions existing between earthquake ground motion features, structural properties, and structural response parameters. The proposed methodology accounts realistically for the inherent variability associated with the ground motion time histories corresponding to a specific source-site pair. Discrete-time ARMA models, nonstationary in both amplitude and frequency content, are used to represent the earthquake loading. They are identified from actual earthquake records using Kalman filtering and a moving time window technique. Idealized nonlinear SDOF systems are used as structural models. Using ARMA Monte Carlo simulation, a comprehensive parametric study is undertaken to investigate the effects of the stochastic earthquake input, the type of structural behavior, and the structural parameters on the stochastic seismic response. Through the results obtained, more insight is gained into the stochastic nature of the seismic response of simple structures.

REFERENCES

1. Box, G. E. P. and Jenkins, G. M., Time Series Analysis: Forecasting and Control, Holden-Day, Revised Edition, San Francisco, 1976.

2. Conte, J. P. Influence of the Earthquake Ground Motion Process and Structural Properties on Response Characteristics of Simple Structures, Report No. UCB/EERC-90/08, Earthquake Engineering Research Center, University of California, Berkeley, 1990.

3. Conte, J. P., Pister, K. S. and Mahin, S. A. Variability of Structural Damage Indices within an Earthquake, Proceedings of the 4th U.S. Natl. Conf. on Earthquake Engineering, Palm Springs, USA, 1990.

4. Chang, M. K., Kwiatkowski, J. W. and Nau, R. F., ARMA Models for Earthquake Ground Motions, Journal of Earthquake Engineering and Structural Dynamics, Vol. 10, pp. 651-662, 1982.

5. Lashkari-Irvani, B., "Cumulative Damage Parameters for Bilinear Systems Subjected to Seismic Excitations," Ph.D. Thesis, Stanford University, Stanford, CA, August 1983.

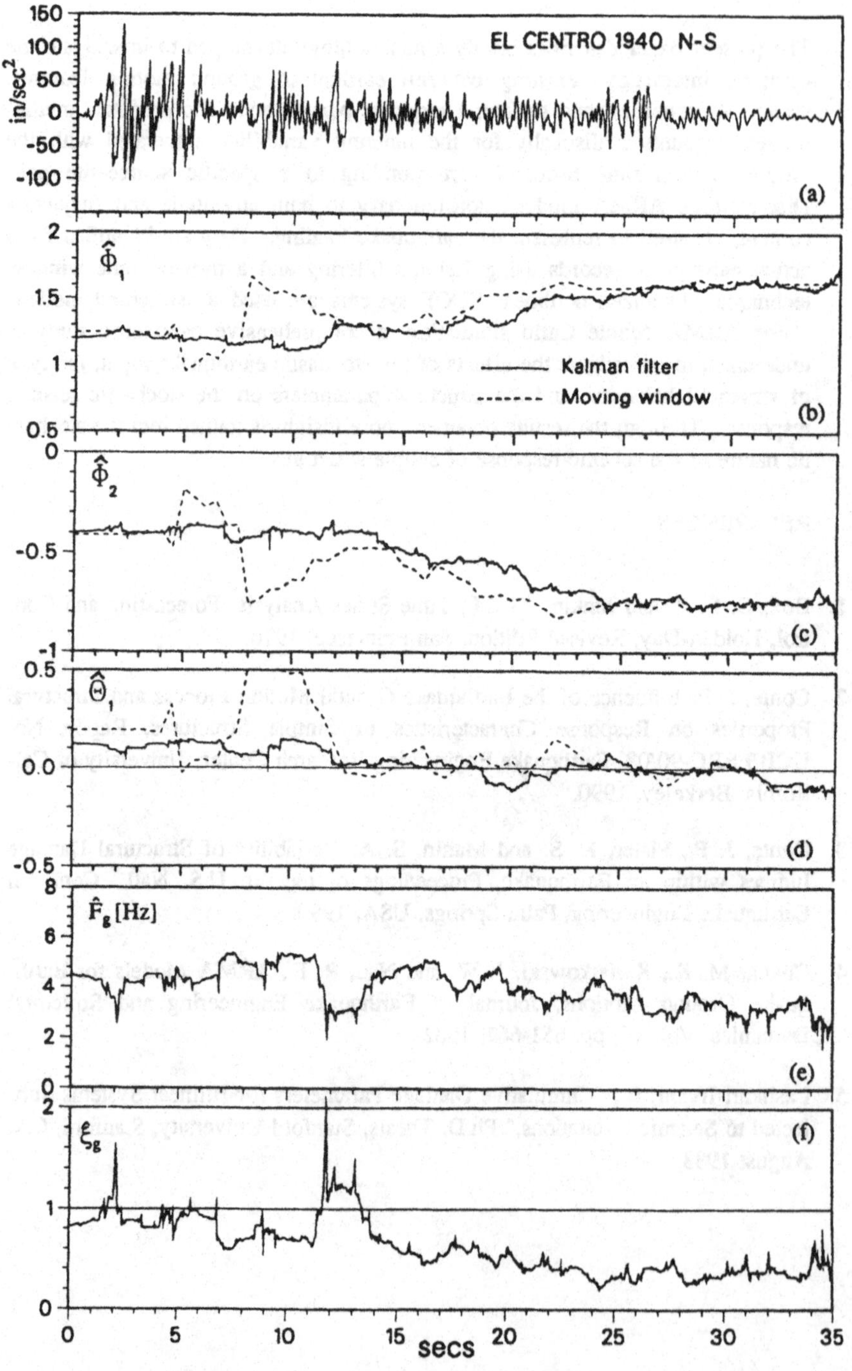

Figure 1. Time-dependent ARMA(2,1) estimation results.

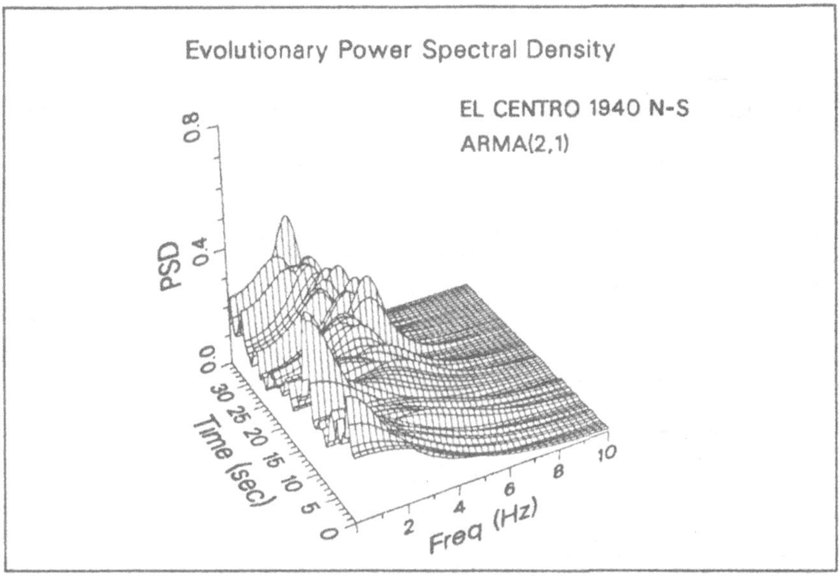

Figure 2. Normalized ARMA(2,1) evolutionary power spectral density estimate.

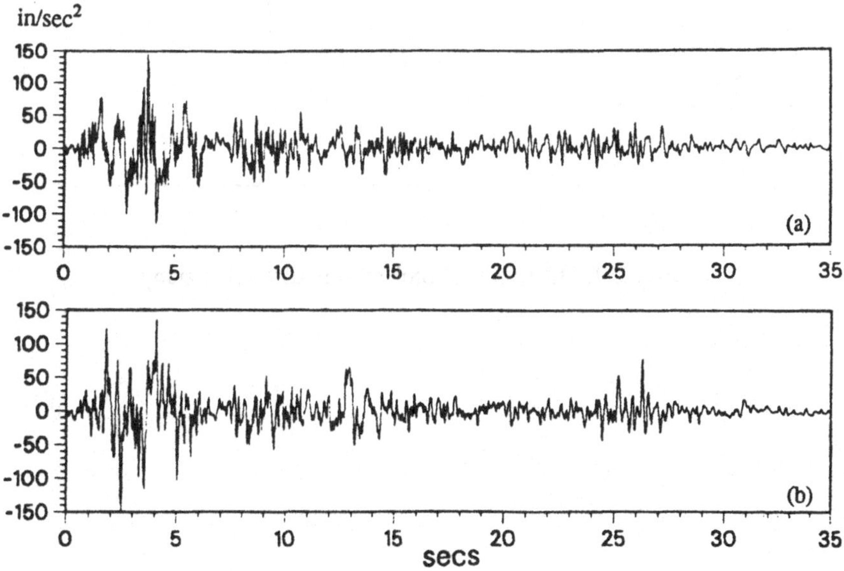

Figure 3. ARMA(2,1) simulated earthquake ground acceleration time histories for the El Centro 1940 target record.

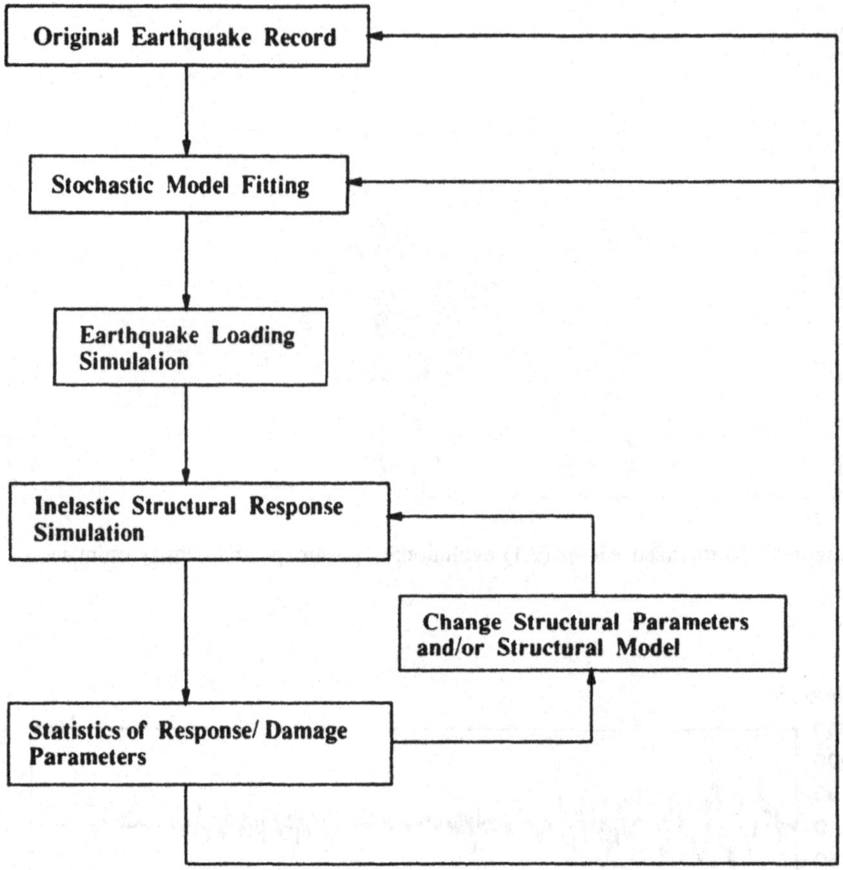

Figure 4. Flow chart of probabilistic parametric study.

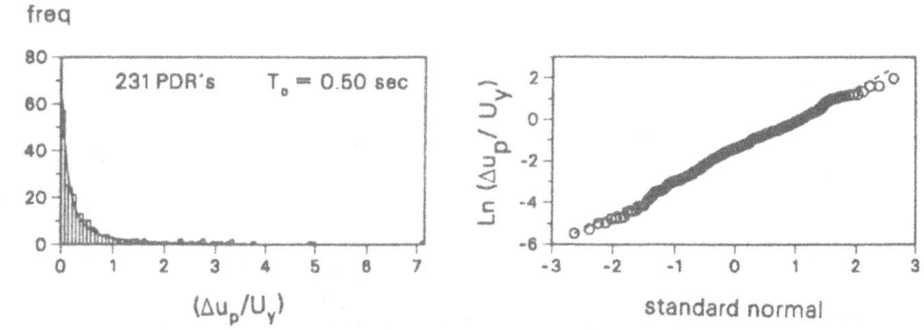

Figure 5. Lognormal distribution of normalized plastic deformation ranges (ARMA(4,2) simulation of SMART 1 - Event 39 and bilinear inelastic structural model with $\xi_0 = 0.05$, $\eta = 0.10$, and $\alpha = 0.00$)

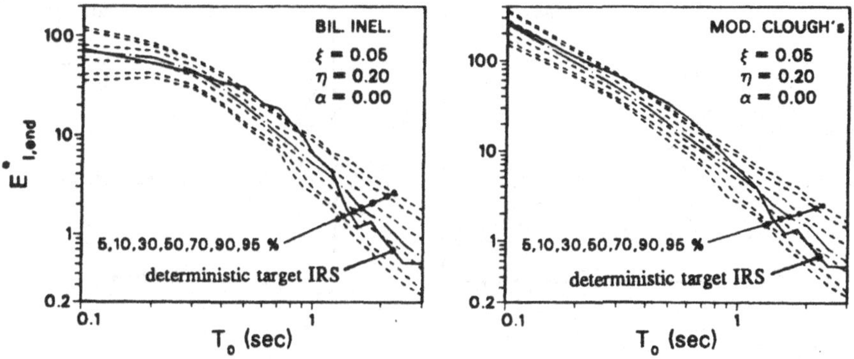

Figure 6. Probabilistic normalized input energy spectra corresponding to the ARMA(2,1) model of El Centro.

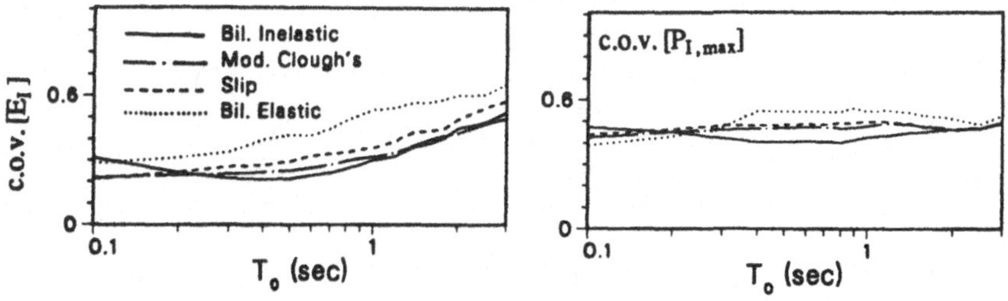

Figure 7. C.O.V. of input energy and peak rate of input energy for different structural models.

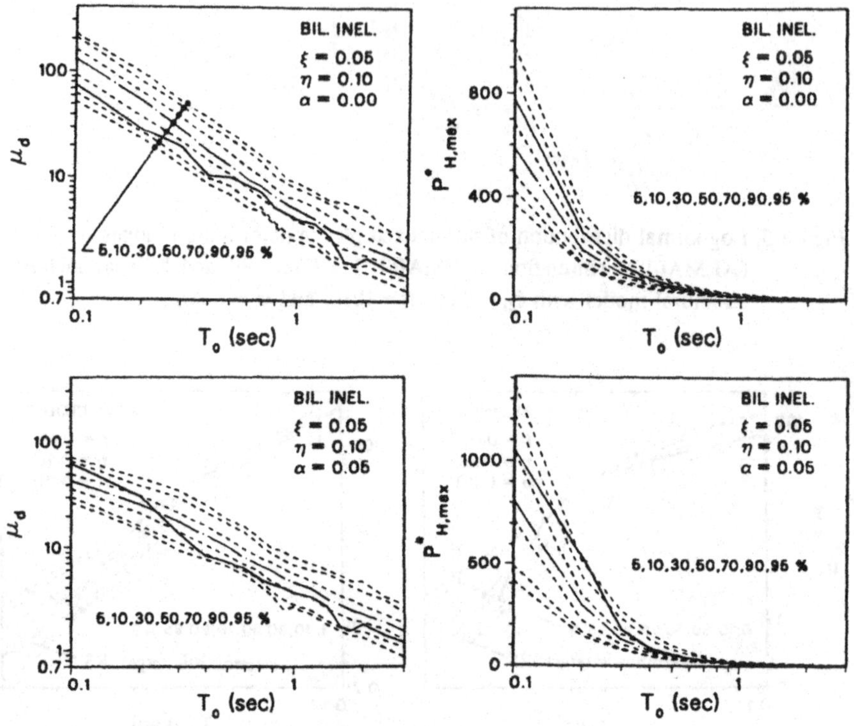

Figure 8. Influence of strain-hardening on maximun displacement ductility and normalized peak rate of hysteretic energy dissipated.

Applications of Stochastic-Adaptive Filters in Earthquake Engineering

E. Safak

U.S. Geological Survey, MS-922, Reston, VA 22092, U.S.A.

ABSTRACT

Stochastic-adaptive filters provide excellent tools to analyze nonstationary and noisy signals, because of their superior time-tracking and noise handling abilities. This paper gives a brief summary of the theory of stochastic-adaptive filtering, and presents three applications in earthquake engineering. The applications are on modeling earthquake ground motions, site amplification, and separation of translational and torsional motions in buildings. An example for each application is presented by using real-life data.

INTRODUCTION

Stochastic-adaptive filters are used to extract nonstationary signals from noisy recordings. Stochastic-adaptive filtering is a digital estimation problem, and involves identifying the parameters of the best filter that passes the signal while suppressing the noise. Because the filtering is equivalent to an identification problem, such filters provide excellent tools to model nonstationary signals and systems.

In this paper we briefly summarize the concept of stochastic-adaptive filtering, and present three applications in earthquake engineering. The applications include time-varying models for earthquake ground motions, estimation of site characteristics, and separation of torsional and translational components of building vibrations.

THEORY

We start with a general input-output relationship in a dynamic system. Let $x(t)$ and $y(t)$ denote the recorded input and output of a dynamic system. Under general conditions we can express the relationship between $x(t)$ and $y(t)$ by the following equation:

$$y(t) = H(q,t)x(t) + G(q,t)e(t) \tag{1}$$

$H(q,t)$ and $G(q,t)$ are time-varying filters representing the dynamics of the system and noise, respectively, and $e(t)$ denotes a white-noise process. For discrete-time recordings, the filters can be represented in terms of rational polynomials of the time-shift operator q, where $q^{-j}x(t) = x(t-j)$.

Stochastic-adaptive filtering tries to find the best estimates of $H(q,t)$ and $G(q,t)$ for given $y(t)$ and $x(t)$. For this, we first isolate the unknown noise term by re-writing Equation 1 in the following form:

$$y(t) = [1 - G^{-1}(q,t)]y(t) + H(q,t)G^{-1}(q,t)x(t) + e(t) \tag{2}$$

where G^{-1} denotes the inverse filter for G. It can be shown that the right-hand side of Equation 2 includes terms up to $y(t-1)$ only [4,5]. Therefore, if we know all the filter terms up to time $t-1$, we can estimate $y(t)$ by taking the expectation of the right-hand side. That is,

$$\hat{y}(t) = [1 - G^{-1}(q,t)]y(t) + H(q,t)G^{-1}(q,t)x(t) \tag{3}$$

where $\hat{y}(t)$ is the expected value of $y(t)$, and the expected value of $e(t)$ is zero. The error, $\epsilon(t)$, in the estimation of $y(t)$ is

$$\epsilon(t) = y(t) - \hat{y}(t) \tag{4}$$

and the total error, $V(t)$, up to time t

$$V(t) = \sum_{s=1}^{t} \beta(t,s)\epsilon^2(s) \tag{5}$$

where $\beta(t,s)$ denote the weighting factors. We need the weighting factors to track the time variations in the system. They are selected such that more weight is given to the current values while gradually discounting the past values. Filters H and G are determined by minimizing V, which can be accomplished recursively by using stochastic approximation techniques. Details of the minimization can be found elsewhere [4]. The filters H and G describe all the characteristics of the system and noise.

MODELING EARTHQUAKE GROUND MOTIONS

Earthquake ground motions are nonstationary both in time and frequency. We can model ground motions as filtered white-noise sequences, where the filters are time-varying. If $y(t)$ denotes the recorded ground motion and $e(t)$ a white-noise sequence with unit variance, we can write

$$y(t) = \frac{1}{A(q,t)} e(t) \tag{6}$$

where

$$A(q,t) = 1 + a_1(t)q^{-1} + \cdots + a_m(t)q^{-m} \tag{7}$$

Equation 6 is the same as Equation 1 with $H = 0$ and $G = 1/A$. For a given earthquake recording, the time-varying parameters, $a_j(t)$, of the filter are determined by using the method described above. The filter order m is selected by observing the variation of $V(t)$ for different values of m [4]. The time-varying spectrum of the record is calculated by evaluating the filter for $q = e^{-i(2\pi f)\Delta}$, where $i = \sqrt{-1}$, f is the cyclic frequency, and Δ is the sampling interval. The equation for the amplitude spectrum, $S_y(f,t)$, is

$$S_y(f,t) = \frac{1}{|1 + a_1(t)e^{-i(2\pi f)\Delta} + \cdots + a_m(t)e^{-im(2\pi f)\Delta}|} \tag{8}$$

To give an example, we will model the ground acceleration given in Figure 1a. The record is the north-south component of the horizontal accelerations recorded at the basement of the Transamerica Building in San Francisco by the U. S. Geological Survey during the magnitude 7.1, October 17, 1989 Loma Prieta earthquake [1]. A 6th-order filter is used for modeling. Time variation of estimated filter parameters are given in Figure 1b. As the figure indicates, the characteristics of the motion vary with time. Figure 1a shows the comparison of the model output with recorded acceleration, and Figure 1c shows the time-varying amplitude spectrum of the record.

SITE AMPLIFICATION

Soft soil layers cause significant increase in the amplitudes of earthquake ground motions. This phenomenon is known as site amplification, and is an important factor in seismic design. The standard method to model site amplification is to use spectral ratios, the ratio of the Fourier amplitude spectrum of the soil-site recording to that of the nearby rock-site recording. Spectral ratios are very sensitive to noise in the records, and are not applicable to nonstationary signals.

We can use the adaptive filtering approach to model site amplification. If $x(t)$ and $y(t)$ denote the rock-site and soil-site recordings, respectively, and $H(q,t)$ the filter representing the site, we can state the site amplification problem by the following equation:

$$y(t) = H(q,t)x(t) + e(t) \tag{9}$$

where it is assumed that the noise is a white-noise, i.e., G=1 in Equation 1. The form selected for $H(q,t)$ is

$$H(q,t) = \frac{B(q,t)}{A(q,t)} = \frac{b_1(t)q^{-1} + \cdots + b_n(t)q^{-n}}{1 + a_1(t)q^{-1} + \cdots + a_m(t)q^{-m}} \tag{10}$$

The parameters $a_j(t)$ and $b_j(t)$ define the filter for site amplification.

To give an example, we consider the transverse (with respect to the epicenter) components of the accelerations, given in Figure 2a, recorded at the Piedmont Junior High School (a rock site) and Oakland Outer Harbor Wharf (a soil site) during the Loma Prieta earthquake [2]. Following the guidelines given in reference 3, we determine the model orders as $m = 30$ and $n = 10$. Time variation of estimated model parameters are plotted in Figure 2b and c. The site amplification is represented by the amplitude of the estimated filter evaluted at $q = e^{-i(2\pi f)\Delta}$, i.e., $|H(e^{-i(2\pi f)\Delta}, t)|$. This value is plotted in Figure 2d. As the figure shows, site amplification has time-varying characteristics.

TORSIONAL AND TRANSLATIONAL MOTIONS IN BUILDINGS

Another application of adaptive filtering techniques is the separation of torsional and translational components of buildings vibrations. This involves first determining the dynamic center of rigidity at a floor level, defined as the point with no torsional component. To formulate the problem, consider the schematic floor plan with three sensors shown in Figure 3. Let $\theta(t)$ be the rotational motion calculated as the difference of two parallel horizontal recordings divided by the distance between the sensors, i.e., $\theta(t) = [u_2(t) - u_1(t)]/d$. If $x_1(t)$ denotes the pure translational part of $u_1(t)$, and c_1 the perpendicular distance from $u_1(t)$ to the center of rigidity, we can write the following equation:

$$u_1(t) = c_1\theta(t) + x_1(t) \tag{11}$$

In this equation, $u_1(t)$ and $\theta(t)$ are known, $x_1(t)$ and c_1 are unknown. It can be shown that as long as torsional and translational frequencies are separated (i.e., $x_1(t)$ and $\theta(t)$ do not have identical dominant frequencies) the problem becomes identical to an adaptive filtering problem, with $u_1(t)$, c_1, and $x_1(t)$ in Equation 11 corresponding to $y(t)$, $H(q,t)$ and $G(q,t)e(t)$ in Equation 1, respectively [6]. We determine c_1 and its time-variation similarly as explained above. If we assume that c_1 is time invariant, the problem becomes much simpler. In this case, c_1 is calculated by the following equation [6]:

$$c_1 = \frac{\sum_{t=1}^{N} u_1^2(t)\theta(t)}{\sum_{t=1}^{N} \theta^2(t)} \tag{12}$$

where N denotes the total number of points in the series. The c_2 and c_3 distances, in Figure 4, can be calculated similarly by replacing u_1 with u_2 and u_3, and observing the right-hand rule sign convention. To reduce the effects of noise and higher modes, it is recommended that the displacement records be used for u_j's and θ, and θ be band-pass filtered around the dominant torsional frequency.

We will give an example for the method by using earthquake recordings from the top floor of the 32-story Wilshire Finance Building in Los Angeles [3]. The building has a rectangular cross-section for the first 12 stories and a triangular one thereafter, and exhibits a large amount of torsion. The cross-section at the top and the sensor layout are shown in Figure 4. We calculate the dynamic center of rigidity by assuming that its location is time invariant. The location of the estimated center of rigidity is also shown, by G, in Figure 4. Recorded displacements and their calculated translational and rotational components are plotted in Figure 5. The figure shows large torsional components for sensor 2.

CONCLUSIONS

Three applications of stochastic-adaptive filtering techniques in earthquake engineering are presented. Stochastic-adaptive filters provide excellent tools to analyze nonstationary and noise signals, because of their superior time-tracking and noise handling abilities. In this paper, stochastic-adaptive filters are used to model earthquake ground motions and site amplification, and to separate translational and torsional motions in buildings. An example for each application is presented by using real-life data.

REFERENCES

1. Brady, A.G. and Mork, P. Loma Prieta, California, Earthquake, October 18 (GMT) 1989, processed strong motion records, volume 1, ***Open File Report 90-247***, U.S. Geological Survey, Menlo Park, CA., 1990.
2. CDMG. The Loma Prieta (Santa Cruz Mountains), California Earthquake of 17 October 1989, ***Special Report 104***, California Division of Mines and Geology (CDMG), Sacramento, California, 1991.
3. Çelebi, M., Şafak, E. and Youssef, N. Torsional response of a unique building, ***Journal of Structural Engineering***, ASCE, May 1991.
4. Şafak, E. Analysis of recordings in structural engineering: Adaptive filtering, prediction, and control, ***Open-File Report 647***, U. S. Geological Survey, Menlo Park, California, 1988.
5. Şafak, E. Adaptive modeling, identification, and control of dynamic structural systems: Part I and Part II, ***Journal of Engineering Mechanics***, ASCE, November, 1989.
6. Şafak, E,. and Çelebi, M. New techniques in record analysis: Torsional vibrations, ***proceedings***, Fourth U. S. National Conference on Earthquake Engineering, Palm Springs, California, May 20-24, 1990.

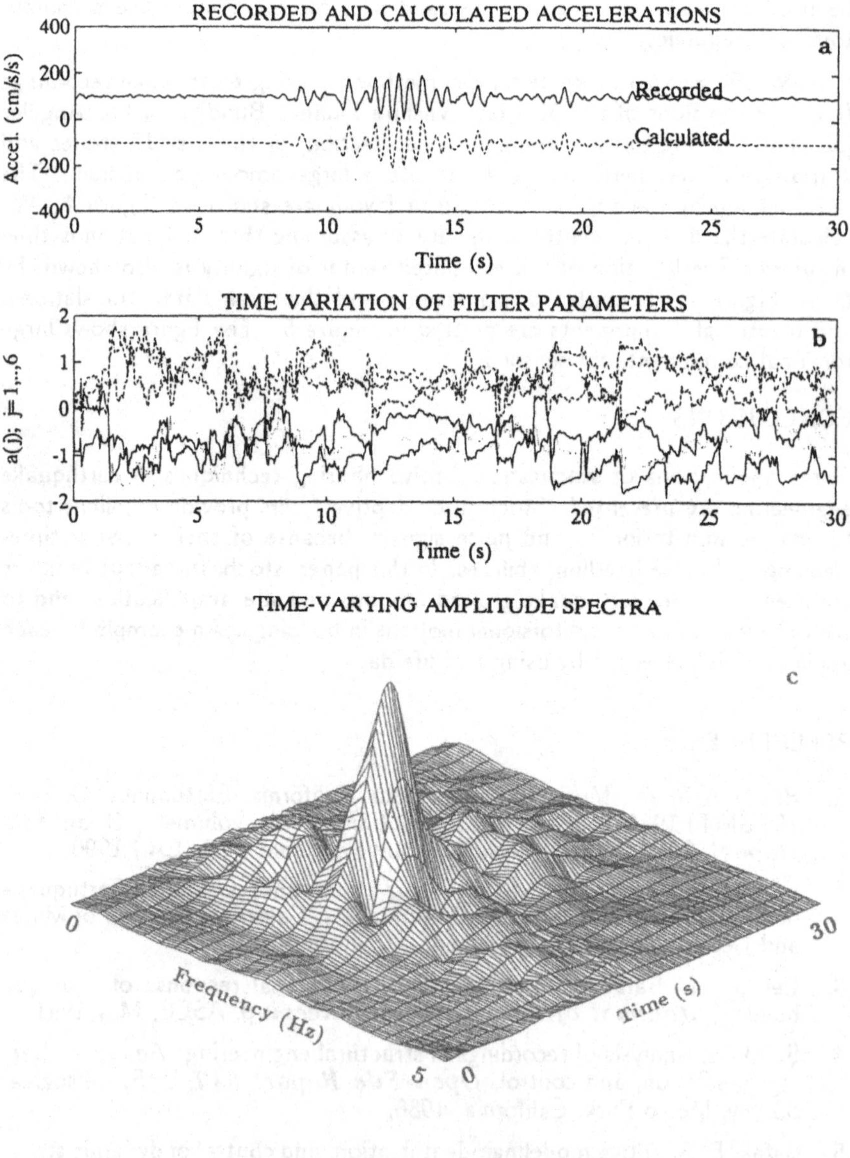

Figure 1. Modeling ground motions by time-varying filters: (a) recorded and calculated accelerations, (b) time variation of filter parameters, and (c) time-varying amplitude spectra.

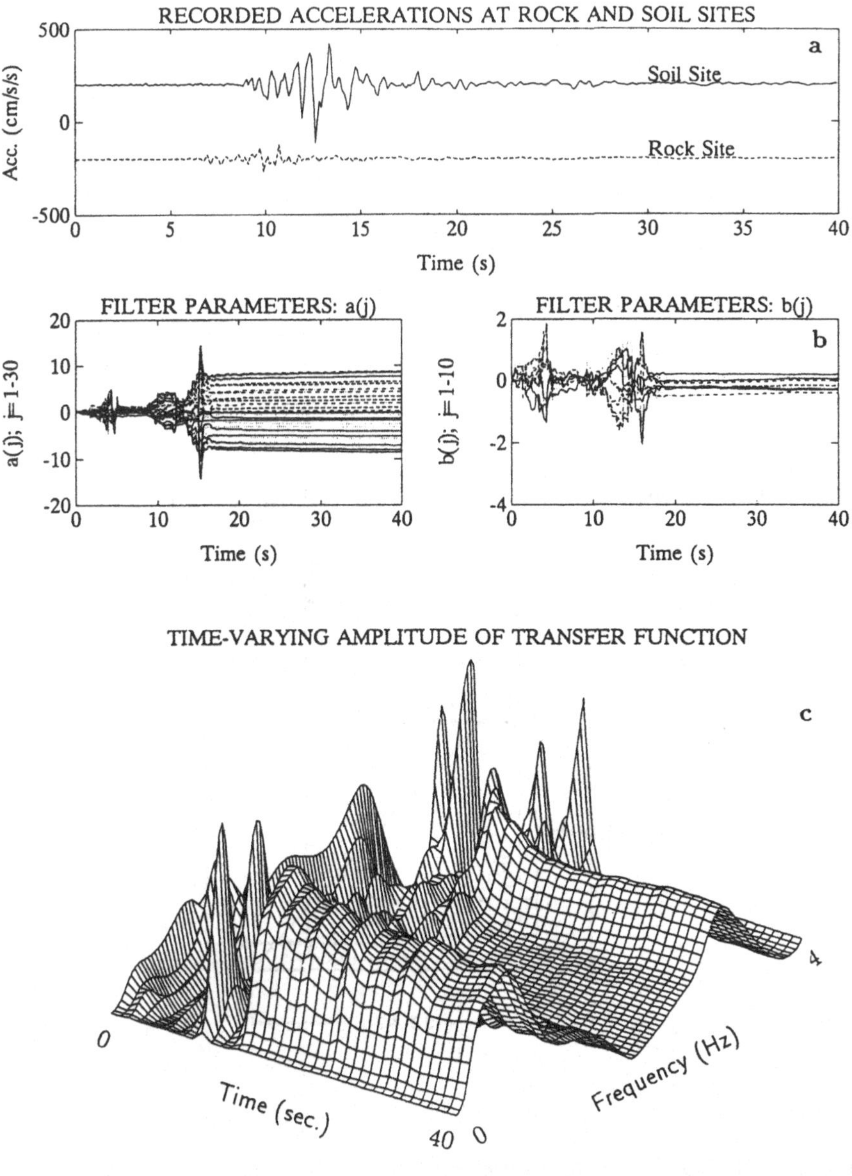

Figure 2. Modeling site amplification by time-varying filters: (a) recorded accelerations at rock and soil sites, (b) time variation of filter parameters, and (c) time-varying amplitude of site amplification.

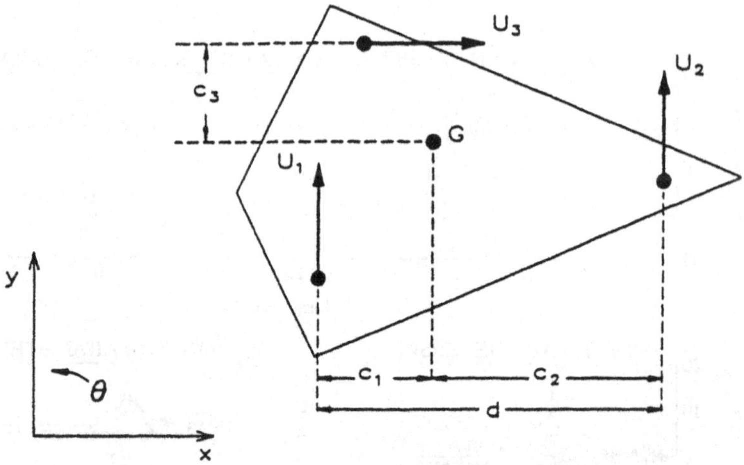

Figure 3. Schematic floor plan of a building with three sensors.

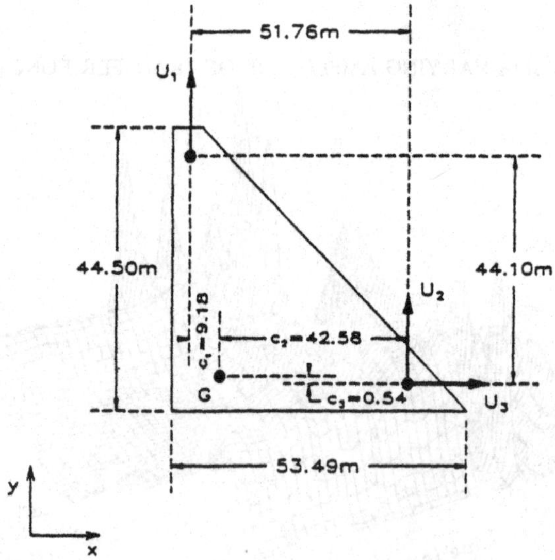

Figure 4. Cross-section and sensor layout at the 32nd floor of Wiltshire Finance Building, and the estimated location (G) of the dynamic center of rigidity.

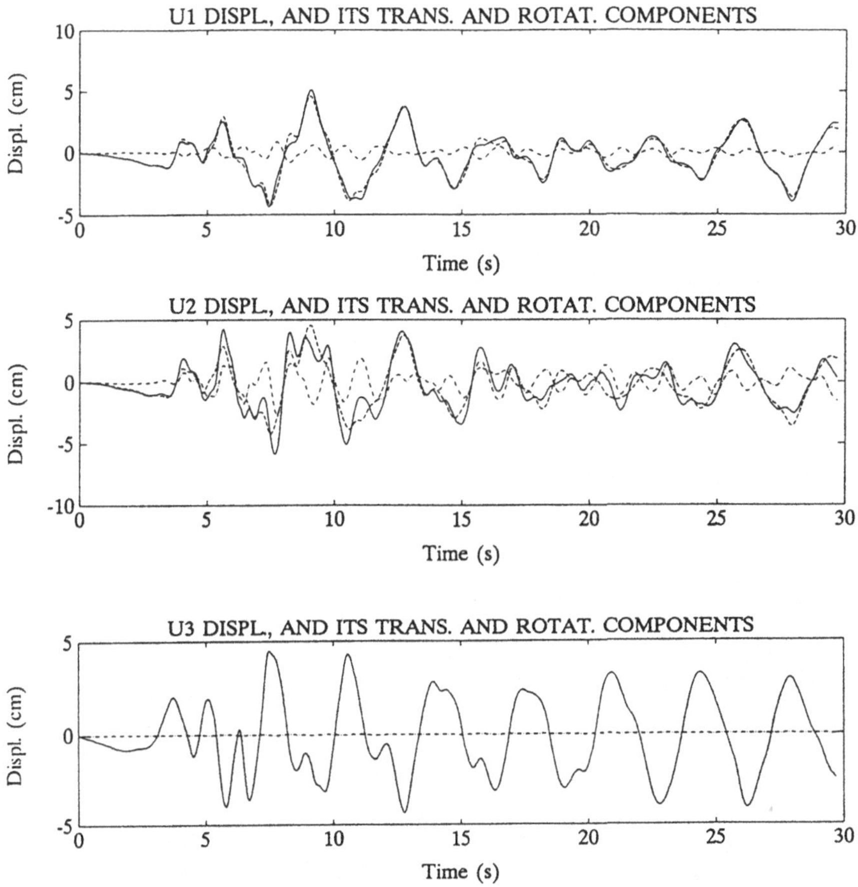

Figure 5. Recorded displacements (solid lines), and their calculated translational (dashed lines) and rotational (dashed-dotted lines) components.

Study of the Propagation and Amplification of Seismic Waves in Caracas Valley with Reference to the July 29, 1967 Earthquake Response

A.S. Papageorgiou, J. Kim

Department of Civil Engineering, Rensselaer Polytechnic Institute, Troy, NY 12180-3590, U.S.A.

ABSTRACT

The response of a 2–D model of the valley of Caracas, Venezuela – a NS cross–section through the Palos Grandes district – to seismic excitations is investigated. It is found that the spectral amplification and time response characteristics of this asymmetric wedge–shaped valley are a function of the direction of incidence of the excitation waves. Furthermore, the differences in the predictions of the 1–D and 2–D models are more pronounced for SV–waves than for SH–waves, especially when SV–waves are incident at (or near) the critical angle i_c. The 2–D model of the valley is used to simulate also the anti–plane response of the valley to the July 29, 1967 Caracas earthquake. The simulations reveal that the intensity of motion on rock and on the sediments is higher by a factor of 3 compared to estimates proposed by previous investigators.

INTRODUCTION

Geologic conditions and topography at or near a site are known to exert a very significant influence on the nature of ground shaking, and their importance on seismic hazard has long been recognized (Milne, 1898; Gutenberg, 1957; Borcherdt, 1970; for recent reviews of the subject see Sanchez–Sesma, 1987 and Aki, 1988).

In particular, sedimentary deposits in the form of sediment–filled valleys or basins very often have a pronounced effect on the intensity of strong ground motion. The finite lateral extent of the sedimentary deposit introduces complex effects through generation of surface waves at the edges and resonance in the lateral direction, and tends to increase both the amplitude and the duration of ground motion (Bard, 1983).

There are several well–documented examples of valleys/basins that were shaken by earthquakes where such effects were clearly observed: The response of La Molina sediment–filled valley to the Lima, Peru earthquake of November 9, 1974 (Zahradnik and Hron, 1987); the response of the deep sedimentary basins of the San Fernando and the Los Angeles areas (Hanks, 1975; Liu and Heaton, 1984; Vidale and Helmberger, 1988); the response of Shirak valley to the December 7, 1988, Armenia, earthquake (Borcherdt et al., 1989); the response of the Mexico City valley to the September 19, 1985 Michoacan earthquake (Bard et al., 1988, Chavez–Garcia and Bard, 1989; Campillo et al., 1990); the response of the Duwamish River valley to the April 29, 1965 Puget Sound earthquake (Mullineaux et al., 1967; Langston and Lee, 1983).

Finally, a case in which the irregularities of the subsurface topography appear to have influenced significantly the motion of the overlying sediments, is the response of the Caracas Valley to the July 29, 1967 earthquake. This event caused extreme damage in the city of Caracas: four buildings collapsed and about three hundred people were killed. A significant feature of the earthquake was the concentration of damage to certain types of structures in localized areas. Specifically, the four buildings which collapsed during this earthquake – as well as all buildings with more than 12 stories which suffered structural damage – were located in the Palos Grandes suburb of east Caracas, while damage to low structures in this same suburb was relatively minor (Whitman, 1969; Seed et al., 1972). Severe earthquake effects in the Palos Grandes district, which may explain this damage concentration in 1967, are associated with site characteristics and they should be expected to repeat in the future. The 1967 earthquake experience in Caracas motivated us to study the site effects in the Palos Grandes area, to try to reconstruct the response of the valley in the aforementioned district to the 1967 event and to compare the simulated response vis–a–vis the documented damage distribution (Papageorgiou and Kim, 1991a). Our effort was particularly facilitated by a recent study (Suárez and Nábelek, 1990) which synthesized evidence from the damage and intensity reports, the epicentral location of the main shock, the distribution of aftershocks, a detailed study of the teleseismic body waves using a formal least–squares inversion algorithm and geologic information, to characterize the time history and geometry of the rupture process of the 1967 Caracas event.

The method of analysis which we employed to study the response of a two–dimensional model of the valley is the frequency domain Boundary Element Method (e.g. Beskos, 1987). For anti–plane motions we used the method as discussed by Wong and Jennings (1975), while for in–plane motions we used the Discrete Wavenumber Boundary Element Method introduced by Kawase (1988) and further refined by Kim and Papageorgiou (1991). The obvious advantages of the method are efficiency, accuracy in the computation and flexibility in the specification of boundary configurations.

RESPONSE TO INCIDENT PLANE WAVES

In order to study the response of the valley in the Palos Grandes district to seismic inputs and in particular to the 1967 event, we adopted a two–dimensional model of a NS cross–section of the valley. The model is based on data obtained from the geophysical seismic surveys (Murphy,1969; Presidential Commission Report, 1978). We analyzed two variations of this 2–D cross–section: (1) Model A, consisting of a uniform sedimentary deposit with no material damping and (2) Model B, consisting of a uniform sedimentary deposit, with a thin softer sedimentary layer at the top, with material damping. For both models, we considered the sediments to be embedded in a uniform half–space with no damping. The material properties of models A and B are listed in Table 1. The material damping (Q_s factor) and shear wave velocities of the sediments in Model B were selected so as to be consistent with estimates of the level of shear strains induced in the 1967 earthquake (Whitman, 1969).

TABLE 1

PHYSICAL PARAMETERS OF CARACAS VALLEY MODELS

		ρ (gm/cm³)	α(m/sec)	β(m/sec)	$Q_s[\xi_s]$*
MODEL A	sediment	1.8	1775	1025	∞ [0.0]
	half–space	2.2	4000	2300	∞ [0.0]
MODEL B	thin, soft† cover layer	1.7	650	310	12 [0.042]
	sediment	1.8	1700	950	12 [0.042]
	half–space	2.2	4000	2300	∞ [0.0]

*$\xi_s[= 1/(2Q_s)]$ = material damping ratio
†The thickness of the soft cover layer varies from 5 m to 35 m

The response of the valley to anti–plane motions (i.e. plane SH–waves and anti–plane sources has been studied by Papageorgiou and Kim (1991a), while its response to in–plane motions (i.e. plane P– and SV–waves, and in–plane sources) has been elaborated in Papageorgiou and Kim (1991 b). Here we summarize results related to the response of the valley only to incident plane waves.

(i) *Amplification Ratios*: Amplification ratio (AR) at a site in the valley is defined to be the ratio of the amplitude of steady–state motion at the site (i.e. free surface of sediment) to the amplitude of motion which would be recorded at the free surface of the half–space that represents the basement rock if the valley did not exist. This definition of amplification is also known as *elastic rock amplification* (Roesset, 1970). Examples of AR's are shown in Figures 1a,b.

The following observations can be made with regard to AR's as predicted by 1–D and 2–D models (Papageorgiou and Kim, 1991b):

- For *vertical incidence* ($i = 0°$), the predictions of the 1–D model compare favorably with the predictions of the 2–D model as far as the overall structure of the amplification ratio vs. frequency across the valley is concerned. However, the 1–D model underpredicts the AR of the 2–D model because it does not account for the lateral interferences. Furthermore, the responses of the 2–D model to vertically propagating SV– and SH–waves exhibit important differences when compared to each other, while for the 1–D model there is no distinction between SV– and SH–waves when $i = 0°$.

- *Oblique incidence; SH–waves*: As the direction of incidence of SH–waves deviates from the vertical direction, the 1–D model provides estimates of site period which are in reasonably good agreement with the predictions of the 2–D model. However the difference in the values of AR's of the two models increases with increasing angle of incidence.

- *Oblique incidence; SV–waves*: The differences in the predictions of the two models (i.e. 1–D vs 2–D) are more pronounced for SV–waves than for SH–waves, especially when SV–waves are incident at (or near) the critical angle i_c ($\sin i_c = \beta_r/\alpha_r$, where β_r, α_r are the S– and P–wave valocities respectively of the half space). This implies that consideration of SH–waves only, even if the model of analysis is 2–D, may not be sufficient for reliable microzonation studies. This conclusion is supported also by the comparative study of observations with numerical simulations presented by Jongmans and Campillo (1991) who point out that 2D–SH modeling alone is not sufficient to explain recorded data.

(ii) Time response to incident plane waves: Next we briefly discuss some important features of the time response of the valley to incident plane waves. The time dependence of the input signal is described by the Ricker wavelet (Ricker, 1945).

Figure 2 shows the velocities du/dt (= in–plane horizontal component) and dw/dt (= vertical component) induced on the surface of the valley by plane SV–waves incident vertically (i.e. $i = 0°$).

As anticipated (Bard 1983) the incident SV–waves induce Rayleigh waves which result from overcritical reflections at the inclined bottom of the sediments and propagate between the two edges of the valley. Two distinct modes of these *locally generated (or valley–induced) Rayleigh waves* can be clearly seen to propagate. One is *the fundamental mode,* with dominant the vertical component of motion, the velocity phase of which is about 0.9 kg/sec. The other is *the first higher mode,* with dominant the horizontal component of motion, which propagates with phase velocity (at frequency $f = 2$ Hz) approximately equal to 1.8 km/sec which is close to the P–wave velocity of the sediment.

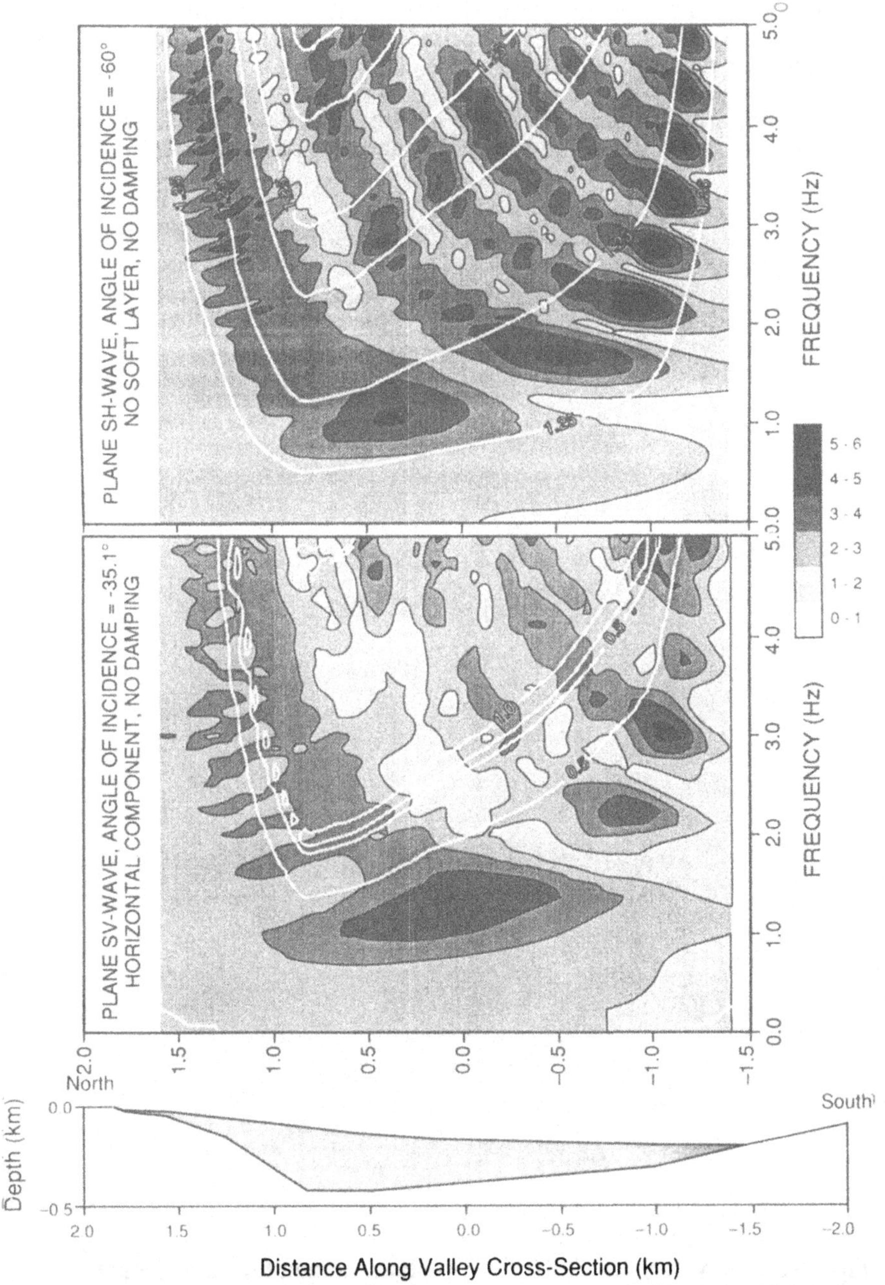

Figure 1. **Amplification ratios (AR) of valley-model A for plane (a) SH and (b) SV waves incident from the north with incidence angles $i = -60°$ and -35.1° respectively. On the AR obtained from the 2-D model are superimposed the AR of the corresponding 1-D model of each site across the valley.**

Considering the case of vertical incidence ($i = 0°$), due to the horizontal particle motion of the incident plane wave, no direct body waves are recorded on the vertical component. Therefore, the vertical component seismograms are comprised of Rayleigh–wave phases and converted or reflected body waves.

Similarly, incident SH waves induce *Love waves* which propagate between the two edges of the valley (Papageorgiou and Kim, 1991a).

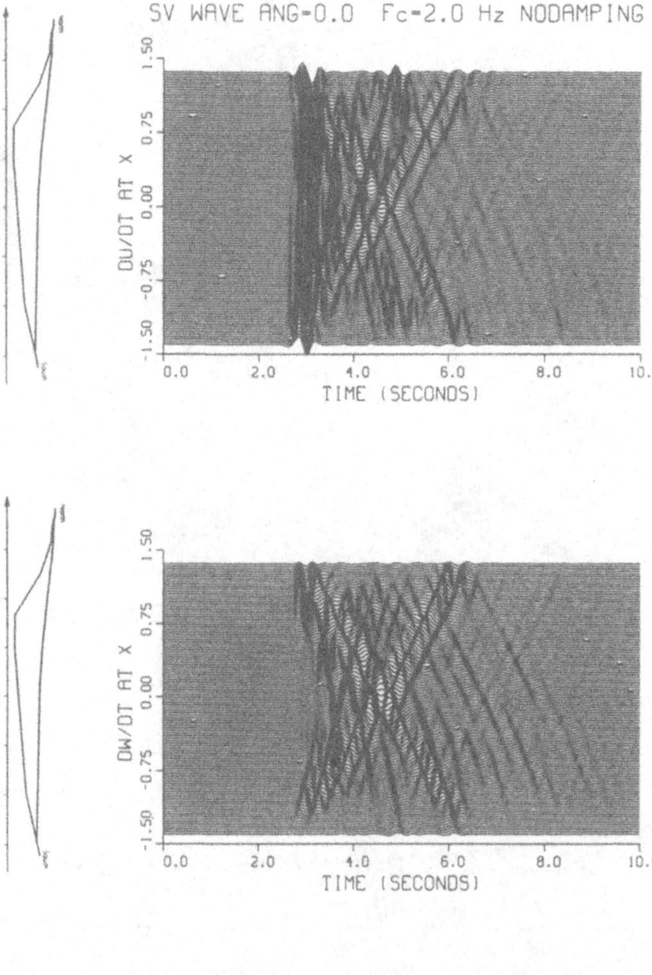

Figure 2. Velocities *du/dt* (= in–plane horizontal component) and *dw/dt* (= vertical component) induced on the surface of the valley (Model A) by plane SV–waves incident vertically (i.e. $i = 0°$).

Finally, as demonstrated by Papageorgiou and Kim (1991b), the valley responds very strongly to the horizontally propagating P–wave (SP–wave) which is induced when SV–waves, incident at the critical angle, interact with the free surface of the half–space.

THE CARACAS EARTHQUAKE OF JULY 29, 1967: SIMULATION OF THE VALLEY RESPONSE

At 8:00 p.m., local time on Saturday night, July 29, 1967, the city of Caracas and several neighboring towns along the Caribbean coast were severely shaken by a moderate–sized earthquake ($M_w = 6.6$), which caused surprisingly heavy damage to multistory buildings in the cities of Caracas and Caraballeda (Figure 3) (Sozen et al., 1968; Whitman, 1969; Hanson and Degenkolb, 1969; Seed et al., 1972).

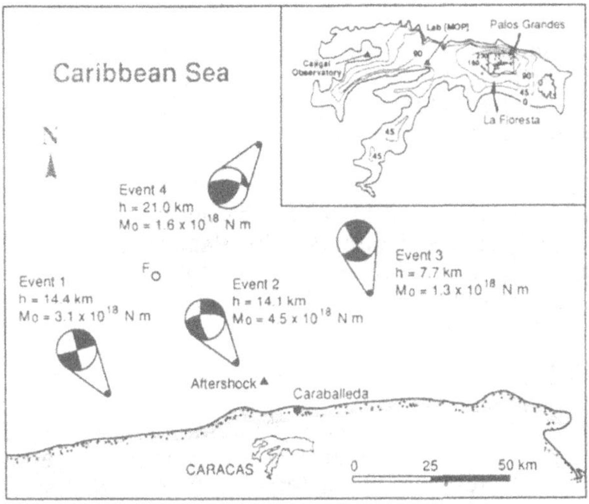

Figure 3. Mechanism and location (solid circles) of the four subevents that form the Caracas earthquake. *Inset:* Map of the valley of Caracas showing depth of alluvium. Dots (·) indicate locations of buildings with more than 6 stories having structural damage.

The fault plane orientation of the 1967 Caracas earthquake had been the source of controversy for several years that was recently resolved by the definitive study of Suárez and Nábelek (1990). These investigators synthesized evidence based on the intensity and damage reports, the distribution of aftershocks, and the results of a joint formal inversion of the teleseismic P and SH waves to show that these data clearly indicate that the rupture of this event occurred on an east–west trending fault system. From the joint inversion of the teleseismic data they inferred that in a time frame of 65 seconds, four distinct bursts of seismic moment release (subevents) occurred with a total seismic moment of 8.6×10^{25} dyn–cm (Figure 3).

No strong motion instrument was operational in the area at the time of the earthquake. Papageorgiou and Kim (1991a), synthesizing all the evidence that is available in the published literature, simulated the anti–plane response of the valley to strong motions equivalent – in amplitude, frequency content and duration – to those that most likely were generated by the July 29, 1967 Caracas earthquake (Shinozuka and Jan, 1972; Boore, 1983). They concluded that the intensity of motions on the floor of the valley is higher at least by a factor of 3 as compared to the estimates of various investigators (Sozen et al., 1968; Whitman, 1969; Seed et al., 1972) who assumed that all the seismic energy was released from a point located at a distance of 55 km from the city of Caracas. In reality, the source that caused the damage in this city was located at a distance of 25 km.

The spectral characteristics of the valley response as described by *spectral acceleration (or spectral velocity)* (Clough and Penzien, 1975) are shown in Figure 4. It is evident that the zone of heavy damage for high rise buildings (no. of stories $N \geq 6$) coincides with the sites where the motions at periods $0.6 \text{ sec} \lesssim T \lesssim 1.4 \text{ sec}$ were strongly amplified (by a factor of 3). This should come as no surprise in view of the amplification characteristics of the valley as described by the AR plot for $i = -60°$ (Figure 1a).

Various pieces of evidence (e.g. seismoscope response recorded on rock) were presented and discussed by Papageorgiou and Kim (1991a) which suggest that the intensity of motion was at least as high as estimated by the simulation.

CONCLUSIONS

The study of the response of the asymmetric, wedge–shaped 2–D model of the Caracas valley that we presented in this paper, revealed that the spectral and time response characteristics of the valley are a function of the direction of incidence of the excitation waves and consequently depend on the position of the seismic source relative to the valley.

From the study of the response of a 2–D model of the cross–section of the Caracas Valley at the Palos Grandes suburb it is concluded that the differences in the predictions of the 1–D and 2–D models are more pronounced for SV–waves than for SH–waves, especially when SV–waves are incident at (or near) the critical angle i_c. The valley responds very strongly to the horizontally propagating P–wave (SP–wave) which is induced when SV–waves, incident at the critical angle, interact with the free surface of the half–space.

The simulated anti–plane response of the valley to strong motions equivalent – in amplitude, frequency content and duration – to those that most likely were generated by the July 29, 1967 Caracas earthquake, revealed that the intensity of motions on the floor of the valley was higher at least by a factor of 3 as compared to the estimates of various investigators (Sozen et al., 1968; Whitman, 1969; Seed et al.,

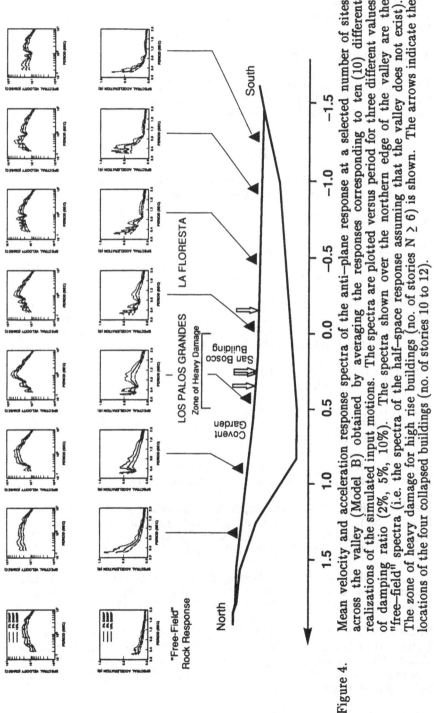

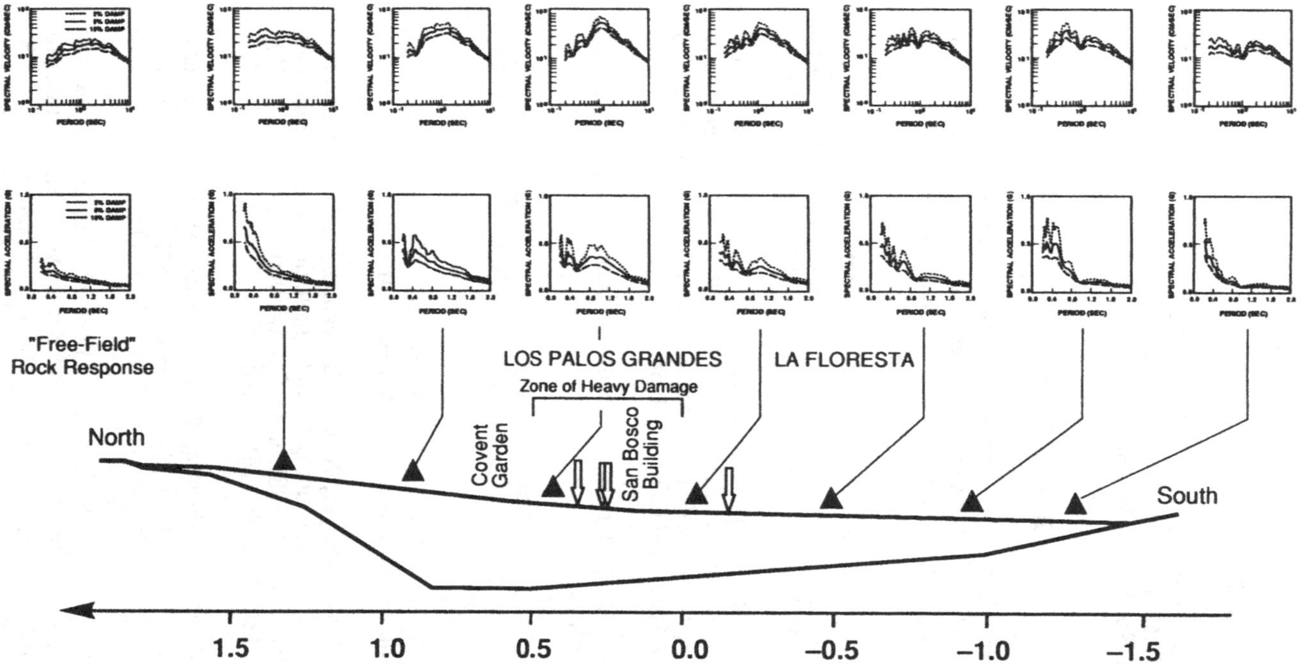

Figure 4. Mean velocity and acceleration response spectra of the anti–plane response at a selected number of sites across the valley (Model B) obtained by averaging the responses corresponding to ten (10) different realizations of the simulated input motions. The spectra are plotted versus period for three different values of damping ratio (2%, 5%, 10%). The spectra shown over the northern edge of the valley are the "free–field" spectra (i.e. the spectra of the half–space response assuming that the valley does not exist). The zone of heavy damage for high rise buildings (no. of stories $N \geq 6$) is shown. The arrows indicate the locations of the four collapsed buildings (no. of stories 10 to 12).

1972) who assumed that all the seismic energy was released from a point located at a distance of 55 km from the city of Caracas. In reality, the source that caused the damage in this city was located at a distance of 25 km.

ACKNOWLEDGMENTS

This work was supported by Contract No. NCEER 89–1305 under the auspices of the National Center for Earthquake Engineering Research under NSF Grant No. ECE–86–07591.

REFERENCES

Aki, K., Local site effect on ground motion, in *Earthquake Engineering and Soil Dynamics II – Recent advances in ground–motion evaluation*, J. Lawrence Von Thun, Editor, ASCE, Geotechnical Special Publication, 20:103–155, 1988.

Bard, P.Y., Les effets de site d'origine structurale en sismologie, modelisation et interpretation, application au risque sismique, *These d'Etat*, Universite Scientifique et Medicale de Grenoble, France, 1983.

Bard, P.Y., M. Campillo, F.J. Chaves–Garcia and F.J. Sanchez–Sesma, The Mexico earthquake of September 19, 1985 – A theoretical investigation of large and small amplification effects in the Mexico City Valley, *Earthquake Spectra*, 4:609–633, 1988.

Beskos, D.E., Boundary element methods in dynamic analysis, *Appl. Mech. Rev.*, 40:1–23, 1987.

Boore, D.M., Stochastic simulation of high–frequency ground motions based on seismological models of the radiated spectra, *Bull. Seism. Soc. Am.*, 73:1865–1894, 1983.

Borcherdt, R.D., Effects of local geology on ground motion near San Francisco Bay, *Bull. Seism. Soc. Am.*, 60:29–61, 1970.

Borcherdt, R.D., G. Glassmoyer, A. Der Kiureghian, and E. Cranswick,, Results and data from seismologic and geologic studies following earthquakes of December 7, 1988 near Spitak, Armenia, S.S.R., *U.S. Geol. Surv. Open–File Rept. 89–163A*, 1989.

Campillo, M., F.J. Sanchez–Sesma, K. Aki, Influence of small lateral variations of a soft surficial layer on seismic ground motion, *Soil Dynamics and Earthquake Engineering*, 9:284–287, 1990.

Chavez–Garcia, F.J. and P.Y. Bard, Effect of Random Thickness Variations on the Seismic Response of a Soft Soil Layer: Applications to Mexico City, in *Engineering Seismology and Site Response*, Eds. A.S. Cakmak and I. Herrera, Computational Mechanics Publications, Southampton Boston, 247–261, 1989.

Clough, R.W. and J. Penzien, *Dynamics of Structures*, McGraw–Hill Book Company, New York, 1975.

Gutenberg, B., Effects of ground on earthquake motion, *Bull. Seism. Soc. Am.*, **47**:221–250, 1957.

Hanks, T.C., Strong ground motion of San Fernando, California, earthquake ground displacements, *Bull. Seism. Soc. Am.*, **65**:193–225, 1975.

Hanson, R.D., and H.J. Degnekolb, *The Venezuela Earthquake, July 29, 1967*, American Iron and Steel Institute, New York, 1969.

Jongmans, D. and M. Campillo, "Site Amplification Effects in the Ubaye Valley (France): Measurements and Modeling," *Proc. Second Int. Conf. on Recent Advances in Geotechnical Earthquake Engineering and and Soil Dynamics*, St. Louis, Missouri, Paper No. 8.8:1171–1180, March 11–15, 1991

Kawase, H., "Time–domain response of a semicircular canyon for incident SV, P and Rayleigh waves calculated by the discrete wavenumber boundary element method, *Bull. Seism. Soc. Am.*, 78:1415–1437, 1988.

Kim, J. and A.S. Papageorgiou, Detailed Presentation of the Boundary Element Discrete Wavenumber Method as Implemented for the Study of Diffraction of Elastic Waves in 2–D Half–Spaces, *Earthquake Engineering and Structural Dynamics* (submitted for publication), 1991.

Langston, C.A., and J.J. Lee, Effect of structure geometry on strong ground motions: the Duwamish River Valley, Seattle, Washington, *Bull. Seism. Soc. Am.*, **73**:1851–1863, 1983.

Liu, H.L. and T.H. Heaton, Array analysis of the ground velocities and accelerations from the 1971 San Fernando California earthquake, *Bull. Seism. Soc. Am.* **74**:1951–1968, 1984.

Milne, J., *Seismology*, 1st ed., Kegan Paul, Trench, Truber, London, pp. 320, 1889.

Mullineaux, D.R., M.G. Bonilla, and J. Schlocker, Relation of building damage to geology in Seattle, Washington, during the April 1965 earthquake, *U.S. Geol. Surv. Profess. Paper, 575–D*, pp. D183–D191, 1967.

Murphy, V., Seismic Investigations, Valley of Caracas and the Litoral Central, *Weston Geophysical Engineers International, Inc.*, Weston, Massachusetts, 1969.

Papageorgiou, A.S. and J. Kim, Study of the Propagation and Amplification of Seismic Waves in Caracas Valley with Reference to the

July 29, 1967 Earthquake: SH–Waves, *Bull. Seism. Soc. Am.* (in press), 1991a.

Papageorgiou, A.S. and J. Kim, Oblique incidence of SV–waves on Sediment–Filled Valleys: Implications for Seismic Zonation, *Proceedings Fourth International Conference on Seismic Zonation*, August 26–29, 1991, San Francisco Bay Region, California, 1991b.

Presidential Earthquake Commission Report, Segunda Fase del Estudio del Sismo ocurrido en Caracas El 29 de Julio de 1967, Volumes A & B, *Fundacion Venezolana de Investigaciones Sismologicas (FUNVISIS)*, 1978.978.

Ricker, N., The computation of output disturbances from amplifiers for true wavelet inputs, *Geophysics*, **10**:207–220, 1945.

Roesset, J.M., Fundamentals of Soil Amplification, *Seismic Design for Nuclear Power Plants*, R.J. Hansen (Editor), M.I.T. Press, Cambridge, Mass., pp. 183–244, 1970.

Sanchez–Sesma, F.J., Site effects on strong ground motion, *Soil Dynamics and Earthquake Engineering*, **6**:124–132, 1987a.

Seed, H.B., R.V. Whitman, H. Dezfulian, R. Dobry and I.M. Idriss, Soil conditions and building damage in 1967 Caracas earthquake, *J. Soil Mech., Found. Div.*, ASCE, **98**:787–806, 1972.

Shinozuka, M. and C.M. Jan, Digital Simulation of Random Processes and its Applications, *J. Sound. Vib.*, **25**:111–128, 1972.

Sozen, M.A., P.C. Jennings, R.B. Mattiesen, G.W. Housner and N.M. Newmark, Engineering Report on the Caracas Earthquake of July 29, National Academy of Sciences, Washington, D.C., 1967.

Suárez, G. and J. Nábelek, The 1967 Caracas Earthquake: Fault Geometry, Direction of Rupture Propagation and Seismotectonic Implications, *J. Geophys. Res.*, **95**:17,459–17,474, 1990.

Vidale, J.E. and D.V. Helmberger, Elastic finite–difference modeling of the 1971 San Fernando, California earthquake, *Bull. Seism. Soc. Am.*, **78**:122–141, 1988.

Whitman, R.V., Effect of Soil Conditions upon Damage to Structures; Caracas Earthquake of 29 July 1967, *Report to Presidential Commission for Study of the Earthquake*, Caracas, Venezuela, 1969.

Wong, H.L. and P.C. Jennings, Effect of canyon topography on strong ground motion, *Bull. Seism. Soc. Am.*, **65**:1239–1257, 1975.

Zahradnik, J., and F. Hron, Seismic ground motion of sedimentary valleys – Example La Molina, Lima, Peru, *J. Geophys.*, **62**:31–37, 1987.

Seismic Response of Stochastic Ground

T. Harada

Department of Civil Engineering, Miyazaki University, Miyazaki, 889-21, Japan

ABSTRACT

The spatial variation of seismic ground motions is an important factor that should be carefully considered in the seismic design of buried lifelines such as tunnels and pipelines. The consideration of the spatial variation of ground motions may also have significant effects on the seismic response of structures with spatially extended foundations or multiple supports. The temporal and spatial variability of seismic ground motions has been inferred experimentally using data from closely spaced seismograph arrays. In this paper, the seismic ground motions varying in time and space domain are described analytically by the stochastic waves filtered by a surface soil layer, with random thickness, resting on rigid bedrock. The seismic wave, being assumed by a single plane wave, is transmitted to the soil layer from the rigid bedrock. The stochastic waves from the model possess the characteristics of both the coherent and incoherent components of ground motions. The analytic expressions of the frequency- wavenumber spectra of the stochastic waves at ground surface are then derived.

INTRODUCTION

The spatial variation of seismic ground motions is an important factor that should be carefully considered in the seismic design of buried lifelines such as tunnels and pipelines. The consideration of the spatial variation of ground motion may also have significant effects on the seismic response of structures with spatially extended foundations or multiple supports. In fact, for buried lifelines, the seismic deformation method was developed (Public Works Research Institute [1]) and is now in practical use in Japan. For the seismic design of the Akashi Kaikyo bridge foundations, a modified response spectrum was used taking account for the spatial variation of ground motions around the foundations (Kashima et al.[2],Kawaguchi et al.[3]).

The temporal and spatial variability of seismic ground motions has been inferred experimentally using data from closely spaced seismograph arrays. The SMART-1 array, for example, located at Lotung in the N E corner of Taiwan has provided valuable data for the analysis of ground motions in time and space domain. Numerous studies using the SMART-1 array data have been reported (Loh et al.[4],Harada[5],Harada and Shinozuka[6],Harichandran[7] ,Abrahamson[8,9]). It is common in these studies that the accelerograms from each seismic event are described as samples from space-time stochastic processes or stochastic fields and eventually the spatial coherence functions or the frequency- wavenumber spectra are estimated.

In this paper, the seismic ground motions varying in time and space domain are described by the stochastic waves filtered by a single surface soil layer, with random thickness caused by the irregular shape of free-surface, resting on rigid bedrock. The seismic wave, being assumed by a single plane wave, is transmitted to the soil layer from the rigid bedrock. The analytic expressions of the frequency-wavenumber spectra of the stochastic waves at ground surface are then derived.

BASIC EQUATIONS OF SEISMIC RESPONSE OF GROUND WITH IRREGULAR INTERFACE

For the wave propagation problem possessing cylindrical symmetry, the displacement wave field (u, v, w) in a Cartesian coordinate system (x, y, z) can be constructed from the solutions (u', v', w') of the two-dimensional problem associated with the P-SV wave propagation and the SH wave propagation in a new coordinate system (x', y', z) (see Fig.1) such that [10],

$$u(x, y, z, t) = \cos\theta\; u'(x', z, t) - \sin\theta\; v'(x', z, t) \tag{1-a}$$

$$v(x, y, z, t) = \sin\theta\; u'(x', z, t) + \cos\theta\; v'(x', z, t) \tag{1-b}$$

$$w(x, y, z, t) = w'(x', z, t) \tag{1-c}$$

Introducing the Fourier transform of u with respect to x, y, t in the following form,

$$u(\kappa_x, \kappa_y, z, \omega) = \frac{1}{(2\pi)^3} \int\int\int u(x, y, z, t) e^{-i(\kappa_x x + \kappa_y y - \omega t)} dx dy dt \tag{2-a}$$

with the inverse transform,

$$u(x, y, z, t) = \int\int\int u(\kappa_x, \kappa_y, z, \omega) e^{i(\kappa_x x + \kappa_y y - \omega t)} d\kappa_x d\kappa_y d\omega \tag{2-b}$$

and similarly for v, w, u', and v', one obtains the relationships in frequency-wavenumber domain such as

$$u(\kappa_x, \kappa_y, z, \omega) = \frac{\kappa_x}{\kappa} u'(\kappa, z, \omega) - \frac{\kappa_y}{\kappa} v'(\kappa, z, \omega) \tag{3-a}$$

$$v(\kappa_x, \kappa_y, z, \omega) = \frac{\kappa_y}{\kappa} u'(\kappa, z, \omega) + \frac{\kappa_x}{\kappa} v'(\kappa, z, \omega) \tag{3-b}$$

$$w(\kappa_x, \kappa_y, z, \omega) = w'(\kappa, z, \omega) \tag{3-c}$$

where κ_x and κ_y are the wavenumbers and ω the frequency. The wavenumber κ denotes:

$$\kappa = \sqrt{\kappa_x^2 + \kappa_y^2} = \kappa_{x'} \tag{4}$$

and the angle θ between the two coordinate systems (x, y, z) and (x', y', z) is given by:

$$\cos\theta = \frac{\kappa_x}{\kappa} \quad ; \quad \sin\theta = \frac{\kappa_y}{\kappa} \tag{5}$$

It is noted here that the special symbols are avoided for simplicity of notations to denote that u (similarly for the other displacements) has been transformed.

The displacement wave field (u', v', w') corresponding to P-SV and SH wave propagations in a two-dimensional homogeneous isotropic medium is known to be expressed as matrix wave equation in the form [11]:

$$\frac{\partial}{\partial z}\mathbf{B}(x', z) = \mathbf{A}_0(z)\mathbf{B}(x', z) \tag{6}$$

where $\mathbf{B}$ is the displacement-stress vector and $\mathbf{A}_0$ is the operator matrix defined such as

$$\mathbf{B}_{SH}(x', z) = col\ [v'\ ,\ \tau_{y'z}], \qquad \mathbf{B}_{P-SV}(x', z) = col\ [u'\ ,\ w\ ,\ \tau_{x'z}\ ,\ \tau_{zz}] \tag{7}$$

$$\mathbf{A}_{0_{SH}}(z) = \begin{pmatrix} 0 & \frac{1}{\mu^*} \\ -\mu^*\partial_{x'x'} - \rho\omega^2 & 0 \end{pmatrix} \tag{8-a}$$

$$\mathbf{A}_{0_{P-SV}}(z) = \begin{pmatrix} 0 & -\partial_{x'} & \frac{1}{\mu^*} & 0 \\ -b\partial_{x'} & 0 & 0 & b/\lambda^* \\ -a\partial_{x'x'} - \rho\omega^2 & 0 & 0 & -b\partial_{x'} \\ 0 & -\rho\omega^2 & -\partial_{x'} & 0 \end{pmatrix} \tag{8-b}$$

with

$$a = \frac{4\mu^*(\lambda^* + \mu^*)}{\lambda^* + 2\mu^*}, \quad b = \frac{\lambda^*}{\lambda^* + 2\mu^*}, \quad \partial_{x'} = \partial/\partial x', \quad \partial_{x'x'} = \partial/\partial x'x' \tag{8-c}$$

where μ^* and λ^* are the complex Lame constants , and ρ the density of medium. It should be noted here that the Fourier transform has been done with respect to time and the frequency ω (time dependence is $exp(-i\omega t)$) is dropped out in equation (6) and will disappear in the following notations for brevity.

Welded boundary conditions at the irregular interface (see Fig.2) require continuity of displacement and traction at each point on the interface. Therefore, a new displacement-stress vector $\mathbf{b}$, being measured with respect to the local tangent plane at each point on the interface, has to be introduced. The new displacement-stress vector takes the form [11]:

$$\mathbf{b}(x', f) = (\mathbf{I} + \frac{\partial f}{\partial x'}\mathbf{Q}_0)\mathbf{B}(x', f) \tag{9}$$

where $\mathbf{b}$ and $\mathbf{B}$ are evaluated along the irregular interface located at the depth $z(x')$ defined by:

$$z(x') = z + f(x') \tag{10}$$

with z being the average depth of the interface and $f(x')$ being the lateral fluctuation of the interface (see Fig.2). In equation (9), $\mathbf{I}$ is the unit matrix and $\mathbf{Q}_0$ is given by:

$$\mathbf{Q}_{0_{SH}} = \begin{pmatrix} 0 & 0 \\ -\mu^*\partial_{x'} & 0 \end{pmatrix} \tag{11-a}$$

$$\mathbf{Q}_{0_{P-SV}} = \begin{pmatrix} 0 & 0 & 0 & 0 \\ 0 & 0 & 0 & 0 \\ -a\partial_{x'} & 0 & 0 & -b \\ 0 & 0 & -1 & 0 \end{pmatrix} \tag{11-b}$$

The irregular interface boundary condition can be expressed in the form:

$$\mathbf{b}_1(x', f) = \mathbf{b}_2(x', f) \tag{12}$$

where subscripts indicate the respective media.

In order to obtain the approximation of equation (9), the scattered wave field (displacement-stress vector in the medium with irregular interface) is approximately related to the background field (displacement-stress vector in the medium with horizontal plane interface) by using a Taylor expansion around the average interface depth $z(x') = z$,

$$\mathbf{B}(x', f) = \mathbf{B}(x', z) + \frac{\partial \mathbf{B}}{\partial z} f \cdots = \{\mathbf{I} + f\mathbf{A}_0(z)\}\mathbf{B}(x', z). \tag{13}$$

In deriving the last term on right hand side of (13), equation (6) has been used. Substituting (13) into (9), and omitting terms of order higher than f and $\partial f/\partial x'$, one obtains:

$$\mathbf{b}(x', f) = \{\mathbf{I} + f\mathbf{A}_0 + \frac{\partial f}{\partial x'}\mathbf{Q}_0\}\mathbf{B}(x', z) \tag{14}$$

Furthermore, introducing the new notations, $\mathbf{b}^0$ and $\mathbf{B}^0$, which represent the background wave fields, and denoting the first-order approximations of $\mathbf{b}$ and $\mathbf{B}$ by $\mathbf{b}^I$ and $\mathbf{B}^I$, respectively, one obtains the first-order approximations of equation (9) as:

$$\mathbf{b}^I(x', f) = \mathbf{B}^I(x', z) + \{f\mathbf{A}_0 + \frac{\partial f}{\partial x'}\mathbf{Q}_0\}\mathbf{B}^0(x', z) \tag{15}$$

with the following condition because $\mathbf{b}$ equals to $\mathbf{B}$ at the interface $z(x') = z$ in the case of horizontal plane interface ($f = 0$):

$$\mathbf{b}^0(x', z) = \mathbf{B}^0(x', z) \tag{16}$$

The Fourier transform of (15) with respect to x', using the result that the Fourier transform of a product is the convolution of the Fourier transform, yields:

$$\mathbf{b}^I(\kappa, f) = \mathbf{B}^I(\kappa, z) + \int_{-\infty}^{\infty} f(\kappa - \kappa')\mathbf{J}(\kappa, \kappa')\mathbf{B}^0(\kappa', z)d\kappa' \tag{17}$$

where

$$\mathbf{J}(\kappa, \kappa') = \mathbf{A}_0(\kappa') + i(\kappa - \kappa')\mathbf{Q}_0(\kappa') \tag{18}$$

with

$$\mathbf{J}_{SH}(\kappa, \kappa') = \begin{pmatrix} 0 & \dfrac{1}{\rho Cs^2} \\ \rho Cs^2\kappa\kappa' - \rho\omega^2 & 0 \end{pmatrix} \tag{19-a}$$

$$\mathbf{J}_{P-SV}(\kappa, \kappa') =$$

$$\begin{pmatrix} 0 & -i\kappa' & \frac{1}{\rho C s^2} & 0 \\ \{2(\frac{Cs}{Cp})^2 - 1\}i\kappa' & 0 & 0 & \frac{1}{\rho C p^2} \\ 4\rho C s^2\{1 - (\frac{Cs}{Cp})^2\}\kappa\kappa' - \rho\omega^2 & 0 & 0 & \{2(\frac{Cs}{Cp})^2 - 1\}i\kappa \\ 0 & -\rho\omega^2 & -i\kappa & 0 \end{pmatrix} \tag{19 - b}$$

In equation (19), C_s and C_p are the complex P-wave and S-wave velocities, respectively, given by:

$$C_s = C_s^0(1 + iD_s), \quad C_p = C_p^0(1 + iD_p) \tag{20}$$

with C_s^0 and C_p^0 being the elastic P-wave and S-wave velocities respectively and D_s, D_p being the ratios of the linear hysteretic damping for P- and S-waves.

The irregular interface boundary condition given by (12) can be written for the first-order approximations in frequency-wavenumber domain as:

$$\mathbf{B}_1^I(\kappa, z) = \mathbf{B}_2^I(\kappa, z) + \int_{-\infty}^{\infty} f(\kappa - \kappa')\mathbf{L}_{21}(\kappa, \kappa')\mathbf{B}_2^0(\kappa', z)d\kappa' \tag{21 - a}$$

where

$$\mathbf{L}_{21} = \mathbf{J}_2 - \mathbf{J}_1 = -\mathbf{L}_{12} \tag{21 - b}$$

Equation (21) indicates the presence of irregular interface results in discontinuity in the scattered wave field $\mathbf{B}(x', z)$ at $z(x') = z$. This discontinuity acts like a seismic source $\mathbf{S}$ which can be evaluated directly from the background wave field such as:

$$\mathbf{S}(\kappa, z) = \int_{-\infty}^{\infty} f(\kappa - \kappa')\mathbf{L}_{21}(\kappa, \kappa')\mathbf{B}_2^0(\kappa', z)d\kappa' \tag{22}$$

RESPONSE OF A SINGLE SOIL LAYER WITH IRREGULAR FREE SURFACE RESTING ON RIGID BEDROCK

Response of a single soil layer with irregular free surface resting on rigid bedrock as shown in Fig.3 is considered. For a free surface the traction has to vanish, so that for an irregular free surface $z(x') = f(x')$, $(z = 0)$, the scattered wave field takes the form:

$$\mathbf{b}(\kappa, f) = \mathbf{B}(\kappa, f) = col\ [\mathbf{U}(\kappa, f), \mathbf{0}] \tag{23}$$

Then, the first-order approximation of the interface boundary condition given by equation (17) can be expressed as:

$$\mathbf{B}^I(\kappa, f) = \mathbf{B}^I(\kappa, 0) + \int_{-\infty}^{\infty} f(\kappa - \kappa')\mathbf{J}(\kappa, \kappa')\mathbf{B}^0(\kappa', 0)d\kappa' \tag{24}$$

Making use of the propagator matrix $\mathbf{P}(\kappa, z, z_0)$ which satisfies [11],

$$\frac{\partial}{\partial z}\mathbf{P}(\kappa, z, z_0) = \mathbf{A}_0(\kappa, z)\mathbf{P}(\kappa, z, z_0), \quad \mathbf{P}^{-1}(\kappa, z, z_0) = \mathbf{P}(\kappa, z_0, z) \tag{25}$$

the displacement-stress vector at the bedrock $z(x') = H$ can be transformed to that at the free-surface $z(x') = 0$ such as:

$$\mathbf{B}(\kappa, 0) = \mathbf{P}(\kappa, 0, H)\mathbf{B}(\kappa, H) \tag{26}$$

It is assumed now that an input seismic motion is specified at the bedrock and propagates in the direction of x' axis with apparent wave speed c. Then, the input motion (represented by displacement-stress vector) at the bedrock is expressed in the form:

$$\mathbf{B}(\kappa', H) = \mathbf{B}(\kappa_0, H)\ \delta(\kappa' - \kappa_0) \tag{27-a}$$

where δ is the Delta function, and κ_0 is given by:

$$\kappa_0 = \frac{\omega}{c}. \tag{27-b}$$

Taking into account for equations (26) and (27) in (24), one obtains:

$$\mathbf{B}^I(\kappa, f) = \mathbf{P}(\kappa, 0, H)\mathbf{B}^I(\kappa, H)\delta(\kappa - \kappa_0) + f(\kappa - \kappa_0)\mathbf{J}(\kappa, \kappa_0)\mathbf{P}(\kappa_0, 0, H)\mathbf{B}^0(\kappa_0, H) \tag{28}$$

By considering the boundary conditions that the traction vanishes at the free-surface and the input motion is specified at the bedrock, equation (28) can be more explicitly expressed in the partitioned form:

$$\begin{pmatrix} \mathbf{U}^I(\kappa, f) \\ \mathbf{0} \end{pmatrix} = \begin{pmatrix} \mathbf{P}_{11} & \mathbf{P}_{12} \\ \mathbf{P}_{21} & \mathbf{P}_{22} \end{pmatrix} \begin{pmatrix} \mathbf{U}^0(\kappa, H) \\ \tau^I(\kappa, H) \end{pmatrix} \delta(\kappa - \kappa_0)$$
$$+ f(\kappa - \kappa_0) \begin{pmatrix} \mathbf{J}_{11} & \mathbf{J}_{12} \\ \mathbf{J}_{21} & \mathbf{J}_{22} \end{pmatrix} \begin{pmatrix} \mathbf{P}^0_{11} & \mathbf{P}^0_{12} \\ \mathbf{P}^0_{21} & \mathbf{P}^0_{22} \end{pmatrix} \begin{pmatrix} \mathbf{U}^0(\kappa_0, H) \\ \tau^0(\kappa_0, H) \end{pmatrix} \tag{29}$$

where $\mathbf{P}^0_{ij} = \mathbf{P}_{ij}(\kappa_0, 0, H)$. In equation (29), the relation, $\mathbf{U}^I(\kappa, H) = \mathbf{U}^0(\kappa, H)$, is used because the input motion displacement is specified at the bedrock. From equation (29), one obtains the scattered wave field such as:

$$\mathbf{U}^I(\kappa, f) = \{(\mathbf{P}_{11} - \mathbf{P}_{12}\mathbf{P}_{22}^{-1}\mathbf{P}_{21})\delta(\kappa - \kappa_0)$$
$$+ f(\kappa - \kappa_0)(\mathbf{J}_{11} - \mathbf{P}_{12}\mathbf{P}_{22}^{-1}\mathbf{J}_{21})(\mathbf{P}^0_{11} - \mathbf{P}^0_{12}(\mathbf{P}^0_{22})^{-1}\mathbf{P}^0_{21})\}\mathbf{U}^0(\kappa_0, H) \tag{30-a}$$

$$\tau^I(\kappa, H)\delta(\kappa - \kappa_0) = [-\mathbf{P}_{22}^{-1}\mathbf{P}_{21}\delta(\kappa - \kappa_0)$$
$$- f(\kappa - \kappa_0)\{\mathbf{P}_{22}^{-1}\mathbf{J}_{21}(\mathbf{P}^0_{11} - \mathbf{P}^0_{12}(\mathbf{P}^0_{22})^{-1}\mathbf{P}^0_{21})\}]\mathbf{U}^0(\kappa_0, H) \tag{30-b}$$

EXAMPLES OF FREQUENCY-WAVENUMBER SPECTRA OF STOCHASTIC WAVES AT THE IRREGULAR FREE-SURFACE

Closed form expressions are presented for the horizontal displacements of stochastic waves at the free-surface of single soil layer with irregular free-surface subjected to the horizontal seismic ground motions at the bedrock. The vertical components in both the input motions and the response of soil layer are not considered, primarily because the horizontal components are much more interested in earthquake engineering problems and also because the closed form expressions corresponding to P-SV wave propagation becomes relatively simple.

From the form given by equation (30-a), the horizontal displacement components (u', v') have the form:

$$u' = \{A\delta(\kappa - \kappa_0) + Bf(\kappa - \kappa_0)\}u'_g \ , \ v' = \{C\delta(\kappa - \kappa_0) + Df(\kappa - \kappa_0)\}v'_g \tag{31}$$

where $u' = u^I(\kappa, f), v' = v^I(\kappa, f)$, $u'_g = u^0(\kappa_0, H), v'_g = v^0(\kappa_0, H)$. By making use of equation (3), the horizontal displacement components (u, v) in a Cartesian coordinate system (x, y, z) can be obtained in the form:

$$\begin{Bmatrix} u \\ v \end{Bmatrix} = \begin{bmatrix} H_{uu} & H_{uv} \\ H_{vu} & H_{vv} \end{bmatrix} \begin{Bmatrix} u_g \\ v_g \end{Bmatrix} \tag{32}$$

where

$$\begin{aligned} H_{uu} &= [A(\frac{\kappa_x}{\kappa})^2 + C(\frac{\kappa_y}{\kappa})^2]\delta(\kappa_x - \kappa_x^0, \kappa_y - \kappa_y^0) \\ &\quad + [B(\frac{\kappa_x}{\kappa})^2 + D(\frac{\kappa_y}{\kappa})^2]f(\kappa_x - \kappa_x^0, \kappa_y - \kappa_y^0) \end{aligned} \tag{33-a}$$

$$\begin{aligned} H_{uv} = H_{vu} &= (A - C)\frac{\kappa_x\kappa_y}{\kappa^2}\delta(\kappa_x - \kappa_x^0, \kappa_y - \kappa_y^0) \\ &\quad + (B - D)\frac{\kappa_x\kappa_y}{\kappa^2}f(\kappa_x - \kappa_x^0, \kappa_y - \kappa_y^0) \end{aligned} \tag{33-b}$$

$$\begin{aligned} H_{vv} &= [A(\frac{\kappa_y}{\kappa})^2 + C(\frac{\kappa_x}{\kappa})^2]\delta(\kappa_x - \kappa_x^0, \kappa_y - \kappa_y^0) \\ &\quad + [B(\frac{\kappa_y}{\kappa})^2 + D(\frac{\kappa_x}{\kappa})^2]f(\kappa_x - \kappa_x^0, \kappa_y - \kappa_y^0) \end{aligned} \tag{33-c}$$

$$\kappa_x^0 = \frac{\omega\cos\theta}{c} \ , \ \kappa_y^0 = \frac{\omega\sin\theta}{c} \tag{33-d}$$

By assuming that the input seismic motion is the stationary stochastic process and the fluctuation of irregular free-surface is the isotropic stochastic field, eventually the response displacement wave fields may be the stationary and homogeneous stochastic wave fields. Then, its frequency- wavenumber spectra are obtained as:

$$S_{uu}(\kappa_x, \kappa_y, \omega) = |u(\kappa_x, \kappa_y, \omega)|^2, \quad S_{vv}(\kappa_x, \kappa_y, \omega) = |v(\kappa_x, \kappa_y, \omega)|^2 \tag{34}$$

If $v_g = 0$ in (34), then, equations (32) to (34) yield:

$$\begin{aligned} S_{uu}(\kappa_x, \kappa_y, \omega) = \ & [|A(\frac{\kappa_x}{\kappa})^2 + C(\frac{\kappa_y}{\kappa})^2|^2\delta(\kappa_x - \kappa_x^0, \kappa_y - \kappa_y^0) \\ & + |B(\frac{\kappa_x}{\kappa})^2 + D(\frac{\kappa_y}{\kappa})^2|^2 S_{ff}(\kappa_x, \kappa_y)]S_{u_g u_g}(\omega) \end{aligned} \tag{35}$$

where S_{ff} is the wavenumber spectrum of the fluctuation of irregular free-surface $f(x')$, being the isotropic stochastic field. The assumption of isotropic characteristics for $f(x')$ may be necessary to maintain cylindrical symmetry of the problem considered in this paper. The quantities A, B, C, and D appearing in (31) to (35) are given by:

$$A = \frac{(\gamma^2 + \kappa^2)\nu\gamma}{R}[(\gamma^2 - \kappa^2)\cos\nu H + 2\kappa^2\cos\gamma H] \tag{36-a}$$

$$B = A_0 \frac{[4\kappa\kappa_0(\gamma^2 - \nu^2) - (\gamma^2 + \kappa^2)^2]\gamma}{R}[\nu\gamma \cos\nu H \sin\gamma H + \kappa^2 \cos\gamma_0 H \sin\nu_0 H] \quad (36-b)$$

$$C = \frac{1}{\cos\gamma H} \quad (36-c)$$

$$D = \frac{\gamma_0 \sin\gamma H}{\cos\gamma H \cos\gamma_0 H}\left(\frac{\kappa\kappa_0 - \kappa_0^2}{\gamma\gamma_0} - \frac{\gamma_0}{\gamma}\right) \quad (36-d)$$

$$R = 4\kappa^2\nu\gamma(\gamma^2 - \kappa^2) + \nu\gamma[4\kappa^4 + (\gamma^2 - \kappa^2)^2]\cos\nu H \cos\gamma H$$
$$-\kappa^2[4\nu^2\gamma^2 + (\gamma^2 - \kappa^2)^2]\sin\nu H \sin\gamma H \quad (36-e)$$

where the vertical wavenumbers ν and γ satisfy:

$$\nu^2 = \left(\frac{\omega}{Cp}\right)^2 - \kappa^2 \ ; \ Re\nu \geq 0 \ ; \ Im\nu \geq 0 \ , \ \gamma^2 = \left(\frac{\omega}{Cs}\right)^2 - \kappa^2 \ ; \ Re\gamma \geq 0 \ ; \ Im\gamma \geq 0 \quad (37)$$

and A_0, ν_0, and γ_0 correspond to the values of A, ν, γ when $\kappa = \kappa_0 = \omega/c$ (see equations (4), (5), and (27-b)). If the vertical displacement w of the response is constrained to be zero for the solutions of P-SV wave propagation, the quantities A and B take the form:

$$A = \frac{1}{\cos\frac{C_p}{C_s}\nu H} \ , \ B = \frac{\frac{C_p}{C_s}\nu_0 \sin\frac{C_p}{C_s}\nu H}{\cos\frac{C_p}{C_s}\nu H \cos\frac{C_p}{C_s}\nu_0 H}\left(\frac{\kappa\kappa_0 - \kappa_0^2}{\nu\nu_0} - \frac{\nu_0}{\nu}\right) \quad (38)$$

Finally, the frequency-dependent auto-correlation function is defined as

$$R(\xi_x, \xi_y, \omega) = \int_{-\infty}^{\infty}\int_{-\infty}^{\infty} S(\kappa_x, \kappa_y, \omega) e^{i(\kappa_x\xi_x + \kappa_y\xi_y)} d\kappa_x d\kappa_y \quad (39)$$

Equation (39) can be evaluated numerically.

NUMERICAL EXAMPLES

Numerical examples are given to visualize the shape of the frequency-wavenumber spectrum given in equation (35) and the frequency-dependent auto-correlation function in (39). In order to visualize the filtering effect of the soil layer with irregular free-surface, the power spectrum of input seismic motion displacement is assumed unity, and the direction and apparent wave speed of input motion is assumed as:

$$S_{u_g u_g}(\omega) = 1 \ ; \ \theta = 45° \ ; \ c = 523 m/s \quad (40)$$

The following values are used for the various constants necessary to evaluate (35):

$$Elastic\ P-wave\ velocity : C_p^0 = 573 m/s \quad (41-a)$$

$$Elastic\ S-wave\ velocity : C_s^0 = 191 m/s \quad (41-b)$$

$$Material\ damping\ ratio : D_p = D_s = 0.3 \quad (41-c)$$

$$Average\ thickness\ of\ layer : H = 100m \quad (41-d)$$

The following analytic expression is assumed for the wavenumber spectrum of $f(x')$:

$$S_{ff}(\kappa_x, \kappa_y) = \begin{cases} \frac{2\sigma_{ff}^2}{\pi\kappa_u^2}\cos^2\left(\frac{m\pi}{\kappa_u}\kappa\right) & 0 \leq \kappa \leq \kappa_u \\ 0 & otherwise \end{cases} \quad (42)$$

where the parameters are assumed as $m = 4, \kappa_u = 4\pi/100, \sigma_{ff} = 20$.

According to the above values, the first natural frequency ω_1 of the soil layer with horizontal plane free-surface, obtained by the vertically propagating S-wave, is estimated as $\omega_1 = 3rad/s(2\pi C_s^0/4H = 2\pi \cdot 191/400)$. Equation (42) is plotted in Fig.4. Figure 5 shows $S_{uu}(\kappa_x, \kappa_y, \omega)$ as a function of wavenumbers κ_x and κ_y for the same region of Fig.4 ($\kappa_{xmax} = \kappa_{ymax} = 0.04rad/m$) at 8 different values of frequency . It is observed from Fig.5 that the relatively sharp mountain appears up to approximately 3 rad/s while widely spreading mountains and troughs over the wavenumber plane tend to appear in high frequency range. Such pattern as shown in Fig.5 is quite similar to the pattern estimated by Abrahamson [8] using the SMART-1 array recordings for event 5.

In order to show the validity of more simple analytic expressions of the frequency-wavenumber spectrum, equation (35), with the quantities given in (36-c), (36-d), and (38) based on the assumption that the vertical displacement of the response is zero for P-SV wave propagation problem, is plotted in Fig.6. Quite similar pattern as shown in Fig.5 is observed from Fig.6, indicating that the validity of equation (35) with the quantities given in (36-c), (36-d), and (38).

Finally, the frequency-dependent auto-correlation function given in (39) is calculated numerically. The normalized $R_{uu}(\xi_x, \xi_y, \omega)/R_{uu}(0, 0, \omega)$, being calculated from the more simple expression of frequency-wavenumber spectrum given in (35) with the quantities given in (36-c), (36-d), and (38), is plotted in Fig.7 as a function of the separation distance ξ_x and ξ_y ($\xi_{xmax} = \xi_{ymax} = 750m$) at 8 different values of frequency used in Figs. 5 and 6.

CONCLUSIONS

In this paper , a closed form analytic expression of the frequency-wavenumber spectrum is established. The corresponding seismic ground motion is produced by the seismic response of a single soil layer, with random thickness, resting on rigid bedrock. The seismic wave, assumed by a single plane wave, is transmitted to the soil layer from the rigid bedrock. The stochastic waves from the model possess the characteristics of both the coherent and incoherent components of ground motions. Although the present paper considers the inhomogeneity of local soil layer caused by the lateral variation of the thickness of soil layer only, the effect of the inhomogeneity caused by the spatial variation of soil properties on the ground motions is also studied (Harada,T., and Fugasa, T [12]).

REFERENCES

1. Public Works Research Institute. A Proposal for Earthquake Resistant Design Method. Technical Memorandum of PWRI, Ministry of Construction, No. 1185, March, 1977 (in Japanese).

2. Kashima, N., Kawashima, K., Harada, T., Isoyama, R., and Masuda, S. Soil-Structure Interaction and Its Implication for Seismic Design of Structures. Proceedings of 9th World Conference on Earthquake Engineering, Prentice Hall, pp.605-612, 1984.

3. Kawaguchi, K., Masuda, S., Isoyama, R., and Saeki, M. Aseismic Design of Akashi

Kaikyo Bridge Foundations. Proceedings of New Zealand- Japan Workshop on Base Isolation of Highway Bridges, Technology Research Center for National Land Development, pp.52-63, 1987.

4. Loh, C. H., Penzien, J., and Tsai, Y. B. Engineering Analyses of SMART-1 Accelerograms. Earthquake Engineering and Structural Dynamics, Vol.10, pp.579-591, 1982.

5. Harada, T. Probabilistic Modeling of Spatial Variation of Strong Earthquake Ground Displacements. Proceedings of 8th World Conference on Earthquake Engineering, Prcentice Hall, pp.605-612, 1984.

6. Harada, T., and Shinozuka, M. Ground Deformation Spectra. Proceedings of 3rd U. S. National Conference on Earthquake Engineering, Earthquake Engineering Reserch Institute, pp2191-2202, 1986.

7. Harichandran, R. Local Spatial Variation of Earthquake Ground Motion. Earthquake Engineering and Soil Dynamics-Recent Advanced in Ground Motion Evaluation, Edited by Lawrence Vom Thun, J., Geotechnical Special Publication, No.20, ASCE, pp.203-217, 1988.

8. Abrahamson, N. A. Estimation of Seismic Wave Coherency and Rupture Velocity Using The SMART-1 Strong-Motion Array Recordings. Report No.UCB/EERC-85/02 Earthquake Engineering Research Center, University of California, Berkley, 1985.

9. Abrahamson, N.A., Schneider, J.F., and Stepp, J.C. Empirical Spatial Coherency Functions for Application to Soil-Structure Interaction Analyses. Earthquake Spectra, Professional Journal of the Earthquake Engineering Research Institute, Vol.7, No.1, pp.1-27, 1991.

10. Bouchon, M. Discrete Wavenumber Representation of Elastic Wave Fields in Three-Space Dimensions. Journal of Geophysical Research, Vol.84, pp.3609-3614, 1979.

11. Kennett, B.L.N. Seismic Wave Scattering by Obstacles on Interfaces. Geophysical Journal of the Royal Astronomical Society, Vol.28, pp.249-266, 1972.

12. Harada, T., and Fugasa, T. Characteristics of Seismic Responses of 3-Dimensional Ground with Stochastic Soil Properties. Memoirs of the Faculty of Engineering, Miyazaki University, No.36, September, 1990.

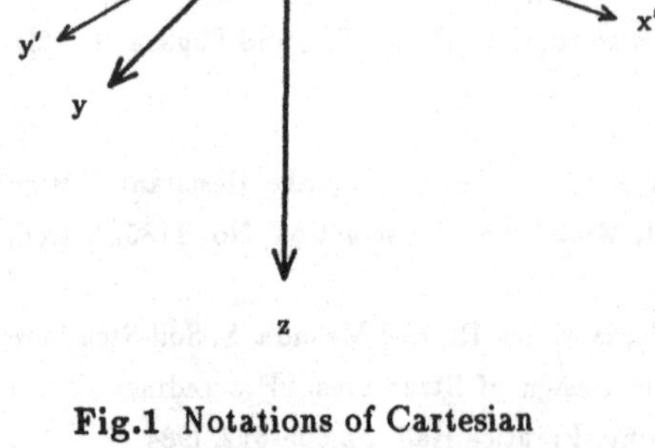

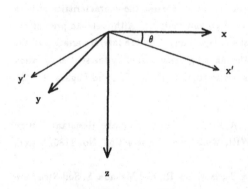

Fig.1 Notations of Cartesian Coordinate Systems

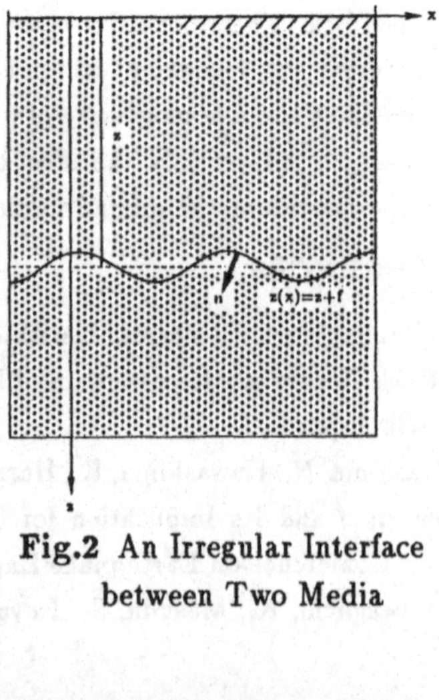

Fig.2 An Irregular Interface between Two Media

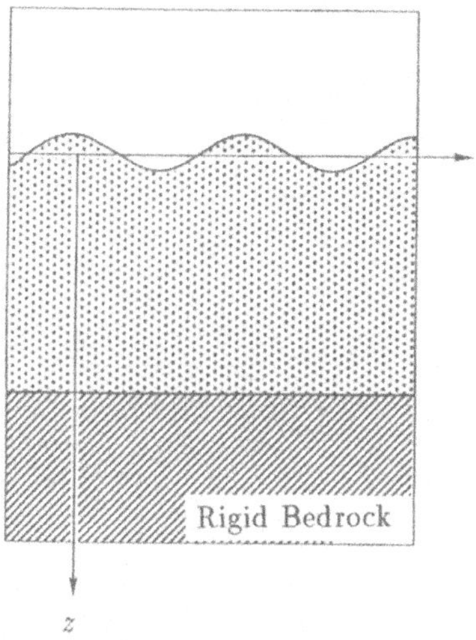

Fig.3 Single Soil Layer with Irregular Free-Surface Resting on Rigid Bedrock

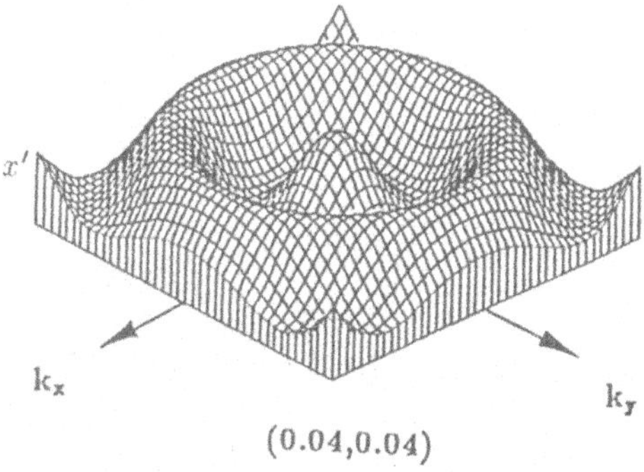

Fig.4 Wavenumber Spectrum of Irregularity of Free-Surface Used in Numerical Example

Fig.5 Frequency-Wavenumber Spectra at 8 Different Frequencies (Rigorous)

(*a*) $\omega = 1$ ($Max = 0.306$)

(*b*) $\omega = 2$ (4.675)

(*c*) $\omega = 3$ (50.346)

(*d*) $\omega = 4$ (26.183)

(*e*) $\omega = 5$ (12.549)

(*f*) $\omega = 6$ (9.191)

(*g*) $\omega = 7$ (6.231)

(*h*) $\omega = 8$ (7.364)

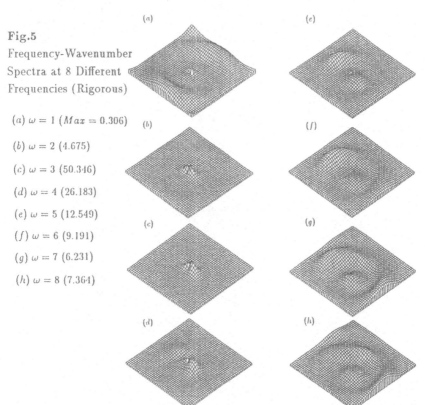

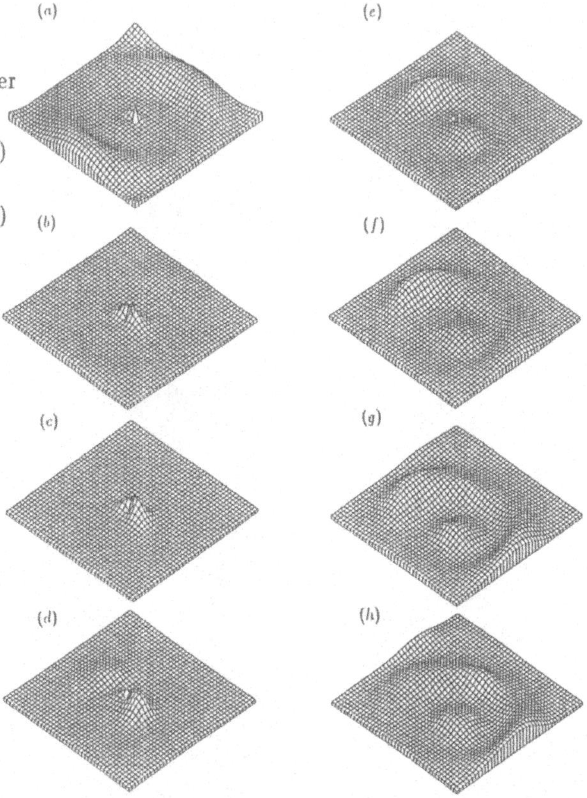

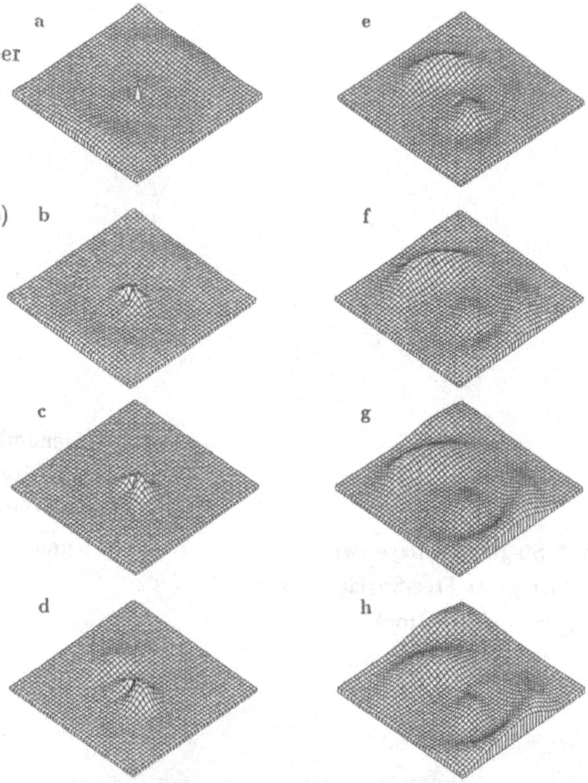

Fig.6
Frequency-Wavenumber Spectra at 8 Different Frequencies (Approximate)

(a) $\omega = 1$ ($Max = 1.135$)

(b) $\omega = 2$ (4.675)

(c) $\omega = 3$ (50.346)

(d) $\omega = 4$ (26.976)

(e) $\omega = 5$ (12.611)

(f) $\omega = 6$ (9.219)

(g) $\omega = 7$ (6.571)

(h) $\omega = 8$ (7.477)

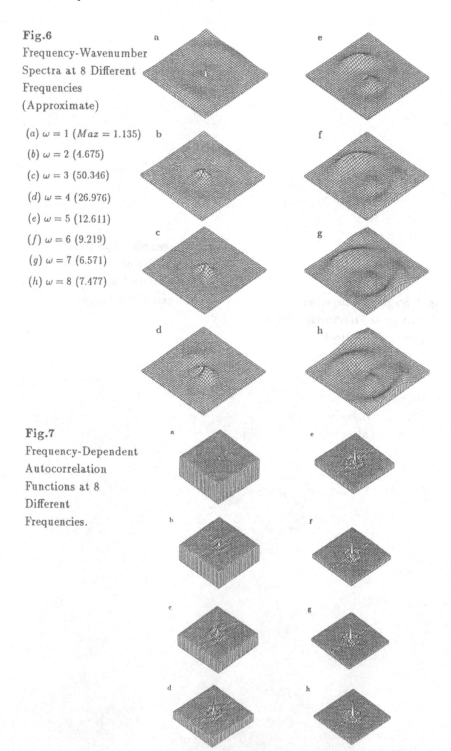

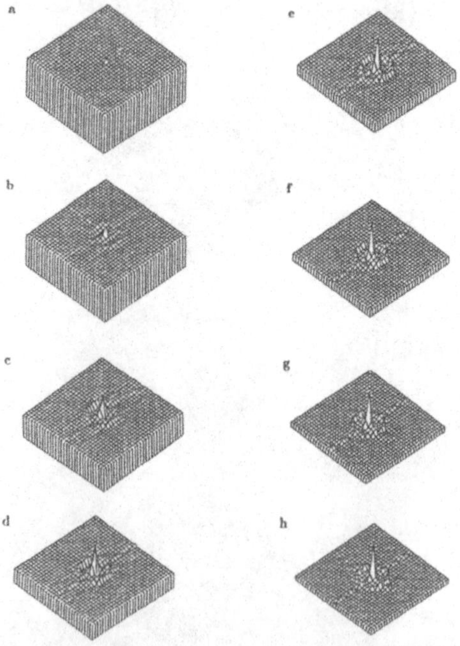

Fig.7
Frequency-Dependent Autocorrelation Functions at 8 Different Frequencies.

An Efficient Treatment of Model Uncertainties for the Dynamic Response of Structures

L.S. Katafygiotis, J.L. Beck

California Institute of Technology, Pasadena, CA 91125, U.S.A.

ABSTRACT

The problem of calculating the uncertainty in the dynamic response of a structure due to uncertainties related to the modeling of its dynamic behavior, is addressed. Based on a Bayesian probabilistic approach, a new efficient and accurate numerical method is proposed to investigate the resulting uncertainties in the structural response, where engineering judgement is used to quantify the uncertainties in the modeling process. The proposed method is useful in design, where dynamic test data is not available to investigate modeling errors through the application of system identification.

INTRODUCTION

The deterministic modeling of the dynamic behavior of a structural system consists of using some method, such as finite elements, to select a particular model out of a class of models. This class is specified by choosing the general mathematical form, such as linear dynamics, which is expected to describe the essential features of the input-output relation of the system. However, uncertainties arising during such modeling should be acknowledged, quantified, and propagated in order to quantify the resulting uncertainty in the structural response.

There are two types of uncertainty introduced when modeling the structural behavior with a model within a specified class. The first type of uncertainty is concerned with which model in the class is the most appropriate to model the system. This type of uncertainty is referred to as uncertainty of the "model parameters ", since certain parameters must be assigned unique values in order to specify a particular model within the given class; therefore, uncertainty in the specification of the most appropriate model within the class can be viewed as uncertainty in the specification of its parameter values. The second type of uncertainty is referred to as uncertainty of the "model error", and is concerned with how well the chosen class of models approximates the dynamic behavior of the given structural system. The existence of "model error", referring to the difference between system and model output, stems from the fact that no mathematical model is good enough to exactly represent the behavior of a real system. To account for model error, a probability model must be chosen out of a specified class of probability models. The uncertainty of the model error is reflected as uncertainty in specifying the model-error parameters that must be assigned unique values in order to specify a particular probability model out of the specified class of probability models.

The uncertainty associated with a quantity, such as the model and model-error parameters, or the system output, is quantified using probability in the Bayesian context, that is, assigning a probability distribution describing how plausible each value is for the uncertain quantity, on the basis of the given information. Note that for most of the applications of interest in this study, the common interpretation of probability, as a relative frequency of occurences in the long run, does not make sense.

It is assumed in this study that no records of measured structural response are available, otherwise the problem becomes one of statistical system identification, which we have treated elsewhere [Beck, 1990; Katafygiotis, 1991]. In particular, this is the case during design, when the structure has not yet been built. In this case, the uncertainties in the modeling process are estimated subjectively, on the basis of any available information and past experience dealing with similar structures. Therefore, the "prior" probability distributions are assigned to the uncertain parameters subjectively, and are chosen to be of a convenient mathematical form, roughly consistent with the engineer's judgement regarding the relative plausibilities of the different values of these parameters.

The probability distribution of the system output, as well as the first two moments of this distribution, after accounting for the uncertainties of the model parameters and the model-error parameters, is given according to the theorem of total probability by appropriate high-dimensional integrals over the domain of all these uncertain parameters. Analytical evaluation of these high-dimensional integrals is not possible, in general.

Different approximate numerical methods have been used in the past in order to estimate the statistics of the uncertain structural response. In some studies, simulation methods were used to investigate the effects of uncertainties in structural properties [Shinozuka 1972], while in other studies perturbation methods were used to compute first and second moment statistics of response quantities [Chen and Saroka, 1973; Contreras 1980; Hisada and Nakagiri, 1982; Vanmarcke and Grigoriou, 1983; Liu et al., 1987]. Simulation techniques are quite powerful but, in general, are very costly and time consuming, since they require a large number of numerical solutions. This disadvantage becomes evident when one deals with large or even medium-sized systems, where numerical simulation becomes unrealistic on conventional digital computers. Numerical integration methods suffer from the same disadvantage, since the number of response solutions to be evaluated increases exponentially with the number of uncertain parameters. Perturbation techniques, on the other hand, are easily integrated into existing computer codes of deterministic structural dynamics, but they suffer from inaccuracy and questions of convergence when dynamic, particularly transient, and wave propagation problems are considered. Liu et al. [1988] came across these problems when using the second moment approach, based on a truncated Taylor series expansion of the model response with respect to the uncertain model parameters. Beck and Katafygiotis [1989] confirmed that the results obtained using the second moment approach when applied to problems of dynamic nature can be very misleading. We showed that much better results are obtained using a truncated Fourier series expansion of the model response with respect to the uncertain model parameters. Jensen [1989] extended this idea to more general orthogonal series expansions. The methodology presented here is an extension of the work presented in Beck and Katafygiotis [1989], and allows for an efficient and accurate estimation of the first two moments of the uncertain response of an uncertain multi-degree of freedom linear structural system.

FORMULATION OF THE PROBLEM

Let $\underline{x}(t;I) \in R^{N_R}$ denote the system output at time t, for given known input I. Let $\mathcal{M}$ denote a chosen class of models with uncertain parameters $\underline{a} \in R^{N_a}$, and $\underline{q}(t;\underline{a},I,\mathcal{M}) \in R^{N_R}$ be the corresponding model output of a particular model $M = M(\mathcal{M},\underline{a})$ of the class $\mathcal{M}$. For the sake of brevity, the symbols t and I will often be assumed without being explicitly written. For example, $\underline{x} \equiv \underline{x}(t) \equiv \underline{x}(t;I)$. Also, often when the parameters $\underline{a}$ are referenced, the symbol $\mathcal{M}$ will be assumed. For example, $\underline{q}(t;\underline{a},I,\mathcal{M}) \equiv \underline{q}(t;\underline{a},I) \equiv \underline{q}(t;\underline{a}) \equiv \underline{q}(\underline{a})$.

As mentioned earlier, because of the simplifications and assumptions used in choosing the particular class of models $\mathcal{M}$, there is an uncertainty concerning how accurately the response of any of its members M can predict the real system output. To account for the model error, that is, for possible differences between system and model output, a class of probability models $\mathcal{P}$, parameterized by the model-error parameters $\underline{\sigma} \in R^{N_\sigma}$ is selected. A particular probability model $P = P(\mathcal{P},\underline{\sigma})$ prescribes the function h giving the probability density function (pdf) of the system output given the model output:

$$p(\underline{x}|\underline{q},\underline{\sigma},\mathcal{P}) = h(\underline{x};\underline{q},\underline{\sigma}) \tag{1}$$

The selection of the classes $\mathcal{M}$ and $\mathcal{P}$ defines the class $\mathcal{M}_P$ parameterized by $[\underline{a}^T,\underline{\sigma}^T]^T \in R^{N_a+N_\sigma}$ prescribing the pdf:

$$p(\underline{x}|\underline{a},\underline{\sigma},\mathcal{M}_P) = h(\underline{x};\underline{q}(\underline{a}),\underline{\sigma}) \equiv f(\underline{x};\underline{a},\underline{\sigma}) \tag{2}$$

In order to account for the uncertainty in the parameters $\underline{a}$ and $\underline{\sigma}$, it is assumed that $\mathcal{M}_P$ also specifies a function $\pi_{\underline{a},\underline{\sigma}}$ corresponding to their prior pdf:

$$p(\underline{a},\underline{\sigma}|\mathcal{M}_P) = \pi_{\underline{a},\underline{\sigma}}(\underline{a},\underline{\sigma}) \tag{3}$$

It is assumed that $\underline{a}$ and $\underline{\sigma}$ are independently distributed, leading to:

$$\pi_{\underline{a},\underline{\sigma}}(\underline{a},\underline{\sigma}) = \pi_{\underline{a}}(\underline{a})\pi_{\underline{\sigma}}(\underline{\sigma}) \tag{4}$$

Both functions $\pi_{\underline{a}}(\underline{a})$ and $\pi_{\underline{\sigma}}(\underline{\sigma})$ are chosen subjectively, on the basis of available information and past experience dealing with similar structures. Denote by $S(\underline{a}) \subseteq R^{N_a}$ and $S(\underline{\sigma}) \subseteq R^{N_\sigma}$ the set of allowable values of $\underline{a}$ and $\underline{\sigma}$. The pdf $p(\underline{x}|\mathcal{M}_P)$ of the response $\underline{x}(t)$, based on the axioms of probability logic and assuming that $\underline{a}$ and $\underline{\sigma}$ are independently distributed, can be expressed as:

$$\begin{aligned} p(\underline{x}|\mathcal{M}_P) &= \int\limits_{S(\underline{a})}\int\limits_{S(\underline{\sigma})} p(\underline{x}|\underline{a},\underline{\sigma},\mathcal{M}_P)p(\underline{a}|\mathcal{M}_P)p(\underline{\sigma}|\mathcal{M}_P)d\underline{a}d\underline{\sigma} \\ &= \int\limits_{S(\underline{a})}\int\limits_{S(\underline{\sigma})} f(\underline{x};\underline{a},\underline{\sigma})\pi_{\underline{a}}(\underline{a})\pi_{\underline{\sigma}}(\underline{\sigma})d\underline{a}d\underline{\sigma} \end{aligned} \tag{5}$$

The "model-error" pdf $h(\underline{x};\underline{q},\underline{\sigma})$ is selected to be Gaussian, so that $f(\underline{x};\underline{a},\underline{\sigma})$ is a joint Gaussian distribution with mean $\underline{q}(\underline{a})$ and covariance matrix $\Sigma(\underline{\sigma})$:

$$\begin{aligned} f(\underline{x};\underline{a},\underline{\sigma}) &= G(\underline{x};\underline{q}(\underline{a}),\Sigma(\underline{\sigma})) \\ &\equiv \frac{1}{(2\pi)^{\frac{N_R}{2}}|\Sigma(\underline{\sigma})|^{\frac{1}{2}}}\exp\left(-\frac{1}{2}[\underline{x}-\underline{q}(\underline{a})]^T\Sigma^{-1}(\underline{\sigma})[\underline{x}-\underline{q}(\underline{a})]\right) \end{aligned} \tag{6}$$

The first two moments of the above distribution can then be shown to be:

$$\bar{\underline{x}} \equiv E[\underline{x}|\mathcal{M}_P] = \int\limits_{S(\underline{a})} \underline{q}(\underline{a})\pi_{\underline{a}}(\underline{a})d\underline{a} \tag{7}$$

$$\mathrm{Cov}(\underline{x}) = \int\limits_{S(\underline{a})} \underline{q}(\underline{a})\underline{q}(\underline{a})^T\pi_{\underline{a}}(\underline{a})d\underline{a} - \bar{\underline{x}}\bar{\underline{x}}^T + \int\limits_{S(\underline{\sigma})} \Sigma(\underline{\sigma})\pi_{\underline{\sigma}}(\underline{\sigma})d\underline{\sigma} \tag{8}$$

Under the additional assumption that the uncertainties of each of the elements x_j of $\underline{x}$ are independent, which leads to a diagonal covariance matrix with elements σ_j^2, the expressions for the first two moments of x_j become:

$$\bar{x}_j \equiv E[x_j|\mathcal{M}_P] = \int\limits_{S(\underline{a})} q_j(\underline{a})\pi_{\underline{a}}(\underline{a})d\underline{a} \tag{9}$$

$$\mathrm{Var}(x_j) = \int\limits_{S(\underline{a})} q_j^2(\underline{a})\pi_{\underline{a}}(\underline{a})d\underline{a} - \bar{x}_j^2 + E[\sigma_j^2] \tag{10}$$

Equations (9) and (10) can be rewritten as:

$$\bar{x}_j = \bar{q}_j \tag{11}$$

$$\mathrm{Var}(x_j) = \mathrm{Var}(q_j) + E[\sigma_j^2] \tag{12}$$

stating that the expected structural response is the mean model response and the variance of the structural response is equal to the variance of the model response due to the uncertain model parameters, plus the mean variance of the model error. Therefore, the problem of evaluating the first two moments of the uncertain structural response reduces to evaluating the first two moments of the model response due to the uncertainties in the model parameters.

MDOF LINEAR STRUCTURAL MODELS

Consider the class $\mathcal{M}_{N_d}$ of N_d-degree of freedom linear structural models, defined by the following equation of motion:

$$M\ddot{\underline{q}} + C\dot{\underline{q}} + K\underline{q} = -M\underline{b}\ddot{z}(t) \quad ; \quad \underline{q}(0) = \underline{q}_0, \dot{\underline{q}}(0) = \dot{\underline{q}}_0 \tag{13}$$

The $N_d \times N_d$ matrices M, C, and K are the mass, the damping and the stiffness matrix, respectively. It is assumed that classically damped modes exist, which requires that [Caughey and O'Kelly, 1965]:

$$CM^{-1}K = KM^{-1}C \tag{14}$$

The vector $\underline{q} = [q_1, q_2, \ldots, q_{N_d}]^T$ consists of the generalized displacements relative to the base of each degree of freedom, while $\underline{q} + \underline{b}z$ represent the corresponding total or absolute generalized displacements. The components of the vector $\underline{b} = [b_1, b_2, \ldots, b_{N_d}]^T$ are called pseudo-static influence coefficients, and they are known from the prescribed geometry of the structural model.

In order to obtain uncoupled equations of motion, modal analysis is used. The following notation is adopted:

$\underline{\phi}^{(r)}$: r^{th} modeshape normalized so that $\underline{\phi}^{(r)T}M\underline{\phi}^{(r)} = 1$

ω_r: r^{th} modal frequency; $\omega_r \geq \omega_s$ for $r > s$

ζ_r: damping ratio of rth mode ; $\zeta_r \geq 0$

α_r:participation factor of the rth mode; $\alpha_r = \underline{\phi}^{(r)T}M\underline{b}$

$\beta_i^{(r)}$: effective participation factor of the rth mode at the ith dof
The value of $\beta_i^{(r)} = \phi_i^{(r)}\alpha_r$ is independent of the normalization chosen for the modeshapes.

Using modal analysis, the equation of motion for the contribution of the rth mode to the response at the ith degree of freedom (dof) is:

$$\ddot{q}_i^{(r)} + 2\zeta_r\omega_r\dot{q}_i^{(r)} + \omega_r^2 q_i^{(r)} = -\beta_i^{(r)}\ddot{z}(t) \tag{15}$$

with initial conditions:

$$q_i^{(r)}(0) = \phi_i^{(r)}\sum_{j=1}^{N_d} m_j\phi_j^{(r)}q_j(0)\ , \ \dot{q}_i^{(r)}(0) = \phi_i^{(r)}\sum_{j=1}^{N_d} m_j\phi_j^{(r)}\dot{q}_j(0) \tag{16}$$

The response at the ith dof can be expressed as a superposition of the first $N_m \leq N_d$ modal contributions, where the higher modal contributions are neglected:

$$q_i(t) \cong \sum_{r=1}^{N_m} q_i^{(r)}(t) \quad ; \quad N_m \leq N_d \tag{17}$$

The parameters of the class of models prescribed by Equation (13) are the elements of M, C, and K, and the components of the initial conditions $\underline{q}_o$ and $\dot{\underline{q}}_o$. The uncertainties associated with some of these parameters can be neglected when compared with the uncertainties associated with the remaining parameters and, therefore, these parameters can be treated as deterministic and need not be included in the vector of uncertain parameters $\underline{a}$. In particular, the elements of the mass matrix M will be assumed to be determinstically known, since they can be estimated accurately enough from the structural drawings. Also, the initial conditions will be treated as being deterministically known, since it is usually assumed that the system starts from rest; in this case, $\underline{q}_o = \dot{\underline{q}}_o = 0$. Since the values of the elements of the damping matrix C cannot be constructed by synthesis from the structural drawings, the uncertainties concerning the damping of a structural model will be accounted for by uncertainties of the modal damping ratios ζ_r. Finally, the uncertainty in the stiffness distribution is assumed to be parameterized through a set of nondimensional positive parameters $\theta_i, i = 1, \ldots, N_\theta$, so that:

$$K = K_0 + \sum_{i=1}^{N_\theta} \theta_i K_i \quad ; \quad \theta_i > 0 \; , \; i = 1, \ldots, N_\theta \tag{18}$$

Each parameter θ_i scales the stiffness contribution K_i of a certain substructure to the total stiffness matrix and K_o accounts for the stiffness contributions of those substructures with deterministic stiffnesses. Usually, the expected value of each scaling parameter θ_i is taken to be unity.

It follows from the above that the vector of original uncertain parameters is:

$$\underline{a} = [\theta_1, \theta_2, \ldots, \theta_{N_\theta}, \zeta_1, \zeta_2, \ldots, \zeta_{N_M}]^T \tag{19}$$

The uncertainties of the remaining modal parameters controlling the dynamic response, that is, the modal frequencies and the effective participation factors, must be evaluated in terms of the uncertainties of the parameters $\underline{\theta}$ on which they depend.

PROBABILISTIC MODELING OF UNCERTAIN PARAMETERS

It is assumed that knowing the value of one of the original uncertain parameters contained in $\underline{a}$ does not influence judgement of the plausibilities of the values of the remaining parameters so that their probability distributions can be specified independently. A Gamma distribution is chosen for the pdf of each parameter a_j:

$$\begin{aligned} \pi_{a_j}(a_j) = g(a_j; \mu_{a_j}, \nu_{a_j}) &= \frac{\mu_{a_j}^{\nu_{a_j}}}{\Gamma(\nu_{a_j})} a_j^{\nu_{a_j}-1} e^{-\mu_{a_j} a_j} \quad ; \quad a_j \geq 0 \\ &= 0 \quad ; \quad a_j < 0 \end{aligned} \tag{20}$$

where $\nu_{a_j} > 0$, $\mu_{a_j} > 0$ and $\Gamma(\nu_{a_j})$ is the Gamma function. Since a Gamma distribution is specified by two parameters μ, ν, it gives the freedom to specify both the expected value and the variance of the uncertain parameter through:

$$\bar{a}_j = \frac{\nu_{a_j}}{\mu_{a_j}} \quad , \quad \sigma^2_{a_j} = \frac{\nu_{a_j}}{\mu^2_{a_j}} \tag{21}$$

Note that a Gamma distribution is defined to be zero for negative values of the parameters, which is consistent with the elements of $\underline{a}$ being allowed only positive values. Another advantage of the Gamma distribution is that it allows a closed form solution for integrals of the following form:

$$\int_0^\infty a_j^\beta e^{-\gamma a_j} \begin{matrix}\sin\\ \cos\end{matrix}(\delta a_j) g(a_j;\mu_{a_j},\nu_{a_j}) da_j \quad ; \quad \beta > -(\nu_{a_j}+1) \quad , \quad \gamma > -\mu_{a_j} \tag{22}$$

The expected values of quantities of interest dependent on $\underline{\theta}$ can be obtained as follows. Let $f(\underline{\theta})$ denote such a quantity of interest. For example, $f(\underline{\theta})$ could be one of the following: $\omega_r(\underline{\theta}), \omega_r(\underline{\theta})^2, \beta_i^{(r)}(\underline{\theta}), \beta_i^{(r)}(\underline{\theta})^2$, $\omega_r(\underline{\theta})\beta_i^{(r)}(\underline{\theta})$, etc. By plotting any such function $f(\underline{\theta})$, it can be seen that it is a smooth slowly-varying function of the parameters $\underline{\theta}$ and, therefore, it can be approximated with a quadratic polynomial in $\underline{\theta}$:

$$f(\underline{\theta}) \simeq c_{0,f} + \sum_{i=1}^{N_\theta} c_{i,f}\theta_i + \sum_{i=1}^{N_\theta}\sum_{j=1}^{N_\theta} c_{ij,f}\theta_i\theta_j = \underline{c}_f^T \underline{\tilde{\theta}}(\underline{\theta}) \tag{23}$$

where

$$\underline{c}_f = \left[c_{0,f}, c_{1,f}, \ldots, c_{N_\theta,f}, c_{11,f}, c_{12,f}, \ldots, c_{1N_\theta,f}, c_{22,f}, c_{23,f}, \ldots, c_{N_\theta N_\theta,f}\right]^T \tag{24}$$

and

$$\underline{\tilde{\theta}}(\underline{\theta}) = [1, \theta_1, \theta_2, \ldots, \theta_{N_\theta}, \theta_1^2, \theta_1\theta_2, \ldots, \theta_1\theta_{N_\theta}, \theta_2^2, \theta_2\theta_3, \ldots, \theta_{N_\theta}^2]^T \tag{25}$$

The coefficient vector $\underline{c}_f$ is evaluated by requiring the above quadratic approximation to pass through a number of points $(\underline{\theta}^{(i)}, f(\underline{\theta}^{(i)}))$, appropriately chosen around the expected value of the parameters $\underline{\theta}$. This requirement results in a system of linear equations from which the coefficient vector can be recovered. After the coefficient vector $\underline{c}_f$ has been evaluated, the expected value of f is approximated as follows.

$$\bar{f}(\underline{\theta}) \equiv E[f(\underline{\theta})] \simeq \underline{c}_f^T \bar{\underline{\tilde{\theta}}} \tag{26}$$

where

$$\begin{aligned}\bar{\underline{\tilde{\theta}}} &\equiv E[\underline{\tilde{\theta}}(\underline{\theta})] \\ &= [1, \bar{\theta}_1, \bar{\theta}_2, \ldots, \bar{\theta}_{N_\theta}, \overline{\theta_1^2}, \bar{\theta}_1\bar{\theta}_2, \ldots, \bar{\theta}_1\bar{\theta}_{N_\theta}, \overline{\theta_2^2}, \bar{\theta}_2\bar{\theta}_3, \ldots, \overline{\theta_{N_\theta}^2}]^T\end{aligned} \tag{27}$$

The variance of $f(\underline{\theta})$ can be calculated from:

$$\text{Var}(f) = E[f(\underline{\theta})^2] - E^2[f(\underline{\theta})] \tag{28}$$

where the expected value of $f(\underline{\theta})^2$ is approximated in a similar way, by assuming a quadratic approximation for the function $f(\underline{\theta})^2$.

The marginal probability distribution of each uncertain modal frequency ω_r is approximated with a Gamma distribution with expected value and variance determined by the above procedure. The validity of this approximation is verified later with a numerical example.

It should be noted that although the quadratic approximation for the modal parameters in terms of the uncertain stiffness parameters $\underline{\theta}$ gives accurate results, a quadratic approximation for the model response in terms of either the modal parameters or the stiffness parameters gives poor results. This is because the latter functional dependencies are not smooth and slowly-varying but instead are highly oscillatory.

STATISTICS OF $q_i(t)$

It follows from Equation (17) that:

$$E[q_i(t)] \simeq \sum_{r=1}^{N_m} E[q_i^{(r)}(t)] \quad , \quad E[q_i(t)^2] \simeq \sum_{r=1}^{N_m}\sum_{s=1}^{N_m} E[q_i^{(r)}(t)q_i^{(s)}(t)] \tag{29}$$

The objective is to evaluate the terms in the above sums. Only the proposed procedure for evaluating the terms of the left sum is described here, but a similar procedure can be used to evaluate the terms in the right sum, as shown in Katafygiotis [1991].

Assuming the system starts from rest, the equation of motion for the rth modal contribution is given by Equation (15) with initial conditions $q_i^{(r)}(0) = \dot{q}_i^{(r)}(0) = 0$. Notice that $q_i^{(r)}(t) = q_i^{(r)}(t;\underline{\theta},\zeta_r)$, $\omega_r = \omega_r(\underline{\theta})$ and $\beta_i^{(r)} = \beta_i^{(r)}(\underline{\theta})$. The response $q_i^{(r)}(t;\underline{\theta},\zeta_r)$ is given by the following Duhamel integral:

$$q_i^{(r)}(t;\underline{\theta},\zeta_r) = -\int_0^t h(\tau;\omega_r(\underline{\theta}),\zeta_r)\beta_i^{(r)}(\underline{\theta})\ddot{z}(t-\tau)d\tau \tag{30}$$

where $h(\tau;\omega_r,\zeta_r) = \frac{1}{\omega_{r,D}}e^{-\zeta_r\omega_r\tau}\sin(\omega_{r,D}\tau)$ and $\omega_{r,D} = \omega_r\sqrt{1-\zeta_r^2}$. The expected value of $q_i^{(r)}(t)$, which is the general term in the left sum of Equation (29), is given by the following convolution integral:

$$\bar{q}_i^{(r)}(t) \equiv E[q_i^{(r)}(t)] = -\int_0^t \bar{g}_i^{(r)}(\tau)\ddot{z}(t-\tau)d\tau \tag{31}$$

$$\text{where} \quad \bar{g}_i^{(r)}(\tau) = \int_{S(\underline{\theta})}\int_0^\infty h(\tau;\omega_r(\underline{\theta}),\zeta_r)\beta_i^{(r)}(\underline{\theta})\pi_{\underline{\theta}}(\underline{\theta})\pi_{\zeta_r}(\zeta_r)d\underline{\theta}d\zeta_r \tag{32}$$

It follows from Equation (31) that $E[q_i^{(r)}(t)]$ can be calculated using FFT's, if the function $\bar{g}_i^{(r)}(t)$ is known. The latter function, as can be seen from Equation (32), is an $(N_\theta + 1)$-dimensional integral, over the domain of all uncertain parameters involved. To evaluate efficiently this high-dimensional integral assume a one-to-one transformation of the parameters $\underline{\theta}$: $\underline{\theta} \rightarrow \underline{\tilde{\eta}}^{(r)} = [\omega_r, (\underline{\eta}^{(r)})^T]^T \in R^{N_\theta}$, then:

$$\bar{g}_i^{(r)}(\tau) = \int\limits_{S(\underline{\eta}^{(r)})} \int\limits_0^\infty \int\limits_0^\infty h(\tau; \omega_r, \zeta_r) \beta_i^{(r)}(\underline{\tilde{\eta}}^{(r)}) \pi_{\underline{\tilde{\eta}}^{(r)}}(\underline{\tilde{\eta}}^{(r)}) \pi_{\zeta_r}(\zeta_r) d\underline{\eta}^{(r)} d\omega_r d\zeta_r \tag{33}$$

Note that:

$$\pi_{\underline{\tilde{\eta}}^{(r)}}(\underline{\tilde{\eta}}^{(r)}) = \pi_{\underline{\tilde{\eta}}^{(r)}}(\omega_r, \underline{\eta}^{(r)}) = \pi_{\omega_r}(\omega_r) \pi_{\underline{\eta}^{(r)}|\omega_r}(\underline{\eta}^{(r)}|\omega_r) \tag{34}$$

where $\pi_{\omega_r}(\omega_r)$ is the marginal distribution of the frequency ω_r, which can be approximated well with a Gamma distribution that can be computed as described earlier.

Substituting (34) into (33), we obtain:

$$\begin{aligned}\bar{g}_i^{(r)}(\tau) &= \int\limits_0^\infty \int\limits_0^\infty h(\tau; \omega_r, \zeta_r) \left[\int\limits_{S(\underline{\eta}^{(r)})} \beta_i^{(r)}(\omega_r, \underline{\eta}^{(r)}) \pi_{\underline{\eta}^{(r)}|\omega_r}(\underline{\eta}^{(r)}|\omega_r) d\underline{\eta}^{(r)} \right] \\ &\quad \pi_{\omega_r}(\omega_r) \pi_{\zeta_r}(\zeta_r) d\omega_r d\zeta_r \\ &= \int\limits_0^\infty \int\limits_0^\infty h(\tau; \omega_r, \zeta_r) E[\beta_i^{(r)}|\omega_r] \pi_{\omega_r}(\omega_r) \pi_{\zeta_r}(\zeta_r) d\omega_r d\zeta_r\end{aligned} \tag{35}$$

where $E[\beta_i^{(r)}|\omega_r]$ is the conditional expected value of $\beta_i^{(r)}$ given ω_r, and is given by:

$$E[\beta_i^{(r)}|\omega_r] = \int\limits_{S(\underline{\eta}^{(r)})} \beta_i^{(r)}(\omega_r, \underline{\eta}^{(r)}) \pi_{\underline{\eta}^{(r)}|\omega_r}(\underline{\eta}^{(r)}|\omega_r) d\underline{\eta}^{(r)} \tag{36}$$

Since $\beta_i^{(r)}(\underline{\theta})$ and $\omega_r(\underline{\theta})$ are slowly varying functions in the θ_i's, $E[\beta_i^{(r)}|\omega_r]$ is also expected to vary slowly with ω_r. Therefore, the following linear approximation in ω_r is assumed:

$$E[\beta_i^{(r)}|\omega_r] \simeq c_{i,0}^{(r)} + c_{i,1}^{(r)} \omega_r \tag{37}$$

The coefficients $c_{i,0}^{(r)}$ and $c_{i,1}^{(r)}$ can be recovered from the following:

$$\begin{aligned}E[\beta_i^{(r)}] &\simeq c_{i,0}^{(r)} + c_{i,1}^{(r)} E[\omega_r] \\ E[\omega_r \beta_i^{(r)}] &\simeq c_{i,0}^{(r)} E[\omega_r] + c_{i,1}^{(r)} E[\omega_r^2]\end{aligned} \tag{38}$$

where the expected value of the quantities in the above relations can be evaluated using quadratic approximations in $\underline{\theta}$, as discussed earlier. The accuracy of the linear approximation of Equation (37) is shown later with a numerical example.

By substituting (37) into (35), the original $(N_\theta + 1)$-dimensional expression for $\bar{g}_i^{(r)}(t)$ reduces to a two-dimensional integral over ω_r and ζ_r. Furthermore, this two-dimensional integral can be evaluated analytically if the transformed variables $\alpha_1 = \omega_r\sqrt{1-\zeta_r^2}$ and $\alpha_2 = \omega_r\zeta_r$ are introduced, and are assumed independently Gamma distributed.

Thus, we finally achieve an evaluation of the expected value of the rth modal contribution without performing any numerical integration, thereby avoiding a problem which is computationally prohibitive for a large number of parameters N_θ.

NUMERICAL EXAMPLE

A three-story shear building is assumed with uniform mass $m_i = m_o$ and interstory stiffness $k_i = k_o\theta_i$, $i = 1,2,3$ ($k_o = 2000 m_o sec^{-2}$). Each θ_i , $i = 1,2,3$ is assumed independently Gamma distributed with expected value $\bar{\theta}_i = 1$ and coefficient of variation $\alpha_{\theta_i} = \sigma_{\theta_i}/\bar{\theta}_i = 0.10$. Similarly, each damping ratio ζ_r , $r = 1,2,3$ is assumed independently Gamma distributed with $\bar{\zeta}_r = 0.05$ and $\alpha_{\zeta_r} = 0.20$. The base acceleration is taken to be the 1940 El Centro earthquake record, NS component. Figure 1 shows the normalized marginal pdfs of the modal frequencies ω_r , $r = 1,2,3$, obtained using simulations (dashed-dotted curves), versus the ones corresponding to the approximations by Gamma distributions described earlier (solid curves). It can be seen that the two sets of curves corresponding to the exact and the approximate case compare very well. Figure 2 shows the conditional expected value $E[\beta_i^{(1)}|\omega_1], i = 1,2,3$, plotted against ω_1. The solid curves are obtained using simulations and the dashed-dotted curves are obtained using the linear approximations of Equation (37). The dashed curve corresponds to the marginal pdf of ω_1, appropriately scaled. It can be seen that the two sets of curves compare very well over the region of interest, that is, over the region where the marginal distribution of ω_1 is significant. Figures 3a and 3b show the expected response at the roof $E[q_3(t)]$ and the corresponding standard deviation $\sigma_{q_3}(t)$, respectively. The solid curves are obtained using numerical integration, while the dashed-dotted curves are obtained using the much more efficient methodology proposed in this paper. Again, it can be seen that the comparison is very good.

CONCLUSIONS

The dominant computational effort required by the proposed method is $\mathcal{O}(N_1^2 N_M^2)$ [Katafytiotis 1991]. In the latter expression, N_1 is the order of the matrices involved in two-dimensional FFT's employed by the proposed methodology when calculating the variance of the response and is given by $N_1 = 2^{\mathrm{INT}(\log_2 N_T)+2}$., where $N_T\Delta t$ is the time interval over which the statistics of the response are to be calculated. N_m is the number of modes included in the response. Note that the amount of computation is practically independent of the number of uncertain parameters N_θ. Therefore, the computational effort required by the proposed method is extremely low when compared with that required using numerical integration, which grows exponentially with the number of uncertain parameters. The proposed approximate method is very accurate, as was demonstrated with a numerical example.

It can be concluded that the proposed methodology is an efficient and accurate method for calculating the statistics of the dynamic response due to model uncertainties. The method provides a tool for the engineer during the design of a structure to investigate the resulting uncertainties in the structural response due to uncertainties in the modeling process.

References

Beck, J.L., "Statistical System Identification of Structures," *Structural Safety and Reliability*, ASCE, II, 1395-1402, 1990.

Beck, J.L. and Katafygiotis, L.S., "Treating Model Uncertainties in Structural Dynamics," *Proceedings 9th World Conference on Earthquake Engineering*, 5, Tokyo-Kyoto, Japan, 1989.

Chen, P.C. and Soroka, W.W., "Impulse Response of a Dynamic System with Statistical Properties," *Journal of Sound and Vibration*, 31, 309-314, 1973.

Contreras, H., "The Stochastic Finite Element Model," *Computers and Structures*, 12, 341-348, 1980.

Hisada, T. and Nakagiri, S., "Stochastic Finite Element Analysis of Uncertain Structural Systems," *4th International Conference on Finite Element Methods*, Melbourne, Australia, 1982.

Jensen, H.A., "Dynamic Response of Structures with Uncertain Parameters," EERL Report No. 89-02, California Institute of Technology, 1989.

Katafygiotis, L.S., "Treatment of Model Uncertainties in Structural Dynamics," EERL Report No. 91-01, California Institute of Technology, 1991.

Liu, W.K., Belytschko, T., Mani, A., and Besterfield, G.A., "A Variational Formulation for Probabilistic Mechanics," *Finite Element Methods for Plate and Shell Structures, Vol. 2: Formulations and Algorithms*, T. Hughes and E. Hinton, Eds., Pineridge Press, U.K., 1988.

Liu, W.K., Mani, A., and Belytschko, T., "Finite Element Methods in Probabilistic Mechanics," *Probabilistic Engineering Mechanics*, 2(4), 201-213, 1987.

Shinozuka, M., "Monte Carlo Solution of Structural Dynamics," *International Journal of Computers and Structures*, 2, 885-874, 1972.

Vanmarcke, E. and Grigoriou, M., "Stochastic Finite Element Analysis of Simple Beams," *Journal of Engineering Mechanics*, 109(5), 1203-1214, 1983.

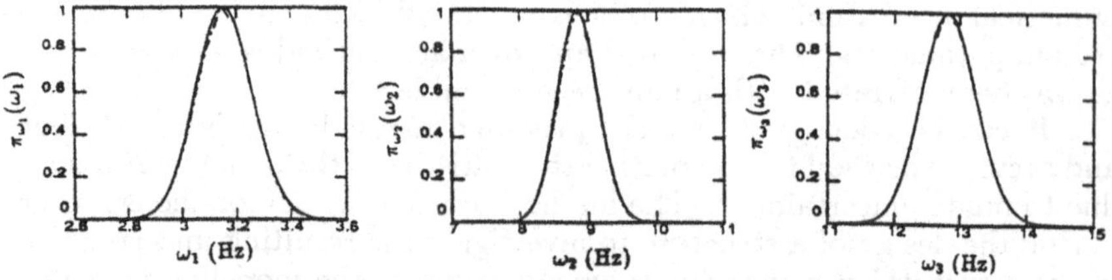

Figure 1. Normalized marginal pdfs of the modal frequencies ω_r , $r = 1, 2, 3$. The solid curves correspond to the approximations by Gamma distributions and the dashed-dotted curves to the exact cases obtained using simulations.

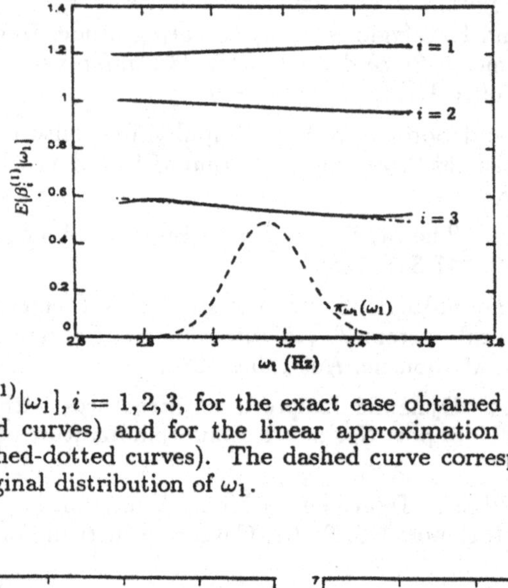

Figure 2. $E[\beta_i^{(1)}|\omega_1], i = 1, 2, 3$, for the exact case obtained using simulations (solid curves) and for the linear approximation of Equation (37) (dashed-dotted curves). The dashed curve corresponds to a scaled marginal distribution of ω_1.

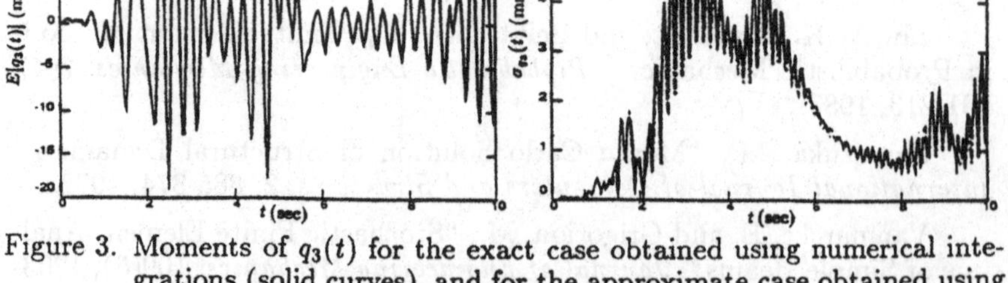

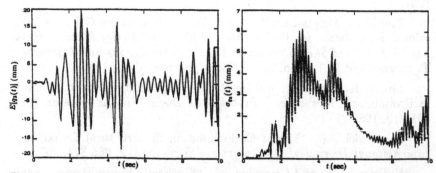

Figure 3. Moments of $q_3(t)$ for the exact case obtained using numerical integrations (solid curves), and for the approximate case obtained using the proposed methodology (dashed-dotted curves). (a) $E[q_3(t)]$ and (b) $\sigma_{q_3}(t)$

Evaluation of Response Statistics of Moment Resisting Steel Frames under Seismic Excitation

D.F. Eliopoulos, Y-K. Wen
University of Illinois at Urbana-Champaign, Urbana, IL 61801, U.S.A.

ABSTRACT

A method of evaluation of the response statistics of a low rise, moment resisting steel frame, located at a high seismicity region is presented in this paper. Available geological and seismological information at the site of interest is used to identify the parameters of a ground motion model for future future seismic excitations. A strong column–weak beam (SCWB) structural model is developed to model the frame with reduced degrees of freedom and hence lower the computational cost of the analysis. The inelastic behavior of the system is modeled by a smooth hysteresis model which can be linearized in closed form and the response statistics to future earthquakes are obtained by the random vibration method. The method yields very good results compared to Monte Carlo simulations with a significant reduction in computational effort. The response statistics obtained can be used in the assessment of the performance and reliability of the building.

INTRODUCTION

The performance and safety of buildings in future earthquakes is a problem of major interest to structural engineers. The knowledge of seismic excitation properties such as the source mechanism and the characteristics of ground motion has expanded considerably in the recent past; future events, however, still can not be predicted with certainty. As a result, probabilistic analysis is needed to quantify the uncertainty due to earthquake ground motion. This study presents a method evaluating the response statistics of a building under ground excitation with specified earthquake magnitude and source–to–site distance, local intensity and duration. The study is part of an ongoing project which evaluates the reliability of steel buildings designed according to current procedures such as UBC–88 and

SEAOC. It concentrates on the modeling of ground motion and structural frame and on statistical response analysis. Other important aspects of the problem, such as uncertainty due to nonstructural elements resistance, damage modeling, reliability evaluation and applications in design will be reported elsewhere. The ground motion is modeled as a nonstationary random process whose amplitude and frequency content vary with time. The ground motion model proposed by Yeh and Wen (1989) is used for this purpose. This model can be used either in generating artificial records for time history response analysis or in an analytical solution for the response statistics by random vibration methods (Wen, 1989). The model parameters are functions of the site and source characteristics.

Low-to medium-rise moment resisting steel frames are considered in this study. A simple analytical model, reflecting the strong column-weak beam design of these frames is developed to reduce the system's degrees of freedom and to facilitate the random vibration analysis. A smooth hysteresis model (Baber and Wen, 1980) is used to describe the inelastic behavior of the system under cyclic loading and the method of statistical equivalent linearization is used to obtain the response statistics.

GROUND MOTION MODELING

Seismic Hazard Analysis

Seismic hazard is dominated by the geological and seismological characteristics of the area of interest. Recent evidence indicates that it is of advantage in hazard analysis to distinguish potential future events presenting a threat to a site as either characteristic (major events which occur in a major fault) or non-characteristic (minor, local events). Non-characteristic events may be more destructive than a distant characteristic earthquake if their epicenter is near the site or because of local geology. On the other hand if the site is located near a major fault, the characteristic earthquake hazard usually governs.

The major parameters of the characteristic earthquake for the analysis are recurrence time T, magnitude M, distance to the fault R and attenuation. Following the characteristic earthquake model (Schwartz and Coppersmith, 1984), recurrence of a characteristic event along a fault segment can be modeled by a renewal process. A lognormal distribution best represents observed recurrence time behavior (Nishenko and Bulland, 1987). The expected moment magnitude of the characteristic event can be evaluated from the dimensions and other physical characteristics of the fault segment (U.S. Geological Survey, 1988). The significant duration t_D, which is associated with the strong phase of the ground motion (Trifunac and Brady, 1975) can then be obtained as a function of magnitude, M, and site-to-source distance, R, using an empirical relation of the form

$\log_{10} t_D = a_0 + a_1 M + a_2 R$ where a_0, a_1 and a_2 are site dependent coefficients, determined by regression analysis based on available data. To identify the spectral properties of the ground motion at the site due to characteristic earthquake, its Fourier amplitude spectrum is scaled in terms of magnitude, site-to-source distance, attenuation and local site geology using the empirical model proposed by Trifunac and Lee (1989). This model uses a frequency (or period) dependent attenuation function, $Att(\Delta, M, T)$, to estimate the Fourier amplitude spectrum, $FS(T)$, as

$$\log_{10} FS(T) = M + Att(\Delta, M, T) + \hat{b}_1(T)M + \hat{b}_2(T)s + \\ + \hat{b}_5(T) + \hat{b}_6(T)M^2 - \epsilon(t) \tag{1}$$

where T is the period (sec), s characterizes the local geology (s=0 for alluvium, s=2 for basement rock, s=1 for intermediate sites), Δ is a measure of the site-to-source distance, $\hat{b}_1(T)$, $\hat{b}_2(T)$, $\hat{b}_3(T)$, $\hat{b}_4(T)$ are regression coefficients and $\epsilon(t)$ is an error (or uncertainty) term following a normal distribution. The Arias intensity, I_A, can be evaluated from the Fourier amplitude spectrum as

$$I_A = \frac{1}{\pi} \int_0^\infty FS^2(\omega)\, d\omega \tag{2}$$

where $\omega = 2\pi/T$ is the cyclic frequency (rad/sec).

The major parameters of non-characteristic (local) earthquakes for the analysis are occurrence rate, N, and modified Mercalli intensity, I_{MM}. Since the source, magnitude and epicentral distance of local events are generally difficult to determine, modified Mercalli intensity is used. An exponential distribution is assumed for I_{MM} and a Poisson process is used to model the occurrence of local events in time (Algermissen et al., 1982). The Poisson assumption seems to be consistent with historical occurrence of intermediate and large events. Small events may depart significantly from a Poisson process but they are only of marginal interest to engineering applications. Based on the above assumptions, the mean occurrence rate, N, of non-characteristic events with modified Mercalli intensity greater than I at a site can be found by the log-linear relation $\log_{10} N = a - bI$, where a and b are constants associated with the site. The significant duration t_D is then determined from an empirical relation of the form $\log_{10} t_D = a_0 - a_1 I_{MM}$ where a_0 and a_1 are site dependent coefficients. To identify the frequency content of the ground motion due to non-characteristic earthquakes, the Fourier amplitude spectrum is scaled in terms of modified Mercalli intensity and local site geology. Following the method proposed by Trifunac and Lee

(1989), the Fourier amplitude spectrum, FS(T), can be expressed as

$$\log_{10} FS(T) = \hat{b}_1(T)I_{MM} + \hat{b}_2(T)s + \hat{b}_4(T) - \epsilon(t) \tag{3}$$

where T is the period (sec), s characterizes the geologic site conditions (s=0 for alluvium, s=2 for basement rock, s=1 for intermediate sites), $\hat{b}_1(T)$, $\hat{b}_2(T)$, $\hat{b}_3(T)$, $\hat{b}_4(T)$ are regression coefficients and $\epsilon(t)$ is an error (or uncertainty) term following a normal distribution. The Arias intensity can again be evaluated from the Fourier amplitude spectrum using Eq. 2.

Identification of Ground Motion Model Parameters

Ground motion is modeled as an amplitude and frequency modulated filtered Gaussian white noise following the approach of Yeh and Wen (1989). Such a process can be characterized by an intensity function, I(t), a frequency modulation function, $\phi(t)$, and a power spectral density function, $S(\omega)$. The intensity function, which is deterministic, controls the amplitude of the process and is given by

$$I^2(t) = A\frac{t^B\, e^{-Ct}}{D + t^E} \tag{4}$$

where parameters A, B, C, D and E are identified from a regression procedure by comparison with data. The frequency modulation function, which controls the change of the frequency content of the process with time is given by

$$\phi(t) = \mu_0(t) / \mu'_0(t_0) \tag{5}$$

where $\mu_0(t)$ is an n-th polynomial approximation of the mean number of zero crossings and $\mu'_0(t)$ its first derivative with respect to time, evaluated at time t_0:

$$\mu_0(t) = r_1 t + r_2 t^2 + \ldots + r_n t^n \tag{6}$$

The coefficients r_1, r_2,..., r_n are identified from data. In most cases, $\phi(t)$ is expressed as a third order polynomial. Finally the power spectral density function (PSD) controls the frequency content of the process. A normalized Clough–Penzien spectrum for this purpose.

$$S_{CP}(\omega) = S_0\left[\frac{\omega_g^4 + 4\zeta_g^2\omega_g^2\omega^2}{(\omega_g^2 - \omega^2)^2 + 4\zeta_g^2\omega_g^2\omega^2}\right]\left[\frac{\omega^4}{(\omega_f^2 - \omega^2)^2 + 4\zeta_f^2\omega_f^2\omega^2}\right] \tag{7}$$

Spectrum parameters, S_0, ω_g, ζ_g, ω_f and ζ_f can be identified from the squared Fourier amplitude spectrum, which is related to the PSD function by

$$S(\omega) = \frac{1}{\pi\, t_F} FS^2(\omega) \tag{8}$$

The intensity envelope parameters are then identified using the Arias intensity and the significant duration of the record. The frequency modulation function coefficients are finally determined from data at the site or at some other site with similar soil conditions (Details can be found in Eliopoulos and Wen, 1991).

STRUCTURAL MODELING AND ANALYSIS

The response of a structural system to a stochastic load is generally described in terms of the first and second order statistics of response variables such as nodal displacements, joint rotations, member forces etc.. These response variables are frequently referred to as state variables, since their values define the state of the structural system. As the number of state variables increases, the size of the covariance matrix (second order statistics) of the state variables increases dramatically (i.e. according to $n(n+1)/2$). Therefore reduction of degrees of freedom of the structural system is essential. Following the strong column–weak beam design philosophy of moment resisting steel frames (International Conference of Building Officials, 1988), a simplified strong column–weak beam model is developed to reduce the computational cost of the analysis.

The Strong Column–Weak Beam (SCWB) Model

In principle, the SCWB model localizes inelastic behavior at the base and the floor levels assuming that all columns of the frame remain elastic and all yielding is concentrated at the base and the beams. A linear beam–column element at each story and a rotational inelastic spring at the base and each floor level represent this behavior (Fig. 1). Masses are lumped at floor levels and rotational inertia is neglected. Rayleigh damping is used, with a 5% critical damping ratio in the first two modes. The moment of inertia of the equivalent beam–column element at a story is assumed to be equal to the sum of the moments of inertia of the columns of the original frame at that story. The restoring moment of the i–th level rotational spring, M_i^{nl}, is given by:

$$M_i^{nl} = a_i G_i \theta_i + (1 - a_i)\, G_i Y_i \quad ; \quad i = 0, ns \tag{9}$$

where θ_i is the joint rotation at the i–th level, G_i the elastic stiffness coefficient and a_i the post–to–pre–yield stiffness ratio of the i–th level rotational hysteretic spring. The smooth differential model developed by Baber and Wen (1980) models the hysteretic nature of the restoring moment. The model is well known for its versatility; it is also computationally suited for random vibration analysis, since it can be easily combined with the method of statistical equivalent linearization. Assuming no stiffness or strength degradation, the hysteretic component of the joint rotation at the i–th level, Y_i, is given by

$$\dot{Y}_i = A_i\dot{\theta}_i - \beta_i\,|\dot{\theta}_i Y_i|Y_i - \gamma_i\,\dot{\theta}_i Y_i^2 \quad ; \quad i = 0, ns \tag{10}$$

where ns is the number of stories. The linear stiffness coefficients of the rotational springs, G_i, are identified to match the first mode of the frame. Inelastic parameters α_i, A_i, β_i, γ_i are then identified for each level, from a simulated quasi–static test. This test is simulated numerically by an incremental static analysis performed by the well known finite element program DRAIN–2DX. A lateral load pattern is applied at floor levels and several cycles of inelastic response are generated using a cyclic load path. The response is then used as input for the identification algorithm (Eliopoulos and Wen, 1991). It is important to notice that the identification can be performed independently for the parameters of each rotational spring since the residuals at different levels are not coupled with respect to the unknown parameters.

The SCWB model has the ability to capture the stiffness coupling effect between adjacent stories of the frame and can therefore reproduce the actual response without significant loss in accuracy. Time history analyses were performed using the SCWB model for different acceleration records and the results were compared to those obtained by DRAIN–2DX. A comparison of interstory drifts for a five story–three bay frame is shown in Fig. 2 for the El Centro Differential Array record of the Imperial Valley, 1979 earthquake.

Random Vibration Analysis

To implement the SCWB model in random vibration analysis, Eqs. 10 are linearized at all levels, using the method of statistical equivalent linearization. The linearized equations are written in matrix form as

$$\dot{\underline{Y}} = [C_{eq}]\dot{\underline{\theta}} + [K_{eq}]\underline{Y} \tag{11}$$

where $[C_{eq}]$ and $[K_{eq}]$ are diagonal matrices, given as functions of the response statistics (Wen, 1980). The equilibrium equations of the model, formulated in terms of interstory drifts, u_i, under base excitation, $\ddot{a}_g$, are:

$$\ddot{\underline{u}} = -a\dot{\underline{u}} - [K](\underline{u} + b\dot{\underline{u}}) - [A](\underline{\theta} + b\dot{\underline{\theta}}) - e_1\ddot{a}_g \tag{12}$$

$$\dot{\underline{\theta}} = -\frac{1}{b}\underline{\theta} - \frac{1}{b}[F]^{-1}[E](\underline{u} + b\dot{\underline{u}}) - \frac{1}{b}[F]^{-1}[D]\underline{Y} \tag{13}$$

where a and b are the Rayleigh damping proportionality coefficients and matrices [A], [D], [E], [F] and [K] are functions of m_i, h_i, EI_i, G_i and α_i. To model $\ddot{a}_g$ as a frequency modulated filtered white noise, the Clough–Penzien filtering equations are added to the system of Eqs. 11, 12 and 13 as follows:

$$\ddot{x}_g + \left[-\frac{\phi''(t)}{\phi'(t)} + 2\zeta_g\omega_g\phi'(t) \right]\dot{x}_g + [\omega_g\phi'(t)]^2 x_g = -[\phi'(t)]^2 I(t)\xi(\phi(t)) \quad (14)$$

$$\ddot{x}_f + \left[-\frac{\phi''(t)}{\phi'(t)} + 2\zeta_f\omega_f\phi'(t) \right]\dot{x}_f + [\omega_f\phi'(t)]^2 x_f =$$

$$= -2\zeta_g\omega_g\phi'(t)\dot{x}_g - [\omega_g\phi'(t)]^2 x_g \quad (15)$$

and

$$\ddot{a}_g = 2\zeta_f\omega_f\frac{\dot{x}_f}{\phi'(t)} + \omega_f^2 x_f + 2\zeta_g\omega_g\frac{\dot{x}_g}{\phi'(t)} + \omega_g^2 x_g \quad (16)$$

The state vector $\underline{y}$ is defined as

$$\underline{y}^T = \left\{ \underline{y}_1^T, \underline{y}_2^T, \underline{y}_3^T, \underline{y}_4^T \right\} \quad (17)$$

where $\underline{y}_1 = \underline{u}$, $\underline{y}_2 = \dot{\underline{u}}$, $\underline{y}_3 = \underline{\theta}$ and $\underline{y}_4^T = \left\{ x_g, \dot{x}_g, x_f, \dot{x}_f \right\}$. Equations 11 to 15 are solved for the highest derivatives of the state variables, and after some algebra, using the state vector representation, the system of equations is written in matrix form as:

$$\dot{\underline{y}} = [G]\underline{y} + \underline{a} \quad (18)$$

Postmultiplying both sides of Eq. 18 by $\underline{y}^T$, adding the resulting equation to its transpose and taking expected values one finally obtains:

$$[\dot{S}] = [G][S] + [S][G]^T + [B] \quad (19)$$

where $\quad [S] = E\left[\underline{y}\,\underline{y}^T\right] \quad ; \quad [\dot{S}] = E\left[\dot{\underline{y}}\,\underline{y}^T\right] + E\left[\underline{y}\,\dot{\underline{y}}^T\right]$

and $\quad [B] = E\left[\underline{a}\,\underline{y}^T\right] + E\left[\underline{y}\,\underline{a}^T\right]$

Since the excitation is a shot noise, the only nonzero term in [B] is B_{kk} (where $k = 4ns + 4$), which in the case of frequency modulated input is

$$B_{kk} = 2\pi\,[\phi'(t)]^3\, I^2(t)\, S_0 \quad (20)$$

Equation 19 is a set of first order linear differential equations which can be solved for the covariance matrix [S] of the state vector $\underline{y}$ by direct integration schemes. For example, an explicit predictor–corrector method or an implicit backward differentiation may be used; the latter is more suitable if the set of covariance matrix differential equations is stiff (Press et al., 1989). Matrix [G] in Eq. 19 contains the linearized coefficient matrices $[C_{eq}]$ and $[K_{eq}]$ which are functions of the response statistics. These matrices are updated continuously during the solution.

The obtained covariance matrix [S] of the state vector contains the response statistics of the structure. These statistics can also be obtained by

Monte Carlo method; this method though is generally computationally expensive as a large number of time history analyses is required.

NUMERICAL EXAMPLE

The response of a five story–three bay special moment resisting space frame (SMRSF), designed according to UBC–88, is considered. The frame provides the required lateral resistance to a generic office building, located in Santa Monica Boulevard, Los Angeles, California, sixty kilometers from the Mojave segment of the southern San Andreas fault. The dimensions and design of the frame and the corresponding SCWB model are shown in Fig. 1. The soil at the site can be characterized as stiff and firm.

The expected magnitude of the characteristic earthquake at the Mojave segment is M = 7.5 (U.S. Geological Survey, 1988). The significant duration, t_D, of the characteristic earthquake can be generated using the empirical relation

$$\log_{10} t_D = -0.14 + 0.2M + 0.002R + \epsilon_D \tag{21}$$

where R is the site-to-source distance in km and ϵ_D is a random variable following a normal distribution N(0., 0.135) which describes the uncertainty in this relationship. Following the procedure of Trifunac and Lee (1989), Fourier amplitude spectra can be scaled for M = 7.5, R = 60 km and s = 1 using Eq. 1 with $E[\epsilon] = 0$ and $\sigma_\epsilon = 0.205$. Similarly, ϵ models the uncertainty in this relationship, which is primarily due to the attenuation. Therefore, given the occurrence of a characteristic earthquake, the local intensity and duration can be described by Eqs. 1 and 21 in which the uncertainty in attenuation and duration are included. For given intensity and duration the ground motion model parameters can be identified. Details can be found in Eliopoulos and Wen (1991).

For non-characteristic events, the coefficients a and b in the log-linear relation $\log_{10} N = a - bI$ given by Algermissen et al. (1982) with specified I_{max} and I_{min}, are used for the site. The probability that $I < i$ given the occurrence can be calculated from

$$F_I = P(I < i) = \frac{N(I_{min}) - N(i)}{N(I_{min}) - N(I_{max})} = \frac{1 - 10^{-b(i - I_{min})}}{1 - 10^{-b(I_{max} - I_{min})}} \tag{22}$$

For the site, $b = 0.37$, $I_{max} = 11$ and $I_{min} = 5$. Eq. 22 can be used to generate modified Mercalli intensity given occurrence. The significant duration, t_D, can be evaluated from the empirical relation

$$\log_{10} t_D = 1.96 - 0.123\, I_{MM} + \epsilon_D \tag{23}$$

in which ϵ_D follows a normal distribution N(0., 0.205). Following the procedure of Trifunac and Lee (1989), Fourier amplitude spectra can be scaled for a given value of I and s = 1 for firm soil using Eq. 3. As in the case of characteristic events, Eqs. 3 and 23 describe the intensity and duration at the site including the uncertainties. For given intensity and duration the ground motion model parameters can be identified.

For example, a non-characteristic earthquake with modified Mercalli Intensity $I = 8.7$ and $t_D = 10.66$ sec, the identified ground motion parameters and the resulting intensity, frequency modulation and power spectral density functions are shown in Fig. 3.

The response statistics of the five story-three bay frame under the excitation of the non-characteristic event specified in the foregoing are evaluated using random vibration analysis. The same statistics are also obtained from 40 samples of Monte Carlo simulations for comparison. The root mean square story drifts are presented in Fig. 4. In the previous example, the computation time required for random vibration analysis was one eighth of that of simulations. Note that the obtained response statistics are conditional on a given combination of source, path and site condition parameters. Other important response statistics such as maximum drift and cumulative damage can also be obtained from the random vibration analysis.

SUMMARY AND CONCLUSIONS

Methods of modeling seismic excitation at a given site due to future earthquakes and steel buildings with moment resisting frames are proposed in this study. Ground motions are treated as nonstationary random processes with time varying intensity and frequency content dependent on source and site characteristics. A random vibration method is also developed to obtain the response statistics. Numerical examples show that the method is accurate and efficient. The SCWB model, used in the analysis, reduced computational effort dramatically without any loss in the quality of the results.

Future research will concentrate on the overall reliability problem. The analytical tools developed in this study will be used in conjunction with the fast integration method for time variant structural reliability (Wen and Chen, 1987) to evaluate the overall reliability of the buildings including consideration of randomness in occurrence time, uncertainties in source and site parameters, and also structural resistance, in particular those due to nonstructural components.

ACKNOWLEDGEMENT

This research is supported by the National Science Foundation under Grant NSF CES-88-22690. Advice and contribution from D. A. Foutch and C-Y. Yu are also gratefully acknowledged.

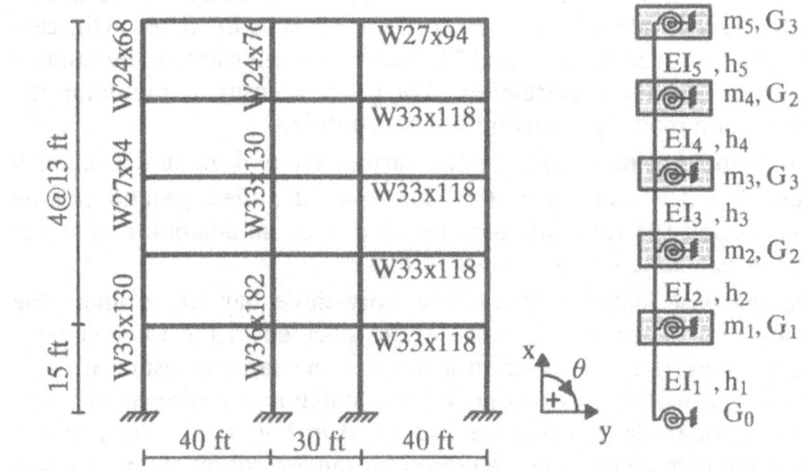

Figure 1 Five story steel frame and corresponding SCWB model

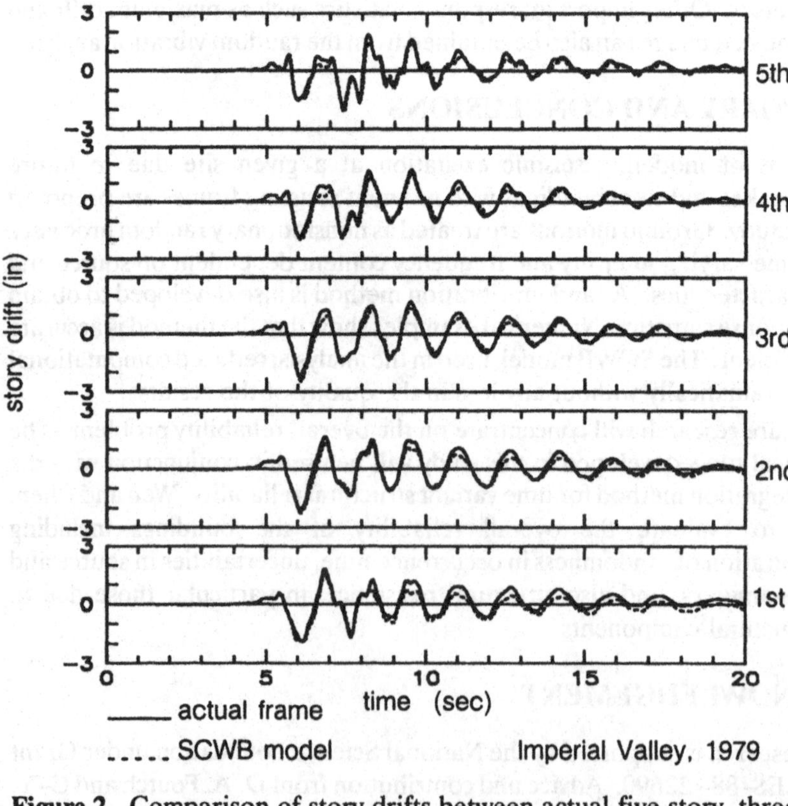

Figure 2 Comparison of story drifts between actual five story, three bay frame (analyzed using DRAIN-2DX) and SCWB model.

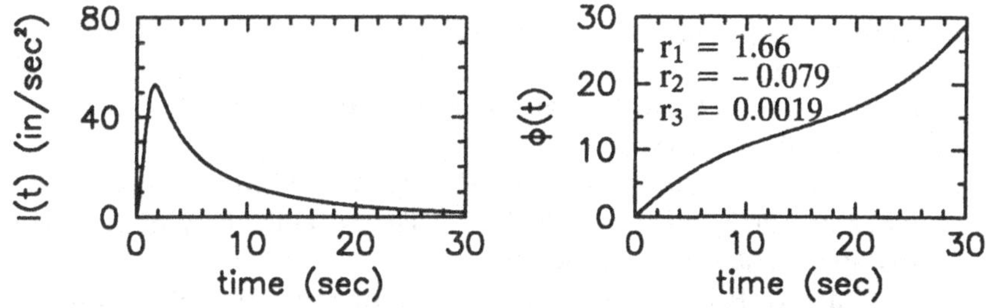

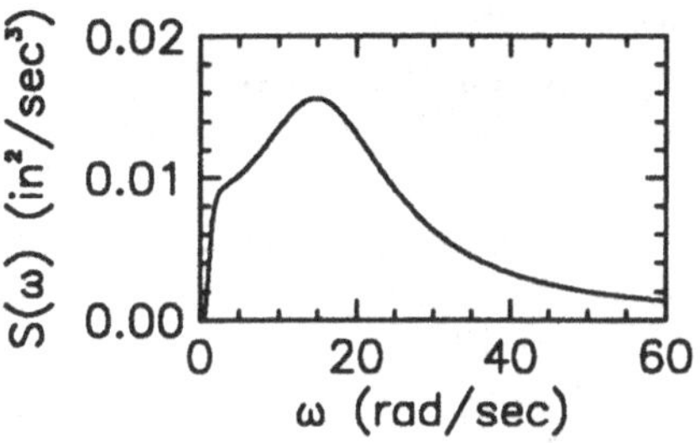

Intensity Function Parameters	Normalized PSD Function Parameters
A = 996.74	S_0 = 0.0087
B = 2.867	ω_g = 18.16
C = 0.124	ζ_g = 0.62
D = 4.274	ω_f = 1.35
E = 4.119	ζ_f = 0.66

Figure 3 Ground motion model functions and parameters

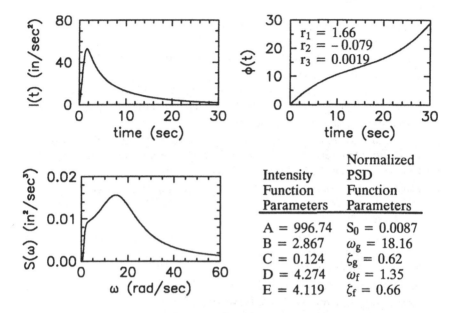

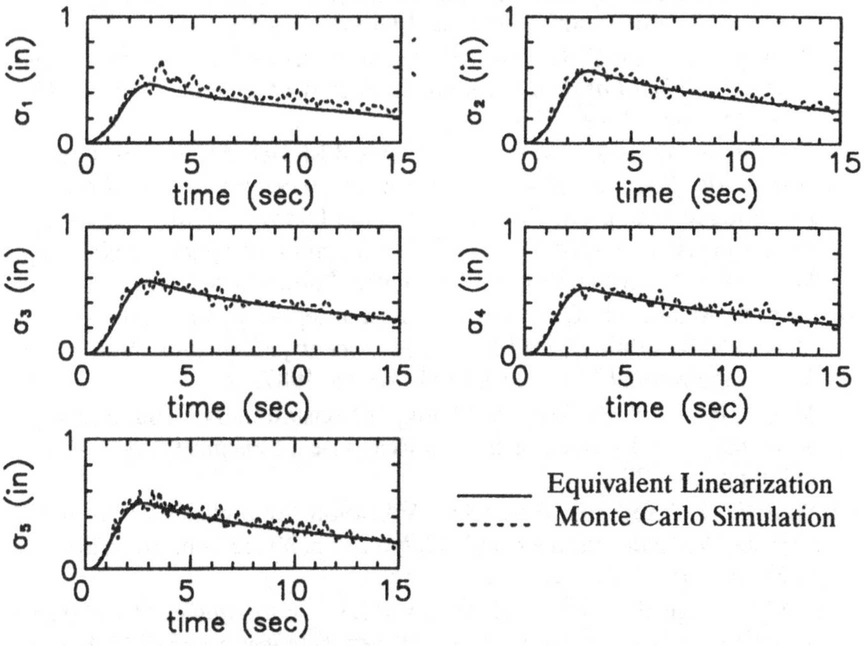

Figure 4 Root–mean–square interstory drifts of five story–three bay special moment resisting space frame

REFERENCES

1. Algermissen, S. T. et al., *Probabilistic Estimates of Maximum Acceleration and Velocity in Rock in the Contiguous United States,* Open–File Report 82–1033. United States Department of the Interior Geological Survey, Denver, Colorado, 1982.
2. Eliopoulos D. F. and Y. K. Wen, *Reliability of Moment Resisting Steel Frames to Earthquake Motion*, Ph.D. Thesis, University of Illinois at Urbana–Champaign, August 1991.
3. Eliopoulos D. F. and Y. K. Wen, "System Identification Techniques for Inelastic Structures", *Proc. Eighth Engineering Mechanics Specialty Conference*, ASCE, Columbus, Ohio, May 1991.
4. International Conference of Building Officials, *Uniform Building Code, 1988 Edition*. International Conference of Building Officials, Whittier, California, 1988.
5. Nishenko, S. P. and R. Buland, "A Generic Recurrence Interval Distribution for Earthquake Forecasting", *Bulletin of the Seismological Society of America*, Vol. 77, pp. 1382–1399, 1987.
6. Press W. H. et al., *Numerical Recipes – The Art of Scientific Computing*, Cambridge University Press, New York, 1989.
7. Scwartz, D. P. and K. J. Coppersmith, "Fault Behavior and Characteristic Earthquakes: Examples from the Wasatch and San Andreas Fault Zones", *Journal of Geophysical Research*, Vol. 89, No. B7, pp. 5681–5698. American Geophysical Union, July 10, 1984.
8. Trifunac M. D. and A. G. Brady, "A Study of Strong Earthquake Ground Motion", *Bulletin of the Seismological Society of America*, Vol. 65, No. 3, pp. 581–626, June 1975.
9. Trifunac M. D. and V. W. Lee, "Empirical Models for Scaling Fourier Amplitude Spectra of Strong Ground Acceleration in Terms of Earthquake Magnitude, Source to Station Distance, Site Intensity and Recording site Conditions", *Soil Dynamics and Earthquake Engineering*, Vol. 8, No. 3. Computational Mechanics Publications, 1989.
10. Wen Y. K. and H.–C. Chen, "On Fast Integration for Time Variant Structural Reliability", *Probabilistic Engineering Mechanics*, Vol. 2, No. 3. Computational Mechanics Publications, 1987.
11. Yeh, C. H. and Y. K. Wen, "Modeling of Nonstationary Ground Motion and Analysis of Inelastic Structural Response", *Structural Safety*, Vol. 8, 1990, pp. 281–298.
12. Wen Y. K. "Methods of Random Vibration for Inelastic Structures", *Applied Mechanics Reviews*, Vol. 42, No. 2, pp. 39–52. American Society of Mechanical Engineering, February, 1989.
13. U.S. Geological Survey, *Probabilities of Large Earthquakes Occurring in California on the San Andreas Fault*, Open–File Report 88–398, United States Department of the Interior Geological Survey, 1988.

SECTION 9: MATERIALS

On Wavefront Propagation in Random Nonlinear Media

M. Ostoja-Starzewski

Department of Metallurgy, Mechanics and Materials Science, Michigan State University, East Lansing, MI 48824-1226, U.S.A.

ABSTRACT

Determination of the effects of material microscale randomness on the wavefront propagation is the focus of this paper. The study is set in the context of one-dimensional microstructures with material randomness, of high signal-to-noise ratio, being present in constitutive moduli and grain lengths. Two different, but related, categories of problems - transient waves in granular microstructures, and acceleration wavefronts in nonlinear elastic/dissipative continua - are discussed. While particular attention is given to a random phenomenon of shock formation in the second case, the common thread of these both problem categories is the use of Markov diffusion processes in description of wave evolutions.

1. INTRODUCTION

Wave propagation in random media is now a subject area dating back several decades [1, 2]. However, due to the inherent mathematical difficulties and relevance in geophysical and radiophysical problems, most studies were confined to harmonic-type solutions of linear wave equations with random index of refraction in continuous or discrete media, and wave scattering at random surfaces. Thus, most of the problems connected with wavefront propagation, and especially in nonlinear setting, are so far understood only in the context of such phenomena taking place in deterministic media.

In this paper we give an account of our research on wavefront propagation in random media. As it is classically the case in stochastic mechanics, the main objective is to study the effects of randomness - that is, material microscale randomness - vis-à-vis deterministic solutions. Two categories of problems are presently being studied: transient waves in granular microstructures, and acceleration wavefronts in nonlinear elastic/dissipative media and nonlinear elastic composites; definitions of these microstructures are given in Section 2. However, due to the lack of space here we only give a short description of the method used in the first problem category (Section 3), and choose to go into more detail in the second one (Sections 4 and 5). A common feature of wave phenomena in both cases is the Markovian character of forward propagating disturbances, and hence the use of Markov diffusion processes in this field.

2. MODELS OF RANDOM NONLINEAR MEDIA

As is commonly done in stochastic mechanics, by a random medium *B* we understand a family $\{B(\omega);\ \omega \in \Omega\}$ where each $B(\omega)$ is a deterministic medium; Ω is a sample space equipped with a σ-algebra and a probability measure P. In this paper we consider all $B(\omega)$ to be semi-infinite one-dimensional media: $x \in [0, \infty]$. Three types of models are introduced.

A. Granular media: every $B(\omega)$ is a sequence of grains - homogeneous continua - of random geometric and physical properties with high signal-to-noise ratio.

A.1 Linear elastic granular media: length l, mass density ρ, and modulus E are random.

A.2 Bilinear elastic granular media: length l, mass density ρ, and two elastic moduli E_0, E_1 are random; see Fig. 1a). The stress level σ_0 separating both linear elastic ranges is assumed deterministic. . The stress-strain law of each grain is

$$\begin{aligned} \sigma &= E_0(\omega)\,\varepsilon \qquad \textit{if} \qquad |\sigma| < |\sigma_0| \\ \sigma &= \sigma_0 + E_1(\omega)\,(\varepsilon - \varepsilon_0) \qquad \textit{if} \qquad |\sigma| \geq |\sigma_0| \end{aligned} \tag{2.1}$$

Thus, *B* is described by a vector random process parametrized by x, that is: $\{l, E_0, E_1\}_x$. Three general types of this bilinear elastic granular medium may be considered:

i) all grains are of a soft characteristic, Fig. 1a),
ii) all grains are of a hard characteristic,
iii) both types of grains are present.

A.3 Nonlinear elastic granular media: length l, mass density ρ, and the elastic modulus E are random; see Fig. 1b). The stochastic stress-strain law is

$$\sigma = E(\omega)\,\varepsilon^n \tag{2.2}$$

where $n > 1$, or < 1. Three general types of this model analogous to those of Model A.2 may be considered here.

A.4 Linear-hysteretic granular media: length l, mass density ρ, and two moduli E_0 and E_1 are random; see Fig. 1c). The stress-strain curve is a straight line OM on initial loading to M; its slope defines the initial modulus E_0. Upon unloading the stress-strain curve is another straight line MN which defines the second modulus E_1

$$\sigma - \sigma_m = E_1(\omega)\,(\varepsilon - \varepsilon_m) \tag{2.3}$$

If the material is reloaded, it follows the line NM to M, and then continues along the initial loading line.

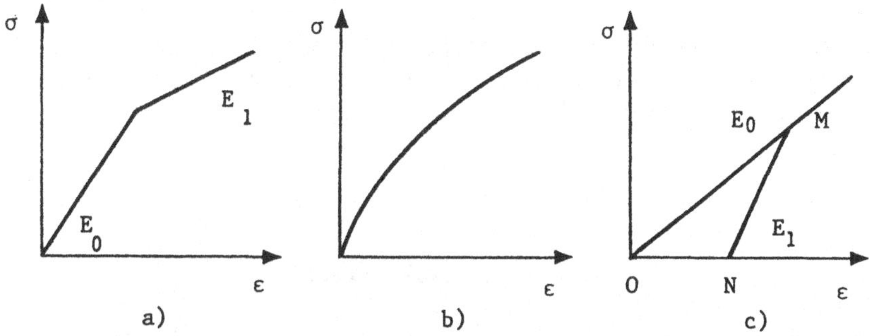

Figure 1. Constitutive laws: a) soft bilinear elastic, b) soft nonlinear elastic, c) linear-hysteretic.

B. Nonlinear elastic/dissipative continuum: every $B(\omega)$ is a deterministic continuum, whose response, with regard to the acceleration waves which we study here, is determined by two random functions of x, namely μ_x and β_x. A physical interpretation of these two functions is given in Section 4.

C. Nonlinear elastic laminated composite: every $B(\omega)$ is a periodic array of alternating layers with parallel plane boundaries; this model is based on a deterministic model analyzed in [3]. Each cell of the composite consists of two layers $\alpha = 1, 2$ with each layer being described by three random variables: mass density ρ_α, tangent modulus E_α, and second-order modulus $\underline{E}_\alpha$.

In all these models material properties are generally assumed to have a Markov property in x, although here, for simplicity, white-noise processes are employed.

3. WAVEFRONT PROPAGATION IN GRANULAR MEDIA

In this section we give a very brief account of the method employed in analysis of wavefronts propagating in granular media (Model A). In accordance with our basic formulation [4, 5], we model disturbance evolution in the x,t space-time

with reference to a mean (ensemble average) characteristic. There are two choices, which we illustrate here for the case of linear elastic grains (Model A.1):
- parametrization by time t, in which case the wavefront evolution is described by a Markov vector process

$$W_t = [(l, \rho, E), \zeta, \xi]_t \tag{3.1}$$

and is referred to a mean characteristic specified by $\langle c \rangle = \langle (E/\rho)^{1/2} \rangle$;
- parametrization by position x, in which case the wavefront evolution is described by a Markov vector process

$$W_x = [(l, \rho, E), \zeta, \tau]_x \tag{3.2}$$

and is referred to a mean characteristic specified by $\langle c^{-1} \rangle^{-1} = \langle (E/\rho)^{-1/2} \rangle^{-1}$. In the above, $\langle . \rangle$ denotes ensemble average, ζ denotes the disturbance amplitude (either stress or velocity), while ξ and τ are defined by

$$\tau(x, \omega) = t(x, \omega) - \langle t(x) \rangle \tag{3.3}$$

$$\xi(t, \omega) = x(t, \omega) - \langle x(t) \rangle \tag{3.4}$$

The evolution of ζ is described by a transmission coefficient ${}^{it}T$ defined by

$${}^{t}\zeta = {}^{it}T\,{}^{i}\zeta \tag{3.5}$$

in which i and t denote the incident and transmitted quantities, respectively. Using a rigid grain boundary model, we derive ${}^{it}T$ as

$${}^{it}T = \frac{2\,{}^{it}\chi}{1 + {}^{it}\chi}, \quad {}^{it}\chi = \left(\frac{{}^{t}\rho\,{}^{t}E}{{}^{i}\rho\,{}^{i}E}\right)^{1/2} \tag{3.6}$$

Markov processes W_t and W_x modelling propagating disturbances may conveniently be approximated by diffusion processes. In view of space limitations, we refer the reader to [6, 7, 8], where transient waves propagating in particular granular media models are analyzed, and only note here the salient features of these studies:
i) in the special case of medium's properties being of δ-correlated type, the vector process $(\zeta, \xi)_t$, or $(\zeta, \tau)_x$, is Markov;
ii) this process is multiplicative in ζ, and additive in ξ, or τ;
iii) in each case the forward characteristics of the deterministic problem (homogeneous medium) provide a reference for W_t and W_x processes - modelling the families of stochastic disturbances and the entire space-time may be described probabilistically.

4. ACCELERATION WAVEFRONT PROPAGATION IN A NONLINEAR ELASTIC/DISSIPATIVE CONTINUUM

An acceleration wavefront is a wavefront on which there is a jump in particle acceleration. It is well known that the evolution of an acceleration wavefront is governed by a Bernoulli (or Riccati) type equation

$$\frac{da}{dx} = -\mu(x)\,a + \beta(x)\,a^2 \qquad (4.1)$$

Here $\mu(x)$ depends upon the state of the medium ahead of the wavefront, while $\beta(x)$ depends on the elastic response of the material. In the special case of a linear elastic material, $\beta(x)$ becomes identically equal to zero, which results in a well-known exponential decay of waves in many dissipative media (e.g. linear viscoelasticity). Chen [9, 10], and McCarhty [11] are good general references on the subject, but Menon et al [12] provide a more systematic and correct study of the global behavior of such waves. However, it has to be pointed out that all the studies were conducted in the setting of deterministic media, that is with $\mu(x)$ and $\beta(x)$ being deterministic functions of x.

The main focus of studies of global behavior of acceleration waves concerns establishing conditions under which the wave damps out ($a(x) \to 0$), becomes stable ($a(x) \to$ const), or blows up to ∞ at a finite distance. The latter case corresponds to a shock wave formation. It is a quite general result that the initial amplitude a_0 has to exceed a so-called critical amplitude a_c in order for the wave to change into a shock wave. In case of a material described by $\mu(x)=\mu=\text{const} > 0$ and $\beta(x)=\beta=\text{const} > 0$, a_c is given by

$$a_c = \frac{\mu}{\beta} \qquad (4.2)$$

while the distance to form a shock is given by

$$\xi = \frac{-1}{\mu} ln\left[1 - \frac{\mu}{\beta a_0}\right] \qquad (4.3)$$

The question we ask is this: "What is the effect of spatial randomness of μ and β on a_c and ξ ?" Henceforth, μ and β are taken as two random processes $\mu_x = \{\mu(x);\, x>0\}$ and $\beta_x = \{\beta(x);\, x>0\}$, i.e. Model B. In particular, we assume

$$\begin{aligned} \mu(x) &= \langle\mu\rangle + \Gamma_1(x) \qquad \forall x \\ \beta(x) &= \langle\beta\rangle + \Gamma_2(x) \qquad \forall x \end{aligned} \qquad (4.4)$$

Where Γ_1 and Γ_2 are zero-mean δ-correlated Gaussian processes, such that

$$\langle \Gamma_i(x)\Gamma_j(x')\rangle = 2g_{ij}\delta(x-x') \qquad g_{ij} = g_{ji} \tag{4.5}$$

By a standard substitution

$$\zeta(x,\omega) = [a(x,\omega)]^{-1} \qquad \forall x \qquad \forall \omega \tag{4.6}$$

the original nonlinear equation (4.1) is transformed into

$$\frac{d\zeta}{dx} = \mu(x)\zeta - \beta(x) \qquad \zeta(0) = z_0 = \frac{1}{a_0} \tag{4.7}$$

so that the problem of shock formation is equivalent to a problem of $\zeta(x)$ going to zero. Henceforth, equation (4.7) will be interpreted as a stochastic differential equation in the sense of Stratonovich.

We introduce a diffusion formulation for the random process $\zeta_x = \{\zeta(x); x>0\}$. Thus, the forward Kolmogorov (or Fokker-Planck) equation is

$$\frac{\partial}{\partial x}P(z,x|z',x') = L^f(z,x)P(z,x|z',x') \tag{4.8}$$

where z and z' are values in the range of the random variable $\zeta(x)$ and

$$L^f(z,x) = -\frac{\partial}{\partial z}A(z,x) + \frac{\partial^2}{\partial z^2}B(z,x) \tag{4.9}$$

in which

$$A(z,x) = az+b \qquad B(z,x) = cz^2+dz+e \tag{4.10}$$

Interpreting the coefficients A(z, x) and B(z, x) according to the Stratonovich rules, we obtain

$$a = \langle\mu\rangle + g_{11}, b = -\langle\beta\rangle, c = g_{11}, d = -2g_{12}, e = g_{22} \tag{4.11}$$

Of course, one could also modify (4.7) to a stochastic differential equation that describes the evolution of $\zeta(x)$ according to the Ito calculus, and arrive at the same diffusion formulation.

The backward Kolmogorov diffusion formulation is given by

$$\frac{\partial}{\partial x}P(z,x|z',x') = L^b(z',x')P(z,x|z',x') \tag{4.12}$$

$$L^b(z',x') = A(z',x')\frac{\partial}{\partial z'} + B(z',x')\frac{\partial^2}{\partial z'^2} \tag{4.13}$$

in which A(z', x') and B(z', x') are given by the same forms as (4.10).

The forward diffusion formulation serves to determine P(z, x | z', x'), or, alternatively, its moments. In what follows, we determine the first and second moments of $\zeta(x)$. Thus, evolution of the first moment is governed by

$$\frac{d}{dx}\langle\zeta\rangle = a\langle\zeta\rangle + b, \ \langle\zeta(0)\rangle = z_0 \tag{4.14}$$

from which we find

$$\langle\zeta(x)\rangle = \left(z_0 + \frac{b}{a}\right)e^{ax} - \frac{b}{a} \tag{4.15}$$

It follows from this and from $(4.11)_{1,2}$ that the ensemble average of a critical inverse amplitude ζ is

$$\langle\zeta_c\rangle = -\frac{b}{a} = \frac{\langle\beta\rangle}{\langle\mu\rangle + g_{11}} \tag{4.16}$$

Next, observing that

$$\langle a_c^{-1}\rangle^{-1} \leq \langle a_c\rangle \tag{4.17}$$

and that the critical amplitude of the deterministic problem is

$$(a_c)_{det} = \frac{\langle\mu\rangle}{\langle\beta\rangle} \tag{4.18}$$

we conclude that the ensemble average critical amplitude of the stochastic problem is higher than $(a_c)_{det}$.

The second moment's evolution is governed by

$$\frac{d}{dx}\langle\zeta^2\rangle = \langle\zeta^2\rangle(2c+2a) + \langle\zeta\rangle(2b+2d) + 2e \tag{4.19}$$

with the initial condition

$$\langle\zeta^2(0)\rangle = 0 \tag{4.20}$$

from which, using (4.15), we find

$$\langle\zeta^2(x)\rangle = C_1\left(e^{(2c+2a)x} - e^{ax}\right) + C_2\left(1 - e^{(2c+2a)x}\right) \tag{4.21}$$

where

$$C_1 = \frac{\left(z_0 + \frac{b}{a}\right)(2b+2d)}{2c+a} \qquad C_2 = \frac{\frac{b(2b+2d)}{a} - 2e}{2c+2a} \tag{4.22}$$

The backward diffusion formulation serves to determine the random distance ξ to form a shock. To that end we observe that the problem of finding the random variable ξ is equivalent to a problem of random exit distances ("random exit times" in the classical terminology) through a boundary at $\zeta = 0$. Now, let $\{X_n(z'); n=0, 1, 2,...\}$ denote the moments $<\xi^n>$ for the initial condition of z' at distance 0. It is well known that $X_n(z')$'s satisfy a hierarchy of equations

$$L^b(z')X_n(z') = -nX_{n-1}(z') \tag{4.23}$$

The first equation, for n=1, yields

$$(az'+b)\frac{dX_1}{dz'} + (cz'^2 + dz'+e)\frac{d^2X_1}{dz'^2} = -1 \tag{4.24}$$

with the boundary conditions

$$X_1(0) = 0 \qquad X_1(\infty) = 0 \tag{4.25}$$

Solution to the above is presently being studied.

Finally we note that the backward diffusion formulation my also be employed to solve for the distribution of this random variable $\zeta(x')$ at an arbitrary preceding distance x', which leads to a shock $\zeta = 0$ at x. This is to be accomplished by solving (4.12) with the initial condition

$$P(0, x'|z', x') = P(0, x|z', x) = \delta(z') \tag{4.26}$$

The mean of $\zeta(x')$ in this problem can be found directly from (4.15) as

$$\langle \zeta(x') \rangle = \frac{b}{a}(e^{a(x'-x)} - 1) \tag{4.27}$$

5. ACCELERATION WAVEFRONT PROPAGATION IN A NONLINEAR ELASTIC COMPOSITE

Here we show how an analogous problem of acceleration wave propagation in a setting different from the one considered in the preceding section - namely in a nonlinear elastic composite - may be studied with the same tools. We follow the model of Chen and Gurtin [3]. Following that reference and Model C of Section 2, it is essential to introduce several more random variables:
$c_\alpha = [E_\alpha/\rho_\alpha]^{1/2}$ - speed in layer α,
$T_\alpha = l_\alpha/c_\alpha$ - transit time in layer α,
$T = T_1 + T_2$ - transit time in one cell.
The evolution of wavefront amplitude after passing through one period of the composite is given by (see equation (19) in [3])

$$\frac{a(T)}{a(0)} = \frac{1-\delta^2}{1-a(0)\,[\varphi_1 + (1+\delta)\,\varphi_2]} \tag{5.1}$$

where

$$\delta = \frac{1-\chi}{1+\chi},\ \chi = \sqrt{\frac{E_2\rho_2}{E_1\rho_1}},\ \varphi_1 = \frac{-E_1T_1}{2c_1E_1},\ \varphi_2 = \frac{-E_2T_2}{2c_2E_2} \tag{5.2}$$

As pointed out in [3], "a careful study of this equation leads to the conclusion that the wave will grow or decay according as a(0) is greater than or less than the number:"

$$a_s = \frac{\delta^2}{\varphi_1 + (1+\delta)\,\varphi_2} \tag{5.3}$$

Since ρ_α, E_α and $\underline{E}_\alpha$ are random (for α=1, 2) in all the layers of the composite, there is no single deterministic value of a_c. Evolution of the acceleration wave is a random process with a blow-up to ∞ - i.e shock formation - occuring at a random time (distance ξ) rather than a single deterministic time. This process is parametrized by a discrete distance: $a_k = \{a(k); k=0,1,2,...\}$. Next, we write (5.1) as

$$a_k = \frac{(1-(\delta_k)^2)}{1-C_k a_{k-1}}, C_k = \varphi_{1_k} + (1+\delta_k)\varphi_{2_k} \tag{5.4}$$

which illustrates the Markov property as well as the nonlinear character of a_k. However, by introducing a random variable

$$\zeta(k,\omega) = [a(k,\omega)]^{-1} \qquad \forall k \qquad \forall\omega \tag{5.5}$$

we arrive at a linear multiplicative Markov process ζ_k, which evolves according to

$$\zeta_k = \zeta_{k-1}\frac{1}{1-(\delta_k)^2} - \frac{C_k}{1-(\delta_k)^2} \tag{5.6}$$

The above may be written in a finite-difference form analogous to (4.7), that is

$$\frac{\zeta_k - \zeta_{k-1}}{l_k} = \mu_k \zeta_{k-1} - \beta_k, \mu_k = \frac{(\delta_k)^2}{(1-(\delta_k)^2)\, l_k}, \beta_k = \frac{C_k}{(1-(\delta_k)^2)\, l_k} \tag{5.7}$$

where we have identified the analogues $\mu(k)$ and $\beta(k)$ of $\mu(x)$ and $\beta(x)$, respectively. It follows now that the same derivations as those of section 4. apply to the determination of a_c and ξ. Thus, we can use a diffusion approximation - in continuous distance x - for the process ζ_k, in which the drift and diffusion coefficients are of the same forms as in (4.10).

6. CONCLUSIONS

Wavefront propagation in random nonlinear media in two broad problem categories has recently been studied. These both categories have been introduced so that the effects of material randomness of either discrete type (piecewise-continuous properties of Model A) or of continuous type (Model B) on propagating wavefronts may be investigated. In the first category (Model A) - we use a combination of the method of characteristics and vector Markov diffusion processes to assess the evolution of forward propagating disturbances. One component of the vector describes the amplitude of the field quantity (e.g. particle velocity) at the disturbance, while another one describes the scatter in arrival times. Microstructures of three different types are considered: bilinear elastic (soft or hard) grains, nonlinear elastic (soft or hard) grains, and linear-hysteretic grains. Various new and heretofore unknown behaviors are discovered. For example, in case of bilinear elastic soft grains an acceleration wavefront forms, which displays a very strong scatter in the arrival times. Results for granular-type media - Models A.1, A.2, A.3 and A.4 - are reviewed in [8].

In the second problem category we investigate propagation of acceleration wavefronts in nonlinear elastic dissipative media. In the deterministic case their evolution is typically described by the Bernoulli equation, see e.g. [11]. One very important aspect of the global behavior of acceleration wavefronts concerns the possibility of formation of shocks, in a finite time, according to the initial amplitude being greater than or less than a certain critical value. We focus on the determination of the probability distributions describing this critical value and the distance to form a shock. Again, Markovian evolution of disturbances provides the basis for analysis. It follows from equations (4.16), (4.17) and (4.18) that the ensemble average critical amplitude in the stochastic problem is greater than its counterpart in the deterministic problem with $<\beta>$ and $<\mu>$ being constant. A similar conclusion follows from (4.27) regarding the amplitude which in a given finite distance would result in a shock. As expected, these are nontrivial results in the sense that the averages of a stochastic problem are different from solutions of the corresponding averaged deterministic problem. Thus, qualitatively similar conclusions are expected in an analogous problem of shock formation from an acceleration wavefront in a nonlinear elastic composite (Model C), which is considered in Section 5. However, not everything is derivable analytically - a recourse to computations has to be made in order to find statistical moments of the random distance to form a shock.

ACKNOWLEDGEMENT

The author benefited from discussions with Dr. B. Rohani of the U.S. Army Corps of Engineers and Prof. K. Sobczyk of the Polish Academy of Sciences. This research was supported in part by the U.S. Army under Contract No. DACA39-90-M-2528 and by the AFOSR under Contract No. AFOSR-89-0423; Lt. Col. G.K. Haritos is the program manager at AFOSR.

REFERENCES

1. Sobczyk, K. Stochastic Wave Propagation, Elsevier, Amsterdam, 1985.

2. Sobczyk, K. Stochastic Waves: The Existing Results and New Problems, Probabilistic Engng. Mech., Vol. 1(3), pp. 167-176, 1986.

3. Chen, P.J. and Gurtin, M.E. On the Propagation of One-Dimensional Acceleration Waves in Laminated Composites, J. Appl. Mech., Vol. 40, pp. 1055-1060, 1973.

4. Ostoja-Starzewski, M. Stress Wave Propagation in Discrete Random Solids, Wave Phenomena: Modern Theory and Applications (Ed. Rogers, C. and Bryant Moodie, T.), North-Holland Mathematics Studies, Vol. 97, pp. 267-278, Elsevier Science Publ., Amsterdam, 1984.

5. Ostoja-Starzewski, M. Wavefront Propagation in Discrete Random Media via Stochastic Huygens' Minor Principle, J. Franklin Inst., Vol. 326 (2), pp. 281-293, 1989.

6. Ostoja-Starzewski, M. Wavefront Propagation in a Class of Random Microstructures - I: Bilinear Elastic Grains, Int. J. Non-Linear Mech., Vol.26, 1991.

7. Ostoja-Starzewski, M. Wavefront Propagation in a Class of Random Micro structures - II: Nonlinear Elastic Grains, forthcoming.

8. Ostoja-Starzewski, M. Transient Waves in a Class of Random Heterogeneous Media, invited paper, Appl. Mech. Rev. (Mechanics Pan-America 1991), Vol. 44 (10, Part 2), 1991, to appear.

9. Chen, P.J. Selected Topics in Wave Propagation, Noordhoff, Leyden, 1976.

10. Chen, P.J. Growth and Decay of Waves in Solids, Encyclopedia of Physics (Ed. Fluegge, S. and Truesdell, C.), Vol. VI a/3, Springer-Verlag, Berlin, 1973.

11. McCarthy, M.F. Singular Surfaces and Waves, Continuum Physics, (Ed. Eringen, A.C.), Vol. II, pp. 450-521, Academic Press, New York, 1975.

12. Menon, V.V., Sharma, V.D. and Jeffrey, A. On the General Behaviour of Acceleration Waves, Applicable Analysis, Vol. 16, pp. 101-120, 1983.

A Stochastic Approach to the Rheology of Fibrous Systems

Y.M. Haddad

Department of Mechanical Engineering, University of Ottawa, Ottawa, KIN 6N5, Canada

ABSTRACT

A new micromechanical approach to the rheological response of a class of fibrous systems is presented. The approach recognizes that the structure of the fibrous system consists of a randomly arranged, approximately two-dimensional array of viscoelastic fibres which are bonded together at regions where they cross. Due to the inherent randomness of the physical and geometrical characteristics of the microstructure, probabilistic concepts are used. In this context, the relevant field quantities are seen as random variables and the associated deformation is treated as a stochastic process. An important notion of the theory is the introduction of a 'Material Operator' which contains the significant microstructural characteristics pertaining to the mechanical response of the fibrous network. The model is presented in a general form and can be applicable to a large class of fibrous systems.

INTRODUCTION

This paper is concerned with a stochastic approach to the rheological behaviour of random fibrous systems with the inclusion of the microstructure. In order to illustrate the type of material investigated in this paper, a micrograph (X170) of a (cellulosic) fibrous system is shown in Figure 1. It represents bleached sulphite paper. The microstructure of the system is seen to be heterogeneous showing a random arrangement of single fibres which are bonded together at certain junctions. Fibrous systems such as introduced above have been classified with respect to their response behaviour as viscoelastic materials [1,2].

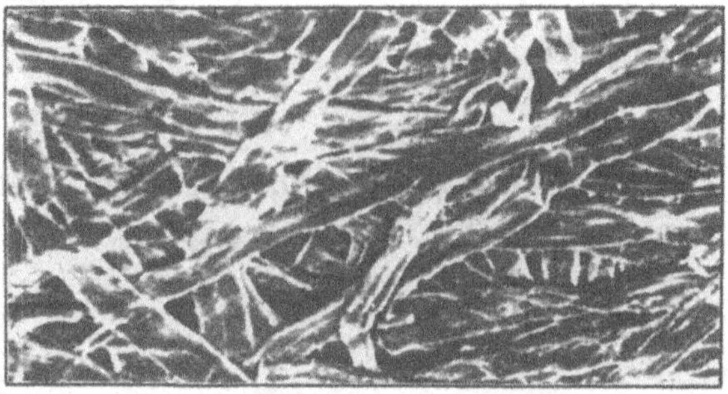

Figure 1. A micrograph (X170) of a natural cellulosic fibrous system (sulpshite; bleached and dried).

Traditionally, models that are based on continuum theories have been used for the prediction of the response behaviour of fibrous systems. These models, in general, refer to a homogeneous medium ignoring thereby the effect of the microstructure. Recently, several attempts have been made to modify the continuum mechanics approach by allowing for microscopic or "local" quantities to enter into the analysis but without removing the main restrictions imposed by continuum physics on such formulations. In this, "modified continuum" models have been proposed by Nissan [3], Onogi and Sasaguri [4], Sternstein and Nissan [5] and Van den Akker [6], amongst others.

The necessity, however, to develop a new approach to the response of fibrous systems that would be based explicitly on microstructural considerations has been frequently discussed in the literature [7-9]. Rance [9], for instance, advanced the argument that the nonrecoverable deformation of a paper sheet is essentially due to the breakage of inter-fibre bonding that occurs at increasing rate leading to the final fracture of the macroscopic specimen. Page et al. [10] verified experimentally that this phenomenon could occur under the effect of elastic tension. Nissan [11] and Corte [12] reported in this regard that hydrogen bonding is the most effective binding mechanism between two adjacent cellulosic fibres.

On the other hand, the failure of fibre-segments during elastic tension of paper samples has been reported by Van den Akker [13]. Thus, the opinion has been put forward that the mechanical response of a fibrous network depends on the response characteristics of inter-fibre bonding as well as those of individual fibres. Experimental investigation by McIntosh and Leopold [14] supported the last statement. Van den Akker [15] highlighted further the importance of the statistical approach to the problem of the prediction of the macroscopic response of fibrous systems with the inclusion of the microstructure. In this context, Corte and Kallmes [16] presented an important study concerning the statistical geometry of a model of fibrous systems.

THE APPROACH

In the present paper, a new microstructural approach to the rheological response of fibrous systems is developed. This approach recognizes that the microstructure of a fibrous system consists of a randomly arranged, approximately two-dimensional array of viscoelastic fibres which are bonded together at regions where they cross. Thus, the analysis takes into account the rheological response of individual fibres as well as the effect of interfibre bonding. Due to the inherent randomness of the physical and geometrical characteristics of the microstructure, probabilistic concepts are used. Moreover, the microsturctural elements forming the fibrous system possess, in general, time-dependent response characteristics. As a consequence, it is appropriate to consider the significant quantities governing the deformation process as stochastic variables and the deformation process itself is seen in this approach as a stochastic process.

In order to describe the mechanical response of a structured fibrous system, it is necessary to consider the response of the individual structural elements which on a local scale could differ considerably from an average response if the phenomenological approach was taken. Such local deviation in the response which are usually neglected by ignoring the microstructure are directly related to basic properties of the nonhomogeneous fibrous network. Accordingly, the present analysis introduces a new definition of the structural element of a fibrous network and deals with the formulation of its rheological response in a probabilistic sense. It is considered in this regard that the rheological response of the fibre-segment as well as the inter-fibre bonding are of significant importance.

In order to extend the analysis to the practical case of a two-dimensional network, it is necessary to make use of "intermediate quantities' arising from the consideration of the existence of a statistical ensemble of structural elements within an intermediate domain of the material specimen. Further, it is equally important to find a connection between the microscopic and the macroscopic response formulations. Thus, the analysis aims at the formulation of a set of "governing response equations" for the structured fibrous system that, in contrast to the classical continuum mechanics formulations, are based on the concepts of statistical theory and probabilistic micromechanics [17, 18]. In this context, it has been found useful to employ operational representation of the various relations. Hence, the notion of a "Material Operator" characteristic of the viscoelastic response of an intermediate domain of the material is introduced. This material operator provides the connection between the stress field and the occurring deformations within the intermediate domain under consideration. It contains in its argument those stochastic variables or functions of such variables distinctive of the microstructure within the intermediate domain. In a very reduced and simplified form, such an operator is expressed by:

$$\Gamma(t) = \Gamma({}^{f}\Gamma, {}^{j}\Gamma, {}^{\alpha}K, p_1, p_2, \ldots, t) \tag{1}$$

where ${}^{f}\Gamma$ and ${}^{j}\Gamma$ are random material operators expressing the response characteristics of elements of the structure, ${}^{\alpha}K$ is a function of one or more geometrical parameters, p_1 and p_2 are geometrical probabilities and t is the time parameter. A comparison between basic concepts of the probabilistic micromechanical approach adopted here and the corresponding postulates of classical continuum mechanics is shown in Table 1.

Table 1. A comparison between basic concepts of the prohabitistic micromechanical approach and those of classical continuum mechanics

	Classical Continuum Mechanics	**Prohabilistic Micromachanics**
Material system	Continuous	Discrete
Local description	Mathematical point	Structural element
Stress and deformation	Continuous	Discontinuous
Analytical approach	- Deterministic. - Constitutive theory	- Stochastic - Operational formalism of a structured material system

In order to extend the formulation to the possibility of prediction of failure in the fibrous system, the evolution of the internal deformation process is considered from a stochastic point of view. Hence, the internal deformation process, corresponding to the steady-state response of an ensemble of structural elements within the intermediate domain is considered to be stochastic of the stationary Markov type [19, 20]. Thus, the fibre-segment deformation process and by dependence, the inter-fibre bonding deformation process are seen to be stochastically interacting processes of Markovian character. In this, a failure criterion of a particular structural element may be conjectured by considering the distributions of the kinematical deformation quantities in connection with their maximum allowable limits. The scope of the present approach is demonstrated in Figure 2.

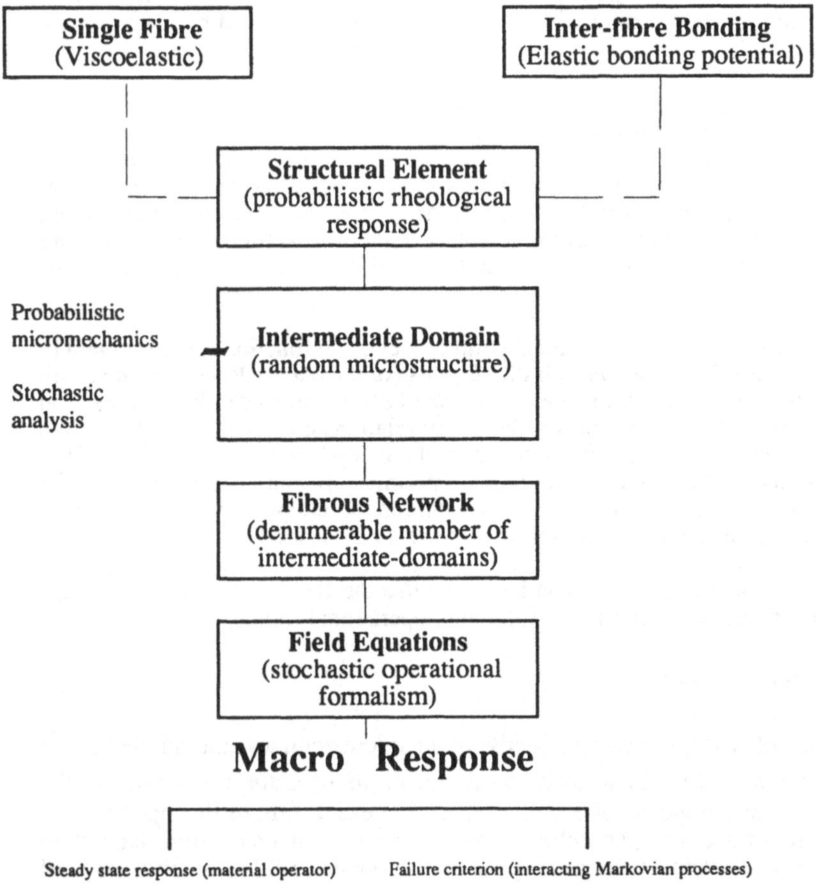

Figure 2. Scope of the probabilistic micromechanical approach

PROBABILISTIC, MICROMECHANICAL RESPONSE

A Structural Element

A structural element (α) is defined as the smallest part of the medium that represents the mechanical and physical characteristics of the microstructure at the "micro" level. In the present analysis, a new model of the structural element is introduced that includes the contribution of a single fibre segment between two neighbouring junctions, as well as one-half of each of the junction areas associated with the actual bonding between the fibres.

Throughout the analysis, a superscript (α) to the left of the symbol will refer in general to a microelement. However, since a distinction is made between the fibre segment and the two half junction areas, the quantities referring to such a segment will be denoted by a superscript "f" while those referring to the bonding within the junction are designated by a superscript "j".

Rheological response of a fibre-segment

In view of the fact that this paper considers the response of "natural fibres" such as cellulose fibres which in most cases exhibit rheological properties, the viscoelastic response of such fibres will be considered. While in this paper, for the simplification of the analysis, the continuum approach is maintained for the response of the fiber segment, it is understood that the effect of fibre substructural parameters will not be considered at the present stage of presentation.

From a phenomersological point of view, two functions are important in the description of the viscoelastic response of the material; they are the creep and relaxation functions. Generally, the two functions can be expressed in the forms of the well-known hereditary relations due to Volterra [21]. We shall consider here the formulation of the creep behaviour of a single fibre with the inclusion of pertaining experimental data. This is with the understanding that the same approach can be adopted for the relaxation behaviour as the two functions are interrelated, Haddad [22].

In this connection, it can be shown that the creep response of the single fibre can be expressed in the following operational form

$$^{f}e(t) = {}^{f}\Gamma(t)\,{}^{f}\xi(t) \tag{2}$$

where **e(t)**, $\xi(t)$ indicate respectively the microstrain and the microstress at time t and ${}^{f}\Gamma(t)$ is a rheological transform operator pertaining to the viscoelastic response of a single fibre. The exact form of this operator can be established for a particular range of the stress input on the fibre and within a specific extent of the time parameter t, Haddad [22]. The fibre-material operator ${}^{f}\Gamma(t)$ will be used subsequently in the expression for the response of the structural element.

Inter-fibre bonding response

The adopted model refers to matching points on the corresponding surfaces of matching fibres α and β. In this connection, a "pair potential" form appears to be most suitable for the description of the binding interaction. From classical considerations, one of the usual forms of such a potential in the one-dimensional case is represented by "Morse function" [23].

$$\Psi = \Psi o\{\exp(-2v \mid d(t) \mid) - 2\exp(v \mid d(t) \mid)\} \tag{3}$$

where $d(t)$ is the bonding displacement at time t, Ψo is the equilibrium value of the Morse potential at the value of $d(t)$ = O and v is the Morse constant. The material properties represented by the constants Ψo and v are obtainable from spectroscopic studies. For the hydrogen bonding in cellulose, Ψo may be taken as 4.5 kcal/mole OH[24]. The constant v may be taken as 2 $\mathring{A}^{-1}$[25].

Based on the bonding potential function (3), the operational equation for the inter-fibre bonding response can be expressed as

$$^{j}e(t) = {}^{j}\Gamma(t)^{j}\xi(t) \tag{4}$$

where the equation for the material operator $^{j}\Gamma(t)$ is given by, Haddad [26],

$$^{j}\Gamma^{-1} = \frac{-2v^{2}\Psi o}{^{B}\mathrm{A}(t)} gkk^{-1}\,{}^{\alpha}\nabla^{-1} \tag{5}$$

In (5), **g** is the unit normal to the junction area, **k** is a unit base vector associated with the local coordinate frame of reference and ∇ is the gradient operator on displacement. In (5), $^{B}\mathrm{A}(t)$ is the time-dependent actual bonded area within the junction.

Probabilistic response of a structural element α

The following geometrical probabilities are introduced; with respect to a scanning line *l,l* that intersects the surface of a macroscopic specimen;

p: the probability that the scanning line *l,l* intersects, at a certain point of the specimen, an individual fibre segment.

q: the probability that *l,l* intersects at the same point, a junction between two fibres.

Thus, at the considered point, one may express the microstrain in a probabilistic manner as

$$^{\alpha}e(t) = p^{f}e(t) + q^{j}e(t) \tag{6}$$

where $o \le (p,q,) \le 1$ and $p+q=1$. In this equation, $^{j}e(t)$ is considered to be totally due to the bonding displacement. That is, the fibre displacement at the point in question is taken as zero. The probability q appearing in (6) can be determined in terms of the physical and geometrical characteristics of the fibrous microstructure; Haddad [26].

Substituting for the microstrains ${}^{f}e(t)$ and ${}^{j}e(t)$ in (6) in terms of the corresponding microstresses from (2) and (4), respectively, it follows that

$$^{\alpha}e(t) = (1-q)^{f}\Gamma(t)^{f}\xi(t) + q\,^{j}\Gamma^{j}\xi(t) \tag{7}$$

However, in order to formulate an operational relation that will serve eventually in establishing a response relation for the structural element, it is necessary to employ a relationship between the fibre microstress and the microstress in the junction. This relation may be expressed as, Haddad [27],

$$^{j}\xi(t) = {}^{\alpha}K(t)^{f}\xi(t)$$

$$^{\alpha}K(t) = \frac{^{f}a}{^{B}\mathrm{A}(t)}\{1 - {}^{f}\lambda(t)\} \tag{8}$$

where ${}^{f}a$ is the fibre cross-sectional area and ${}^{f}\lambda(t)$ is a time-dependant parameter associated with the equilibrium of the junction between two fibres α and β. In ref [27], the parameter ${}^{f}\lambda(t)$ is determined in terms of the orientation of the fibrous elements forming a particular junction.

Now, substituting for ${}^{j}\xi(t)$ from (8) into (7), the latter equation can be written as

$$^{\alpha}e(t) = [(1-q)^{f}\Gamma(t) + q\,^{j}\Gamma^{\alpha}K(t)]^{f}\xi(t) \tag{9}$$

Thus, if we introduce the transform operator ${}^{\alpha f}\Gamma(t)$ associated with the mechanical response of the structural element such that the value of this operator corresponds to the expression between square brackets in (9), i.e.

$$^{\alpha f}\Gamma(t) = [(1-q)^{f}\Gamma(t) + q\,^{j}\Gamma^{\alpha}K(t)] \tag{10}$$

then, one would be able to express the response equation (9) in the following operational form

$$^{\alpha}e(t) = {}^{\alpha f}\Gamma(t)^{f}\xi(t) \tag{11}$$

Further, it is possible by the inversion of (11), to determine the microstress ${}^{f}\xi(t)$ from the microstrain ${}^{\alpha}e(t)$ as

$$^{f}\xi(t) = {}^{\alpha f}\Gamma^{-1}(t)^{\alpha}e(t) \tag{12}$$

In the same manner, given (8), one can express (11) as a function of the bonding microstress ${}^{j}\xi(t)$ as

$$ {}^{\alpha}e(t) = {}^{\alpha j}\Gamma(t)\,{}^{j}\xi(t) \tag{13}$$

where

$$ {}^{\alpha j}\Gamma(t) = {}^{\alpha f}\Gamma(t)\,{}^{\alpha}K^{-1}(t) \tag{14}$$

Transition to the Macroscopic Response Behaviour

Following the concepts of the micromechanical theory of structured media, Axelrad [17], all microscopic field quantities within the intermediate domain are considered to be stochastic functions of premitive random variables. Thus, the components of the microstrain, for instance, are seen as stochastic functions ${}^{\alpha}e(r,t)$ that can be regarded as a family of random variables ${}^{\alpha}e_{t}(r)$ within the intermediate domain depending on the time parameter t, or a family of curves ${}^{f}e_{r}(t)$ depending on the structural element position vector ${}^{\alpha}r$.

Letting ${}^{M}P\{.\}$ denotes the probability distribution of a random variable within an intermediate domain M, then, in view of the structural element response equation (11), the probability distribution of the microstrain within the mesodomain may be expressed as

$$ {}^{M}P\{{}^{\alpha}e(t)\} = {}^{M}P\{{}^{\alpha f}\Gamma(t)\}\,{}^{M}P\{{}^{f}\xi(t)\} \tag{15}$$

At the same time, with reference to (12), one can express the probabilistic distribution of the fibre-microstress as

$$ {}^{M}P\{{}^{f}\xi(t)\} = {}^{M}P\{{}^{\alpha f}\Gamma^{-1}(t)\}\,{}^{M}P\{{}^{\alpha}e(t)\} \tag{16}$$

Further, following (13), the distribution of the bonding microstress becomes

$$ {}^{M}P\{{}^{j}\xi(t)\} = {}^{M}P\{{}^{\alpha j}\Gamma^{-1}(t)\}\,{}^{M}P\{{}^{\alpha}e(t)\} \tag{17}$$

In addition, the distribution of the bonding microstress is associated, in view of (8), with the distribution of the fibre-stress via the relation

$$ {}^{M}P\{{}^{j}\xi(t)\} = {}^{M}P\{{}^{\alpha}K(t)\}\,{}^{M}P\{{}^{f}\xi(t)\} \tag{18}$$

Time evolution of the Internal Deformation Process

The internal deformation process, corresponding to the steady state response of an ensemble of structural elements within the intermediate domain is considered to be stochastic of the stationary Markov type [19, 20]. Denoting the time duration of such process by the $t_1 \le t \le t_2$, the internal deformation process $\{{}^{\alpha}e(t); t_1 \le t \le t_2\}$ is regarded to be timewise continuous of Markovian character. Accordingly, the probability distribution of ${}^{\alpha}e(t)$ is considered to be completely defined for all $t > \tau, t_1 \le (\tau, t) \le t_2$, by the value assumed at $t = \tau$ and, in particular, is independent of the history of such process for all $t_1 \le t < \tau$.

Analytically, a Markov process is completely determined by its transition probabilities. Letting $\underline{P}(t)$ denote the transition probabilities of the stochastic process $\{{}^{\alpha}e(t); t_1 \le t \le t_2\}$, then the time history of the probability distribution $P\{{}^{\alpha}e(t)\}$ can be described by

$$P\{{}^{\alpha}e(t)\} = \underline{P}(t-\tau)P\{{}^{\alpha}e(\tau)\} \tag{19}$$

where $\underline{P}(t-\tau)$, under the restriction that the process $\{{}^{\alpha}e(t), t_1 \le t \le t_2\}$ is represented by a stationary Markov process, is governed by the Chapman-Kolmogorov equation:

$$\underline{P}^{e}(t) = \underline{P}(\Delta t)\underline{P}^{e}(\tau); \qquad \Delta t = t - \tau \tag{20}$$

and the backward and forward Kolmogorov equations given, respectively, by

$$\begin{aligned} \frac{d\underline{P}^{e}(\Delta t)}{dt} &= \underline{P}^{e}(\Delta t)\underline{Q}^{e}(\Delta t) \\ \frac{d\underline{P}^{e}(\Delta t)}{dt} &= \underline{Q}^{e}(\Delta t)\underline{P}^{e}(\Delta t) \end{aligned} \tag{21}$$

subject to the initial condition $\underline{P}^{e}(t_1) = \underline{I}$ ($\underline{I}$ is the identity matrix). In eqns (21), $\underline{Q}^{e}(\Delta t)$ is a time-dependent transition probability matrix, given by

$$Q^{e}(\Delta t) = \lim_{\Delta t \to 0} \left(\frac{\underline{P}^{e}(t, t+\Delta t)}{\Delta t} \right)$$

However, for simplification of the analysis, one may assume that this transition probability matrix is time-independent, i.e.

$$\underline{Q}^{e}(\Delta t) = \underline{Q}^{e}$$

This suggests that the solution of the Kolmogorov equations given in eqns (21) can be written as [19]

$$\underline{P}^{e}(\Delta t) = \exp(\underline{Q}^{e}\Delta t) \tag{22}$$

Thus, by combining eqns (19) and (22), the former becomes

$$P\{{}^{\alpha}e(t)\} = \exp(\underline{Q}^{e}\Delta t)P\{{}^{\alpha}e(\tau)\} \qquad (\Delta t = t - \tau) \tag{23}$$

i.e. the probabilistic distribution of the microstrain within the intermediate domain at any time $t > \tau$ can be found from the corresponding distribution at $t = \tau$. On the other hand, if two successive distributions of microstrain can be assessed experimentally, a value for the transition probability matrix $\underline{Q}^{e}$ can be obtained.

Having established the time evolution of the internal deformation process via equation (23), a local failure criterion ${}^{\alpha}S(t)$ of the microstructure, within the intermediate domain, may be conjectured [28] by setting

$$^{\alpha}S(t) = 1 - \int_{^{\alpha}e_{max}}^{\infty} d^{M}P\{{}^{\alpha}e(t)\} \tag{24}$$

In this connection, ${}^{\alpha}S(t)$ may be interpreted, in a probabilistic sense, to be associated with the failure of the fibrous element or the interfibre bonding through the relation.

$$^{\alpha}S(t) = (1-q)^{f}S(t) + q^{j}S(t) \tag{25}$$

where the probability q is used.

In (25), ${}^{f}S(t)$ and ${}^{j}S(t)$ may be expressed, respectively, in terms of the statistical distributions ${}^{M}P\{{}^{f}e(t)\}$ and ${}^{M}P\{{}^{j}e(t)\}$ in an analogous manner to (24) as,

$$^{f}S(t) = 1 - \int_{^{f}e_{max}}^{\infty} d^{M}P\{{}^{f}e(t)\} \tag{26}$$

and

$$^{j}S(t) = 1 - \int_{^{j}e_{max}}^{\infty} d^{M}P\{{}^{j}e(t)\} \tag{27}$$

This is with the understanding that the two processes $\{{}^{f}e(t), {}^{j}e(t);\ t_1 \le t \le t_2\}$ can be considered as stationery Markovian and both processes are interacting via equations (14) and (18).

REFERENCES

1. Brezinski, J.P., Tappi, 39, 116, 1956.
2. Schulz, J.H., Tappi 44, 736, 1961.
3. Nissan, A.H., Trans. Faraday Soc., 55, 2048, 1959.
4. Onogi, S. and Sasaguri, K., Tappi, 44(12), 874, 1961.
5. Sternstein, S.S. and Nissan, A.H., Br. Paper Board Makers' Assoc., Proc. Tech. Sec., 1, 319, 1962
6. Van den Akker, J.A., Br. Paper Board Makers' Assoc, Proc. Tech. Sec., 1, 205, 1962.
7. Rance, H.F., Br. Paper Board Makers' Assoc., Proc. Tech. Sec., 1, 1, 1962.
8. Haddad, Y.M., Ph.D Thesis, McGill University, Montreal, 1975.
9. Rance, H.F., Paper Makers' Assoc. G.B. Ireland, Proc. Tech. Sec., 29, 448, 1948.
10. Page, D.H., Tydeman, P.A. and Hunt, M., Br. Paper Board Makers' Assoc., Proc. Tech. Sec., 1, 171, 249, 1962.
11. Nissan, A.H., in Surf. coat. relat. pap. wood symp., R.H. Marchessault and C. Skaar, eds., Syracuse University Press, New York, 221, 1967.
12. Corte, H. Composite Materials, (Ed. Holiday, L). Elsevier, New York, 475, 1966.
13. Van den Akker, J.A., Lathrop, A.L., Voelker, M.H. and Dearth, L.R., Tappi, 41(8), 416, 1958.
14. McIntosh, D.C. and Leopold, B., Br. Paper Board Maker's Assoc., Proc. Tech. Sec., 1, 265, 1962.
15. Van den Akkar, J.A., Tappi, 53(3), 388, 1970.
16. Corte, H and Kallmes, O.J., Br. Paper Board Makers' Assoc., Proc. Tech. Sec., 1, 13, 1962.
17. Axelrad, D.R., Foundations of the probabilistic mechanics of discrete media, Pergamon Press, Oxford, 1984.
18. Haddad, Y.M., Mater. Sci. Engg., 72, 135, 1985.
19. Bharucha Reid, A.T. Elements of the Theory of Markov Processes and their Applications, McGraw-Hill, New York, 1960.
20. Kendall, D.G., Ann. Math. Statist, 19, 1, 1948.
21. Volterra, V., "Fonctions de Lignes". Gauthier-Villard, Paris, 1913.
22. Haddad, Y.M. and Tanary, S., Journal of Rheology, 31(7), 515-526, 1987.

23. Morse, P.M., Phys. Rev. 34, 57, 1929.

24. Corte, H. and Schashek, H., Papier 9, 319, 1955.

25. Sokolov, N.D., "Symposium on Hydrogen Bonding, Ljubljana (1957)" (Eds. Hadzi, D. and Thomson, W.H.), p.385, Pergamon, London, 1959.

26. Haddad, Y.M., Res Mechanics, 22, 243-265, 1987.

27. Haddad, Y.M., J. Colloid and Interface Science, (100)1, 143-166, 1984.

28. Haddad, Y.M., J. Materials Science 21, 3767-3776, 1986.

Stability of Composite Structures with Random Material Imperfections

A. Tylikowski

Institute of Machine Design Fundamentals, Warsaw University of Technology, Narbutta 84, 02-524 Warsaw, Poland

ABSTRACT

The effect of stochastically varying material properties on an asymptotic behaviour of viscoelastic laminated structures is investigated in the paper using a Liapunov type method. The random stiffnesses of structures are described by random fields in the form of truncated harmonic series with random coefficients. The Galerkin method is employed to solve an auxiliary problem. The conditions for a mean stability of trivial solution are derived for a laminated plate in cylindrical bending and a cross-ply laminated circular cylindrical shell in an axisymmetrical state. A Monte Carlo analysis is carried to find a mean value of critical force dor different types of both imperfection mode and the Galerkin expansion. It is shown that the critical loading of laminated structures decreases with increasing variances of stiffnesses.

INTRODUCTION

It is well known that processing imperfections have great effect on the buckling behaviour of plate and shell structures. Randomly arranged disturbances of fiber reinforcements can lead to two basic problems: initial geometric imperfections and space-dependent stiffnesses. In the first case the resultant effect of imperfections is described by assumption that the middle plate plane or middle shell surface in their natural state contain geometric imperfections, which are random fields with known statistics. Numerous papers are available on free and forced vibrations [3] as well as dynamic stability [9] of laminated structures with geometrical imperfections. The imperfections lead to forcing terms in the equations of motion and no equilibrium state exists in both linear and non-linear cases.

The presence of processing imperfections sometimes does not eliminate the existence of the equilibrium state if it merely introduces space-varying coefficients in the equations of motion and it does not deviate the mean surface from the perfect plane and the perfect cylinder, respectively. An example would be a perfectly straight column with a space-dependent cross-section subjected to a time-dependent axial load, for which the undeflected shape is a possible state irrespective of the axial force. In this case one may examine the stability of the equilibrium state. Such problems have received considerable attention [10, 11, 12] for elastic structures under constant forces.

The present paper is concerned with the problem of second type, in which the buckling of perfect laminated structures with space-dependent stiffnesses is investigated. Contrary to the previous common treatment, in which the analyzed structures are purely elastic, the viscous damping described by the Voigt-Kelvin model is taken into account. The problem should be described and solved using an approach and method of stochastic media dynamics [4]. In order to avoid difficulties caused by need of solving partial differential equations with space-dependent coefficients (cf.[7]) one can simplify the problem associated with the buckling of viscoelastic bodies using recent results relating to asymptotic behaviour of a creep buckling analysis. Comparing more recent investigations carried out by Knauss and Menahim [5] and Tylikowski [8] we can demonstrate the small effect of the inertial term on an asymptotic estimation of a column displacement. Following this idea we omit the inertial force in equations of viscoelastic structure compressed by constant forces. The corresponding principle is used to derive the equations for viscoelastic continuous systems. In the analysis a Liapunov type method is used in order to derive the mean stability condition of long laminated plate in cylindrical bending long laminated shell in axisymmetrical state. In the important case when a variability of structure stiffnesses is described by a small harmonic function of space, simple asymptotic formulae are obtained for the critical loading.

DYNAMIC STABILITY OF VISCOELASTIC LAMINATED PLATE WITH MATERIAL IMPERFECTIONS

Let us consider a thin laminated plate in the cylindrical bending compressed by a constant force uniformly distributed along the longer edge. The damping properties of plate are described by a Voigt-Kelvin model. Assuming material nonhomogenities the bending stiffness EI is a random field. Using results concerning the creep buckling of viscoelastic bodies [5], [8] we omit an inertial term in equation governing the transverse displacement of the plate

$$\lambda(EI(\gamma,x)\, v_{,xx})_{,xx} + (EI(\gamma,x)\, w_{,xx})_{,xx} + F\, w_{,xx} = 0, \qquad x \in (0,1) \tag{1}$$

where $\gamma \in (\Omega, B, P)$ - probability space, λ - retardation time. The plate is assumed to be simply supported along the edges, i.e.

$$w = 0, \quad M_x = EI(\gamma,x)\, w_{,xx} = 0. \tag{2}$$

We are interested in deriving almost sure stability criteria of the trivial solution of equation(1) (an equilibrium state). As the (random) Liapunov functional we choose the following expression

$$V = \int_0^l EI(\gamma,x)\, w_{,xx}^2\, dx, \tag{3}$$

which is positive-definite with probability one. Thus we choose the measure of distance as a square root of functional $\| \; \| = V^{\frac{1}{2}}$. As realizations of the random field $EI(\gamma,x)$ are sufficiently smooth and the classical solutions exist with probability one we can use the ordinary differentiation rule in order to obtain the time derivative of functional (3)

$$\frac{dV}{dt} = 2 \cdot \int_0^l EI(\gamma,x)\, w_{,xx}\, v_{,xx}\, dx \tag{4}$$

Integrating by parts and using boundary conditions (2) we can write the time-derivative in the form

$$\frac{dV}{dt} = 2 \cdot \int_0^l w \left(EI(\gamma,x)\, v_{,xx} \right)_{,xx} dx$$

After substituting equation(1) we have

$$\frac{dV}{dt} = -\frac{2}{\lambda}\int_o^l \left[\left(EI(\gamma,x)\, w_{,xx}\right)_{,xx} + F w_{,xx}\right] w\, dx \tag{5}$$

Once again integrating by parts equation(5) yields

$$\frac{dV}{dt} = -\frac{2}{\lambda}(V - U) \tag{6}$$

where

$$U = F\int_o^l w_{,x}^2\, dx. \tag{7}$$

Now we look for an upper bound of the second functional in equation(6)

$$U \leq \kappa V, \tag{8}$$

where the random variable κ is to be determined. In another words we look for number κ defined by

$$\kappa = \max_w \frac{U}{V}\ . \tag{9}$$

Substituting equation(8) into (6) we have the inequality

$$\frac{dV}{dt} = -\frac{2}{\lambda}((1 - \kappa(\gamma))V$$

Therefore after solving the above inequality we can write the upper estimation

$$V(t) \leq V(0)\ \ \exp\left[-\frac{2}{\lambda}(1 - \kappa(\gamma))t\right] \tag{10}$$

If $E\kappa < 1$ (11)

then the trivial solution of equation(1) is stable in the mean [2] with respect to measure $\|\ \|$.

The number κ defined in equation(9) can be calculated using the calculus of variations

$$\delta\left(\frac{U}{V}\right) = \frac{1}{V}\left(\delta U - \frac{U}{V}\delta V\right) = \frac{1}{V}\delta(U - \kappa V) = 0. \tag{12}$$

Substituting equations(3) and (7) yields

$$\delta(U - \kappa V) = 2\int_o^l \left(F w_{,x}\,\delta w_{,x} - \kappa EI(\gamma,x) w_{,xx}\,\delta w_{,xx}\right) dx = 0$$

Integrating by parts and taking into account an arbitrariness of variation δw we obtain the following associate Euler equation

$$\kappa(EI(\gamma,x)w_{,xx})_{,xx} + F w_{,xx} = 0 \qquad x \in (0,1) \tag{13}$$

There is no general solution to equation(13), as $EI(\gamma,x)$ is an arbitrary random field. Thus we have to restrict ourselves to a particular random field. Let us assume that $EI(\gamma,x)$ is a sum of a constant mean value and a space dependent harmonic component of the form

$$EI(\gamma,x) = EI + e(\gamma)\cos\frac{n\pi x}{l} \quad , \tag{14}$$

where $e(\gamma)$ is a random variable with known probabilistic characteristics. Applying the Galerkin method with base functions in the form

$$\sin\frac{m\pi x}{l}$$

to equation(13) yields

$$\int_o^l \left\{ \kappa \left[\left(EI + e(\gamma)\cos\frac{n\pi x}{l} \right) \sum_{m=1}^{N} \left(\frac{m\pi}{l} \right)^2 a_m \sin\frac{m\pi x}{l} \right]_{xx} + \right.$$

$$\left. F \sum_{m=1}^{N} a_m \left(\frac{m\pi}{l} \right)^2 \sin\frac{m\pi x}{l} \right\} \sin\frac{i\pi x}{l}\, dx = 0, \qquad i = 1,...,N \tag{15}$$

Elementary integrations lead to the system of uniform linear equations with respect to coefficients a_i

$$\left[\kappa EI \left(\frac{i\pi}{l} \right)^4 - F \left(\frac{i\pi}{l} \right)^2 \right] a_i +$$

$$+ \frac{\kappa e}{2} \sum_{m=1}^{N} \left(\frac{i\pi}{l} \right)^2 \left(\frac{m\pi}{l} \right)^2 a_m \left(-\delta_{i,n-m} + \delta_{i,m-n} + \delta_{i,n+m} \right) = 0, \quad i = 1,...,N \tag{16}$$

where $\delta_{i,j}$ is the Kronecker delta.

For n=1 and N=2 the changing bending stiffnesses of the plate is described by function

$$EI(\gamma,x) = EI + e(\gamma)\cos\frac{\pi x}{l}. \tag{17}$$

System of equations (16) reduces to the form

$$\left[\kappa\, EI \left(\frac{\pi}{l} \right)^4 - F \left(\frac{\pi}{l} \right)^2 \right] a_1 + 2\kappa \left(\frac{\pi}{l} \right)^4 e(\gamma) a_2 = 0,$$

$$2\kappa \left(\frac{\pi}{l} \right)^2 e(\gamma) a_1 + \left[16\kappa EI \left(\frac{\pi}{l} \right)^4 - 4F \left(\frac{\pi}{l} \right)^2 \right] a_2 = 0. \tag{18}$$

Denoting $\chi = \kappa EI \left(\frac{\pi}{l} \right)^2$, $\varepsilon = \frac{e(\gamma)}{EI}$ we notice that equations(18) have nontrivial solution, if the determinant is equal to zero.

$$\begin{bmatrix} \chi - F & 2\chi\varepsilon \\ 2\chi\varepsilon & 4(4\chi - F) \end{bmatrix} = 0.$$

Finally the number χ is given by the following expression

$$\chi = \frac{5 + \left(9 + 4\varepsilon^2\right)^{\frac{1}{2}}}{2(4-\varepsilon^2)} F.$$

For small values of ε^2 we can expand the square root to obtain the following estimation of number κ

$$\kappa = \frac{1}{EI}\left(\frac{l}{\pi}\right)^2 \left(1 + \frac{1}{3}\varepsilon^2\right) F.$$

Therefore stability condition (11) is fulfilled if the axial force satisfies inequalities independent on the damping coefficient

$$1 = \frac{1}{EI}\left(\frac{l}{\pi}\right)^2 \left(1 + \frac{1}{3}E(\varepsilon^2)\right) F. \tag{19}$$

$$\frac{F}{F_E} < \frac{1}{1 + \frac{1}{3}E\left(\varepsilon^2\right)}$$

where F_E denotes the Euler critical force

$$F_E = \frac{EJ\,\Pi^2}{l^2},$$

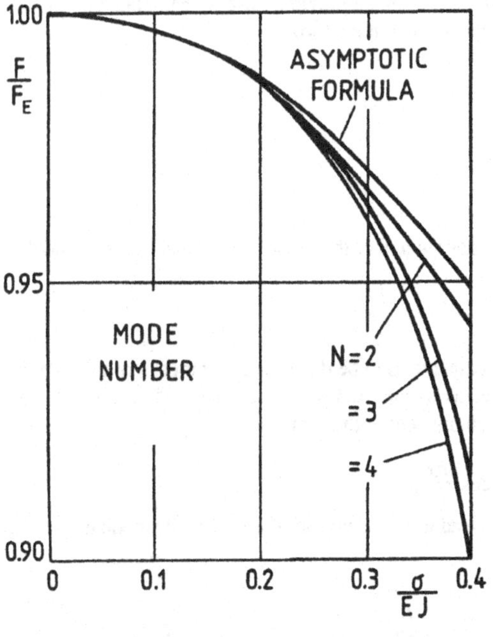

Figure 1. Influence of imperfections standard deviation on mean critical force for laminated plate in cylindrical bending

The calculations are carried out for different numbers of modes in the Galerkin expansion N = 2,3,4 and for the first imperfection mode n=1. The Monte Carlo method is used to generate 10000 realizations of Gaussian variable e and to calculate the mean value of critical force, for which the equilibrium state w = 0 is stable in the mean. When the ratio of standard deviation to the mean bending stiffness is less then 0.2 the asymptotic formula (19) provides mean critical loads in good agreement with the numerical simulations.

In order to avoid the random functional introduced by equation(3) we can examine the stability of laminated plate lowering the order of equation(1)

$$\lambda EI(\gamma,x)\, v_{,xx} + EI(\gamma,x)\, w_{,xx} + Fw = 0. \qquad x \in (0,1) \tag{20}$$

Dividing by the random bending stiffness $EI(\gamma,x)$ (positive with probability one) and denoting the reciprocal by $e(\gamma,x)$ we have the following equation

$$\lambda v_{,xx} + w_{,xx} + Fe(\gamma,x) = 0. \qquad x \in (0,1) \tag{21}$$

Assuming the same boundary conditions (2) we choose the following deterministic expression as the Liapunov functional

$$V = \int_o^l w_{,x}^2 dx, \tag{22}$$

which is positive definite. Thus we choose the measure of distance as a square root of functional $\| \; \| = \sqrt{2}$.

Proceeding similarly as in the previous case we calculate the time derivative of functional along equation (21) and write it in the form

$$\frac{dV}{dt} = -\frac{2}{\lambda}(V - U), \tag{23}$$

where $$U = F\int_o^l e(\gamma,x)\, w_{,x}^2 dx. \tag{24}$$

Now we look for an upper bound of the second functional in equation(24)

$$U \le \kappa^* V, \tag{25}$$

where the random variable κ^* is to be determined by means of the calculus of variation and the Galerkin method. Let us assume that $e(\gamma,x)$ is a sum of a constant mean value and a space dependent harmonic component of the form

$$e(\gamma,x) = e_o + e(\gamma)\cos\frac{n\pi x}{l}, \tag{26}$$

Standard calculations lead us to the system of uniform linear equations with respect to coefficients a_i

$$\left[\kappa^*\left(\frac{i\pi}{l}\right)^2 - Fe_o\right] + \frac{Fe(\gamma)}{2}\sum_{m=1}^{N} a_m\left(-\delta_{i,n-m} + \delta_{i,m-n} + \delta_{i,n+m}\right) = 0. \quad i = 1,\dots,N \tag{27}$$

For n=1 and N=2 system of equations (16) reduces to the form

$$\left[\kappa^*\left(\frac{\pi}{l}\right)^2 - Fe_o\right]a_1 - \frac{Fe(\gamma)}{2}a_2 = 0, \tag{28}$$

$$-\frac{Fe(\gamma)}{2}a_1 + \left[4\kappa^*\left(\frac{\pi}{l}\right)^2 - Fe_o\right]a_2 = 0.$$

Denoting $\chi = \frac{\kappa^*}{F e_o}\left(\frac{\pi}{l}\right)^2$, $\varepsilon = \frac{e(\gamma)}{e_o}$ we finally find the number χ given by the following expression

$$\chi = \frac{5 + \left(9 + \frac{4}{9}\varepsilon^2\right)^{\frac{1}{2}}}{8}.$$

For small values of ε^2 we obtain the following estimation of number κ^*

$$\kappa^* = Fe_o\left(\frac{l}{\pi}\right)^2\left(1 + \frac{1}{3}\varepsilon^2\right).$$

CROSS -PLY LAMINATED CIRCULAR CYLINDRICAL SHELL

Let us consider a closed viscoelastic simply supported cylindrical shell of radius a, length l and total thickness h. The shell consists of an even number of equal thickness orthotropic layers antisymmetrically laminated with respect to its midsurface from both the geometric and the material property standpoint. The Kirchhoff-Love hypothesis on nondeformable normal elements is taken into account. All inertias are neglected. For the shell subjected to a concentrated load P, the initial membrane load can be determined by assuming that the shell remains circular and undergoes a uniform compression circumferentially

$$\mathcal{N} = \frac{P}{2\pi a}.$$

Assuming that the shell buckles in an axisymmetrical mode we can write the equations of equilibrium in the form

$$N_{x,x} = 0, \tag{29}$$

$$M_{x,xx} - \frac{N_\theta}{a} - \mathcal{N}w_{,xx} = 0, \qquad x \in (0,l)$$

where an axial force N_x, and a bending moment are given as follows

$$\mathcal{N}_x = A_{11}u_{,x} + A_{12}\frac{w}{a} - B_{11}w_{,xx}, \tag{30}$$

$$N_\theta = A_{12}u_{,x} + A_{22}\frac{w}{a},$$

$$M_x = B_{11}u_{,x} - D_{11}w_{,xx}.$$

We also assume boundary conditions corresponding to simply supported edges

$$w=0, \quad M_x=0, \quad N_x=0 \qquad \text{for } x=0,l. \tag{31}$$

Applying the corresponding principle to equations(29) we write equations for the viscoelastic shell made of the Voigt - Kelvin material

$$N_{x,x}+\lambda N_{x,xt}=0, \tag{32}$$
$$M_{x,xx}+\lambda M_{x,xxt}-\frac{N_\theta}{a}-\lambda\frac{N_{\theta,t}}{a}+\mathcal{N}w_{,xx}=0, \qquad x\in(0,l)$$

In order to investigate the mean stability of trivial solution of equations(32) we construct the Liapunov functional in the form

$$V=\frac{1}{2}\int_o^l\left[-M_x w_{,xx}+N_x u_{,x}+N_\theta\frac{w}{a}\right]d\Omega. \tag{33}$$

Displacements dependent form of functional (33), which will be useful in the further analysis, is obtained by using equations(30) in equations(33)

$$V=\frac{1}{2}\int_o^l\left[D_{11}(x)w_{,xx}^2-2B_{11}(x)w_{,xx}u_{,x}+A_{11}(x)u_{,x}^2+A_{12}(x)u_{,x}\frac{w}{a}+A_{11}\left(\frac{w}{a}\right)^2\right]d\Omega. \tag{34}$$

Differentiating functional (33) with respect to time yields

$$\frac{dV}{dt}=\frac{1}{2}\int_o^l\left(-M_{x,t}w_{,xx}-M_x w_{,xxt}-N_{x,t}u_{,x}+N_x u_{,xt}+N_\theta\frac{w}{a}+N_\theta\frac{w_{,t}}{a}\right)d\Omega. \tag{35}$$

Integrating by parts and using boundary conditions (31) we prove the following formulae

$$\int_o^l\left(-M_{x,t}w_{,xx}\right)dx=\int_o^l\left(-M_{x,xxt}w\right)dx,$$
$$\int_o^l N_{x,t}u_{,x}dx=\int_o^l N_{x,xt}u\,dx, \tag{36}$$
$$\int_o^l\left(-M_x w_{,xxt}+N_x u_{,xt}+N_\theta\frac{w_{,t}}{a}\right)dx=\int_o^l\left(-\left(B_{11}u_{,x}-D_{11}w_{,xx}\right)w_{,xxt}+N_x u_{,xt}+\left(A_{12}u_{,x}+A_{22}\frac{w}{a}\right)\frac{w_{,t}}{a}\right)d\Omega=$$
$$=\int_o^l\left(-M_{x,t}w_{,xx}+N_\theta\frac{w}{a}+\left(N_x+B_{11}w_{,xx}-A_{12}\frac{w}{a}-A_{11}u_{,x}\right)u_{,xt}+\left(A_{11}u_{,xt}+A_{12}\frac{w_{,t}}{a}-B_{11}w_{,xxt}\right)u_{,x}\right)d\Omega=$$
$$\int_o^l\left(-M_{x,t}w_{,xx}+N_{x,t}u_{,}x+N_{\theta,t}\frac{w}{a}\right)d\Omega.$$

Substituting equations(36) into time derivative (35) yields

$$\frac{dV}{dt}=\int_o^l\left(-M_{x,t}w_{,xx}+N_{x,t}u_{,x}+N_{\theta,t}\frac{w}{a}\right)d\Omega.$$

Integrating by parts and using equations of motion in order to eliminate terms with time derivatives we can rewrite the time derivative of functional in the form

$$\frac{dV}{dt}=\frac{1}{\lambda}\int_o^l\left[\left(-M_{x,xx}-\frac{N_\theta}{a}-\mathcal{N}w_{,xx}\right)w+N_{x,x}u\right]d\Omega. \tag{37}$$

Finally after integration by parts we have

$$\frac{dV}{dt}=-\frac{1}{\lambda}(V-U), \tag{38}$$

where the auxiliary functional U has the form

$$U=\mathcal{N}\int_o^l w_{,x}^2dx.$$

Proceeding similarly to the analysis presented in section 2 we look for a random number κ satisfying inequality (8) by means of the calculus of variation. But contrary to the previous considerations (cf.equation(9)) the number κ is defined in another way

$$\kappa=\max_{u,w}\frac{U}{V}$$

In order to derive Euler equations we calculate variation δU and δV using the formulae similar to equation(36). Comparing with zero variation delta $\delta(U-\kappa V)$ and coefficients of δu and δw we obtain two Euler equations

$$N_{x,x}=0, \tag{39}$$

$$\kappa M_{x,xx}-\frac{\kappa N_\theta}{a}-\mathcal{N}w_{,xx}=0.$$

Eliminating the axial displacement u we write the appropriate Euler equation in the form

$$\kappa\left\{\left[\left(D_{11}-\frac{B_{11}^2}{A_{11}}\right)w_{,xx}\right]_{,xx}+\frac{B_{11}A_{12}}{aA_{11}}w_{,xx}+\left(\frac{B_{11}A_{12}}{aA_{11}}w\right)_{,xx}+\right.$$

$$\left.\left(A_{11}-\frac{A_{12}^2}{A_{11}}\right)\frac{w}{a^2}+\mathcal{N}w_{,xx}\right\}=0, \qquad x\in(0,1), \tag{40}$$

We assume that coefficients in equation(40) can be presented as a sum of constant mean values and relatively small space-dependent parts in the following form

$$D_{11}-\frac{B_{11}^2}{A_{11}}=\alpha_o+\alpha(x), \quad \frac{B_{11}A_{12}}{aA_{11}}=\beta_o+\beta(x),$$

$$A_{11}-\frac{A_{12}^2}{A_{11}}=\gamma_o+\gamma(x),$$

where the space dependent parts are harmonics with random independent amplitudes ε_1, ε_2, ε_3

$$\alpha(x)=\varepsilon_1\cos\frac{m\pi x}{l},$$

$$\beta(x)=\varepsilon_2\cos\frac{n\pi x}{l},$$

$$\gamma(x)=\varepsilon_3\cos\frac{p\pi x}{l},$$

The system with respect to coefficients of Galerkin expansion can be written as follows

$$\left[\alpha_o\left(\frac{i\pi}{l}\right)^4-\frac{2}{a}\beta_o\left(\frac{i\pi}{l}\right)^2+\frac{\gamma_o}{a^2}\right]w_i+\frac{1}{2}\sum_{j=1}^{N}\varepsilon_1\left(\frac{i\pi}{l}\right)^2\left(\frac{j\pi}{l}\right)^2\left(-\delta_{i,m-j}+\right.$$

$$\left.+\delta_{i,j-m}+\delta_{i,j+m}\right)-\frac{\varepsilon_2}{a}\left[\left(\frac{i\pi}{l}\right)^2+\left(\frac{j\pi}{l}\right)^2\right]\left(-\delta_{i,n-j}+\delta_{i,j-n}+\delta_{i,j+n}\right)+$$

$$\left.\frac{\varepsilon_3}{a^3}\left(-\delta_{i,p-j}+\delta_{i,j-p}+\delta_{i,j+p}\right)\right]w_j-\left(\frac{i\pi}{l}\right)^2\frac{N}{\kappa}w_i=0,\qquad i=1,\ldots,N. \tag{41}$$

Specifying system of equations for $N=2$ and $m=n=p=1$ we have two linear equations with respect to w_1, w_2

$$\left[\alpha_o\left(\frac{i\pi}{l}\right)^4-\left(\frac{2}{a}\beta_o+\frac{N}{\kappa}\right)\left(\frac{\pi}{l}\right)^2+\frac{\gamma_o}{a^2}\right]w_1+\frac{1}{2}\left[4\varepsilon_1\left(\frac{\pi}{l}\right)^4-\frac{5\varepsilon_2}{a}\left(\frac{\pi}{l}\right)^2+\frac{\varepsilon_3}{a^2}\right]w_2=0 \tag{42}$$

$$\frac{1}{2}\left[4\varepsilon_1\left(\frac{\pi}{l}\right)^4-\frac{5\varepsilon_2}{a}\left(\frac{\pi}{l}\right)^2+\frac{\varepsilon_3}{a^2}\right]w_1+\left[16\alpha_o\left(\frac{\pi}{l}\right)^4-4\left(\frac{2}{a}\beta_o+\frac{N}{\kappa}\right)\left(\frac{\pi}{l}\right)^2+\frac{\gamma_o}{a^2}\right]w_2=0.$$

In order to obtain a number κ satisfying inequality (8) we equal to zero the determinant of equations(42). Then we have an asymptotic formula for the number κ, assuming the imperfections are small

$$\kappa=\mathcal{N}_o\left(1+c_1E(\varepsilon_1^2)+c_2E(\varepsilon_2^2)+c_3E(\varepsilon_3^2)\right) \tag{43}$$

where $\mathcal{N}_o$ denotes the critical loading of shell with constant stiffnesses given by the following formula

$$\mathcal{N}_o=\frac{\alpha_o\left(\frac{\pi}{l}\right)^4-\frac{2}{a}\beta_o\left(\frac{\pi}{l}\right)^2+\frac{\gamma_o}{a^2}}{\left(\frac{\pi}{l}\right)^2}. \tag{44}$$

and c_1, c_2, c_3 are given constants depending on the shell geometry and material properties.Therefore we can conclude that the trivial solution of equation(32) is mean stable if the mean value of number κ is smaller than 1. This condition shows us that the critical loading of shell with the space-dependent stiffnesses described by equation(32) is smaller than that of constant stiffnesses

$$N < N_e\left(1 - c_1 E(\varepsilon_1^2) - c_2 E(\varepsilon_2^2) - c_3 E(\varepsilon_3^2)\right) \tag{45}$$

The calculations are performed for modes in the Galerkin expansion, which are closest to the lowest vibration mode i = 3,4,5,6 or i = 3,4,5,6,7 and for different types of imperfection modes.

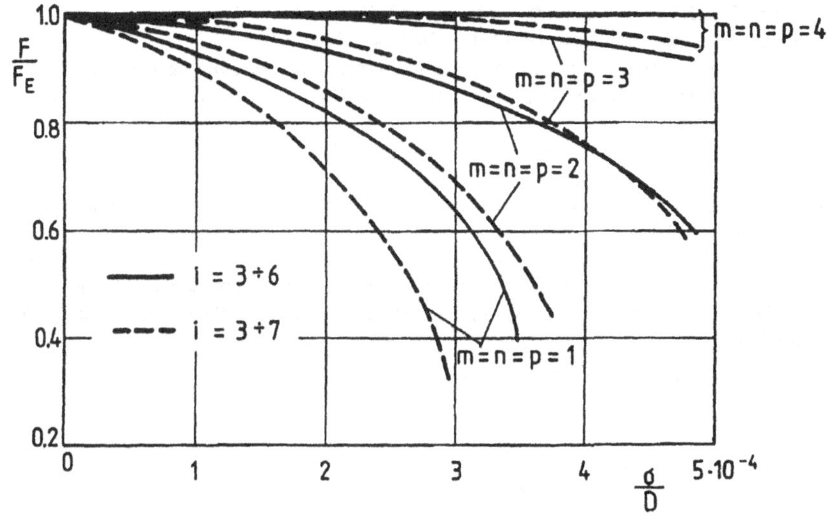

Figure 2. Mean critical force for laminated axisymmetrical shell versus imperfection standard deviation.

The stiffness D corresponds to the i=5, for which we have the lowest buckling force. The stability regions for laminated shells are much more sensitive to the imperfection mode and the dependence on the standard deviation is more pronounced, in comparison to the laminated plates.

Acknowledgements The author would like to acknowledge support from CPBP 02.02 coordinated by the Institute of Fundamental Technological Research of Polish Academy of Sciences.

References

1. Bert, C.W., Baker, J.L., and Eagle, D.M., Free vibrations of multilayer anisotropic cylindrical shells, J. Comp. Mat., Vol.3 pp. 480-499, 1969 .
2. Bertram, J.E., Sarachik, P.E., Stability of circuits with random time-varying parameters, IRE Trans. on Circuit Theory, Spec. Supl. Vol.6, pp. 260-270, 1959 .
3. Fei-Yue Wang, Monte Carlo analysis of nonlinear vibration of rectangular plates with random geometric imperfections, Int. J. Solids Struct., Vol.26 pp. 99-109, 1990.
4. Howe, M.S., Wave propagation in random media, J.Fluid Mech., Vol.45, pp. 769, 1971.
5. Knauss,W.G., Minahem, T.M., Creep buckling in viscoelastic materials, Fourth IUTAM Symp. Creep in Structures, Abstracts, Cracow, 1990. (in print)
6. Schultz, A.B., Tsai, S.W., Measurements of complex moduli for laminated fiber-reinforced composites, J. Comp. Mat., Vol. 3 pp. 434-442, 1969.
7. Sobczyk, K., Fale stochastyczne, PWN, Warszawa,1982.
8. Tylikowski, A., Vibration and bounds on motion of viscoelastic columns with imperfections and time-dependent forces, Fourth IUTAM Symp. Creep in Structures, Abstracts, Cracow, 1990. (in print)
9. Tylikowski, A., Dynamic stability of symmetrical cross-ply plate with random geometric imperfections, ZAMM, Vol.71 1991 (in print)
10. Wanin, G.A., Semenyuk, N.P., Stability of composite materials with imperfections, Naukowa Dumka, Kiew, 1987 (Rusian).
11. Gulyaew, V.I., Bazhenov, W.A., Gotsulyak, E.A., Dehtaryuk, E.S., Lizunow, P.P., Stability of periodic process in nonlinear mechanical systems, Lvov, Wisha Shkola, 1983 (Rusian).
12. Potapow, V.D., Stability of viscoelastic construction elements, Stroyizdat, Moskva, 1985 (Rusian).

SECTION 10: APPLIED RANDOM VIBRATIONS

Thermally Forced Random Vibrations of Thick Plates

R. Heuer, H. Irschik, F. Ziegler

Department of Civil Engineering, Technical University of Vienna, A-1040 Vienna, Austria

ABSTRACT

Nonstationary random vibrations of thermally forced moderately thick plates are studied. Influence of plate shear and rotatory inertia is taken into account according to Mindlin's theory. In case of polygonal platform and hard hinged support conditions, the deflections are shown to be analogous to the deflections of a thin plate with effective loading and stiffness. The frequency response function of the undamped plate due to time-harmonic thermal curvature is calculated by an advanced boundary element method. Subsequently, light damping is built in by applying the quadrature type of elastic-visco-elastic correspondence. Random vibrations are considered by employing the evolutionary power spectral method where the plate structure is excited by a nonstationary envelope-type stochastic process of the thermal curvature. Particularly, stationary spectral relations between an imposed plate surface temperature process and the thermal curvature of thick plates are derived following Fourier's law of heat conduction.

INTRODUCTION

When considering linear elastic plates of moderate thickness the influence of shear and rotatory inertia is to be taken into account according to Mindlin's theory [1] or in the approximations of Reissner [2]. The original equations are generalized to include thermal strain effects. It has been shown by Irschik,

Heuer and Ziegler [3] that the isothermal case of a polygonally shaped plate with hinged supports can be reduced to two Helmholtz-Klein-Gordon boundary value problems. Especially the sixth-order system of differential equations of the Mindlin theory reduces properly to a fourth-order equation for the deflection, see [1] and Irschik [4] and [5] for special cases. That effectively loaded "Kirchhoff"-plate is then projected onto two membrane problems. The thickness-shear mode becomes separated. Irschik [5] has treated the natural vibrations and found the analogies to hold even in the more general case of a linear Pasternak foundation and in-plane forces according to the Brunnelle and Robertson theory, [6]. Static thermal bending of thick rectangular plates is considered by Das and Rath [7]. Thermally forced random vibrations are considered by the method of nonstationary spectra, see, eg Ziegler, Irschik and Heuer [8]. Ie the dynamic properties of the plate are summarized in the frequency response functions where we assume the surface temperature to vary in a time-harmonic manner such that the mean temperature is zero. With increasing forcing frequency the temperature waves become more and more concentrated to a small surface layer. Numerically, the undamped solution is found by an advanced boundary element method, Irschik, Heuer and Ziegler [9]. Before entering the spectral method, light damping must be considered to limit the resonance amplitudes. Dasgupta and Sackman [10] presented a quadrature type of the elastic-viscoelastic correspondence which has been properly extended to include singularities by Ziegler [11]. Thus, the classical correspondence applies to the resonance singularities and a regular portion of the undamped frequency response functions remains to be subject to a weighted integration over the real frequency axis. A version of the nonstationary spectral method which requires just one more weighted integration of the damped frequency response function is discussed by Ziegler [12]. Thus the spectral densities and their evolution in time of the deflections at preselected points are easily derived even in the case of the non-simple geometries of skew and parallelogram plates.

Effects of nonlinear foundation even with random parameters have been discussed by Shinozuka, Deodatis and Wu [13]. A generalization of the present method to include such nonlinearities seems to be possible.

REDUCTION OF THE EQUATIONS OF THERMALLY FORCED MINDLIN PLATES

Mindlin [1] derived a widely accepted sixth-order system of differential equations for the deflection w and for the cross-sectional rotations ψ_x, ψ_y, which, when generalized to include the imposed thermal curvature κ^{th} takes on the form

$$\frac{K}{2}\left[(1-\nu)\Delta\psi_x+(1+\nu)\Phi_{,x}\right]-(\psi_x+w_{,x})/s \;-K(1+\nu)\kappa^{th}_{,x}=r\ddot{\psi}_x \tag{1}$$

$$\frac{K}{2}\left[(1-\nu)\Delta\psi_y+(1+\nu)\Phi_{,y}\right]-(\psi_y+w_{,y})/s \;-K(1+\nu)\kappa^{th}_{,y}=r\ddot{\psi}_y \tag{2}$$

$$\left(\Delta\overline{w}+\phi\right)/s=\rho h\ddot{w} \tag{3}$$

$$\Phi=\psi_{x,x}+\psi_{y,y}\;,\quad K=Gh^3/6\,(1-\nu)\,,\quad s=1/\kappa^2Gh\,,\quad r=\rho h^3/12\;. \tag{4}$$

The material is assumed to be isotropic and homogeneous. Pao and Kaul [14] reported on the best choice of the shear factor κ^2. The moderately large but constant thickness of the plate is denoted h, and ρ is the mass density, G is the shear modulus and ν is Poisson's constant. The parameter r accounts for the influence of rotatory inertia. Differentiation with respect to the coordinates x and y of equations (1) and (2), respectively, adding the results and substituting the potential Φ yields, the time factor $w=w\exp(i\omega t)$ is suppressed,

$$\overline{K}\Delta\Delta w-\overline{n}\,\Delta w-\overline{\mu}w=-\overline{K}(1+\nu)\Delta\kappa^{th} \tag{5}$$

$\overline{K}=K$, $\overline{n}$ and $\overline{\mu}$ denote the effective stiffness, hydrostatic in-plane force and inertia parameter of a Kirchhoff-plate loaded by the thermal time-harmonic curvature. The parameters with an overbar depend on the frequencly ω,

$$\overline{n}=-K(\rho hs+r)\,\omega^2\;,\quad \overline{\mu}=\left(1-sr\omega^2\right)\rho h\omega^2\;. \tag{6}$$

The boundary conditions along any simply supported straight edge take on the simple forms, (n,s) are local coordinates with normal n,

$$\begin{aligned}\Gamma:\ & w=0\ ,\\ & m_n = K(\psi_{n,n}+\nu\psi_{s,s}) - K(1+\nu)\kappa^{th} = 0\ ,\\ & \psi_s = \psi_{s,s} = 0,\ \ \psi_{n,n} = (1+\nu)\kappa^{th}\ ,\\ & \Delta w + \psi_{n,n} + \psi_{s,s} = 0 \rightarrow \Delta w = -(1+\nu)\kappa^{th}\ ,\end{aligned} \tag{7}$$

ie an inhomogeneous Navier's condition results. The latter is typical for classical thermal problems of thin plates.

The work of Federhofer [15] on natural vibrations of prestressed Kirchhoff plates suggests an extension to the inhomogeneous boundary value problem, equations (5) and (7): Putting the deflection

$$w = w_1 + w_2\ , \tag{8}$$

equation (5) is quite naturally decomposed into two membrane problems,

$$\Delta w_j + \alpha_j w_j = -\alpha_j\, \vartheta_j\, \kappa^{th},\ \ j = 1,2\ . \tag{9}$$

When substituting these set of equations into equation (5), the coefficients of w_j must vanish identically. These conditions render the characteristic values in terms of the parameters of equation (6),

$$\alpha_j = -\frac{1}{2K}\left[\bar{n} + (-1)^j\,(\bar{n}^2 + 4K\bar{\mu})^{1/2}\right]\ ,\ \ j = 1,2\ . \tag{10}$$

Furthermore, the coefficient of κth is to be compared to render

$$\alpha_j\,\vartheta_j = -(-1)^j\,\alpha_j\,(1-\nu)/(\alpha_1-\alpha_2)\ ,\ \ j = 1,2\ . \tag{11}$$

The boundary conditions (7) are transformed to render Dirichlet boundary value problems of the effectively loaded membranes of same platform,

$$\Gamma:\ \ w_j = 0,\ \ j = 1,2\ . \tag{12}$$

Thus, the analogy to vibrating, prestressed membranes is complete, and the two potentials (deflections) w_j are uncoupled, even in the case of rather generally shaped planforms under conditions of forced vibrations. The natural mode shapes of the flexural motion, w_1, and of the thickness shear vibration, w_2, are affine to that of the deflection of the associated membranes with fixed edges. Irschik [5] considered also the third type of eigenmotions, the thickness twist motion.

The frequency response functions w_1 and w_2 are easily and efficiently determined by means of the boundary element method BEM and by sweeping the forcing frequency. Green's function of a rectangular basic domain is applied, Irschik, Heuer and Ziegler [3].

FREQUENCY RESPONSE FUNCTION OF DEFLECTION AND LIGHT DAMPING

The frequency response functions of the membrane boundary value problems are determined by means of the BEM and the Green's functions of rectangular domains, especially for the forcing $\kappa^{th} = 1.e^{i\omega t}$, the time factor is understood,

$$F_{w_j} = \alpha_j \vartheta_j \int_0^B \int_0^L G_j(x,y;\, \xi,\eta)\, d\xi d\eta + \int_\Gamma G_j(x,y;\, \sigma)\mu_j^F(\sigma)\, d\sigma \tag{13}$$

see Fig.1. The distribution of forces μ_j^F enforces F_{w_j} to vanish along Γ. If some edges coincide with those for the basic domain the integral is to be extended over $C \subset \Gamma$, only. At the boundary the Fredholm integral equation of the first kind is established

$$\Gamma:\ F_{w_j}(s) = 0\ . \tag{14}$$

Green's function is determined by the single series, convergence is proved by Heuer and Irschik [17]

$$G_j = \frac{1}{L}\sum_{n=1}^{\infty} \frac{\sin n\pi x/L \sin n\pi\xi/L}{\bar{k}_n\left[1 - \exp\left(-2\bar{k}_n B\right)\right]} \left\{\exp\left[-\bar{k}_n|y-\eta|\right] - \exp\left[-\bar{k}_n(y+\eta)\right]\right.$$
$$\left. + \exp\left[-\bar{k}_n\left(2B - |y-\eta|\right)\right] - \exp\left[-\bar{k}_n\left(2B - (y+\eta)\right)\right]\right\}\ ,\quad \bar{k}_n = \left[(n\pi/L)^2 - \alpha_j\right]^{1/2}\ . \tag{15}$$

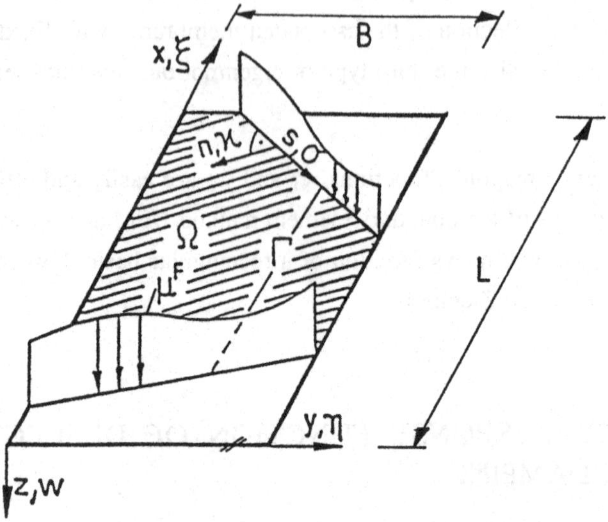

Fig. 1 Polygonal plate domain, embedded into the basic (rectangular) domain; line load density μ^F enforcing boundary conditions at Γ

Numerical evaluations have been made for trapezoidal and parallelogram plates. The singular part of Green's function is to be integrated analytically,

$$G_S = -\frac{1}{2\pi} \ln\left[(\sigma - s)^2 + (\nu - n)^2\right]^{1/2} \quad . \tag{16}$$

The result of integration over a straight boundary element of length ϑ is

$$\int_{s_i - \vartheta/2}^{s_i + \vartheta/2} G_S d\sigma \Big|_{n \to 0} = \frac{\vartheta}{2\pi}\left(1 - \ln\frac{\vartheta}{2}\right) \quad . \tag{17}$$

The regular part is approximated by linear interpolation within the singular interval. Superposition gives

$$F_w(x,y;\omega) = F_{w_1} + F_{w_2} \quad . \tag{18}$$

Light damping is considered by putting the shear modulus complex,

$$G^* = G(1 + i\,\varepsilon) \quad . \tag{19}$$

The shear wave speed becomes complex in that single parameter model,

$$c^* = [G(1+ i\,\varepsilon)/\rho]^{1/2} \quad . \tag{20}$$

The undamped Mindlin plate has resonance singularities,

$$F_w = F_w^{(a)} + F_w^{(r)} \;, \tag{21}$$

$$F_w^{(a)} = \bar{\lambda}^{2} \sum_k f_k / \left(\nu^2 - \omega_k^2\right) \tag{22}$$

and the mode-dependent coefficients are determined numerically from the discrete values of (18),

$$f_k \doteq 2\omega_k \lim_{N\to\infty} \frac{1}{N\lambda^2} \sum_{i=1}^{N} (\nu_i - \omega_k)\, F_w(\nu_i) \quad . \tag{23}$$

The regular part is determined from equation (21).

Equation (22) is subject to the classical viscoelastic correspondence principle, frequency is considered of dimension one,

$$F_w^{(a)*} = \lambda^{*2} \sum_k f_k / \left(\nu^{*2} - \omega_k^2\right), \;\; \nu^* = \omega L / c^* \quad . \tag{24}$$

The regular part is extended into the lower half of the complex frequency plane by the weighted integration

$$\begin{aligned} F_w^{(r)*} &= \lambda^{*2} \int_{-\nu_n}^{\nu_n} F_w^{(r)}(\xi)\; K(\xi;\nu^*)\, d\xi \,, \\ K &= \frac{1}{\pi} \frac{\omega_I}{\omega_I^2 + (\xi - \omega_R)^2} \;, \qquad \nu^* = \omega_R - i\omega_I \quad . \end{aligned} \tag{25}$$

Errors from the approximations (23) are somewhat corrected by (25).

POWER SPECTRAL DENSITY OF THE FLEXURAL RESPONSE

Putting the thermal loading process

$$\kappa^{th} = \sqrt{\psi(t)}\ \kappa_s(t) \tag{26}$$

where $\psi(t)$ is a deterministic time function and $\kappa_s(t)$ is a stationary stochastic process with zero mean and with the given power spectral density S_{th} of the thermal curvature, the power spectral density of a deflection w(x,y,t) becomes

$$S_{ww}(t,\nu) = \phi^*(t,\nu)\ \overline{\phi^*(t,\nu)}\ S_{th}(\nu)\ , \tag{27}$$

where

$$\phi^*(t,\nu) = \frac{1}{2\pi}\int_{-\infty}^{+\infty} F_w^*(\xi)\ \varphi(t,\nu,\xi)\ d\xi \tag{28}$$

is determined by numerical integration with the weighting function

$$\varphi(t,\nu,\xi) = \int_0^t \psi\left(t-\tau\right) \exp\left[i(\xi-\nu)\tau\right] d\tau\ . \tag{29}$$

Equation (29) in simple cases can be analytically integrated:

$$\begin{aligned} \varphi &= -i\left\{\exp\left[i(\xi-\nu)\,t\right] - \exp\left(\beta t\right)\right\} / \left(\xi-\nu+i\beta\right) + \\ &\quad +i\left\{\exp\left[i(\xi-\nu)\,t\right] - \exp\left(\delta t\right)\right\} / \left(\xi-\nu+i\delta\right)\ , \\ \psi(t) &= e^{\beta t} - e^{\delta t}\ ,\quad \beta,\delta<0\ . \end{aligned} \tag{30}$$

Switching by a simple Heaviside function is included by taking the first part with $\beta=0$.

A further numerical integration renders the autocorrelation function

$$R_w(t_1,t_2;x,y) = \int_{-\infty}^{+\infty} \exp\left[i\nu(t_1-t_2)\right]\ \phi^*(t_1,\nu)\ \overline{\phi^*(t_2,\nu)}\ S_{th}(\nu)\ d\nu\ . \tag{31}$$

RANDOM TEMPERATURE WAVES

Skew symmetric stationary temperature waves produce, ie such imposed stationary thermal curvatures. The temperature waves in a half-space $z \geq 0$ when forced by a prescribed surface temperature $\theta_o(t)=e^{i\omega t}$ with assigned frequency w are given by the solution of the linear head conduction equation, see Carslaw and Jaeger [16], the time factor is suppressed,

$$F(z,\omega) = e^{-kz}\, e^{-ikz}, \quad k = \sqrt{\omega/2a}\,, \quad \lambda=2\pi/k \;, \tag{32}$$

a is the temperature conductivity, k the wave number and λ is the wave length. The latter in a metal is typically 3,5 cm at 1 Hertz. In case of thick plates the influence of the second surface is negligible and the temperature frequency response function in case of an unsymmetrical thermal loading of the two faces at $z=\pm h/2$ becomes approximately,

$$F(z,\omega) = e^{-k(-z+h/2)}\, e^{-ik(-z+h/2)} \;, \quad 0 \leq z \leq h/2 \;. \tag{33}$$

Hence, the time-reduced imposed thermal curvature is sufficiently well approximated even for small frequencies by, α is the linear coefficient of thermal expansion,

$$\kappa^{th} = 2\,\frac{12}{h^3}\int_0^{h/2} \alpha\, z\, F(z,\omega)\, dz = \frac{12}{h^3}\, F_{th}(\omega) \;, \tag{34}$$

$$F_{th}(\omega) = 2\alpha\left\{\left[\frac{h}{2k}\frac{\sqrt{2}}{2}\, e^{-i\pi/4} + \frac{1}{k^2}\frac{1}{2}\, e^{-i\pi/2}\right] - \frac{1}{k^2}\frac{1}{2}\, e^{-kh/2}\, e^{-i(kh/2+\pi/2)}\right\} . \tag{35}$$

The squared amplitude becomes

$$F_{th}\,\overline{F}_{th} = 4\alpha^2\left\{\frac{h^2}{8k^2} + \frac{1}{4k^4}\left(1+e^{-kh}\right) - \frac{1}{2k^4}\, e^{-kh/2}\cos kh/2 + \frac{h}{4k^3} - \frac{h}{4k^3}\, e^{-kh/2}\left[\cos kh/2 - \sin kh/2\right]\right\} . \tag{36}$$

Given the power spectral density of the surface temperature, $S_{\phi\phi} = S_0$ =const in the frequency band $\omega_a \leq |\omega| \leq \omega_b$, see Fig. 2 for positive ω, and switching on the thermal loading at t=0, its stationary power spectrum is given by

$$S_{th}(\omega) = \left(\frac{12}{h^3}\right)^2 F_{th}\,\overline{F}_{th}\,S_{\phi\phi}\ , \tag{37}$$

see Fig. 3. A good approximation results if $l_{max} < h/2$, ie

$$\omega_a/2\pi > 16\pi a/h^2\ . \tag{38}$$

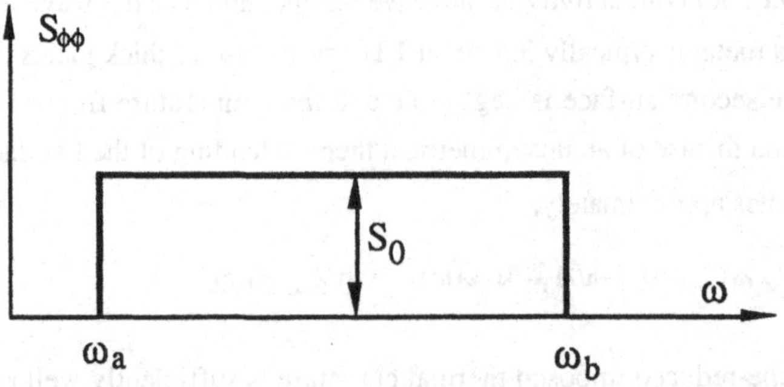

Fig. 2 Broad-band power spectral density of the surface temperature (symmetric)

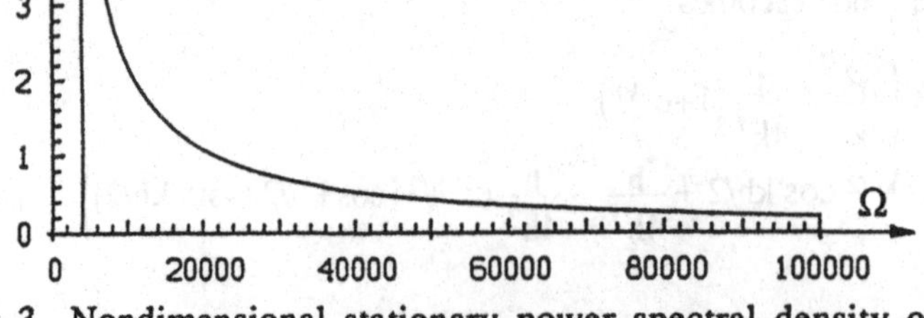

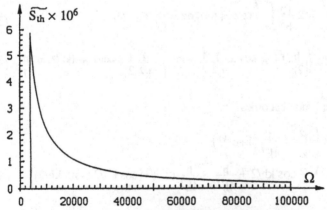

Fig. 3 Nondimensional stationary power spectral density of the thermal curvature $\widetilde{S_{th}}(\Omega) = \frac{S_{th}}{S_0}(h^3/12)^2 L^4$, $\Omega = \omega L^2/2a$, $h/L = 0.2$;

L...characteristic length of the plate

CONCLUSIONS

An efficient numerical algorithm for nonstationary random vibrations of thermally loaded, moderately thick plates is developed. The analysis is performed in the frequency domain. An analogy with respect to deflections of thin plates with effective stiffness and loading parameters is derived. Combination of an advanced BEM with the elastic-visco-elastic correspondence of quadrature type renders the damped frequency response functions that contains all the information of the dynamics of the linear elastic structure. In case of random excitations of envelope type it enters directly the evolutionary power spectrum of the thermal curvature. The time dependent spectral densities and autocorrelation functions of the plate deflections are determined by numerical integration. The nonstationary spectral method introduced in this paper puts probabilistic analysis at one´s fingertips.

REFERENCES

1. Mindlin, R. D. Influence of rotatory inertia and shear on flexural motions of isotropic, elastic plates, ASME Journal Applied Mechanics, Vol. 18 , pp. 31-38, 1951.

2. Reissner, E. On Bending of Elastic Plates, Quarterly of Applied Mathematics, Vol. 5, pp. 55-68, 1947.

3. Irschik, H., Heuer, R. and Ziegler, F. Dynamic analysis of polygonal Mindlin plates on two-parameter foundation using classical plate theory and an advanced BEM, Computational Mechanics, Vol. 4, pp. 293-300, 1989.

4. Irschik, H. Eine Analogie zwischen Lösungen für schubstarre und schubelastische Platten, ZAMM, Vol. 62, pp. T129-T131, 1982.

5. Irschik, H. Membrane-type eigenmotions of Mindlin plates, Acta Mechanica, Vol. 55, pp. 1-20, 1985.

6. Brunnelle, E.J. and Robertson, S.R. Initially stressed Mindlin plates, AIAA J., Vol. 12, pp. 1036-1045, 1974.

7. Das, Y.S. and Rath, B.K. Thermal Bending of Moderately Thick Rectangular Plate, ASME Journal of Applied Mechanics, Vol. 10, pp 1349-1351, 1972.

8. Ziegler, F., Irschik, H. and Heuer, R. Nonstationary Response of Polygonally Shaped Membranes to Random Excitation, in: Random Vibration-Status and Recent Developments (R. Lyon, I. Elishakoff, eds.), Elsevier, pp. 555-565, 1986.

9. Irschik, H., Heuer, R. and Ziegler, F. Free and forced vibrations of polygonal Mindlin-plates by an advanced BEM, Proc. IUTAM-Symp. on Advanced Boundary Element Methods, San Antonio 1987 (T.A.Cruse, ed.), Springer-Verlag Berlin, pp. 179-188, 1988.

10. Dasgupta, G. and Sackman, J. An alternative representation of the elastic-viscoelastic correspondence principle for harmonic oscillation. Transact, ASME Journal of Applied Mechanics, Vol. 44, pp. 57-60, 1977.

11. Ziegler, F. The elastic-viscoelastic correspondence in case of numerically determined discrete elastic response spectra, ZAMM, Vol. 63, pp. T135-T137, 1983.

12. Ziegler, F. Random Vibrations, A spectral method for linear and nonlinear structures, Prob. Engng. Mech., Vol. 2, pp. 92-99, 1987.

13. Shinozuka, M., Deodatis, G. and Wu, W.F. Nonlinear Dynamic Response and System Stochasticity, in: Proc. IUTAM Symp. on Nonlinear Stochastic Dynamic Engineering Systems, Innsbruck 1987, (F.Ziegler, G.I. Schueller, eds.) Springer-Verlag Berlin, pp. 255-268, 1988.

14. Pao, Y.-H. and Kaul, R.K. Waves and vibrations in isotropic and anisotropic plates," In: R. D. Mindlin and Applied Mechanics (G.Herrmann, ed.), Pergamon Press, New York, pp. 149-195, 1974.

15. Federhofer, K. Biegeschwingungen der in ihrer Mittelebene belasteten Kreisplatte, Ing.-Arch., Vol. 6, pp. 68-74, 1935.

16. Carslaw, H.S. and Jaeger, J.C. Conduction of Heads in Solids, Oxford, At the Clarendon Press, 1959.

17. Heuer, R. and Irschik, H. A Boundary Element Method for Eigenvalue Problems of Polygonal Membranes and Plates, Acta Mechanica, Vol. 66, pp. 9-20, 1987.

Random Vibration of the Rigid Block

R. Giannini (*), R. Masiani (**)
(*) Department of Engineering Structures, University of L'Aquila, Italy
(**) Department of Structural and Geotechnical Engineering, University of Rome "La Sapienza", Italy

ABSTRACT

The dynamic response of the rigid block oscillator to a Gaussian white noise excitation process is studied. The motion equation is linearized with the assumption of small displacements. The energy dissipation due to the inelastic impact is modeled as an impulsive process with arrivals occurring at displacement equal zero. The solution of the associated *Fokker-Plank* equation is obtained with two techniques of approximation: Gaussian closure and non-Gaussian closure. The non-Gaussian equations are written in terms of moments of arbitrary order, while the numerical application is limited to the fourth order.

INTRODUCTION

The analytical model of the rigid body oscillator was formulated by G.W. Housner in 1963 [7] to study the seismic response of block-like structures. It is characterized by a force-displacement relationship with a step-discontinuity at the rest position, and it is based on the hypotheses of infinite stiff, no bouncing and absence of sliding. These hypotheses are well verified in the case of "slender" block with friction, and recently the same equation was used to model some multi-bodies problems in the field of the seismic response of masonry structures (Baggio and al.[2]).

The exact solution for the rigid-block equation is known only for the case of stationary response to harmonic base motion (Spanos and Koh [15], Giannini and Masiani [6]); in all the other cases it is necessary to integrate step-by-step the equation in the time domain. This may be an heavy task, especially when the excitation is defined in non-deterministic terms, such as in the case of seismic shacking (Aslam and al. [1], Yim and al. [18], Muto and al. [11], Giannini [5]). For these problems, it is evident the usefulness of defining the excitation as a stochastic process and to evaluate the statistical properties of the response by means of a non-linear *random vibrations* approach.

In this work, as a first approximation, the excitation is assumed to be a white noise process, and the properties of the response process are obtained using two different techniques of closure: the classic Gaussian closure and the more recent non-Gaussian closure technique, first proposed by Crandall [3] and, for example, successfully applied to the Duffing oscillator by Liu and Davis [10].

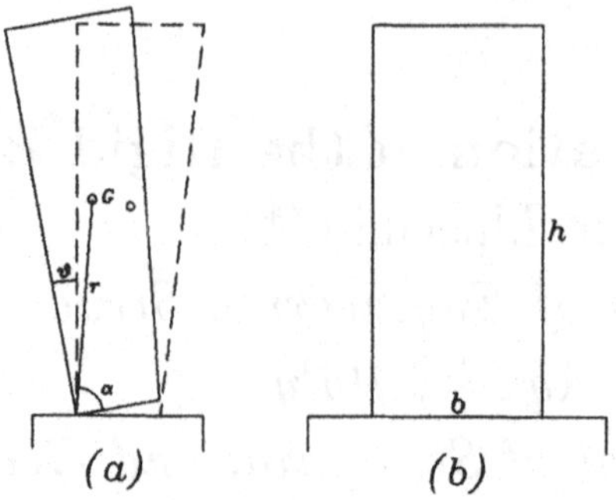

Figure 1: Sketch of the block.

THE RIGID BLOCK OSCILLATOR

The Housner's model assumes that the only motion allowed to a slender block-like structure is the rotation around hinges located at the boundary of the rest surface. According to this hypothesis, which is valid when the friction is sufficient to avoid sliding, the dynamic equilibrium equation is:

$$I\ddot{\theta} + mgr\cos(\alpha + |\theta|)\mathrm{sig}(\theta) = mr\sin(\alpha + |\theta|)A(t) \tag{1}$$

where I is the mass moment of inertia around the rotation axis, m is the mass, $A(t)$ is the acceleration of the horizontal dragging motion, g is the gravitational acceleration and the other symbols are explained in fig. 1a.

Equation (1) may be linearized: assuming $\theta \ll 1$, the second order infinitesimals on the left side and the first order on the right side vanish and equation (1) becomes:

$$\ddot{\theta} + \frac{mgr\cos\alpha}{I}\mathrm{sig}(\theta) - \frac{mgr\sin\alpha}{I}\theta = \frac{mr\sin\alpha}{I}A(t) \tag{2}$$

Letting:

$$x = \frac{I}{mgr\cos\alpha}\theta \tag{3}$$

equation (2) may be written as:

$$\ddot{x} + \mathrm{sig}(x) - Kx = F(t) \tag{4}$$

where:

$$K = \frac{mgr\sin\alpha}{I} \qquad F(t) = \tan\alpha\frac{A(t)}{g} \tag{5}$$

and, in the particular case of the prismatic body as in figure 1b:

$$K = \frac{3}{2}\frac{g\lambda}{b(1+\lambda^2)} \qquad F(t) = \lambda\frac{A(t)}{g} \tag{6}$$

letting $\lambda = h/b$ the *slender ratio* of the block.

Equations (1) and (4) are not valid in the instant when the body impacts on the rest surface and the kinetic energy of the system is reduced because of the inelastic shock. This energy loss is measured by the *restitution coefficient* ϵ, defined as the ratio between the velocity after and before the impact. The value of ϵ must be experimentally evaluated: nevertheless, simple considerations on the balance of the angular momentum give a lower bound for ϵ: for example, in the case of prismatic bodies it is:

$$\epsilon \geq \frac{\lambda^2 - 0.5}{\lambda^2 + 1} \tag{7}$$

These values are confirmed by the weakly damping measured experimentally for the slender block (Wong and Tso [16]).

The effect of the velocity reduction caused by the inelastic impact is modeled by an impulsive force equal to:

$$-2\frac{1-\epsilon}{1+\epsilon}\dot{x}|\dot{x}|\delta(x(t)) \tag{8}$$

where $\delta(x)$ is the Dirac function.

Taking into account the dissipative term, the motion equation becomes:

$$\ddot{x} + \pi\zeta\dot{x}|\dot{x}|\delta(x) + \text{sig}(x) - Kx = F(t) \tag{9}$$

with:

$$\zeta = \frac{2}{\pi}\frac{1-\epsilon}{1+\epsilon} \tag{10}$$

RESPONSE TO A RANDOM EXCITATION

Let $F(t)$ be a white noise process whose spectral density function is:

$$S_F = \frac{D_F}{\pi} \tag{11}$$

If $\boldsymbol{X} = [X,\ \dot{X}]$ is the vector of the state variables of the response process, equation (9) may be written in a standard form as:

$$\frac{d}{dt}\begin{pmatrix} X_1 \\ X_2 \end{pmatrix} = \begin{pmatrix} X_2 \\ -\pi\zeta|X_2|X_2\delta(X_1) - \text{sig}(X_1) + KX_1 \end{pmatrix} + \begin{pmatrix} 0 \\ F(t) \end{pmatrix} \tag{12}$$

and the corresponding *Fokker-Plank* equation (Soong [14]) is:

$$\frac{\partial f}{\partial t} = -\frac{\partial}{\partial x_1}(x_2 f) - \frac{\partial}{\partial x_2}[(-\pi\zeta|x_2|x_2\delta(x_1) - \text{sig}(x_1) + Kx_1)f] + D_F\frac{\partial^2 f}{\partial x_2^2} \tag{13}$$

where $f = f(x_1, x_2, t)$ is the probability density function of the response process.

The exact solution of equation (13) is not known. Nevertheless, it is easy to demonstrate that a "softening" system, like the rigid block, does not have stationary solutions (Roberts and Spanos [13]): to have information about the dynamic response it is necessary to study the transient state. Many approximate techniques are available to solve this problem, several of which are based on the formulation of the *Fokker-Plank* equation in terms of the moments of the components of the process $\boldsymbol{X}$. For equation (12) this formulation gives (Soong [14]):

$$\begin{aligned} \frac{dE[h]}{dt} &= E\left[X_2\frac{\partial h}{\partial X_1}\right] - \pi\zeta E_2\left[|X_2|X_2\frac{\partial h}{\partial X_2}\middle| X_1 = 0\right] - \\ &- E\left[\text{sig}(X_1)\frac{\partial h}{\partial X_2}\right] + E\left[KX_1\frac{\partial h}{\partial X_2}\right] + D_F E\left[\frac{\partial^2 h}{\partial X_2^2}\right] \end{aligned} \tag{14}$$

where h is a generic function of a random vector $\boldsymbol{X}$ and $E_2[\ ]$ is the expected value of the variable X_2 conditioned to $X_1 = 0$:

$$E_2[\psi(X_1, X_2)\big| X_1 = 0] = \int_{-\infty}^{\infty} \psi(0, x_2) f(0, x_2) dx_2 \tag{15}$$

The explicit formulation of the terms holding $\text{sig}(X_1)$ or $|X_2|$ requires some hypotheses on the distribution function of $\boldsymbol{X}$. The simplest one assumes $\boldsymbol{X}$ to be a Gaussian process (Iyengar and Dash [8], Nigam [12]); this is certainly not exact because of the non-linearity of equation (9), but it may be assumed as the first approximation level when the excitation intensity is low.

GAUSSIAN CLOSURE

If $\boldsymbol{X}$ is a Gaussian process, its distribution function is completely defined by five parameters: $\mu_1, \mu_2, \sigma_{11}^2, \sigma_{22}^2, \sigma_{12}$, being the mean values and the three non-zero terms of the covariant matrix respectively.

Two further hypotheses are assumed: i) the body shall be initially in the rest position and ii) it has a vertical plane of symmetry, so that the coefficients of the equation (9) are constant in time. Consequently, the odd moments of $\boldsymbol{X}$ vanish: $\mu_1(t) = \mu_2(t) \equiv 0$, and the covariance matrix of the response may be evaluated from equations (14) relative only to the 2th moments, that is:

$$\begin{aligned} \dot{m}_{20} &= 2m_{11} \\ \dot{m}_{11} &= m_{02} - E[|X_1|] + Km_{20} \\ \dot{m}_{02} &= -2\left(\pi\zeta E_2[|X_2|X_2^2\big|X_1 = 0] + E[\text{sig}(X_1)X_2] - Km_{11}\right) + 2D_F \end{aligned} \tag{16}$$

where dot represents derivative with respect to time, and:

$$m_{hk} = E[X_1^h X_2^k] \tag{17}$$

with (in particular): $m_{20} = \sigma_{11}$, $m_{11} = \sigma_{12}^2$ and $m_{02} = \sigma_{22}^2$.

According to the Gaussian hypothesis on $\boldsymbol{X}$, the terms $E_2[|X_2|X_2^2\big|X_1 = 0]$, $E[|X_1|]$ and $E[\text{sig}(X_1)X_2]$ may also be written in terms of the moments m_{hk}:

$$E_2[|X_2|X_2^2\big|X_1 = 0] = \frac{2}{\pi}m_{20}\left[\frac{m_{02}}{m_{20}} - \left(\frac{m_{11}}{m_{20}}\right)^2\right]^{3/2}$$

$$E[|X_1|] = \sqrt{\frac{2}{\pi}}\sqrt{m_{20}} \qquad E[\text{sig}(X_1)X_2] = \sqrt{\frac{2}{\pi}}\frac{m_{11}}{\sqrt{m_{20}}} \tag{18}$$

Substituting these expressions into equations (16), three ordinary differential equations are obtained:

$$\begin{aligned} \dot{m}_{20} &= 2m_{11} \\ \dot{m}_{11} &= m_{02} - \sqrt{\frac{2}{\pi}}\sqrt{m_{20}} + Km_{20} \\ \dot{m}_{02} &= -2\left(2\zeta m_{20}\left[\frac{m_{02}}{m_{20}} - \left(\frac{m_{11}}{m_{20}}\right)^2\right]^{3/2} + \sqrt{\frac{2}{\pi}}\frac{m_{11}}{\sqrt{m_{20}}} - Km_{11}\right) + 2D_F \end{aligned} \tag{19}$$

the solution of which represents the Gaussian closure of the problem.

Unlike the *Fokker-Plank* equation (13), equations (19) allow stationary solutions for small values of the excitation D_F; this corresponds to situations for which the probability that the solution diverges is so small that it vanishes in the Gaussian approximation (Roberts and Spanos [13]).

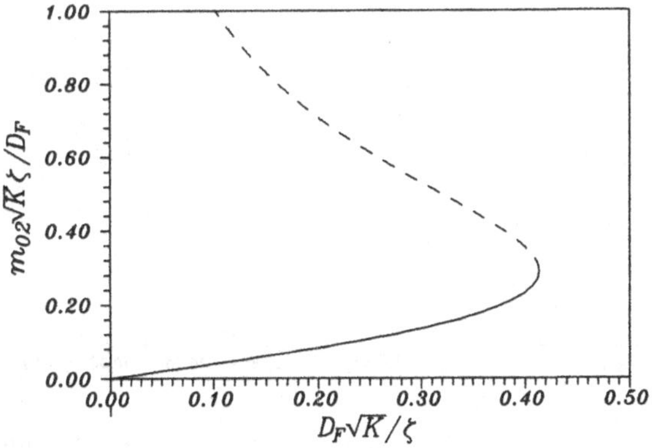

Figure 2: Existence domain for the stationary solution (Gaussian closure)

The stationary solution of equations (19) is easily obtained letting $\dot{m}_{20} = \dot{m}_{11} = \dot{m}_{02} \equiv 0$:

$$\begin{aligned} & m_{11} = 0 \\ & m_{02} - \sqrt{\frac{2}{\pi}}\sqrt{m_{20}} + K m_{20} = 0 \\ & 2\zeta\frac{(m_{02})^{3/2}}{\sqrt{m_{20}}} - D_F = 0 \end{aligned} \tag{20}$$

The first equation indicates that, after the transient state, X and $\dot{X}$ are orthogonal and, then, independent.

Letting:

$$v = \zeta\frac{\sqrt{K}}{D_F}m_{02} \tag{21}$$

the solution in terms of m_{02} is:

$$\frac{\sqrt{K}D_F}{\zeta} = \frac{8v}{\pi(1+4v^2)^2} \tag{22}$$

In figure 2 the normalized response v is plotted as a function of the normalized excitation intensity $\sqrt{K}D_F/\zeta$. The diagram shows that equation (22) admits real solutions only if D_F is less than a given threshold: of the two real solutions, only the lower one has physical meaning, because it satisfies the condition that $v \to 0$ when $D_F \to 0$. The existence threshold is given in terms of dragging motion intensity by:

$$D_{F_s} = \frac{3\sqrt{3}}{4\pi}\frac{\zeta}{\sqrt{K}} \approx 0.4135\frac{\zeta}{\sqrt{K}} \tag{23}$$

Equations (19) are easy to integrate in the time domain. In figures 3 and 4 the variance of the displacement is plotted as a function of time. In figure 3 are also reported the results of Monte Carlo simulations obtained averaging over 500 samples of integrations of equation (1). When $D_F > D_{F_s}$ (see on the right side of figure 3) equations (19) do not admit stationary solutions: the solution diverges and $m_{hk} \to \infty$ as $t \to \infty$.

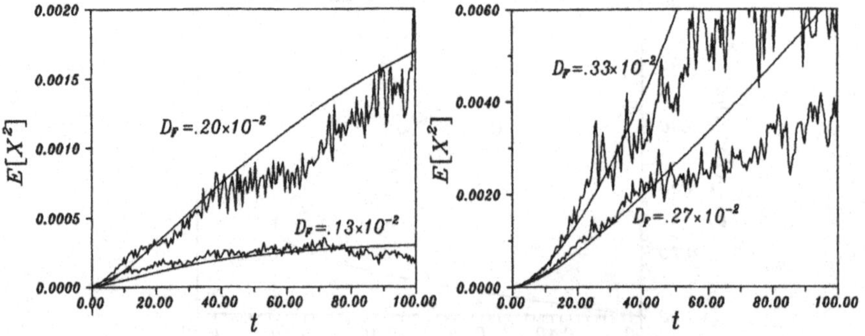

Figure 3: Variance of the response displacement vs. time: Gaussian closure and simulations: $b = 1m$ $\lambda = 8$ ($K = 1.811, \zeta = 0.743 \times 10^{-2}, D_{F_s} = 0.228 \times 10^{-2}$)

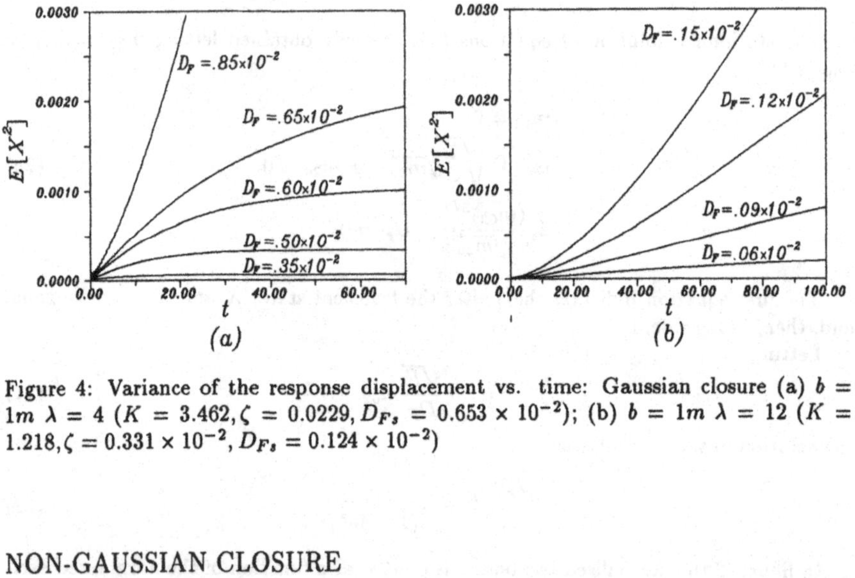

Figure 4: Variance of the response displacement vs. time: Gaussian closure (a) $b = 1m$ $\lambda = 4$ ($K = 3.462, \zeta = 0.0229, D_{F_s} = 0.653 \times 10^{-2}$); (b) $b = 1m$ $\lambda = 12$ ($K = 1.218, \zeta = 0.331 \times 10^{-2}, D_{F_s} = 0.124 \times 10^{-2}$)

NON-GAUSSIAN CLOSURE

The Gaussian closure is the first level of approximation available to solve the equations (13), and it is equivalent to the stochastic linearization. Recently, other techniques have been proposed in order to overcome the limit of the Gaussian approximation (Wu and Lin [17], Lin [9], Crandall [3, 4]). In this paper, the characteristics of the problem have suggested to adopt the so called *non-Gaussian closure*, based on the choice of a *richer* probability density function, i.e. depending on a number of parameters greater than three.

Several formulation are possible (see, for example, Liu and Davies [10]); in this paper the probability density is assumed to be a Gauss function modulated by a polynomial:

$$f(x_1, x_2) = \frac{1}{\sigma_1 \sigma_2} \phi(\xi, \eta, \rho) \sum_{n=0}^{\infty} \sum_{m=0}^{\infty} c_{nm} \xi^n \eta^m \tag{24}$$

where $\xi = (x_1 - \mu_1)/\sigma_1$, $\eta = (x_2 - \mu_2)/\sigma_2$, and $\phi(\xi, \eta, \rho)$ is the standard normal bivariate density with correlation ρ:

$$\phi(\xi,\eta,\rho)=\frac{1}{2\pi\sqrt{1-\rho^2}}\exp\left(-\frac{\xi^2-2\xi\eta\rho+\eta^2}{2(1-\rho^2)}\right) \tag{25}$$

while $\mu_1,\mu_2,\sigma_1,\sigma_2,\rho$ e c_{nm} are the time dependant parameters which characterize the distribution; in general, the greater is the number of these coefficients the better is the solution. It must be noted immediately that, because of the symmetry of the problem, function $f_X(x)$ is symmetric, and then:

$$c_{nm}=0 \quad (n+m\in\mathbb{O})$$

$$\mu_1=\mu_2=0$$

being $\mathbb{O}$ the set of odd integer numbers.

Let be $h(\boldsymbol{X})=X_1^lX_2^k$: from (14) it is:

$$\begin{aligned}\dot{m}_{lk}&=l(m_{l-1,k+1})-k\Big(2\zeta E_2[|X_2|X_2^k\Big|X_1=0]\delta_{0l}+\\&+E[\mathrm{sig}(X_1)X_1^lX_2^{k-1}]-Km_{l+1,k-1}\Big)+k(k-1)D_Fm_{l,k-2}\end{aligned} \tag{26}$$

where $\delta_{00}=1$ and $\delta_{0l}=0$ as $l\neq 0$.

Using (24) it is possible to write explicitly the moment $E[X_1^nX_2^m]$:

$$m_{nm}=E[X_1^nX_2^m]=\begin{cases}0 & (n+m\in\mathbb{O})\\ \sigma_1^n\sigma_2^m\displaystyle\sum_{h=0}^{\infty}\sum_{k=0}^{\infty}c_{hk}M(n+h,m+k) & (n+m\in\mathbb{E})\end{cases} \tag{27}$$

where $\mathbb{E}$ is the set of even integer numbers, and $M(n,m)$ is the moment of order n,m of two standard Gaussian variables. Taking into account the symmetry it is:

$$M(n,m)=\begin{cases}0 & (n+m\in\mathbb{O})\\ \displaystyle\frac{m!\rho^m}{2^{(n+m)/2}}\sum_{k=0}^{(m/2)}\frac{(n+m-2k)!}{k!(m-2k)!(\frac{n+m}{2}-k)!}\left(\frac{1-\rho^2}{\rho^2}\right)^k & (n+m\in\mathbb{E})\end{cases} \tag{28}$$

The normalization condition ($m_{00}=1$) and equation (27) give the expression for the coefficient c_{00}:

$$c_{00}=1-\sum_{h=0}^{\infty}\sum_{k=0}^{\infty}c_{hk}M(h,k) \quad (h+k\in\mathbb{E}^+) \tag{29}$$

being $\mathbb{E}^+$ the set of even natural numbers greater than zero.

Substituting this expression into equation (27) we can obtain the moments as an explicit function of the unknown coefficients of the distribution:

$$m_{n,m}=\sigma_1^n\sigma_2^m\left\{M(n,m)+\sum_{h=0}^{\infty}\sum_{k=0}^{\infty}c_{hk}\Big[M(n+h,m+k)-M(h,k)M(n,m)\Big]\right\}$$

$$(n+m\in\mathbb{O},\quad h+k\in\mathbb{E}^+) \tag{30}$$

The set of algebraic equations (30) may be solved to evaluate the unknowns σ_1,σ_2,ρ and c_{hk} $(h+k\in\mathbb{E}^+)$ as a function of the terms m_{nm}, and therefore defining completely the distribution function f_X and the others expected values (the terms with $\mathrm{sig}(X_1)$ and $|X_2|$) of the equations (26). In fact, it is:

$$\begin{aligned}E[\mathrm{sig}(X)X^nY^m]&=\sigma_1^n\sigma_2^m\Big\{M_S(n,m)+\\&\sum_{h=0}^{\infty}\sum_{k=0}^{\infty}c_{hk}\Big[M_S(n+h,m+k)-M(h,k)M_S(n,m)\Big]\Big\}\\&(n+m\in\mathbb{O},\quad h+k\in\mathbb{E}^+)\end{aligned} \tag{31}$$

where $M_S(n,m)$ is the joint moment of two Gaussian variables with the term sig(ξ):

$$M_S(n,m) = \int_{-\infty}^{\infty} \mathrm{sig}(\xi)\xi^n \eta^m \phi(\xi,\eta,\rho)d\xi d\eta \tag{32}$$

and it is:

$$M_S(n,m) = \begin{cases} 0 & (n+m \in \mathbb{E}) \\ \dfrac{2^{(n+m)/2}}{\sqrt{\pi}} \rho^m m! \displaystyle\sum_{k=0}^{(m/2)} \frac{((n+m-1)/2-k)!}{k!(m-2k)!}\left(\frac{1-\rho^2}{4\rho^2}\right)^k & (n+m \in \mathbb{O}) \end{cases} \tag{33}$$

Finally, the terms containing $|X_2|$ are:

$$E_2[|X_2|X_2^m \big| X_1 = 0] = 2\frac{\sigma_2^{m+1}}{\sigma_1} \sum_{k=0}^{\infty} c_{0,2k}\; M_0(0,2k) \tag{34}$$

$$M_0(m) = \int_0^{\infty} \eta^{m+1}\phi(0,\eta,\rho)d\eta = \frac{1}{\pi}(1-\rho^2)^{\frac{(m+1)}{2}} 2^{\frac{m}{2}-1}\left(\frac{m}{2}\right)! \tag{35}$$

FOURTH ORDER NON-GAUSSIAN CLOSURE

As an example of application of this procedure, equations (26) are developed until the fourth order moments. In such a way, we can write eight ordinary differential equations:

$$\begin{aligned}
\dot{m}_{20} &= 2m_{11} \\
\dot{m}_{11} &= m_{02} - E[|X_1|] + Km_{20} \\
\dot{m}_{02} &= -2\left(\pi\zeta E_2[|X_2|X_2^2 \big| X_1 = 0] + E[\mathrm{sig}(X_1)X_2] - Km_{11}\right) + 2D_F \\
\dot{m}_{40} &= 4m_{31} \\
\dot{m}_{31} &= 2m_{22} - E[\mathrm{sig}(X_1)X_1^3] + Km_{40} \\
\dot{m}_{22} &= 2m_{13} - 2E[\mathrm{sig}(X_1)X_1^2X_2] + 2Km_{31} + 2D_F m_{20} \\
\dot{m}_{13} &= m_{04} - 3E[\mathrm{sig}(X_1)X_1X_2^2] + 3Km_{22} + 6D_F m_{11} \\
\dot{m}_{04} &= -4\left(\pi\zeta E_2[|X_2|X_2^4 \big| X_1 = 0] + E[\mathrm{sig}(X_1)X_2^3] - Km_{13}\right) + 12D_F m_{02}
\end{aligned} \tag{36}$$

The number of the unknown coefficients of the distribution function (24) must be the same as the numbers of equations (36). An effective choice for $f_X(x)$ is, for example:

$$f(x_1,x_2) = \frac{1}{\sigma_1\sigma_2}\phi(\xi,\eta,\rho)\left(c_{00} + c_{20}\xi^2 + c_{11}\xi\eta + c_{02}\eta^2 + c_{31}\xi^3\eta + c_{13}\xi\eta^3\right) \tag{37}$$

and then the explicit expressions for the moment $m_{n,m}$ are:

$$\begin{aligned}
m_{20} &= \sigma_1^2\left[1 + 2c_{20} + 2\rho c_{11} + 2\rho^2 c_{02} + 12\rho c_{31} + 6\rho(\rho^2+1)c_{13}\right] \\
m_{11} &= \sigma_1\sigma_2\left[\rho + 2\rho c_{20} + (\rho^2+1)c_{11} + 2\rho c_{02} + 3(3\rho^2+1)c_{31} + 3(3\rho^2+1)c_{13}\right] \\
m_{02} &= \sigma_2^2\left[1 + 2\rho^2 c_{20} + 2\rho c_{11} + 2c_{02} + 6\rho(\rho^2+1)c_{31} + 12\rho c_{13}\right] \\
m_{40} &= 3\sigma_1^4\left[1 + 4c_{20} + 4\rho c_{11} + 4\rho^2 c_{02} + 32\rho c_{31} + 4\rho(5\rho^2+3)c_{13}\right] \\
m_{31} &= 3\sigma_1^3\sigma_2\Big[\rho + 4\rho c_{20} + (3\rho^2+1)c_{11} + 2\rho(\rho^2+1)c_{02} +
\end{aligned}$$

$$+ (27\rho^2 + 5)c_{31} + (8\rho^4 + 21\rho^2 + 3)c_{13}\Big] \tag{38}$$

$$m_{22} = \sigma_1^2\sigma_2^2\Big[1 + 2\rho^2 + 2(5\rho^2 + 1)c_{20} + 4\rho(\rho^2 + 2)c_{11} +$$

$$+ 2(5\rho^2 + 1)c_{02} + 6\rho(9\rho^2 + 7)c_{31} + 6\rho(9\rho^2 + 7)c_{13}\Big]$$

$$m_{13} = 3\sigma_1\sigma_2^3\Big[\rho + 2\rho(\rho^2 + 1)c_{20} + (3\rho^2 + 1)c_{11} + 4\rho c_{02} +$$

$$+ (8\rho^4 + 21\rho^2 + 3)c_{31} + (27\rho^2 + 5)c_{13}\Big]$$

$$m_{04} = 3\sigma_2^4\Big[1 + 4\rho^2 c_{20} + 4\rho c_{11} + 4c_{02} + 4\rho(5\rho^2 + 3)c_{31} + 32\rho c_{13}\Big]$$

while the moments of the non-linear terms are:

$$E[|X_1|] = \sigma_1\sqrt{\tfrac{2}{\pi}}\Big[1 + c_{20} + \rho c_{11} + \rho^2 c_{02} + 5\rho c_{31} + \rho(2\rho^2 + 3)c_{13}\Big]$$

$$E[\mathrm{sig}(X_1)X_2] = \sigma_2\sqrt{\tfrac{2}{\pi}}\Big[\rho + \rho c_{20} + c_{11} + \rho(2 - \rho^2)c_{02} + (3\rho^2 + 2)c_{31} +$$

$$+ (3 + 3\rho^2 - \rho^4)c_{13}\Big]$$

$$E[\mathrm{sig}(X_1)X_1^3] = \sigma_1^3\sqrt{\tfrac{2}{\pi}}2\Big[1 + 3c_{20} + 3\rho c_{11} + 3\rho^2 c_{02} + 21\rho c_{31} + 3\rho(4\rho^2 + 3)c_{13}\Big]$$

$$E[\mathrm{sig}(X_1)X_1^2X_2] = \sigma_1^2\sigma_2\sqrt{\tfrac{2}{\pi}}2\Big[\rho + 3\rho c_{20} + (2\rho^2 + 1)c_{11} + \rho(\rho^2 + 2)c_{02} +$$

$$+ (17\rho^2 + 4)c_{31} + 3(\rho^4 + 5\rho^2 + 1)c_{13}\Big] \tag{39}$$

$$E[\mathrm{sig}(X_1)X_1X_2^2] = \sigma_1\sigma_2^2\sqrt{\tfrac{2}{\pi}}\Big[1 + \rho^2 + (5\rho^2 + 1)c_{20} + \rho(\rho^2 + 5)c_{11} +$$

$$+ (2 + 5\rho^2 - \rho^4)c_{02} + 21\rho(\rho^2 + 1)c_{31} + \rho(27 + 17\rho^2 - 2\rho^4)c_{13}\Big]$$

$$E[\mathrm{sig}(X_1)X_2^3] = \sigma_2^3\sqrt{\tfrac{2}{\pi}}3\Big[\rho(1 - \rho^2/3) + \rho(\rho^2 + 1)c_{20} + (\rho^2 + 1)c_{11} +$$

$$+ \rho(\rho^4 - 3\rho^2 + 4)c_{02} + (3\rho^4 + 9\rho^2 + 2)c_{31} + (\rho^6 - 4\rho^4 + 12\rho^2 + 5)c_{13}\Big]$$

and

$$E_2[|X_2|X_2^2\Big|X_1 = 0] =$$

$$= \frac{2}{\pi}\frac{\sigma_2^3}{\sigma_1}(1 - \rho^2)^{3/2}(1 - c_{20} - \rho c_{11} + (3 - 4\rho^2)c_{02} - 3\rho c_{31} - 3\rho c_{13})$$

$$E_2[|X_2|X_2^4\Big|X_1 = 0] = \tag{40}$$

$$= \frac{2}{\pi}\frac{\sigma_2^5}{\sigma_1}(1 - \rho^2)^{5/2}4(1 - c_{20} - \rho c_{11} + (5 - 6\rho^2)c_{02} - 3\rho c_{31} - 3\rho c_{13})$$

Equations (36) are a set of ordinary differential equation with time-dependent coefficients. Its numerical integration is very difficult, probably because of the strong nonlinearity of the coefficients. A less complex problem is to evaluate the stationary solution, when it exists. In figure 5 the intensity of the stationary response is plotted as a function of the intensity of the dragging motion acceleration, for the Gaussian closure (dashed lines) and non-Gaussian closure (solid lines): beyond the vertical dashed line $D_F = D_{F_s}$ the Gaussian solutions diverge to infinity, while non-Gaussian solutions still exist.

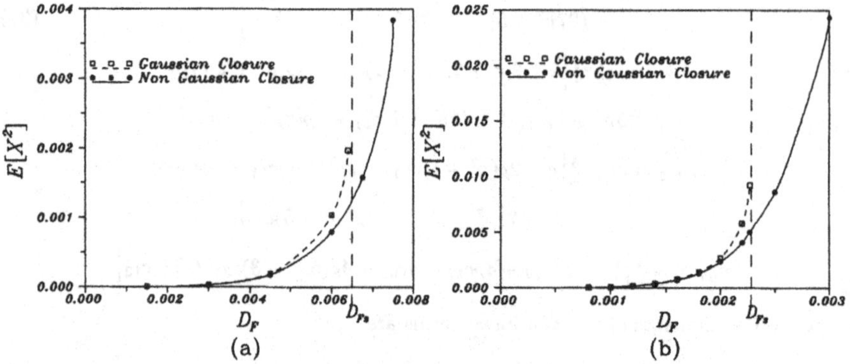

Figure 5: Stationary solutions, response vs. exciting intensity: comparison between Gaussian and non-Gaussian closure; (a) $b = 1m$ $\lambda = 4$ ($K = 3.462, \zeta = 0.0229$); (b) $b = 1m$ $\lambda = 8$ ($K = 1.811, \zeta = 0.743 \times 10^{-2}$)

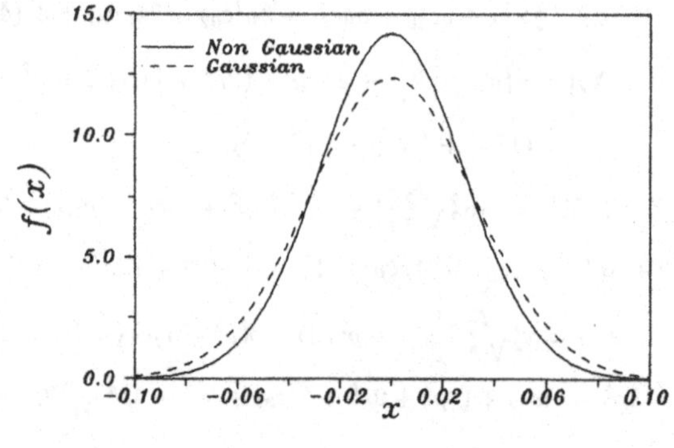

Figure 6: Comparison between Gaussian and non-Gaussian marginal p.d.f. of X: $b = 1m$ $\lambda = 4, D_F = 0.006$

The results from the non-Gaussian closure are lower than those from the Gaussian one, and are closer to the Monte Carlo simulations showed in figure 3. This is evident also in figure 6, where the p.d.f. obtained from the Gaussian closure approximation (dashed line) and from the non-Gaussian closure (solid line) are compared for the case of $\lambda = 4$ and $D_F = 0.006$.

CONCLUSIONS

The dynamic behavior of block-like structures has been studied adopting a random vibration approach. The excitation was modeled as a white noise process, and the associated *Fokker-Plank* equation was solved using the technique of closure. Two level of approximation were used: Gaussian and non-Gaussian closure. The first one gives a good approximation of the transient and the stationary solution when the intensity of the excitation is low, but it overestimates the response for stronger excitations. Non-Gaussian closure better approximates the stationary response intensity, but, at the present, the numerical integration of the differential equations involves some difficulties. In the future the response to a filtered white noise excitation process will be studied in order to better simulate an earthquake dragging motion.

This research was partially supported by the *Ministero dell'Università e della Ricerca Scientifica.*

References

[1] M. Aslam, W. Godden, and D. T. Scalise. Earthquakes rocking response of rigid bodies. *J. Struct. Div. ASCE*, 106(2):377–392, 1980.

[2] C.Baggio, A. Giuffrè, and R. Masiani. Seismic response of masonry assemblages. In *Proc. 9nd Europ. Conf. on Earthquake Eng.*, pages 221–230, Mosca,USSR, 1990.

[3] S. H. Crandall. Non-gaussian closure for random vibration of non-linear oscillators. *Int. J. Non-Linear Mech.*, 15:303–313, 1980.

[4] S. H. Crandall. Non-gaussian closure techniques for stationary random vibration. *Int. J. Non-Linear Mech.*, 20(1):1–8, 1985.

[5] R. Giannini. Dynamic analysis of systems of staked blocks. In *Proc. 2nd Nat. Conf. on Seismic Eng. in Italy*, pages 47–62, Rapallo, Italy, 1984. (in Italian).

[6] R. Giannini and R. Masiani. Frequency domain response of the rigid-block. In *Proc. 10th Conf. AIMETA*, pages 169–174, Pisa, Italy, 1990. (in Italian).

[7] G. W. Housner. The behaviour of inverted pendulum structures during earthquakes. *Bull. Seism. Soc. America*, 53(2):403–417, 1963.

[8] R. N. Iyengar and P. K. Dash. Study of random vibration of nonlinear systems by gaussian closure technique. *J. Appl. Mech. ASME*, 45:393–399, 1978.

[9] Y. K. Lin. Application of Markov processes theory to nonlinear random vibration problems. In Schueller and Shinozuka, editors, *Stochastic Methods in Structural Dynamics*, M. Nijhoff Publisher, 1987.

[10] Q. Liu and H. G. Davies. Application of non-gaussian closure to the nonstationary response of a Duffing oscillator. *Int. J. Non-Linear Mech.*, 23(3):241–250, 1988.

[11] K. Muto, H. Takase, K. Horikoshi, and H. Ueno. 3-D non linear dynamic analysis of staked blocks. In *Proc. 2nd Special Conf. Dyn. Response Struct.*, pages 917–930, Atlanta, USA, 1981.

[12] N. C. Nigam. *Introduction to Random Vibrations.* MIT-Press, Cambridge MA., 1983.

[13] J. B. Roberts and P. D. Spanos. *Random Vibration and Statistical Linearization.* J. Wiley & Sons, Chichester, 1990.

[14] T. T. Soong. *Random Differential Equations in Science and Engineering.* Academic Press, New York, 1973.

[15] P. D. Spanos and A. S. Koh. Rocking of rigid blocks due to harmonic shaking. *J. Eng. Mech. ASCE*, 110(11):1627–1642, 1984.

[16] C. M. Wong and W. K. Tso. Steady state rocking response of rigid blocks part 2: experiment. *Earth. Eng. Struct. Dyn.*, 18:107–120, 1989.

[17] W. F. Wu and Y. K. Lin. Cumulant neglect closure for non-linear oscillators under random parametric and external excitations. *Int. J. Non-Linear Mech.*, 19(4):349–362, 1984.

[18] C. S. Yim and J. Penzien A. K. Chopra. Rocking response of rigid blocks to earthquakes. *Earth. Eng. Struct. Dyn.*, 8(6):565–587, 1980.

Response of Non-linear Trusses with Spatial Variability under Random Excitations

R-H. Cherng, Y-K. Wen

Department of Civil Engineering, University of Illinois at Urbana-Champaign, Urbana, IL 61801, U.S.A.

ABSTRACT

This study considers the response of nonlinear plane truss structures with spatially uncertain system parameters subjected to random excitations. The nonlinearities due to both large deflection and inelastic deformation are considered. Within the framework of a finite element formulation, a stochastic equivalent linearization method in conjunction with a perturbation method is developed to solve for the response statistics. Comparison of responses of a structure under Gaussian filtered white noise ground excitation with Monte-Carlo simulation results indicates that the stochastic equivalent linearization method yields reasonably accurate results with much less computational time. It is found that the contribution of the system spatial uncertainties to the total response statistics can be significant. A non-dimensional response sensitivity coefficient is also defined to identify the most dominant system parameters in terms of the contribution to the response variability.

INTRODUCTION

Structural system parameters such as stiffness, damping, and those of the material model are seldom completely known and loading environments are generally random and often dynamic in nature. In the assessment of the reliability of structures, stochastic response analyses are often necessary. Under severe loadings such as those caused by earthquakes, sea waves, or winds, structures generally exhibit nonlinear behaviors due to inelastic and/or large deformation. The exact solutions for response statistics of such nonlinear, uncertain structures under random excitations are difficult to obtain except via Monte-Carlo method which can be very costly. Many approximate methods have been proposed. As far as the loading uncertainty is concerned, the stochastic equivalent linearization method has been successfully applied to a large class of nonlinear discrete (Wen [9]) and continuous (e.g., Simulescu, Mochio, and Shinozuka [8]) systems under random excitations. For systems with spatial uncertainties, the mean-centered perturbation method (Liu, Besterfield, and Belytschko [7]) has been proven to be most effective.

This study concentrates on the response of uncertain nonlinear plane trusses subjected to random excitations. The constitutive law is based on the differential equation model in Wen [9] with the parameters determined in terms of the material property constants. The nonlinearities due to both large deflection and inelasticity are treated by the stochastic equivalent linearization method, and the system uncertainties are included by the perturbation method.

ELEMENT EQUATIONS OF MOTION

Constitutive Law

The constitutive law in this study is described explicitly by the following differential equations:

$$\sigma = \hat{a}E\varepsilon + (1-\hat{a})Ez \tag{1}$$

$$\dot{z} = \kappa\dot{\varepsilon} - \gamma|\dot{\varepsilon}|\,|z|^{n-1}z - \beta\dot{\varepsilon}|z|^{n} \tag{2}$$

where σ is the second Piola-Kirchhoff axial stress , ε is the Green-Lagrange axial strain, E is the Young's modulus, $\hat{a}$ is the hardening ratio, z is the element auxiliary variable, and n , κ , β , and γ are control parameters of the hysteresis loops. A modification of the one dimensional version of Karray and Bouc's [6] constitutive law of plasticity shows that (Cherng and Wen [2]) $\kappa = 1$ and $\beta = \gamma = E^n/2\sigma_y^n$ where σ_y is the yielding stress of the material. The value of n controls the smoothness in transition from elastic to plastic state.

Total Lagrangian Formulation

A prismatic plane truss element consisting of two end nodes (each with two translational degree-of-freedoms) is considered. It is assumed that the element may undergo large displacement and large rotation, and its constitutive law is given by Eqs. (1) and (2). The element equations of motion are obtained with the general total Lagrangian formulation.

STOCHASTIC EQUIVALENT LINEARIZATION METHOD

Linearized Global Equations of Motion

Consider the stationary response of the element under zero-mean stationary Gaussian excitations. The mean value is generally not zero because of the geometric nonlinearity. It has been shown that (Cherng and Wen [2]), one can derive the linearized equations of motion in global coordinates by applying the stochastic linearization method on the element equations of motion and following a standard assembling procedure used in finite element analyses. The equations are given in matrix forms as follows:

$$\mathbf{M}\ddot{\underline{\mathbf{U}}} + \mathbf{C}\dot{\underline{\mathbf{U}}} + \mathbf{K}\underline{\mathbf{U}} + \mathbf{P}\underline{\mathbf{Z}} + \underline{\mathbf{Q}} = \underline{\mathbf{f}}(\mathbf{t}) \tag{3}$$

$$\dot{\underline{\mathbf{Z}}} = \mathbf{A}\underline{\mathbf{Z}} + \mathbf{B}\dot{\underline{\mathbf{U}}} \tag{4}$$

in which $\underline{\mathbf{U}}$ is the relative displacement vector to the ground, $\underline{\mathbf{Z}}$ is the element auxiliary variable vector, $\mathbf{M}$ is the mass matrix, $\mathbf{C}$ is the viscous damping matrix, $\underline{\mathbf{f}}(\mathbf{t})$ is the ground excitation vector, and $\mathbf{K}$, $\mathbf{P}$, $\underline{\mathbf{Q}}$, $\mathbf{A}$, and $\mathbf{B}$ are explicit functions of system parameters and response statistics.

Governing Equations for Response Statistics

It is assumed that the ground acceleration can be modeled as Kanai-Tajimi (Kanai [5]) filtered Gaussian white noise, i.e., $2\zeta_g\omega_g\dot{x}_g + \omega_g^2 x_g$ where x_g satisfies the following equation:

$$\ddot{x}_g + 2\zeta_g\omega_g\dot{x}_g + \omega_g^2 x_g = w(t) \tag{5}$$

in which ω_g and ζ_g are the natural frequency and damping ratio of the filter and $w(t)$ is a Gaussian white noise with power spectral density S_0. By defining a response state vector $\underline{\mathbf{Y}} = (x_g\ \dot{x}_g\ \underline{\mathbf{U}}\ \dot{\underline{\mathbf{U}}}\ \underline{\mathbf{Z}})^T$, we can rewrite Eqs. (3), (4), and (5) as:

$$\frac{d\underline{\mathbf{Y}}}{dt} = \mathbf{G}\underline{\mathbf{Y}} + \underline{\mathbf{F}} \tag{6}$$

Taking the first and second statistical moments of Eq. (6) respectively and considering stationary responses, we obtain one governing equation for $\overline{\underline{\mathbf{U}}}$, the mean stationary nodal displacement vector, and one equation for $\mathbf{S}$, the one-time covariance matrix of the state vector, as follows:

$$\mathbf{K}\underline{\bar{\mathbf{U}}} + \underline{\mathbf{Q}} = \underline{\mathbf{o}} \tag{7}$$

$$\mathbf{GS} + \mathbf{SG}^T + \mathbf{D} = \mathbf{O} \tag{8}$$

where $\mathbf{K}$, $\underline{\mathbf{Q}}$, $\mathbf{G}$ are functions of $\underline{\bar{\mathbf{U}}}$ and $\mathbf{S}$ and the components of $\mathbf{D}$ are all zeros except $\mathbf{D}(2,2) = 2\pi S_0$.

Solution for Response Statistics

Applying a iterative solution method (Cherng and Wen [2]) to Eqs. (7) and (8) we obtain $\underline{\bar{\mathbf{U}}}$ and $\mathbf{S}$, which contains the first two statistical moments of displacements, velocities, and element auxiliary variables. The statistical moments of other response quantities, such as element stresses, can be derived from $\underline{\bar{\mathbf{U}}}$ and $\mathbf{S}$.

The extreme statistics of a response can be also calculated (Davenport [4]) based on the first two moments of the response and its time derivative.

PERTURBATION METHOD

Perturbed Governing Equations for Response Statistics

In Eqs. (7) and (8) the system parameters are assumed to be deterministic. The spatial uncertainties of system parameters, however, may be substantial and contribute significantly to the total response statistics and the probability of structural failure. They can be incorporated into the foregoing formulation by applying the mean-centered perturbation method as follows.

The system spatial uncertainties are represented by a vector, $\underline{\lambda}$, whose mean vector and covariance matrix are assumed to be either known or derivable by discretizing the prescribed random fields. $\underline{\lambda}$ is then transformed into a non-dimensional zero-mean basic random variable vector, $\underline{\epsilon}$, via the following:

$$\epsilon_i = \frac{\lambda_i}{\bar{\lambda}_i} - 1 \; ; \; i = 1, \; NB \tag{9}$$

in which the subscript i denotes the i th component of a vector, the bar over a variable denotes the variable's mean value, and NB is the dimension of $\underline{\lambda}$. Now Eqs. (7) and (8) are rewritten below to indicate that they are conditional on $\underline{\epsilon}$:

$$\mathbf{K}|_{\underline{\epsilon}}\ \underline{\overline{\mathbf{U}}}|_{\underline{\epsilon}} + \underline{\mathbf{Q}}|_{\underline{\epsilon}} = \underline{\mathbf{o}} \tag{10}$$

$$\mathbf{G}|_{\underline{\epsilon}}\ \mathbf{S}|_{\underline{\epsilon}} + \mathbf{S}|_{\underline{\epsilon}}\ \mathbf{G}^T|_{\underline{\epsilon}} + \mathbf{D}|_{\underline{\epsilon}} = \mathbf{O} \tag{11}$$

in which $(.)|_{\underline{\epsilon}}$ designates the dependence of a matrix (or vector) on $\underline{\epsilon}$.

The coefficient matrices of Eqs. (10) and (11) are then expanded into Taylor's series of $\underline{\epsilon}$ at $\underline{\epsilon} = 0$ respectively as follows:

$$\mathbf{K}|_{\underline{\epsilon}} = \mathbf{K}|_{\underline{\epsilon}=0} + \sum_{i=1}^{NB} \frac{\partial \mathbf{K}}{\partial \epsilon_i}\bigg|_{\underline{\epsilon}=0} \epsilon_i + \frac{1}{2}\sum_{i=1}^{NB}\sum_{j=1}^{NB} \frac{\partial^2 \mathbf{K}}{\partial \epsilon_i \partial \epsilon_j}\bigg|_{\underline{\epsilon}=0} \epsilon_i\epsilon_j + \ldots \tag{12}$$

$$\underline{\mathbf{Q}}|_{\underline{\epsilon}} = \underline{\mathbf{Q}}|_{\underline{\epsilon}=0} + \sum_{i=1}^{NB} \frac{\partial \underline{\mathbf{Q}}}{\partial \epsilon_i}\bigg|_{\underline{\epsilon}=0} \epsilon_i + \frac{1}{2}\sum_{i=1}^{NB}\sum_{j=1}^{NB} \frac{\partial^2 \underline{\mathbf{Q}}}{\partial \epsilon_i \partial \epsilon_j}\bigg|_{\underline{\epsilon}=0} \epsilon_i\epsilon_j + \ldots \tag{13}$$

$$\mathbf{G}|_{\underline{\epsilon}} = \mathbf{G}|_{\underline{\epsilon}=0} + \sum_{i=1}^{NB} \frac{\partial \mathbf{G}}{\partial \epsilon_i}\bigg|_{\underline{\epsilon}=0} \epsilon_i + \frac{1}{2}\sum_{i=1}^{NB}\sum_{j=1}^{NB} \frac{\partial^2 \mathbf{G}}{\partial \epsilon_i \partial \epsilon_j}\bigg|_{\underline{\epsilon}=0} \epsilon_i\epsilon_j + \ldots \tag{14}$$

$$\mathbf{D}|_{\underline{\epsilon}} = \mathbf{D}|_{\underline{\epsilon}=0} + \sum_{i=1}^{NB} \frac{\partial \mathbf{D}}{\partial \epsilon_i}\bigg|_{\underline{\epsilon}=0} \epsilon_i + \frac{1}{2}\sum_{i=1}^{NB}\sum_{j=1}^{NB} \frac{\partial^2 \mathbf{D}}{\partial \epsilon_i \partial \epsilon_j}\bigg|_{\underline{\epsilon}=0} \epsilon_i\epsilon_j + \ldots \tag{15}$$

in which $(.)|_{\underline{\epsilon}=0}$ designates a matrix (or vector) evaluated at $\underline{\epsilon} = 0$. The solutions of Eqs. (10) and (11), $\underline{\overline{\mathbf{U}}}|_{\underline{\epsilon}}$ and $\mathbf{S}|_{\underline{\epsilon}}$, are also represented by Taylor's series of $\underline{\epsilon}$ at $\underline{\epsilon} = 0$ respectively as:

$$\underline{\overline{\mathbf{U}}}|_{\underline{\epsilon}} = \underline{\overline{\mathbf{U}}}|_{\underline{\epsilon}=0} + \sum_{i=1}^{NB} \underline{\overline{\mathbf{U}}}^{1i}|_{\underline{\epsilon}=0}\ \epsilon_i + \frac{1}{2}\sum_{i=1}^{NB}\sum_{j=1}^{NB} \underline{\overline{\mathbf{U}}}^{2ij}|_{\underline{\epsilon}=0}\ \epsilon_i\epsilon_j + \ldots \tag{16}$$

$$\mathbf{S}|_{\underline{\epsilon}} = \mathbf{S}|_{\underline{\epsilon}=0} + \sum_{i=1}^{NB} \mathbf{S}^{1i}|_{\underline{\epsilon}=0}\ \epsilon_i + \frac{1}{2}\sum_{i=1}^{NB}\sum_{j=1}^{NB} \mathbf{S}^{2ij}|_{\underline{\epsilon}=0}\ \epsilon_i\epsilon_j + \ldots \tag{17}$$

in which $(.)^{1i}$ denotes the first derivative of a matrix (or vector) with respect to ϵ_i and $(.)^{2ij}$ denotes the second derivative of a matrix (or vector) with respect to ϵ_i and ϵ_j.

Substituting Eqs. (12), (13), (14), (15), (16) and (17) into Eqs. (10) and (11) and balancing the zero order terms, we obtain a set of zero order nonlinear simultaneous equations:

$$\mathbf{K}|_{\underline{\varepsilon}=0}\ \underline{\mathbf{U}}|_{\underline{\varepsilon}=0} + \underline{\mathbf{Q}}|_{\underline{\varepsilon}=0} = \underline{\mathbf{o}} \tag{18}$$

$$\mathbf{G}|_{\underline{\varepsilon}=0}\ \mathbf{S}|_{\underline{\varepsilon}=0} + \mathbf{S}|_{\underline{\varepsilon}=0}\ \mathbf{G}^T|_{\underline{\varepsilon}=0} + \mathbf{D}|_{\underline{\varepsilon}=0} = \mathbf{O} \tag{19}$$

The first order terms must satisfy the following sets of simultaneous equations:

$$\mathbf{K}|_{\underline{\varepsilon}=0}\ \underline{\mathbf{U}}^{1i}|_{\underline{\varepsilon}=0} + \frac{\partial \mathbf{K}}{\partial \epsilon_i}\bigg|_{\underline{\varepsilon}=0}\ \underline{\mathbf{U}}|_{\underline{\varepsilon}=0} + \frac{\partial \underline{\mathbf{Q}}}{\partial \epsilon_i}\bigg|_{\underline{\varepsilon}=0} = \underline{\mathbf{o}} \tag{20}$$

$$\mathbf{G}|_{\underline{\varepsilon}=0}\ \mathbf{S}^{1i}|_{\underline{\varepsilon}=0} + \mathbf{S}^{1i}|_{\underline{\varepsilon}=0}\ \mathbf{G}^T|_{\underline{\varepsilon}=0} + \frac{\partial \mathbf{G}}{\partial \epsilon_i}\bigg|_{\underline{\varepsilon}=0} \mathbf{S}|_{\underline{\varepsilon}=0}$$

$$+ \mathbf{S}|_{\underline{\varepsilon}=0} \frac{\partial \mathbf{G}^T}{\partial \epsilon_i}\bigg|_{\underline{\varepsilon}=0} + \frac{\partial \mathbf{D}}{\partial \epsilon_i}\bigg|_{\underline{\varepsilon}=0} = \mathbf{O} \tag{21}$$

$i = 1, 2, \ldots, NB$

Similarly, the equalization of the second order terms leads to

$$\mathbf{K}|_{\underline{\varepsilon}=0}\ \underline{\mathbf{U}}^{2ij}|_{\underline{\varepsilon}=0} + \frac{\partial \mathbf{K}}{\partial \epsilon_i}\underline{\mathbf{U}}^{1j}|_{\underline{\varepsilon}=0} + \frac{\partial \mathbf{K}}{\partial \epsilon_j}\underline{\mathbf{U}}^{1i}|_{\underline{\varepsilon}=0} + \frac{\partial^2 \mathbf{K}}{\partial \epsilon_i \partial \epsilon_j}\bigg|_{\underline{\varepsilon}=0}\ \underline{\mathbf{U}}|_{\underline{\varepsilon}=0}$$

$$+ \frac{\partial^2 \underline{\mathbf{Q}}}{\partial \epsilon_i \partial \epsilon_j}\bigg|_{\underline{\varepsilon}=0} = \underline{\mathbf{o}} \tag{22}$$

$$\mathbf{G}|_{\underline{\varepsilon}=0}\ \mathbf{S}^{2ij}|_{\underline{\varepsilon}=0} + \mathbf{S}^{2ij}|_{\underline{\varepsilon}=0}\ \mathbf{G}^T|_{\underline{\varepsilon}=0} + \frac{\partial \mathbf{G}}{\partial \epsilon_i}\bigg|_{\underline{\varepsilon}=0}\ \mathbf{S}^{1j}|_{\underline{\varepsilon}=0}$$

$$+ \mathbf{S}^{1j}|_{\underline{\varepsilon}=0}\ \frac{\partial \mathbf{G}^T}{\partial \epsilon_i}\bigg|_{\underline{\varepsilon}=0} + \frac{\partial \mathbf{G}}{\partial \epsilon_j}\bigg|_{\underline{\varepsilon}=0}\ \mathbf{S}^{1i}|_{\underline{\varepsilon}=0} + \mathbf{S}^{1i}|_{\underline{\varepsilon}=0}\ \frac{\partial \mathbf{G}^T}{\partial \epsilon_j}\bigg|_{\underline{\varepsilon}=0}$$

$$+ \frac{\partial^2 \mathbf{G}}{\partial \epsilon_i \partial \epsilon_j}\bigg|_{\underline{\varepsilon}=0}\ \mathbf{S}|_{\underline{\varepsilon}=0} + \mathbf{S}|_{\underline{\varepsilon}=0}\ \frac{\partial^2 \mathbf{G}^T}{\partial \epsilon_i \partial \epsilon_j}\bigg|_{\underline{\varepsilon}=0} + \frac{\partial^2 \mathbf{D}}{\partial \epsilon_i \partial \epsilon_j}\bigg|_{\underline{\varepsilon}=0} = \mathbf{O} \tag{23}$$

$i,j = 1, 2, \ldots, NB$

The higher order perturbed equations may be included in the formulation as well. Nevertheless, if the coefficients of variation of system parameters are not large (say, less than 40%), the computational effort involved which generally increases dramatically in higher order computation may not be justified.

Series Expansion Solution of Conditional Response Statistics

The zero order equations, i.e., Eqs. (18) and (19), are solved first by the aforementioned iterative method (Cherng and Wen [2]). Based on the zero order solutions, we can proceed to solve for the first order equations, i.e., Eqs. (20) and (21). The second order equations, i.e., Eqs. (22) and (23), are similarly solved by using the zero and first order solutions. It is important to recognize that the time consuming decompositions of matrices **K** and **Q** are not necessary in solving the first and second equations since their decomposed forms have already become available when solving the zero order equations.

The substitution of the above solutions into Eqs. (16) and (17) gives us the second order series expansions for $\underline{\overline{\mathbf{U}}}|_{\underline{\epsilon}}$ and $\mathbf{S}|_{\underline{\epsilon}}$, which can be further used to obtain the second order series expansions for other response quantities.

Total Response Statistics

The first two moments of a response R are obtained according to Ang and Tang [1]:

$$\overline{R} = E_{\underline{\epsilon}}\left[\overline{R}|_{\underline{\epsilon}}\right] \tag{24}$$

$$Var[R] = E_{\underline{\epsilon}}\left[Var[R]|_{\underline{\epsilon}}\right] + Var_{\underline{\epsilon}}\left[\overline{R}|_{\underline{\epsilon}}\right] \tag{25}$$

where $E_{\underline{\epsilon}}$ [.] and $Var_{\underline{\epsilon}}$ [.] denote the mean and variance of the quantity taken with respect to $\underline{\epsilon}$, and the conditional first two moments of R, $\overline{R}|_{\underline{\epsilon}}$ and $Var[R]|_{\underline{\epsilon}}$, can be approximated by their truncated series expansions obtained in the preceding section. For example, the total mean and variance of the displacement U_k, the kth component of $\underline{\mathbf{U}}$, are

$$\overline{U_k} \approx \overline{U_k}|_{\underline{\epsilon}=0} + \frac{1}{2}\sum_{i=1}^{NB}\sum_{j=1}^{NB}\overline{U_k^{2ij}}|_{\underline{\epsilon}=0}\; E[\epsilon_i\epsilon_j] \tag{26}$$

$$Var[U_k] \approx Var[U_k]|_{\underline{\epsilon}=0} + \sum_{i=1}^{NB}\sum_{j=1}^{NB}\left(\frac{1}{2}Var[U_k]^{2ij}|_{\underline{\epsilon}=0} + \overline{U_k^{1i}}\,\overline{U_k^{1j}}\right) \cdot E[\epsilon_i\epsilon_j] \tag{27}$$

Non-dimensional Response Sensitivity Coefficient

To provide a universal measure of the sensitivity of a response moment to a basic random variable, we define a non-dimensional sensitivity coefficient of a response moment μ to a system parameter λ_i as shown below [3]:

$$\Omega_i(\mu)\Big|_{\underline{\lambda}=\bar{\underline{\lambda}}} = \frac{\bar{\lambda}_i}{\bar{\mu}}\frac{\partial \mu}{\partial \lambda_i}\Big|_{\underline{\lambda}=\bar{\underline{\lambda}}} \tag{28}$$

$$= \frac{1}{\bar{\mu}}\frac{\partial \mu}{\partial \epsilon_i}\Big|_{\underline{\epsilon}=0} = \frac{1}{\bar{\mu}}\mu^{1i}\Big|_{\underline{\epsilon}=0} \tag{29}$$

in which $\bar{\mu}$ and $\mu^{1i}|_{\epsilon=0}$ are already computed in solving the perturbed equations. Note that $\Omega_i(\mu)|_{\underline{\lambda}=\bar{\underline{\lambda}}}$ can be approximately interpreted as the percentage change in μ due to one percent change in λ_i.

Eq. (29) allows us to compare systematically the influence of basic random variables on the response statistics and determine the less significant variables which may be treated as deterministic in subsequent reliability analyses.

NUMERICAL EXAMPLE

The response of a three-tier plane truss structure is considered subjected to a horizontal ground excitation, modeled as a Kanai-Tajimi filtered Gaussian white noise. It is assumed that each plane truss member can be adequately represented by one single element and there is no viscous damping. The basic random variables include material constants of each element (E, σ_y, and $\hat{\alpha}$) and parameters of the ground excitation (S_0, ω_g, and ζ_g). They are assumed to be statistically independent of one another. Their mean values are given in Fig. 1. The coefficients of variation are assumed to be 0.10 for all material constants, 0.30 for S_0, 0.43 for ω_g, and 0.36 for ζ_g.

To examine the accuracy of the stochastic equivalent linearization method, Monte-Carlo simulations are first performed on a deterministic system (all system parameters are set equal to their mean values) under random excitation. For one run of simulation the displacement time history at the top of the structure and the stress-strain history of the first story column are shown in Figs. 2 and 3 respectively. The time histories of ensemble r.m.s. displacement and velocity of forty simulations are also shown in Figs. 4 and 5. The stationary response statistics by Monte-Carlo method and the stochastic linearization method are summarized in Table 1. It is seen that the stochastic linearization method gives results which are lower by 4 to 20%. It requires, however, only a small fraction of computational time of Monte-Carlo simulations (approximately 1/30 of one run of 40 sec time history in simulations).

When the uncertainties of system parameters are included , the total variance of the maximum displacement increases from 1.31 in^2 (the zero order terms) to 2.94 in^2 (the zero order plus the first order terms). Therefore the system parameter uncertainties account for 55.4% of the total variance. The results of non-dimensional sensitivity analysis of the mean maximum displacement are shown in Table 2. It indicates that the response of this given structure is insensitive to the material constant variations in beams and braces but quite sensitive to those in columns and variabilities of ground excitation parameters. In terms of the total variance (Eq. (27)), the ground excitation parameters have the most significant contribution because of their larger coefficients of variation.

CONCLUSION

A method of stochastic finite element analysis of nonlinear plane trusses is proposed. The emphasis is on nonlinearities due to large deflection and inelasticity and on effect of system uncertainties and random excitation. The constitutive law is based on a differential equation model; the finite element discretization is according to the total Lagrangian formulation; and a stochastic equivalent linearization method in conjunction with a perturbation method are used to obtain response statistics. Numerical examples indicate that the stochastic linearization method gives reasonably accurate results at a small fraction of computational cost of Monte-Carlo method, and system uncertainties have a significant contribution to the total response statistics. Future research will consider the related reliability problem.

ACKNOWLEDGEMENT

This study is supported by National Science Foundation under Grants NSF ECS 85-11972 and CES 88-22690. The support is gratefully acknowledged. Advises and suggestions from D. A. Pecknold are also greatly appreciated.

	40 Simulations	Linearization
R.M.S. Top Displacement (in)	3.32	2.65
R.M.S. Top Velocity (in/sec)	32.92	31.63
Mean Maximum Top Displacement in 20 sec. (in)	8.96	8.22
Standard Deviation of Maximum Top Displacement in 20 sec. (in)	1.42	1.14
R.M.S. First Story Column Stress (ksi)	22.81	21.82

Table 1 Comparison of Results by Monte-Carlo Simulations and Stochastic Linearization Method

	First Story Column			First Story Brace			First Story Beam			Ground Excitation		
	E	σ_y	$\hat{a}$	E	σ_y	$\hat{a}$	E	σ_y	$\hat{a}$	S_0	ω_g	ζ_g
Sensitivity Coefficients	−.2	.3	−.3 E−1	−.2 E−1	−.1 E−3	.9 E−2	.5 E−3	.2 E−7	.2 E−3	.2	−.3	−.7 E−1

Table 2 Non-dimensional Sensitivity Coefficients of Mean Maximum Top Displacement in 20 sec. to system parameters

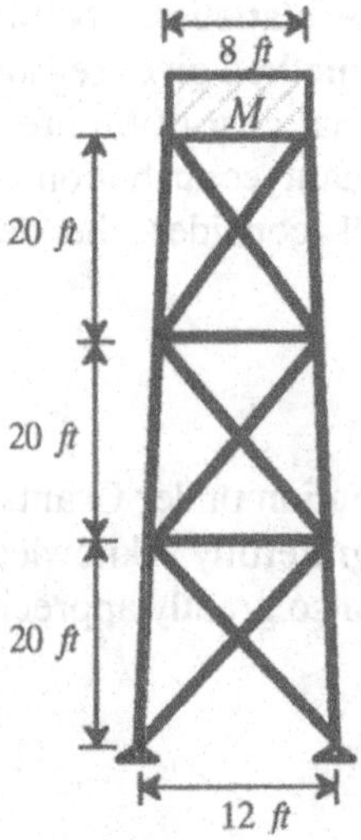

$M = .6\ kp.sec^2/in$

Material Constants (mean values)

$E = 29000\ ksi\ ,\ A = 48\ in^2$

$\sigma_y = 58\ ksi$

$\kappa = 1\ ,\ n = 2\ ,\ \hat{a} = .001$

ϱ *(Density)* $= 7.3 \text{x} 10^{-7}\ kp.sec^2/in^4$

Horizontal Ground Excitation Parameters

$S_0 = 150\ in^2/\sec^3$

$\omega_g = 19\ rad/sec\ ,\ \zeta_g = .32$

Figure 1 Three-tier Plane Truss Structure

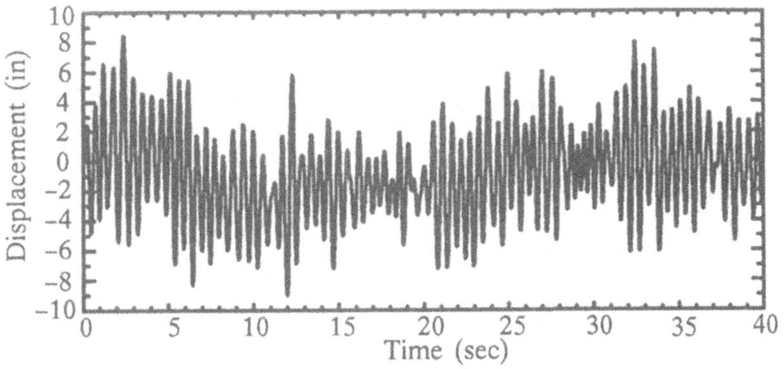

Figure 2 Time History of Displacement at the Top of the Structure

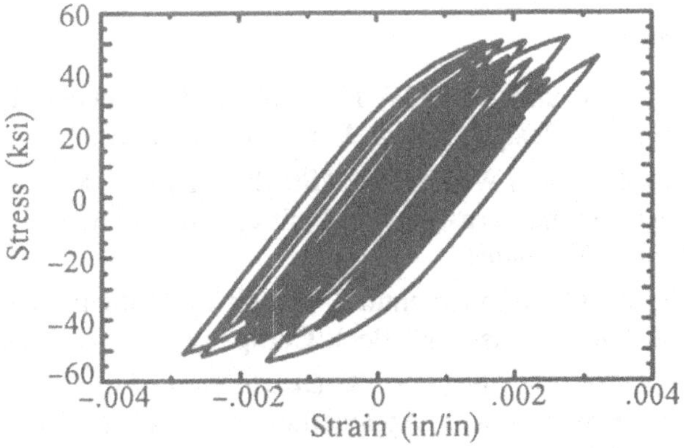

Figure 3 Hysteretic Stress-Strain History of First Story Column

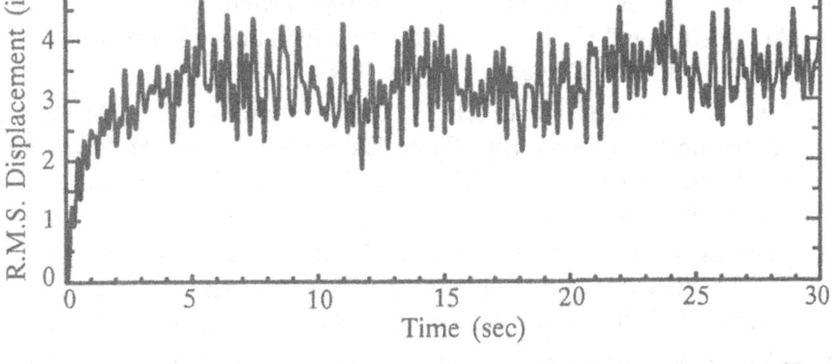

Figure 4 Time History of Ensemble R.M.S. Displacement at the Top of the Structure

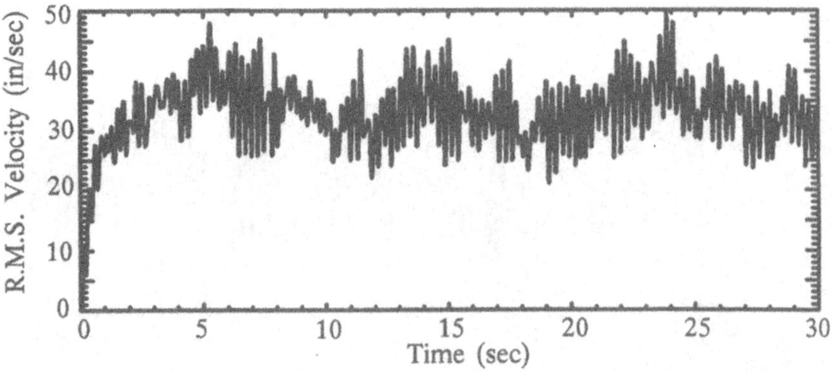

Figure 5 Time History of Ensemble R.M.S. Velocity at the Top of the Structure

REFERENCES

1. Ang, A. H-S. and Tang, W. H., Probability Concepts in Engineering Planning and Design, Vol. I, John Wiley & Sons, New York, 1975.
2. Cherng, R.-H. and Wen, Y. K., "Stochastic Finite Element Analysis of Nonlinear Plane Trusses," submitted to International Journal of Non-Linear Mechanics.
3. Chiang, A. C., Fundamental Methods of Mathematical Economics, McGraw-Hill International Book Company, Tokyo, 1984.
4. Davenport, A. G., "Note on the Distribution of the Largest Value of a Random Function with Application to Gust Loading," Proceedings Institute of Civil Engineers, London, Vol. 28, 1964.
5. Kanai, K., "Semiempirical Formula for the Seismic Characteristics of the Ground Motion," Bulletin Earthquake Research Institute, University of Tokyo, Vol. 35, pp. 309–325, 1957.
6. Karray, M. A. and Bouc, R., "Etude Dynamique d'un Systeme Anti-Sismique," (in French) Accepted for Publication in Annales de l'Ecole National d'Ingenieur Tunis (ENIT), 1986.
7. Liu, W. K., Besterfield, G. H., and Belytschko, T., "Variational Approach to Probabilistic Finite Elements" Journal of Engineering Mechanics Division, ASCE, Vol. 114, No. 12, pp. 2115–2133, December 1988.
8. Simulescu, I., Mochio, T., and Shinozuka, M., "Equivalent Linearization Method in Nonlinear FEM," Journal of the Engineering Mechanics Division, ASCE, Vol. 115, No. 3, pp. 475–492, March 1989.
9. Wen, Y. K., "Methods of Random Vibration for Inelastic Structures," Applied Mechanics Reviews, Vol. 42, No. 2, pp. 39-52, February 1989.

Nonlinear Effects on TLP Springing Response and Reliability

S.R. Winterstein (*), T. Marthinsen (**)
(*) Civil Engineering Department, Stanford University, Stanford, CA 94305-4020, U.S.A.
(**) Saga Petroleum a.s., Oslo, Norway

Abstract

Vertical "springing" motions of a tension leg platform are studied. First- and second-order forces are found from wave diffraction theory, and corresponding first- and second-order responses from a linear structural model. Tether tensions and platform motions—particularly accelerations—show clear second-order effects. In predicting extreme springing responses, it is found useful to consider several design seastates: one with extreme H_S ("wave height driven"), and another with resonant second-order effects ("wave period driven"). These are found to provide good agreement with critical seastates found by full FORM analysis.

Introduction

The tension leg platform (TLP), a floating structure tethered to the sea bottom, has become an increasingly popular design for deep-water oil production. We focus here on its vertical "springing" motions, which contribute to tether fatigue and to platform accelerations. Extreme tether tensions and platform motions are estimated, including uncertainty both in seastate parameters, such as significant wave height and average period, and in structural properties such as natural frequencies. Numerical results are shown for the Snorre platform, currently under construction for use in the North Sea.

Several features complicate the problem at hand. Chief among these is the effect of nonlinear forces, whose frequencies are sums and differences of wave elevation frequencies. The sum-frequency terms are of interest here as they drive resonant springing response; horizontal slow-drift motions are instead governed by difference-frequency contributions. An additional feature of the problem is noteworthy: the resulting force power spectrum is found to oscillate rather rapidly near typical heave, roll, and pitch natural frequencies. This suggests that these frequencies and associated damping ratios should be selected with care, and that uncertainty in these quantities may have significant impact.

We use wave diffraction theory here to find first- and second-order TLP forces, and a linear structural model for corresponding first- and second-order responses. The resulting Volterra response model is similar to that used in various slow-drift studies (Neal, 1972; Vinje, 1984; Naess, 1985; Naess, 1986). As in those studies, we use eigenvalue analysis to re-cast the second-order springing response in terms of independent normal variables (Kac and Siegert, 1947). Our analysis of this model is somewhat simpler, however. Response moments are found directly from these eigenvalues, and extreme respones are estimated from moment-based non-Gaussian models (Winterstein, 1988; Winterstein and Ness, 1989). Non-Gaussian effects on fatigue can also be predicted with these models.

First-order reliability methods (FORM) are used here to directly estimate design seastates for various responses. In general prediction of extreme responses, it is found useful to consider several design seastates: one with extreme H_S ("wave height driven"), and another with resonant second-order effects ("wave period driven"). These may be directly estimated from the wave height-period scattergram for the region, and provide good agreement with critical seastates found by full FORM analysis.

Wave Climate Model

The sea surface elevation $\eta(t)$ at a fixed location is modelled as a stationary Gaussian process over a seastate duration of T_{ss}=3 hours. The JONSWAP model is adopted for the spectral density of $\eta(t)$, with constant peak factor γ=3.3 in all seastates. Variations in γ, either deterministic or stochastic, can be readily included as well.

Note that non-Gaussian models of $\eta(t)$ can also be constructed, from either observed wave statistics (Winterstein, 1990) or nonlinear wave theory (Marthinsen, 1991a; Winterstein et al, 1991). These nonlinear wave effects are included, however, in the force transfer functions to be used subsequently. That is, they relate forces to the *first-order* wave elevation $\eta(t)$; hence, a Gaussian model of $\eta(t)$ is used here.

With our model, each seastate is characterized by its significant wave height, $H_S=4\sigma_\eta$, and its spectral peak period T_P. Following common practice, H_S is assigned a three-parameter Weibull distribution:

$$P[H_S < h] = 1 - \exp\left\{-\left[\frac{(h-\gamma)}{h_0}\right]^\beta\right\} \tag{1}$$

Conditional on H_S, T_P is assumed lognormally distributed with parameters of the following form:

$$E[\ln T_P|H_S] = a_0 + a_1 \ln(H_S + a_2)\,; \quad Var[\ln T_P|H_S] = b_0 + b_1 \exp(-b_2 {H_S}^{b_3}) \tag{2}$$

For the Snorre platform we consider a wave climate representative of the Northern North Sea, for which γ=0, β=1.547, h_0=2.822, a_0=1.59, a_1=0.42, a_2=2.0, b_0=.005, b_1=.085, b_2=0.13 and b_2=1.34 (Haver and Nyhus, 1986). Following this reference an alternate lognormal distribution is used for H_S values < 3.27 m.; however, all design H_S values found here are above this value.

Figure 1 shows the variation of "typical" (50% fractile) seastate period T_P with H_S, as well as somewhat higher (p=0.84) and lower (p=0.16) T_P fractiles. Also shown is the 100-year contour, outside which seastates fall with probability p=.01 per year = 3.43×10^{-6} per seastate. Seastates along this contour are approximately equally likely, with roughly constant probability density. In fact they become equally likely after transforming H_S and T_P to standard normal variables, as required by FORM.

For wave-dominated structures such as jackets and jackups, it is common to choose a design seastate with the 100-year H_S value and typical associated T_P. This would represent the intersection of the 100-year contour and the 50% T_P–fractile. Figure 1 shows this "wave-height-driven" seastate to be

$$H_S = H_{100} = 14.5 \text{ m.}\,; \quad \text{associated 50\%-fractile } T_P = 15.9 \text{ sec.} \tag{3}$$

Force and Response Model

Volterra series have become an increasingly common way to model nonlinear systems, both in offshore structural engineering and other areas. They provide a natural extension of linear models. They are also particularly convenient for systems with a single input: here, the wave elevation $\eta(t)$ at the center of the platform. The corresponding output $x(t)$, which may be a structural force or motion, is then modelled in the time domain as

$$x(t) = x_1(t) + x_2(t)\,; \quad x_1(t) = \int_{-\infty}^{+\infty} h_1(\tau)\eta(t-\tau)d\tau \tag{4}$$

$$x_2(t) = \int_{-\infty}^{+\infty}\int_{-\infty}^{+\infty} h_2(\tau_1,\tau_2)\eta(t-\tau_1)\eta(t-\tau_2)d\tau_1\tau_2 \tag{5}$$

Thus, in addition to the first-order impulse response function $h_1(\tau)$ that defines a linear system, quadratic nonlinear effects are included through the second-order impulse response $h_2(\tau_1,\tau_2)$. The nonlinear system is then completely described by either h_1 and h_2 or their Fourier transforms, the first- and second-order transfer functions $H_1(\omega)$ and $H_2(\omega_1,\omega_2)$. Note that this second-order model cannot represent all nonlinear systems; exact models of cubic and higher-order nonlinearites would require still higher-order transfer functions.

The transfer functions $H_1(\omega)$ and $H_2(\omega_1, \omega_2)$ can be estimated in various ways. In nonlinear system identification, they may be estimated empirically from simultaneous histories of $\eta(t)$ and $x(t)$ (Kim et al, 1990). Alternatively, they may be estimated analytically from an underlying nonlinear theory. This latter approach is used here, using wave diffraction theory to find force transfer functions and a linear structural model to obtain corresponding response transfer functions.

Specifically, we consider here the Snorre TLP. This platform has four circular columns with 25 m. diameters at 76 m. center-to-center distance, and four square pontoons with 11.3 m. sides and 2.5 m. corner radii. Wave diffraction analysis has been used to estimate the first- and second-order force transfer functions (Molin and Chen, 1990). We denote these by $F_{1i}(\omega)$ and $F_{2i}(\omega_1, \omega_2)$, where i=1,...,6 to include the three translational and three rotational degrees of freedom.

Motion transfer functions, H_{1i} and H_{2i}, corresponding to the force transfer functions F_{1i} and F_{2i}, are found by solving the simultaneous equations of motion in the 6 degrees of freedom (Marthinsen, 1991b). This is done for each frequency ω and frequency pair (ω_1, ω_2) of interest. Wave radiation damping is used for the first-order response, while drag damping is important for both slow-drift and springing second-order responses. We use here a generalized Morison drag model in the 6 degrees of freedom; the linear damping coefficient is used to control low-amplitude springing oscillations (Marthinsen, 1989).

Note that the transfer function H_2 weights a term of the form $\exp[i(\omega_1 + \omega_2)t]$. Therefore, if ω_1 and ω_2 are of the same sign ($\omega_1 \cdot \omega_2 > 0$), this transfer function truly reflects a higher, "sum" frequency contribution. Alternatively, H_2 shows difference frequency contributions for $\omega_1 \cdot \omega_2 < 0$. Because of our interest in sum-frequency forces and springing response, we set force and response transfer functions to zero for $\omega_1 \cdot \omega_2 < 0$.

Response Moments and Extremes

In TLP reliability analysis, response quantities of interest include both extreme values and fatigue damage. Unfortunately, exact statistics of these quantities are not known even for linear Gaussian responses, let alone for the second-order Volterra model. Hence we must approximate these quantities, using limited response statistics from Eqs. 4–5. We choose here to characterize the physical response model by its first four statistical moments.

It is convenient first to recast Eqs. 4–5, which involve correlated $\eta(t)$ values at various t, in terms of standard normal processes $u_i(t)$ that are independent at fixed t:

$$x(t) = \sum_i c_i u_i(t) + \sum_i \lambda_i u_i^2(t) \tag{6}$$

The coefficients λ_i are eigenvalues of the integral equation (Kac and Siegert, 1947):

$$\int_{-\infty}^{+\infty} \Gamma(\omega_1, \omega_2)\Phi(\omega_2)d\omega_2 = \lambda\Phi(\omega_1) \tag{7}$$

in which $\Gamma(\omega_1, \omega_2) = H_2(\omega_1, -\omega_2)\sqrt{S_\eta(|\omega_1|)S_\eta(|\omega_2|)}/2$. Once these second-order coefficients λ_i have been found, the corresponding first-order coefficients c_i in Eq. 6 can be recovered:

$$c_i = \int_{-\infty}^{+\infty} H_1(\omega)[\frac{1}{2}S_\eta(|\omega|)]^{1/2}\Phi_i^*(\omega)d\omega \tag{8}$$

This eigenvalue solution for the coefficients c_i and λ_i comprises the bulk of the computational effort. The remaining moment-based response analysis is straightforward. The moments of $x(t)$ are found directly from simple algebraic expressions (Winterstein and Marthinsen, 1991):

$$m_x = \sum_i \lambda_i \tag{9}$$

$$\sigma_x^2 = \sum_i c_i^2 + 2\lambda_i^2 \tag{10}$$

$$\alpha_{3x} = \mathrm{E}[(x - m_x)^3]/\sigma_x^3 = \sum_i (6c_i^2\lambda_i + 8\lambda_i^3)/\sigma_x^3 \tag{11}$$

$$\alpha_{4x} = \mathrm{E}[(x - m_x)^4]/\sigma_x^4 = 3 + \sum_i 48\lambda_i^2(c_i^2 + \lambda_i^2)/\sigma_x^4 \tag{12}$$

The Gaussian model, with α_{3x}=0 and α_{4x}=3, is approached either when the second-order coefficients λ_i become small relative to the first-order factors c_i, or when the number of significant λ_i terms grows. For example, if $\lambda_1 = \cdots = \lambda_n$ and all c_i=0, $x(t)$ is proportional to a chi-square variable with n degrees of freedom. The Gaussian model is then recovered as $n \to \infty$.

With these response moments, the Hermite transformation model can be used to relate $x(t)$ to a standard Gaussian process $u(t)$ (Winterstein, 1988):

$$x = m_x + \kappa\sigma_x[u + c_3(u^2 - 1) + c_4(u^3 - 3u)] \tag{13}$$

in terms of the coefficients $c_4 = (\sqrt{1 + 1.5(\alpha_{4x} - 3)} - 1)/18$, $c_3 = \alpha_{3x}/(6 + 36c_4)$, and $\kappa = 1/\sqrt{1 + 2c_3^2 + 6c_4^2}$. Eq. 13 applies for $\alpha_{4x} \geq 3$, and reduces to the Gaussian model if α_{3x}=0 and α_{4x}=3. Alternative models are available when $\alpha_{4x} < 3$, although such cass do not arise here.

The p-fractile extreme response in seastate duration T_{ss} can be estimated from Eq. 13, taking u as the corresponding Gaussian extreme fractile:

$$u_{max,p} = [2\ln(n\frac{T_{ss}}{T_0\ln(1/p)})]^{1/2} \tag{14}$$

Here T_0 is the average response period and n=1 or 2 for max $x(t)$ or max $|x(t)|$, respectively. We consider here the *median* extreme (p=0.5) throughout, assuming that the uncertainty in the seastate parameters dominates that in maximum response given these parameters. The additional uncertainty in the actual extreme can also be readily included, introducing the variable u_{max} in the FORM analysis. Finally, note that fatigue life estimates consistent with Eq. 13 can also be constructed.

Computational Notes

In practice we discretize Eq. 7 over a discrete set of frequencies, and truncate Eq. 6 when these eigenvalues become sufficiently small. This discrete frequency mesh should include both first-order wave and second-order response frequencies, with sufficient resolution to sample the lightly damped response mode. Alternatively, moments of $x(t)$ can be directly found from multiple integrals involving $H_1(\omega)$ and $H_2(\omega_1, \omega_2)$. This approach has been used, for example, to estimate slow-drift response skewness under linear potential forces and nonlinear Morison drag (Spanos and Donley, 1990). The eigensolution may be an attractive alternative, however, particularly for the 4-fold integration required for the kurtosis value.

Finally, we may also use the second-order system model to find still higher response moments, full probability densities, and more detailed descriptors such as mean upcrossing rates. Such quantities have been estimated, for example, for the slow-drift problem (Naess, 1985; Naess, 1986). A possible danger, however, is to seek more information than the second-order system representation can accurately provide. Indeed, even the kurtosis value in Eq. 12 omits third-order effects, which may be of comparable importance. Thus, in addition to their simplicity, the moment-based models used here may capture most of the useful information in the approximate second-order Volterra model.

Numerical Results

Figure 2 shows how various response statistics change with significant wave height H_S. Response statistics for tether tension are shown at left, and for heave acceleration at right. In both cases, we show results for the three T_P fractiles in Figure 1. As in Figure 1, the solid line shows results for a seastate with "typical" period T_P at the median fractile p=0.5 for the given H_S value. The other curves show the effects of somewhat higher (p=0.84) and lower (p=.16) periods T_P. The RMS and kurtosis values are found from Eqs. 10 and 12. Associated skewness values are nearly zero, and hence not shown. The extreme response is predicted from these moments with Eqs. 13–14, with n=1 for one-sided (tensile) tether tensions and n=2 for two-sided (absolute) extreme accelerations. As noted previously, p=0.5 in

Eq. 14 to focus on representative seastate extremes.

The tension and acceleration responses share several features. Chief among these is the association of larger responses—both rms and max values—with *smaller* periods. This can be explained by noting that the relevant structural periods (heave and pitch) are roughly 2.5 sec. Sum-frequency responses will thus be governed by the wave energy content in the 5-sec. range. All seastates in Figure 2 have $T_P > 5$ sec.; those at the lowest (16%) T_P-fractile have the most energy near 5 sec. and give the largest sum-frequency effect. Further, seastates closest to T_P=5 sec. correspond to 16% T_P-fractiles *at small* H_S values. Figure 2 shows the kurtosis to be greatest in these seastates, reflecting that they indeed show the largest relative contribution of nonlinear effects.

Differences between tension and acceleration results are equally notable. In essence, they show that second-order sum-frequency effects are greater for the heave acceleration. Evidence for this includes its greater sensitivity to period T_P, and its considerably greater kurtosis value. Physically, this is because the higher (sum) frequencies are amplified in these acceleration responses. Note also that although the system is lightly damped, relatively high kurtosis values can be found in springing responses.

Variations with Return Period

In reliability analysis and design, one often seeks the response level with a specified mean return period, such as 100 years. For fixed offshore structures, the seastate described by Eq. 3 might be expected to produce this 100-year response. From Figure 2, however, we might anticipate critical seastates with somewhat lower H_S values, as they are associated with lower, more damaging periods T_P. We may also expect Eq. 3 to better model the design seastate for extreme tether tension than for heave acceleration, because the latter is more sensitive to T_P values.

Figure 3 confirms these expectations. Extremes with various mean return periods are estimated by FORM, together with the design seastate parameters that produce them. The dashed lines show analogous results for wave-dominated seastates; i.e., using H_S values with the desired return period and representative, 50%-fractile associated periods. For tether tensions (at left), the relative agreement between solid and dashed lines suggests that this response is indeed driven largely by wave height H_S. The actual seastate most likely to produce the 100-year response is roughly H_S=13.0 m., T_P=13.1 sec.; however, the seastate in Eq. 3 predicts the maximum response to within 10%.

In contrast, the wave height dominated seastate from Eq. 3 underpredicts the maximum heave acceleration by more than an order of magnitude. In fact, the actual design seastate for this acceleration has relatively constant period T_P over a range of return periods. This suggests the alternative, "wave period driven" 100–year seastate:

$$T_P = 2T_n = 4.7 \text{ sec.}; \quad \text{associated } H_S \text{ from 100-year contour } = 4.6 \text{ m.} \tag{15}$$

This seastate, in terms of the relevant natural period T_n, is chosen to emphasize second-order effects. For heave acceleration, we choose T_n as the vertical natural period T_y=2.35 sec. Eq. 15 provides excellent agreement with the 100-year heave acceleration predicted by FORM, for which H_S=4.7 m. and T_P=4.8 sec.

Variations with Structural Periods

Finally, Figure 4 shows the effect of varying the governing natural frequency—the heave frequency ω_y for heave acceleration and the pitch frequency ω_θ for tether tension—on the 100-year extreme responses. Results in Figure 3 are recovered when ω_y=2π/2.35=2.67 and ω_θ=2π/2.60=2.42 [rad/sec]. Note first that over a broad range of ω_y, the period driven seastate in Eq. 15 well approximates the heave acceleration design seastate. For example, Figure 4 shows design seastate periods between T_P=5.0 sec. and 8.0 sec. for T_y between about 2.5 sec. and 4.0 sec. Note also that while the design seastate parameters H_S and T_P vary smoothly with ω_y, the resulting extreme heave acceleration does not. This is due to similar irregularities in the heave force transfer function (Winterstein and Marthinsen, 1991).

Results for tether tension are somewhat less expected. The base case ω_θ=2.67, which is largely wave height dominated, is not typical of seastates with smaller pitch frequencies. When $\omega_\theta < 2.25$ the critical seastates become closer to those of heave acceleration; i.e., more nearly driven by wave period. In this range the extreme tether tension increases, exceeding the wave height dominated result by factors of 2-3. Again, the irregular variation of extreme response with ω_θ mirrors that of the pitch moment transfer function (Winterstein and Marthinsen, 1991). Note that because of limited transfer function data, second-order effects are not included for frequencies above 2.8 [rad/sec]. Also, the tension results neglect set-down effects, which were found relatively small in some trial calculations.

Summary and Conclusions

- Tether tensions and platform motions show clear effects of force nonlinearity, both on response variance and higher moments. These effects are most pronounced on acceleration responses, which amplify sum-frequency contributions.

- Many springing responses combine negligible skewness with considerably non-Gaussian values of kurtosis. In this regard they differ from other hydrodynamic effects studied through second-order models, such as the wave elevation and TLP slow-drift motions. As these second-order models may be less adequate to estimate kurtosis than skewness, results should be interpreted with care.

- TLP force transfer functions, which reflect local geometry, oscillate rather rapidly near typical heave, roll, and pitch natural frequencies. Hence, extreme springing responses show similar oscillations with these natural frequencies. The inherent uncertainty both in these frequencies and in force transfer functions should also be noted. FORM analysis of this uncertainty may be hindered by the non-monotonic behavior of response rms and extremes.

- Design seastates for extreme TLP heave acceleration are generally driven by wave period T_P. These are well approximated by Eq. 15. For extreme tether tension, the corresponding design seastate varies with ω_θ, the pitch natural frequency. The period driven choice in Eq. 15 is critical when $\omega_\theta < 2.25$, while the wave dominated choice (Eq. 3) is more nearly critical for larger ω_θ values. This suggests that second-order effects may be important for both responses, and that both seastates (Eqs. 3 and 15) should generally be checked to predict extremes.

ACKNOWLEDGEMENTS

Financial support for the first author has been provided by the Office of Naval Research, Contract No. N00014-87-K-0475, and by the Reliability of Marine Structures Program of Stanford University. This work was undertaken during the second author's visit to Stanford; the authors would like to thank Saga Petroleum for making this visit possible.

REFERENCES

Haver, S. and K.A. Nyhus (1986). A wave climate description for long term response calculations. *Proc., 5th Intl. Offshore Mech. Arctic Eng. Symp.*, ASME, **IV**, 27–34.

Kac, M. and A.J.F. Siegert (1947). On the theory of noise in radio receivers with square law detectors. *J. Appl. Phys.*, **18**, 383–400.

Kim, S.B., E. Powers, R.K. Miksad, F.J. Fischer and J.Y. Hong (1990). Spectral estimation of second order wave forces on a TLP subject to nongaussian irregular seas. *Proc., 9th Intl. Offshore Mech. Arctic Eng. Symp.*, ASME, **I**(A), 159–164.

Marthinsen, T. (1989). Hydrodynamics in TLP design. *Proc., 8th Intl. Offshore Mech. Arctic Eng. Symp.*, ASME, **I**, 127–133.

Marthinsen, T. (1991a). The statistics of irregular second-order waves. *Proc., 9th Intl. Workshop on Water Waves and Floating Bodies*, Woods Hole, Apr. 14–17.

Marthinsen, T. (1991b). Second-order hydrodynamic load and response statistics. *Proc., 1st Intl. Workshop on Very Large Floating Structures*, Honolulu, 209–220.

Molin, B. and X.B. Chen (1990). *Calculation of second-order sum-frequency loads on TLP hulls*, Institute Français du Petrole Report.

Naess, A. (1985). Statistical analysis of second-order response of marine structures. *J. Ship Res.*, **29**, 270–284.

Naess, A. (1986). The statistical distribution of second-order slowly-varying forces and motions. *Appl. Ocean Res.*, **8**(2), 110–118.

Neal, E. (1972). Second-order hydrodynamic forces due to stochastic excitation. *Proc., 10th ONR Symp.*, Cambridge, MA, 517–539.

Spanos, P.D. and M.G. Donley (1990). Stochastic response of a tension leg platform to viscous drift forces. *Proc., 9th Intl. Offshore Mech. Arctic Eng. Symp.*, ASME, **II**, 107–114.

Vinje, T. (1984). On the statistical distribution of second-order forces and motions. *Int. Shipbuilding Prog.*, **30**, 58–68.

Winterstein, S.R. (1988). Nonlinear vibration models for extremes and fatigue. *J. Engrg. Mech.*, ASCE, **114**(10), 1772–1790.

Winterstein, S.R. and O.B. Ness (1989). Hermite moment analysis of nonlinear random vibration. *Computational mechanics of probabilistic and reliability analysis*, ed. W.K. Liu and T. Belytschko, Elme Press, Lausanne, Switzerland, 452–478.

Winterstein, S.R. (1990). Random process simulation with the Fast Hartley Transform. *J. Sound Vib.*, **137**(3), 527–531.

Winterstein, S.R., E.M. Bitner-Gregersen, and K.O. Ronold (1991). Statistical and physical models of nonlinear random waves. *Proc., 10th Int. Offshore Mech. Arc. Eng. Sym.*, Stavanger.

Winterstein, S.R. and T. Marthinsen (1991). Second-order springing effects on TLP extremes and fatigue. *J. Engrg. Mech.*, ASCE, submitted for possible publication.

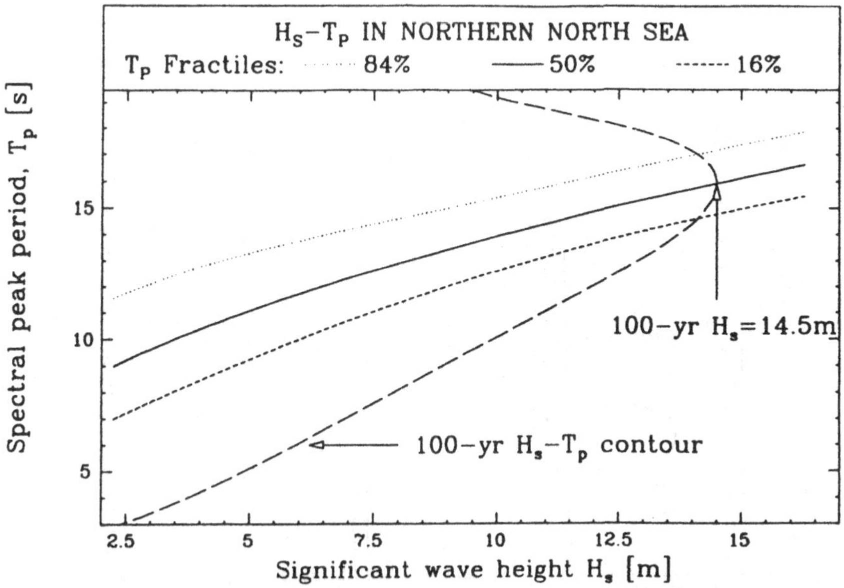

Figure 1: H_S-T_P fractiles and 100-year contour.

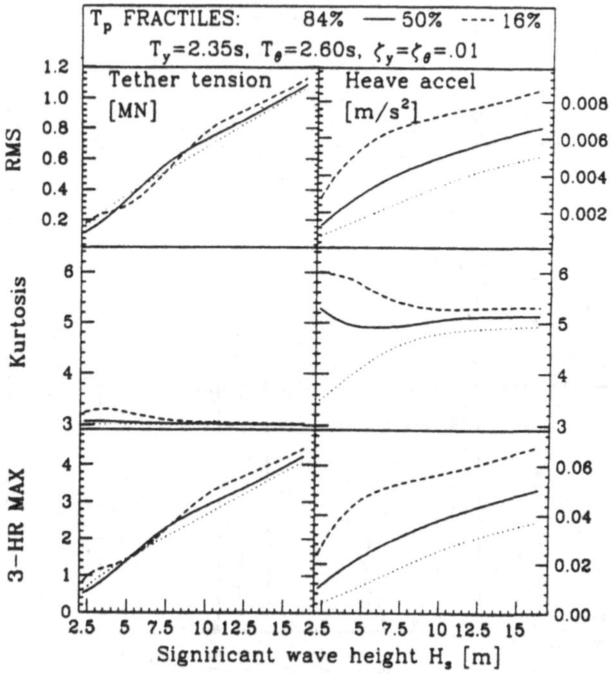

Figure 2: Response for various T_p fractiles.

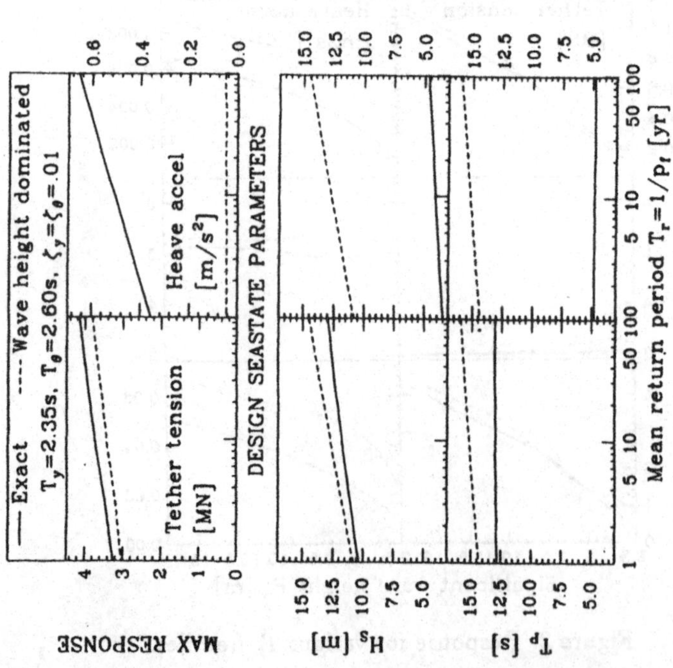

Figure 3: Extreme responses with various return periods.

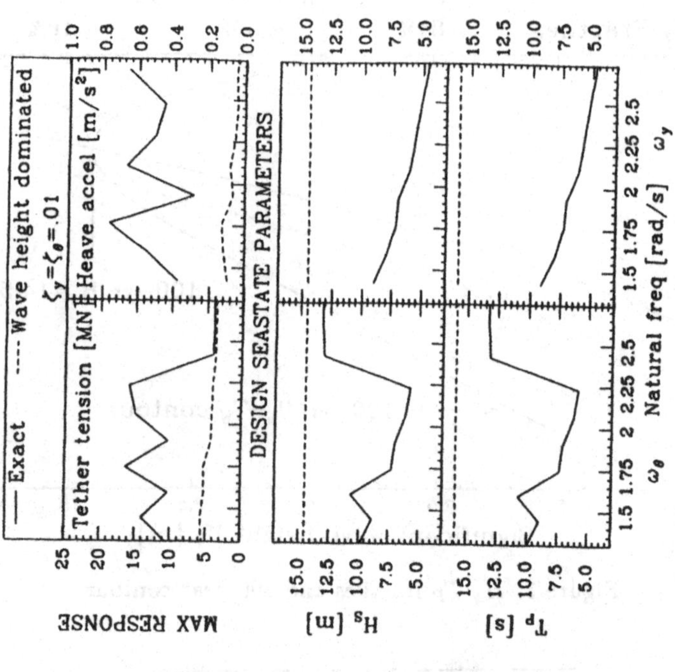

Figure 4: Extreme responses for various structural periods.

Disordered Periodic Structures

Y.K. Lin, G.Q. Cai

Center for Applied Stochastics Research, Florida Atlantic University, Boca Raton, FL 33431, U.S.A.

ABSTRACT

A structure designed to be spatially periodic in configuration cannot be exactly periodic due to material, geometric and manufacturing variabilities. The departure from exact periodicity is known as disorder, and its effects are investigated in this paper in terms of additional decay in wave propagation, and possible localized higher response.

INTRODUCTION

An ideal periodic structure is one composed of identical units which are connected end-to-end to form a spatially periodic array. Considerable progress has been made for the analysis of such structures in recent years; e.g. Brillouin [1], Miles [2], Lin [3], Lin and McDaniel [4], and Mead [5]. One of the distinctive properties of a periodic structure is its alternate wave-passage frequency bands and wave-stoppage frequency bands. If a periodic structure is undamped, then disturbances at a frequency within a wave-passage band will propagate indefinitely, whereas those at a frequency within a wave-stoppage band will attenuate in a short distance. However, an ideal periodic structure does not exist due to geometric and material variabilities. The departure from perfect periodicity is called disorder, and it has been investigated by Soong and Bagdanoff [6,7], Lin and Yang [8,9], Kissel [10], Pierre [11], Cai and Lin [12], and others.

In this paper, two effects of disorder will be examined. One is the attenuation of wave propagation due to disorder even when the frequency is within a wave-passage band. Another is the variation of the structure response near where an excitation is applied. Due to the length limitation, only a generic mono-coupling disordered structure, in which only one generalized displacement is permitted at each cell-to-cell interface, will be discussed in detail, and the frequency is assumed to lie within a wave-passage band. Furthermore, structural damping will be neglected since our objective is to examine the effects due to disorder alone. The more general case of a damped structure with multiple cell-to-cell coupling will be treated in another paper, Cai and Lin [13].

WAVE TRANSFER MATRIX

Consider a disordered periodic structure, consisting of N cell-units. Stations n and n+1 mark the boundaries of cell n. The state vectors at stations n and n+1 are related as follows

$$\begin{Bmatrix} w(n+1) \\ f(n+1) \end{Bmatrix} = [T(n)] \begin{Bmatrix} w(n) \\ f(n) \end{Bmatrix} \tag{1}$$

where w is a generalized displacement, f is the corresponding generalized force, and $[T(n)]$ is a 2×2 transfer matrix associated with the nth cell. It is implied in (1) that the motion is time-wise sinusoidal with a frequency ω. Then, w and f are interpreted as complex amplitudes of the displacement and force, respectively, and matrix $[T(n)]$ is frequency dependent. Without disorder, the transfer matrices for all cells would be identical, denoted by $[T]$.

The eigenvalues of $[T]$ are a reciprocal pair, denoted by λ and λ^{-1} where $|\lambda| = 1$ for the present undamped structure if the frequency is within a wave-passage band. The eigenvectors corresponding to these eigenvalues constitute a transformation matrix $[D]$, with which the state vectors in (1) are transformed to wave vectors as follows [14]:

$$\begin{Bmatrix} w(j) \\ f(j) \end{Bmatrix} = [D] \begin{Bmatrix} \mu^r(j) \\ \mu^\ell(j) \end{Bmatrix} \; ; \quad j = 1, 2, \ldots, N+1 \tag{2}$$

where superscript r or ℓ indicates the direction of wave propagation, to be either right-going or left-going. Substituting (2) into (1), one obtains

$$\begin{Bmatrix} \mu^r(n+1) \\ \mu^\ell(n+1) \end{Bmatrix} = [D]^{-1}[T(n)][D] \begin{Bmatrix} \mu^r(n) \\ \mu^\ell(n) \end{Bmatrix} = [Q(n)] \begin{Bmatrix} \mu^r(n) \\ \mu^\ell(n) \end{Bmatrix} \tag{3}$$

where $[Q(n)]$ is a transfer matrix for wave vectors, instead of state vectors. Without disorder, $[Q(\mathrm{n})]$ would become independent of n; i.e.

$$[Q(n)] = [Q] = [D]^{-1}[T][D] = \begin{bmatrix} \lambda & 0 \\ 0 & \lambda^{-1} \end{bmatrix} \tag{4}$$

The diagonal form of the wave transfer matrix given in (4) for an ideal periodic structure implies that no reflection occurs when a wave motion passes through the interface of two adjoining units. However, in a disordered periodic structure, off-diagonal elements appear in the wave transfer matrix $[Q(n)]$, indicating reflections at the interfaces.

WAVE PHASE DIFFERENCE

We shall now investigate the wave motions μ^r and μ^ℓ in detail. Without damping, the average energy in-flow and energy out-flow of a typical inner cell, say the nth cell of a N-cell chain, must be equal within one period of time $2\pi/\omega$; i.e.

$$|\mu^r(n)|^2 + |\mu^\ell(n+1)|^2 = |\mu^\ell(n)|^2 + |\mu^r(n+1)|^2 \tag{5}$$

If, furthermore, no energy dissipation mechanism exists at the right boundary of the chain, then $|\mu^r(N+1)|^2 = |\mu_B^r|^2 = |\mu_B^\ell|^2 = |\mu^\ell(N+1)|^2$. It follows $|\mu^r(n)| = |\mu^\ell(n)|$; namely, the right-going and left-going waves at the same station are always equal in magnitude for an undamped periodic structure. However, they may differ in phase; i. e.

$$e^{i\theta(n)} = \frac{|\mu^r(n)|}{|\mu^\ell(n)|} \tag{6}$$

where $i = \sqrt{-1}$, and $\theta(n)$ is purely real. The relationship between $\theta(n)$ and $\theta(n+1)$ at two neighboring stations can be deduced from (3) to yield

$$e^{i\theta(n)} = \frac{q_{12}(n) - e^{i\theta(n+1)}q_{22}(n)}{e^{i\theta(n+1)}q_{21}(n) - q_{11}(n)} \; ; \qquad n = N-1, \ldots, 2, 1 \tag{7}$$

where $q_{ij}(n)$ are elements of the wave transfer matrix $[Q(n)]$. Equation (7) shows that $\theta(n)$ depends on $\theta(n+1)$ as well as the physical properties of the nth cell. For the last cell at the right end of the chain,

$$e^{i\theta(N)} = \frac{q_{12}(N) - e^{i\theta_B}q_{22}(N)}{e^{i\theta_B}q_{21}(N) - q_{11}(N)} \tag{8}$$

where θ_B is the phase difference of the right-going and left-going waves at the right boundary. However, θ_B can be determined from the boundary condition. For example, $e^{i\theta_B} = -d_{22}/d_{21}$, if the right boundary of the chain is clamped, and $e^{i\theta_B} = -d_{12}/d_{11}$, if the boundary is free. Here, d_{ij} are elements of the tranformation matrix $[D]$. It can be shown that $|d_{22}| = |d_{21}|$ and $|d_{12}| = |d_{11}|$ using the fact $|\lambda| = 1$.

Equations (7) and (8) provide a simple way to calculate the wave phase different θ at any station in a disordered chain, given the physical properties of every cell. We begin with computing θ_B for a specific boundary condition at the right end. We then determine progressively θ(N), θ(N-1), ..., until the desired θ(n). The scheme is numerically efficient; it can be applied to a large number of such chains for the purpose of Monte-Carlo simulation.

For a randomly disordered periodic structure, each of the disordered parameter in a cell-unit is described by a random variable, which may be represented by a mean plus a random variable with zero mean. For cell n, these random variables may be denoted by a vector $x(n)=\{x_1(n), x_2(n), ..., x_k(n)\}^T$. Being functions of the disordered parameters, the elements of the wave transfer matrix, $q_{ij}(n)$, may be denoted by $q_{ij}(n)=q_{ij}[x(n)]$. Also, it is reasonable to assume that $x_j(n)$ for each j, but different n, are independent and identically distributed random variables, and that they form an ergodic sequence if the disordered chain is infinitely long. In this case, the wave phase differences θ(n) are also random variables, and their probability density functions can be obtained as follows:

$$p[\theta(N)] = \int p[x(N)] \left| \frac{\partial x_1(N)}{\partial \theta(N)} \right| dx_2(N)...dx_k(N) \tag{9}$$

$$p[\theta(n)] = \int p[\theta(n+1)]\, p[x(n)] \left| \frac{\partial \theta(n+1)}{\partial \theta(n)} \right| dx(n) \tag{10}$$

where each $p[\ ]$ denotes a probability density. In obtaining (10), use has been made of the independence of the two random variables $\theta(n+1)$ and $x(n)$ due to the fact that $\theta(n+1)$ is only related to random vectors $x(n+1)$, $x(n+2)$, ..., $x(N)$, as can be seen from the recursive relationship, (7). For a finite disordered chain, we can evaluate $p[\theta(n)]$ by numerically integrating (9) and (10) recursively.

It is conceptually illuminating to consider the case of a semi-infinite disordered chain; namely, when $N\to\infty$. In this case, $p[\theta(n)]$ becomes independent of n, for a finite n. The existence of such an invariant probability density has been proved by Furstenberg [15]. A closed form solution for the invariant $p(\theta)$ is difficult to obtain; however, it must satisfy the relation

$$\{p[\theta(n)]\}_{\theta(n)=\theta} = \left\{ \int p[\theta(n+1)]\, p[x(n)] \left| \frac{\partial \theta(n+1)}{\partial \theta(n)} \right| dx(n) \right\}_{\theta(n)=\theta} \tag{11}$$

$$= \{p[\theta(n+1)]\}_{\theta(n+1)=\theta}$$

For specific cases, it can be calculated approximately by using a perturbation approach, as shown later in an example. Of course, a semi-infinite disordered chain does not actually exist. However, the invariant probability distribution, obtained for a semi-infinite chain, is a good approximation for at least the first several cells, if the entire chain is long enough.

LOCALIZATION FACTOR

The term localization factor refers to a semi-infinite disordered chain and is defined as the exponential decay rate per unit cell of the wave amplitude as a wave motion propagates along the chain. The subject has been investigated in an earlier paper [12]. It will now be re-examined using the framework of the present paper. We begin by modifying equation (3) as follows:

$$\begin{Bmatrix} \mu^r(n) \\ \mu^\ell(n) \end{Bmatrix} = [V(n)] \begin{Bmatrix} \mu^r(n+1) \\ \mu^\ell(n+1) \end{Bmatrix} \tag{12}$$

where $[V(n)] = [Q(n)]^{-1}$. The first row of (12) reads

$$\mu^r(n) = v_{11}(n)\mu^r(n+1) + v_{12}(n)\mu^\ell(n+1) \tag{13}$$

where v_{ij} are elements of $[V(n)]$. Dividing both sides of (13) by $\mu^r(n+1)$, we obtain

$$\frac{\mu^r(n)}{\mu^r(n+1)} = v_{11}(n) + v_{12}(n)e^{-i\theta(n+1)} \tag{14}$$

The exponential decay rate of wave amplitude due to passage through cell n is given by

$$\begin{aligned} \gamma(n) &= ln\left|\frac{\mu^r(n)}{\mu^r(n+1)}\right| \\ &= \frac{1}{2} ln\left\{ |v_{11}(n)|^2 + |v_{12}(n)|^2 + 2Re[v_{11}(n)v_{12}^*(n)\, e^{i\theta(n+1)}]\right\} \end{aligned} \tag{15}$$

where $Re[\ ]$ denotes the real part of a complex quantity, and an asterisk denotes the complex conjugate. Since random variables x(n) and θ(n+1) are independent, the localization factor can be obtained as

$$\begin{aligned} \gamma &= \lim_{N\to\infty}\frac{1}{N}\sum_{n=1}^{N}\gamma(n) = E[\gamma(n)] \\ &= \frac{1}{2}\int ln\left\{ |v_{11}(x)|^2 + |v_{12}(x)|^2 + 2Re[v_{11}(x)v_{12}^*(x)\, e^{i\theta}]\right\} p(x)p(\theta)\,dx\,d\theta \end{aligned} \tag{16}$$

in which spatial averaging has been replaced by ensemble averaging on the basis of the ergodic assumption for the random sequence $x(n)$, n=1,2,....

Equation (16) is applicable to any randomly disordered mono-coupling chain, whether the localization is weak, moderate or strong, and its accuracy depends only on the accuracy of the computed invariant $p(\theta)$.

FREQUENCY RESPONSE

Let the input to the disordered structure be a sinusoidal force of frequency ω with a complex amplitude $f(1)$, acting at the left boundary of the first cell, and let the output of interest be the steady-state displacement at the same location with a complex amplitude $w(1)$. It can be shown, by using (2), that the frequency response function $H(\omega)$ is given by

$$H(\omega) = \frac{w(1)}{f(1)} = \frac{d_{11}\mu^r(1) + d_{12}\mu^l(1)}{d_{21}\mu^r(1) + d_{22}\mu^l(1)} = \frac{e^{i\theta(1)}d_{11} + d_{12}}{e^{i\theta(1)}d_{21} + d_{22}} \tag{17}$$

where d_{ij} are elements of matrix $[D]$. For a randomly disordered structure, the probability of the magnitude $z=|H|$ of the frequency response function can be obtained as follows:

$$p[z] = p[\theta(1)]\left|\frac{d\theta(1)}{dz}\right| \tag{18}$$

where $p[\theta(1)]$ can be calculated recursively from the right boundary as discussed previously. If a randomly disordered structure is very long, the invariant $p(\theta)$ of the wave phase difference for a semi-infinite chain is a good approximation for $p[\theta(1)]$.

NUMERICAL EXAMPLE

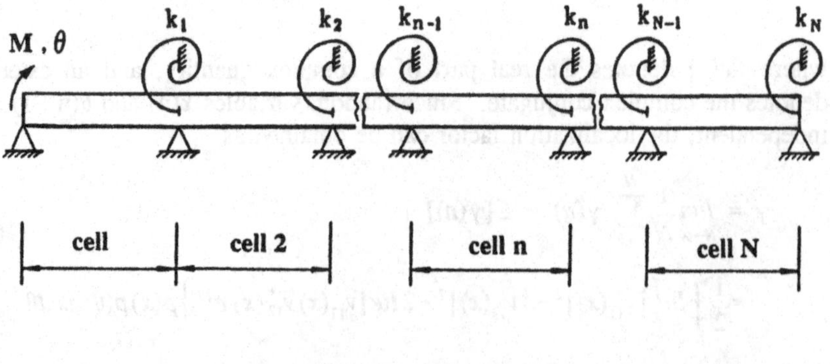

Fig. 1 A multi-span beam

The theory will now be applied to an Euler-Bernoulli beam on evenly spaced supports, shown in Fig. 1. With the exception of the first support, a torsional spring is added at each support. The torsional spring stiffnesses are assumed to be random, and they are described by

$$k_n = k_0[1 + x(n)] \qquad n = 1, 2, \ldots \tag{19}$$

where k_0 is the average k_n, and $x(n)$ are independent and identically distributed random variables with a zero mean. For numerical computation, $x(n)$ are further assumed to be uniformly distributed between $-\sqrt{3}\sigma$ and $\sqrt{3}\sigma$ where σ is the standard deviation. The other physical parameters are treated as being deterministic, including the distance between two neighboring supports l, the mass of the beam per unit length M, and the bending rigidity of the beam EI. The following physical constants were used in the numerical calculations: l=0.1051 m, M=1.8043 kg/m, E_0I=0.3134 N-m^2, and $\delta_0 = E_0I/lk_0 = 0.1$.

A typical cell-unit is chosen to include a beam element between two neighboring supports and the entire torsional spring on its right. The torsional spring on the left is treated as belonging to the preceding cell. The generalized displacement and the corresponding generalized force at each cell-to-cell interface are the rotation angle and the bending moment, respectively. Accordingly, the transfer matrix for the nth cell is given by

$$[T(n)] = \begin{bmatrix} \beta & \dfrac{\alpha}{\nu} \\ -\dfrac{\nu(1-\beta^2)}{\alpha} + \dfrac{\beta}{\delta}[1+x(n)] & \beta + \dfrac{\alpha}{\delta\nu}[1+x(n)] \end{bmatrix} \tag{20}$$

where $\delta = \dfrac{EI}{lk_0}$, $\nu = l\left(\dfrac{\omega^2 M}{EI}\right)^{\frac{1}{4}}$,

$$\alpha = \frac{\cosh\nu\,\cos\nu - 1}{\sinh\nu - \sin\nu}, \qquad \beta = \frac{\sinh\nu\,\cos\nu - \cosh\nu\,\sin\nu}{\sinh\nu - \sin\nu}.$$

The ideal transfer matrix $[T]$ is obtained by letting $x(n)=0$ in (20). Its two eigenvalues can be represented by $e^{\pm i\psi}$, where ψ is purely real and is obtainable from

$$\cos\psi = \beta + \frac{\alpha}{2\delta_0\nu_0}, \qquad \left|\beta + \frac{\alpha}{2\delta_0\nu_0}\right| \le 1 \tag{21}$$

where $\nu_0 = l\left(\dfrac{\omega^2 M}{E_0 I}\right)^{\frac{1}{4}}$. If the multi-span disordered beam is semi-infinite, then (11) has the following specific form

$$[p(\theta')]_{\theta'=\theta} = \left\{ \frac{1}{2\sqrt{3}\sigma} \int_{-\sqrt{3}\sigma}^{-\sqrt{3}\sigma} \left| \frac{\partial\theta}{\partial\theta'} \right| p[\theta(\theta',x)]\,dx \right\}_{\theta'=\theta} = p(\theta) \tag{22}$$

where $\theta(\theta',x)$ represents an implicit relation given by

$$sin\theta = \left\{ sin(\theta'+2\psi) - 2A_0 x[1+cos(\theta'+2\psi)] \right\} G(\theta',x) \tag{23}$$

$$cos\theta = \left\{ cos(\theta'+2\psi) + 2A_0 x\, sin(\theta'+2\psi) - 2A_0^2 x^2[1+cos(\theta'+2\psi)] \right\} G(\theta',x)$$

where $A_0 = a/(2\delta_0\nu_0 sin\psi)$,

$$\left| \frac{\partial\theta}{\partial\theta'} \right| = G(\theta',x) = \left\{ 1 - 2A_0 x\, sin(\theta'+2\psi) + 2A_0^2 x^2[1+cos(\theta'+2\psi)] \right\}^{-1} \tag{24}$$

If $|A_0\sigma|$ is small, namely, if the localization is weak, then an approximate invariant probability density may be obtained as follows

$$p(\theta) = \frac{1}{2\pi} \left\{ 1 + A_0^2\sigma^2 \frac{sin(\theta+\psi)}{sin\psi} [1 + \frac{cos(\theta+\psi)}{cos\psi}] \right\} + 0(A_0\sigma)^4 \; ; \tag{25}$$

$$cos(2n\psi) \neq 1, \quad -\pi \leq \theta \leq \pi$$

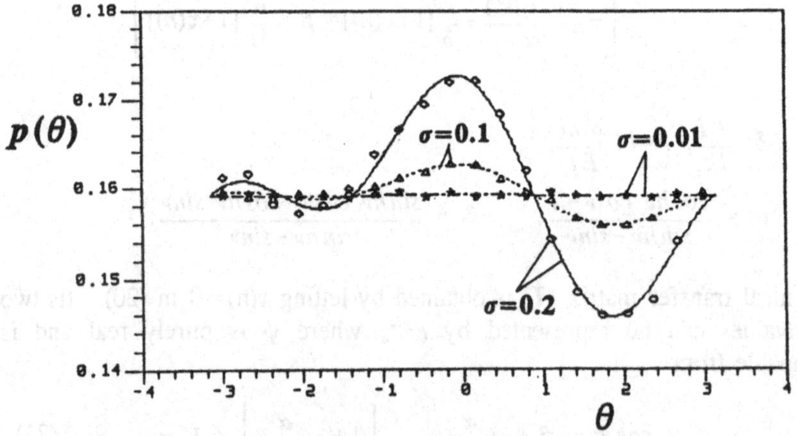

Fig. 2 Invariant probability density of wave phase difference θ, undamped semi-infinite disordered beams, ω=250 rad/s

Fig. 2 depicts the computed approximate invariant $p(\theta)$ for three different cases corresponding to three standard deviations σ = 0.01, 0.1 and 0.2 for the disorder parameter x. The results are in good agreement with those obtained from Monte-Carlo simulations, also shown in the figure as circles, triangles and stars. The simulations were carried out according to (7) and (8), starting from the right boundary. The distribution of θ becomes essentially invariant after 1000 cells, and the simulation results shown in Fig. 2 correspond to the relative frequencies calculated for the subsequent 10^6 cells.

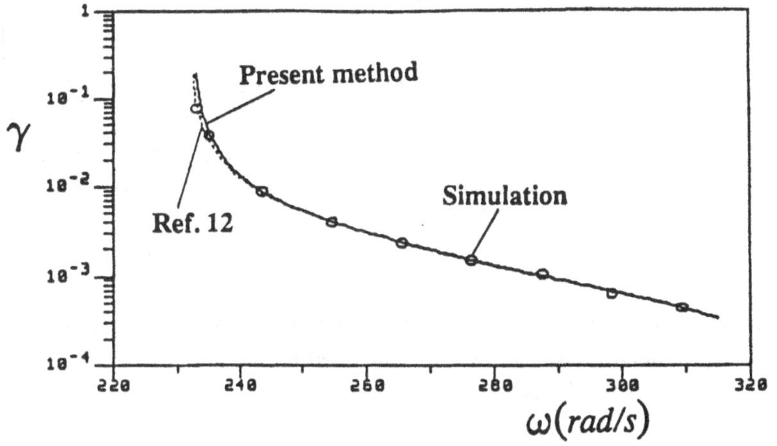

Fig. 3 Localization factors of a disordered beam

The localization factor of a disordered beam was calculated from (16), using the invariant measure $p(\theta)$ obtained from equation (25). The results are shown in Fig. 3 for the case of $\sigma = 0.1$. The frequency range was chosen to be within the first wave-passage frequency band. Also depicted are the simulation results and our earlier results from a different procedure [12]. The present method yields accurate results in comparison with the simulations, except for frequencies very close to the lower boundary of the first wave-passage band, where a strong localization occurs and the invariant measure $p(\theta)$ obtained from perturbation (25) is not very accurate.

The probability densities of the frequency response amplitude, computed from (9), (10) and (18) and normalized with respect to the reference value z_p of the corresponding ideal periodic system, are shown in Figs. 4 and 5. In Fig. 4, four different cases corresponding to four total cell numbers, N = 5, 15, 30 and 100 are plotted, with the standard deviation of the disorder parameter x fixed at $\sigma=0.1$. In Fig. 5, the cell number is fixed at N=30, but the standard deviation σ is chosen to be either 0.05, 0.1 or o.2. It is seen that an increase of either the cell number or the level of disorder causes $p(z/z_p)$ to become more extended and flatter. In both the cases of N=100 in Fig. 4 and $\sigma=0.2$ in Fig. 5, the distribution of the wave phase difference $\theta(1)$ becomes almost uniform within $(-\pi, \pi)$.

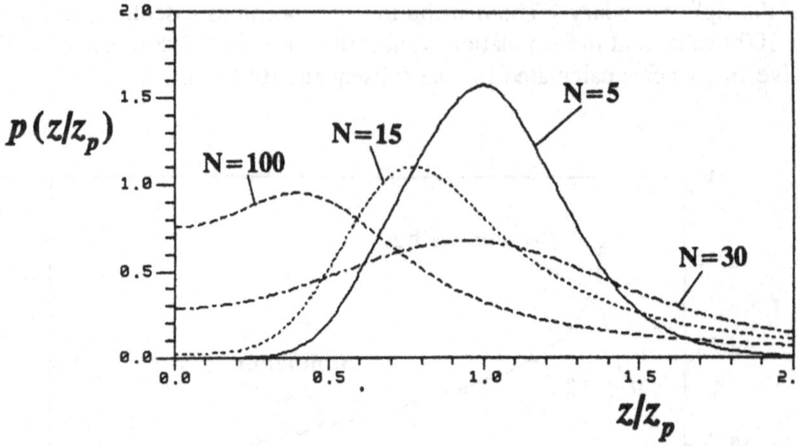

Fig. 4 Probability density of non-dimensional frequency response magnitude z/z_p, undamped disordered beams with different number of spans, $\sigma=0.1$, $\omega=250$ rad/s

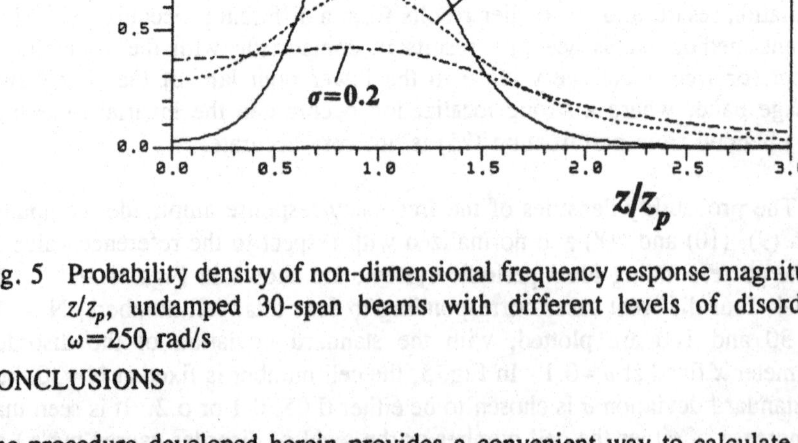

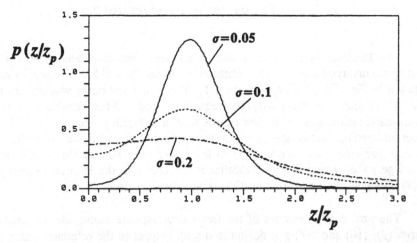

Fig. 5 Probability density of non-dimensional frequency response magnitude z/z_p, undamped 30-span beams with different levels of disorder, $\omega=250$ rad/s

CONCLUSIONS

The procedure developed herein provides a convenient way to calculate the localization factor and the probability density of the frequency response for a randomly disordered structure both analytically and by simulation. Some important characteristics of a randomly disordered structure have been observed from the computed results. The shape of the probability distribution of the frequency response magnitude of a randomly disordered structure indicates the relative likelihood that the magnitude can be either higher or lower than that of the corresponding ideally periodic structure. The number of disordered cells and the level of disorder play a similar role in spreading the frequency response.

ACKNOWLEDGEMENT

This paper is prepared under grant AFOSR-91-0073 from the Air Force Office of Scientific Research, Air Force Systems Command, USAF, monitored by Dr. Spencer Wu. The U.S. Government is authorized to reproduce and distribute reprints for governmental purposes, notwithstanding any copyright notation thereon.

REFERENCES

1. Brillouin, L. Wave Propagation in Periodic Structures, Dover, New York, 1953.
2. Miles, J. M. Vibration of Beams on Many Supports, Journal of Engineering Mechanics, ASCE, Vol. 82, No. 1, pp. 1-9, 1956.
3. Lin, Y. K. Free Vibration of Continuous Beam on Elastic Supports, Journal of Mechanical Science, Vol. 4, pp. 409-423, 1962.
4. Lin, Y. K. and McDaniel, T. J. Dynamics of Beam-Type Periodic Structures, Journal of Engineering for Industry, Vol. 91, pp. 1133-1141, 1969.
5. Mead, D. J. Wave Propagation and Natural Modes in Periodic Systems: I. Mono-Coupled Systems, II. Multi-Coupled Systems, with and without Damping, Journal of Sound an Vibration, Vol. 40, No. 1, pp. 1-39, 1975.
6. Soong, T. T. and Bogdanoff, J. L. On the Natural Frequencies of a Disordered Linear Chain of N-Degrees of Freedom, International Journal of Mechanical Sciences, Vol. 5, pp. 237-265, 1963.
7. Soong, T. T. and Bogdanoff, J. L. On the Impulsive Admittance and Frequency response of a Disordered Linear Chain of N-Degrees of Freedom, International Journal of Mechanical Sciences, Vol. 6, pp. 225-237, 1964.
8. Lin, Y. K. and Yang, J. N. Free Vibration of a Disordered Periodic Beam, Journal of Applied Mechanics, Vol. 41, No. 2, pp. 383-391, 1974.
9. Yang, J. N. and Lin, Y. K. Frequency Response Functions of a Disordered Periodic Beam, Journal of Sound and Vibration, Vol. 38, No. 3, pp. 317-340, 1975.
10. Kissel, G. J. Localization in Disordered Periodic Structures, Ph.D.

Thesis, Massachusetts Institute of Technology, Boston, MA, 1988.

11. Pierre, C. Weak and Strong Vibration Localization in Disordered Structures: A Statistical Investigation, Journal of Sound and Vibration, Vol.139, No. 1, pp. 549-564, 1990.
12. Cai, G. Q. and Lin, Y. K. Localization of Wave Propagation in Disordered Periodic Structures, AIAA Journal, Vol. 29, No. 3, pp. 450-456, 1991.
13. Cai, G. Q. and Lin, Y. K. Statistical Distribution of Frequency Response of Disordered Periodic Structures, to appear.
14. Yong, Y. and Lin, Y. K. Propagation of Decaying Wave in Periodic and Piecewise Periodic Structures of Finite Length, Journal of Sound and Vibration, Vol. 129, No. 2, pp. 99-118, 1989.
15. Furstenberg, H. Noncommuting Random Products, Transaction of the American Mathematical Society, Vol. 108, No. 3, pp. 377-428, 1963.

Response Statistics of a Nonlinear Coupled Oscillator Under Parametric Excitation

Y.J. Yoon, R.A. Ibrahim

Department of Mechanical Engineering, Wayne State University, Detroit, MI 48202, U.S.A.

ABSTRACT

The nonlinear random response of a cantilever beam to a band limited parametric excitation is investigated numerically and experimentally. The beam response is represented by the first two modes which are found to be coupled through nonlinear inertia and parametric terms. The excitation is modeled by a filtered white noise whose center frequency is close to either twice the first, or twice the second, or the sum of the two modal frequencies. The response statistics are predicted by using Monte Carlo simulation. The results are compared qualitatively with the experimental measurements. It is found that both predicted and measured response statistics are nonstationary. When one mode is parametrically excited the other mode is also indirectly excited through nonlinear modal interaction. In addition the response probability density functions are characterized by multi-maxima curves.

1. INTRODUCTION

Parametric instability takes place when the external excitation appears as a coefficient to the system parameters. Under periodic excitation instability occurs when the excitation frequency is close to twice the system natural frequency. In recent years extensive research activities have been devoted to examine various modes of stochastic stability of linear systems [1]. However, few attempts have been directed to predict the nonlinear random response of one mode excitation [2-4]. These studies model the excitation by a wide band random process. In practice one should generate a band limited random process whose center frequency is close to the mode under question. The band limited random process can be generated from a linear filter excited by a white noise. In this case the response of the system can be represented by a Markov process whose probability density function can be described by the Fokker-Planck equation. Wedig [5,6] obtained parametric excitation by passing Gaussian white noise either through a low-pass filter or band-pass filter and examined the limit of instability regions by a perturbation technique. He also obtained three regions of instability of two-degree-of-freedom system in the mean square sense in the neighborhood of $2\omega_1$, $2\omega_2$ and $\omega_1+\omega_2$ where ω_1 and ω_2 are the system natural frequencies.

The experimental investigation of the response of nonlinear systems under band-limited random excitation is very valuable not only to verify the analytical results but also to provide physical insight into complex response characteristics. It is believed that the earliest fundamental experimental investigation was conducted by Bogdanoff and Citron [7]. The purpose of their tests was to verify the theoretical conclusion that the inverted pendulum may be stabilized under second order stationary random parametric excitation having a discrete-power spectra. They found that all differences among excitation frequencies must be large to ensure the stability of the pendulum in the inverted position. Dalzell [8] conducted exploratory studies to measure the statistics of nonlinear liquid motion in cylindrical tanks subjected to parametric excitation. Baxter and Evan-Iwanowski [9] reported the results of random excitation of an elastic column subjected to random axial force. Ibrahim and Heinrich [10] conducted a series of random parametric excitation tests of liquid tanks and compared the measured results with those predicted by Gaussian and non-Gaussian closure schemes and Stratonovich stochastic averaging. The time history response records reveal four response characteristic regimes such as zero motion, uncertain zero motion, partially developed motion and fully developed motion. The parametric random excitation of a nonlinear hanging articulated pipe system was examined by Kawashima and Shimogo [11,12] as a technique of experimental modal analysis and to identify modal parameters such as damping coefficients and natural frequencies. This paper presents the response statistics of a cantilever beam subjected to parametric random excitation. Two different approaches are adopted. These are the Monte Carlo simulation and experimental investigation.

2. Analytical Background

2-1. Mathematical Modeling

The system considered in this study is a cantilever beam of length L, width b, thickness h and mass m per unit length. The fixed end is subjected to a prescribed vertical support motion Z(t). The governing equations of motion of the first two modes derived by using Lagrange's equation. Generally the beam deflection is represented by the summation of n-modes

$$u(x,t) = \sum_{i=1}^{n} q_i(t)\, \phi_i(x) \tag{1}$$

where $q_i(t)$ are the generalized coordinates and $\phi_i(x)$ are the corresponding mode shape functions given by the relationship

$$\phi_i(x) = A_i\left[\cos(\beta_i x) - \cosh(\beta_i x)\right] + \left[\sin(\beta_i x) - \sinh(\beta_i x)\right] \tag{2}$$

where $A_i = \dfrac{\cos(\beta_i L) + \cosh(\beta_i L)}{\sin(\beta_i L) - \sinh(\beta_i L)}$

The axial drop w(x,t) is expressed in terms of the lateral displacement u(x,t) as

$$w(x,t) = \frac{1}{2}\int_0^x \left(\frac{\partial u(\xi,t)}{\partial \xi}\right)^2 d\xi \tag{3}$$

The kinetic energy T and elastic potential energy V of the beam are

$$T = \frac{1}{2}\int_0^L \left(\frac{\partial u(x,t)}{\partial t}\right)^2 m\,dx + \frac{1}{2}\int_0^L \left(\dot{Z}(t) - \frac{\partial w(x,t)}{\partial t}\right)^2 m\,dx \tag{4a}$$

$$V = \frac{1}{2}EI\int_0^L \left(\frac{\partial^2 u(x,t)}{\partial x^2}\right)^2 dx \tag{4b}$$

where E is Young's modulus and I is the area moment of inertia about the bending axis.
The Rayleigh's dissipation function D is

$$D = \frac{1}{2}\sum_{i=1}^{n} c_i \dot{q}_i^2 \tag{5}$$

where c_i is a linear viscous modal damping coefficient.
Applying Lagrange's equation yields the coupled nonlinear equations of motion of the first two modes

$$\begin{aligned} X_1'' + 2\zeta_1 X_1' + X_1 = {} & \varepsilon^2 (-\alpha_{13}X_1 + \alpha_{14}X_2)(X_1'^2 + X_1X_1'') \\ & + \varepsilon^2 (\alpha_{14}X_1 - \alpha_{15}X_2)(X_1''X_2 + 2X_1'X_2' + X_1X_2'') \\ & + \varepsilon^2 (-\alpha_{16}X_1 + \alpha_{17}X_2)(X_2'^2 + X_2X_2'') + \varepsilon (\alpha_{11}X_1 - \alpha_{12}X_2)\, r_f^2\, U(\tau) \end{aligned} \tag{6.a}$$

$$\begin{aligned} X_2'' + 2\zeta_2 r X_2' + r^2 X_2 = {} & \varepsilon^2 (\alpha_{23}X_1 - \alpha_{24}X_2)(X_1'^2 + X_1X_1'') \\ & + \varepsilon^2 (-\alpha_{25}X_1 + \alpha_{26}X_2)(X_1''X_2 + 2X_1'X_2' + X_1X_2'') \\ & + \varepsilon^2 (\alpha_{26}X_1 - \alpha_{27}X_2)(X_2'^2 + X_2X_2'') + \varepsilon (-\alpha_{21}X_1 + \alpha_{22}X_2)\, r_f^2\, U(\tau) \end{aligned} \tag{6.b}$$

where U(t) is generated from the linear filter equation

$$U'' + 2\zeta r_f U' + r_f^2 U = W(\tau) \tag{7}$$

where

$$X_1 = \frac{q_1}{h} \quad X_2 = \frac{q_2}{h} \quad U = \frac{Z}{h} \quad r = \frac{\omega_2}{\omega_1} \quad \omega_1^2 = \frac{12.36EI}{mL^4} \quad \omega_2^2 = \frac{485.55EI}{mL^4}$$
$$\zeta_1 = \frac{0.27\,c_1}{mL\omega_1} \quad \zeta_2 = \frac{0.52\,c_2}{mL\omega_2} \quad \tau = \omega_1 t \quad r_f = \frac{\Omega}{\omega_1} \quad \varepsilon = \frac{h}{L} \tag{8}$$

a prime denotes differentiation with respect to the dimensionless time τ, and α_{ij} are constant coefficients.

It is seen that some nonlinear terms in equations (6) include non-linear inertia coupling. In equation (7) $W(\tau)$ is a white noise with autocorrelation:

$$R_w(\tau') = E\left[w''(\tau)\, w''(\tau+\tau')\right] = 2D\,\delta(\tau')$$

where $\delta(t)$ is the Dirac delta function.

After successive elimination of coupled nonlinear acceleration terms, the system equations are written in the state vector form:

$$
\begin{aligned}
X_1' &= X_3 \\
X_2' &= X_4 \\
X_3' &= -2\zeta_1 X_3 - X_1 + \zeta_2 r \varepsilon^2 \Psi_{11} + \zeta_1 \varepsilon^2 \Psi_{12} + r^2 \varepsilon^2 \Psi_{13} + \varepsilon^2 \Psi_{14} + \varepsilon \Psi_{15} \\
X_4' &= -2\zeta_2 r X_4 - r^2 X_2 + \zeta_2 r \varepsilon^2 \Psi_{21} + \zeta_1 \varepsilon^2 \Psi_{22} + r^2 \varepsilon^2 \Psi_{23} + \varepsilon^2 \Psi_{24} + \varepsilon \Psi_{25} \\
U' &= U_1 \\
U_1' &= -2\zeta r_f U_1 - r_f^2 U + W(\tau)
\end{aligned}
\qquad (9)
$$

where Ψ_{ij} are nonlinear functions of the coordinates.

To estimate the response statistics such as mean squares power spectral density functions and probability density functions, three hundred records of the white noise $W(\tau)$ were simulated and a sequence of 12800 random numbers were sampled for the generation of a single record by using the IMSL subroutine DRNNOR (random number generation from a normal distribution using an inverse Cumulative Distribution Function (CDF) method). Then the 6 coupled differential equations (9) are numerically solved by the IMSL subroutine DIVPRK (Runge-Kutta-Verner fifth-order and six-order method). The numerical integration is performed with double precision on the Sun 3/260 computer using time increment $\Delta\tau = 0.2$ and integration accuracy 0.1D-03. The response statistics are then estimated from the response records X_i.

2-2. Numerical Results

Systems with parametric filtered white noise can exhibit instability if the excitation central frequency is close to either twice the first mode frequency, or twice the second mode frequency, or the sum of the two.

Figure 1 shows the time history records of mean squares of excitation and responses of first and second modes for three cases of excitation center frequency Ω. These are $\Omega=2\omega_1$, $\Omega=2\omega_2$, and $\Omega=\omega_1+\omega_2$, respectively. The mean square responses were plotted against the mean square of the filtered noise. These plots indicated that the mean square stability boundary of the first, second and combination modes correspond to excitation levels (mean square value) 5, 220, and 50, respectively. It is seen that the parametric responses under stationary filtered white noise is a non-stationary random process. When the first mode is parametrically excited the second mode also is excited due to cubic nonlinear inertia coupling terms. Furthermore, the mean square response of the excited mode is much larger than the other mode which is indirectly excited.

Figure 2 shows the corresponding power spectral density functions. These are obtained by using the fast Fourier transform algorithm for the filtered excitation, response records of the three cases. The FFT order 8 together with rectangular window are used. Figure 3(a) shows power spectra for the case of first mode parametric excitation. The second mode appears to have entrained oscillations at two frequencies corresponding to one and three times the first mode natural frequency. Figure 3(b) shows also the power spectra of second mode parametric excitation case. Again, the first mode exhibits smaller level than the second mode which is parametrically excited. Under combination parametric resonance the energy is shared by the two modes and a large portion is transferred to the first mode.

Figure 3 shows estimated probability density function of the first and second mode responses under Gaussian filtered white noise excitation. The degree of deviation of response processes from normality can be inferred by plotting the corresponding Gaussian curves shown by dotted graphs which are determined from the mean and variance of each process. The estimated probability density functions of the first and second mode parametric responses are found to be non-Gaussian while those of other mode responses due to the nonlinear couplings are Gaussian, see figure 3(a, b). Depending on the level of power spectral density we obtain different shapes of probability density function from the first mode response in combination resonance. For the case of combination resonance the probability density function the first mode exhibits multi maxima as shown in figure 3(c). In this case $X_1(\tau)$ is known as a stochastic process with multifurcation [13].

3. Experimental Investigation

3-1. Experiment Model and Equipment

The experimental model emulates the analytical model. The beam is spring steel with cross section 0.0254 x 0.001 m^2 and the length L = 0.337 m. The measured natural frequencies, the damping ratios and bandwidths of excitations are:

$f_1 = 5.75$ Hz, $\zeta_1 = 0.003$, $f_2 = 36$ Hz, $\zeta_2 = 0.005$
$\Delta f = 0.4$ Hz, $f_c = 11.5$ Hz for the first mode excitation
$\Delta f = 0.16$ Hz, $f_c = 72$ Hz for the second mode excitation
$\Delta f = 12.5$ Hz, $f_c = 41.75$ Hz for two-mode excitation

The base of the model is allowed to move vertically and is directly connected to the head of a VTS-600 electrodynamic shaker. The shaker thrust is 600-lb and provides up to 0.0254 m (1 in) peak-to-peak stroke. The shaker receives a filtered random signal from either Hp 3565S hardware or Scientific Atlanta SD1715 Sine/Random Vibration Controller. The filtered signal is amplified via two Techron Model 7560 Power Amplifiers. The excitation is measured by using a Bruel & Kjaer model 4383 accelerometer which is mounted on the shaker platform. A Bruel & Kjaer type 2626 conditioning charge amplifier is used to amplify the accelerometer signal from the shaker. The response signals of the cantilever beam are detected by 350 ohm strain gauges in a two arm bridge and amplified by using Micro-Measurements Model 2310 signal conditioning amplifier. The excitation and response signals are read and converted into binary numbers via a Data Translation Model 2828 Analog/Digital converter board which can read signals from 4 channels (+/- 10V) simultaneously at up to 25k points per second per channel. The board is installed in an expansion slot of an AST Computer (40 mb hard disk and 4 mb Ram).

The data processing is performed at equally spaced intervals. The sampling rate is selected to be at least two times the maximum frequency of the model to avoid the occurrence of aliasing. An Interactive Laboratory system (ILS) software is used to estimate a number of statistical parameters for the response and excitation. In addition to the subroutines available in the ILS software other programs are developed to determine mean squares and probability densities.

3-2. Experimental Results

The cantilever beam is excited by a band-limited parametric random excitation at different levels of excitation power spectral density. Based on preliminary tests and computer capacity it is decided to set the time duration of each test to 45 minutes which is found to be enough to exhibit all possible dynamic characteristics of the beam response. The mean square of any signal X(t) is calculated at every data point using the formula

$$\Phi^2(t_n) = \frac{1}{n}\sum_{i=1}^{n} X^2(t_i)$$

where $\Phi^2(t_n)$ is the mean square value of n data points.

Figure 4 is the measured response statistics for the case of the first mode parametric excitation. Figure 4(a) shows time histories of mean squares of the shaker and beam response. It is seen that the shaker signal is stationary while the cantilever beam response is non-stationary. The degree of non-stationarity of cantilever beam response depends on the level of excitation power spectral density. The power spectra of the shaker and beam are plotted in figure 4(b). The probability density functions as estimated from the signals of the shaker and cantilever beam are plotted in figure 4(c). The mean value of the shaker signal is not zero produced from an offset in the shaker amplifier. The shaker signal is nearly Gaussian while the cantilever beam response has significant deviation from normality and tends to have a spike shape form in the undeveloped parametric beam oscillation.

Figure 5 shows that the measured power spectral density functions for the cases of second mode and two-mode excitations. The observed spectra at zero frequency are due to offsets of the measured signals. It is seen that under second mode excitation, the level of its power spectrum is higher than the first. For the case of two mode excitation the level of the first mode power spectral density is much greater than that of the second mode.

4. Conclusions

The parametric random excitation of two normal modes of a cantilever beam is investigated numerically and experimentally. A Monte Carlo simulation is used to predict the response statistics and the predicted results exhibit stochastic bifurcation for the case of two mode excitation as reflected from the multi-maxima of the first mode probability density. In both numerical simulations and experimental measurements the response is found non-Gaussian under relatively higher excitation levels. In addition, both approaches show the common statistical properties such as non-stationarity of mean square responses, shape of power spectral density function, and deviation of normality.

Acknowledgement

This research is supported by a grant from the National Science Foundation under grant No. 87-96342 and by additional funds from the Institute for Manufacturing Research of Wayne State University.

5. References

1. Ibrahim, R. A. *Parametric Random Vibration*, John Wiley & Sons, New York, 1985.
2. Wu, C. M. and Lin, Y. K. Cumulant - Neglect Closure for Nonlinear Oscillators under Parametric and External Excitations, *Int. J. Nonlinear Mech.* 19, pp. 349-362, 1984.
3. Ibrahim, R. A., Soundararajan, A. and Heo, H. Stochastic Response of Nonlinear Dynamic Systems Based on a Non-Gaussian Closure, *Journal of Applied Mech.* 52(4), pp. 965-970, 1985.
4. Ibrahim, R. A. and Soundararajan, A. An Improved Approach for Random Parametric Response of Dynamic Systems with Nonlinear Inertia, *Int. J. Nonlinear Mech.* 20(4), pp. 309-323, 1985.
5. Wedig, W. Stability Conditions for Oscillations with Parametric Filtered Noise Excitation, *Z. Angew. Math. Mech.*52(3), pp. 161-166, 1972.
6. Wedig, W. Instability regions of First and Second Type for Vibrating Systems with Random Excitation, *Z. Angew. Math. Mech.* 53(4), pp.T248-T250, 1973.
7. Bogdanoff, J. L. and Citron, S. J. Experiments with an Inverted Pendulum Subjected to Random Parametric Excitation, *J. Acoust. Soc. Amer.* 38, pp. 447-452, 1965.
8. Dalzell, J. F. Exploratory Studies of Liquid Behavior in Randomly Excited Tanks: Longitudinal Excitation, Tech. Report No. 1, Southwest Research Inst., San Antonio, May 1967
9. Baxter, G. K. and Evan-Iwanowski, R. M. Response of a Column in Random Vibration Tests, *J. Struct. Mech. Div.* 101, pp. 1749-1761, 1975.
10. Ibrahim, R. A. and Heinrich, R. Experimental Investigation of Liquid Sloshing Under Parametric Random Excitation, *J. Applied Mechanics* 55, pp. 467-473, 1988.
11. Kawashima, T. and Shimogo, T. The response of a hanging articulated pipe under vertical random excitation in water (the response of a system with random parametric excitation and nonlinear damping, *JSME International Journal* 3-262, pp. 608-613, 1987.
12. Kawashima, T. and Shimogo, T. Experimental modal analysis of a multi-degree-of-freedom system with random parametric excitation and nonlinear damping," *Mechanical systems and Signal Processing* 4(3), pp. 257-268, 1990.
13. Kapitaniak, T. *Chaos in Systems with Noise*, World Scientific, Singapore, 1988.

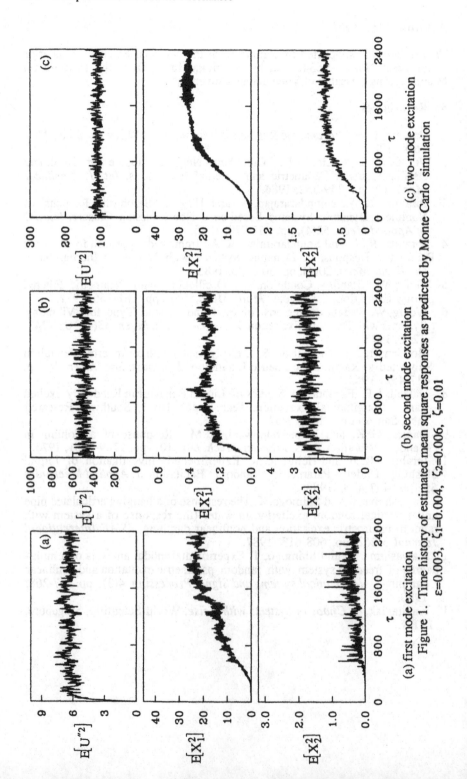

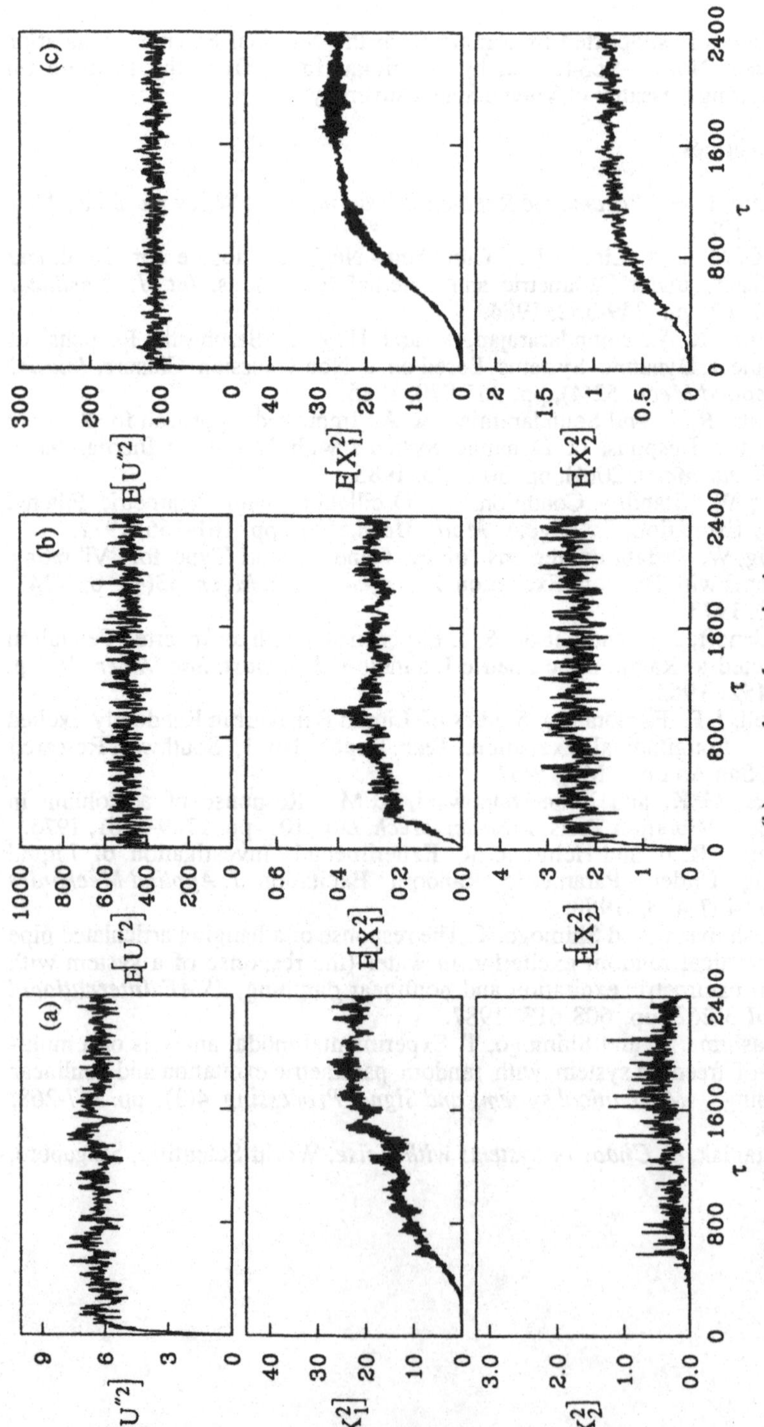

(a) first mode excitation (b) second mode excitation (c) two-mode excitation

Figure 1. Time history of estimated mean square responses as predicted by Monte Carlo simulation $\varepsilon=0.003$, $\zeta_1=0.004$, $\zeta_2=0.006$, $\zeta=0.01$

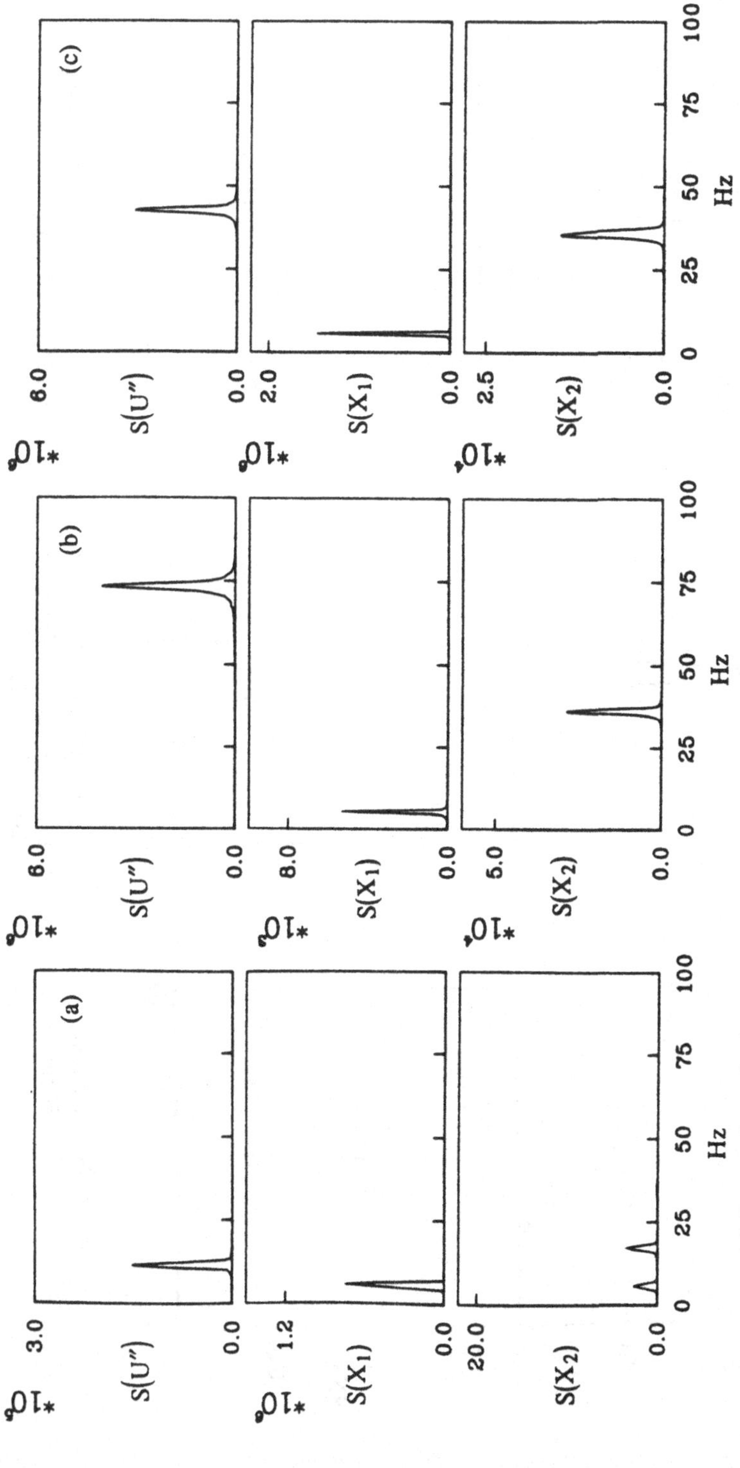

(a) first mode excitation

(b) second mode excitation

(c) two mode excitation

Figure 2. The estimated power spectral density functions from Monte Carlo simulation
ε=0.003, ζ_1=0.004, ζ_2=0.006, ζ=0.01

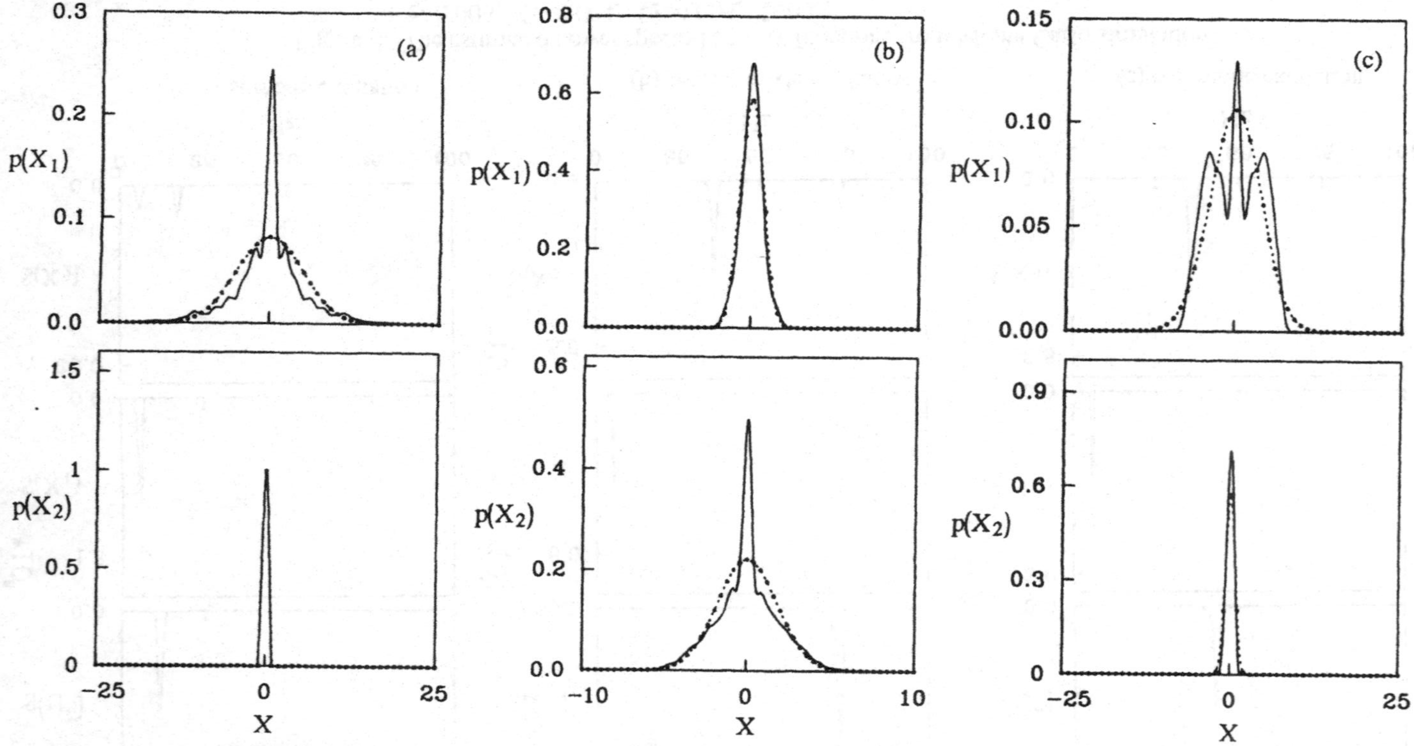

(a) first mode excitation (b) second mode excitation (c) two-mode excitation

Figure 3. The estimated probability density functions from Monte Carlo simulation $\varepsilon=0.003$, $\zeta_1=0.004$, $\zeta_2=0.006$, $\zeta=0.01$

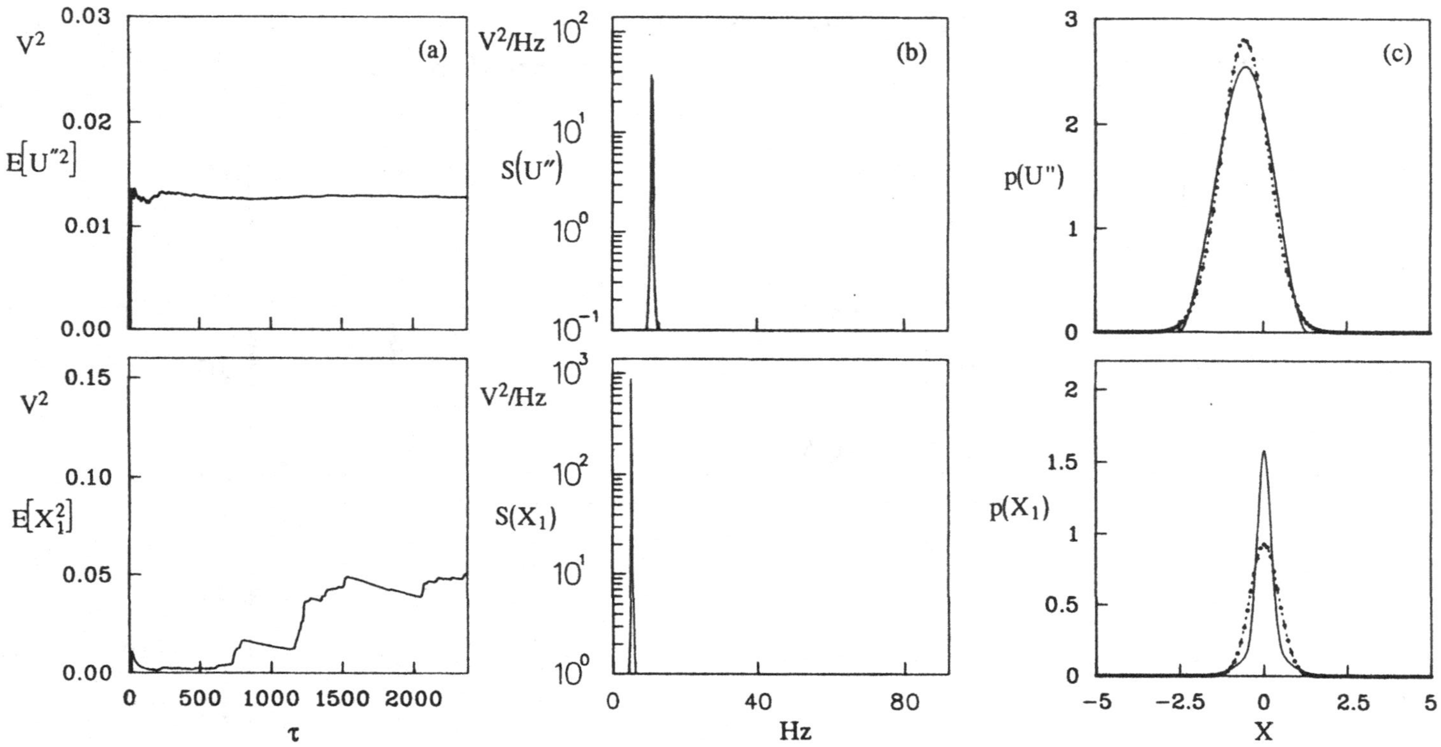

(a) time history of mean square response (b) power spectral density function (c) probability density function

Figure 4. The measured response statistics of the first mode excitation

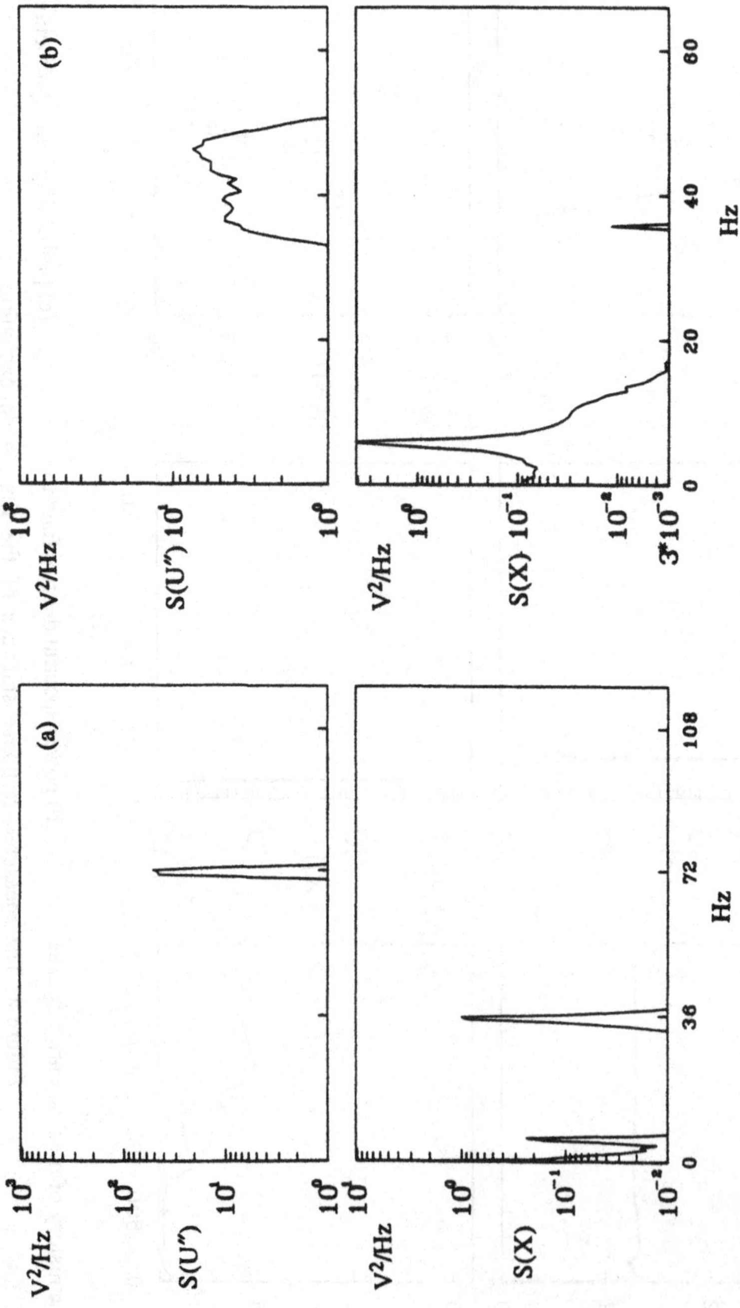

(a) second mode excitation

(b) two-mode excitation

Figure 5. The measured power spectral density functions

Random Field Interpolation Between Point by Point Measured Properties

O. Ditlevsen

Department of Structural Engineering, Technical University of Denmark, Building 118, DK 2800, Denmark

Abstract Computational practicability imposes strong restrictions on the choice of a random field model for stochastic interpolation (kriging) between measured values of a material property, say, at a large set of points in a material body. With background in an actual example of a large number of cone tip resistance measurements in clay till it is demonstrated that pragmatic considerations lead to an almost unique mathematical structure of the model. The problem of pragmatism originates from the necessity of being able to invert a very large covariance matrix a large number of times. In order to appreciate this a general introduction to random field interpolation is given.

Introduction

Interpolation between point by point measured values of a spatially distributed property has been based on statistical principles for decades within the field of mining and mineral prospecting. In this field of mining geostatistics the method is called "kriging" alluding the works of D.G. Krige quoted in Journel and Huijbregts [1]. In a framework of reasoning very much different from the one considered in the following G. Matheron [2,3] has formulated the philosophy and the principles of geostatistics that includes the principles of kriging.

In modern reliability analysis in particular in geotechnical engineering [4] there is a need to make interpolation that includes measures of uncertainty of the interpolation. This paper first reconsiders the classical problem of interpolation and the difficulty of error estimation in the classical theory except under conditions that only rarely are satisfied in practice. The modern interpretation of the kriging procedure as the formulation of a conditional random field model is next introduces as a pragmatic alternative to the classical interpolation technique. Inherently it contains a probabilistic measure of the interpolation error. The random field method of interpolation as such has been considered and applied in several papers on engineering applications [4] but generally without a discussion of a proper interpretation of the method. Herein it is attempted to present the method with a linking to classical interpolation. This philosophical basis may ease the understanding of the method as a stochastic interpolation method which is not tied to some physical process that behaves according to a random mechanism. It is merely a rational model of the uncertain engineering guess on interpolated values. It rests on the idea that continuity and differentiability principles make the variation among the measured values representative for what should be expected with respect to the uncertainty of the interpolation values.

Finally it is shown that pragmatic principles of computational practicability of random field interpolation in large regularly organized tables lead to a very narrow class of applicable correlation functions.

Classical interpolation

The solution to the problem of interpolation in a table (x_0,y_0), (x_1,y_1), ..., (x_{n-1},y_{n-1}) of values of the n times differentiable function $y = f(x)$ is in classical mathematical analysis given by Lagrange on the form

$$f(x) = y_0\, Q_0(x) + y_1\, Q_1(x) + \ldots + y_{n-1}\, Q_{n-1}(x) + R_n(x) \qquad (1)$$

in which $Q_i(x)$, $i = 0, 1, \ldots, n-1$, is the uniquely defined polynomial of (n–1)th degree that takes the value 1 for $x = x_i$ and the value 0 for $x = x_0, \ldots, x_{i-1}, x_{i+1}, \ldots, x_{n-1}$. The Lagrangian remainder $R_n(x)$ is

$$R_n(x) = (x-x_0) \cdot \ldots \cdot (x-x_{n-1}) \frac{f^{(n)}(\zeta)}{n!} \qquad (2)$$

in which ζ is some number contained within the smallest interval I that contains $x, x_0, \ldots, x_{n-1}$. Interpolation to order $n-1$ then consists of using (1) with the remainder neglected. The error committed by the interpolation is bounded in absolute value by

$$|R_n(x)| \le \frac{|(x-x_0)\cdot\ldots\cdot(x-x_{n-1})|}{n!} \max_{\zeta\in I} \left|f^{(n)}(\zeta)\right| \qquad (3)$$

Thus the error evaluation in classical interpolation is not offering a model of the error in terms of a random variable but only an upper bound that can be highly conservative. Moreover, the evaluation of the bound (3) requires that the nth derivative can be obtained. This, however, is in contrast to most situations in practice where interpolation is needed. Often only the table (x_0,y_0), (x_1,y_1), ..., (x_{n-1},y_{n-1}) is given and the values of the function f(x) are unknown between these points. This is the typical situation when the table is obtained by point by point measurement of some physical quantity that varies with x in an unknown way. The table may also be the result of a lengthy computation. In principle f(x) may be computed for any needed value of x but it may be more economical to accept even a considerable error in the evaluation of f(x) as obtained by interpolation between already computed values. The same may hold in the case of values obtained by measurements. In some situations it may even be impossible to measure intermediate values (as in the obvious case of x being the time but sometimes also in cases where x is a spatial coordinate). With only the table of points given the interpolation procedure is usually guided by a principle of simplicity perhaps supported by physical and geometrical arguments. To appreciate the philosophy of the principle of simplicity behind practical interpolation it is only needed to note that there is an infinite set of interpolation functions passing through the given points. Besides satisfying differentiability requirements to some specified order this set is obviously characterized by the property that the difference between any two functions of the set is a function with zero points at $x_0, x_1, \ldots, x_n$. In practice the choice of the interpolation function from the set is usually made in view of choosing an interpolation function that in itself and in all the derivatives up to the specified order has as small a variability as possible. Even though this principle of simplicity of the interpolation may be given a precise mathematical definition and it thereafter may be demonstrated that it leads to a unique choice of the interpolation function, it is still just an arbitrary principle that does not give any indication of the interpolation error.

A solution to this problem can be obtained by introducing an alternative but somewhat similar principle of simplicity. It is based on the principles of statistical reasoning.

Random field interpolation. Maximum likelihood principle of choice

Let $J(y_0,\ldots,y_{n-1})$ be the set of interpolation functions and consider the union

$$\mathcal{J} = \bigcup_{z \in \mathbb{R}^n} J(z) \tag{4}$$

in which $z = (z_0,\ldots,z_{n-1})$. In case it is only required that any interpolation function is n times continuously differentiable, then $\mathcal{J}$ is the class of n times continuously differentiable functions. Define a probability measure over $\mathcal{J}$ that depends on a number of parameters $\theta_1, \ldots, \theta_q$. For each value set of $\theta_1, \ldots, \theta_q$ we then have a random field over the range of values of x. Let the probability measure have such properties that there is a probability density at the set J(z) for each $z \in \mathbb{R}^n$ in the sense of being the unique limit

$$\lim_{d \to 0} \left\{ \frac{P\left[\bigcup_{\zeta \in \mathcal{N}(z)} J(\zeta) \middle| \theta_1,\ldots,\theta_q\right]}{\text{Vol}[\mathcal{N}(z)]} \right\} \tag{5}$$

where d = diameter of $\mathcal{N}(z)$, $\mathcal{N}(z)$ is an arbitrary neighborhood of z with volume $\text{Vol}[\mathcal{N}(z)]$, and $P[\cdot \mid \theta_1,\ldots,\theta_q]$ is the probability measure.

The principle of simplicity now concerns a principle about how to choose the values of the parameters $\theta_1, \ldots, \theta_q$ that fix the probability measure on $\mathscr{F}$. Instead of the deterministic principle of least variability it is reasonable to choose $\theta_1, \ldots, \theta_q$ so that the probability density has its maximum for $\mathbf{z} = \mathbf{y} = (y_o, y_1, \ldots, y_{n-1})$, that is, at the actual set of interpolation functions. This is the well–known principle of maximum likelihood estimation in the theory of mathematical statistics. When the parameter values are chosen for example according to the principle of maximum likelihood, the probability measure on $\mathscr{F}$ induces a probability measure in the relevant set of interpolation functions J($\mathbf{y}$) simply as the conditional probability measure given J($\mathbf{y}$), that is, given $\mathbf{z} = \mathbf{y}$. Thus a conditional random field is defined that possesses the desired properties of expressing the interpolated value at x of the unknown function as a random variable. At the points of the given table (x_o, y_o), ..., (x_{n-1}, y_{n-1}) the random variable degenerate to be atomic, but at a point x different from the points of the table it gets a mean and a standard deviation that depend on x. The detailed nature of this dependency is laid down in the mathematical structure of the probability measure introduced in $\mathscr{F}$.

From an operational point of view the most attractive probability measure to choose is the Gaussian measure. Let the Gaussian measure on $\mathscr{F}$ have mean value function $\mu(x)$, $x \in \mathbb{R}$, and covariance function $c(\xi,\eta)$, $\xi, \eta \in \mathbb{R}$. Then the conditional measure on J($\mathbf{y}$) is Gaussian with mean value function given by the linear regression

$$E[Y(x)\,|\,\mathbf{z}=\mathbf{y}] = \mu(x)+[c(x,x_o)\ldots c(x,x_{n-1})]\{c(x_i,x_j)\}^{-1}\begin{bmatrix} y_o & -\mu(x_o) \\ \vdots & \\ y_{n-1} & -\mu(x_{n-1}) \end{bmatrix} \tag{6}$$

and covariance function given by the residual covariance function corresponding to the linear regression. It is

$$\mathrm{Cov}[Y(\xi),Y(\eta)\,|\,\mathbf{z}=\mathbf{y}] = c(\xi,\eta)-[c(\xi,x_o)\ldots c(\xi,x_{n-1})]\{c(x_i,x_j)\}^{-1}\begin{bmatrix} c(\eta,x_o) \\ \vdots \\ c(\eta,x_{n-1}) \end{bmatrix} \tag{7}$$

Comparison with classical interpolation

The conditional mean (6), in particular, is written out explicitly in order to display an interpretation of the mean value function $\mu(x)$ and the function $c(\xi,\eta)$ that relates them to usual deterministic interpolation practice. In fact, if (1) and (6) are compared, it is seen that (6) is obtained from (1) if the polynomial $Q_i(x)$, $i = 0, \ldots, n-1$, is replaced by the linear combination

$$a_{oi}\, c(x,x_o) + a_{1i}\, c(x,x_1) + \ldots + a_{(n-1)i}\, c(x,x_{n-1}) \tag{8}$$

where the coefficients a_{oi}, ..., $a_{(n-1)i}$ are the elements of the ith column in the inverse to the covariance matrix $\{c(x_i,x_j)\}$. It is seen that this function like $Q_i(x)$ is 1 for $x = x_i$ and 0 for $x = x_o, ..., x_{i-1}, x_{i+1}, ..., x_{n-1}$. Indeed, if (8) is identified with $Q_i(x)$ for $i = 0, ..., n-1$, the functions $c(x,x_o), ..., c(x,x_{n-1})$ are uniquely determined by their values at x_o, $x_1, ..., x_{n-1}$ as

$$c(x,x_i) = c(x_o,x_i)\, Q_o(x) + ... + c(x_{n-1},x_i)\, Q_{n-1}(x) \tag{9}$$

As function of x this is a valid covariance, that is, there exists a non–negative definite function $c(\xi,\eta)$ such that (9) is obtained for $(\xi,\eta) = (x,x_i)$. This is seen directly by computing the covariance function of the random field

$$Z_o\, Q_o(x) + ... + Z_{n-1}\, Q_{n-1}(x) \tag{10}$$

in which $(Z_o,...,Z_{n-1})$ is a random vector with covariance matrix $\{c(x_i,x_j)\}$. The covariance function is

$$c(\xi,\eta) = [Q_o(\xi) \;...\; Q_{n-1}(\xi)]\, \{c(x_i,x_j)\} \begin{bmatrix} Q_o(\eta) \\ \vdots \\ Q_{n-1}(\eta) \end{bmatrix} \tag{11}$$

which gives (9) for $\xi = x$ and $\eta = x_i$.

It follows from this that (1) is a special case of (6). The remainder becomes replaced by the term

$$\mu(x) - [c(x,x_o) \;...\; c(x,x_{n-1})]\, \{c(x_i,x_j)\}^{-1} \begin{bmatrix} \mu(x_o) \\ \vdots \\ \mu(x_{n-1}) \end{bmatrix} \tag{12}$$

plus a Gaussian zero mean random field with covariance function given by (7). This interpretation of the mean value function $\mu(x)$ and the covariance function $c(\xi,\eta)$ of the random field as essentially being interpolation functions makes it easier to appreciate the consequence of a specific mathematical form of the covariance function. For example, if the general appearance of the values in the table suggests the choice of a homogeneous Gaussian field, the choice of a correlation function like $\exp[-\alpha|\xi-\eta|]$ implies that the mean interpolation function given by (6) will be non–differentiable at $x = x_o, ..., x_{n-1}$ while a correlation function like

$\exp[-\beta(\xi-\eta)^2]$ will give a differentiable interpolation function also at $x = x_0, \ldots, x_{n-1}$. In fact, the sample functions corresponding to the first correlation function will with probability 1 be continuous but not differentiable at any point while the second correlation function corresponds to sample functions that are differentiable of any order.

Example 1 Consider the case where the points $x_0, \ldots, x_{n-1}$ are equidistant with $x_{i+1} - x_i = h$. Let the random interpolation field on the line be a homogeneous Gaussian field with mean μ and covariance function

$$c(\xi,\eta) = \begin{cases} \left[1 - \frac{|\xi-\eta|}{h}\right] \sigma^2 & \text{for } |\xi-\eta| \leq h \\ 0 \text{ otherwise} \end{cases} \tag{13}$$

Then (6) gives

$$E[Y(x)\,|\,z=y] = \frac{1}{h}\left[(x_1-x)\,y_0+(x-x_0)\,y_1\right] \tag{14}$$

for $x_0 \leq x \leq x_1$. This is equivalent with linear interpolation in the mean. The covariance function (7) becomes

$$\mathrm{Cov}[Y(\xi),Y(\eta)\,|\,z=y] =$$

$$\begin{cases} \left[\frac{\sigma}{h}\right]^2 [h^2-h|\xi-\eta|-(x_1-\xi)(x_1-\eta)-(x_0-\xi)(x_0-\eta)] & \text{for } |\xi-\eta| \leq h \\ 0 \text{ otherwise} \end{cases} \tag{15}$$

from which the standard deviation

$$D[Y(x)\,|\,z=y] = \frac{\sigma}{h}\sqrt{2(x-x_0)(x_1-x)} \tag{16}$$

is obtained. It is interesting to compare this measure of the interpolation error with the Lagrangian remainder obtained from (2) for $n = 2$. It is seen that (16) varies like the square root of $R_2(x)$, that is, it predicts larger errors close to x_0 or x_1 than $R_2(x)$ does.

While the interpolation functions in the classical procedure are broken at the points of the table implying that no way is indicated of reasonable extrapolation outside the range from x_0 to x_{n-1}, the homogeneous random field procedure represents a solution to both the interpolation and the extrapolation problem in terms of a set of sample functions defined everywhere. The points of the table only play the role that all the sample functions of the set pass through the points of the table. The homogeneity of the field implies that the variability of the points within the range from x_o to x_{n-1} is reflected in the field description outside this range. Specifically we have for $x \leq x_o$

$$E[Y(x)|\mathbf{z}=\mathbf{y}] = \begin{cases} \mu & \text{for } x \leq x_o - h \\ \frac{1}{h}[(x_o-x)\,\mu+(x-x_o+h)\,y_o] & \text{for } x_o - h < x \leq x_o \end{cases} \tag{17}$$

and similarly for $x > x_{n-1}$. The standard deviation is

$$D[Y(x)|\mathbf{z}=\mathbf{y}] = \begin{cases} \sigma & \text{for } x \leq x_o - h \\ \frac{\sigma}{h}\sqrt{(x_o-x)(x-x_o+2h)} & \text{for } x_o - h < x \leq x_o \end{cases} \tag{18}$$

and similarly for $x > x_{n-1}$. Outside the interval from $x_o - h$ to $x_{n-1} + h$ the field is simply given as the homogeneous field. The variation among $y_o, \ldots, y_{n-1}$ determines the values of μ and σ^2 as the well–known maximum likelihood estimates

$$\bar{y} = \frac{1}{n}\sum_{i=0}^{n-1} y_i \quad \text{and} \quad s^2 = \frac{1}{n}\sum_{i=0}^{n-1}(y_i-\bar{y})^2 \tag{19}$$

corresponding to a Gaussian sample of independent outcomes $y_o, \ldots, y_{n-1}$.

Bayesian principle of choice

The maximum likelihood principle leads to a specific choice of the distributional parameters $\theta_1, \ldots, \theta_q$, that is, it leaves no room for doubt about what values to choose. Such doubt can be included in the stochastic interpolation procedure by assigning not just the distribution family to the interpolation problem but also assigning a joint probability distribution to $\boldsymbol{\theta} = (\theta_1,\ldots,\theta_q)$. The mathematical technique then becomes exactly the same as in the Bayesian statistical method: The joint probability density of the parameter vector $\boldsymbol{\theta}$ and any vector of field random variables that includes the random variables $Y(x_o), \ldots, Y(x_{n-1})$ as a subvector is obtained as the product of the given relevant conditional probability density of the field random variables given the parameters and the assigned joint density of the parameter vector (the prior). With only the interpolation function set $J(\mathbf{y})$ being relevant, the conditional joint density of the parameter vector and the field random variables given $(Y(x_o),\ldots,Y(x_{n-1})) = (y_o,\ldots,y_{n-1})$ is obtained except for proportionality by only considering the joint density at points for which $Y(x_0)$, ..., $Y(x_{n-1})$ are fixed at y_0, ..., y_{n-1} respectively. Integration over $\mathbb{R}^q$ with respect to the parameters then finally gives the joint distribution of the field random variables (the predictive distribution). The family of all predictive distributions obtained in this way defines the random field of interpolation and extrapolation functions such that it includes the statistical uncertainty effect of the size of the given table (x_0,y_0), ..., (x_{n-1},y_{n-1}).

Example 2 The parameter vector (μ,σ) in example 1 may be considered as an outcome of the random vector (M,Σ). Assuming a non–informative prior of $(M,\log \Sigma) \in \mathbb{R}^2$ (degenerate uniform density over $\mathbb{R}^2$), the special random field model of example 1 leads to the well–known Bayesian standard results in Gaussian statistics. For example, in the interpolation case the predictive distribution of

$$\frac{Y(x) - \frac{1}{h}\,[(x_1-x)y_0+(x-x_0)y_1]}{\frac{s}{h}\,\sqrt{2(x-x_0)(x_1-x)}}\,\sqrt{1-\frac{1}{n}} \tag{20}$$

for $x \in \,]x_0,x_1[$ is the t–distribution with $n-1$ degrees of freedom.

Computational practicability

Both for the maximum likelihood principle and the Bayesian principle of stochastic interpolation on the basis of a Gaussian field assumption the governing mathematical function is the joint Gaussian density

$$f_{\mathbf{Y}}(y_0,\dots,y_{n-1};\mu,\sigma,\mathbf{P}) \propto \frac{1}{\sqrt{\det(\mathbf{P})}}\,\frac{1}{\sigma^n}\exp\left[-\frac{1}{2\sigma^2}(\mathbf{y}-\mu\,\mathbf{e})'\,\mathbf{P}^{-1}(\mathbf{y}-\mu\,\mathbf{e})\right] \tag{21}$$

written here for the case of a homogeneous field with mean μ, standard deviation σ, and a correlation function that defines the correlation matrix $\mathbf{P}$ of the random vector $(Y(x_0),\dots,Y(x_{n-1}))$. The correlation matrix $\mathbf{P}$ contains in its elements the unknown parameters of the correlation function. As function of μ, σ and the correlation parameters the right side of (21) defines the likelihood function. Generally neither $\mathbf{P}^{-1}$ nor the determinant $\det(\mathbf{P})$ can be expressed explicitly in terms of the correlation parameters. Thus in order to maximize the likelihood with respect to the parameters both $\mathbf{P}^{-1}$ and $\det(\mathbf{P})$ must be evaluated several times during an iteration procedure.

The Bayesian principle of stochastic interpolation has particular relevance for structural reliability evaluations of structures with carrying capacities that depend on the strength variation throughout a material body. For example, a direct foundation on saturated clay may fail due to undrained failure extended over a part of the clay body. The evaluation of the failure probability corresponding to a specific rupture figure requires a specifically weighted integration of the undrained shear strength across the rupture figure. If the undrained shear strength is measured only at a finite number of points in the clay body, it is required to make interpolation and or extrapolation in order to obtain the value of the integrand in any relevant point of the body. Irrespective of whether the maximum likelihood principle is applied or whether the Bayesian principle is applied in order to properly take statistical uncertainty into account, it is required that $\mathbf{P}^{-1}$ be computed iteratively for different values of the correlation parameters. For example, for the Bayesian principle of choice this is the case when the reliability is evaluated by a first or second order reliability method (FORM or SORM) in a space of basic variables that among its dimensions includes the correlation parameters. In the search for the most central limit state point by some gradient method, say, the inverse $\mathbf{P}^{-1}$ has to be computed iteratively several times for different parameter values.

From this discussion it follows that computational practicability put limits to the order of $\mathbf{P}$ or it requires that the mathematical structure of $\mathbf{P}$ is such that $\mathbf{P}$ can be inverted analytically or such that $\mathbf{P}^{-1}$ can be obtained in terms of matrices of considerably lower order than the order of $\mathbf{P}$. A possibility of breaking down to lower order is present in case of what here is called factorized correlation structure:

Let $\mathbf{P}$ and $\mathbf{Q} = \{q_{ij}\}_{mm}$ be correlation matrices. Then the matrix defined as

$$[\mathbf{Q}] \circ \mathbf{P} \equiv \begin{bmatrix} q_{11}\mathbf{P} & q_{12}\mathbf{P} & \cdots \\ q_{21}\mathbf{P} & q_{22}\mathbf{P} & \cdots \\ \cdots & & q_{mm}\mathbf{P} \end{bmatrix} \tag{22}$$

is a correlation matrix. The proof is as follows: Let $\mathbf{x}' = [\mathbf{x}_1' \ldots \mathbf{x}_m']$ be an arbitrary vector and let $\mathbf{A}$ be an orthogonal matrix such that $\mathbf{A}'\mathbf{P}\mathbf{A} = \mathbf{\Lambda} = \lceil \lambda_1 \ldots \lambda_n \rfloor$ where $\lambda_1, \ldots, \lambda_n$ are the non–negative eigenvalues of $\mathbf{P}$. Then

$$\mathbf{x}'([\mathbf{Q}] \circ \mathbf{P})\,\mathbf{x} = \sum_{i=1}^{m} \sum_{j=1}^{m} q_{ij}\,\mathbf{x}_i'\,\mathbf{P}\,\mathbf{x}_j = \sum_{i=1}^{m} \sum_{j=1}^{m} q_{ij}\,\mathbf{y}_i'\,\mathbf{\Lambda}\,\mathbf{y}_j =$$

$$\lambda_1 \sum_{i=1}^{m} \sum_{j=1}^{m} q_{ij}\, y_{ij}\, y_{j1} + \ldots + \lambda_n \sum_{i=1}^{m} \sum_{j=1}^{m} q_{ij}\, y_{in}\, y_{jn} \tag{23}$$

in which $\mathbf{y}_i = \mathbf{A}\,\mathbf{x}_i$, $i = 1, \ldots, m$. Since $\mathbf{Q}$ is a correlation matrix, the right hand side of (23) is non–negative. Thus it follows that $[\mathbf{Q}] \circ \mathbf{P}$ is a correlation matrix.

It follows directly by the multiplication test that if **P** and **Q** are both regular, then

$$([\mathbf{Q}]\circ\mathbf{P})^{-1} = [\mathbf{Q}^{-1}] \circ \mathbf{P}^{-1} \tag{24}$$

By application of simple row operations first diagonalizing **P** to **D** in all places in (22) to obtain

$$\begin{bmatrix} q_{11}\mathbf{D} & q_{12}\mathbf{D} & \cdots \\ q_{21}\mathbf{D} & q_{22}\mathbf{D} & \cdots \\ \cdots & & \end{bmatrix} \tag{25}$$

and next diagonalizing (25) by exactly those row operations that diagonalize **Q**, it follows that

$$\det([\mathbf{Q}]\circ\mathbf{P}) = \det(\mathbf{Q})[\det(\mathbf{P})]^m \tag{26}$$

Isotropic factorized correlation structure in the plane

Assume that the table of points corresponds to points in the plane arranged in a square mesh of equidistant points in both directions with mesh width L. Let there be k points in the first direction and ℓ points in the second direction in total giving $k\ell$ points with given values of the otherwise unknown function. The field random variables $Y_{11}, Y_{21}, \ldots, Y_{k1}, Y_{12}, Y_{22}, \ldots, Y_{k2}, \ldots, Y_{1\ell}, Y_{2\ell}, \ldots, Y_{k\ell}$ corresponding to the points (i,j) of the mesh are collected in a vector of dimension $k\ell$ and in the order as indicated. Then it is easily shown that the correlation matrix of order $k\ell$ for an arbitrary choice of L has the factorized structure as in (22) if and only if the random field is correlation homogeneous and the correlation function of the field can be written as the product of a correlation function solely of the coordinate along the first mesh direction and a correlation function solely of the coordinate along the second direction. In the class of such product correlation functions the correlation functions that correspond to isotropic fields are uniquely given as

$$\exp[-\beta r^2] \tag{27}$$

where r is the distance between the two points at which the random variables are considered and β is a free positive parameter. Thus the assumption of isotropy of the interpolation field and the requirement of computational practicability for large $k\ell$ make it almost a "must" to adopt the correlation function defined by (27). Writing (27) for r = L as the correlation coefficient κ, the correlation matrix of the vector of Y–values becomes $[S_1] \circ S_2$ where

$$S_i = \begin{bmatrix} 1 & \kappa & \kappa^4 & \dots & \kappa^{(\nu-1)^2} \\ \kappa & 1 & \kappa & \dots & \kappa^{(\nu-2)^2} \\ \dots \end{bmatrix} \tag{28}$$

i = 1 (ν = k), i = 2 ($\nu = \ell$). These considerations generalize directly to 3 dimensions for a spatial equidistant mesh of kℓm points. Then the correlation matrix becomes $[[S_1]\circ S_2] \circ S_3$ with each of the matrices S_1, S_2, S_3 of form as in (28) in case of spatial isotropy.

For a soil body it may be reasonable to have isotropy in any horizontal plane but not necessarily in the 3–dimensional space. Then S_3 may be any other correlation matrix obtained from the correlation function in the vertical direction. When soil strength measurements are made for example by the socalled CPT method (Cone Penetration Test), the distance h between measurement points is much smaller in the vertical direction than the mesh width L in the horizontal direction. Then S_3 can be of impracticable high order. However, there is at least one case of such a matrix S corresponding to a homogeneous field for which S^{-1} is known explicitly. This case is

$$S = \begin{bmatrix} 1 & \rho & \rho^2 & \dots & \rho^m \\ \rho & 1 & \rho & \dots & \rho^{m-1} \\ \dots \end{bmatrix}; \; S^{-1} = \frac{1}{1-\rho^2} \begin{bmatrix} 1 & -\rho & 0 & \dots & 0 \\ -\rho & 1+\rho^2 & -\rho & \dots & 0 \\ \dots \end{bmatrix} \tag{29}$$

which corresponds to a homogeneous Markov field on the line with $\rho = \exp[-\alpha h]$ where α is a free positive parameter.

This homogeneous Gaussian 3–dimensional field interpolation model with (27) defining the horizontal correlation properties and with Markov properties in the vertical direction has recently been used in a reliability study of the anchor blocks for the future suspension bridge across the eastern channel of Storebælt in Denmark [5]. The size of the interpolation problem in that investigation was k = 6, ℓ = 10, m = 150 giving a table of 9000 points. Without the unique structuring of the random field correlation given here, the likelihood analysis would have been totally impracticable.

Acknowledgement

This work has been financially supported by the Danish Technical Research Council.

References

1. Journel, A.B. and Huijbregts, C.J., Mining Geostatistics, Academic Press, New York, 1978.

2. Matheron, G., Principles of Geostatistics, Economic Geology, Vol. 58, pp. 1246–1266, 1963.

3. Matheron, G., Estimating and Choosing (transl. from French by A.M. Hasofer), Springer Verlag, Berlin–Heidelberg, 1989.

4. Keaveny, J.M., Nadim, F. and Lacasse, S., Autocorrelation Functions for Offshore Geotechnical Data, in Structural Safety & Reliability, Vol. 1 (Ed. Ang, A.H.–S., Shinozuka, M. and Schuëller, G.I.), pp. 263–270, Proceedings of ICOSSAR '89, the 5th International Conference on Structural Safety and Reliability, San Francisco, August 7–11, 1989. ASCE, New York, 1990.

5. Ditlevsen, O. and Gluver, H., Parameter Estimation and Statistical Uncertainty in Random Field Representations of Soil Strengths, in Proceedings of CERRA–ICASP6, the Sixth International Conference on Applications of Statistics and Probability in Civil Engineering, México City, México, June, 1991.

Application of Dynamic Response Analysis in Calculation of Natural Frequencies

S. Moossavi Nejad (*), G.D. Stefanou (**)
(*) Polytechnic of Central London, U.K.
(**) University of Patras, Greece

ABSTRACT

The application of dynamic response analysis in calculation of the natural frequencies and mode shapes of linear and nonlinear structures are presented and an example is given for a nonlinear structure, with distinct frequencies which may change with amplitudes.

INTRODUCTION

Calculation of natural frequencies by solving the eigenvalue problem normally requires setting up of large matrices and the solution of the characteristic equations. The dynamic response analysis by minimisation of total potential work however is a vector based analysis and therefore requires relatively smaller computer time and space. This method is used to calculate the response of structures to an impulse load in the desired direction, the results are then analysed to calculate the frequency components and the mode shapes. An example is given for a cable net, solved by both conventional eigenvalue analysis and the dynamic response analysis.

THEORETICAL BACKGROUND

The dynamic response analyses due to the impulse loading are carried out using a step by step response calculation in the time domain for which equilibrium of the dynamic forces at the end of each time increment is established by minimisation of the total potential dynamic work. This method was developed by Bucholdt Moossavi-Nejad (1) and further developed to implement various types of dynamic loading and in particular turbulent wind loading (2).

In order to achieve maximum performance and convergence, a scaling technique and a nonlinear steplength control are built into the algorithm. Stability is ensured for larger time steps by using an appropriate set of difference equations. The size of the time step, however, is limited by the accuracy required and the frequency components of the dynamic load.

Although the algorithm does not require the natural frequencies of the structure, it is desirable to calculate those for wind or earthquake loading, since it can be established whether or not the frequency range of frequency components of dynamic loading covers the dominant natural frequencies of the structure.

THE NET

The numerical model used for this work is a 7x5 flat net, shown in Fig(1). The flat net was chosen since it is both nonlinear and can have a highly accurate mathematical model.

The specification of the net is given bellow.

Overall dimensions	3000x4000 mm
Spacing of cables in plan	500 mm
Effective cross sectional area of cable	2 mm^2
Equivalent Young Modulus of cable	105 kN/mm^2
Weight of cable	.167E-3 N/mm
Number of cable links	72
Number of free cable joints	35
Number of boundary joints	24
Number of degrees of freedom	105
Logarithmic damping ratio	3.0%

DESCRIPTION OF THE NUMERICAL WORK

In order to compare the natural frequencies, calculated by the eigenvalue analysis and those obtained by dynamic response analysis, the eigenvalues of the net were calculated and the mode shapes were obtained. Figs(2 and 3) show modes 1 and 2 of the net respectively. The frequencies appear in bands and one value for each band bellow 10 Hz is shown in Table 1 together with those obtained by dynamic analysis.

The value shown by dashes did not appear in spectrum due to the direction of the load release.

Table 1, frequencies of the net.

Mode No.	By Eigenvalue Analysis	By Dynamic Analysis
1	2.571	2.56
2	3.195	3.20
3	3.668	----
4	4.108	4.10
7	5.008	5.00
12	5.889	5.78
18	6.709	6.70
24	7.105	7.05
30	7.898	7.85
34	9.193	9.15

The net was subjected to a load release of 50 N from joints 18, the centre joint having the largest amplitude in mode 1, and joint 16 having the largest amplitude in mode 2. For each case the response of the net was calculated for a period of 10 seconds with a time step of 0.005 seconds.

Frequency spectrum for the response of joints 16, 17 and 18 were calculated for each case in the range of 0 to 10 Hz. The response of each joint and its frequency spectrum are shown in Figs(4,5,6) for the release from the centre joint and in Figs(7,8,9) for the release from joint 16.

In each case the spectrum contains a number of distinct frequencies bellow 10 Hz with one or two frequencies having significantly large spectrum values. This indicates that the net has vibrated in a mixture of modes with the dominant mode related to the direction of the load release. The number of dominant frequencies in the spectrum depend on the joint and the contribution of different modes to its amplitude. This can be observed in Figs(7 and 8) where joints 16 and 17 respond to the release of load from joint 16 with a mixture of the modes 1 and 2, but joint 16 shows larger spectrum value for the frequency of mode 2 and joint 17 shows larger spectrum value for mode 1. The response of joint 18 contains no contribution from the second mode as can be expected by observing the mode shapes. Although in each case one or two modes are dominant, most of the other frequencies are preset.

CONCLUSIONS

The response of the net was calculated for the release of a load, an impulse load, using minimisation of total potential work. The frequencies were calculated by two methods and compared. The results show close comparison and indicate that frequencies can be obtained, for the desired mode, by applying an impulse load to the structure on the joints having largest amplitude of that mode and in the direction of that amplitude. These frequencies also show good correlation with those obtained experimentally, which are given in (3). This method is especially appropriate for lower modes of vibration where the mode shapes are related to the deformation of the structure under static loading.

Although it is desirable to make comparison between the computer time and space required by each method to carry out these analyses, the availability and the range of computers and algorithms to perform these analyses are formidable and may not lead to a clear conclusion. However for large scale space frame structures with rigid joints, setting up of the characteristic matrix and its solution may require significantly larger computer space than those required by the minimisation method. The above analyses were carried out on a microcomputer with an 80-386 processor.

REFERENCES

1- Buchholdt H. A. and Moossavi Nejad S. Nonlinear Dynamic Analysis Using Conjugate Gradients, pp 44-52, Eng. Structures, Vol 4, 1982.

2- Moossavi Nejad, S. Implementation of Wind Loading in Time Domain Dynamic Response Analysis by Minimisation of Total Potential Work, pp 261 to 268, Proceedings of the Asia Pacific Symposium on Wind Engineering, Roorkee, India, 1985.

3- Moossavi Nejad S. A Gradient Method for Dynamic Analysis of Tension Structures, PhD Thesis, CNAA 1982.

Fig(1)

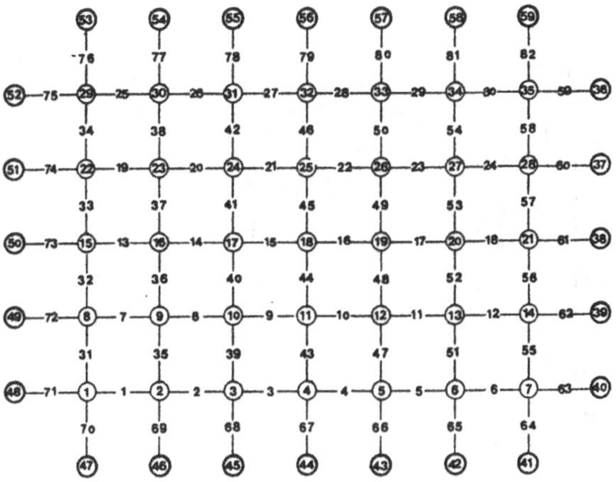

Numbering of links and joints

Fig(2)

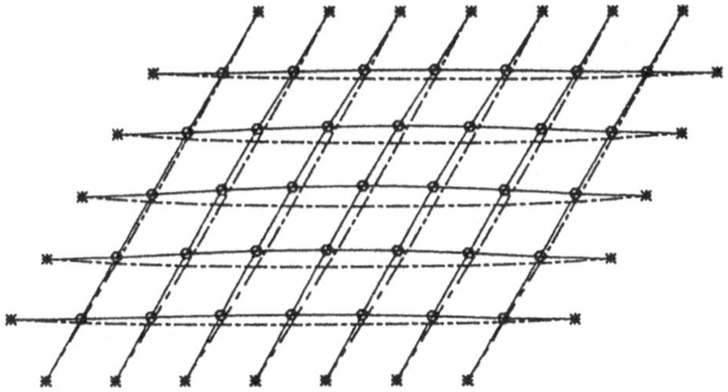

7x5 Net Mode 1

Fig(3)

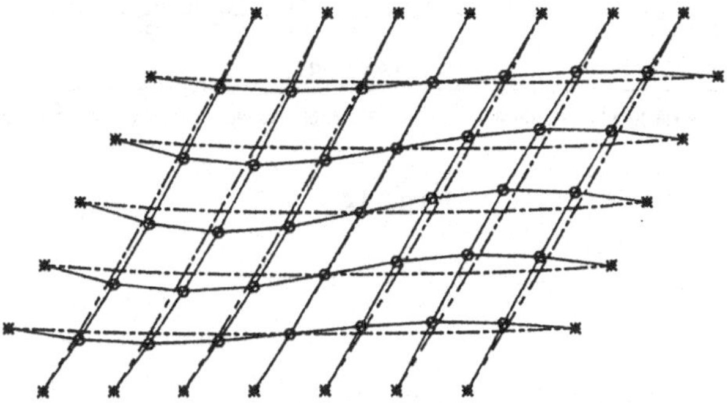

7x5 Net Mode 2

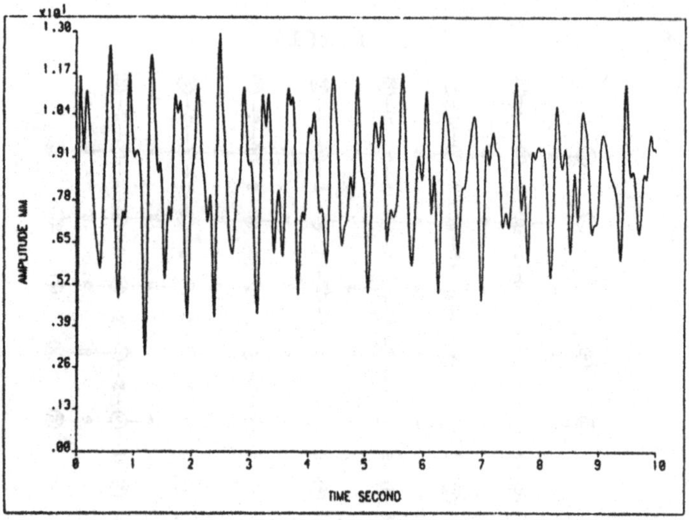

RESPONSE OF 7x5 NET TO 50N RELEASED FROM J18 3% DAMP. J 18

Fig(4a)

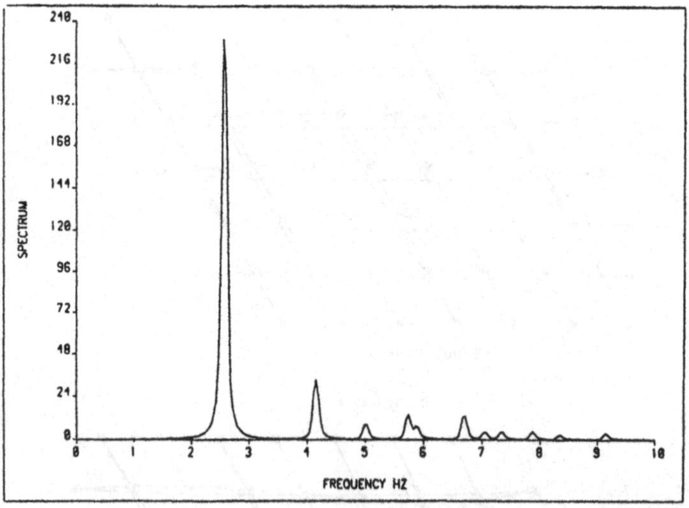

REFREQUENCY SPECTRUM OF THE RESPONSE OF 7x5 NET TO 50N R. J18 3% J 18

Fig(4b)

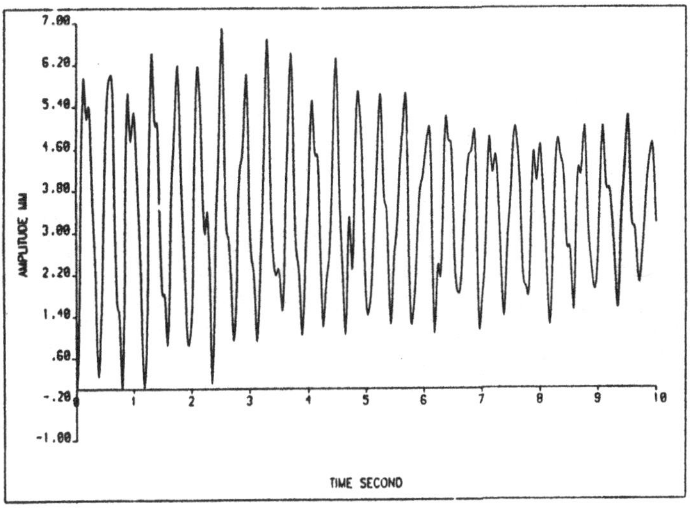

RESPONSE OF 75NET TO 50N RELEASED FROM J18 3% DAMP. J 17

Fig(5a)

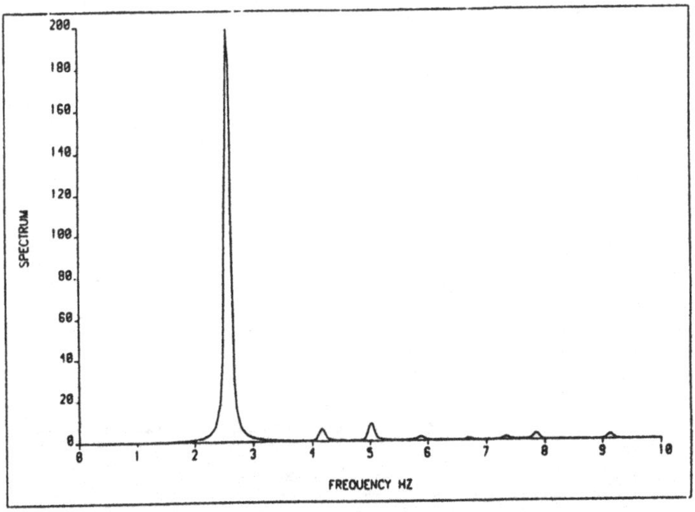

REFREQUENCY SPECTRUM OF THE RESPONSE OF 7x5 NET TO 50N R. J18 3% J17

Fig(5b)

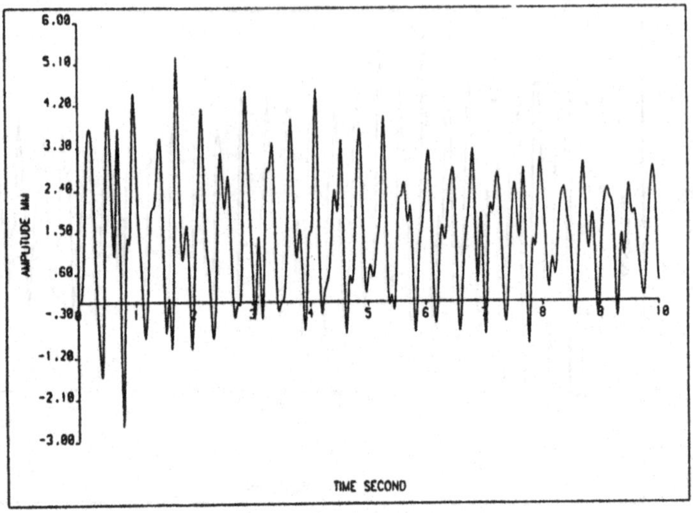

RESPONSE OF 7x5 NET TO 50N RELEASED FROM J18 3% DAMP. J 16

Fig(6a)

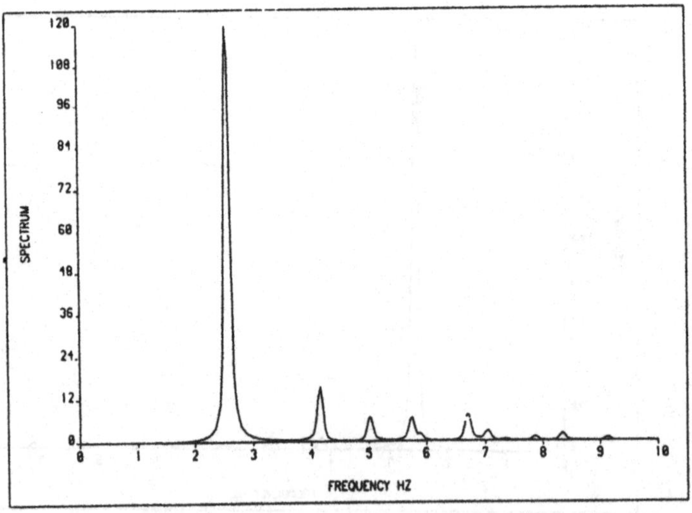

REFREQUENCY SPECTRUM OF THE RESPONSE OF 7x5 NET TO 50N R. J18 3% J 16

Fig(6b)

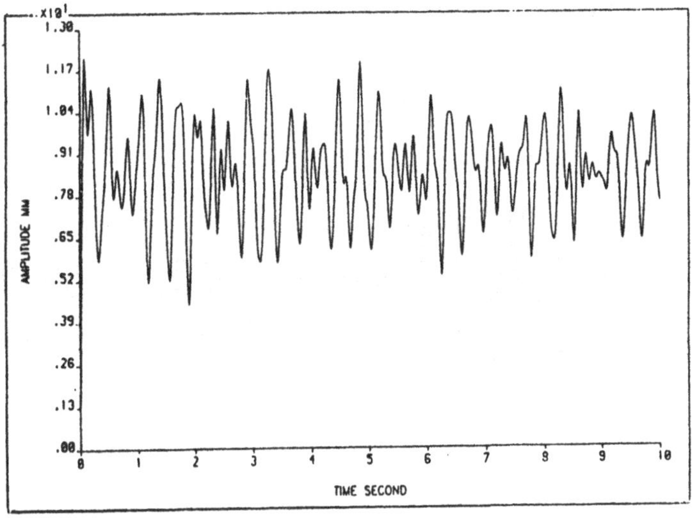

RESPONSE OF 7x5 NET TO RELEASE OF 50 N FROM J16 3% DAMP J 16

Fig(7a)

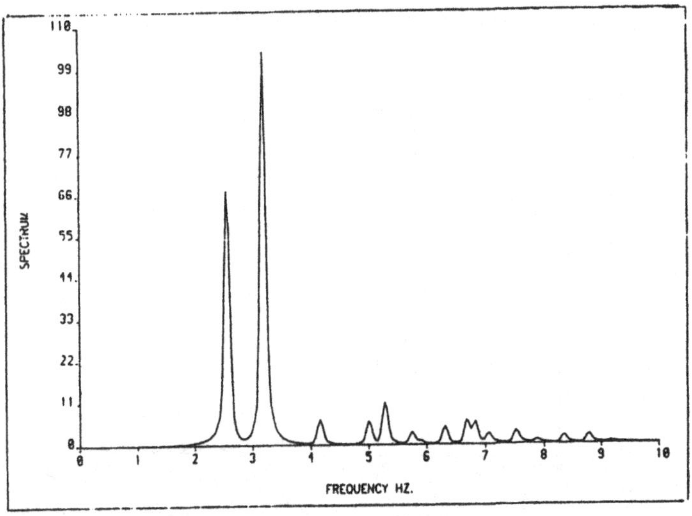

FREQUENCY SPECTRUM OF 7x5 NET RELEASE OF 50 N FROM J16 J 16

Fig(7b)

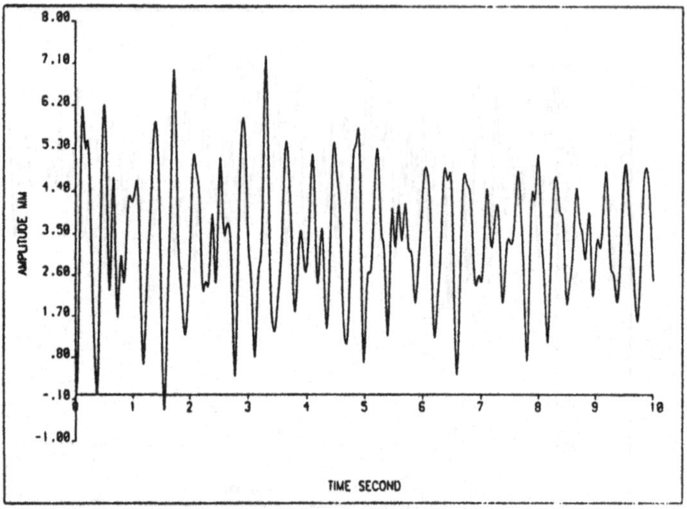

RESPONSE OF 7x5 NET TO RELEASE OF 50 N FROM J16 3% DAMP J 17

Fig(8a)

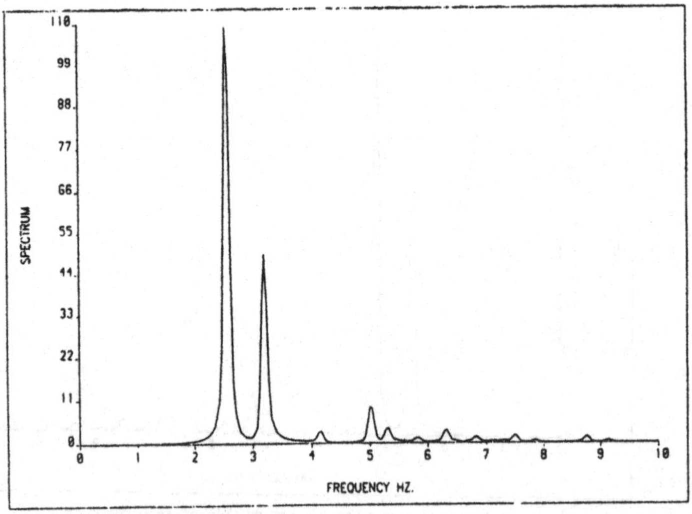

FREQUENCY SPECTRUM OF 7x5 NET RELEASE OF 50 N FROM J16 J 17

Fig(8b)

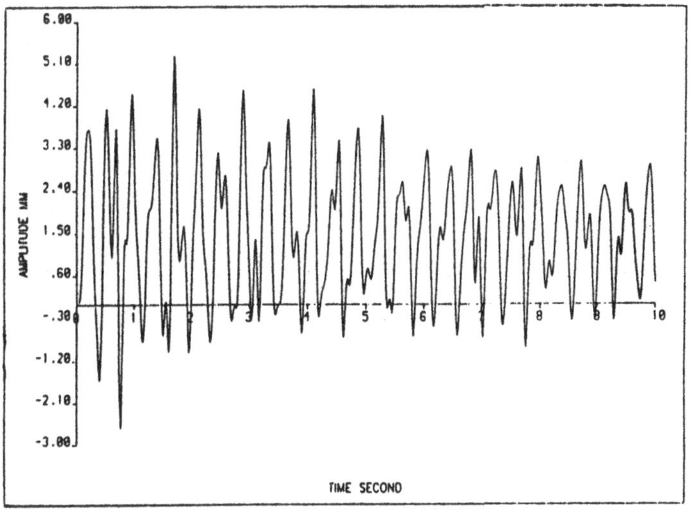

RESPONSE OF 7x5 NET TO RELEASE OF 50 N FROM J16 3% DAMP J 18

Fig(9a)

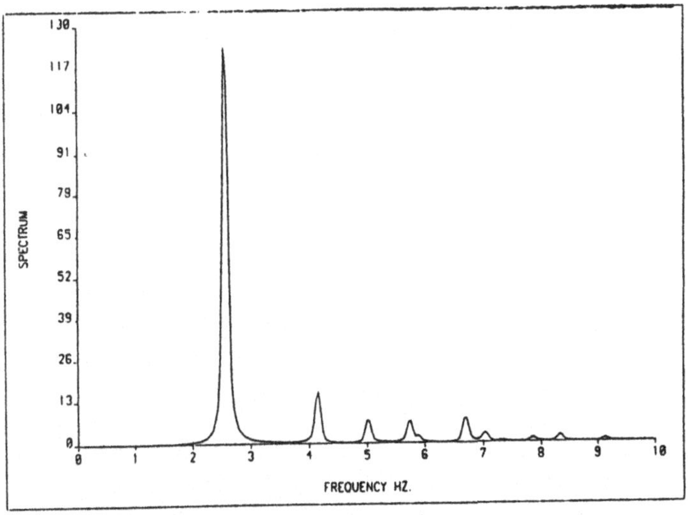

FREQUENCY SPECTRUM OF 7x5 NET RELEASE OF 50 N FROM J16 J 18

Fig(9b)

SECTION 11: APPLIED STOCHASTIC FINITE ELEMENT ANALYSIS

Stochastic Eigenvalue Analysis by SFEM Code "SAINT"

T. Nagashima, T. Tstusumi

Advanced Engineering Department, Science & Technology Division, Mitsubishi Research Institute, Inc., Time and Life Building, 2-3-6 Otemachi, Chiyoda-ku, Tokyo 100, Japan

ABSTRACT

A formulation and its example of perturbation eigenvalue analysis for a structure which has doubly degenerated eigenvalues are presented. In such cases, the zero-th order eigenvectors (i.e. the baseline for perturbation analysis) cannot be determined only by ordinary deterministic analysis. They should be calculated from original vectors so that they diagonalize first order perturbation. Statistical disposition can be done if all the considered perturbation result in the same zero-th order eigenvectors. A numerical example is carried out on a cylindrical shell structure which has a seam along the axial direction.

1. INTRODUCTION

Stochastic Finite Element Method (SFEM) is one of the efficient procedures to analyze structures which contain uncertainties of shape, material properties and/or boundary conditions etc. A general purpose SFEM code "SAINT" (Stochastic finite element Analysis code for INTegrated design) was developed to facilitate the design of reliable and safe structures. Elastic static analysis, eigenvalue analysis and heat transfer analysis for structures which may have a variety of uncertainties can be executed by using "SAINT". Deterministic evaluation by conventional FEM and sensitivity analysis with respect to specified design variables (i.e. probablistic variables) are done successively in "SAINT". The sensitivities are used for statistical disposition under the assumption of the first order approximation.

The structural system which has a cyclic symmetry such as a bladed-disk assembly generally has doubly degenerated natural frequencies.[1] In SFEM based on the perturbation method, the degenerated eigenvalue problems should be dealt with more carefully than ordinary cases. The eigenvectors which are used as the baseline of the perturbation (so-called the "zero-th order eigenvectors") cannot be decided directly by the deterministic analysis. They should be re-arranged to diagonalize the first order perturbation.

In this paper, the formulation of the first order perturbation method for doubly degenerated eigenvalue problem is discussed. As a numerical example, the natural frequency analysis of a cylindrical shell by SFEM is presented.

2. FORMULATION

2.1 Perturbation Method in simple (non-degenerated) undamped eigenvalue problems

The eigenvalue problem can be written in the matrix form

$$[K]\{u\}=\lambda[M]\{u\} \tag{1}$$

where [K], [M], {u} represent the stiffness matrix, the mass matrix and the displacement vector respectively and may be formulated by FEM. The eigenvalue λ equals to the square of the natural angular frequency ω.

If the structure contains uncertainty of the design variables, the stiffness matrix and the mass matrix have fluctuations. Such fluctuations may change the eigenvalues and the eigenvectors. If the fluctuations can be assumed to be small, the first order perturbation method is efficient to establish relationship between fluctuation of design variables and those of eigenvectors and eigenvalues.

The i-th (mode's) eigenvalue and the corresponding eigenvector satisfy the equation

$$[K]\{\phi^{(i)}\}=\lambda^{(i)}[M]\{\phi^{(i)}\} \tag{2}$$

Assume that the stiffness matrix and the mass matrix have fluctuations around the unperturbed values and that they can be expressed as follows,

$$[K]=[K_0]+[\delta K] \tag{3}$$
$$[M]=[M_0]+[\delta M] \tag{4}$$

where $[K_0]$ $[M_0]$ represent the unperturbed stiffness matrix and mass matrix. Because the eigenvalues and the eigenvectors will vary continuously from the corresponding unperturbed values, one can write (up to the first order),

$$\lambda^{(i)}=\lambda_0^{(i)}+\delta\lambda^{(i)} \tag{5}$$
$$\{\phi^{(i)}\}=\{\phi_0^{(i)}\}+\{\delta\phi^{(i)}\} \tag{6}$$

where $\lambda_0{}^{(i)}$ $\{\phi_0{}^{(i)}\}$ represent the i-th unperturbed eigenvalue and eigenvector and $\delta\lambda^{(i)}$ $\{\delta\phi^{(i)}\}$ represent the first order perturbation of i-th eigenvalue and eigenvector, respectively.

Introducing equations (3)(4)(5)(6) into equation (2) and equating the zero-th and the first order terms respectively, one obtains

$$\left([K_0]-\lambda_0^{(i)}[M_0]\right)\left\{\phi_0^{(i)}\right\}=0 \tag{7}$$

$$\left([K_0]-\lambda_0^{(i)}[M_0]\right)\left\{\delta\phi^{(i)}\right\}=-\left([\delta K]-\lambda_0^{(i)}[\delta M]-\delta\lambda^{(i)}[M_0]\right)\left\{\phi_0^{(i)}\right\} \tag{8}$$

In case of simple (non-degenerated) eigenvalues, the unperturbed eigenvalues and the eigenvectors can be calculated by using equation (7).
Premultiply both sides of (8) by $\{\phi_0^{(i)}\}^T$ to obtain

$$\left\{\phi_0^{(i)}\right\}^T\left([K_0]-\lambda_0^{(i)}[M_0]\right)\left\{\delta\phi^{(i)}\right\}=-\left\{\phi_0^{(i)}\right\}^T\left([\delta K]-\lambda_0^{(i)}[\delta M]-\delta\lambda^{(i)}[M_0]\right)\left\{\phi_0^{(i)}\right\} \tag{9}$$

Note that the eigenvectors are normalized with respect to $[M_0]$ by setting

$$\left\{\phi_0^{(i)}\right\}^T[M_0]\left\{\phi_0^{(i)}\right\}=1 \tag{10}$$

For any simple eigenvalues, introducing equations (7) (10) into equation (9), one obtains

$$\delta\lambda^{(i)}=\left\{\phi_0^{(i)}\right\}^T\left([\delta K]-\lambda_0^{(i)}[\delta M]\right)\left\{\phi_0^{(i)}\right\} \tag{11}$$

Substituting equation (11) into (8) and imposing an extra condition of constraint to avoid the singularity associated with the scale arbitrariness in eigen space,the perturbation of the eigenvectors $\{\delta\phi^{(i)}\}$ is determined.

2.2 Calculation of the unperturbed eigenvectors for the multiple (degenerated) eigenvalues

If the objective eigenvalue $\Lambda_0^{(i)}$ is degenerated, it is impossible to determine the eigenvectors associated with $\Lambda_0^{(i)}$ uniquely which is to be the starting point of perturbation (zero-th order eigenvectors) only by using equation (7). We concentrate on doubly degenerated cases here, but the same methodology can be easily applied to higher order degeneracy. Suppose that we used an arbitrary combination of degenerated eigenvectors to be the zero-th order vectors (for example, originally calculated vectors from the deterministic analysis). In that case, perturbed eigenvectors cannot be expressed in the form of equation (6). There are two certain deterministic eigenvectors which are parallel to the projection of perturbed vectors within the degenerated eigen subspace, but the vectors are not always the selected zero-th order vectors. They will be a certain linear combination of original vectors. The linear combination alters according to the nature of the perturbation. This fact makes it difficult to use equation (8) and to perform the first order perturbation analysis. To avoid the difficulty, the zero-th order vectors must be determined by taking information of perturbations into consideration.

A zero-th order vector can be expressed as

$$\left\{\phi_0^{(i)}\right\}=c_1^{(i)}\left\{\Phi_0^{(i)}\right\}+c_2^{(i)}\left\{\Phi_0^{(i+1)}\right\} \tag{12}$$

where $c_1^{(i)}, c_2^{(i)}$ represent the unknown coefficients to be determined and $\{\Phi_0^{(i)}\}$, $\{\Phi_0^{(i+1)}\}$ represent arbitrary eigenvectors derived from equation (7). $\{\Phi_0^{(i)}\}$ satisfies the orthogonality condition

$$\{\phi_0^{(i)}\}^T[M_0]\{\phi_0^{(j)}\}=\delta_{ij} \tag{13}$$

where δij is the Kronecker's delta.

Put equation (12) into equation(8), and premultiply both sides of (8) by $\{\Phi_0^{(i)}\}$, $\{\Phi_0^{(i+1)}\}$ to obtain

$$\{\Phi_0^{(i)}\}^T\{[\delta K]-\Lambda_0^{(i)}[\delta M]-\delta\Lambda^{(i)}[M_0]\}\{\phi_0^{(i)}\}=0 \tag{14.1}$$

$$\{\Phi_0^{(i+1)}\}^T\{[\delta K]-\Lambda_0^{(i)}[\delta M]-\delta\Lambda^{(i)}[M_0]\}\{\phi_0^{(i)}\}=0 \tag{14.2}$$

After reordering equation (14), the 2-dimensional eigenvalue problem is derived as

$$\begin{bmatrix} a_{11} & a_{12} \\ a_{21} & a_{22} \end{bmatrix}\begin{Bmatrix} c_1^i \\ c_2^i \end{Bmatrix}=\delta\Lambda^{(i)}\begin{Bmatrix} c_1^i \\ c_2^i \end{Bmatrix} \tag{15}$$

where

$$a_{11}=\{\Phi_0^{(i)}\}^T\{[\delta K]-\Lambda_0^{(i)}[\delta M]\}\{\Phi_0^{(i)}\}$$
$$a_{12}=a_{21}=\{\Phi_0^{(i)}\}^T\{[\delta K]-\Lambda_0^{(i)}[\delta M]\}\{\Phi_0^{(i+1)}\}$$
$$a_{22}=\{\Phi_0^{(i)}\}^T\{[\delta K]-\Lambda_0^{(i)}[\delta M]\}\{\Phi_0^{(i)}\}$$

Solving equation (15), the eigenvalues $\delta\Lambda^{(i)}$ and the coefficients $c_1^{(i)}, c_2^{(i)}$ are determined. By using equation (12), one can determine the zero-th order perturbed eigenvectors. They diagonalize perturbation so that the actual perturbed eigenvectors differ reasonably little from them.

The first order perturbative eigenvectors can be derived by using the same method for the simple eigenvalues.

2.3 Calculation of the fluctuations of stiffness matrix and mass matrix

In FEM analysis for thin shell structures, the plate and shell elements are generally used. The isoparametric shell element with the reduced integration method is utilized in a lot of general purpose FEM codes as an efficient technique to avoid so-called "shear locking phenomenon". But sometimes problems are aroused and produce bad results with element distortion and spurious zero energy modes (so called hourglass mode). Bathe et al. had developed isoparametric shell element avoiding the numerical assumptions.[2][3] This is by using the new technique in interpolation of out-of-plane strain components. This technique enables full integration for estimating element stiffness matrix. The outline of methodology to calculate the fluctuations of stiffness matrix and mass matrix is shown here in brief. Element stiffness/mass matrices which are used in assemblage of global matrices are calculated from design variables such as Young's modulus, Poisson's ratio, shell thickness and density. The stiffness matrix and the mass matrix of an element are described as

$$[K^e]=\int_{-1}^{1}\int_{-1}^{1}\int_{-1}^{1}[B]^T[D][B]\det[J]dr_1dr_2dr_3 \tag{16}$$

$$[M^e]=\int_{-1}^{1}\int_{-1}^{1}\int_{-1}^{1}\rho[N]^T[N]\det[J]dr_1dr_2dr_3 \tag{17}$$

and

$$[B]=[B(h,r_1,r_2,r_3)] \tag{18.1}$$
$$[D]=[D(E,\nu,h,r_1,r_2,r_3)] \tag{18.2}$$
$$[J]=[J(h,r_1,r_2,r_3)] \tag{18.3}$$
$$[N]=[N(h,r_1,r_2,r_3)] \tag{18.4}$$

where [B],[D],[J] and [N] represent the strain-displacement matrix, the elastic matrix, the Jacobi matrix and the interpolation matrix respectively.

In this method, elastic matrix [D] is defined according to the natural coordinate system, so [D] is a function of thickness h, natural coordinate r1,r2,r3. If the fluctuation of E, ν, h and ρ are specified, one can calculate $[\delta B][\delta D][\delta J][\delta N]$ by using equation (18). And the perturbation of the stiffness matrix and the mass matrix of an element can be described as

$$[\delta K^e]=\int_{-1}^{1}\int_{-1}^{1}\int_{-1}^{1}([\delta B]^T[D][B]\det[J]+[B]^T[\delta D][B]\det[J]$$
$$+[B]^T[D][\delta B]\det[J]+[B]^T[D][B]\delta\det[J])dr_1dr_2dr_3 \tag{19}$$

$$[\delta M^e]=\int_{-1}^{1}\int_{-1}^{1}\int_{-1}^{1}(\delta\rho[N]^T[N]\det[J]+\rho[\delta N]^T[N]\det[J]$$
$$+\rho[N]^T[\delta N]\det[J]+\rho[N]^T[N]\delta\det[J])dr_1dr_2dr_3 \tag{20}$$

Merging the perturbed stiffness matrix and mass matrix, the perturbative global stiffness matrix and mass matrix can be derived.

3.NUMERICAL EXAMPLE

By using the procedure shown in the previous section, the perturbation eigenvalue analysis procedure for shell structures is introduced into the SFEM code "SAINT". In this procedure,a 4-node isoparametric shell element as shown in the previous section is used. Matrices are calculated by using Gauss-Legendre integration in the second order with 2x2x2 integration points (Full integration). The skyline method based on the modified Cholesky Method is utilized for solving simultaneous equations and inverse power method is used for calculating eigenvalues and eigenvectors.

The numerical example is carried out with regard to vibrational characteristics of a cylindrical shell structure which has a seam portion along axial direction where some properties can have small deviation from the definite value. The cylindrical shell as shown in Figure 1 is made of the film of h in thickness. The material properties of the film are Young's modulus E, Poisson's ratio ν and mass per unit volume ρ. It is glued to form its shape with a lap of d in width. The radius and the length of the cylindrical shell are R and L respectively. The values of the model parameters are shown in Table 1. The FEM model used in this analysis is shown in Figure 2.

Firstly, assuming homogeneity for the structure, the deterministic analysis is performed. The first (lowest) six eigenvalues are calculated. The result is shown in Table 2, being compared with those of analytical evaluation. The summary of this FEM analysis is shown in table 3. Next, considering the fluctuation of Young's modulus and shell thickness in the porsion of the seam, the sensitivity of the eigenvalues are calculated by the perturbation method. The result is shown in Table 4.

In this case, the zero-th order vectors according to Young's modulus and the ones according to shell thickness are the same pair. One is the mode in which the seam vibrates with a large amplitude, and the other is that in which it moves little. Note that this fact enables us to combine them in one statistical calculation.

The relationship between natural frequencies and the fluctuation of the properties in the seam are estimated by using calculated sensitivity. The results are shown in Figure 3,4,5 being compared with those of the direct method. Using the SFEM procedure, the standard deviations of the natural frequencies can be evaluated if those of the design variable are given and if the zero-th order vectors are common regardless of the combination of perturbation parameters in consideration. As an example ,the standard deviation of the fundamental frequency (the lowest natural frequency) is calculated under the condition that the shell thickness and the Young's modulus of the seam has the fluctuation of which standard deviations are specified. Two properties' fluctuations are assumed to be uncorrelated. The standard deviation of the natural frequency σf can be estimated by

$$\sigma_f=\sqrt{(\partial f/\partial h)^2\sigma_h{}^2+(\partial f/\partial E)^2\sigma_E{}^2} \tag{21}$$

where σh, σE are the standard deviations of the thickness and Young's modulus respectively. Sensitivities $\partial f/\partial E$, $\partial f/\partial h$ are given in Table 4. If σh and σE are assumed 1% respectively, the σf is estimated 0.03% (the seam vibrates most) and 0.005% (the seam vibrates least).

4.CONCLUDING REMARKS

The first order perturbation method was applied to the eigenvalue problem of the structure system which has doubly degenerated eigenvalues. The methodology was introduced into the SFEM code "SAINT" to improve the accuracy of perturbation eigenvalue analysis. By using "SAINT", the vibrational behavior of a cylindrical shell which has inhomogeneous seam was investigated.

ACKNOWLEDGEMENT

The authors are indebted to Mr. Hirohisa Noguchi, Research Associate of University of Tokyo Research Center for Advanced Science and Technology, for his guidance during this study.

REFFERENCES

[1]Wei,S.T,and Pierre C. 'Localization Phenomena in Mistuned Assemblies with Cyclic Symmetry Part 1:Free Vibrations',ASME journal of Vibration,Acoustics,Stress,and Reliability in Design,Vol 106,429-438(1988)

[2]Dvorkin,E.N. and Bathe,K.J.'A continuum mechanics based four-node shell element for general nonlinear analysis' ,Eng.Comput.,1,77-88(1984)

[3]Bathe,K.J. and Dvorkin,E.N.'Short communication a four-node plate bending element based on Mindlin/Reissner plate theory and a mixed interpolation.',Int.j.numer.methods eng.,21,367-383(1985)

Table 1 Properties of shell structure

thickness	h=0.25mm
radius	R=80mm
length	L=480mm
lap width	d=13.96mm
Young's modulus	$E=600kgf/mm^2$
Poisson's ratio	$\nu=0.3$
density	$\rho=1.4X10^{-10}kgfsec/mm$

Table2 Natural frequencies

mode	FEM	Analytical
1st	98.137	93.974
2nd	98.137	(m=1,n=4)
3rd	115.145	108.100
4th	115.145	(m=1,n=5)
5th	128.270	126.994
6th	128.270	(m=1,n=3)

Unit :Hz

According to the Donnel shell theory,the natural frequencies are caluculated as

$$\omega^2=\frac{Eh^2}{12(1-\nu^2)\rho R^4}\left[\left(n^2+k^2\right)^2+\frac{\alpha k^4}{\left(n^2+k^2\right)^2}\right] \qquad k=\frac{m\pi R}{L},\alpha=12\left(1-\nu^2\right)\left(\frac{R}{h}\right)^2$$

where m,n are number of half waves in the axial direction and number of waves in the circumferential direction respectively.

Table 3 Summary of FEM analysis

Number of nodes	396
Number of elements	360
Number of degree of freedoms	2376
Number of subspace vectors	12
Number of eigenvalues to get	6
Number of iterations	7
*CPU time(sec)/Direct method	180

*Use IBM RISC6000/320

Table 4 Sensitivity of natural frequencies

mode	sensitivity $h_0 \partial f/\partial h$	sensitivity $E_0 \partial f/\partial E$
1st	0.192626	0.421969
2nd	1.92441	2.30406
3rd	-0.0196430	0.251754
4th	5.03021	2.94671
5th	0.205396	0.645273
6th	0.221938	2.91777

Unit :Hz

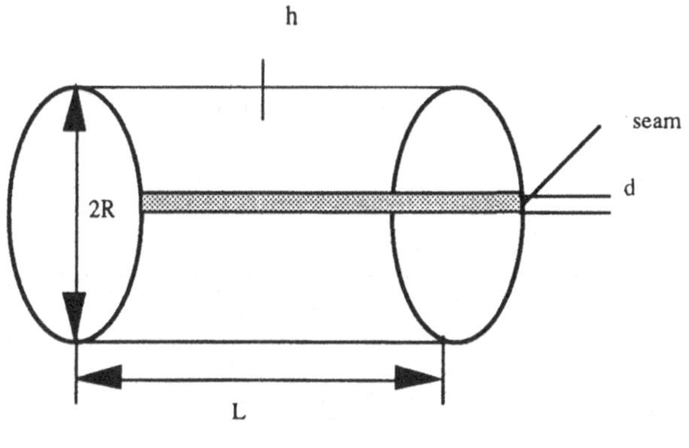

Figure 1 Shell structure

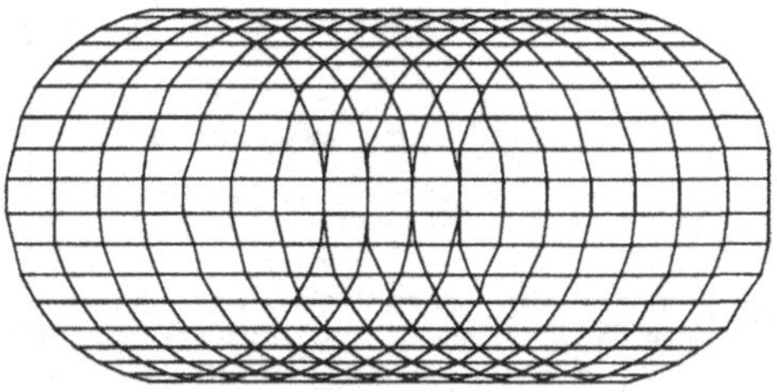

Figure2 FEM model

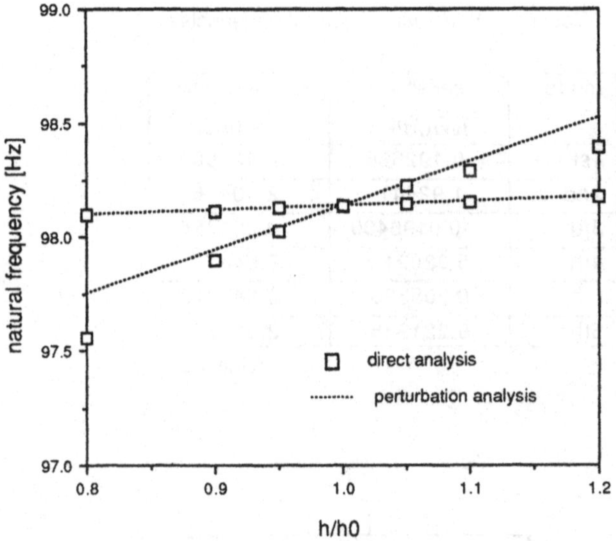

(a) fluctuation of thickness

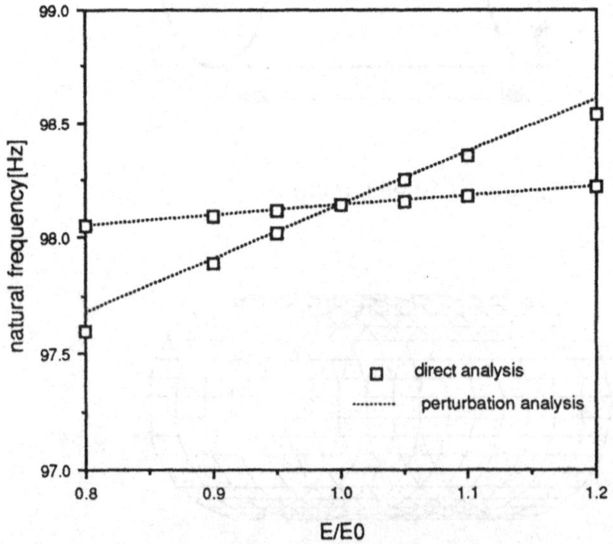

(b)fluctuation of Young's modulus

Figure3 1st and 2nd natural frequency vs. fluctuation

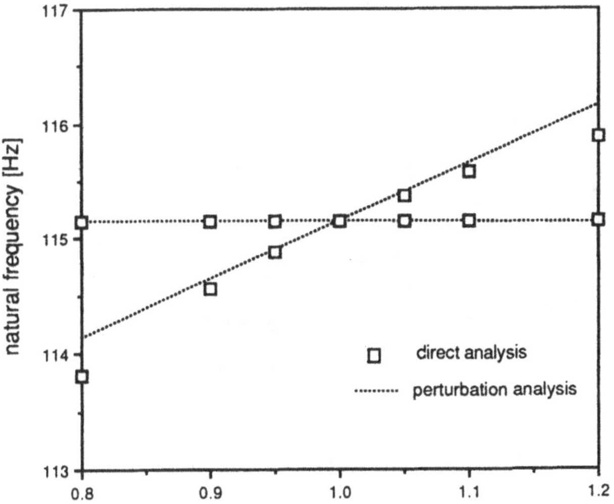

h/h0
(a)fluctuation of thickness

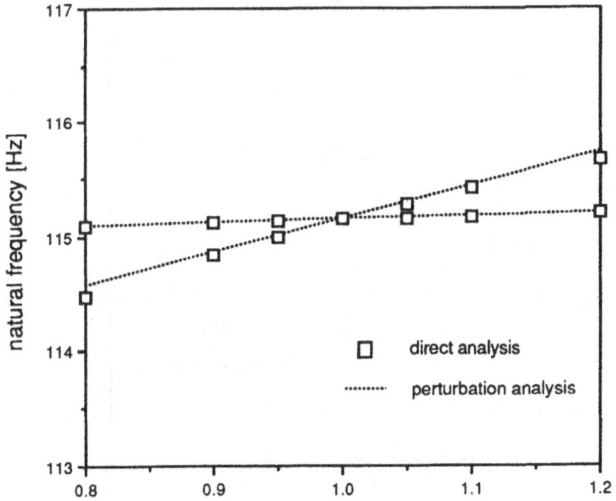

E/E0
(b) fluctuation of Young's modulus

Figure4 3rd and 4th natural frequency vs fluctuation

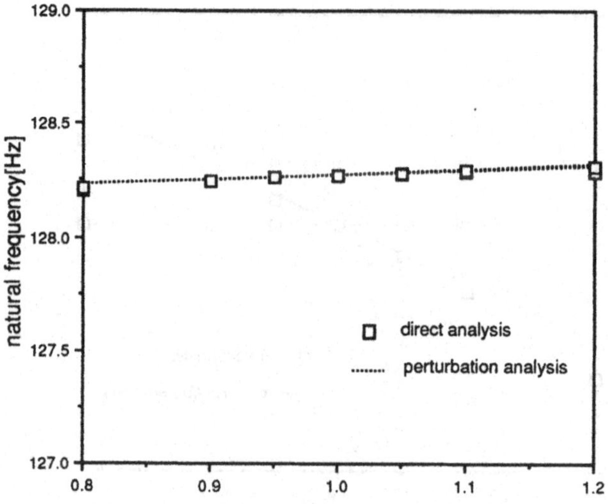

(a)fluctuation of thickness

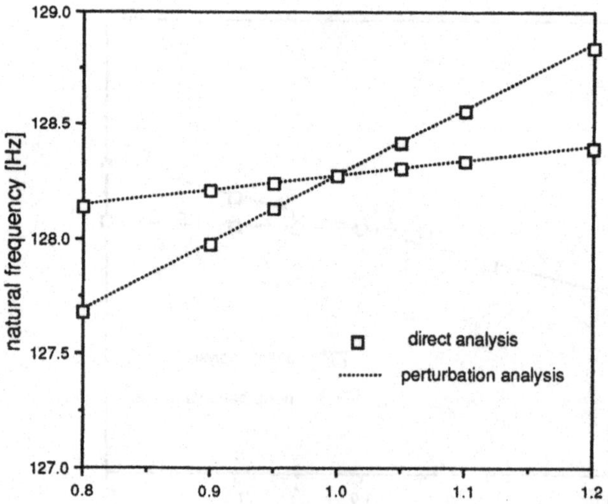

(b)fluctuation of Young's modulus

Figure5 5th and 6th natural frequency vs. fluctuation

A Probabilistic Postbuckling Analysis of Composite Shells

S.P. Engelstad, J.N. Reddy

Department of Engineering Science and Mechanics, Virginia Polytechnic Institute and State University, Blacksburg, VA 24061, U.S.A.

ABSTRACT

A probabilistic finite element analysis procedure for postbuckling analysis of laminated composite shells is developed. The total Lagrangian formulation, employing a degenerated 3–D laminated composite shell finite element with the full Green–Lagrange strains and first–order shear deformable kinematics is used. The first–order second–moment technique for probabilistic finite element analysis of random fields is employed and results are presented in the form of mean and variance of the structural response. Reliability calculations are made by using the first–order reliability method combined with sensitivity derivatives from the finite element analysis. Two examples are solved to illustrate the variability of shell and panel postbuckling problems. In general, the procedure is quite effective in determining the response statistics and reliability for postbuckling behavior of laminated composite shells.

INTRODUCTION

The use of fiber reinforced composite materials in modern engineering structural design has become a common practice. For example, organic matrix composite materials such as graphite–epoxy have been used extensively with substantial weight savings. This savings is partly due to the larger number of composite design variables, allowing the material to be tailored to suit the loading environment. However, these extra design variables result in increased variability as compared to the same structure produced with conventional materials. By compensating for this variability, it is possible to lose the original advantages of the material. In addition, application of these materials to airframe structures often requires that they survive even after buckling has occurred. As a result, more accurate methods for predicting the postbuckling response and failure are necessary. Thus the motivation of this paper is to use the probabilistic finite element technique to quantify the variability and reliability of postbuckled composite structures.

Several authors have developed probabilistic finite element methods for random fields or variables. The most commonly used method for nonlinear analysis is the second–moment method, as outlined by Nakagiri et al. [1], and Liu et al. [2]. Nakagiri applied the method to the analysis of linear composite plates using classical lamination theory. Liu and his coworkers have done extensive work applying the method to geometric and material nonlinear analysis, but only for isotropic materials.

This paper focuses on the application of the second–moment method to probabilistic finite element analysis of postbuckling behavior of composite shells. The procedures developed by Liu et al. [2] for determining first–order mean and variance of the deflection, strain, and stress response are utilized. The sensitivity derivatives computed in the second–moment method are also used to perform the first–order reliability optimization. In order to more accurately model the postbuckling response and failure of composite plates and shells, the shell element with shear deformation theory is used in the present study. The computational difficulties in analyzing actual laminated composite panels involving models with large numbers of layers and degrees of freedom, and also that proceed deep into the postbuckling range, are investigated. Probabilistic sensitivity studies are performed, and comparisons are made with experimental results.

FINITE ELEMENT FORMULATION

The incremental equations of a continuous medium are formulated based on the principle of virtual displacements and the total Lagrangian description. The detailed description can be found in Reference 3, and is thus omitted here for brevity.

The degenerated 3–D shell element is obtained from the three dimensional solid element by imposing two constraints: (1) straight lines normal to the midsurface before deformation remain straight but not normal after deformation; (2) the transverse normal components of strain and hence stress are ignored in the development. The resulting nonlinear formulation admits arbitrarily large displacements and rotations of the shell element and small strains, since the thickness does not change and the normal does not distort.

PROBABILISTIC FINITE ELEMENT METHOD

The second–moment perturbation method as developed by Liu et al. [2] is summarized in this section for geometric nonlinear, time independent behavior. Discussions of importance to composite random variables are presented, and developments are applied to the degenerated 3–D shell element. The first–order reliability method incorporated into the computational scheme is described as well.

Consider the assembled form of the finite element equations,

$$\{F(\{\Delta\},\{b\})\} = \{R(\{b\})\} \tag{1}$$

where $\{F\}$ is the internal force vector, $\{\Delta\}$ is the displacement vector, $\{R\}$ is the external force, and $\{b\}$ is a discretized vector of the random function $b(x)$, where x is a spatial coordinate $\{x\}$. As in typical finite element analysis, the random function $b(x)$ is expanded using shape functions $\psi_i(x)$

$$b(x) = \sum_{i=1}^{n} \psi_i(x) b_i \tag{2}$$

where b_i are the nodal values of $b(x)$. Generally the random quantity b can be a material property, geometric dimension, or a load.

The probabilistic finite element method for moment evaluation proceeds by applying second–moment analysis. The vectors in equation (1) are expanded about a value of the random function b via Taylor's series, and substituting the result into equation (1) yields the following perturbation equations:

Zeroth–order equation:

$$\{\bar{F}\} = \{\bar{R}\} \tag{3}$$

First–order equations:

$$[\bar{K}^T]\{\bar{\Delta}\}_{b_i} = \{\bar{R}\}_{b_i} - \{\bar{F}\}_{b_i} \quad , \ i = 1,\ldots,n \tag{4}$$

where the bar over the symbols represent evaluation of that term at a particular value of b, and the subscript b_i represents the derivative with respect to b_i. As is common in the literature, $[K^T]$ represents the tangent stiffness matrix.

Composite Random Variables

Sources of randomness can be material properties, geometric dimensions, or loads. For the present study, the only geometric dimensions selected as random variables are the ply thickness and ply angle. The loading is considered to be deterministic throughout this study. All material properties are treated as random variables, which could be any of the engineering material properties E_{11}, E_{22}, ν_{12}, G_{12}, G_{13}, G_{23}. These properties along with the ply angle θ define the coefficients of the constitutive matrix.

Spatial Correlation

The probabilistic finite element procedure developed herein has the ability to model the correlation involved in spatial fields. For an orthotropic ply, we assume an orthotropic autocorrelation coefficient function in the following form

$$A(b_i,b_j) = \exp\left[-\left[\frac{\xi_i' - \xi_j'}{\lambda_{\xi'}}\right]^2\right]\exp\left[-\left[\frac{\eta_i' - \eta_j'}{\lambda_{\eta'}}\right]^2\right] \quad i,j = 1...,n \quad (5)$$

ξ and η are the curvilinear coordinates of the shell which are aligned with the body coordinates, ξ' and η' are the curvilinear coordinates in the principal material coordinate system, aligned with the fibers, and $\lambda_{\xi'}$, $\lambda_{\eta'}$ are input correlation lengths.

Reliability Estimation

Probability of failure estimation or reliability is a very important aspect of the computational procedure. The method used here is the first–order reliability method developed by Hasofer and Lind [4] and extended by Rackwitz and Fiessler [5]. Various optimization schemes are used to solve the reliability constrained optimization problem; the one chosen here is a Lagrange multiplier method [6]. In the optimization scheme, gradients of the limit state function with respect to the random variables are required. Since the limit state function is typically a function of the structural response (e.g., displacement, strain, stress), then by the chain rule of differentiation, derivatives of the response with respect to the basic random variables are needed. These are determined in the perturbation probabilistic finite element method. In this way an efficient procedure for reliability estimation incorporating the probabilistic finite element method is achieved.

APPLICATIONS

Postbuckling of Spherical Shell Under External Pressure

A shallow spherical shell with simply supported boundary conditions and uniform external pressure is studied in this example. The problem description is given in Figure 1, and all input material properties, variances and levels of spatial correlation are given in Table 1. All examples in this study used a coefficient of variation (COV) of .05 except for the case of ply orientation angle, where the standard deviation was chosen to be 2 degrees. All variables were treated as random fields, with typical correlation lengths of one half of the domain of the problem in each principal material direction. Only one quadrant of the shell was modeled in the interest of computational savings, even though the spatial correlation assumptions may not be symmetric across the shell. Of course, this is not a limitation of the procedures or computational model developed herein.

A postbuckling analysis of the graphite–epoxy shell is performed next. The modified–Riks method [7] is used to trace the nonlinear path past the snap through point (or limit point) and into the postbuckling range. Using the first–order second–moment method combined with the modified–Riks technique, the w–displacement response throughout the postbuckling range, including the limit points, is calculated and presented in Figure 1. Here the squares indicate mean response while (*)

Table 1 Material Properties and Statistics for Spherical Shell Problem

Random Variable	Mean	Standard Deviation	Coef. of Variation	$\lambda_{\xi'}$ (mm)	$\lambda_{\eta'}$ (mm)
E_{11}	109.0 GPa	5.45 GPa	0.05	635	381
E_{22}	6.27 GPa	.314 GPa	0.05	381	635
G_{12}	3.09 GPa	.155 GPa	0.05	508	508
ν_{12}	0.222	1.112×10^{-2}	0.05	508	508
G_{13}	3.09 GPa	.155 GPa	0.05	508	508
G_{23}	2.41	.121 GPa	0.05	508	508
Fiber Angle	0°,90°	2°	–	635	635
Ply Thickness	12.7 mm	.636 mm	0.05	635	635

the stars are the mean plus or minus one standard deviation (one sigma) points. It is evident that at the limit points the COV is very large, in fact reaching a value of 0.5.

The increase in variance at the limit points occurs since the buckling behavior of the structure is more sensitive to any changes in stiffness or load at these points. It should be noted that while the displacement COV begins at about .05 before the limit points, it becomes quite small after the limit points, settling to a value of about .01.

Postbuckling of Flat Panel Under Axial Compression

The problem under consideration here is a flat, rectangular, 50.8 cm long by 17.8 cm wide graphite–epoxy panel loaded in axial compression. An experimental study was performed on a series of these panels by Starnes and Rouse [8]. Figures 2a and b show a typical panel with fixture and the resulting failure mode. The loaded ends of the panels were clamped by fixtures and the unloaded edges were simply supported by knife–edge restraints to prevent the panels from buckling as wide columns. The material properties, layup, and statistics are as follows: E_{11} = 131.0 GPa (19,000 ksi), E_{22} = 13.0 GPa (1,890 ksi), G_{12} = G_{13} = 6.4 GPa (930 ksi), ν_{12} = 0.38, G_{23} = 1.72 GPa (250 ksi), and lamina thickness = .14 mm (.0055 in); the layup was a 24–ply orthotropic laminate with a $[\pm45/0_2/\pm45/0_2/\pm45/0/90]_s$ stacking sequence; the random variables were E_{11}, E_{22}, G_{12}, ν_{12}, G_{13}, G_{23}, and ply thickness each with a coefficient of variation of 0.05, also ply angle with a standard deviation of 2 degrees. The model used 72 nine–node Lagrange elements uniformly spaced, with 12 along the longitudinal (x–direction) and 6 along the width (y–direction), with 325 nodes and 1625 degrees of freedom. The boundary conditions are shown in Fig. 3.

A previous deterministic analytical study was performed on the panel (denoted as Panel C4 in [8]) by this author in Reference 9. Comparisons were made between analytical and experimental results. In general the comparisons were very good, even deep in the postbuckling range. In order to pass the critical buckling load, a geometric imperfection of a small percentage of the plate thickness (typically 1 to 5) times the normalized linear buckling mode was added to the original geometry of the panel. The purpose of the previous analysis was to study the effect of shear deformation on postbuckling response and failure prediction. The purpose of this analysis is to study the variability of the panel results.

In order to understand the failure mode, it is necessary to see the nonlinear buckling mode shape. An analytical contour plot of the out–of–plane deflection at an applied load of 2.1 P_{cr} is shown in Figure 2c. A moire fringe pattern photograph from Reference 8 of the out–of–plane deflections at the same load is shown in Figure 2d. Both patterns indicate two longitudinal half–waves with a buckling–mode nodal line at panel midlength.

The probabilistic analysis of this panel assumed a fully correlated random field for each random function in each layer so that each degenerated to a single random variable per layer. An attempt was made to employ the random field techniques; however, the computational expense was too high due to eight random fields in 24 layers, 1625 degrees of freedom in the model, and a nonlinear analysis.

The postbuckling response is shown in Figure 3. The load P is normalized by the analytical buckling load P_{cr} and the end shortening deflection u by the analytical end shortening u_{cr} at buckling. A 1% plate thickness geometric imperfection was used. The analytical results compare favorably with the experimental results. In addition, the plus or minus one standard deviation points indicate the variation in the data. It should be noted that in [9], it was shown that one of the reasons for the good agreement here is the inclusion of the shear deformation in the element formulation. Reference 10 contains more analytical and experimental comparisons.

In References 8 and 9, it was determined that the failure mode was primarily due to transverse shear stress τ_{13} (τ_{xz} in 0° ply) along the midlength (nodal line) of the panel. Other stresses such as σ_{11} and τ_{12} also contribute to failure along the nodal line region in other layers.

The distribution of the transverse shear stress τ_{13} in the third layer of the laminate (a 0° ply) is shown in Figure 4a for a load of 2.1 P_{cr}. It is observed that high transverse shear stress develops along the nodal line of the panel. Figure 4b shows the redistribution of the τ_{13} stress for three different applied loads. At the experimental failure load of 2.1 P_{cr}, the τ_{13} stress approaches the material allowable value of 62 MPa (9 ksi). The COV for the τ_{13} stress was typically .055 except at the critical buckling load when it reached a value of .12.

It is of interest to determine the most significant random variables in the variance for the τ_{13} stress. Figure 5 contains the peak mean τ_{13} stress, along with the COV for the combined and individual random variables for increasing load values. It is observed that the τ_{13} stress variance is most influenced by G_{13}, with the ply thickness effect increasing as the buckling load is passed and bending effects become more important.

Even though for this graphite–epoxy composite panel a first–ply failure does not represent overall panel failure, a reliability analysis was performed for first–ply failure. In Reference 9, it was shown that although the failure mechanism was primarily due to τ_{13} transverse shear stress, the Tsai–Wu failure criterion which accounts for interaction of stresses did a better job of predicting failure than the maximum stress criterion. The Tsai–Wu criterion (limit state function) can be stated as follows:

$$g = 1 - (F_1\sigma_1 + F_2\sigma_2 + 2F_{12}\sigma_1\sigma_2 + F_{11}\sigma_1^2 + F_{22}\sigma_2^2 + F_{44}\sigma_4^2$$

$$+ F_{55}\sigma_5^2 + F_{66}\sigma_6^2) = 0$$

where $F_1 = \frac{1}{X_t - X_c}$; $F_2 = \frac{1}{Y_t - Y_c}$; $F_{11} = \frac{1}{X_t X_c}$; $F_{22} = \frac{1}{Y_t Y_c}$;

$$F_{44} = \frac{1}{R^2}; \; F_{55} = \frac{1}{S^2}; \; F_{66} = \frac{1}{T^2}; \; F_{12} = -\frac{1}{2}\sqrt{\frac{1}{X_t X_c Y_t Y_c}}$$

and σ_1, σ_2 are the normal stresses along the fiber and normal to the fiber, respectively, σ_4, σ_5, σ_6 are the shear stresses in the 23, 13, 12 planes respectively; X and Y correspond to strengths in the 1 and 2 directions, and the subscripts t and c denote tension and compression, respectively; R, S, and T are the shear strengths. The values of the strengths used are as follows: X_t = 1400 MPa (203 ksi), X_c = 1138 MPa (165 ksi), Y_t = 80.9 MPa (11.7 ksi), Y_c = 189 MPa (27.4 ksi), R = S = 62 MPa (9 ksi), and T = 69.0 MPa (10.0 ksi).

All random variables were assumed to have normal distributions in this example. Since no experimental distribution data was available, then this "Normal" assumption reduced the reliability computation time in the optimization routine (as extra iterations due to equivalent-Normal approximations were unnecessary). While all stiffness random variables were given a COV of 0.05, the strengths were assumed to have COV's of 0.10. In addition, zero correlation was assumed between each random variable. Of course, none of these simplifications are limitations of the method.

Figure 6a shows the deterministic comparison of the Tsai–Wu analytical prediction of first ply failure to the experimental failure. The failure location is indicated by an "x" in the insert and the thickness location is the first layer of the laminate. Figure 6b contains the reliability analysis. The probability of failure (P_f) was found to be quite small (.0021) at the load P/P_{cr} = 1.39, which is still quite deep in the postbuckling range. The number of iterations for each numbered data point in Figure 6b is as follows: point 1 required 5 iterations, point 2 required 6 iterations, point 3 required 8 iterations. The numerical tolerance for convergence in the reliability index β was set to .001. In terms of CPU time (on a Convex computer), the following expenditures were required for each data point: point 1 required 1.59 hours, point 2 required 1.75 hours, point 3 required 2.05 hours.

CONCLUDING REMARKS

A geometric nonlinear finite element analysis procedure has been applied to the postbuckling analysis of laminated composite plates and shells. The first–order second–moment perturbation method was used to compute statistical moments and quantify the variability of response. The first–order reliability method was employed to compute probability of failure. The degenerated 3–D laminated composite shell element proved to be accurate in modeling the geometric nonlinear behavior of the postbuckled panel. The inclusion of transverse shear deformation was found to be critical, especially in the postbuckling range. It was demonstrated that the modified–Riks arc length method works quite well with the second–moment probabilistic method and allowed mean and variance calculations to be made beyond zero–slope limit points, which often exist in shell structures.

ACKNOWLEDGEMENT

The support of this research by NASA Lewis Research Center through Grant NAG–3–933 is gratefully acknowledged.

REFERENCES

1. Tani, S. and Nakagiri, S., "Reliability Synthesis of CFRP Laminated Plate," *Proc. ICOSSAR '89, 5th Int. Conf. on Structural Safety and Reliability*, San Francisco, CA, p. 2079, 1989.

2. Liu, W. K., Belytschko, T., and Mani, A., "Random Field Finite Elements," *Int. J. for Numerical Methods in Engineering*, Vol. 23, p. 1831, 1986.

3. Chao, W. C. and Reddy, J. N., "Analysis of Laminated Composite Shells Using a Degenerated 3–D Element," *Int. J. Numerical Methods in Engineering*, Vol. 20, p. 1991, 1984.

4. Hasofer, A. M. and Lind, N. C., "Exact and Invariant Second–Moment Code Format," *J. of the Engineering Mechanics Division*, ASCE, Vol. 100, No. EM1, p. 111, 1974.

5. Rackwitz, R. and Fiessler, B., "Structural Reliability Under Combined Random Load Sequences," *J. of Computers and Structures*, Vol. 9, p. 489, 1978.

6. Ang, A. H. S. and Tang, W. K., *Probability Concepts in Engineering Planning and Design, Volumes I and II, Basic Principles*, Wiley, New York, 1984.

7. Crisfield, M. A., "A Fast Incremental/Iterative Solution Procedure that Handles "Snap–Through"," *Computers and Structures*, Vol. 13, p. 55, 1980.

8. Starnes, J. H., Jr., and Rouse, M., "Postbuckling and Failure Characteristics of Selected Flat Rectangular Graphite–Epoxy Plates Loaded in Compression," AIAA Paper No. 81–0543, 1981.

9. Engelstad, S. P., Knight, Jr., N. F., and Reddy, J. N., "Postbuckling Response and Failure Prediction of Flat Rectangular Graphite–Epoxy Plates Loaded in Axial Compression," Proceedings of the 32nd Structures, Structural Dynamics and Materials Conference, Baltimore, Maryland, A Collection of Technical Papers, Part 2, p. 888, 1991.

10. Engelstad, S. P., "Nonlinear Probabilistic Finite Element Modeling of Composite Shells," Ph.D. Dissertation, Virginia Polytechnic Institute and State University, Blacksburg, VA, December 1990.

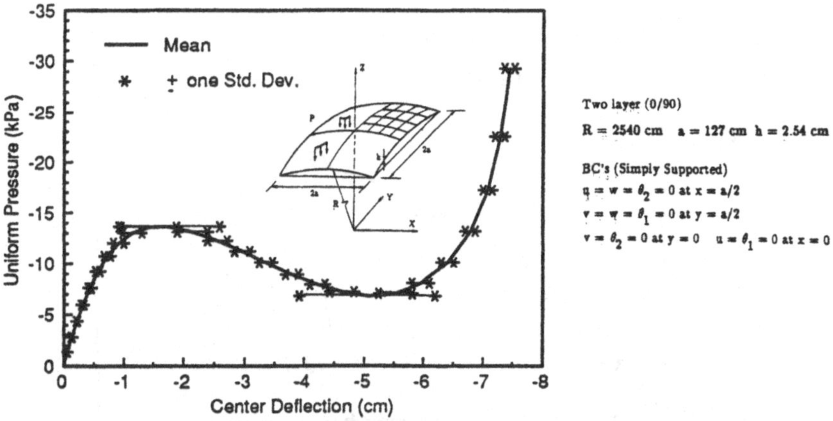

Figure 1 Mean center displacement w and plus or minus one standard deviation points versus load throughout postbuckling range of spherical shell.

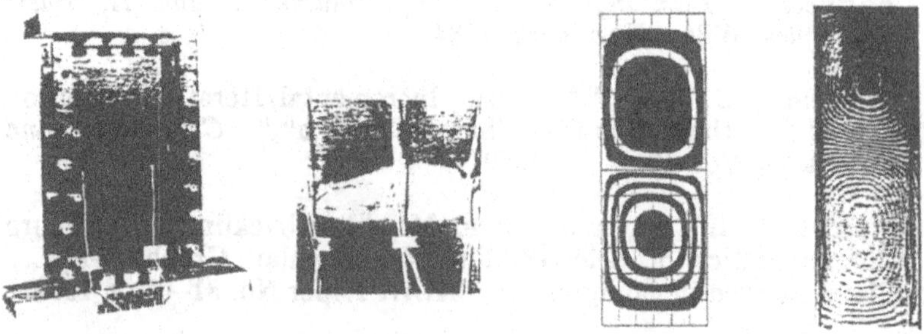

(a) Panel fixture

(b) Panel failure mode

(c) Contour plot of out-of-plane analytical displacement

(d) Photograph of moire-fringe pattern

Figure 2 **Flat rectangular graphite-epoxy panel under axial compression [8].**

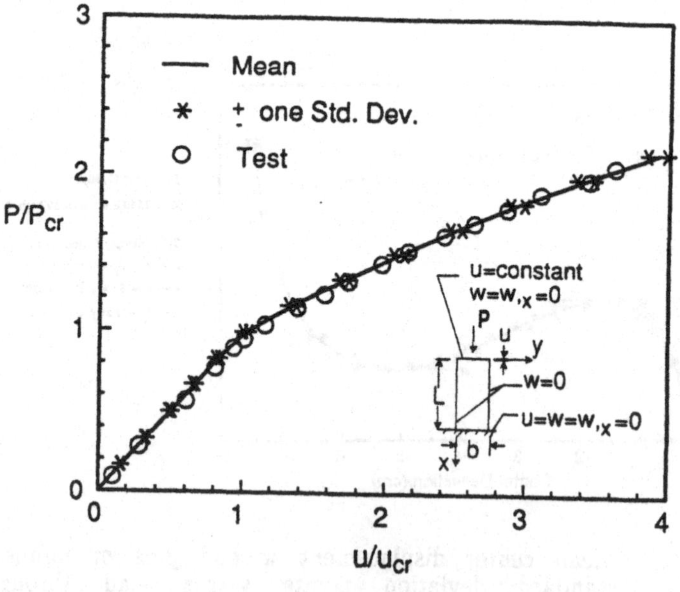

Figure 3 **End-shortening postbuckling response of graphite-epoxy panel.**

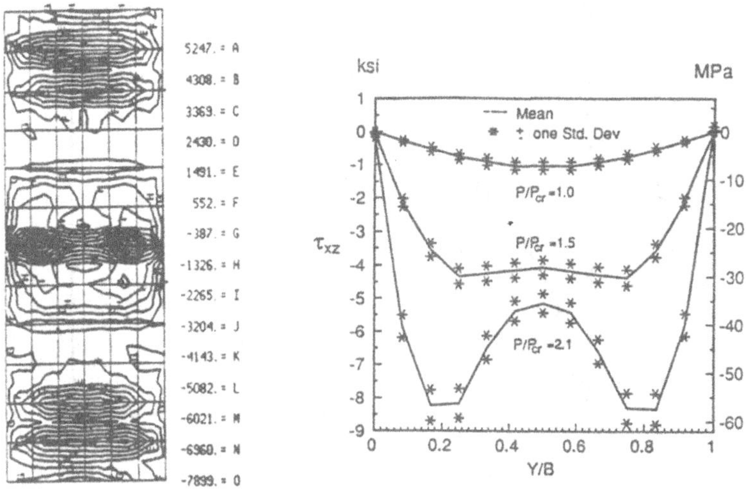

(a) Contour plot of transverse shear stress distribution

(b) Stress distributions across panel midlength

Figure 4 Transverse shear stress τ_{xz} distributions in the third layer (0° ply) from the surface of the graphite–epoxy panel.

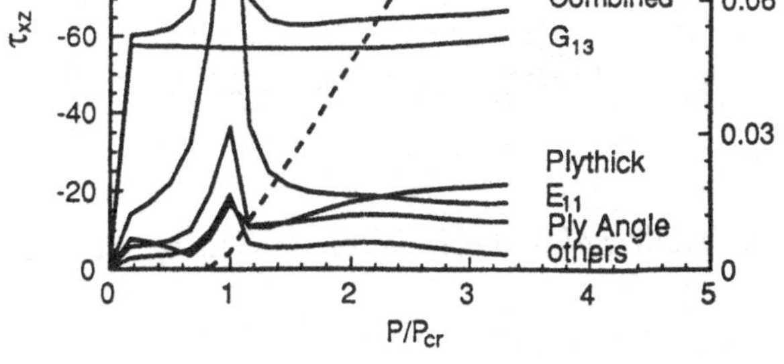

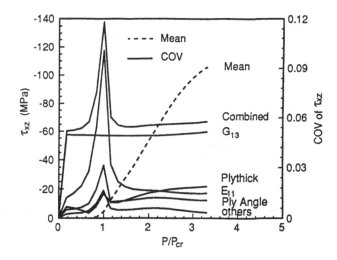

Figure 5 Mean and coefficient of variation (COV) of τ_{xz} stress versus load throughout postbuckling range showing the combined and individual effects of the ply–level random variables for the graphite–epoxy panel.

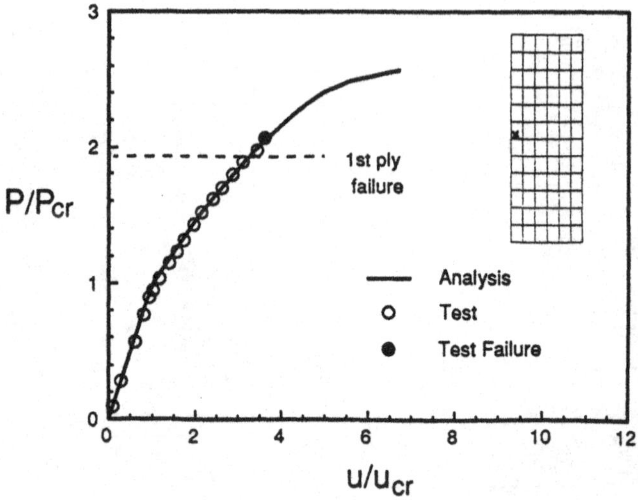

(a) End–shortening postbuckling response of panel showing deterministic Tsai–Wu first–ply failure.

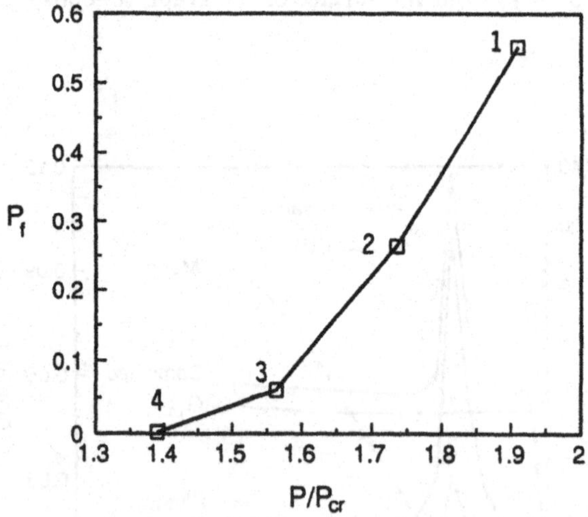

(b) Probability of first–ply failure (P_f) using the Tsai–Wu criterion.

Figure 6 First–ply Tsai–Wu deterministic failure and reliability analysis.

An Efficient SFEM Algorithm for Nonlinear Structures

A. Haldar, Y. Zhou

Department of Civil Engineering & Engineering Mechanics, University of Arizona, Tucson, AZ 85721, U.S.A.

ABSTRACT

An efficient SFEM method is proposed for reliability analysis of highly redundant elastic-perfectly-plastic large deformed frame structures under proportional loading. Any frame or truss structures can be handled using this method, and the geometric and material nonlinearities can be considered at the same time. Uncertainties in the load and resistance-related parameters are considered. The proposed method has several advantages over the other available methods. The method is verified with two examples. The results match very well with Monte Carlo simulation.

INTRODUCTION

The nonlinear behavior of structures can be attributed to geometric and material nonlinearities. A class of problems, where the material behavior can be modeled as elastic-perfectly-plastic, received a considerable amount of attention in the past. Both deterministic and probabilistic methods were proposed. The reliability analysis of such elasto-plastic frame structures is the object of this paper.

To estimate the reliability of such frame structures, some of the most commonly used methods are the plastic mechanism, stable configuration and failure surface approaches.

According to the plastic mechanism approach, an upper bound of the structural reliability can be evaluated on the basis of a set of plastic mechanisms. The success of this approach depends on the completeness of the mechanisms identified. However, it may be impractical to identify all failure mechanisms for complicated structures. Also, the failure mechanisms are correlated through material properties, plastic hinge formations and loads. The correlation among different failure mechanisms makes this approach extremely complicated. Several attempts have been made to set up search procedures to identify the most likely failure modes. Murotsu et al (1984) used the method of automatic generation of failure mechanisms. An upper bound of the structural failure probability is evaluated by the systematic selection of the stochastically dominant mechanism through branching and bounding operations. Bjerager (1989) proposed a method based on the lower bound theorem of plasticity theory by using the force method formulation to consider different equilibrium states of the structure. The failure mechanism in his method is generated in random order. Siões (1990) proposed an upper bound theorem fomulation. He used a mathematical programming-based procedure for identifying the stochastically most relevant failure mechanisms. The mechanisms are found by minimizing a quadratic concave function. The interaction of bending and axial forces is considered instead of the predominant bending action alone.

The stable configuration approach (Quek and Ang 1990) examines how a structure or any of its damaged states can carry the loads. Each configuration in which the loads can be carried without collapse is considered to be undamaged and the union of corresponding probabilities would give the probability of survival of a structure. In general, both the plastic mechanism and the stable configuration techniques approximately evaluate the failure probability of the structures. Failure to identify some failure mechanisms could lead to an underestimation of the probability of failure, and failure to consider some stable configurations would overestimate the probability of failure.

In order to avoid dealing with the complicated failure mechanisms and stable configurations, the failure surface approach was proposed (Wang et al 1990). In this method, the limit state function is defined by performing overall structural analyses. In order to obtain the limit state surfaces, the load space formulation or the shakedown theory is used. However, the limit surface obtained by this method is a group of discrete points, and every point is computed by assuming the external loads are only random variables. The limit surface obtained in this way is approximate. Since the limit surface is based on the analysis in the load space, with the increase in the number of loads, the description of the limit surface will be more difficult. Using this technique, only the results obtained by Monte Carlo simulation are reported in the literature.

An efficient method is proposed in this paper for the reliability analysis of elasto-plastic frame structures using the stochastic finite element method (SFEM). The limit state function is expressed in terms of the limit load of plastic collapse instead of failure mechanisms involving internal forces or displacements and threshold values. In the multiple load case, the virtual work principle is used to describe the limit state function. There is no need to specify individual failure modes a priori and it is easy to describe the limit state function. Since the nonlinear finite element method is used to compute the response of the elasto-plastic structure, any complexity of frame or truss structures can be handled using this method, and the geometric and material nonlinearities can be considered at the same time. Unlike the mechanism analysis method in which the failure probability of the structure has no relation to the history of the structural deformation, the result obtained by the proposed method can reflect the influence of the nonlinear deformation process and the redistribution of generalized forces at the plastic hinges, thus improving the structural reliability estimation. The first order second moment method (AFOSM) is used in the reliability analysis. Thus, the randomness in the external loading and the resistance-related parameters of the structure can be considered simultaneously. The proposed methodology has several advantages over all the other available methods to estimate reliability of elasto-plastic frame structures.

ELASTO-PLASTIC ANALYSIS OF NONLINEAR FRAME STRUCTURES

Most of the deterministic analyses of elasto-plastic frame structures under increasing loads are based on small deformation theory. However, in a lot of structural problems, the occurrence of plastic collapse is usually accompanied by geometrically large deformations. An efficient finite element analysis method for the elasto-plastic large deformed frame structures based on the assumed stress approach is used in this study for the structural analysis. Details of this method can be found elsewhere (Nee and Haldar 1988; Shi and Atluri 1988; Haldar and Nee 1989). Some key points of this method are discussed below.

For the elasto-plastic analysis, the following assumptions are made for the plastic properties of the frame structures:

1) The plastic deformations are developed only at the plastic hinges of an element, and the material is still linearly elastic except at plastic hinges; and

2) An elastic-perfectly-plastic material model is assumed.

For plane frames with solid rectangular cross-sections, the generalized yield function can be expressed as:

$$f = |\frac{\mathbf{M}}{\mathbf{M}_0}| + (\frac{\mathbf{N}}{\mathbf{N}_0})^2 - 1 = 0 \tag{1}$$

Where N_0 is the full plastic axial force and M_0 is the plastic bending moment. Other forms of yield functions can be used in place of equation (1).

In the loading process, the stress-strain relationship is elastically linear at the beginning, and the tangent stiffness is the same as in the elastic case. When the combination of the axial force and the bending moment at any node of the element satisfies the prescribed yield function, a plastic hinge occurs instantly at that node; the elasto-plastic tangent stiffness matrix $\mathbf{K}_p$ and the elasto-plastic internal force $\mathbf{R}_p$ of this element should be used to compute the response of the whole structure. By using the assumed stress approach, $\mathbf{K}_p$ and $\mathbf{R}_p$ can be expressed as:

$$\mathbf{K}_p = \mathbf{K} - \mathbf{A}_{\sigma do}^T \cdot \mathbf{A}_{\sigma\sigma}^{-1} \cdot \mathbf{V}_p \cdot \mathbf{C}_p^T \cdot \mathbf{A}_{\sigma do} \tag{2}$$

$$\mathbf{R}_p = \mathbf{A}_{\sigma do}^T(\mathbf{A}_{\sigma\sigma}^{-1} \cdot \mathbf{V}_p \cdot \mathbf{C}_p^T - \mathbf{A}_{\sigma\sigma}^{-1}) \cdot \tilde{\mathbf{R}}_\sigma + \mathbf{R}_{do} \tag{3}$$

where $\mathbf{K}$ is the elastic tangent stiffness matrix, $\mathbf{A}_{\sigma do}, \mathbf{A}_{\sigma\sigma}^{-1}, \mathbf{R}_{do}$ are the same as in the elastic case (Shi and Atluri 1988; Nee and Haldar 1988), and $\mathbf{V}_p, \mathbf{C}_p^T$ and $\tilde{\mathbf{R}}_\sigma$ are shown below:

$$\mathbf{V}_p = \{ \ -\frac{\partial f}{\partial \mathbf{N}}, \quad -\frac{\partial f}{\partial \mathbf{M}}(1 - \mathbf{x}_1/l), \quad -\frac{\partial f}{\partial \mathbf{M}}(\mathbf{x}_1/l) \ \}^t \tag{4}$$

$$\mathbf{C}_p^T = (\mathbf{V}_p^T \cdot \mathbf{A}_{\sigma\sigma}^{-1} \cdot \mathbf{V}_p)^{-1} \cdot \mathbf{V}_p^T \cdot \mathbf{A}_{\sigma\sigma}^{-1} \tag{5}$$

and

$$\tilde{\mathbf{R}}_\sigma = \mathbf{R}_\sigma + \begin{bmatrix} \mathbf{H}_p \\ \theta_p^*(1 - \mathbf{x}_1/l) \\ \theta_p^*(\mathbf{x}_1/l) \end{bmatrix} \tag{6}$$

Where $\mathbf{H}_p$ is the additional axial elongation of an element due to the presence of plastic hinges, it can be expressed as:

$$\mathbf{H}_p = \sum \Delta\lambda \frac{\partial f}{\partial \mathbf{N}}; \quad at \ \ \mathbf{x}_1 = l_p \tag{7}$$

θ_p^* is the additional relative rotation due to the presence of plastic hinges, and can be expressed as:

$$\theta_p^* = \sum \Delta\lambda \frac{\partial f}{\partial \mathbf{M}}; \quad at \ \ \mathbf{x}_1 = l_p \tag{8}$$

where $\Delta\lambda$ is a scalar factor of proportionality or plastic multiplier, and l_p is the location of the plastic hinge along the length of an element.

In the loading process, when plastic hinges formed in some elements, the corresponding element tangent stiffness matrices and internal forces can be computed from equations (2) and (3), respectively. The global tangent stiffness matrix $\tilde{\mathbf{K}}$ and the internal force matrix $\tilde{\mathbf{R}}$ are formed using appropriate values for $\mathbf{K}_p$, $\mathbf{R}_p$, $\mathbf{K}$, and $\mathbf{R}$ depending on whether the element is in the elastic or plastic state. The global displacement increment $\Delta\mathbf{D}$ can be obtained using an iteration strategy as given below:

$$\tilde{\mathbf{K}}^{(k)} \Delta\mathbf{D} = \mathbf{F}^{(k)} - \tilde{\mathbf{R}}^{(k-1)} \tag{9}$$

COMPUTATION OF PLASTIC LIMIT LOAD

Equation (9) is usually solved by applying the iterative Newton-Raphson (N-R) method. In the iteration solution procedure, the current tangent stiffness matrix is always computed based on the responses of the previous step. For the elasto-plastic structure, with the increase in applied external loadings, plastic hinges will occur at some critical points of the structure and will continue to form at other locations if the loadings continue to increase. When the presence of the plastic hinges increase the nodal displacements considerably, the corresponding element tangent stiffness matrices will be in ill condition. The global tangent stiffness matrix will deteriorate and equation (9) will not converge with the increased load step. The corresponding point is referred to as the plastic limit point. At the plastic limit point, the total load factor reaches its maximum value, and the displacements at some points of the structure increase with no change in the applied loading or even when the applied loading is decreased.

When the N-R method is used to solve equation (9) for the elasto-plastic geometrically nonlinear problems, numerical difficulties will be encountered when a plastic limit point is approached. A robust method of carrying out the analysis around or beyond the plastic limit point is the arc-length method (Nee and Haldar 1988). In this method, the N-R scheme is still used to obtain a convergent solution; however, the size of the load step within each increment will vary depending on the incremental arc-length Δl. For the displacement control arc-length algorithm, the incremental arc-length can be expressed as:

$$\Delta l^2 = (\delta \mathbf{D})^T \cdot (\delta \mathbf{D}) \tag{10}$$

and the increment of the load step can be computed as:

$$\Delta p = -\frac{a_1}{a_2} \tag{11}$$

where

$$a_1 = 2.\Delta \mathbf{D}_e^T(\delta \mathbf{D}_u^{(k-1)} + \Delta \mathbf{D}_u^{(k)}) \tag{12}$$

$$\begin{aligned} a_2 &= (\Delta \mathbf{D}_u^{(k)})^T \cdot (\Delta \mathbf{D}_u^{(k)}) + 2(\delta \mathbf{D}^{(k-1)})^T \cdot \Delta \mathbf{D}_u^{(k)} \\ &\quad + [(\delta \mathbf{D}^{(k-1)})^T \cdot (\delta \mathbf{D}^{(k-1)}) - \Delta l^2] \end{aligned} \tag{13}$$

In equations (10), (12) and (13), $\delta \mathbf{D}$ is the total displacement increment vector, $\Delta \mathbf{D}_u$ is the displacement vector due to the unbalanced force which can be expressed as:

$$\Delta \mathbf{D}_u^{(k)} = (\tilde{\mathbf{K}})^{-1}(p^{(k-1)} \cdot \mathbf{F} - \tilde{\mathbf{R}}^{(k-1)}) \tag{14}$$

and $\Delta \mathbf{D}_e$ is the displacement vector due to the external force which can be expressed as:

$$\Delta \mathbf{D}_e = (\tilde{\mathbf{K}})^{-1} \cdot \mathbf{F} \tag{15}$$

Since the load factor reaches its maximum value at the plastic limit point, this maximum load factor can be located by tracing the loading path of the structure. If the incremental load is positive at the $(k-1)$th step and is negative at the kth step, it implies that the maximum load factor must be between the $(k-1)$th and kth iteration steps. The correct maximum load factor can be computed using an interpolation method. In this study, a quadratic interpolation shown in Fig.1 is used to compute the maximum load factor. In the N-R iteration solution procedure, there is a particular load factor increment Δp_i corresponding to every load factor level p_i. Therefore, the load factor increment is a function of the load factor level, and the quadratic interpolation is based on this quadratic relationship. Three pieces of information are needed in this interpolation procedure: Δp_k, the load factor increments at the kth step, represents the first time when the load factor increment becomes negative; $\Delta p_{(k-1)}$ and $\Delta p_{(k-2)}$ are the two positive load factor increments before step k. p is the total load factor, and p_0 is the total load factor at the $(k-2)$th step. The load factor increment Δp can be found using the following equation:

$$\Delta p = a + bp + cp^2 \tag{16}$$

where

$$a = A \tag{17}$$

$$b = \frac{1}{AB(A+B)}(2AB^2 - A^3 - BA^2 + B^3 + B^2C) \tag{18}$$

$$c = \frac{1}{AB(A+B)}(AA - BB - AB - CB) \tag{19}$$

and

$$A = \Delta p_{(k-2)} \tag{20}$$

$$B = \Delta p_{(k-1)} \tag{21}$$

$$C = \Delta p_k \tag{22}$$

Once the constants in equations (17)-(19) are known, a positive solution p^* can be obtained from the following quadratic equation:

$$a + bp^* + cp^{*2} = 0 \tag{23}$$

The plastic limit load factor can then be obtained as:

$$p_m = p_0 + p^* \tag{24}$$

Because the occurrence of the plastic limit load is due to the deterioration of the global stiffness matrix, in the elasto-plastic geometrically nonlinear frame structure analysis, the plastic limit load will indicate the beginning of the plastic failure of the structure corresponding to a particular failure mechanism. It can be considered as the minimum plastic failure load. It should be noted that, for different structural configurations and loading conditions, a plastic limit load can correspond to a global failure or a local failure state.

SFEM-BASED RELIABILITY ANALYSIS

The reliability analysis of a highly redundant elastic-perfectly-plastic structure by the proposed plastic limit load method is based on the concept of the Advanced First Order Second Moment (AFOSM) reliability analysis method (Der Kiureghian and Ke 1985; Mahadevan and Haldar 1989). The plastic limit load method proposed here locates the very probable failure point in the transformed standard normal space using the linear programming technique. Thus, the corresponding failure probability, design point and the failure mechanism of the structure can be obtained very easily.

In the basic random variable space, there is a boundary between the safe and unsafe regions of the structure, and it can be described as:

$$g(\mathbf{x}) = g(x_1, x_2, ..., x_n) \tag{25}$$

The function in equation (25) is called the limit state function. The structure can be considered safe when the following inequality is satisfied:

$$g(\mathbf{x}) = g(x_1, x_2, ..., x_n) \leq 0 \tag{26}$$

In the context of the AFOSM, the starting point of the structural reliability analysis is the limit state function. By transforming the vector of basic random variables $\mathbf{x}$ into reduced variables $\mathbf{y}$ in a standard normal space, the limit state function G($\mathbf{y}$) becomes another n-dimensional surface in the standard normal space. The minimum distance from the origin to this limit state surface is defined as the reliability index β, which is a measure of reliability of the structure. The probability of failure can be obtained as $p_f = \Phi(-\beta)$. The reliability index β can be found by solving the following optimization problem:

$$\beta = \sqrt{\mathbf{y}^T\mathbf{y}} \tag{27}$$

under the condition:

$$G(\mathbf{y}) = 0 \tag{28}$$

For simple structures, the response can be expressed in a closed form in terms of all the basic random variables, and equations (27) and (28) can be solved in closed forms. However, for complicated structures, the response can only be obtained through a numerical algorithm, and the following iteration formula can be used to find the solution (Mahadevan and Haldar 1989):

$$\mathbf{y}_{i=1} = (\mathbf{y}_i^t \alpha_i + \frac{G(\mathbf{y}_i)}{|\nabla G(\mathbf{y}_i)|})\alpha_i \tag{29}$$

where $\nabla G(\mathbf{y})$ is the gradient vector of the limit state function and α_i is the unit vector normal to the limit state surface away from the origin. They can be expressed as:

$$\nabla G(\mathbf{y}) = (\frac{\partial G(\mathbf{y})}{\partial y_1}, \ldots \frac{\partial G(\mathbf{y})}{\partial y_N})^t \tag{30}$$

and

$$\alpha_i = -\frac{\nabla G(\mathbf{y}_i)}{|\nabla G(\mathbf{y}_i)|} \tag{31}$$

In the proposed plastic limit load method, the total structural failure probability needs to be computed only once. As discussed earlier, a plastic limit load can be computed for a deterministic structure corresponding to a unique failure mechanism. For a structure with random parameters and random external load $\mathbf{P}$, the plastic limit load $\mathbf{P}_m$ will be a function of basic random variables and every realization of the plastic limit load corresponds to a realization of random failure mechanisms. The region where all of the realizations of plastic limit load P_m are greater than the applied load P is the safe region, and the following equation should be the limit state function for the plastic failure problem:

$$g = P_m - P \tag{32}$$

Equation (32) is applicable when only one external load is applied to the structure. When more than one load is present, the plastic limit load will be a random vector. According to the plasticity theory, the critical collapse condition can be found when the work done by the plastic limit loads is equal to the work done by the internal forces. That is,

$$\sum_{i=1}^{N} \mathbf{P}_{m,i}\mathbf{D}_i = W \tag{33}$$

where N is the number of the applied forces, $\mathbf{P}_m$ is the plastic limit load vector, $\mathbf{D}$ is the displacement vector corresponding to $\mathbf{P}_m$, and W is the work done by the internal forces. If the work done by the load vector $\mathbf{P}$ is greater than W, the structure will collapse. So the limit state function of plastic collapse for the structure can be expressed as:

$$g = \sum_{i=1}^{N} (\mathbf{P}_{m,i} - \mathbf{P}_i)\mathbf{D}_i = 0 \tag{34}$$

It is obvious that when $N = 1$, equation (34) becomes equation (32).

In order to use equation (29) to find the design point, the gradient of the limit state function should be computed first, and it has the following expression:

$$\nabla G(\mathbf{y}) = \{\frac{\partial G}{\partial y_1}, \ \ldots, \ \frac{\partial G}{\partial y_k}, \ \ldots, \ \frac{\partial G}{\partial y_K}\} \tag{35}$$

where K is the number of basic random variables. The common term in equation (35) can be expressed as:

$$\frac{\partial G}{\partial y_k} = \sum_{i=1}^{N}\{\frac{\partial P_{m,i}}{\partial y_k}D_i - \frac{\partial P_i}{\partial y_k}D_i + (P_{m,i} - P_i)\frac{\partial D_i}{\partial y_k}\} \tag{36}$$

where

$$\frac{\partial P_i}{\partial y_k} = \begin{cases} 0 & \text{if } P_i = y_k; \\ 1 & \text{elsewhere.} \end{cases} \tag{37}$$

$\partial D_i/\partial y_k$ are elements of the gradient of the global displacement vector about the basic random variables and can be obtained by solving the following matrix equation:

$$\tilde{\mathbf{K}}\frac{\partial \mathbf{D}}{\partial y_k} = \frac{d\mathbf{P}}{dy_k} - \frac{\partial \tilde{\mathbf{R}}}{\partial y_k} \tag{38}$$

where $\tilde{\mathbf{K}}$ is the tangent stiffness matrix; $d\mathbf{P}/dy_k$ is an N-dimensional vector in which all elements are zero except when $P_i = y_k$; and $\tilde{\mathbf{R}}$ is the internal force matrix as discussed earlier. $\partial\tilde{\mathbf{R}}/\partial y_k$ and $\partial P_{m,i}/\partial y_k$ can be computed by the difference method. The forward difference scheme is used in this study.

After computing the gradient of the limit state function $\nabla_y G$, the unit vector α_i can also be obtained. Substituting them in equation (29), the new iteration point can be computed. The design point will be obtained when the iteration is convergent. Equation (27) then can be used to compute the reliability index. The methodology discussed here can be better explained with the help of examples.

NUMERICAL EXAMPLES

Example 1

A one-story plane frame structure is considered first, and its reliability is estimated using the proposed method. The geometry of the frame is shown in Fig.2. The Young's modulus E and the vertical load P are considered as normal random variables. E has a mean of 29,000 ksi and a coefficient of variation (COV) of 0.1; the COV for P is assumed to be 0.2, and its mean is subject to change. The frame consists of members with solid rectangular cross-sections. Equation (1) is used to express the yield function of the sections, where N_0 and M_0 are considered to be deterministic, with 100 kips and 500 kip-in, respectively. The limit state function for the collapse of the structure is represented by equation (32). Since there is only one load acting on the structure, no force interaction is involved and the plastic limit load has no relation to the applied load. The reliability indices are computed with respect to different mean values of P, and the results are shown in Fig.2. The result is compared with 10,000 cycles of Monte Carlo simulation.

Example 2

The same frame considered in Example 1 but subjected to different loading conditions is considered in this example. Equation (1) is also used as the yield function; N_0 and M_0 are considered to be random variables. N_0 and M_0 are normally distributed with means of 100 kips and 500 kip-in, respectively. The COVs for both variables are assumed to be 0.1. The frame is subjected to a vertical load P and a horizontal load H. P and H are considered to be normal random variables. The mean and COV of H are 2 kips and 0.15, respectively. The COV for P is 0.2, and its mean is subject to change. The limit state function for the collapse of the structure can be expressed as:

$$g = (H_{m,B} - H_B)u_B + (P_{m,A} - P_A)v_A \tag{39}$$

where v_A is the vertical displacement at point A and u_B is the horizontal displacement at point B; P_m and H_m are the vertical and horizontal plastic limit loads. Reliability indices of the structure are computed for different mean values of P, and the results are shown in Fig.3. This example is a general case in terms of random structural response, since the structural parameters as well as the loads are considered to be random and the load interaction effects are involved. It can be seen that the result by the proposed method matches very well with the Monte Carlo simulation results.

CONCLUSIONS

An assumed stress-based SFEM is proposed for the reliability analysis of elasto-plastic large deformed frame structures. Any frame or truss structures can be handled using this method, and the geometric and material nonlinearities can be considered at the same time. Uncertainties in the load and resistance-related parameters are considered. The proposed method has several advantages over the other available methods. The method is verified with two examples. The results match very well with Monte Carlo simulation.

ACKNOWLEDGEMENT

This paper is based upon work partly supported by the National Science Foundation under Grant No. MSM-8896267. Any opinions, findings and conclusions or recommendations expressed in this publication are those of the authors and do not necessarily reflect the views of the sponser.

REFERENCES

1. Bjerager, P. Plastic system reliability by LP and FORM, Computers & Structures, Vol. 31, No. 2, pp.187-196, 1989.
2. Der Kiureghian, A. and Ke, J-B. FEM-based reliability analysis of framed structures, Proc. 4th International conference on structural safety and reliability, ICOSSAR'85, Kobe, Japan, Vol.1, pp.375-381, 1985.

3. Haldar, A. and Nee, K.M. Elasto-plastic large deformation analysis of PR steel frames for LRFD, Comput. & Struct., Vol.31, No.5, pp.811-823, 1989.
4. Mahadevan, S. and Haldar, A. Stochastic finite element-based structure reliability analysis and optimization, J. of Struct. Eng., ASCE, Vol.115, No.7, pp.1579-1598, 1989.
5. Murotsu, Y., Okada,H., Taguchi,K., Grimmelt,M., and Yonezawa, M. Automatic generation of stochastically dominant failure modes of frame structures, Structural Safety, No. 2, pp.17-25, 1984.
6. Nee, K.M. and Haldar, A. Elastoplastic nonlinear post-buckling analysis of partially restrained space structures, Comp. Meth. in App. Mech. and Engrg., No.71, pp.69-97, 1988.
7. Quek, S.T. and Ang, A.H-S. Reliability analysis of structural systems by stable configurations, J. Struct. Engrg., ASCE, Vol.116, No.10, pp.2656-2670, 1990.
8. Shi, G. and Atluri, S.N. Elasto-plastic large deformation anslysis of space-frames: A plastic-hinge and stress-based explicit derivation of tangent stiffness, Int. J. Num. Meth. Engrg., Vol.26, pp.589-615, 1988.
9. Siões, L. M. C. Reliability of portal frames with interacting stress resultants, J. Struct. Engrg., Vol.116, No.12, pp.3475-3497, 1990.
10. Wang, T.Y., Corotis, R.B. and Ellingwood, B. Limit state sensitivity of structural frames subjected to cycled forces, J. Struct. Engrg., ASCE, Vol.116, No.10, pp.2824-2841, 1990.

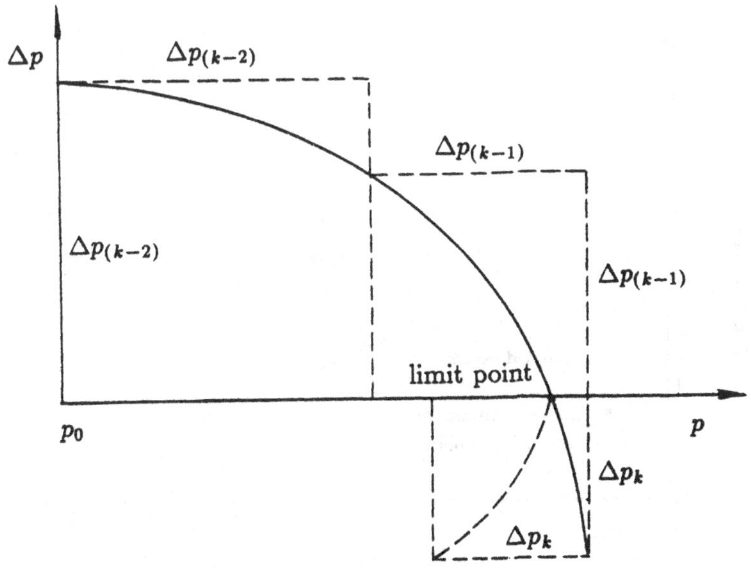

Fig.1 Interpolation to find plastic limit load

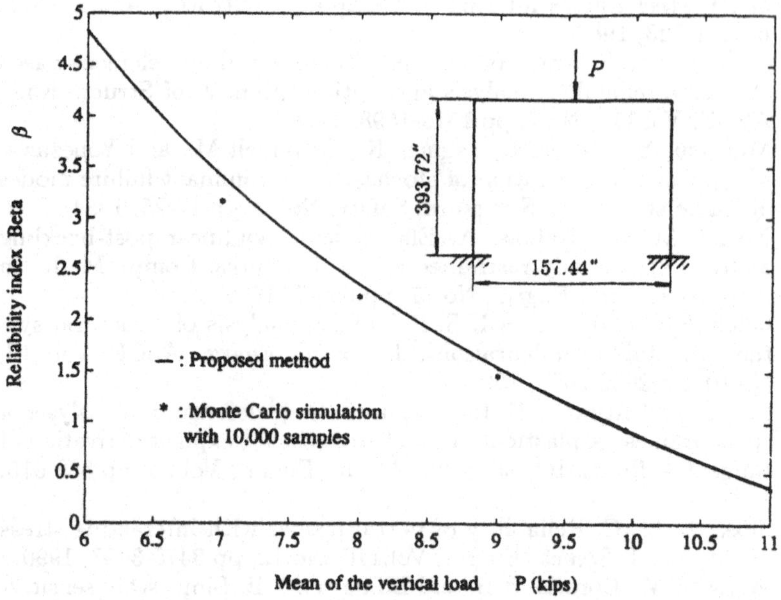

Fig.2 Reliability indices changed with the vertical load (one load case)

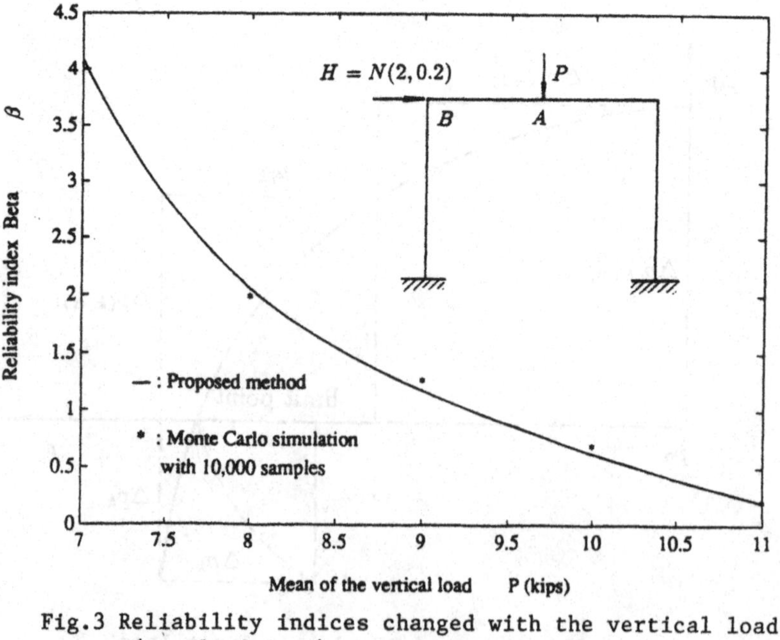

Fig.3 Reliability indices changed with the vertical load (one load case)

Stochastic Methods for Systems with Uncertain Properties under Plane Stress/Strain

C.G. Bucher, W.A. Wall
Institute of Engineering Mechanics, University of Innsbruck, Technikerstr. 13, A-6020 Innsbruck, Austria

ABSTRACT

For the analysis of systems with stochastic properties such as random variability of the elastic modulus various methods have been developed. The present paper aims at a comparison of two particular approaches to this type of analysis. The methods compared are the (analytical) perturbation method and the recently developed (numerical) Stochastic Finite Element Method (SFEM) based on so-called weighted integrals. The potential advantages of the SFEM are clearly shown by the extension of this recent method from one-dimensional structures to two-dimensional systems under plane stress/strain conditions. The comparison is performed with the aid of numerical examples dealing with circular plates.

INTRODUCTION

Stochastic analysis of systems with uncertain properties became increasingly important in the last few years. While the methodology for deterministic systems under stochastic loading has progressed considerably in to the range of practical applicability there is still lack of competitive procedures for the analysis of systems with stochastic parameters. One of the major problems arises from the fact that the system properties generally must be described in terms of continuous-type random fields. This, at least in principle, introduces an uncountable number of random variables into the analysis. Typically, however, modern structural analysis methods are based on discrete type formulations, i.e. Finite Element Methods. Although this method and its accuracy are well established by now its extension into problems of system stochasticity must be accompanied by - preferably closed-form - alternative solutions for verification purposes. This enables the evaluation of different discretization strategies with respect to both accuracy and efficiency.

Several early approaches in Stochastic Finite Elements utilized a rather crude approximation method by pointwise describing the randomness of the system properties in terms of discrete random variables. These methods imply restriction of the element size to be small as compared to the correlation distance of the random field. This might, at least in some cases, be a severe drawback when comparing to deterministic FE analysis.

An alternative way of describing the random fields is given in terms of weighted integrals, i.e. in terms of (weighted) averages over a particular domain, generally an element of the FE mesh. This method has been developed by Takada and Shinozuka, 1989, Deodatis, 1991 as well as Deodatis and Shinozuka, 1991 for primarily one-dimensional truss and frame structures. Extensions to 2D problems like plates under plane stress/strain conditions have been suggested by Deodatis et al., 1991 as well as Takada, 1991. This method is utilized in the subsequent comparative study in order to show the advantages and possible disadvantages of Stochastic Finite Element Methods.

STATEMENT OF PROBLEM

The static analysis of continuous elastic structures usually requires the solution of a boundary value problem (equilibrium equations)

$$\mathbf{D}_1 E\, \mathbf{D}_2 \mathbf{u}(x,y,z) = \mathbf{X}(x,y,z) \tag{1}$$

where $\mathbf{u}=(u, v, w)$ is the vector of displacements in 3D (x, y, z)-space, $\mathbf{D}_1$ and $\mathbf{D}_2$ represent matrix valued linear spatial differential operators, E is Young's modulus and $\mathbf{X}$ is the vector of distributed volume forces. For a plane stress/strain problem neglecting the effect of Poisson's ratio the above equation reduces to

$$\mathbf{D}_1 E(x,y)\, \mathbf{D}_2 \begin{bmatrix} u(x,y) \\ v(x,y) \end{bmatrix} = \begin{bmatrix} X(x,y) \\ Y(x,y) \end{bmatrix} \tag{2}$$

where

$$\mathbf{D}_1 = \begin{bmatrix} \frac{\partial}{\partial x} & \frac{\partial}{\partial y} & 0 \\ 0 & \frac{\partial}{\partial x} & \frac{\partial}{\partial y} \end{bmatrix}; \ \mathbf{D}_2 = \begin{bmatrix} \frac{\partial}{\partial x} & 0 \\ \frac{1}{2}\frac{\partial}{\partial y} & \frac{1}{2}\frac{\partial}{\partial x} \\ 0 & \frac{\partial}{\partial y} \end{bmatrix} \tag{3}$$

with appropriate boundary conditions on u, v, and their derivatives. In the following, it will be assumed that the elastic properties as described by Young's modulus $E(x, y)$ are given in terms of a homogeneous random field with a mean value function

$$E_0 = \in\left[E(x,y)\right] \tag{4}$$

and an autocovariance function

$$\kappa_{EE}(\sqrt{(x_2 - x_1)^2 + (y_2 - y_1)^2}) = \in\left[E(x_1,y_1)E(x_2,y_2)\right] - E_0^2 \tag{5}$$

A closed-form solution for the above boundary-value problem appears to be impossible unless the spatial variation of $E(x, y)$ is of very special form. Hence approximate solutions must be sought, either analytically by means of a perturbation approach or numerically by employing Finite Element Methods.
It should be mentioned that frequently the result of such an analysis should be given in terms of stresses rather than displacements or strains since this type of response quantity provides more useful information for design purposes such as reliability analyses or structural optimization.

PERTURBATION APPROACH

If the random field $E(x, y)$ can be written in terms of

$$E(x,y) = E_0 + \varepsilon E_1(x,y) \tag{6}$$

where ε is a small parameter then the solution vector $\mathbf{u}(x, y)$ may be expanded into a power series with respect to ε

$$\mathbf{u}(x,y) = \mathbf{u}_0(x,y) + \varepsilon\mathbf{u}_1(x,y) + \varepsilon^2\mathbf{u}_2(x,y) + \ldots \tag{7}$$

Convergence of this series is expected for $|\varepsilon| < 1$. By inserting eq.(7) into eq.(2) and ordering with respect to powers of ε the following hierarchy of equations is obtained

$$\begin{aligned}
\varepsilon^0: &\quad \mathbf{D}_1 E_0\, \mathbf{D}_2 \mathbf{u}_0 = \mathbf{X} \\
\varepsilon^1: &\quad \mathbf{D}_1 E_0\, \mathbf{D}_2 \mathbf{u}_1 = -\mathbf{D}_1 E_1\, \mathbf{D}_2 \mathbf{u}_0 \\
\varepsilon^2: &\quad \mathbf{D}_1 E_0\, \mathbf{D}_2 \mathbf{u}_2 = -\mathbf{D}_1 E_1\, \mathbf{D}_2 \mathbf{u}_2 \\
&\quad \vdots
\end{aligned} \tag{8}$$

This set of equations clearly shows that successive solutions of the deterministic system with random loads describe the behavior of the stochastic system. In the following, the analysis will be restricted to the terms of first order in ε.
The most general formulation of the solution to the boundary value problem eq.(8) is formally given in terms of the respective Green's function $\mathbf{h}(x, y)$. This matrix valued function allows an integral representation of the solution $\mathbf{u}_0(x, y)$

$$\mathbf{u}_0(x,y) = \iint_A \mathbf{h}(x,y,\xi,\eta)\mathbf{X}(\xi,\eta)\,d\xi\,d\eta \tag{9}$$

where the region of integration A is the area of the plate structure. By introducing the (fictitious) load vector $\mathbf{X}_1(x, y)$ given by

$$\mathbf{X}_1(x,y) = -\mathbf{D}_1 E_1(x,y)\, \mathbf{D}_2 \mathbf{u}_0(x,y) \tag{10}$$

the first order perturbation term $\mathbf{u}_1(x, y)$ can be written analogously as

$$\mathbf{u}_1(x,y) = \iint_A \mathbf{h}(x,y,\xi,\eta)\mathbf{X}_1(\xi,\eta)\,d\xi\,d\eta \tag{11}$$

where $\mathbf{h}(.)$ denotes the same Green's function as given in eq.(9).
The stresses $\mathbf{S} = (\sigma_x, \tau_{xy}, \sigma_y)$ (to first order in ε) can be calculated from

$$\mathbf{S} \equiv \mathbf{S}_0 + \varepsilon \mathbf{S}_1 = E_0\, \mathbf{D}_2 \mathbf{u}_0 + \varepsilon(E_1\, \mathbf{D}_2 \mathbf{u}_0 + E_0\, \mathbf{D}_2 \mathbf{u}_1) \tag{12}$$

The perturbation term in the last part of eq.(12) obviously contains both local (pointwise) and global contributions of $E(x, y)$ to the variability of the stresses. In contrast, the displacement field $\mathbf{u}_1(x, y)$ is influenced by the global properties of $E(x, y)$ only.

Finally, the statistical moments of displacements **u** and stresses S can be calculated from eqs.(11) and (12) by taking appropriate mathematical expectations.

This procedure is completely analogous to the analytical approach as suggested by Bucher and Shinozuka, 1988 as well as Kardara et al., 1989.

STOCHASTIC FINITE ELEMENT METHOD

Stochastic Element Stiffness Matrix

Consider the constant stress/strain triangular element shown in Fig. 1 with six degrees of freedom and linear shape functions in the displacement field. It is assumed that the elastic modulus of the element may be described by a homogeneous stochastic field with autocorrelation function $R(\xi,\eta)$. Therefore, the distribution of Young's modulus can be written as

$$E_{(x,y)}^{(e)} = E_0^{(e)}\left[1+f_{(x,y)}^{(e)}\right] \tag{13}$$

where $E_0{}^{(e)}$ = the mean value of the elastic modulus; x, y denote the global cartesian coordinates; and $f_{(x,y)}{}^{(e)}$ = two-dimensional, univariate, zero-mean, homogeneous stochastic field.

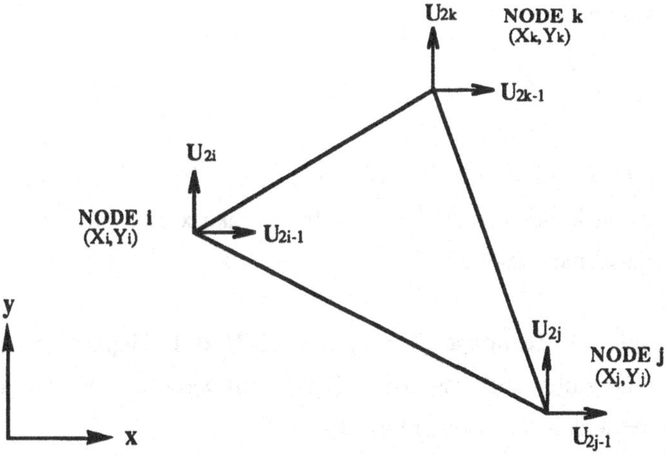

FIG. 1. Triangular finite element with six degrees of freedom

The volume integral defining the element stiffness matrix is given by

$$\mathbf{K}^{(e)} = \int_{V^{(e)}} \mathbf{B}^{(e)T}\mathbf{D}^{(e)}\mathbf{B}^{(e)}\, dV^{(e)} \tag{14}$$

where $\mathbf{B}^{(e)}$ denotes the gradient matrix, containing the derivatives of the shape functions and $\mathbf{D}^{(e)}$ denotes the materials property matrix containing the randomly distributed Young's modulus. For a CST-element, the gradient matrix is constant within one element and therefore eq.14 can be written as

$$\mathbf{K}^{(e)} = t^{(e)}\mathbf{B}^{(e)T} \int_{A^{(e)}} \mathbf{D}^{(e)} dA\ \mathbf{B}^{(e)} \tag{15}$$

Using the formulation for the distribution of Young's modulus from eq.(1) and integrating with respect to $A^{(e)}$, the stochastic element stiffness matrix is found to be

$$\mathbf{K}^{(e)} = \mathbf{K}_0^{(e)} + X_0^{(e)}\Delta\mathbf{K}_0^{(e)} \tag{16}$$

where the matrix $\mathbf{K}_0^{(e)}$, which is the mean value of matrix $\mathbf{K}^{(e)}$, and $X_0^{(e)}\Delta\mathbf{K}_0^{(e)}$ can be denoted as deterministic and stochastic part of the element stiffness matrix $\mathbf{K}^{(e)}$, respectively.

Finally, after introducing the appropriate boundary conditions, the equations of equilibrium are given by

$$\mathbf{K}\ \mathbf{U} = \mathbf{P} \tag{17}$$

where $\mathbf{U}$ = global nodal displacement vector, $\mathbf{P}$ = deterministic global force vector and $\mathbf{K}$ denotes the global stiffness matrix calculated as the sum of the element stiffness matrices.

Analysis of Response Variability of Nodal Displacements

Since the global stiffness matrix $\mathbf{K}$ contains weighted integrals which in this case are random variables of the form

$$X_{e0} = \int_{A^{(e)}} f_{(x,y)}^{(e)}\, dA^{(e)} \tag{18}$$

(cf. eq.(13, 15, 16)) with mean values $\overline{X}_{e0}$ all equal to zero the global nodal displacement vector **U** is also a function of these random variables. The first-order Taylor expansion of function **U** around their mean values can be derived as

$$\mathbf{U} \cong \mathbf{U}_0 + \sum_{e=1}^{N_e} X_{e0}\left[\frac{\partial \mathbf{U}}{\partial X_{e0}}\right]_E \tag{19}$$

where the symbol $[\;]_E$ denotes evaluation at the mean values of random variables X_{e0}. Introducing the notations $\mathbf{U}_0$ and $\mathbf{K}_0$ for the values of the respective matrices evaluated at the mean values of X_{e0}, and using eq.17, the expressions appearing in eq.19 are found to be

$$\mathbf{U}_0 = \mathbf{K}_0^{-1}\;\mathbf{P} \tag{20}$$

$$\left[\frac{\partial \mathbf{U}}{\partial X_{e0}}\right]_E = -\mathbf{K}_0^{-1}\left[\frac{\partial \mathbf{K}}{\partial X_{e0}}\right]_E \;\mathbf{U}_0 \tag{21}$$

Hence, the first-order approximation of the mean value and the covariance matrix of global nodal displacement vector **U** are easily evaluated as

$$\epsilon[\mathbf{U}] = \mathbf{U}_0 \tag{22}$$

$$\mathrm{Cov}\,[\mathbf{U},\mathbf{U}] = \sum_{e_1=1}^{N_e}\sum_{e_2=1}^{N_e} \mathbf{K}_0^{-1}\Delta\mathbf{K}_0^{(e_1)}\mathbf{U}_0\mathbf{U}_0^T\Delta\mathbf{K}_0^{(e_2)}\left(\mathbf{K}_0^{-1}\right)^T \cdot \int_{A^{(e_1)}}\int_{A^{(e_2)}} R_{ff}(\xi,\eta)\; dA^{(e_1)}dA^{(e_2)} \tag{23}$$

Analysis of Response Variability of Stresses

Utilizing the internal nodal forces given by

$$\mathbf{S}^{(f)} = \mathbf{K}^{(f)}\;\mathbf{U}^{(f)} \tag{24}$$

to calculate the averages of the stress field of finite element (f), $\sigma^{(f)}$, the mean value and the variance vector of $\sigma^{(f)}$ are evaluated as

$$\epsilon\left[\ \sigma^{(f)}\ \right] = \mathbf{G}^{(f)}\ \tilde{\mathbf{S}}_0^{(f)} \tag{25}$$

$$\begin{aligned}
\mathrm{Var}\left[\ \sigma^{(f)}\ \right] &= \\
&= \mathrm{diag}\left(\mathbf{G}^{(f)}\Delta\tilde{\mathbf{K}}_0^{(f)}\mathbf{U}_0^{(f)}\right)\ \mathbf{G}^{(f)}\Delta\tilde{\mathbf{K}}_0^{(f)}\mathbf{U}_0^{(f)} \cdot \int_{A^{(f)}}\int_{A^{(f)}} R_{ff}(\xi,\eta)\ dA^{(f)}dA^{(f)} \\
&- 2\sum_{e=1}^{N_e} \mathrm{diag}\left(\mathbf{G}^{(f)}\tilde{\mathbf{K}}_0^{(f)}\left\{\mathbf{K}_0^{-1}\Delta\mathbf{K}_0^{(e)}\mathbf{U}_0\right\}_{(f)}\right)\ \mathbf{G}^{(f)}\Delta\tilde{\mathbf{K}}_0^{(f)}\mathbf{U}_0^{(f)} \\
&\cdot \int_{A^{(e)}}\int_{A^{(f)}} R_{ff}(\xi,\eta)\ dA^{(e)}dA^{(f)} \\
&+ \sum_{e_1=1}^{N_e}\sum_{e_2=1}^{N_e} \mathrm{diag}\left(\mathbf{G}^{(f)}\tilde{\mathbf{K}}_0^{(f)}\left\{\mathbf{K}_0^{-1}\Delta\mathbf{K}_0^{(e_1)}\mathbf{U}_0\right\}_{(f)}\right)\ \mathbf{G}^{(f)}\tilde{\mathbf{K}}_0^{(f)}\left\{\mathbf{K}_0^{-1}\Delta\mathbf{K}_0^{(e_2)}\mathbf{U}_0\right\}_{(f)} \\
&\cdot \int_{A^{(e_1)}}\int_{A^{(e_2)}} R_{ff}(\xi,\eta)\ dA^{(e_1)}dA^{(e_2)}
\end{aligned} \tag{26}$$

where $\{\ \}_{(f)}$ denotes a vector, consisting only of the components of $\{\ \}$ corresponding to the degrees of freedom of element (f) and $\mathbf{G}^{(f)}$ is given by

$$\mathbf{G}^{(f)} = \frac{1}{t^{(f)}A^{(f)}}\left(\tilde{\mathbf{B}}^{(f)\ T}\right)^{-1} \tag{27}$$

where $\tilde{}$ denotes a submatrix of the respective matrix, as can be seen in Wall, 1991.

NUMERICAL EXAMPLES

Verification of Perturbation Approach

A circular plate with uniform axial loading is considered in the following (cf. Fig. 2)

Let the elastic properties be described by

$$E(r,\varphi) = E_0\left[1+\varepsilon\frac{r}{R}\cos 2\varphi\right] \tag{28}$$

where $\varepsilon = 0.2$ is assumed.

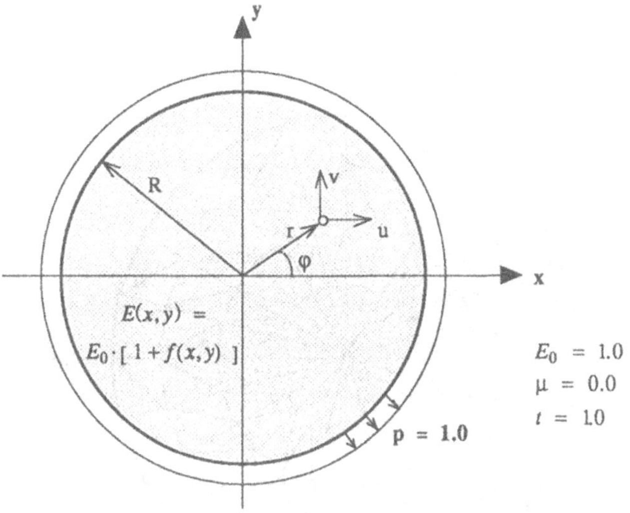

FIG. 2. Circular Plate Under Uniform Radial Loading

Fig. 3 compares the resulting radial stresses along the axis $\varphi = 0$ obtained from a first order perturbation to those from a Finite Element analysis based on 168 triangular CST elements. The Green's function for this particular problem is derived according to Muskhelishvili (1971). The Finite Element mesh is shown in Fig.4.

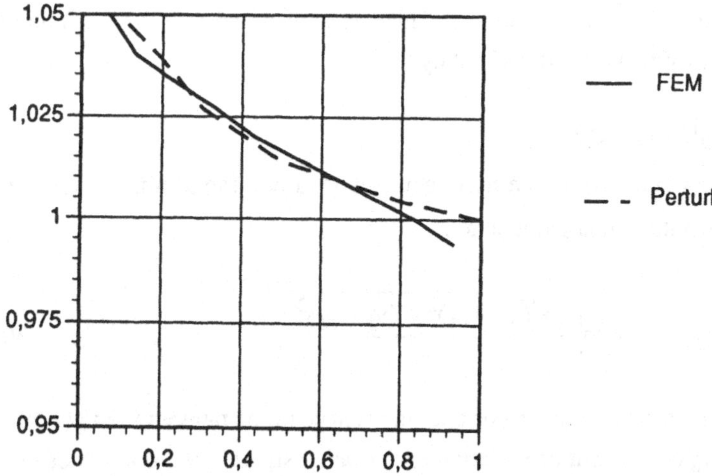

FIG. 3. Radial Stress from Perturbation Approach Compared to FEM result

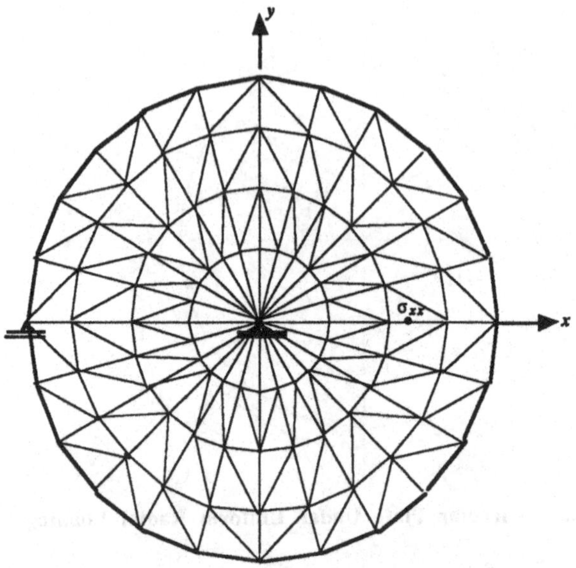

FIG. 4. FE mesh for Circular Plate Under Uniform Radial Loading

The results clearly show that a first order perturbation may yield accurate results for systems with spatially varying elastic properties.

Circular plate with Stochastic Young's modulus

Again a circular plate with uniform load is considered (cf. Fig.2). Young's modulus is assumed to be described by

$$E(x,y) = E_0\left[1+\varepsilon g(x,y)\right] \tag{29}$$

where $\varepsilon = 0.1$ and $g(x, y)$ is a homogeneous isotropic random field with zero mean and an autocorrelation function

$$R_{gg}(x_1,x_2,y_1,y_2) = \exp\left(-\frac{\sqrt{(x_2-x_1)^2+(y_2-y_1)^2}}{b}\right) \tag{30}$$

In this equation, b denotes the correlation distance of the random field $g(x, y)$. The resulting coefficient of variation of the radial stress in the point A of Fig.4 from the perturbation apprach based on Green's functions is plotted in Fig.5. In the same figure, the results from SFEM based on weighted integrals are plotted as well.

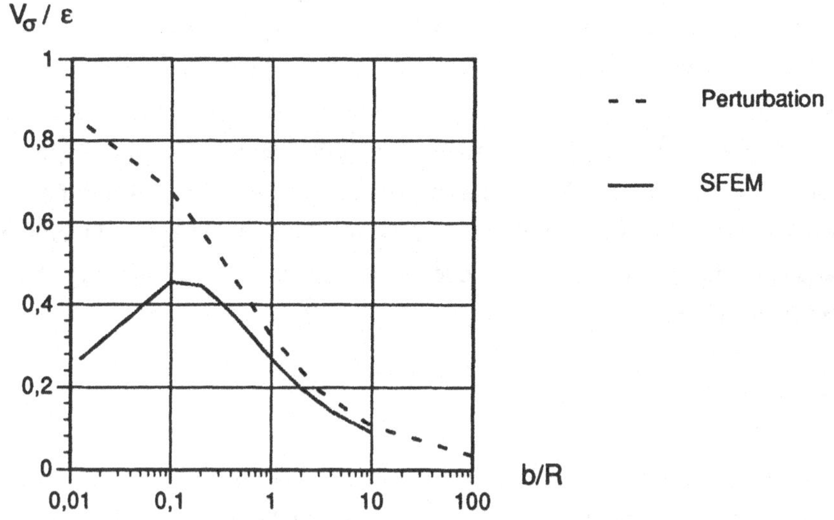

FIG. 5. Coefficient of Variation of Radial Stress in Circular Plate Under Uniform Radial Loading

The results agree quite well for values of the correlation length b > 0.3R. At very small correlation lengths the local effects present in the continuum-based perturbation approach become dominant while these effects are averaged out in the SFEM. Since for most problems the stress in a finite region is much more important than stress at a particular point (e.g. for crack propagation) the above difference is not a significant shortcoming of the SFEM. On the other hand, the Green's functions as utilized above are not available for general structures so that there is a considerable advantage in the SFEM.

CONCLUSIONS

The numerical results indicate that the perturbation method can be utilized as a tool for calibration and verification in the context of the development of Stochastic Finite Element Procedures. The inherent differences between the analytical method based on Green's functions for a boundary value problem and the numerical procedure based on weighted integrals become evident when cases of low spatial correlation are investigated. Otherwise, the agreement in the statistical moments of the stresses is found to be excellent. This indicates that, particularly due to its versatility in modeling complex system geometries, the SFEM is a very efficient method of analysis. It should be noted, however, that both methods as presented here primarily aim at a probabilistic description in terms of first and second moments. They are not suitable for a more sophisticated model e.g. in terms of probability densities of the response quantity under investigation. A natural extension of the SFEM would be the development of efficient and accurate Monte Carlo techniques (including variance reduction) in order to provide the desired probabilistic information by statistical methods.

ACKNOWLEDGEMENT

This research has ben partially supported by the Austrian Industrial Research Promotion Fund under Grant No. 6/546 to the Institute of Engineering Mechanics (Director: G.I. Schuëller) at the University of Innsbruck.

REFERENCES

Bucher, C.G., Shinozuka, M. (1988) "Response Variability II", *Journal of Engineering Mechanics*, ASCE, Vol.114, No.12, Dec. 1988, pp 2035 - 2054.

Deodatis, G. (1991). "The weighted integral method.I:Stochastic stiffness matrix." to appear in *Journal of Engineering Mechanics*, ASCE, August 1991

Deodatis, G., Wall, W.A., Shinozuka, M. (1991). "Analysis of two-dimensional stochastic systems by the weighted integral method". Proc, 1st Int. Conf. Computational Stochastic Mechanics, 17-19 Sept 1991, Corfu, Greece.

Deodatis, G., Shinozuka, M. (1991). "The weighted integral method.II:Response variability and Reliability." to appear in *Journal of Engineering Mechanics*, ASCE, August 1991

Kardara, A., Bucher, C.G., Shinozuka, M.: "Response Variability III", *Journal of Engineering Mechanics*, ASCE, Vol.115, No.8, Aug.1989, pp 1726 - 1747.

Muskhelishvili, N.I. (1971). "Einige Grundaufgaben zur mathematischen Elastizitätstheorie" (translated from Russian), Carl Hanser Verlag, München.

Takada, T., Shinozuka M. (1989). "Local Integration Method in Stochastic Finite Element Analysis", Proc. ICOSSAR '89, Vol. II, pp. 1073-1080.

Takada, T. (1991). Fundamental Study on Application of Stochastic Field Theory to Engineering", ORI Report 91-02, Ohsaki Research Institute, Shimizu Corp, March 1991.

Wall, W. A. (1991). "Entwicklung eines stochastischen finite Elemente Verfahrens für dynamische Analyse." Masters Thesis, University of Innsbruck, June 1991.

SECTION 12: FLOW RELATED APPLICATIONS AND CHAOTIC DYNAMICS

Nonlinear Response and Sonic Fatigue of High Speed Aircraft

R. Vaicaitis, P. Kavallieratos

Department of Civil Engineering and Engineering Mechanics, Columbia University, New York, NY 10027, U.S.A.

ABSTRACT

The surface thermal protection systems of advanced high speed aircraft such as the National Aerospace Plane (NASP) will be constructed from high temperature resistant composite and intermetallic materials. The various dynamic response studies indicate that under severe aerodynamic, acoustic and thermal loads the response of these surface panels is nonlinear. The equations of motion for nonlinear response of composite surface panels are developed in space-time domain. A time domain Monte Carlo type approach is used to determine solutions and fatigue live predictions. The formulation includes surface loads due to nonsteady aerodynamic flow, cavity and static pressure, parametric excitations, turbulent boundary layer and engine exhaust noise, oscillating surface shocks and aerodynamic heating. Numerical results are presented for a simplified laminated composite panel.

INTRODUCTION

Dynamic response characteristics and acoustic fatigue are important factors in the structural design and safety of modern high speed aircraft. In many cases, sonic fatigue is the controlling factor in assessing reliability of high speed aircraft. For the NASP configuration, the surface temperatures exceed 3000° F and noise levels reach 185 decibels.[1-4] To withstand these severe environments, new composite and intermetallic materials are being developed.[4-6] However, the information on dynamic response and fatigue characteristics of these materials is limited and new research efforts are required in order that the advanced high speed aircraft could become a reality.

The main objective of the present study is to develop a time domain analytical model for nonlinear response and fatigue of composite surface panels. The governing nonlinear equations of motion are solved utilizing simulation techniques of stationary random processes, a Galerkin like modal methods and a numerical integration procedure. The required statistical quantities such as moments, probability density function, peak distribution and crossing rates are calculated directly from the response time history. A preliminary fatigue model based on constant amplitude S-N data, cumulative damage rule and theoretical histogram of stress peaks is constructed.

PROBLEM FORMULATION AND SOLUTION PROCEDURE

Consider a thin rectangular composite panel exposed to surface flow, high intensity acoustic noise and thermal heating. Following the procedures developed in Refs. 7-11, the governing nonlinear equations of motion of thin laminated composite panels can be derived. If rotary inertia, inplane inertia, transverse shear and bending coupling due to midplane stretching are retained, the coupled nonlinear equations of motion are very cumbersome and require lengthy numerical solutions. For the thin composite panel considered in this study, these effects are neglected and the governing equations of motion can be written as

$$\begin{aligned} &D_{11}w_{,xxxx}+4D_{16}w_{,xxxy}+2\,(D_{12}+2D_{66})\,w_{,xxyy}+\\ &4D_{26}w_{,xyyy}+D_{22}w_{,yyyy}-F_{,yy}w_{,xx}+2F_{,xy}w_{,xy}\\ &-F_{,xx}w_{,yy}-N_x^b w_{,xx}-N_{y,yy}^b+M_{x,xx}^T+2M_{xy,xy}^T\\ &+M_{y,yy}^T+cw_{,t}+\rho w_{,tt}=p^r(x,y,t)-p^c(x,y,t)\\ &+p^a(x,y,t)+p_{shock}(x^*,y^*,t)+\Delta p \end{aligned} \tag{1}$$

$$\begin{aligned} &A_{11}F_{,yyyy}+A_{22}F_{,xxxx}-2A_{26}F_{,xxxy}+(2A_{12}+A_{66})F_{,xxyy}\\ &-2A_{16}F_{,xyyy}=w_{,xy}^2-w_{,xx}w_{,yy}-K(N_x^T)-L(N_y^T)-M(N_{xy}^T) \end{aligned} \tag{2}$$

in which w is the normal displacement to the panel surface, a comma followed by subscripts indicates differentiation with respect to independent variables x and y, D_{ij} are bending stiffnesses, F is the Airy stress function, $N^b{}_x$ and $N^b{}_y$ are inplane loads applied at the boundaries, c is viscous damping coefficient, ρ is mass of the composite panel per unit area, p^r is the random surface pressure due to turbulent flow and/or engine exhaust noise, p^c is the cavity back-up pressure acting on the lower face of the panel, p^a is the nonsteady aerodynamic pressure due to convected flow and panel motions, p_{shock} is the localized pressure at $x=x^*$ and $y=y^*$ induced by impinging shocks, Δp is the static pressure differential and A_{ij} are the membrane compliances of the laminated panel. The operators K,L and M are defined as

$$K = A_{11}\partial^2/\partial y^2 + A_{12}\partial^2/\partial x^2 - A_{16}\partial^2/\partial x\partial y \tag{3}$$

$$L = A_{12}\partial^2/\partial y^2 + A_{22}\partial^2/\partial x^2 - A_{26}\partial^2/\partial x\partial y \tag{4}$$

$$M = A_{16}\partial^2/\partial y^2 + A_{26}\partial^2/\partial x^2 - A_{66}\partial^2/\partial x\partial y \tag{5}$$

The thermal parameters M^T_x, M^T_y, M^T_{xy}, N^T_x, N^T_y and N^T_{xy} can be determined from

$$\begin{Bmatrix} M_x^T \\ M_y^T \\ M_{xy}^T \end{Bmatrix} = \sum_{k=1}^{N} \int_{z_{k-1}}^{z_k} (\bar{Q}_{ij})_k \begin{Bmatrix} \alpha_x \\ \alpha_y \\ \alpha_{xy} \end{Bmatrix} \Delta T \cdot z dz \tag{6}$$

$$\begin{Bmatrix} N_x^T \\ N_y^T \\ N_{xy}^T \end{Bmatrix} = \sum_{k=1}^{N} \int_{z_{k-1}}^{z_k} (\bar{Q}_{ij})_k \begin{Bmatrix} \alpha_x \\ \alpha_y \\ \alpha_{xy} \end{Bmatrix} \Delta T dz \tag{7}$$

in which α_x, α_y, α_{xy} are the directional thermal expansion coefficients, $\Delta T(x,y,z)$ is the temperature distribution in the panel, z_k denotes the distance from the neutral surface to the k-th layer of the laminated composite and $(\bar{Q}_{ij})_k$ are the transformed reduced directional stiffness coefficients. The mass per unit area of a composite panel can be calculated from

$$\rho = \sum_{k=1}^{N} \rho_0^{(k)} (z_k - z_{k-1}) \tag{8}$$

where $\rho_0^{(k)}$ is the material density of the k-th layer.

The nonsteady aerodynamic pressure p^a and the cavity pressure p^c are functions of the panel motions. Depending on surface flow and cavity conditions, solutions for p^a and p^c have been developed.[12-14] When the surface panel vibrations are highly nonlinear, a non-linear piston theory might need to be utilized for the nonsteady aerodynamic pressure p^a.[15] The random pressure p^r and localized pressures induced by impinging shocks can be obtained utilizing the statistical information of their cross-spectral densities and simulation techniques of random processes.[16-20] These simulation procedures could be extended to certain classes of nonstationary and non-Gaussian random processes.[20]

To solve the nonlinear equations (1) and (2), panel boundary conditions need to be prescribed. Exact boundary conditions for the Airy stress function F are very complicated and, for the present study, the inplane boundary conditions are satisfied on the average.[13,14] Furthermore, the panel is assumed to be simply supported on all four edges. Then, panel deflections are expanded in terms of panel modes as

$$w(x,y,t)=\sum_{m=1}^{\infty}\sum_{n=1}^{\infty} q_{mn}(t)\, X_{mn}(x,y) \tag{9}$$

in which q_{mn} are modal amplitudes and X_{mn} are the corresponding modes. For a simply supported panel

$$X_{mn}=\sin(m\pi x/a)\cdot\sin(n\pi y/b) \tag{10}$$

The solution for F consists of a homogeneous and a particular solution. These solutions were presented in Refs. 13,21 for homogeneous plates and in Ref. 22 for an orthotropic plate. Similar procedure can be implemented to equation (2). However, the final result is very lengthy and it is not presented in this paper. After the solution for Airy stress function F is known, these results and equation (9) are substituted into equation (1) and then solved using the Galerkin method by computing the integral average of equation (1) weighted by each term of equation (9). The result is a system of nonlinear coupled differential equations in the generalized coordinate $q_{mn}(t)$. The solution to these nonlinear differential equations can be developed utilizing a step by step numerical integration procedure.

The nonlinear stresses at each lamina layer can be calculated from

$$\begin{Bmatrix}\sigma_x\\ \sigma_y\\ \tau_{xy}\end{Bmatrix}_k=[\bar{Q}]_k\begin{Bmatrix}A_{11}(N_x+N_x^T)+A_{12}(N_y+N_y^T)-z_k w_{,xx}\\ A_{12}(N_x+N_x^T)+A_{22}(N_y+N_y^T)-z_k w_{,yy}\\ A_{66}(N_{xy}+N_{xy}^T)-2z_k w_{,xy}\end{Bmatrix} \tag{11}$$

where

$$N_x=\partial^2F/\partial y^2,\ N_y=\partial^2F/\partial x^2,\ N_{xy}=-\partial^2F/\partial x\partial y \tag{12}$$

and $[\bar{Q}]_k$ is the matrix of transformed stiffness coefficients.

The displacement and stress response time histories calculated from equations (9) and (11) are obtained for each realization of the simulated random pressure $p^r(x,y,t)$ and/or $p_{shock}(x^*,y^*,t)$. These solutions would need to be repeated for a number of different realizations and then the response statistics calculated using ensemble averages in a Monte Carlo sense. However, by assuming the input pressures to be ergodic random processes, it is sufficient to obtain a solution for only one realization and then use temporal averaging to calculate the required response statistics. Following this procedure, the various statistical moments, histograms of probability density function, peak distribution and crossing rates can be determined directly from the nonlinear response time history of displacement or stress.

SONIC FATIGUE

The prediction of fatigue life in metal structures has been based either on a cumulative damage rule[23,24] or fracture mechanics. The fracture mechanics theories are nonlinear and deal with stress intensity at the crack tip rather than the continuum state stress in the cumulative damage approach. However, for composite and intermetallic materials, the fracture mechanics models seem to fail in reliable prediction of fatigue damage. In the present study, a preliminary fatigue damage model based on experimental data from constant amplitude coupon tests and theoretical histogram of stress peak distribution is suggested.

The nonlinear nature of the stress response gives a non-Gaussian distribution of stress amplitudes. Fatigue damage estimates for non-Gaussian and nonlinear response are available only for few limited cases. In developing fatigue damage prediction for nonlinear and non-Gaussian stress response that is consistent with the present time domain approach, we consider the expected total damage accumulated from $t=0$ to $t=\tau$[25]

$$E[D(\tau)] = \frac{E[M_T]\,\tau}{B}\int_{-\infty}^{\infty}|S|^{\lambda}p_I(S)\,dS \qquad (13)$$

where $E[M_T]$ is the expected total number of peaks in the selected time interval, B and λ are experimental constants from constant amplitude tests, S is the stresses and p_I is the probability density function of the stress peak magnitude. In developing equations (13), it was assumed that the stress response process is stationary. The histogram of stress peak magnitude, p_I, is calculated directly from stress response time history and the integral in equation (13) is evaluated numerically. The fatigue life of a surface panel is the time duration of the total damage to reach unity.

NUMERICAL EXAMPLE

As an example of time domain approach, the nonlinear response and fatigue life of rectangular panels made from 6Al-4V titanium material and laminate composite are presented.[22] The random input is assumed to be uniformly distributed Gaussian white noise with lower and upper cut-off frequencies of 0 Hz and 500 Hz, respectively. The random pressure is taken as the sole input to the panel. In addition, in-plane loads and thermal effects are neglected.

Isotropic Panel

The numerical results were obtained for a panel with the following material and geometric properties: a=20 in., b=8.2 in., h=0.06 in., $E=16 \times 10^6$ psi, $\nu=0.34$, $\rho=h \times .000414$lb-sec^2/in^4. The modal damping coefficients were selected as $\zeta_{mn} = 0.02\ (\omega_{mn}/\omega_{11})$ where ω_{mn} are the natural frequencies of a linear panel.

The root-mean-square (rms) normal stress σ_{yy} evaluated at z=h/2 and the middle of the panel is plotted in Fig. 2 versus the rms of the input pressure. For comparison, results of linear time domain and PSD solutions are also included in this figure. It can be seen that for high input levels the linear response predictions overestimate the actual nonlinear response by a large amount. A probability density and peak distribution histograms of stress component σ_{yy} are presented in Fig. 3. As can be observed from these results, the nonlinear response is no longer Gaussian and peaks do not follow the Rayleigh distribution. Fatigue life predictions were obtained for $\lambda=6$ and $B=1.518 \times 10^6$. When using these parameters, the stress amplitude is in units of ksi. By selecting the histogram of p_I as shown in Fig. 3 and setting $E[D(\tau)] = 1$ in equation 13, the fatigue life at 140 dB (167 dB overall) input is 18.7 hrs (nonlinear) and 2.393 hrs (linear). The total number of expected stress peaks per second is 556 (nonlinear) and 124 (linear). Thus, fatigue life predictions based on a linear theory underestimate the fatigue of a nonlinear panel by a large amount. The fatigue life estimates presented in this paper should be viewed as preliminary merely to illustrate the time domain analysis.

Composite Panel

The panel shown in Fig. 1 is assumed to be an orthotropic panel composed of eight lamina layers with lay-up [0/+45/-45/90] and overall thickness of 0.0416 inches. Each layer is made from A-S/3501 Graphite/Epoxy.[22]

The time histories of linear and nonlinear stress response taken from Ref. (22) are shown in Figs. 4 and 5. Note that the mean value of the nonlinear stress is not zero. These results indicate that the use of linear strain-displacement relationship is not warranted for applications involving high levels of acoustic loading.

CONCLUSIONS

The time domain analysis presented in this paper indicates that for severe input pressure levels acting on thermal protection systems of advanced high speed aircraft, the various linearized theories used to predict response and sonic fatigue life might not be realistic. The time domain approach could provide a viable and practical alternative for response and fatigue life predictions of flight structures.

REFERENCES

1. Mixon, J.S. and Roussos, L.A. Acoustic Fatigue: Overview of Activities at NASA Langley, NASA TM-89143, 1987.

2. Murrow, H.N. and Powell, C.A. Hypersonic Loads and Response Workshop, NASA Langley Research Center, Hampton, VA, 1989.

3. Ellis, P.A. Overview-Design of an Efficient Lightweight Airframe Structure for the National Aerospace Plane, Proceedings of the AIAA/ASME/ASCE/AHS/ASC 30th SDM Conference, Paper No. 89-1046, Mobile, AL, 1989.

4. Cazier, F.W., Jr., Doggett, V., Jr., and Ricketts, R.H. Structural Dynamic and Aeroelastic Considerations for Hypersonic Vehicles, Proceedings of the AIAA/ASME/ASCE/-AHS/ASC 32nd SDM Conference, Paper No. 91-1255, Baltimore, MD, 1991.

5. Prescott, C. Hypersonic Vehicle Requirements, Proceedings of the AISS/ASME/ASCE/AHS/ASC 32nd SDM Conference, Paper No. 91-1252, Baltimore, MD, 1991.

6. Hjelm, L. Results of NASP Materials and Structures Program, Proceedings of the AIAA/ASME/ASCE/AHS/ASC 32nd SDM Conference, Paper No. 91-1254, Baltimore, MD, 1991.

7. Jones, R.M. Mechanics of Composite Materials, McGraw-Hill, New York.

8. Vinson, J. and Sierakowski, R. The Behavior of Structures Composed of Composite Materials, Martinus Nijhoft.

9. Chia, C.C. Nonlinear Analysis of Plates, McGraw-Hill, New York, 1960.

10. Sathyamoorthy, M. and Chia, C.Y. Nonlinear Vibration of Anisotropic Rectangular Plates Including Shear and Rotatory Inertia, Fibre Science and Technology, Vol. 13, pp. 337-361, 1980.

11. Prasad, C. and Mei, C., Effects of Transverse Shear on Large Deflection Random Response of Symmetric Composite Laminates with Mixed Boundary Condition, Proceedings of AIAA/ASME/ASCE/AHS/ASC 30th SDM Conference, Paper No. 89-1356, Mobile, AL, 1989.

12. Dowell, E.H. Aeroelasticity of Plates and Shells, Noordhoft International Publishing, Leydon, 1975.

13. Dowell, E.H. Transmission of Noise from a Turbulent Boundary Layer Through a Flexible Plate into a Closed Cavity, Journal of the Acoustical Society of America, Vol. 46, No. 1, Pt. 2, 1969.

14. Vaicaitis, R., Dowell, E.H. and Ventres, C.S. Nonlinear Panel Response by a Monte Carlo Approach, AIAA Journal, Vol. 12, No. 5, pp. 685-691, 1974.

15. Vaicaitis, R. Time Domain Approach for Nonlinear Response and Sonic Fatigue of NASP Thermal Protection Systems, Proceedings of the AIAA/ASME/ASCE/AHS/ASC 32nd SDM Conference, Paper No. 91-1177, Baltimore, MD, 1991.

16. Langanelli, A.L. Prediction of the Pressure Fluctuations on Maneuvering Re-Entry Weapons, AFWAL-TR-83-3133, 1984.

17. Maestrello, L. Radiation from and Panel Response to a Supersonic Turbulent Boundary Layer, Journal of Sound and Vibration, Vol. 10, No. 2, pp. 261-295, 1969.

18. Coe, C.F. and Chyu, W.J. Pressure Fluctuation Inputs and Response of Panels Underlying Attached and Separated Supersonic Turbulent Boundary Layers, NASA TM X-62, p. 189, 1972.

19. Shinozuka, M. and Jan, C.M. Digital Simulation of Random Processes and its Applications, Journal of Sound and Vibration, Vol. 25, No. 1, pp. 111-128, 1972.

20. Shinozuka, M. (Ed.). Stochastic Mechanics, Vol. I and II, Department of Civil Engineering and Engineering Mechanics, New York, 1987.

21. Hong, H.-K. and Vaicaitis, R. Nonlinear Response of Double Wall Sandwich Panels, Journal of Structural Mechanics, 12(4), pp. 483-503, 1985.

22. Vaicaitis, R. and Arnold, R.R. Time Domain Monte Carlo Approach for Nonlinear Response and Sonic Fatigue, Proceedings of AIAA 13th Aeroacoustics Conference, Paper no. 90-3938, Tallahassee, FL, 1990.

23. Jacobson, M.J. Advanced Composite Joints: Design and Acoustic Fatigue Characteristics, AFFDL-TR-71-126, Air Force Flight Dynamics Laboratory, Wright Patterson Air Force Base, Ohio, 1972.

24. Swanson, S.R. Random Load Fatigue Testing: A State of the Art Survey, Materials Research and Standards, Vol. 8, No. 4, pp. 10-44, 1968.

25. Lin, Y.K. Probabilistic Theory of Structural Dynamics, McGraw-Hill, New York, 1967.

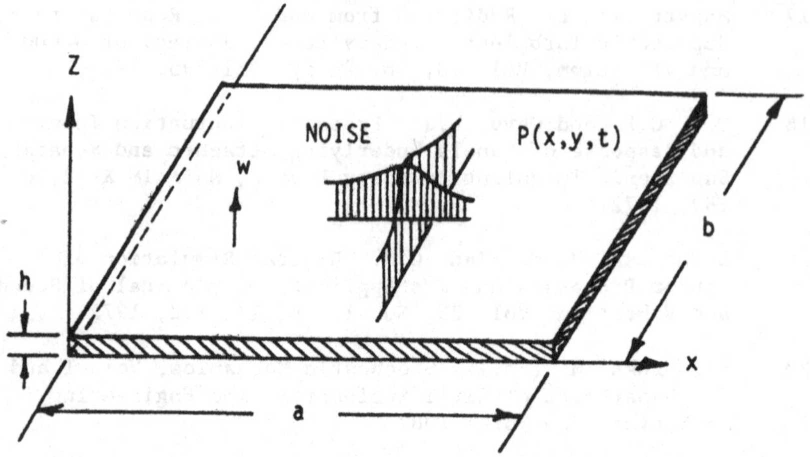

Fig. 1 A rectangular panel exposed to random pressure

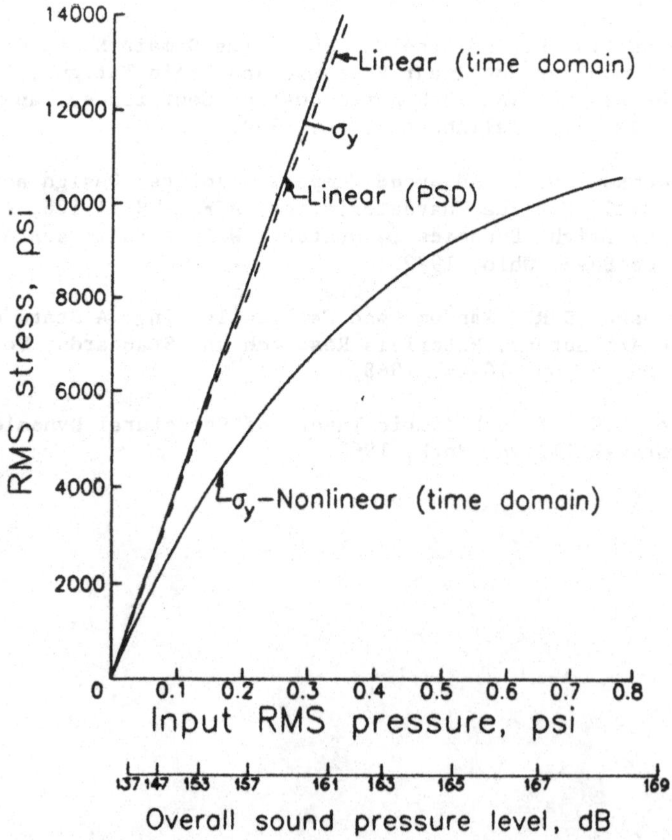

Fig. 2 Linear and nonlinear root-mean-square response (rms) of normal stress component σ_{yy}

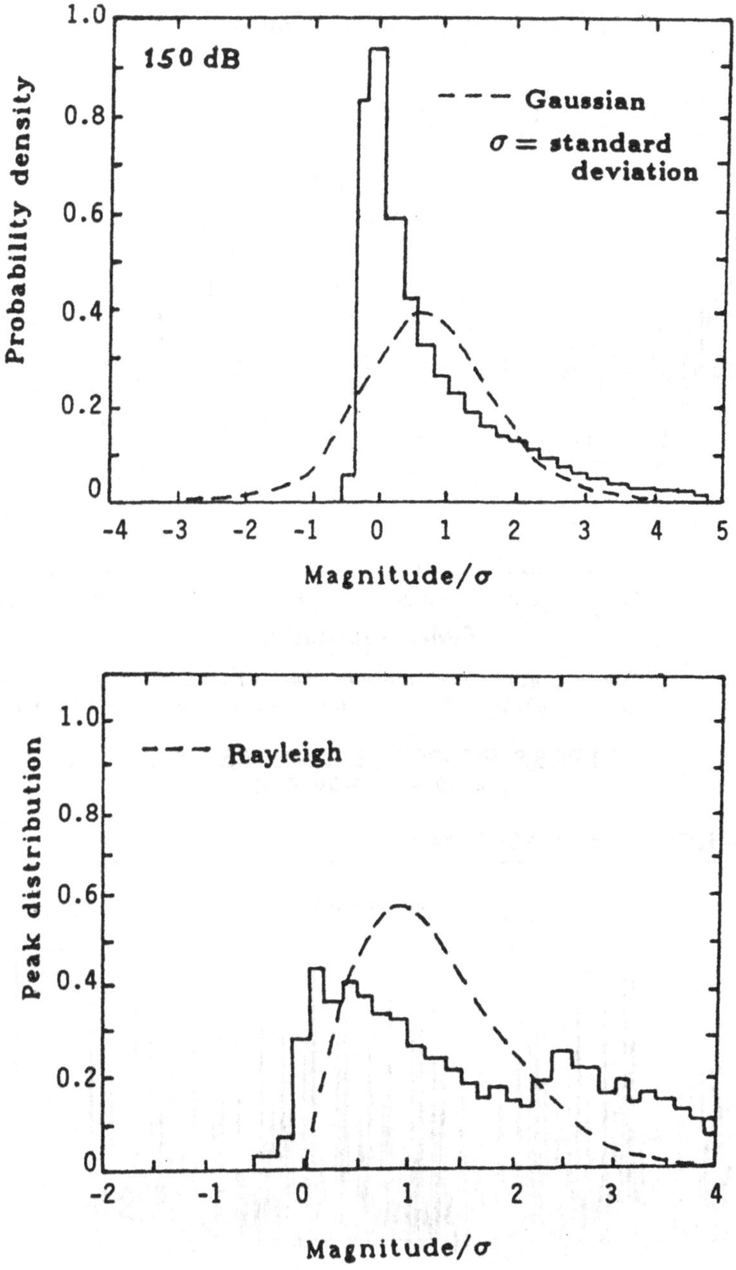

Fig. 3 Probability density and peak distribution histograms of normal stress σ_{yy} for 150 dB input

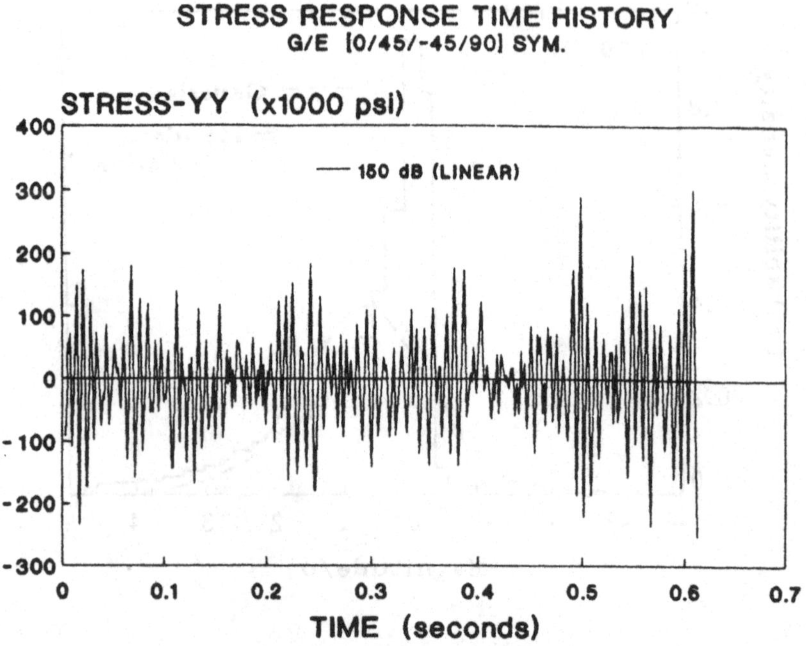

Fig. 4 The σ_{yy} stress response time history for a composite panel at 150 dB input (linearized solution)

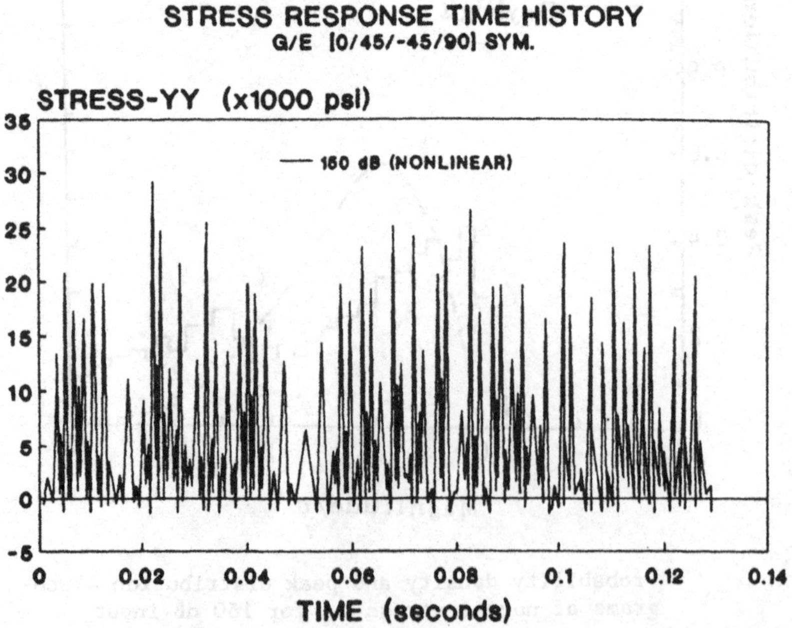

Fig. 5 The σ_{yy} stress response time history for a composite panel at 150 dB input (nonlinear solution)

Necessary Condition for Homoclinic Chaos Induced by Additive Noise

E. Simiu (*), M. Frey (**), M. Grigoriu (***)
(*) Building and Fire Research Laboratory, and (**) Computing and Applied Mathematics Laboratory, National Institute of Standards and Technology, Gaithersburg, MD 20899, U.S.A.
(***) Department of Civil Engineering, Cornell University, Ithaca, NY 14853, U.S.A.

ABSTRACT

The effect of noise on the possible occurrence of chaos in systems with a homoclinic orbit was recently investigated in the literature on the basis of a redefinition of the Melnikov function. The purpose of this note is to show that, even in the case of deterministic equations, this redefinition is not consistent with the geometry of the perturbed orbits and would therefore lead to incorrect solutions. The possibility is then explored of developing a necessary condition for the occurrence of homoclinic chaos in forced systems perturbed additively by a commonly used approximate representation of white noise.

INTRODUCTION

The effect of noise on the possible occurrence of chaos in systems with a homoclinic orbit (e.g., the Duffing equation) was recently the object of studies by Kapitaniak[1], who based his work on numerical solutions of Ito equations, and by Bulsara, Schieve, and Jacobs[2], and Schieve and Bulsara[3], who adopted an approach based on a redefinition of the Melnikov function. In this note we analyze this redefinition, and show that it is inconsistent with the original concept of the Melnikov function, and that it yields incorrect results even in the case of deterministic equations. We then explore an alternative approach, and obtain a necessary condition for the occurrence of chaos in forced systems perturbed additively by a commonly used approximate representation of white noise (see, e.g., Rice[4]).

MELNIKOV FUNCTION: CASE OF TWO SMALL PERTURBATIONS

Effect of Single Perturbation and of Sum of Two Perturbations

Consider the two-dimensional conservative system

$$\ddot{x} + f(x) = 0 \tag{1}$$

where $f(x)$ is a nonlinear function of the dynamic variable $x(t)$, such that Equation 1 possesses a homoclinic trajectory $x_s(t)$. We now introduce in the right-hand side of Equation 1 the perturbation $\epsilon[\gamma\cos\omega t - k\dot{x}(t)]$, $\epsilon > 0$. For ϵ sufficiently small, the saddle point of Equation 1 perturbs to a nearby hyperbolic invariant manifold. The Melnikov function is related to the separation between those manifolds (e.g., Arrowsmith[5], pp. 172, 177), and has the expression

$$M(t_o, \theta_o) = \int_{-\infty}^{\infty} \dot{x}_s(t)[\gamma\cos(\omega t + \omega t_o + \theta_o) - k\dot{x}_s(t)]dt \tag{2}$$

(Wiggins[6], p. 507). A clear explanation of the geometric meaning of t_o and θ_o is available in Wiggins, p. 487. For the purposes of this note it is convenient to keep constant the time t_o which defines the particular orbit being considered and vary the angle $\theta_o = \omega\tau$ (where τ is a running parameter with the dimension of time) at which the Poincaré section is carried out, see Wiggins[6], p. 507.

We now consider the perturbation

$$\epsilon[\gamma\cos\omega t - k\dot{x}(t)] + F(t) \tag{3}$$

where $F(t) = \epsilon\gamma_1\cos\omega_1 t$. The Melnikov function will then be

$$M(t_o, \theta_o, \theta_1) = M(t_o, \theta_o) + \int_{-\infty}^{\infty} \dot{x}_s(t)\gamma_1\cos(\omega_1 t + \omega_1 t_o + \theta_1)dt \tag{4}$$

It is seen that, for any set of values t_o, θ_o, θ_1, the corresponding correction $\Delta M = M(t_o, \theta_o, \theta_1) - M(t_o, \theta_o)$, which accounts for the effect of the perturbation $F(t)$, involves only the parameters of Equation 1 and those of the perturbing function $F(t)/\epsilon$. In particular, <u>ΔM does not depend upon the parameter k</u>. A similar observation can be made if we introduce additional perturbations $\epsilon\gamma_n\cos\omega_n t$ (see, e.g, Wiggins[7], pp. 461-467).

<u>Example</u>. We consider for definiteness the case of the Duffing oscillator [$f(x) = -x + x^3$] with the perturbation given by Equation 3, where $F(t)/\epsilon = \gamma_1\cos\omega_1 t$. We assume that $\gamma > 0$, $\gamma_1 \neq 0$ (we do not impose the restriction $\gamma_1 > 0$), and that ω, ω_1 are incommensurate. Following Wiggins[6] (p. 463) and Wiggins[7] (p. 516), the necessary condition for the occurrence of chaos is

$$-4/3k + \gamma S(\omega) + |\gamma_1| S(\omega_1) > 0 \tag{5}$$

where

$$S(\omega) = \sqrt{2}\omega \operatorname{sech}(\pi\omega/2) \tag{6}$$

It is clear that, for any given parameters k and ω, the presence of the perturbation F(t) lowers the minimum (threshold) value of γ for which the occurrence of chaos is possible. If, instead of confining ourselves to one amplitude γ_1, we consider an ensemble of values γ_1 (such as for Gaussian-distributed γ_1), then the average effect of the perturbation F(t) is still to lower the threshold for γ. This is because $E[|\gamma_1|] > 0$.

Critique of a Proposed Alternative Approach

We now describe briefly an approach proposed by Bulsara, Schieve, and Jacobs[2]. A small perturbation F(t) is introduced in Equation 1, and the hyperbolic invariant manifold resulting from the corresponding perturbation of the homoclinic orbit is denoted by

$$x(t) = x_s(t) + \delta x(t) \tag{7}$$

A Melnikov function, corresponding to the perturbation $\epsilon[\gamma\cos\omega t - k\dot{x}(t)] + F(t)$, is then redefined by analogy with Equation 2, and written (in a slightly different notation) as follows:

$$M_b = \int_{-\infty}^{\infty} \dot{x}(t)[\gamma\cos(\omega t + \theta_o) - k\dot{x}(t)]dt \tag{8}$$

Neglecting higher order terms, the correction to the redefined Melnikov function is then obtained from Equations 8, 7, and 2 as follows:

$$\Delta M_b = -2k \int_{-\infty}^{\infty} \dot{x}_s(t)\delta\dot{x}(t)dt + \gamma\int_{-\infty}^{\infty} [\cos(\omega t + \theta_o)]\delta\dot{x}(t)dt - k\int_{-\infty}^{\infty} \delta\dot{x}^2(t)dt \tag{9}$$

As in the earlier example, we consider again an ensemble of smooth perturbations F(t) with zero mean. The corresponding ensemble average of ΔM_b would: (1) depend on k, and (2) be negative, that is, on the average, the threshold for chaos would be raised by the presence of the perturbation F(t), since the total value of the redefined Melnikov function would, on the average, be reduced. These two incorrect results yielded by Equation 9 are due to the redefinition of the Melnikov function implicit in Equation 8. That redefinition can be interpreted as meaning that it is permissible to apply the original Melnikov approach with reference to a set of manifolds that is already separated. In fact the total Melnikov distance should be referenced with respect to the homoclinic separatrix of the unperturbed equation.

MELNIKOV FUNCTION AND APPROXIMATE REPRESENTATION OF NOISE

Approximate Representation of Noise

Band-limited white noise $W_{\omega_o}(t)$ with spectral density

$$f(\omega) = \begin{cases} \sigma^2, & 0 \leq \omega \leq \omega_o \\ 0, & \omega > \omega_o \end{cases}$$

has the Wiener integral representation

$$W_{\omega_o}(t) = \sigma \int_0^{\omega_o} \cos\omega t \, dU_\omega + \sigma \int_0^{\omega_o} \sin\omega t \, dV_\omega \qquad (10)$$

where U, V are independent standard Brownian motions. It is easily verified using properties of the Wiener integral[8] that

$$\lim_{\omega_o \to \infty} E[W_{\omega_o}(t) W_{\omega_o}(t + t_1)] = \sigma^2 \delta(t_1) \qquad (11)$$

Approximating the integrals in Equation 10 by sums, a common representation of noise is obtained (Rice[4]):

$$Z(t) = \sigma\left[\frac{\omega_o}{N}\right]^{1/2} \sum_{n=1}^{N} (A_n \cos \omega_n t + B_n \sin\omega_n t) \qquad (12)$$

where $\omega_n = n\omega_o/N$ and $\{A_n, B_n; n = 1, 2, \ldots, N\}$ are independent Gaussian random variables with zero mean and unit variance. Independence of the random variables is a consequence of the independent increment property of the Brownian motions U_ω, V_ω. The noise representation in Equation 12 is attributed to Nyquist and is used, e.g., in electrical, seismic, and wind engineering applications.

Application to the Duffing Oscillator

It can be shown (Wiggins[6], p. 516) that the Melnikov function corresponding to the perturbation of Equation 1 by the function

$$\epsilon[-k\dot{x}(t) + \gamma\cos\omega t + Z(t)] \qquad (13)$$

is

$$M(t_o, \theta_o, \theta_1, \theta_2, \ldots, \theta_N) = -4k/3 + \gamma S(\omega)\sin(\omega t_o + \theta_o)$$
$$+ \mu_1(t_o, \theta_1, \theta_2, \ldots, \theta_N; \omega_o) \qquad (14)$$

where

$$\mu(t_o,\theta_1,\ldots,\theta_N;\omega_o) =$$

$$\left[\frac{\omega_o}{N}\right]^{1/2} \sum_{n=1}^{N} S(\omega_n)[A_n\sin(\omega_n t_o+\theta_n) + B_n\cos(\omega_n t_o+\theta_n)] \qquad (15)$$

A necessary condition for the occurrence of chaos is that the right-hand side of Equation 15 have simple zeroes. Such is the case provided that

$$4k/3 < \max[\gamma S(\omega)\sin(\omega t_o+\theta_o) + \mu(t_o,\theta_1,\ldots,\theta_N;\omega_o)] \qquad (16)$$

In Equation 16 and in subsequent expressions the maximum is taken over all possible values of θ_o, $\theta_1,\ldots,\theta_N$ [recall that $\theta_o = \omega\tau$, $\theta_j = \omega_j\tau$ $(j = 1, 2,\ldots,N)$]. If ω and ω_o are incommensurate, then this maximum can be slightly simplified to give the following condition for chaos:

$$4k/3 < \gamma S(\omega) + \sigma \max[\mu(t_o,\theta_1,\ldots,\theta_N;\omega_o)] \qquad (17)$$

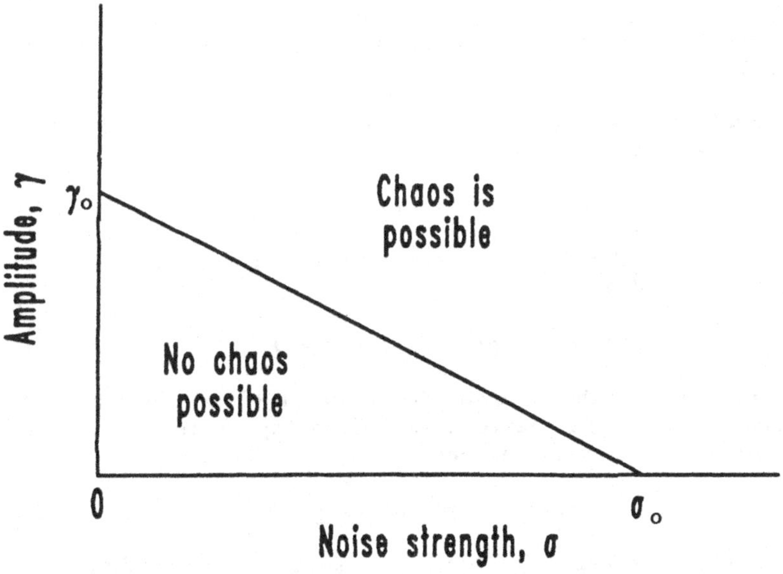

Figure 1. γ-threshold as a function of noise strength, σ.

The expected value of the maxand in Equation 17 is zero, so the expectation of the maximum is positive. Hence the γ-threshold for chaos is lowered, on the average, by the presence of the approximate white noise perturbation Z(t). Indeed, as shown by Inequality 17, the γ-threshold for chaos has a negative linear relationship to σ. This is illustrated in Figure 1. The area above the line in Figure 1 is the region of the γ-σ parameter space where, on the average, chaotic dynamics are possible. Let

$$\gamma_o = \frac{4k}{3S(\omega)}, \qquad \sigma_o = \frac{4k}{3E\{\max[\mu(t_o,\theta_1,\ldots,\theta_N;\omega_o)]\}} \tag{18,19}$$

Equations 18 and 19 give, respectively, the minimum forcing amplitude γ necessary for the occurrence of chaos in the absence of noise, and the noise strength σ necessary for the occurrence of chaos in the absence of the perturbation $\gamma\cos\omega t$. The linear relationship between γ and σ pointed out here is a result of the linearity with respect to effects of a sum of perturbations inherent in the Melnikov method.

We have found that, on the average, stochastic perturbation of the Duffing equation by Z(t) lowers the threshold condition for homoclinic chaos. In fact, a stronger statement can be made. The γ-threshold is lowered with probability one. To see this, observe that the average of the function μ with respect to τ is zero for any realization of the random coefficients in Z(t). Thus, the maximum of μ is positive with probability one. Hence, barring zero-probability events, the presence of Z(t) always lowers the γ-threshold.

We evaluate the maximum in Inequality 17 for the particular case N = 1. We note that

$$A_1\sin(\omega_1 t_o + \theta_1) + B_1\cos(\omega_1 t_o + \theta_1) = R\cos(\omega_1 t_o + \theta_1 + \Phi) \tag{20}$$

where $R_1 = (A_1{}^2 + B_1{}^2)^{1/2}$ is a standard Raleigh random variable with probability density

$$f(r) = r\exp(-r^2/2), \quad r > 0 \tag{21}$$

Φ is uniformly distributed over the interval $[0,2\pi]$, and R_1 and Φ are independent[9]. The maximum in Inequality 17 is now simply $S(\omega_1)R_1$ and Inequality 17 becomes

$$4k/3 < \gamma S(k) + \sigma\surd\omega_1\ S(\omega_1)R_1 \tag{22}$$

Thus, for N = 1 the effect of Z(t) is to shift the γ-threshold for chaos by an amount $-\sigma\surd\omega_1 R_1\ S(\omega 1)/S(\omega)$. Since $R_1 > 0$ with probability 1, the presence of Z(t) does indeed always lower the γ-threshold. Since $E[R_1] = (\pi/2)^{1/2}$, on the average the threshold is lowered by the amount $\sigma\surd\omega_1(\pi/2)^{1/2}S(\omega_1)/S(\omega)$.

The form of the maxand in Inequality 17 is quite suggestive. Considering that the sums composing Z(t) are derived from a Wiener integral representation of white noise, one might make the conjecture that for the Duffing oscillator perturbed by band-limited noise, $W_{\omega o}(t)$, the sums in the maxand of Inequality 17 should be replaced by Wiener integrals. For a Duffing oscillator with the perturbation

$$\epsilon[-kx(t) + \gamma\cos\omega t + W_{\omega_o}(t)] \tag{23}$$

a necessary condition for the occurrence of chaos might formally be written

$$\frac{4k}{3} < \gamma S(\omega) + \sigma \max_{t_o,\theta(\omega)} \{\int_o^{\omega_o} S(\omega)\sin[\omega t_o+\theta(\omega)]dU_\omega + \int_o^{\omega_o} S(\omega)\cos[\omega t_o+\theta(\omega)]dV_\omega\} \tag{24}$$

However, while the validity of Inequality 17 as a condition for chaos in the presence of Z(t) is clear, the validity of Inequality 24 as a condition for chaos in the presence of band-limited noise has not been established.

Finally, we note that the relation between results obtained here for systems subjected to a perturbation Z(t) approximating white noise on the one hand, and similar results for systems subjected to ideal white noise on the other, also remains to be studied.

SUMMARY AND CONCLUSIONS

A simple analysis of an existing approach to the development of a stochastic Melnikov function for systems with additive noise showed it does not describe correctly the geometry of the stable and unstable manifolds in the perturbed system. The possibility was then explored of developing a necessary criterion for the occurrence of homoclinic chaos in forced systems perturbed additively by a commonly used approximate representation of noise. Our results indicate that the effect of perturbations approximating noise is to lower the forcing parameter threshold which must be exceeded for homoclinic chaos to occur. This suggests that for certain regions of the system parameter space to which there would correspond non-chaotic behavior in the deterministic system, additive white noise can induce chaotic behavior.

The conditions for the occurrence of chaos discussed here are necessary but not sufficient. We note, however, that numerical simulations reported by Kapitaniak[1] and by Bulsara, Schieve and

Jacobs[2] have shown that homoclinic chaos can occur in systems forced by weak noise.

ACKNOWLEDGMENTS

The work reported in this paper was supported in part by the Minerals Management Service (MMS), US Department of the Interior. Mr. Charles E. Smith of the Technology Assessment Branch, MMS, was Project Research Manager.

REFERENCES

1. Kapitaniak, T. Chaos in Systems with Noise, Second Edition, World Scientific, Singapore and New Jersey, 1990.
2. Bulsara, A.R., Schieve, W.C., and Jacobs, E.W. Homoclinic chaos in systems perturbed by homoclinic noise, Physical Review A, Vol. 41, pp. 668-681, 1990.
3. Schieve, W.C. and Bulsara, A.R. Multiplicative noise and homoclinic crossing: Chaos, Physical Review A, Vol. 41, pp. 1172-1174, 1990.
4. Rice, C. O. Mathematical Analysis of Random Noise. Selected Papers on Noise and Stochastic Processes, (ed. Wax, N.), pp. 133-294, Dover Publications, New York, 1954
5. Arrowsmith, D.K. and Place, C.M. An Introduction to Dynamical Systems, Cambridge University Press, Cambridge and New York, 1990.
6. Wiggins, S. Introduction to Applied Nonlinear Dynamical Systems and Chaos, Springer-Verlag, New York and Berlin, 1990.
7. Wiggins, S. Global Bifurcations and Chaos, Springer-Verlag, New York and Berlin, 1988.
8. Hoel, P., Port, S., and Stone, Introduction to Stochastic Processes, Houghton-Mifflin, New York, 1972.
9. Benjamin, J.R. and Cornell, C.A. Probability, Statistics and Decision for Civil Engineers, McGraw-Hill, New York, 1970.

AUTHORS' INDEX

Printed in the United States
By Bookmasters